GENES

SECOND EDITION

GENES

SECOND EDITION

BENJAMIN LEWIN
Editor, *Cell*

JOHN WILEY & SONS
New York Chichester Brisbane
Toronto Singapore

Book supervised by Linda Indig
Book and cover designed by Laura Ierardi
Illustrations by John Balbalis with the assistance of the Wiley Illustration Department

Library of Congress Cataloging in Publication Data:

Lewin, Benjamin M.
Genes.

Includes index.
1. Genetics. I. Title.
QH430.L487 1985 575.1 84-15350
ISBN 0-471-80789-3

Printed in Singapore

10 9 8 7 6 5 4 3 2 1

PREFACE

Another title for this book would be *Modern Genetics*. Even in the two years since the first edition of *Genes* was published, the technology for dealing with DNA has become more powerful; an increasing wealth of information about the genome continues to overwhelm classical genetics. The brute force of DNA technology has to some degree obscured the elegance of traditional genetics. Studying genetics now essentially means dealing with DNA. Indeed, the major part of most introductory books on genetics usually focuses on direct analysis of the genome, the ground covered in *Genes*.

With the thought in mind that this book will increasingly be treated as part of an introduction to genetics, I have changed the perspective and considerably extended the introductory chapters. They now focus on the biochemical basis for heredity, as seen through the structure of DNA. It should be possible for a student with even only a casual knowledge of biochemistry to use this book as a text.

The pace of advance in molecular biology seems scarcely to have slackened, and traditional means of publishing seem incapable of keeping up. Books and reviews of the field are often substantially out of date even before they appear in print. Short reviews in some journals are up to date on individual topics, but of course they cannot provide an overview of the field as a whole. (Innumerable volumes of "timely" research articles appear, but all they represent are hastily assembled articles, disreputable for their lack of peer review, and in any case unintelligible to readers outside the immediate field.) Certainly there are all too few wide-ranging and intelligible reviews.

The discrepancy between the rapidity of scientific advance and the inefficiency of the traditional means for reviewing is shown by an amusing sidelight: reviews of the first edition of this book began to appear in major journals only as the second edition was being prepared for press. Yet a major part of the purpose of a text such as this is to offer an up-to-date assessment of the field.

I decided upon a short revision cycle for *Genes* because a surprising amount has changed in two years and much of the fun of the field lies in being au courant with present opinion. In addition to some changes of perspective and general updating, a substantial number of new figures have been introduced, not merely to explain new points, but also to clarify old issues; a large correspondence about the first edition has made it evident that accurate illustrations of conceptual points are very useful, and this feature has therefore been expanded. By keeping the current version of the text up to date with the latest facts and thoughts in the field, I hope to make it possible for both teachers and students to keep abreast of this central area in modern genetics.

Benjamin Lewin
Cambridge, Massachusetts

PREFACE TO THE FIRST EDITION

The title of this book almost makes a preface superfluous. *Genes* is simply about genes, recognizing what amounts to a new field whose extraordinary progress has all but overwhelmed the traditional discipline of genetics. My aim is to cut through the enormous mass of information that has accumulated recently, to discern general principles and describe the state of the art in this exciting area. This text asks: what is a gene, how is it reproduced, how is it expressed, what controls its expression?

The underlying theme of this book is that the gene has at its disposal a vast repertoire of strategies for survival, different examples of which are displayed in various systems. Reflecting the perspective of current research, procaryotic and eucaryotic molecular biology are given equal weight. Both are now part of the same story. The starting point is the issue of how a gene is represented in protein; and from the protein, we work backward at the molecular level, as it were, to the DNA itself.

As a comprehensive introduction to the molecular biology of the gene, this book assumes no prior knowledge and is up to date with current research. I hope that this overview will make this rapidly advancing subject more accessible and readily allow readers to proceed to more advanced works. In view of the size of the relevant literature, it would only be confusing to rely on the citation of individual research articles. Each chapter therefore concludes with a bibliography to suggest useful reviews and some research articles that lead more deeply into the subject.

I have tried to illustrate all important points diagramatically, and where appropriate the illustrations attempt to give some feeling for the scale and relationships of the elements involved. This preface would certainly be incomplete without acknowledgment of the considerable artistic endeavors of John Balbalis to realise this aim.

One of the pleasures of writing this book has been the ensuing discussions with my friends and colleagues who have commented on it. Many improvements have resulted from the generous efforts of Sankar Adhya, Sidney Altman, French Anderson, David Clayton, Nicholas Cozzarelli, Bernard Davis, Igor Dawid, Arg Efstratiadis, Nina Fedoroff, Alice Fulton, Joe Gall, Nicholas Gillham, Philip Hanawalt, Ira Herskowitz, Lee Hood, Joel Huberman, George Khoury, Nancy Kleckner, Marilyn Kozak, Charles Kurland, Art Landy, Jeffrey Miller, Masayasu Nomura, Charles Radding, Jeff Roberts, Rich Roberts, Gerry Rubin, Robert Schimke, David Schlessinger, David Shafritz, Phil Sharp, Allen Smith, Phang-C. Tai, Susumu Tonegawa, Harold Varmus, Alex Varshavsky, and Hal Weintraub. Finally, it hardly needs saying that the book would have been much less fun to write without the enthusiastic participation of my family.

Benjamin Lewin

CONTENTS

PART 7 REACHING MATURITY: RNA PROCESSING 393

CHAPTER 25 CUTTING AND TRIMMING STABLE RNA 395

CHAPTER 26 RNA AS CATALYST: MECHANISMS OF RNA SPLICING 405

CHAPTER 27 CONTROL OF RNA PROCESSING 429

PART 8 THE PACKAGING OF DNA 445

CHAPTER 28 ABOUT GENOMES AND CHROMOSOMES 447

CHAPTER 29 CHROMATIN STRUCTURE: THE NUCLEOSOME 468

CHAPTER 30 THE NATURE OF ACTIVE CHROMATIN 489

GENES

SECOND EDITION

PART 1
DNA AS A STORE OF INFORMATION

When alcohol reaches a concentration of about 9/10 volume there separates out a fibrous substance which on stirring the mixture wraps itself about the glass rod like thread on a spool and the other impurities stay behind as a granular precipitate. The fibrous material is redissolved and the process repeated several times. In short, this substance is highly reactive and on elementary analysis conforms *very* closely to the theoretical values of pure DNA (who could have guessed it).

Oswald Avery, 1944

CHAPTER 1
DNA IS THE GENETIC MATERIAL

Ever since it was realized that an organism does not pass on a simulacrum of itself to the next generation, but instead provides it with **genetic material** containing the **information** needed to construct a progeny organism, we have wanted to define the nature of this information and the manner in which it is utilized. Now that we know the physical structure of the genetic material, we may define the aim of modern biology as accounting for the complexity of living organisms in terms of the properties of their constituents.

The **gene** is the unit of genetic information, and the critical step toward defining its nature was the realization that it is a distinct entity. The era of the molecular biology of the gene began with Schrödinger's review in 1945 of the properties to be expected from the perspective of the physicist. "Incredibly small groups of atoms, much too small to display exact statistical laws, do play a dominating role in the very orderly and lawful events within a living organism. . . . The gene is much too small . . . to entail an orderly and lawful behavior according to physics."

From this perspective, Schrödinger went on to develop the concept that the laws of physics might be inadequate to account for the properties of the genetic material, in particular its stability during untold generations of inheritance. The gene was expected to obey the laws of physics so far established, but it was thought that characterizing the genetic material might lead to the discovery of new laws of physics, a prospect that brought many physicists into biology.

Schrödinger summarized the anticipated properties of the gene in terms that, with modest redefinition, correspond remarkably well with our present view. "We shall assume the structure of the gene to be that of a huge molecule, capable only of discontinuous change, which consists in a rearrangement of the atoms and leads to an isomeric molecule. The rearrangement [mutation] may affect only a small region of the gene, and a vast number of different rearrangements may be possible." At that time, the gene was visualized as an *aperiodic crystal*, which seemed to provide the only physical form consistent with the demands placed on the genetic material.

Now, of course, we know that the gene is indeed a huge molecule, but one that does not function autonomously; instead it relies upon other cellular components for its perpetuation and function. All of its activities obey the known laws of physics and chemistry and it has not, in the end, been necessary to invoke new regulations.

HEREDITY WORKS THROUGH MACROMOLECULES

All living organisms are composed of cells. **Unicellular** organisms consist of individual cells, able to survive, reproduce, and (in some cases) mate. **Multicellular** organisms exist by virtue of cooperation between many cells, which may have different specialties to contribute to the survival of the individual.

Both types of organism may be classified in terms of the type of cell. Bacteria (generally unicellular) are the **prokaryotes**, organisms in which nominally there is only a single cell compartment, bounded by a membrane or membranes that give security against the outside world. **Eukaryotes** may be either unicellular or multicellular and are defined by the division of each cell into a **nucleus** that contains the genetic material, surrounded by a **cytoplasm**, which in turn is bounded by the **plasma membrane** that marks the periphery of the cell.

Both the nature of the genetic information and the general means (although not every detail) of its utilization are the same in prokaryotes and eukaryotes. In describing how the inheritance of genetic information accounts for the perpetuation of both prokaryotic and eukaryotic cells, we may regard their genetic instructions as being of the same kind, although the process of reading the instructions involves extra steps in eukaryotes.

We may view the molecules of which cells are made as falling into two classes. The **small molecules** are the substrates and products of metabolic pathways, providing the energy needed for cell survival; they also provide the subunits from which larger molecules are assembled. Thus **polysaccharides** are assembled from **sugars** and **lipids** are assembled from **fatty acids**.

Genetic information is conveyed through **macromolecules**, very large molecules. **Nucleic acids**, the largest macromolecules, are assembled from **nucleotides**; **proteins** are assembled from **amino acids**. In each case, *the sequence in which the individual building blocks are joined together is the critical factor that determines the property of the resulting macromolecule.*

The macromolecules are responsible for maintaining the structure of the cell. Both the physical characteristics and the location of a macromolecule are important. A cell, after all, does not consist merely of protoplasm contained in a membraneous bag: each structural component of the cell occupies a specific location. This is true not just to the extent that membranes contain lipids while the nucleus contains the genetic material, but also in the sense that perpetuation of the structure of the cell from one generation to the next depends upon the maintenance of a highly ordered structure, in which particular functions can be exercised only at particular places.

The physical and functional properties of macromolecules can be accounted for in terms of their structures. The ultimate responsibility for these structures lies with the genetic material.

All biological reactions are accomplished by **enzymes**, proteins that possess specific catalytic activities enabling them to bring together two moieties in an environment that makes it possible for reaction to occur. Enzymes perform the series of reactions by which metabolic pathways convert raw material into compounds that the organism needs. They are responsible for the assembly of fatty acids into lipids or of sugars into polysaccharides.

All proteins are assembled from amino acids by a complex apparatus that works under the direction of genetic information provided by a nucleic acid. Each segment of nucleic acid consists of a sequence of nucleotides that directly determines the sequence of amino acids in a corresponding protein. The amino acid sequence in turn determines the function of the protein. All the macromolecular structures of the cell are therefore determined by the sequence of the genetic material.

Is the structure in terms of macromolecules sufficient to account for the properties of the cell? This question asks whether the structure of a macromolecule alone can explain its location in the cell, or whether some extraneous form of information is necessary. At least in some cases, a macromolecule can be located properly only at the time when it is made. A profound influence during the development of a multicellular organism from the fertilized egg is the need for macromolecules to be located at the right place. We do not understand this effect at all well, but the effect that location may have on function is sometimes called **positional information**, a phrase that implies the importance of location as well as structure per se.

PROTEINS FUNCTION THROUGH THEIR CONFORMATION

Each protein consists of a unique sequence of amino acids. Just twenty amino acids are commonly found in proteins. (Some others are found in special circumstances, but they are generated by modifications made to one of the common twenty after the protein has been constructed.)

Each free amino acid has the general structure

$$\begin{array}{c} NH_2 \\ | \\ R\text{–}CH \\ | \\ HO\text{–}C{=}O \end{array}$$

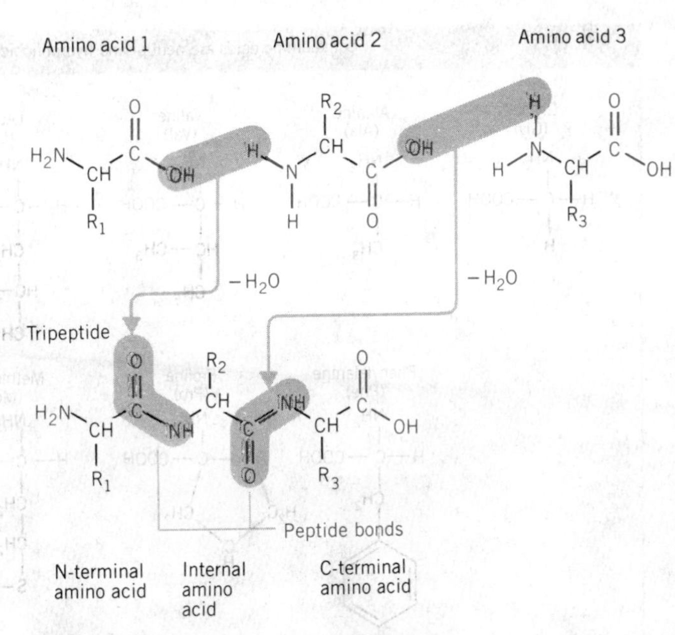

Figure 1.1
A peptide bond is formed by a condensation reaction (so-called because water is lost).

Amino acids are joined together to form a protein by **peptide bonds**, which are created by the *condensation* of the amino (NH_2) group of one amino acid with the carboxyl (COOH) group of the next, as illustrated in **Figure 1.1**. A chain of amino acids joined in this manner is called a **polypeptide**. Since peptide bonds are covalent, they are relatively stable, and in the living cell are broken only rarely, usually as the result of a specific enzymatic action.

As illustrated in two dimensions, the polypeptide chain forms a zig-zag **backbone**, from which the **side groups** (denoted **R**) stick out. The term **protein** usually is used to describe the functional unit, which sometimes consists of a single polypeptide chain and sometimes consists of more than one.

We can define the direction of a polypeptide chain according to the orientation of the peptide bonds. The amino acid at one end of the chain must have a free NH_2 group and thus defines the **amino-** or **N-terminal** end, while the amino acid at the other end must have a free COOH group and thus defines the **carboxyl-** or **C-terminal** end. Protein sequences are conventionally written from N-terminus (at the left) to C-terminus (at the right).

The side group **R** is different for each amino acid and determines the nature of its contribution to the overall protein structure. **Figure 1.2** shows the structures of the twenty common amino acids. Classified by their ionic charge, lysine, arginine and histidine are basic; aspartic acid and glutamic acid are acidic; and the remainder are neutral. Some of the neutral amino acids are polar (electrically charged), while others are hydrophobic (water-repelling). The polar amino acids may react with other (nonprotein) groups bearing electric charges. On the other hand, clusters of the hydrophobic amino acids in a protein may create "pockets" with distinct properties that welcome uncharged compounds.

Certain unusual (modified) amino acids are found in particular proteins or types of cell. For example, hydroxyproline and hydroxylysine are variants of proline and lysine that each have an additional —OH group; they are found in the collagens, which are proteins of connective tissue. More common modifications include **phosphorylation** (addition of a phosphate group, usually to the R group of serine or tyrosine) and **acetylation** or **methylation** (addition of acetyl or methyl groups, usually to the basic amino acid lysine). These modifications may be important in protein function because they change the properties of the **R** group.

The critical feature of a protein is its ability to fold into a **conformation** that creates **active sites** or other structural features that allow it to play its role in the cell. Each protein exists in a unique conformation (or

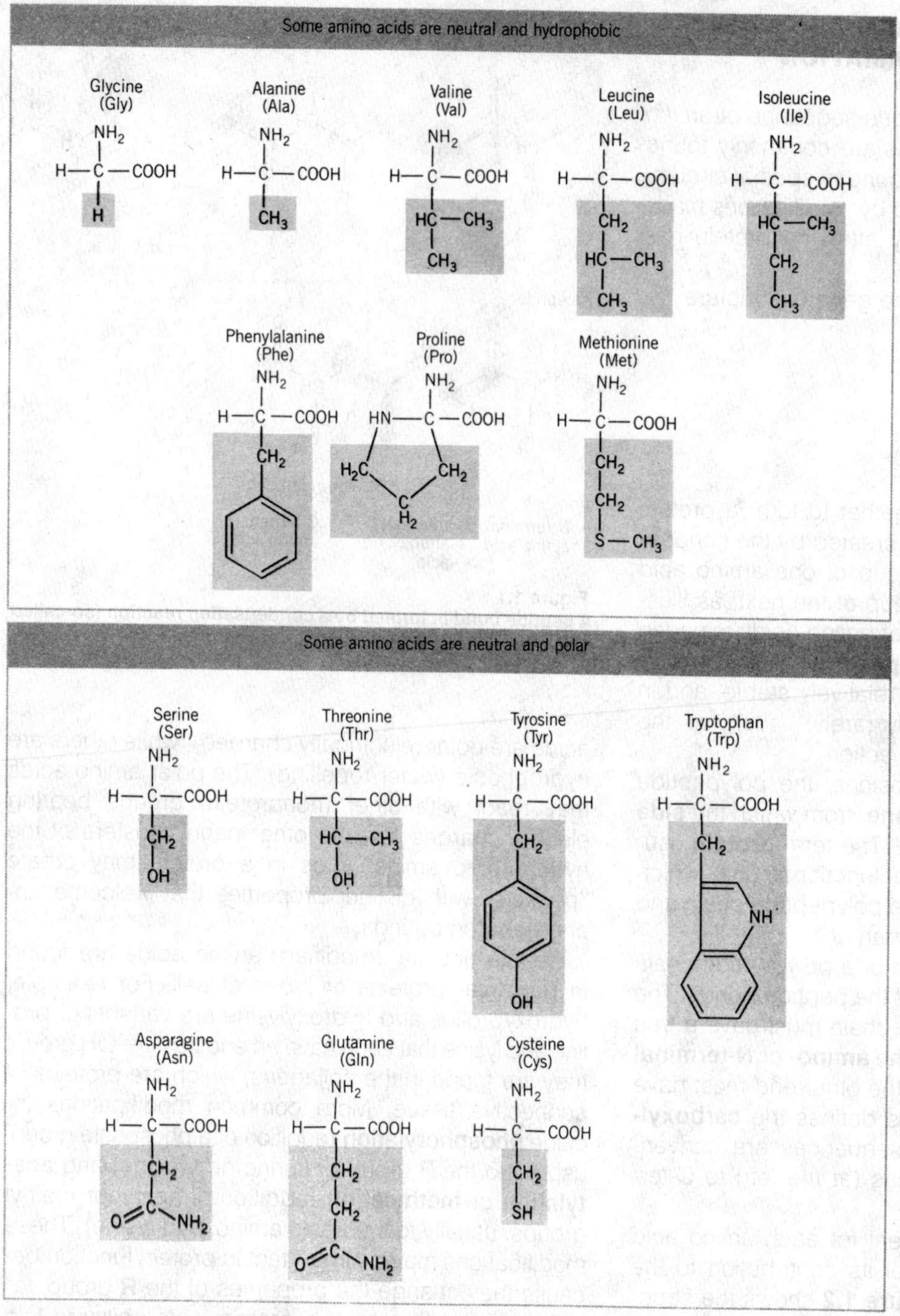

Some amino acids are neutral and hydrophobic

Glycine (Gly)
Alanine (Ala)
Valine (Val)
Leucine (Leu)
Isoleucine (Ile)
Phenylalanine (Phe)
Proline (Pro)
Methionine (Met)

Some amino acids are neutral and polar

Serine (Ser)
Threonine (Thr)
Tyrosine (Tyr)
Tryptophan (Trp)
Asparagine (Asn)
Glutamine (Gln)
Cysteine (Cys)

Figure 1.2
Amino acids are classified according to the nature of their side groups.
Each amino acid is known by a three letter abbreviation.

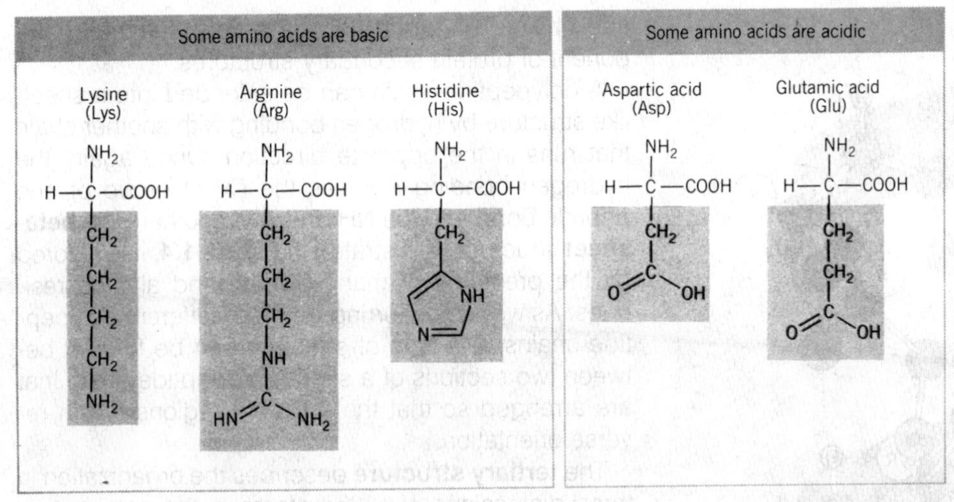

Figure 1.2 (continued)

sometimes a series of alternate conformations). The conformation is described in terms of several levels of structure.

The series of amino acids that constitutes a polypeptide chain comprises its **primary structure.** The **secondary structure** is generated by the folding of the primary sequence, which is made possible by the ability to move freely about bonds of free rotation. Secondary structure refers to the path that the polypeptide backbone of the protein follows in space.

Several types of interactions between amino acids contribute to the acquisition of secondary structure. Both covalent and noncovalent bonds are involved.

In many proteins, one of the important features responsible for establishing the secondary structure is the formation of S—S disulfide "bridges" between two cysteine residues (forming cystine). Each cysteine is separately placed into the polypeptide chain at the appropriate location; the condensation of the two —SH groups into the S—S bridge occurs later, when they are brought into apposition as the chain begins to fold into the correct conformation.

In all proteins, a major force underlying the acquisition of conformation is the formation of noncovalent bonds, in particular ionic interactions between oppositely charged groups (such as acidic and basic amino acids), hydrophobic interactions between amino acids with apolar side-chains, and hydrogen bonds.

Hydrogen bonds are weak electrostatic bonds that occur between a partially negatively charged oxygen atom and a partially positively charged hydrogen atom, as in the examples

$$
\begin{array}{lcl}
\overset{\delta^+\ \delta^-}{>C=O} & \cdots\cdots\cdots & \overset{\delta^+\ \delta^-}{H—N<} \\
\overset{\delta^+\ \delta^-}{>C=O} & \cdots\cdots\cdots & \overset{\delta^+\ \delta^-}{H—O—}
\end{array}
$$

(The δ indicates the partial nature of the electric charge.) The polarization of the C═O and the N—H or O—H bonds in effect allows the formation of a hydrogen bond between the two groups. The bond takes its name from the fact that the hydrogen atom is to some degree shared between the reacting groups.

Hydrogen bonds are much weaker than covalent bonds, by a factor of more than 10. However, because a large number of hydrogen bonds can be formed in a macromolecule, their overall contribution to the stability of the conformation can be substantial. But the weakness of the individual hydrogen bond (and of other noncovalent bonds) allows them to be broken relatively easily under physiological conditions. This is an important aspect of the function of both proteins and nucleic acids.

Proteins show enormous diversity of form as a result of their ability to generate a huge range of conformations. Certain types of secondary structure are relatively common. Two particularly well known types are the alpha helix and beta sheet.

Hydrogen bonding between groups on the same polypeptide chain may cause the backbone to twist

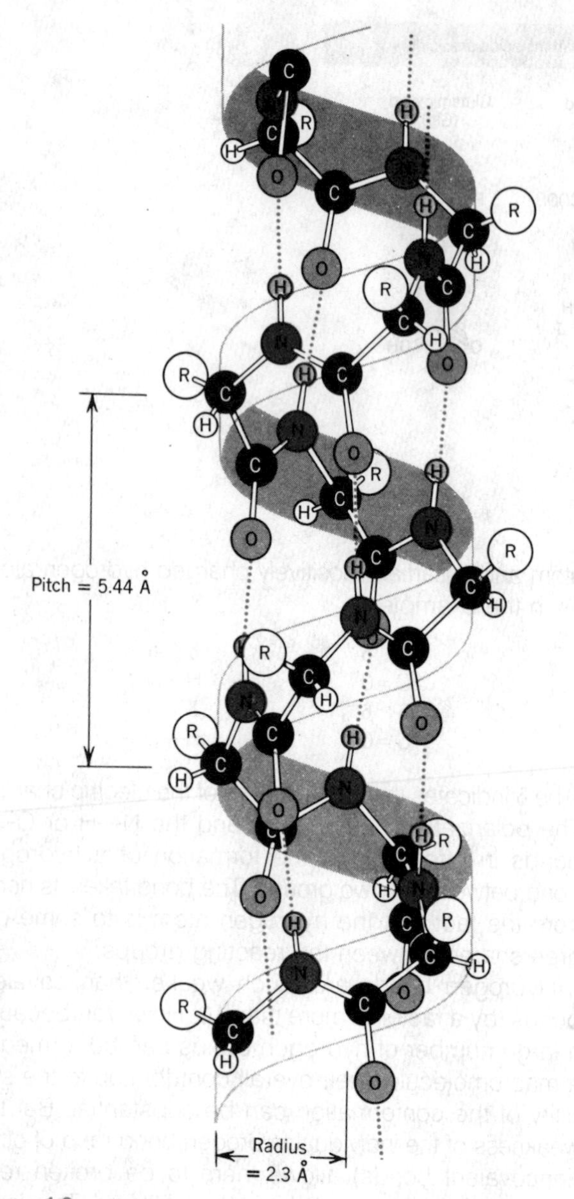

Figure 1.3
A polypeptide alpha helix is stabilized by hydrogen bonding between amino acids four residues apart in the same polypeptide. (H= bonds are shown by dots.)

into a helix, most often the form known as the **alpha helix**, illustrated in **Figure 1.3**. The helix is stabilized in part by hydrogen bonds formed between the C═0 group of one peptide bond and the NH group of the peptide bond four residues farther along the polypeptide chain. The alpha helix is a very common component of protein secondary structures.

A polypeptide chain can be extended into a sheet-like structure by hydrogen bonding with another chain that runs in the opposite direction. Once again, the hydrogen bonding involves the C═O group of one peptide bond and the NH group of another. The **beta-sheet** structure is illustrated in **Figure 1.4**; it is favored by the presence of many glycine and alanine residues. As well as occurring between different polypeptide chains, this type of structure can be formed between two sections of a single polypeptide chain that are arranged so that the adjacent regions are in reverse orientation.

The **tertiary structure** describes the organization in three dimensions of all the atoms in the polypeptide chain, including the side (R) groups as well as the polypeptide backbone. In a protein consisting of a single polypeptide chain, this level describes the complete structure. Proteins can be divided into two general classes (which represent extremes of structure) on the basis of their tertiary structure.

Fibrous proteins have elongated structures, with the polypeptide chains arranged in long strands. Depending on the protein, the tertiary structure may be based on an α-helix or β-sheet. The particular form of secondary structure may stretch through a large part of the protein. The fibrous proteins are major structural components of the cell or tissue; for example, they are prominent in connective tissues. Thus their role tends to be static in providing a structural framework.

Globular proteins have more compact structures. The tertiary structure is generally rather irregular, often containing many different types of secondary structure. For example, although a part of the structure often takes the form of an alpha helix, it is rare for it to comprise the entire structure. Many globular proteins consist of partly helical and partly non-helical secondary structures. Relatively short stretches of α-helix may be held in a particular relationship to one another. Globular proteins include most enzymes and most of the proteins involved in gene expression and regulation with which we shall be involved. Their roles tend to be dynamic, involving the ability to catalyze reactions or change conformation.

Yet a higher level of organization is recognized in **multimeric proteins**, which consist of aggregates of

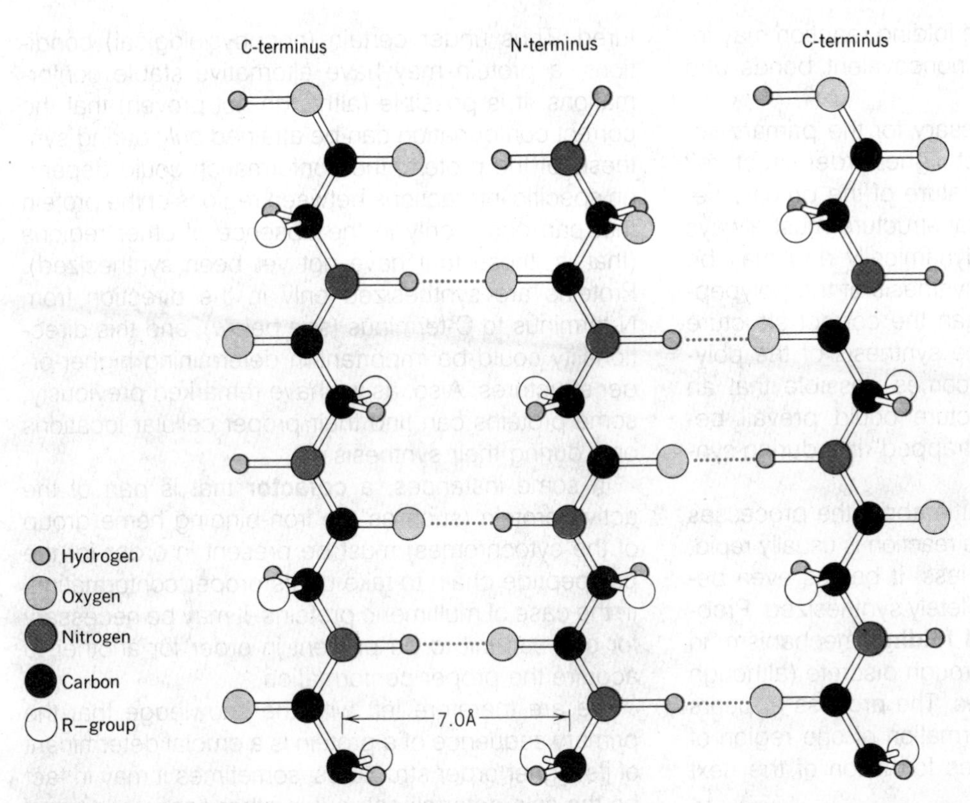

Figure 1.4
A beta sheet is stabilized by hydrogen bonding between amino acids in different polypeptide chains (or in parts of the same chain that run in opposite directions).

more than one polypeptide chain. The **quaternary structure** describes the conformation assumed by the multimeric protein. The individual polypeptide chains that make up a multimeric protein are often described as the **protein subunits**.

The versatility of protein structure can be illustrated by considering the alternative forms possible for a polypeptide chain of (say) 300 amino acids. As a fully extended array of amino acids, it would stretch for 100 nm; coiled in an alpha helix, it would extend for 45 nm. But organized as a beta-sheet it could become a flat box of 7 × 7 nm only 0.8 nm deep; and as a sphere utilizing various forms of secondary structure, it could have a diameter of only 4.3 nm.

The relationship between structure and function depends on the type of protein. Some proteins, especially enzymes that catalyze metabolic reactions, may have an **active site** at which the catalytic reaction occurs. The structure of this site, comprised by the juxtaposition of a handful of amino acids, may be absolutely crucial, and admitting of no variation. It may be possible to change other regions of the protein without abolishing its catalytic function. Yet other proteins, for example some structural proteins, may have a conformation large parts or even all of which is involved directly in its function, and any change at all may therefore be likely to prevent its function. Proteins therefore differ in the extent to which their structure can vary without preventing their function.

HOW DO PROTEINS ACQUIRE THE CORRECT CONFORMATION?

A fundamental principle is that higher-order structures are determined directly by lower-order structures. This means that the primary sequence of amino acids carries the information that is necessary for folding into the correct conformation. The folding reaction may involve the formation of both noncovalent bonds and covalent bonds.

What conditions are necessary for the primary sequence to fold into the correct higher-order structure? Is this process an intrinsic feature of the primary sequence? In this case, the final structure must always be the most stable thermodynamically and may be generated at any time after synthesis of the polypeptide chain is complete. Or can the correct structure be generated only during the synthesis of the polypeptide? In this case, it becomes possible that an intrinsically less stable structure could prevail because the protein becomes "trapped" in it during synthesis.

We still know remarkably little about the processes involved in protein folding. The reaction is usually rapid, occurring within seconds or less. It begins even before a protein has been completely synthesized. Probably it involves a **sequential folding** mechanism, in which the reaction passes through discrete (although highly transient) intermediates. The process appears to be cooperative, so that formation of one region of secondary structure enhances formation of the next region, and so on.

The relationship between higher-order structures and the primary structure may be revealed when a protein is **denatured** by heating or chemical treatments that cause it to lose its conformation. Most denaturing events involve the breakage of hydrogen and other noncovalent bonds. An exception is the disruption of S—S bridges that results from treatment with reducing agents. However, all of these changes affect the *conformation*; the primary sequence of amino acids in the polypeptide chain remains unaltered.

In at least some cases, the higher-order structure follows ineluctably from the primary sequence. This situation was first identified by experiments in which the protein *ribonuclease* was denatured. On reversing the denaturing procedure, the active conformation is regained, showing that *all* the necessary information resides in the primary sequence. Thus the production of active ribonuclease occurs automatically whenever the intact primary chain is placed in the appropriate conditions.

In other cases, proteins can be irreversibly denatured. Thus under certain (nonphysiological) conditions, a protein may have alternative stable conformations. It is possible (although not proven) that the correct conformation can be attained *only* during synthesis of the protein; the conformation could depend on specific interactions between regions of the protein that can occur only in the absence of other regions (that is, those that have not yet been synthesized). Proteins are synthesized only in the direction from N-terminus to C-terminus (see below), and this directionality could be important in determining higher-order structures. Also, as we have remarked previously, some proteins can find their proper cellular locations only during their synthesis.

In some instances, a **cofactor** that is part of the active protein (such as the iron-binding heme group of the cytochromes) must be present in order for the polypeptide chain to take up its proper conformation. In the case of multimeric proteins, it may be necessary for one subunit to be present in order for another to acquire the proper conformation.

We are therefore left with the knowledge that the primary sequence of a protein is a crucial determinant of its higher order structures; sometimes it may in fact be the sole determinant, but in other cases additional interactions may be involved in acquisition of the final conformation. In each case, however, if the primary sequence is synthesized within the appropriate environment, it will inevitably acquire the proper higher-order structures.

Although higher order structure follows from primary sequence, there may be several ways to make a particular type of tertiary structure. For example, the globin (red blood cell) proteins of different species vary substantially in sequence, but acquire the same general tertiary structure.

Some proteins exist in alternate conformations and have different biological properties in each conformation. Proteins with this ability are called **allosteric**. The transition between conformations may be influenced by the interaction of the protein with a cofactor or with another protein. **Figure 1.5** illustrates the consequences of an allosteric transition. The change in conformation brought about by an interaction at one site may alter the structure and thus the function of another active site. Allosteric transitions affect *only* the

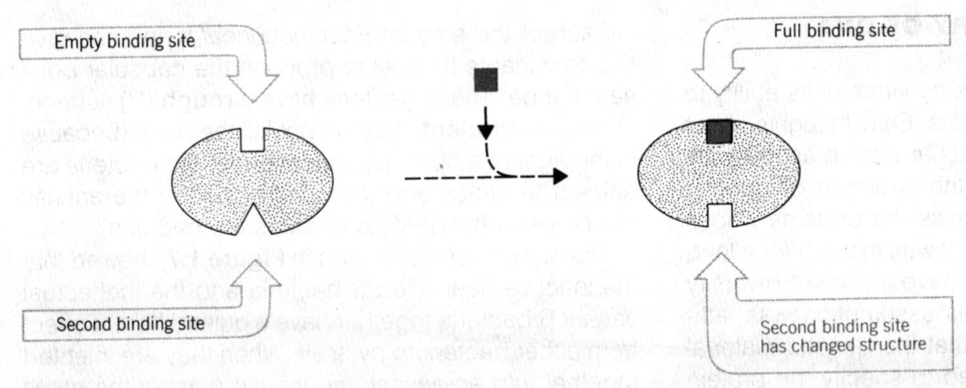

Figure 1.5
When a small molecule binds to its active site, an allosteric protein undergoes a transition to an alternative conformation in which the structure and function of a second active site are altered.

conformation—they do not change the primary sequence—and they rely heavily on making and breaking hydrogen bonds.

Allosteric proteins play an important role in both metabolic and genetic regulation. Often the product of a metabolic pathway can bind to an enzyme catalyzing an early step in the pathway to prevent it from sending further small molecules through the pathway. This interaction is called **feedback inhibition**. In genetic regulation, the activity of a protein that controls gene expression may respond to a small molecule. In either case, the critical interaction is the ability of a small molecule to bind to a specific site on the protein and thereby to change the conformation in such a way that the activity of the protein is altered.

Some allosteric proteins are multimers of identical subunits. In this case, the conformation of one subunit can alter the conformation of the other subunit(s). For example, when a small molecule that controls the protein's activity binds to its site on one subunit, it may become much easier for the other subunits to bind the small molecule. This effect is illustrated in **Figure 1.6**. It amplifies the effect of binding the first small molecule in such a way that the protein characteristically flips very rapidly from one state to another. This is an important ability in a regulatory protein.

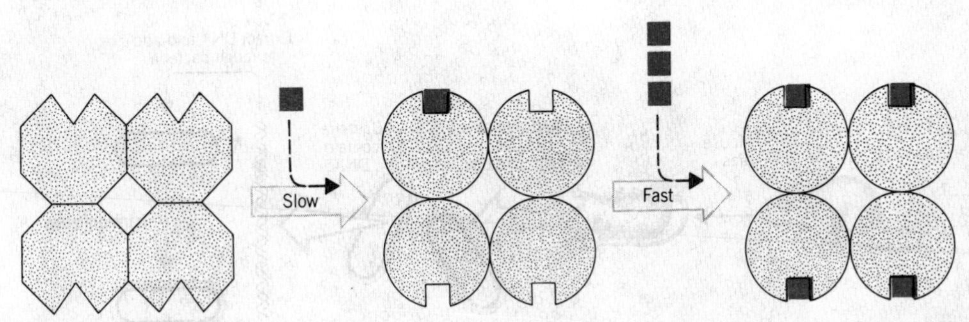

Figure 1.6
When an allosteric protein is a multimer of identical subunits, binding to one subunit may enhance binding at other subunits.

THE DISCOVERY OF DNA

The genetic material functions by virtue of its ability to specify a large variety of proteins. Early thoughts about the nature of the genetic material were biased by an erroneous assumption: that the structure of genetic material must be as complex as the proteins whose production it specifies. In fact, it was thought for a long time that *only* proteins could have sufficient diversity to specify other proteins. This assumption was jettisoned when it was realized that the genetic material carries the information needed to specify the protein in an enciphered form, a code.

The demonstration that genetic material is nucleic acid had its roots in the discovery of **transformation** by Griffith in 1928. The bacterium *Pneumococcus* kills mice by causing pneumonia. The virulence of the bacterium is determined by the **capsular polysaccharide**, a component of the surface. Several **types** (I, II, III) of *Pneumococcus* have different capsular polysaccharides. All of these types have a **smooth** (S) appearance, due to the presence of the capsular polysaccharide on the surface.

Naturally, the smooth bacteria are effective in killing mice only when the bacteria are able to multiply. If the bacteria are killed by heat treatment, they lose their ability to harm the animal.

Each of the smooth *Pneumococcal* types can give rise to variants that fail to produce the capsular polysaccharide. These bacteria have a **rough** (R) surface. They are **avirulent**; they do not kill the mice (because in the absence of the polysaccharide, the bacteria are subject to attack and then destruction in the animal; the polysaccharide coat protects the bacteria).

The experiment illustrated in **Figure 1.7** showed that the inactive heat-killed S bacteria and the ineffectual variant R bacteria together have a quite different effect from either bacterium by itself. When they are injected together into an animal, the mouse dies as the result of a *Pneumococcal* infection. Virulent S bacteria can be recovered from the mouse postmortem.

In the first of these experiments, the dead S bacteria were of type I. The live R bacteria had been derived from type II. The virulent bacteria recovered from the mixed infection had the smooth coat of type I. Thus some property of the dead type I S bacteria can **transform** the live R bacteria so that they make the type I capsular polysaccharide, and as a result become virulent.

The component of the dead bacteria responsible for transformation was called the **transforming principle.** Purifying this component required the development of a cell-free system, in which extracts of the dead S bacteria could be added to the live R bacteria before

Figure 1.7
The transforming principle is DNA.
Neither heat-killed smooth bacteria nor live rough (mutant) bacteria can kill mice. But the mixture kills mice; and live smooth bacteria can be recovered from them. The transformation of inactive rough bacteria into virulent smooth bacteria can be accomplished *in vitro* by the addition of DNA extracted from smooth bacteria.

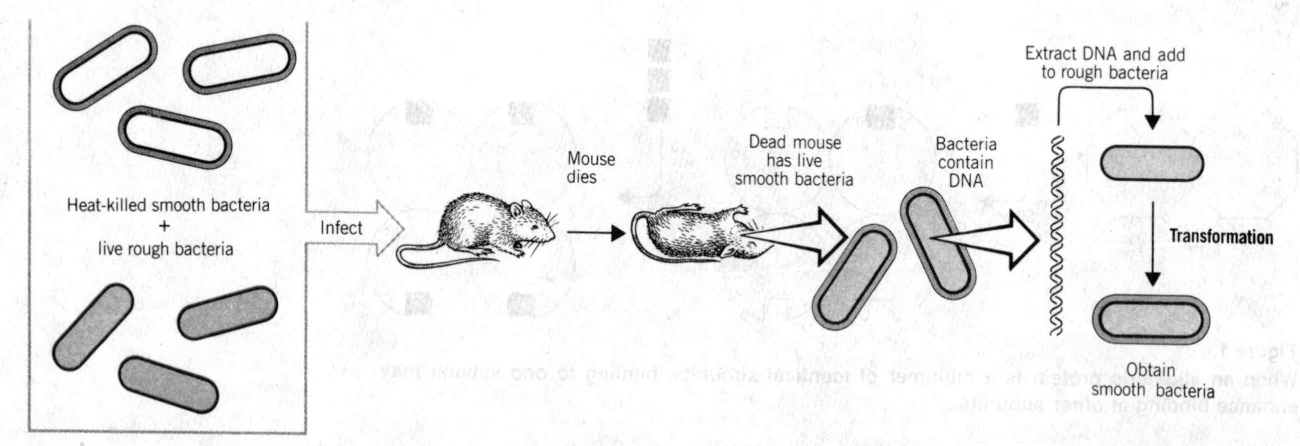

injection into the animal. The classic studies of Avery and his colleagues showed chemically in 1944 that the isolated transforming principle is **deoxyribonucleic acid (DNA).**

The role of DNA was confirmed shortly after when the enzyme deoxyribonuclease (DNAase) was purified. This enzyme specifically destroys DNA. Addition of DNAase to the preparation of transforming principle abolished transformation. The possibility that transformation might be due to the presence of some protein contaminant in the DNA preparation was excluded by showing that treatment with the enzyme trypsin (which degrades most proteins) had no effect on the activity.

The surprise of this result is indicated by the fact that, at this time, DNA was not even known to be a component of *Pneumococcus*, although of course it had been recognized for many decades as a major component of eukaryotic chromosomes. In showing that the genetic material of a prokaryote is DNA, this result therefore offered a unifying view for the basis of heredity in bacteria and in higher organisms.

The implications of the result are precisely captured by the discussion in the original paper reporting the research. "The inducing substance, on the basis of its chemical and physical properties, appears to be a highly polymerized and viscous form of DNA. On the other hand, the type III capsular polysaccharide, the synthesis of which is evoked by this transforming agent, consists chiefly of a nonnitrogenous polysaccharide. . . . Thus it is evident that the inducing substance and the substance produced in turn are chemically distinct and biologically specific in their action and that both are requisite in determining the type specificity of the cells of which they form a part." This discussion marked the introduction of a distinction between the genetic material and the products of its expression, a view that remained implicit in subsequent studies.

In spite of the clarity of these results, considerable reluctance remained in accepting DNA as the genetic material, on the grounds that no way could be imagined for it to represent the variety of protein structures. However, the identification of DNA as the transforming principle was buttressed by showing that other capsular types and, indeed, entirely different properties can be transferred not only in *Pneumococcus*, but also in other bacteria. (The next property to be transformed was that of resistance to penicillin.)

These further results disposed of arguments that transformation might reflect some particular role of DNA in formation of the capsular polysaccharide rather than a general genetic function. (In fact, we now know that the DNA transferred in the original experiments carried the ability to produce an enzyme—UDPG dehydrogenase—which catalyzes one of the steps of capsular polysaccharide synthesis.)

DNA IS THE (ALMOST) UNIVERSAL GENETIC MATERIAL

After the transforming principle had been shown to consist of DNA, the next step was to demonstrate that DNA provides the genetic material in a quite different system. Phage T2 is a virus that infects the bacterium *E. coli*. When phage particles are added to bacteria, they adsorb to the outside surface, some material enters the bacterium, and then about 20 minutes later each bacterium bursts open (lyses) to release a large number of progeny phage particles.

In 1952, Hershey and Chase infected bacteria with T2 phages that had been radioactively labeled *either* in their DNA component (with ^{32}P) *or* in their protein component (with ^{35}S). **Figure 1.8** illustrates the results of this experiment. The infected bacteria were agitated in a blender, and two fractions were separated by centrifugation. One contained the empty phage coats that were released from the surface of the bacteria; these consist of protein and therefore carried the ^{35}S radioactive label. The other fraction consisted of the infected bacteria themselves.

Most of the ^{32}P label was present in the infected bacteria. The progeny phage particles produced by the infection contained about 30% of the original ^{32}P label. The progeny received very little—less than 1%—of the protein contained in the original phage population. This experiment therefore showed directly that the DNA of parent phages enters the bacteria and then becomes part of the progeny phages, exactly the pattern of inheritance expected of genetic material.

Whether viruses are to be classified as "living" has been a debating point. The classic example of phage T2 (remember that phages are simply viruses that infect bacteria) shows that, on infection, the virus is able to commandeer the machinery of the host cell to man-

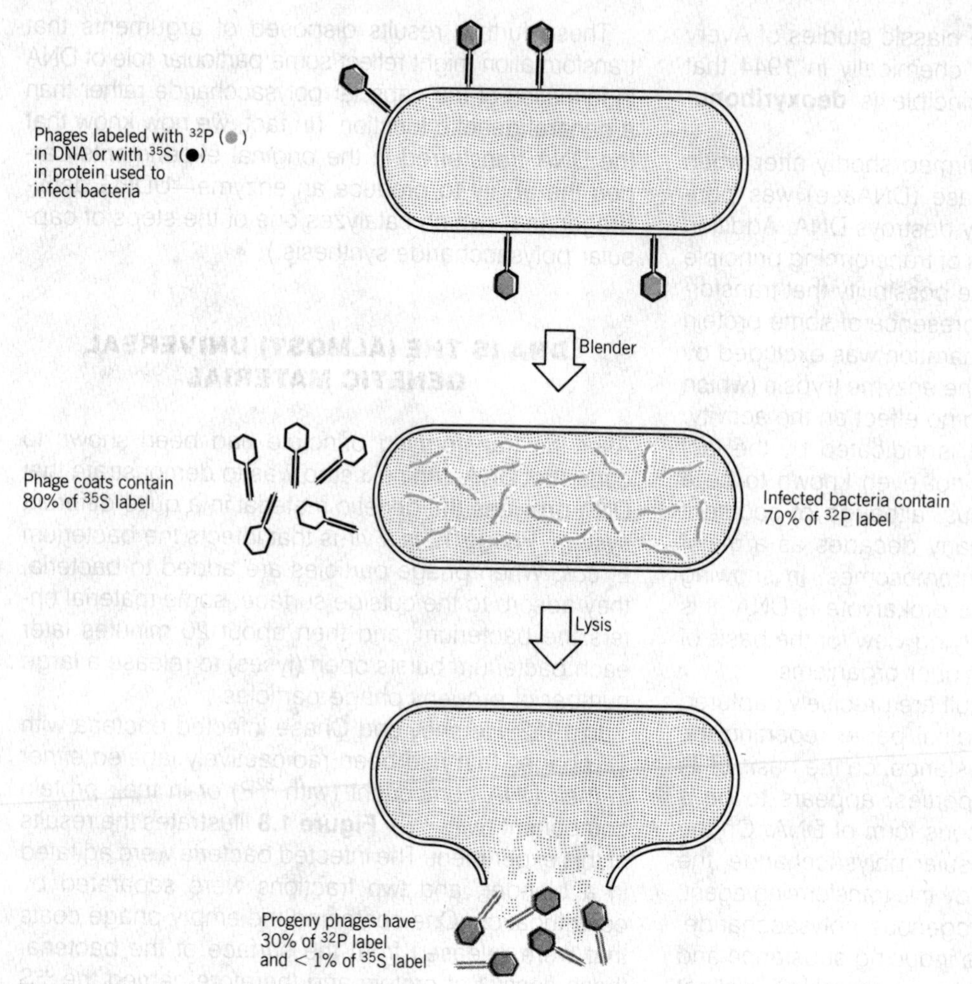

Figure 1.8
The genetic material of phage T2 is DNA.
When phages are labeled with ^{32}P in DNA or with ^{35}S in protein, the ^{32}P label enters the infected bacteria and can be recovered from the progeny phages. But the ^{35}S label is not inherited by the progeny.

ufacture more copies of itself. The phage possesses genetic material whose behavior is analogous to that of cellular genomes: its traits are faithfully reproduced, and they are subject to the same rules that govern inheritance. Thus the case of T2 reinforces the general conclusion that the genetic material is DNA.

Although clearly demonstrating the role of DNA as genetic material, the phage infection experiments actually are not quite as precise as bacterial transformation in excluding protein as a contaminant of the DNA. Increased purification of transforming DNA reduced to ridiculously small proportions the amount of contaminant that could be present. During phage infection, however, a small amount of protein actually *is* transferred into the bacteria together with the DNA; but, of course, it is only the DNA that is recovered from the progeny. The climate of opinion had much to do with the immediate acceptance of the Hershey-Chase experiment compared with the disbelief in Avery's work.

Bacteria and bacteriophages clearly have DNA as their genetic material. But what about eukaryotes? For a long time the evidence was only inferential. DNA is present in the right location and behaves in the appropriate manner. Direct evidence became available only rather recently, long after the matter was regarded as settled.

In discussing his results on transformation, Avery made a comment with wider-ranging implications than he could have known. "If we are right," he wrote, "it means that nucleic acids are not merely structurally important but functionally active substances in determining the biochemical activities and specific characteristics of cells and that by means of a known chemical substance it is possible to induce predictable and hereditary changes in cells." He had in mind the bacterial system, but similar results now have been obtained with eukaryotes as well.

When DNA is added to populations of single eukaryotic cells growing in culture, the nucleic acid enters the cells, and in some of them results in the production of new proteins. At first performed with DNA extracted en masse, these experiments now can be routinely performed with purified DNA whose incorporation leads to the production of a particular protein. **Figure 1.9** depicts one of the standard systems.

Although for historical reasons these experiments are described as **transfection** when performed with eukaryotic cells, they are a direct counterpart to bacterial transformation. The DNA that is introduced into the recipient cells may become part of its genetic material, inherited in the same way as any other part. At first, these experiments were successful only with individual cells adapted to grow in a culture medium. Since then, however, DNA has been introduced in mouse eggs by microinjection; and it becomes a stable part of the genetic material of the mouse (see Chapter 38). Such experiments show directly not only that DNA is the genetic material in eukaryotes, but they demonstrate also that it can be transferred between different species and yet remain functional.

The genetic material of all known organisms and many viruses is DNA. However, there are also some viruses in which an alternative nucleic acid, **ribonucleic acid (RNA)**, is found as the genetic material. Although its chemical formula is slightly different from that of DNA, in these circumstances RNA exercises the same role. The general principle of the nature of the genetic material, then, is that it is always nucleic acid; in fact, it is DNA except in some viruses that use RNA instead.

THE COMPONENTS OF DNA

One of the reasons for skepticism about the genetic function of DNA was the misunderstanding of its struc-

Figure 1.9
Eukaryotic cells can acquire a new phenotype as the result of transfection by added DNA.

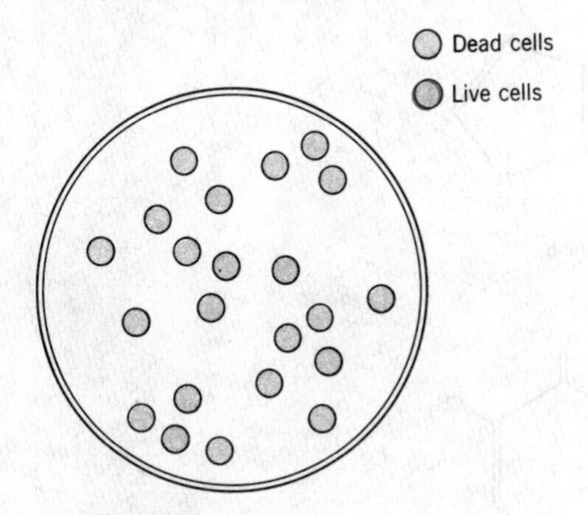

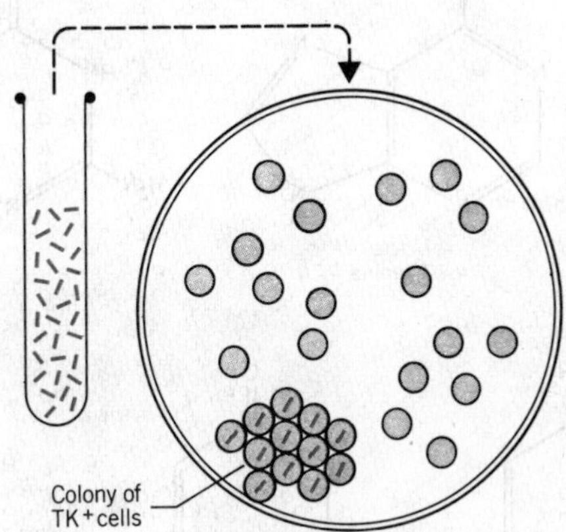

ture. Although known to contain four different bases, DNA was thought to consist of a regularly repeating unit containing all the four bases. By the 1930s Caspersson had shown that DNA consists of extremely large molecules—much larger than proteins—but the monotony of its structure appeared to preclude any possible genetic role, for which, after all, variability of form would be expected.

But then the continuing analysis carried out by Chargaff demonstrated that the bases found in DNA are present in amounts that are different in each species. This observation led to the concept that the *sequence* of bases in DNA might be the form in which **genetic information** is carried, this sequence in some way specifying, or **coding** for, the sequence of amino acids in a protein. By the 1950s, the concept of genetic information was common: the twin problems that it posed were working out the structure of the nucleic acid and explaining how a sequence in DNA could represent the sequence of a protein.

A nucleic acid consists of a chemically linked sequence of nucleotides. Each nucleotide contains a heterocyclic ring of carbon and nitrogen atoms (the **nitrogenous base**), a five-carbon sugar in ring form (a **pentose**), and a **phosphate** group.

The nitrogenous bases fall into the two types shown in **Figure 1.10**: **pyrimidines** and **purines**. Pyrimidines have a six-member ring; purines have fused five- and six-member rings. Each nucleic acid is synthesized from only four types of base. The same two purines, adenine and guanine, are present in both DNA and RNA. The two pyrimidines in DNA are cytosine and thymine; in RNA uracil is found instead of thymine. The only difference between uracil and thymine is the presence of a methyl substituent at position C_5. The bases are usually referred to by their initial letters; so DNA contains A, G, C, T, while RNA contains A, G, C, U.

Two types of pentose are found in nucleic acids. They distinguish DNA and RNA and give rise to the general names for the two types of nucleic acid. **Figure 1.11** shows their structures. In DNA the pentose is **2-deoxyribose**; whereas in RNA it is **ribose**. The difference lies in the absence/presence of the hydroxyl group at position 2 of the sugar ring.

The nitrogenous base is linked to position 1 on the pentose ring by a glycosidic bond from N_1 of pyrimidines or N_9 of purines. To avoid ambiguity between the numbering systems of the heterocyclic rings and the sugar, positions on the pentose are given a prime (′).

Figure 1.10
Purines and pyrimidines provide the nitrogenous bases in nucleic acids.
The "purine" and "pyrimidine" rings show the general structures of each type of base; the numbers identify the positions on the ring.

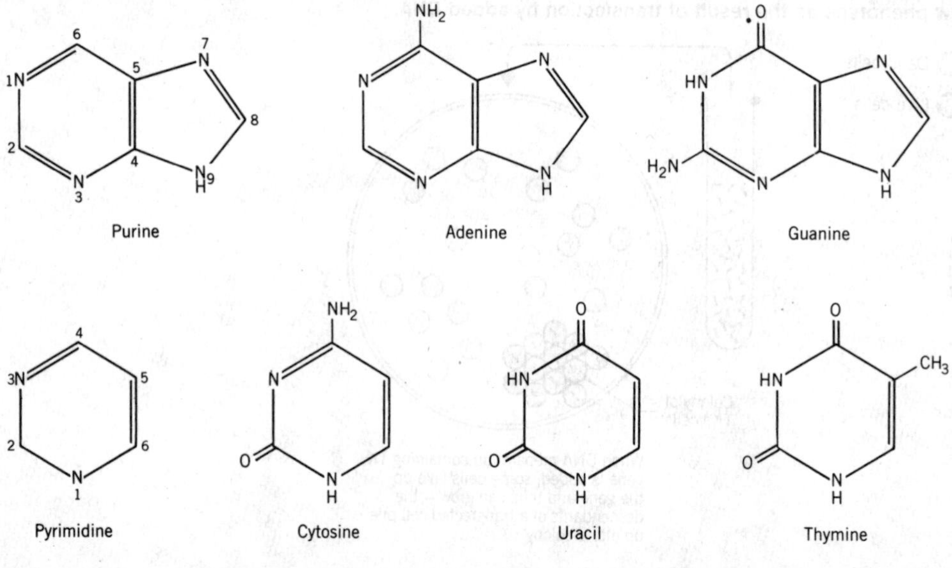

Purine

Adenine

Guanine

Pyrimidine

Cytosine

Uracil

Thymine

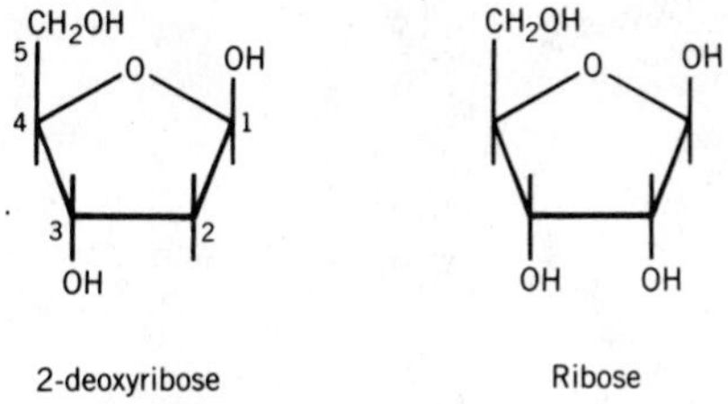

Figure 1.11
2-Deoxyribose is the sugar in DNA and ribose is the sugar in RNA.
The carbon atoms are numbered as indicated for deoxyribose.

A base linked to a sugar is called a **nucleoside**; when a phosphate group is added, the base-sugar-phosphate is called a **nucleotide**. The nomenclature of the individual units is described in **Table 1.1**.

Nucleotides provide the building blocks from which nucleic acids are constructed. The nucleotides are linked together into a **polynucleotide chain** by a backbone consisting of an alternating series of sugar and phosphate residues. The 5′ position of one pentose ring is connected to the 3′ position of the next pentose ring via a phosphate group, as shown in **Figure 1.12**. Thus the phosphodiester-sugar backbone is said to consist of 5′–3′ linkages. The nitrogenous bases "stick out" from the sugar-phosphate backbone.

The terminal nucleotide at one end of the chain has a free 5′ group; the terminal nucleotide at the other end has a free 3′ group. It is conventional to write nucleic acid sequences in the 5′–3′ direction—that is, from the 5′ terminus at the left to the 3′ terminus at the right.

When DNA or RNA is broken into its constituent nucleotides, the cleavage may take place on either side of the phosphodiester bonds. Depending on the circumstances, nucleotides may have their phosphate group attached to either the 5′ or the 3′ position of the pentose, as shown in **Figure 1.13**. The two types of nucleotide released from nucleic acids are therefore the nucleoside-3′-monophosphates and nucleoside-5′-monophosphates.

All the nucleotides can exist in a form in which there is more than one phosphate group linked to the 5′ position. An example is shown in Figure 1.12. The bonds between the first (α) and second (β) and between the second (β) and third (γ) phosphate groups are **energy-rich** and are used to provide an energy source for various cellular activities. The abbreviation for a triphosphate takes the form NTP; the abbreviation for a diphosphate is NDP.

The 5′ triphosphates are the precursors for nucleic acid synthesis. **Figure 1.14** shows the reaction, in which the 5′ end of the triphosphate reacts with a 3′—OH group at the end of the polynucleotide chain. The two terminal phosphate groups (γ and β) of the triphosphate are released, and a bond is formed from the α phosphate to the sugar.

DNA IS A DOUBLE HELIX

The discovery of the double helix was a dramatic event. Two notions converged in the construction of this model by Watson and Crick in 1953.

First, X-ray diffraction data showed that DNA has the form of a regular helix, making a complete turn every 34Å (3.4 nm), and with a diameter of about 20Å (2 nm). Since the distance between adjacent nucleotides is 3.4Å, there must be 10 nucleotides per turn.

Table 1.1
Bases, nucleosides, and nucleotides have related names.

Base	Nucleoside	Nucleotide	Abbreviation RNA	Abbreviation DNA
Adenine	Adenosine	Adenylic acid	AMP	dAMP
Guanine	Guanosine	Guanylic acid	GMP	dGMP
Cytosine	Cytidine	Cytidylic acid	CMP	dCMP
Thymine	Thymidine	Thymidylic acid		dTMP
Uracil	Uridine	Uridylic acid	UMP	

Abbreviations of the form NMP stand for nucleoside monophosphate; "d" is used to indicate the deoxy form and its absence implies the presence of a 2′-OH group.

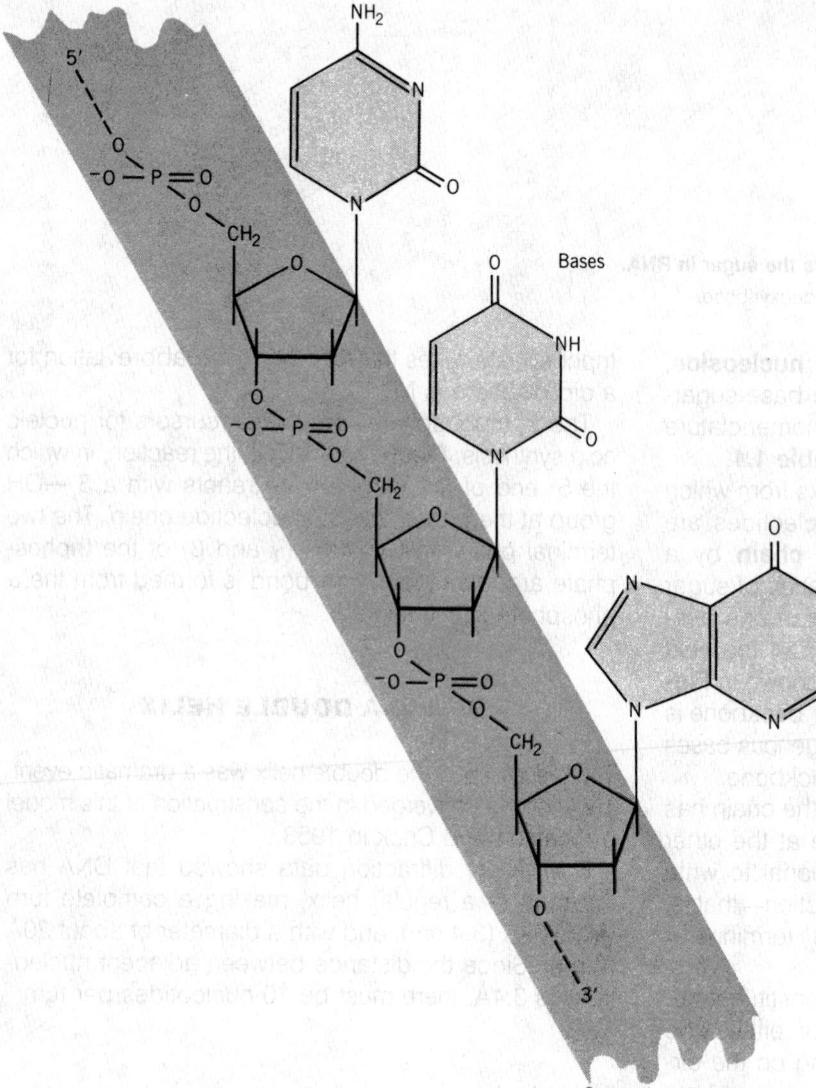

Figure 1.12
A polynucleotide chain consists of a series of 5′–3′ sugar-phosphate links that form a backbone from which the bases protrude.

The density of DNA suggested that the helix must contain two polynucleotide chains. The constant diameter of the helix can be explained if the bases in each chain face inward and are restricted so that a purine is always opposite a pyrimidine, avoiding purine-purine (too thick) or pyrimidine-pyrimidine (too thin) partnerships.

The second critical feature lay in the earlier observation of Chargaff that, irrespective of the actual amounts of each base, the proportion of G and C is always the same in DNA, and the proportion of A and T is always the same. (Thus the DNA of any species can be characterized by the proportion of its bases that is G + C, which ranges from 26% to 74%, a clear refutation of the repeating-tetranucleotide theory.)

Watson and Crick proposed that the two polynucleotide chains in the double helix are not connected by covalent bonds, but associate by virtue of **hydrogen bonding** between the nitrogenous bases. **Figure 1.15** demonstrates that, in their usual forms, G can

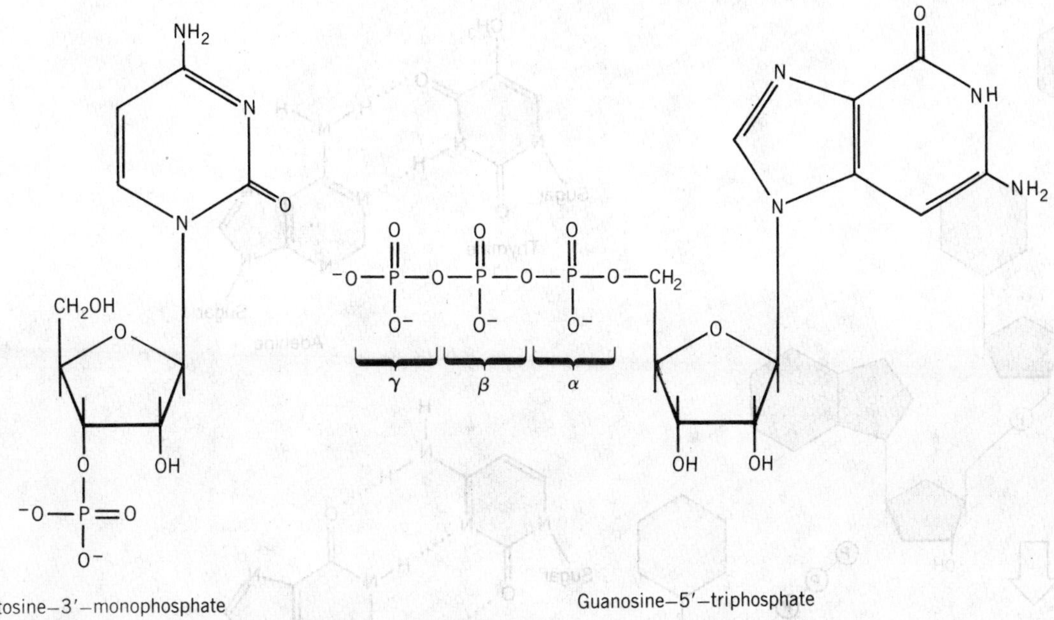

Cytosine–3′–monophosphate

Guanosine–5′–triphosphate

Figure 1.13
Nucleotides may carry phosphate in the 5′ or 3′ position.

hydrogen bond specifically with only C, while A can bond specifically only with T. These reactions are described as **base pairing**, and the paired bases (G with C or A with T) are said to be **complementary**.

To permit specificity in base pairing, the use of the appropriate form of the base is crucial. The movement of a hydrogen atom allows each base to exist in **tautomeric** forms. The forms present in the double helix have amino groups (NH_2) and keto groups (C═O), as opposed to the tautomeric alternative of imino groups (NH) and enol groups (COH).

The model requires the two polynucleotide chains to run in opposite directions (**antiparallel**), as illustrated in **Figure 1.16**. Looking along the helix, therefore, one strand runs in the 5′–3′ direction, while its partner runs 3′–5′.

The sugar-phosphate backbone is on the outside and carries negative charges on the phosphate groups. When DNA is in solution *in vitro*, the charges are neutralized by the binding of metal ions; usually Na^+ is provided. In the natural state *in vivo*, positively charged proteins provide some of the neutralizing force. The properties of these proteins may play an important role in determining the organization of DNA in the cell.

The bases lie on the inside. They are flat structures, lying in pairs perpendicular to the axis of the helix. (If we think of the double helix in terms of a spiral staircase, the base pairs form the treads.) The base pairs contribute to the thermodynamic stability of the double helix in two ways.

First, energy is released by the hydrogen bonding between the bases in each pair. Second, proceeding along the helix, bases are **stacked** above one another, in a sense like a pile of plates. Energy is released by hydrophobic **base-stacking**, resulting from interactions between the electron systems of the stacked base pairs.

Each base pair is rotated ~36° around the axis of the helix relative to the next base pair. Thus ~10 base pairs make a complete turn of 360°. The twisting of the two strands around one another forms a double helix that has a **narrow groove** (about 12Å across) and a **wide groove** (about 22Å across), as can be seen from the scale model presented in **Figure 1.17**. The double helix is **right-handed**; the turns run clockwise looking along the helical axis. These features represent the accepted model for what is known as the **B-form** of DNA.

P
P
P
OH
P
P
P
OH
P
P
P
+
P
P—P
OH

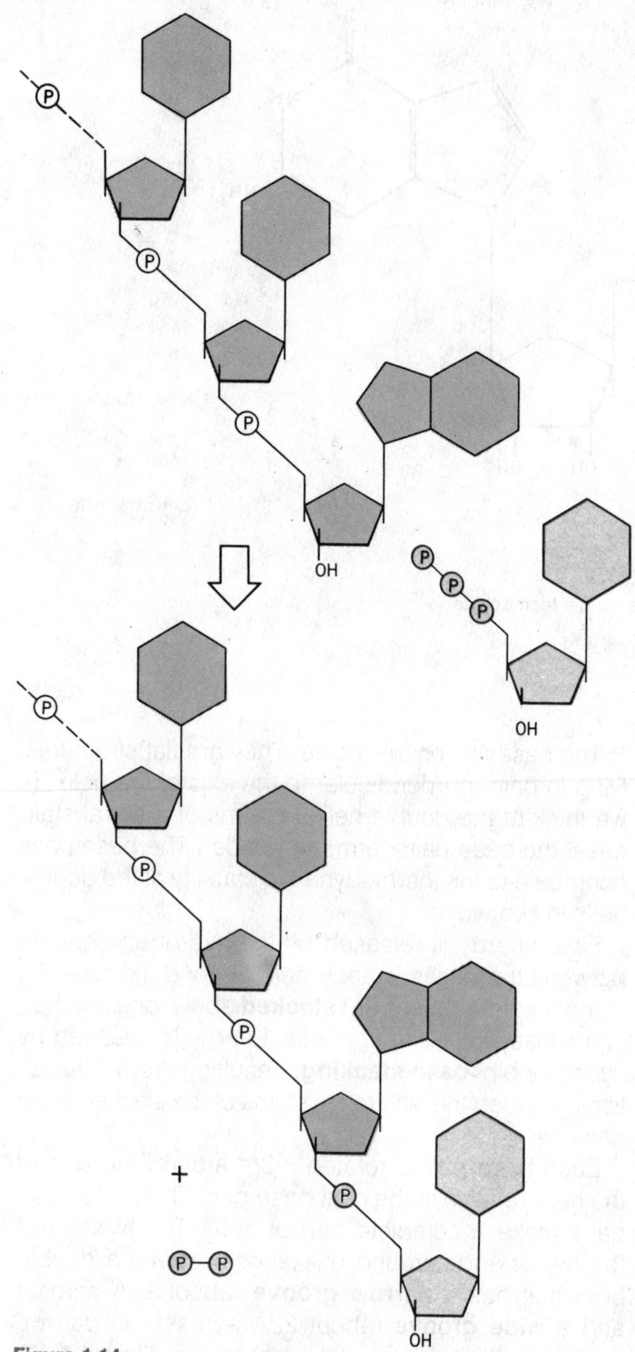

Figure 1.14
Nucleic acid synthesis occurs by adding the nucleoside-5′-monophosphate moiety of a nucleoside triphosphate to the 3′-OH end of the polynucleotide chain.

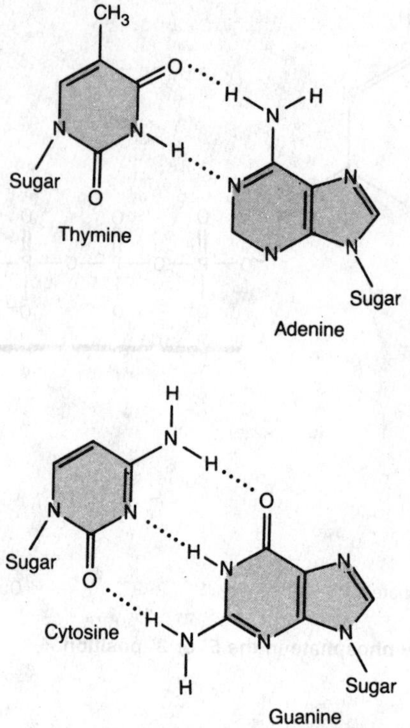

CH3
O
H
N
H
N
N
H
Sugar
O
N
N
N
N
Thymine
Sugar
Adenine
H
N
H
N
N
O
Sugar
H
N
O
N
N
H
N
N
Cytosine
H
Sugar
Guanine

Figure 1.15
Complementary base pairing involves the formation of two hydrogen bonds between A and T, and of three hydrogen bonds between G and C. No other pairs form in DNA.

DNA REPLICATION IS SEMICONSERVATIVE

One of the crucial requirements for the genetic material is that it must be reproduced accurately. In a classic understatement, Watson and Crick concluded their paper by saying that "it has not escaped our notice that the specific pairing we have postulated immediately suggests a possible copying mechanism for the genetic material."

This idea was taken up in a subsequent paper, where they pointed out that, because the two polynucleotide

Figure 1.16
The double helix maintains a constant width because purines always face pyrimidines in the complementary A-T and G-C base pairs.

Reading down the page, the strand on the left runs 5′–3′ and the strand on the right runs 3′–5′. The figure is diagrammatic and does not show the winding of the two strands about one another.

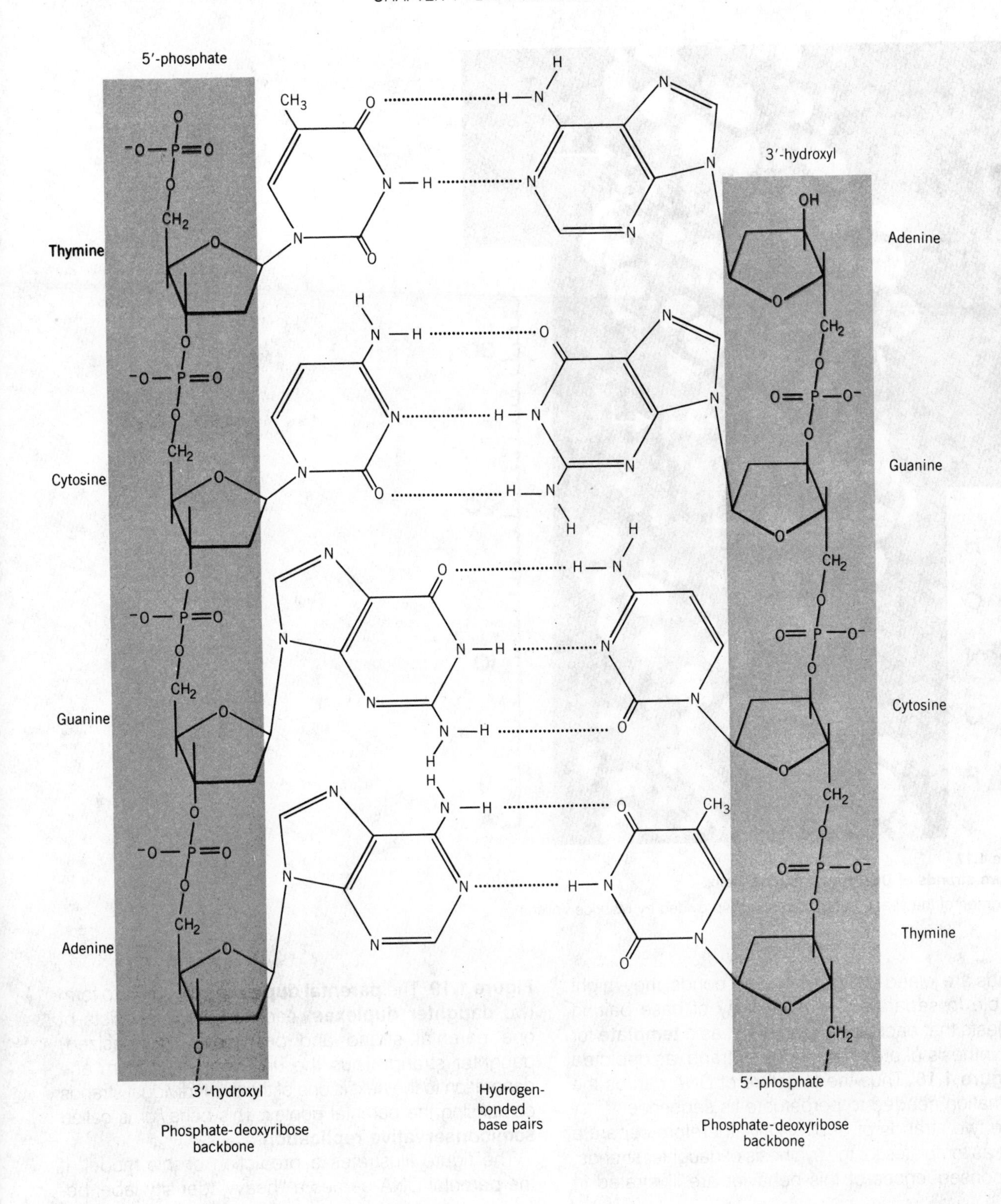
5′-phosphate
Thymine
Cytosine
Guanine
Adenine
3′-hydroxyl
Phosphate-deoxyribose backbone
Hydrogen-bonded base pairs
3′-hydroxyl
Adenine
Guanine
Cytosine
Thymine
5′-phosphate
Phosphate-deoxyribose backbone

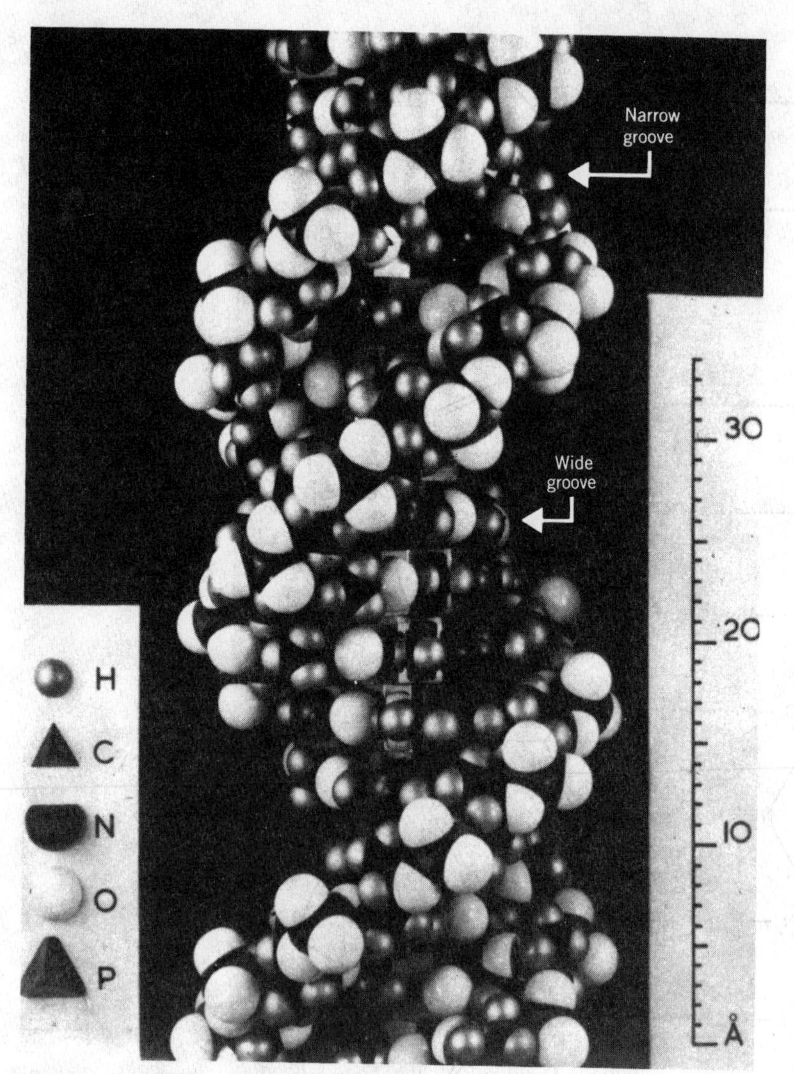

Figure 1.17
The two strands of DNA form a double helix.
Photograph of the space-filling model kindly provided by Maurice Wilkins.

strands are joined only by hydrogen bonds, they might be able to separate. The specificity of base pairing suggests that each strand could act as a **template** for the synthesis of a complementary strand, as depicted in **Figure 1.18**. Thus the structure of DNA carries the information needed to perpetuate its sequence.

The two strands of duplex DNA therefore separate to act as templates for the synthesis of daughter strands. The consequences of this behavior are illustrated in **Figure 1.19**. The **parental duplex** is replicated to form two **daughter duplexes**, each of which consists of one parental strand and one (newly synthesized) daughter strand. Thus the unit conserved from one generation to the next is one of the two individual strands comprising the parental duplex. This behavior is called **semiconservative replication**.

The figure illustrates a prediction of this model. If the parental DNA carries a "heavy" density label be-

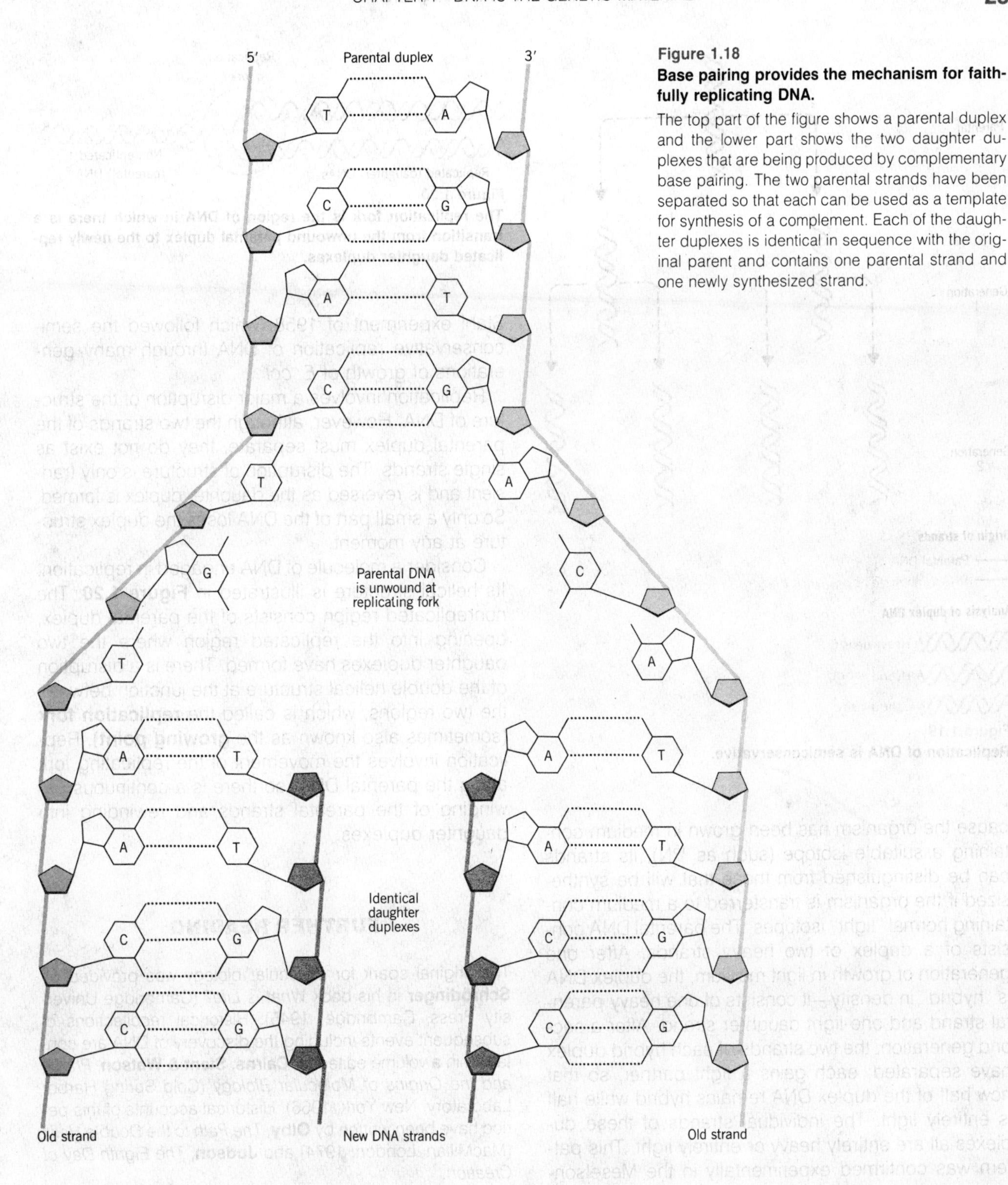

Figure 1.18

Base pairing provides the mechanism for faithfully replicating DNA.

The top part of the figure shows a parental duplex and the lower part shows the two daughter duplexes that are being produced by complementary base pairing. The two parental strands have been separated so that each can be used as a template for synthesis of a complement. Each of the daughter duplexes is identical in sequence with the original parent and contains one parental strand and one newly synthesized strand.

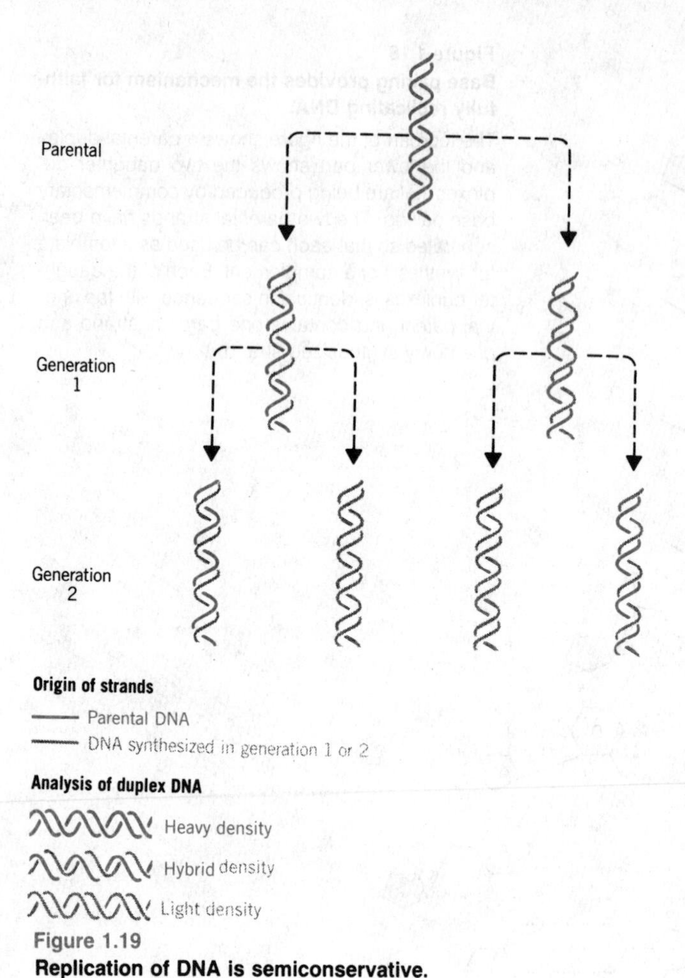

Figure 1.19
Replication of DNA is semiconservative.

cause the organism has been grown in medium containing a suitable isotope (such as ^{15}N), its strands can be distinguished from those that will be synthesized if the organism is transferred to a medium containing normal "light" isotopes. The parental DNA consists of a duplex of two heavy strands. After one generation of growth in light medium, the duplex DNA is "hybrid" in density—it consists of one heavy parental strand and one light daughter strand. After a second generation, the two strands of each hybrid duplex have separated; each gains a light partner, so that now half of the duplex DNA remains hybrid while half is entirely light. The individual strands of these duplexes all are entirely heavy or entirely light. This pattern was confirmed experimentally in the Meselson-Stahl experiment of 1958, which followed the semiconservative replication of DNA through many generations of growth of *E. coli*.

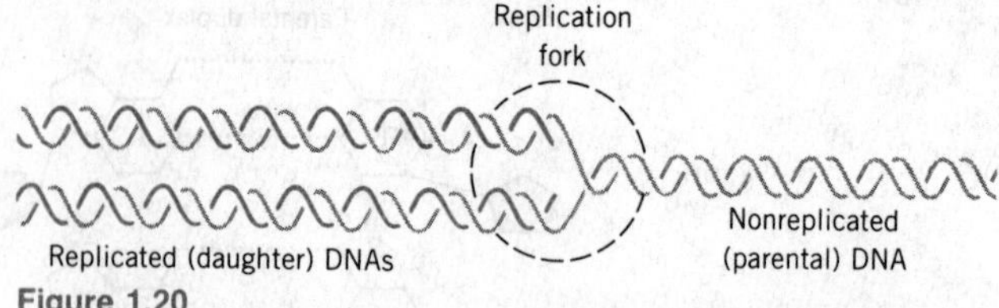

Figure 1.20
The replication fork is the region of DNA in which there is a transition from the unwound parental duplex to the newly replicated daughter duplexes.

Replication involves a major disruption of the structure of DNA. However, although the two strands of the parental duplex must separate, they do not exist as single strands. The disruption of structure is only transient and is reversed as the daughter duplex is formed. So only a small part of the DNA loses the duplex structure at any moment.

Consider a molecule of DNA engaged in replication. Its helical structure is illustrated in **Figure 1.20**. The nonreplicated region consists of the parental duplex, opening into the replicated region where the two daughter duplexes have formed. There is a disruption of the double helical structure at the junction between the two regions, which is called the **replication fork** (sometimes also known as the **growing point**). Replication involves the movement of the replicating fork along the parental DNA, so there is a continuous unwinding of the parental strands and rewinding into daughter duplexes.

FURTHER READING

The original spark for molecular biology was provided by **Schrödinger** in his book *What is Life?* (Cambridge University Press, Cambridge, 1945). Historical recollections of subsequent events including the discovery of DNA are contained in a volume edited by **Cairns, Stent & Watson**, *Phage and the Origins of Molecular Biology* (Cold Spring Harbor Laboratory, New York, 1966). Historical accounts of this period have been written by **Olby**, *The Path to the Double Helix* (MacMillan, London, 1974) and **Judson**, *The Eighth Day of Creation*.

CHAPTER 2
GENES ARE CARRIED ON CHROMOSOMES

Inherited traits are defined by their ability to be passed from one generation to the next in a predictable manner. Before we consider the nature of the unit of inheritance, it is important to realize the distinction between the appearance of the organism (which is what we observe) and the underlying genetic constitution (which we must infer). Visible or otherwise measurable properties are called the **phenotype**, while the genetic factors responsible for creating the phenotype are called the **genotype**.

As seen by their effects on the phenotype, inherited traits vary widely in complexity. Some appear in principle to be relatively limited—for example, humans may have either brown or blue eyes. Some are apparently more complex; consider, for example, the inheritance of the shape of a nose. Through the concept of the gene, genetics finds common ground for the inheritance of traits ranging from the ability to perform a simple metabolic reaction to the construction of a complex shape.

The gene is the unit of inheritance. Each gene constitutes a sequence of DNA that carries the information representing a particular protein. The effect of a gene on the organism is revealed by the change(s) in the phenotype that occur when the DNA sequence is altered. A change in the sequence of a gene is called a **mutation**. The organism carrying the altered gene is called a **mutant**; an organism carrying the normal (unaltered) gene is called **wild type**. "Wild type" may be used to describe either the genotype or phenotype. Since most mutations damage the function of a protein, the study of mutations is biased toward examining situations in which a gene is not functioning properly.

The effects of a change in a protein depend on its functions in the organism, and may be relatively minor or may be serious. Complex changes in the phenotype can result from single mutations. A classic example of a deleterious mutation that exerts its effect by interfering with a metabolic pathway was discovered by Garrod at the turn of the century. His phrase "inborn errors of metabolism" carries the concept that a genetic defect may cause a metabolic failure that creates a mutant phenotype.

As an example, phenylketonuria is an inherited disease of man, among other effects impeding brain function. The disease results from the absence of the enzyme that usually converts the amino acid phenylalanine into another amino acid, tyrosine. The failure results in an accumulation of phenylalanine, which is then converted into the toxic compound phenylpyruvic acid, and a variety of defects ensues.

As revealed by the effects of mutations, some phenotypic traits are determined by single genes, while others are determined by several genes. Thus some features are altered only when a specific gene is mutated, while others can be affected by mutation in any one of several genes. When a phenotypic trait can be identified with the function of a particular protein, we can relate the phenotype and genotype (as, for example, in the case of phenylketonuria). However, the basis for other phenotypic traits is not yet known (for example, how the shape of a nose is determined), and in these cases we have as guide only the working assumption that the interactions of proteins represented by as yet unidentified genes will be found to explain the phenotypic differences between individuals.

DISCOVERY OF THE GENE

The essential attributes of the gene were defined by Mendel's work more than a century ago. Summarized in his two laws, the gene was recognized as a "particulate factor" that passes unchanged from parent to progeny. Although we usually regard genetics from the perspective of the organism, we should recognize the sense in which the organism is the gene's way of expressing and perpetuating itself. This point is made by the alternation of generations depicted in **Figure 2.1**.

The figure illustrates the life cycle of **diploid** organisms, which have two copies of each gene. One of the two copies is passed from the parent to a **gamete** (sex cell: egg or sperm). The alternate types of gamete produced by parents of different sexes unite to form a **zygote** (fertilized egg). Thus the zygote gains one copy of each gene from each parent, restoring the situation in which every organism has one copy of paternal origin and one of maternal origin.

A gene may exist in alternate forms that determine the expression of some particular characteristic (for example, the color of a flower may be red or white). These forms are called **alleles**. Mendel's first law describes the **independent segregation of alleles:** *alleles do not affect each other when present in the same plant, but segregate unchanged by passing into different gametes at the formation of the next generation.*

When both alleles are of the same type, an organism is said to be **homozygous** (or true-breeding). If the alleles are of different types, the organism is **heterozygous** (or hybrid). The phenotype of a homozygote directly reflects the genotype of the allele, but the phenotype of a heterozygote depends on the relationship between the types of allele that are present. Often one allele is **dominant** and the other is **recessive**; in this case, the phenotype is determined by the dominant allele (the presence of the recessive allele is in effect irrelevant), and so the appearance of the heterozygote is indistinguishable from that of the homozygous dominant parent.

Mendel's first law in effect recognizes that the genotype of the hybrid includes both alleles, even though only one contributes to the phenotype. **Figure 2.2** illustrates the experiment with the color of garden peas that revealed this situation. When a dominant homozygote is crossed with a recessive homozygote, all the progeny of the first (F1) generation are heterozygotes whose phenotype is the same as that of the dominant homozygous parent. But when the heterozygotes are crossed with one another to generate a second (F2) generation, the recessive phenotype reappears. The critical point is that the alleles must consist of discrete physical entities whose interaction is at the level of expression.

Although the characteristics studied by Mendel (fortunately) showed complete dominance, this is not necessarily always the case. Alleles may exhibit **incomplete (partial) dominance** or no dominance (sometimes known as **codominance**). In the latter case, the phenotype of the heterozygote is intermediate between that of the homozygotes. In the snapdragon, for example, a cross between red and white generates heterozygotes with pink flowers. Although there is no dominance, the same rule is observed that the first hybrid (F1) generation is uniform in phenotype; and the same ratios are generated in the second (F2) hybrid generation, except that three phenotypes can be distinguished instead of two.

Mendel's second law summarizes the **independent assortment of different genes**. When a homozygote that is dominant for two different characters is crossed with a homozygote that is recessive for both characters, as before the F1 consists of plants whose pheno-

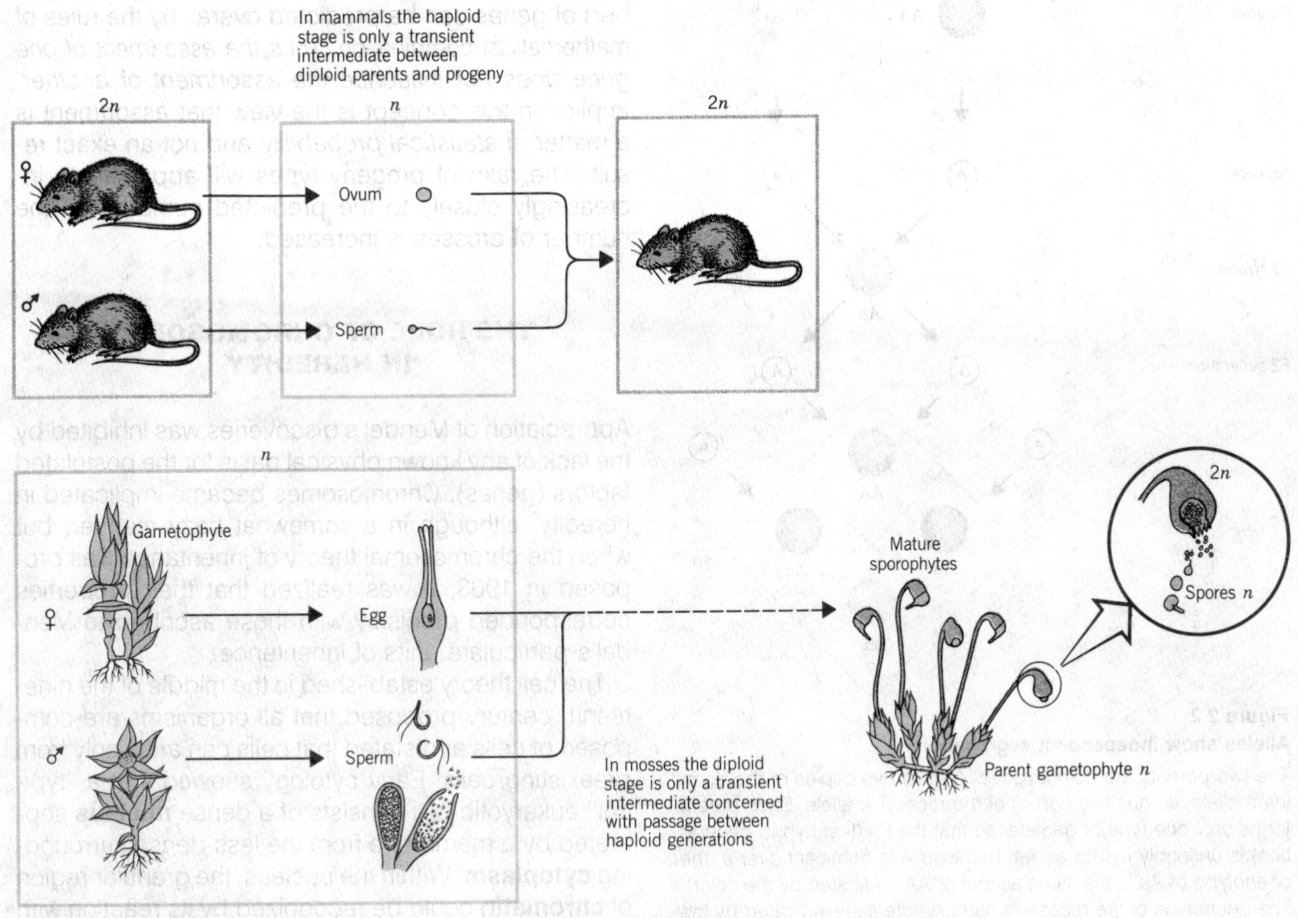

Figure 2.1

When eukaryotes perpetuate their genes through an alternation of diploid and haploid states, either type of state may provide the predominant "adult" type while the other is concerned solely with gamete and zygote formation. The mammals and mosses are extreme examples in which the adult generations are diploid and haploid, respectively.

Red indicates diploid (2n) tissue; grey indicates haploid (n) tissue.

type is the same as the dominant parent. But in the next (F2) hybrid cross, two general classes of progeny are found. One class consists of the two **parental types**. The other class consists of *new* phenotypes, representing plants with the dominant feature of one parent and the recessive feature of the other. These are called **recombinant types;** and they occur in both possible (**reciprocal**) combinations.

Figure 2.3 shows that the ratios of the four phenotypes comprising the F2 can be explained by supposing that gamete formation involves an entirely random association between one of the alleles for the first character and one of the alleles for the second character. All four possible types of gamete are formed in equal proportion; and then they associate at random to form the zygotes of the next generation.

Once again, the characteristic ratio of phenotypes conceals a greater variety of genotypes. The classic technique to confirm this conclusion consists of making a **backcross** to the recessive parent, whose al-

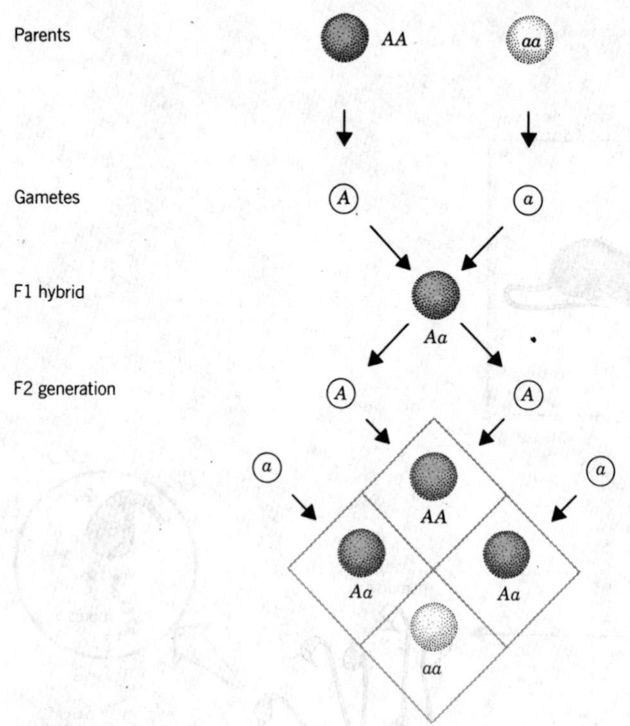

Figure 2.2
Alleles show independent segregation.
The two parents are homozygous. *AA* has two copies of the dominant allele; *aa* has two copies of the recessive allele. Each parent forms only one type of gamete, so that the F1 (first hybrid generation) is uniformly hybrid as *Aa*. Because *A* is dominant over *a*, the *phenotype* of *Aa* is the same as that of *AA* (indicated by the color). The phenotype of the recessive homozygote *aa* is indicated by the lack of color.

The Predictions of Mendel's First Law

Each F1 hybrid forms both *A* and *a* gametes in equal amounts. On mating, these unite randomly to generate an F2 (second hybrid generation) consisting of 1 *AA*: 2 *Aa*: 1 *aa*. Since the *AA* and *Aa* have the same phenotype, this gives the classic 3:1 ratio of dominant : recessive types.

In cases in which the heterozygote *Aa* has a phenotype intermediate between the parental *AA* and *aa*, the F1 would be distinct from either parent; and in the F2 the ratio of phenotypes would be 1 dominant : 2 intermediate : 1 recessive.

leles make no contribution to the phenotype of the progeny. **Figure 2.4** shows that the backcross essentially makes it possible to look directly at the genotype of the organism that is being examined.

The law of independent assortment establishes the principle that the behavior of any pair (or greater number) of genes can be predicted overall by the rules of mathematical combination. *Thus the assortment of one gene does not influence the assortment of another.* Implicit in this concept is the view that assortment is a matter of *statistical probability* and not an exact result. The ratio of progeny types will approximate increasingly closely to the predicted numbers as the number of crosses is increased.

THE ROLE OF CHROMOSOMES IN HEREDITY

Appreciation of Mendel's discoveries was inhibited by the lack of any known physical basis for the postulated factors (genes). Chromosomes became implicated in heredity, although in a somewhat hazy manner, but when the chromosomal theory of inheritance was proposed in 1903, it was realized that their properties corresponded precisely with those ascribed to Mendel's particulate units of inheritance.

The cell theory established in the middle of the nineteenth century proposed that all organisms are composed of cells and stated that cells can arise only from preexisting cells. Early cytology showed that a "typical" eukaryotic cell consists of a dense **nucleus** separated by a membrane from the less-dense surrounding **cytoplasm**. Within the nucleus, the granular region of **chromatin** could be recognized by its reaction with certain stains. Not long after Mendel's work, it was found that the chromatin consists of a discrete number of thread-like particles, the **chromosomes**.

Chromosomes can be visualized in most cells only during the process of cell division. The two types of division in sexually reproducing organisms explain both the perpetuation of the genetic material within an organism and the process of inheritance as predicted by Mendel's laws.

Starting as a fertilized egg, an organism develops through many cell divisions, each division generating two cells from one. The **cell cycle** comprises the period between the release of a cell as one of the progeny of a division and its own subsequent division into two daughter cells.

The cell cycle falls into two parts. A relatively long **interphase** represents the time during which there is no visible change, but the (single) cell engages in its

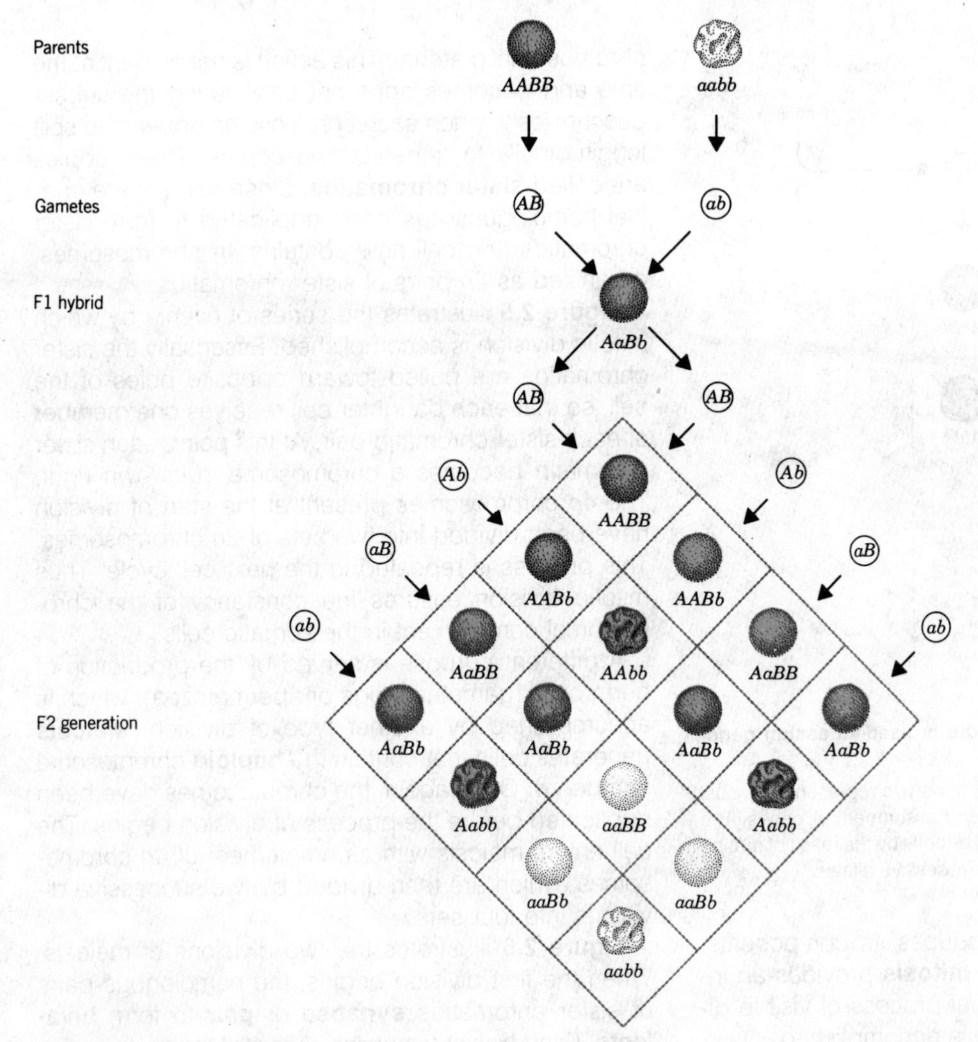

Figure 2.3
Different genes assort independently.
One parent is homozygous for two dominant genes, *A* determining color, and *B* determining shape (shown by round structure). The other parent is homozygous for the recessive alleles *a* and *b* (characteristics shown by no color and wrinkled shape). The F1 is uniform for the dominant characteristics.

The Predictions of Mendel's Second Law

Each F1 plant produces gametes in which alleles segregate independently and genes assort independently, so that equal numbers are produced of each of the four possible types of gamete. The gametes unite randomly to form 9 genotypic classes. Because of the dominance of *A* and *B*, there are only four phenotypic classes: 9 colored-smooth: 3 colored-wrinkled: 3 noncolored-smooth: 1 noncolored-wrinkled.

Note that each reciprocal genotype is present in the same number; for example, the two parents (one each of *AABB* and *aabb*) or any type of recombinant (such as one each of *AAbb* and *aaBB*). The 3:1 ratios are maintained for each individual segregating character.

The number of phenotypic classes will be greater if either or both of the characters are not dominant (so that heterozygotes appear different from either homozygote). It will be less if two genes affect a single characteristic of the phenotype. So if both *A* and *B* were needed to produce color, the F2 would display a ratio of 9 colored : 7 noncolored.

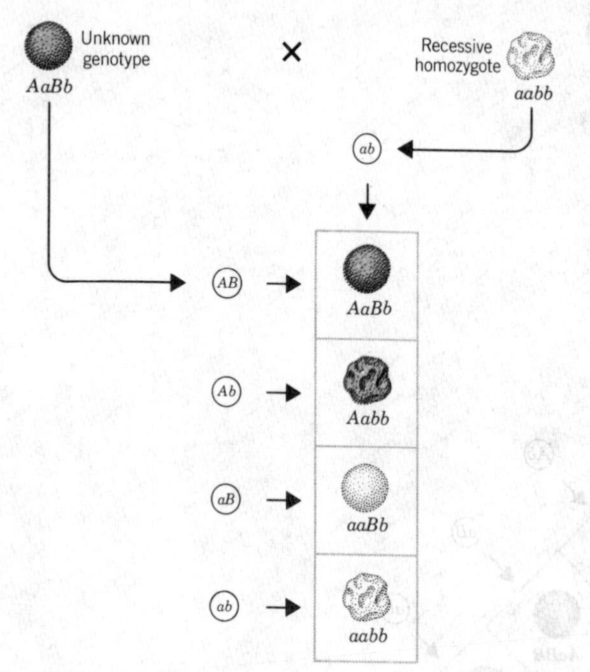

Figure 2.4
Backcross to recessive homozygote is used to assign genotypes.
The gametes formed by an unknown genotype represent the usual independent segregation of alleles and assortment of genes. The genotype of each gamete is revealed directly by the result of mating with a gamete that brings in only the recessive alleles.

synthetic activities and reproduces its components. The relatively short period of **mitosis** provides an interlude during which the actual process of visible division into two daughter cells is accomplished.

The products of the series of mitotic divisions that generate the entire organism are called the **somatic cells**. During embryonic development, many or most of the somatic cells will be proceeding through the cell cycle. In the adult organism, however, many cells are **terminally differentiated** and no longer divide; they remain in a perpetual interphase.

At the end of mitosis, each daughter cell can be seen to start its life with two copies of each chromosome. These copies are called **homologues**. The *total number* of chromosomes is called the **diploid set** and has **2*n*** members. The typical somatic cell exists in the diploid (2*n*) state (except when it is preparing for or is actually engaged in the act of division).

During interphase, a growing cell duplicates its chromosomal material. This action is not evident at the time and becomes apparent only during the subsequent mitosis, when each chromosome appears to split longitudinally to generate two copies. These copies are called **sister chromatids**. Since each of the original homologues has been duplicated to form sister chromatids, the cell now contains 4*n* chromosomes, organized as 2*n* pairs of sister chromatids.

Figure 2.5 illustrates the series of events by which mitotic division is accomplished. Essentially the sister chromatids are pulled toward opposite poles of the cell, so that each daughter cell receives one member of each sister chromatid pair. At this point, each sister chromatid becomes a chromosome in its own right. The 4*n* chromosomes present at the start of division have been divided into two sets of 2*n* chromosomes. This process is repeated in the next cell cycle. Thus mitotic division ensures the constancy of the chromosomal complement in the somatic cells.

A different purpose is served by the production of germ cells (gametes: eggs or spermatozoa), which is accomplished by another type of division. **Meiosis** generates cells that contain the **haploid** chromosome number, ***n***. Once again, the chromosomes have been duplicated before the process of division begins. The cell enters meiosis with a complement of 4*n* chromosomes, which are then divided by two successive divisions into four sets.

Figure 2.6 illustrates the two divisions of meiosis. When the first division begins, the homologous pairs of sister chromatids **synapse** or **pair** to form **bivalents**. *Each bivalent contains all four of the cell's copies of one homologue*. The first division causes each bivalent to segregate into its constituent sister chromatid pairs. This generates two sets of 2*n* chromosomes, each set consisting of *n* sister chromatid pairs.

Now the second meiotic division follows, in which both of the sets of 2*n* divide again. This division is formally like a mitotic division, since one member of each sister chromatid pair segregates to a different daughter cell.

The overall result of meiosis is therefore to divide the starting number of 4*n* chromosomes into four haploid (*n*) cells. These cells may then give rise to mature eggs or sperm.

In forming the gametes, homologues of paternal and maternal origin are separated, so that each gamete

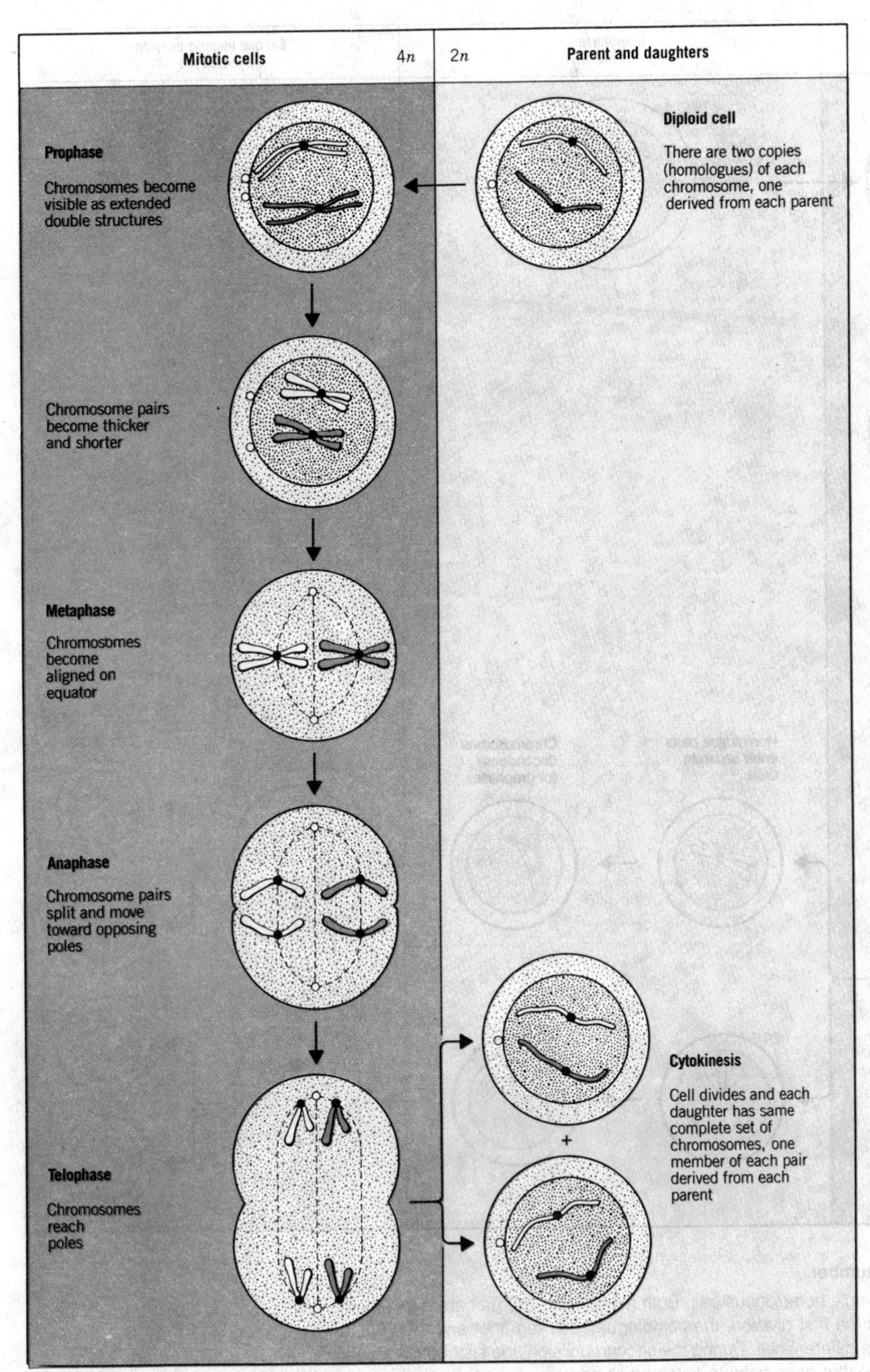

Figure 2.5

Mitosis perpetuates the chromosome constitution of the cell.

The figure shows the behavior of a single pair of homologous chromosomes (actually a eukaryotic cell has many such pairs). The source of each homologue is indicated by its color (the white homologue is derived from one parent, the colored homologue from the other parent). Each chromosome is duplicated before the start of mitosis. During mitosis, the duplicates separate and are segregated into different progeny cells. Each daughter cell has the same complement of chromosomes as the parent cell.

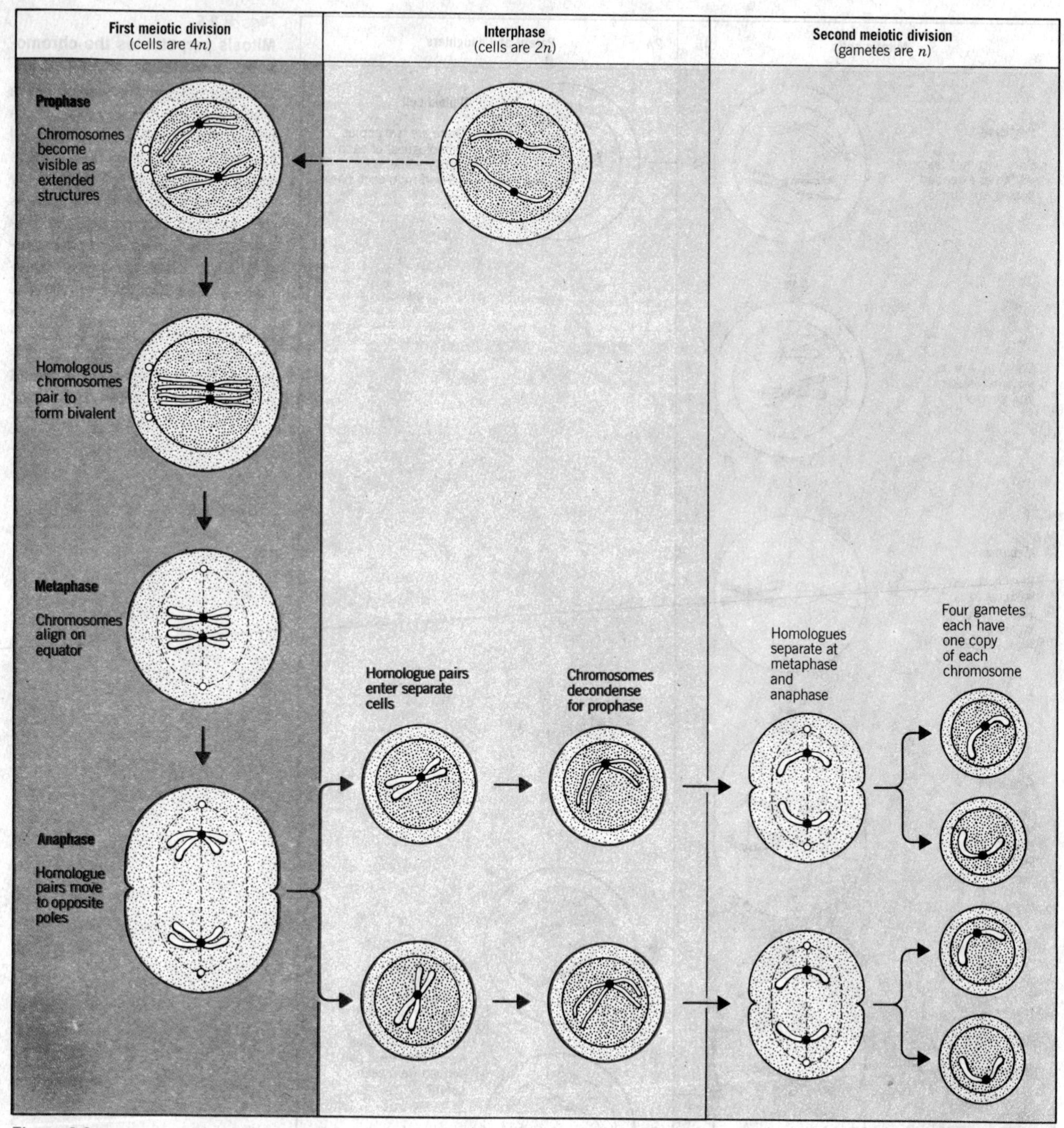

Figure 2.6

Meiosis halves the chromosome number.

The figure shows the behavior of a single homologous pair. Both members of the pair are duplicated before the start of prophase. During the first division, the homologues pair together and then each pair of sister chromatids moves into a different cell. During the second division, the sister chromatids are segregated into different cells, so that each gamete obtains one copy.

gains only *one of the two homologues* of its parent. A critical feature in relating this process to the predictions of Mendel's laws was the realization that *nonhomologous* chromosomes segregate *independently*, so that either member of one homologous pair enters the gamete at random with either member of a different homologous pair.

To summarize the parallels between chromosomes and Mendel's units of inheritance: Genes occur in allelic pairs, one member of each pair having been contributed by each parent; so the diploid set of chromosomes results from the contribution of a haploid set by each parent. The assortment of nonallelic genes into gametes should be independent of origin; correspondingly, nonhomologous chromosomes undergo independent segregation. The critical proviso is that each gamete obtains a complete haploid set, and this condition is fulfilled whether viewed in terms of Mendel's factors or chromosomes.

EACH GENE LIES ON A SPECIFIC CHROMOSOME

The correlation between the behavior of genes and chromosomes immediately suggests a further question: can we identify particular genes with particular chromosomes? Proof that a specific gene always lies on a certain chromosome was provided by the properties displayed by a mutant of the fruit fly *Drosophila melanogaster* obtained by Morgan in 1910. This white-eyed male appeared spontaneously in a line of flies of the usual (wild-type) red eye color.

The *white* mutation could be located on a particular chromosome because of its association with sexual type. In many sexually reproducing organisms, there is an exception to the rule that chromosomes occur in homologous pairs whose separation (disjunction) at meiosis produces identical haploid sets. Male and female sets may differ visibly in chromosome constitution, the most common form of difference being the replacement of one member of a homologue pair with a different chromosome in *one* of the sexes.

This pair is referred to as the **sex chromosomes**, and the remaining homologous pairs are called the **autosomes**. The chromosome complements of the two sexes can be described as $2A + XX$ and $2A + XY$, where A represents the haploid set of autosomes and X and Y are the individual sex chromosomes. The sex with the (homogametic) complement $2A + XX$ forms gametes only of the type $A + X$. The sex with the (heterogametic) complement of $2A + XY$ forms equal proportions of gametes of the types $A + X$ and $A + Y$. The union of gametes from one sex with gametes from the other sex perpetuates the equal sex ratio at zygote formation. In *Drosophila*, the homogametic sex happens to be the female.

A critical prediction of Mendel's laws is that the results of a genetic cross should be the same regardless of orientation—that is, irrespective of which parent introduces which allele. But the reciprocal crosses with white eye in *Drosophila* give different results, as shown in **Figure 2.7**.

The cross of white male x red female gives the entirely red-eyed F1 expected if red is dominant and white is recessive. But all the white-eyed flies that reappear in the F2 are *males*. In the reciprocal cross of red male x white female, all the F1 males are white-eyed and all the females are red-eyed. Crossing these flies gives an F2 with equal proportions of white and red eyes in each sex.

This pattern of inheritance exactly follows that of the sex chromosomes. If the alleles for red and white eyes reside on the X chromosome, the phenotype of a female will be determined in the usual way by the alleles present on the two X chromosomes. However, if there is no locus for eye color present on the Y chromosome, the phenotype of a male will be determined by the single allele present on its X chromosome. This allele will be transmitted to all of its daughters and none of its sons, constituting the typical pattern of **sex linkage**.

GENES LIE IN A LINEAR ARRAY

The independent segregation of chromosomes at meiosis explains the independent assortment of genes that are carried on different chromosomes. But the number of genetic factors is much greater than the number of chromosomes. If all genes lie on chromosomes, many genes must be present on each chromosome. What is the relationship between these genes?

Soon after the discovery of *white* eye, several other

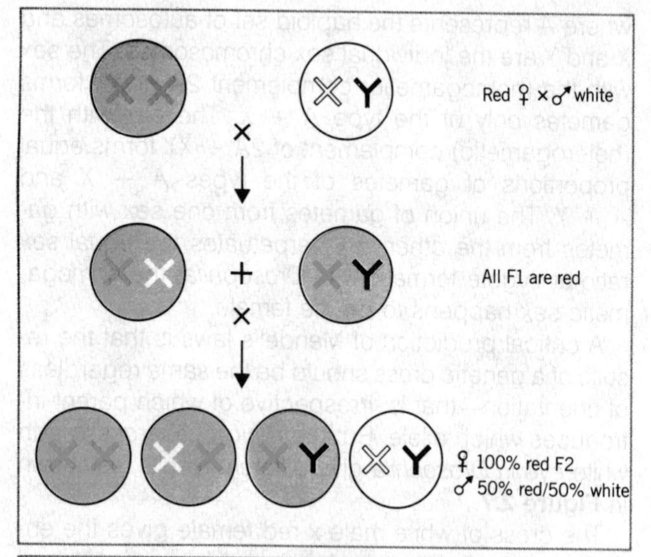

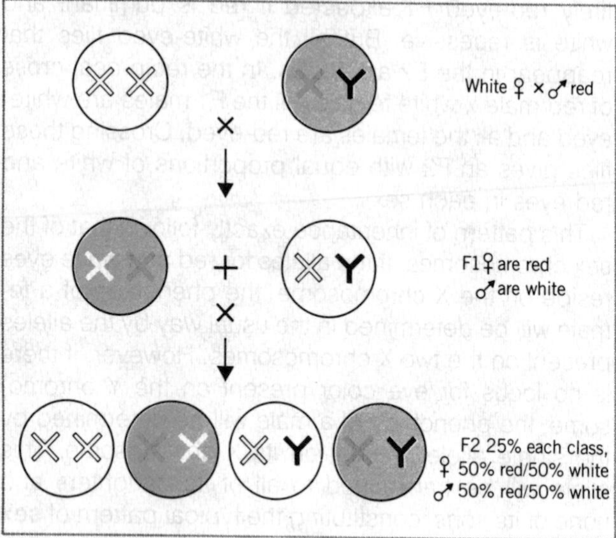

Indicates chromosome with red allele (dominant)

Indicates chromosome with white allele (recessive)

Chromosome has no allele (=recessive)

Red eye fly White eye fly

Figure 2.7
Genes on the X chromosome show sex-linked inheritance.
Red/white eye color of a male fly depends only on the X chromosome received from its mother. The phenotype of the female is determined by whether it receives a dominant allele from *either* parent. This generates the characteristic 'crisscross' pattern of sex-linked inheritance.

factors were found to show the same sex-linked pattern. By the same logic, each of these genes must reside on the X chromosome. When their inheritance is followed in pairs, their behavior deviates from the predictions of Mendel's second law. Instead of generating the proportions depicted in Figure 2.3, the proportion of parental genotypes in the F2 is greater than expected, *because there is a reduction in the formation of recombinant genotypes*. The propensity of some characters to remain associated instead of assorting independently is given the name **linkage**.

Figure 2.8 shows that linkage is measured by the proportion of recombinants in the progeny; the smaller the number, the tighter the linkage. Morgan proposed that the cause of genetic linkage is the "simple mechanical result of the location of the (genes) in the chromosomes." He suggested that the production of recombinant classes can be equated with the process of **crossing-over** that is visible during meiosis. Early in meiosis, at the stage when all four copies of each chromosome are organized in a bivalent, pairwise exchanges of material occur between the closely associated (synapsed) homologue pairs. This exchange is called a **chiasma**; it is illustrated diagramatically in **Figure 2.9**.

A chiasma represents a site at which two of the chromatids in a bivalent have been broken at corresponding points. The broken ends have been rejoined crosswise, generating new chromatids each of which consists of material derived from one chromatid on one side of the junction point, with material of the other chromatid on the opposite side. The two recombinant chromatids have reciprocal structures. The mechanism is described as **breakage and reunion**.

Formal proof that crossing-over is responsible for recombination was not obtained until some years after the original suggestion. In maize and in *Drosophila*, suitable **translocations** had occurred in which part of one chromosome had broken off and become attached to another. The translocation chromosome can therefore be distinguished by its appearance from the normal chromosome. In suitable crosses, as indicated in **Figure 2.10**, it is possible to show that the formation of genetic recombinants occurs only when there has been a physical crossing-over between the appropriate chromosome regions.

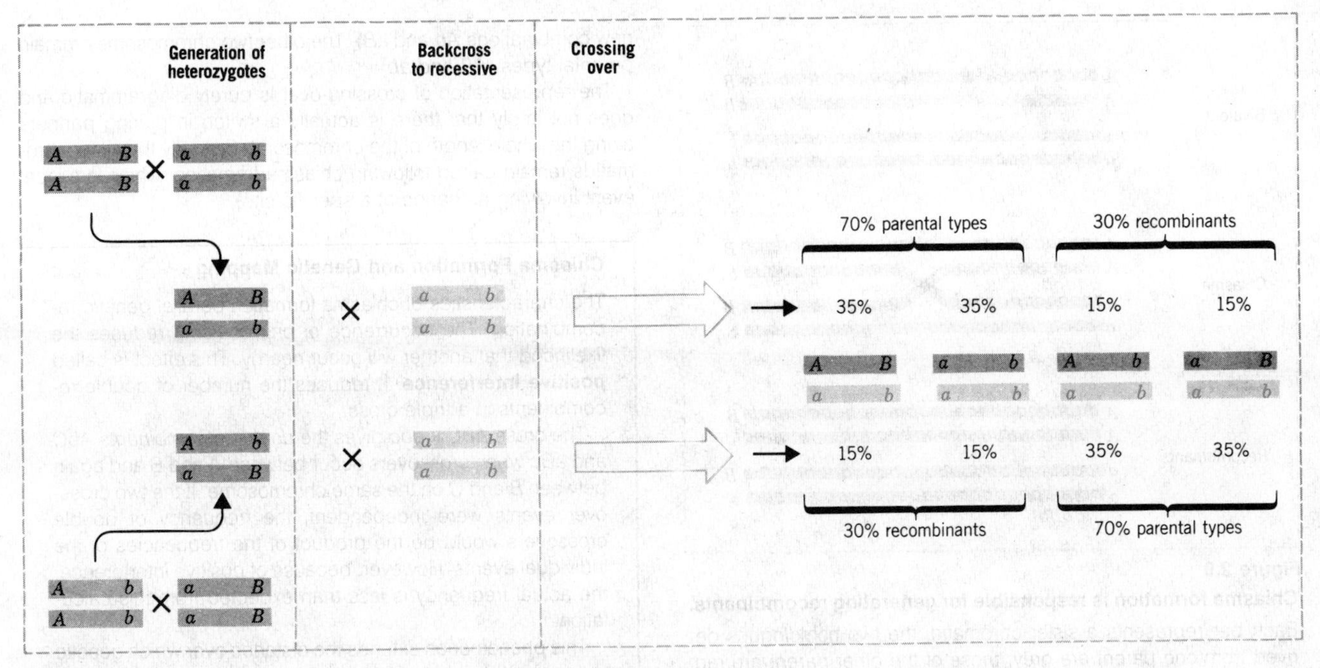

Figure 2.8
Linkage can be measured by a backcross with a double homozygote.
The chromosomes are indicated by shaded bars. Black represents material from the heterozygote; color represents material from the double homozygote.

In the upper cross, one chromosome of the heterozygote carries both dominant alleles; the other carries both recessive alleles. Thus the parental types are *AB* and *ab*. The recombinant types are *Ab* and *aB*.

In the lower cross, one chromosome of the heterozygote carries one dominant and one recessive allele; the other chromosome carries the reverse combination. This makes *Ab* and *aB* the parental types; the recombinants are *AB* and *ab*.

In each cross the progeny contain an increase in the proportion of parental types (70%) and a decrease in the proportion of recombinant types (30%) compared with the 50% of each type that is expected from independent assortment. Note that both parental types are present in the same amount, and both recombinant types are present in the same amount. The linkage between *A* and *B* is measured as 30% or 30 map units.

If the likelihood that a chiasma will form between two points on a chromosome depends on their distance apart, genes located near each other will tend to remain together. As the distance increases, the probability of crossing-over between them will increase. Thus if crossing-over is responsible for recombination, the closer genes lie to one another, the more tightly they will be linked. Reversing the argument, genetic linkage can be taken to be a measure of physical distance.

The concept that genes on the same chromosome are linked was extended by Sturtevant with the proposal that the extent of recombination between them can be used as a **map distance** to measure their relative locations. This is expressed as the percent recombination; that is,

$$\text{Map distance} = \frac{\text{Number of recombinants} \times 100}{\text{Total number of progeny}}$$

Distance measured in this way is given as **map units**, defined by 1 map unit (or centiMorgan) equals 1% recombination.

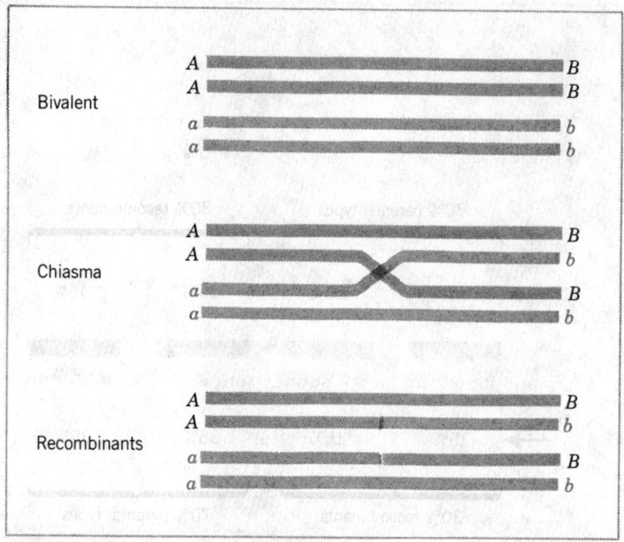

Figure 2.9

Chiasma formation is responsible for generating recombinants.

Each bar represents a sister chromatid; the two homologues derived from one parent are grey, those of the other parent are red. They associate to form a bivalent at the meiotic prophase. (Because the bivalent contains four sister chromatids, this is sometimes referred to as the "four strand stage" of meiosis.)

Crossing over occurs between two of the sister chromatids, when both are broken at the same site, and the opposite ends are joined. This generates the two recombinant chromosomes (carrying the new combinations *Ab* and *aB*). The other two chromosomes remain parental types (*AB* and *ab*).

The representation of crossing-over is purely diagrammatic and does not imply that there is actually a switch in pairing partners along the whole length of the chromosome (actually the sister chromatids remain paired following chiasma formation, which is a local event involving a change at a specific site).

Chiasma Formation and Genetic Mapping

The characteristics of chiasma formation parallel genetic recombination. The occurrence of one chiasma reduces the likelihood that another will occur nearby. This effect is called **positive interference**. It reduces the number of double recombinants in a triple cross.

The cross *ABC* x *abc* gives the *double recombinants AbC* and *aBc* when crossovers occur between *A* and *B* and again between *B* and *C* on the same chromosome. If the two crossover events were independent, the frequency of double crossovers would be the product of the frequencies of the individual events. However, because of positive interference, the actual frequency is less than expected from this calculation.

This phenomenon extends the distance over which genetic maps can be constructed by direct linkage, because it reduces the actual recombination frequencies between distant markers.

When adjacent crossovers occur, which two of the four sister chromatids are involved in one crossover is independent of which two are involved in the other crossover.

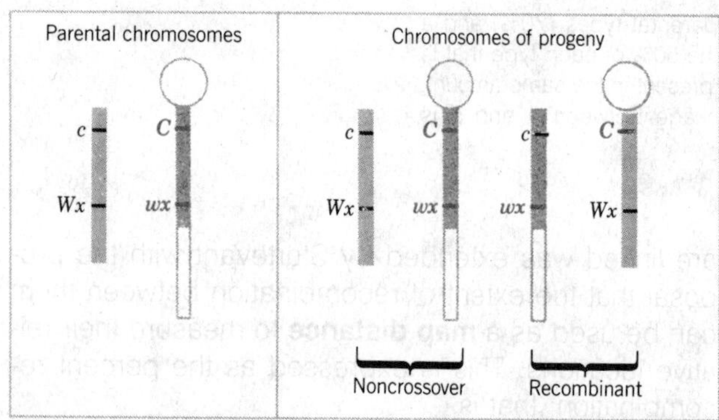

Figure 2.10

Genetic recombination is caused by physical crossing-over.

In a mutant of maize, the chromosome carrying the alleles *C* and *wx* has gained a "knob" at one end, while the other end has a long translocation of material from another chromosome. In a heterozygote with a normal chromosome carrying the alleles *c* and *Wx*, recombination between genetic markers is always associated with the formation of a new type of chromosome.

A critical feature was observed when the six sex-linked characters in *D. melanogaster* were followed together. *Individual map distances are additive.* Thus if two genes *A* and *B* are 10 units apart, and gene *C* lies a further 5 units beyond *B*, the direct measure of distance between *A* and *C* will be close to 15. The genes can therefore be placed in a **linear order**.

A crucial concept in the construction of a genetic map is that the distance between genes does not depend on the particular *alleles* that are used, but only on the gene **loci**. The locus defines the *position* occupied on the chromosome by the gene representing a particular trait. *The various alternate forms of the gene*—that is, the alleles used in mapping—*all reside at the same location.*

So genetic mapping is concerned with identifying the positions of gene loci, which are fixed and lie in a linear order. In a mapping experiment, the same result is obtained irrespective of the particular combination of alleles (see Figure 2.8). Sturtevant concluded that his results "form a new argument in favor of the chromosome view of inheritance, since they strongly indicate that the factors investigated are arranged in a linear series, at least mathematically."

This last qualification is interesting. Although a genetic map can be constructed to represent a chromosome as a linear array of genes, this does not prove that the genetic content of the chromosome is *physically* a continuous array of genes. Many other models were considered before it became clear that a chromosome contains a single thread of genetic material.

Linkage is not displayed between all pairs of genes located on a single chromosome. The maximum recombination that occurs between two loci is the 50% that corresponds to the independent segregation predicted by Mendel's second law (see Figure 2.3). (Although there is a high probability that recombination will occur between two genes lying far apart on a chromosome, each individual recombination event involves only two of the four associated chromatids, thus generating 50% crossover between the genes.)

So in spite of their presence on the same chromosome, genes that are well separated may assort independently. However, although they show no direct linkage, each can be linked to genes that lie between them, and so a genetic map can be extended beyond the limit of 50 map units directly measurable between any pair of genes. In fact, the 50 map unit limit does not come into play abruptly, but is reached as a plateau after linkage measurements have become distorted between relatively distant genes. Thus a genetic map is usually based on measurements involving fairly close genes, and may be subject to corrections from the simple percent recombination.

The **linkage group** includes all those genes that can be connected either directly or indirectly by linkage relationships. Genes lying close together show direct linkage; those more than 50 units apart in practice assort independently. As linkage relationships are extended, the genes of any organism fall into a discrete number of linkage groups. Each gene identified in the organism can be placed into one of the linkage groups. Genes in one linkage group always show independent assortment with regard to genes located in other linkage groups.

The number of linkage groups is the same as the number of chromosomes. The relative lengths of the linkage groups are similar to the actual relative sizes of the chromosomes. The example of *D. melanogaster* (where it happens to be particularly easy to measure the relative chromosome lengths) is illustrated in **Figure 2.11**.

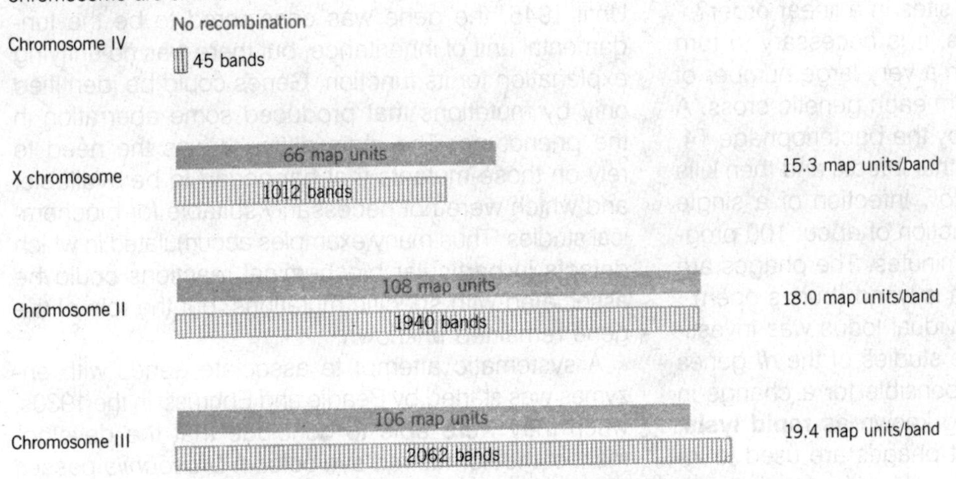

Figure 2.11
The linkage map can be related to the physical size of the chromosomes.
Each chromosome in *D. melanogaster* can be visualized as a number of "bands." The relative number of bands reflects the physical length of each chromosome. The relative numbers of map units per chromosome are similar.

Mendel's concept of the gene as a discrete particulate factor can therefore be extended into the concept that the chromosome constitutes a linear unit, divided into many genes, whose physical arrangement may underlie their genetic behavior.

THE GENETIC MAP IS CONTINUOUS

On the genetic maps of higher organisms established during the first half of this century, the genes are arranged like beads on a string. They occur in a fixed order, and genetic recombination involves the transfer of corresponding portions of the string between homologous chromosomes. The gene is to all intents and purposes a mysterious object (the bead), whose relationship to its surroundings (the string) is not clear.

The resolution of the recombination map of a higher eukaryote is restricted by the small number of progeny that can be obtained from each mating. Recombination occurs so infrequently between nearby points that it is rarely observed between different mutations in the same gene. This poses a general question: is the structure of the gene the same as the structure of the regions between genes?

In biochemical terms, we approach this issue by considering the structure of DNA. A chromosome consists of a length of DNA whose secondary structure is the same whether it is part of a gene or part of a region between genes. Even given this knowledge, in genetic terms we still need to ask: can recombination occur within a gene and can its frequency at these close quarters be used to arrange sites in a linear order?

To answer these questions, it is necessary to turn to microbial systems in which a very large number of progeny can be obtained from each genetic cross. A suitable system is provided by the bacteriophage T4, a virus (related to phage T2) that infects and then kills the bacterium *Escherichia coli*. Infection of a single bacterium leads to the production of about 100 progeny phages in less than 30 minutes. The phages are released when the bacterium is lysed (bursts open).

The constitution of an individual locus was investigated in a series of intensive studies of the *rII* genes of the phage, which are responsible for a change in the pattern of bacterial killing known as **rapid lysis**. When two different *rII* mutant phages are used to infect a bacterium simultaneously, it is possible to arrange the conditions so that progeny phages will be produced *only* if recombination has occurred between the two mutations to generate a wild-type recombinant. The frequency of recombination depends on the distance between sites, just as in the eukaryotic chromosome.

The selective power of this technique in distinguishing recombinants of the desired type allows even the rarest recombination events to be quantitated, so that the map distance between *any* pair of mutations can be measured. (A limit is set by the frequency at which mutants "revert" spontaneously to wild type. The reversion frequency is about 0.0001%, so to achieve a minimum level of measurement 10-fold above this "background," only recombination frequencies greater than 0.001% can be assayed.)

About 2400 mutations were analyzed by Benzer, and they identified 304 different mutant sites. (When two mutations failed to recombine, they were assumed to represent independent and spontaneous occurrences at the *same* genetic site.) The mutations could be arranged into a linear order, showing that the gene itself has the same linear construction as the array of genes on a chromosome. Thus the genetic map is linear within as well as between loci; that is, it consists of an unbroken sequence within which the genes reside.

ONE GENE—ONE PROTEIN

Until 1945, the gene was considered to be the fundamental unit of inheritance, but there was no unifying explanation for its function. Genes could be identified only by mutations that produced some aberration in the phenotype. The main difficulty was the need to rely on those mutants that happened to be available, and which were not necessarily suitable for biochemical studies. Thus many examples accumulated in which defects in particular biochemical reactions could be associated with specific mutations, but the role of the gene remained unknown.

A systematic attempt to associate genes with enzymes was started by Beadle and Ephrussi in the 1930s, when they were able to conclude that the development of the normal red eye color in *Drosophila* passes through a discrete series of stages. Blockage at different stages results in the production of different mutant colors. But it was not until much later that the complete pathway could be worked out. Beadle and Tatum thus attempted to approach the problem from the other direction. As Beadle recollected, "it suddenly occurred to me that it ought to be possible to reverse the procedure we had been following and instead of attempting to work out the chemistry of known genetic differences we should be able to select mutants in which known chemical reactions were blocked. *Neurospora* was an obvious organism on which to try this approach, for it probably could be grown in a culture medium of known composition."

So mutants of the fungus *Neurospora crassa* were generated (by irradiation with X-rays) and selected for their inability to grow on a minimal medium. The biochemical nature of the defect could be identified by finding which growth factor allowed the mutant strain to grow. Each mutant proved to be blocked in a particular metabolic step, undertaken in the wild type strain by a single enzyme. Blockage at each step leads to accumulation of the metabolic intermediate immediately prior to the step. **Figure 2.12** illustrates the pathway subsequently elucidated for production of brown eye pigment in *D. melanogaster*, in which blockage at different steps causes different phenotypes that depend on the metabolic intermediate that accumulates.

By 1945 the results of the analysis had become known in common parlance as the **one gene : one enzyme hypothesis**. This proposed that each metabolic step is catalyzed by a particular enzyme, whose production is the responsibility of a single gene. A mutation in the gene may alter the activity of the protein for which it is responsible. Since a mutation is a random event with regard to the structure of the gene, the greatest probability is that it will damage gene function. Thus the majority of mutations create a nonfunctional gene; but this is by no means the only consequence of mutation, which sometimes alters rather than abolishes gene function.

These ideas explain the nature of recessive mutations: they represent an absence of function, because the mutant gene has been prevented from producing its usual enzyme. As illustrated in **Figure 2.13**, however, in a heterozygote containing one wild type and one mutant gene, the wild-type gene is able to direct

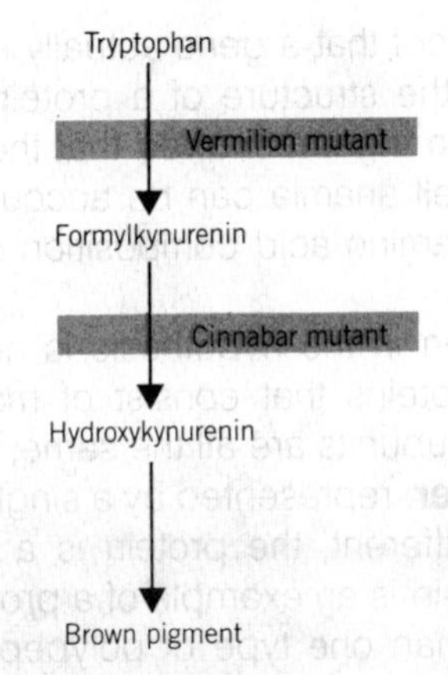

Figure 2.12
Genes control metabolic steps.
Some mutations affecting eye color of *D. melanogaster* act by blocking different stages in the pathway for converting tryptophan to brown pigment. Each mutation results in the absence of a particular enzyme, causing accumulation of the intermediate on which it acts. Eye color is influenced by the effect (or lack thereof) of this intermediate.

production of the enzyme. The wild type gene is therefore dominant. (This assumes that an adequate *amount* of protein is made by the single wild-type allele. When this is not true, the smaller amount made by one allele as compared to two may result in the intermediate phenotype of a partially dominant or codominant wild-type gene.)

Figure 2.13
Genes code for proteins.
In the wild type, both alleles (solid bars) are active and produce protein. In the heterozygote, the dominant allele (solid bar) is active and produces protein, but the recessive allele (open bar) does not prouce any active protein. In a homozygote with two recessive alleles, no protein is produced; so the organism lacks the function.

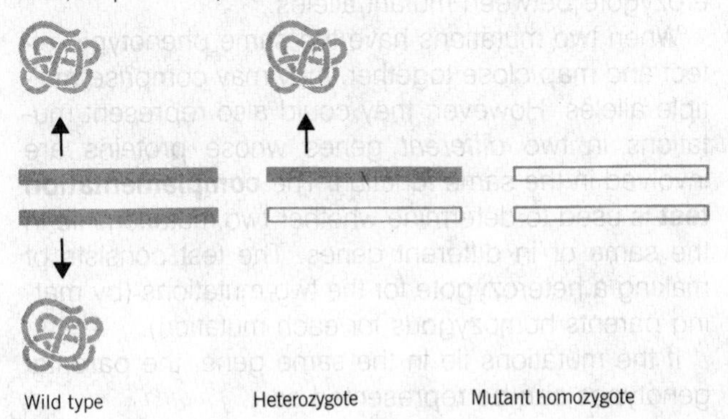

The direct proof that a gene actually is responsible for controlling the structure of a protein had to wait until 1957, when Ingram showed that the single-gene trait of sickle-cell anemia can be accounted for by a change in the amino acid composition of the protein hemoglobin.

A modification in the hypothesis is needed to accommodate proteins that consist of more than one subunit. If the subunits are all the same, the protein is a **homomultimer**, represented by a single gene. If the subunits are different, the protein is a **heteromultimer**. Hemoglobin is an example of a protein that consists of more than one type of polypeptide chain; a heme (iron-binding) group is associated with two α subunits and two β subunits. Each type of subunit comprises a different polypeptide chain and is represented by its own gene. Thus the function of hemoglobin may be inhibited by mutation in the genes coding for either the α or β polypeptides.

Stated as a more general rule applicable to any heteromultimeric protein, the one gene : one enzyme hypothesis becomes more precisely expressed as **one gene : one polypeptide chain**.

A DETAILED DEFINITION: THE CISTRON

If a recessive mutation is produced by every change in a gene that prevents the production of an active protein, there should be a large number of such mutations in any one gene; many amino acid replacements may be able to change the structure of the protein sufficiently to impede its function. Different variants of the same gene are called **multiple alleles**, and their existence makes it possible to have a heterozygote between mutant alleles.

When two mutations have the same phenotypic effect and map close together, they may comprise multiple alleles. However, they could also represent mutations in two *different* genes whose proteins are involved in the same function. The **complementation test** is used to determine whether two mutations lie in the same or in different genes. The test consists of making a heterozygote for the two mutations (by mating parents homozygous for each mutation).

If the mutations lie in the same gene, the parental genotypes can be represented as

$$\frac{m_1}{m_1} \text{ and } \frac{m_2}{m_2}$$

The first parent provides an m_1 mutant allele and the second parent provides an m_2 allele, so that the heterozygote has the constitution

$$\frac{m_1}{m_2}$$

in which *no wild type gene is present*.

If the mutations lie in different genes, the parental genotypes can be represented as

$$\frac{m_1\ +}{m_1\ +} \text{ and } \frac{+\ m_2}{+\ m_2}$$

where each has a wild type copy of one gene (represented by the plus sign) and a mutant copy of the other. Then the heterozygote has the constitution

$$\frac{m_1\ +}{+\ m_2}$$

in which the two parents between them have provided a wild-type copy of each gene. The heterozygote has wild phenotype; the two genes are said to **complement**.

Figure 2.14 provides a more elaborate description of the complementation test. If we consider just the individual sites of mutation (without regard to whether they lie in the same or in different genes), the double heterozygote may have either of two configurations. In the **cis** configuration, both mutations are present on the *same* chromosome. In the **trans** configuration, they are present on opposite chromosomes. The relative effects of these configurations are determined by whether the mutations lie in the same or in different genes.

First consider the situation in which the mutations lie in the same gene. The *trans* configuration corresponds to the test we have just described. Both copies of the gene are mutant. In the *cis* configuration, however, one genome elaborates a protein that has two mutations, while the other has none and is therefore wild type. Thus when two mutations lie in the same gene, the phenotype of a heterozygote is determined by the configuration. It is mutant when the mutations lie in *trans* and wild type when they lie in *cis*.

By contrast, when the mutations lie in different genes, the configuration is irrelevant. In either case there is

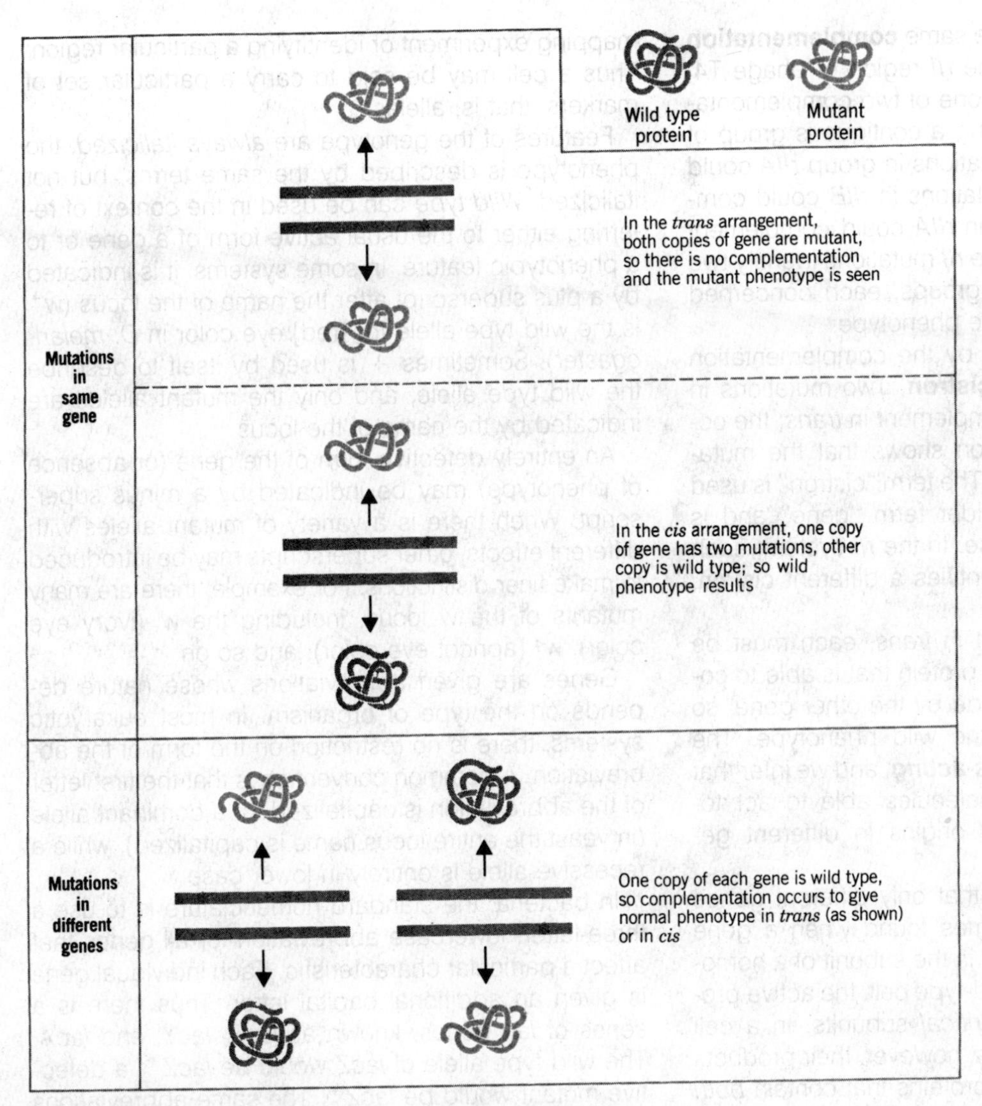

Figure 2.14
The cistron is defined by the *cis/trans* complementation test.
Genes are represented by bars; asterisks identify sites of mutation.

one copy of each mutant gene and one copy of each wild type gene.

Thus when two mutations are tested in the *cis* configuration, the same result occurs whether they lie in the same or in different genes. So the *cis* configuration is used as a control to demonstrate the presence of wild-type function (and against which the failure of complementation in *trans* can be judged). This gives rise to the full name of the procedure as the **cis/trans complementation test**.

Complementation is tested in practice by determining whether the *trans* heterozygote shows wild phenotype (the mutations lie in different genes) or mutant phenotype (the mutations lie in the same gene). For eukaryotes the experiment is performed by constructing the appropriate double heterozygote. For viruses a host cell is simultaneously infected with the two mutant types.

When two mutations *fail* to complement in *trans*, the inference is that both affect the same function. They are therefore assigned to the same **complementation group**. In the analysis of the *rII* region of phage T4, all of the mutations fell into one of two complementation groups, each constituting a contiguous group of mutant sites. No pair of mutations in group *rIIA* could complement, no pair of mutations in *rIIB* could complement; but any mutation in *rIIA* could complement any mutation in *rIIB*. Thus the *rII* mutations identify two adjacent complementation groups, each concerned with the same function of the phenotype.

The genetic unit defined by the complementation test is formally called the **cistron**. Two mutations in the same cistron *cannot* complement in *trans*; the occurrence of complementation shows that the mutations lie in *different* cistrons. The term "cistron" is used interchangeably with the older term "gene" and is generally passing into disuse. In the *rII* system, each complementation group identifies a different cistron, or gene.

If two genes complement in *trans*, each must be responsible for producing a protein that is able to cooperate with the protein made by the other gene, so that together they create the wild phenotype. The products are said to be ***trans*-acting**, and we infer that they represent diffusible molecules able to act together irrespective of their origins in different genomes.

An exception to the rule that only different genes can complement is sometimes found when a gene codes for a polypeptide that is the subunit of a homomultimeric protein. In the wild-type cell, the active protein consists of several *identical* subunits. In a cell containing two mutant alleles, however, their products can mix to form multimeric proteins that contain *both types* of subunit. Sometimes the two mutations compensate, so that the mixed-subunit protein is active, even though the proteins consisting solely of either type of mutant subunit are inactive. This effect is called **interallelic** complementation.

A NOTE ABOUT TERMINOLOGY

The terminology used to describe genetic loci and their effects varies somewhat between different organisms, but there are some common terms and features that we will depend on.

Genetic marker is sometimes used to describe a gene of interest, for example, one being used in a mapping experiment or identifying a particular region. Thus a cell may be said to carry a particular set of markers, that is, alleles.

Features of the genotype are *always italicized*; the phenotype is described by the same terms, but not italicized. *Wild type* can be used in the context of referring either to the usual active form of a gene or to a phenotypic feature. In some systems, it is indicated by a plus superscript after the name of the locus (w^+ is the wild type allele for [red] eye color in *D. melanogaster*). Sometimes + is used by itself to describe the wild type allele, and only the mutant alleles are indicated by the name of the locus.

An entirely defective form of the gene (or absence of phenotype) may be indicated by a minus superscript. When there is a variety of mutant alleles with different effects, other superscripts may be introduced to make finer distinctions. For example, there are many mutants of the *w* locus, including the w^i (ivory eye color), w^a (apricot eye color), and so on.

Genes are given abbreviations whose nature depends on the type of organism. In most eukaryotic systems, there is no restriction on the form of the abbreviation. A common convention is that the first letter of the abbreviation is capitalized for a dominant allele (in yeast the entire locus name is capitalized), while a recessive allele is entirely in lower case.

In bacteria, the standard nomenclature is to use a three-letter, lowercase abbreviation for all genes that affect a particular characteristic. Each individual gene is given an additional capital letter. Thus there is a series of *lac* genes, known as *lacZ*, *lacY*, and *lacA*. The wild type allele of *lacZ* would be $lacZ^+$; a defective mutant would be $lacZ^-$. The same abbreviations are used to describe the phenotype, but the first letter is capitalized. Thus wild-type or mutant *lac* alleles would give rise to the respective bacterial phenotypes of Lac^+ or Lac^-. The product of a gene is a feature of the phenotype and is referred to accordingly; for example, the LacZ protein is the enzyme β-galactosidase.

FURTHER READING

The modern view of the gene/cistron started with **Benzer's** paper (*Proc. Nat. Acad. Sci. USA* **41**, 344, 1955). The work with bacteriophages is reviewed in the edited volume by **Cairns, Stent & Watson**, *Phage and the Origins of Molecular Biology* (Cold Spring Harbor Laboratory, New York, 1966).

CHAPTER 3
MUTATIONS CHANGE THE SEQUENCE OF DNA

The concept that the gametes provide the *information* for development of the embryo rather than comprising miniature versions of the body parts actually was introduced by Aristotle 2500 years ago (although he appreciated the significance only of the egg, not of the sperm). It became possible to use this idea as a principle of modern work only when Mendel's analysis showed that genes are particulate factors. The next step was to show that each gene represents a particular polypeptide chain. This faces us with the issue of how a sequence of nucleotides in DNA represents a sequence of amino acids in protein.

A crucial feature of the structure of DNA is that *it is independent of the particular sequence of base pairs*. Thus the sequence of bases in the polynucleotide chain is important not in the sense of structure per se, but because it **codes** for the sequence of amino acids that constitutes the corresponding polypeptide. The concept that each protein consists of a particular series of amino acids dates from Sanger's characterization of insulin in the 1950s. The relationship between a sequence of DNA and the sequence of the corresponding protein is called the **genetic code.**

The structure and/or enzymatic activity of each protein follows from its primary sequence of amino acids. Thus by determining the sequence of amino acids in each protein, the gene is able to carry all the information needed to specify an active polypeptide chain. In this way, a single type of structure—the gene—is able to represent itself in innumerable polypeptide forms.

Together the various protein products of a cell undertake the catalytic and structural activities that are responsible for establishing its phenotype. Of course, in addition to gene sequences that code for proteins, DNA also contains certain sequences whose function is to be recognized by a regulator molecule, usually a protein. Here the function of the DNA is determined by its sequence directly, not via any intermediary code. Both types of sequence, genes expressed as proteins and regions recognized as such, constitute the genetic information of the organism.

The isolation of mutants has been the pivot of the analysis of the gene. Until very recently, genes have been identifiable only by characterizing mutations that either are lethal or at least block the development of some visible or otherwise measurable feature of the phenotype. The mapping of mutations leads to the central concept that a group of mutations all affecting the same characteristic may lie close together; and the complementation test shows that each such group constitutes a functional genetic unit. What is the nature of these mutations?

From the discovery that DNA is the genetic material,

the concept that a mutation is a change in the sequence of base pairs follows naturally. Through the genetic code, a change in the nucleotide sequence may lead to a change in the amino acid sequence, thus altering or abolishing the activity of the protein.

THE GENETIC CODE IS READ IN TRIPLETS

A nucleotide sequence must contain sufficient coding units to represent 20 different amino acid residues. But there are only 4 bases in DNA. This implies that the **coding ratio** (the number of bases required to specify each amino acid) must be greater than 1. If 2 bases coded for each amino acid, the DNA could specify 16 (i.e., 4^2) different amino acids. Since this is inadequate, the coding ratio must be at least 3.

If the genetic code is triplet, three adjacent bases must represent each amino acid. Since there are 64 possible triplet combinations (i.e., 4^3), a triplet code implies that either only some of the triplets actually specify amino acids or some amino acids must be specified by more than one triplet combination.

Our first ideas about the nature of the genetic code envisaged some form of direct relationship between the nucleic acid and protein sequences, such as a stereochemical fit between an amino acid and a nucleotide triplet. Such ideas still form the basis of some proposals for the origin of the genetic code, but in its current operation the code is deciphered by a complex apparatus that stands between nucleic acid and protein (and which is therefore necessary if the information carried in DNA is to have meaning). Only one of the two strands of DNA codes for protein, so we write the genetic code as a sequence of bases (rather than base pairs).

The genetic code is read in groups of three nucleotides, each group representing one amino acid. Each trinucleotide sequence is called a **codon.** A gene includes a series of codons that is read in series from a starting point at one end to a termination point at the other end. Written in the conventional 5′–3′ direction, the nucleotide sequence of the DNA strand that codes for protein corresponds to the amino acid sequence of the protein written in the direction from N-terminus to C-terminus.

The general basis of the code was discovered by genetic analysis of mutants of the *rII* region of phage T4 (the same region used previously to analyze the complementation group). In 1961, Crick and his colleagues showed that the code must be read in *nonoverlapping triplets from a fixed starting point.*

Nonoverlapping implies that each codon consists of three nucleotides and that successive codons are represented by successive trinucleotides.

The use of a *fixed starting point* means that assembly of a protein must start at one end and work to the other, so that different parts of the coding sequence cannot be read independently.

If the genetic code is read in nonoverlapping triplets, there are three possible ways of translating a nucleotide sequence into protein, depending on the starting point. These are called **reading frames.** For the sequence

ACGACGACGACGACGACGACG

the three possible frames of reading are

ACG ACG ACG ACG ACG ACG
CGA CGA CGA CGA CGA CGA
GAC GAC GAC GAC GAC GAC

A mutation that inserts or deletes a single base will change the reading frame for the entire subsequent sequence. A change of this sort is called a **frameshift.** Since the new reading frame is completely different in sequence from the old one, the entire amino acid sequence of the protein is altered beyond the site of mutation. Thus the function of the protein is likely to be lost completely.

This type of mutation is induced by the **acridines**, compounds that bind to DNA and distort the structure of the double helix. The result of their action is that an additional base may be incorporated or a base may be omitted during replication. Each mutagenic event sponsored by an acridine involves the addition or removal of a single base pair.

If an acridine mutant is produced by, say, addition of a nucleotide, it should revert to wild type by deletion of the nucleotide. But reversion can also be caused by the deletion of a different base, at a site close to the first. The second mutation is described as a **suppressor**. (In the context of this work, "suppressor" is used in an unusual sense, to describe an *intragenic* event, whereas more usually it describes an *extragenic* relationship, when a mutation in one gene suppresses a mutation in another gene, as described later.)

Figure 3.1 shows that the combination of an acridine mutation and its suppressor causes the code to be read in the incorrect frame only between the two sites of mutation; but on either side, the correct reading frame is used. When genetic recombination is used to separate acridine mutants and their suppressors, the suppressor can be characterized as a mutation by itself. Stated as a general rule, all acridine mutations can be classified into one of two sets, described as (+) and (−). Either type of mutation by itself causes a frameshift, the (+) type by virtue of a base addition, the (−) type by virtue of a base deletion. Double mutant combinations of the types (+ +) and (− −) continue to show mutant behavior. But combinations of the types (+ −) or (− +) suppress one another.

The genetic code must be read as a sequence in a reading frame that is fixed by the starting point, so that additions or deletions compensate for each other, whereas double additions or double deletions remain mutant. But this does not reveal how many nucleotides make up each codon. However, when triple mutants are constructed, only (+ + +) and (− − −) combinations show the wild phenotype, while other combinations remain mutant. If we take three additions or three deletions to correspond respectively to the addition or omission overall of a single amino acid, this implies that the code is read in triplets. As with the compensating double mutants, an incorrect amino acid sequence is found between the two outside sites of mutation, and the sequence on either side remains wild type, as indicated in Figure 3.1.

The prediction that active double or triple mutants should have the wild type amino acid sequence on either side of the mutant region could not be confirmed with the *rII* system used for the genetic analysis. Unfortunately, at this time it was not possible to obtain the protein products for biochemical analysis. However, comparable analysis with other systems was able to confirm the prediction.

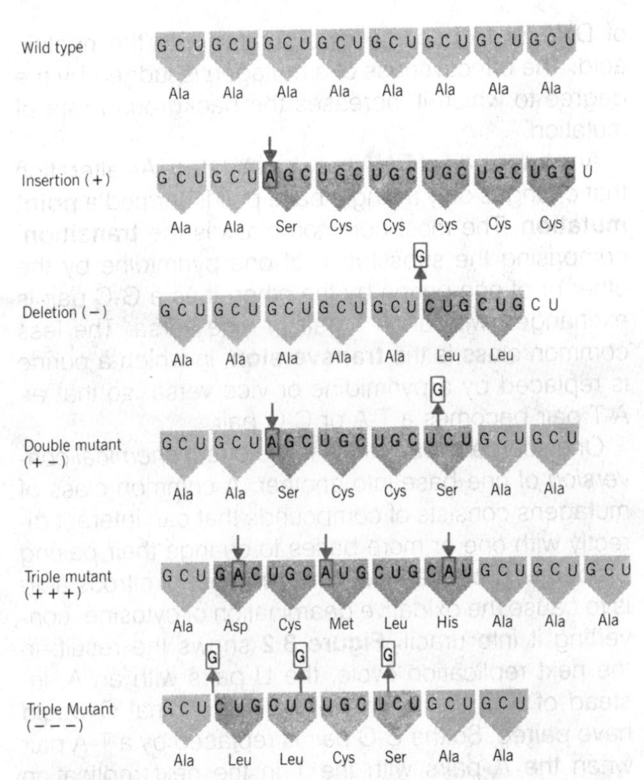

Figure 3.1
Frameshift mutations show that the genetic code is read in triplets from a fixed starting point.
The triplet reading frame is indicated by pentagons; grey indicates wild-type codons and color indicates mutant codons. Bases inserted into or deleted from the wild-type sequence are indicated by boxes and arrows.

POINT MUTATIONS CHANGE SINGLE BASE PAIRS

All organisms suffer a certain number of mutations as the result of normal cellular operations or random interactions with the environment. Such mutations are called **spontaneous**; the rate at which they occur is characteristic for any particular organism and sometimes is called the **background level**.

In the normal replication cycle, an error may occasionally occur when the wrong base is inserted. Also, sometimes DNA is damaged, activating "repair" systems that synthesize a replacement stretch of polynucleotide chain. Repair systems too may make errors during the synthesis of DNA. We do not yet have much accurate information on the relative contributions of these types of event to the overall spontaneous mutation rate.

The occurrence of mutations can be increased by treatment with certain compounds. These are called **mutagens**, and the changes they cause are referred to as **induced mutations**. Most mutagens act directly by virtue of an ability either to act on a particular base of DNA or to become incorporated into the nucleic acid. The effectiveness of a mutagen is judged by the degree to which it increases the background rate of mutation.

Any base pair of DNA may be mutated. An alteration that changes only a single base pair is termed a **point mutation**. The most common class is the **transition**, comprising the substitution of one pyrimidine by the other or of one purine by the other; thus a G-C pair is exchanged with an A-T pair or vice versa. The less common class is the **transversion**, in which a purine is replaced by a pyrimidine or vice versa, so that an A-T pair becomes a T-A or C-G pair.

One source of transitions is the direct *chemical conversion* of one base into another. A common class of mutagens consists of compounds that can interact directly with one or more bases to change their pairing properties. For example, a major effect of nitrous acid is to cause the oxidative deamination of cytosine, converting it into uracil. **Figure 3.2** shows the result: in the next replication cycle, the U pairs with an A, instead of with the G with which the original C would have paired. So the C-G pair is replaced by a T-A pair when the A pairs with the T in the next replication cycle. Nitrous acid also deaminates adenine, causing the reverse transition from A-T to G-C.

Another cause of transitions is **base mispairing**, when unusual partners pair in defiance of the usual restriction to Watson-Crick pairs. Base mispairing may occur either as an aberration involving the usual bases or as a consequence of the introduction of an abnormal base.

Some mutagens consist of analogs of the usual bases that have ambiguous pairing properties, and whose mutagenic action results from their incorporation into DNA in place of one of the regular bases. **Figure 3.3** shows the example of bromouracil (BrdU), which is incorporated into DNA by mistake for thymine. But because BrdU can also mispair reasonably well with guanine (instead of with adenine), the presence of the bromine leads to substitution of the original A-T pair by a G-C pair.

The mistaken pairing may occur either during the original incorporation of the base or in a subsequent replication cycle. The transition is induced with a certain probability in each replication cycle, so the incorporation of BrdU has continuing effects on DNA.

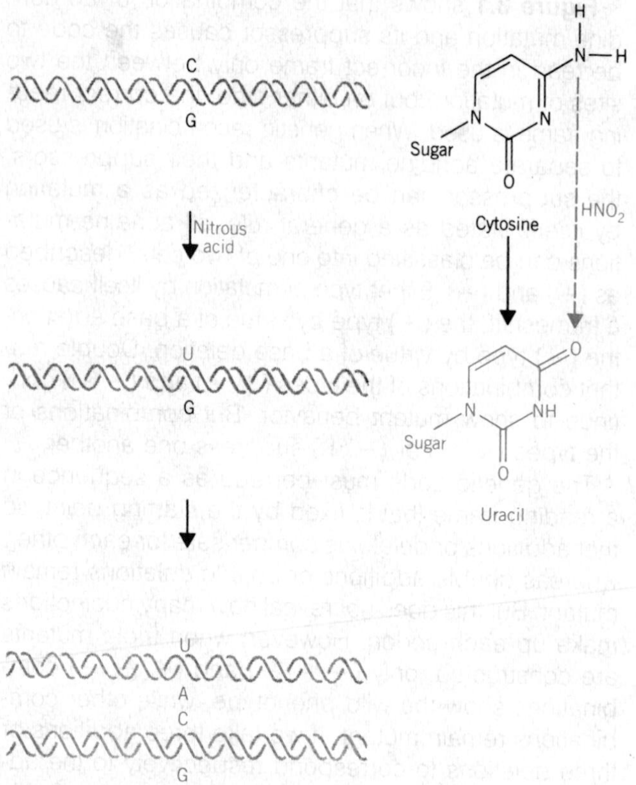

Figure 3.2
Mutations can be induced by chemical modification of a base. Nitrous acid oxidatively deaminates cytosine to uracil. Replication generates one daughter duplex with the wild-type C-G pair; but the other has a U-A pair, which is replicated to give A-T pairs in subsequent generations.

Mutations induced by base substitution quite often are **leaky**: the mutant has some residual function. This situation arises when the change in sequence of the corresponding protein does not entirely abolish its activity. The nature of the genetic code explains how this occurs. A point mutation that alters only a single base will change only the one codon in which that base is located. So only one amino acid is affected in the protein. While this substitution may reduce the activity of the protein, it does not necessarily abolish it entirely. This contrasts with the mutations induced by acridines, where a long series of codons may be altered by a shift of reading frame, completely abolishing protein function.

Point mutations were thought for a long time to be

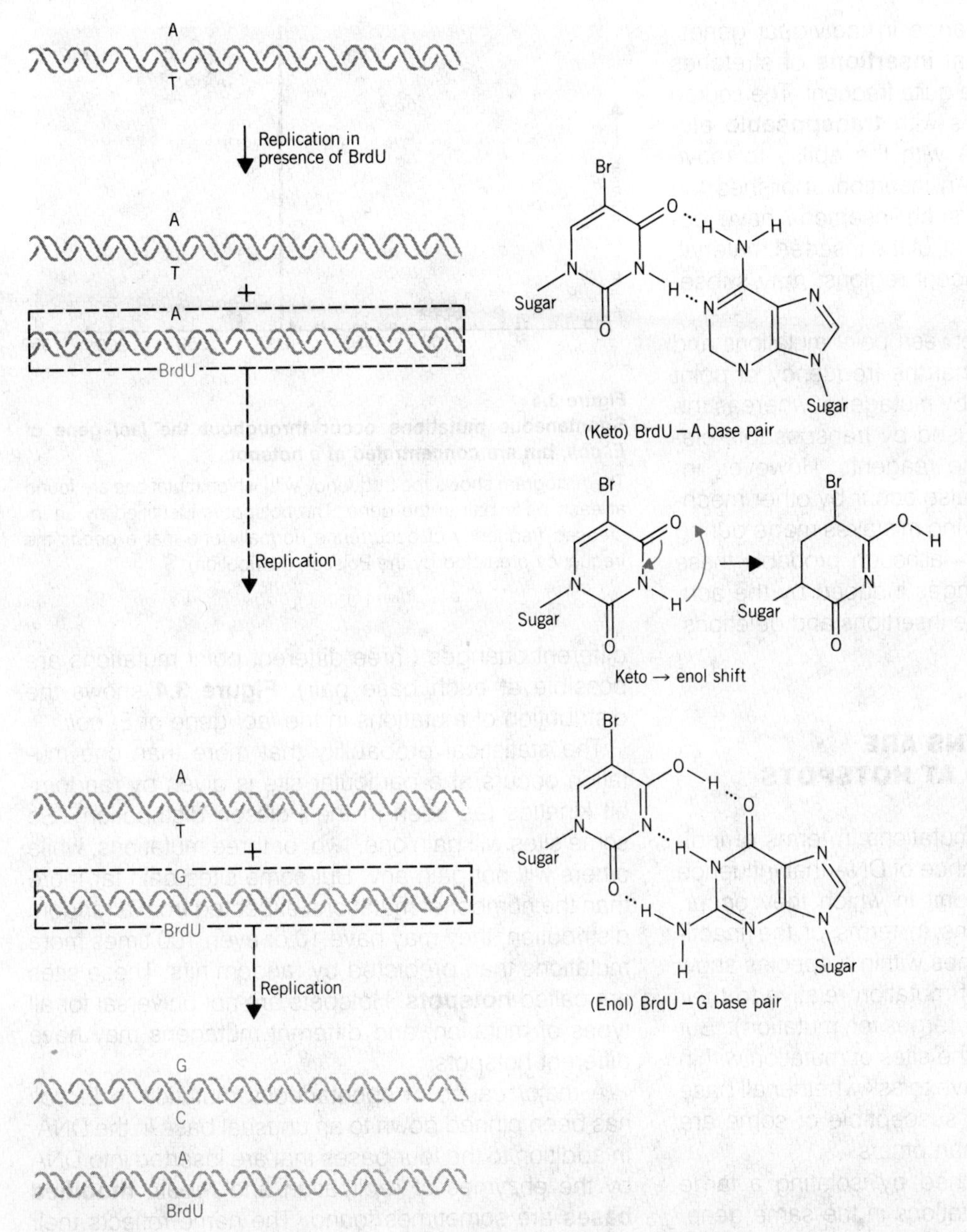

Figure 3.3
Mutations can be induced by the incorporation of base analogs into DNA.
Bromouracil contains a bromine atom in place of the methyl group of thymine. Its presence permits the keto - enol shift to occur more frequently, allowing BrdU to pair with guanine as well as with adenine. Thus if BrdU is incorporated in DNA in place of thymine, it may cause the replacement of an A by a G at a certain frequency in each generation. The result is that A-T pairs are replaced by G-C pairs.

the principal means of change in individual genes. However, we now know that **insertions** of stretches of additional material may be quite frequent. The source of the inserted material lies with **transposable elements**, sequences of DNA with the ability to move from one site to another. An insertion abolishes the activity of a gene. Where such insertions have occurred, **deletions** of part or all of the inserted material, and sometimes of the adjacent regions, may subsequently occur.

A significant difference between point mutations and the insertions/deletions is that the frequency of point mutation can be increased by mutagens, whereas the occurrence of changes caused by transposable elements is indifferent to these reagents. However, insertions and deletions can also occur by other mechanisms—for example, involving mistakes made during replication or recombination—although probably these are less common. The changes induced by the acridines formally also constitute insertions and deletions.

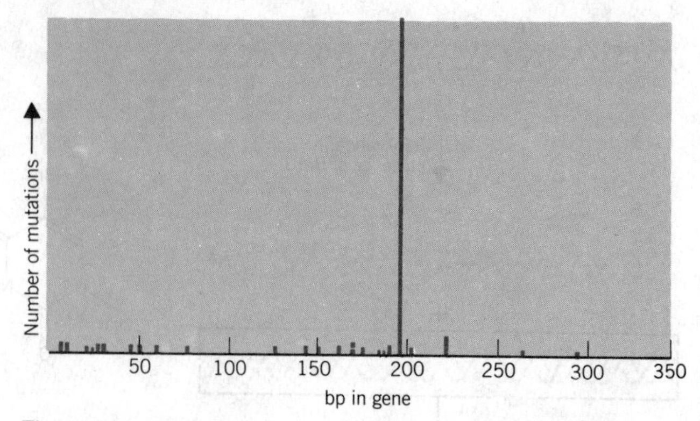

Figure 3.4
Spontaneous mutations occur throughout the *lacI* gene of *E. coli,* but are concentrated at a hotspot.
The histogram shows the frequency with which mutations are found at each base pair in the gene. The hotspot is identified by an increased frequency of occurrence (formally one that exceeds the frequency predicted by the Poisson distribution).

MUTATIONS ARE CONCENTRATED AT HOTSPOTS

So far we have dealt with mutations in terms of individual changes in the sequence of DNA that influence the activity of the genetic unit in which they occur. When we consider mutations in terms of the inactivation of the gene, most genes within a species show more or less similar rates of mutation relative to their size (that is, relative to the target for mutation). But when we come to consider the sites of mutation within the sequence of DNA, we have to ask whether all base pairs in a gene are equally susceptible or some are more likely to be mutated than others.

This question is approached by isolating a large number of independent mutations in the same gene. Many mutants are obtained, each of which has suffered an individual mutational event. Then the site of each mutation is determined. Most mutations will lie at different sites, but some will lie at the same position. Two independently isolated mutations at the same site may constitute exactly the same change in DNA (in which case the same mutational event has happened on more than one occasion), or they may constitute different changes (three different point mutations are possible at each base pair). **Figure 3.4** shows the distribution of mutations in the *lacI* gene of *E. coli*.

The statistical probability that more than one mutation occurs at a particular site is given by random-hit kinetics (as seen in the Poisson distribution). So some sites will gain one, two, or three mutations, while others will not gain any. But some sites gain far more than the number of mutations expected from a random distribution; they may have 10 or even 100 times more mutations than predicted by random hits. These sites are called **hotspots**. Hotspots are not universal for all types of mutation; and different mutagens may have different hotspots.

A major cause of spontaneous mutation in *E. coli* has been pinned down to an unusual base in the DNA. In addition to the four bases that are inserted into DNA by the enzymes of replication and repair, **modified bases** are sometimes found. The name reflects their origin; they are produced by chemically modifying one of the four bases already present in DNA. The most common modified base is 5-methylcytosine, which is generated by a methylase enzyme that adds a methyl group to a small proportion of the cytosine residues (at specific sites in the DNA).

Sites containing 5-methylcytosine provide hotspots for spontaneous point mutation. In each case, the mutation takes the form of a G-C to A-T transition. In strains of *E. coli* that are unable to perform the methylation reaction, these hotspots do not exist.

The reason for the existence of the hotspots is that 5-methylcytosine suffers a spontaneous deamination at an appreciable frequency; replacement of the amino group by a keto group converts 5-methylcytosine to thymine (the structures of the pyrimidines are shown in Figure 1.10). The conversion creates a mispaired G.T partnership, whose separation at the subsequent replication produces one wild type G-C pair and one mutant A-T pair.

Figure 3.5 compares the effect of deaminating the (rare) 5-methylcytosine or the more common cytosine. The deamination of cytosine generates uracil. However, *E. coli* contains an enzyme, uracil-DNA-glycosidase, that removes uracil residues from DNA. This action leaves an unpaired G residue, and a repair system then inserts a C base to partner it. The net result of these reactions is to restore the original sequence of the DNA. Presumably this system serves to protect DNA against the consequences of spontaneous deamination of cytosine (although it is not active enough to prevent nitrous acid from acting as a mutagen; see Figure 3.2). In contrast, however, the deamination of 5-methylcytosine leaves thymine; because this base is a perfectly respectable constituent of DNA in its own right, the system does not operate in these circumstances, and a mutation results.

The operation of this system casts an interesting light on the use of T in DNA compared with U in RNA. Perhaps it relates to the need of DNA for stability of sequence; the use of T means that any deaminations of C are immediately recognized, because they generate a base (U) not usually present in the DNA.

Another major hotspot in the *lacI* gene occurs at a site where the sequence $\begin{matrix}\text{CTGG}\\\text{GACC}\end{matrix}$ is repeated three times in succession. Mutations result from the addition or deletion of one unit of four bases from this sequence, thus:

	Sequence of Upper Strand
Deletion mutant	C T G G C T G G
	↑
Wild type	C T G G C T G G C T G G
	↓
Insertion mutant	C T G G C T G G C T G G C T G G

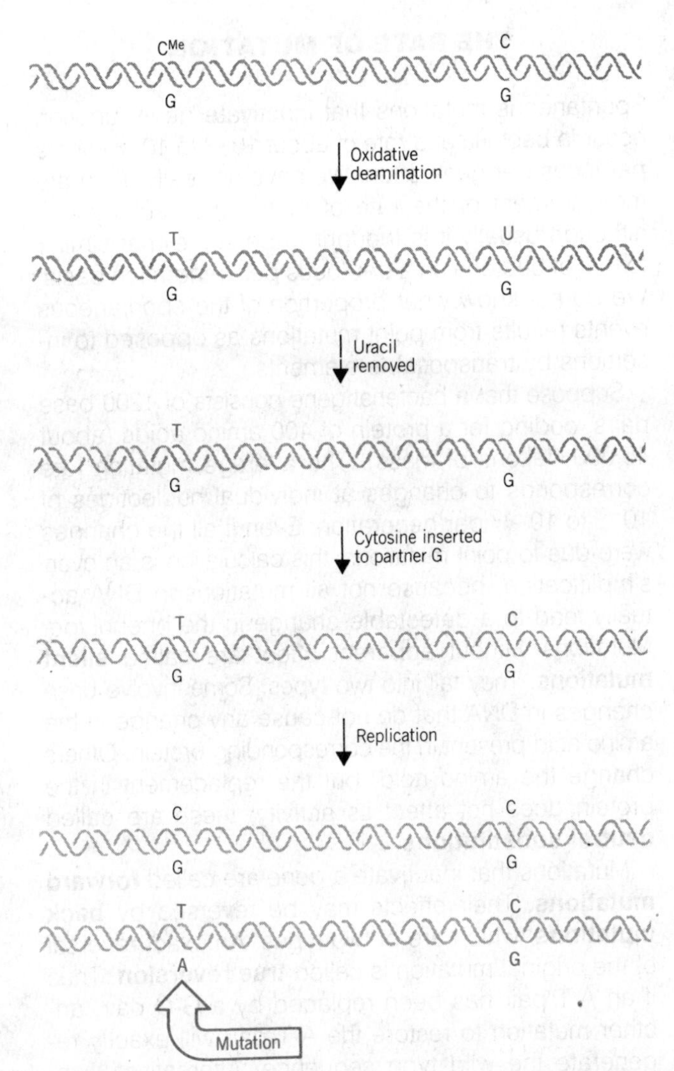

Figure 3.5
The deamination of 5-methylcytosine produces thymine (causing C-G to A-T transitions), while the deamination of cytosine produces uracil (which is removed and then replaced by cytosine).

The origin of these mutations may lie in a "slippage" in which one strand of DNA becomes misaligned with the other during replication; for example, the first CTGG of one strand could become misaligned with the second GACC of the other strand during replication. This could lead to additions or losses of individual repeats of the CTGG sequence.

THE RATE OF MUTATION

Spontaneous mutations that inactivate gene function occur in bacteria at a rate of about 10^{-5} to 10^{-6} events per locus per generation. We have no really accurate measurement of the rate of mutation in eukaryotes, although usually it is thought to be somewhat similar to that of bacteria on a per-locus per-generation basis. We do not know what proportion of the spontaneous events results from point mutations as opposed to insertions by transposable elements.

Suppose that a bacterial gene consists of 1200 base pairs, coding for a protein of 400 amino acids (about 45,000 daltons of mass). The average mutation rate corresponds to changes at individual nucleotides of 10^{-9} to 10^{-10} per generation. Even if all the changes were due to point mutations, this calculation is an oversimplification, because not all mutations in DNA actually lead to a detectable change in the phenotype. Mutations without apparent effect are called **silent mutations**. They fall into two types. Some involve base changes in DNA that do not cause any change in the amino acid present in the corresponding protein. Others change the amino acid, but the replacement in the protein does not affect its activity; these are called **neutral substitutions**.

Mutations that inactivate a gene are called **forward mutations**. Their effects may be reversed by **back mutations** which are of two types. An exact reversal of the original mutation is called **true reversion**. Thus if an A-T pair has been replaced by a G-C pair, another mutation to restore the A-T pair will exactly regenerate the wild type sequence. Alternatively, another mutation may occur elsewhere in the gene, and its effects may compensate for the first mutation. This is called **second-site reversion**. For example, one amino acid change in a protein may abolish gene function, but a second alteration may compensate for the first and restore protein activity. (And the compensating acridine frameshift mutations fall into the category of second site revertants.)

A forward mutation results from any change that inactivates a gene, whereas a back mutation must restore function to a protein damaged by a particular forward mutation. Thus the demands for back mutation are much more specific than those for forward mutation. The rate of back mutation is correspondingly lower than that of forward mutation, typically by a factor of about 10.

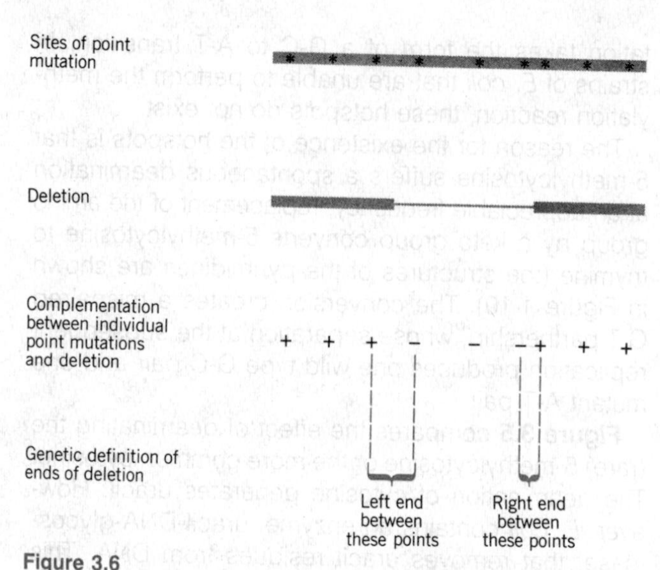

Figure 3.6
A deletion will complement a series of point mutations on either side of the deleted region, but will not complement point mutations that lie within the deleted region. The boundaries of the deletion are indicated by the switch from complementation to noncomplementation.

Mutations may also occur in other genes to circumvent the effects of mutation in the original gene. This effect is called **suppression** (or, more formally, extragenic suppression; this is the usual use of the term *suppression*, as opposed to its use to describe the behavior of acridine-induced mutations). The isolation of revertants is an important characteristic that distinguishes point mutations and insertions from deletions. A point mutation can revert by restoring the original sequence or by a second-site mutation; an insertion can revert by deletion of the inserted material.

However, a deletion of part of a gene cannot revert. Such deletions have a critical use in genetic mapping. The genetic extent of a deletion is defined by its inability to complement with a series of adjacent mutations (because it lacks the entire corresponding stretch of wild type DNA). The deletion can complement with point mutations on either side. **Figure 3.6** shows that a deletion is visualized genetically by the transitions between complementation and noncomplementation with a series of point mutations.

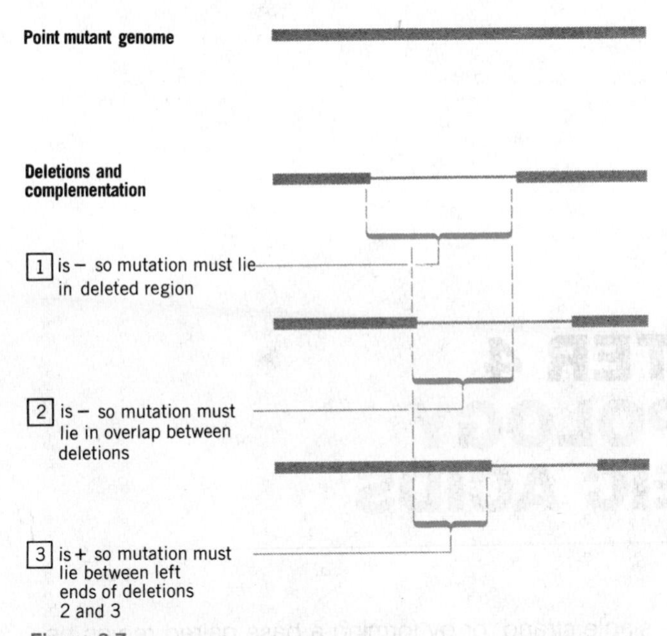

Figure 3.7

Deletion mapping can be used to locate point mutations between the ends of overlapping deletions.

If a point mutation cannot be complemented by a deletion, the site of mutation must lie within the region that has been deleted. If a point mutation is complemented by a deletion, the site of mutation must lie outside the deleted region. By comparing the ability of a series of deletions to complement a point mutation, the site of mutation can be identified.

By obtaining a series of partially overlapping deletions, we can map any point mutation by testing its ability to complement with them. **Figure 3.7** illustrates the protocol. When two deletions both fail to complement a point mutation, the site of mutation must lie in the region common to the deletions. When one deletion complements and one does not, the site of mutation must lie in the region in which the deletions do not overlap.

In characterizing a genome, we should like ideally to be able to mutate every single gene of the organism. But the length to which we can apply this approach is restricted in two ways.

First, we need a criterion for distinguishing the mutant from the wild type. The simplest criterion is visible change, for example, in eye color. Other criteria can be established by devising suitable selective procedures; for example, by adjusting the conditions of growth so that the absence or presence of some enzyme is required for survival. The problem in all of this is that there may be functions of whose existence we are ignorant, and for which we therefore fail to devise suitable tests. Taken to an extreme, there is the possibility that genes exist whose products are unnecessary and whose absence therefore has no effect (at least under the conditions that we use). How are they to be detected?

Our second difficulty concerns functions that are *essential* for viability. A mutation in such a gene is likely to kill the organism. To isolate mutants in these genes, they must be obtained in the form of **conditional lethal** mutations. These mutations are lethal under one set of conditions, but exhibit no deficiency, or only a much reduced deficiency, under alternative conditions that allow the cell to be perpetuated.

Thus the same (mutant) organism can be studied under two conditions. In **permissive** conditions, it does not display the mutant phenotype and may therefore be perpetuated. In **nonpermissive** conditions, it dies or becomes severely ill, but it can be studied during the transition from permissive to nonpermissive conditions. (Of course, conditional mutations are not restricted to essential functions, but in principle can be found in any gene.)

FURTHER READING

Perspectives on the original work on mutations can be obtained from the review by **Drake & Balz** (*Ann. Rev. Biochem* **45**, 11–37, 1976). The discovery of hotspots was reported by **Benzer** (*Proc. Nat. Acad. Sci. USA* **47**, 403–416, 1961) and (nearly twenty years later) they were equated with modified bases by **Coulondre et al.** (*Nature* **274**, 775–780, 1978). The basis of frameshift mutations has been reviewed by **Roth** (*Ann. Rev. Genet.* **8**, 319–346, 1974). Their use to define the nature of the genetic code was reported in the classic paper by **Crick et al.** (*Nature* **192**, 1227–1232, 1961).

CHAPTER 4
THE TOPOLOGY OF NUCLEIC ACIDS

There is more to DNA than just a sequence of base pairs organized into an invariant double helical structure. Although genetic information is represented by the sequence rather than the structure per se, for the information to be utilized, the strands of the double helix must separate. Strand separation can be influenced by the overall structure of the DNA molecule; and the ability of certain regions to undergo changes in the double helical structure is an important aspect of the function of DNA.

The double helix does not exist as a long straight rod, but is coiled in space in order to fit into the dimensions of the cell or virus whose genome it provides. A consequence of this further level of organization is that the duplex may be placed under stress in such a way that its own structure is affected. The result may be to change the winding of the two partner strands around each other, at times causing their separation into single strands, or even generating an alternative form of the double helix.

Although DNA usually is found in the form of a double helix, the genomes of some viruses consist of single-stranded DNA. Within the cell there are several forms of the other nucleic acid, RNA, usually present as single strands. However, single strands of DNA or RNA may generate double helical regions, either by base pairing with an independent but complementary single strand, or by forming a base paired region between two complementary regions within the same otherwise single-stranded molecule.

The concept of base pairing is central to all processes involving nucleic acids. Disruption of the base pairs is a crucial aspect of the function of a double-stranded molecule, while the ability to form base pairs is essential for the activity of a single-stranded nucleic acid.

SINGLE STRANDED NUCLEIC ACIDS MAY HAVE SECONDARY STRUCTURE

The stability of the double helix results from the hydrogen bonding between the complementary A-T and G-C pairs and also from interactions between the bases as they are "stacked" above each other along the axis of the helix. These forces can be used to predict the stability of a double helix between two complementary sequences, and this is an important technique in analyzing the structure of a single-stranded nucleic acid sequence. Because RNA is the predominant single-stranded nucleic acid, the formation of double-stranded regions from a single strand is usually analyzed in terms of RNA, but the technique is equally valid for single-stranded DNA.

The primary structure of RNA is the same as that of DNA: a polynucleotide chain with 5′–3′ sugar-phosphate links. (However, the base composition of a single-stranded nucleic acid is not restricted to the overall G = C, A = U [or T] equalities.) Considered just as a single strand, the molecule can follow a random path in space, but base pairing within it can fix the location of one region relative to another.

When a sequence of bases is followed by a complementary sequence nearby in the same molecule, the chain may fold back on itself to generate an antiparallel duplex structure. This is called a **hairpin**. It consists of a base paired, double helical region, the **stem**, often with a **loop** of unpaired bases at one end. **Figure 4.1** shows an example. When the complementary sequences are relatively distant in the molecule, their juxtaposition to form a double-stranded region essentially creates a stem with a very long single-stranded loop.

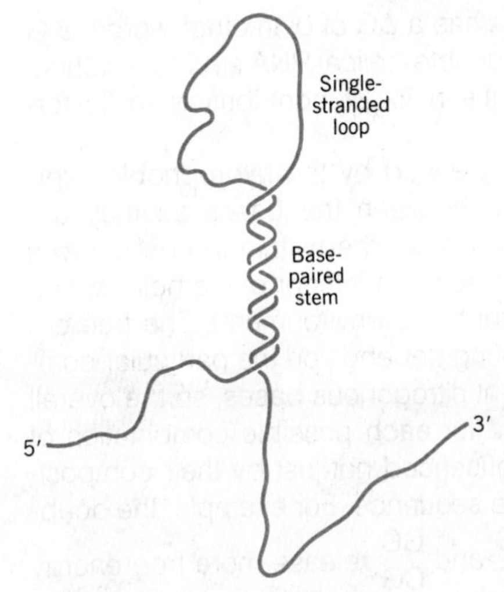

Figure 4.1

A single-stranded nucleic acid may fold back on itself to form a duplex hairpin by base pairing between complementary sequences.

In view of the irregular occurrence of secondary structure in single-stranded nucleic acids, how is it to be predicted from the sequence or measured in practice? Sometimes a single-stranded molecule may have several regions that are able potentially to base pair with one another in alternate arrangements, so it is necessary to resolve which (if any) actually occurs.

The *overall* extent of base pairing is reflected in the biophysical properties of the molecule. However, this does not reveal which individual regions are involved. Single-stranded and double-stranded regions have different susceptibilities to some **nucleases** (enzymes that degrade nucleic acids), and this provides a test for analyzing which particular regions are involved in base pairing. However, such data have limited applicability, and in general are effective usually only with relatively short molecules. For the most part, the analysis of RNA secondary structure still depends more on theory than experimental information.

The *plausibility* of a particular base-paired structure can be predicted by **rules** that describe the interaction of the base pairs. When alternative structures exist, their relative stabilities can be assessed by these rules. Of course, this approach treats the RNA as an isolated and stable structure, ignoring any other factors that may intervene to influence the structure (for example, binding by proteins). However, the application of these rules provides a first step to seeing whether a particular structure is able in principle to exist.

The basis of these rules is to calculate the **free energy** for the formation of each structure. The free energy is (briefly) a thermodynamic constant that gives the amount of energy released by a reaction. It is measured in kcal/mole as the parameter ΔG, which has a *negative* value corresponding to the available energy. Thus if alternate structures have free energies of formation of $\Delta G = -21$ kcal/mole and $\Delta G = -35$ kcal/mole, the latter is more likely to be formed, other things being equal.

The overall free energy of a double-stranded structure is calculated by summing the free energies of the individual base pairing reactions that are involved. The formation of each hydrogen bond releases rather a small amount of energy. Since G-C base pairs have three hydrogen bonds, they are the more stable, with a ΔG of -2.4 kcal/mole; the two hydrogen bonds of an A-U pair confer a ΔG of -1.2 kcal/mole. Thus the stability of a double helix increases with the proportion of G-C base pairs. Another base pair that can form in RNA is the irregular partnership G.U (and G.T can form in double-stranded regions of single-stranded DNA, although it is excluded from a regular double helix of DNA). This has a ΔG of 0; in other words, a G may face a U in a double-helical RNA structure without either disrupting the helix or contributing to its formation.

Free energy is released by the hydrophobic interactions that occur between the bases as they are stacked on top of one another within the helical axis (water is excluded from the interior of the helix, which thus forms a hydrophobic environment). The free energy of base stacking depends on the particular combination of adjacent nitrogenous bases, so the overall free energy differs for each possible combination of base pairs. It is influenced not just by their composition but also by the sequence. For example, the doublet sequences $\begin{matrix}GG\\CC\end{matrix}$ and $\begin{matrix}GC\\CG\end{matrix}$ release more free energy ($\Delta G = -5.0$ kcal/mole) than the sequence $\begin{matrix}CG\\GC\end{matrix}$ ($\Delta G = -3.2$ kcal/mole), although they all consist of two G-C pairs. Similarly, the sequence $\begin{matrix}AA\\UU\end{matrix}$ releases less energy ($\Delta G = -3.2$ kcal/mole) than the alternatives of the same composition, $\begin{matrix}AU\\UA\end{matrix}$ and $\begin{matrix}UA\\AU\end{matrix}$ ($\Delta G = -1.6$ kcal/mole). (The sequences of the doublets are written following the convention that the top row represents a strand running 5′ to 3′ from left to right; the lower strand is the complement running 3′ to 5′ from left to right.)

The free energy of formation for a duplex region is usually calculated by summing the free energies for each doublet of adjacent base pairs. Each base pair within the duplex region is involved in the calculation for two doublets, one for the base pair on each side. (A base pair that lies at one end of the duplex participates in the formation of only one doublet, with its single neighbor.) The exact free energy of formation is determined by the sequence of the duplex, as illustrated in the example of **Figure 4.2**.

Unlike double-stranded DNA, which is maintained as a perfect duplex structure, double helical regions of RNA usually form between single strands that are not perfectly complementary. This means that there may be interruptions in the integrity of a duplex region. They take the form of:

hairpin loops (occurring at the end of the duplex region when the double helix is formed by looping back to pair with an immediately adjacent complementary sequence);

interior loops (when there are corresponding regions in each potential complement that do not base pair);

bulge loops (when one of the potential complements contains an additional sequence that does not pair).

All of these unpaired regions *hinder* the formation of the double helix. Their effect is taken into account in the free energy calculation by assigning each type of interruption a **positive** free energy value that must be included in the sum total. In other words, holding these regions in a particular structure requires the *input* of energy, and this must be offset against the free energy that is released from the base pairing and stacking. Overall, the free energy calculation must produce a sufficiently negative value or the secondary structure will be unable to form.

INVERTED REPEATS AND SECONDARY STRUCTURE

When a single-stranded nucleic acid contains two sequences that are complementary, they may base pair to form a hairpin as described in Figure 4.2. The sequence in a double-stranded DNA that corresponds to such a structure consists of two copies of an identical sequence present in the reverse orientation. They are called **inverted repeats**.

The sequence of both inverted repeats is called a **palindrome**. It is defined as a sequence of duplex DNA that is the same when either of its strands is read in a defined direction. An example of a palindrome is

5′ GGTACC 3′
3′ CCATGG 5′

because when either strand is read in the direction 5′ to 3′, it generates the sequence GGTACC (that is, reading left to right on the upper strand and reading right to left on the lower strand).

A palindromic sequence is formally described as a **region of dyad symmetry**, in which the **axis of symmetry** identifies the central point about which the sequence is the same on either side (when read in op-

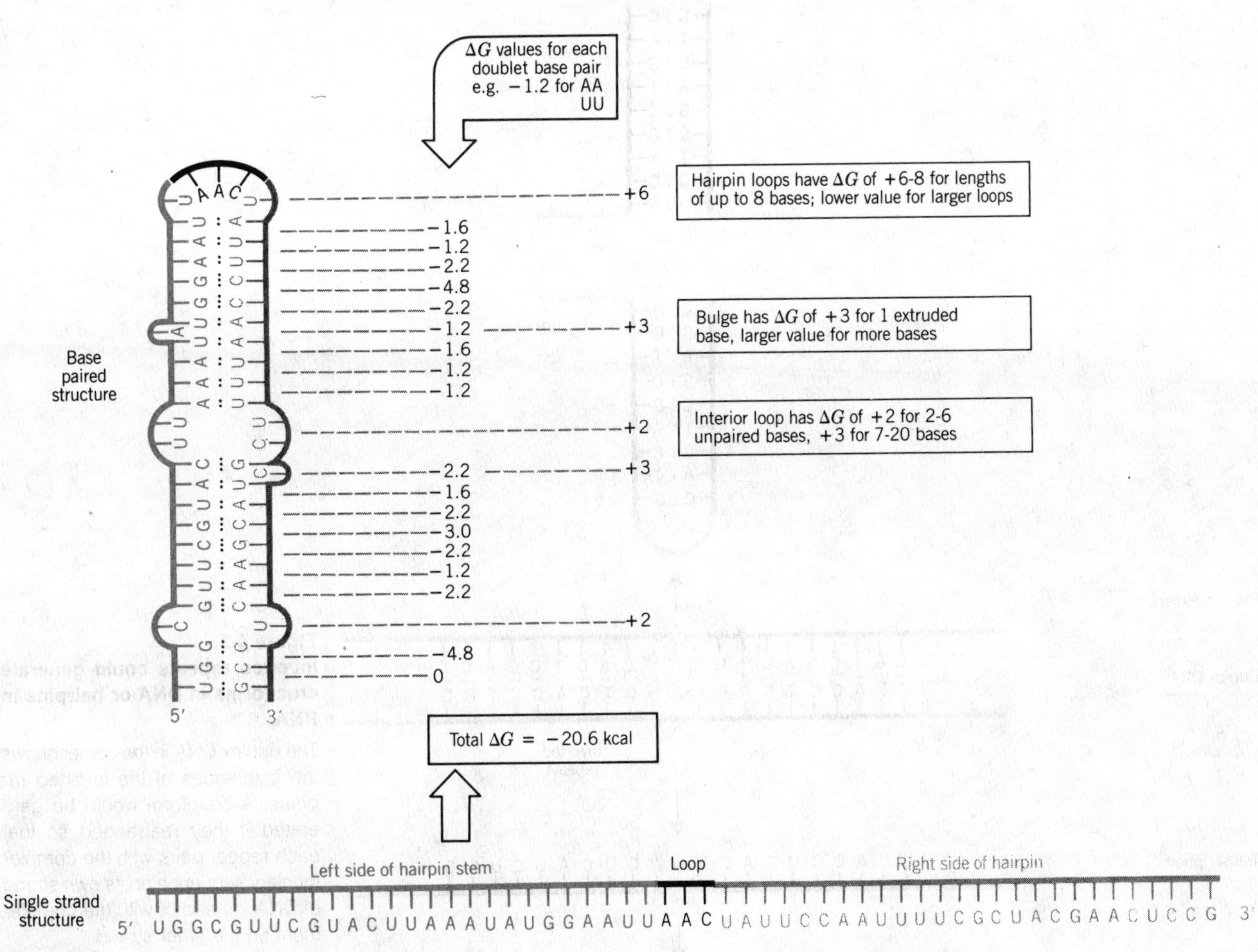

Figure 4.2
The free energy of formation for a potential base-paired region can be calculated by summing the ΔG values released (by base pair formation) or demanded (to maintain single-stranded regions in a restricted conformation).

posite directions). By drawing a line through the axis of symmetry,

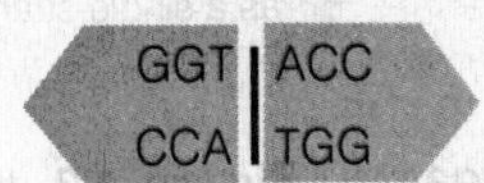

we see that the same sequence $\begin{matrix}\text{GGT}\\\text{CCA}\end{matrix}$ is present on either side of the axis of symmetry, in inverse orientation, as indicated by the arrows. Each arrow identifies one of the inverted repeats.

The two copies of an inverted repeat need not necessarily be contiguous. Consider the same inverted repeat in a different situation

We still have an inverted repeat of the same triplet sequence, although the axis of symmetry has been shifted into the center of the additional four base pairs that separate the two copies of the repeat.

An inverted repeat can be reflected in the second-

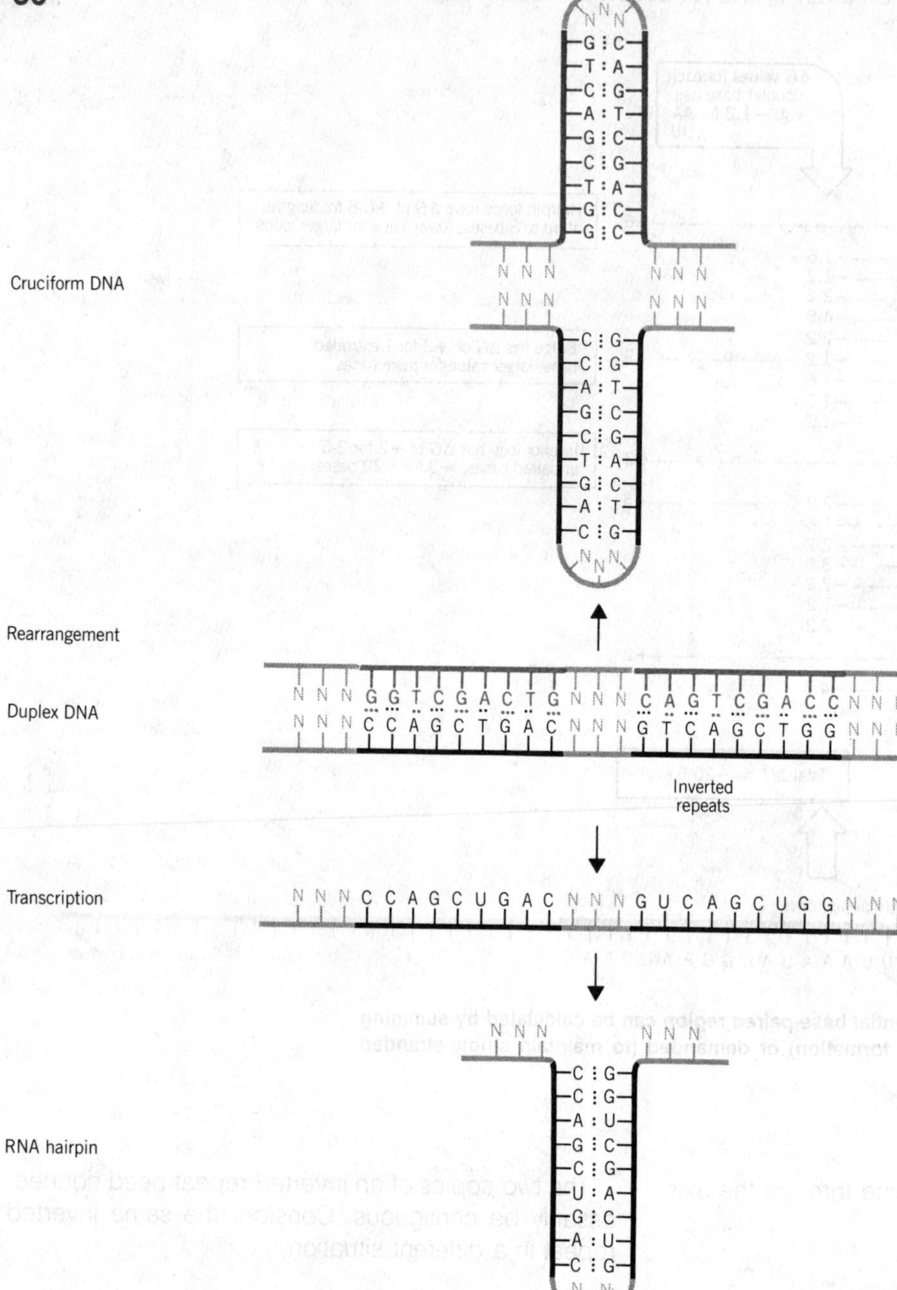

Figure 4.3
Inverted repeats could generate cruciforms in DNA or hairpins in RNA.

The duplex DNA in the center shows the sequences of the inverted repeats. A cruciform would be generated if they rearranged so that each repeat pairs with the complementary sequence on its *own* strand of DNA instead of with the complement on the other strand.

Transcription of the upper DNA strand would generate an RNA with the same sequence as the lower DNA strand. The repeats can base pair to generate a hairpin, which has the same sequence as the lower part of the cruciform.

The inverted repeats themselves form the base-paired stem of the hairpin. The three bases that separate the inverted repeats appear as a single-stranded loop at the end of the stem.

ary structure of either single-stranded or double-stranded nucleic acid. Considering just one strand, the copies of the repeat are complementary when read in opposite directions. Reading away from the axis of symmetry, we obtain TGG going left and ACC going right. This situation is the same as that illustrated in Figure 4.2 and again in **Figure 4.3**; the complementary sequences may base pair to form a hairpin.

In a double-stranded DNA, the complementary sequences on one strand have the opportunity to base pair only if the strand separates from its partner. In that case, the situation is in fact the same on each strand: a hairpin could be formed. The formation of the two apposed hairpins creates a **cruciform**, so called because it represents the junction of four duplex regions. The original duplex extends on either side, and the intrastrand hairpins protrude from it.

The conditions under which cruciforms might form *in vivo* are controversial. A double helix is unlikely to abandon its continuous duplex structure to form the protruding hairpin loops spontaneously. A considerable amount of energy would be required to separate the two strands ($\Delta G \sim +50$ kCal/mol), and only some of it would be regained by the formation of the cruciform, which is energetically less stable than a continuous duplex (by about $\Delta G = +18$ kCal/mol). However, cruciforms can be generated by providing appropriate conditions *in vitro*, as shown in the example of **Figure 4.4**.

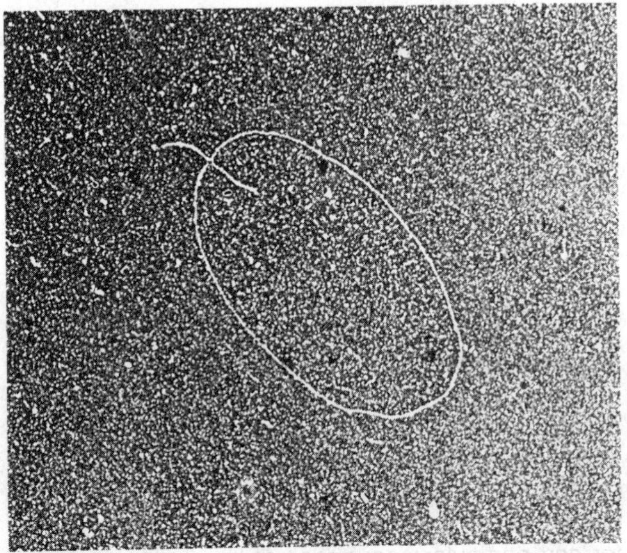

Figure 4.4
An electron microscope of a cruciform generated in DNA *in vitro.*
Kindly provided by Martin Gellert.

We do not know whether cruciforms occur *in vivo*. An interesting discrepancy has been noted between the abilities of DNA molecules with short and long palindromic sequences to survive in bacteria. A short palindromic sequence presents no problem, but DNA that has long palindromes either fails to survive or has its sequences grossly rearranged. This could indicate that cruciforms tend to form in the longer palindromes, but cannot survive in cell (possibly because they resemble the intermediates formed in recombination, and are therefore targets for certain enzymes).

Current opinion therefore favors the view that probably cruciforms do not form naturally in cells; and if they are induced by providing appropriate DNA sequences, they are likely to be unstable.

DUPLEX DNA HAS ALTERNATIVE DOUBLE-HELICAL STRUCTURES

It has been known for a long time that DNA can exist in more than one type of double-helical structure, although under physiological conditions it appeared always to take the form of the now classic B-type duplex. We still think that duplex DNA is almost entirely in the B-form within the cell, although some of the parameters that describe the precise structure of the B-type duplex have been adjusted, in particular the number of base pairs per turn. However, it seems also that, under certain conditions, duplex DNA can make a transition from the B-form to another form. Probably this affects a rather small proportion of the DNA.

The problem in assigning specific values to describe the parameters of the double helix is that most of the structural evidence rests on X-ray diffraction studies of fibers. These data indicate basic features, such as the number of residues per turn or the spacing of residues along the helical axis, but do not assign positions to specific atoms in the way that is possible with X-ray crystallography. This leaves it to model building to provide structures consistent with the experimental data.

One consequence is that (in theory) it is always possible that another model may be constructed to explain the same data. Although other models have been proposed—a brief notoriety was enjoyed by one suggesting that two antiparallel strands lie "side by side" instead of being wound into a helix—none of them has stood up to other experimental predictions.

The existence of DNA as a double helix has been confirmed by experiments to measure directly the number of base pairs per turn. This proves to be 10.4 instead of the 10.0 predicted by the classic B-model. The change requires a slight adjustment in the angle of rotation between adjacent base pairs along the helix, to 34.6°, so that it takes slightly longer to accomplish the full 360° turn.

The B-duplex provides a model to fit what may be *average* data; variations could occur in particular regions, either as the result of a particular base sequence or because of conditions imposed by the environment. We know that the precise structure can indeed be altered; for example, the structure of a particular 12 base pair molecule has been shown by X-ray crystallography to have 10.1 base pairs per turn, achieved by a slight twist of each base pair that improves base stacking relative to the original model.

In vivo, the tight coiling and compaction that is necessary to fit DNA into the cell may change details of its structure. Thus we take the value of 10.4 base pairs per turn to represent an average for DNA as a whole under certain conditions, and we assume that the value for any particular sequence will be close to, but not necessarily identical with, this estimate.

To accommodate variations, the idea that the DNA double helix has a single structure has been replaced by the view that there are families of structures. Each family represents a characteristic type of double helix, as described by the parameters *n* (the number of nucleotides per turn) and *h* (the distance between adjacent repeating units). The variation is achieved by changes in the rotation of groups about bonds with rotational freedom. Within each family, the parameters can vary slightly; for example, for B-DNA *n* could be 10.0–10.6.

Table 4.1 summarizes the average properties of four families of structure. The existence of the A, B and C forms has been known for a considerable time, and transitions between them occur when appropriate changes are made in the conditions. All of them are right-handed structures, differing in the governing parameters. However, the Z form represents a rather different structure that was discovered more recently.

Table 4.1
DNA can exist in several types of structure families.

Helix Type	Base Pairs per Turn	Rotation per Base Pair	Vertical Rise per Base Pair	Helical Diameter
A	11	+32.7°	2.56Å	23Å
B	10	+36.0°	3.38Å	19Å
C	9.33	+38.6°	3.32Å	19Å
Z	12	−30.0°	3.71Å	18Å

The rotation per base pair is indicated as (+) for a right-handed duplex and (−) for a left-handed duplex.

The **B-form** for which Watson and Crick constructed their model is found in fibers of very high (92%) relative humidity and in solutions of low ionic strength. The living cell provides appropriate conditions for this form.

The **A-form** is found in fibers at 75% relative humidity and requires the presence of sodium, potassium or cesium as the counterion. Instead of lying flat, the bases are tilted with regard to the helical axis; and there are more base pairs per turn.

The A-form is biologically interesting because it is probably very close to the conformation of double-stranded regions of RNA, which may take this form because the presence of the 2′ hydroxyl group prevents RNA from lying in the B-form. Hybrid duplexes with one strand of DNA and one strand of RNA also probably lie in the A-form.

Figure 4.5 compares structural forms of double-stranded nucleic acids. B-DNA has a major (wide) groove and a minor (narrow) groove; differences between the bases are most easily recognized in the wide groove, which is therefore a major point of contact for proteins that bind specific sequences of DNA. In the relatively compact structure of A-form RNA, the major groove is much less accessible, because it is much deeper and phosphate groups overhang it. The result is that the features of individual bases are buried deep in the groove. The minor groove, however, is superficial. So the structural basis for recognizing a double-stranded nucleic acid sequence is likely to be different in B-form DNA and A-form RNA.

The **C-form** occurs when DNA fibers are maintained in 66% relative humidity in the presence of lithium ions. It has fewer base pairs per turn than B-DNA. There is no evidence for its existence *in vivo*.

These three forms are available to all DNAs, irrespective of nucleotide sequence. Some further forms have been found that appear to represent options open only to DNA molecules with particular quirks in their base composition.

The **D-form** and **E-form** (actually possible extreme variants of the same form) have the fewest base pairs per turn (only 8 and 7½) and are taken up only by certain DNA molecules that lack guanine.

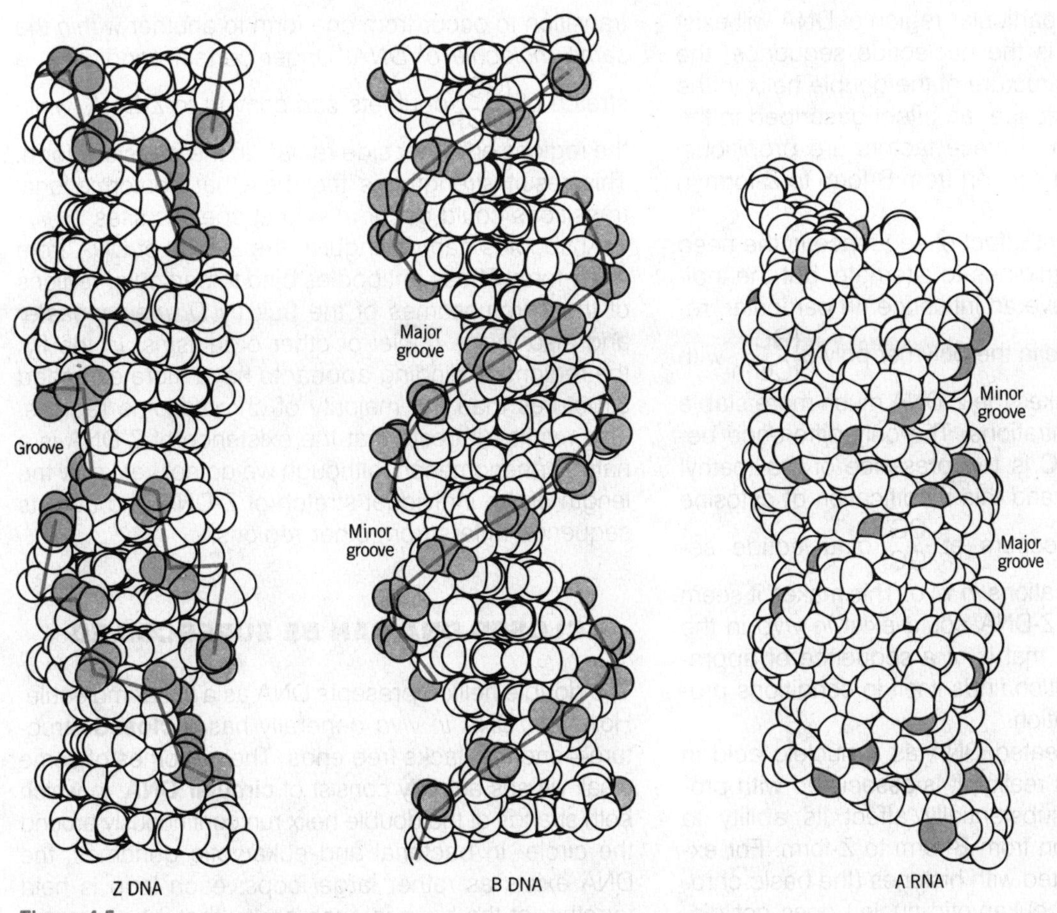

Figure 4.5
Nucleic acids can form several types of double helix.

A LEFT-HANDED FORM OF DNA

The **Z-form** provides the most striking contrast with the classic structure families. In contrast with all the others, it is a **left-handed helix**. Its structure is illustrated in **Figure 4.5**.

Z-DNA has the most base pairs per turn of any structure for DNA, and so has the least twisted structure; it is very skinny, and its name is taken from the zigzag path that the sugar-phosphate backbone follows along the helix, quite different from the smoothly curving path of the backbone of B-form DNA. Z-DNA has only a single groove, with a greater density of negative charges than either of the two grooves in B-DNA.

The Z-form double helix has been found in polymers that have a sequence of alternating purines and pyrimidines. Two that have been examined have a simple repeating dinucleotide sequence:

$$\text{poly d}\begin{Bmatrix}GC\\CG\end{Bmatrix} \text{ and poly d}\begin{Bmatrix}AC\\TG\end{Bmatrix}.$$

(The *d* indicates that these are the deoxy forms, that the sequence is DNA and not RNA.)

Z-form DNA was discovered *in vitro* under somewhat unusual conditions (using a high salt concentration to counter the increased electrostatic repulsion between the nucleotides compressed into the slimmer double helix of Z-DNA). Under what conditions might it exist *in vivo*?

Two factors intrinsic to the double helix determine

the likelihood that a particular region of DNA will exist in the Z-form. One is the nucleotide sequence; the other is the overall structure of the double helix in the sense of its path in space, an effect described in the next section. If both of these factors are propitious, DNA may be able to convert from B-form to Z-form in the natural state.

The most prominent effect of sequence is the need for purines and pyrimidines to alternate, but the individual bases also have an influence. In particular, replacing the C residue in the polymer poly d$\begin{matrix}\{GC\}\\ \{CG\}_n\end{matrix}$ with 5-methylcytosine makes the Z-DNA much more stable at lower salt concentrations. The only difference between 5-Me-C and C is the presence of the methyl group at position 5; and this modification of cytosine is a reaction that occurs at $\begin{matrix}CG\\ GC\end{matrix}$ dinucleotide sequences at some locations *in vivo*. This makes it seem more plausible that Z-DNA could exist *in vivo* in the right circumstances, that is, if a sequence of appropriate base composition finds itself in conditions promoting Z-DNA formation.

So far we have treated DNA as a nucleic acid in splendid isolation. In reality, it is associated with proteins. These may substantially affect its ability to undergo the transition from B-form to Z-form. For example, DNA associated with histones (the basic chromosomal proteins of eukaryotic nuclei) does not display the transition under conditions when free DNA can do so.

On the other hand, there could be proteins that bind specifically to Z-DNA. Attempts have been made to isolate such proteins by separating those associating with Z-DNA from those that bind to any (that is, B-form) DNA. Such experiments may lead to the characterization of particular proteins that stabilize DNA in the Z-form *in vivo*.

Z-DNA was discovered as an alternate form taken by the entire sequence of a particular DNA molecule consisting of a simple repetitive sequence. But this is not a realistic situation. The majority of cellular DNA almost certainly exists in the B-form (although possibly with regional variations in the values for the helical parameters). Only rather short stretches are likely to be involved in transitions to other forms.

We must therefore ask whether it is possible for a transition to occur from one form to another *within the same molecule of DNA*. Under certain conditions, a stretch of $\begin{matrix}\{GC\}\\ \{CG\}\end{matrix}$ doublets *can* convert to Z-DNA, while the regions on either side remain in the classic B-form. This result strengthens the idea that conformational transitions could occur *in vivo* at specific sites.

Antibodies can distinguish the Z-form of DNA from the B-form. These antibodies bind to particular regions of the chromosomes of the fruit fly *D. melanogaster* and also to the nuclei of other organisms. In the fly, the regions of binding appear to have more extended structures than the majority of chromosome regions. This result suggests that the existence of Z-DNA is a natural phenomenon, although we do not yet know the length of the individual stretch of Z-DNA and how its sequence differs from other regions.

CLOSED DNA CAN BE SUPERCOILED

The double helix represents DNA as a linear molecule. However, DNA *in vivo* generally has a **closed** structure—one that lacks free ends. The genomes of some small viruses actually consist of **circular DNA**, in which both strands of the double helix run continuously around the circle. In bacterial and eukaryotic genomes, the DNA exists as rather large loops; each loop is held together at the base in such a way that it becomes a simulacrum of a circle. This form of organization is important because it allows an additional structural constraint to be superimposed on the double helix.

Supercoils are introduced into DNA when a duplex is twisted in space around its own axis. The usual analogy is to consider a rubber band twisted about itself to generate a tightly coiled structure in which the rubber band (the double helix) crosses over itself in space. The possible forms of a DNA molecule are compared in **Figure 4.6**, which is an electron micrograph of a linear molecule, a nonsupercoiled circular version of the same DNA, and a supercoiled circular molecule.

The twisting introduced by supercoiling places a DNA molecule under torsion and gives the type of structure depicted in **Figure 4.7**. Supercoiling can occur *only* in closed structures, because an **open** molecule can release the torsion simply by untwisting. (A

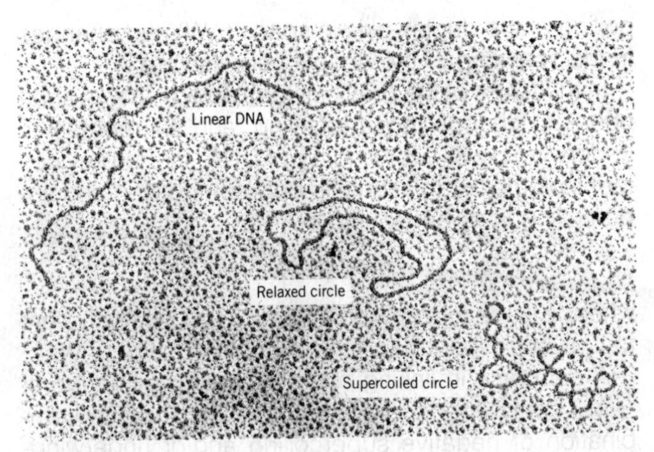

Figure 4.6
Supercoiled DNA has a compact twisted structure compared with a nonsupercoiled circle of linear molecule.
Kindly provided by Svend O. Freytag.

closed molecule must have no breaks on *either* strand of DNA; a break even in one strand of a circular molecule allows untwisting.) A molecule that lacks supercoiling, whether closed or open, is said to be **relaxed**.

The principle by which supercoiling is maintained is that, if a duplex is wound around itself in space *and then the ends are fixed*, the structure cannot unravel. One supercoil (or more formally, supercoiled turn) is introduced every time that the duplex thread is twisted about its axis; and the greater the number of supercoils, the greater the torsion in a closed molecule.

For a rubber band, it does not matter in which *direction* we apply the twist that generates the supercoils. (The two edges of the rubber band are equivalent.) But because the double helix is itself a twisted structure (as seen in the intertwining of the two strands), its response to torsion depends on the direction of the supercoiling.

Negative supercoils twist the DNA about its axis in the *opposite* direction from the clockwise turns of the right-handed double helix. This allows the DNA in principle to relieve the torsional pressure by adjusting the structure of the double helix itself. Relief generally takes the form of reducing the rotation per base pair, that is, of loosening the winding of the two strands about each other. Thus DNA with negative supercoils is said to be **underwound**. If the torsion is great enough, it may even lead to a limited disruption of base pairing, as illustrated in Figure 4.6.

The opposite types of effect are caused if DNA is supercoiled in the *same* direction as the intrinsic winding of the double helix. The introduction of **positive supercoils** tightens the structure, applying torsional pressure to wind the double helix even more tightly. Positively supercoiled DNA is said to be **overwound**. This state can be created by treatment *in vitro*, but does not occur naturally.

Figure 4.7
Supercoiling causes a DNA duplex to be twisted about itself in space; negative supercoiling can be relieved by disrupting base pairs.

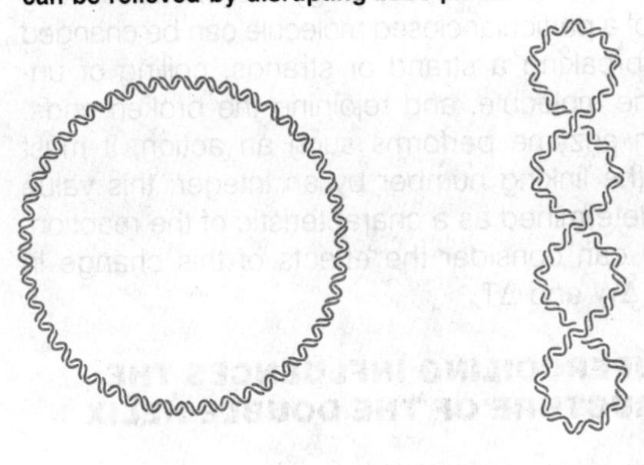

Circular DNA with zero supercoiling

Negatively supercoiled DNA

Negative supercoiling may be converted into strand separation

The supercoiling of DNA is often described in terms of the superhelical density, the number of superhelical turns per turn of the double helix. It is described as the parameter $\sigma = \tau/\beta$, where τ is the number of superhelical turns and β is the number of turns of the duplex, approximately the number of base pairs divided by 10. Thus σ corresponds to the number of superhelical turns per ~10 base pairs. It is negative for negative supercoiling and positive for positive supercoiling.

Since the original definition of supercoiling, a new nomenclature has been introduced to describe the situation. The concept is based on the **linking number**, which specifies the number of times that the two strands of the double helix of a closed molecule cross each other *in toto*. The linking number is the number of revolutions that one strand makes around the other when the DNA is considered (hypothetically) to lie flat on a plane surface. The convention is to count the linking number so that it is positive for each crossover in a right-handed double helix. It is necessarily an integer.

The linking number has two components, the twist and the writhe, defined by the equation

$$L = W + T$$

The **twisting number**, T, is a property of the double helical structure itself, representing the rotation of one strand about the other. It represents the *total number of turns of the duplex*. Thus it is determined by the number of base pairs per turn. For a relaxed closed circular DNA lying flat in a plane, the twist is the total number of base pairs divided by the number of base pairs per turn.

The **writhing number**, W, represents the *turning of the axis of the duplex in space*. It corresponds to the intuitive concept of supercoiling, but does not have exactly the same quantitative definition or measurement. For a relaxed molecule, W = 0. In this case, the linking number equals the twist.

The parameter measured in many experiments is the *change* in linking number, ΔL. It is given by the equation

$$\Delta L = \Delta W + \Delta T$$

The equation states that any change in the total number of revolutions of one DNA strand about the other can be expressed as the sum of the changes of the coiling of the duplex axis in space (ΔW) and changes in the screwing of the double helix itself (ΔT).

In the terms of the previous terminology, ΔW is analogous to change in supercoiling, and ΔT represents underwinding or overwinding of DNA. In the previous terminology, the relationship between twisting and supercoiling is given by

$$\alpha = \beta + \tau$$

where α is the *topological winding number*, equivalent to L, β corresponds to T, and τ corresponds to W.

A decrease in linking number, that is, a change of $-\Delta L$, corresponds to the introduction of some combination of negative supercoiling and/or underwinding. An increase in linking number, measured as a change of $+\Delta L$, corresponds to a decrease in negative supercoiling/underwinding.

The critical feature about the use of the linking number is that *this parameter is an invariant property of any individual closed DNA molecule.* The linking number cannot be changed by any deformation short of one that involves the breaking and rejoining of strands. So a circular molecule with a particular linking number can express this in terms of different combinations of T and W, but cannot change their sum so long as the strands are unbroken. (In fact, the partition of L between T and W prevents the assignment of fixed values for the latter parameters for a DNA molecule in solution.)

We now see the utility of the linking number. *It is related to the actual enzymatic events by which changes are made in the topology of DNA.* The linking number of a particular closed molecule can be changed only by breaking a strand or strands, coiling or uncoiling the molecule, and rejoining the broken ends. When an enzyme performs such an action, it must change the linking number by an integer; this value can be determined as a characteristic of the reaction. Then we can consider the effects of this change in terms of ΔW and ΔT.

SUPERCOILING INFLUENCES THE STRUCTURE OF THE DOUBLE HELIX

Changes in the structure of DNA do not occur spontaneously *in vivo*—they happen in an environment in which DNA is under a variety of constraints. Some of these changes are assisted by negative supercoiling.

The introduction of negative supercoiling requires energy; the supercoiled molecule has more energy through its possession of the supercoils. We might regard them as a store of energy. So when a change is to be made in the structure of DNA that *requires* energy, *less input of energy* will be needed to make the change if the DNA is negatively supercoiled. In practice, this means that negative supercoiling may influence the equilibrium of a structural change, and thus supercoiled DNA may undergo structural transitions in conditions in which relaxed DNA would be unable to do so.

One such transition is strand separation, which is assisted by the effect of negative supercoiling in underwinding the molecule. The excess energy *possessed* by a negatively supercoiled molecule can be used to help provide the energy *needed* to separate the strands of DNA.

Can we quantitate the effect of negative supercoiling? All genomes that have been examined exhibit some negative supercoiling. A typical level *in vivo* is about 1 negative turn for every 200 base pairs, which is described as a superhelical density of -0.05. This confers an energy on the molecule of about -9 kcal/mole.

The energy needed to unwind DNA depends on the sequence of base pairs; it is in effect the reverse of the energy released when a double helix forms as described in Figure 4.2. Thus we need to provide 12–50 kcal/mole to separate 10 base pairs. So the actual level of supercoiling probably corresponds to enough energy to assist the unwinding of just a very few base pairs.

The extreme case of unwinding a right-handed double helix would be to rewind it into a left-handed duplex. This is precisely what is involved in converting B-DNA to Z-DNA, and, indeed, negatively supercoiled DNA has a greater propensity to take the Z-form than relaxed DNA. In a circular DNA molecule that contains blocks of $\genfrac{}{}{0pt}{}{\text{CG}}{\text{GC}}$ repeats, their transition to the Z-form occurs at physiological salt concentrations only when the density of negative supercoiling reaches a critical value. When the C residues are methylated, fewer negative supercoils are needed to sponsor the transition.

The longer the stretch of $\genfrac{}{}{0pt}{}{\text{CG}}{\text{GC}}$ doublets, the more readily it can be converted to Z-form. Thus a stretch of 32 doublets converts at a superhelical density of -0.05, but a density of -0.07 is needed to convert a stretch of 14 doublets. These values imply that a relatively large amount of energy is needed to *initiate* the conversion of B-DNA to Z-DNA, but a relatively small amount of energy is needed to *extend* the region of Z-DNA within the $\genfrac{}{}{0pt}{}{\text{CG}}{\text{GC}}$ doublet stretch.

According to these results, the free energy needed to convert B-DNA to Z-DNA is the sum of two parameters, a ΔG per junction of 7.7 kcal/mole and a ΔG per additional base pair of 0.45 kcal/mole. The required energy is of the same order of magnitude as the energy made available by negative supercoils, so we see that the introduction of supercoiling is potentially a potent force in influencing the structure of the double helix.

What happens at the junction between the region of Z-DNA and B-DNA? Apparently it consists of at least one base pair (and perhaps several) whose members have been separated and are no longer in double helical form. The need to maintain the junctions as separated strands partly explains the large energy requirement for generating each junction.

Another possible effect of supercoiling is to generate cruciforms in palindromic DNA. A considerable amount of energy is needed to maintain a cruciform, of the order of $\Delta G = +18$ kCal/mol. However, an even greater energy of activation is required to disrupt the double helix to generate the cruciform initially, about $\Delta G = +50$ kCal/mol. Also, the reaction occurs very slowly under physiological conditions. In fact, it seems that the degree of supercoiling needed to generate a cruciform may be much greater than is actually found in the living cell.

DNA CAN BE DENATURED AND RENATURED

From the perspective of its functions in replicating itself and being expressed as protein, the central property of the double helix is the ability of the two strands to separate without needing to disrupt covalent bonds. This makes it possible for the strands to separate and to reform under physiological conditions at the (very rapid) rates needed to sustain genetic functions. The specificity of the process is determined by complementary base pairing. The same features that allow DNA to fulfill its biological role make it possible for us to manipulate the nucleic acid *in vitro*, and ultimately to isolate the segment of DNA that represents a particular protein.

The noncovalent forces that stabilize the double helix may be disrupted by heating, by exposure to high salt concentration, or by combination of such treatments. The two strands of a double helix separate entirely when all the hydrogen bonds between the two strands are broken. The process of strand separation is called **denaturation** or (more colloquially) **melting**. (Since "denaturation" is also used to describe loss of authentic protein structure, it is a general term implying that the natural conformation of a macromolecule has been converted to some other form.)

The denaturation of DNA occurs over a narrow temperature range and results in striking changes in many of the physical properties of DNA. A particularly useful change occurs in the optical density. The heterocyclic rings of nucleotides adsorb light strongly in the ultraviolet range (with a maximum close to 260 mu that is characteristic for each base). But the adsorption of DNA itself is some 40% less than would be displayed by a mixture of free nucleotides of the same composition.

This is called the **hypochromic** effect; it results from interactions between the electron systems of the bases, made possible by their stacking in the parallel array of the double helix. Any departure from the duplex state is immediately reflected by a decline in this effect—that is, by an increase in optical density toward the value characteristic of free bases. The denaturation of DNA can therefore be followed by this **hyperchromicity**.

The midpoint of the temperature range over which the strands of DNA separate is called the **melting temperature**, denoted $\mathbf{T_m}$. An example of a melting curve determined by change in optical adsorbance is shown in **Figure 4.8**. The curve always takes the same form, but its absolute position on the temperature scale (that is, its T_m) is influenced by both the base composition of the DNA and the conditions employed for denaturation.

When DNA is in solution under approximately physiological conditions, the T_m usually lies in a range of

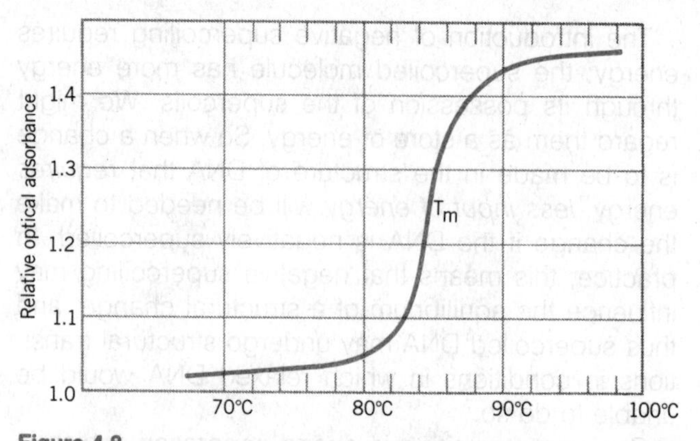

Figure 4.8
Denaturation of DNA can be followed by the increase in optical density and is described by the T_m.

85–95°C. Thus without intervention from cellular systems, the duplex DNA is stable at the temperature prevailing in the cell.

The exact value of the T_m depends on the proportion of G-C base pairs; because each G-C base pair has three hydrogen bonds, it is more stable than an A-T base, which has only two hydrogen bonds. Thus the more G-C base pairs are contained in a DNA, the greater the energy that is needed to separate the two strands.

The dependence of T_m on base composition is linear, increasing about 0.4°C for every percent increase in G-C content. So a DNA that is 40% G-C (a value typical of mammalian genomes) denatures with a T_m of about 87°C under the usual conditions, whereas a DNA that is 60% G-C has a T_m of about 95°C under the same conditions.

A major effect on T_m is exerted by the ionic strength of the solution. The T_m increases 16.6°C for every tenfold increase in monovalent cation concentration. The most commonly used condition is to perform manipulations of DNA in 0.12 M phosphate buffer, which provides a monovalent Na^+ concentration of 0.18 M, and a T_m of the order of 90°C.

The T_m can be greatly varied by performing the reaction in the presence of reagents, such as formamide, that destabilize hydrogen bonds. This allows the T_m to be reduced to as low as 40°C, with the advantage that the DNA does not suffer damage (such as strand breakage) that can result from exposure to high temperatures.

NUCLEIC ACIDS HYBRIDIZE BY BASE PAIRING

The denaturation of DNA is reversible under appropriate conditions, when the two separated complementary strands reform into a double helix. The reaction is called **renaturation** and is illustrated in **Figure 4.9**.

Renaturation depends on specific base pairing between the complementary strands. The reaction takes place in two stages. First, DNA single strands in the solution continually encounter one another by chance; if (and only if) their sequences are complementary, the two strands pair by forming a series of individual G-C and A-T pairs that generate a short double-helical region. Then the region of base pairing extends along the molecule by a zipper-like effect to form a lengthy duplex molecule. Renaturation of the double helix restores the original properties that were lost when the DNA was denatured.

Renaturation describes the reaction between two complementary sequences that were separated by denaturation. However, the technique can be extended to allow any two complementary nucleic acid sequences to **anneal** with each other to form a duplex structure. The reaction is generally described as **hybridization** when nucleic acids from different sources are involved, as in the case when one preparation consists of DNA and one of RNA. The ability of two nucleic acid preparations to hybridize constitutes a precise test for their sequence complementarity.

The principle of the hybridization reaction is to expose two single-stranded nucleic acid preparations to each other and then to measure the amount of double-stranded material that is formed. There are two common ways of performing the reaction: **solution** (or **liquid) hybridization** and **filter hybridization**.

Liquid hybridization is described by its name: the two preparations of single-stranded DNA are mixed together in solution. When large amounts of material are involved, the reaction may be followed by the change in optical density. With smaller amounts of material, one of the preparations may carry a radioactive label, whose entry into duplex form is followed by determining directly the amount of double-stranded DNA containing the label. Double-stranded DNA can be assayed either directly by using chromatography to separate duplex DNA from single strands or by degrading all the single strands that have not reacted and then measuring the amount of material that remains.

Solution hybridization is not an appropriate technique for investigating the relationship of two preparations one or both of which consist of duplex DNA. The problem is that if two duplex DNA preparations are denatured and then the single strands are mixed, two types of reaction occur. The *original* complementary single strands can renature. Or each single strand can hybridize with a complementary sequence in the *other* DNA. The competition between the two reactions makes it difficult to assess the extent of hybridization.

This difficulty can be overcome by immobilizing one

Figure 4.9
Denatured single strands of DNA can renature to give the duplex form.

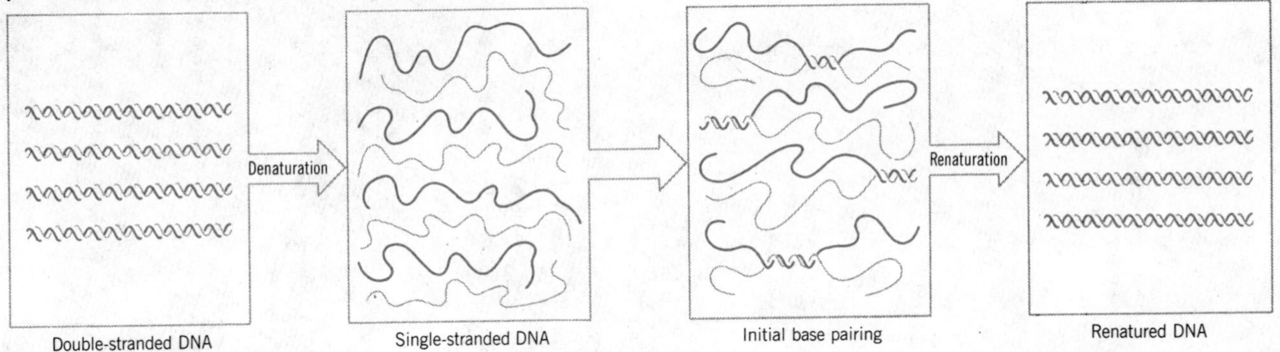

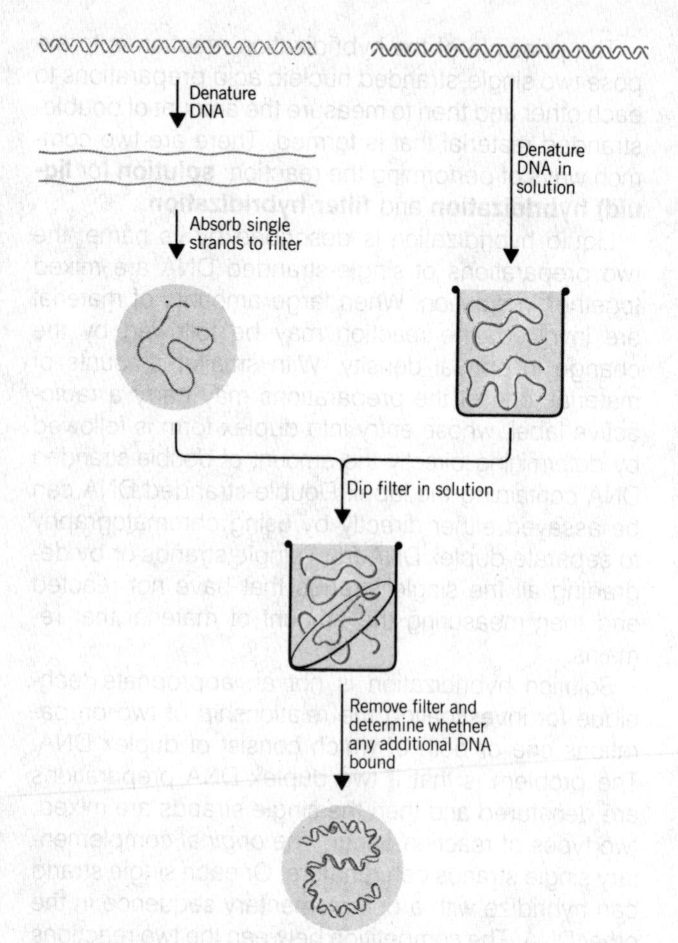

Figure 4.10
Filter hybridization establishes whether a solution of denatured DNA (or RNA) contains sequences complementary to the strands immobilized by the filter.

of the DNA preparations so that it cannot renature. Nitrocellulose filters have the useful property of adsorbing single strands of DNA but not RNA; and once a filter has been used to adsorb DNA, any further adsorption of single strands can be prevented by certain treatments. **Figure 4.10** illustrates the resulting procedure in which a DNA preparation is denatured and the single strands are adsorbed to the filter. Then a second denatured DNA (or RNA) preparation is added. This material adsorbs to the filter only if it is able to base pair with the DNA that was originally adsorbed. The usual form of the experimental procedure is to add a radioactively labeled RNA or DNA preparation to the filter, allowing the extent of reaction to be measured as the amount of radioactive label retained by the filter.

FURTHER READING

Alternative structures for DNA have been reviewed by **Cantor** (*Cell* **26**, 293–295, 1981). The structure of Z-DNA has been discussed in detail by **Rich, Nordheim & Wang** (*Ann. Rev. Biochem.*, **53**, 791–846, 1984). Conditions for cruciform generation have been analyzed by **Wang** (*Cell* **33**, 817–829, 1983).

CHAPTER 5
ISOLATING THE GENE

A chromosome contains a long, uninterrupted thread of DNA along which lie many genes. Identifiable sites in the genetic material are provided by mutations that create changes in the sequence of base pairs. On the linkage map of the chromosome, distances are represented by the frequencies of recombination. A eukaryotic chromosome may contain thousands of genes, so at the molecular level, a map unit is a very large measure indeed.

By comparing the total map length of all the linkage groups in a genome with the total amount of DNA, we can obtain an idea of how frequently genetic exchange occurs on the average relative to length of DNA. We may take 50 map units (50% recombination) to correspond to the average distance between independent recombination events. Values for a range of genomes are summarized in **Table 5.1**.

In bacteriophages, there are *exceedingly* frequent genetic exchanges (although any quantitation of the frequency is only *very* approximate). Also, it is possible to generate enormous numbers of progeny so that even the rarest recombination events (between mutations in adjacent base pairs) can be measured. These features allow the detailed mapping within a gene that we have described in Chapter 2. The rate of genetic exchange in bacteria is also high, and again it is possible to identify recombinants from very large numbers of progeny, so that here too mapping within a gene is possible.

In lower eukaryotes, at least as typified by yeast and other fungi, the frequency of recombination remains high, but in higher eukaryotes it is roughly a hundred times lower. We do not yet have enough information to establish a range for the higher eukaryotes, but the two examples of fruit fly and mouse illustrate the difficulties of detailed genetic mapping. Recombination events occurring at a frequency of 0.1% are likely to be 50,000 base pairs apart in the fruit fly and 180,000

Table 5.1
The frequency of genetic recombination is much lower in eukaryotes than in prokaryotes.

Species	Average Distance (in base pairs) Between Independent Crossovers	Base Pairs per Map Unit
Bacteriophage (T4)	1.0×10^4	200
Bacterium (*E. coli*)	1.2×10^5	2,400
Yeast (*S. cerevisiae*)	2.5×10^5	5,000
Fungus (*N. crassa*)	8.2×10^5	16,000
Fruit fly (*D. melanogaster*)	2.5×10^7	500,000
Mammal (*M. musculus*)	9.0×10^7	1,800,000

base pairs distant in the mouse. These distances could correspond to many genes.

Scrutinized from the perspective of the genetic map, the outlook for characterizing individual genes is therefore somewhat poor. Suppose a gene in the fruit fly consists of 5000 base pairs. This distance corresponds to 0.01% recombination. To detect recombination between mutations located at opposite ends of the gene, it would be necessary to isolate 1 fly in 10,000 progeny. The result is that genetic mapping within genes of the fruit fly is barely possible. It is not at all practical with mice, in which the frequency of recombination is lower and the number of progeny from each mating is less.

How accurately does the genetic map represent the actual length of the genetic material? At the gross level of the chromosome, the genetic map usually provides a reasonable representation of the underlying molecular structure (as illustrated in Figure 2.11). However, the map unit depends on the relative frequency of recombination; since this may be strikingly different in each species, the genetic maps of different organisms cannot be compared. And at the molecular level, the genetic map is not informative about the real relationship between sites of mutation.

Even in prokaryotes, the information to be gained from fine structure mapping is limited by the nature of the recombination event. At the level of mapping within a gene, recombination frequencies are not entirely independent of the particular mutations used in each cross, but may be influenced quite strongly by the local sequence of DNA. In other words, they do not have the ideal property of allele-independence that we described in Chapter 2, but actually show allele-specific effects. Our view of the gene in terms of the genetic map is therefore distorted by the characteristics of the recombination systems.

So at the molecular level, the map distances between mutations do not necessarily correspond with the distance that actually separates them on DNA. The resolution of the genetic map also is limited by the availability of mutations; a gap in the map could represent a region in which no mutations have been found, or it could be caused by a local increase in recombination frequency.

We have no reason to suppose that the situation would be any different in eukaryotes if we were able to perform fine structure mapping. So for both theoretical and practical reasons, we need another source of information. How are we to establish the molecular structure of the gene and relate it to the structure of the protein product? How far apart are adjacent genes and how are we to recognize the regions between them?

RESTRICTION ENZYMES CLEAVE DNA INTO SPECIFIC FRAGMENTS

The ultimate aim of genetic mapping is to establish the nucleotide sequence of the gene and to extend the sequence to the neighboring genes on either side. By working back from a protein, we can use nucleic acid hybridization techniques to isolate the corresponding DNA. It is in fact now almost routine to isolate the cellular DNA for which a protein product is available (although this can be a lengthy process if the product is scarce).

Having isolated the DNA, a crucial step en route to obtaining its sequence is to map the nucleic acid at the molecular level. A physical map of any DNA molecule can be obtained by breaking it at defined points whose distance apart can be accurately determined. The necessary specific breakage is made possible by the ability of certain **restriction enzymes** to recognize rather short, specific sequences of DNA as targets for cleavage. A map of DNA obtained by identifying these points of breakage is known as a **restriction map**.

A restriction map represents a linear sequence of the sites at which particular restriction enzymes find their targets. Distance along such maps is measured directly in base pairs (abbreviated **bp**) for short distances; longer distances are given in **kb**, corresponding to kilobase pairs in DNA or to kilobases in RNA.

From the restriction map, we can proceed to determine the sequence of DNA between points that are sufficiently close together (within about 300 bp for practical purposes). By suitable choice of points, these regions can be connected into a sequence of an entire gene and its environs. By comparing the sequence of the DNA with the sequence of the protein that the gene represents, we can delineate the regions that code for the polypeptide; and by extending the sequence in either direction, the distance to the next gene can be determined.

By comparing the sequence of a wild type DNA with that of a mutant allele, we can determine the nature of the mutation and its exact site of occurrence. This defines the relationship between the genetic map (based entirely on sites of mutation) and the physical map (based on or even comprising the sequence of DNA). Ultimately, the map of a region of the genome can be expressed in base pairs of DNA rather than in the relative map units of formal genetics. Indeed, genes can now be identified by virtue of their protein product, or even sometimes by their sequence alone, so we are no longer entirely dependent on mutations to provide the raw material for constructing a map of the genome. Of course, mutants remain essential for identifying the functions of gene products.

Each restriction enzyme has a particular target in duplex DNA, usually a specific sequence of between 4 and 6 base pairs. The enzyme cuts the DNA at every point at which its target sequence occurs. Different restriction enzymes have different target sequences, and a large range of these activities (obtained from a wide variety of bacteria) now is available. (They are discussed in the context of their natural habitat in Chapter 33.)

If we take a particular DNA molecule and cut it with a suitable restriction enzyme, a limited number of breaks are made, producing a number of distinct fragments. These fragments can be separated on the basis of their size by **gel electrophoresis**. In this technique, the cleaved DNA is placed on top of a gel made of agarose or polyacrylamide. When an electric current is passed through the gel, fragments move down it at a rate that depends on their length. The smaller a fragment, the more rapidly it moves. (The distance moved depends on the log of the fragment length.)

This movement produces a series of **bands**, each band corresponding to a fragment of particular size, the size decreasing down the gel. The length of any particular fragment can be determined by **calibrating** the gel. This is done by running a control in parallel in another slot of the same gel. The control has a mixture of standard fragments all of known size (often called **markers**). The migration of the markers defines the relationship between fragment length and distance moved for the particular gel.

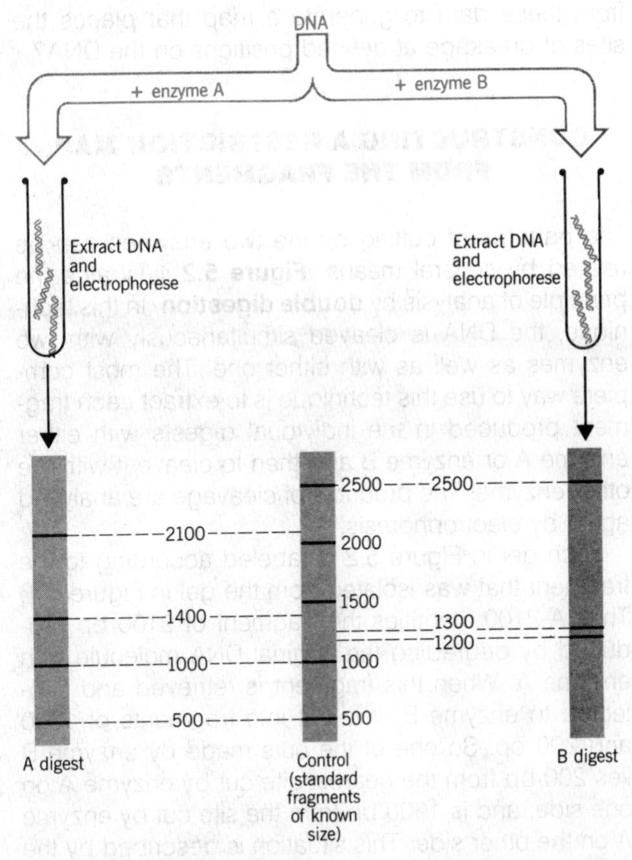

Figure 5.1
DNA can be cleaved by restriction enzymes into fragments that can be separated by agarose gel electrophoresis.
The sizes of the individual fragments generated by enzyme A (left) or enzyme B (right) are determined by comparison with the positions of fragments of known size, such as the control shown in the center.

Figure 5.1 shows an example of this technique. A DNA molecule of length 5000 bp is incubated separately with two restriction enzymes, A and B. After cleavage the DNA is electrophoresed, revealing that enzyme A has cut the substrate DNA into four fragments (of lengths 2100, 1400, 1000 and 500 bp), while enzyme B has generated three fragments (of lengths 2500, 1300 and 1200 bp). Can we proceed further from these data to generate a map that places the sites of breakage at defined positions on the DNA?

CONSTRUCTING A RESTRICTION MAP FROM THE FRAGMENTS

The patterns of cutting by the two enzymes can be related by several means. **Figure 5.2** illustrates the principle of analysis by **double digestion**. In this technique, the DNA is cleaved simultaneously with two enzymes as well as with either one. The most complete way to use this technique is to extract each fragment produced in the individual digests with either enzyme A or enzyme B and then to cleave it with the other enzyme. The products of cleavage are analyzed again by electrophoresis.

Each gel in Figure 5.2 is labeled according to the fragment that was isolated from the gel in Figure 5.1. Thus A-2100 identifies the fragment of 2100 bp produced by degrading the original DNA molecule with enzyme A. When this fragment is retrieved and subjected to enzyme B, it is cut into fragments of 1900 and 200 bp. So one of the cuts made by enzyme B lies 200 bp from the nearest site cut by enzyme A on one side, and is 1900 bp from the site cut by enzyme A on the other side. This situation is described by the map

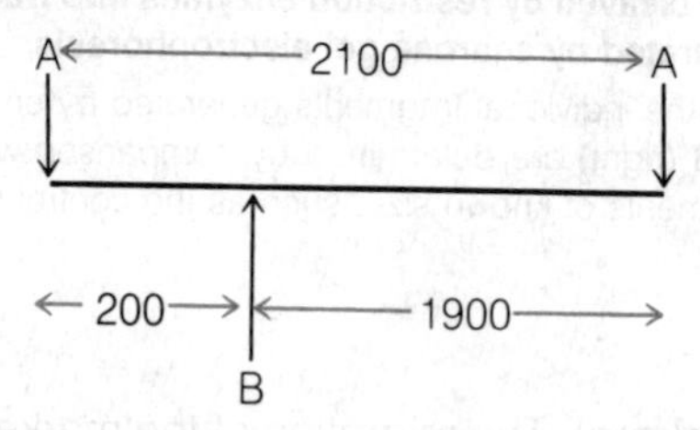

The vertical arrows indicate the sites of cleavage by enzyme A or enzyme B; the horizontal arrows and numbers indicate the lengths of individual fragments.

A related pattern of cuts is seen when we examine the susceptibility of fragment B-2500 to enzyme A. It is cut into fragments of 1900 and 600 bp. So the 1900 bp fragment is generated by double cuts, with an A site at one end and a B site at the other end. It can be released from either of the single-cut fragments (A-2100 or B-2500) that contain it. These single-cut fragments must therefore **overlap** in the region of the 1900

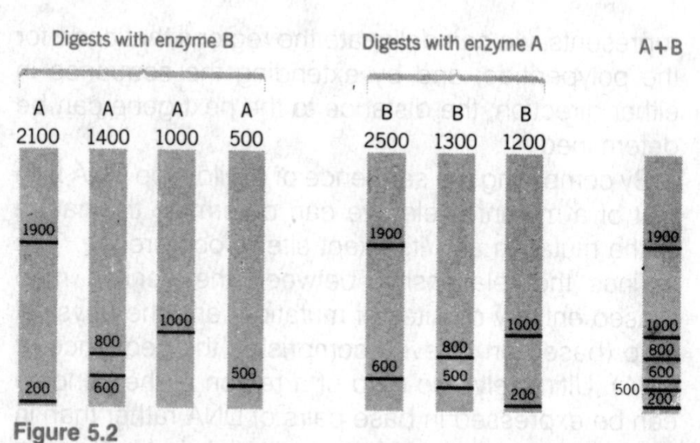

Figure 5.2

Double digests define the cleavage positions of one enzyme with regard to the other.

The four fragments obtained by digestion with enzyme A can be eluted from the gel shown in Figure 5.1; then they are degraded by digestion with enzyme B (four gels on left). Similarly, the three bands originally produced by enzyme B can be obtained and then digested with enzyme A (three bands in center). A complete double digest is obtained by subjecting the intact DNA to both enzymes simultaneously (band at right).

bp of the common fragment that can be generated from them. This is described by extending our map to the right to add a cleavage site for enzyme B.

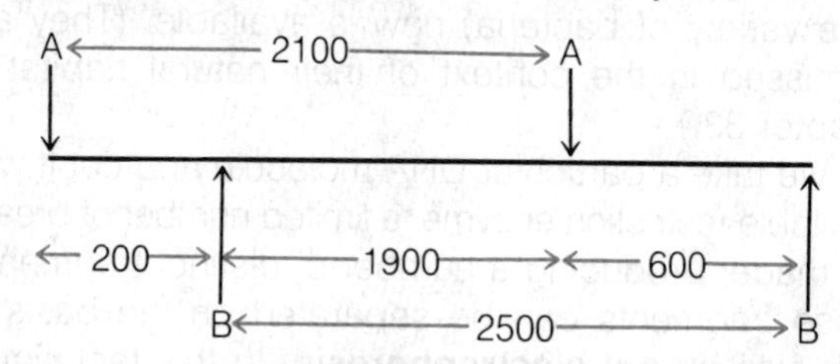

This map demonstrates an important principle of restriction mapping. When we consider the construction of larger fragments from smaller fragments, we can rely on the **complete additivity** of lengths (within experimental limits). Thus fragment A-2100 consists of fragments of 200 bp and 1900 bp, while fragment B-2500 consists of fragments of 1900 bp and 600 bp.

When all of the fragments are analyzed in this manner, we can see that every fragment produced by cutting an original A fragment with the B enzyme *also* is found in one of the double digests in which an original B fragment was cut with the A enzyme. The entire pattern can be seen in the complete double-digest (the gel at the right of Figure 5.2), in which every double-cut fragment occurs once. These data allow the sites of cutting to be placed into an unequivocal map.

The key to restriction mapping is the use of overlapping fragments. Because of the overlap of A-2100 and B-2500 in the central region of 1900 bp, we can relate the A site 200 bp to the left of the 1900 bp region with the B site 600 bp to the right. In the same way, we can now extend the map farther on either side. The 200 bp fragment at the left is also produced by cutting B-1200 with enzyme A, so the next B site must lie 1000 bp to the left. The 600 bp fragment at the right is also produced by cutting A-1400 with enzyme B, so the next A site must lie 800 bp to the right. This gives the map

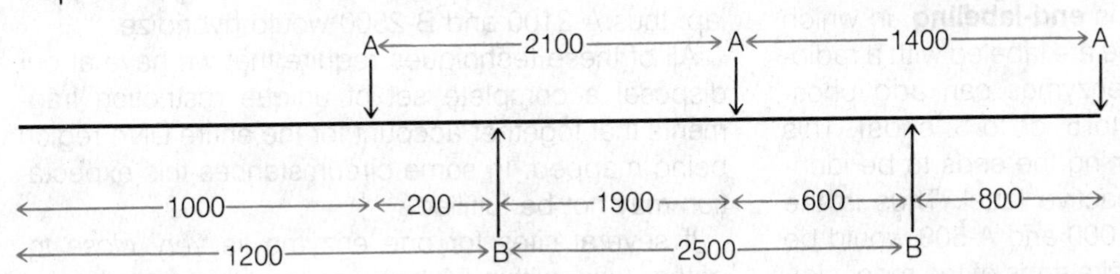

We can now complete the map by identifying the source of the two fragments at each end. At the left end, the 1000 bp fragment arises from B-1200 or in the form of A-1000, which is not cut by enzyme B. Thus A-1000 lies at the end of the map; in other words, proceeding from the left end of the complete 5000 bp region, it is 1000 bp to the first A site and 1200 bp to the first B site. (This is why a B cut is not shown at the left end of the map above, although formally we treated the end as a B-cutting site in the analysis.)

At the right end of the map, the 800 bp double-cut fragment is generated by cutting B-1300 with enzyme A, so we must add a fragment of 500 bp to the right. This is the terminal fragment, as seen by its presence as A-500 in the single-cut A digest. Thus our completed map takes the form

We have now produced a restriction map of the entire 5000 bp region. This is recapitulated in its more usual form in **Figure 5.3**. The map shows the positions at which particular restriction enzymes cut DNA; the distances between the sites of cutting are measured in base pairs. Thus the DNA is divided into a series of regions of defined lengths that lie between sites recognized by the restriction enzymes.

The actual construction of a restriction map usually requires recourse to several enzymes, so it becomes necessary to resolve quite a complex pattern of the overlapping fragments generated by the various enzymes. Several further techniques are used to assist the mapping.

One approach is to make **partial digests**, by using conditions in which an enzyme does not recognize every target in every DNA molecule, but instead sometimes fails to cleave some targets. (Formally, the conditions can be set so that the enzyme has a specified probability, for example, 50%, of cutting at any particular one of its recognition sites.) With this technique, enzyme A might generate partial cleavage fragments of 3500, 3100 and 1400 bp with the DNA segment mapped in Figure 5.3. By comparing the partial products with the complete digest pattern, the 1000 and 2100 bp fragments can be assigned as adjacent components of the 3100 partial fragment. Similarly, the 2100 and 1400 bp fragments could be placed as adjacent components of the 3500 partial fragment. Thus this technique allows the sites cut by a single enzyme to be ordered.

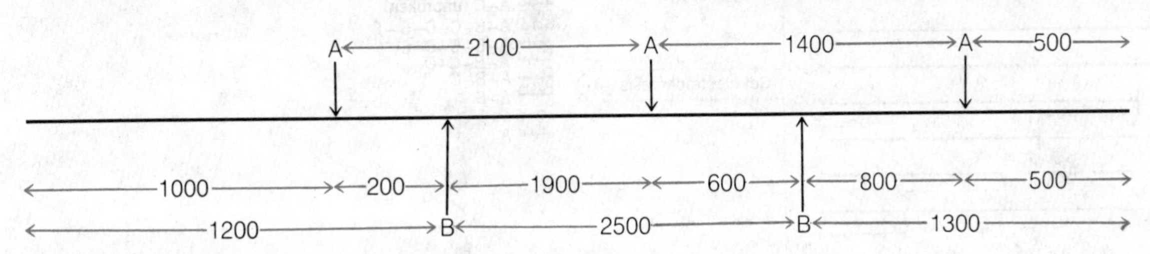

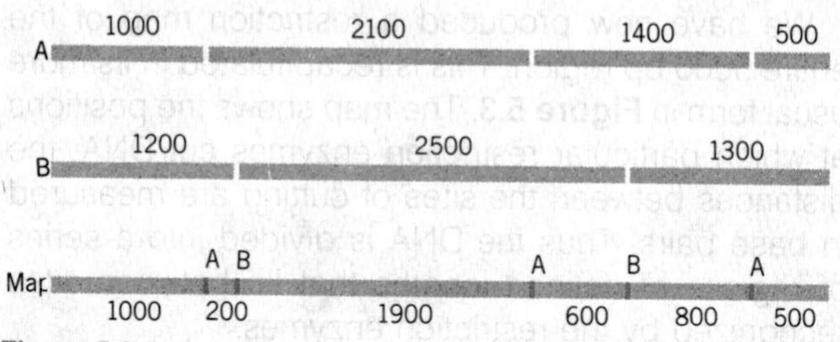

Figure 5.3

A restriction map is a linear sequence of sites separated by defined distances on DNA.

The map identifies the sites cleaved by enzymes A and B, as defined by the individual fragments produced by the single and double digests.

Another useful technique is **end-labeling**, in which the ends of the DNA molecule are labeled with a radioactive phosphate (certain enzymes can add phosphate moieties specifically to 5' or to 3' ends). This allows the fragments containing the ends to be identified directly by their radioactive label. Thus in the fragment A preparation, A-1000 and A-500 would be placed immediately at opposite ends of the map; similarly, fragments B-1200 and B-1300 would be picked out as ends.

By combining partial digestion with end-labeling, a series of sites cut by one enzyme can be mapped directly relative to the end. **Figure 5.4** shows that if fragments are identified only by their radioactively labeled ends, those cut from within the molecule are ignored, while a series of fragments identifies the distance of each cutting site from the labeled end. If the left end of the 5000 bp fragment of Figure 5.3 was labeled, partial cleavage with enzyme A would immediately identify cutting sites at 1000, 3100, and 4500 bp from the end.

In analyzing the products of cleavage by multiple enzymes, fragments are often compared by the technique of nucleic acid hybridization. This greatly strengthens the analysis. For example, in constructing the map of Figure 5.3, we inferred that the A-2100 and B-2500 fragments must overlap because they both generate a fragment of 1900 bp in double digests. However, in an analysis of a longer DNA, there could be several fragments of the same size. Hybridization allows us to test directly whether two fragments overlap; thus A-2100 and B-2500 would hybridize.

All of these techniques require that we have at our disposal a complete set of unique restriction fragments that together account for the entire DNA region being mapped. In some circumstances this expectation may not be fulfilled.

If several sites for one enzyme lie *very close together* (say, within 50 bp), very small fragments may be generated that are lost from the agarose gel. Of course, this will result in a discrepancy when the molecular weights of the other fragments are totalled, but this would not necessarily be considered sinister by itself, since there are always modest experimental discrepancies when fragment sizes are compared. This difficulty is dealt with by using several restriction enzymes to construct a map, so that there are sufficient overlaps to be sure that all of the DNA is contained.

Figure 5.4

When restriction fragments are identified by their possession of a labeled end, each fragment directly shows the distance of a cutting site from the end. Successive fragments increase in length by the distance between adjacent restriction sites.

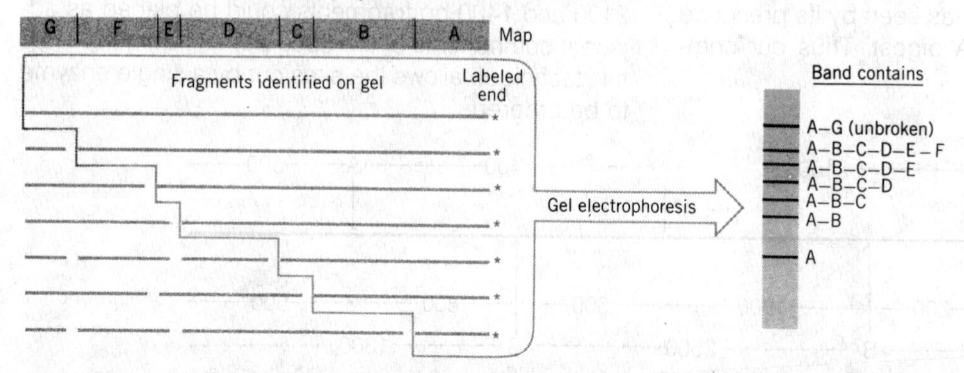

A complication is provided by the assumption that every restriction fragment is seen as an individual band on the gel. This will not be true when more than one fragment has the same molecular weight (or, more practically, whenever fragments exist with sizes so similar that they cannot be resolved.) To check whether each band is unique, its intensity can be quantitated. Bands that contain more than one fragment should be correspondingly more intense. Multiple fragments of the same size can be resolved by using further restriction enzymes.

A final point about restriction enzymes concerns nomenclature. The enzymes are named by a three letter (sometimes four letter) abbreviation that identifies their origin. Roman numerals are added to distinguish multiple enzymes that have the same origin. Examples are HpaI (derived from *Haemphilus parainfluenzae*) and EcoRI (derived from *E. coli*).

RESTRICTION SITES CAN BE USED AS GENETIC MARKERS

A restriction map and a genetic map provide different views of the same material. The two types of map can be related, however, to reveal the physical basis underlying the genetic map. Also, restriction cleavage patterns can be integrated with the genetic map.

A restriction map identifies a linear series of sites in DNA, separated from one another by actual distance along the nucleic acid. *A restriction map can be obtained for any sequence of DNA, irrespective of whether mutations have been identified in it, or, indeed, whether we have any knowledge of its function.*

A genetic map identifies a series of sites at which mutations occur. *The existence of the sites depends on the fact that changes in base sequence have altered the phenotype. The distance between the sites is determined by recombination frequencies,* which are influenced by the local conditions.

To relate the restriction map to the genetic map, we must compare the restriction maps of wild type and corresponding mutant DNAs. The relationship between the genetic and physical maps can be determined on the basis of restriction mapping alone when there are mutants that make large changes in the genetic map. For example, a small deletion or insertion

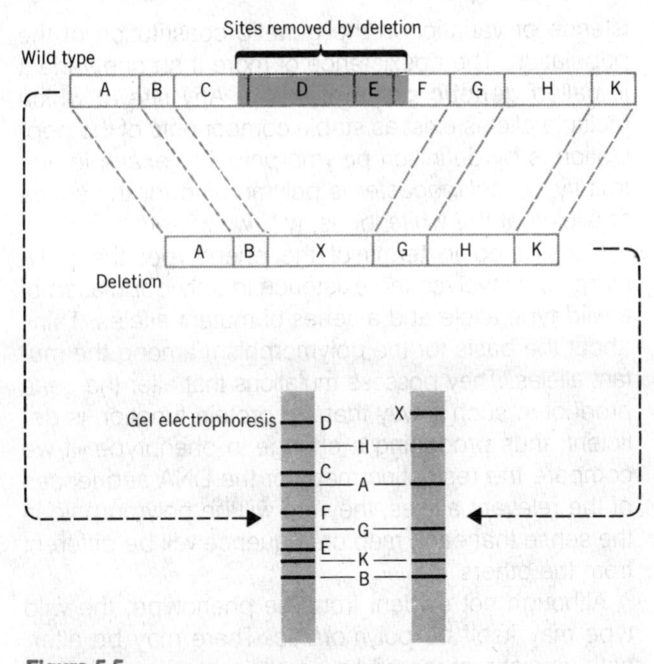

Figure 5.5

A deletion removes a series of restriction sites, and fuses the restriction fragments containing its ends.

The deletion entirely eliminates fragments D and E, which are present in the restriction fragments of the wild type but not the deletion. Fragments C and F of the wild type also are absent from the deletion, which has a new fragment, X, that contains part of their sequence. The fragments representing regions on either side of the deletion, A and B on the left, and G, H and K on the right, remain exactly the same in both wild type and deletion digests.

An insertion would have precisely the reverse effect; a mutant would contain additional fragments that are not present in the wild type.

that affects the genetic map should cause a corresponding reduction or increase in size of the restriction fragment in which it lies. A longer deletion (or insertion) may remove (or introduce) a series of restriction sites. **Figure 5.5** shows an example.

Locating point mutations on the restriction map is more difficult. Occasionally they may change target sites for restriction enzymes, but otherwise they remain undetectable, since the sizes of the restriction fragments remain the same in wild type and mutant DNAs. To locate these base substitutions, it is therefore often necessary to determine the sequence of the DNA.

Construction of a genetic map is based on the existence of variation in the genetic constitution of the population. The coexistence of more than one variant is called **genetic polymorphism**. Any site at which multiple alleles exist as stable components of the population is by definition polymorphic. For example, the fruit fly *D. melanogaster* is polymorphic for the series of alleles at the *white* locus, w^+, w^i, w^a, etc.

Considered in terms of the phenotype, the polymorphism involves the existence in a fly population of a wild-type allele and a series of mutant alleles. Think about the basis for the polymorphism among the mutant alleles. They possess mutations that alter the gene product in such a way that the protein function is deficient, thus producing a change in phenotype. If we compare the restriction maps or the DNA sequences of the relevant alleles, they too will be polymorphic in the sense that each map or sequence will be different from the others.

Although not evident from the phenotype, the wild type may itself be polymorphic. There may be alternate versions of the wild-type allele, distinguished by differences in sequence that do not affect their function, and which therefore are not detected in the form of phenotypic variants. Considered in terms of genotype, a fly population may have extensive polymorphism; many different sequence variants may exist at the *w* locus, some of them evident because they affect the phenotype, others hidden because they have no visible effect.

Polymorphisms in the genome can be detected by comparing the restriction maps of different individuals. The criterion is a change in the pattern of fragments produced by cleavage with a restriction enzyme. **Figure 5.6** shows that when a target site is present in the genome of one individual and absent from another, the extra cleavage in the first genome will generate two fragments corresponding to the single fragment in the second genome.

Because the restriction map is independent of gene function, a polymorphism at this level can be detected *irrespective of whether the sequence change affects the phenotype*. In fact, probably only a minority of the restriction polymorphisms in a genome actually affect the phenotype. The majority may involve sequence changes that have no effect on the production of proteins (for example, because they lie between genes).

A difference in restriction maps between two individuals can be used as a genetic marker in exactly the same way as any other marker. The sole difference is that instead of examining some feature of the phenotype, we directly assess the genotype, as revealed by the restriction map. **Figure 5.7** illustrates

Figure 5.6
A change in DNA that affects a restriction site is detected by a difference in restriction fragments.

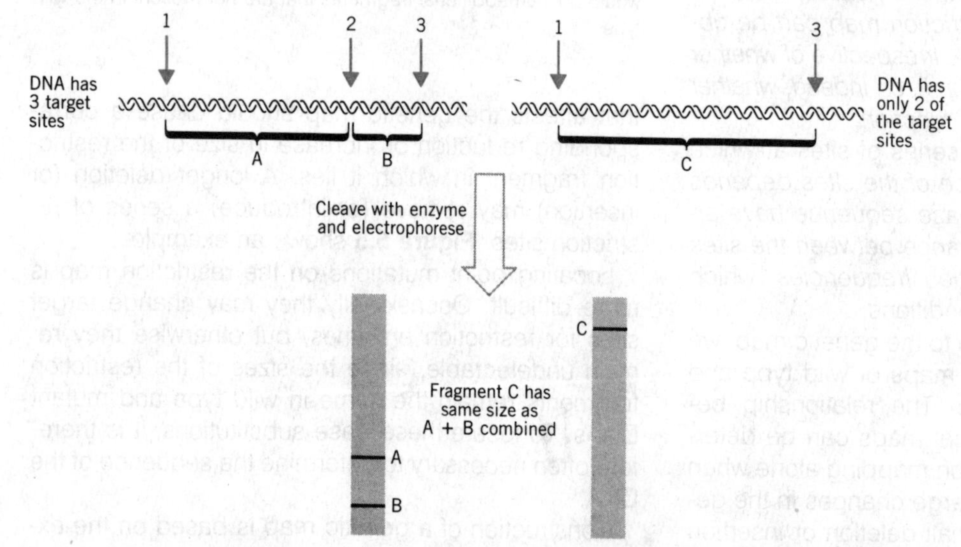

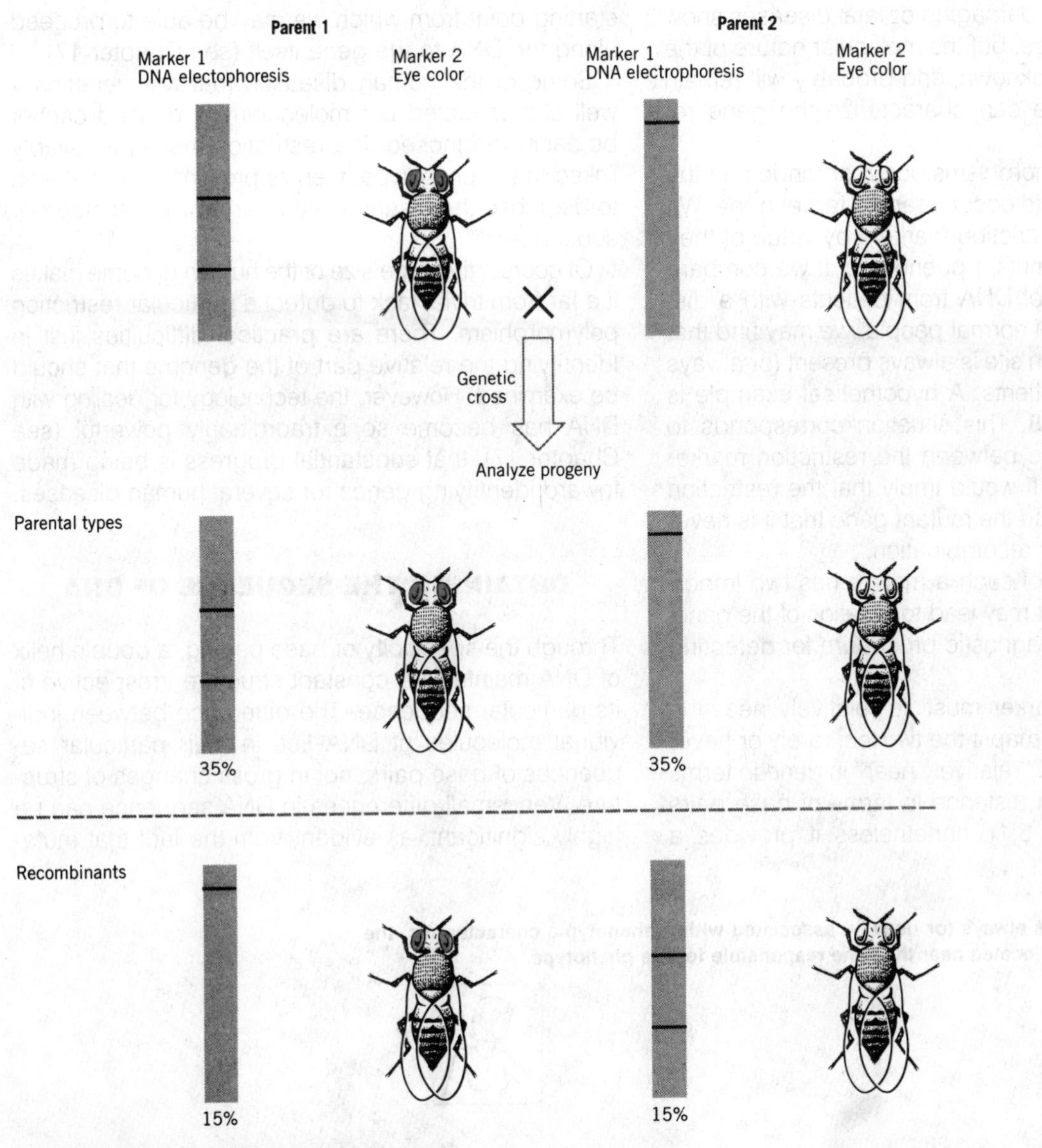

Figure 5.7
A restriction polymorphism can be used as a genetic marker to measure recombination distance from a phenotypic marker (such as eye color).

the measurement of recombination frequency between a restriction marker and a phenotypic marker.

Because restriction markers are not restricted to those genome changes that affect the phenotype, they provide the basis for an extremely powerful technique for identifying genetic loci at the molecular level. A typical problem concerns a mutation with known effects on the phenotype, where the relevant genetic locus can be placed on a genetic map, but for which we have no knowledge about the corresponding gene or protein. Some important human diseases fall into this category. For example, cystic fibrosis, Huntington's chorea, and many other damaging or fatal diseases show Mendelian inheritance, but the molecular nature of the mutant function is unknown, and probably will remain unidentified until we can characterize the gene responsible.

If restriction polymorphisms occur at random in the genome, some should occur near the target gene. We can identify such restriction markers by virtue of their tight linkage to the mutant phenotype. If we compare the restriction map of DNA from patients with a disease with the DNA of normal people, we may find that a particular restriction site is always present (or always absent) from the patients. A hypothetical example is shown in **Figure 5.8**. This situation corresponds to finding 100% linkage between the restriction marker and the phenotype. It would imply that the restriction marker lies so close to the mutant gene that it is never separated from it by recombination.

The identification of such a marker has two important consequences. It may lead to isolation of the gene. And it may offer a diagnostic procedure for detecting the disease.

The restriction marker must lie relatively near the gene on the genetic map if the two loci rarely or never recombine. Although "relatively near" in genetic terms can be a substantial distance in terms of base pairs of DNA (see Table 5.1), nonetheless it provides a starting point from which we may be able to proceed along the DNA to the gene itself (see Chapter 17).

Some of the human diseases that are genetically well characterized but molecularly ill defined cannot be easily diagnosed. If a restriction marker is reliably linked to the phenotype, then its presence can be used to diagnose the disease, either at a prenatal stage or subsequently.

Of course, the large size of the human genome makes it a far from trivial task to detect a particular restriction polymorphism. There are practical difficulties just in identifying the relative part of the genome that should be examined. However, the technology for dealing with DNA has become so extraordinarily powerful (see Chapter 17), that substantial progress is being made toward identifying genes for several human diseases.

OBTAINING THE SEQUENCE OF DNA

Through the specificity of base pairing, a double helix of DNA maintains a constant structure irrespective of its particular sequence. The difference between individual molecules of DNA lies in their particular sequences of base pairs, not in gross changes of structure. Very small differences in DNA sequence can be highly significant, as evident from the fact that muta-

Figure 5.8
If a restriction marker is always (or usually) associated with a phenotypic characteristic, the restriction site must be located near the gene responsible for the phenotype.

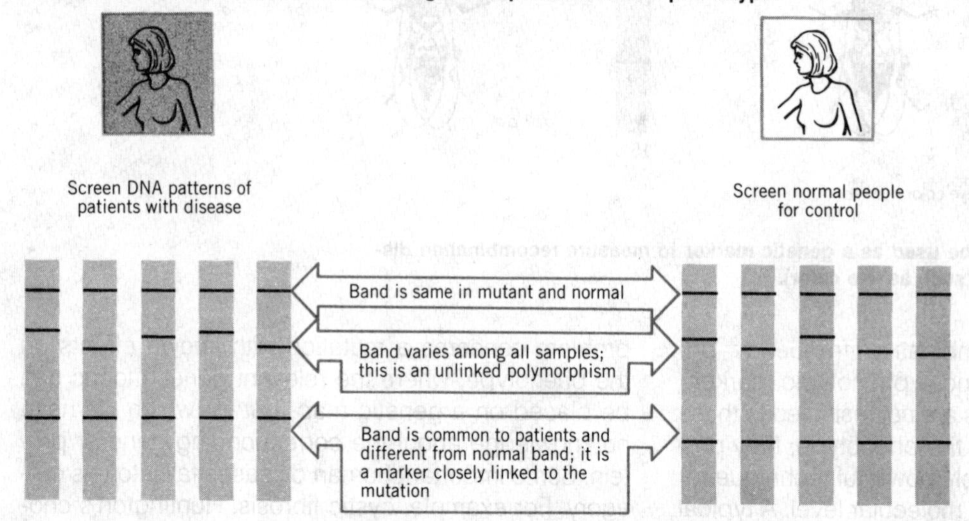

tions consist of changes in DNA sequence as small as one base pair.

How can the exact nucleotide pair sequence of DNA be determined? The first approach to nucleic acid sequencing followed that of protein sequencing: break the molecule into small fragments, determine their base composition, and by obtaining overlapping fragments deduce the exact sequence. With proteins this is practical because of the variety offered by the 20 amino acids; with nucleic acids, it poses a problem because there are only four bases. The old techniques were therefore restricted to determining the sequences of very short nucleic acids.

But now it is possible to determine DNA sequences *directly*. In fact, DNA sequences can be obtained much more rapidly than protein sequences.

All sequencing protocols share the same principle of approach. Each protocol generates a series of single-stranded DNA molecules, each molecule one base longer than the last. DNA molecules of the *same sequence*, but differing in length by as little as one base at one end, can be separated by electrophoresis on acrylamide gels. Bands corresponding to molecules of increasing length form a "ladder" that can be followed for up to about 300 nucleotides.

Two types of approach have been used to obtain these sets of bands. One is to use chemical reactions that cleave DNA at individual bases. The other is to synthesize the DNA *in vitro* in such a way that the reaction terminates specifically at the position corresponding to a given base.

To determine the sequence of the molecule by either approach, it is subjected to the appropriate protocol in four separate reactions, each reaction specific for one of the bases. The products are electrophoresed in four parallel slots of the same gel. The band corresponding to a particular length appears in only one of the four slots. The slot identifies the nucleotide that is present at the corresponding position in DNA. By proceeding from each band to the band of the next size, the series of nucleotides can be "read" according to which slots the bands appear in. This gives the DNA sequence.

In the chemical method, a DNA fragment previously labeled at one (unique) end with radioactive ^{32}P is broken specifically at one of the four types of base. The key to using the cleavage for sequencing is to ensure that there is only partial breakage at any susceptible position, occurring with a probability of, say, 1–2%.

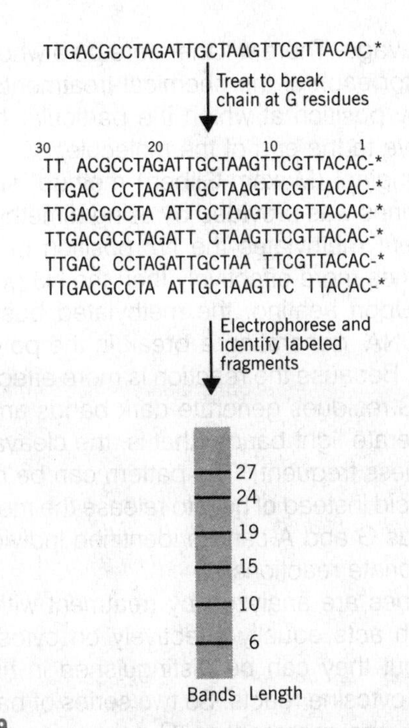

Figure 5.9

The locations of a particular base in single-stranded DNA can be determined by electrophoresing end-labeled fragments after treatment with a suitable reagent.

The chain is broken at G residues in this example. Some chains are broken only once, some more than once, but in each case the fragment that is identified is the one with the labeled end (indicated by the asterisk). Altogether, the labeled molecules extend from the common end to every position at which a G was present in the original DNA. When they are electrophoresed, each labeled molecule gives a band whose position on the gel indicates its length.

So we start with a preparation that contains a large number of identical molecules, all labeled at the same end. When subjected to the chemical treatment, each molecule is likely to be cleaved at only a small number of the sites at which the appropriate base occurs. But in the entire preparation of DNA molecules, every position will be attacked on some occasion, as illustrated in **Figure 5.9**.

The treated molecules are electrophoresed, and those carrying the radioactive label are identified (by autoradiography). Each will form a band whose length depends on the distance from the labeled end to the site of cleavage. There will therefore be a whole series of bands for each of the chemical treatments, identifying every position at which the particular base occurs relative to the end of the molecule.

In the original (Maxam-Gilbert) method, specificity for the purines was provided by using dimethylsulfate. This reagent methylates the N7 position of guanine about 5 times more effectively than the N3 position of adenine. Upon heating, the methylated base is lost from the DNA, generating a break in the polynucleotide chain. Because the reaction is more effective with guanine, G residues generate dark bands and A residues generate light bands (that is, the cleavage of A has been less frequent). The pattern can be reversed by using acid instead of heat to release the methylated bases. Thus G and A can be identified individually in the appropriate reactions.

Pyrimidines are analyzed by treatment with hydrazine, which acts equally effectively on cytosine and thymine; but they can be distinguished in high salt, when only cytosine reacts. So two series of bands are generated, one representing C only and one representing the combination of C + T.

A set of four reaction mixtures is set up as indicated in **Figure 5.10**. The products of the reactions are analyzed in parallel by electrophoresis. Each of the four contains a series of bands identifying molecules in which breakage has occurred at the various positions at which the target base was located. The sequence of DNA is obtained by proceeding from each band to the band that is one base longer. The shortest bands are present at the bottom of the gel, so the DNA sequence is read upward.

How accurate is DNA sequencing? The set of reactions gives the sequence of one strand of the DNA. There is always a risk of an occasional error. However, the accuracy can be improved by independently sequencing *both* strands of DNA. Any sites at which they are not complementary are identified as possible errors that can be corrected. With this precaution, errors in the sequence should be very rare.

DNA molecules of up to about 300 nucleotides can be separated on acrylamide gels. How do we proceed from determining the sequences of these relatively short fragments to that of a longer region? Just as with restriction mapping itself, the key concept is the use of

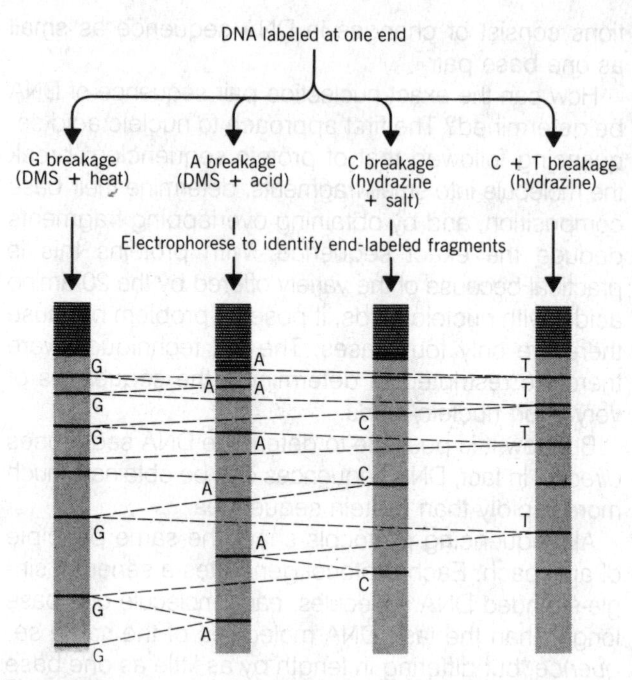

Figure 5.10
DNA can be sequenced by a set of reactions that together are specific for all four bases.
Four reactions are performed in parallel and the products are electrophoresed. The sequence is read up from the bottom of the gel, following the bands successively across as indicated by the zigzag. G, A, and C are identified by the left three gels, both C and T appear on the right, so the difference between the C and the C+T bands identifies the positions of T. The DNA sequence is

end-N_{20}GAGCATGACGGTAGCTAGAGTA. . . .

At the bottom of the gel, the shortest fragments are usually about 20 bases long, which is the closest we can come to the labeled end. The bands become progressively closer together proceeding up the gel, until at the top it becomes impossible to read the sequence.

overlapping fragments. Consider a series of adjacent fragments, ordered on a restriction map as

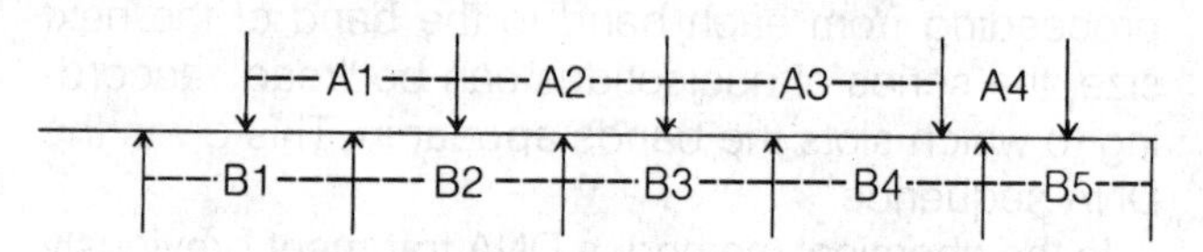

Suppose that we determine the sequence of each member of the A series of fragments. We know that

they are adjacent from the restriction map, so their individual sequences should in theory abut directly and give us the entire sequence. But to be sure that there are no small missing fragments generated by closely adjacent cleavages of the same enzyme (suppose that there were two cleavage sites, 10 bp apart, between A2 and A3!), it is essential to sequence *across each junction*. Thus we require another set of fragments, one in which all the junctions between the A fragments are intact. This is provided by the B series of fragments, each of which contains the sequence on either side of the junction between two A fragments.

Just as we can construct the restriction map from overlapping fragments, we can construct the overall DNA sequence by overlapping the ends of each individually sequenced fragment. Suppose, for example, that the ends of fragments A1 and A2 have the sequences

A1. . .CGTAGGGTCAAGTCAT
AATGCTGCCCAAA. . .A2

Then in fragment B2 we should find the sequence

CGTAGGGTCAAGTCATAATGCTGCCCAAA

in which the two ends run continuously from one to the other.

PROKARYOTIC GENES AND PROTEINS ARE COLINEAR

By comparing the nucleotide sequence of a gene with the amino acid sequence of a protein, we can determine directly whether the gene and the protein are **colinear**: whether the sequence of nucleotides in the gene corresponds exactly with the sequence of amino acids in the protein. In bacteria and their viruses, there

Figure 5.11
The recombination map of the tryptophan synthetase gene corresponds with the amino acid sequence of the protein.
Black bars indicate the sites of mutation, as identified by amino acid substitutions in the protein sequence or by mapping of mutations in the gene. The distance between bars indicates their percent separation on the map (as detailed by the numbers). The recombination map expands the distances between some mutations, but otherwise corresponds well with the physical structure of the gene and protein.

Note that in two cases there are mutations that can be separated on the genetic map, but that affect the same amino acid on the upper map (the connecting lines converge). In one case Gly is changed to Arg or Val; in the other Gly is changed to Cys or Asp. Formally this indicates that the unit of genetic recombination (actually one base pair) is smaller than the unit coding for the amino acid (actually three base pairs).

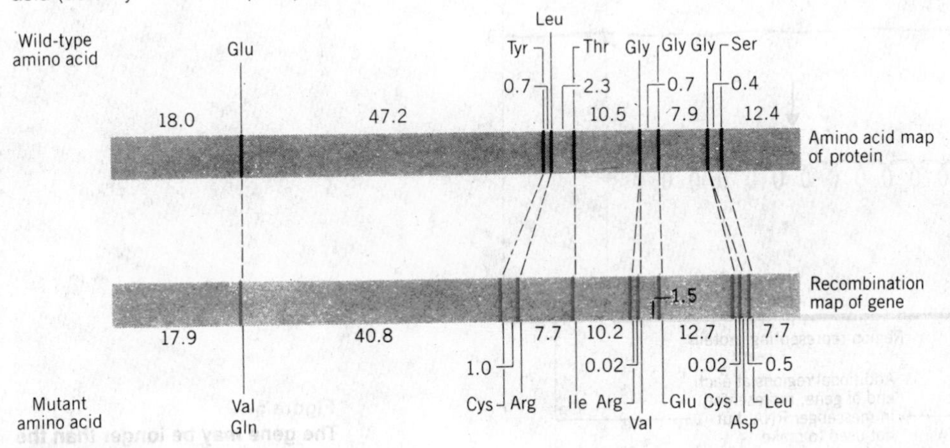

is an exact equivalence. Each gene contains a continuous stretch of DNA whose length is directly related to the size of the protein that it represents. A sequence of 3N base pairs is needed to code for a protein of N amino acids.

The equivalence of the prokaryotic gene and its product means that a restriction map of DNA will exactly match an amino acid map of the protein. How well do these maps fit with the recombination map?

The colinearity of gene and protein was originally investigated in the tryptophan synthetase gene of *E. coli*. Genetic distance was measured by the percent recombination between mutations; protein distance was measured by the number of amino acids separating sites of replacement. **Figure 5.11** compares the two maps. The order of seven sites of mutation is the same as the order of the corresponding sites of amino acid replacement. And the recombination distances are relatively similar to the actual distances in the protein (so in this case, there is little distortion of the recombination map relative to the physical map).

In comparing gene and protein, we are restricted to dealing with the sequence of DNA stretching between the points corresponding to the ends of the protein. However, a gene is not directly translated into protein, but is expressed via the production of a **messenger RNA**, a nucleic acid intermediate actually used to synthesize a protein (as we shall see in detail in Part 2). Messenger RNA is synthesized by the same process of complementary base pairing that is used in the replication of DNA, with the important difference that it corresponds to only one strand of the DNA double helix. Thus the sequence of messenger RNA is complementary with the sequence of one strand of DNA and is identical (apart from the replacement of T by U) with the other strand of DNA.

A messenger RNA includes a sequence of nucleotides that corresponds with the sequence of amino acids in the protein. This part of the nucleic acid is called the **coding region**. But the messenger RNA may include additional sequences on either end; these sequences do not directly represent protein. However, the "gene" usually is considered to include the entire sequence represented in messenger RNA. Sometimes mutations impeding gene function are found in the additional, noncoding regions, confirming the view that these comprise a legitimate part of the genetic unit.

Figure 5.12 illustrates this situation, in which the gene is considered to comprise a continuous stretch of DNA, needed to produce a particular protein and including the sequence coding for that protein, but also including sequences on either side of the coding region.

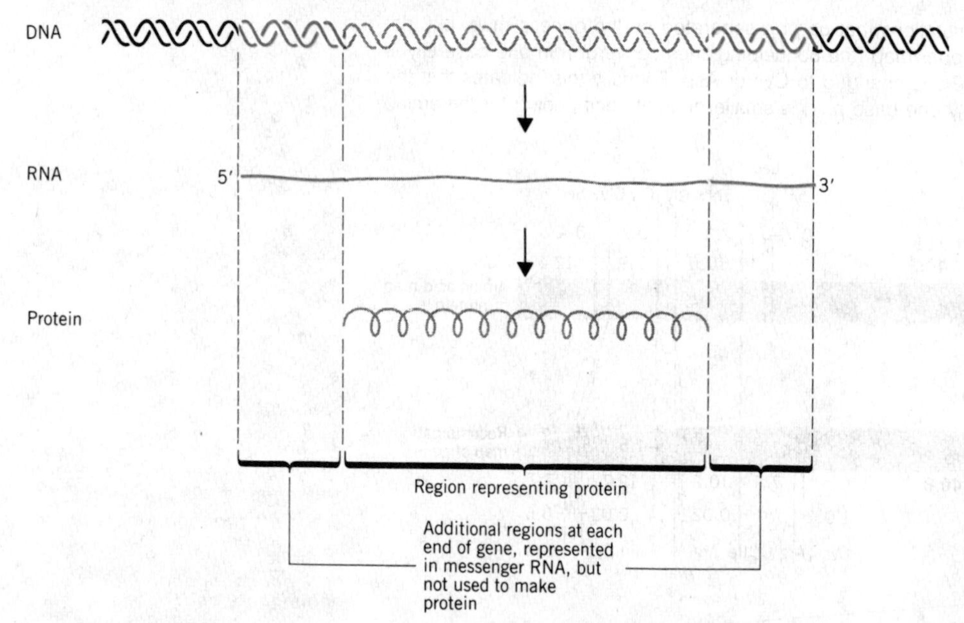

Figure 5.12
The gene may be longer than the sequence coding for protein.

EUKARYOTIC GENES CAN BE INTERRUPTED

This simple view of the gene was upset in 1977 by the discovery of **interrupted genes**. The primary evidence for the existence of these interruptions was a comparison between the structure of DNA and the corresponding messenger RNA. The messenger RNA always includes a nucleotide sequence that corresponds exactly with the protein product according to the rules of the genetic code. However, the gene may include additional sequences that lie within the coding region, interrupting the sequence that represents the protein. This discrepancy between the structure of the DNA and messenger RNA is quite common in eukaryotes, has been found in an archaebacterium (representing another evolutionary branch of the bacteria), but has never been found in the eubacteria (which provide the typical prokaryotes).

The sequences of DNA comprising an interrupted gene are divided into the two categories depicted in **Figure 5.13**. The **exons** are the regions that are represented in the messenger RNA. The **introns** are missing from the messenger RNA. The process of gene expression requires a new step, one that does not occur in eubacteria. The DNA gives rise to an RNA copy that exactly represents the genome sequence. But this RNA is only a precursor; it cannot be used for producing protein. First the introns must be removed from the RNA to give a messenger RNA that consists only of the series of exons. This process is called **RNA splicing**.

How does this change our view of the gene? The exons are always joined together in the same order in which they lie in DNA. Thus the colinearity of gene and protein is maintained between the individual exons and the corresponding parts of the protein chain. The *order* of mutations in the gene remains the same as the order of amino acid replacements in the protein. But the *distances* in the gene may not correspond at all with the distances in the protein.

All the exons are represented on the same molecule of RNA, and their splicing together occurs only as an *intra*-molecular reaction. Thus there is no joining of exons carried by *different* RNA molecules. The mechanism therefore excludes any splicing together of sequences representing different alleles. So mutations located in different exons of a gene cannot complement one another; thus they continue to be defined as members of the same complementation group.

If mutations that affect the sequence of a higher eukaryotic protein could be mapped with sufficient resolution, they should be found in clusters. Each cluster should correspond to an exon, and should be separated from the next cluster on the genetic map by a distance corresponding to the relative length of the intron. Thus recombination frequencies could not be taken as a guide to relative distance within a gene.

What is the nature of mutations in the introns? Since the introns are not part of the messenger RNA, mu-

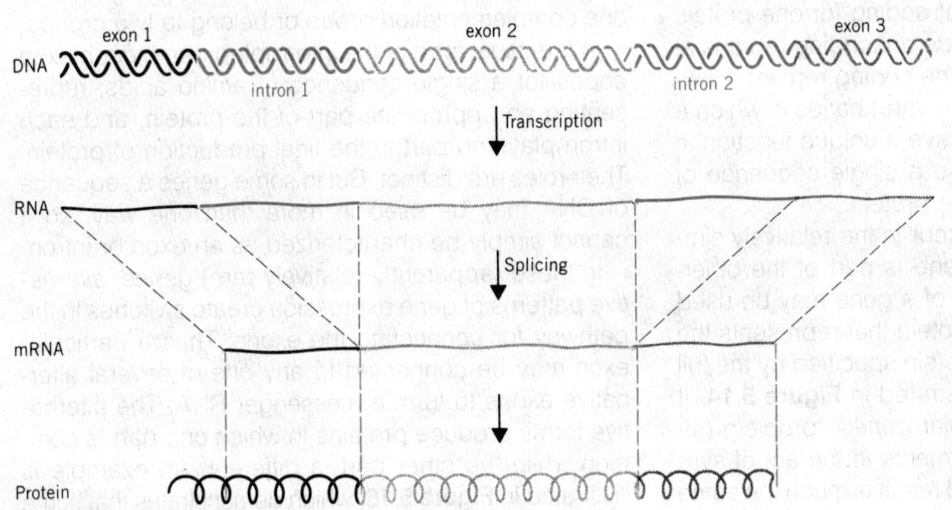

Figure 5.13
Interrupted genes are expressed via a precursor RNA, from which the introns are removed when the exons are spliced together to give the messenger RNA.

tations in them cannot directly affect protein structure. However, they could prevent the production of the messenger RNA—for example, by inhibiting the splicing together of exons. A mutation of this sort would act only on the allele that carries it, would therefore fail to complement any other mutation in that allele, and so must constitute part of the same complementation group as the exons.

In the interrupted genes of eukaryotes, most introns appear to serve no function other than to be removed during gene expression. However, there are some exceptions, most notably in the yeast mitochondrion, in which an intron itself codes for the production of a protein that functions independently from the protein coded by the exons. In this case, mutations in the intron fall into a different complementation group from that represented by mutations in the exons. Thus one complementation group may lie between clusters of mutations that comprise another group.

Eukaryotic genes are not necessarily interrupted. Some correspond directly with the protein product in the same manner as eubacterial genes. We do not yet know the relative proportions of interrupted and uninterrupted genes in eukaryotes, although it seems that the former may be in the majority.

SOME DNA SEQUENCES CODE FOR MORE THAN ONE PROTEIN

Most genes consist of a sequence of DNA that is devoted solely to the purpose of coding for one protein (although the gene may include noncoding regions at either end and introns within the coding region). However, there are some (relatively rare) cases in which a sequence of DNA does not have a unique function in representing protein, because a single sequence of DNA codes for more than one protein.

Overlapping genes may occur in the relatively simple situation in which one gene is part of the other. The first half (or second half) of a gene may be used independently to specify a protein that represents the first (or second) half of the protein specified by the full gene. This relationship is illustrated in **Figure 5.14**. It does not present any particular *genetic* problem (although it does require adjustments in the act of synthesizing the protein). The end result is much the same as though a partial cleavage took place in the protein

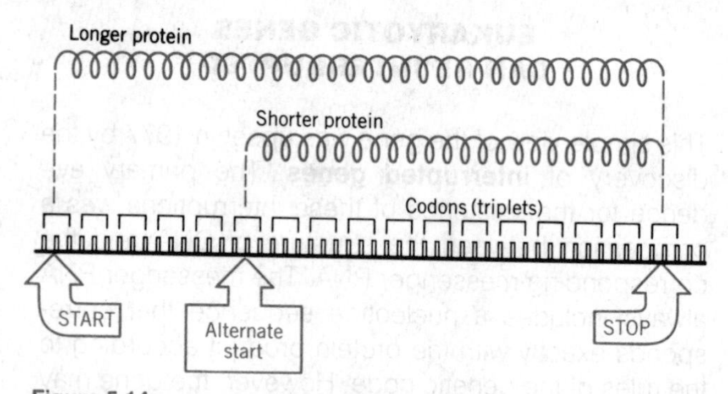

Figure 5.14
Two proteins can be generated from a single gene by starting (or terminating) expression at different points.

product to generate part-length as well as full-length forms.

Two genes may overlap in a more subtle manner when the same sequence of DNA is shared between two *nonhomologous* proteins. This situation arises when the same sequence of DNA is translated in more than one reading frame. In cellular genes, a DNA sequence usually is read in only one of the three potential reading frames, but in some viral and mitochondrial genes, two adjacent genes that are read in different reading frames may overlap. This situation is illustrated in **Figure 5.15**. The distance of overlap is usually relatively short, so that most of the sequence representing the protein retains a unique coding function. A mutation in the shared sequence may, depending on its type, affect either gene or both; thus it could be part of only one complementation group or belong to two groups.

In the usual form of the interrupted gene, each exon codes for a single sequence of amino acids, representing an appropriate part of the protein, and each intron plays no part in the final production of protein. Their roles are distinct. But in some genes a sequence of DNA may be used in more than one way, so it cannot simply be characterized as an exon or intron.

In these (apparently relatively rare) genes, *alternative* patterns of gene expression create switches in the pathway for connecting the exons. Thus a particular exon may be connected to any one of several alternative exons to form a messenger RNA. The alternative forms produce proteins in which one part is common while the other part is different. An example is illustrated in **Figure 5.16**, which demonstrates that some regions may behave as exons when expressed via

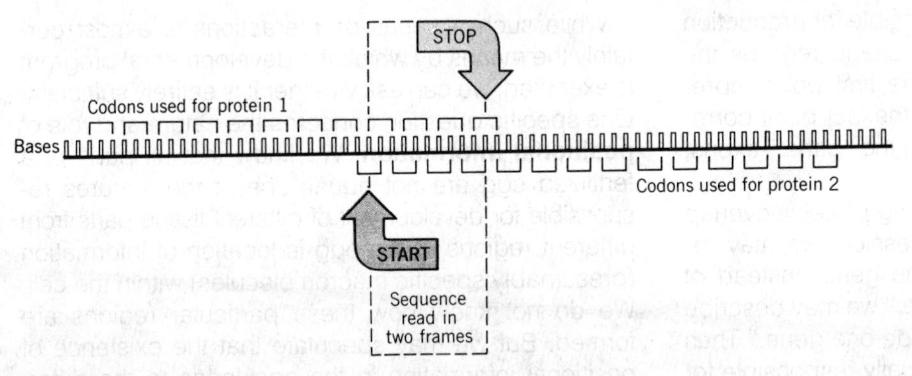

Figure 5.15
Two genes may share the same sequence by reading the DNA in different frames.

one pathway, but are introns when expressed via another.

Sometimes both pathways operate simultaneously, a certain proportion of the RNA being spliced in each way; sometimes the pathways are alternatives that are expressed under different conditions, one in one cell type and one in another cell type. The complementation properties of such mutations can in principle be complex, and it is not necessarily possible to assign all mutations to one or another independent complementation group.

THE SCOPE OF THE PARADIGM

The question of what's in a name is especially appropriate for the gene. Clearly we can no longer say that a gene is a sequence of DNA that continuously and uniquely codes for a particular protein. For situations

Figure 5.16
Genes may be difficult to define when there are alternative pathways for expression.
In one pathway, *exon 1* is joined to *exon 2* by removing from RNA the regions marked *segment* and *intron*. In the other pathway, the region of *exon1—segment* is joined directly to *exon 2*, removing only the intron. Thus in the first pathway the *segment* region is an intron, but in the second pathway it is an exon. The pathways produce two proteins that are the same at their ends, but one of which has an additional sequence in the middle. Thus the region of DNA may code for more than one protein.

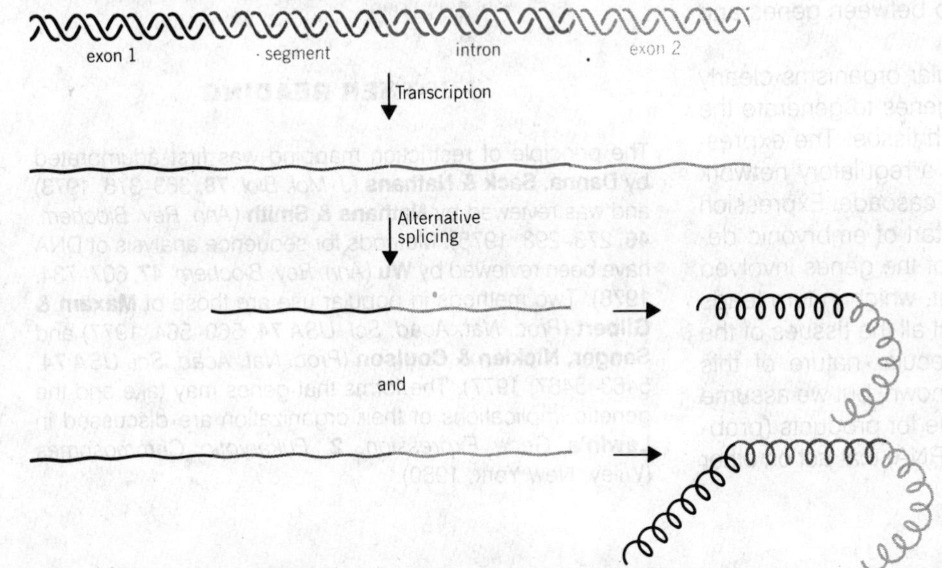

in which a stretch of DNA is responsible for production of one particular protein, current usage regards the entire sequence of DNA, from the first point represented in the messenger RNA to the last point corresponding to its end, as comprising the "gene," exons, introns, and all.

When the sequences representing proteins overlap or have alternative forms of expression, we may reverse the usual description of the gene. Instead of saying "one gene-one polypeptide," we may describe the relationship as "one polypeptide-one gene." Thus we may regard the sequence actually responsible for production of the polypeptide (including introns as well as exons) as the gene, while recognizing that from the perspective of another protein, part of this same sequence may also belong to *its* gene. This allows the use of descriptions such as "overlapping" or "alternative" genes.

We can now see how far we have come from the original hypothesis that Beadle and Tatum proposed. Up to that time, the driving question was the nature of the gene. Once it was discovered that genes represent proteins, the paradigm became fixed in the form of the concept that every genetic unit functions through the synthesis of a particular protein. This view remains the central paradigm of molecular biology: a sequence of DNA functions either by directly coding for a particular protein or by being necessary for the use of an adjacent segment that actually codes for the protein. How far does this paradigm take us beyond explaining the basic relationship between genes and proteins?

The development of multicellular organisms clearly rests upon the use of different genes to generate the different cell phenotypes of each tissue. The expression of genes is determined by a regulatory network that probably takes the form of a cascade. Expression of the first set of genes at the start of embryonic development leads to expression of the genes involved in the next stage of development, which in turn leads to a further stage, and so on until all the tissues of the adult are functioning. The molecular nature of this regulatory network is largely unknown, but we assume that it consists of genes that code for products (probably protein, perhaps sometimes RNA) that act on other genes.

While such a series of interactions is almost certainly the means by which the developmental program is executed, we can ask whether it is entirely sufficient. One specific question concerns the nature and role of **positional information**. We know that all parts of a fertilized egg are not equal; one of the features responsible for development of different tissue parts from different regions of the egg is location of information (presumably specific macromolecules) within the cell. We do not know how these particular regions are formed. But we may speculate that the existence of positional information in the egg leads to the differential expression of genes in the cells subsequently formed in these regions, which leads to the development of the adult organism, which leads to the development of an egg with the appropriate positional information. . .

This possibility prompts us to ask whether some information needed for development of the organism may be contained in a form that we cannot directly attribute to a sequence of DNA (although the expression of particular sequences may be needed to perpetuate the positional information). Put in a more general way, we might ask: if we could read out the entire sequence of DNA comprising the genome of some organism and interpret it in terms of proteins and regulatory regions, could we then construct an organism (or even a single living cell) by controlled expression of the proper genes?

FURTHER READING

The principle of restriction mapping was first adumbrated by **Danna, Sack & Nathans** (*J. Mol. Biol.* **78**, 363–376, 1973) and was reviewed by **Nathans & Smith** (*Ann. Rev. Biochem.* **46**, 273–293, 1975). Methods for sequence analysis of DNA have been reviewed by **Wu** (*Ann. Rev. Biochem.* **47**, 607–734, 1978). Two methods in popular use are those of **Maxam & Gilbert** (*Proc. Nat. Acad. Sci. USA* **74**, 560–564, 1977) and **Sanger, Nicklen & Coulson** (*Proc. Nat. Acad. Sci. USA* **74**, 5463–5467, 1977). The forms that genes may take and the genetic implications of their organization are discussed in **Lewin's** *Gene Expression,* **2**, *Eukaryotic Chromosomes* (Wiley, New York, 1980).

PART 2
TURNING GENES INTO PROTEINS

The central dogma states that once "information" has passed into protein it cannot get out again. The transfer of information from nucleic acid to nucleic acid, or from nucleic acid to protein, may be possible, but transfer from protein to protein, or from protein to nucleic acid, is impossible. Information means here the precise determination of sequence, either of bases in the nucleic acid or of amino acid residues in the protein.

Francis Crick, 1958

CHAPTER 6
BREAKING THE GENETIC CODE

The **central dogma** defines the paradigm of molecular biology: that genes are units perpetuating themselves and functioning through their expression in the form of proteins. **Figure 6.1** illustrates the restrictions imposed by the dogma.

Within the cell, genetic information is carried by the sequence of DNA. The information is perpetuated by the act of **replication**, a process involving duplication of the nucleic acid to give identical copies of it. Information is expressed by a two stage process. **Transcription** generates a single-stranded RNA identical in sequence with one of the strands of the duplex DNA. Then **translation** converts the nucleotide sequence of the RNA into the sequence of amino acids comprising a protein.

The expression of *cellular* genetic information is *irreversible*. A sequence in the form of RNA can be used further *only* to generate a protein sequence; it cannot be retrieved for use as genetic information. Originally we thought this restriction was absolute, but now we know that it does not apply to some viruses, whose genetic material consists of RNA rather than DNA.

Some RNA viruses replicate purely through the RNA form, using complementary base pairing, sometimes with and sometimes without the formation of a stable RNA duplex. Thus replication of viral RNA is possible in the infected cell. This is a relatively minor extension of the central dogma.

A more important extension is presented by the RNA tumor viruses. The genomes of these **retroviruses** consist of single-stranded RNA molecules. During infection of a cell, the RNA is converted by the process of **reverse transcription** into a single-stranded DNA, which in turn is converted into a double-stranded DNA. This duplex DNA may become part of the genome of

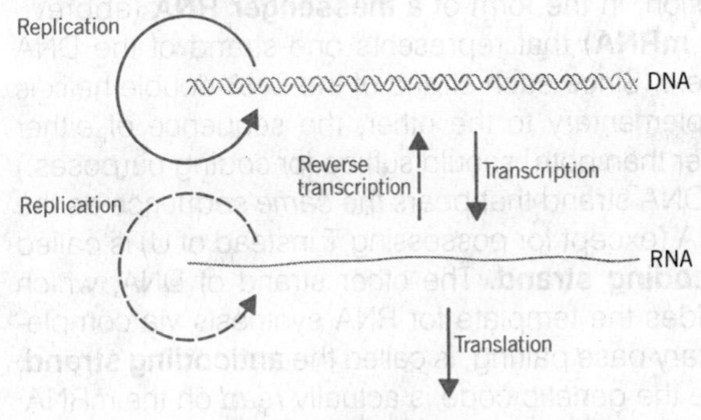

Figure 6.1
The central dogma states that information in nucleic acid can be perpetuated or transferred, but the transfer of information into protein is irreversible.
The processes used by the cell are indicated as continuous lines, while those regularly used only in viral infection are broken.

the cell, and then is inherited like any other gene. Thus reverse transcription allows a sequence of RNA to be retrieved and used as genetic information.

The existence of RNA replication and reverse transcription establishes the general principle that information in the form of either type of nucleic acid sequence can be converted into the other type. In the usual course of events, however, neither of these mechanisms is used by the cell itself, which relies on the processes of replication, transcription, and translation. But on rare occasions (possibly mediated by infection by an RNA virus), information in the form of a cellular RNA may be retrieved and inserted into the genome. Although reverse transcription plays no role in the regular operations of the cell, it therefore becomes a mechanism of potential importance when we consider the evolution of the genome.

PROTEINS ARE SYNTHESIZED IN ONE DIRECTION

The breaking of the genetic code showed that genetic information is stored in the form of nucleotide triplets, but this did not reveal how each codon specifies its corresponding amino acid.

The concept that there must be a code evolved pari passu with the idea that the process of translation must involve a **template**. Because the genetic material in the nucleus is physically separated from the site of protein synthesis in the cytoplasm of a eukaryotic cell, it was clear that the DNA could not *itself* provide the template.

The template is generated by the process of transcription, in the form of a **messenger RNA** (abbreviated **mRNA**) that represents one strand of the DNA duplex. (Since each strand of the DNA double helix is complementary to the other, the sequence of either [rather than both] should suffice for coding purposes.) The DNA strand that bears the *same* sequence as the mRNA (except for possessing T instead of U) is called the **coding strand**. The other strand of DNA, which provides the template for RNA synthesis via complementary base pairing, is called the **anticoding strand**. Since the genetic code is actually *read* on the mRNA, usually it is described in terms of the four bases present in RNA: U, C, A, and G.

The description of the template as *messenger* RNA reflects its ability to move from the nucleus where it is synthesized to the cytoplasm where it functions. Translation of mRNA into protein is accomplished by reading the genetic code: each triplet of nucleotides is converted into one amino acid. (The process is called translation to reflect the role of the mRNA. Because mRNA provides the template on which the amino acids are actually assembled into a polypeptide chain, it is at this step that the nucleotide sequence is deciphered as representing individual amino acids.)

If the genetic code is read as a series of adjacent triplets, the corresponding protein should be assembled sequentially from one end to the other. In which direction does this process proceed?

This point was resolved in 1961, when Dintzis investigated the synthesis of the red blood cell protein, hemoglobin. In cells synthesizing hemoglobin, one of the amino acids was replaced with a radioactively labeled form. Then completed protein chains were extracted, and the level of radioactivity in each part of the chain was measured.

Figure 6.2 shows that the radioactivity initially enters the polypeptide chains at all possible positions, corresponding to whatever point the synthesis of each individual molecule has reached. But in *completed* chains, the radioactivity should appear *first* at the end that is synthesized *last*. After a longer period, the radioactivity is found in the parts of the chain that were made earlier.

Thus the radioactivity works its way back, as it were, from the end to the start. This experiment showed that proteins are assembled only in the direction from the N-terminus to the C-terminus.

THE SEARCH FOR MESSENGER RNA

Translation is undertaken by the **ribosome**, a compact **ribonucleoprotein particle** consisting of two subunits. They are generally called the large and small subunits, and each consists of several proteins associated with a long RNA molecule; the RNAs are known as **ribosomal RNA** (abbreviated **rRNA**). A ribosome attaches to mRNA at the 5′ end; moving along the RNA toward the 3′ end, it translates each triplet codon into an amino acid en route.

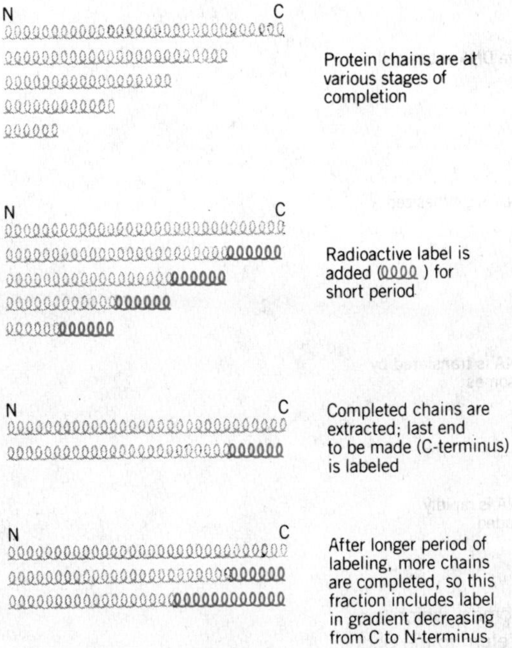

N C

Protein chains are at various stages of completion

N C

Radioactive label is added ( ) for short period

N C

Completed chains are extracted; last end to be made (C-terminus) is labeled

N C

After longer period of labeling, more chains are completed, so this fraction includes label in gradient decreasing from C to N-terminus

Figure 6.2

Proteins are synthesized from the N-terminus toward the C-terminus.

A radioactive label enters all parts of the growing polypeptide chains simultaneously. When completed chains are extracted, they contain a label in the *last* region to be completed. So the gradient of label from C-terminus toward N-terminus indicates that synthesis proceeds in the opposite direction.

Ribosomes were first identified as the site of protein synthesis by following the fate of radioactively labeled amino acids as they are incorporated into proteins. Within a couple of minutes, the radioactive label is found in the fraction containing the ribosomes. Then it is released into the general cytoplasm. This result suggests that amino acids are assembled into polypeptides at the ribosomes; and the polypeptides are released as soon as synthesis has been completed.

When ribosomes were discovered, it was thought that they might actually provide the template for protein synthesis. According to this view, different ribosomes might contain different RNAs, used for synthesizing different proteins. This idea actually was almost correct, missing the point in quite a subtle way (by confusing the rRNA with the mRNA). Now we know that all ribosomes are identical; but they undertake the synthesis of different proteins by associating with the different mRNAs that provide the actual templates.

The search for the messenger was originally conducted in bacteria, where the mRNA was hard to identify. We now understand that the bacterial mRNA proved so elusive because it represents only a very small proportion of the mass of the bacterial RNA and is unstable. In bacteria, mRNA is synthesized, translated by the ribosomes, and degraded, all in rapid succession, as illustrated in **Figure 6.3**. A given molecule of mRNA may survive for only a matter of minutes or even less.

There were also difficulties in identifying mRNA in eukaryotes. As with bacteria, mRNA constitutes only a small proportion of the total cellular RNA (roughly 3% of the mass). However, it does have some distinctive properties, and now the mRNA for any particular protein can be isolated almost routinely from the fraction of a cell extract that contains ribosomes and mRNA. Eukaryotic mRNA is relatively stable, often surviving for a period of some hours in the cell.

A eukaryotic mRNA usually can be isolated intact and translated *in vitro* when ribosomes and other necessary components are added. In fact, the proof that a given mRNA represents a particular protein now is provided by showing that the protein is synthesized under its direction *in vitro*. The proof can be taken further by determining the nucleotide sequence of the RNA and comparing it with the amino acid sequence of the protein.

TRANSFER RNA IS THE ADAPTOR

The incongruity of structure between a trinucleotide sequence and an amino acid immediately raises the question of how each codon is matched to its particular amino acid. Even before the exact form of the code had been discovered, Crick suggested that translation might be mediated by an "adaptor" molecule. We now know that this adaptor is **transfer RNA** (abbreviated **tRNA**), a small molecule whose polynucleotide chain is only 75–85 bases long.

Each tRNA has two crucial properties. First, it is able to recognize only one amino acid, to which it is *covalently linked*. Second, each tRNA contains a trinucleotide sequence, the **anticodon**, which is com-

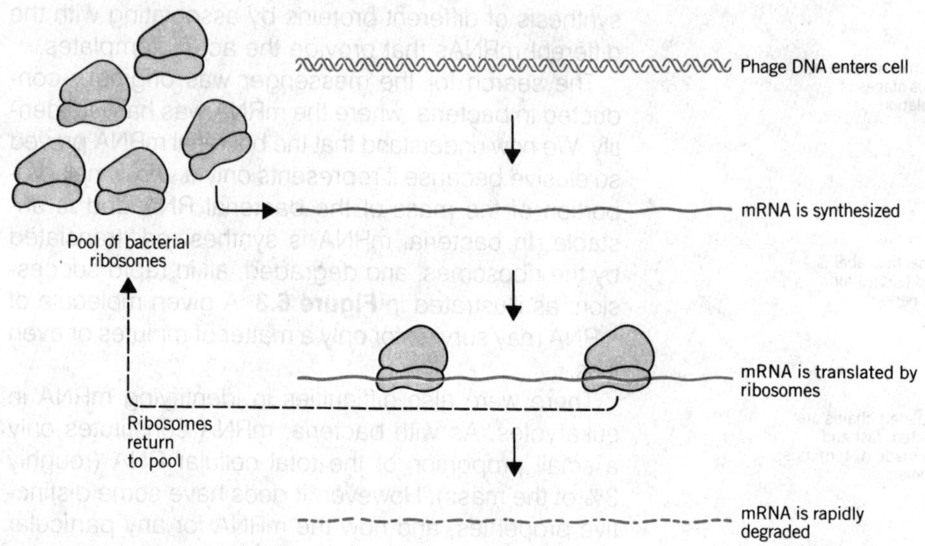

Figure 6.3
Messenger RNA has only a transient existence in bacteria.
An mRNA is synthesized by transcription from DNA and is translated by the ribosomes, which are drawn from a pool of free subunits. The mRNA is rapidly degraded; the ribosomes return to the pool and are available for translating further mRNAs.

plementary to the codon representing its amino acid. The anticodon enables the tRNA to recognize the codon via complementary base pairing.

Transfer RNA has a characteristic secondary structure. The nucleotide sequence of every tRNA can be written in the form of a **cloverleaf**, as shown in **Figure 6.4**, in which complementary base pairing forms **stems** for single-stranded **loops**. The stem-loop structures are called the **arms** of tRNA.

When a tRNA is **charged** with the amino acid corresponding to its anticodon, it becomes **aminoacyl-tRNA.** The amino acid is linked by an ester bond from its carboxyl group to a hydroxyl group of the ribose of the last base of the tRNA (which is always adenine). The structure of the link is illustrated in **Figure 6.5**.

The process of charging the tRNA to form aminoacyl-tRNA is catalyzed by a specific enzyme, **aminoacyl-tRNA synthetase**. There are (at least) 20 aminoacyl-tRNA synthetases. Each recognizes a single amino acid and also *all the tRNAs* on to which it can legitimately be placed. (There may be many such tRNAs. In addition to the several tRNAs that may be needed to respond to synonym codons, sometimes there are multiple species of tRNA reacting with the same codon.)

The charging process involves a two stage reaction, as illustrated in **Figure 6.6**. First, the amino acid reacts with ATP to form aminoacyl~adenylate, releasing pyrophosphate. Energy for the reaction is provided by cleaving the high energy bond of the ATP. Then the activated amino acid is transferred to the tRNA, releasing AMP.

Does the anticodon sequence alone allow aminoacyl-tRNA to recognize the correct codon? An early experiment showed that, *once aminoacyl-tRNA has been formed, its behavior is dictated solely by the anticodon.* **Figure 6.7** shows that reductive desulfuration can be used to convert the amino acid of cysteinyl-tRNA into alanine. Thus the product is alanyl-$tRNA^{Cys}$. (The convention for naming tRNAs is to use the three letter abbreviation for the amino acid as a superscript; if there is more than one tRNA for the same amino acid, subscript numerals are used to distinguish them.) The alanine residue is incorporated

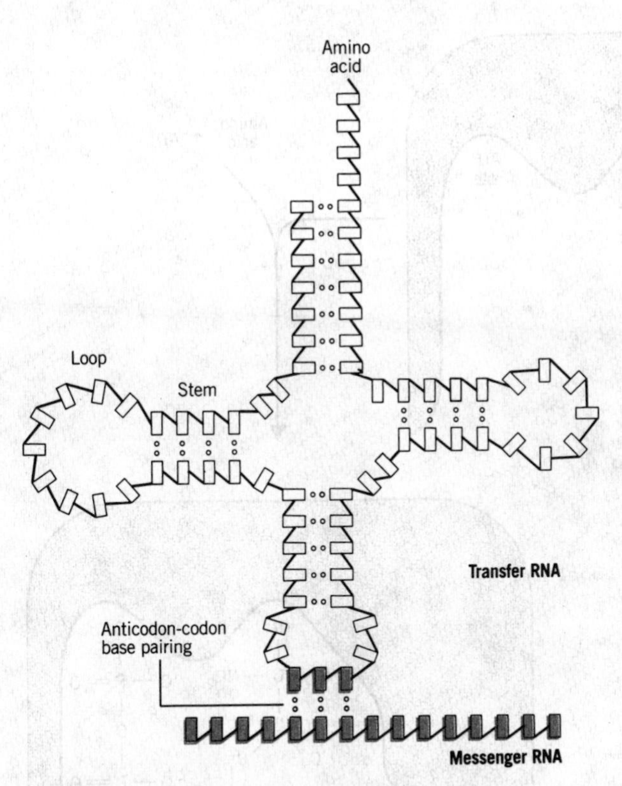

Figure 6.4
Transfer RNA has the dual properties of an adaptor.
The sugar-base moieties are represented by the boxes; the lines connecting them are phosphodiester bonds. The stems of each arm are maintained by complementary base pairing (indicated by small open circles); the bases at the end of each arm form a single-stranded loop. At the loop shown at the bottom, a triplet anticodon sequence is complementary to the codon for the amino acid represented by the tRNA. At the arm shown at the top, the amino acid is linked to the nucleotide at the 3′ terminus of the tRNA.

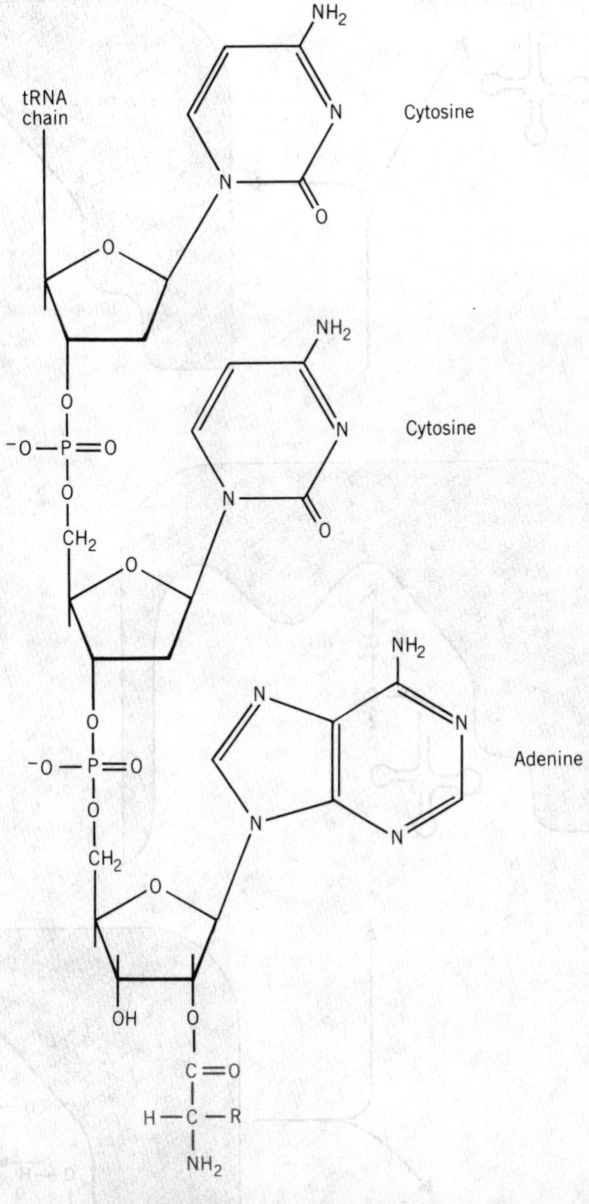

Figure 6.5
Aminoacyl-tRNA consists of an amino acid linked through an ester bond to (usually) the 2′-OH terminal position of the tRNA, which always ends in the three base sequence -CCA.
In some cases, the amino acid may be placed on the 3′-OH group, but the question of which hydroxyl group is used in any particular case is moot, because the 2′ and 3′ forms of aminoacyl-tRNA are in rapid equilibrium with each other in solution—and therefore presumably in the cell.

into protein in place of cysteine. Thus once a tRNA has been charged, the amino acid plays no further role in its specificity.

RIBOSOMES TRAVEL IN CONVOY

The interaction between tRNA and mRNA is sponsored by the ribosome, which provides an environment that controls the recognition between codon and anticodon. To accomplish the sequential synthesis of a protein, the ribosome **moves** along the mRNA. At

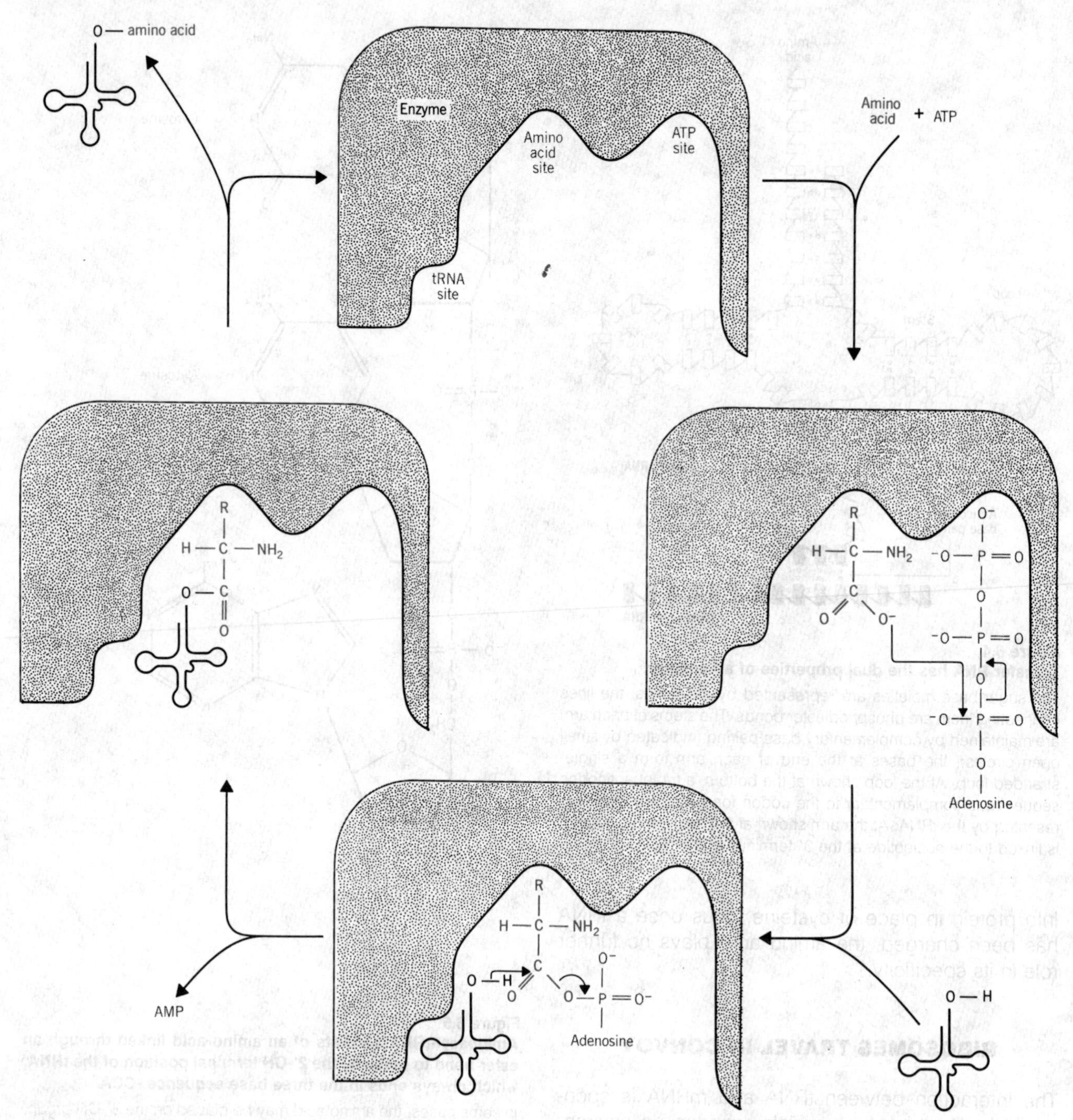

Figure 6.6

An aminoacyl-tRNA synthetase charges tRNA with an amino acid.

The enzyme has different sites for tRNA, an amino acid, and ATP. The first reaction requires the amino acid and ATP; it generates aminoacyl-AMP. Then tRNA is bound and the amino acid is transferred to it. When the aminoacyl-tRNA is released, the enzyme is ready for another cycle.

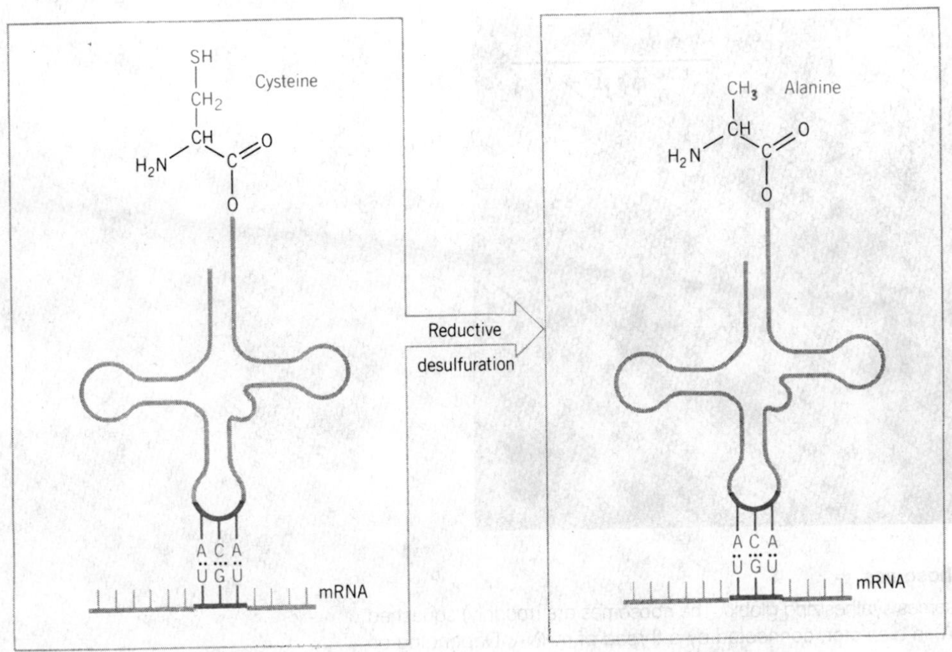

Figure 6.7
The meaning of tRNA is determined by its anticodon and not by its amino acid.
Cysteinyl-tRNA has an anticodon that responds to the codon UGU. When the cysteinyl residue is chemically converted to an alanyl moiety, the tRNA continues to respond to the UGU codon, but now places alanine instead of cysteine in the protein.

any given moment, the ribosome accommodates the two species corresponding to successive codons, thus allowing a peptide bond to form between the two appropriate amino acids. As a ribosome moves along an mRNA, the appropriate aminoacyl-tRNAs associate with it, donating their amino acids to the polypeptide chain. At each step, the growing polypeptide chain becomes longer by one amino acid.

Ribosomes are traditionally described in terms of their (approximate) rate of sedimentation. (The unit of sedimentation is the Svedberg, denoted as **S**. The greater the mass, the faster the rate of sedimentation, and the higher the S value. The rate is also influenced by shape, because more compact bodies sediment more rapidly.) Bacterial ribosomes generally sediment at about 70S. The ribosomes of the cytoplasm of higher eukaryotic cells are larger, usually sedimenting at about 80S.

When active ribosomes are isolated in the form of the fraction associated with radioactive amino acids, they sediment more rapidly than 70S (or 80S). The rapidly sedimenting unit consists of an mRNA associated with several ribosomes. This is the **polyribosome** or **polysome**.

A classic characterization of polysomes is shown in the electron micrograph of **Figure 6.8**, which shows pentasomes of red blood cells engaged in the synthesis of globin protein. Each pentasome consists of five ribosomes connected by a thread of mRNA. The ribosomes are located at various positions along the messenger. Those at one end have just started protein synthesis; those at the other end are about to complete production of a polypeptide chain.

The polypeptide chain lengthens progressively with each amino acid that is added. Each ribosome synthesizes a single polypeptide during its traverse of the message sequence. Thus the mRNA has a series of ribosomes that carry increasing lengths of the protein product, moving from the 5′ to the 3′ end, as illustrated in **Figure 6.9**.

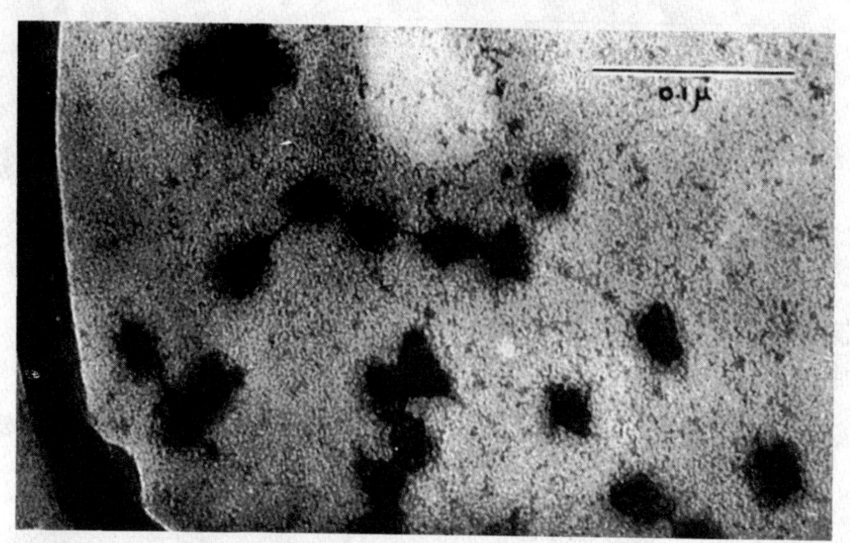

Figure 6.8
Protein synthesis occurs on polyribosomes.
The electron micrograph shows pentasomes synthesizing globin. The ribosomes are (roughly) squashed spherical objects of about 7 nm (70 Å) in diameter, connected by a thread of mRNA. Depending on the conditions, the polysomes may appear compact or elongated.
Photograph kindly provided by Alex Rich.

Roughly the last 30–35 amino acids added to a growing polypeptide chain are protected by the structure of the ribosome from the environment. Probably all of the preceding part of the polypeptide protrudes and is free to start folding into its proper conformation. (Thus proteins can display parts of the mature conformation even before synthesis has been completed.)

The size of the polysome depends on a variety of factors. In bacteria, it may be very large, with tens of ribosomes simultaneously engaged in translation. Partly the size is due to the length of the mRNA (which actually may code for several proteins); partly it is due to the high efficiency with which the ribosomes translate the mRNA. Since ribosomes attach to bacterial mRNA even before its transcription has been completed, the polysome is likely still to be attached to DNA.

In eukaryotic cytoplasm the mRNA must be transported from the nucleus to reach the ribosomes in the cytoplasm. The polysomes are likely to be smaller than those in bacteria; again, their size is a function both of the length of the mRNA (representing only a single protein in eukaryotes) and of the characteristic frequency with which ribosomes attach. One thinks of eukaryotic mRNAs generally as having fewer than 10 ribosomes attached at any one time.

The number of ribosomes on each mRNA molecule synthesizing a particular protein is not precisely determined, in either bacteria or eukaryotes, but is a matter of statistical fluctuation, determined by the variables of mRNA size and efficiency.

An overall view of the attention devoted to protein synthesis in the intact bacterium is given in **Table 6.1**. The 20,000 or so ribosomes account for an appreciable proportion of the cell mass. The tRNA molecules outnumber the ribosomes by almost tenfold; most of them are present as aminoacyl-tRNAs, that is, ready to be used at once in protein synthesis. Because of their instability, it is difficult to calculate the number of mRNA molecules, but a reasonable guess would be 2000–3000, in varying states of synthesis and decomposition.

MOST CODONS REPRESENT AMINO ACIDS

There are 64 combinations of trinucleotide sequences but only 20 amino acids. When the triplet nature of the

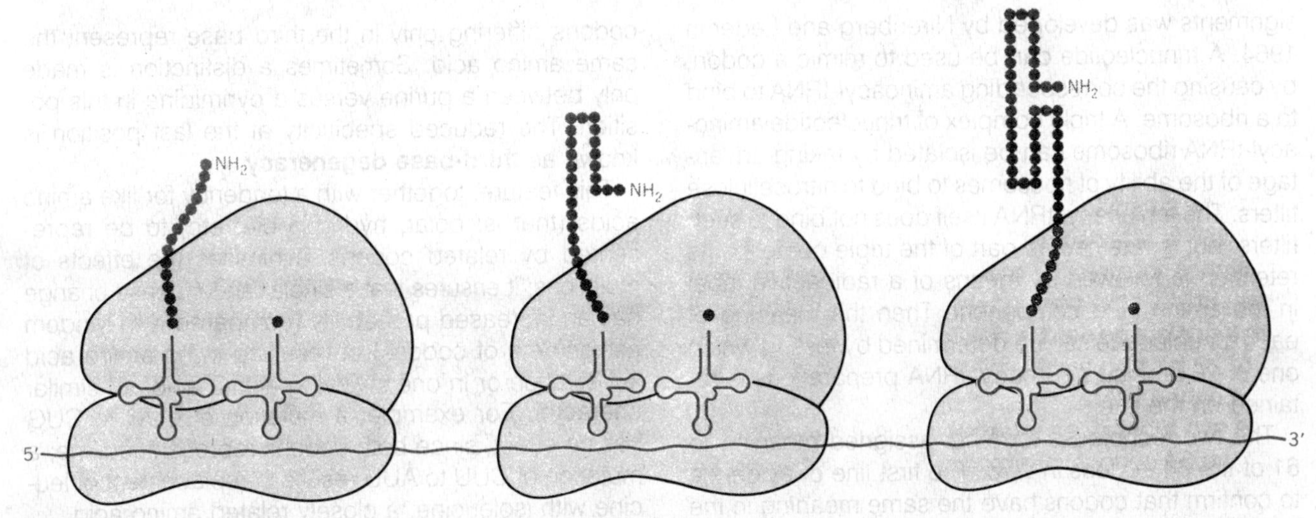

Figure 6.9
Polyribosomes consist of several ribosomes translating a messenger by moving in the direction from 5′ to 3′.
Each ribosome has two tRNA molecules, the first carrying the last amino acid added to the chain (that is, in the form of the polypeptide chain synthesized so far), the second carrying the next amino acid to be added.

Table 6.1
Considering *E. coli* in terms of the components of the bacterium.

Component	Proportion of Cell Dry Weight, %	Number/Cell
Wall	10	
Membrane	10	
DNA	2	1
mRNA	2	2,500?
tRNA	3	160,000
rRNA	21 } 30	20,000
Ribosomal protein	9 } 30	
Soluble protein	42	10^6
Small metabolites	1	6.5×10^6

code was discovered, no approach was available to allow codons to be assigned systematically to amino acids. But it was only a very short time before two methods were able to assign meanings to most of the codons. Both methods relied on the use *in vitro* of purified components of the apparatus that synthesizes proteins.

Protein-synthesizing systems can be prepared in the form of cell-free extracts. The original procedure consisted essentially of breaking cells of the bacterium *E. coli* and centrifuging the mixture to remove cell wall and membrane fragments. When provided with a suitable source of energy and precursors, the supernatant can translate any mRNA that is added to it. (Similar procedures have since been used with other organisms, and today many sophisticated and highly purified systems are available.)

The *E. coli* system was used to translate *synthetic* polynucleotides. The first report of success with such a system was made by Nirenberg in 1961 (dramatically at a Biochemical Congress in Moscow). This was the demonstration that polyuridylic acid [poly(U)] can act as an mRNA to direct the assembly of phenylalanine into polyphenylalanine. This result means that UUU must be a codon for phenylalanine. Subsequently, many other synthetic polynucleotides, consisting of known sequences of different bases, were used by Khorana; they allowed the meaning of about half of the 64 codons to be assigned.

The **ribosome binding assay** for making codon assignments was developed by Nirenberg and Leder in 1964. A trinucleotide can be used to mimic a codon, by causing the corresponding aminoacyl-tRNA to bind to a ribosome. A triple complex of trinucleotide·aminoacyl-tRNA·ribosome can be isolated by taking advantage of the ability of ribosomes to bind to nitrocellulose filters. The aminoacyl-tRNA itself does not bind to such filters, but is retained as part of the triple complex. Its retention is followed by means of a radioactive label in the amino acid component. Then the meaning of each trinucleotide can be determined by testing which one of 20 labeled aminoacyl-tRNA preparations is retained on the filter.

The two techniques together assigned meaning to 61 of the 64 codons *in vitro*. The first line of evidence to confirm that codons have the same meaning in the living cell was provided by analyzing mutations. When a point mutation causes one amino acid to be substituted for another, their codons must be related by the appropriate type of base change. More productively, a frameshift mutation causes one series of amino acids to be changed into another; again, the series of codons for the wild type and mutant amino acid sequences must be related by a single base insertion or deletion.

Such experiments tended generally to confirm the codon assignments, but fell short of complete proof of their meaning. But since then, the sequencing of DNA has made it possible to compare a nucleotide sequence and amino acid sequence directly. *The sequence of the coding strand of DNA, read in the direction from 5′ to 3′, consists of triplets corresponding to the amino acid sequence of the protein read from N-terminus to C-terminus*. The entire genetic code has been confirmed in overwhelming detail from such analysis.

The code is summarized in **Figure 6.10**. A striking feature is its **degeneracy**: 61 codons represent 20 amino acids. Almost every amino acid is represented by several codons. Codons that have the same meaning are called **synonyms**. **Figure 6.11** shows that the number of codons representing each amino acid accords quite well with the frequency with which it is used in proteins.

Codons tend to be clustered in groups representing a single amino acid. Often the base in the third position of a codon is not significant, because the four codons differing only in the third base represent the same amino acid. Sometimes a distinction is made only between a purine versus a pyrimidine in this position. The reduced specificity at the last position is known as **third-base degeneracy**.

This feature, together with a tendency for like amino acids (that is, polar, hydrophobic, etc) to be represented by related codons, minimizes the effects of mutations; it ensures that a single random base change has an increased probability (compared with random assignment of codons) of resulting in no amino acid substitution or in one involving amino acids of similar character. For example, a mutation of CUC to CUG has no effect, since both codons represent leucine; a mutation of CUU to AUU results in replacement of leucine with isoleucine, a closely related amino acid.

Three codons (UAA, UAG and UGA) do not represent amino acids. They are used specifically to terminate protein synthesis; one of these codons marks the end of every gene.

Figure 6.10

All the triplet codons have meaning.

Sixty-one of the codons represent amino acids. All of the amino acids except tryptophan and methionine are represented by more than one codon. The synonym codons usually form groups in which the base in the third position has the least meaning. Three codons cause termination (TERM). The order of bases in a codon is written in the same way as other sequences, in the direction from 5′ to 3′.

FIRST BASE	SECOND BASE: U	C	A	G
U	UUU, UUC: Phe UUA, UUG: Leu	UCU, UCC, UCA, UCG: Ser	UAU, UAC: Tyr UAA, UAG: TERM	UGU, UGC: Cys UGA: TERM UGG: Trp
C	CUU, CUC, CUA, CUG: Leu	CCU, CCC, CCA, CCG: Pro	CAU, CAC: His CAA, CAG: Gln	CGU, CGC, CGA, CGG: Arg
A	AUU, AUC, AUA: Ile AUG: Met	ACU, ACC, ACA, ACG: Thr	AAU, AAC: Asn AAA, AAG: Lys	AGU, AGC: Ser AGA, AGG: Arg
G	GUU, GUC, GUA, GUG: Val	GCU, GCC, GCA, GCG: Ala	GAU, GAC: Asp GAA, GAG: Glu	GGU, GGC, GGA, GGG: Gly

IS THE CODE UNIVERSAL?

The genetic code was originally elucidated in the bacterium *E. coli*. We must therefore ask whether it is the same in all other living organisms.

Messenger RNA from one species can be correctly translated *in vitro* or *in vivo* by the protein synthetic apparatus of a different species. Many **heterologous translation systems** now have been developed. Their successful operation indicates that the codons used in the mRNA of one species have the same meaning for the ribosomes and tRNAs of another species. Comparisons of DNA sequences with the corresponding protein sequences have confirmed that the *identical set of codon assignments* is used in all bacteria and eukaryotic nuclei.

Because there are drastic variations in the base compositions of different genomes, contrasted with a general stability in the amino acid contents of proteins, different species must use the various synonym codons for each amino acid to different and characteristic extents. Indeed, it is only the degeneracy of the code that permits it to be maintained unchanged in such circumstances.

The universality of the code argues that it must have been established very early in evolution. Some models suppose that originally there may have been a stereochemical relationship between amino acids and the codons representing them. Then the system now used for protein synthesis may have evolved by selection for features such as greater efficiency or accuracy.

Perhaps the code started in a primitive form in which a small number of codons were used to represent comparatively few amino acids, possibly even with one codon corresponding to any member of a group of amino acids. More precise codon meanings and more amino acids could have been introduced later. One possibility is that at first only two of the three bases in each codon were used; discrimination at the third position could have evolved later.

The present code could have become "frozen" at some point because the system was becoming so sophisticated that any changes in codon meaning would disrupt existing proteins by substituting amino acids. Its universality implies that this must have happened at such an early stage that all living organisms are descended from a single pool of primitive cells in which this occurred.

Exceptions to the universal genetic code have been found only in the mitochondria from several species. The only common change is that in each case UGA has the same meaning as UGG, and therefore represents tryptophan instead of termination. Other changes are characteristic for each species, as summarized in **Table 6.2.**

Figure 6.11
The number of codons for each amino acid correlates with its frequency of use in proteins.
Arginine is an exception, because there is a reduction in the occurrence of the doublet sequence CG in eukaryotic DNA. The four arginine codons that start with this doublet are therefore present in DNA at lower frequency than would be predicted by their base composition per se.

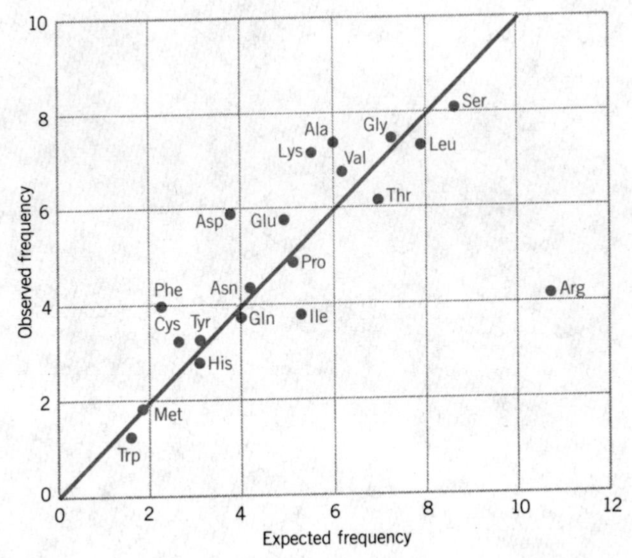

Table 6.2
Changes occur in the mitochondrial genetic code.

Organism	Codon	Probable Meaning in Mitochondrion	Usual Meaning
All	UGA	Tryptophan	Termination
Yeast	CUA	Threonine	Leucine
Fruit fly	AGA	Serine	Arginine
Mammal	AG^{A}_{G}	Termination	Arginine
"	AUA	Methionine	Isoleucine

Some of these changes make the code simpler, by replacing two codons that had different meanings with a pair that has a single meaning. Thus in mammalian mitochondria, AUA has the same meaning as AUG (methionine), instead of a different meaning. Many of the changes affect codons involved in either initiating or terminating protein synthesis (see Chapter 7).

Why has the mitochondrion been able to evolve changes in the code? Since the mitochondrial codon assignments are simplified (relative to the general code), they could represent a more primitive pattern. Alternatively, the simplification may have evolved as a response to the special needs of the system for mitochondrial protein synthesis (see Chapter 8).

The existence of species-specific changes implies that there must be more flexibility in the mitochondrion than in the nucleus. Because the mitochondrion synthesizes only a small number of proteins (about 10), the problem of disrupting them by changes in meaning is much less severe. Probably the codons that are altered were not used extensively in locations where amino acid substitutions would have been deleterious.

FURTHER READING

The breaking of the genetic code is well known. An extensive historical record was given by **Ycas** in *The Biological Code* (North Holland, Amsterdam, 1969). A briefer but complete account was presented in **Lewin's** *Gene Expression*, **1**, *Bacterial Genomes* (Wiley, New York, 1974), which also gives a general overview of the processes involved in protein synthesis.

CHAPTER 7
THE ASSEMBLY LINE FOR PROTEIN SYNTHESIS

Synthesis of proteins involves an assembly line in which the ribosomes proceed inexorably along the messenger, bringing in the aminoacyl-tRNAs that provide the actual building blocks of the protein product. The ribosome itself constitutes a small migratory factory, in which a compact package of proteins and rRNAs forms several active centers able to undertake various catalytic activities. Different sets of accessory factors assist the ribosome in each of the three stages of protein synthesis: initiation, elongation, and termination. Energy for protein synthesis is provided by hydrolysis of GTP.

Initiation involves the reactions that precede formation of the peptide bond between the first two amino acids of the protein. It requires the ribosome to bind to the mRNA, sponsoring an initiation complex that contains the first aminoacyl-tRNA. This is a relatively slow step in protein synthesis.

Elongation includes all the reactions from synthesis of the first peptide bond to addition of the last amino acid to the chain. Amino acids are added one at a time, and the addition of an amino acid is the most rapid step in protein synthesis.

Termination encompasses the steps that are needed to release the completed polypeptide chain; at the same time, the ribosome dissociates from the mRNA. Termination is slow compared with the time required to add an amino acid to the chain.

Protein synthesis overall is a rapid process (although the rate depends strongly on temperature). In bacteria at 37°C, about 15 amino acids are added to a growing polypeptide chain every second. So it takes only about 20 seconds to synthesize an average protein of 300 amino acids. About 80% of the bacterial ribosomes are engaged in protein synthesis, so the pool of free units is only a minor proportion. In eukaryotes, the rate of protein synthesis is slower; in red blood cells at 37°C, elongation typically sees about 2 amino acids added to the chain per second.

Most of the experiments to define the stages of protein synthesis have been performed with *in vitro* systems, consisting of ribosomes, aminoacyl-tRNAs, accessory factors, and an energy source. In these systems, the rate of protein synthesis may be slower by an order of magnitude or so than the rate *in vivo*.

THE RIBOSOMAL SITES OF ACTION

A ribosome consists of two subunits, which dissociate *in vitro* when the concentration of Mg^{2+} ions is reduced. Bacterial (70S) ribosomes have subunits that sediment at roughly 50S and 30S. The subunits of eukaryotic cytoplasmic (80S) ribosomes sediment at about 60S and 40S. In both cases, the larger subunit has a more globular shape, and is about twice the size of the smaller subunit, which is flatter. The two subunits work together as part of the complete ribosome, but each undertakes distinct reactions in protein synthesis.

Messenger RNA is associated with the small subunit, about 30 bases of the mRNA being bound at any time. But only two molecules of tRNA can be in place on the ribosome at any moment. So protein synthesis involves reactions taking place at just two of the (roughly) ten codons covered by the ribosome.

Figure 7.1 shows that each tRNA lies in a distinct site. The two sites have different features.

The only site that can be entered by an incoming aminoacyl-tRNA is the **A site** (or **entry site**). Prior to the entry of aminoacyl-tRNA, the site exposes the codon representing the next amino acid that is due to be added to the chain.

The codon representing the *last* amino acid to have been added to the chain lies in the **P site** (or **donor site**). This site is occupied by **peptidyl-tRNA**, a tRNA carrying the entire polypeptide chain synthesized up to this point (see *step 1*).

When both sites are occupied, peptide bond formation occurs by a reaction in which the polypeptide carried by the peptidyl-tRNA is transferred to the amino acid carried by the aminoacyl-tRNA. This transfer takes place on the large subunit of the ribosome.

The end of the tRNA that carries an amino acid is located on the large subunit, while the anticodon at the other end interacts with the mRNA bound by the small subunit. So the P and A sites must extend across both ribosomal subunits, as drawn in Figure 7.1.

Transfer of the polypeptide generates a ribosome in which the **deacylated tRNA** now devoid of any amino acid lies in the P site, while a new peptidyl-tRNA has been created in the A site (see *step 2*). This peptidyl-tRNA is one amino acid residue longer than the peptidyl-tRNA that had been in the P site in *step 1*.

Then the ribosome moves one triplet along the messenger. Its movement expels the deacylated tRNA and moves the peptidyl-tRNA into the P site (see *step 3*). The next codon to be translated now lies in the A site, ready for a new aminoacyl-tRNA to enter, when the cycle will be repeated.

STARTING THE POLYPEPTIDE CHAIN

What initiates protein synthesis: how is the first codon of the gene recognized as providing the starting point for translation? The signal that marks the start of the reading frame is a special codon, the triplet AUG (in bacteria, occasionally GUG is used instead).

The first clue that initiation cannot take place at *any* codon, but requires a special mechanism, was provided by the observation that the amino acid composition of *E. coli* proteins at their N-terminal ends is highly biased. Methionine occurs as the first amino acid in almost 50% of the proteins. When the species **N-formyl-methionyl-tRNA** was discovered, it seemed to be a good candidate for the special initiator, because the blocked amino acid group would prevent it from participating in chain elongation, but would permit it to initiate a protein.

Two types of tRNA can carry methionine in *E. coli*. One is used for initiation, the other for recognizing AUG codons during elongation.

The initiator tRNA is known as $\mathbf{tRNA_f^{Met}}$. First, it is charged with the amino acid to generate Met-$tRNA_f$; then the formylation reaction shown in **Figure 7.2** blocks the free NH_2 group. The name of the product is usually abbreviated to fMet-$tRNA_f$. It is used *only* for initiation, and recognizes either of the codons AUG or GUG.

The species responsible for recognizing AUG codons in internal locations is $\mathbf{tRNA_m^{Met}}$. This tRNA responds *only* to internal AUG codons; and its methionine cannot be formylated. Thus there are two differences between the initiating and elongating Met-tRNAs: the tRNA moieties themselves are different; and the amino acids differ in the state of the amino group.

The meaning of the AUG and GUG codons depends on their **context**. When the AUG codon is used for initiation, it is read as formyl-methionine; when used within the coding region, it represents methionine. The meaning of the GUG codon is even more dependent on its location. When present as the *first* codon, it is read via the initiation reaction as formyl-methionine.

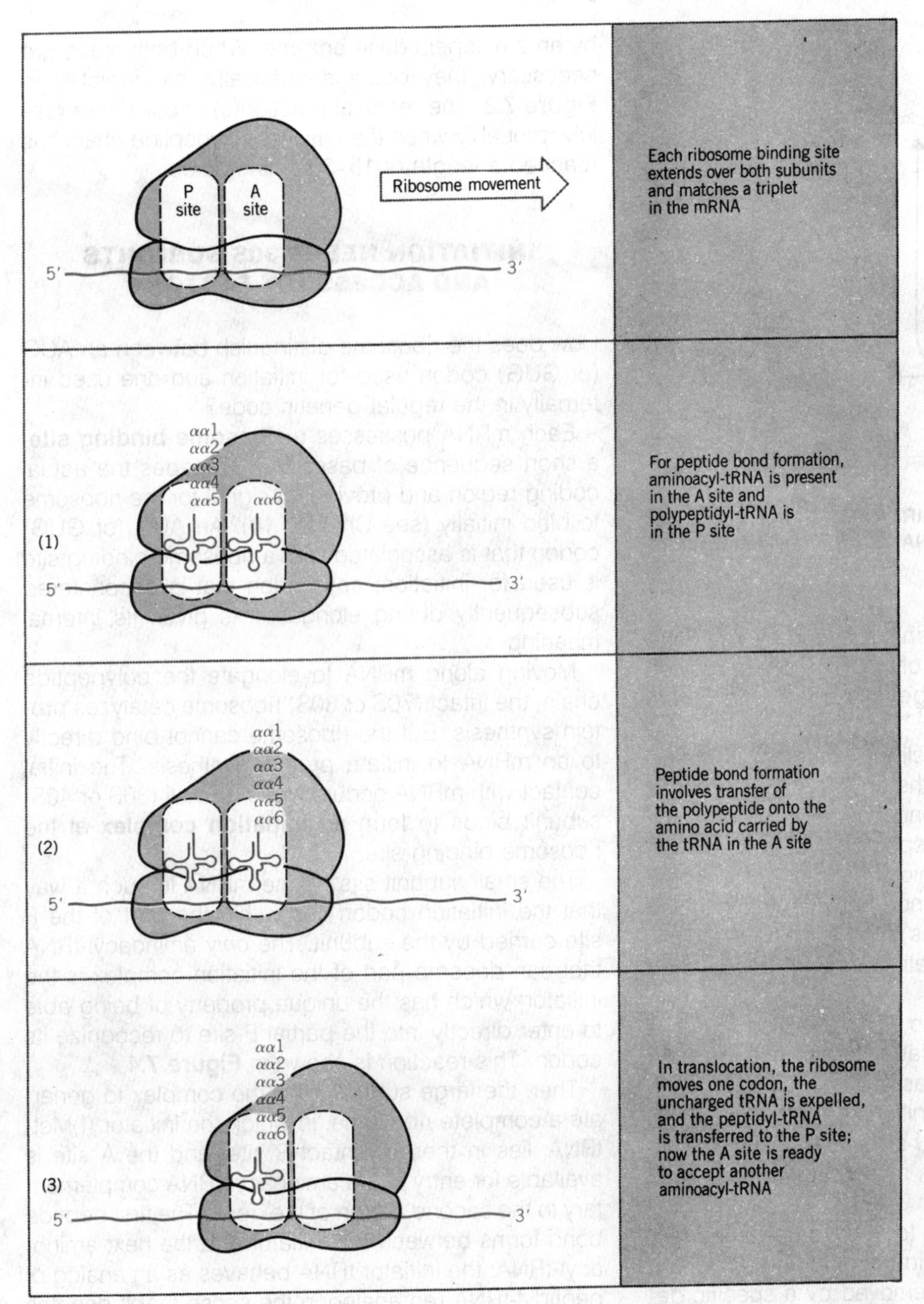

Figure 7.1
The ribosome has two sites for binding tRNA.

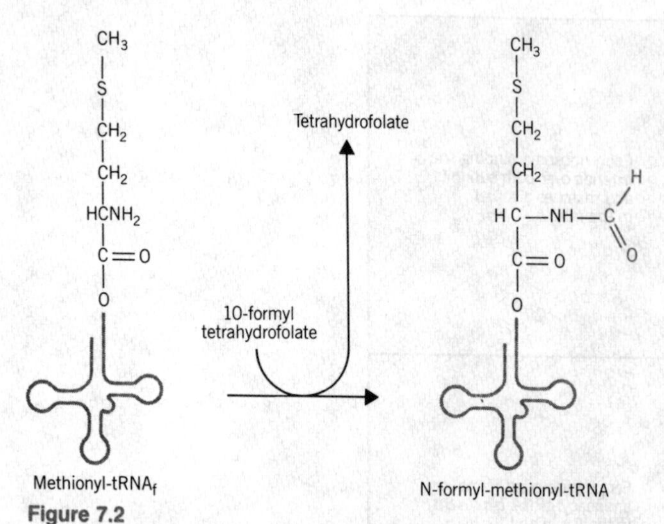

Figure 7.2
The initiator N-formyl-methionyl-tRNA (fMet-tRNA$_f$) is generated by formylation of methionyl-tRNA, using formyl-tetrahydrofolate as cofactor.

Yet when present *within* a gene, it is read by Val-tRNA, one of the regular members of the tRNA set, to provide valine as required by the genetic code (see Figure 6.10).

The initiation reaction is slightly different in eukaryotic cytoplasm. First, only the codon AUG is used as an initiator; GUG is not found naturally. Second, the initiator tRNA is a distinct species, but is known as **tRNA$_i^{Met}$**, because its methionine does *not* become formylated. Thus the difference between the initiating and elongating Met-tRNAs lies solely in the tRNA moiety, with Met-tRNA$_i$ used for initiation and Met-tRNA$_m$ used for elongation.

Two tRNAs responding to AUG are also found in mitochondria, where the initiator species may be used with formylated methionine as in bacteria. Changes in the pattern of codon recognition seem to allow AUA, and possibly AUC and AUU also, to be used as initiation codons in mammalian mitochondria (see Table 6.2).

There is always a system to remove unwanted residues from the N-terminus. In bacteria and mitochondria, the formyl residue is removed by a specific deformylase enzyme. If methionine is to be the N-terminal amino acid of the protein, this is the only necessary step. For any protein that has some other amino acid at the N-terminus, the methionine must be removed by an aminopeptidase enzyme. When both steps are necessary, they occur sequentially, as depicted in **Figure 7.3**. The removal reaction(s) occur rather rapidly, probably when the nascent polypeptide chain has reached a length of 15–30 amino acids.

INITIATION NEEDS 30S SUBUNITS AND ACCESSORY FACTORS

How does the ribosome distinguish between an AUG (or GUG) codon used for initiation and one used internally in the regular genetic code?

Each mRNA possesses a **ribosome binding site**, a short sequence of bases that precedes the actual coding region and provides a signal for the ribosome to bind initially (see Chapter 10). An AUG (or GUG) codon that is associated with a ribosome binding site is used for initiation; any codon that is encountered subsequently during elongation is given its internal meaning

Moving along mRNA to elongate the polypeptide chain, the intact (70S or 80S) ribosome catalyzes protein synthesis. But the ribosome cannot bind directly to an mRNA to initiate protein synthesis. The initial contact with mRNA occurs when a small (30S or 40S) subunit binds to form an **initiation complex** at the ribosome binding site.

The small subunit sits on the mRNA in such a way that the initiation codon lies within the part of the P site carried by the subunit. The *only* aminoacyl-tRNA that can become part of the initiation complex is the initiator, which has the unique property of being able to enter directly into the partial P site to recognize its codon. This reaction is shown in **Figure 7.4**.

Then the large subunit joins the complex to generate a complete ribosome, in which the initiator (f)Met-tRNA$_f$ lies in the now-intact P site; and the A site is available for entry of the aminoacyl-tRNA complementary to the second codon of the gene. The first peptide bond forms between the initiator and the next aminoacyl-tRNA; the initiator tRNA behaves as an analog of peptidyl-tRNA (an analog in the sense that it donates a peptidyl chain consisting of only one amino acid).

So initiation prevails when an AUG (or GUG) codon is juxtaposed with a signal to indicate that small subunits should bind *de novo* to the mRNA, because only

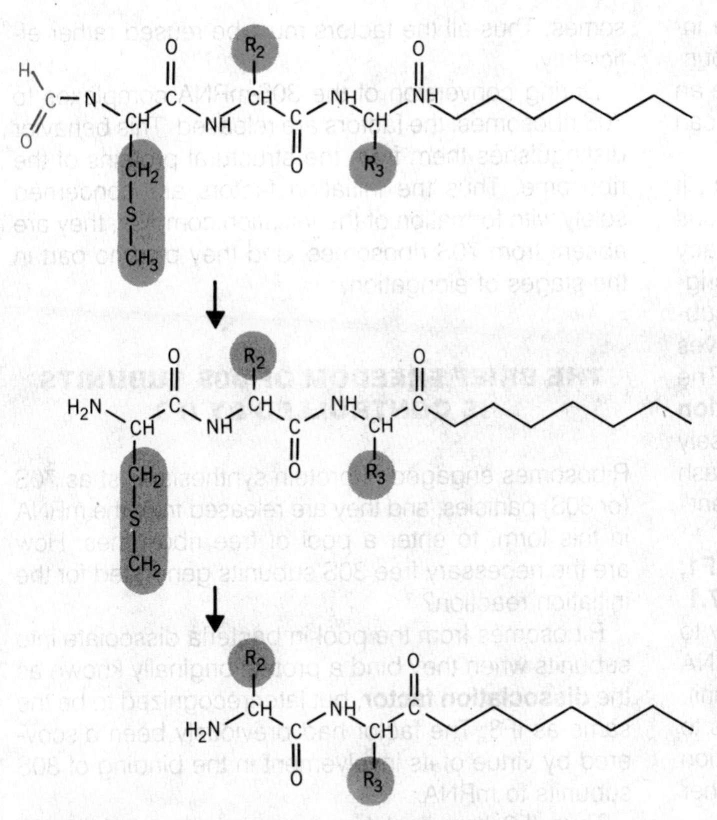

Figure 7.3
Newly synthesized proteins start with formyl-methionine.
The formyl group is removed by a deformylase to generate a normal NH_2 terminus. In about half the proteins, the methionine then at the terminus is removed by an aminopeptidase, creating the new terminus of R_2 (originally the second amino acid incorporated into the chain).

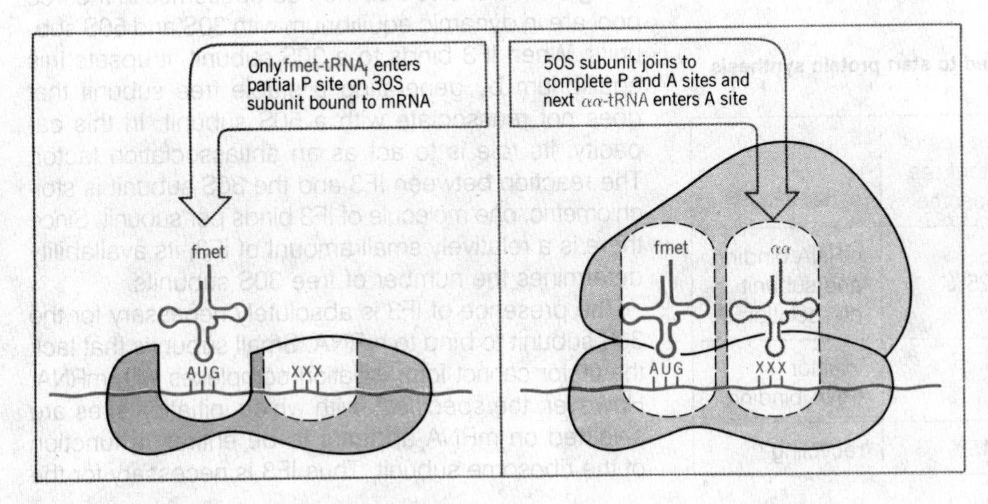

Figure 7.4
Only fMet-tRNA$_f$ can be used for initiation by 30S subunits; only other aminoacyl-tRNAs can be used for elongation by 70S ribosomes.

the initiator tRNA can enter the partial P site. The internal reading prevails when the codons are encountered by a ribosome that is *continuing* to translate an mRNA, because only the regular aminoacyl-tRNAs can enter the (complete) A site.

Although the 30S subunit is involved in initiation, it is not by itself competent to undertake the reactions of binding mRNA and initiator tRNA. The inadequacy of ribosome subunits in the initiation reaction was originally discovered by the effects of "washing" 30S subunits with ammonium chloride. This treatment leaves them unable to sponsor initiation of new chains. The reason is that active 30S subunits contain **initiation factors (IF)**. The factors are proteins bound loosely enough to be released by the ammonium chloride wash (which does not have any effect on the "permanent" ribosomal proteins).

Bacteria use three initiation factors, numbered **IF1, IF2**, and **IF3**. Their roles are summarized in **Table 7.1**. IF3 is needed for 30S subunits to bind specifically to initiation sites in mRNA. IF2 binds the initiator tRNA and places it in the partial P site of the 30S subunit. The role of IF1 has yet to be pinned down; it binds to 30S subunits only as a part of the complete initiation complex, and could be involved in stabilizing it, rather than in recognizing any specific component.

The initiation factors are found only on "free" 30S subunits, which are derived from the 20% of ribosomes not currently engaged in protein synthesis. The number of copies of each factor is sufficient to serve only a small proportion of the number of free ribosomes. Thus all the factors must be reused rather efficiently.

Table 7.1
Three initiation factors are needed to start protein synthesis in *E. coli*.

Factor	Mass (daltons)	Molecules of Factor/Free Ribosome	Function
IF3	23,000	25%	mRNA binding and subunit dissociation
IF2	100,000	?	initiator tRNA binding
IF1	9,000	15%	recycling?

During conversion of the 30S·mRNA complexes to 70S ribosomes, the factors are released. This behavior distinguishes them from the structural proteins of the ribosome. Thus the initiation factors are concerned solely with formation of the initiation complex, they are absent from 70S ribosomes, and they play no part in the stages of elongation.

THE BRIEF FREEDOM OF 30S SUBUNITS IS CONTROLLED BY IF3

Ribosomes engaged in protein synthesis exist as 70S (or 80S) particles, and they are released from the mRNA in this form, to enter a pool of free ribosomes. How are the necessary free 30S subunits generated for the initiation reaction?

Ribosomes from the pool in bacteria dissociate into subunits when they bind a protein originally known as the **dissociation factor**, but later recognized to be the same as IF3. The factor had previously been discovered by virtue of its involvement in the binding of 30S subunits to mRNA.

Thus IF3 has dual functions: it is needed first to generate 30S subunits; and then it must be present for them to bind to mRNA. IF3 essentially controls the freedom of 30S subunits, which lasts from their dissociation from the pool to their reassociation with a 50S subunit at initiation.

Figure 7.5 shows that the 70S ribosomes in the free pool are in dynamic equilibrium with 30S and 50S subunits. When IF3 binds to a 30S subunit, it upsets this equilibrium by generating a stable free subunit that does not reassociate with a 50S subunit. In this capacity, its role is to act as an antiassociation factor. The reaction between IF3 and the 30S subunit is stoichiometric: one molecule of IF3 binds per subunit. Since there is a relatively small amount of IF3, its availability determines the number of free 30S subunits.

The presence of IF3 is absolutely necessary for the 30S subunit to bind to mRNA. Small subunits that lack the factor cannot form initiation complexes with mRNA. However, the *specificity* with which initiation sites are selected on mRNA appears to be entirely a function of the ribosome subunit. Thus IF3 is necessary for the

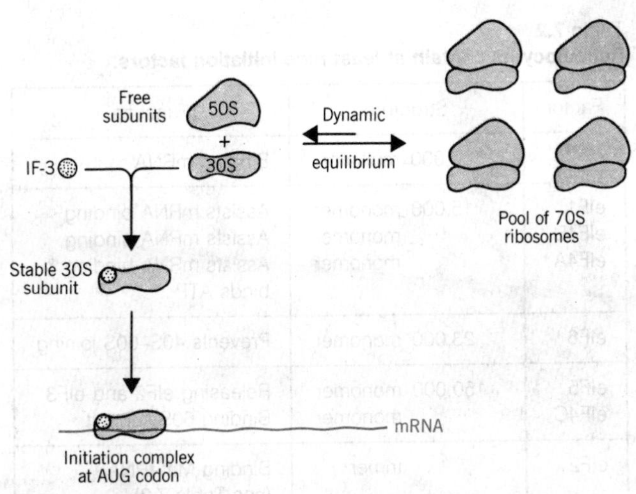

Figure 7.5
IF3 upsets the equilibrium of 70S ribosomes with subunits by generating stable 30S subunits with the ability to bind at initiation sites on mRNA.

reaction, but does not seem to be involved in selecting the site.

IF3 is released before the small subunit is joined by the 50S subunit. A 30S subunit cannot simultaneously bind IF3 and a 50S subunit. The options open to the 30S subunit are therefore either to be "free" by virtue of the presence of IF3, or to be part of a 70S ribosome.

The 30S subunit is driven around its life cycle by these alternatives. When a ribosome is a member of the free pool, its dynamic equilibrium with the subunits allows IF3 to replace the 50S subunit. When an initiation complex has been formed, the IF3 is released, and the 30S subunit is joined by a 50S subunit. On its release, IF3 immediately recycles by finding another 30S subunit.

IF2 PICKS OUT THE INITIATOR tRNA

A role for IF2 in binding the fMet-$tRNA_f$ initiator to the 30S·mRNA complex is illustrated in **Figure 7.6**. By forming a complex specifically with fMet-$tRNA_f$, IF2 ensures that only the initiator tRNA, and none of the regular aminoacyl-tRNAs, participates in the initiation reaction. It is not clear which features of the initiator tRNA are involved in distinguishing it from other

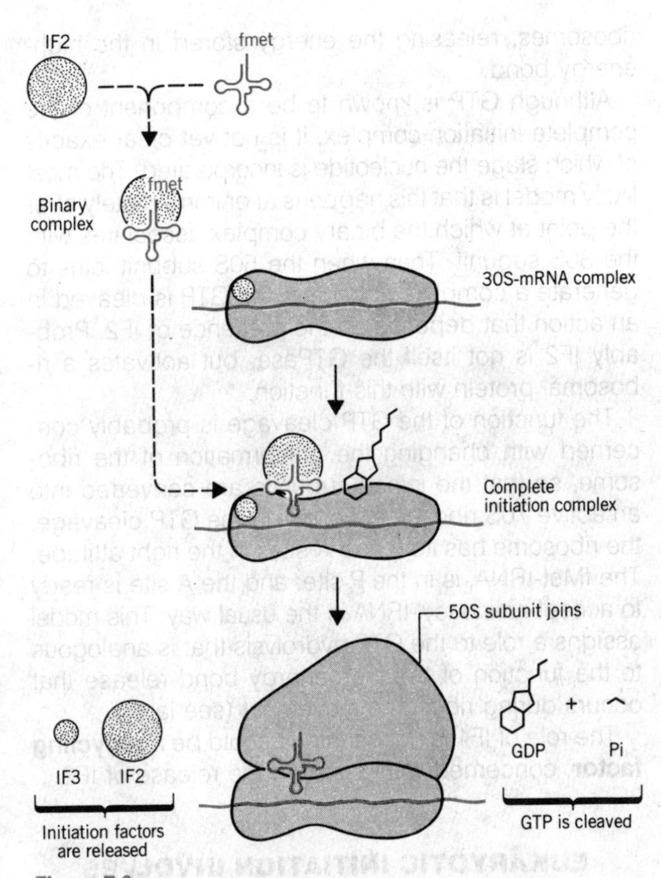

Figure 7.6
Bacterial initiation factors are needed for 30S subunits to bind mRNA and for fMet-$tRNA_f$ to bind to the 30S-mRNA complex; after 50S binding, the factors are released and GTP is cleaved.

aminoacyl-tRNAs, except that the presence of a blocked NH_2 group is essential.

The binary complex of IF2·fMet-$tRNA_f$ could represent the first stage in the entry of the initiator tRNA into the initiation complex. Then the 30S·mRNA complex binds the IF2·fMet-$tRNA_f$ complex, placing the tRNA in the partial P site. This order of events is not proven; IF2 can bind directly to 30S subunits, so it is possible that the fMet-$tRNA_f$ is recognized by IF2 already present on the 30S subunit.

Whatever the order of events, IF2 remains part of the 30S subunit at this stage; it has a further role to play. It has been known for a long time that IF2 has a **ribosome-dependent GTPase activity**: the factor sponsors the hydrolysis of GTP in the presence of ribosomes, releasing the energy stored in the high-energy bond.

Although GTP is known to be a component of the complete initiation complex, it is not yet clear exactly at which stage the nucleotide is incorporated. The most likely model is that this happens at or immediately after the point at which the binary complex associates with the 30S subunit. Then when the 50S subunit joins to generate a complete ribosome, the GTP is cleaved in an action that depends on the presence of IF2. Probably IF2 is not itself the GTPase, but activates a ribosomal protein with this function.

The function of the GTP cleavage is probably concerned with changing the conformation of the ribosome, so that the joined subunits are converted into an active 70S ribosome. Following the GTP cleavage, the ribosome has its P and A sites in the right attitude. The fMet-$tRNA_f$ is in the P site; and the A site is ready to accept aminoacyl-tRNA in the usual way. This model assigns a role to the GTP hydrolysis that is analogous to the function of the high-energy bond release that occurs during ribosome movement (see later).

The role of IF1 is not certain. It could be a **recycling factor**, concerned with assisting the release of IF2.

EUKARYOTIC INITIATION INVOLVES MANY FACTORS

There is a difference in the way that bacterial 30S and eukaryotic 40S subunits find their binding sites for initiating protein synthesis on mRNA. In bacteria, the initiation complex forms directly at a sequence surrounding the AUG initiation codon. In eukaryotes, small subunits first recognize the 5' end of the mRNA, and then move to the initiation site, where they are joined by large subunits. We shall return to this issue in Chapter 10.

Apart from this difference, the process of initiation in eukaryotes appears to be generally analogous to that in *E. coli*. There are more initiation factors—at least nine already have been found in reticulocytes (immature red blood cells), in which the most work has been done. The factors are named similarly to those in bacteria, but with a prefix "e" to indicate their eukaryotic origin. The present set of factors is summarized in **Table 7.2**. The eIF2 and eIF3 fractions each

Table 7.2
Reticulocytes contain at least nine initiation factors.

Factor	Structure	Function
eIF3	>500,000 multimer	Binding mRNA
eIF1 eIF4B eIF4A	15,000 monomer monomer monomer	Assists mRNA binding Assists mRNA binding Assists mRNA binding & binds ATP
eIF6	23,000 monomer	Prevents 40S–60S joining
eIF5 eIF4C	150,000 monomer monomer	Releasing eIF2 and eIF3 Binding 60S subunit
eIF2	trimer	Binding Met-$tRNA_i$ (see Table 7.3)
eIF4D	monomer	Unknown

contain multiple chains. The other factors are mostly single polypeptides with roles that are not well characterized.

Eukaryotic initiation proceeds through the formation of a **ternary complex** containing Met-$tRNA_i$, eIF2, and GTP. The complex is formed in two stages. First, GTP binds to eIF2. Second, the binding increases the factor's affinity for Met-$tRNA_i$, which then is bound. **Figure 7.7** shows that the ternary complex associates directly with free 40S subunits, in a reaction that is independent of the presence of mRNA. In fact, the Met-$tRNA_i$ initiator must be present in order for the 40S subunit to bind to mRNA.

Binding of the 40S-ternary complex to mRNA depends on eIF3. Unlike its bacterial counterpart (IF3), eIF3 may be concerned *solely* with mRNA binding, a role in which several other factors also may be involved. A high energy bond (in ATP) is hydrolyzed.

The role of maintaining subunits in their dissociated state may belong to another factor, eIF6; in another difference from the bacterial reaction, eIF6 may act on the large ribosomal subunit.

Junction of the 60S subunits with the initiation complex cannot occur until eIF2 and eIF3 have been released from the initiation complex, a function mediated by eIF5. The 40S-60S joining reaction may depend directly on eIF4C. These two factors therefore fulfill a role that is not necessary in bacteria. Probably

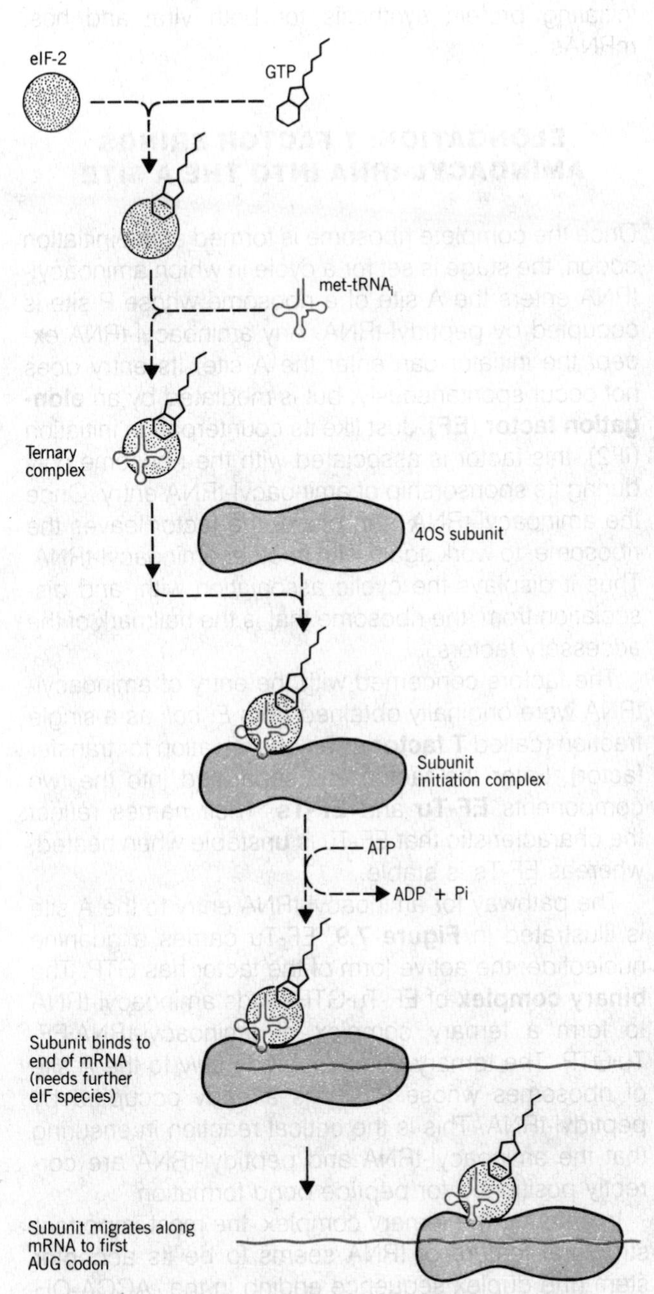

Figure 7.7
In eukaryotic initiation, eIF2 forms a ternary complex with Met-tRNA$_f$. The ternary complex binds to free 40S subunits, which attach to the 5′ end of mRNA and migrate to the initiation site.

all of the remaining factors are released when the complete 80S ribosome is formed.

It is important to realize that, in the context of characterizing initiation factors, "eukaryotic" refers to only a small number of systems. Others could follow a pathway that is different in detail. (Similarly, there are some variations in the initiation pathway in bacteria other than *E. coli*.)

Of the eukaryotic initiation factors, the most is known about eIF2, which has been implicated as a possible control point for protein synthesis in two circumstances. eIF2 contains three subunits; their general properties appear to be similar in mammals and wheat germ, which suggests that the function of the factor may be similar in many eukaryotes. **Table 7.3** summarizes the characteristics of the factor.

When reticulocytes are deprived of hemin, the cofactor needed to assemble the red blood cell protein, hemoglobin, the initiation of protein synthesis ceases. The reason is that an inhibitor of protein synthesis becomes active when the hemin concentration falls below a critical level. The inhibitor is a protein kinase (an enzyme that catalyzes the addition of phosphate groups to proteins); eIF2α is the substrate for the enzyme.

The phosphorylation of eIF2α inhibits its activity by an indirect mechanism. After playing its part in the initiation reaction, eIF2 is released in the form of a complex with the GDP generated by hydrolyzing the GTP that was part of the ternary complex. The factor cannot function again until the GDP is replaced with GTP. **Figure 7.8** illustrates this pathway. The replacement reaction requires an additional factor, known by various names, but presently called eIF2B. This factor cannot function if the α subunit is phosphorylated.

Table 7.3
eIF2 consists of three subunits.

Subunit	Mass (daltons)	Function in Initiation
α	35,000	Binds GTP Controlled by phosphorylation
β	38,000	May be a recycling factor
γ	55,000	Binds Met-tRNA$_i$

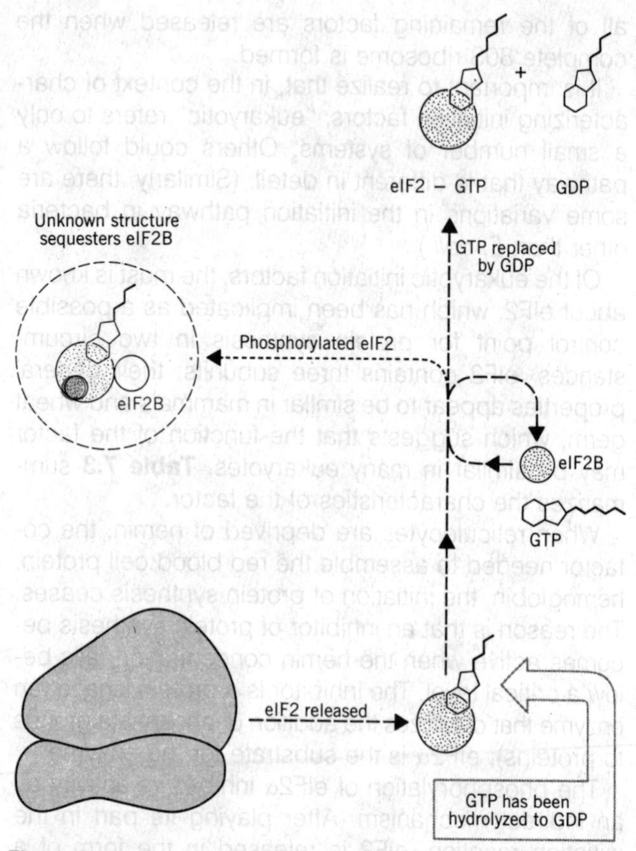

Figure 7.8
eIF2 is released from the eukaryotic initiation reaction as a binary complex containing GDP; the factor eIF2B is needed to replace the GDP with GTP so that eIF2 can be used again.

Phosphorylation of only 20–30% of the eIF2 is sufficient to halt initiation of protein synthesis, probably because the phosphorylated eIF2 binds all the eIF2B (there are only about 10–20% as many eIF2B molecules as eIF2 molecules) and prevents it from functioning on unphosphorylated eIF2.

The same phosphorylation of eIF2 occurs in a variety of cells when double-stranded RNA (dsRNA) is added. The dsRNA activates another kinase, which makes exactly the same phosphorylation of eIF2α. Synthesis of the kinase may be induced by the antiviral agent, interferon (which is itself induced very potently by dsRNA). Interferon specifically inhibits the protein synthetic activity of infecting viruses, while leaving host cell protein synthesis unaffected. However, we do not understand how the phosphorylation of eIF2 may be related to this discriminatory inhibition, because eIF2 itself appears to be equally involved in initiating protein synthesis for both viral and host mRNAs.

ELONGATION: T FACTOR BRINGS AMINOACYL-tRNA INTO THE A SITE

Once the complete ribosome is formed at the initiation codon, the stage is set for a cycle in which aminoacyl-tRNA enters the A site of a ribosome whose P site is occupied by peptidyl-tRNA. Any aminoacyl-tRNA *except* the initiator can enter the A site. Its entry does not occur spontaneously, but is mediated by an **elongation factor** (**EF**). Just like its counterpart in initiation (IF2), this factor is associated with the ribosome only during its sponsorship of aminoacyl-tRNA entry. Once the aminoacyl-tRNA is in place, the factor leaves the ribosome, to work again with another aminoacyl-tRNA. Thus it displays the cyclic association with, and dissociation from, the ribosome that is the hallmark of the accessory factors.

The factors concerned with the entry of aminoacyl-tRNA were originally obtained from *E. coli* as a single fraction (called **T factor** as an abbreviation for transfer factor). Later the factor was separated into the two components **EF-Tu** and **EF-Ts**. Their names reflect the characteristic that EF-Tu is **u**nstable when heated, whereas EF-Ts is **s**table.

The pathway for aminoacyl-tRNA entry to the A site is illustrated in **Figure 7.9**. EF-Tu carries a guanine nucleotide; the active form of the factor has GTP. The **binary complex** of EF-Tu·GTP binds aminoacyl-tRNA to form a ternary complex of aminoacyl-tRNA·EF-Tu·GTP. The ternary complex binds *only* to the A site of ribosomes whose P site is already occupied by peptidyl-tRNA. This is the critical reaction in ensuring that the aminoacyl-tRNA and peptidyl-tRNA are correctly positioned for peptide bond formation.

In entering the ternary complex, the most important structural feature of tRNA seems to be its acceptor stem (the duplex sequence ending in the -ACCA-OH to which the amino acid is attached). It does not matter whether the amino acid is attached to the 2′ or 3′ position of the sugar ring, but at some stage before

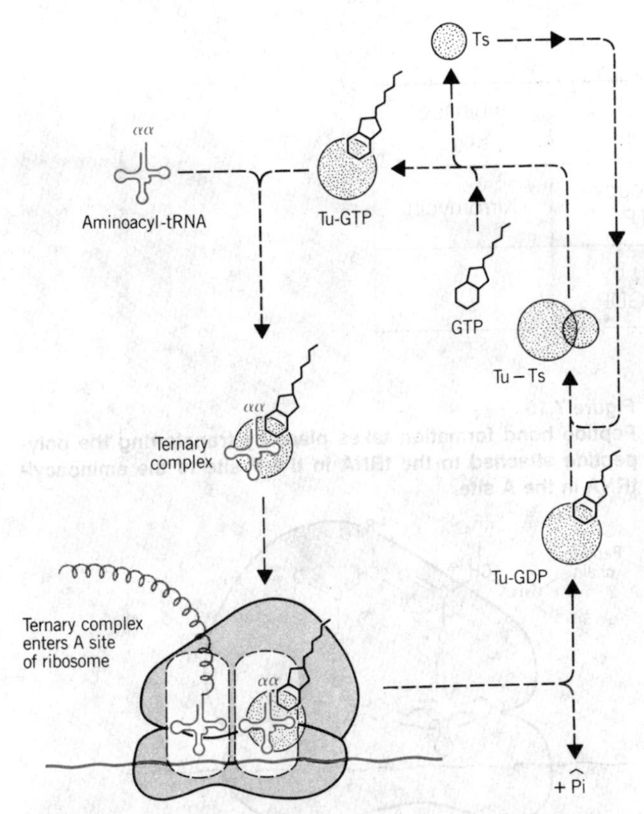

Figure 7.9
EF-Tu·GTP places aminoacyl-tRNA on the ribosome and then is released as EF-Tu·GDP. The binary complex requires EF-Ts to mediate the replacement of GDP by GTP.

binding to the ribosome, the ternary complex stabilizes the ester bond of aminoacyl-tRNA in the 2′ configuration.

The only aminoacyl-tRNA that cannot be recognized by the binary complex is fMet-$tRNA_f$, whose failure to bind prevents it from responding to internal AUG or GUG codons. One reason for the lack of reaction may be the unusual structure of the acceptor stem, which differs from other aminoacyl-tRNAs in ending in an unpaired A·C instead of an A·U pair. The blocking of the NH_2 group is also important. (The exact features responsible for maintaining this discrimination may be peculiar to bacteria, although the same principle applies in eukaryotes.)

After aminoacyl-tRNA has been placed in the A site, the GTP is cleaved; and then the binary complex EF-Tu·GDP is released. This form of EF-Tu does not bind aminoacyl-tRNA effectively, probably because the type of guanine nucleotide controls the conformation of the factor.

The factor EF-Ts mediates the regeneration of the used form, EF-Tu·GDP, into the active form, EF-Tu·GTP. First, EF-Ts displaces the GDP from EF-Tu, forming the combined factor EF-Tu·EF-Ts. (This is the original complete T factor.) Then the EF-Ts is in turn displaced by GTP, reforming EF-Tu·GTP. The active binary complex binds aminoacyl-tRNA; and the released EF-Ts can recycle.

The interactions between EF-Ts, EF-Tu, and GTP are reversible *in vitro*. The reaction with aminoacyl-tRNA is irreversible; and it is this step that drives the reaction sequence only in the forward direction. From the perspective of the elongation factors, EF-Ts cycles between freedom and binding EF-Tu, while EF-Tu oscillates between binding GDP or GTP in the binary and ternary complexes or being associated with EF-Ts in the T factor.

Table 7.4 summarizes some properties of EF-Tu and EF-Ts. The amount of EF-Tu approaches the number of aminoacyl-tRNA molecules. This implies that most aminoacyl-tRNAs are likely to be present in ternary complexes rather than free. Coded by two genes, EF-Tu provides about 5% of the total protein. Most of the EF-Tu must be in the form of binary or ternary complexes, because there is much less EF-Ts; its amount approaches the number of ribosomes.

The role of GTP in the ternary complex has been studied by substituting an analog that cannot be hydrolyzed. The compound **GMP-PCP** has a methylene bridge in place of the oxygen that links the β and γ phosphates in GTP. In the presence of GMP-PCP, a ternary complex can be formed that binds aminoacyl-tRNA to the ribosome. But the peptide bond cannot be formed. Thus the *presence* of GTP is needed for aminoacyl-tRNA to be bound at the A site; but the *hydrolysis* is not required until later.

Much information about the individual steps of bacterial protein synthesis has been obtained by using antibiotics that inhibit the process at particular stages. **Kirromycin** inhibits the function of EF-Tu. When EF-Tu is bound by kirromycin, it remains able to bind aminoacyl-tRNA to the A site. But the EF-Tu·GDP complex cannot be released from the ribosome. Its continued presence prevents formation of the peptide bond between the peptidyl-tRNA and the aminoacyl-tRNA.

Table 7.4
Aminoacyl-tRNA utilization requires two protein factors.

Factor	Coded by	Mass	Molecules per Cell	Function	Inhibited by
EF-Tu	*tufA* *tufB*	43,225	70,000	Binds aminoacyl--tRNA & GTP	Kirromycin
EF-Ts	*tsf*	74,000	10,000	Binds EF-Tu by displacing GDP	

As a result, the ribosome becomes "stalled" on mRNA, bringing protein synthesis to a halt.

This effect of kirromycin demonstrates that inhibiting one step in protein synthesis blocks the next step. In this case, the release of EF-Tu·GDP is needed for the ribosome to acquire the right conformation to sponsor peptide bond formation. The same principle is seen at other stages of protein synthesis: one reaction must be completed properly before the next can occur.

In eukaryotes, as typified by reticulocytes, there are two elongation factors, **eEF1** and **eEF2**, which are equivalent respectively to the bacterial factors EF-T and EF-G (see later).

The role of eEF1 is to bring aminoacyl-tRNA to the ribosome, again in a reaction that involves cleavage of a high-energy bond in GTP. The active factor consists of aggregates of polypeptide chains of various sizes. The details of how GTP is regenerated after cleavage are not known, but the factor may have components analogous to EF-Tu and EF-Ts (which is why it is described as a counterpart to the combined factor EF-T). Quantitatively, the situation may be similar in eukaryotes and prokaryotes, with eEF1 (like EF-Tu) constituting a major protein of the cell.

TRANSLOCATION MOVES THE RIBOSOME

The ribosome remains in place while the polypeptide chain is elongated by transferring the polypeptide attached to the tRNA in the P site to the aminoacyl-tRNA present in the A site. The reaction is shown in **Figure 7.10**. Synthesis of the peptide bond is catalyzed by **peptidyl transferase**, an enzyme activity which prefers 3′ aminoacyl-tRNA as substrate. Possibly the amino

Figure 7.10
Peptide bond formation takes place by transferring the polypeptide attached to the tRNA in the P site to the aminoacyl-tRNA in the A site.

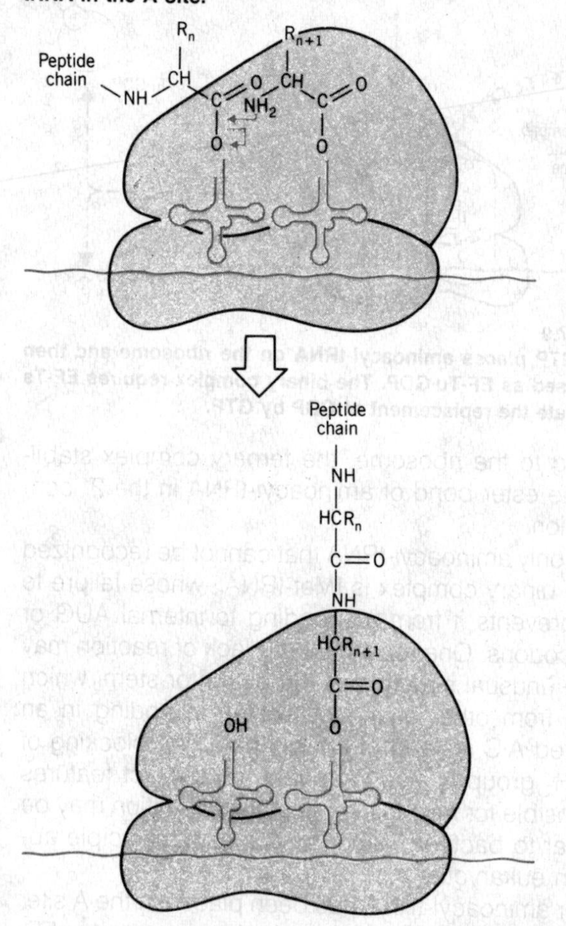

acid is transferred from the 2′ position favored in the ternary complex before the peptide bond is formed.

Peptidyl transferase is a function of the large (50S or 60S) ribosomal subunit. The transferase must be part of a ribosomal site at which the ends of the peptidyl-tRNA and aminoacyl-tRNA are brought close together. One of the problems still unsolved in protein synthesis is how two quite bulky tRNA molecules can have their anticodon ends base paired at adjacent codons on mRNA, while their acceptor ends are close enough for peptide bond formation. We do not know how the geometry of the P and A sites avoids the expected steric hindrance.

Following peptide bond formation, the ribosome carries uncharged tRNA in the P site and peptidyl-tRNA in the A site. The cycle of addition of amino acids to the growing polypeptide chain is completed by the **translocation** illustrated in **Figure 7.11**, in which the ribosome advances three nucleotides along the mRNA. In a concerted action, translocation simultaneously expels the uncharged tRNA and moves the peptidyl-tRNA into the P site. The ribosome then has an empty A site ready for entry of the aminoacyl-tRNA corresponding to the next codon.

The need for an additional factor in translocation was revealed by some of the original experiments that

Figure 7.11
After peptide bond formation, EF-G and GTP are needed for translocation. Usually, EF-G and GDP are released, but fusidic acid can stabilize the ribosome in the posttranslocation state.

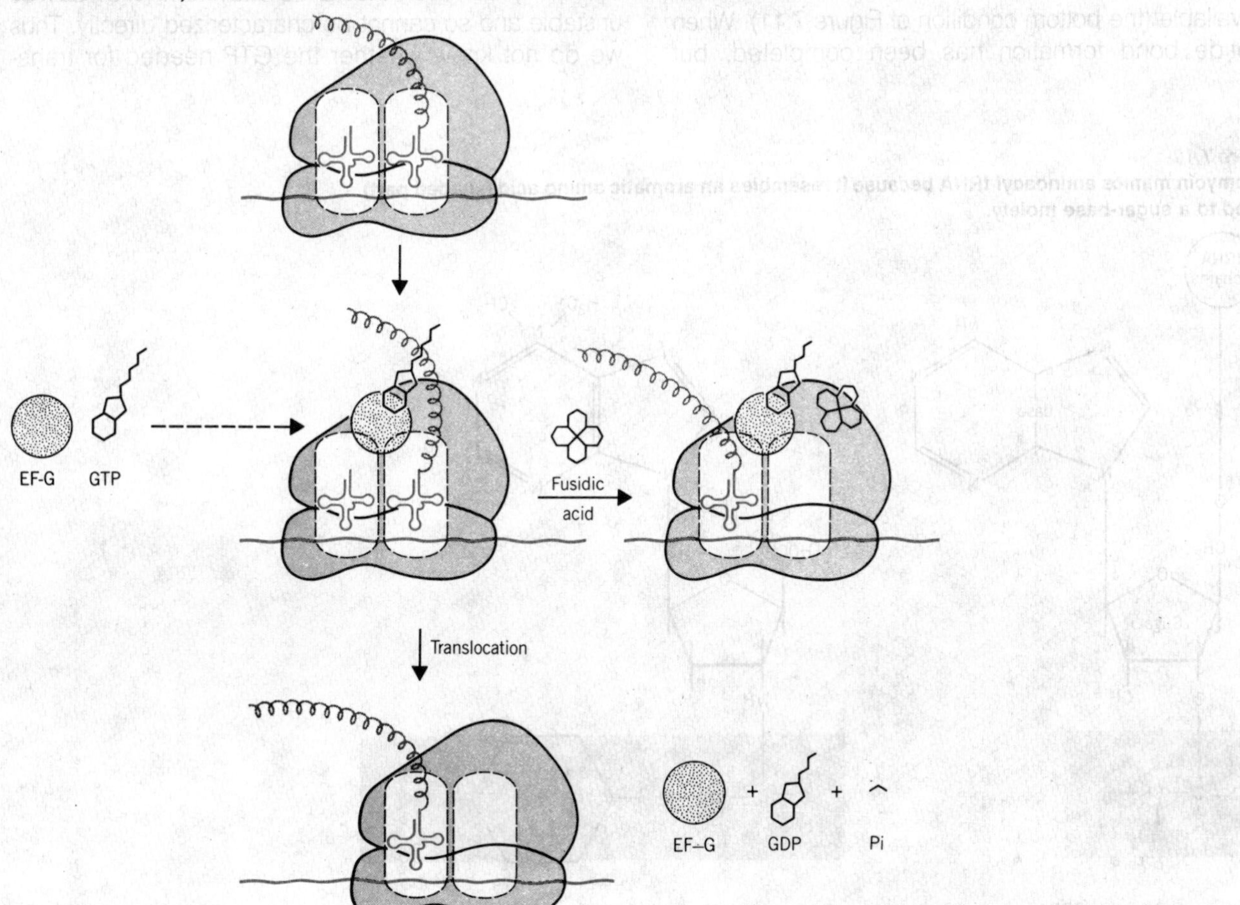

distinguished the A and P sites. This work used the antibiotic **puromycin**, which closely resembles an amino acid attached to the terminal adenosine of tRNA.

Figure 7.12 shows that puromycin has an N at the position of the O that joins the amino acid to tRNA. The antibiotic is treated by the ribosome as though it were an incoming aminoacyl-tRNA; the polypeptide attached to peptidyl-tRNA is transferred to the NH_2 group of the puromycin. But the polypeptidyl-puromycin link cannot be cleaved, so the polypeptide cannot be transferred to the next aminoacyl-tRNA. Instead it is released from the ribosome in the form of polypeptidyl-puromycin. This premature termination of protein synthesis is responsible for the lethal action of the antibiotic.

Because puromycin enters the ribosome by the same route as aminoacyl-tRNA, it can react with a peptidyl-tRNA only when the P site is occupied and the A site is available (the bottom condition of Figure 7.11). When peptide bond formation has been completed, but translocation has not yet occurred, puromycin cannot enter, because the A site remains blocked by the peptidyl-tRNA (the top state of Figure 7.11).

Translocation is needed to make ribosomes in the latter condition sensitive to puromycin. The acquisition of sensitivity *in vitro* requires GTP and another elongation factor, **EF-G**. (The name of the protein reflects its isolation as a **G**TP-requiring factor.) EF-G is a major constituent of the cell; as summarized in **Table 7.5**, it is present at a level of approximately one copy per ribosome.

EF-G binds to the ribosome to sponsor translocation; and then it is released when GTP hydrolysis occurs. Binding can still occur when GMP-PCP is substituted for GTP; thus the *presence* of a guanine nucleotide is needed for binding, but its *hydrolysis* is not needed until the actual translocation reaction occurs. Any complex that EF-G may form with GTP is unstable and so cannot be characterized directly. Thus we do not know whether the GTP needed for trans-

Figure 7.12
Puromycin mimics aminoacyl-tRNA because it resembles an aromatic amino acid (shaded part) linked to a sugar-base moiety.

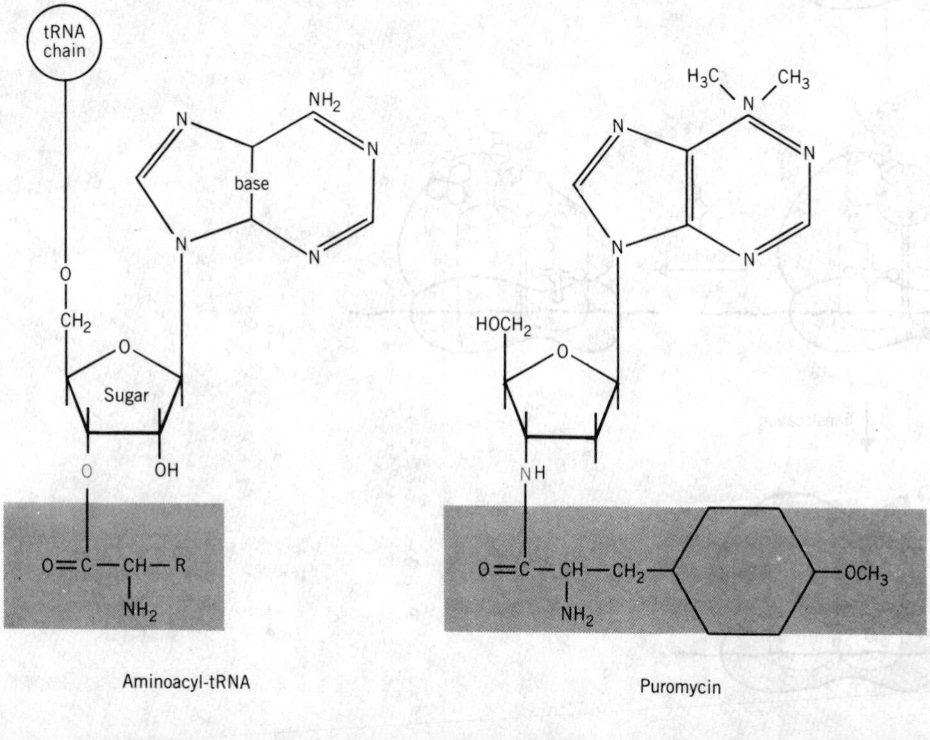

tRNA chain
NH2
N
N
base
N
N
O
CH2
O
Sugar
O
OH
O=C—CH—R
NH2
Aminoacyl-tRNA

H3C CH3
N
N
N
N
N
HOCH2
O
NH
O=C—CH—CH2—
—OCH3
NH2
Puromycin

Table 7.5
Translocation requires an elongation factor.

Factor	Coded by	Mass	Molecules per Cell	Function	Inhibited by
EF-G	*fus*	77,444	20,000	Binds ribosome and GTP	Fusidic acid

location is brought to the ribosome by EF-G or is provided in some other way. The hydrolysis of GTP is not catalyzed by the factor, but is a ribosomal function.

Ribosomes cannot bind EF-Tu and EF-G simultaneously, so protein synthesis follows a cycle in which the factors are alternately bound to, and released from, the ribosome. Thus EF-Tu·GDP must be released before EF-G can bind; and then EF-G must be released before aminoacyl-tRNA·EF-Tu·GTP can bind.

The need for EF-G release was discovered by the effects of the steroid antibiotic fusidic acid, which "jams" the ribosome in its posttranslocation state. In the presence of fusidic acid, one round of translocation occurs: EF-G binds to the ribosome, GTP is hydrolyzed, and the ribosome moves three nucleotides. But fusidic acid stabilizes the ribosome·EF-G·GDP complex, so that EF-G and GDP remain on the ribosome instead of being released (see Figure 7.11). Because the ribosome then cannot bind aminoacyl-tRNA, no further amino acids can be added to the chain.

We do not know whether the ability of each elongation factor to exclude the other is mediated via an effect on the overall conformation of the ribosome or by direct competition for overlapping binding sites. (The drug thiostrepton prevents both EF-Tu and EF-G from binding.) The need for each factor to be released before the other can bind ensures that the events of protein synthesis proceed in an orderly manner.

Both factors use GTP in a similar manner, requiring its presence for binding to the ribosome, but actually needing to hydrolyze it only later. Since neither factor can use GDP to support its binding to the ribosome, the triphosphate form may be needed for the factor to acquire the right conformation. This mechanism ensures that factors obtain access to the ribosome only in the company of the GTP that they will need later to fulfill their function. Hydrolysis of the GTP provides energy to change the conformation of the ribosome; with EF-Tu, it is needed for the aminoacyl-tRNA in the A site to be reactive; with EF-G it is needed for ribosome movement.

The eukaryotic counterpart to EF-G is the protein eEF2, which seems to function in a similar manner, as a translocase dependent on GTP hydrolysis. Its action also is inhibited by fusidic acid, although less effectively than occurs with bacterial ribosomes. Unlike the bacterial factor, a stable complex of eEF2 with GTP can be isolated. The complex can bind to ribosomes with consequent hydrolysis of its GTP.

A unique reaction of eEF2 is its susceptibility to diphtheria toxin. The toxin uses NAD (nicotinamide adenine dinucleotide) as a cofactor to transfer an ADPR moiety (adenosine diphosphate ribosyl) on to the eEF2. The ADPR-eEF2 conjugate is inactive in protein synthesis. The substrate for the attachment is an unusual amino acid, produced by modifying a histidine; it is common to the eEF2 of many species.

The ADP-ribosylation is responsible for the lethal effects of diphtheria toxin. The reaction is extraordinarily effective: a single molecule of toxin can modify sufficient eEF2 molecules to kill a cell.

FINISHING OFF: THREE CODONS TERMINATE PROTEIN SYNTHESIS

Only 61 triplets are assigned to amino acids. The other three triplets are **termination codons** that end protein synthesis. They have casual names that reflect the history of their discovery. The UAG triplet is called the **amber** codon; UAA is the **ochre** codon; and UAG is sometimes called the **opal** codon.

The nature of these triplets was originally shown by a genetic test that distinguished two types of point mutation.

A point mutation that changes a codon to represent

a different amino acid is called a **missense** mutation. One amino acid replaces the other in the protein; the effect on protein function depends on the site of mutation and the nature of the amino acid replacement.

But when a point mutation creates one of the three termination codons, it causes **premature termination** of protein synthesis at the mutant codon. This is likely to abolish protein function, since only the first part of the protein is made in the mutant cell. A change of this sort is called a **nonsense** mutation. (Sometimes the term *nonsense codon* is used to describe these triplets. "Nonsense" is really a misnomer, since the codons actually do have meaning, albeit an unpalatable one for a mutant gene.)

Analysis of nonsense mutants showed that any one of the three codons UAG, UAA, or UGA is sufficient to terminate protein synthesis within a gene. Are the same signals used to terminate protein synthesis naturally at the end of the gene?

In every gene that has been sequenced, one of the termination codons lies immediately after the codon representing the C-terminal amino acid of the wild-type sequence. These triplet sequences are therefore necessary and sufficient to end protein synthesis, whether occurring naturally at the end of a gene or created by mutation within a coding sequence.

None of the termination codons is represented by a tRNA. They function in an entirely different manner from other codons, and are recognized directly by protein factors. (Since the reaction does not depend on codon-anticodon recognition, there seems to be no particular reason why it should require a triplet sequence. Presumably this reflects some aspect of the evolution of the genetic code.)

In *E. coli* two proteins catalyze termination. They are called **release factors** (**RF**), and are specific for different sequences. **RF1** recognizes UAA and UAG; **RF2** recognizes UGA and UAA. The factors act at the ribosome A site and require polypeptidyl-tRNA to be present in the P site. A third factor, **RF3**, appears to stimulate the action of both the others.

In eukaryotic systems, there is only a single release factor, **eEF**. GTP is needed for eEF to bind to ribosomes (it is not involved in bacteria). Probably the GTP is cleaved after the termination step has occurred; the hydrolysis may be needed to allow eEF to dissociate from the ribosome.

The termination reaction involves release of the completed polypeptide from the last tRNA, expulsion of the tRNA from the ribosome, and dissociation of the ribosome from mRNA. The cleavage of polypeptide from tRNA could take place by a reaction analogous to the usual transfer from peptidyl-tRNA to aminoacyl-tRNA during elongation; perhaps the release factor diverts the reaction. We do not yet understand how the ribosome dissociates from mRNA. The dissociation may result from a conformational change triggered by the RF factor, but it is possible that another protein factor(s) could be involved.

FURTHER READING

Since the basic pathway for protein synthesis was worked out, most reviews have concentrated on the roles of the factors that control the reactions. The pattern of factors has been described by **Maitra et al**. (*Ann. Rev. Biochem.* **51**, 869–900, 1982). Eukaryotic initiation and its regulation have been analyzed by **Jagus, Anderson & Safer** (*Prog. Nucleic Acids. Res.* **25**, 127–185, 1981). Progress in working out the control of eIF2 was reviewed by **Safer** (*Cell* **33**, 7–8, 1983). The processes involved in prokaryotic, eukaryotic and organelle initiation have been compared by **Kozak** (*Microbiol. Rev.* **47**, 1–45, 1983).

CHAPTER 8
TRANSFER RNA: THE TRANSLATIONAL ADAPTOR

Transfer RNA occupies a pivotal position in protein synthesis, providing the adaptor molecule that accomplishes the translation of each nucleotide triplet into an amino acid. In Crick's phrase, tRNA represents nature's attempt to make a nucleic acid fulfill the sort of role usually performed by a protein. For tRNAs are involved in a multiplicity of reactions in which it is necessary for them to share certain characteristics in common, yet be distinguished by others. The crucial feature that confers this capacity is the ability of tRNA to fold into a specific tertiary structure.

Working backward from the final action of tRNA in protein synthesis, all tRNAs are able to sit in the P and A sites of the ribosome, where at one end they are associated with mRNA via codon-anticodon pairing, while at the other end the polypeptide is being transferred. For the P and A sites to welcome all tRNAs, the entire set must conform to the same general dictates for size and shape.

Similarly, all tRNAs (except the initiator) share the ability to be recognized by the translation factors (EF-Tu or eEF1) for binding to the ribosome. The initiator tRNA is recognized instead by IF2 or eIF2. So the tRNA set must possess common features for interaction with elongation factors, but the initiator tRNA must lack this feature or possess some contradictory feature.

On the other hand, there must be critical differences between small groups of tRNAs. Usually, each amino acid is represented by more than one tRNA. Multiple tRNAs representing the same amino acid are called **isoaccepting tRNAs**. A group of isoaccepting tRNAs must be charged only by the single aminoacyl-tRNA synthetase specific for their amino acid. Thus isoaccepting tRNAs must share some common feature(s) enabling the enzyme to pick them out from the other tRNAs. The entire complement of tRNAs is divided into 20 isoaccepting groups, each group being able to identify itself to its particular synthetase.

We still have very little idea of what features identify tRNAs to the proteins with which they interact. Early attempts to deduce common features focused on seeking out short nucleotide stretches that are common to all tRNAs or to a particular group of tRNAs. However, although some features in the primary sequence of tRNA are highly conserved, and are presumably essential for one or more of its functions, attempts to correlate individual regions or sequences with particular protein recognition reactions generally have not been successful.

The common (or distinctive) features of tRNAs may transcend the primary and secondary structures and be conveyed at least in part by the tertiary structure. To analyze the function of tertiary structure in any detail requires comparison of several (perhaps all) of the complement of tRNAs. This will be a long job, since at present only a few individual tRNA molecules have been analyzed at the level of tertiary structure.

THE UNIVERSAL CLOVERLEAF

The sequences of several hundred tRNAs from a wide variety of bacteria and eukaryotes have been determined, and they all conform to a general secondary structure that was one of the possibilities deduced from the sequence of the very first tRNA to be sequenced. Each tRNA sequence can be written in the form of a **cloverleaf**, which is maintained by base pairing between short regions that are complementary (see Figure 6.4). The reality of this conformation, rather than some alternatives, was authenticated by showing that in several tRNAs the bases expected to be in the more exposed (unpaired) positions indeed are those more susceptible to attack by nucleases.

A general form of the cloverleaf is illustrated in **Figure 8.1**. There are four major **arms**, named for their structure or function. The **acceptor arm** consists of a base-paired stem that ends in an unpaired sequence whose free 2′/3′-OH group is aminoacylated. The other arms consist of base-paired **stems** and unpaired **loops**. The **TψC arm** is named for the presence of this triplet sequence, the **anticodon arm** always contains the anticodon triplet in the center of the loop, and the **D arm** is named for its content of the base dihydrouridine. (ψ stands for pseudouridine and D stands for dihydrouridine, two of the "unusual" bases in tRNA that are discussed later.)

The numbering system for tRNA illustrates the constancy of the structure. Positions are numbered from 5′ to 3′ according to the most common tRNA structure, which has 76 residues. The overall range of tRNA lengths is from 74 to 95 bases. The variation in length is caused by differences in the structure of only two of the arms.

In the D loop, there is variation of up to four residues. The extra nucleotides relative to the most common structure are denoted 17:1 (lying between 17 and 18) and 20:1 and 20:2 (lying between 20 and 20:1). However, in the smallest D loops, residue 17 as well as these three may be absent.

The most variable feature of tRNA is the so-called **extra arm**, which lies between the TψC and anticodon arms. Depending on the nature of the extra arm, tRNAs can be divided into two classes. **Class 1 tRNAs** have a small extra arm, consisting of only 3–5 bases. They represent about 75% of all tRNAs. **Class 2 tRNAs** have a large extra arm—it may even be the longest in the tRNA—with 13–21 bases, and about 5 base pairs in the stem. The additional bases are numbered from 47:1 through 47:16.

The base pairing that maintains the secondary structure is virtually invariant. Going clockwise around the cloverleaf, there are always 7 base pairs in the acceptor stem, 5 in the TψC arm, 5 in the anticodon arm, and usually 3 (sometimes 4) in the D arm. Within a given tRNA, most of the base pairings will be conventional partnerships of A-U and G-C, but occasional G-U, G-ψ, or A-ψ pairs will be found. The additional types of base pairs are less stable than the regular pairs, but still allow a double-helical structure to form in RNA.

When the sequences of tRNAs are compared, the bases found at some positions are **invariant** (or **conserved**); there is almost always a particular base found at the position. Actually, as more tRNAs are sequenced, positions that seemed entirely invariant do display occasional exceptions. So for practical purposes, the description of any position as invariant means that the specified base is present in more than 90–95% of tRNAs. Sometimes the exceptions are individual; sometimes they fall into groups representing some peculiarity of a particular cell.

Some positions are described as **semiinvariant** (or **semiconserved**) because they seem to be restricted to one type of base (purine versus pyrimidine), but either base of that type may be present.

THE TERTIARY STRUCTURE IS L-SHAPED

The cloverleaf form in which the secondary structure of tRNA is written should not be allowed to convey a misleading view of the tertiary structure, which actually

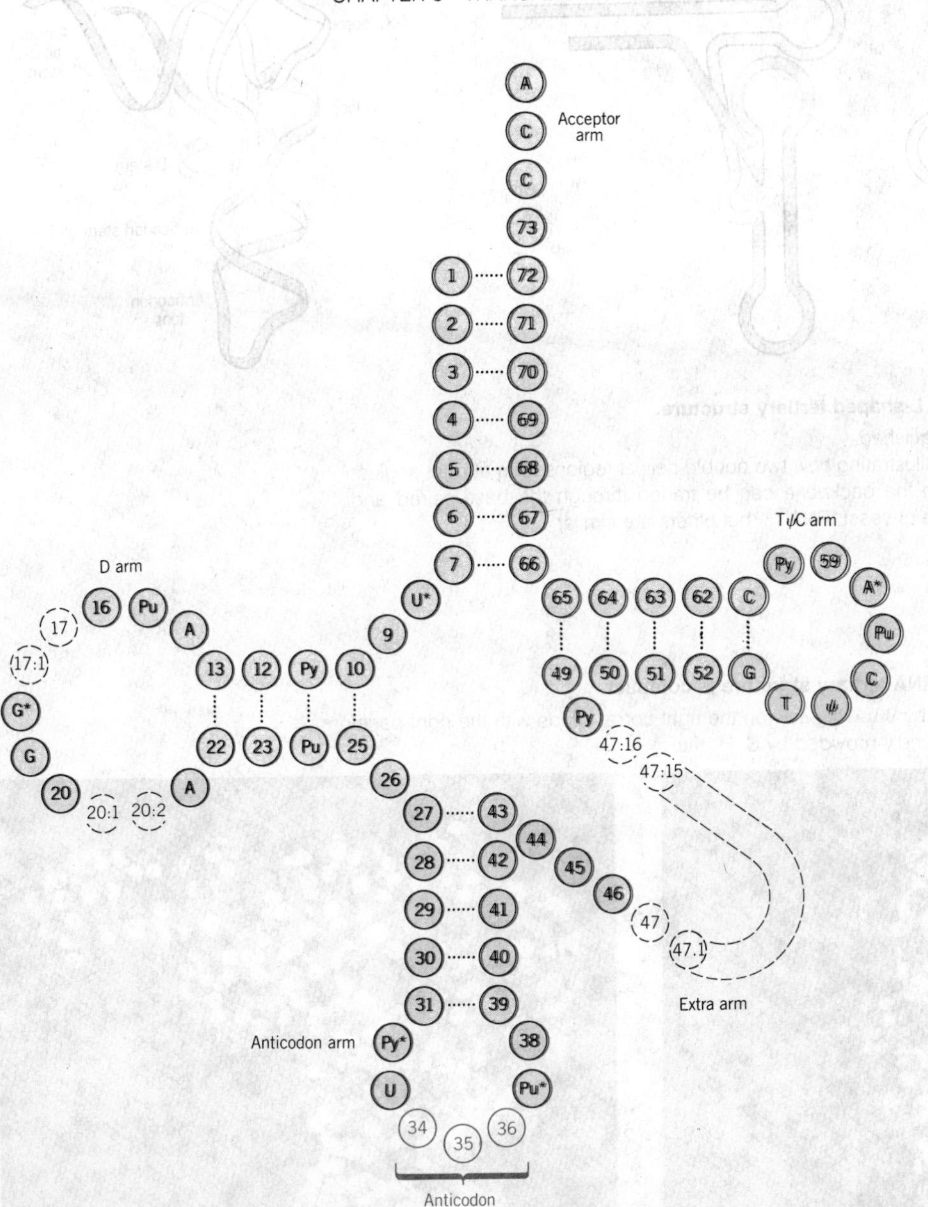

Figure 8.1
Transfer RNA forms a cloverleaf.
Circles indicate base positions. Numbers in shaded circles indicate positions that are always present but may be filled by any base. When a position is always occupied by a particular base, the base is indicated in place of the number. For invariant bases, the actual base is indicated; for semiinvariant bases, Py and Pu indicate the presence of either pyrimidine or either purine. An asterisk indicates that the base is modified, but that the form of the modification may vary. Open circles identify positions present in some but not all tRNAs. Dots indicate base pairing.

is rather compact. To determine the tertiary structure, it is necessary to grow crystals of a tRNA for X-ray crystallography. This is difficult, and actually has been achieved for only a few tRNAs, most from yeast. However, the similarities of the structures, and the nature of the interactions that maintain the tertiary conformation, suggest that a common tertiary theme may be honored by all tRNAs, although each will present its own variation.

The base paired double-helical stems of the secondary structure are maintained in the tertiary structure, but their arrangement in three dimensions essentially creates two double helices at right angles to each other, as illustrated in **Figure 8.2**. The acceptor stem and the TψC stem form one continuous double helix with a single gap; the D stem and anticodon stem form another continuous double helix, also with a gap. The region between the double helices, where the turn

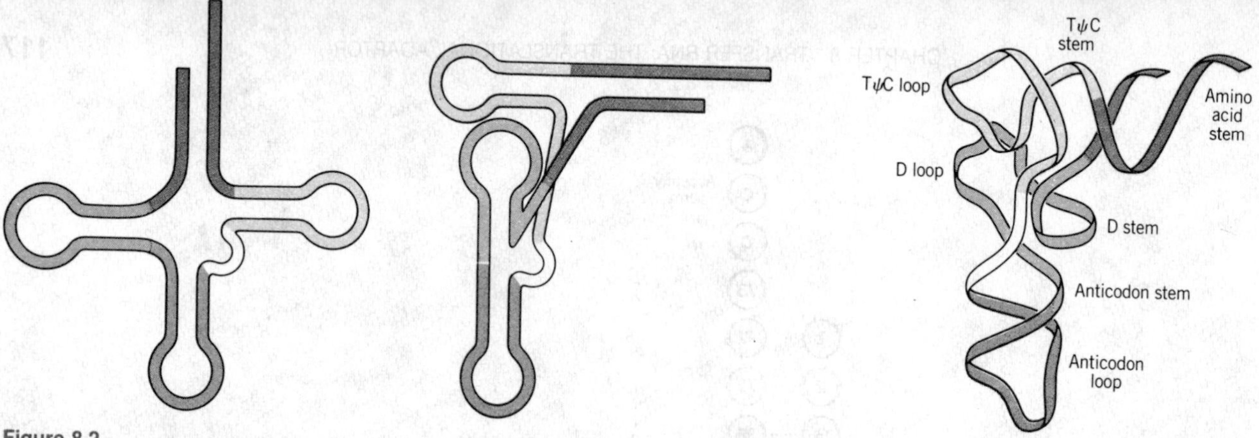

Figure 8.2
Transfer RNA folds into a compact L-shaped tertiary structure.
Left: a cloverleaf with the arms distinguished.
Center: a two-dimensional projection illustrating how two double-helical regions form at right angles.
Right: a schematic drawing in which the backbone can be traced through the base-paired and unpaired regions. This is the structure of yeast $tRNA^{Phe}$, but others are similar.

Figure 8.3
A space-filling model shows that tRNA tertiary structure is compact.
The two views of $tRNA^{Phe}$ are rotated by 90°. The view on the right corresponds with the right panel in the preceding figure. Photograph kindly provided by S. H. Kim.

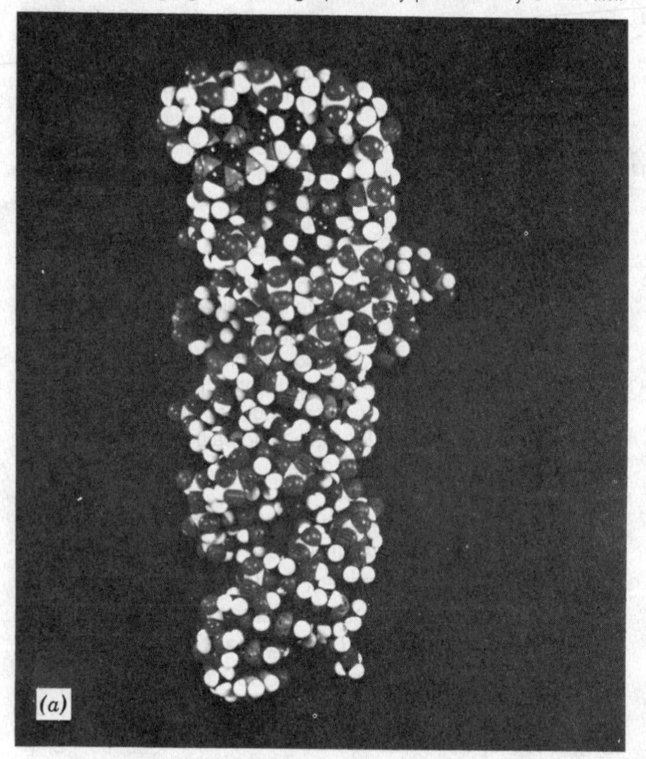

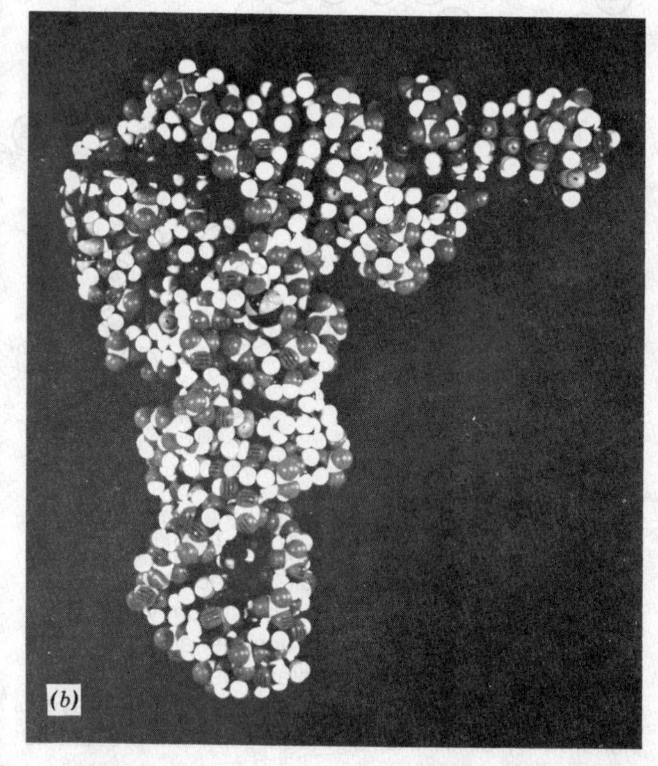

in the L is made, contains the TψC loop and the D loop. Thus the amino acid resides at the extremity of one arm of the L, and the anticodon loop forms the other end.

This model was worked out in detail for $tRNA^{Phe}$ of yeast, the first tRNA to be crystallized. A molecular model of the structure is shown in **Figure 8.3**. In another case, $tRNA^{Asp}$, the angle between the two axes is slightly greater, so the molecule has a similar, but slightly more open, conformation. Detailed differences affecting one region or another are found in each tRNA, thus accommodating the dilemma that all tRNAs must have a similar shape, yet it must be possible to recognize differences between them.

The tertiary structure is created by hydrogen bonding, mostly involving bases that are unpaired in the secondary structure. The bonds of the cloverleaf are described as **secondary H bonds**; the additional bonds of the tertiary structure are called **tertiary H bonds**. Many of the invariant and semiinvariant bases are involved in the tertiary H bonds, which explains their conservation and also suggests that the general form of the tertiary structure may be common to all tRNAs.

The pattern of additional associations for $tRNA^{Phe}$ is summarized in **Figure 8.4**, which shows that some of the bonds involve unusual bases or arrangements for triple pairing. The greatest concentration of the tertiary interactions occurs between the D loop and the TψC loop. Not every one of these interactions is universal—for example, some are absent from $tRNA^{Asp}$—but probably they identify the *general* pattern for establishing tRNA structure.

Figure 8.4
Most tertiary interactions in tRNA structure involve invariant or semiinvariant bases.
The molecule is drawn as a cloverleaf (left) or L-shape (right), with tertiary bonds indicated by lighter lines; colored circles indicate invariant or semiinvariant bases. Heavy lines indicate the covalent bonds joining the arms of the L-shape.

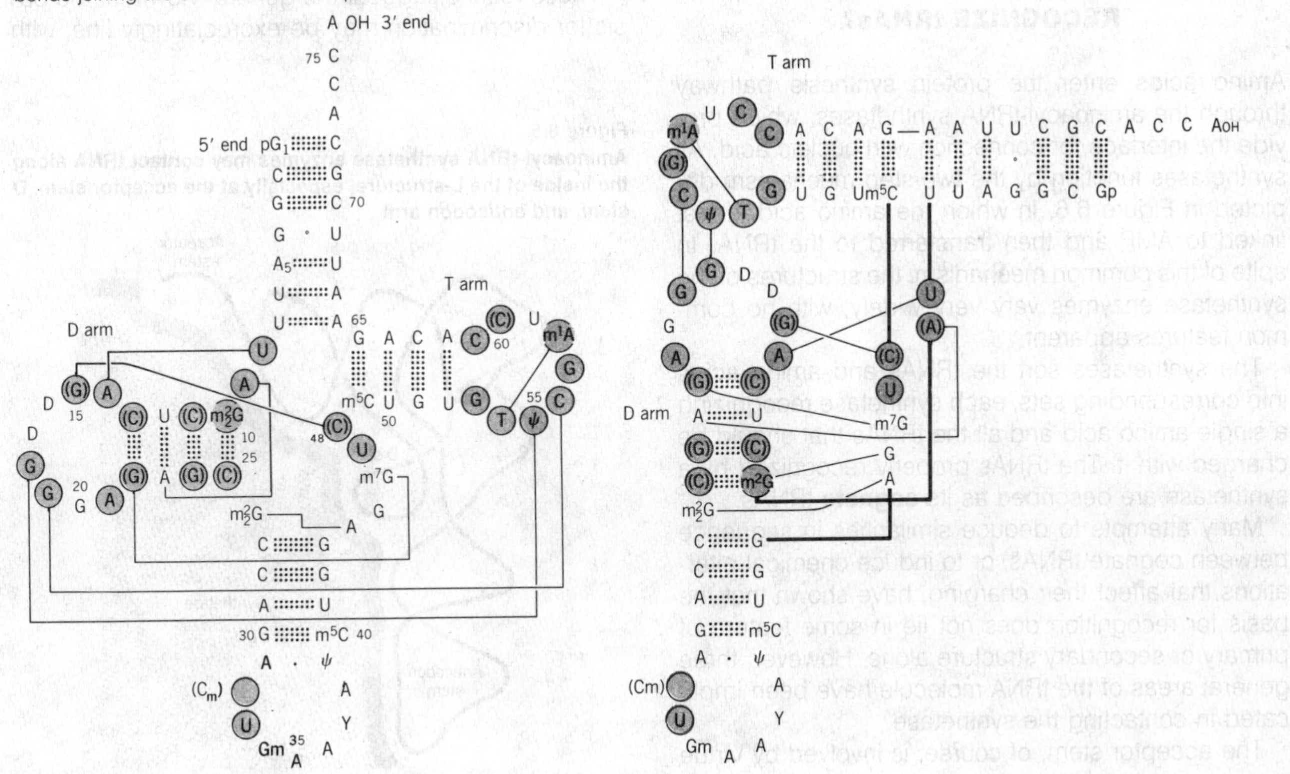

The structure visualized by X-ray crystallography represents a stable organization of the molecule. In solution, or in association with proteins, some flexibility may be shown. The conformation of tRNA can change in both circumstances, but we do not know yet how these changes are related to the crystal structure.

The structure suggests some general conclusions about the function of tRNA. First, its sites for exercising particular functions are maximally separated. The amino acid is as far distant from the anticodon as possible, which is consistent with the need for the aminoacyl group to be near the peptidyl transferase site on the large subunit of the ribosome, while the anticodon pairs with mRNA on the small subunit. The TψC sequence lies at the junction of the arms of the L, possibly a critical location for controlling changes in the tertiary structure. The structure may therefore accommodate the various demands on tRNA by providing several different sites, analogous to the active sites of a protein.

HOW DO SYNTHETASES RECOGNIZE tRNAs?

Amino acids enter the protein synthesis pathway through the aminoacyl-tRNA synthetases, which provide the interface for connection with nucleic acid. All synthetases function by the two-step mechanism depicted in Figure 6.6, in which the amino acid is first linked to AMP and then transferred to the tRNA. In spite of this common mechanism, the structures of the synthetase enzymes vary very widely, with no common features apparent.

The synthetases sort the tRNAs and amino acids into corresponding sets, each synthetase recognizing a single amino acid and all the tRNAs that should be charged with it. The tRNAs properly recognized by a synthetase are described as its **cognate** tRNAs.

Many attempts to deduce similarities in sequence between cognate tRNAs, or to induce chemical alterations that affect their charging, have shown that the basis for recognition does not lie in some feature of primary or secondary structure alone. However, three general areas of the tRNA molecule have been implicated in contacting the synthetase.

The acceptor stem, of course, is involved by virtue of its role in being charged on the 3′ terminus with amino acid. The D stem is implicated by reactions in which enzymes are cross-linked photochemically to their tRNAs; this area is always one of the regions of contact. Often the anticodon stem also is linked to the enzyme. This suggests the general model for tRNA·synthetase binding depicted in **Figure 8.5**, in which the protein binds the tRNA along the "inside" of the L-shaped molecule.

The recognition reaction is extraordinarily precise. Mutations that alter only single bases may change the recognition of a tRNA, preventing it from being recognized by its synthetase, and/or allowing recognition by another synthetase. These mutations are clustered in the acceptor stem and the anticodon arm. (Of course, not all mutations have this effect; for example, the "suppressor" mutations discussed later in this chapter change a base in the anticodon, and therefore the codons to which a tRNA responds, without altering its charging with amino acids.) The reverse mutational experiment is to create substitutions in the synthetase enzyme that alter its discrimination of tRNA.

These results suggest the general view that the basis for discrimination may be excruciatingly fine, with

Figure 8.5
Aminoacyl-tRNA synthetase enzymes may contact tRNA along the inside of the L-structure, especially at the acceptor stem, D stem, and anticodon arm.

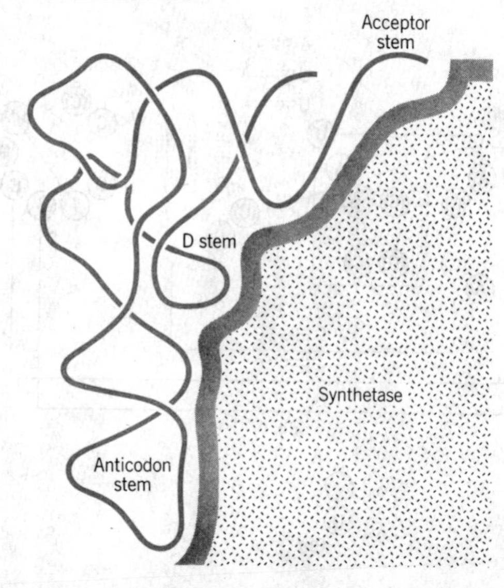

single points of contact distinguishing one tRNA from another. Recognition may depend on an interaction between a few points of contact in the tRNA, concentrated at the extremities and in the D stem, and a few amino acids constituting the active site in the protein.

DISCRIMINATION IN THE CHARGING STEP

The nature of discriminatory events is a general issue raised by several steps in gene expression. How do synthetases recognize just the corresponding tRNAs and amino acids? How does a ribosome recognize only the tRNA corresponding to the codon in the A site? How do DNA or RNA polymerases recognize only the base complementary to the template? Each case poses a similar problem: how to distinguish one particular member from the entire set, all of which share the same general features.

Probably there is a random-hit process in which any member initially can contact the active center, but then the wrong members are rejected and only the appropriate one is accepted. The appropriate member is always in a minority (1 of 20 amino acids, 1 of ~40 tRNAs, 1 of 4 bases), so the criteria for discrimination must be strict. We can imagine two general ways in which the decision whether to reject or accept might be taken.

First, the cycle of admittance, scrutiny, rejection/acceptance could represent a single binding step that *precedes all other stages* of whatever reaction is involved. This is tantamount to saying that the affinity of the binding site is controlled in such a way that only the appropriate species is comfortable there. In the case of synthetases, this would mean that only the cognate tRNAs could form a stable attachment at the site.

Alternatively, the reaction may *proceed through some of its stages*, after which a decision is reached on whether the correct species is present. If it is present, the reaction is reversed, or a bypass route is taken, and the wrong member is expelled. This sort of post-binding scrutiny is generally described as **proofreading**. In the example of synthetases, it would constitute allowing the charging reaction to proceed through certain stages even if the wrong tRNA or amino acid were present.

Some synthetases are highly specific for initially binding a single amino acid, but others can also activate amino acids closely related to the proper substrate. Although the analog amino acid can sometimes be converted to the adenylate form, in none of these cases is a misactivated amino acid actually used to form a stable aminoacyl-tRNA. At what *stage* is a false aminoacyl-adenylate rejected? **Figure 8.6** summarizes the enzyme's options.

One possibility is that the structure of the aminoacyl-adenylate is simply subjected to closer scrutiny than was the structure of the amino acid itself. Then the wrong aminoacyl-adenylate is hydrolyzed. This could happen as an intrinsic process on closer examination (**kinetic proofreading**); or because tRNA binding induces a conformational change in the enzyme, establishing more exacting requirements for the aminoacyl-adenylate (**conformational proofreading**).

However, present opinion favors the idea that selection occurs at a subsequent step by **chemical proofreading**. In this process, the wrong amino acid is actually transferred to tRNA, is then recognized as incorrect by its structure in the tRNA binding site, and so is hydrolyzed and released. The process requires a continual cycle of linkage and hydrolysis until the correct amino acid is transferred to the tRNA.

A classic example in which discrimination between amino acids depends on the presence of tRNA is provided by the Ile-tRNA synthetase of *E. coli*. The enzyme can charge valine with AMP, but hydrolyzes the valyl-adenylate when $tRNA^{Ile}$ is added. The overall error rate depends on the specificities of the individual steps. In the example, the erroneous activation of valine occurs with a frequency of about 1/225. Chemical proofreading allows only about 1/270 of the $Val\text{-}tRNA^{Ile}$ molecules to be released. The overall error rate is the product, $1/60{,}000 = 1.5 \times 10^{-5}$. The measured rate of substitution of valine for isoleucine (in rabbit globin) is $2\text{–}5 \times 10^{-4}$, which suggests that mischarging probably provides only a small fraction of the errors that actually occur.

Binding of tRNA to the synthetase involves the steps depicted in **Figure 8.7**. The initial binding is rapid; then, if the correct tRNA is present, binding is stabilized by a conformational change in the enzyme. This allows aminoacylation to occur rapidly. If the wrong tRNA is present, the failure to make the conformational

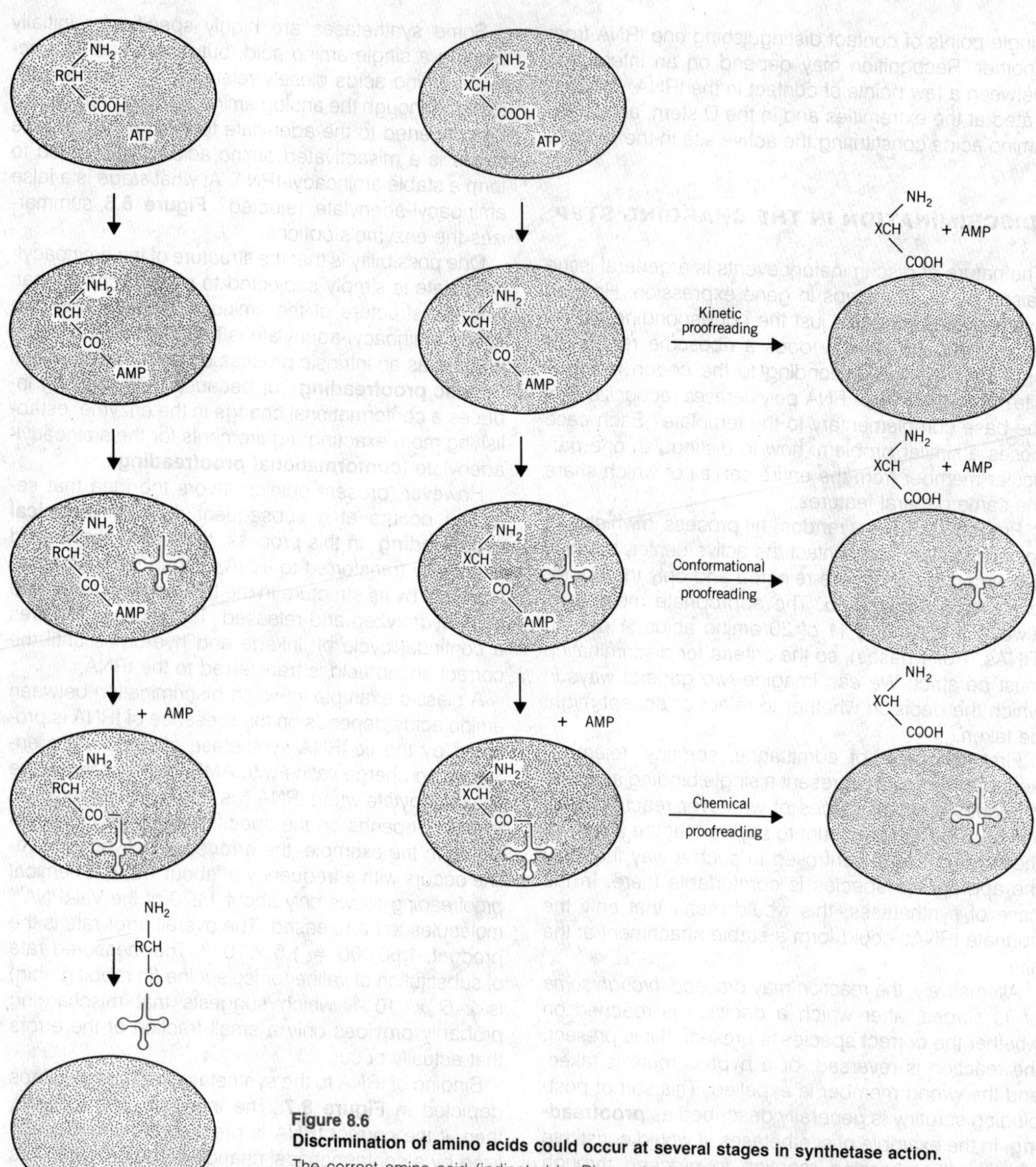

Figure 8.6
Discrimination of amino acids could occur at several stages in synthetase action.
The correct amino acid (indicated by R) may be bound in preference to most (perhaps all) other amino acids (indicated by X). If an incorrect amino acid is bound, it may be proofread kinetically, by a conformational change induced by tRNA, or chemically via an abortive transfer to tRNA.

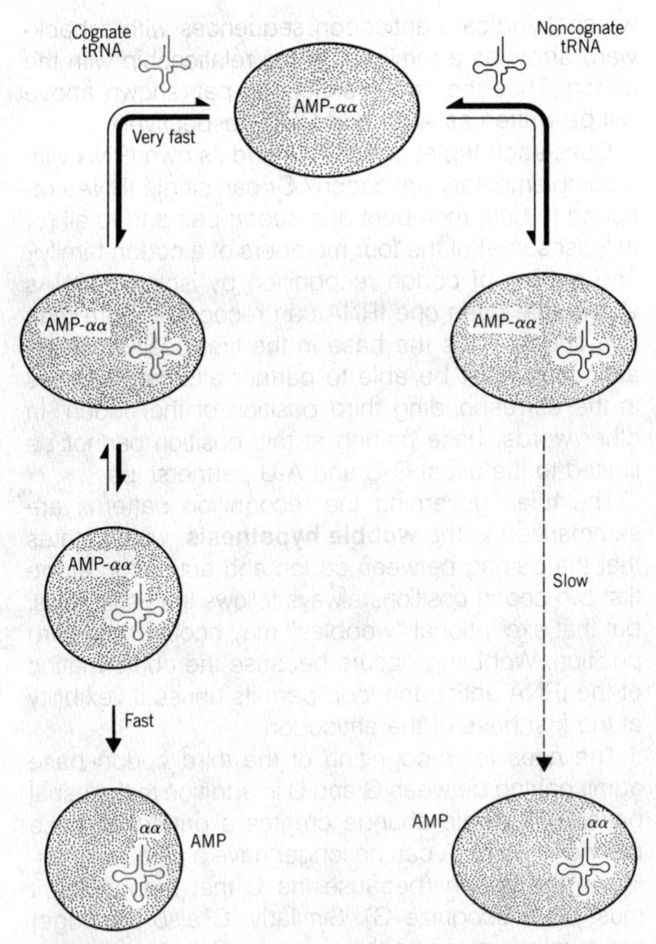

Figure 8.7
Recognition of the correct tRNA by synthetase is controlled at two stages.
First, the enzyme has a greater affinity for its cognate tRNA, whose binding is stabilized by a conformational transition in the enzyme. *Second*, the aminoacylation of the incorrect tRNA is very slow, which further increases the probability the tRNA will dissociate.

change ensures that the reaction proceeds much more slowly; this increases the chance that the tRNA will dissociate from the enzyme before it is charged.

So there may be a dual system for discrimination, in which first the accomplishment of stable binding depends on making proper contacts with the tRNA; and then the catalytic reaction itself is influenced by the tRNA present.

CODON-ANTICODON RECOGNITION INVOLVES WOBBLING

The function of tRNA in protein synthesis is fulfilled when it recognizes the codon in the ribosome A site. The interaction between anticodon and codon takes place by base pairing, but under rules that relax the usual restriction to G-C and A-T partnerships.

Experimental data about the forces involved in codon-anticodon recognition are indecisive; there is even some doubt about the extent to which G-C pairing stabilizes the reaction compared with A-U pairing. Also, interactions measured in solution may not be appropriate, because the structure of tRNA and the environment of the A site play an important role in establishing the conditions for the association of tRNA with mRNA.

We must therefore deduce the rules governing the interaction from information about the sequences of anticodons corresponding to particular codons. The ability of any tRNA to respond to a given codon can be measured by the trinucleotide binding assay or by its use in an *in vitro* protein synthetic system (the procedures used originally to define the genetic code, discussed in Chapter 6).

The genetic code itself yields some important clues about the process of codon recognition. The pattern of third-base degeneracy is drawn in **Figure 8.8**, which shows that in almost all cases either the third base is irrelevant or a distinction is made only between purines and pyrimidines. There are eight **codon families** in which all four codons sharing the same first two bases have the same meaning, so that the third base has no role at all in specifying the amino acid. There are seven **codon pairs** in which the meaning is the same whichever pyrimidine is present; and there are five codon pairs in which either purine may be present without changing the amino acid that is coded.

It may be more significant to look at the code from the reverse perspective. There are only three cases in which a unique meaning is conferred by the presence of a particular base at the third position: AUG (for methionine), UGG (for tryptophan), and UGA (nonsense). This means that C and U never have a unique meaning in the third position, and A never signifies a unique amino acid.

Because the anticodon is complementary to the co-

UUU UUC UUA UUG	UCU UCC UCA UCG	UAU UAC UAA UAG	UGU UGC UGA UGG
CUU CUC CUA CUG	CCU CCC CCA CCG	CAU CAC CAA CAG	CGU CGC CGA CGG
AUU AUC AUA AUG	ACU ACC ACA ACG	AAU AAC AAA AAG	AGU AGC AGA AGG
GUU GUC GUA GUG	GCU GCC GCA GCG	GAU GAC GAA GAG	GGU GGC GGA GGG

Figure 8.8
Third bases have the least influence on codon meanings.
Boxes indicate groups of codons within which third-base degeneracy ensures that the meaning is the same. These can be divided into four classes.

Third Base Relationship	Third Bases with Same Meaning	Number of Codons
Third base irrelevant	U, C, A, G	32 (8 families)
Purines distinguished from Pyrimidines	U or C A or G	14 (7 pairs) 12 (6 pairs)
Three out of four	U, C, A	3 (AUX = Ile)
Unique definitions	G only	2 (AUG = Met UGG = Trp)
	A only	1 (UGA = term)

don, it is the *first* base in the anticodon sequence written conventionally in the direction from 5′ to 3′ that pairs with the *third* base in the codon sequence written by the same convention. Thus the combination

Codon	5′	A	C	G	3′
Anticodon	3′	U	G	C	5′

is usually written as codon ACG/anticodon CGU, where the anticodon sequence must be read backward for complementarity with the codon.

To avoid confusion, we shall retain the usual convention in which all sequences are written 5′–3′, but we shall indicate anticodon sequences with a backward arrow as a reminder of the relationship with the codon. Thus the codon/anticodon pair shown above will be written as ACG and CGU, respectively.

Does each triplet codon demand its own tRNA with a complementary anticodon? Or can single tRNAs respond to both members of a codon pair and to all (or at least some) of the four members of a codon family? The pattern of codon recognition by isolated tRNAs shows that often one tRNA can recognize more than one codon. Thus the base in the first position of the anticodon must be able to partner alternative bases in the corresponding third position of the codon. In other words, base pairing at this position cannot be limited to the usual G-C and A-U partnerships.

The rules governing the recognition patterns are summarized in the **wobble hypothesis**, which states that the pairing between codon and anticodon at the first two codon positions always follows the usual rules, but that exceptional "wobbles" may occur at the third position. Wobbling occurs because the conformation of the tRNA anticodon loop permits unusual flexibility at the first base of the anticodon.

The rules for recognition of the third codon base admit pairing between G and U in addition to the usual pairs. This single change creates a pattern of base pairing in which A can no longer have a unique meaning in the codon (because the U that recognizes it must also recognize G). Similarly, C also no longer has a unique meaning (because the G that recognizes it also must recognize U). **Table 8.1** summarizes the pattern of recognition.

It is therefore possible to recognize unique codons only when the third bases are G or U; this option is

Table 8.1
Codon-anticodon pairing involves wobbling at the third position.

Base in First Position of Anticodon	Base(s) Recognized in Third Position of Codon
U	A or G
C	G only
A	U only
G	C or U

not used very often, since UGG and AUG are the only examples of the first type, and there is none of the second type.

tRNA CONTAINS MANY MODIFIED BASES

Transfer RNA is unique among nucleic acids in its content of "unusual" bases. An unusual base is any purine or pyrimidine ring except the usual A, G, C and U from which all RNAs are synthesized. All other bases are produced by **modification** of one of the four bases after it has been incorporated into the polyribonucleotide chain. All classes of RNA display some degree of modification, but in all cases except tRNA this is confined to rather simple events, such as the addition of methyl groups. Only in tRNA is there a vast range of modifications, ranging from simple methylations to wholesale restructuring of the purine rings. These modifications confer on tRNA a much greater range of structural versatility, which may be important in its various functions.

Figure 8.9 shows some of the more common modified bases. There is a tendency for the modifications of pyrimidines (C and U) to be less complex than those of purines (A and G). In addition to the modifications of the bases themselves, methylations at the 2′-O position of the ribose ring also may be found.

The most common modifications of uridine are straightforward. Methylation at position 5 creates ribothymidine (T). The base is the same commonly found in DNA; but here it is attached to ribose, not deoxyribose. In RNA, thymine constitutes an unusual base, originating by modification of U. (And in DNA, uracil is an unusual base, generated by some mutagenic treatments, as described in Chapter 3.)

Dihydrouridine (D) involves the saturation of a double bond, changing the ring structure. Pseudouridine (ψ) has an interchange of the positions of N and C atoms. And 4-thiouridine has sulfur substituted for oxygen. Other changes that occur to uridine may involve combinations of these changes, or the addition of more complex groups; for example, oxyacetic acid may be added at position 5.

The most common alterations of cytidine are methylations; also acetylation at position 4, and the generation of 2-thiocytidine.

The base inosine is found normally in the cell as an intermediate in the purine biosynthetic pathway. However, it is not incorporated directly into RNA, where instead its existence depends on deamination of A. Other modifications of A include the addition of complex groups.

Two complex series of nucleotides depend on modification of G. The Q bases, of which queuosine is the simplest example, have an additional pentenyl ring added via an NH linkage to the methyl group of 7-methylguanosine. The pentenyl ring may carry various further groups. The Y bases, of which wyosine is the simplest example, have an additional ring fused with the purine ring itself, and this carries a long carbon chain, again to which further groups are added in different cases.

In all cases except one, the modification reaction involves the alteration of, or addition to, existing bases in the tRNA. The exception is the synthesis of Q bases, where an enzyme has been found that can exchange free queuosine with a guanosine residue in the tRNA. The reaction involves breaking and remaking bonds on either side of the nucleoside, a unique event in nucleic acid synthesis.

The list of modified nucleosides found in tRNA extends to about 50 residues at present. They are synthesized by specific tRNA-modifying enzymes. The original nucleoside present at each position can be determined either by comparing the sequence of tRNA with that of its gene or (less efficiently) by isolating precursor molecules that lack some or all of the modifications. The sequences of precursors show that different modifications are introduced at different stages during the maturation of tRNA.

Some modifications are constant features of all tRNA molecules—for example, the D residues that give rise to the name of the D arm, and the ψ found in the TψC sequence. On the 3′ side of the anticodon there is always a modified purine, although the modification varies widely.

Other modifications are specific for particular tRNAs or groups of tRNAs. For example, wyosine bases are characteristic of $tRNA^{Phe}$ in bacteria, yeast, and mammals. There are also some species-specific patterns. Overall, modifications are found in all parts of the tRNA molecule.

The features recognized by the tRNA-modifying en-

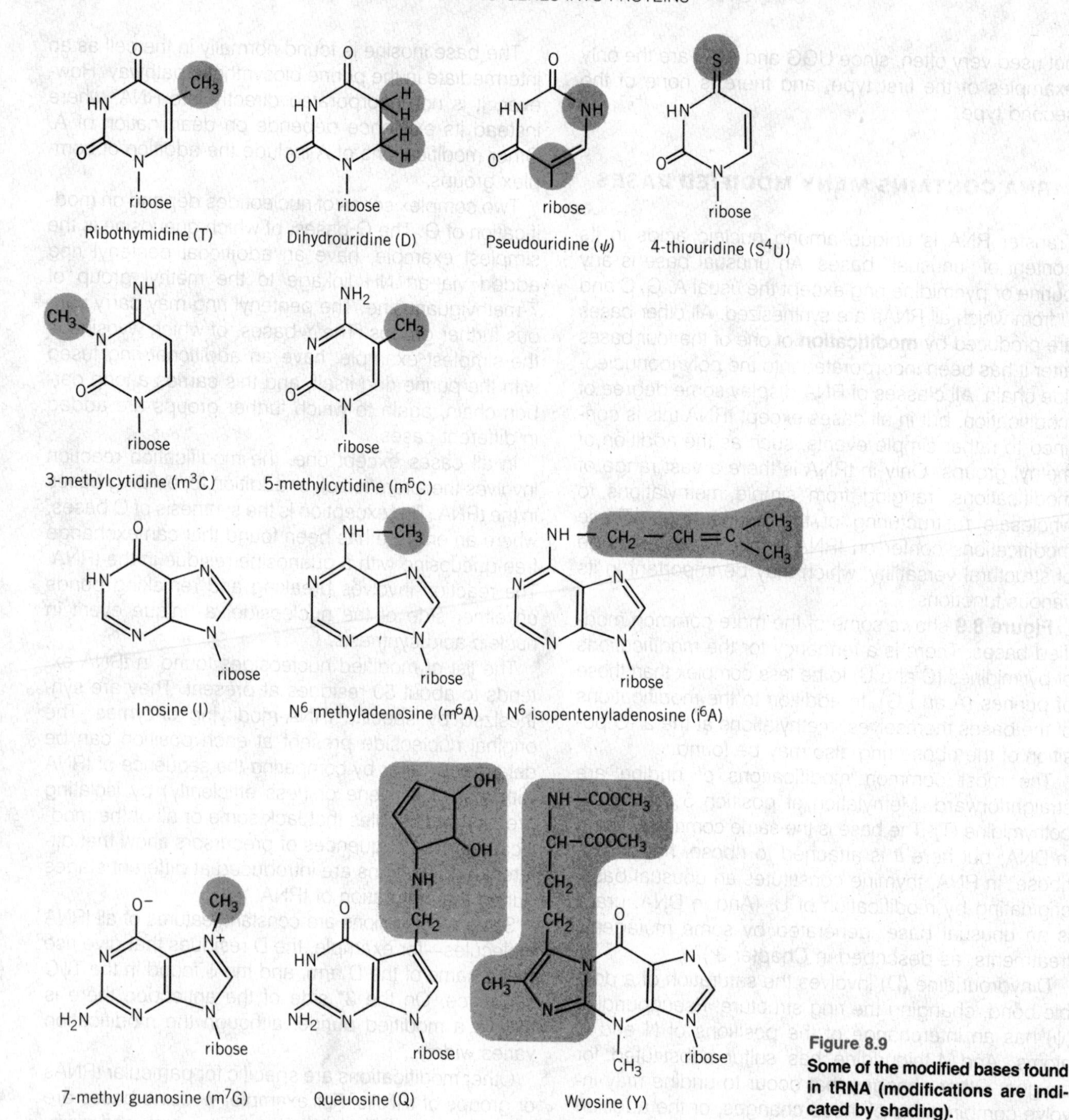

Figure 8.9
Some of the modified bases found in tRNA (modifications are indicated by shading).

zymes are unknown. When a particular modification is found at more than one position in a tRNA, the same enzyme does not necessarily make all the changes; for example, a different modification enzyme may be needed to synthesize the pseudouridine at each location. We do not know what controls the specificity of the modifying enzymes, but it is clear that there are many enzymes, with varying specifities. Some enzymes may undertake single reactions with individual tRNAs; others may have a range of substrate molecules. Some modifications require the successive actions of more than one enzyme.

BASE MODIFICATION MAY CONTROL CODON RECOGNITION

We certainly do not yet understand all the functions of modification. The most direct effect is seen in the anticodon itself, in which modification may influence the ability to pair with the codon, thus determining the meaning of the tRNA. Other modifications in the vicinity of the anticodon also may influence its pairing.

When bases in the anticodon are modified, further pairing patterns become possible in addition to those predicted by the regular and wobble pairing involving A, C, U, and G. **Figure 8.10** illustrates some of the variations in codon-anticodon pairing that are allowed by the combination of wobbling and modification.

Actually, some of the regular combinations do not occur, because some bases are *always* modified; in particular, U and A are not employed at the first position of the anticodon. Usually, U at this position is converted to a modified form that may have altered pairing properties. There seems to be an absolute ban on the employment of A, which has never been found; usually it is converted to I.

Inosine (I) is often present at the first position of the anticodon, where it is able to pair with any one of three bases, U, C, and A. Note that this ability is especially important in the isoleucine codons, where AUA codes for isoleucine, while AUG codes for methionine. Remembering that with the usual bases it is not possible to recognize A alone in the third position, any tRNA with U starting its anticodon would have to recognize AUG as well as AUA. So AUA must be read together with AUU and AUC, a problem that is solved by the existence of tRNA with I in the anticodon.

Two other modified nucleosides that can pair with three bases in the third codon position are uridine-5-oxyacetic acid and 5-methoxyuridine. They recognize A and G efficiently, and recognize U less efficiently.

Another case in which multiple pairings can occur, but with some preferred to others is provided by the series of queuosine and its derivatives. These modified G bases continue to recognize both C and U, but pair with U more readily.

A restriction not allowed by the usual rules can be achieved by the employment of 2-thiouridine in the anticodon. This modification allows the base to continue to pair with A, but prevents it from indulging in wobble pairing with G.

These and other pairing relationships make the general point that there is more than one way to construct a set of tRNAs able to recognize all the 61 codons representing amino acids. No particular pattern predominates in any given organism, although the absence of a certain pathway for modification can prevent the use of some recognition patterns. Thus a particular codon family may be read by tRNAs with different anticodons in different cells. Often the tRNAs will have overlapping responses, so that a particular codon may be read by more than one tRNA. In such cases there may be differences in the efficiencies of the alternative recognitions. And in addition to the construction of a set of tRNAs able to recognize all the codons, there will often be multiple tRNAs that respond to the same codons.

The predictions of wobble pairing accord very well with the observed abilities of almost all tRNAs. But there are exceptions in which the codons recognized by a tRNA differ from those predicted by the wobble rules. Such effects probably result from the influence of neighboring bases and/or the conformation of the anticodon loop in the overall tertiary structure of the tRNA. Indeed, the importance of the structure of the anticodon loop is inherent in the idea of the wobble hypothesis itself. Further support for the influence of the surrounding structure is provided by the isolation of occasional mutants in which a change in a base in some other region of the molecule alters the ability of the anticodon to recognize codons (see later).

Another unexpected pairing reaction is presented by the ability of the bacterial initiator, fMet-$tRNA_f$, to recognize both AUG and GUG. This misbehavior involves the third base of the anticodon.

MITOCHONDRIA HAVE MINIMAL tRNA SETS

According to the wobble hypothesis, a minimum of 31 tRNAs (excluding the initiator) are required to recog-

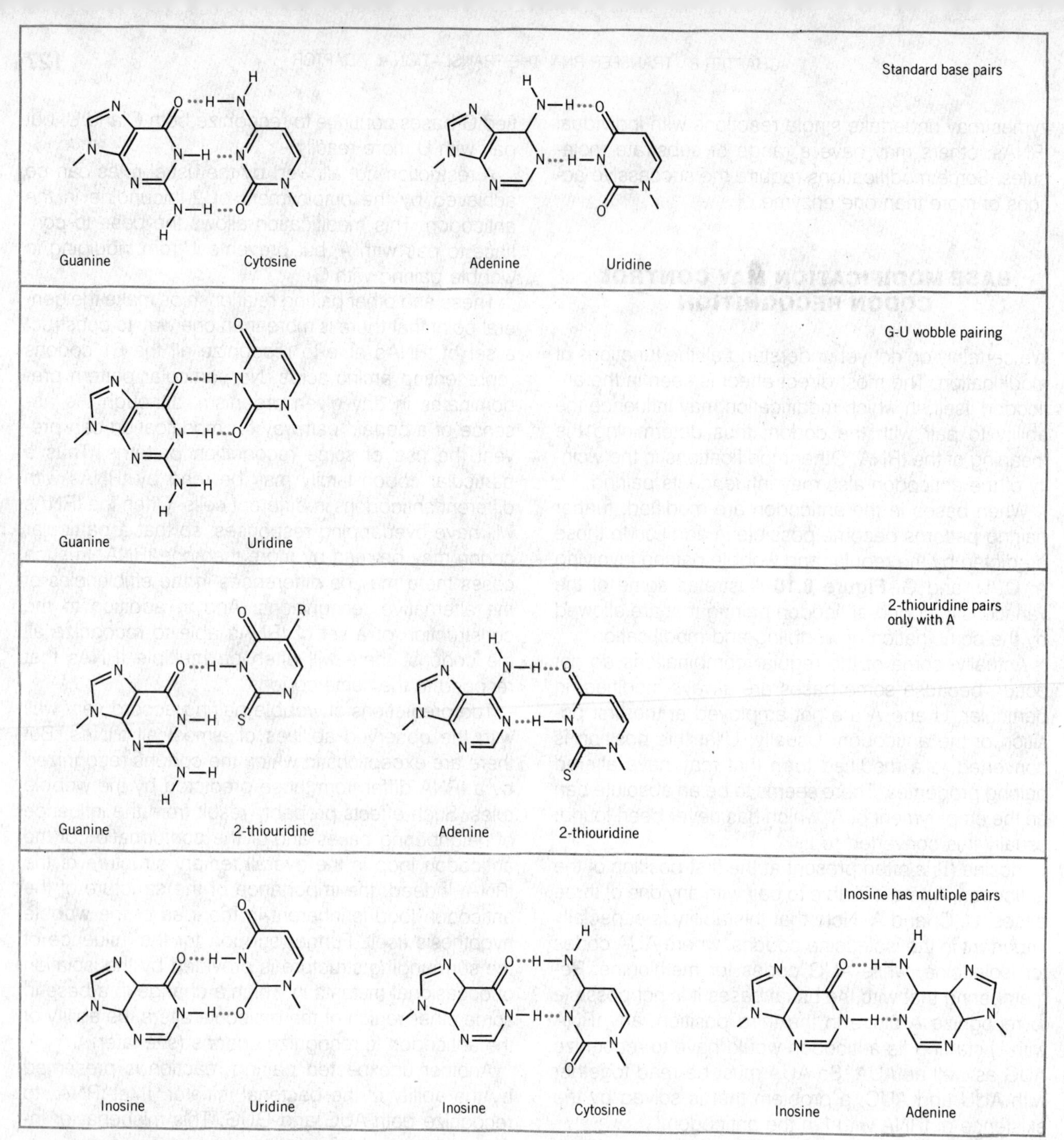

Figure 8.10

Wobble in base pairing allows some bases at the first position of the anticodon to recognize more than one base in the third position of the codon.

Pairing between standard bases is extended from the G-C and A-U pairs to the G-U wobble pair. Base modifications may restrict or extend the pattern. Modification to 2-thiouridine restricts pairing to A alone because only one H-bond could form with G. Modification to inosine allows pairing with U, C, and A.

nize all 61 codons (at least 2 tRNAs are required for each codon family and 1 tRNA is needed per codon pair or single codon). But an unusual situation exists in at least some mitochondria in which there may be as few as 23–24 different tRNAs. How does this limited set of tRNAs accommodate all the codons?

The critical feature lies in a simplification of codon-anticodon pairing, in which one tRNA recognizes all four members of a codon family. This reduces to 23 the minimum number of tRNAs required to respond to all internal codons. In all eight codon families, the sequence of the tRNA contains an unmodified U at the first position of the anticodon.

The remaining codons are grouped into pairs in which all the codons ending in pyrimidines are read by G in the anticodon, and all the codons ending in purines are read by U in the anticodon, as predicted by the wobble hypothesis. The complication of the single codons UGG and AUG is avoided completely in mammalian mitochondria, where the reading pattern changes the genetic code, so that UGA is read with UGG as tryptophan, while AUA ceases to represent isoleucine and instead is read with AUG as methionine (see Table 6.2). This allows all the nonfamily codons to be read as 14 pairs.

The 23 identified tRNA genes therefore code for an initiator, 14 tRNAs representing pairs, and 8 tRNAs representing families. This leaves the two usual nonsense codons UAG and UAA unrecognized by tRNA, together with an arginine codon pair (AG^{A}_{G}) that also may be used for termination (another change in the code). Similar rules are followed in the mitochondria of fungi.

A paradox is presented by the use of U in the anticodons responding both to the entire codon families and to the purine-ending pairs. The difficulty seems to be overcome by modifying the U in the tRNAs that respond to codon pairs only. An unmodified U in the anticodon may be able to wobble with U and C as well as with the A and G shown in Figure 8.10. If this is an inherent ability, it may explain why U is always modified in the first position of anticodons in bacteria and eukaryotic cytoplasm (although this leaves us wondering why U is never used to recognize entire codon families there).

MUTANT tRNAs READ DIFFERENT CODONS

Isolation of mutant tRNAs has been one of the most potent tools for analyzing the ability of a tRNA to respond to its codon(s) in mRNA, and for determining the effects that different parts of the tRNA molecule may have on codon-anticodon recognition.

Mutant tRNAs are isolated by virtue of their ability to overcome the effects of mutations in genes coding for proteins. We have already described the terminology in which a mutation that is able to overcome the effects of another is called a **suppressor** (see Chapter 3).

In tRNA suppressor systems, the primary mutation changes a codon in an mRNA so that the protein product is no longer functional; the secondary, suppressor mutation changes the anticodon of a tRNA, so that it recognizes the mutant codon instead of (or as well as) its original target codon. The amino acid that is now inserted restores protein function. The suppressors are named as **nonsense** or **missense**, depending on the nature of the original mutation.

This type of effect was originally discovered by analyzing the ability of nonsense mutants of phage T4 to grow on different strains of *E. coli*. On a nonpermissive (wild-type) strain, a nonsense phage mutant cannot grow, because the nonsense mutation terminates protein synthesis prematurely. But the *same phage* can grow on permissive strains of the bacterium, because they possess suppressor mutations that overcome the effects of the nonsense mutation.

In the normal (nonpermissive) cell, the nonsense mutation is recognized only by a release factor, terminating protein synthesis. The suppressor mutation creates an aminoacyl-tRNA that can recognize the termination codon; by inserting an amino acid, it allows protein synthesis to continue beyond the site of nonsense mutation. This new capacity of the translation system allows a full-length protein to be synthesized and is illustrated in **Figure 8.11**. If the amino acid inserted by suppression is different from the amino acid that was originally present at this site in the wild-type protein, the activity of the protein may be reduced.

Nonsense suppressors fall into three classes, one for each type of termination codon. **Table 8.2** de-

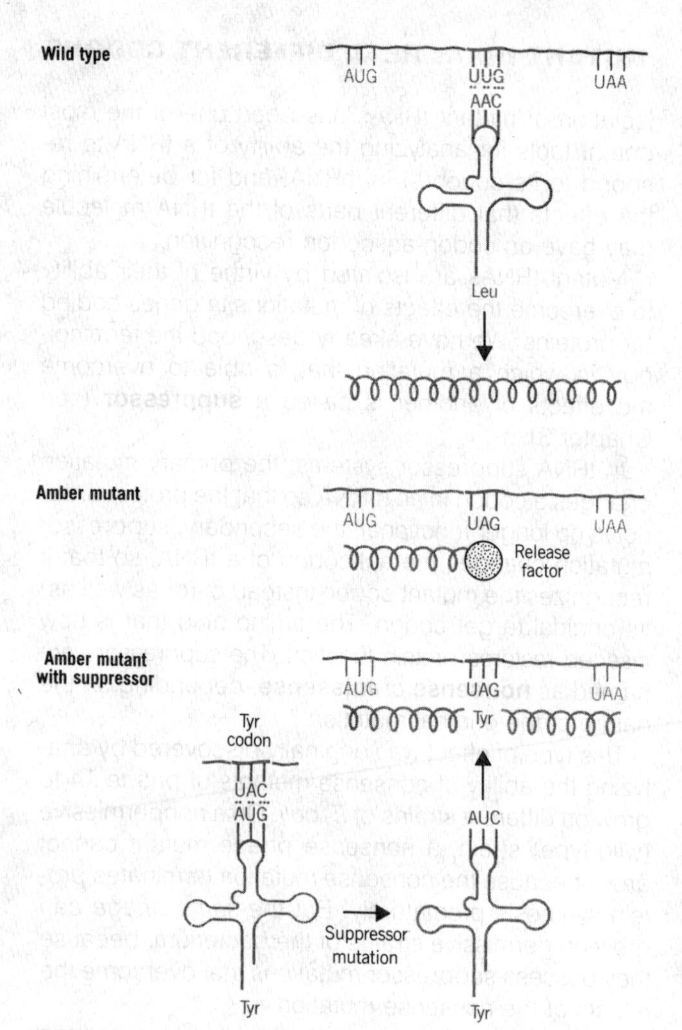

Figure 8.11

Nonsense mutations can be suppressed by a tRNA with a mutant anticodon.

The wild-type gene contains a UUG codon specifying leucine, which is recognized by a tRNA with the anticodon CAA. A mutation changes the codon to UAG, so it is now recognized by release factor. A second mutation changes the anticodon of tyrosine tRNA from GUA to CUA, so that it recognizes the amber codon. Suppression results in synthesis of a full length protein in which the original Leu residue has been replaced by Tyr.

scribes the properties of some of the best characterized suppressors.

The easiest to characterize have been amber suppressors. In *E. coli*, at least 6 tRNAs have been mutated to recognize UAG codons. All of the amber suppressor tRNAs have the anticodon CUA, in each case derived from wild type by a single base change. The site of mutation can be any one of the three bases of the anticodon, as seen from *supD*, *supE*, and *supF*. Each suppressor tRNA recognizes *only* the UAG codon, instead of its former codon(s). The amino acids inserted are serine, glutamine, or tyrosine, the same as those carried by the corresponding wild-type tRNAs.

Ochre suppressors also arise by mutations in the anticodon. The best known are *supC* and *supG*, which insert tyrosine or lysine in response to *both* ochre (UAA) and amber (UAG) codons. This conforms exactly with the prediction of the wobble hypothesis that UAA cannot be recognized alone. All ochre suppressors in bacteria act on both ochre and amber codons. (In yeast, ochre suppressors have been found that act only on UAA; probably this ability is conferred by the use of a modified base in the anticodon.)

The suppressors of UGA are interesting. One type *(supU)* is a mutant of tryptophan tRNA. The anticodon of the wild-type $tRNA^{Trp}$ is CCA, which recognizes only the UGG codon. The suppressor tRNA has the mutant anticodon UCA, which recognizes *both* its original codon and the termination codon UGA.

Another UGA suppressor has an entirely unexpected property. It is derived from the same $tRNA^{Trp}$, but its only mutation is the substitution of A in place of G *at position 24*. This change replaces a G-U pair in the D stem with an A-U pair, increasing the stability of the helix. The sequence of the anticodon remains the same as the wild type, CCA. So the mutation in the D stem must in some way alter the conformation of the anticodon loop, allowing CCA to pair with UGA in an unusual wobble pairing of C with A. The suppressor tRNA continues to recognize its usual codon, UGG.

Actually, this ability of the $tRNA^{Trp}$ to recognize the UGA termination codon may not be entirely unique to the suppressor form, because UGA mutants are always somewhat "leaky"—there is some level of residual expression even in the absence of a mutant tRNA suppressor. So perhaps the wild-type $tRNA^{Trp}$ reacts (rather poorly) with UGA as well as UGG; the mutation at position 24 may enhance this effect rather than create it de novo. In this case, a partial suppression of UGA might occur in the bacterium, so that any genes ending with UGA sometimes are extended to the next termination codon.

A related response is seen with a eukaryotic tRNA.

Table 8.2
Nonsense suppressor tRNAs are generated by mutations in the anticodon.

		Wild Type		Suppressor	
Locus	tRNA	Codons Recognized	Anticodon	Anticodon	Codons Recognized
supD (su1)	Ser	UCG	$\overleftarrow{CGA}$	$\overleftarrow{C\mathbf{U}A}$	UAG
supE (su2)	Gln	CAG	$\overleftarrow{CUG}$	$\overleftarrow{CU\mathbf{A}}$	UAG
supF (su3)	Tyr	UA^{C}_{U}	$\overleftarrow{GUA}$	$\overleftarrow{\mathbf{C}UA}$	UAG
supC (su4)	Tyr	UA^{C}_{U}	$\overleftarrow{GUA}$	$\overleftarrow{\mathbf{U}UA}$	UA^{A}_{G}
supG (su5)	Lys	AA^{A}_{G}	$\overleftarrow{UUU}$	$\overleftarrow{UU\mathbf{A}}$	UA^{A}_{G}
supU (su7)	Trp	UGG	$\overleftarrow{CCA}$	$\overleftarrow{\mathbf{U}CA}$	UG^{A}_{G}

Boldface indicates the base in the mutant anticodon that has changed.
Nomenclature. The loci were originally known as *su1*, *su2*, etc., as indicated in parentheses, with su^{+} indicating the suppressor mutant form of the gene, and su^{-} indicating the wild-type form. Now the loci have been renamed *sup* and lettered from *A* to *V*. All the loci described here are structural genes coding for tRNAs.

Bovine liver contains a $tRNA^{Ser}$ with the anticodon $^{m}\overleftarrow{CCA}$. The wobble rules predict that this tRNA should respond to the tryptophan codon UGG; but in fact it responds to the termination codon UGA. Thus it is possible that UGA is suppressed naturally in this situation.

The general importance of these observations lies in the demonstration that *codon-anticodon recognition of either wild-type or mutant tRNA cannot be predicted entirely from the relevant triplet sequences, but may well be influenced by other features of the molecule.*

All of these nonsense suppressors are created by point mutations in the structural gene coding for the tRNA. However, in principle suppressor tRNAs also could be caused by changes in the modification of bases in the anticodon. Such changes could result from mutation of the gene coding for a tRNA-modifying enzyme.

Missense mutations alter a codon representing one amino acid into a codon representing another amino acid, one that cannot function in the protein in place of the original residue. (Formally, any substitution of amino acids constitutes a missense mutation, but in practice it is detected only if it changes the activity of the protein.) The mutation can be suppressed by the insertion either of the original amino acid or of some other amino acid that is acceptable to the protein.

Figure 8.12 demonstrates that missense suppression can be accomplished in the same way as nonsense suppression, by mutating the anticodon of a tRNA carrying an acceptable amino acid so that it responds to the mutant codon. Thus missense suppression involves a change in the meaning of the codon from one amino acid to another.

Missense suppressors have been isolated from several tRNAs that insert glycine in response to one of the codons GGN. In one case, a mutation in an *E. coli* gene changes a glycine codon to an arginine codon (AGA). Suppressor mutations in either of two glycine tRNAs change the original anticodon from $\overline{CCC}$ or $\overline{UCC}$ to $\overline{UCU}$, which now recognizes the arginine codons AG^{A}_{G}.

An alternative way to create missense suppressors is for a mutation in the tRNA (or synthetase) to change the amino acid with which a tRNA is charged. Although the tRNA would continue to respond to the same codon(s), the insertion of a different amino acid might cause suppression. This type of suppression is rather unusual, probably because it demands a mutation that *creates* a new recognition site.

Most suppressors have been characterized in *E. coli*, although some comparable mutants have been obtained in other bacteria (notably *S. typhimurium*). Much less is known about the occurrence and pos-

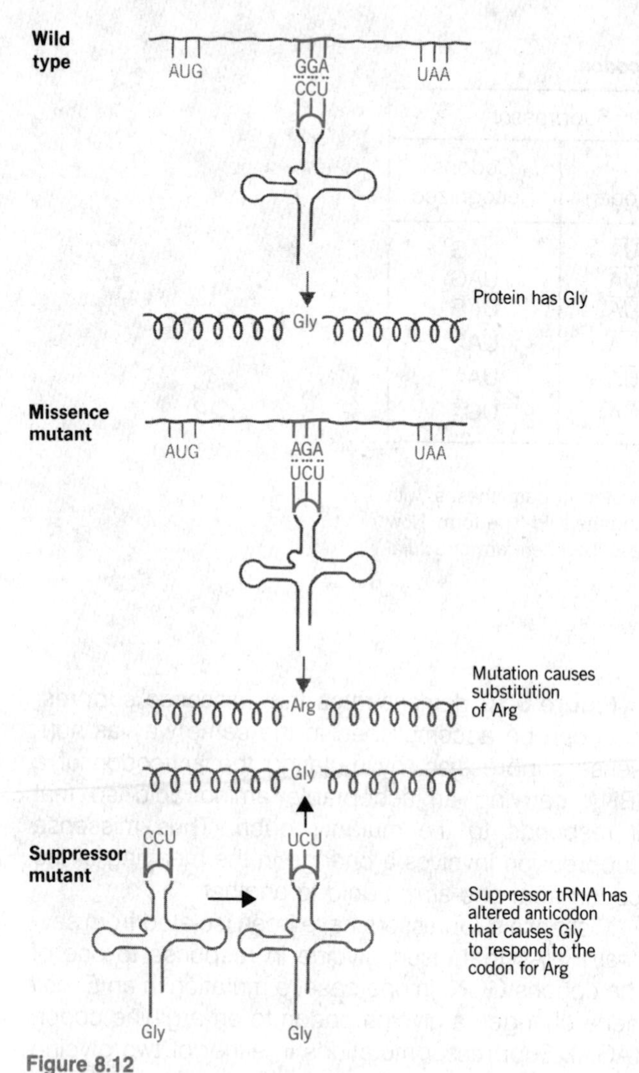

Figure 8.12

Missense suppression occurs when the anticodon of tRNA is mutated so that it responds to the "wrong" codon.

Mutation of a codon from GGA to AGA causes the wild-type Gly to be replaced by Arg in the mutant protein. If the anticodon in a $tRNA^{Gly}$ is mutated from $\overleftarrow{UCC}$ to $\overleftarrow{UCU}$, it inserts Gly in response to the codon AGA that should represent Arg. Note that both the wild-type $tRNA^{Arg}$ and the suppressor $tRNA^{Gly}$ will respond to AGA, so that suppression is only partial.

sible suppression of nonsense and missense mutants in eukaryotes, but generally the picture is probably similar for ochre and amber codons, although there are hints that UAG sometimes may be partially suppressed.

SUPPRESSOR tRNAs COMPETE FOR THEIR CODONS

There is an interesting difference between the usual recognition of a codon by its proper aminoacyl-tRNA and the situation in which mutation allows a suppressor tRNA to recognize a new codon. In the wild-type cell, *only one meaning* can be attributed to a given codon, which represents either a particular amino acid or a signal for termination. But in a cell carrying a suppressor mutation, the mutant codon may have the *alternatives* of being recognized by the suppressor tRNA or of being read with its usual meaning.

A nonsense suppressor tRNA must compete with the release factors that recognize its target termination codon(s). A missense suppressor tRNA must compete with the tRNAs that respond properly to its new codon. The extent of competition will influence the efficiency of suppression; so the effectiveness of a particular suppressor may depend not only on the affinity between its anticodon and the target codon, but also on its concentration in the cell, and the comparable parameters governing the competing termination or insertion reactions.

The efficiency with which any particular codon is read may be influenced by its location. Thus the efficiency of nonsense suppression by a given tRNA can vary quite widely, depending on the **context** of the codon. We do not understand the effect that neighboring bases in mRNA have on codon-anticodon recognition, but it can change the frequency with which a codon is recognized by a particular tRNA by more than an order of magnitude. The base on the 3′ side of a codon appears to have a particularly strong effect.

A nonsense suppressor is isolated by its ability to respond to a mutant nonsense codon. But the same triplet sequence constitutes one of the normal termination signals of the cell! The mutant tRNA that suppresses the nonsense mutation must in principle be able to suppress natural termination at the end of any gene that uses this codon. **Figure 8.13** shows that this **readthrough** results in the synthesis of a longer protein, with additional C-terminal material. The extended protein will end at the next termination triplet sequence found in the phase of the reading frame. Any extensive suppression of termination is likely to be deleterious

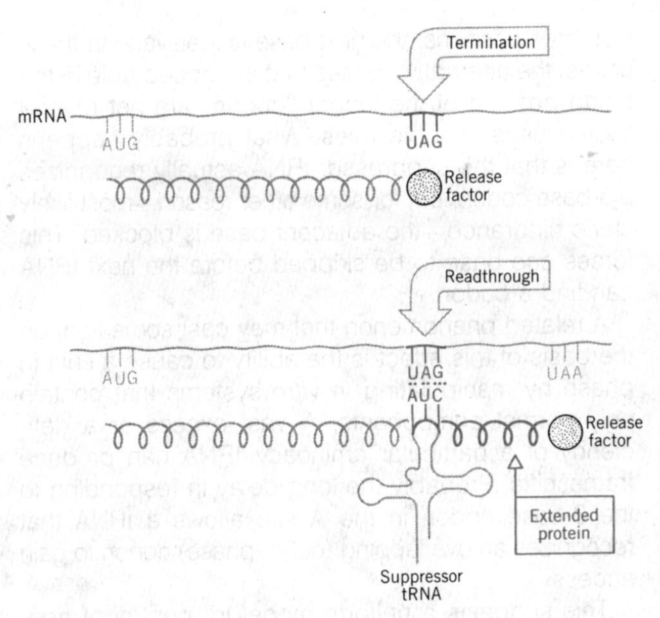

Figure 8.13
Nonsense suppressors also read through natural termination codons, synthesizing proteins longer than wild-type.

to the cell by producing extended proteins whose functions are thereby altered.

Amber suppressors tend to be relatively efficient, working at rates measured from 10% to 50%, depending on the system. If we suppose that the cell could not tolerate such a high level of termination of natural proteins, this may indicate that amber codons are used less frequently than the other termination codons at the ends of genes.

Ochre suppressors are difficult to isolate. They are always much less efficient, usually with activities below 10%. All ochre suppressors grow rather poorly, which indicates that suppression of both UAA and UAG is damaging to *E. coli*, probably because the ochre codon is used most frequently as a natural termination signal.

In the same way, one gene's missense suppressor is likely to be another gene's mutator. If a suppressor corrects a mutation by substituting one amino acid for another at the mutant site, it will have the same effect elsewhere. But in other locations it will be introducing a new amino acid instead of the normal wild-type residue. The change may inhibit normal protein function. This poses a dilemma for the cell: it must suppress what is a mutant codon at one location, while failing to change too extensively its normal meaning at other locations. The absence of any strong missense suppressors is therefore explained by the damaging effects that would be caused by a general and efficient substitution of amino acids.

A mutation that creates a suppressor tRNA can have two consequences. First, it allows the tRNA to recognize a new codon. Second, it may sometimes *prevent* the tRNA from recognizing the codons to which it previously responded. It is significant that all the high-efficiency amber suppressors are derived by mutation of one copy of a redundant tRNA set. In these cases, the cell has several tRNAs able to respond to the codon originally recognized by the wild-type tRNA. Thus the mutation does not abolish recognition of the old codons, which continue to be served adequately by the tRNAs of the set.

In contrast with this situation, another amber suppressor tRNA is a **recessive lethal**. It can act as a suppressor when the cell has two copies of the gene, one wild type and one mutant. But if there is only a single, mutant copy, the cell dies. The relevant gene is *supU*, the only gene coding for tryptophan tRNA in *E. coli*. A mutation changes its anticodon from $\overleftarrow{\text{CCA}}$ to $\overleftarrow{\text{CUA}}$. But this alteration leaves the cell without any tRNA that can respond to UGG, a lethal change in its capacity. (By contrast, when this tRNA is mutated to recognize *both* UGA and UGG, as described in Table 8.2, the change is not lethal.) Limitations of this sort may explain why it has not been possible to isolate all possible nonsense suppressors—that is, derivatives of every tRNA recognizing a codon differing by just one base from a termination codon.

tRNA MAY INFLUENCE THE READING FRAME

The reading frame of a messenger usually is invariant. Translation starts at an AUG codon and continues in triplets to a termination codon. Reading takes no notice of sense: insertion or deletion of a base causes a frameshift mutation, in which the reading frame is changed beyond the site of mutation. Ribosomes and tRNAs continue ineluctably in triplets, synthesizing an entirely different series of amino acids.

Frameshift suppression restores the original reading frame. We have already discussed how this can be achieved by compensating base deletions and insertions (see Chapter 3). However, *extragenic* frameshift suppressors also can be found.

One type of external frameshift suppressor corrects the reading frame when a mutation has been caused by inserting an additional base within a stretch of identical residues. For example, a G may have been inserted in a run of several contiguous G bases. The frameshift suppressor is a $tRNA^{Gly}$ that has an extra base inserted in its anticodon loop, converting the anticodon from the usual triplet sequence $\overleftarrow{CCC}$ to the quadruplet sequence $\overleftarrow{CCCC}$. The suppressor tRNA recognizes a 4-base "codon" with the usual wobble at the last (fourth) position.

A related type of frameshift suppression has been found in the yeast mitochondrion, in which mutations that insert or delete a single T residue in a run of five T residues all are leaky. The reason seems to be that the $tRNA^{Phe}$ responding to the UUU codon sometimes allows the ribosome to "slip" a base in either direction, thus suppressing the mutation. This slippage seems to be a normal, if infrequent, occurrence (less than 5%).

Another situation in which an occasional frameshift is a normal event is presented by the RNA phage MS2. A frameshift (whose mechanism is unknown) causes the ribosome to recognize a termination codon at an early position in its new reading frame. The terminating ribosome then can recognize the initiation codon of the lysis gene, which lies just a few bases farther along. When the ribosome does not terminate, it reads right over the lysis gene initiation codon. So the frameshift-dependent termination event is a prerequisite for initiation of lysis gene expression. This makes the important point that the rare occurrence of "misreading" events can be relied on as a necessary step in natural translation.

Some frameshift suppressors can recognize more than one 4-base "codon." For example, a bacterial $tRNA^{Lys}$ suppressor can respond to either AAAA or AAAU, instead of the usual codon AAA. Another suppressor can read any 4-base "codon" with ACC in the first three positions; the next base is irrelevant. In these cases, the alternative bases that are acceptable in the fourth position of the longer "codon" are not related by the usual wobble rules. What probably happens here is that the suppressor tRNA actually recognizes a 3-base codon, but for some other reason—most likely steric hindrance—the adjacent base is blocked. This forces one base to be skipped before the next tRNA can find a codon.

A related phenomenon that may cast some light on the basis of this effect is the ability to cause a shift in phase by manipulating *in vitro* systems that contain only normal components. A vast excess or a deficiency of a particular aminoacyl-tRNA can produce frameshifts. Probably the long delay in responding to the 3-base codon in the A site allows a tRNA that recognizes an overlapping (out-of-phase) codon to gain access.

This suggests a uniform model for control of ribosome translocation. When the ribosome moves, it shifts the tRNA that was in the A site so that now it properly occupies the P site. When a tRNA is bound to a 4-base "codon," or if it extends sterically beyond the usual 3-base codon, the next aminoacyl-tRNA binds to a triplet that is out of phase in the A site. But when translocation occurs, the ribosome places this species properly into the P site, so that subsequent aminoacyl-tRNAs encounter a triplet codon in the A site in the usual way. Thus the geometry of the tRNA/mRNA interaction is used to count distance for the ribosome.

FURTHER READING

The literature on tRNA has been well served by some review books. In **Altman's** (Ed.) *Transfer RNA* (MIT Press, Cambridge, 1978), primary and secondary structure were discussed by **Clark** (pp. 14–47) and tertiary structure by **Kim** (pp. 248–293). Modified bases were discussed by **Nishimura** (pp. 168–195). One viewpoint of aminoacyl-tRNA synthetases was given by **Igloi & Cramer** (pp. 294–340), another by **Schimmel & Soll** (*Ann. Rev. Biochem.* **48**, 601–648, 1979). A two volume set of reviews and research papers also gives a broad view of tRNA: **Schimmel, Soll & Abel-**

son, *Transfer RNA: Structure, Properties and Recognition*, and **Soll, Abelson & Schimmel**, *Transfer RNA: Biological Aspects* (Cold Spring Harbor Lab., New York, 1979). Original work on the coding properties of tRNA and their change via mutation have been dealt with in Chapter 5 of **Lewin's** *Gene Expression*, **1**, *Bacterial Genomes* (Wiley, New York, 1974). Suppression has been reviewed by **Korner, Feinstein & Altman** (pp. 105–135 of Altman's book), and more recent developments were briefly covered by **Roth** (*Cell* **24**, 601–602, 1981).

CHAPTER 9
THE RIBOSOME TRANSLATION FACTORY

The ribosome behaves like a small migrating factory that travels along the template engaging in rapid cycles of peptide bond synthesis. Aminoacyl-tRNAs shoot in and out of the particle at a fearsome rate, depositing amino acids, and elongation factors cyclically associate and dissociate. Together with the accessory factors, the ribosome provides the full range of activities required for all the steps of protein synthesis.

Ribosomes are a major cellular component. In an actively growing bacterium, there may be roughly 20,000 ribosomes per genome (see Table 6.1). They contain about 10% of the total bacterial protein and account for 80% or so of the total mass of cellular RNA. In eukaryotic cells, their proportion of total protein is less, but their absolute number is greater, and still they contain most of the mass of RNA of the cell. The number of ribosomes (in prokaryotes or eukaryotes) is directly related to the protein-synthesizing activity of the cell.

Bacterial ribosomes are attached to mRNAs that are themselves still connected with the DNA. In eukaryotic cytoplasm, the ribosomes commonly are associated with the cytoskeleton, a fibrous matrix. In some eukaryotic cells, some of the ribosomes are associated with membranes of the endoplasmic reticulum. The common feature is that ribosomes engaged in translation are not free in the cell, but are associated, directly or indirectly, with cellular structures.

All ribosomes can be dissociated into two subunits, one roughly the size of the other. Each subunit contains a major rRNA component and many different protein molecules, almost all present in only one copy per subunit. The larger subunit may contain smaller RNA molecule(s) in addition to the major rRNA. A pointer to the importance of rRNA is that its sequence remains almost invariant within a cell, although it is coded by many genes. This conservation argues that there is selection against changes in its sequence.

The role of the major rRNAs has been a subject of debate. A major function is clearly structural. Proteins bind to the major rRNA at particular sites and in a specific order that is required to assemble each subunit. Indeed, ribosomes are interesting not only for their function, but also for the process by which they self-assemble from the constituent RNA and protein molecules.

The ribosome possesses several active centers, each of which is constructed from a particular group of proteins. Some catalytic functions can be identified with individual proteins, but none of the activities work with isolated proteins or groups of proteins. The ribosome represents a collection of many enzymes, each active only in the context of the proper overall structure, whose coordinated activities together accomplish the act of translation. Many of the proteins (and the rRNA) may be concerned principally with establishing the overall structure that brings the various active sites into the right relationship; they need not necessarily participate directly in the synthetic reactions.

RIBOSOMES ARE COMPACT RIBONUCLEOPROTEIN PARTICLES

Ribosomes can be constructed in several ways. There are appreciable variations in the overall size and proportions of RNA and protein in the ribosomes of bacteria, eukaryotic cytoplasm, and organelles. **Table 9.1** summarizes the components of the bacterial ribosome, as exemplified by *E. coli*.

All ribosomes in a bacterium are exactly the same. The structure of every component is known for *E. coli* ribosomes. The ribosomal RNAs have been sequenced, and the amino acid sequences of most of the proteins have been determined. The small (30S) subunit consists of the 16S rRNA associated with 21 proteins (named S1–S21). The large (50S) subunit contains 23S rRNA, the small 5S RNA, and 31 proteins (numbered L1–L34 because of some early misassignments). With the exception of one protein present at four copies per ribosome, there is one copy of each protein. The rRNA provides most of the mass of the particle.

The ribosomes of higher eukaryotic cytoplasm are larger than those of bacteria. **Table 9.2** summarizes the properties of a mammalian example. The total content of both RNA and protein is greater; the major RNA molecules are longer, and the number of proteins is greater. The small (40S) subunit consists of 18S rRNA associated with about 33 polypeptides; the large (60S) subunit contains the major 28S rRNA, two smaller RNAs, and about 49 polypeptides. Probably most or all of the proteins are present in stoichiometric amounts. RNA is still the predominant component by mass.

Ribosomes take varied forms in organelles. Some examples are summarized in **Table 9.3**. In chloroplasts, they are roughly the same size as in bacteria, although with a greater proportion of nucleic acid. In plant mitochondria, they are only a little smaller than the ribosomes of the surrounding cytoplasm. In the mitochondria of lower eukaryotes (such as fungi), their overall size is slightly larger than in *E. coli*. In mammalian or amphibian mitochondria, however, they are much smaller, with a total size of 60S, and a low proportion of RNA by mass (<30%).

The shapes of bacterial and eukaryotic cytoplasmic ribosomes are generally similar (organelle ribosomes have not been characterized). **Figure 9.1** is a diagrammatic representation of the small bacterial subunit, which has a somewhat flat shape. **Figure 9.2**

Table 9.1
The components of the *E. coli* ribosome.

	Ribosome	Small Subunit	Large Subunit
Sedimentation rate Mass (daltons)	70S 2,520,000	30S 930,000	50S 1,590,000
RNA Major Smaller RNA mass RNA proportion	66%	16S = 1541 bases 560,000 60%	23S = 2904 bases 5S = 120 bases 1,104,000 70%
Protein number Protein mass Protein proportion	34%	21 polypeptides 370,000 40%	31 polypeptides 487,000 30%

Table 9.2
The components of the rat liver cytoplasmic ribosome

	Ribosome	Small Subunit	Large Subunit
Sedimentation rate	80S	40S	60S
Mass (daltons)	4,420,000	1,400,000	2,820,000
RNA			
Major		18S = 1900 bases	28S = 4700 bases
Smaller			5.8S = 160 bases 5S = 120 bases
RNA mass		700,000	1,820,000
RNA proportion	60%	50%	65%
Protein number		33 polypeptides	49 polypeptides
Protein mass		700,000	1,000,000
Protein proportion	40%	50%	35%

shows that the large subunit has a more spherical structure.

The complete 70S ribosome has not been characterized directly in as much detail as the subunits, but one model for its construction is illustrated in **Figure 9.3**. This supposes that the two subunits are attached principally by an association between the cleft of the 30S subunit and the central protuberance of the 50S subunit. There could be a space or sort of tunnel between them.

Some electron micrographs of subunits and complete bacterial ribosomes are shown in **Figure 9.4**, together with models in the corresponding orientation.

The apparent structure of mammalian cytoplasmic ribosomes is similar, although they are larger. There is a suggestive similarity between the structures of bacterial and eukaryotic small subunits; the large subunits seem more divergent. Although a detailed model has not yet been constructed, the relationship between the mammalian subunits looks similar to that of prokaryotic ribosomes.

Much attention has been paid to the location of mRNA in the ribosome. One popular idea is that it fits between or close to the junction of the subunits, possibly in the "tunnel" between them. Some models depict the mRNA as "threaded" through the ribosome, but its location is just as likely to be superficial. As is evident from **Figure 9.5**, the two tRNAs are quite large relative to the ribosome; they seem more likely to be inserted into large "clefts" opening from the surface than to be inserted completely into holes in the interior.

Table 9.3
Ribosomes vary in size in organelles.

Organelle	Size of Ribosome
Euglena chloroplast	70S
Maize mitochondrion	78S
Yeast mitochondrion	73S
Xenopus mitochondrion	60S
Human mitochondrion	60S

RIBOSOMAL PROTEINS INTERACT WITH rRNA

The RNAs constitute the major part of the mass of the ribosome. Their presence is pervasive, and probably most or all of the ribosomal proteins associate with rRNA. Thus the major rRNAs form what is sometimes thought of as the backbone of each subunit, a continuous thread whose presence dominates the structure, and which determines the positions of the ribosomal proteins.

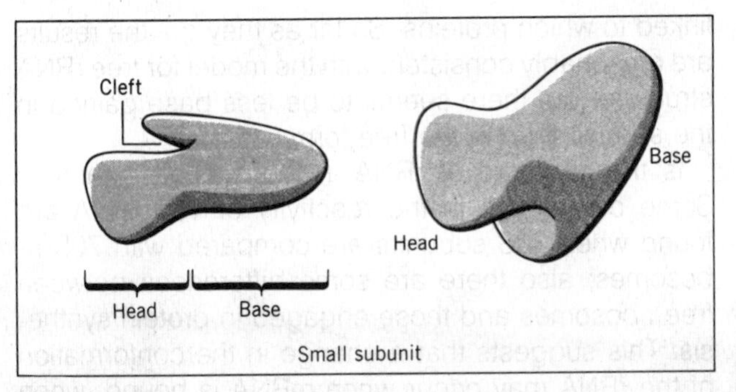

Figure 9.1
The 30S subunit has an elongated and asymmetrical shape, about 55 × 220 × 220 Å, with a constricted region and a "cleft" that divides the "head" from the "base."
The two views display the subunit rotated by 90°.

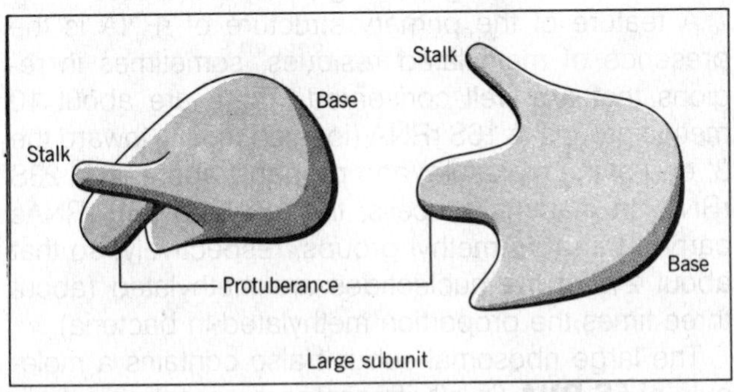

Figure 9.2
Two views show that the 50S subunit has a fairly compact body, about 150 × 200 × 200 Å, from which a "central protuberance" and a "stalk" stick out.

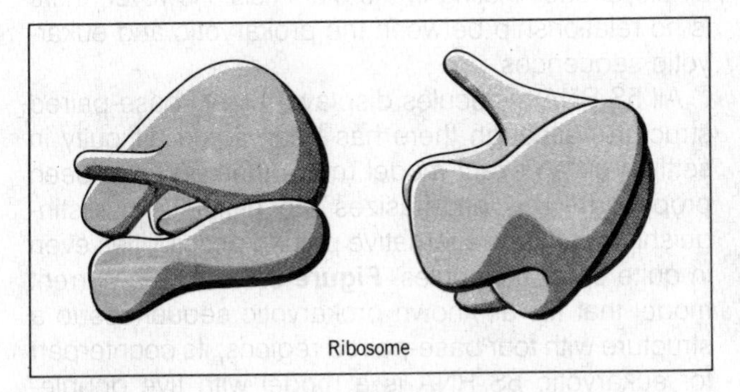

Figure 9.3
The 70S ribosome may be held together by associations between discrete areas of the two subunits, as indicated in these two views.

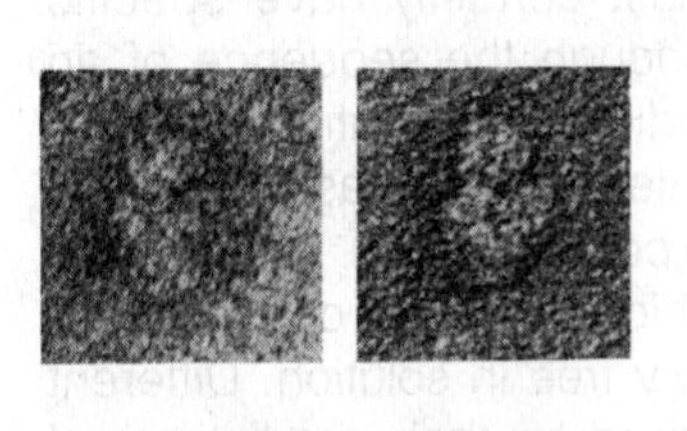
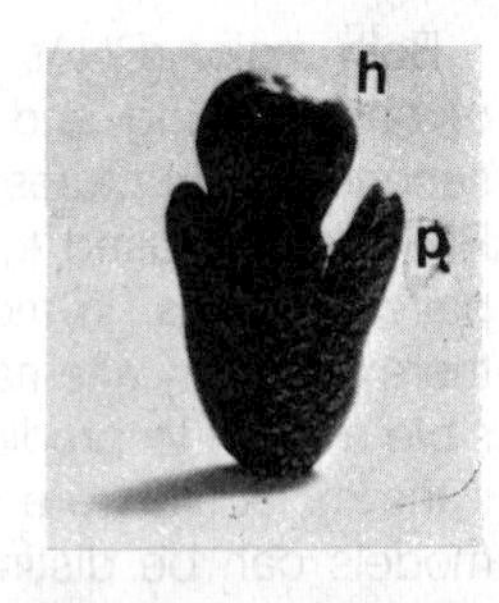

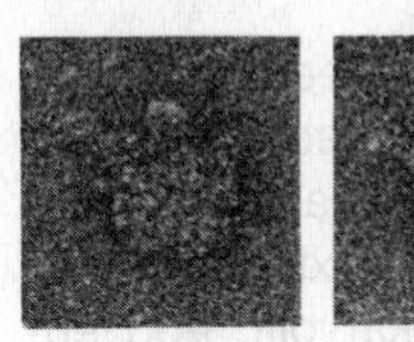
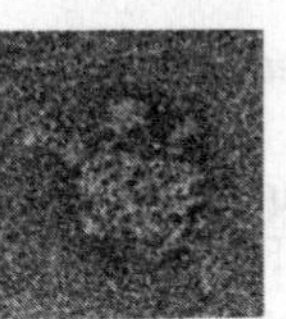
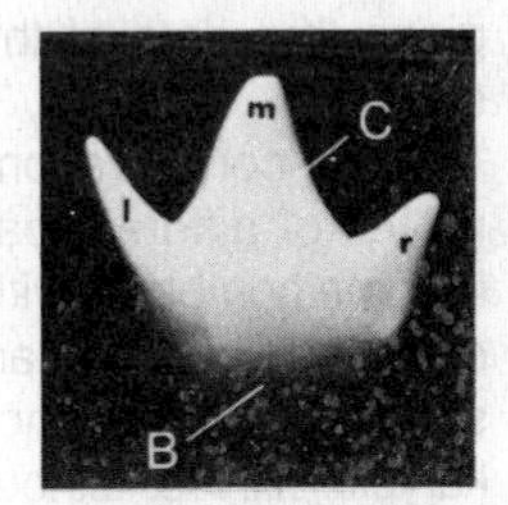

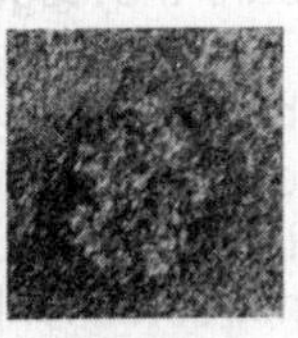
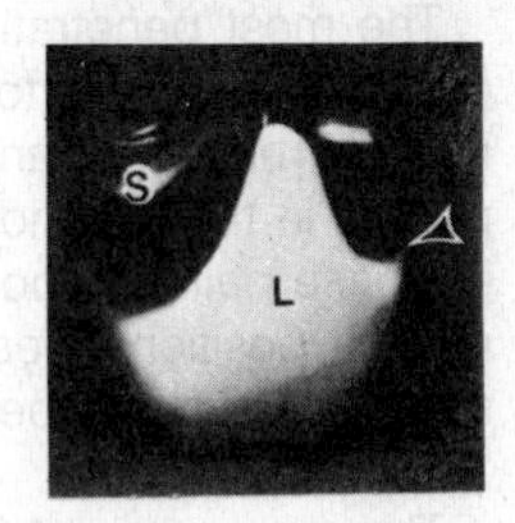

Figure 9.4
Electron microscopic images of bacterial ribosomes and subunits reveal their shapes.
An example of each structure is shown together with a model. Photographs kindly provided by Miloslav Boublik.

Figure 9.5
Size comparisons show that the ribosome is large enough to bind two tRNAs (as well as 40 bases of mRNA).

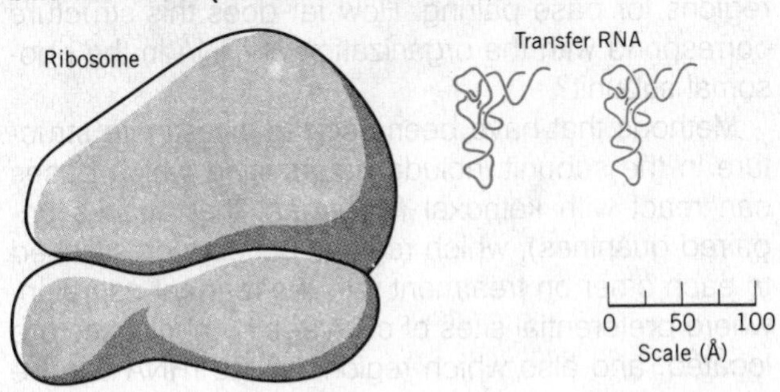

Both major rRNAs contain a considerable amount of base pairing and almost certainly have specific secondary structures. Although the sequence of an RNA can be used to predict the formation of base-paired regions, in molecules as large as the rRNAs there are many alternative conformations. It is not possible a priori to predict which would be chosen even if the molecule were simply free in solution. Different models can be distinguished by their predictions of the relative stabilities of particular base-paired regions; but data of this type are somewhat limited in scope.

Some conservation of sequence is seen when the rRNAs of different bacteria are compared; and there is some homology with organelle rRNAs (a relationship of uncertain significance). There is no extensive conservation of sequence between prokaryotic and eukaryotic rRNAs, although some conserved regions can be found in eukaryotic rRNAs.

The most penetrating approach to analyzing secondary structure is to compare the sequences of an rRNA in related organisms. Those regions that are important in the secondary structure are conserved; so if a base pair is important, one can form at the same relative position in each rRNA. From such comparisons, models have been constructed for both 16S and 23S rRNA.

The current model for *E. coli* 16S rRNA is illustrated in diagrammatic form in **Figure 9.6**. The molecule forms four general domains, in which just under half of the sequence is base paired. The individual double-helical regions tend to be short (less than 8 base pairs long). Often the duplex regions are not perfect, but contain bulges of unpaired bases.

This model represents a structure for free rRNA; it does not take account of the ribosomal proteins, whose binding to the RNA could influence the availability of regions for base pairing. How far does this structure correspond with the organization of rRNA in the ribosomal subunit?

Methods that have been used to investigate structure in the subunit include determining which bases can react with kethoxal (a reagent that attacks unpaired guanines), which regions become crosslinked to each other on treatment with the reagent psoralen, where preferential sites of cleavage by nucleases are located, and also which regions of the rRNA can be linked to which proteins. So far as they go, the results are reasonably consistent with the model for free rRNA structure, but there seems to be less base pairing in the subunit than in the free form.

Is the structure of rRNA in the subunit invariant? Some differences in the reactivity of 16S rRNA are found when 30S subunits are compared with 70S ribosomes; also there are some differences between free ribosomes and those engaged in protein synthesis. This suggests that a change in the conformation of the rRNA may occur when mRNA is bound, when the subunits associate, or when tRNA is bound. However, we do not know whether this change reflects a direct interaction of the rRNA with mRNA or tRNA, or is caused indirectly by some other change in ribosome structure. The main point is that ribosome conformation may change during protein synthesis.

A feature of the primary structure of rRNA is the presence of methylated residues, sometimes in regions that are well conserved. There are about 10 methyl groups in 16S rRNA (located mostly toward the 3′ end of the molecule) and probably about 20 in 23S rRNA. In mammalian cells, the 18S and 28S rRNAs carry 43 and 74 methyl groups, respectively, so that about 2% of the nucleotides are methylated (about three times the proportion methylated in bacteria).

The large ribosomal subunit also contains a molecule of **5S RNA** (in all ribosomes except those of mitochondria). Prokaryotic 5S RNAs show some conservation of sequence, especially in the regions that bind to ribosomal proteins. Similarly, there is conservation of eukaryotic 5S RNAs, with one sequence (for example) predominating in the mammals. However, there is no relationship between the prokaryotic and eukaryotic sequences.

All 5S RNA molecules display a highly base-paired structure, although there has been some difficulty in settling on an exact model (more than 20 have been proposed). This emphasizes the difficulty in distinguishing between alternative pairing possibilities even in quite small molecules. **Figure 9.7** shows a current model that fits all known prokaryotic sequences to a structure with four base-paired regions; its counterpart for eukaryotic 5S RNA is a model with five double-helical regions. As with tRNA, the secondary structure is unlikely to convey an accurate impression of the form the molecule actually takes in three dimensions.

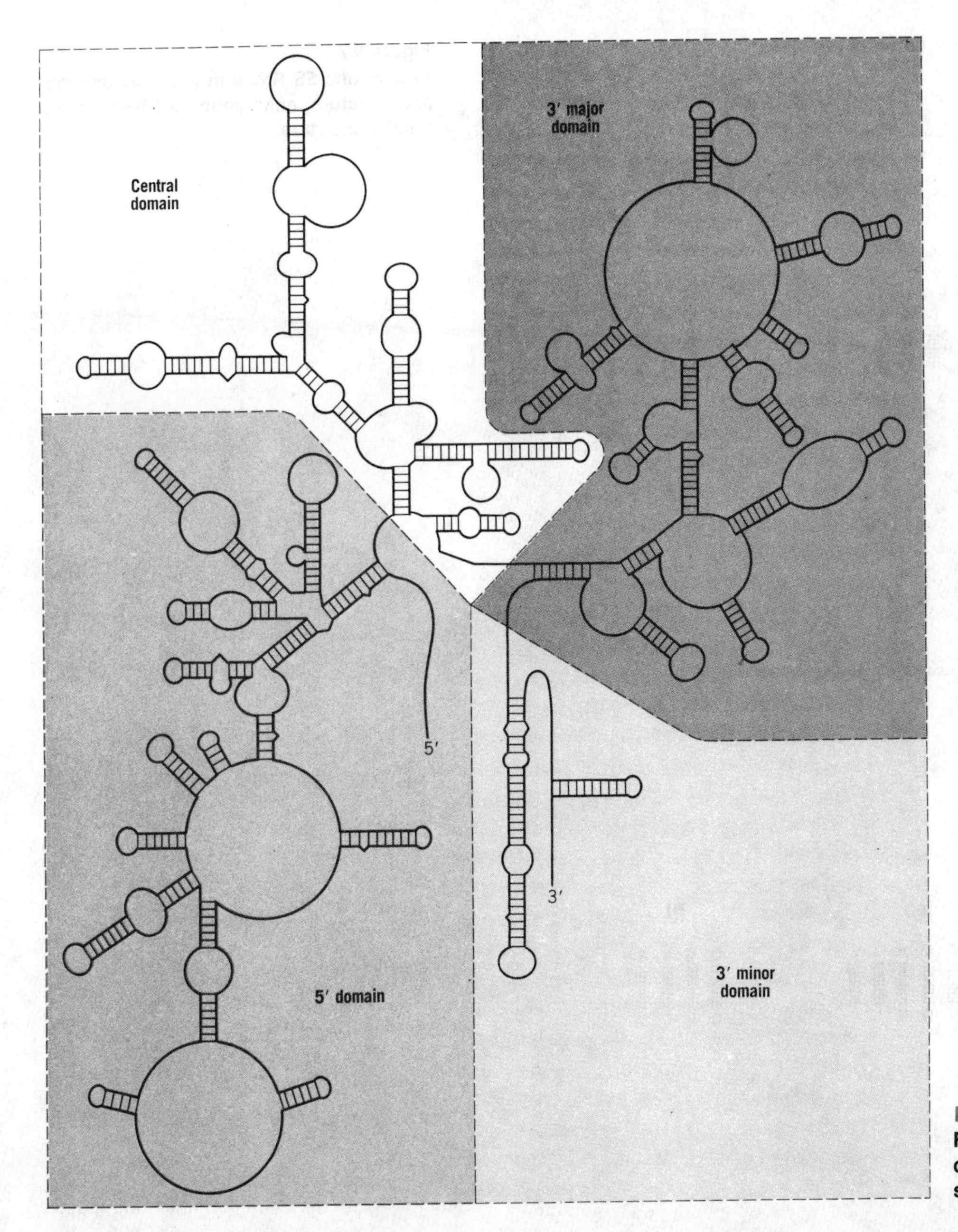

Figure 9.6
Four domains of 16S rRNA each contain many short double-stranded regions.

In eukaryotic cytoplasmic ribosomes, another small RNA is present in the large subunit. This is the **5.8S RNA**. Its sequence may correspond to the 5′ end of the prokaryotic 23S rRNA.

Both the major and smaller rRNA molecules bind proteins at specific sites. Some of the ribosomal proteins bind strongly to the isolated major rRNAs; these proteins include those that are assembled first into the particle *in vitro*. The positions on the rRNA at which these proteins bind can be defined by characterizing the parts of the nucleic acid that they protect against degradation by nucleases. Such experiments give a linear map of sites on the rRNA. A common feature of the binding sites is the presence of secondary struc-

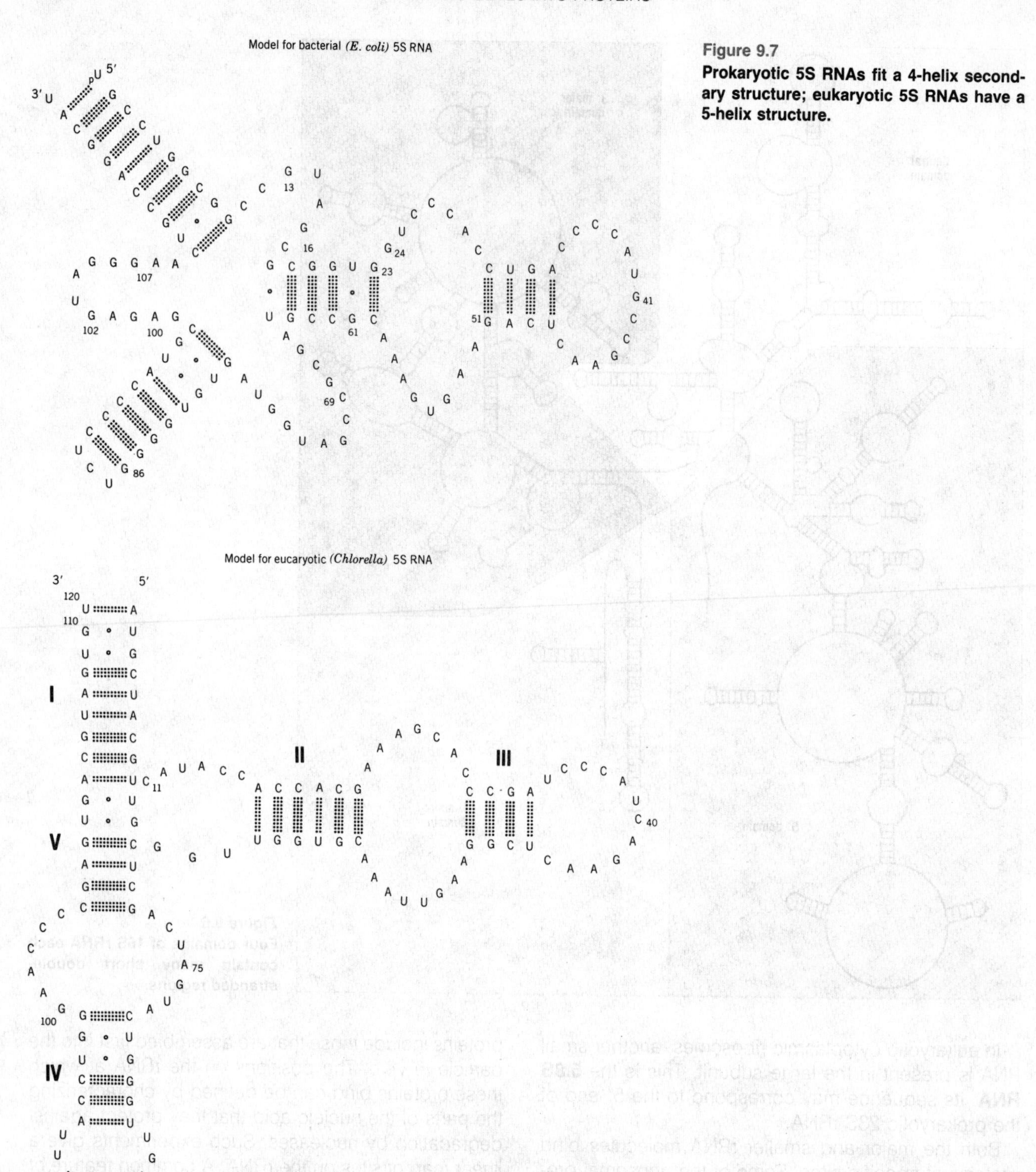

Figure 9.7

Prokaryotic 5S RNAs fit a 4-helix secondary structure; eukaryotic 5S RNAs have a 5-helix structure.

ture, often hairpins whose duplex stem contains unpaired bulges.

Other proteins do not bind strongly or even at all to free rRNA isolated in the usual way, but may be able to bind to rRNA prepared in other ways or to rRNA to which other proteins are already bound. This suggests that the conformation of the rRNA (which is influenced by the method of isolation), is important in determining whether binding sites exist for some proteins. As each protein binds, it may induce changes in the conformation of rRNA that make it possible for other proteins to bind. In *E. coli*, it now seems that almost or perhaps even all of the ribosomal proteins can bind (albeit with varying affinities) to one of the rRNAs.

Two approaches have gone hand in hand in analyzing the construction of the ribosomal subunits: the study of assembly (see next section); and the direct analysis of structure.

Biochemical studies have concentrated on identifying the locations of individual ribosomal proteins by immune electron microscopy, determining the relationships of adjacent proteins by crosslinking them, analyzing the interactions between particular ribosomal proteins and rRNA, and using more biophysical approaches such as neutron scattering of subunits containing specifically deuterated proteins.

Immune electron microscopy is performed by preparing antibodies against particular ribosomal proteins, binding the antibodies to intact subunits, and then visualizing the sites of binding by electron microscopy. The locations of virtually all the proteins of the 30S subunit have been determined in this way, and many of the 50S.

When pairwise relationships are analyzed by crosslinking, the closeness of two proteins can be ascertained by using reagents with varying crosslinking spans. Intact subunits are treated with the crosslinking reagent, and then the proteins present in crosslinked pairs are analyzed. Crosslinking of proteins to rRNA allows the binding sites on the nucleic acid to be determined.

A somewhat similar picture emerges from all the techniques. Each ribosomal protein can be located at a particular position in the structure. Some particular proteins can even be equated with particular features of ribosomal appearance. Probably the organization of each subunit is highly defined.

RECONSTITUTION *IN VITRO* MIMICS ASSEMBLY *IN VIVO*

When bacterial ribosomal subunits are centrifuged in CsCl, a discrete group of proteins (the **split proteins**) is lost from each subunit. The dissociation generates 23S and 42S particle **cores** from the 30S and 50S subunits, respectively. In fact, when ribosomes are exposed to increasing concentrations of either CsCl or LiCl, groups of proteins are lost in successive disruptions. This suggests that the ribosome contains groups of cooperatively organized proteins, so that disruption of a group effectively causes loss of all its members together.

The stepwise dissociation can be reversed by removing the CsCl and providing Mg^{2+} ions. The reconstituted subunits are active in protein synthesis. **Figure 9.8** summarizes the reconstitution procedure, which involves a two stage reaction, relying on a temperature-dependent transition *in vitro*.

Figure 9.8
30S subunits can be dissociated into subparticles that are related to assembly intermediates.

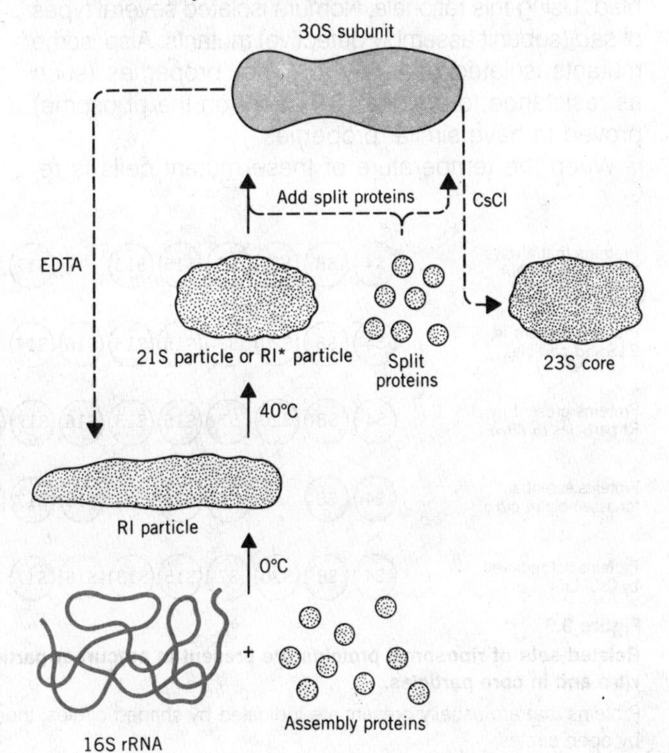

The first stage in reconstituting a 30S subunit is to assemble the 16S rRNA with a group of about 15 proteins. They react in the cold to form **RI particles**. Then these particles must be heated to allow a unimolecular rearrangement to take place, generating **RI* particles**. Finally, the remaining 6 proteins can join.

The nature of the unimolecular rearrangement is indicated by the effects of EDTA on 30S subunits. The chelating agent removes magnesium ions and causes the subunit to unfold and lose some proteins. The resulting particles are similar to the RI particles in requiring heating before they can reassociate with the proteins that have been lost. So the limiting step in assembly appears to be a temperature-dependent conversion in which the assembling particle folds into a more compact structure.

The dependence of reconstitution *in vitro* on incubation at elevated temperature suggested that mutations blocking assembly *in vivo* might be detected as **cold-sensitive** mutants. Such mutants should be recovered as bacteria in which protein synthesis is normal at the usual temperature, but is prevented at low temperature, because ribosomes cannot be assembled. Using this rationale, Nomura isolated several types of *sad* (subunit assembly defective) mutants. Also, some mutants isolated originally for other properties (such as resistance to antibiotics that act on the ribosome) proved to have similar properties.

When the temperature of these mutant cells is reduced, production of the affected class of subunit ceases. Some of the mutations constitute alterations in individual ribosomal proteins; others may lie in genes whose products directly or indirectly affect assembly.

The *sad* mutants accumulate smaller particles that are precursors in the assembly of the affected ribosome subunit. The precursor particles contain rRNA, associated with some, but not all, of the subunit proteins. A deficiency in a particular protein always blocks assembly at a specific stage.

When assembly of 30S subunits is blocked, a precursor sediments at 26S or 21S, depending on the conditions. It has also been identified as a minor (and transient) component of wild-type bacteria.

Blocking 50S subunit assembly leads to the accumulation of 32S or 43S particles. The 32S particle is the first identified stage of assembly. The 43S particle represents a later stage; it has all the proteins present in the 32S particle plus some additional species.

SUBUNIT ASSEMBLY IS LINKED TO TOPOLOGY

The small subunit precursor found *in vivo* and the RI particles reconstituted *in vitro* have very similar (or identical) protein components, as implied in Figure 9.8. The RNA molecules are slightly different, because the *in vivo* precursor has a precursor form of the rRNA,

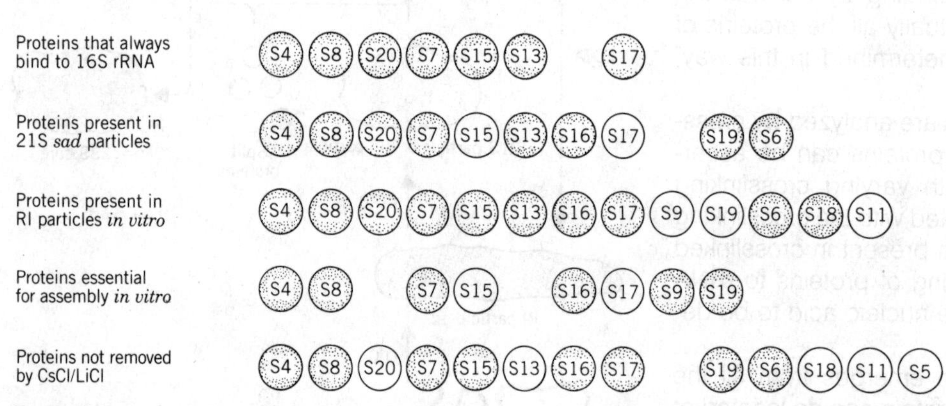

Figure 9.9

Related sets of ribosomal proteins are present in precursor particles assembled *in vivo* or *in vitro* and in core particles.

Proteins that are usually present are indicated by shaded circles, those that are sometimes present by open circles.

while the *in vitro* particle has mature 16S rRNA. It is interesting (and fortunate) that mature rRNA can be used for assembly.

The single precursor to the 30S subunit found *in vivo* is the only discrete intermediate that can be found. The precursor RNA is a little longer than the mature rRNA and is only slightly methylated. Probably the conformational change in the precursor is associated with removal of the surplus sequences and methylation of the rRNA; then the remaining proteins are added. Free 16S rRNA cannot be methylated *in vitro*, but the core particle is a substrate for methylase activity, which is consistent with this scheme.

Figure 9.9 shows that the proteins present in the precursors and RI particles also are very similar to the proteins that are the last to be removed by treatment with LiCl or CsCl. This suggests that the topology of the 30S subunit reflects the assembly process. Those proteins that are part of the precursor (which include all the species binding most strongly to the rRNA) tend to be the more secure components of the 30S subunit, and are harder to remove by abusive treatment.

The roles of particular components in ribosome subunit assembly can be analyzed by attempting to reconstitute subunits in the absence of individual proteins. When some proteins are omitted from the assembly mixture, no reconstitution occurs. These proteins are therefore essential for assembly. Other proteins *can* be omitted; these are the species added last, whose presence is not needed to determine the overall structure. The subunits that assemble may appear biophysically normal, but when tested in protein synthesis they prove to be deficient in some function (thus identifying the role of the missing protein).

Partial reconstitution can be used to examine whether the omission of one protein affects the ability of other proteins to assemble into the particle. Essentially the protocol is to leave out one protein and then to see which other proteins fail to bind. (This can be done irrespective of whether the partial assembly that occurs generates a functional subunit or not.)

The *in vitro* assembly map of **Figure 9.10** was constructed by determining which proteins must be bound to 16S rRNA for another particular protein to bind. The results show that ribosomal proteins must bind to the assembling particle in a certain order. Some of the proteins that show dependent relationships in the assembly map in fact lie physically close to one another. This is consistent with a scheme in which dependence largely involves physical interactions; but clearly this is not the whole story, since (for example) the binding of one protein may open a site elsewhere on rRNA for another to bind. All the proteins defined as essential for assembly are located in the early part of the assembly map.

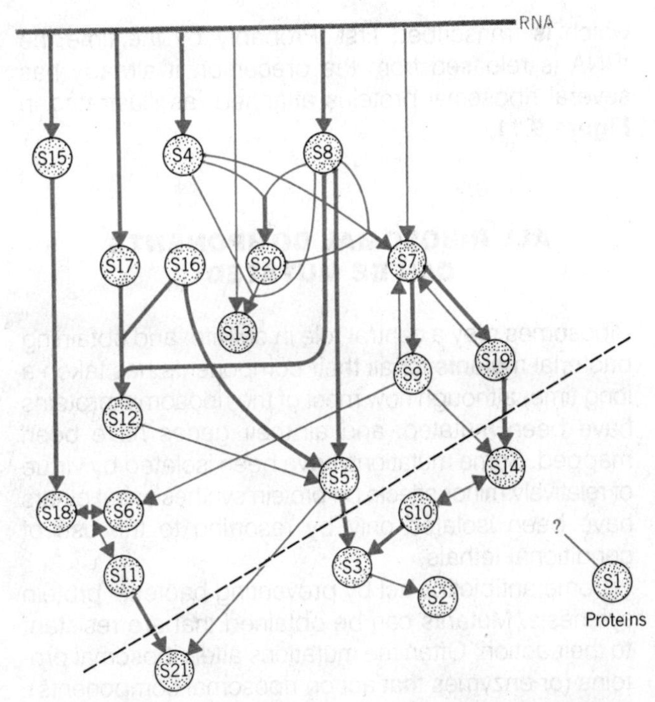

Figure 9.10
Ribosome assembly involves a series of sequential reactions.
An arrow from one protein to another protein indicates that the first assists binding of the second (double arrows indicate mutual effects). Major effects are indicated by heavy lines and lesser effects by thinner lines. Arrows from rRNA identify the most strongly binding proteins, but others also bind.

Proteins closest to the rRNA are those found most persistently in precursor particles; those farther away are assembled into the particle later and are removed more easily by salt. The dashed line separates those proteins usually or often found in precursor particles from those never found (which are equivalent to the split proteins).

Assembly of the 30S subunit probably starts while the rRNA is being transcribed. In this context, it may be significant that most of the strongest rRNA binding sites are clustered in the 5′ region of the molecule, which is transcribed first. Probably by the time the rRNA is released from the precursor, it already has several ribosomal proteins attached, as illustrated in **Figure 9.11**.

ALL RIBOSOMAL COMPONENTS CAN BE MUTATED

Ribosomes play a central role in cell life, and obtaining bacterial mutants for all their components has taken a long time, although now most of the ribosomal proteins have been mutated, and all their genes have been mapped. Some mutations have been isolated by virtue of relatively minor effects on protein synthesis; but others have been isolated only by resorting to the use of conditional lethals.

Some antibiotics act by preventing bacterial protein synthesis. Mutants can be obtained that are resistant to their action. Often the mutations alter ribosomal proteins (or enzymes that act on ribosomal components). Five 30S proteins and four 50S proteins have been mutated to confer antibiotic resistance. These variants may be useful in investigating the role of the protein in the ribosome.

Direct attempts to isolate mutants in protein synthesis arise from changes in the protein-synthetic apparatus that alter the accuracy of translation. Nonsense and missense mutations may be suppressed by tRNA mutants (see Chapter 8); another type of suppression is conferred by mutation in the ribosome. Thus several ribosomal mutants have been obtained in the form of second-site revertants that overcome defects caused by mutations elsewhere. Sometimes a known step in translation is affected, so the role of a particular ribosomal protein can be determined. About six proteins have been mutated in this way.

We have already seen that mutants can be obtained in ribosome assembly by virtue of their cold-sensitivity. This phenotype also is shown by some antibiotic-resistant mutants. So far, four of the 30S proteins and one 50S protein have been identified in assembly mutants. This type of mutation is a conditional lethal, since lack of ribosomes kills the cell.

Some ribosomal proteins have been identified simply by virtue of their alteration in mutants that are temperature-sensitive. This means that the bacteria show

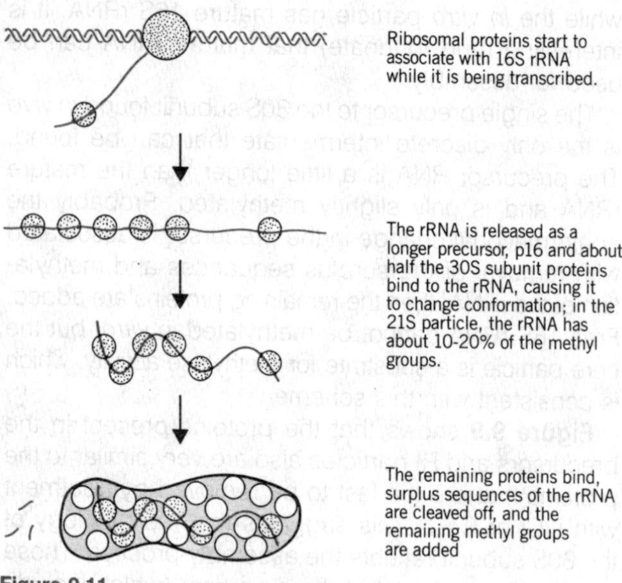

Figure 9.11
Ribosome subunit assembly is an ordered process in which changes in conformation occur as each protein is added.

reduced growth at higher temperature, and the effect is due to an alteration in a ribosomal protein that (presumably) malfunctions when the temperature is raised. The generality of the defect does not leave any clues about the function of the protein.

There are still a few ribosomal proteins whose genes have not yet provided sites in which mutations are damaging. It remains possible that these proteins are dispensable for ribosome assembly and activity.

The properties of kasugamycin-resistant mutants show that 16S rRNA plays a direct role in protein synthesis. Kasugamycin blocks the initiation reaction. Resistant mutants of the type *ksgA* lack a methylase enzyme that introduces four methyl groups into two adjacent adenines at a site near the 3′ terminus of the 16S rRNA. The methylation generates the highly conserved sequence G—$m_2{}^6$A—$m_2{}^6$A, found in both prokaryotic and eukaryotic small rRNA.

In the wild-type, therefore, the presence of the four methyl groups renders the ribosome sensitive to kasugamycin. Why does the absence of the methyl groups confer resistance to the drug? Both wild-type and mutant 30S subunits are equally sensitive to inhibition by kasugamycin of the binding of fMet-$tRNA_f$. But in the presence of 50S subunits, the fmet-tRNA$_f$ is released from the sensitive (wild-type) ribosomes; whereas it is retained by the mutants, thus allowing protein synthesis to continue.

This suggests that methyl groups are involved in the joining of the 30S and 50S subunits; and the association reaction must be connected also with the retention of initiator tRNA. The influence of the large subunit on the reaction is shown by the isolation of some 50S protein mutations that make the ribosome dependent on kasugamycin.

Another indication of the importance of the 3′ end of 16S rRNA is given by its susceptibility to the lethal agent colicin E3. Produced by some bacteria, the colicin cleaves about 50 nucleotides from the 3′ end of the 16S rRNA of *E. coli*. The cleavage entirely abolishes initiation of protein synthesis.

Resistance to chloramphenicol arises in quite different ways in bacteria and in mitochondria, where it blocks protein synthesis by interfering with the peptidyl transferase reaction. In bacteria, resistant mutants have altered 50S proteins. In mitochondria, they possess single-base changes at sites about 243 nucleotides from the 3′ end of the large rRNA.

RIBOSOMES HAVE SEVERA ACTIVE CENTERS

We can distinguish several ribosomal activities on the basis of function or location. The 30S subunits bind mRNA and the initiator-tRNA·initiation factor complex; then they bind 50S subunits. The 70S ribosomes possess the functionally distinct P and A sites at which tRNA is bound; the peptidyl transferase center is carried on the 50S subunit. The EF-G binding site, and hence responsibility for translocation, is carried on the 50S subunit. A simplified view of these sites is drawn in **Figure 9.12**. How are these sites related to the actual topology of the ribosome? Which ribosomal components are involved in its various functions? Several approaches have been used to analyze the relationship between structure and function.

One way to identify particular sites is affinity labeling. This technique uses analogs of components that bind to the ribosome; the analogs are either themselves chemically reactive or can be activated *in situ*.

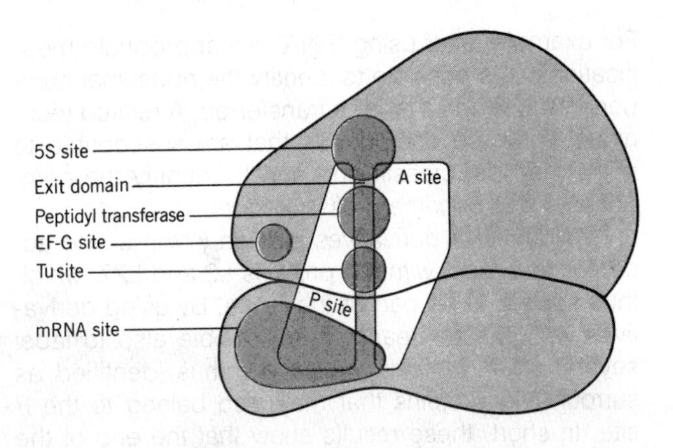

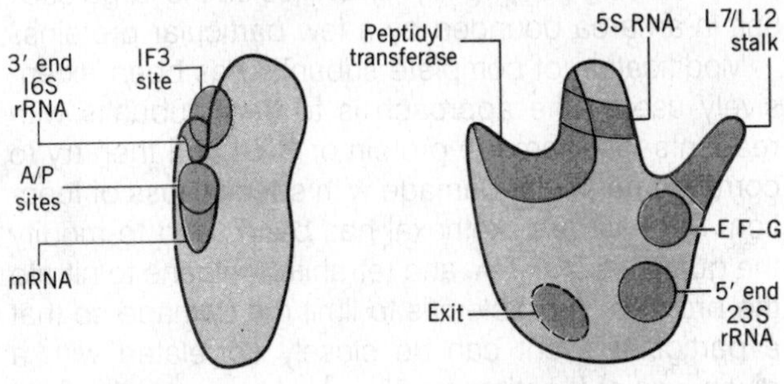

Figure 9.12
The ribosome has several separate active centers.

The subunits are shown at a rotation of 90° relative to the intact ribosome. The locations of the various sites are strictly diagrammatic and cannot yet be related to the precise three dimensional structure of the ribosome.

mRNA binding site is located on the 30S subunit and includes proteins S1, S18 and S21, as well as S3, S4, S5, S12, and the 3′ terminal region of the 16S rRNA; also, this is the site where IF3 binds. It is close to the P site.

P site is probably located mostly on the 30S subunit and is able to bind initiator tRNA. The 16S rRNA is a component, probably involving a region not far from its 3′ end. On the 50S subunit, the P site includes L2 and L27, as well as L14, L18, L24, and L33.

A site must lie close to the P site, but the exact geometry is unknown. The 16S rRNA influences the A site and could be a component. Most of the proteins identified with the A site are on the 50S subunit; they include L1, L5, L7/L12, L20, L30, and L33.

Peptidyl transferase site should lie somewhere in the region connecting the A and P sites, close to the terminus of tRNA. Proteins needed for this activity are L2, L3, L4, L15, and L16.

5S RNA site may lie near the peptidyl transferase site, including L5, L8, and L25.

Tu site includes several proteins lying close together that can be crosslinked to EF-Tu; these are L5, L1, L20, and L7/L12.

L7/L12 site forms the stalk of the 50S subunit, which consists of two dimers.

EF-G site may be located on the large subunit close to the interface with the small subunit, near S12.

For example, after using tRNA with appropriate modifications, it is possible to identify the ribosomal components to which a label is transferred. A related technique is to use compounds that are susceptible to photochemical crosslinking and to identify the components that become crosslinked.

Peptidyl-tRNA derivatives labeled in the amino acceptor end react with the proteins L2 and L27, which thus appear to be part of the P site. By using derivatives with greater reach, it is possible also to label several other proteins, which are thus identified as surrounding proteins that may also belong to the P site. In short, these results show that the end of the tRNA carrying the peptidyl chain lies on the large subunit in an area bounded by a few particular proteins.

Modification of complete subunits has been extensively used. One approach is to treat subunits with reagents that damage protein or RNA and then try to correlate particular damage with specific loss of function. For example, kethoxal has been used to modify the guanines of rRNA, and tetranitromethane to nitrate the proteins. A problem is to limit the damage so that a particular event can be closely correlated with a given loss of function.

Initial binding of 30S subunits to mRNA requires protein S1, which has a strong affinity for single-stranded nucleic acid. It may be responsible for maintaining mRNA in the single-stranded state upon binding to the 30S subunit. This action may be necessary to prevent the mRNA from taking up a base-paired conformation that would be unsuitable for translation.

Analyzing the location of S1 in the ribosome is complicated by its extremely elongated structure. Crosslinking studies show that it is closely related to S18 and S21, which are among the proteins that react in affinity labeling experiments when initiator tRNA is bound to an AUG codon. They are located in the head of the small subunit. The three proteins may constitute a relatively small domain that is involved in both the initial binding of mRNA and binding initiator tRNA.

The 3′ end of the 16S rRNA also is located near this domain and is involved in the binding reaction. It possesses a short sequence that is complementary to a sequence found just prior to each AUG initiation codon in bacterial mRNA. It is likely that complementary base pairing between these sequences is involved in the initial binding of ribosomes to initiation sites on mRNA (see Chapter 10).

Initiation factor IF3 binds in the same region of the ribosome. IF3 can be crosslinked to the 3′ end of the rRNA, as well as to several ribosomal proteins, including those probably involved in binding mRNA. If this area of the small subunit is the same region involved in binding the large subunit, the role of IF3 could be to bind there to stabilize mRNA·30S subunit binding, then to be displaced when the 50S subunit joins.

Not very much is known about the mechanism of subunit joining. There is some complementarity between a part of 23S rRNA and (again) the 3′ region of 16S rRNA. One possibility is that base pairing is involved in bringing the subunits together. In this case, the ability of IF3 to bind to 16S rRNA might be incompatible with subunit association.

Treatment with kethoxal prevents ribosomes from binding aminoacyl-tRNA (although it leaves them able to bind mRNA, which provides a control to show that the treatment has not simply irreversibly damaged the entire function of the particle). About 6 guanine residues of the 16S rRNA are modified. This suggests that the 16S rRNA may be part of the A site. More direct evidence is provided by crosslinking studies, which show that peptidyl-tRNA can be linked to a point not far from the 3′ end of 16S rRNA. The rRNA has a conserved sequence that may interact with tRNA. However, crosslinking does not occur between rRNA and aminoacyl-tRNA in the A site.

Crosslinking studies involving EF-Tu show that the A site lies in the vicinity of a group of several proteins, but we do not at present know much about how these (and other) proteins interact to provide the tRNA-binding site.

The incorporation of 5S RNA into 50S subunits that are assembled *in vitro* depends on the ability of three proteins, L5, L8, and L25, to form a stoichiometric complex with it. The complex can bind to 23S rRNA, although none of the isolated components can do so. It lies in the vicinity of the P and A sites.

The conserved sequence in the TψC loop of tRNA is complementary to a sequence in 5S RNA. There have been suggestions that base pairing between 5S RNA and tRNA might occur, but experiments *in vitro* have shown that removal of the relevant segment from 5S RNA does not impede ribosome function.

A group of several proteins is needed for peptidyl transferase activity. Attempts to identify a particular one of these proteins with the enzymatic activity have not been successful. The best two candidates are L2 and L16.

The growing polypeptide chain appears to be extruded from the 50S subunit at a point some distance from the peptidyl transferase site. It probably extends through the ribosome as an unfolded polypeptide chain until it leaves the exit domain, when it is free to start folding.

The only exception to the rule that there is one copy of each ribosomal protein is presented by L7/L12. L7 differs from L12 only in the presence of an acetyl group on the N-terminus; there are two copies of the dimer per ribosome. The L7/L12 aggregate forms the stalk of the large subunit. When it is removed, the particles become unable to perform GTP hydrolysis at the behest of any accessory factor. This does not necessarily mean that L7/L12 is the GTPase; it may instead be necessary for the activity of another protein.

The translocation factor EF-G binds to the 50S subunit; but translocation involves movement of the mRNA through the 30S subunit. The binding site for EF-G is close to S12, one of the proteins of the mRNA binding site on the 30S subunit. This places EF-G at the interface between subunits, in the vicinity of the L7/L12 dimers.

The P and A sites must lie close together, since their tRNAs are bound to adjacent triplets on mRNA. The P site influences the activity of the A site, since peptidyl-tRNA must be present for aminoacyl-tRNA to bind. It has always been a problem to see how two tRNA molecules might fit into the ribosome next to each other. The distance between the anticodons cannot be greater than about 10Å, yet the diameter of the tRNA is about 20Å.

The conformation of the tRNA seems to be the same in both the P site and A site (although it is not certain that it is the same as the tertiary structure displayed in a tRNA-crystal). One solution to the stereochemical problem would be to have a twist or kink in the mRNA between the codons, so that the two tRNAs fit onto the tRNA from different directions. Then ribosome movement might be a matter of flicking the tRNA out of the A site, as it were, around the corner into the P site. This would allow mRNA advancement to be a function of tRNA movement, which would let anticodon-codon pairing measure the distance for translocation, as discussed in Chapter 8.

The functional relationships between the various ribosomal sites have yet to be defined. The ribosome may well be a highly interactive structure, in which a change at one point could greatly affect the activity of another site elsewhere.

THE ACCURACY OF TRANSLATION

The lack of detectable variation when the sequence of a protein is analyzed demonstrates that protein synthesis must be extremely accurate: very few mistakes are apparent in the form of substitutions of one amino acid for another. There are two stages in protein synthesis at which errors might be made.

Charging a tRNA only with its correct amino acid clearly is critical. We have seen in Chapter 8 that this is a function of the aminoacyl-tRNA synthetase enzyme. Probably the error rate will vary with the particular enzyme, but current estimates are that mistakes occur in less than 1 in 10^5 aminoacylations.

Correct recognition of a codon by its anticodon is crucial. In a case in which all misreadings are likely to occur at codon recognition, the substitution of arginine by cysteine (in bacterial flagellin), there are roughly 1 in 10^4 misreadings per codon.

One of the puzzling features of protein synthesis is its accuracy. Although binding constants vary with the individual codon-anticodon reaction, the specificity is always much too low to provide an error rate of 10^{-4} When free in solution, tRNAs bind to their trinucleotide codon sequences only relatively weakly; and related, but erroneous triplets (with two correct bases out of three) are recognized 10^{-1} to 10^{-2} times as efficiently as the correct triplets. Thus codon-anticodon base pairing per se is not nearly specific enough to account for the accuracy of translation.

This suggests that the ribosome has some function that directly or indirectly acts as a "proofreader," distinguishing correct and incorrect codon-anticodon pairs, and thus amplifying the rather modest intrinsic difference. Now suppose that there is no specificity in the initial collision between the aminoacyl-tRNA·EF-Tu·GTP complex and the ribosome. If any complex, irrespective of its tRNA, can enter the A site, the number of incorrect entries must far exceed the number of correct entries. So there must be some mechanism for stabilizing the correct aminoacyl-tRNA, allowing its amino acid to be accepted as a substrate for receipt of the polypeptide chain; contacts with an incorrect aminoacyl-tRNA must be rapidly broken, so that the complex leaves without reacting. How does a ribosome assess the codon-anticodon reaction in the A site to determine whether a proper fit has been achieved?

The ability of the ribosome to influence the accuracy of translation was first shown by the effects of mutations that confer resistance to streptomycin. One of the effects of streptomycin is to increase the level of misreading of the pyrimidines U and C (usually one is mistaken for the other, occasionally for A). The site at which streptomycin acts may be the S12 protein; in resistant mutants, it is this protein whose sequence is altered. Ribosomes with an S12 protein derived from resistant bacteria show a reduction in the level of misreading compared with wild-type ribosomes. This compensates for the effect of streptomycin on misreading.

Mutations at two other loci, coding for proteins S4 and S5, influence misreading, since revertants showing the usual level of misreading can be isolated at either locus. Thus the accuracy of translation is controlled by the interactions of these three proteins. The level of misreading and the nature of the response to streptomycin both depend on the versions of these proteins that are present. (Some combinations even make the ribosome **dependent** on the presence of streptomycin for correct translation.)

A direct mechanism for controlling accuracy would be to establish the stereochemistry of the A site, which could change so as to determine the latitude of codon-anticodon recognition. The original results obtained with streptomycin were interpreted in terms of ribosome geometry, supposing that codon-anticodon binding is scrutinized in such a way that the criteria for accepting aminoacyl-tRNA could be made more or less precise.

An alternative is that the effect could be indirect. The speed of ribosome function could determine the time available for tRNA recognition, and thus the efficiency of the process. This model explains the effects of streptomycin by adjusting the kinetics of chain elongation. The relevant parameter is the speed of ribosome action relative to the time required to make and break contacts. If the velocity of peptide bond formation is increased, incorrect aminoacyl-tRNAs are more likely to be trapped by bond formation before the aminoacyl-tRNA escapes. Slowing the rate of protein synthesis gives more time to correct errors. There is some evidence that the rate of polypeptide chain elongation may be related to the level of misreading.

One idea is that the making of a correct contact between codon and anticodon could be signalled to the other end of the tRNA by a change in conformation. The change could be needed to place the amino acid in the appropriate location to accept the polypeptide from the peptidyl-tRNA.

A question that is important in calculating the cost of protein synthesis is the stage at which the decision is taken on whether to accept a tRNA. If the decision occurs immediately to release the aminoacyl-tRNA·EF-Tu·GTP complex, there is little extra cost for rejecting the large number of incorrect tRNAs that are likely (statistically) to enter the A site before the correct tRNA is recognized. But if the GTP is hydrolyzed when the complex binds, an additional high-energy bond will be cleaved for every incorrectly associating tRNA. This would increase the cost of protein synthesis well above the three high-energy bonds that are used in adding every (correct) amino acid to the chain. There is some evidence that the use of GTP *in vivo* is greater than had been expected, possibly involving an extra 3–4 GTP cleavages per amino acid.

FURTHER READING

A major reference work on ribosomes is still the book edited by **Nomura, Tissieres, & Lengyel:** *Ribosomes* (Cold Spring Harbor Lab., New York, 1974). Although now outdated, it provides a useful general view of both structural and functional aspects. The various models for ribosome structure have been reviewed and partially reconciled by **Wittman** (*Ann. Rev. Biochem.* **52**, 35–65, 1983). Ribosomal RNA has been analyzed in minute detail by **Woese et al**. (*Microbiol. Rev.* **47**, 621–669, 1983 and **Noller** (*Ann. Rev. Biochem.* **53**, 119–162, 1984). The structure and function of ribosome sites have been described by **Pongs** (pp. 78–104 in Altman (Ed), *Transfer RNA*, MIT Press, Cambridge, 1978). The parameters involved in the accuracy of translation have been summarized briefly by **Kurland** (*Cell* **28**, 201–202, 1982).

CHAPTER 10
THE MESSENGER RNA TEMPLATE

The existence of mRNA was first suspected because of the evident need for an intermediary in eukaryotic cells to carry genetic information from the nucleus where DNA resides into the cytoplasm where proteins are synthesized. Conceived as providing a template on which amino acids would be assembled into polypeptides, the messenger was first sought in bacteria. The ribosomes, established as the site of protein synthesis, were excluded from the role of providing the template, and then mRNA was found as a transient species that associates with the ribosomes to be translated into proteins (see Chapter 6).

Although the transient existence of bacterial mRNA prevented isolation of the molecule, the properties of the messenger were deduced in detail from the features of its translation into protein. It was some time before mRNA could be isolated from eukaryotes, but then it proved to be a relatively more stable component of the cytoplasm, found again as part of the polyribosomes. Because of its greater stability, eukaryotic mRNA can be isolated with some facility; these days, it is possible in principle to isolate the mRNA for any particular protein. By an irony of circumstance, therefore, the wheel has come full circle, and mRNA has been characterized directly in the greatest detail in eukaryotes, where its existence was originally postulated.

Messenger RNA is fixed of purpose in all living cells: to be translated via the genetic code into protein. Yet there are differences in the details of the synthesis and structure of prokaryotic and eukaryotic mRNA. The most evident is that in eukaryotes the mRNA is synthesized as a large precursor molecule in the nucleus. After an involved process of maturation, often involving a considerable reduction in size as well as other modifications, the mRNA is exported to the cytoplasm. Its synthesis and expression thus occur in different cellular compartments. In bacteria, on the other hand, mRNA is transcribed and translated in the single cellular compartment; and the two processes are so closely linked that they occur simultaneously. The principal difference in the process of translation is that a bacterial mRNA may code for several proteins, whereas a eukaryotic mRNA invariably is translated into only one polypeptide chain.

THE TRANSIENCE OF BACTERIAL MESSENGERS

The instability of bacterial mRNAs is very pronounced. It may be measured in two ways. Both rely in halting the transcription of mRNA and then following the fate of the existing mRNA molecules in the cell.

The **functional half-life** measures the ability of the mRNA to serve as template for synthesis of its protein product. It is usually about 2 minutes. In other words, the amount of new protein that an individual mRNA can synthesize is halved about every 2 minutes.

The **chemical half-life** is determined by measuring the decline in the amount of mRNA able to hybridize with the DNA of its gene. Generally the chemical decline lags slightly behind the functional decline.

This (slight) discrepancy suggests that the degradation of mRNA involves an initial step sufficient to prevent its use as a template, but not detectable as a structural alteration sufficient to impede its hybridization. For example, a single cleavage in the mRNA would not affect its ability to hybridize with DNA, but could prevent its translation.

The initial step is followed by breakdown of the mRNA into its constituent nucleotides. The details of the molecular events responsible for the breakdown are still elusive. All we know is that breakdown occurs more or less sequentially along the mRNA from the 5′ end to the 3′ end—that is, in the same direction in which the mRNA is transcribed and translated.

Transcription and translation are intimately related in bacteria. The exact rates vary with the temperature, but usually are similar to one another. For example, at 37°C, transcription of mRNA occurs at about 2500 nucleotides/minute, which corresponds to synthesis of about 14 codons/second. This is very close to the rate of protein synthesis, roughly 15 amino acids/second. When expression of a new gene is initiated, its mRNA typically will appear in the cell within about 2.5 minutes. The corresponding protein will appear within perhaps another 0.5 minute.

The coincidence of these rates, and the close timing of the initial appearance of mRNA and protein, implies that transcription and translation occur simultaneously. Gene expression is initiated when an **RNA polymerase** enzyme binds to DNA and starts transcribing the mRNA. Ribosomes attach to the 5′ end of the mRNA and start translation, even before the rest of the message has been synthesized. In effect, a bunch of ribosomes follow closely behind the RNA polymerase.

Electron microscopy of *E. coli* cells has identified the (unknown) transcription units shown in **Figure 10.1**, in which several mRNAs are under synthesis simultaneously, and each carries many ribosomes engaged in translation.

The degradation of mRNA in turn very closely follows its translation. Probably it begins within 1 minute of the start of transcription. Again, the 5′ end of the mRNA may have started to decay before the 3′ end has been synthesized or translated. Degradation seems

Figure 10.1
Transcription units can be visualized in bacteria.
The thin central line is DNA, and the lines extending from it are RNA molecules being transcribed. The ~15 RNAs being synthesized from the transcription unit increase in length from left to right, defining the direction of synthesis. Each RNA is covered in ribosomes.
Photograph kindly provided by Oscar Miller.

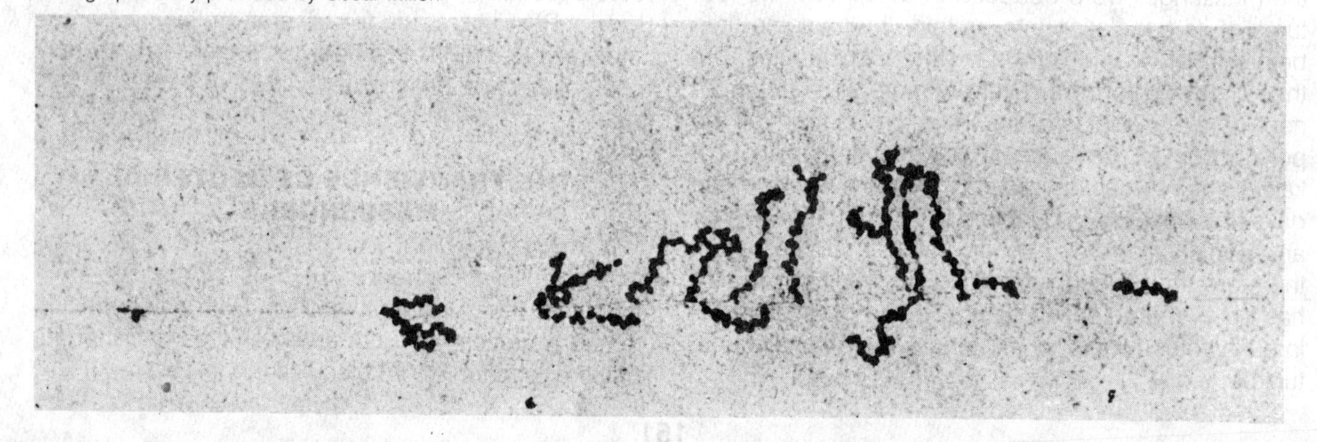

to follow the last ribosome of the convoy along the mRNA. But the degradation proceeds more slowly, probably at about half the speed of transcription or translation.

This series of events is only possible, of course, because transcription, translation, and degradation all occur in the same direction. **Figure 10.2** gives an idea of the timing of events for the expression of a typical bacterial unit of transcription.

We should not allow our need to visualize the "average" case obscure the harsh reality that every mRNA is in statistical jeopardy at all times, with a constant probability that its decay will begin. Thus "young" mRNAs are as likely to be attacked as "old" mRNAs. Some copies of an mRNA may be translated many times, while others function hardly at all. This random life expectancy is a feature of individual mRNA molecules of both prokaryotes and eukaryotes. But the overall translational yield of any messenger sequence is predictable.

MOST BACTERIAL mRNAs ARE POLYCISTRONIC

Because of their instability *in vivo*, bacterial mRNAs can rarely be isolated intact. Information about their constitution can be deduced in some detail, however,

Figure 10.2

Transcription, translation, and degradation of mRNA all proceed simultaneously in bacteria.

The figure shows a hypothetical transcription unit of about 12,000 bp, whose expression starts at time 0. For simplicity, the unit is drawn as though it consisted of a single gene (an actual unit of this length would probably represent several genes).

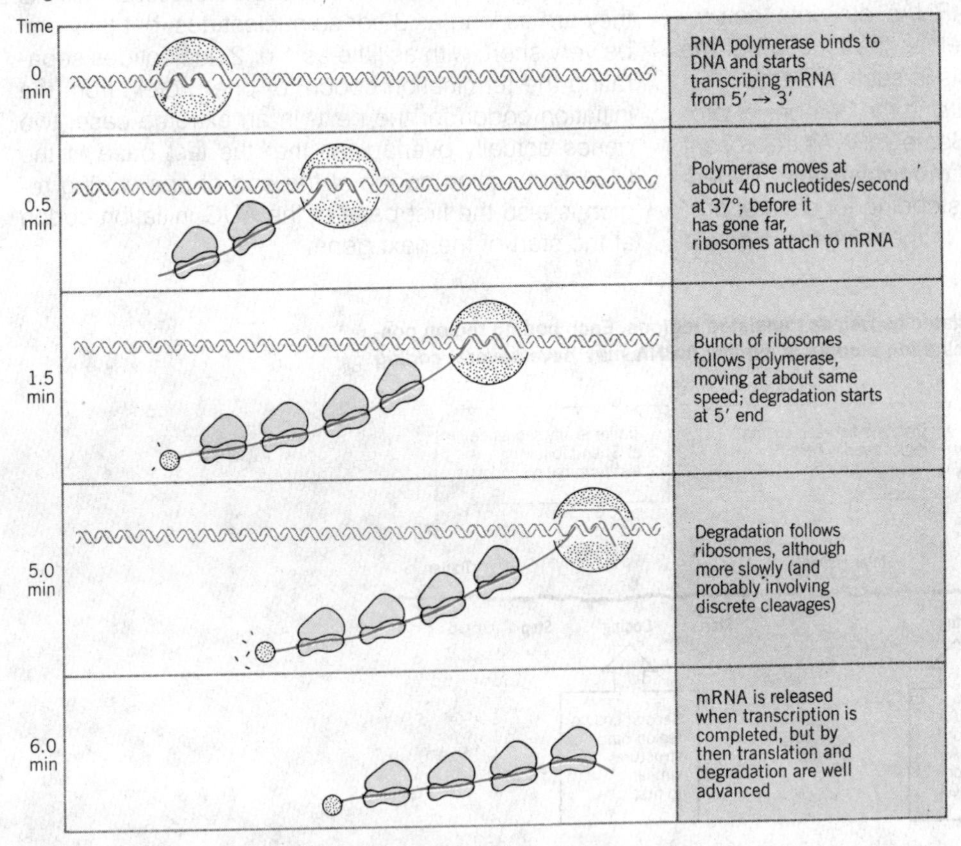

from the sequences of their genes. The region of DNA represented in the mRNA can be defined by hybridization assays, which measure the ability of radioactively labeled mRNA to hybridize with DNA fragments representing specific parts of the gene. In this way, a detailed picture of the structure and function of bacterial mRNA can be constructed without characterizing the mRNA molecules directly.

Useful exceptions where the mRNA can be obtained are provided by phage infections. In some cases, mRNA is more stable in the phage-infected cell than in the bacterial host, and this allows physical characterization of the molecule. The example par excellence is provided by the RNA phages, where the mRNA also serves as the genome, and therefore is not only stable but also can be obtained in large amounts.

An important approach for working directly with bacterial mRNA is the use of cell-free systems for both transcription and translation. The mRNA can be transcribed *in vitro* from the appropriate template, usually a "cloned" copy of the gene. It can be translated by *E. coli* ribosomes, aminoacyl-tRNAs, etc., into the protein. Thus the structure as well as the function of the mRNA is amenable to analysis in such systems.

Bacterial mRNAs vary greatly in the number of proteins for which they code. Some mRNAs represent only a single gene: they are **monocistronic**. Others (the majority) carry sequences coding for several proteins: they are **polycistronic**. In these cases, a single mRNA is transcribed from a group of adjacent genes. (As we shall see in Chapter 14, such a cluster of genes constitutes an *operon* that is controlled as a single genetic unit.)

All mRNAs contain two types of region. The **coding region** consists of a series of codons representing the amino acid sequence of the protein, starting (usually) with AUG and ending with a termination codon. But the mRNA is always longer than the coding region. In a monocistronic mRNA, extra regions may be present at both ends. An additional sequence at the 5′ end, preceding the start of the coding region, is described as a **leader**. An additional sequence following the termination signal, forming the 3′ end, is described as a **trailer**. Although part of the transcription unit, these sequences are not used to code for protein.

The structure of a polycistronic mRNA is illustrated in **Figure 10.3**. The **intercistronic regions** that lie between the various coding regions vary greatly in size. In the example of the RNA phages, they may be quite large, up to 100 or so bases. In some bacterial mRNAs they are as long as 30 or so nucleotides, but they can be very short, with as little as 1 or 2 nucleotides separating the termination codon for one protein from the initiation codon for the next. In an extreme case, two genes actually overlap, so that the last base of the UGA termination codon at the end of one coding region is also the first base of the AUG initiation codon at the start of the next gene.

Figure 10.3
Bacterial mRNA includes nontranslated as well as translated regions. Each coding region possesses its own initiation and termination signals. A typical mRNA may have several coding regions.

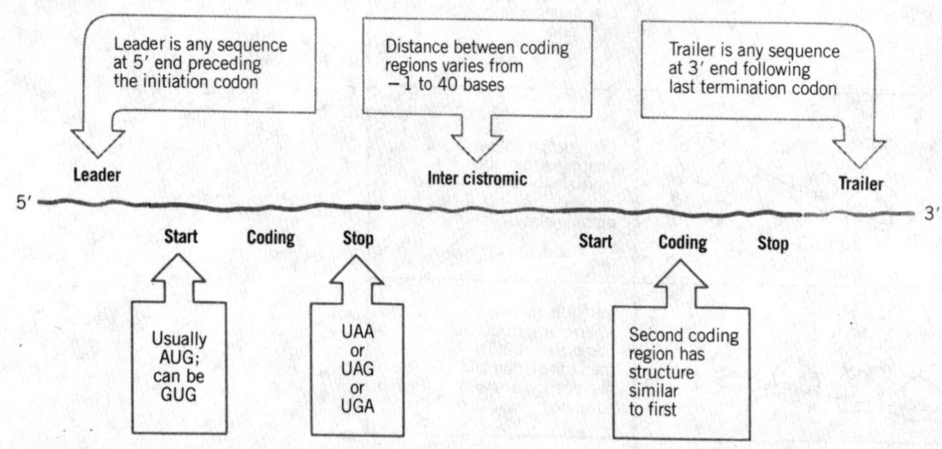

TRANSLATION OF POLYCISTRONIC MESSENGERS

The rates at which RNA and protein chains are extended are more or less independent of the individual gene, so the amount of protein that is synthesized depends primarily on the efficiencies with which transcription and translation are initiated. The frequency of transcription depends on the affinity of the RNA polymerase for the gene, while the number of ribosomes engaged in translating a particular cistron depends on the efficiency of its initiation site.

In the example of the tryptophan genes of *E. coli*, about 15 transcriptional initiation events occur every minute. Each of the 15 mRNAs probably is translated by about 30 ribosomes in the interval between its transcription and degradation. So roughly 150 molecules of protein are translated per minute in the steady state.

There are potentially more opportunities to control translation in a polycistronic mRNA. Are the several coding regions in a polycistronic mRNA translated independently or is their expression connected? Is the mechanism of initiation the same for all cistrons, or is it different for the first cistron and the internal cistrons?

The timing of events indicates that translation of bacterial mRNAs must proceed sequentially through the cistrons. At the time when ribosomes attach to the first coding region, the subsequent coding regions may not yet even have been transcribed. By the time the second ribosome site is available, translation is well under way through the first cistron.

What happens between the coding regions may depend on the individual mRNA. Probably in most cases the ribosomes bind independently at the beginning of each cistron. The most likely series of events is illustrated in the upper part of **Figure 10.4**. When synthesis of the first protein terminates, the ribosomes dissociate into subunits and leave the mRNA. Then a new 30S subunit must attach at the next initiation codon, be joined by a 50S subunit, and set out to translate the next cistron.

But this does not mean that translation of one cistron is always without effect on translation of the next cistron. In some units, a nonsense mutation in one gene prevents expression of a gene located farther along a

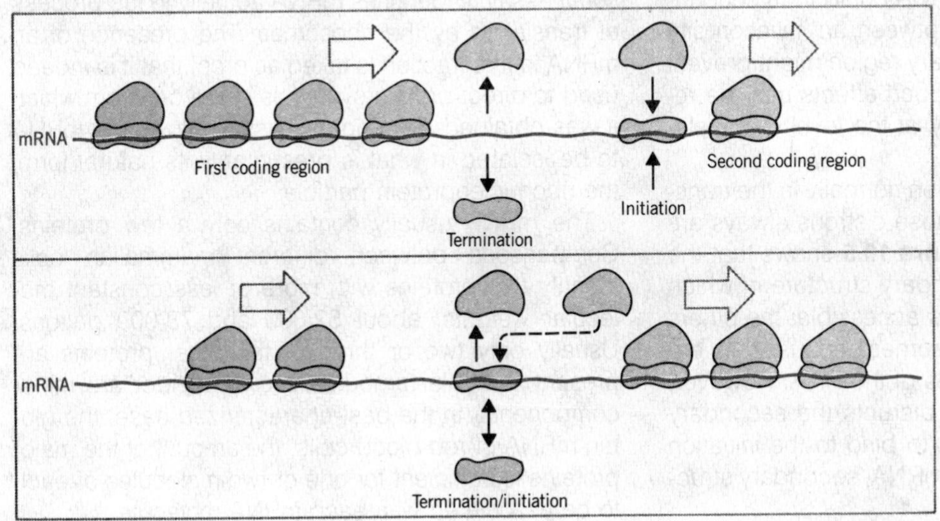

Figure 10.4
Reinitiation within polycistronic mRNAs may be influenced by the intercistronic region.
Upper. When the intercistronic region is longer than the span of the ribosome, dissociation at the termination site is followed by independent reinitiation at the next cistron.
Lower. When the intercistronic region is very short, the 30S subunit might dissociate transiently or even remain on the mRNA during termination and reinitiation.

polycistronic mRNA. This effect is called **polarity**. The principal cause of polarity is an (indirect) effect on transcription, which stops soon after the site of mutation (see Chapter 13). However, polarity may also result from a relationship between the translation of the two cistrons in mRNA.

A direct relationship between the translation of two cistrons may occur when the space between the coding regions is small. A ribosome physically spans about 35 bases of mRNA, so that it may simultaneously contact a termination codon and the next initiation site even if they are separated by up to a few base pairs. This juxtaposition could allow some of the usual intercistronic events to be bypassed. For example, the 30S subunit of a terminating ribosome might fail to dissociate from mRNA, because it is instantly attracted to the initiation site of which it is already virtually in possession. As illustrated in the lower part of Figure 10.4, this would mean that, when the 50S subunits and the completed polypeptide chain are released, the 30S subunit would remain *in situ* to reinitiate translation of the next cistron.

A less direct relationship between cistrons in a polycistronic mRNA is mediated by secondary structure. Usually this will not apply in bacterial mRNAs, because the ribosomes follow the RNA polymerase closely, and the regions ahead of them are not offered the option to base pair into stable double-stranded regions. However, when the ribosomes dissociate prematurely at a nonsense mutation in an early cistron, the subsequent regions of mRNA could form secondary structure. Base pairing between an initiation site and some other, complementary region might prevent a 30S subunit from binding. Such effects may be responsible for creating polarity at the level of translation.

An effect of this nature is seen normally in the translation of the RNA phages, whose cistrons always are expressed in a set order. **Figure 10.5** shows that the phage RNA takes up a secondary structure in which only one initiation sequence is accessible; the others cannot be recognized by ribosomes because they are base paired with other regions of the RNA. However, translation of the first cistron disrupts the secondary structure, allowing ribosomes to bind to the initiation site of the next cistron. In this mRNA, secondary structure controls translatability.

A FUNCTIONAL DEFINITION FOR EUKARYOTIC mRNA

Eukaryotic polyribosomes are reasonably stable and can be isolated by centrifuging the cellular components. But because the mRNA is only a minor component of the total mass of the RNA in this fraction (the overwhelming proportion is rRNA), it cannot be isolated directly by the usual fractionation techniques. Attempts to label the mRNA preferentially by using radioactive nucleotide precursors (the approach used in bacteria) proved unsuccessful, because some contaminating cytoplasmic RNA fractions also are labeled just as rapidly and efficiently. This problem was responsible for the delay in isolating eukaryotic mRNAs after the discovery of bacterial mRNA.

The first technique to distinguish mRNA from other RNAs took advantage of the results of treating polysomes with the chelating agent EDTA (whose main effect in these circumstances is to remove Mg^{2+} ions from ribosomes). As a result, the ribosomes dissociate into individual subunits, releasing the mRNA as a **ribonucleoprotein** (**mRNP**) fraction sedimenting at ~18S. This fraction consists of mRNA associated with proteins. The nontranslated material that was contaminating the polysomes is not disrupted by EDTA and continues to sediment rapidly at ~200S.

In spite of considerable advances since this technique was developed, the release of mRNA by EDTA remains important for two reasons. First, it is a *functional* assay: it identifies mRNA actually in the process of translation by the ribosomes. The presence of an mRNA in this fraction is taken as proof that it is indeed used to direct protein synthesis in the cell from which it was obtained. Second, the assay allows the mRNA to be isolated in what is presumably its natural form, the ribonucleoprotein particle.

The mRNP usually contains only a few proteins. Comparisons between different mammalian cells identify two proteins with more or less constant molecular weights, about 52,000 and 78,000 daltons. Usually only two or three of the other proteins are present in similar amounts; the remainder are minor components. In the best-characterized case, the globin mRNA of red blood cells, the amount of the major proteins is sufficient for one or two molecules of each to be associated with each mRNA molecule.

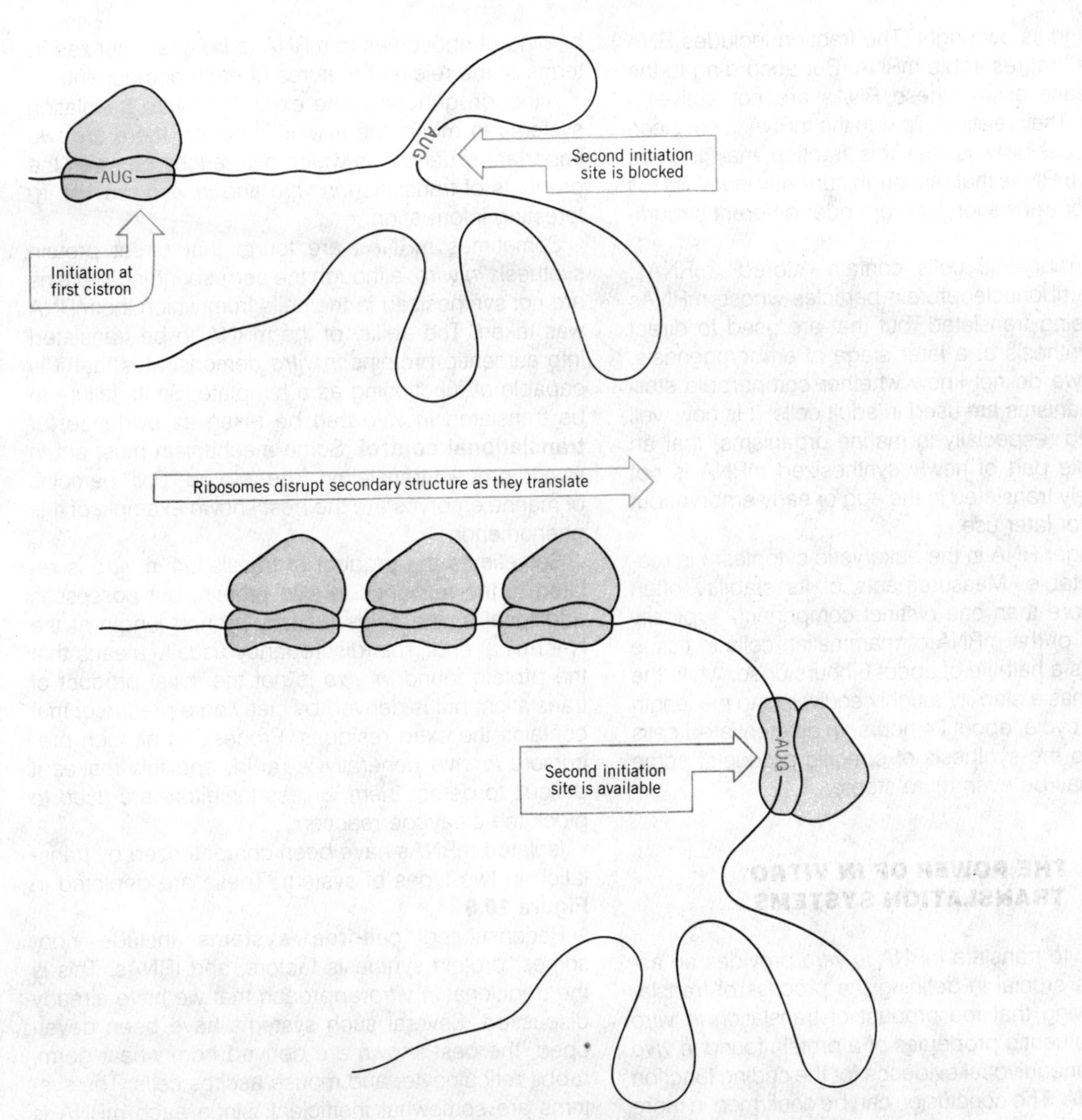

Figure 10.5
Secondary structure can control initiation. Only one initiation site is available in the RNA phage, but translation of the first cistron changes the conformation of the RNA so that other initiation site(s) become available.

The functional significance of these proteins is not known. They could be concerned with transporting the mRNA from nucleus to cytoplasm or possibly with influencing its translation.

The ribonucleoprotein fraction that contaminates the polysomes is "contaminating" only in the sense that it is an unwanted component when mRNA is being isolated. The RNP is presumably a legitimate component of the cell in its own right. The fraction includes RNA molecules that resemble mRNA. But according to the EDTA-release assay, these RNAs are not active in translation. Their relationship with the mRNA is not clear, but one possibility is that this fraction may include legitimate mRNAs that (although currently inactive) will be used for translation later or under different circumstances.

Some embryonic cells contain "stored" mRNAs. These are ribonucleoprotein particles whose mRNAs are not being translated, but that are used to direct protein synthesis at a later stage of embryogenesis. Although we do not know whether comparable storage mechanisms are used in adult cells, it is now well established (especially in marine organisms) that an appreciable part of newly synthesized mRNA is not immediately translated in the egg or early embryo, but is stored for later use.

Messenger RNA in the eukaryotic cytoplasm is reasonably stable. Measurements of its stability often identify more than one distinct component. Typically about half of the mRNA of mammalian cells in tissue culture has a half-life of about 6 hours or so, while the other half has a stability roughly equivalent to the length of the cell cycle, about 24 hours. In differentiated cells devoted to the synthesis of specific products, some mRNAs may be even more stable.

THE POWER OF *IN VITRO* TRANSLATION SYSTEMS

The ability to translate mRNA *in vitro* provides an assay that is crucial in defining the process of translation. Showing that the product of translation *in vitro* has the authentic properties of a protein found *in vivo* provides unequivocal evidence for the coding function of an mRNA. The conclusion can be confirmed in more detail by showing that the sequence of the mRNA includes a coding region corresponding with the sequence of the protein. (Sometimes the nucleotide sequence of the mRNA can be used to predict the amino acid sequence of an unsequenced protein.)

In vitro systems provide the only approach for defining the *mechanism* of either translation or transcription. In both cases, the critical step in expression occurs at initiation. The translation systems allow the binding of ribosomes to mRNA to be characterized in terms of the relevant features of each component.

Although generally one expects *in vitro* translation systems to mimic the *in vivo* process, there are two important situations in which *differences* between the products of translation *in vitro* and *in vivo* provide interesting information.

Sometimes mRNAs are found that direct protein synthesis *in vitro*, although the corresponding proteins are not synthesized in the cells from which the mRNA was taken. The ability of the mRNA to be translated into authentic proteins *in vitro* demonstrates that it is capable of functioning as a template. So its failure to be translated *in vivo* can be taken as evidence for **translational control**. Some mechanism must act *in vivo* to prevent translation. The "stored" RNP particles of marine embryos are the best known example of this phenomenon.

Sometimes the product of translation *in vitro* is related to the authentic *in vivo* protein, but possesses additional amino acids, usually a short length at the N-terminal end. This discrepancy usually means that the protein found *in vivo* is not the initial product of translation, but is derived by cleaving a precursor that contains the extra residues. Processing of such precursors *in vivo* generally is rapid, and this makes it difficult to detect them, unless inhibitors are used to block the cleavage reaction.

Isolated mRNAs have been characterized by translation in two types of system. These are depicted in **Figure 10.6**.

Reconstituted **cell-free systems** include ribosomes, protein synthesis factors, and tRNAs. This is the traditional *in vitro* approach that we have already discussed. Several such systems have been developed; the best known are derived from wheat germ, rabbit reticulocyte, and mouse ascites cells. The systems are somewhat inefficient, since each mRNA is translated a relatively low number of times at a rate much below that found *in vivo*. At best the systems function for 90–120 minutes before stopping. In all cases, there is some background due to translation of endogenous mRNAs that were not removed; its level varies with the system.

An alternative translation system is presented by the intact **Xenopus oocyte**. Injected mRNAs are translated by the natural protein synthetic apparatus. The

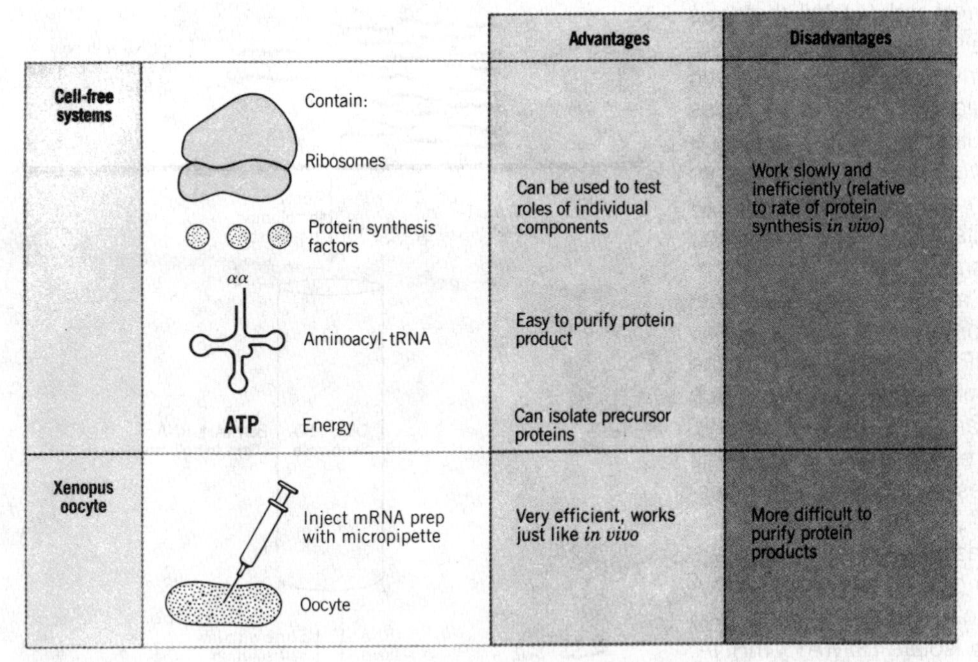

Figure 10.6
Exogenous mRNAs can be translated by cell-free systems or by injection into *Xenopus* oocytes.

system is efficient and treats the injected mRNAs as though they were endogenous mRNAs, so that they are used for repeated rounds of translation. The only limit is that too much mRNA may saturate the translation system (which is present in great excess relative to the demands placed on it by endogenous mRNAs). Generally the system continues to be active for 24–48 hours.

Neither type of system displays any tissue or species specificity. This indicates that mRNAs and the protein-synthetic apparatus of (perhaps) any cytoplasm are interchangeable. There does not seem to be any translational control in the sense of factors or ribosomes that function with one set of mRNAs but not with another. The translational machinery is not preprogrammed, but will accept any mRNA as template. Thus any translational control must take the form of preventing mRNAs from reaching the protein-synthesizing apparatus, most probably by sequestering them in a form in which they are physically unavailable.

The oocyte system is able to process at least some proteins that usually are cleaved during or soon after synthesis. In some cases, it can even direct protein products to enter facsimiles of the appropriate cell compartment. The processing ability implies that signals for processing may be common in different cell types and species; and as a practical consequence, it means that the isolation of precursor proteins must be undertaken in cell-free translation systems.

MOST EUKARYOTIC mRNAs ARE POLYADENYLATED AT THE 3′ END

Most eukaryotic mRNAs have a sequence of polyadenylic acid at the 3′ end. This terminal stretch of A residues is often described as the **poly(A) tail**; and mRNA with this feature is denoted **poly(A)$^+$**.

The poly(A) sequence is not coded in the DNA, but is added to the RNA in the nucleus after transcription. The addition of poly(A) is catalyzed by the enzyme poly(A) polymerase, which adds about 200 A residues to the free 3′-OH end of the mRNA. We do not know how the length of the added stretch is controlled.

When mRNA first enters the cytoplasm, it has approximately the same length of poly(A) tail that was added in the nucleus. The tail gradually is shortened, perhaps in discrete steps involving endonucleolytic cleavage. The cellular population of mRNA includes both "new" and "old" molecules, with relatively longer and shorter stretches of poly(A). However, the length of the poly(A) tail present on any particular molecule does not seem to influence its ability to be translated or its stability in the cytoplasm.

The poly(A) of mammalian mRNA is associated with the common 78,000 dalton protein that is a predominant component of the mRNP. This suggests that the structure of the 3′ end of the mRNA consists of a stretch of poly(A) bound to roughly an equal mass of protein. (This location for the protein explains why at least this component of the mRNP does not interfere with translation.)

The presence of poly(A) has an extremely important practical consequence. The poly(A) region of mRNA can bind by base pairing to oligo(U) or oligo(dT); and this reaction can be used to isolate poly(A)$^+$ mRNA. The most convenient technique is to immobilize the oligo(U or dT) on a solid support material, for example, by chemical linkage to Sepharose. Then when an RNA population is applied to the column, as illustrated in **Figure 10.7**, only the poly(A)$^+$ RNA is retained. It can be retrieved by treating the column with a solution that breaks the bonding to release the RNA.

The only drawback in this procedure is that it does isolate *all* the RNA that contains poly(A). If RNA of the whole cell is used, for example, both nuclear and cytoplasmic poly(A)$^+$ RNA will be retained. If preparations of polysomes are used (a common procedure), most of the isolated poly(A)$^+$ RNA will be active mRNA; but some of the RNA in the contaminating RNP fraction also carries poly(A) and therefore will be included. This makes it necessary to use the EDTA-release assay as a functional test when it is important to isolate precisely the active mRNA population.

The "cloning" approach for purifying mRNA also makes use of the poly(A) tail. First a small oligo(dT) fragment is hybridized to the poly(A) stretch, providing a starting point or "primer" for the enzyme **reverse transcriptase**. The enzyme copies the mRNA to make a complementary DNA strand (known as **cDNA**). Then the cDNA can be used as a template to synthesize a DNA strand that is identical with the original mRNA

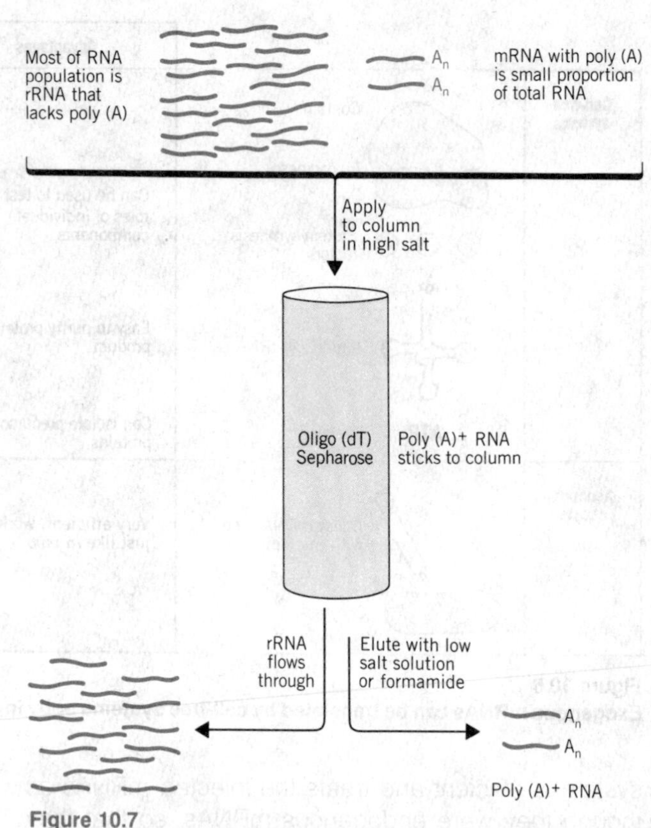

Figure 10.7
Poly(A)$^+$ RNA can be separated from the other RNAs by fractionation on Sepharose-oligo(dT).

sequence. The product of these reactions is a double-stranded DNA corresponding to the sequence of the mRNA. This DNA can readily be cloned by the techniques described in Chapter 17.

The availability of large amounts of any cloned DNA makes it easy to isolate the corresponding mRNA by hybridization techniques. Even mRNAs that are present in only a very few copies per cell can be isolated by this approach. Indeed, only mRNAs that are present in relatively large amounts can be isolated directly without using a cloning step.

When poly(A) was discovered, we thought that all cellular mRNAs might possess it. Although the fraction of mRNA that contains poly(A) always appears to be less than 100%, the discrepancy was attributed to breakage during preparation. A single break in a poly(A)$^+$ mRNA would generate a 5′ end lacking poly(A); this fragment could be mistaken for an authentic mRNA. However, using extremely careful techniques to prepare mRNA in which breakage is minimized, a fraction of the mRNA still lacks poly(A). Typically this constitutes up to one third of the total mRNA.

A predominant component of this **poly(A)$^-$** fraction is provided by the mRNAs that code for the histone proteins of the chromosomes. The remaining (roughly two thirds) part of the poly(A)$^-$ fraction is identical with the poly(A)$^+$ fraction in all respects except for the presence of poly(A). Length, stability, translational efficiency, and nucleocytoplasmic transport all appear the same for poly(A)$^-$ and poly(A)$^+$ mRNA. (The presence of poly(A) does seem to increase the stability of some mRNAs when they are injected into the Xenopus oocyte; but the effect is neither general nor apparently applicable to the normal cell.)

Do the poly(A)$^+$ and poly(A)$^-$ mRNAs code for the same or for different proteins? The histone mRNAs are unique to the poly(A)$^-$ fraction, but the other components of the poly(A)$^-$ fraction show considerable overlap with those of the poly(A)$^+$ fraction. All of these mRNAs may also exist in the polyadenylated form. Thus a particular gene may be represented in transcripts some possessing poly(A) and some not. There may be differences in the proportion of any particular mRNA that is polyadenylated, but at the present we have no idea about the significance of polyadenylation (or its absence).

ALL EUKARYOTIC mRNAs HAVE A METHYLATED CAP AT THE 5′ END

As with other nucleic acids, only the usual four ribonucleotides are incorporated into mRNA during its synthesis. Modifying enzymes are responsible for introducing additional groups at specific locations. Two types of modification event occur at the 5′ end of mRNA found in the eukaryotic cytoplasm (but not in the mitochondrion or chloroplast). Probably these reactions are common to all eukaryotes.

Transcription starts with a nucleoside triphosphate (usually a purine, A or G). The first nucleotide retains its 5′ triphosphate group and makes the usual phosphodiester bond from its 3′ position to the 5′ position of the next nucleotide. The initial sequence of the transcript can be represented as:

$$5' \qquad ppp^{A}_{G}pNpNpNp\ldots$$

But when the mature mRNA is treated *in vitro* with enzymes that should degrade it into individual nucleotides, the 5′ end does not give rise to the expected nucleoside triphosphate. Instead it contains *two* nucleotides, connected by a *5′–5′ triphosphate linkage* and also bearing methyl groups. The terminal base is always a guanine that is added to the original RNA molecule *after transcription*.

Addition of the 5′ terminal G is catalyzed by a nuclear enzyme, guanylyl transferase. The reaction occurs so soon after transcription has started that it is not possible to detect more than trace amounts of the original 5′ triphosphate end in the nuclear RNA. The overall reaction can be represented as a condensation between GTP and the original 5′ triphosphate terminus of the RNA. Thus

$$\overset{5'}{\mathbf{Gppp}} + \overset{5'}{pppApNpNp\ldots}$$
$$\downarrow$$
$$\overset{5'5'}{\mathbf{Gp}ppApNpNp\ldots} + \mathbf{pp} + p$$

The new **G** residue now present at the end of the RNA is in the reverse orientation from all the other nucleotides.

This structure is called the **cap** of mRNA. It is a substrate for several methylations. **Figure 10.8** shows the full structure of a cap after all possible methyl groups have been added.

The first methylation occurs in all eukaryotes and consists of the addition of a methyl group to the 7 position of the terminal guanine. A cap that possesses this single methyl group is known as a **cap 0**. This is as far as the reaction proceeds in unicellular eukaryotes. The enzyme responsible for this modification, guanine-7-methyltransferase, is present in the cytoplasm.

The next step is to add another methyl group, to the 2′-O position of the penultimate base (which was ac-

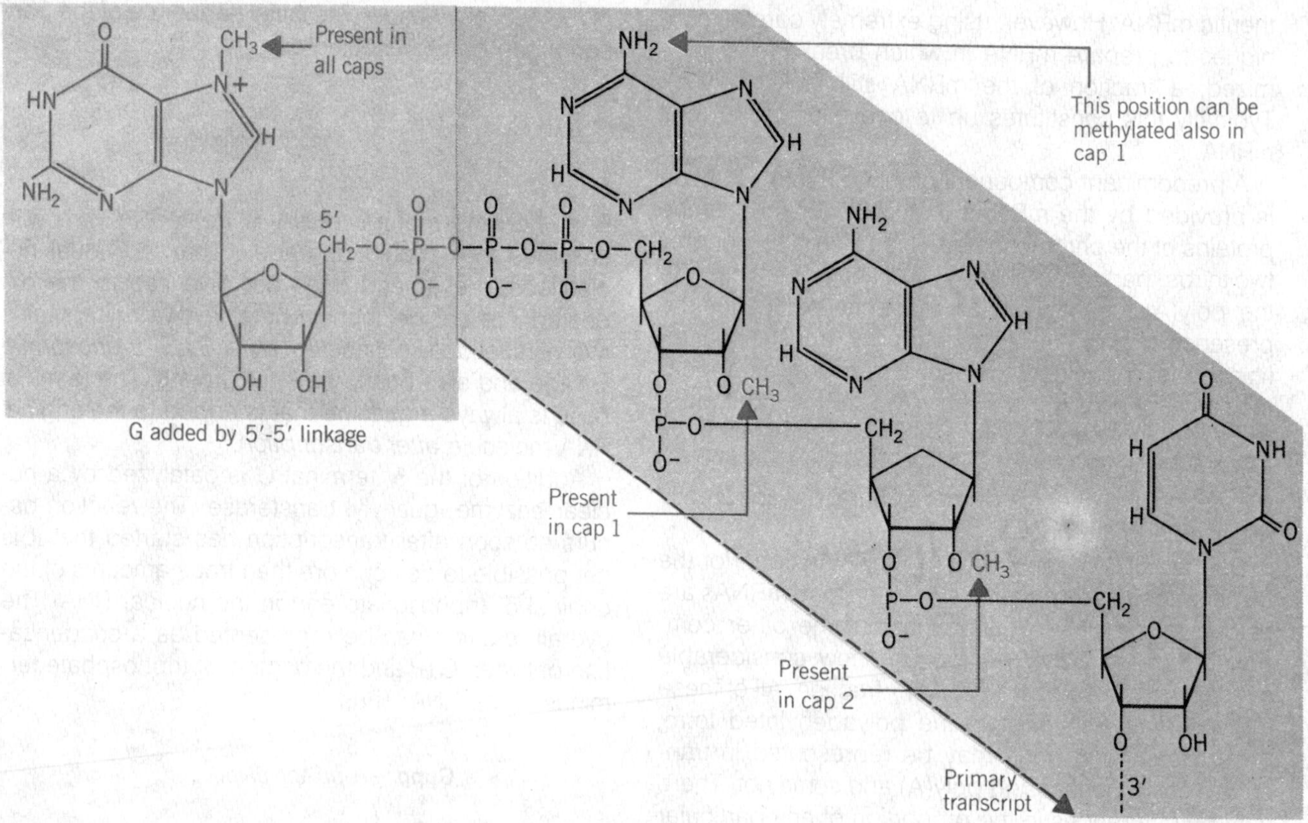

Figure 10.8
The cap blocks the 5′ end of mRNA and may be methylated at several positions.

tually the original first base of the transcript before any modifications were made). This reaction is catalyzed by another enzyme (2′-O-methyl-transferase). A cap with the two methyl groups is called **cap 1**. This is the predominant type of cap in all eukaryotes except the unicellular organisms.

In a small minority of cases in higher eukaryotes, another methyl group is added to the second base. This event happens only when the position is occupied by adenine; the reaction involves addition of a methyl group at the N^6 position. The enzyme responsible acts only on an adenosine substrate that already has the methyl group in the 2′-O position.

In some species, a methyl group may be added to the third base of the capped mRNA. The substrate for this reaction is the cap 1 mRNA that already possesses two methyl groups. The third-base modification is always a 2′-O ribose methylation. This creates the **cap 2** type. This cap usually represents less than 10–15% of the total capped population.

In a population of eukaryotic mRNAs, every molecule is capped. The proportions of the different types of cap are characteristic for a particular organism. We do not know whether the structure of a particular mRNA is invariant or can have more than one type of cap.

In addition to the methylation involved in capping, a low frequency of internal methylation occurs in the mRNA only of higher eukaryotes. This is accomplished by the generation of N^6 methyladenine residues at a frequency of about one modification per thousand bases. There may be 1–2 methyladenines in a typical higher eukaryotic mRNA, although their presence is not obligatory, since some mRNAs (for example, globin) do not have any.

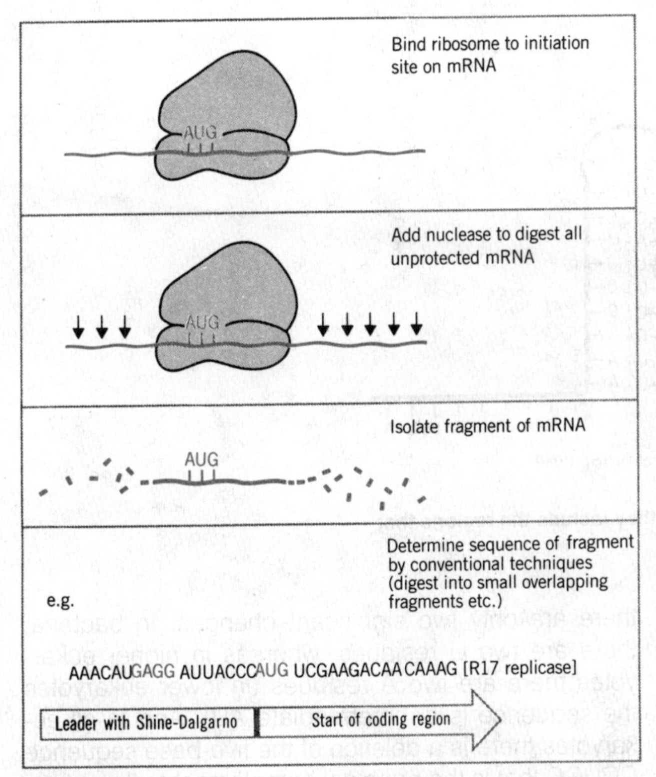

Figure 10.9
Ribosome binding sites on mRNA can be recovered from initiation complexes.

INITIATION MAY INVOLVE BASE PAIRING BETWEEN mRNA AND rRNA

The sites on mRNA where protein synthesis is initiated can be identified by binding ribosomes to the mRNA under conditions that block elongation. Then the ribosomes remain at the initiation site. When ribonuclease is added to the blocked initiation complex, all the regions of mRNA outside the ribosome are degraded, but those actually bound to it are protected, as illustrated in **Figure 10.9**. The protected fragments can be recovered and characterized.

The initiation sequences protected by bacterial ribosomes are 35–40 bases long. Very little homology exists between the ribosome binding sites of different bacterial mRNAs. They display only two common features. First, the AUG (or GUG) initiation codon is always included within the protected sequence. Second, they contain a short sequence that is complementary to part of a sexamer that lies close to the 3′ end of the 16S rRNA.

General evidence for the involvement of the 3′ terminal region of rRNA in initiation is provided by its identification as the target for the antibiotic kasugamycin (an inhibitor of initiation) and its cross-linkage to initiation factors. The 3′ region of 16S rRNA is highly conserved among bacteria. It is self-complementary and could form the base-paired hairpin drawn in **Figure 10.10**. The sequence complementary to mRNA is part of this potential hairpin, as highlighted in the figure. Written in reverse direction, the conserved sexamer is

3′ . . . U C C U C C . . . 5′

All but one of the known *E. coli* mRNA initiation sites include a sequence complementary to at least a trinucleotide part of this sexamer, and more usually to 4–5 bases. Thus bacterial mRNA contains part or all of the oligonucleotide

5′ . . . A G G A G G . . . 3′

This polypurine stretch is known as the **Shine-Dalgarno** sequence. It lies 4–7 bases before the AUG codon. Does the Shine-Dalgarno sequence pair with its complement in rRNA during mRNA-ribosome binding?

A principal reason for believing that the Shine-Dalgarno sequence is involved in initiation is its ubiquity, but there is direct evidence in one case. Ribosome binding to the mRNA of gene *0.3* of phage T7 is destabilized by a mutation that changes the sequence from GAGG to GAAG. We do not know whether the AUG codon itself is recognized by the small subunit in the binding reaction.

The complementary sequence in the rRNA cannot simultaneously be part of the intramolecular hairpin and bind to mRNA. These partnerships could be alternatives, in which case the initiation reaction may involve disrupting the terminal hairpin to allow base pairing between mRNA and rRNA. After initiation, the mRNA-rRNA duplex might be broken by reconstitution of the rRNA base-paired hairpin. This mechanism could reconcile the need to form a stable initiation complex

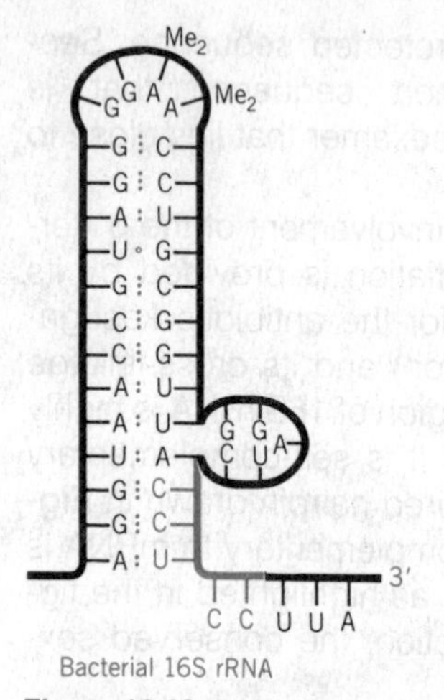

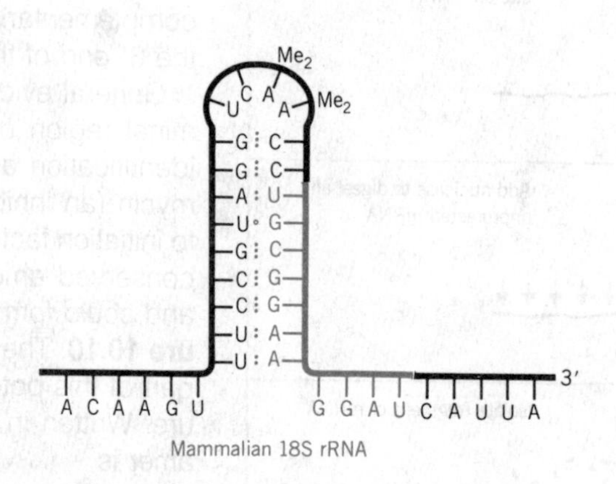

Figure 10.10
Hairpins can be formed by base pairing near the 3′ ends of rRNA; they include the regions that might be involved in recognizing mRNA (indicated in black).

at a particular site with the need later to move off along the mRNA.

When eukaryotic 40S subunits bind to mRNA, they protect a region of up to 60 bases; when the 60S subunits join the complex, the protected region contracts to about the same length of 30–40 bases seen in prokaryotes. Again the protected region surrounds the AUG codon. The reduction in the length of the region protected by 80S ribosomes could be caused by a conformational change triggered in the small subunit by association with the large subunit, or it could represent the loss of initiation factors that directly protect some additional region of the mRNA.

The sequence at the 3′ end of rRNA is well conserved between prokaryotes and eukaryotes, as shown in **Figure 10.11**. In the 20 nucleotides between the adjacent double-methylated adenines and the 3′ end, there are only two significant changes. In bacteria, there are two U residues, whereas in higher eukaryotes there are two A residues (in lower eukaryotes the sequence is an intermediate AU). And in all eukaryotes there is a deletion of the five-base sequence CCUCC that is the principal complement to the Shine-Dalgarno sequence. (So if binding to the Shine-Dalgarno sequence is essential for initiation, eukaryotic ribosomes should be unable to initiate translation of bacterial mRNAs.)

Within the highly conserved 3′ terminal region (and present in bacterial as well as in eukaryotic rRNA), there is a purine-rich sequence,

3′. . U A G G A A G G C G U . . .5′

that in eukaryotes is about the same distance from the 3′ end as the CCUCCU sequence is in bacteria. Again,

Figure 10.11
The 3′ ends of bacterial and mammalian small rRNAs are well conserved. Differences are indicated in boxes; regions that may be involved in recognizing mRNA are shown in red.

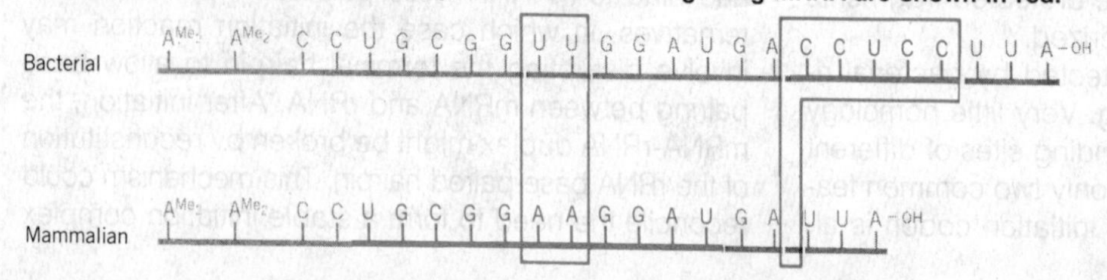

it could participate in hairpin formation, as shown previously in Figure 10.10. Some eukaryotic mRNAs have a tetra- or pentanucleotide sequence able to pair with the purine-rich sequence (if the rRNA hairpin were disrupted). However, the distance of this potential site from the AUG codon is variable; often it lies outside the expected span of the ribosome binding site. We do not know whether this sequence is involved in initiation, but if it is, its use cannot be obligatory, since some eukaryotic mRNAs entirely lack it.

In one case, the general mechanism for initiation cannot involve pairing between mRNA and the 3′ terminal region of rRNA. In mammalian mitochondrial 12S–13S rRNA, the 3′ terminal region does resemble that found in eukaryotic cytoplasm and bacteria, but is less well conserved. It remains able to form a terminal hairpin, but no sequence complementarity with any region near the AUG codons on mRNA has been noticed.

SMALL SUBUNITS MAY MIGRATE TO INITIATION SITES ON EUKARYOTIC mRNA

Virtually all eukaryotic mRNAs are monocistronic, but each mRNA is usually substantially longer than is necessary just to code for its protein. The coding region therefore occupies only a part of the messenger. The average mRNA in eukaryotic cytoplasm is between 1000 and 2000 bases long, has a methylated cap at the 5′ terminus, and carries 100–200 bases of poly(A) at the 3′ terminus. The nontranslated 5′ leader is relatively short, usually (but not always) less than 100 bases. The length of the coding region is determined by the size of the protein. The nontranslated 3′ trailer is often rather long, sometimes 1000 bases or so. By virtue of its location, the leader is not ignored during initiation, but we do not know of any function for the trailer in translation.

The ribosomes of eukaryotic cytoplasm do not bind directly to the initiation site at the start of the coding region. Instead, the first feature to be recognized is the methylated cap that marks the 5′ end. Messengers that lack caps (because synthesis of the cap has been prevented or because it has been removed enzymatically from the mRNA) are not translated efficiently in the *in vitro* systems. Ribosome binding at the cap requires additional factors called **cap binding proteins**, whose current status is not quite clear, but which seem to include several polypeptides. These constitute an additional group of initiation factors.

With the sole exception of a few viral mRNAs (such as poliovirus) that are not capped, the modification is made to the 5′ end of *all* mRNAs in eukaryotic cytoplasm (although not in organelles). Only the exceptional viral mRNAs can be translated *in vitro* without caps. Thus for most mRNAs, the cap is obligatory for translation; the viral mRNAs must have some alternative feature that renders it unnecessary.

Some viruses take advantage of this difference. Poliovirus infection inhibits the translation of host mRNAs. Apparently this is accomplished by interfering with the cap binding proteins that are needed for initiation of cellular mRNAs, but that are superfluous for the noncapped poliovirus mRNA.

Which features of the cap are necessary for its binding by 40S subunits? Introduction of the first methyl group onto the 7 position of the terminal G is essential. The subsequent methylations (to progress from cap 0 to cap 2) probably improve the efficiency of ribosome binding, but are not of primary importance. The 40S subunits can bind to capped poly(U) or other synthetic polynucleotides, which suggests that the cap per se (not the following sequence) is the main recognition feature; but 60S subunits cannot associate with the initiation complex without some sequence information from the mRNA itself.

Sometimes the AUG initiation codon lies within 40 bases of the 5′ terminus of the mRNA, so that both the cap and AUG lie within the span of ribosome binding. But many mRNAs are known in which the cap and AUG are farther apart, even 200–300 bases distant. Yet the presence of the cap still is necessary for a stable complex to be formed at the initiation codon. How can the ribosome rely on two sites so far apart?

A current model supposes that the 40S subunit initially recognizes the 5′ cap and then "migrates" along the mRNA until it encounters the AUG initiation codon (see Figure 7.7). Usually, although not always, this will be the first AUG triplet sequence it meets. The AUG triplet by itself does not seem sufficient to halt migration; probably it is recognized as an initiation codon only when it is in the right context. However, the additional information that is needed may consist of merely two or three specific additional bases near by.

Binding is stabilized at the initiation site. When the 40S subunit is joined by a 60S subunit, the intact ribosome is located at the site identified by the protection assay. When the leader sequence is long, a second 40S subunit could recognize the 5′ end before the first has left the initiation site, so there could be a queue of subunits proceeding along the leader to the initiation site.

The idea that ribosomes must start at the 5′ end is consistent with the monocistronic nature of eukaryotic mRNAs. Some exceptional viral mRNAs contain more than one coding region. But in these cases, only the coding region nearest the 5′ end is translated from the intact mRNA. The other(s) can be read only after a cleavage has occurred in the mRNA to generate a new 5′ end in the vicinity of the initiation codon. This behavior supports the idea that internal initiation sites cannot be recognized directly.

PROTEIN SYNTHESIS IS LINKED TO CELLULAR LOCATION

The eukaryotic cell is a well-ordered structure, whose functions are exercised at specific locations. This general view applies in particular to protein synthesis.

Figure 10.12
Polyribosomes are associated with the fibrous network of the cytoskeleton.
Photograph kindly provided by Sheldon Penman.

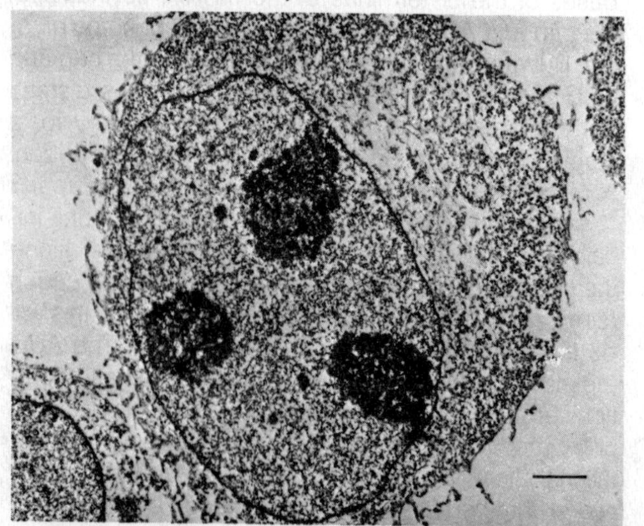

Polyribosomes can be divided into two classes: "free" and membrane-bound. They are engaged in the synthesis of different groups of proteins. The "free" class of polysomes is responsible for synthesizing proteins that do not interact with membranes. The membrane-bound polysomes synthesize proteins whose specific locations within the cell depend on their ability to bind to or pass through a membrane.

The "free" polysomes are somewhat misnamed, since in fact they are not able to diffuse freely through the cytoplasm, but are associated with the **cytoskeleton**. When cells are extracted with a nonionic detergent (Triton) in hypertonic buffer, most of the lipids and soluble proteins are removed, leaving a "cytoskeletal framework" anchored on a remnant nucleus. This framework is a complex network of fibers. The entire class of "free" polysomes is associated with the cytoskeleton, as seen visibly in the electron micrograph of **Figure 10.12**, or as detected by biochemical fractionation.

The polysomes tend to be clustered near the nucleus, where the mRNA enters the cytoplasm. Most of the products of translation are soluble proteins that rapidly diffuse away from the site of synthesis when they are released from the polysomes. Proteins that are components of the cytoskeleton tend to be incorporated into it at points close to their sites of synthesis.

In contrast with the polysomes, monomeric ribosomes are genuinely free in the cell. Association with the cytoskeleton may be a function of the mRNA. Translation of the mRNA may be connected with its association with the cytoskeleton, as seen the most clearly by the events involved in viral infection. In the example of VSV (vesicular stomatitis virus) infection, the viral mRNAs associate with the cytoskeleton shortly after their synthesis; they are translated only during this period. Then they are released from the cytoskeleton.

The existence of *in vitro* systems for translating mRNA shows that the act of protein synthesis can be disengaged from extraneous cellular structures. However, this does not diminish the importance of the observation that the natural course of events is for mRNA to be associated with the cytoskeleton during translation. This may be important in the metabolism of mRNA *in vivo* as well as being necessary for the proper location of newly synthesized cytoskeletal proteins.

Proteins synthesized by membrane-bound polysomes have several destinations. Some are sequestered into compartments, such as the mitochondrion or lysosome. Some are membrane components, of which the most attention has been paid to those residing in the plasma membrane (which circumscribes the cytoplasm). Other proteins may be **secreted** from the cell into the surrounding milieu. How do these proteins find their ultimate sites?

For many proteins that must be inserted in membranes, the sequence of the mature polypeptide is not itself sufficient to direct membrane insertion. Additional information is needed; this most often takes the form of a **leader sequence** at the N-terminal end of the protein. The protein carrying this leader is called a **preprotein**. It is a transient precursor to the mature protein, since the leader is cleaved as part of the process of membrane insertion.

The **pre** sequence is distinct from the **pro** sequence that describes the additional regions present on proteins that exist as *stable* precursors. Some proteins may have both. For example, insulin is initially synthesized as **preproinsulin**; the **pre** sequence is cleaved during secretion, generating **proinsulin**, which is the substrate for processing to mature insulin.

The role of the leader sequence is tailored to fit the circumstances. For certain proteins synthesized within the cytoplasm, but destined to reside within the chloroplast or mitochondrion, the process is **post-translational**. The product of cytoplasmic protein synthesis is a precursor roughly 45 amino acids longer than the mature protein. The completed precursor is released from polysomes. If it is added to intact organelles *in vitro*, it can be incorporated into the compartment.

As illustrated in **Figure 10.13**, the leader sequence leads the precursor through the organelle membrane. During this passage, the leader sequence is cleaved, probably by a protease located on the outside of the envelope. The function of the leader is to be recognized by the organelle membrane, triggering the passage and cleavage reaction. (A single passage is involved for proteins that reside between the outer and inner mitochondrial membranes; two passages are involved for proteins that live within the inner compartment.) Note that a cleavable leader is not the only acceptable form of such information; some mitochondrial proteins are recognized as such in their mature

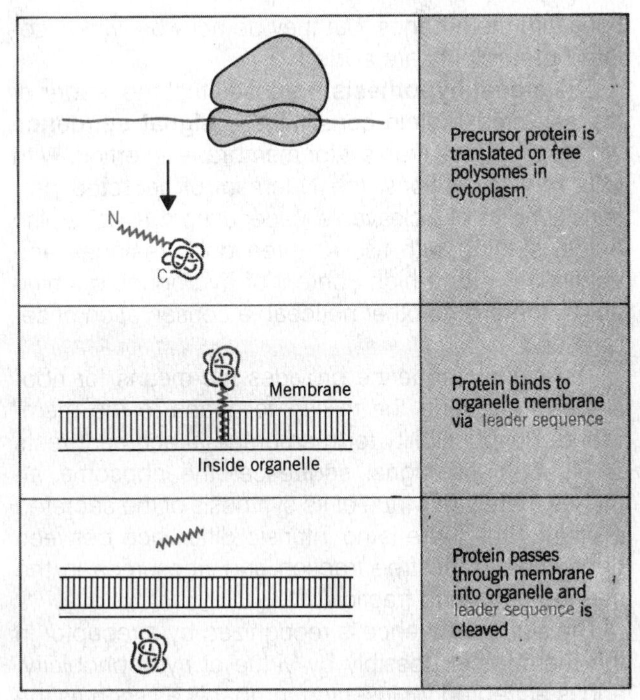

Figure 10.13
Leader sequences allow proteins to recognize mitochondrial or chloroplast surfaces.

form, and may have an internal sequence that can sponsor membrane passage without cleavage.

For proteins that are secreted through, or inserted into, other cellular membranes, the process of association most often starts during translation. The polysomes synthesizing these proteins are associated with the membranes of the endoplasmic reticulum. The preproteins are not released into the cytoplasm to form a precursor pool, but instead pass directly from the ribosome to the membrane. From the membrane, the proteins enter the Golgi apparatus, and then are directed to their ultimate destination, such as the lysosome or plasma membrane. The mechanism by which proteins are synthesized in direct juxtaposition with the membrane is called **cotranslational transfer**. It applies to a variety of secreted proteins, including immunoglobulins and many hormones.

A model for the mechanism of membrane insertion has been based on work with eukaryotic microsomal systems (which contain ribosomes and endoplasmic reticulum). These systems can package *nascent* proteins into membranes; but they do not work when isolated preproteins are added.

The **signal hypothesis** proposes that the leader of the secreted protein constitutes a **signal sequence** whose presence marks it for membrane insertion. With only rare exceptions, the N-termini of secreted proteins consists of a cleavable leader of from 16–29 amino acids, starting with two or three polar residues and continuing with a high content of hydrophobic amino acids; there is no other noticeable conservation of sequence.

The signal sequence provides the means for ribosomes translating the mRNA to attach to the membrane. Responsibility for membrane attachment rests solely with the signal sequence; the ribosome attaches merely by virtue of its synthesis of the secreted protein. Thus there is no intrinsic difference between ribosomes in the free fraction and ribosomes in the membrane-bound fraction.

The signal sequence is recognized by a receptor in the membrane, possibly by virtue of hydrophobicity, and is accepted for insertion, probably as soon as the signal sequence and a few additional amino acids have been synthesized. **Figure 10.14** shows that, as synthesis of the protein chain continues, a time comes when it is well inserted into the membrane, and the signal sequence can be cleaved. When the ribosomes complete translation, the protein is already well on its way through the membrane. The critical feature of this mechanism is that a protein can be inserted into the membrane *only* as it is synthesized.

A route to characterizing the receptor has been opened by the discovery that salt-washed membranes cannot sponsor ribosomal attachment; but this ability can be recovered by adding back the salt wash. The active component is called the **signal recognition particle (SRP)**. It consists of a complex of 6 proteins with a small (305 base, 7S) RNA. On the one hand it can bind to the N-terminal region of nascent secretory proteins; on the other it can bind to a receptor in the membrane. Its activity can be reconstituted *in vitro* from the individual components, so we may expect soon to see a detailed definition of its function.

How does a secreted protein pass through the membrane? The original idea was that the N-terminus would lead the way, as illustrated in Figure 10.14. However, another possibility is that the N-terminus enters, but does not necessarily pass through; in fact, the signal sequence could be anchored at the SRP, with the rest of the protein looping into the membrane. At all events, the signal sequence can exercise its function only during translation, and not as part of a mature preprotein.

How are proteins that are secreted *through* the membrane distinguished from those that reside *within* it? The mechanism of insertion appears to be the same, via a signal sequence, but proteins that are not to pass through the membrane possess a second, internal **halt-transfer** signal. This may take the form of a cluster of hydrophobic amino acids adjacent to some ionic residues. The cluster serves as a "hook" that latches onto the membrane and stops the protein from passing right through.

Some proteins do not have leader sequences, but manage to be inserted into membranes anyway. Probably they possess some other sequence that promotes cotranslational passage, although in this case the mechanism cannot involve cleavage.

Although originally conceived for eukaryotic cells, the signal hypothesis applies to bacteria. Proteins that are to be exported may have N-terminal leader sequences, with a hydrophilic N-terminus and an adjacent hydrophobic core. Exportable proteins may be synthesized on membrane-bound ribosomes. The development of an *in vitro* system for transporting proteins through bacterial membranes suggests that the cytoplasmic face of the inner membrane contains protein(s) that are needed for the transport process, and which could recognize the signal sequence.

Mutations in N-terminal leaders may prevent secretion; they may be suppressed by mutations in other genes, which are thus defined as components of the protein export apparatus. As with eukaryotes, secondary signals may be necessary for proper location after the protein has entered the membrane. For example, the C-terminal region of the *E. coli* β-lactamase is necessary for the protein to leave the membrane to enter the periplasmic space on the other side.

Some bacterial proteins are secreted by posttranslational mechanisms. The best characterized example is the coat protein of phage M13, which is synthesized in the form of a procoat that can be inserted in the membrane. The procoat contains a leader sequence that is cleaved by an enzyme called **leader peptidase** that recognizes precursor forms of several exported proteins.

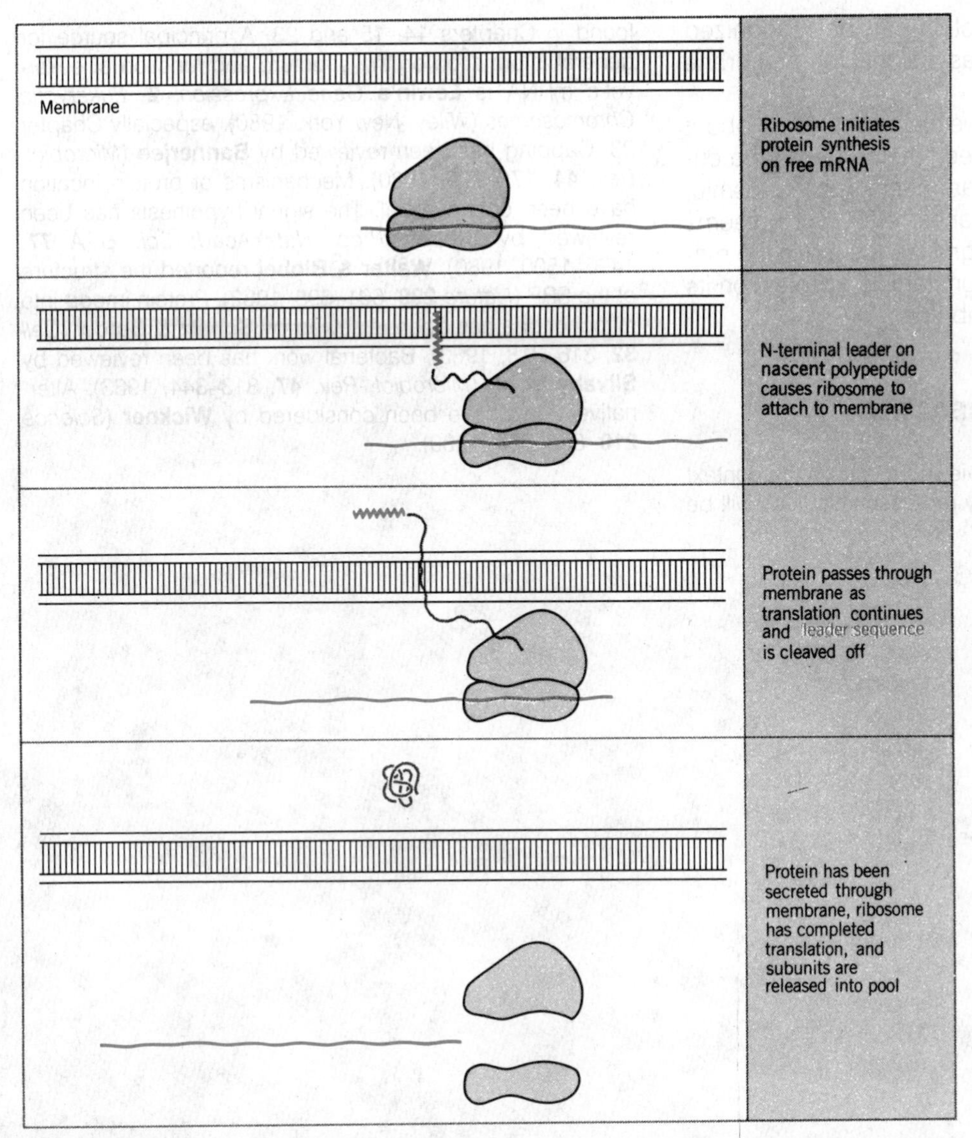

Figure 10.14
The signal hypothesis proposes that ribosomes synthesizing secretory proteins are attached to the membrane via the leader sequence on the nascent polypeptide.

A possible role for posttranslational cleavage has been summarized in the **membrane trigger** hypothesis. The model argues that the role of the leader sequence is to alter the pathway for protein folding. With the leader, the protein exists in a water-soluble conformation that is appropriate for the cytoplasm in which it is synthesized. Loss of the leader converts it into a water-insoluble conformation that is appropriate for membrane insertion.

In at least some cases, the information used to establish the location of a protein is neither species- nor tissue-specific. For example, *X. laevis* oocytes translate injected rabbit globin mRNA to produce free globin protein, but translate rat albumin mRNA into a protein product that is incorporated into membrane vesicles. The signal on mammalian preproinsulin may be adequate to ensure its secretion when its mRNA is translated in the bacterium *E. coli*. The leader sequence on M13 procoat protein can be recognized by mammalian microsomes as a signal for membrane insertion.

There is a great deal that we have yet to learn about the basis for selectivity between membranes (in a eukaryotic cell, after all, there are nuclear, cytoplasmic, Golgi, mitochondrial membranes, etc.), but it seems probable that the use of a signal sequence is the major mechanism responsible for starting a protein on its route to or through any membrane.

FURTHER READING

Bacterial mRNAs have been reviewed mostly in the context of the individual genes that they represent; citations will be found in Chapters 14, 15 and 23. A principal source for general information on the structure and function of eukaryotic mRNA is **Lewin's** *Gene Expression*, **2**, *Eukaryotic Chromosomes* (Wiley, New York, 1980), especially Chapter 23. Capping has been reviewed by **Bannerjee** (*Microbiol. Rev.* **44**, 175–205, 1980). Mechanisms of protein location have been controversial. The signal hypothesis has been reviewed by **Blobel** (*Proc. Nat. Acad. Sci. USA* **77**, 1496–1500, 1980); **Walter & Blobel** reported the structure of the SRP (*Nature* **299**, 691–698, 1982). Protein import into mitochondria has been reviewed by **Schatz & Butow** (*Cell* **32**, 316–318, 1983). Bacterial work has been reviewed by **Silvahy et al**. (*Microbiol. Rev.* **47**, 313–344, 1983). Alternative views have been considered by **Wickner** (*Science* **210**, 861–868, 1980).

PART 3
PRODUCING THE TEMPLATE

In calling the structure of the chromosome fibers a code-script we mean that the all-penetrating mind could tell from their structure whether the egg would develop, under suitable conditions, into a black cock or into a speckled hen, into a fly or a maize plant, a beetle, a mouse or a woman... But the term code-script is, of course, too narrow. The chromosome structures are at the same time instrumental in bringing about the development they foreshadow. They are law-code and executive power—or, to use another simile, they are architect's plan and builder's craft—in one.

Erwin Schrödinger, 1945

CHAPTER 11
RNA POLYMERASES: THE BASIC TRANSCRIPTION APPARATUS

How RNA came to be the intermediary between DNA and protein is unknown. Perhaps the first primitive cells made no distinction between types of nucleic acid, so that what passed for the genome was involved directly in both replication and translation. At some point, it must have become advantageous to separate translation from the genome, so that proteins were synthesized on messengers distinct from the genetic material itself. It is impossible to say how this related in time to the development of the other ribonucleic acid components involved in translation, but it is striking that RNA is present in the ribosome as well as constituting the tRNA adaptor. (It would not be surprising if the role of RNA in the ribosome was more prominent in the past than is apparent today.) Perhaps RNA was the original nucleic acid, and its ubiquitous presence today is but a recollection of its former activity.

All this speaks to the fact that RNA plays central roles in gene expression, not merely in constituting the messenger, but also in providing the means for its translation into protein. In a sense, these roles represent the various specific interests of an RNA conglomerate generally concerned with gene expression, but with several different and apparently independent functions. The production of each type of RNA has a common origin: transcription of DNA. In the case of mRNA, the product is an intermediate whose function requires translation. In the case of tRNA and rRNA, the transcriptional product itself fulfills the final function.

Transcription is the principal stage at which gene expression is controlled. The first (and sometimes the only) step in control is the decision on whether or not to transcribe the gene. In considering the various stages of transcription, we should therefore keep in mind the opportunities that they offer for regulating gene activity.

(Transcription is not the only means by which RNA can be synthesized. Viruses whose genomes are RNA specify enzymes able to synthesize RNA on a template itself consisting of RNA. Such reactions both produce mRNAs coding for proteins needed in the infective cycle (RNA transcription) and provide genomic RNAs to perpetuate the infective cycle (RNA replication). Yet a further reaction is possible with the retroviruses, in which viral RNA serves as a template for reverse transcription to produce a DNA complement.)

TRANSCRIPTION IS CATALYZED BY RNA POLYMERASE

Transcription involves synthesis of an RNA chain representing one strand of a DNA duplex. By "representing" we mean that the RNA is identical in sequence with one strand of the DNA; it is complementary to the other strand, which provides the template for its synthesis.

Transcription takes place by the usual process of complementary base pairing, catalyzed by the enzyme **RNA polymerase**. The stages of the reaction are illustrated in **Figure 11.1.**

First, RNA polymerase must bind to the double-stranded helix. To make the template strand available for base pairing with ribonucleotides, the strands of DNA must be separated. The unwinding is a local event. It begins at the site to which RNA polymerase has bound. The **initiation** stage is usually regarded as including the initial recognition of duplex DNA, unwinding to generate a short single-stranded region, and incorporation of the first nucleotide of the RNA chain. The entire sequence of DNA that is necessary for these reactions is called the **promoter**. The site at which the first nucleotide is incorporated is called the **startsite** or **startpoint**.

Elongation starts when a few bases are incorporated into an RNA chain, forming an RNA-DNA hybrid. To continue synthesis, the enzyme moves along the DNA, unwinding the double helix to expose a new segment of the template in single-stranded condition. As it moves, the RNA that was made previously is displaced from the DNA template strand, which pairs with its original partner to reform the double helix. Thus elongation involves the movement along DNA of a short segment that is transiently unwound, existing as a hybrid RNA-DNA duplex and a displaced single strand of DNA. In principle, any sequence of DNA that does not contain particular signals for initiation or termination can be transcribed in this manner.

Termination involves the ability to recognize the point at which no further bases should be added to the chain. To terminate transcription, the formation of phosphodiester bonds must cease, and the transcription complex must come apart. Thus when the last base is added to the RNA chain, the RNA-DNA hybrid is disrupted, the DNA reforms in duplex state, and the enzyme and RNA are both released from it. The sequence of DNA required for these reactions is called the **terminator**.

Originally defined simply by its ability to incorporate nucleotides into RNA under the direction of a DNA template, the enzyme RNA polymerase now is seen as part of a more complex apparatus involved in transcription. The ability to catalyze RNA synthesis defines the *minimum* component that can be described as RNA polymerase. It supervises the base pairing of the substrate ribonucleotides with DNA and catalyzes the formation of phosphodiester bonds between them.

But ancillary activities may be needed both to initiate and to terminate the synthesis of RNA, when the enzyme must associate with, or dissociate from, a specific site on DNA. The analogy with the division of labors between the ribosome and the protein synthesis factors is obvious. Sometimes it is difficult to decide whether a particular protein that is involved in transcription at one of these stages should be considered a part of the "RNA polymerase" or an ancillary factor.

Initiation requires more functions than elongation. All the components involved in elongation are necessary for initiation, but additional polypeptides also are involved. Whether they are recognized as subunits of RNA polymerase tends to rest on the generality of their involvement. If they are needed to recognize all promoters, they are likely to be classified as part of the enzyme. If they are needed only for the transcription of particular genes, they are likely to be classified as ancillary control factors.

Termination also involves additional activities that are not needed for elongation. In bacteria, there are different classes of terminator sites. At least some of them are not recognized by the form of the enzyme that is involved in elongation. Again, there is variation in the degree to which the necessary additional components are regarded as part of the enzyme. In eukaryotes, the issue of termination has been cloudy, but now it is clear that in at least certain cases additional factors are necessary.

With bacterial enzymes, it is possible to begin to define the roles of individual polypeptides in the stages of transcription. With eukaryotes, the enzymes are less well purified, and the actual enzymatic activities have yet to be resolved from the crude preparations. Ironically enough, in eukaryotes we have begun to isolate ancillary factors needed to initiate or terminate particular genes, while the basic polymerase preparation itself remains rather poorly characterized.

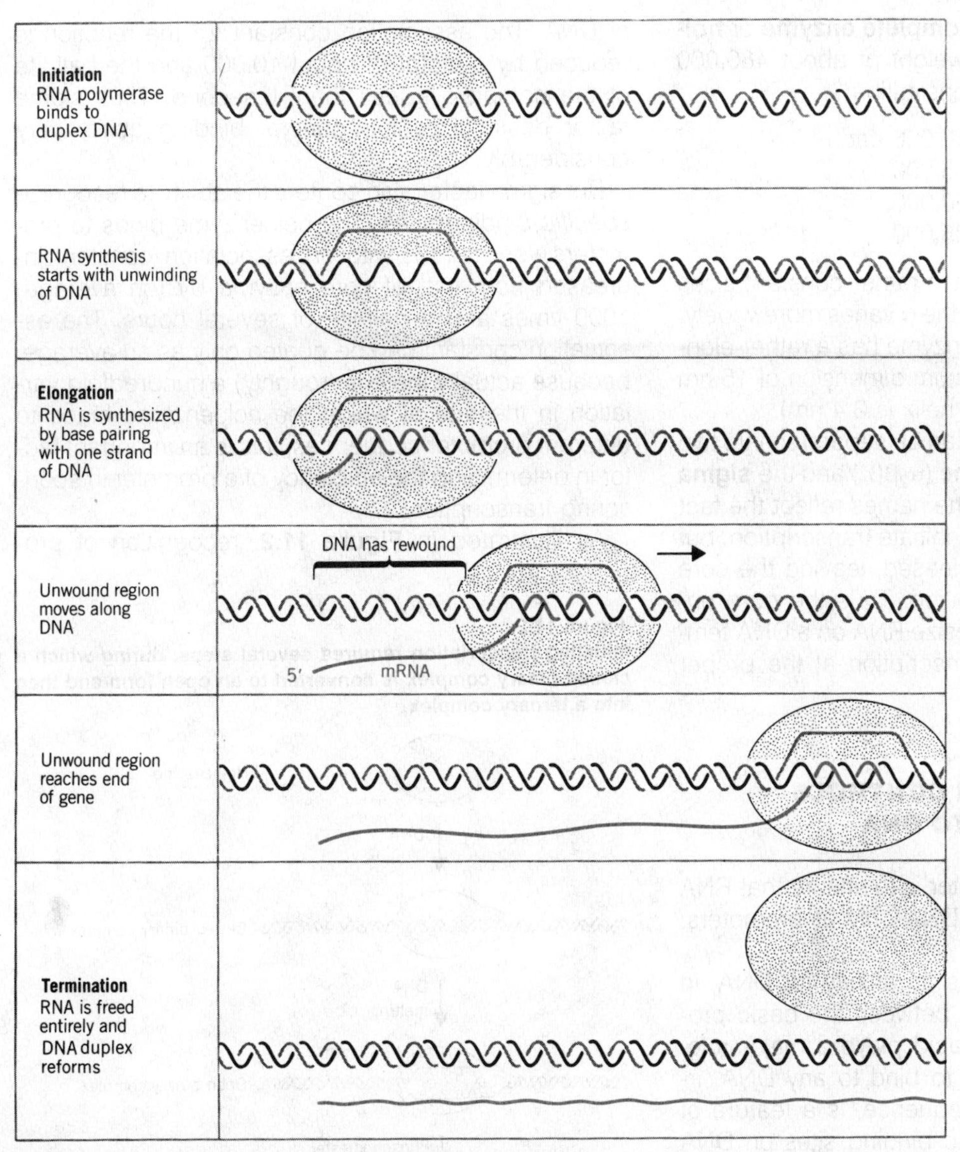

Figure 11.1
RNA is synthesized by base pairing with one strand of DNA in a region that is transiently unwound. As the region of unwinding moves, the DNA duplex reforms behind it, displacing the RNA in the form of a single polynucleotide chain.

A single type of RNA polymerase is responsible for all synthesis of mRNA, rRNA and tRNA in bacteria. The total number of RNA polymerase molecules present in an *E. coli* cell is around 7000. Many of them are actually engaged in transcription; probably between 2000 and 5000 enzymes are synthesizing RNA at any one time, the number depending on the growth conditions.

The RNA polymerase that has been characterized best is that of *E. coli*, but its structure is similar in all other bacteria studied. The **complete enzyme** or **holoenzyme** has a molecular weight of about 480,000 daltons. It has the subunit constitution

2	α	40,000 each
1	β	155,000
1	β'	160,000
1	σ	85,000

The α, β, and β' subunits have rather constant sizes in different bacterial species; the σ varies more widely, from 44,000 to 92,000. The enzyme has a rather elongated structure, with a maximum dimension of 15 nm (one turn of the DNA double helix is 3.4 nm).

The holoenzyme ($\alpha_2\beta\beta'\sigma$) can be separated into two components, the **core enzyme** ($\alpha_2\beta\beta'$) and the **sigma factor** (the σ polypeptide). The names reflect the fact that only the holoenzyme can initiate transcription; but then the sigma "factor" is released, leaving the core enzyme to undertake elongation. Thus the core enzyme has the ability to synthesize RNA on a DNA template, but cannot initiate transcription at the proper sites.

SIGMA FACTOR CONTROLS BINDING TO DNA

The function of the sigma factor is to ensure that RNA polymerase binds stably to DNA *only* at promoters, not at other sites.

The core enzyme itself has an affinity for DNA, in which electrostatic attraction between the basic protein and the acidic nucleic acid plays a major role. Probably this general ability to bind to any DNA, irrespective of its particular sequence, is a feature of all proteins that have specific binding sites on DNA (see Chapter 14). Any sequence of DNA that is bound by RNA polymerase in this reaction is described as a **loose binding site**; the enzyme·DNA complex is described as **closed**, because the DNA remains strictly in the double-stranded form. A closed complex is pretty stable; the half-life for dissociation of the enzyme from DNA is about 60 minutes.

Sigma factor introduces a major change in the affinity of RNA polymerase for DNA. The holoenzyme has a drastically *reduced* ability to recognize loose binding sites—that is, to bind to any general sequence of DNA. The association constant for the reaction is reduced by a factor of about 10,000 and the half-life of the complex is less than 1 second. Thus sigma factor destabilizes the general binding ability very considerably.

But sigma factor also confers the ability to recognize *specific* binding sites. The holoenzyme binds to promoters very tightly, with an association constant increased from that of core enzyme by (on average) 1000 times and a half-life of several hours. The association constant can be quoted only as an average, because actually there is (roughly) a hundredfold variation in the rate at which the holoenzyme binds to different promoter sequences; this is an important factor in determining the efficiency of a promoter in sponsoring transcription.

As illustrated in **Figure 11.2**, recognition of pro-

Figure 11.2
Initiating transcription requires several steps, during which a closed binary complex is converted to an open form and then into a ternary complex.

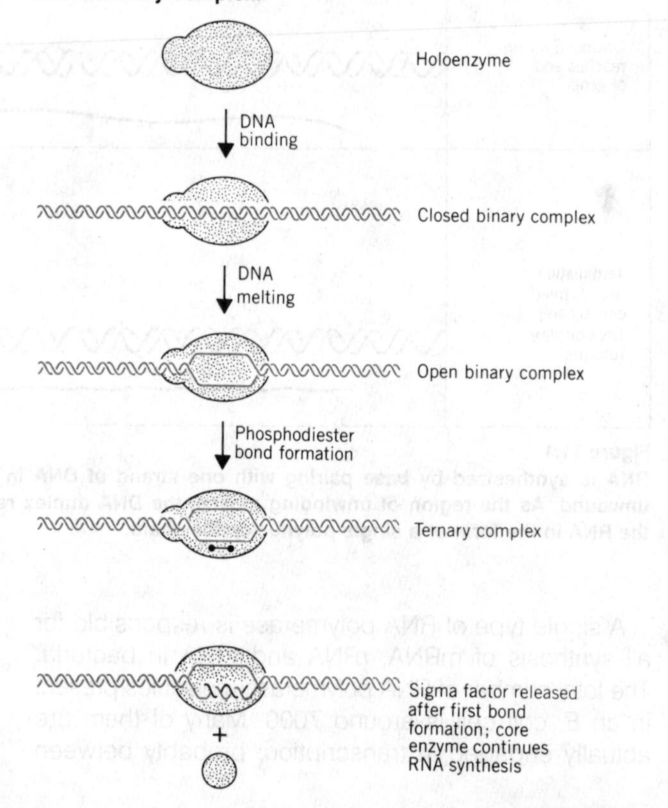

moters by holoenzyme is very different from the reaction of core enzyme with loose binding sites. The holoenzyme·promoter reaction starts in the same way by forming a closed complex. But then this is converted into an **open complex** by the "melting" of a short region of DNA within the sequence bound by the enzyme. The series of events leading to formation of an open complex is called **tight binding**.

Consistent with the need to disrupt the structure of DNA, the RNA polymerase forms open complexes and transcribes negatively supercoiled DNA more readily than linear DNA. The stress placed on a double helix by supercoiling makes it easier to unwind the two strands.

RNA polymerase may find promoters on DNA by the process of trial and error illustrated in **Figure 11.3**.

If any core enzyme is free in the cell, it would exist largely in the form of closed loose complexes, because the enzyme would enter into them rapidly and would leave them slowly. (We don't know for sure how much of the free RNA polymerase in the cell is in the form of core enzyme and how much exists as holoenzyme.) By contrast, the holoenzyme very rapidly associates with, and dissociates from, loose binding sites. So it is likely to continue to make and break a series of closed complexes in an agitated manner until (by chance) it encounters a promoter. Then its recognition of the specific sequence will allow tight binding to occur by formation of an open complex.

Three steps are needed for RNA polymerase to move from one binding site to another on DNA. It must dissociate from the first binding site, find the second site, and associate with it. Movement from one site to another is limited by the speed of diffusion through the medium. The rate constant for binding promoters is very close to this limit, so close, in fact, that insufficient time is left for association and dissociation from loose binding sites during a random search cycle.

RNA polymerase may therefore use another means to seek its binding sites, possibly the direct displacement of one bound sequence by another. Instead of moving about by leaving one binding site and diffusing to another, the enzyme is likely to take hold of one sequence of DNA, exchange it very rapidly for another, and continue to exchange sequences in this promiscuous manner until a promoter is found. Then the enzyme forms a stable, open complex, after which

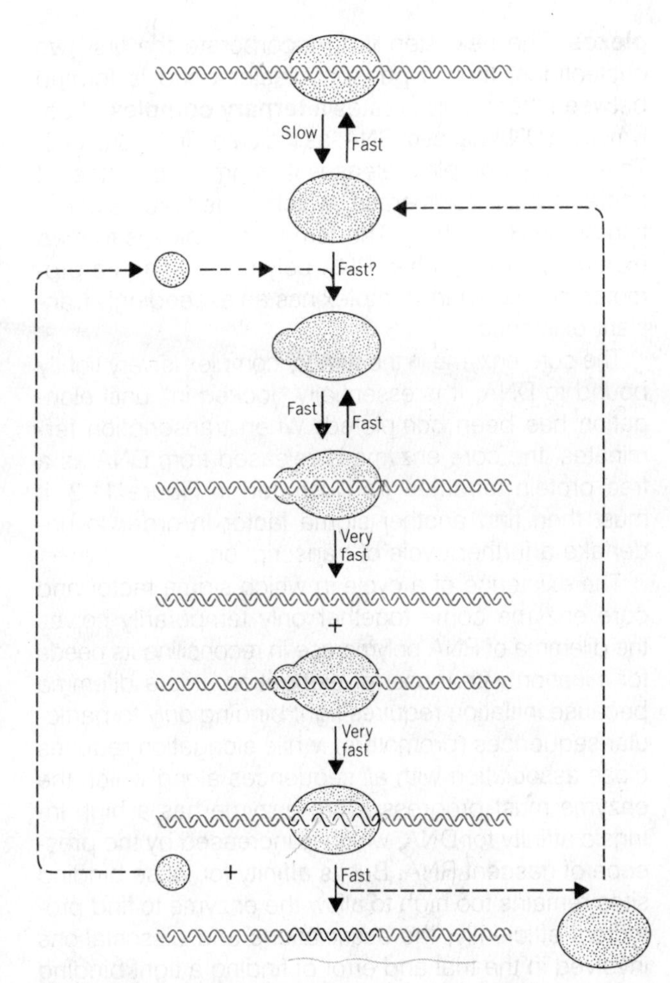

Figure 11.3
Sigma factor and core enzyme recycle at different points in transcription.
Sigma factor is released as soon as a ternary complex has formed at an initiation site; it becomes available for use by another core enzyme. The core enzyme is released at termination; it must either find a sigma and form a holoenzyme that can bind stably only at promoters or it must bind to loose sites on DNA.

initiation occurs. The search process becomes much faster because association and dissociation are virtually simultaneous, and time is not spent commuting between sites.

Sigma factor is involved *only* in initiation. It is released from the core enzyme when RNA synthesis has been initiated.

Both the closed and open associations of RNA polymerase with DNA are described as **binary com-**

plexes. The next step is to incorporate the first two nucleotides; then a phosphodiester bond is formed between them. This creates a **ternary complex** of polymerase·DNA·nascent RNA, as indicated in Figure 11.2. The ternary complex releases its sigma factor; then it contains core polymerase, which undertakes elongation of the RNA chain. The ternary complex is formed extremely rapidly when RNA polymerase finds a promoter, so the binary complex has an exceedingly transient existence.

The core enzyme in the ternary complex is very tightly bound to DNA. It is essentially "locked in" until elongation has been completed. When transcription terminates, the core enzyme is released from DNA as a free protein tetramer. As illustrated in Figure 11.3, it must then find another sigma factor in order to undertake a further cycle of transcription.

The existence of a cycle in which sigma factor and core enzyme come together only temporarily solves the dilemma of RNA polymerase in reconciling its needs for initiation with those for elongation. It is a dilemma because initiation requires tight binding *only* to particular sequences (promoters), while elongation requires close association with *all* sequences along which the enzyme must progress. Core enzyme has a high intrinsic affinity for DNA, which is increased by the presence of nascent RNA. But its affinity for loose binding sites remains too high to allow the enzyme to find promoters efficiently; the associations and dissociations involved in the trial and error of finding a tight binding site could take many hours.

By reducing the stability of the loose complexes, sigma allows the process to occur much more rapidly; and by stabilizing the association at tight binding sites, the factor drives the reaction irreversibly into the formation of open complexes. But then the holoenzyme would be paralyzed by its specific affinity for the promoter, so by releasing sigma, the enzyme reverts to a general affinity for all DNA, irrespective of sequence, that suits it to continue transcription.

CORE ENZYME SYNTHESIZES RNA

Core enzyme starts transcription at the separated DNA strands of an open promoter complex. As the enzyme moves along the template extending the RNA chain, the region of local unwinding moves with it. The enzyme covers about 60 bp of DNA; the unwound segment comprises only a small part of this stretch, < 17 bp according to the overall extent of unwinding.

As the DNA unwinds to free the template, each of its strands probably enters a separate site in the enzyme structure. As **Figure 11.4** indicates, the template strand will be free just ahead of the point at which the ribonucleotide is being added to the RNA chain, and it will exist as a DNA-RNA hybrid in the region where RNA has just been synthesized. The length of the hybrid region may be a little shorter than the stretch of unwound DNA. Probably the RNA-DNA hybrid is about 12 bp long.

As the enzyme leaves the area, the DNA duplex reforms, and the RNA is displaced as a free polynucleotide chain. About the last 50 ribonucleotides added to a growing chain are complexed with DNA and/or enzyme at any moment.

We still do not really understand the topology of unwinding and rewinding during transcription. But the ability of purified RNA polymerase to transcribe double-stranded DNA *in vitro* implies that the reaction depends on an intrinsic property of the enzyme. The formal possibility that the enzyme revolves around the axis of the DNA as it proceeds seems unlikely; for this

Figure 11.4
Bacterial RNA polymerase covers ~60 bp of DNA and has several active centers.
The length of the region of unwinding is exaggerated (about 4 fold) for the purposes of illustration.

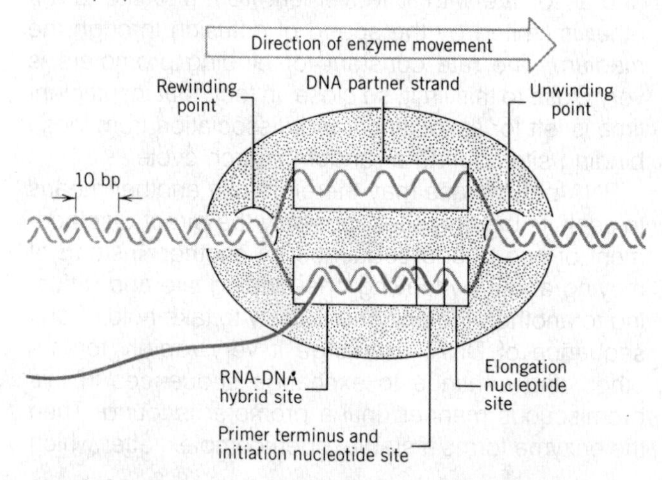

would involve rotation of the entire RNA chain and its associated ribosomes. Another possibility is that the DNA revolves in the unwinding sense ahead of the enzyme, and revolves in the opposite sense behind it. This could require assistance *in vivo* from other enzymatic activities to adjust the topology of the DNA.

All nucleic acids are synthesized from nucleoside 5′ triphosphate precursors. **Figure 11.5** shows the condensation reaction between the 5′ triphosphate group of the incoming nucleotide and the 3′-OH group of the last nucleotide to have been added to the chain. The incoming nucleotide loses its terminal two phosphate groups (γ and β); its α group is used in the phosphodiester bond linking it to the previous nucleotide.

The core enzyme must hold the two reacting groups in the proper apposition for phosphodiester bond formation; then, once they are covalently linked, it moves one base farther along the DNA template so that the reaction can be repeated. The reaction rate is fast, about 40 nucleotides/second at 37°C (see Chapter 10).

Figure 11.5
Phosphodiester bond formation involves a hydrophilic attack by the 3′ – OH group of the last nucleotide of the chain on the 5′ triphosphate of the incoming nucleotide, with release of pyrophosphate.

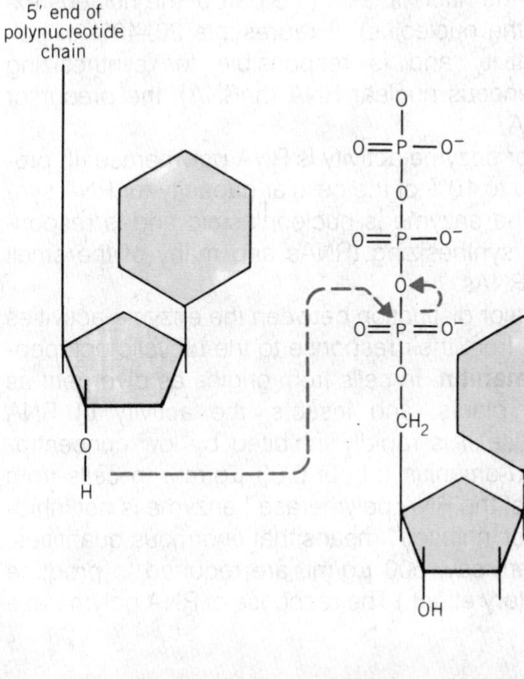

5′ end of polynucleotide chain

O=P—O⁻

O=P—O⁻

O=P=O⁻

CH_2

O

H

OH

The reacting groups are held in two sites. The location occupied by the incoming nucleoside triphosphate is the **elongation nucleotide site**. The position of the last nucleotide added to the chain defines the **primer terminus site**. When transcription is initiated, of course, there is no primer terminus, and the very first nucleotide (usually a purine) enters the **initiation nucleotide site**, which must largely overlap with the primer terminus site. The first nucleotide incorporated into the chain retains all its 5′ triphosphate residues.

All four nucleoside triphosphates can enter the elongation nucleotide site. The acceptability of an incoming precursor is judged by its base pairing with the template strand of DNA, an action apparently supervised by the enzyme. Probably the site has a structure that allows phosphodiester bond formation to proceed only when the nucleotide is properly base paired with DNA. Presumably the nucleotide is expelled if its ability to base pair is deemed inadequate; then another can enter.

Our knowledge of the topology of the core enzyme is really very primitive, and the best we can do at present is to make a diagrammatic representation of the sites defined by the various enzymatic functions, as illustrated in Figure 11.4. None of these sites has yet been physically located on the polypeptide subunits. However, there is some general information about the roles of individual subunits.

Two types of antibiotic both act on the β subunit, as defined by the location of mutations conferring resistance. The **rifamycins** (of which rifampicin is the most used) prevent initiation, acting prior to formation of the first phosphodiester bond, after which the ternary complex becomes resistant to inhibition. **Streptolydigins** inhibit chain elongation. The β subunit is implicated as their target by reconstitution experiments, in which the source of the β subunit determines the response to the antibiotic. Also the β subunit is labeled by certain affinity analogs of the nucleoside triphosphates. Together these results suggest that the β subunit may be involved in binding the nucleotide substrates.

Heparin is a polyanion that binds to the β′ subunit and inhibits transcription *in vitro*. Heparin competes with DNA for binding the polymerase in a binary complex. The β′ subunit is the most basic, which would fit with a role in template binding.

The β and β′ subunits both appear to be contacted by the RNA chain on its way out of the enzyme. The σ subunit is contacted only by the first couple of nucleotides, consistent with its role in initiation. As an independent polypeptide, sigma does not seem to bind DNA, but when holoenzyme forms a tight binding complex, it does contact the DNA in the region of the initial melting. Addition of sigma changes the conformation of core enzyme so that its ability to recognize DNA is altered, but no one of the core subunits can be singled out as the binding site for sigma. All of them contribute, directly or indirectly, to the conversion of core enzyme to holoenzyme.

The α subunit has no known role. However, when phage T4 infects *E. coli*, the α subunit is modified by ADP-ribosylation of an arginine. The modification is associated with a reduced affinity for the promoters formerly recognized by the holoenzyme, so the α subunit might play a role in promoter recognition.

Why does bacterial RNA polymerase require a large, multimeric structure? The existence of much smaller RNA polymerases, comprising single polypeptide chains coded by certain phages, demonstrates that the apparatus required for RNA synthesis can be much smaller than that of the host enzyme.

These enzymes give some idea of the "minimum" apparatus necessary for transcription. They recognize a very few promoters on the phage DNA; and they have no ability to change the set of promoters to which they respond. Thus they are limited to the intrinsic ability to recognize a very few specific DNA binding sequences and to synthesize RNA. How complex are they?

The RNA polymerases coded by the related phages T3 and T7 are single polypeptide chains of about 11,000 daltons each. They synthesize RNA very rapidly (at rates of about 200 nucleotides/second at 37°C). The initiation reaction shows very little variation.

By contrast, the enzyme of the host bacterium can transcribe any one of many (more than 1000) transcription units. Some of these units are transcribed directly, with no further assistance. But many units can be transcribed only in the presence of further protein factors. Some of these factors are specific for a single transcription unit; others are involved in coordinating transcription from many units. Certain phages induce general changes in the affinity of host RNA polymerase, so that it stops recognizing host genes and instead initiates at phage promoters.

So the host enzyme requires the ability to interact with a variety of host and phage functions that modify its intrinsic transcriptional activities. The complexity of the enzyme may therefore at least in part reflect its need to interact with a multiplicity of other factors, rather than any demand inherent in its catalytic activity.

COMPLEX EUKARYOTIC RNA POLYMERASES

The transcription apparatus of eukaryotic cells is more complex and less well defined than that of bacteria. There are three nuclear RNA polymerases, occupying different locations, each with a very large number of polypeptide subunits. Also, different RNA polymerase activities are found in mitochondria and chloroplasts.

The most prominent RNA-synthesizing activity is the enzyme RNA polymerase I, which resides in the nucleolus and is responsible for transcribing the genes coding for rRNA. It accounts for 50–70% of cellular RNA synthesis.

The other major enzyme is RNA polymerase II, located in the nucleoplasm (the part of the nucleus excluding the nucleolus). It represents 20–40% of cellular activity and is responsible for synthesizing heterogeneous nuclear RNA (hnRNA), the precursor for mRNA.

A minor enzyme activity is RNA polymerase III, providing up to 10% of the cellular capacity for RNA synthesis. The enzyme is nucleoplasmic and is responsible for synthesizing tRNAs and many of the small nuclear RNAs.

The major distinction between the enzyme activities is drawn from their response to the bicyclic octapeptide **α-amanitin**. In cells from origins as divergent as animals, plants, and insects, the activity of RNA polymerase II is rapidly inhibited by low concentrations of α-amanitin (about 0.03 μg/ml). In cells from all origins, the RNA polymerase I enzyme is not inhibited. ("Not inhibited" means that enormous quantities, more than, say, 500 μg/ml, are required to produce an inhibitory effect.) The response of RNA polymerase

III to α-amanitin has not been so well conserved; in animal cells it is inhibited by high levels (20 μg/ml), but in yeast and insects it is not inhibited.

The crude enzyme activities all are large proteins, appearing as aggregates of 500,000 daltons or more. Their subunit compositions are complex. Each enzyme has two large subunits, generally one about 200,000 daltons and one about 140,000 daltons. There are up to 10 smaller subunits, ranging in size from 10,000 to 90,000 daltons. We do not know whether any of the subunits found in the different enzymes are the same.

Because it is not yet possible to reconstitute active RNA polymerase from the subunits of any of these enzymes, we have no evidence as to whether all of the protein subunits are integral parts of each enzyme. We do not know which subunits may represent catalytic activities and whether others may be involved in regulatory functions.

Do the enzyme preparations represent the basic transcription apparatus, essentially similar in all cells and subject to regulation by further protein factors? Or do they include such factors as well as a basic catalytic apparatus?

The route to investigating this question is to use isolated enzyme preparations to transcribe defined templates *in vitro*. Several heterologous reactions have been characterized, in which an RNA polymerase II preparation from one cell type and species is used to transcribe a gene that is active in a different cell type and species. The success of such experiments indicates that neither tissue- nor species-specific features are involved in promoter recognition per se. Thus there is some fundamental and conserved feature that identifies a sequence as a promoter.

Of course, this conclusion does not exclude the possibility that further protein factors or other sequences are involved in modulating the reaction (especially increasing its efficiency) in the natural situation. For example, a protein factor has been found in human cells that is necessary for RNA polymerase II to transcribe the genes of the virus SV40. The factor is not necessary for the enzme to transcribe other genes under the same conditions. Since the factor is a normal component of the cell, we may infer it is needed for the natural transcription of some (at present unidentified) host genes. Thus it is possible that there are sets of factors needed to transcribe particular groups of genes.

RNA polymerase III of *X. laevis* can specifically transcribe the genes for 5S RNA only when provided with an additional protein factor, a 37,000 dalton polypeptide that is found complexed with the 5S RNA in oocytes. The protein appears to be an ancillary factor necessary for the basic transcription apparatus to work with the 5S RNA genes; it is not needed with other genes.

The overall complexity of the eukaryotic transcription apparatus is therefore being defined only system by system; all we can say at present is that multimeric enzymes are needed for each of the three classes of RNA synthesis. Whether all the components of these enzymes are essential, and how many other proteins are needed, remains to be seen.

Because of these uncertainties, relatively crude, "dirty" systems may offer more chance of characterizing transcription *in vitro* than purified "clean" systems; too much purification may remove the very factors that we need to characterize!

The RNA polymerase activities of mitochondria and chloroplasts appear to be smaller and distinct from the nuclear enzymes. Of course, the organelle genomes are much smaller, the resident polymerase needs to transcribe only a few genes, and the control of transcription is likely to be very much simpler (if existing at all). So these enzymes may be analogous to the phage enzymes that have a single fixed purpose and do not need the ability to respond to a more complex environment.

FURTHER READING

A source for many reviews and original research articles is the volume edited by **Losick & Chamberlin**, *RNA Polymerase* (Cold Spring Harbor Laboratory, New York, 1976). Two chapters that cover the general matters dealt with here were written by **Chamberlin**, giving first a general overview (pp. 17–68) and then an account of the interactions of the bacterial enzyme with its template (pp. 159–192).

CHAPTER 12
PROMOTERS: THE SITES FOR CONTROLLING INITIATION

To transcribe or not to transcribe: that is the question? Often the major decision on whether a gene is to be expressed is taken by controlling the initiation of transcription. Certainly this is not the only stage at which gene expression can be controlled; in both prokaryotes and eukaryotes there are many examples of genes that are transcribed, but whose ultimate expression depends on further mechanisms that influence utilization of the RNA. However, the initiation of transcription represents a major step toward expressing a gene, and often does represent the only point of decision: for many genes, if transcription is initiated, the transcript inevitably proceeds through all the subsequent stages up to translation into protein.

Initiation occurs when an RNA polymerase binds to a promoter. What controls the ability of the enzyme to initiate at a particular promoter? We can start by thinking about promoters in two general classes.

Some promoters can be recognized by RNA polymerase alone; in these cases, an accessible promoter will always be transcribed. Promoter availability may be determined by extraneous proteins, which may either act directly at the promoter to block access by RNA polymerase, or may function indirectly by controlling the structure of the genome in the region.

Other promoters are not by themselves adequate to support transcription; ancillary protein factors are needed for initiation to occur. The additional protein factors act by recognizing sequences of DNA that may be close to, or overlap with, the sequence bound by RNA polymerase itself.

As sequences of DNA whose function is to be *recognized* by proteins, a promoter and any adjacent control sites differ from other sequences whose role is exerted by being transcribed or translated. The information for promoter function is provided directly by the DNA sequence: its structure is the signal. By contrast, expressed regions gain their meaning only after the information is transferred into the form of some other nucleic acid or protein.

The key question in examining the interaction between an RNA polymerase and its promoter is how a protein can recognize a specific sequence of DNA. Does the enzyme have an active site that distinguishes the chemical structure of a particular sequence of bases in the DNA double helix? How specific are its requirements? Promoters vary in their affinities for RNA polymerase, and this can be an important factor in controlling the frequency of initiation and thus the extent of gene expression.

Binding at the promoter is rapidly followed by initiation at the startpoint; then RNA polymerase continues along the template until it reaches a terminator sequence. This action defines a **transcription unit** that extends from the promoter to the terminator. The critical feature of the transcription unit, depicted in **Figure 12.1**, is that it constitutes a stretch of DNA that is *expressed via the production of a single RNA molecule*. A transcription unit may include only one or several genes. Sometimes regions close to the promoter are described as **proximal**, while those toward the terminator are described as **distal**.

Sequences prior to the startpoint are described as **upstream**; those after the startpoint (within the transcribed sequence) are **downstream**. Sequences are conventionally written so that transcription proceeds from left (upstream) to right (downstream). This corresponds to writing the mRNA in the usual 5′ to 3′ direction. Often the DNA sequence is written just to show the strand whose sequence is the same as the RNA. Base positions are numbered in both directions away from the startpoint, which is assigned the value +1; numbers are increased going downstream. The base before the startpoint is numbered −1, and the negative numbers increase going upstream.

The product of transcription, before any changes have been made to it, is called the **primary transcript**. It would consist of an RNA extending from the promoter to the terminator, possessing its original 5′ and 3′ ends. However, the primary transcript is almost always unstable and therefore very difficult to characterize *in vivo*. In prokaryotes, it is rapidly degraded (mRNA) or cleaved to give mature products (rRNA and tRNA). In eukaryotes, it is modified at the ends (mRNA) or cleaved to give mature products (all RNA).

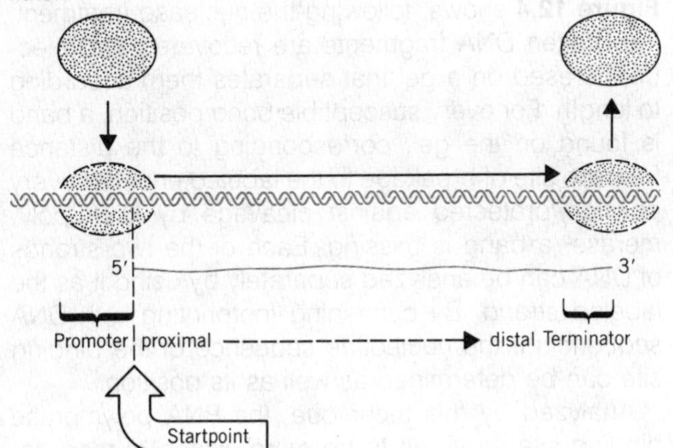

Figure 12.1
A transcription unit is a sequence of DNA transcribed into a single RNA, starting at the promoter and ending at the terminator.

The startpoint is defined as the base pair in DNA that corresponds to the first nucleotide incorporated into RNA. A common method for identifying the startpoint by hybridizing the transcript with its DNA template is illustrated in **Figure 12.2**. In principle, it consists of degrading all the DNA that cannot hybridize with the RNA, and then determining the sequence of the surviving DNA.

To define the startpoint, it is necessary to examine the 5′ end of the primary transcript. Because of the difficulty of isolating primary transcripts *in vivo*, most information about startpoints is provided by *in vitro* studies. However, in those cases in which the authentic 5′ end *has* been identified *in vivo* (by its possession of a triphosphate terminus), it is identical with the 5′ end generated by transcription *in vitro*. In particular, the capped terminus of eukaryotic mRNA appears to coincide with the startpoint.

Usually the startpoint is a unique base pair, sometimes it consists of either of two adjacent base pairs, and occasionally it may involve any one of several adjacent positions. It always represents a defined part of the promoter, usually within the sequence bound by RNA polymerase, in some cases at one extremity.

BINDING SITES FOR *E. COLI* RNA POLYMERASE

The tight binding sites at which RNA polymerase forms stable initiation complexes lie within promoters. These sequences of DNA can be recovered by the protocol summarized in **Figure 12.3**, in which RNA polymerase is bound *in vitro* to a DNA fragment containing a promoter. Then digestion with the enzyme DNAase is used to degrade all the regions of DNA that are not protected by the RNA polymerase.

The recovered fragments vary from 41 to 44 bp in length. The polymerase actually can initiate transcription if left attached to the protected fragment, to synthesize a short RNA of 17–20 bases, terminated when the polymerase "runs off" the end of the fragment. This shows that the protected fragment corresponds to the sequence at which transcription is initiated, and lo-

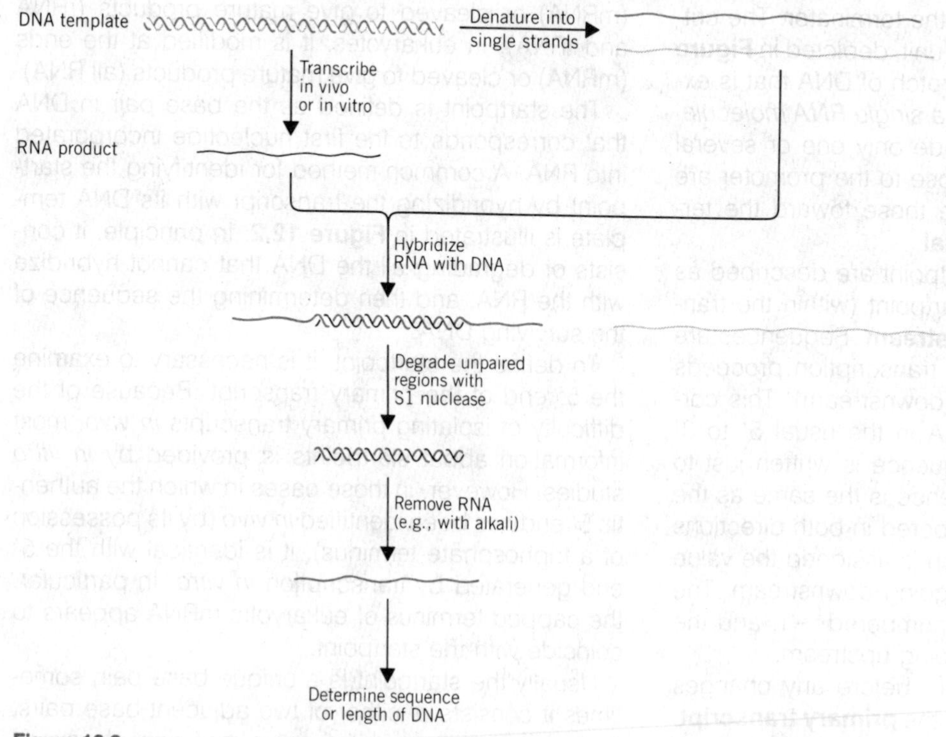

Figure 12.2
Startpoints can be determined by comparing the RNA product with the DNA template.
When the RNA is hybridized with the denatured strands of DNA, it forms an RNA-DNA hybrid with the strand that acted as its template; the other strand of DNA remains unpaired. Treatment with the enzyme S1 nuclease, which specifically degrades single-stranded DNA, destroys both the unpaired strand and regions of the template strand beyond the transcription unit. The RNA of the RNA-DNA hybrid can be removed. Then the sequence of the DNA can be determined or its length can be used to locate the position of the RNA on the template.

cates the startpoint in the center of this binding site. The protected fragment extends from about −20 upstream to +20 downstream.

The ability of RNA polymerase to recognize DNA can be characterized in more detail by a technique known as **footprinting**. A sequence of DNA bound to RNA polymerase (or any other protein) is *partially* digested with an endonuclease, an enzyme that attacks individual phosphodiester bonds *within* a nucleic acid. Under appropriate conditions, every phosphodiester bond that is in principle accessible to the nuclease is broken in some, but not in all, molecules of DNA. Only if RNA polymerase blocks access of the nuclease to DNA will a particular bond fail to be broken at all.

The positions that are cleaved are recognized by using DNA labeled on one strand at one end only. As **Figure 12.4** shows, following the nuclease treatment, the broken DNA fragments are recovered and electrophoresed on a gel that separates them according to length. For every susceptible bond position, a band is found on the gel, corresponding to the distance from the site of breakage to the labeled end. For every position protected against cleavage by RNA polymerase, a band is missing. Each of the two strands of DNA can be analyzed separately by using it as the labeled strand. By combining footprinting with DNA sequencing, the nucleotide sequence of the binding site can be determined as well as its position.

Analyzed by this technique, the RNA polymerase binding site turns out to be more extensive than de-

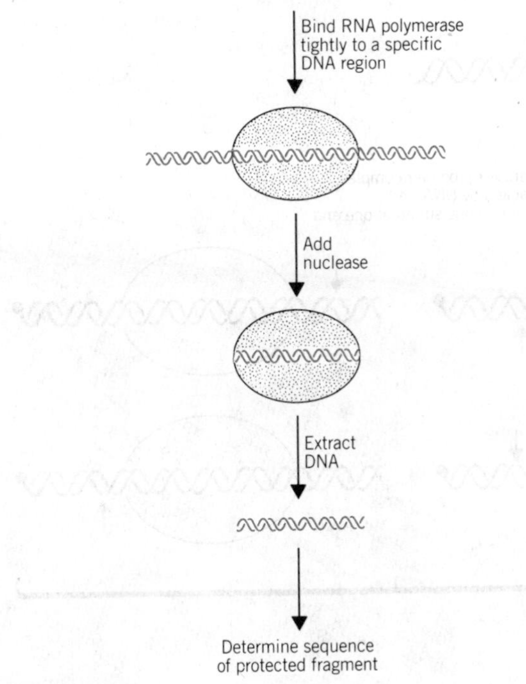

Figure 12.3
DNA binding sites for RNA polymerase can be recovered because they are protected by the enzyme against degradation by nucleases.

tected by the recovery of protected fragments. It extends for roughly an extra 30 bp upstream, that is, from about −50 to +20. The inability of RNA polymerase to *rebind* to the protected (−20 to +20) fragments shows that the extra sequences are needed for binding. The two strands of DNA are not equally well protected in all regions of the promoter, especially at the ends of the binding site. This implies that RNA polymerase has an asymmetric conformation when bound to DNA, which accords with the need to transcribe only one strand of DNA.

Somewhat similar results can be obtained with digestion using an **exonuclease**, an enzyme that attaches to an end of DNA and degrades it continuously. If a protein is bound to the DNA, digestion stops when the nuclease encounters the protein. The limits of the promoter can be defined in terms of the point on each side at which the exonuclease is blocked from proceeding, about −44 upstream and +20 downstream.

So RNA polymerase binds asymmetrically to a region of DNA stretching from about 44–50 bp upstream to a point about 20 bp downstream. The terminal 25–30 bp of the binding site upstream must be more loosely associated with the enzyme. This region is bound sufficiently well to be protected against gentle endonucleolytic or exonucleolytic cleavage (in footprinting), but cannot withstand the stronger conditions used to retrieve intact fragments. These data argue in favor of a model in which RNA polymerase binds to a single, continuous sequence of DNA where (as we shall see) some points of contact are more important than others.

CONSENSUS SEQUENCES IN *E. COLI* PROMOTERS

An obvious way to design a promoter would be for an invariant sequence of DNA to be recognized by RNA polymerase. In the bacterial genome, the minimum length that could provide an adequate signal is 12 bp. (Any shorter sequence is likely to occur—just by chance—a sufficient number of additional times to provide false signals.) The 12 bp sequence need not be contiguous; and, in fact, if a specific number of base pairs separates two constant shorter sequences, their combined length could be less than 12 bp, since the *distance* of separation itself provides a part of the signal (even if the intermediate *sequence* is itself irrelevant).

Attempts to identify the features in DNA that are necessary for RNA polymerase binding started by comparing the sequences of different promoters. Any essential nucleotide sequence should be present in all the promoters. Such a sequence is said to be **conserved**. However, a conserved sequence need not necessarily be conserved at every single position; some variation may be permitted. How do we analyze a sequence of DNA to determine whether it is sufficiently conserved to constitute a recognizable signal?

Putative DNA recognition sites can be defined in terms of an idealized sequence that represents the base most often present at each position. A **consensus** or **canonical sequence** is defined by aligning all known examples so as to maximize their homology. For a sequence to be accepted as a consensus, each particular base must be reasonably predominant at its

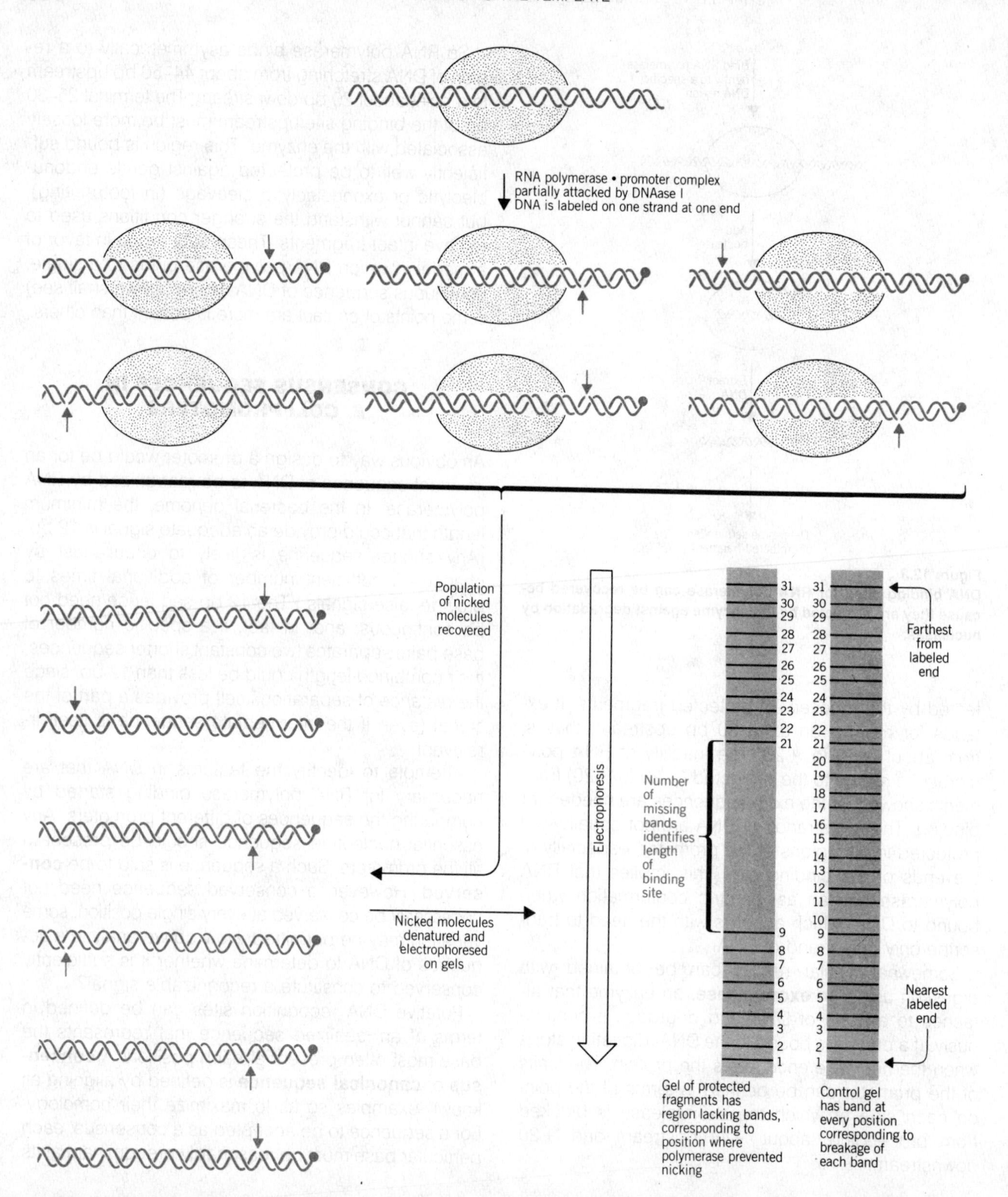
RNA polymerase • promoter complex
partially attacked by DNAase I
DNA is labeled on one strand at one end
Population
of nicked
molecules
recovered
Nicked molecules
denatured and
electrophoresed
on gels
Electrophoresis
Number
of
missing
bands
identifies
length
of
binding
site
31
30
29
28
27
26
25
24
23
22
21
20
19
18
17
16
15
14
13
12
11
10
9
8
7
6
5
4
3
2
1
Farthest
from
labeled
end
Nearest
labeled
end
Gel of protected
fragments has
region lacking bands,
corresponding to
position where
polymerase prevented
nicking
Control gel
has band at
every position
corresponding to
breakage of
each band

position, and most of the sequences must be related to the consensus by only a small number of substitutions, say, 1 or 2.

More than 100 promoters have been sequenced in *E. coli*, and a striking feature is the *lack of any extensive conservation of sequence* over the 60 bp associated with RNA polymerase. The sequence of much of the binding site may actually be irrelevant. But some short stretches within the promoter are conserved.

The startpoint is usually (> 90% of the time) a purine. It is quite common for the startpoint to be the central base in the sequence CAT, but the conservation of this triplet is not great enough to regard it as an obligatory signal.

Just upstream from the startpoint, a 6 bp region is recognizable in almost all promoters. The consensus sequence is **TATAAT**; sometimes it is called the **Pribnow box**. The conservation of the base at each position of the Pribnow box varies from 45% to 100%. The consensus can be summarized in the form

$$T_{80}\ A_{95}\ t_{45}\ A_{60}\ a_{50}\ T_{96}$$

where the subscript denotes the percent occurrence of the most frequently found base. Capital letters are used to indicate bases conserved >54% and lower case letters are used to indicate bases not so well conserved, but nonetheless present more often than predicted from a random distribution. A position at which there is no discernible preference for any base would be indicated by N.

If the frequency of occurrence indicates likely importance in binding RNA polymerase, we would expect the initial highly conserved TA and the final almost completely conserved T in the Pribnow box to be the most important bases. The center of the Pribnow box generally is close to 10 bp upstream from the startpoint. Sometimes it is therefore called the **−10 sequence**. The actual location of the hexamer varies from centering on position −18 to −12.

Similarities of sequence also occur at another location, centered about 35 bp upstream from the startpoint. This is called the **−35 sequence**; sometimes it is known as the "recognition region," because it is part of the sequence that must be recognized by RNA polymerase, but is not included in the tightly bound protected region. The consensus is **TTGACA**; in more detailed form, the conservation is

$$T_{82}\ T_{84}\ G_{78}\ A_{65}\ C_{54}\ a_{45}$$

The distance separating the −35 and −10 sites is between 16 and 18 bp in 90% of the promoters; in the exceptions, it may be as little as 15 or as great as 20 bp. Although the actual sequence in the intervening region may be unimportant, the distance may be critical in holding the two sites at the appropriate separation for the geometry of RNA polymerase.

Figure 12.4
Footprinting identifies DNA binding sites for proteins by their protection against nicking.
The principle is the same as that involved in DNA sequencing; partial cleavage of an end-labeled molecule at a susceptible site creates a fragment of unique length. In a free DNA, *every* susceptible bond position is broken in one or another molecule. But when the DNA is complexed with a protein, the region covered by the DNA-binding protein is protected in every molecule. So two reactions are run in parallel: a control of pure DNA; and an experimental mixture containing the protein.

When the strands are separated and electrophoresed, a radioactive band is produced by each fragment that retains a labeled end. The position of the band corresponds to the number of bases in the fragment. The shortest fragments move the fastest, so distance from the labeled end is counted up from the bottom of the gel (see Figure 5.9). In the control, every bond is broken, generating a series of bands, one representing each base. In the figure, 31 bands can be counted. In the protected fragment, bonds cannot be broken in the region bound by the protein, so bands representing fragments of the corresponding sizes are not generated. The absence of bands 10-20 in the figure identifies a protein-binding site covering the region located 10-20 bases from the labeled end of the DNA.

Instead of using a single gel that analyzes DNA simply by length, both the control and experimental mixtures can be treated to generate four sequencing gels (see Figure 5.10). Comparison of the two gel sets allows the sequence to be "read off" directly, thus identifying the nucleotide sequence of the binding site.

A very few promoters lack a recognizable version of one of the consensus sequences. In at least some of these cases, the promoter cannot be recognized by RNA polymerase alone, for the reaction requires the intercession of ancillary proteins. Possibly their reaction with adjacent sequences overcomes the deficiency in the promoter. We do not yet know all the details of the recognition reaction; all that can be said firmly now is that the "typical" promoter can use the −35 and −10 sequences to be recognized by RNA polymerase. From their absence from exceptional promoters, we realize that other means also can be used for recognition, but we do not know how many alternatives there are, nor exactly how they substitute for the absence of the consensus sequences.

A major source of information about promoter function is provided by mutations. Mutations in promoters affect the level of expression of the gene(s) they control, without altering the gene products themselves. Most are identified in the form of bacterial mutants that have lost, or have very much reduced, transcription of the adjacent genes. These are known as **down** mutations. Less often, mutants are found in which there is increased transcription from the promoter. These are called **up** mutations.

Almost all of the point mutations that affect promoter function fall within the two consensus sequences. Occasionally a mutation is found just upstream of either consensus sequence (see Figure 12.5). The bases present at other positions in the vicinity are clearly much less important or even irrelevant in most promoters. In addition to these mutations, deletions or insertions between the consensus sequences may alter their separation.

The usual assumption is that the function of a recognition site is likely to be fulfilled most effectively by the consensus sequence itself. This expectation is borne out by the simple rule that up mutations usually increase homology with one of the consensus sequences or bring the distance between them nearer to 17 bp. Down mutations usually decrease the resemblance of either site with the consensus or make the distance between them more distant from 17 bp. Down mutations tend to be concentrated in the most highly conserved positions, which confirms their particular importance as the main determinant of promoter efficiency. Only a few mutations substitute one base for another where neither conforms with the consensus.

Occasional exceptions to these rules demonstrate that promoter efficiency cannot be predicted entirely from homology with the consensus. Virtually all actual promoters vary from the consensus, so the neighbors of any particular base may differ from promoter to promoter, even if the base itself is conserved. We cannot predict the effects of context; it is possible that a nonconsensus base at some position can function effectively in one promoter but not in another. It is also possible that the important feature for a promoter may be the *exclusion* of some base from a particular position, rather than the presence of one particular nucleotide. With these caveats, however, *we can define the optimal promoter as a sequence consisting of the −35 septamer, separated by 17 bp from the −10 hexamer, lying 7 bp upstream of the startpoint.*

In all of this analysis, it is important to remember that "up" and "down" mutations are defined relative to the *usual* efficiency with which a particular promoter functions. This varies widely. So a change that is recognized as a down mutation in one promoter might never have been isolated in another (which in its wild-type state could be even less efficient than the mutant form of the first promoter). Thus information gained from studies *in vivo* simply identifies the overall nature of the change caused by mutation.

FUNCTIONS OF THE CONSENSUS SEQUENCES

To analyze promoter function at the molecular level, and to determine the absolute effects of promoter mutations, we must measure the affinity of RNA polymerase for wild-type and mutant promoters *in vitro*. There is roughly a hundredfold variation in the rate at which RNA polymerase binds to different promoters *in vitro*, and this correlates well with the frequency of transcription when their genes are expressed *in vivo*. Taking this analysis further, we can investigate the stage at which a mutation influences the capacity of the promoter. Does it change the affinity of the promoter for binding RNA polymerase? Does it leave the enzyme able to bind but unable to initiate? Is the influence of an ancillary factor altered?

By measuring the rate constants for formation of a closed complex and its conversion to an open complex, we can dissect the two stages of the initiation reaction. Down mutations in the −35 sequence reduce the rate of closed complex formation, but do not inhibit the conversion to an open complex. On the other hand, down mutations in the −10 sequence do not slow the initial formation of a closed complex, but they slow its conversion to the open form. These results suggest that *the function of the −35 sequence is to provide the signal for recognition by RNA polymerase, while the −10 sequence allows the complex to convert from closed to open form.*

The consensus sequence of the −10 site consists exclusively of A-T base pairs. Its content arouses suspicion that it may be concerned with the initial melting of DNA into single strands. The lower energy needed to disrupt A-T pairs compared with G-C pairs means that a stretch of A-T pairs demands the minimum amount of energy for strand separation. Most of the actual −10 sites do have one G-C pair, so more than the bare minimum energy is required for their activation.

The points at which RNA polymerase contacts the promoter can be identified by treating RNA polymerase·promoter complexes with reagents that modify particular bases. The presence of the enzyme may either increase or decrease the availability of a particular base (relative to a control consisting of the DNA by itself). Changes in sensitivity reflect the geometry of the complex and can be used to deduce its overall conformation.

The reagent dimethylsulfate (DMS) reacts specifically with purines, to methylate G residues in the wide groove and A residues in the narrow groove of DNA. The methylation renders the purines labile to heat; so after heating, the base is lost from DNA, leaving an exposed position at which the phosphodiester backbone can be broken by alkali. (This reaction is used in DNA sequencing, as described in Chapter 5).

When bromouracil (BUdR) is substituted for thymine in DNA, ultraviolet irradiation releases the bromine. This creates a free radical that then breaks the DNA chain.

The backbone of DNA is attacked directly by ethylnitrosourea, which converts the phosphodiester bond linking two nucleotides to a phosphotriester (by an alkylation). The phosphotriesters are labile to alkali and provide positions at which the DNA chain can be broken.

The common feature of all these modifications is that they allow a breakage to be made at the corresponding bond in the polynucleotide chain. This site of breakage can be identified by the same approach used in footprinting experiments (see Figure 12.4). By labeling DNA at one end of one strand, each breakage generates an electrophoretic band of corresponding length. When the susceptibility of an RNA polymerase·DNA complex is compared with free DNA, some bands disappear, identifying sites at which the enzyme has protected the promoter against modification. Other bands may increase in intensity, identifying sites at which the DNA must be held in a conformation in which it is more exposed.

The reverse experiment can be performed with DMS or with ethylnitrosourea. The DNA is modified *first*; and then it is bound to RNA polymerase. Actually, the DNA consists of a population of molecules, each modified at some, but not all, of the susceptible sites. Those DNA molecules that cannot bind RNA polymerase are recovered and treated in the usual way to generate strand breakages whose positions can be identified. This locates points at which prior modification *prevents* RNA polymerase from binding to DNA.

Such experiments show that the regions at −35 and −10 contain most of the contact points for the enzyme. Within these regions, the same sets of positions tend both to prevent binding if previously modified, and to show increased or decreased susceptibility to modification after binding. **Figure 12.5** compares the points of contact with sites of mutation; although they do not coincide completely, they occur in the same limited region. At both consensus sites, the region of contact extends for 12–15 bp, somewhat longer than the conserved region.

It is noteworthy that the same *positions* in different promoters may provide the contact points, even though a different base is present. This indicates that there may be a common mechanism for RNA polymerase binding, although the reaction does not depend on the presence of particular bases at some of the points of contact. This model may explain why some of the points of contact are not sites of mutation. Also, not every mutation lies in a point of contact; could some

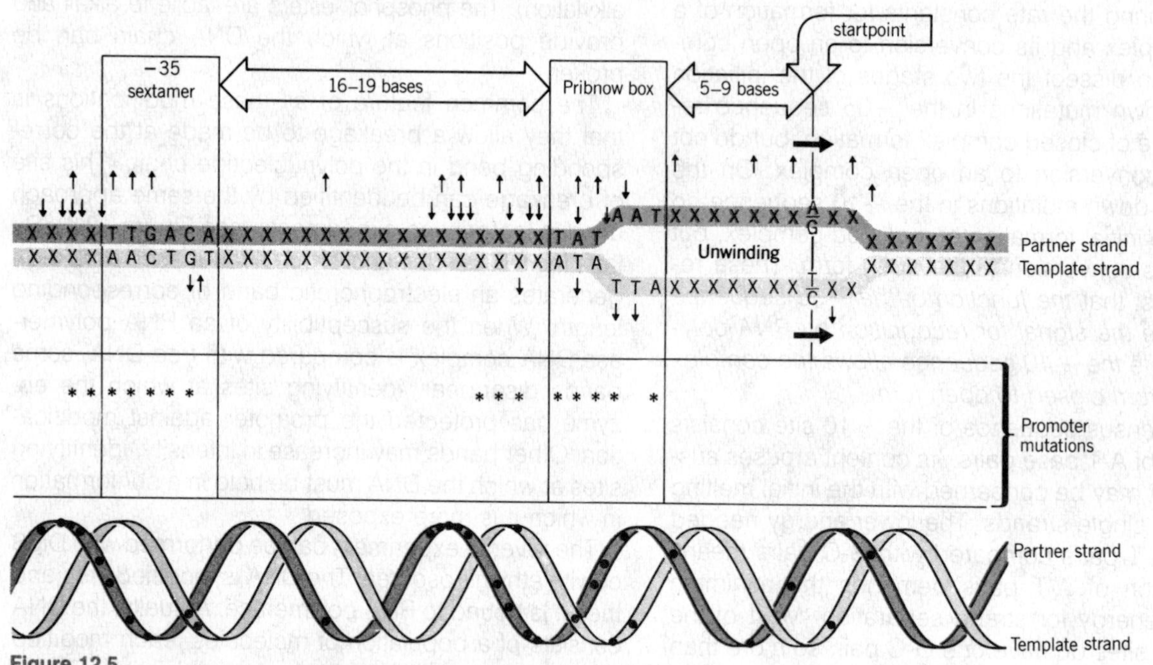

Figure 12.5
One face of the promoter contains the contact points for RNA
The DNA sequence shows a typical promoter, with consensus sequences at −35 and −10, and a region for initial unwinding extending from within the Pribnow box to just past the startpoint.
Upper: sites at which modification prevents RNA polymerase binding are shown by arrows pointing toward the double helix. Sites at which the DNA is protected by RNA polymerase against modification are shown by the arrows pointing away from the double helix. Arrows pointing at bases indicate modification of the base itself; arrows pointing between bases indicate modification of the connecting phosphodiester bond.
Center: sites at which mutations affect promoter function are indicated by asterisks.
Lower: when the DNA is drawn as a double helix viewed from one side, as indicated diagrammatically, all the contact points lie on one face. Most lie on the partner strand (that is, not on the template strand).

influence the neighborhood without actually being touched by the enzyme?

It is especially significant that the experiments with prior modification identify *only* sites in the same region that is protected by the enzyme against subsequent modification. These two experiments measure different things. The first identifies all those sites that the enzyme must recognize in order to bind to DNA. The second recognizes all those sites that actually make contact in the binary complex. The protected sites include all the recognition sites and also some additional positions, which suggests that the enzyme first recognizes a set of sites necessary for it to "touch down," and then extends its points of contact to further sites.

A modification experiment allows the region of DNA that is unwound in the binary complex to be identified directly. When the strands of DNA are separated, the unpaired bases may become susceptible to reagents that cannot reach them in the double helix. The susceptibility of sites in the RNA polymerase·DNA binary complex therefore indicates that they lie in an unpaired region. Experiments using methylation of adenine or cytosine have implicated positions between −9 and +3 in the initial melting reaction. The region unwound during initiation therefore includes the right end of the −10 sequence and extends just past the startpoint. (This measure for the extent of strand separation is less than the 17 bp estimated by the overall degree of unwinding.)

When viewed in three dimensions, the points of contact upstream from the −10 sequence all lie on one face of DNA, as illustrated in Figure 12.5. These bases could be recognized in the initial formation of a closed binary complex. This would make it possible for RNA polymerase to approach DNA from one side and recognize that face of the DNA. As DNA unwinding commences, further sites that originally lay on the other face of DNA might be recognized and bound.

The importance of strand separation in initiating transcription is emphasized by the effects of supercoiling. Both prokaryotic and eukaryotic RNA polymerases can initiate transcription more efficiently *in vitro* when the template is supercoiled, presumably because the supercoiled structure requires less free energy for the initial melting of DNA in the initiation complex.

The involvement of this effect in controlling promoter activity in bacteria is shown by the effects of interfering with enzymes that influence the degree of supercoiling. Among the relevant enzymes are DNA gyrase, which *introduces* negative supercoils, and topoisomerase I, which *relaxes* (removes) negative supercoils (see Chapter 34). Inhibitors of DNA gyrase reduce transcription; mutations in topoisomerase I may increase transcription. Both effects are seen only at some promoters.

The nature of these effects is not entirely clear; bacteria evidently endeavor to set the degree of supercoiling between certain limits, because mutations in one enzyme that alter it may be compensated by mutations in another to restore the balance. However, it does seem that the efficiency of some promoters may be dependent on a certain degree of supercoiling.

Why should some promoters be influenced by the extent of supercoiling while others are not? One possibility is that every promoter has a characteristic dependence on supercoiling, determined by its sequence. This would predict that some promoters have sequences that are easier to melt (and are therefore less dependent on supercoiling), while others have more difficult sequences (and have a greater need to be supercoiled). An alternative is that the location of the promoter might be important, if different regions of the bacterial chromosome have different degrees of supercoiling.

SUBSTITUTION OF SIGMA FACTORS MAY CONTROL INITIATION

The RNA polymerase of *E. coli* is inviolate; the same basic enzyme transcribes all genes, and must therefore recognize a wide spectrum of promoters, whose associated genes are transcribed at different levels and on different occasions. The ability to initiate at particular promoters is controlled by a seemingly endless series of ancillary factors, assisting or interfering with the enzyme. In none of these instances is any change made in the subunits of the enzyme itself. Presumably such a mechanism is not favored, because a change that allowed the RNA polymerase specifically to recognize one type of promoter might prevent it from finding another. This would reduce the flexibility of the enzyme.

Yet the division of labors between a core enzyme that undertakes chain elongation and a sigma factor involved in site selection immediately raises the question of whether there could be more than one type of sigma, each specific for a different class of promoters. In normal growth of *E. coli* there is only a single sigma factor. However, *Bacillus subtilis* contains multiple sigma factors with different specificities. And in certain circumstances, when a drastic change occurs in the life style of the cell, there is a massive switch in gene expression. Then there is evidently less impediment to introducing permanent changes in the RNA polymerase itself, and *B. subtilis* utilizes a wholesale substitution of its original sigma factor by others with different promoter specificities.

The principal RNA polymerase enzyme found in *B. subtilis* cells engaged in normal vegetative growth has the same structure as that of *E. coli*, $\alpha_2\beta\beta'\sigma$. The sigma factor has a mass of 55,000 daltons and is described as σ^{55}. It recognizes promoters by the same criteria as does the *E. coli* enzyme. Variants of the enzyme that contain other sigma factors are found in much smaller amounts. These factors are denoted σ^{00}, where "00" indicates the molecular weight of the factor. The variant enzymes recognize different promoters from the major enzyme.

One occasion on which substitutions of sigma factors cause a transition from expression of one set of genes to expression of another set occurs during bacteriophage infection. **Lytic infection** is the process by which a phage (bacterial virus) takes over a bacterium and destroys it in the process of reproducing more phage particles. In all but the very simplest cases, the development of the phage involves shifts in the pattern of transcription. In *E. coli*, these shifts are accomplished by the synthesis of a phage-coded RNA polymerase or by the efforts of phage-coded ancillary factors that control the bacterial RNA polymerase. During infection of *B. subtilis* by phage SPO1, however, two new sigma factors are elaborated.

The infective cycle of SPO1 passes through three stages of gene expression. Immediately on infection, the **early** genes of the phage are transcribed. After 4–5 minutes, the early genes cease transcription and the **middle** genes are transcribed. Then at 8–12 minutes, middle gene transcription is replaced by transcription of **late** genes.

The early genes are transcribed by the holoenzyme of the host bacterium. They are essentially indistinguishable from host genes whose promoters have the intrinsic ability to be recognized by the RNA polymerase $\alpha^2\beta\beta'\sigma^{55}$.

Expression of phage genes is required for the transitions to middle and late gene transcription. Three regulatory genes, named *28*, *33*, and *34*, control the course of transcription. Their functions are summarized in **Figure 12.6**. The pattern of regulation creates a **cascade**, in which the host enzyme transcribes an early gene whose product is needed to transcribe the middle genes; and then two of the middle genes code for products that are needed to transcribe the late genes.

Mutants in the early gene *28* cannot transcribe the middle genes. The product of gene *28* (called gp28) is a protein of 26,000 daltons that replaces the sigma factor on the core enzyme. This substitution is the sole event required to make the transition from early to middle gene expression. It creates a complete enzyme that can no longer transcribe the host genes, but instead specifically transcribes the middle genes. How gp28 displaces σ^{55}, and what happens to the host sigma polypeptide, is not known. Probably gp28 has a greater affinity for the core enzyme.

Two of the middle genes are involved in the next transition. Mutation in either gene *33* or *34* prevents transcription of the late genes. The products of these genes are proteins of 13,000 and 24,000 daltons, respectively, that replace gp28 on the core polymerase. Again, we do not know how gp33 and gp34 exclude gp28 (or any residual host σ^{55}), but once they have bound to the core enzyme, it is able to initiate transcription only at the promoters for late genes.

The successive replacements of sigma factor have dual consequences. Each time the subunit is changed, the RNA polymerase becomes able to recognize a new class of genes, *and* no longer recognizes the previous class. These switches therefore constitute global changes in the activity of RNA polymerase. Probably all or virtually all of the core enzyme becomes associated with the sigma factor of the moment; and the change is irreversible.

New sigma factors are utilized during **sporulation**, an alternative life style available to some bacteria. At the end of the **vegetative phase**, logarithmic growth ceases because nutrients in the medium become depleted. This triggers sporulation; DNA is replicated, a genome is segregated at one end of the cell, and eventually it is surrounded by the tough spore coat. The process takes about 8 hours.

Sporulation involves a drastic change in the biosynthetic activities of the bacterium, and many genes are involved. The basic level of control lies at transcription. Some of the genes that functioned in the vegetative phase are turned off during sporulation, but most continue to be expressed. In addition, the genes specific for sporulation are expressed only during this period. At the end of the period of sporulation, about 40% of the bacterial mRNA is sporulation-specific.

Genes that are necessary specifically for sporulation can be identified by *spo* mutations, which allow normal vegetative growth but block sporulation. These asporogenous mutants can be classified by the stage at which sporulation is blocked; they are described as *spo0* (sporulation cannot start at all), *spoI*, *spoII*, etc., as later stages are defective. Some constitute mutations in enzymes or structural products that are needed for spore development; but some identify genes concerned with regulating the transition in lifestyles.

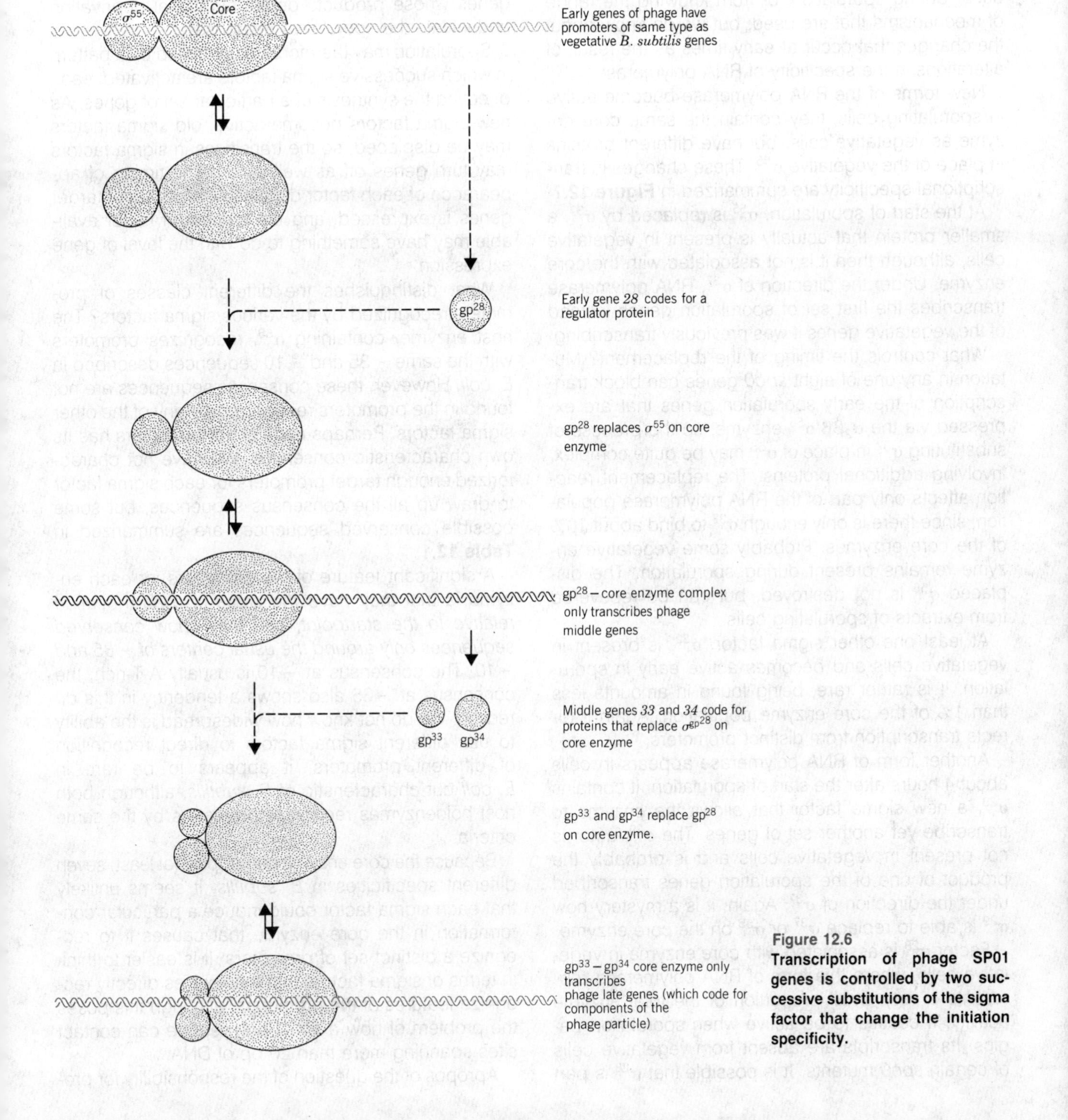

Figure 12.6
Transcription of phage SP01 genes is controlled by two successive substitutions of the sigma factor that change the initiation specificity.

We are a long way from defining all the changes that occur during sporulation, or from knowing the range of mechanisms that are used; but we do know about the changes that occur at early times as the result of alterations in the specificity of RNA polymerase.

New forms of the RNA polymerase become active in sporulating cells; they contain the same core enzyme as vegetative cells, but have different proteins in place of the vegetative σ^{55}. These changes in transcriptional specificity are summarized in **Figure 12.7**.

At the start of sporulation, σ^{55} is replaced by σ^{37}, a smaller protein that actually is present in vegetative cells, although then it is not associated with the core enzyme. Under the direction of σ^{37}, RNA polymerase transcribes the first set of sporulation genes instead of the vegetative genes it was previously transcribing.

What controls the timing of the replacement? Mutation in any one of eight *spo0* genes can block transcription of the early sporulation genes that are expressed via the $\alpha_2\beta\beta'\sigma^{55}$ enzyme, so the process of substituting σ^{37} in place of σ^{55} may be quite complex, involving additional proteins. The replacement reaction affects only part of the RNA polymerase population, since there is only enough σ^{37} to bind about 10% of the core enzymes. Probably some vegetative enzyme remains present during sporulation. The displaced σ^{55} is not destroyed, but can be recovered from extracts of sporulating cells.

At least one other sigma factor, σ^{32}, is present in vegetative cells and becomes active early in sporulation. It is rather rare, being found in amounts less than 1% of the core enzyme population. Again it directs transcription from distinct promoters.

Another form of RNA polymerase appears in cells about 4 hours after the start of sporulation. It contains σ^{29}, a new sigma factor that allows the enzyme to transcribe yet another set of genes. The σ^{29} factor is not present in vegetative cells and is probably the product of one of the sporulation genes transcribed under the direction of σ^{37}. Again, it is a mystery how σ^{29} is able to replace σ^{37} or σ^{55} on the core enzyme.

Factor σ^{28} is associated with core enzyme in vegetative cells, where this form of RNA polymerase represents a very small proportion of the total enzyme activity. It *ceases* to be active when sporulation begins. Its transcripts are absent from vegetative cells of certain *spo0* mutants. It is possible that σ^{28} is part of some signalling system, responsible for expressing genes whose products detect nutritional deprivation and initiate the sporulative response.

Sporulation may therefore be controlled by a pattern in which successive sigma factors are activated, each directing the synthesis of a particular set of genes. As new sigma factors become active, old sigma factors may be displaced, so the transitions in sigma factors may turn genes off as well as on. The timing of appearance of each factor dictates when its set of target genes is expressed; and the amount of factor available may have something to do with the level of gene expression.

What distinguishes the different classes of promoters recognized by the various sigma factors? The host enzyme, containing σ^{55}, recognizes promoters with the same -35 and -10 sequences described in *E. coli*. However, these consensus sequences are not found in the promoters recognized by any of the other sigma factors. Perhaps each set of promoters has its own characteristic consensus. We have not characterized enough target promoters for each sigma factor to draw up all the consensus sequences, but some possible conserved sequences are summarized in **Table 12.1**.

A significant feature of the promoters for each enzyme is that *they have the same size and location relative to the startpoint, and they show conserved sequences only around the usual centers of -35 and -10*. The consensus at -10 is usually A-T-rich; the consensus at -35 also shows a tendency in this direction. We do not know how widespread is the ability to use different sigma factors to direct recognition of different promoters; it appears to be rare in *E. coli* but characteristic of *B. subtilis*, although both host holoenzymes recognize promoters by the same criteria.

Because the core enzyme can display (at least) seven different specificities in *B. subtilis*, it seems unlikely that each sigma factor could induce a particular conformation in the core enzyme that causes it to recognize a distinct set of promoters. It is easier to think in terms of sigma factors that themselves directly recognize features of the promoters, although this poses the problem of how a small polypeptide can contact sites spanning more than 20 bp of DNA.

Apropos of the question of the responsibility for pro-

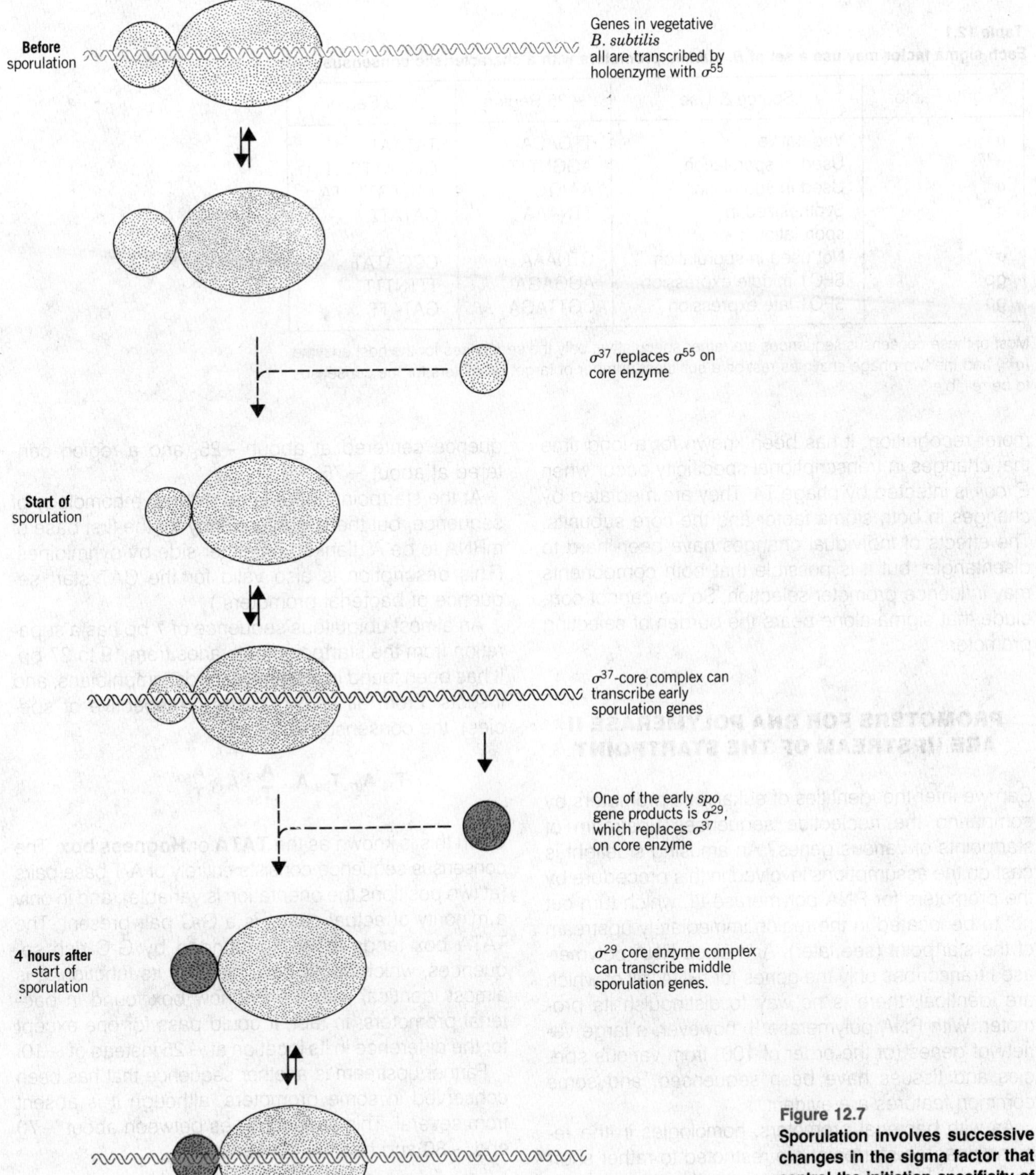

Figure 12.7
Sporulation involves successive changes in the sigma factor that control the initiation specificity of RNA polymerase.

Table 12.1
Each sigma factor may use a set of *B. subtilis* promoters with a characteristic consensus.

Sigma Factor	Source & Use	−35 Region	−10 Region
σ^{55}	Vegetative	TTGACA	TATAAT
σ^{37}	Used in sporulation	AGGNTTT	GGNATTGNT
σ^{32}	Used in sporulation	AAATC	TANTGTTNTA
σ^{29}	Synthesized in sporulation	TTNAAA	CATATT
σ^{28}	Not used in sporulation	CTNAAA	CCGATAT
gp^{28}	SPO1 middle expression	AGGAGA	TTTNTTT
gp^{33-34}	SPO1 late expression	CGTTAGA	GATATT

Most of these consensus sequences are rather speculative; only the sequences for the host enzyme (σ^{55}) and the two phage enzymes rest on a sufficient number of target promoters for the consensus to be reliable.

moter recognition, it has been known for a long time that changes in transcriptional specificity occur when *E. coli* is infected by phage T4. They are mediated by changes in both sigma factor and the core subunits. The effects of individual changes have been hard to disentangle, but it is possible that both components may influence promoter selection. So we cannot conclude that sigma alone bears the burden of selecting promoters.

PROMOTERS FOR RNA POLYMERASE II ARE UPSTREAM OF THE STARTPOINT

Can we infer the identities of eukaryotic promoters by comparing the nucleotide sequences upstream of startpoints of various genes? An amusing sidelight is cast on the assumptions involved in this procedure by the promoters for RNA polymerase III, which turn out not to be located in the region immediately upstream of the startpoint (see later). And since RNA polymerase I transcribes only the genes for rRNA, all of which are identical, there is no way to distinguish its promoter. With RNA polymerase II, however, a large variety of genes (of the order of 100) from various species and tissues have been sequenced, and some common features are evident.

As with bacterial promoters, homologies in the regions near the startpoint are restricted to rather short sequences. Homologies of potential significance have been noted in three regions: the startpoint, a sequence centered at about −25, and a region centered at about −75.

At the startpoint, there is no extensive homology of sequence, but there is a tendency for the first base of mRNA to be A, flanked on either side by pyrimidines. (This description is also valid for the CAT start sequence of bacterial promoters.)

An almost ubiquitous sequence of 7 bp has a separation from the startpoint that varies from 19 to 27 bp. It has been found in mammals, birds, amphibians, and insects. From all known cases (irrespective of species), the consensus is

$$\mathbf{T}_{82}\ \mathbf{A}_{97}\ \mathbf{T}_{93}\ \mathbf{A}_{85}\ \genfrac{}{}{0pt}{}{\mathbf{A}_{63}}{\mathbf{T}_{37}}\ \mathbf{A}_{83}\ \genfrac{}{}{0pt}{}{\mathbf{A}_{50}}{\mathbf{T}_{37}}$$

Often this is known as the **TATA** or **Hogness box**. The consensus sequence consists entirely of A-T base pairs (at two positions the orientation is variable), and in only a minority of actual cases is a G-C pair present. The TATA box tends to be surrounded by G-C rich sequences, which could be a factor in its function. It is almost identical with the Pribnow box found in bacterial promoters; in fact, it could pass for one except for the difference in its location at −25 instead of −10.

Farther upstream is another sequence that has been conserved in some promoters, although it is absent from several. This sequence lies between about −70 and −80 and has the consensus

$$\mathbf{G\,G}\genfrac{}{}{0pt}{}{\mathbf{T}}{\mathbf{C}}\mathbf{C\,A\,A\,T\,C\,T}$$

It is sometimes known as the **CAAT box**. We do not yet know whether it is a general feature of promoters or is specific for certain classes.

Attempts to define the promoters for eukaryotic RNA polymerases have taken advantage of the precedents established with bacterial RNA polymerase. Eukaryotic systems suffer from two particular limitations. The first is that virtually no promoter mutations have been identified *in vivo*, so we start without any prior information on the location of the promoter. The second is that we have not yet been able to retrieve the DNA sequences that are bound by any of the RNA polymerases. Recovery of protected fragments is rendered difficult by the complexity of the enzyme preparations and the lack of information on exactly what constitutes the active structure of the enzyme. Of course, it is only a matter of time until a successful system is developed that will allow us to recover the DNA binding site from an initiation complex.

A less direct assay, but one that actually is more informative in some ways, is to define the promoter in terms of its ability to initiate transcription in a suitable test system. Three types of system have been used.

The ***in vitro* system** takes the classic approach of purifying all the components and manipulating conditions until faithful initiation is seen. "Faithful" initiation is defined as production of an RNA starting at the site corresponding to the 5′ end of mRNA (or rRNA or tRNA precursors). Systems for each of the three RNA polymerases are now in various stages of purification; ultimately each will consist of a preparation in which all of the components have been defined. Then we shall be able to compare the activities *in vitro* of RNA polymerases from different tissues or species.

The **oocyte system** follows the principles established for translation, and relies on injection of a suitable DNA template, this time into the nucleus of the *X. laevis* oocyte. The RNA transcript can be recovered and analyzed. The main limitation of this system is that it is entirely restricted to the conditions that prevail in the oocyte.

***In vivo* systems** do not (as the name might imply) necessarily involve characterizing the ability of a cell to transcribe one of its usual products. What is done here is to follow the ability of a cultured cell to transcribe a template introduced by transfection (a procedure that allows an exogenous DNA to enter a cell and be expressed; the events involved are discussed in Chapter 38). The system is genuinely *in vivo* in the sense that transcription is accomplished by the same apparatus responsible for expressing the cell's own genome; but it differs from the natural situation in the sense that the template may consist of a gene that would not usually be transcribed in the host cell.

The approach to characterizing the promoter is the same in all three systems. We seek to manipulate the template *in vitro* before it is submitted to the system for transcription. Sometimes this is called "surrogate genetics."

When a particular fragment of DNA can be used to initiate transcription, it is taken to include a functional promoter. Then the boundaries of the sequence constituting this promoter can be determined by reducing the length of the fragment from either end, until at some point it ceases to be active. The type of protocol is illustrated in **Figure 12.8**. The boundary upstream can be identified by progressively removing material from this end until promoter function is lost. For the boundary downstream, it is necessary to reconnect the shortened promoter to the sequence to be transcribed (since otherwise there is no product to assay).

We have to take several precautions to avoid extraneous effects. To ensure that the promoter is always in the same context, the same long upstream sequence is always placed next to it. Because termination may not occur properly in the *in vitro* systems, the template may be cut at some distance from the promoter (usually about 500 bp downstream), to ensure that all polymerases "run off" at the same point, generating an identifiable transcript.

Once the boundaries of the promoter have been defined, the importance of particular bases within it can be determined by introducing point mutations or other rearrangements in the sequence. As with bacterial RNA polymerase, these can be characterized as *up* or *down* mutations. Some of these rearrangements may affect only the *rate* of initiation; others may influence the *site* at which initiation occurs, as seen in a change of the startpoint. To be sure that we are dealing with comparable products, in each case it is necessary to characterize the 5′ end of the RNA, as described previously in Figure 12.2.

Results with the *in vitro* systems are mostly fairly straightforward. Proceeding from the upstream direc-

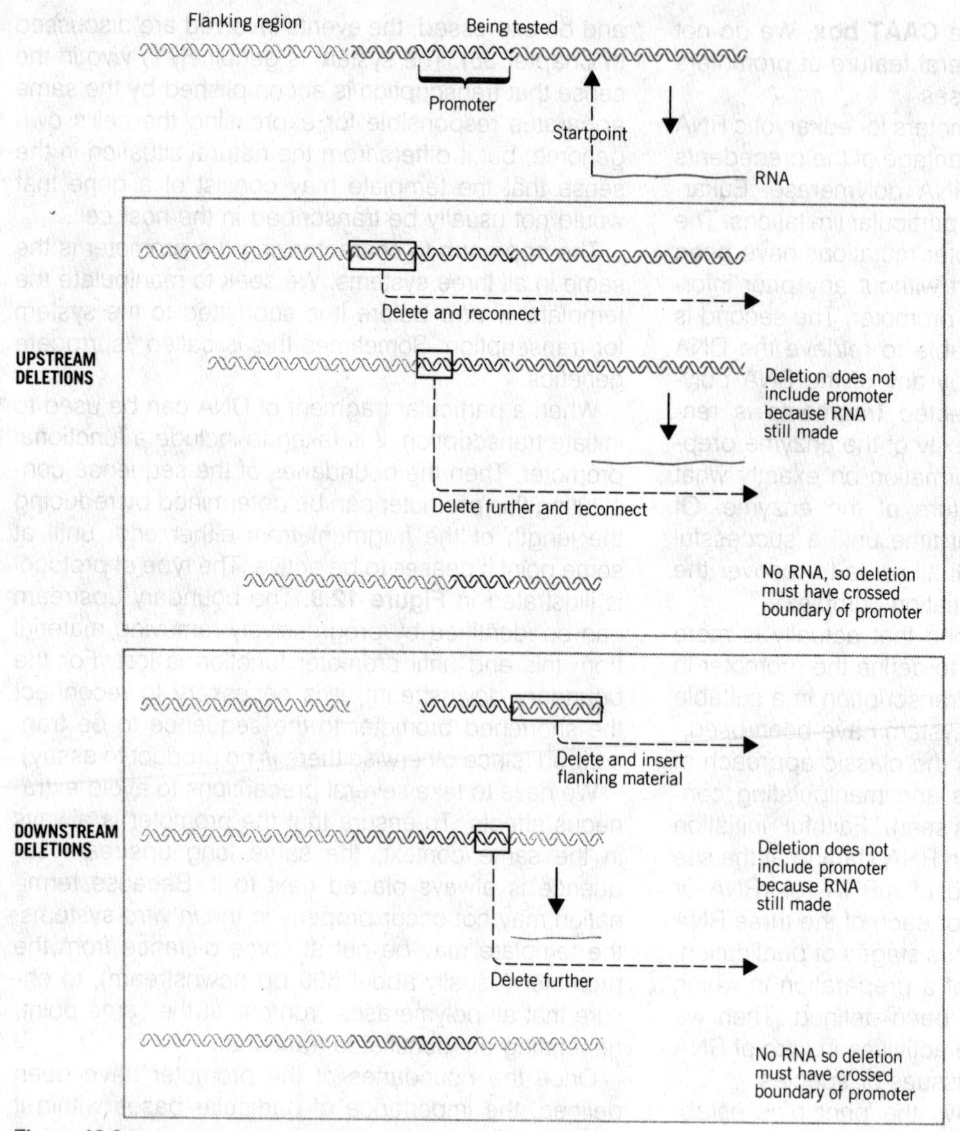

Figure 12.8

Promoter boundaries can be determined by deletions that approach from either side.

Each deletion removes the material on one side of the unit being tested and extends farther into the unit than the last. When one deletion fails to prevent RNA synthesis but the next stops transcription, the boundary of the promoter must lie between them.

tion, sequences can be progressively removed without any effect, until reaching a point located somewhere between −45 and −30. The limit defines the left boundary of the promoter. It is always upstream of the TATA box. Once the limit is transgressed, transcription is reduced twentyfold or more. The importance of the TATA box is confirmed by the fact that single base substitutions in it act as strong down mutations. One such mutation reversed the orientation of an A-T pair, so the base composition alone is not sufficient to fulfill the function of the sequence.

Proceeding from the downstream direction, the right boundary of the promoter is usually close to the startpoint. The farthest location to the right seen so far is +6. In two well-characterized cases, it lies at about −10 or −12, *upstream of the startpoint itself*. When the startpoint is deleted, initiation occurs at a point in the template that is the *same distance* from the promoter as the original startpoint, except that it may be adjusted by a base or two in order to find a purine (usually A) with which to initiate. The apparent failure always to include the startpoint could mean that the geometry of the complex may determine where initiation occurs, presumably because the enzyme stretches a certain distance downstream from its binding site. However, the efficiency of initiation may be somewhat reduced by the absence of the usual startpoint.

RNA POLYMERASE II PROMOTERS ARE MULTIPARTITE

Analysis *in vitro* identifies an element whose behavior is analogous to our concept of the bacterial promoter: a short and well-defined sequence just upstream from the startpoint. But with *in vivo* or *in oocyte* systems, we see instead a strong dependence on sequences farther upstream from the TATA region.

Their exact location varies, but they lie more than 40 bp upstream of the TATA box. When this region is deleted, initiation at the usual startpoint is reduced to about 2% of its previous level. By contrast, if the TATA box is removed, initiation continues to occur, but the startpoint varies from its usual precise location.

More detailed analysis of the sequences needed for promoter function is provided by the technique of **linker scanning**, which allows clusters of mutations to be introduced at particular sites. **Figure 12.9** illustrates the protocol. Deletion mutants are made from both sides, removing the regions either 5′ or 3′ to the relevant site. A "linker sequence" is added to the end of each deletion; the linker consists of a short synthetic oligonucleotide that includes the sequence recognized by some restriction enzyme. Both fragments are cleaved with the enzyme, and then they are joined

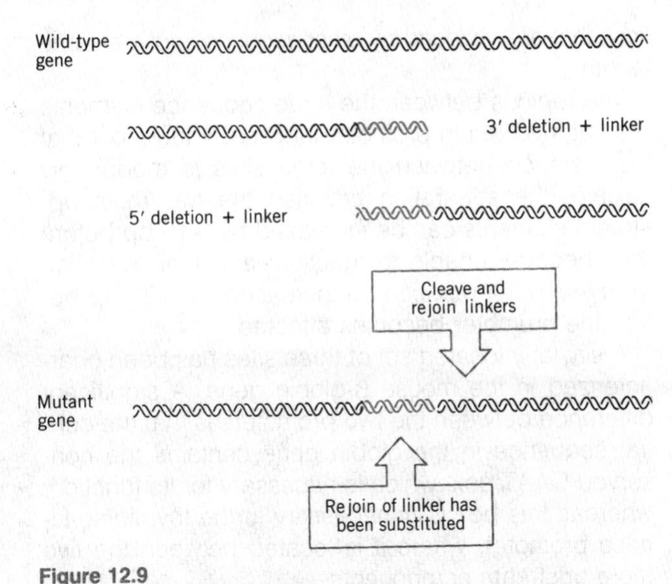

Figure 12.9
The linker-scanning technique allows a short sequence of wild-type gene to be replaced by the linker sequence at the point corresponding to the ends of the matching deletions.

crosswise (a reaction that is characteristic of ends cleaved by some restriction enzymes; see Chapter 17).

The result is to insert the sequence of the linker in place of the sequence originally located at the point where the deletions meet. By using matching 5′ and 3′ deletions that end at a series of sites along the wild-type sequence, the entire sequence can be "scanned" for its sensitivity to mutation.

By using this technique, three distinct sequence elements have been identified in the thymidine kinase gene of herpes virus. The region around the TATA box is needed for accurate initiation. Two other separate regions, located between −47 and −61 and between −80 to −105, are needed for efficient initiation. The introduction of double mutations shows that these latter two regions are concerned with the same function, for mutations in either one alone have as much effect as mutations in both together.

This multipartite structure means that we cannot investigate the internal structure of the promoter simply by making deletions within it; deletions could abolish function by changing critical distances between components, even though the deleted sequences are themselves irrelevant. So we need to make small deletions that are replaced by other sequences of the same length.

The regions between the three sequence elements are not involved in promoter function. It turns out that the distance between the three sites is moderately flexible; the separation between the two most upstream elements can be increased by >15 bp before they become unable to function; and their distance from the TATA box can be increased by >30 bp before the promoter becomes affected.

A similarly located set of three sites has been characterized in the mouse β-globin gene. A significant difference between the two promoters is that the central sequence in the globin gene contains the conserved CAAT box, which is necessary for its function, whereas this box is unnecessary in the thymidine kinase promoter, where it is located between the two more upstream components.

In vivo the promoter therefore seems to have the two separate types of component illustrated in **Figure 12.10**.

The most upstream component may consist of two sequence elements, one lying between −80 and −110, the other lying between −50 and −70. The second may or may not include the CAAT box sequence. Together these elements have a strong influence on the *frequency of initiation*. This component probably has a major effect on the binding of RNA polymerase. However, the residual transcription that occurs in its absence does initiate at the proper startpoint.

Close to the startpoint is an element that surrounds the highly conserved TATA box, and whose deletion introduces a more erratic quality into the *choice of startpoint*, although any overall reduction in transcription is relatively small. The role of this sequence could be to align the RNA polymerase so that it initiates at the proper site. This idea is consistent with the fact that some promoters naturally lack the TATA sequence, and their mRNAs start at more than one point instead of being restricted to the usual single startpoint.

How can a promoter consist of separated elements that stretch over a distance of DNA greater than RNA polymerase could contact? Two models are illustrated in **Figure 12.11**. One possibility is to think in terms of models that suppose RNA polymerase initially contacts the site farther upstream and then moves to the site nearer the startpoint. Another view requires us to remember that *in vivo* DNA is not stretched out in linear fashion; its compact organization could bring into juxtaposition sites that are separated on the duplex molecule (see Chapter 29). The binding site for RNA polymerase could consist of DNA sequences that are not contiguous, but are held together by proteins that bind DNA. This model implies further that the *spacing* (rather than the exact sequence) between these components of the promoter may be important, which is somewhat contradicted by the analysis of the thymidine kinase promoter.

In either type of model, we can see that the region farther upstream could be the most important for initially binding RNA polymerase, but the region nearer the startpoint could be required to hold the enzyme in a configuration that allows it to recognize the exact

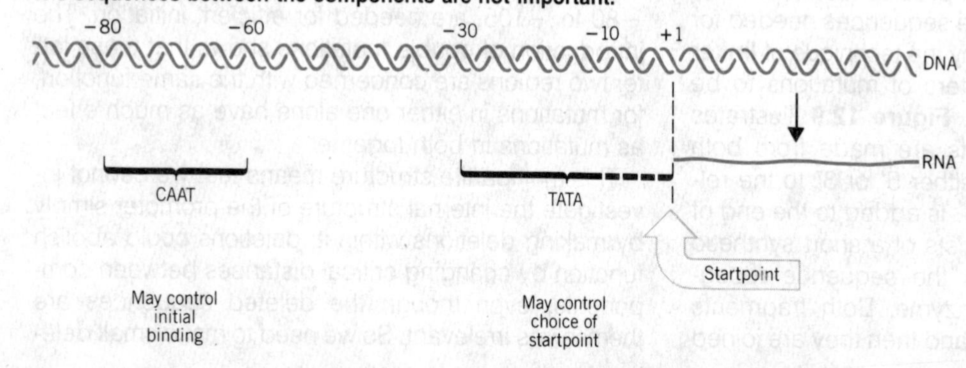

Figure 12.10
Promoters for RNA polymerase II contain separate sequence components with different roles. The sequences between the components are not important.

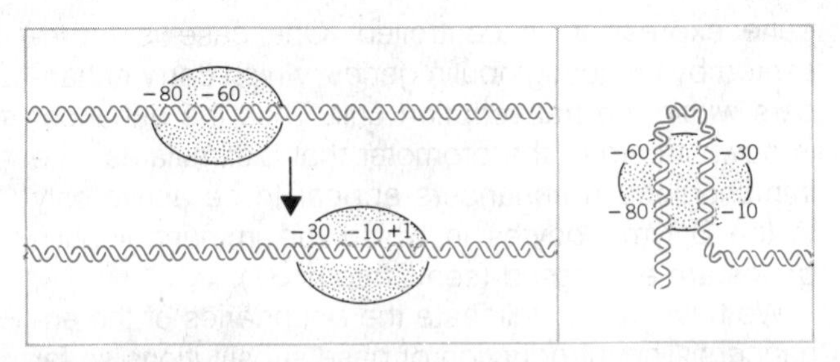

Figure 12.11
One model to reconcile the size of the promoter with the size of the RNA polymerase supposes that the enzyme moves. Another supposes that the DNA is compactly organized.

startpoint. Thus when the sequence farther upstream is absent, the ability of RNA polymerase to bind to the DNA is much reduced; but those enzyme molecules that do bind can initiate accurately because the TATA box is present. On the other hand, when the TATA box is deleted, RNA polymerase remains able to bind efficiently to the upstream sequence, but its contacts in the region around the startpoint are less precise, allowing initiation to occur at more than one point.

How are we to explain the discrepancy between the need *in vivo* for the sequences upstream from −50 and their apparent lack of importance *in vitro*? A major responsibility for this effect may lie with the differing efficiencies of transcription in the two circumstances. The *in vitro* system is relatively inefficient; less than 1% of the templates actually are transcribed. We do not know just what proportion of the templates are utilized *in vivo* or *in oocyte*, but it could very well be much higher. The *in vitro* level could even correspond to the residual expression that we see *in vivo* in the absence of sequences upstream from −60.

At all events, the simpler structure of the template *in vitro*, as a DNA molecule not organized in the usual proteinaceous structure, may mean that its recognition by RNA polymerase is different, certainly less efficient, perhaps because it uses only some features of the promoter.

We have scarcely begun to characterize the protein components with which the elements of the promoter interact. Immediately this brings us back to the question of which polypeptides are subunits of RNA polymerase and which are accessory factors.

At least two factors are needed for *D. melanogaster* RNA polymerase to initiate transcription *in vitro*, and one of them binds to a region containing the TATA box. Some of the characteristic features of promoters therefore may interact directly with factors (as distinguished chromatographically) rather than with RNA polymerase II itself.

A factor in HeLa (human) cells that is needed to transcribe only some promoters (including SV40) binds to the DNA in the upstream region (between −70 and −110). Thus the factor could be involved in allowing RNA polymerase initially to recognize a certain class of promoters. Factors with this type of activity could be involved in regulating transcription.

ENHANCERS ARE BIDIRECTIONAL ELEMENTS THAT ASSIST INITIATION

Even in considering the demands of the *in vivo* systems, we have thought of the promoter essentially as an isolated region responsible for binding RNA polymerase. But eukaryotic promoters do not necessarily function alone. In at least some cases, the activity of a promoter is enormously increased by the presence of another sequence, known as an **enhancer**.

An enhancer is distinguished from the promoter itself by two characteristics: its position relative to the promoter need not be fixed, but can vary substantially; and it can function in either orientation. An enhancer is not restricted to assisting a particular promoter, but can stimulate any promoter placed in its vicinity.

The first enhancer to be discovered lies in the genome of the virus SV40. The DNA contains two identical sequences of 72 bp each, repeated in tandem about 200 bp upstream of the startpoint of a transcription unit. These **72 bp repeats** lie in a region that has an unusual nucleoprotein structure, apparently one that is more exposed than usual (see Chapter 30). Deletion mapping shows that either one of these repeats is adequate to support normal transcription; but removal of both repeats greatly reduces transcription *in vivo*.

By this type of criterion, we might argue that the repeated region constitutes an upstream component of the promoter. But reconstruction experiments in which the 72 bp sequence is removed from the DNA and then is inserted elsewhere show that normal transcription can be sustained so long as it is present *anywhere* on the DNA molecule. In fact, if a β-globin gene is placed on a DNA molecule that contains a 72 bp repeat, its transcription is increased *in vivo* more than 200-fold, even when the 72 bp sequence is as much as 1400 bp upstream or 3300 bp downstream from the startpoint. And these are simply the limits that have been tested so far; we have yet to discover at what distance the 72 bp sequence fails to work. The sequence can be inverted and replaced; and it still works.

The moral of these results is clear. We need to be careful in defining the components of the promoter. It is not enough just to show that deletion of a particular sequence reduces transcription; we must also investigate the importance of the *location* of the deleted sequence. So what is a promoter? *If we use a working definition that it constitutes a sequence or sequences of DNA that must be in a fixed location relative to the startpoint, the TATA box and other upstream elements are included, but the enhancer is excluded.*

Several viral genomes have elements that bear a structural resemblance to the SV40 72 bp repeats, although there is no homology of sequence. One of particular interest is carried by retroviruses, viruses whose insertion into a host genome may activate genes in the vicinity, possibly because they provide an enhancer. Some viral enhancers show specificity for cell type in their function. The ability of the enhancer to function in a particular cell may be partly responsible for determining the host range of the virus. In the case of polyoma, a mutation in the enhancer extends the range of cells that the virus can infect.

In one particularly striking case, an enhancer is responsible for a hormonal response. Transcription of the mouse mammary tumor virus DNA is stimulated by steroid hormones. The element responsible for the hormone response is located about 100 bp upstream of the startpoint, binds a complex consisting of the hormone and its protein receptor, and can stimulate the function of other genes when placed in either orientation at variable distances from their promoters. In short, it behaves as an enhancer associated with a hormone-binding sequence.

Cellular enhancers have been discovered in the form of elements in the genome that stimulate the use of a nearby promoter in a specific tissue. Such enhancers may provide part of the regulatory network by which gene expression is controlled. One case is represented by immunoglobulin genes, which carry enhancers *within* the transcription unit. Thus the enhancer is downstream of the promoter that it stimulates. The immunoglobulin enhancers appear to be active only in the B lymphocytes in which the immunoglobulin genes are expressed (see Chapter 37).

We have yet to delineate the boundaries of the enhancers. The introduction of base substitutions so far has created only down mutations, at a variety of sites. Some homologies of sequence have been noticed between various enhancers, but we do not have sufficient data to deduce a convincing consensus sequence.

How can an enhancer stimulate initiation at a promoter that can be located on either side of it at apparently any distance? Some possibilities are that it might be concerned with structure, location, or enzyme binding.

One possibility is that it might be responsible for locating the template at a particular place within the cell—for example, attaching it to the nuclear matrix.

A more popular idea is that the enhancer might change the overall structure of the template—for example, by influencing the DNA-protein organization of chromatin, or by changing the density of supercoiling. All the enhancers so far characterized include a stretch of alternating pyrimidine-purine residues, just the sort of sequence likely to form Z-DNA (see Chapter 4). One possibility is that enhancers function by forming a short length of Z-DNA, although we remain mystified as to what effect the Z-DNA has on local transcription. However, a structural effect of this nature would explain why the enhancer can affect a promoter on either side.

Enhancers could be responsible for directly binding RNA polymerase (because of altered structure, location, or some other feature), although then the enzyme would have to move to the promoter proper.

The fact that at least some enhancers show tissue specificity implies that whatever feature is responsible for its function may require activation in the particular cell type, presumably by binding of some regulator protein. The generality of enhancement is not yet clear. We do not know what proportion of cellular promoters usually rely on an enhancer to achieve their customary level of expression.

A DOWNSTREAM PROMOTER FOR RNA POLYMERASE III

Before the promoter of the genes coding for 5S RNA in *X. laevis* was identified, all attempts to identify promoter sequences assumed that they would lie upstream from the startpoint. But in the 5S RNA genes, transcribed by RNA polymerase III, the promoter lies well *within* the transcription unit, more than 50 bases downstream from the startpoint.

A 5S RNA gene can be transcribed when a plasmid carrying it is used as template for a nuclear extract obtained from *X. laevis* oocytes, the tissue in which the gene usually is expressed. The promoter was located by using plasmids in which deletions extended into the gene from either direction. The 5S RNA product continues to be synthesized when the entire sequence upstream of the gene is removed.

When the deletions continue into the gene, a product very similar in size to the usual 5S RNA continues to be synthesized so long as the deletion ends before about base +55. The first part of the RNA product represents plasmid DNA; the second part represents whatever segment remains of the usual 5S RNA sequence. But when the deletion extends past +55, transcription does not occur. Thus the promoter lies *downstream of position +55*, but causes RNA polymerase III to initiate transcription a more or less fixed distance away. The wild-type startpoint is unique; in deletions that lack it, transcription initiates at the purine base nearest to the position 55 bp upstream of the promoter.

When deletions extend into the gene from its distal end, transcription is unaffected so long as the first 80 bp remain intact. Once the deletion cuts into this region, transcription ceases. This places the downstream boundary position of the promoter at about position +80.

So the promoter for 5S RNA transcription lies between positions +55 and +80 within the gene. A fragment containing this region can sponsor initiation of any DNA in which it is placed, at a position about 55 bp upstream. How does RNA polymerase initiate transcription upstream of its promoter? The most likely explanation is that the enzyme binds to the promoter, but is large enough to contact regions 55 bp away.

As with RNA polymerase II, the geometry of binding to the promoter must dictate the position of the startpoint, subject to the reservation that pyrimidines cannot be used for initiation. The difference between the enzymes is that RNA polymerase II reaches forward to the startpoint from its promoter, whereas RNA polymerase III reaches backward.

Are other promoters for RNA polymerase III similarly located within their transcription units? The boundaries of some of them are indicated in **Figure 12.12**. In the VA1 gene of adenovirus, the promoter lies between +9 and +72. In the $tRNA_1^{Met}$ gene of *X. laevis*, the promoter lies in two separate parts, both within the gene. Deletion mapping shows that both of the sequences from +8 to +30 and from +51 to +72 must be present. Any deletion between them that reduces their separation prevents initiation, but changes in the intermediate sequence, or insertions that increase the separation of the two components (at least up to 30 bp) have no effect. The promoter therefore consists of two separate regions of about 20 bp each, which must be separated by a minimum distance of 20 bp.

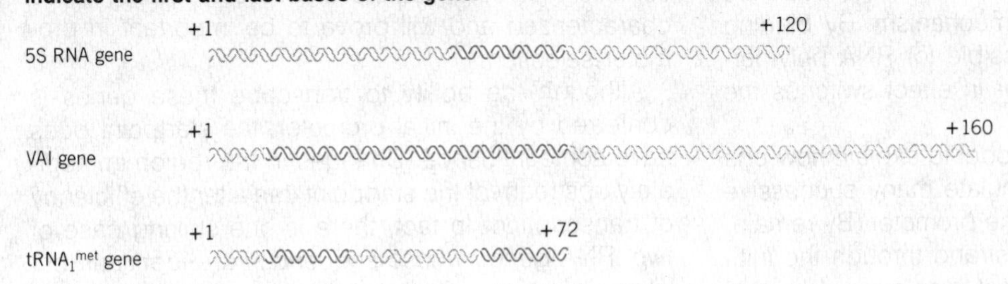

Figure 12.12
Three genes for RNA polymerase III all contain internal promoters (regions in red). The numbers indicate the first and last bases of the gene.

Some other tRNA genes have comparable promoters. The sequences of these regions are highly conserved in eukaryotic tRNAs, a fact that had been interpreted solely in terms of tRNA function, but that now also may be attributed to the needs of the promoter.

The internal location of the promoter poses an important question. When a promoter lies outside the transcription unit itself, presumably it can evolve freely just to meet the needs of the enzyme. But the sequences needed for initiation of the RNA polymerase III transcription units lie within rather different types of genes, and are therefore constrained to meet the needs of products as diverse as 5S RNA, VA RNA, tRNA, and small nuclear RNAs. How then are they able also to provide whatever features are needed for recognition by RNA polymerase III?

An alternative to imposing a requirement that the enzyme can recognize a wide variety of promoter sequences is to suppose that it acts via an ancillary factor. In this case, there could be a different factor responsible for allowing the enzyme to bind to each type of promoter. In fact, it turns out that RNA polymerase III can transcribe the 5S RNA genes only in the presence of an added factor, a 37,000 dalton protein that binds to the region from +45 to +96. (The protein serves a dual purpose; it also binds the 5S product in the oocyte.)

A single copy of the protein is bound to the control region, forming a "preinitiation complex" (which also includes another protein, a factor needed for transcription of all class III genes). The existence of the complex signals that the gene is in an "active" state, ready to be transcribed. RNA polymerase can bind to the gene only when the factor has previously bound. The complex is stable, and may remain in existence through many cycles of replication.

The ability to form a preinitiation complex could be a generally used regulatory mechanism. By binding to a promoter to make it possible for RNA polymerase in turn to bind, the factor in effect switches the gene on.

Figure 12.13 presents a model to explain how one molecule of factor could stimulate many successive rounds of transcription from the promoter. By remaining bound to the noncoding strand through the initiation process, the factor is not displaced. We may surmise that other factors of this sort remain to be characterized and will prove to be important in promoter selection.

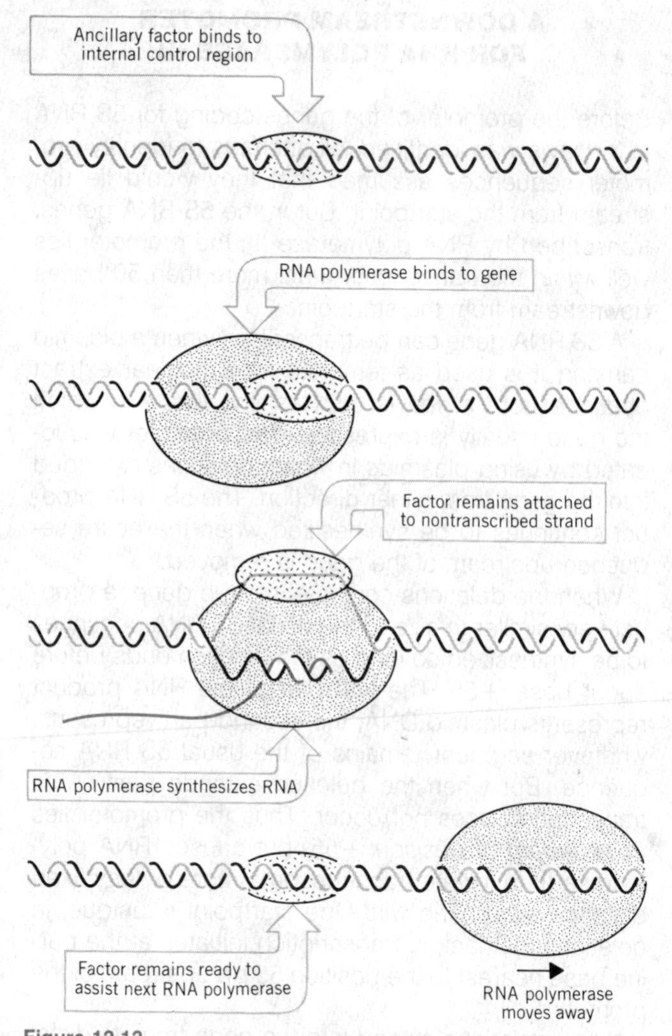

Figure 12.13
The ancillary control factor for the 5S gene binds to the noncoding strand, so it can remain attached to the DNA through many cycles of transcription.

Although the ability to transcribe these genes is conferred by the initial promoter, the startpoint does have some influence. Changes in the region immediately upstream of the startpoint can alter the efficiency of transcription. In fact, there is one striking case of two tRNA genes whose sequences are identical, but whose upstream flanking regions are different; the genes are transcribed at very different rates. Thus the primary responsibility for recognition lies with the internal promoter; but some responsibility for establishing the frequency of initiation lies with the region at the startpoint.

FURTHER READING

The molecular interaction of bacterial RNA polymerase with its promoters has been well described by **Gilbert** in *RNA Polymerase* (Losick & Chamberlin, Eds., Cold Spring Harbor Laboratory, New York, 1976, pp. 193–206) and by **Siebenlist, Simpson & Gilbert** (*Cell* **20**, 269–281, 1980). The stages of the interaction have been reviewed by **Von Hippel et al.** (*Ann. Rev. Biochem.* **53**, 389–449, 1984). The analysis of consensus sequences has been dealt with by **Rosenberg & Court** (*Ann. Rev. Genet.* **13**, 319–353, 1979). Bacterial promoter sequences have been compiled by **Hawley & McClure** (*Nucleic Acids Res.* **11**, 2237–2255, 1983). Involvement of supercoiling in promoter function has been reviewed by **Smith** (*Cell* **24**, 599–601, 1981). Changes induced by sporulation have been reviewed by **Losick & Pero** (*Cell* **25**, 582–584, 1981). An account of phage development via synthesis of new RNA polymerases has been provided by **Bautz** in *RNA Polymerase* (op. cit., pp. 273–284).

Techniques in characterizing eukaryotic RNA polymerase II promoters were discussed by **Corden et al.** (*Science* **209**, 1406–1414, 1981). The discrepancy between promoter function *in vivo* and *in vitro* was brought to light by **McKnight et al.** (*Cell* **25**, 385–398, 1981), who also dissected promoter components (*Science* **217**, 316–324, 1982; *Cell* **31**, 355–365, 1982), as did **Dierks et al.** (*Cell* **32**, 695–706, 1983). A thoughtful analysis of the SV40 enhancer has been provided by **Banerji, Rusconi & Schaffner** (*Cell* **27**, 299–308, 1981); the steroid-regulated enhancer of MTV was discovered by **Chandler, Maler & Yamamoto** (*Cell* **33**, 489–499, 1983). Transcription of 5S genes has been reviewed by **Korn** (*Nature* **295**, 101–105, 1982). The nature of tRNA split promoters has been reviewed by **Hall et al** (*Cell* **29**, 3–5, 1982).

CHAPTER 13
TERMINATION AND ANTITERMINATION

Once RNA polymerase has started transcription, the enzyme continues to move along the template, synthesizing RNA, until it meets a signal to cease activity. At this point, the enzyme stops adding nucleotides to the growing RNA chain, releases the completed product, and dissociates from the DNA template. (We do not know in which order the last two events occur.) Termination requires that all hydrogen bonds holding the RNA-DNA hybrid together must be broken, after which the DNA duplex reforms.

The DNA sequence that provides the signal to cease transcription is called a **terminator** (abbreviated ***t***). Not very much is known about eukaryotic terminators, but terminators in bacteria and their phages have been characterized in detail. They vary widely in both their efficiencies of termination and their dependence on ancillary proteins, at least as seen *in vitro*.

At some terminators, the termination event can be *prevented* by specific ancillary factors that interact with RNA polymerase. **Antitermination** causes the enzyme to continue transcription past the terminator sequence, an event called **readthrough** (the same term used to describe a ribosome's suppression of termination codons).

Antitermination provides a sophisticated control mechanism. Different ancillary proteins (**antitermination factors**) allow RNA polymerase to bypass specific terminator sequences. Such interactions control the ability of the enzyme to read past a terminator into genes lying beyond. This mechanism for controlling gene expression is a feature of phage regulatory circuits.

In approaching the termination event, we must therefore regard it not simply as a mechanism for generating the 3′ end of the RNA molecule, but as an opportunity to control gene expression. Thus the stages when RNA polymerase associates with DNA (initiation) or dissociates from it (termination) are both subject to specific control. There are interesting parallels between the systems employed in initiation and termination. Both require breaking of hydrogen bonds (initial melting of DNA at initiation, RNA-DNA dissociation at termination), and both require additional proteins to interact with the core enzyme. In fact, they may be accomplished by forms of the enzyme that are alternatives to one another.

It is difficult to define the termination point of an RNA molecule that has been synthesized in the living cell. Whether in prokaryotes or eukaryotes, it is always possible that the 3′ end of the molecule has been generated by *cleavage* of the primary transcript, and therefore does not represent the actual site at which RNA polymerase terminated. Unfortunately the 3′ end looks the same whether generated by termination or cleavage; there is no marker comparable to the triphosphate that should identify the original 5′ end.

The best identification of termination sites is provided by systems in which RNA polymerase terminates *in vitro*. Because the ability of the enzyme to terminate is strongly influenced by parameters such as the ionic strength, its termination at a particular point *in vitro* does not prove that this same point is a natural terminator. But now several cases are known in which the same 3′ end can be identified on a particular (prokaryotic) species of RNA, irrespective of whether it has been synthesized *in vivo* or *in vitro*. This is good evidence that we have found a natural termination sequence.

TWO TERMINATION MODES IN *E. COLI* INVOLVE PALINDROMES

The sequences at prokaryotic terminators show no similarities beyond the point at which the last base is added to the RNA. So the responsibility for termination may lie with the *sequences already transcribed* by RNA polymerase. Thus the polymerase does not stop in anticipation of some forthcoming sequence; it relies exclusively on scrutiny of the region it is currently transcribing.

All prokaryotic terminators contain a palindromic sequence just prior to the point of termination. In each case, the palindrome consists of inverted repeats, separated by a short distance. Pairing between the inverted repeats in RNA can generate hairpins consisting of loop and stem structures, as illustrated in the examples of **Figure 13.1**.

The palindromic sequence is implicated directly as a functional site by the locations of point mutations that prevent termination. The mutations occur in a region of about 35 bp prior to the actual site of termination. All of them reside within the stem region of the hairpin. Most would disrupt base pairs, but some do not, so the stability of the hairpin stem cannot be the sole factor involved in its contribution to termination. (However, mutations that enhance pairing usually *increase* termination.)

What is the effect of a hairpin on transcription? Probably all hairpins that form in the RNA product cause the polymerase to slow or pause in RNA synthesis. At a typical terminator, the polymerase pauses for about 60 seconds. (If continuing to transcribe, the enzyme would move about 2000 bp during 1 minute.) The exact delay varies from case to case.

Termination cannot depend simply on encountering a hairpin; there is too much secondary structure in the RNA product for this view to be tenable. The events that follow pausing at a palindrome must depend on the sequences in the vicinity. If no terminator sequences are present, usually the enzyme moves off again to continue transcription.

Two types of terminator have been distinguished in *E. coli* according to whether RNA polymerase requires any additional factors to terminate. The core enzyme can terminate *in vitro* at certain sites in the absence of any other factor. These sites are called **rho-independent** (or **simple**) **terminators**.

Simple terminators have two structural features: the hairpin common to all terminators; and a run of about 6 U residues at the very end of the unit. Both features are needed for termination.

The string of U residues probably provides the signal that allows RNA polymerase to dissociate from the template when it pauses at the hairpin. The rU-dA RNA-DNA hybrid has an unusually weak base-paired structure, so it requires the least energy of any RNA-DNA hybrid to break the association between the two strands. When the polymerase pauses, the RNA-DNA hybrid may unravel from the weakly bonded rU-dA terminal region. Often the actual termination event takes place at any one of several positions toward or at the end of the U-run, as though the enzyme "stutters" during termination.

The importance of the stretch of U bases is confirmed by making deletions that remove some of them; although the polymerase still pauses at the hairpin, it no longer terminates. The series of U bases, of course, corresponds to an A-T-rich region in DNA, so we see that A-T-rich regions may be important in rho-independent termination as well as initiation.

When certain transcription units are used as templates for RNA synthesis *in vitro*, proper termination does not occur; the polymerase pauses at the terminator, but then resumes RNA synthesis. **Rho (ρ) factor** was discovered as a protein whose addition to the *in vitro* system allows RNA polymerase to terminate at certain sites, generating RNA molecules with unique

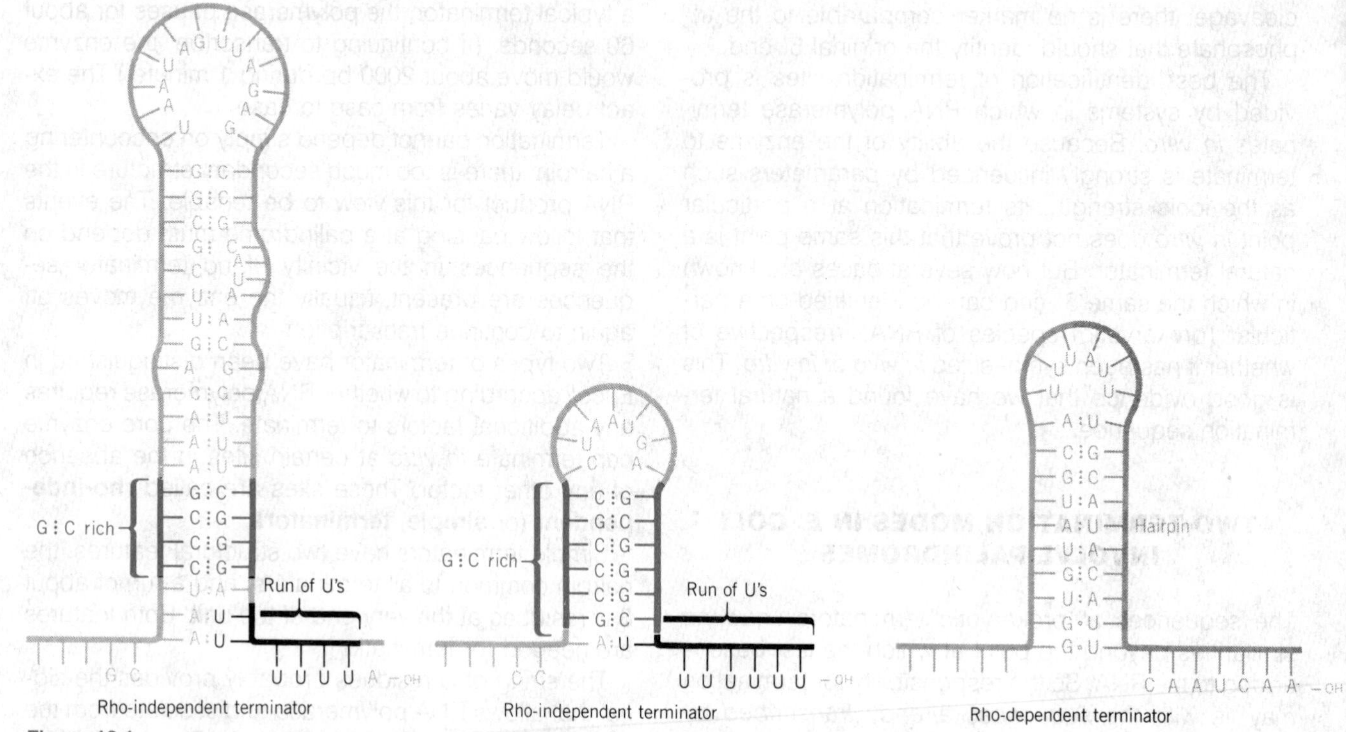

Figure 13.1
Terminator sequences include palindromic regions that form hairpins varying in length from 7 to 20 bp. The stem-loop structures are followed by a run of U residues in rho-independent but not in rho-dependent sites.

3′ ends. This action is called **rho-dependent termination**.

The need for rho factor *in vitro* is somewhat variable. Some terminators require relatively high concentrations of rho, while others function just as well at lower levels. We do not know whether this type of effect represents an artefact of the *in vitro* system or a genuine difference in the efficiency of termination *in vivo*.

Rho-dependent terminators have hairpins that are not particularly G-C-rich; no common sequence has been identified within or near to the hairpin. There is no U-run or other A-T-rich region. The distinction between the rho-dependent and -independent terminators is supported by the results of mutations in the rho factor, which prevent termination at rho-dependent sites, but do not affect termination at the independent sites.

HOW DOES RHO FACTOR WORK?

Rho factor functions solely at the stage of termination. It is a protein of about 55,000 daltons, probably active as a multimer. It displays two activities *in vitro*. It functions catalytically as a termination factor; typically its maximum activity is displayed when it is present at about 10% of the concentration of the RNA polymerase. And it has an NTPase activity—this means that it hydrolyzes free nucleoside triphosphates—which is necessary for termination.

Does rho factor act via recognizing DNA, RNA or RNA polymerase? Its NTPase activity is RNA-dependent; it requires the presence of a polyribonucleotide, preferably of more than 50 or so residues. This suggests that rho binds RNA. A model for rho action supposes that it binds to the 5′ end of a nascent RNA chain and then moves along the RNA, using the hydrolysis of NTPs to provide the necessary energy.

Figure 13.2 suggests that the pausing of RNA polymerase at a terminator site provides an opportunity for rho to catch up. Then the factor interacts with RNA polymerase, causing it to release RNA, and the two proteins dissociate from the nucleic acids. This model predicts that the critical feature of the terminator site is its ability to make RNA polymerase pause. The differences between the various terminators could be explained if they vary in the efficiency with which they cause a pause, for example, because they delay the polymerase for different characteristic periods. The delay could depend on the stability of the hairpin, determined by its length and G-C content. We usually assume that rho will not act just at any site where RNA polymerase has paused, but needs some particular additional sequence.

The idea that rho moves along RNA leads to an

Figure 13.2
Rho factor pursues RNA polymerase along the RNA and can cause termination when it catches the enzyme pausing at a rho-dependent terminator.

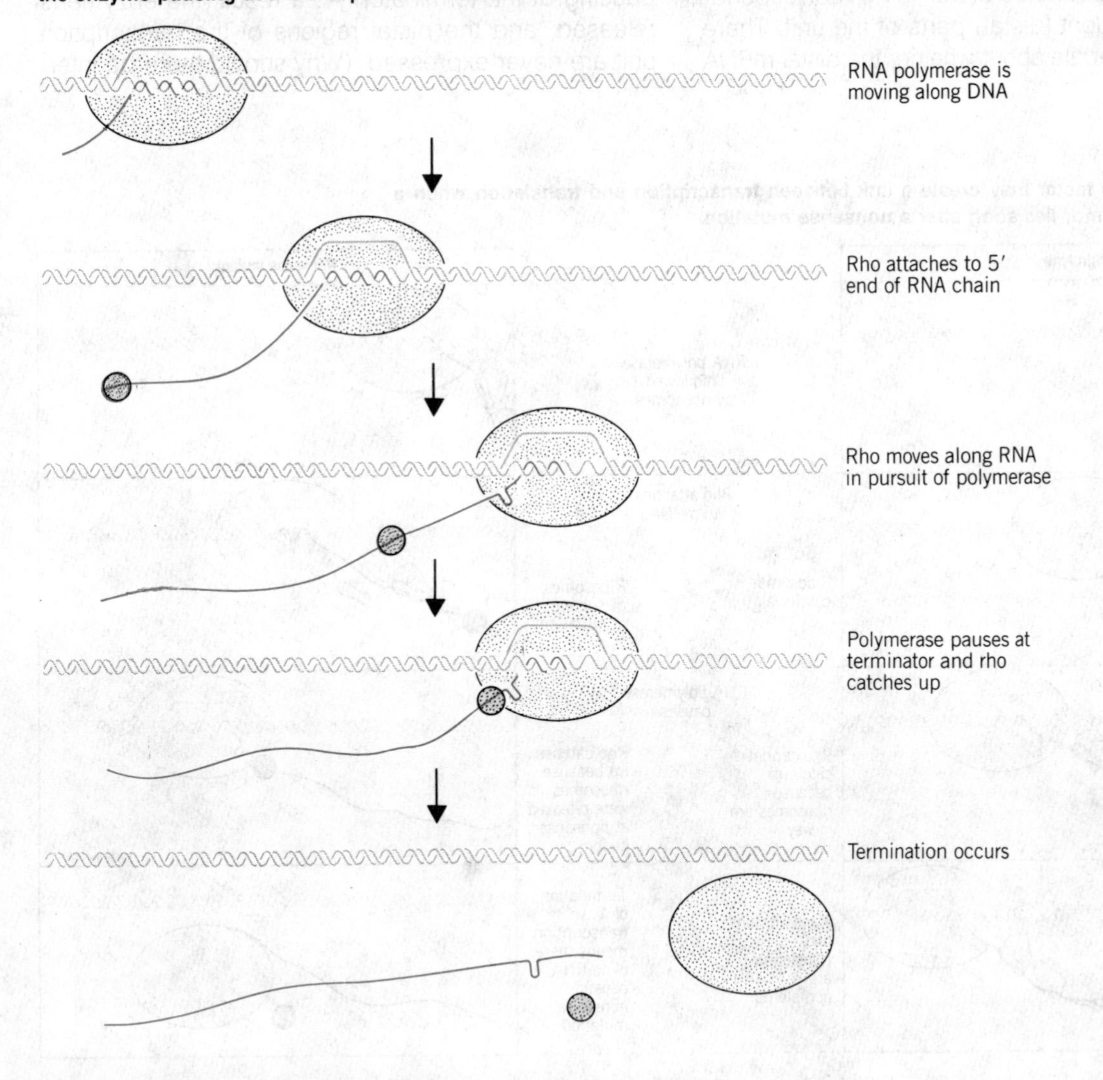

important prediction about the relationship between transcription and translation. Rho must first have access to a free 5′ end and then be able to move along the RNA. Either or both of these conditions may be prevented if ribosomes are translating an RNA. Thus the ability of rho factor to reach the terminator *while RNA polymerase is pausing there* may depend on what is happening in translation.

This model may explain a phenomenon that has been puzzling for a long time. In some cases, a nonsense mutation in one gene of a transcription unit prevents the expression of subsequent genes in the unit. This effect is called **polarity** (see Chapter 10). A common cause is the disappearance of the mRNA corresponding to the subsequent (distal) parts of the unit. There has been a long debate about whether the distal mRNA sequences disappear because they are synthesized but then rapidly degraded, or because the premature termination of translation prevents RNA synthesis from continuing.

Suppose that there may be rho-dependent terminators *within* the transcription unit, that is, before the terminator that *usually* is used. The consequences are illustrated in **Figure 13.3**. Normally these earlier terminators are not used, because the ribosomes prevent rho from catching up with RNA polymerase. But a nonsense mutation releases the ribosomes, so that rho is free to move along the mRNA, enabling it to catch up with RNA polymerase while the enzyme is pausing at the terminator. As a result, the enzyme is released, and the distal regions of the transcription unit are never expressed. (Why should there be inter-

Figure 13.3
The action of the rho factor may create a link between transcription and translation when a rho-dependent terminator lies soon after a nonsense mutation.

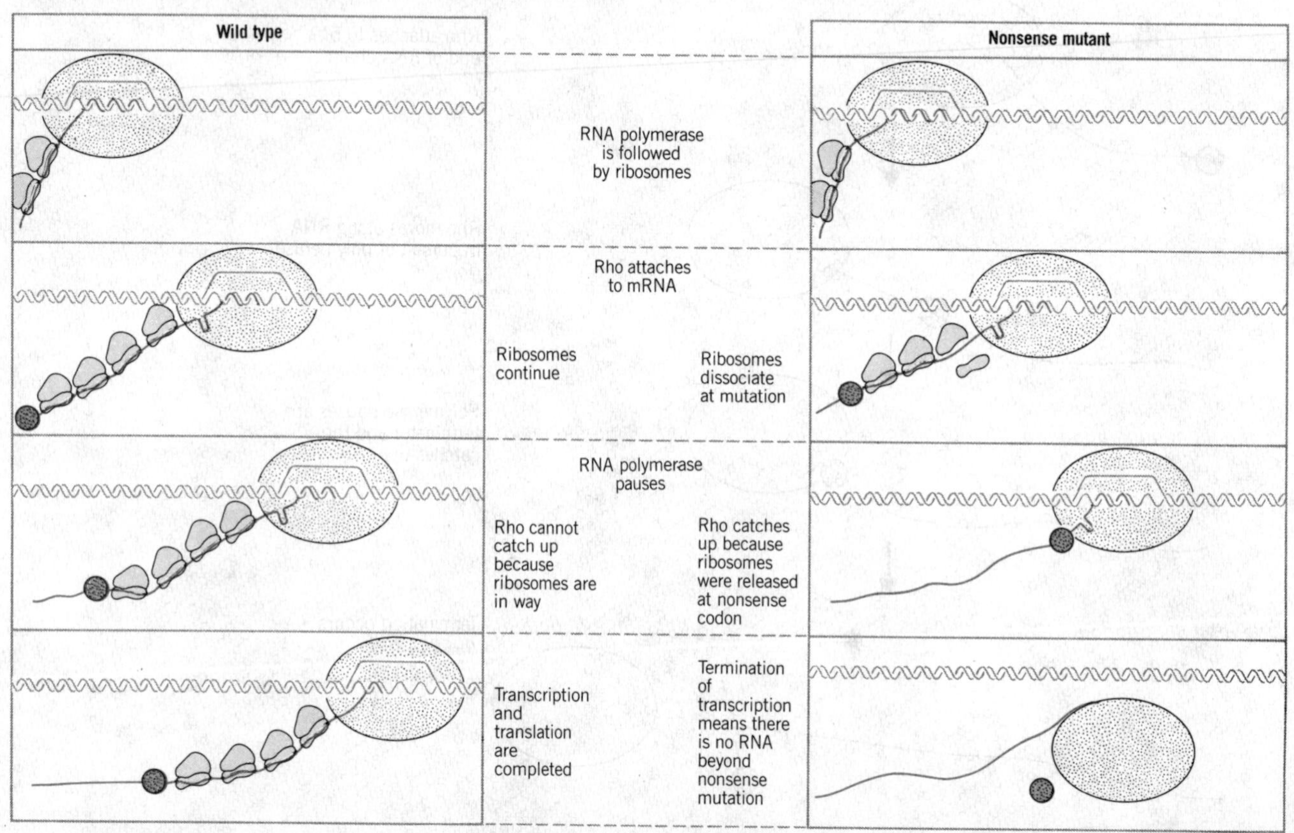

nal terminators? Perhaps they are simply hairpins that mimic the usual rho-dependent terminator by a fortuitous coincidence of sequence.)

Mutations that inhibit rho factor activity have identified the gene that codes for the factor. It is now known as *rho*. Because of the various effects of these mutations, different mutants originally were isolated by a variety of phenotypes and were given different names, of which the best known is *suA* (reflecting the ability to suppress polarity).

Rho mutations show wide variations in their influence on termination. The basic nature of the effect is a failure to terminate. But the magnitude of the failure, as seen in the percent of readthrough *in vivo*, depends on the particular target locus. This effect is probably due to the leakiness of most *rho* mutations, which generally allow the production of a relatively substantial amount of activity. If different terminators require different levels of rho factor for termination, the amount of residual rho activity may allow some terminators to function (those active at low rho concentrations), while others fail to function and instead allow RNA polymerase to readthrough.

The ability of *suA* mutations to suppress polarity is explained if they reduce the probability that rho will act on the internal terminator that follows the nonsense codon. Then the termination of translation does not cause transcription also to terminate; and the regions of mRNA beyond the mutation can be translated by ribosomes that reattach farther along.

Some *rho* mutations can be suppressed by mutations in other genes. This approach provides an excellent way to identify proteins that interact with rho. The β subunit of RNA polymerase is implicated by two types of mutation. First, mutations in the *rpoB* gene coding for it can reduce termination at a rho-dependent site. Second, mutations in *rpoB* can restore the ability to terminate transcription at rho-dependent sites in *rho* mutant bacteria.

ANTITERMINATION DEPENDS ON SPECIFIC SITES

Termination provides an opportunity for certain phages to control the switch from early genes to the next stage of expression. **Figure 13.4** compares the use of antitermination as a control mechanism with the use of new promoters.

We saw in the previous chapter that one mechanism for switching from early gene transcription to the next stage of expression is to replace the sigma factor of the host enzyme with another factor that redirects its specificity in initiation. An alternative is to synthesize a new phage RNA polymerase. In either case, the critical feature that distinguishes the new set of genes is their possession of *different promoters from those originally recognized by host RNA polymerase.*

In this case, expression of the new set of genes does not depend on expression of the early genes after the critical sigma factor or new polymerase has been synthesized. The two sets of transcripts are independent. Early gene expression can cease when the switch to the next stage is made.

By contrast, the use of antitermination depends on a particular arrangement of genes. The early genes lie adjacent to the genes that are to be expressed next, but are separated from them by terminator sites. If termination is prevented at these sites, the polymerase reads through into the genes on the other side. Thus in antitermination, the *same promoters* continue to be recognized by RNA polymerase. So the new genes are expressd only via extension of the RNA chain to form molecules that contain the early gene sequences at the 5′ end and the new gene sequences at the 3′ end. Since the two types of sequence remain linked, early gene expression inevitably continues.

The best characterized example of antitermination is provided by phage lambda. The overall design of the control network is very similar to that of other phages: the host RNA polymerase transcribes the early genes, and one of the early gene products is needed to transcribe the next set of phage genes. In lambda, the genes transcribed immediately by the host RNA polymerase are called the **immediate early** genes, and those expressed at the next stage are called the **delayed early** genes.

The regulator gene that controls the switch from immediate early to delayed early expression was identified by mutations that prevent the transition. Lambda mutants in gene *N* can transcribe *only* the immediate early genes; they proceed no further into the infective cycle. The effect is the same as seen when gene *28* of phage SPO1 is mutated to prevent the production

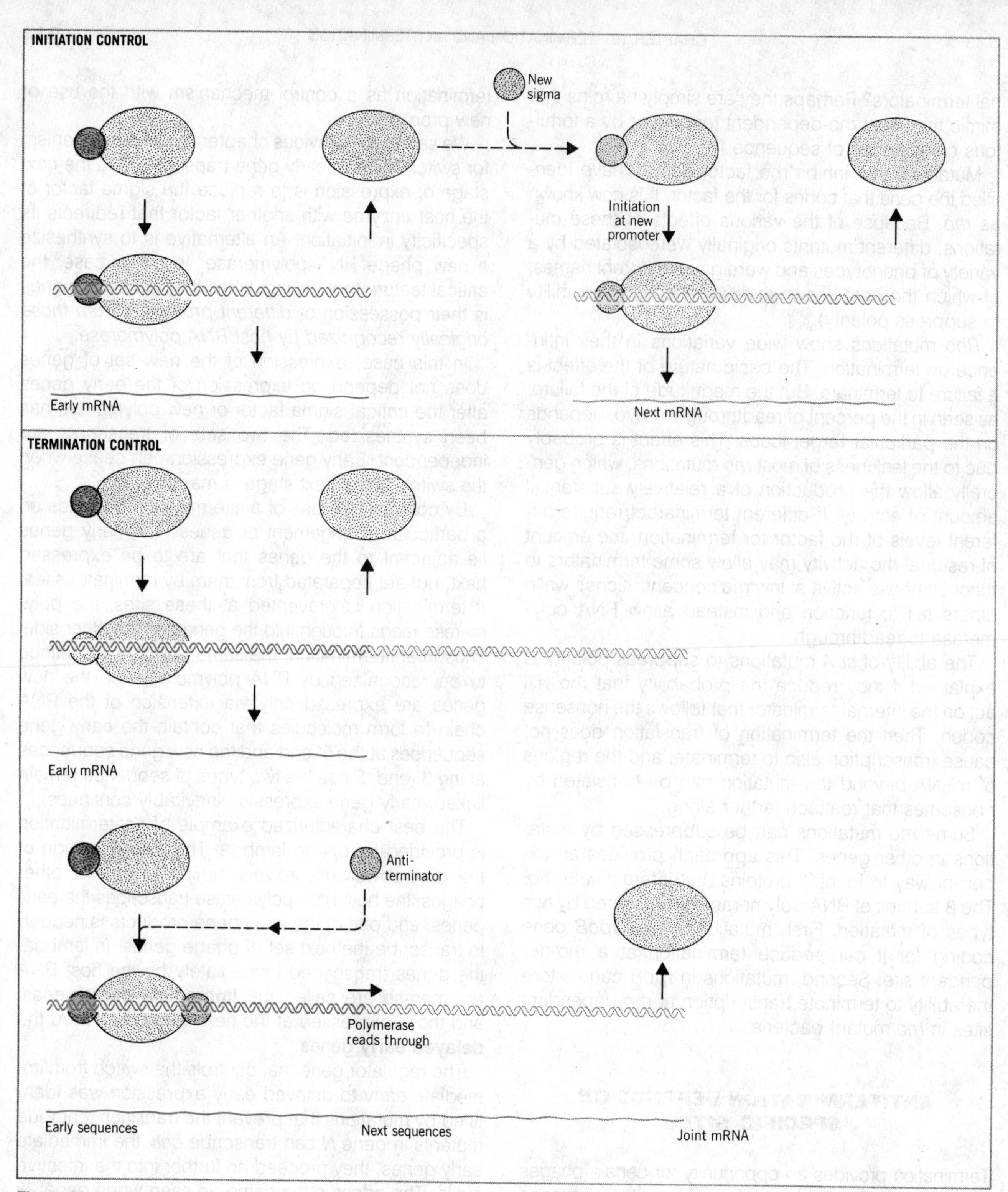

Figure 13.4
Switches in transcriptional specificity can be controlled at initiation or termination.

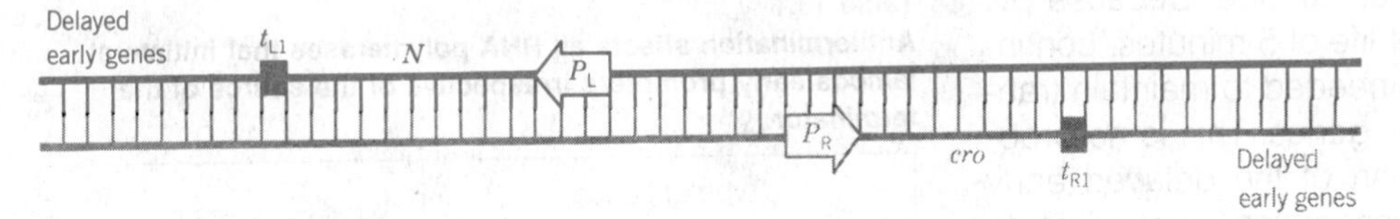

Figure 13.5
Phage lambda has two early transcription units.
The map of lambda represents the two strands of DNA by parallel lines. The "upper" strand is transcribed toward the left; while the "lower" strand is transcribed toward the right. The promoters are indicated by the arrowheads. The terminators are indicated by the solid boxes. Genes *N* and *cro* are the immediate early functions, and are separated from the delayed early genes by the terminators.

of σ^{gp28}. From the genetic point of view, the mechanisms of new initiation and antitermination are similar. Both are **positive controls** in which *an early gene product must be made by the phage in order to express the next set of genes.*

A map of the early region of phage lambda is drawn in **Figure 13.5**. The immediate early genes, *N* and *cro*, are transcribed respectively to the left and to the right from the promoters indicated as P_L and P_R by the *E. coli* RNA polymerase. This means that different strands of DNA are used as the template in the **leftward** and **rightward** transcription units. **Figure 13.6** shows that transcription by *E. coli* RNA polymerase itself stops at the ends of genes *N* and *cro*, at the terminators t_{L1} and t_{R1}, respectively. Both terminators depend on rho; in fact, these were the terminators with which rho was originally identified.

The situation is changed by expression of the *N* gene. The product pN is an antitermination protein that allows RNA polymerase to readthrough t_{L1} and t_{R1} into

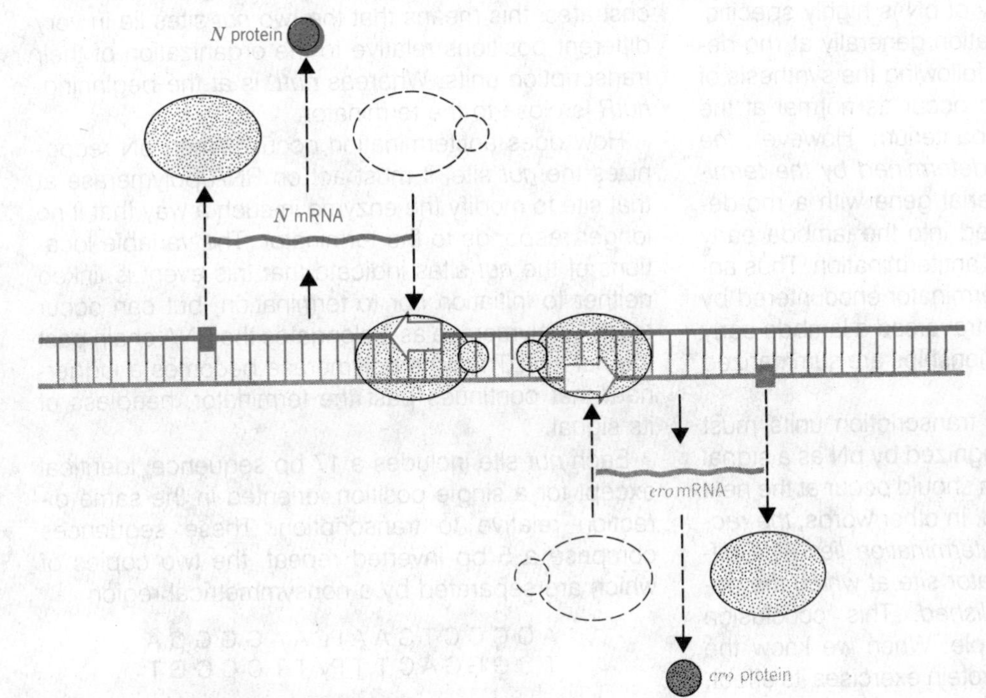

Figure 13.6
Host RNA polymerase transcribes the immediate early genes of lambda.
The enzyme binds at the two promoters, P_L and P_R, and transcribes genes *N* and *cro* up to the terminators t_{L1} and t_{R1}, respectively. Termination requires the rho factor at both sites (this is not shown in the illustration). Transcription of these units produces two immediate early mRNA species, coding for pN and Cro.

the delayed early genes on either side. Because pN is highly unstable, with a half-life of 5 minutes, continued expression of gene *N* is needed to maintain transcription of the delayed early genes. This is no problem, since the *N* gene is part of the delayed early transcription unit, and its transcription must precede that of the delayed early genes.

Like other phages, still another control is needed to express the late genes that code for the components of the phage particle. This switch is regulated by gene *Q*, itself one of the delayed early genes. Its product, pQ, is another antitermination protein, one that specifically allows RNA polymerase initiating at another site, the late promoter $P_{R'}$, to readthrough a terminator that lies between it and the late genes. Thus by employing antitermination proteins with different specificities, a cascade for gene expression can be constructed.

The different specificities of pN and pQ establish an important general principle: RNA polymerase interacts with transcription units in such a way that an ancillary factor can sponsor termination of one unit but not another. Termination can in fact be controlled with the same sort of precision as initiation. What sites are involved in controlling the specificity of termination?

The antitermination activity of pN is highly specific. It does not suppress termination generally at rho-dependent sites. For example, following the synthesis of pN, termination continues to occur as normal at the ends of genes of the host bacterium. However, *the antitermination event is not determined by the terminators t_{L1} and t_{R1}*; if a bacterial gene with a rho-dependent terminator is inserted into the lambda early transcription unit, pN causes antitermination. Thus antitermination occurs at *any* terminator encountered by an RNA polymerase that has traversed a lambda early transcription unit. These relationships are summarized in **Table 13.1**.

Both of the lambda early transcription units must contain some site that is recognized by pN as a signal indicating that antitermination should occur at the next terminator to be encountered. In other words, *the recognition site needed for antitermination lies at a different place from the terminator site at which the action eventually is accomplished*. This conclusion establishes a general principle. When we know the site on DNA at which some protein exercises its effect,

Table 13.1
Antitermination affects all RNA polymerases that initiate at lambda early promoters irrespective of the source of the terminator.

Type of Locus		What Happens in Lambda-Infected Cells
Promoter →	Terminator	
Bacterial	Bacterial	Termination occurs
Early lambda	Early lambda	pN prevents termination
Early lambda	Bacterial	pN prevents termination

we cannot assume that this coincides with the DNA sequence that it initially recognizes. They may be far apart.

The recognition sites for pN are called *nut* (for *N* *u*tiliza*t*ion). The sites responsible for determining leftward and rightward antitermination are described as *nutL* and *nutR*, respectively, and we know that they must lie between P_L and t_{L1} on one side, and between P_R and t_{R1} on the other. Where exactly are they located? Mapping of nut^- deletions and point mutations locate *nutL* between the startpoint of P_L and the beginning of the *N* coding region; and *nutR* lies between the end of the *cro* gene and t_{L1}. As **Figure 13.7** demonstrates, this means that the two *nut* sites lie in very different positions relative to the organization of their transcription units. Whereas *nutL* is at the beginning, *nutR* is close to the terminator.

How does antitermination occur? When pN recognizes the *nut* site, it must act on RNA polymerase at that site to modify the enzyme in such a way that it no longer responds to the terminator. The variable locations of the *nut* sites indicate that this event is linked neither to initiation nor to termination, but can occur to RNA polymerase as it elongates the RNA chain past the *nut* site. Then the polymerase becomes a juggernaut that continues past the terminator, heedless of its signal.

Each *nut* site includes a 17 bp sequence, identical except for a single position, oriented in the same direction relative to transcription. These sequences comprise a 5 bp inverted repeat, the two copies of which are separated by a nonsymmetrical region

A G C C C T G A A Pu A A G G G C A
T C G G G A C T T Py T T C C C G T

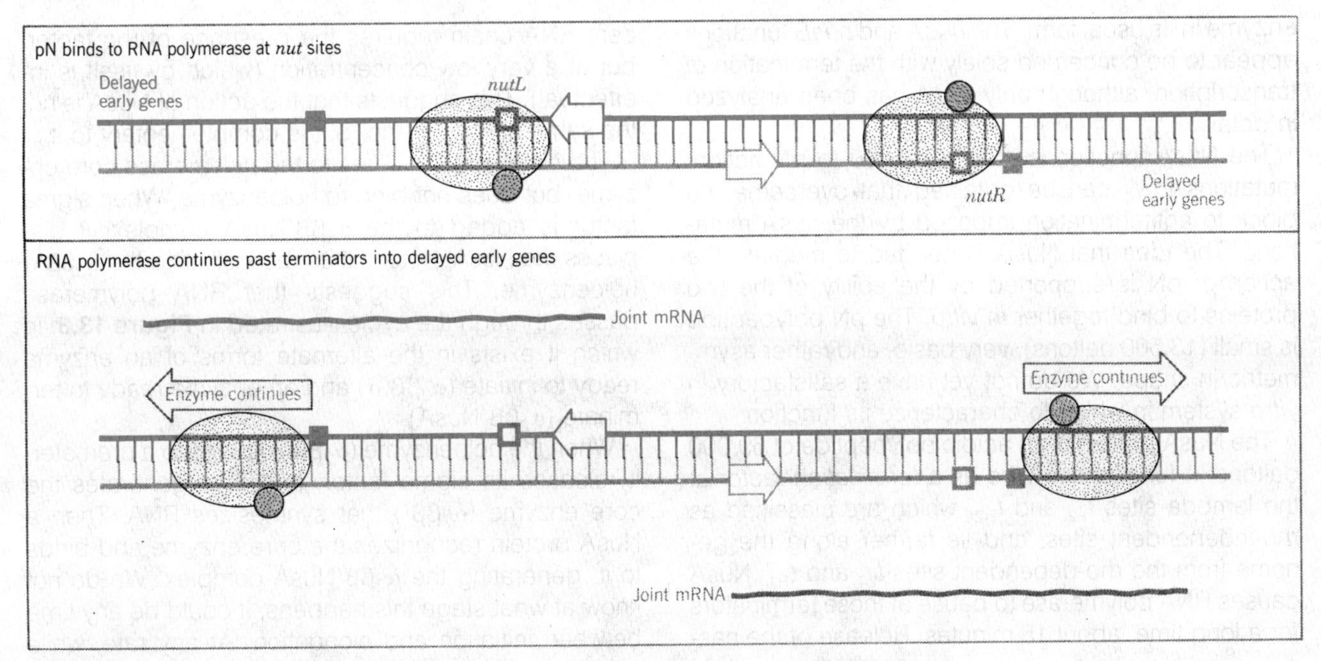

Figure 13.7
The pN protein binds to RNA polymerase as it passes a *nut* site; its presence allows the enzyme to readthrough the terminators t_{L1} and t_{R1}, producing a joint mRNA that contains delayed early as well as immediate early gene sequences.

where the shaded box shows the inverted repeats, and the base pair Pu-Py indicates the single difference between *nutL* and *nutR*. The sequences needed for recognition by N may extend beyond this central core, since some *nut* mutations have been found outside this region.

Is the ability of pN to recognize a short sequence within the transcription unit an example of a more widely used mechanism for antitermination? Other phages, related to lambda, have different *N* genes and different antitermination specificities. The region of the phage genome in which the *nut* sites lie has a different sequence in each of these phages, and presumably each phage has characteristic *nut* sites recognized specifically by its own pN. Each of these pN products must have the same general ability to interact with the transcription apparatus in an antitermination capacity, but has a different specificity for the sequence of DNA that activates the mechanism.

MORE SUBUNITS FOR RNA POLYMERASE?

The discovery of antitermination as a phage control mechanism has led to the identification of further components of the transcription apparatus. The bacterial proteins with which pN interacts can be identified by isolating mutants of *E. coli* in which pN is ineffective. The mutants cannot be infected successfully by lambda, because the phage is limited to expressing only its immediate early genes. Several of these mutations lie in the *rpoB* gene. This argues that pN (like rho factor) interacts with the β subunit of the core enzyme.

Other *E. coli* mutations that prevent pN function identify the *nus* loci, of which there are at least three: *nusA*, *nusB*, and *nusE*. (The term "*nus*" is an acronym for *N* utilization substance.) The *nus* loci must code for proteins that form part of the transcription apparatus, but are not isolated with the RNA polymerase enzyme in its usual form. The *nusA* and *nusB* functions appear to be concerned solely with the termination of transcription, although only *nusA* has been analyzed in detail.

The NusA function is closely related to pN action; mutations in *N* can be obtained that overcome the block to antitermination imposed by the *nusA* mutations. The idea that NusA is needed to mediate the action of pN is supported by the ability of the two proteins to bind together *in vitro*. The pN polypeptide is small (13,500 daltons), very basic, and rather asymmetric in shape. We do not yet have a satisfactory *in vitro* system in which to characterize its function.

The NusA protein is an acidic polypeptide of 69,000 daltons. It functions *in vitro* as a *termination factor* at the lambda sites t_{L2} and t_{R2}, which are classified as rho-independent sites, and lie farther along the genome from the rho-dependent sites t_{L1} and t_{R1}. NusA causes RNA polymerase to pause at these terminators for a long time, about 15 minutes. Release of the nascent RNA chain requires the presence of rho factor, but at a very low concentration (which by itself is ineffectual). This suggests that the action of NusA is not the same as that of rho, but is complementary to it.

The NusA protein binds to the polymerase core enzyme, but does not bind to holoenzyme. When sigma factor is added to the $\alpha_2\beta\beta'$NusA complex, it displaces the NusA protein, thus reconstituting the $\alpha_2\beta\beta'\sigma$ holoenzyme. This suggests that RNA polymerase passes through the cycle illustrated in **Figure 13.8**, in which it exists in the alternate forms of an enzyme ready to initiate ($\alpha_2\beta\beta'\sigma$) and an enzyme ready to terminate ($\alpha_2\beta\beta'$NusA).

When the holoenzyme ($\alpha_2\beta\beta'\sigma$) binds to a promoter, it releases its sigma factor, and thus generates the core enzyme ($\alpha_2\beta\beta'$) that synthesizes RNA. Then a NusA protein recognizes the core enzyme and binds to it, generating the $\alpha_2\beta\beta'$NusA complex. We do not know at what stage this happens; it could be any time between initiation and elongation. At any rate, while

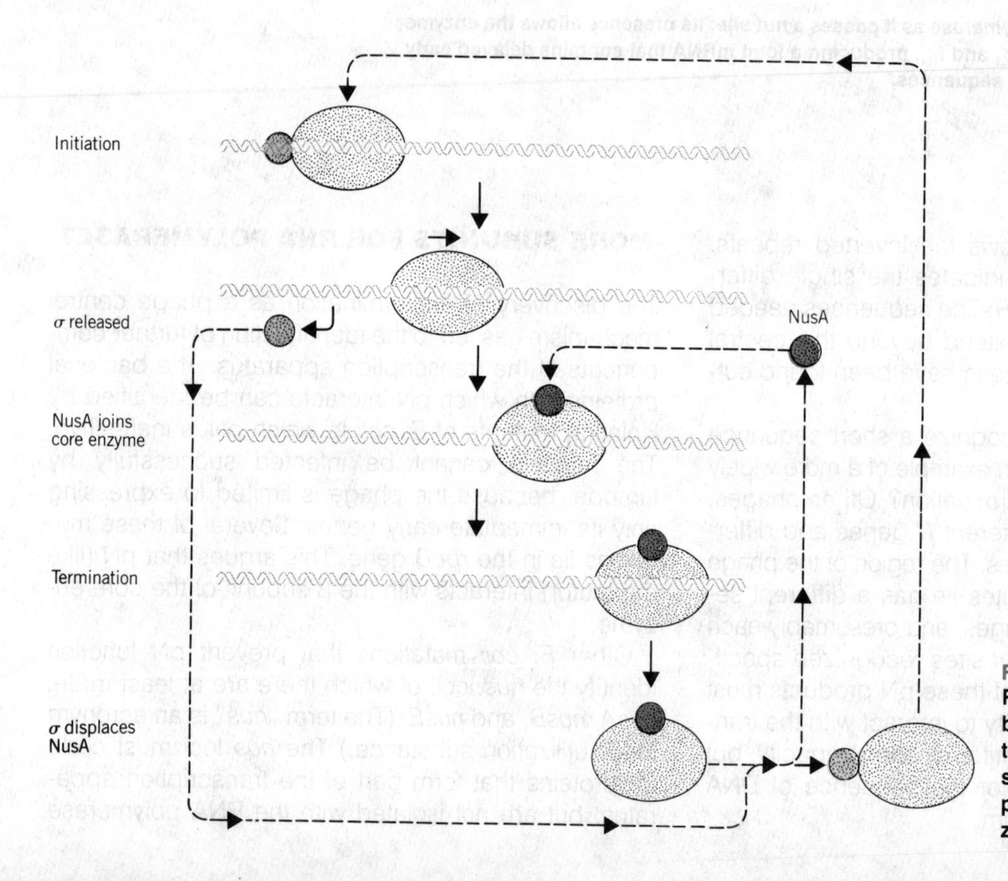

Figure 13.8
RNA polymerase may alternate between initiation-competent and termination-competent forms as sigma and NusA alternately replace one another on the core enzyme.

the $\alpha_2\beta\beta'$NusA polymerase is bound to DNA, the NusA component cannot be displaced. But when termination occurs, the enzyme is released in a state in which NusA either is automatically freed or can be displaced by sigma factor.

From the perspective of the core enzyme, life consists of an alternation between association with sigma for initiation and association with NusA for termination. The two factors are mutually incompatible associates of the core. There seems no reason to regard one as any more a component than the other; we may regard them as alternate subunits. Thus the core enzyme represents a minimal form of RNA polymerase, competent to engage in the basic function of RNA synthesis, but lacking subunits necessary for other functions. As we have remarked, it is a moot point where RNA polymerase ends and the wider transcription apparatus begins.

The NusA component seems to provide the link between pN function and the core enzyme. This makes sense if NusA is a termination subunit, whose incorporation into the enzyme confers specificity toward certain terminators. Thus pN may act as an antiterminator by preventing NusA from exercising its function. Immediately this raises the possibility that the host bacterium may contain other proteins, analogous to NusA (or to pN), whose interactions with the core enzyme control its recognition of termination sites.

This conclusion would explain a curious finding about NusA. It is identical with a protein previously identified as the L factor, which is *necessary* for transcription of some genes *in vitro*. This behavior is not at all what would be expected of a termination factor! But if it is the presence of NusA that allows these hypothetical bacterial functions to suppress termination at particular sites, NusA could be necessary to transcribe gene(s) lying beyond such a site. Alternatively, the NusA subunit itself might give the core enzyme the ability to read past some *t* sites while acknowledging others.

We do not know exactly how core enzyme, NusA protein, and rho factor cooperate at termination. Nor is it clear how their activities relate to the classification of terminators into rho-dependent and rho-independent sites. The existence of NusA implies that core enzyme usually may never terminate by itself; so *in vitro* experiments with purified core enzyme could reflect an aberrant activity. Rho-dependent and -independent sites are recognized under somewhat different conditions *in vitro* (low versus high ionic strength), which has always raised concern that one set of conditions might not reflect the natural situation. The use of these types of *t* sites needs to be reinvestigated with *in vitro* systems that contain the other possible termination components.

DIFFICULTIES IN EUKARYOTES

Little is known about either the signals for termination or the process involved for most eukaryotic RNA polymerases. The major difficulty in analyzing transcripts is the lack of certainty about the actual site of termination. Although the RNA molecules identified *in vivo* have defined 3′ ends, how are we to know whether they were produced by a termination event or by cleavage of a longer original transcript?

For the products of RNA polymerase II, the problem is exacerbated by the processing that occurs at the 3′ end, in which a length of polyadenylic acid is added. In at least some cases, there is evidence that the 3′ terminus for polyadenylation is indeed generated by cleavage and not by termination (see Chapter 27). We therefore have some information about the signals needed to generate the end for addition of poly(A), but the 3′ region of the original primary transcript generally remains uncharacterized. A termination site has been identified in DNA of the virus SV40; this sequence is like a rho-independent bacterial *t* site, with a hairpin followed by a stretch of U bases.

The importance of secondary structure in the generation of 3′ ends has been shown by analyzing the ability of various templates to give rise to histone (nonpolyadenylated) mRNAs when templates are injected into *Xenopus* oocytes. The RNA terminates in a stem-loop structure, and mutations that prevent formation of the duplex stem prevent formation of the end of the RNA. Secondary mutations that restore duplex structure (though not necessarily the original sequence) behave as revertants. This suggests that *formation of the secondary structure is more important than the exact sequence.*

Either or both of the DNA strands could in principle be involved in forming secondary structure. They can be distinguished by using templates consisting of heteroduplex molecules, in which the two strands of DNA are not identical. It turns out that it is important to be able to write a duplex structure for the *coding strand*, not the strand used as template. This strongly suggests that the secondary structure exerts its effect by forming in the RNA as it is transcribed.

For RNA polymerase I, the sole product is a large precursor that contains the sequences of the major rRNA. The precursor is subjected to extensive processing, and it has not yet been proved that the 3′ end actually is generated by termination. However, the terminus does correspond to a run of U bases.

With RNA polymerase III, transcription *in vitro* generates molecules with the same 5′ and 3′ ends as those synthesized *in vivo*. This suggests that proper initiation and termination have occurred. The system can be manipulated by introducing changes in the sequence of the template around the termination region.

When the 5S genes of *X. laevis* are transcribed by the homologous enzyme *in vitro*, termination occurs within a run of 4 U bases. Termination usually occurs at the second, but there is heterogeneity, with some molecules ending in 3 or even 4 U bases. The same heterogeneity is seen in molecules synthesized *in vivo*, so it seems to be a bona fide feature of the termination reaction.

Just like the prokaryotic terminators, the U run is embedded in a G-C-rich region. Although sequences of dyad symmetry are present, they are not needed for termination, since mutations that abolish the symmetry do not prevent the normal completion of RNA synthesis. Nor are any sequences beyond the U run necessary, since all nucleotides to its right can be replaced by other sequences without any effect on termination.

The U run itself is not sufficient for termination, because regions of 4 successive U residues exist within transcription units read by RNA polymerase III. (However, there are no internal 5-U runs, which fits with the greater efficiency of termination when the terminator is a U_5 rather than U_4 sequence.) The critical feature in termination must therefore be the recognition of a U_4 sequence in a context that is rich in G-C base pairs.

How does the termination reaction occur? It cannot rely on the weakness of the rU-dA RNA-DNA hybrid region that lies at the end of the transcript, because often only the first two U residues are transcribed. Perhaps the G-C-rich region plays a role in slowing down the enzyme, but there does not seem to be a counterpart to the hairpin involved in prokaryotic initiation. We remain puzzled how the enzyme can respond so specifically to such a short signal. And in contrast with the initiation reaction, which RNA polymerase cannot accomplish alone, termination seems to be a function of the enzyme itself.

FURTHER READING

The mechanisms of bacterial termination have been reviewed by **Adhya & Gottesman** (*Ann. Rev. Biochem.* **47**, 967–996, 1978) and **Platt & Bear** in *Gene Function in Prokaryotes* (Beckwith et al., Eds., Cold Spring Harbor Laboratory, New York, 1984, pp. 123–161). Attenuation has been reviewed by **Bauer et al**. in *Gene Function in Prokaryotes* (pp. 65–89). The nature of the terminator signals has been reviewed in the context of consensus sequences by **Rosenberg & Court** (*Ann. Rev. Genet.* **13**, 319–353, 1979). The discovery of the wider bacterial transcription apparatus has been put together by **Greenblatt** (*Cell* **24**, 8–9, 1981). Too little is known about eukaryotic termination for reviews yet to be available.

PART 4
CONTROLLING PROKARYOTIC GENE EXPRESSION

According to the strictly structural concept, the genome is considered as a mosaic of independent molecular blueprints for the building of individual cellular constituents. In the execution of these plans, however, coordination is evidently of absolute survival value. The discovery of regulator and operator genes, and of repressive regulation of the activity of structural genes, reveals that the genome contains not only a series of blueprints, but a coordinated program of protein synthesis and the means of controlling its execution.

François Jacob & Jacques Monod, 1961

CHAPTER 14
THE OPERON: THE LACTOSE PARADIGM

Bacteria need to be able to respond swiftly to changes in their environment. Because of the conditions under which they live, capricious fluctuations in the supply of nutrients may occur at any time; survival depends on the ability to switch from metabolizing one substrate to another. Unicellular eukaryotes may share this subjection to an incessantly changing world; but more complex, multicellular organisms are restricted to a more constant set of metabolic pathways, and may not have the same need to respond to external circumstances.

Flexibility is therefore at a premium in the bacterial world. Yet economy also is demanded, since a bacterium that indulges in energetically expensive ways to meet the demands of the environment may be at a disadvantage. Certainly it would be expensive to produce unnecessarily all the enzymes for a metabolic pathway that cannot be used because the substrate is absent. So the bacterial compromise is to avoid synthesizing the enzymes of a pathway in the absence of their substrate; but to be ready at all times to produce the enzymes if the substrate should appear.

This line of reasoning explains the central features of the organization of bacterial genes. They are grouped in clusters, so that all the enzymes needed to make up a particular pathway are represented by genes adjacent to one another. The entire gene cluster may be transcribed into a single polycistronic mRNA, sequentially translated by the ribosomes into each of the proteins. We have already alluded in previous chapters to the fact that few genes are individually transcribed in *E. coli;* most are part of larger transcription units. This form of organization allows all the genes in the unit to be coordinately regulated by the interaction of a regulator protein with a site that lies close to the promoter.

INDUCTION AND REPRESSION ARE CONTROLLED BY SMALL MOLECULES

It has been known since the beginning of this century that certain enzymes of yeast are present only when the cells are grown on the appropriate substrates. This effect now is known as enzyme **induction.** It has been best characterized in bacteria, and in particular with *E. coli*, where the lactose system provides the paradigm for this sort of control mechanism. When cells of *E. coli* are grown in the absence of a β-galactoside, they contain very few molecules of the enzyme β-galactosidase—say, fewer than five. The function of the enzyme is to break the β-galactoside into its component sugars. For example, lactose is cleaved into glucose and galactose, which are then further metabolized. There is no need for the enzyme in the absence of the substrate.

When a suitable substrate is added, the enzyme activity appears very rapidly in the bacteria, as the result of synthesis of new enzyme molecules. Within 2–3 minutes some enzyme is present, and soon there may be up to 5000 molecules of enzyme per bacterium. (Under suitable conditions, β-galactosidase can account for 5–10% of the total soluble protein of the bacterium.) If the substrate is removed from the medium, the synthesis of enzyme stops as rapidly as it originally started.

This type of rapid response to changes in nutrient supply is shown not only in dealing with the need to metabolize new substrates, but is used also to shut off synthesis of compounds that may suddenly appear in the medium. For example, *E. coli* synthesizes the amino acid tryptophan through the action of the enzyme tryptophan synthetase. But if tryptophan is provided in the medium on which the bacteria are growing, immediately the production of the enzyme is halted. This effect is called **repression.** It allows the bacterium to avoid devoting its resources to unnecessary synthetic activities.

Induction and repression represent the same phenomenon seen from different perspectives. In one case the bacterium adjusts its ability to use a given substrate for growth; in the other it adjusts its ability to synthesize a particular metabolic intermediate. The trigger for either type of adjustment is the small molecule that is the substrate for the enzyme, or the product of the enzyme activity, respectively. Small molecules that cause the production of enzymes able to metabolize them are called **inducers.** Those that prevent the production of enzymes able to synthesize them are called **corepressors.**

The ability to act as inducer or corepressor is highly specific. Only the substrate/product or a closely related molecule can serve. But in both cases, the activity of the small molecule does *not* depend on its interaction with the target enzyme. Some inducers resemble the natural inducers for β-galactosidase, but cannot be metabolized by the enzyme. The example par excellence is isopropylthiogalactoside (IPTG), one of several thiogalactosides with this property. Although it is not recognized by β-galactosidase, IPTG is a very efficient inducer.

Molecules that induce enzyme synthesis but are not metabolized are called **gratuitous inducers.** They are extremely useful because they remain in the cell in their original form. (A real inducer would be metabolized, interfering with study of the system.) The existence of gratuitous inducers reveals an important point. *The system must possess some component, distinct from the target enzyme, that recognizes the appropriate substrate; and its ability to recognize related potential substrates is different from that of the enzyme.*

STRUCTURAL GENE CLUSTERS ARE CONTROLLED BY REGULATOR GENES

Induction or repression is not usually confined to a single enzyme. We have mentioned that often all the enzymes of a metabolic pathway are regulated together. And in addition to the enzymes actually involved in the pathway, other related activities may be included in the unit of coordinate control; an example is the protein responsible for transporting the small molecule substrate into the cell.

Genes that code for the proteins required by the cell, for either enzymatic or structural functions, are called **structural genes.** The overwhelming majority of bacterial genes fall into this category, which therefore represents an enormous variety of protein structures and functions.

Structural genes also include the genes coding for rRNA and tRNA. All types of structural gene tend to be organized into clusters that may be coordinately controlled, as in the example of the three *lac* structural genes, *lacZYA*.

Three genes map in the cluster drawn in **Figure 14.1.** First, *lacZ* codes for the enzyme β-galactosidase, whose active form is a tetramer of about 500,000 daltons. Second, *lacY* codes for the β-galactoside permease, a 30,000 dalton membrane-bound protein constituent of the transport system. Third, *lacA* codes for β-galactoside transacetylase, an enzyme that transfers an acetyl group from acetyl-CoA to β-galactosides.

Mutations in either *lacZ* or *lacY* can create the lac^- genotype, in which cells cannot utilize lactose. The $lacZ^-$ mutations abolish enzyme activity, directly preventing metabolism of lactose. The $lacY^-$ mutants cannot take up lactose from the medium. No defect is

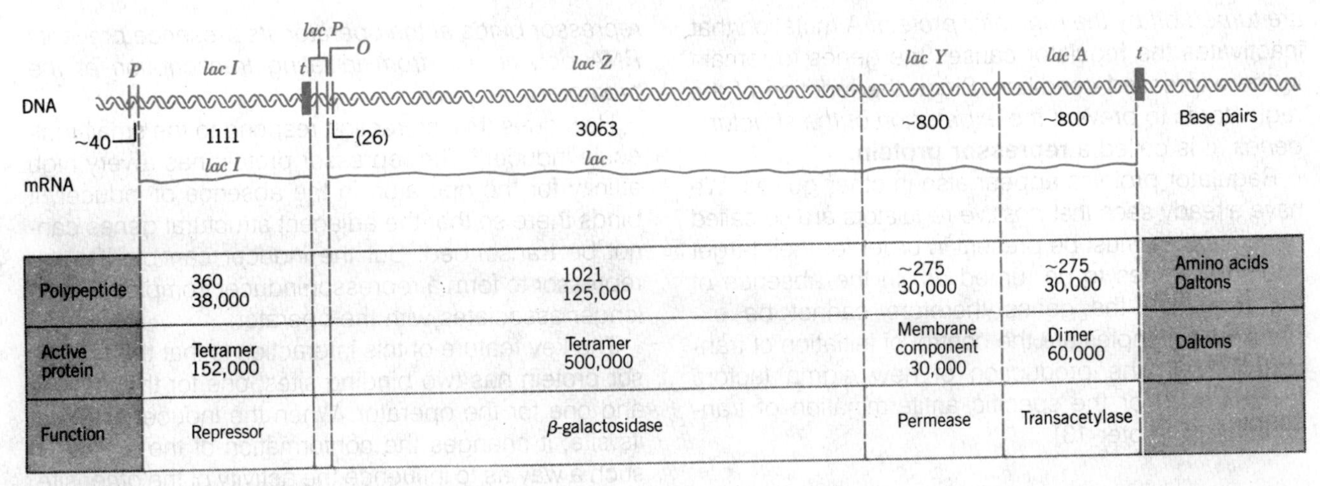

Figure 14.1
The *lac* operon occupies about 6000 bp of DNA.
At the left the *lacI* gene has its own promoter and terminator. The end of the *lacI* coding region is adjacent to the *lacP* promoter. The *lacO* operator occupies the first 26 bp of the *lacZ* gene, which is extremely long and is followed by the *lacY* and *lacA* genes and a terminator.

identifiable in *lacA*$^-$ cells, which is a puzzle. The significance of the acetylation reaction is uncertain, but it is possible that it may give an advantage when the bacteria grow in the presence of certain nonmetabolizable analogs of β-galactosides, because acetylation results in detoxification and excretion.

The cluster of three genes, *lacZYA,* is transcribed into a single mRNA from a promoter just upstream from *lacZ*. Their induction is controlled at the level of transcription. In the absence of an inducer, the gene cluster is not transcribed. When an inducer is added, transcription starts at the single promoter, *lacP,* and proceeds through the genes to a terminator located somewhere beyond *lacA*. This accomplishes a **coordinate regulation,** in which *all the genes are expressed (or not expressed) in unison.*

The mRNA is translated sequentially from its 5′ end, which explains why induction always causes the appearance of β-galactosidase, β-galactoside permease, and β-galactoside transacetylase, in that order. The expression of the three genes via a common mRNA explains why the relative amounts of the three enzymes always remain the same under varying conditions of induction. Induction essentially represents a switch that causes the genes to be expressed. Inducers may vary in their effectiveness, and other factors may influence the absolute level of transcription or translation, but the relationship between the three genes is predetermined by their organization.

The mRNA is extremely unstable, and decays with a half-life of only ~3 minutes. It is this feature that allows induction to be reversed so rapidly. Transcription ceases as soon as the inducer is removed; and in a very short time all the lactose mRNA has been destroyed, and the cell stops producing the enzymes.

What is the target for the small-molecule inducer? We know that it is distinct from the structural gene products. It is another protein, whose sole function is to control the expression of the structural genes. The gene that codes for this protein is called the **regulator gene.** Regulator genes are responsible for controlling the expression of the structural gene clusters, usually via the synthesis of proteins that act to control transcription. The regulator proteins exercise this function by binding to particular sites on DNA.

We can distinguish between structural genes and regulator genes by the effects of mutations. A mutation in a structural gene simply deprives the cell of the particular protein for which that gene codes. But a mutation in a regulator gene influences the expression of all those structural genes that it controls. The nature of this influence reveals the type of regulation.

The *lac* genes are controlled by **negative regulation.** This means that they are transcribed *unless they* *are turned off by the regulator protein.* A mutation that inactivates the regulator causes the genes to remain in the expressed condition. Since the function of the regulator is to *prevent the expression of the structural genes,* it is called a **repressor protein.**

Regulator proteins appear also in other guises. We have already seen that *positive* regulators are so called because they must be present in order for their target structural genes to be turned *on*. In the absence of the regulator, the genes therefore cannot be expressed. Examples are the control of initiation of transcription via the production of new sigma factors (Chapter 12) or the specific antitermination of transcription (Chapter 13).

THE CONTROL CIRCUIT OF THE OPERON

The concept that two different classes of genes might be distinguished by their functions was proposed by Jacob and Monod in 1961 in their classic formulation of the operon model. The **operon** is a unit of gene expression, including structural genes and elements that control their expression. The activity of the operon is controlled by regulator gene(s), whose protein products interact with the control elements. The original model was constructed for the *lac* operon, which remains the best-characterized example, but similar arguments have been used to construct analogous circuits for other operons.

The basic control circuit of the lactose operon is illustrated in **Figure 14.2.** The cluster of three structural genes, *lacZYA,* forms a transcription unit that is conventionally drawn from left to right. Transcription starts at the promoter *lacP,* just to the left (upstream) of the first gene, *lacZ*. The regulator gene, *lacI*, lies a little farther to the left, and forms an independent transcription unit. It is transcribed from its own promoter into a monocistronic mRNA that codes for a repressor protein. The crucial features of the control circuit reside in the dual properties of this repressor: it can prevent transcription; and it can recognize the small-molecule inducer.

How does the repressor prevent transcription? It binds to a sequence of DNA called the **operator,** denoted *lacO*. This site lies between the promoter *lacP* and the cluster of structural genes *lacZYA. When the repressor binds at the operator, its presence prevents RNA polymerase from initiating transcription at the promoter.*

How does this interaction respond to the small-molecule inducer? The repressor protein has a very high affinity for the operator; in the absence of inducer, it binds there so that the adjacent structural genes cannot be transcribed. But the inducer can bind to the repressor to form a repressor•inducer complex that no longer associates with the operator.

The key feature of this interaction is that the repressor protein has two binding sites, one for the inducer and one for the operator. When the inducer binds at its site, it changes the conformation of the protein in such a way as to influence the activity of the *other* site. This type of relationship is called **allosteric control.** The result in this case is that, *when inducer is added, the repressor is converted to a form that leaves the operator.* Then RNA polymerase can initiate transcription of the structural genes.

The lactose repressor is a tetramer of identical subunits of 38,000 daltons each. There are about 10 tetramers per cell (this means that there must be 40 translations of the *lacI* gene in each cell cycle). The regulator gene is transcribed at a rate that appears to be governed simply by the affinity of its promoter for RNA polymerase.

CONSTITUTIVE MUTATIONS DEFINE REPRESSOR ACTION

The *lacI* regulator gene was originally identified by the isolation of mutations that *affect the expression of all three structural genes and map outside them*. Since the *lacI* mutations complement mutations in the structural genes, they can be taken to identify another gene coding for a diffusible product.

The *lacI*$^-$ genotype can arise from either point mutations or deletions. The latter indicates that it represents *loss* of the usual regulation. The *lacI*$^-$ mutants express the structural genes all the time, irrespective of whether the inducer is present or absent. This behavior is described as **constitutive** gene expression. This conforms to our expectation for a negative control system. The *lacI*$^+$ gene codes for a repressor protein that is able to turn off the transcription of the *lacZYA*

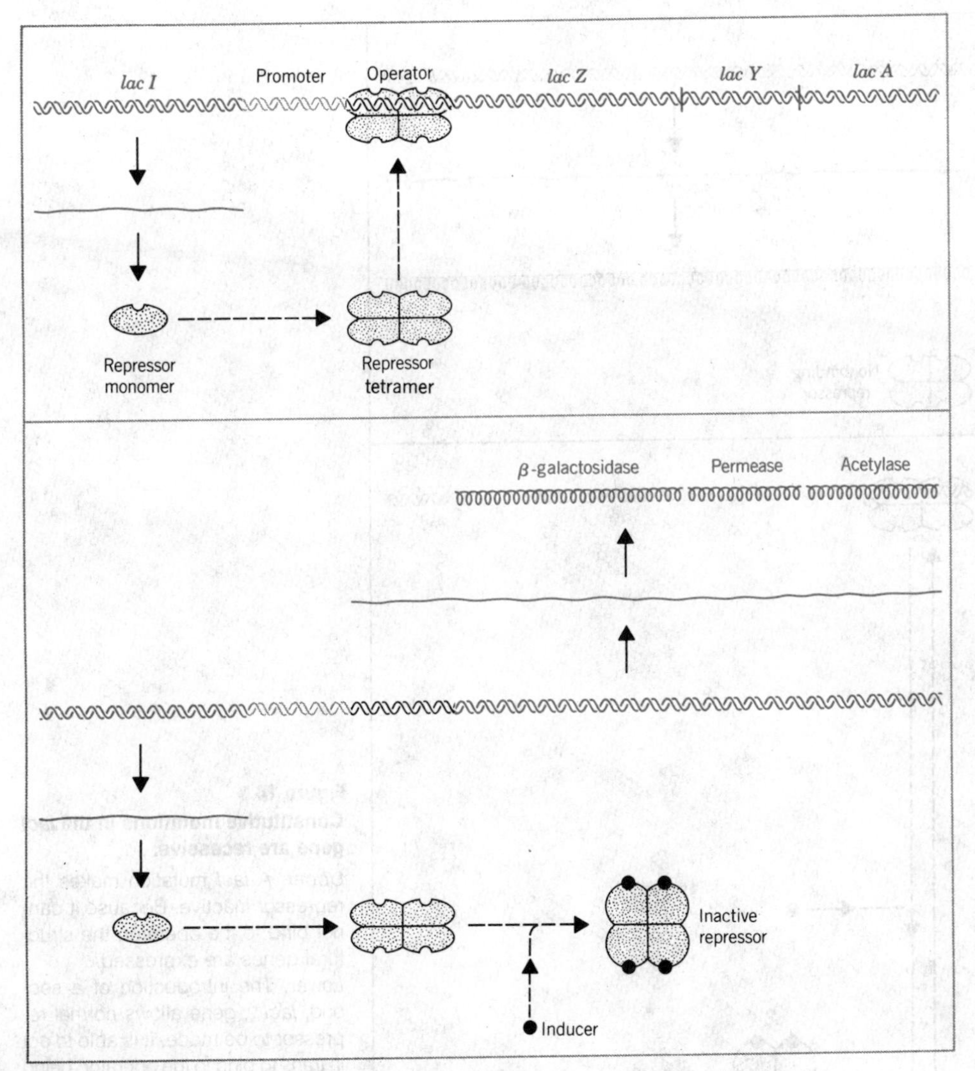

Figure 14.2
The *lac* operon is turned on by an inducer.
Upper. The *lacI* gene synthesizes a repressor whose tetramer binds to the operator and prevents transcription of the structural genes.
Lower. Addition of inducer converts the repressor into an inactive form that cannot bind to the operator; so transcription starts at the promoter and the three enzymes are synthesized.

cluster. Mutation of the gene to form the *lacI*$^-$ type allows the genes to be constitutively expressed because the repressor now is inactive.

We can confirm this conclusion by determining the relationship between the *lacI*$^-$ constitutive mutant gene and the wild-type *lacI*$^+$ gene when both are present in the same cell. This is accomplished by forming a **partial diploid,** when one copy of the operon is present on the bacterial genome itself, and the other is introduced via an independent self-replicating DNA molecule that carries only a few genes, including a copy of the operon or part of it. This additional DNA molecule is called a **plasmid.**

In cells in which one regulator gene is *lacI*$^+$ and the

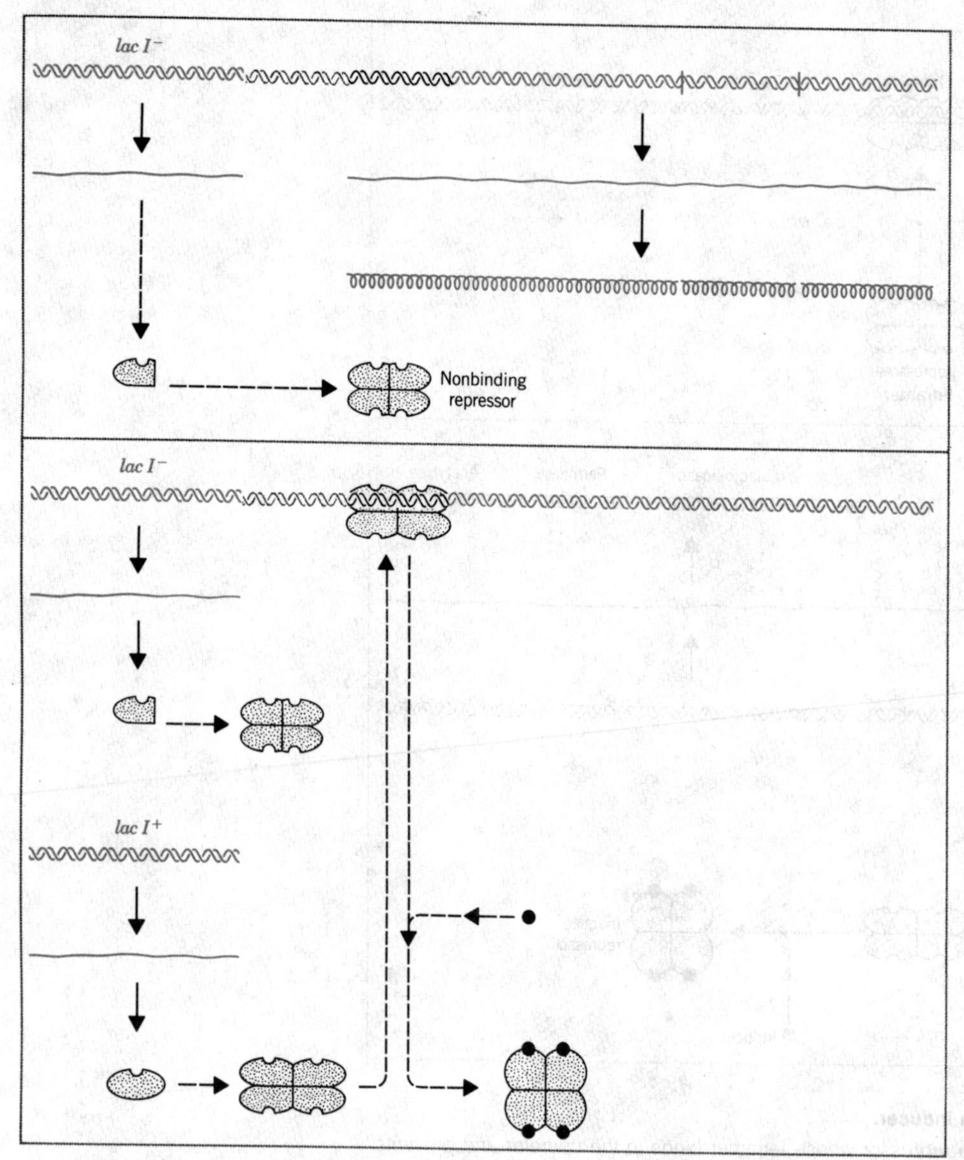

Figure 14.3
Constitutive mutations in the *lacI* gene are recessive.
Upper. A *lacI* mutation makes the repressor inactive. Because it cannot bind to the operator, the structural genes are expressed.
Lower. The introduction of a second, *lacI*$^+$, gene allows normal repressor to be made. It is able to act in *trans* to bind to the operator, being displaced from the DNA when inducer is added. This restores normal control of the structural genes.

other is *lacI*$^-$, normal regulation is restored. The structural genes again are repressed when the inducer is removed. We can explain this effect in the way depicted in **Figure 14.3.** Constitutive mutations in the repressor abolish its ability to bind to the operator. The result is that transcription can initiate freely at the promoter. But the introduction of a second, wild-type regulator gene restores the presence of normal repressor. So once again the operon is turned off in the absence of inducer. In genetic parlance, the wild-type **inducibility** is *dominant* over the mutant constitutivity. This is the hallmark of negative control.

The *lacI* regulator gene lies just to the left of the cluster of structural genes. However, since it specifies a *trans*-diffusible product, it need not be located near the structural genes. Indeed, as we have seen, a *lacI* gene on an independent plasmid is able to control a *lacZYA* cluster on the bacterial chromosome (or the orientation can be reversed). In other operons in *E. coli*, the regulator gene actually is located at some distance from the structural genes. However, we might speculate that an advantage to keeping a regulator near its structural genes is that it has a useful purpose only together with them.

THE OPERATOR IS *CIS*-DOMINANT

The operator was originally identified by constitutive mutations, denoted o^c, whose distinctive properties demonstrate that this region is not represented in a diffusible product. The structural genes contiguous with the o^c mutation are expressed constitutively. The reason is that the operator has suffered a mutation that prevents the repressor from binding to it. Thus the repressor is no longer able to prevent RNA polymerase from initiating transcription. So the operon is continuously expressed, as illustrated in **Figure 14.4.**

The operator can control *only the lac genes that are adjacent to it:* If a second *lac* operon is introduced into the bacterium on an independent molecule of DNA, it has its own operator. Neither operator is influenced by the other. Thus if one operon has a wild-type operator, it will be repressed under the usual conditions, while a second operon with an o^c mutation will be expressed in its characteristic fashion.

The ability of a site to control adjacent genes irrespective of the presence in the cell of other alleles of the site defines the phenomenon of ***cis*-dominance.** It indicates that the mutant site does not specify any product (usually protein, but in principle also RNA) that can diffuse through the cell to exercise its effect at the other allele. Sometimes a *cis*-dominant site is referred to as ***cis*-acting,** to make the contrast with a ***trans*-acting** function, which, because it specifies a diffusible product, can act on all relevant sites in the cell, whether they are present on the same or different molecules of DNA (see Chapter 2).

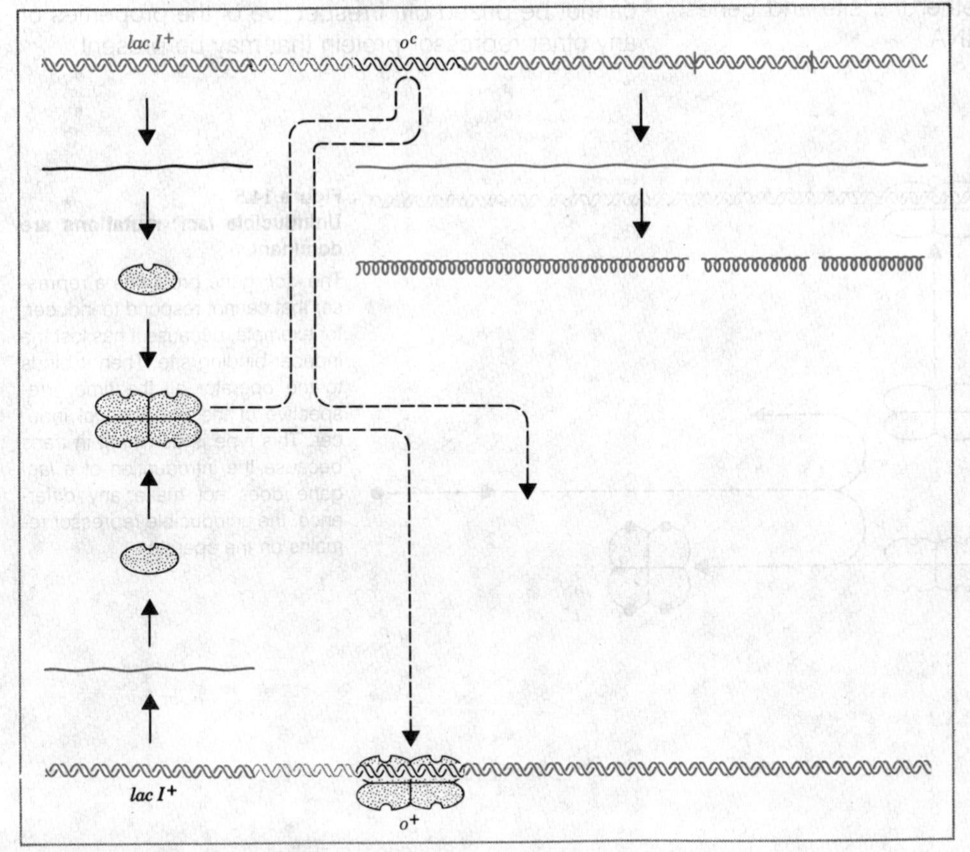

Figure 14.4
Operator-constitutive (o^c) mutations are *cis*-dominant.
Upper: the operon has a mutant operator that cannot bind repressor, so the adjacent structural genes are expressed constitutively.
Lower: the operon has a wild-type operator that responds to repressor and is therefore expressed only on induction.
The behavior of each set of structural genes is controlled solely by the contiguous operator, whose properties are not influenced by the other operator.

The concept of *cis*-dominance applies to *any sequence of DNA that functions by being recognized rather than by being converted into a diffusible product.* Thus mutations in a promoter or terminator also are *cis*-acting. A consequence of *cis*-dominance is that a mutation cannot be assigned to a complementation group. (For the ability to complement is characteristic only of genes expressed as diffusible products.) When two *cis*-acting sites lie close together—for example, a promoter and an operator—we cannot classify the mutations by a complementation test. We are restricted to distinguishing them by their effects on the phenotype.

Finally, we should point out that *cis*-dominance is a characteristic of any site that is *physically contiguous with the sequences it controls*. Thus if a control site functions as part of a polycistronic mRNA, mutations in it will display *exactly the same pattern* of *cis*-dominance as they would if functioning in DNA. The critical feature is that the control site cannot be physically separated from the genes. From the genetic point of view, it does not matter whether the site and genes are together on DNA or on RNA.

UNINDUCIBLE MUTATIONS CAN OCCUR IN THE PROMOTER OR REPRESSOR

Mutants of the operon that are **uninducible** cannot be expressed at all. They fall into the same two types of genetic classes as the constitutive mutants. The dominance relationships of each type can be used in the same way to define the nature of the locus.

Promoter mutations, as we have mentioned, are *cis*-acting. Those that prevent RNA polymerase from binding at *lacP* render the operon nonfunctional because it cannot be transcribed.

Mutations in *lacI* that abolish the ability of repressor to bind the inducer can cause the same phenotype. The repressor is "locked in" to the active form that recognizes the operator and prevents transcription. The addition of inducer is to no avail. Such mutants are described as *lacI*^{s}. **Figure 14.5** shows that they are *trans*-acting and dominant over wild type. This happens because the mutant repressor binds to all operators in the cell to prevent their transcription, and cannot be prized off, irrespective of the properties of any other repressor protein that may be present.

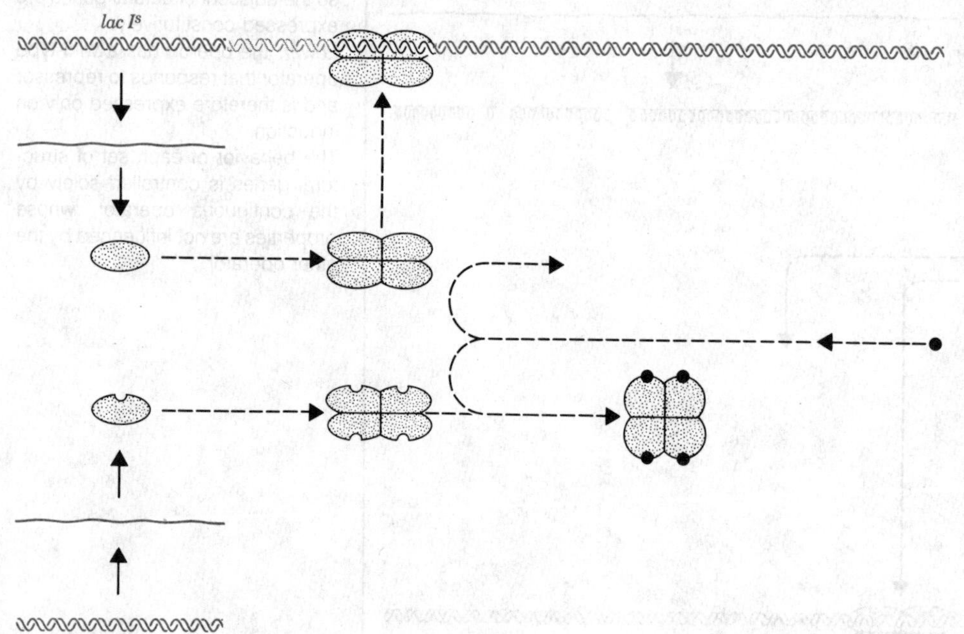

Figure 14.5
Uninducible *lacI*^{s} mutations are dominant.
The *lacI*^{s} gene produces a repressor that cannot respond to inducer; for example, because it has lost the inducer-binding site. Then it binds to the operator all the time, irrespective of the presence of inducer. This type is dominant in *trans* because the introduction of a *lacI* gene does not make any difference; the uninducible repressor remains on the operator.

HOW DOES REPRESSOR BLOCK TRANSCRIPTION?

The repressor was originally isolated by purifying from extracts of *E. coli* the component able to bind the gratuitous inducer IPTG. (Because the amount of repressor in the cell is so small, in order to obtain enough material it was necessary to use a promoter up mutation to increase *lacI* transcription, and to place this *lacI* locus on a DNA molecule present in many copies per cell. This results in an overall overproduction of 100 to 1000-fold.) The binding of the purified repressor protein to DNA was first characterized by a **filter-binding assay.** This takes advantage of the ability of nitrocellulose filters to retain protein but not double-stranded DNA. However, DNA complexed with protein is retained; so sequences of DNA that can bind to the repressor are specifically retained on the filter.

The repressor binds to DNA containing the sequence of the wild-type *lac* operator. The DNA must be in double-stranded form. The repressor does not bind if the DNA has been obtained from an o^c mutant. The addition of IPTG releases the repressor from *lacO* DNA *in vitro*. The *in vitro* reaction between repressor protein and *lacO* DNA therefore displays the characteristics of control inferred *in vivo;* so it can be used to establish the basis for repression.

What is the precise location of the operator in relation to the promoter? Defined by the length of DNA protected by repressor *in vitro* against nuclease attack, *lacO* comprises a region of about 26 bp. It stretches from position −5 just upstream from the mRNA startpoint to position +21 within the transcription unit. Thus it overlaps the right end of the promoter.

Does this overlap prevent repressor and RNA polymerase from binding simultaneously to *lac* DNA? Competition experiments show that the binding of either protein to the DNA prevents the binding of the other. This situation is illustrated in **Figure 14.6.** Similar results have been obtained in several other operons, although the overlap does not always occur in the same region of the promoter (see Figure 15.13). For example, in phage lambda, the operator lies in the upstream region of the promoter (see Chapter 16). But the general rule is that when the repressor is bound at its operator, its presence denies the RNA polymerase access to its promoter, and therefore prevents transcription from being initiated.

Figure 14.6
Repressor binding and RNA polymerase binding are mutually exclusive at the *lac* operator-promoter region.

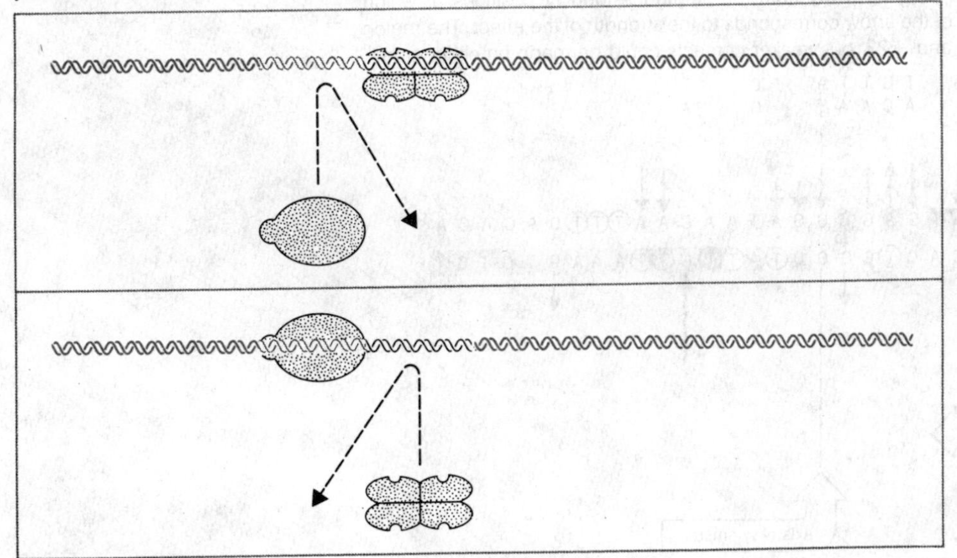

CONTACTS IN THE OPERATOR

How does the repressor recognize the specific sequence of operator DNA? The interaction between lactose repressor and operator is often taken as a paradigm for sequence-specific DNA-binding reactions. It is the best characterized from the aspects of both the DNA sequence and the protein structure.

The operator has a feature common to many recognition sites for proteins: it has an axis of dyad symmetry. The inverted repeats are highlighted in **Figure 14.7.** (The prominent exception to the occurrence of symmetry in recognition sites, of course, is provided by the promoter, which must convey information about the direction of transcription.)

What use is made of the symmetry in recognizing the operator? The possibility that there is a rearrangement of DNA to form a cruciform has been excluded. When the repressor binds to the DNA, there is a change in the structure of the duplex equivalent to a rather small unwinding (but not enough to create free base pairs or a cruciform).

The approaches used to define the points that the repressor contacts in the operator are the same as those we have already discussed for the polymerase-promoter interaction (see Chapter 12). Constitutive point mutations identify individual base pairs that must be crucial; deletions of material on either side define the end points of the region. Experiments in which DNA bound to repressor is compared with unbound DNA for its susceptibility to methylation or UV cross-linking identify bases that are either protected or more susceptible when associated with the protein.

When the DNA bound to repressor is isolated by degrading unprotected sequences with nuclease, it does not extend over the entire region of symmetry.

Figure 14.7
The *lac* operator has a symmetrical sequence.
The sequence is numbered relative to the startpoint for transcription at +1. The regions of dyad symmetry are indicated by the shaded blocks; the axis passes through the base pair at position +11.

Substitutions that result in constitutive expression are shown by the base pairs above the sequence. The two mutations on the right are exact symmetrical counterparts of mutations found on the left. At position +17 T-A replaces G-C, while at position +5 A-T replaces G-C. Similarly, at position +14 C-G replaces T-A, while at position +8 G-C replaces A-T.

Thymines at positions where the substitution of bromouracil permits cross-linking to repressor are indicated by circles. Arrows pointing into the sequence identify purines that are protected by repressor against methylation. Arrows pointing away from the sequence identify positions at which methylation is enhanced. The length of the arrow corresponds to the strength of the effect. The region of strong contacts lies between +1 and +23, but weaker contacts could be made on either side.

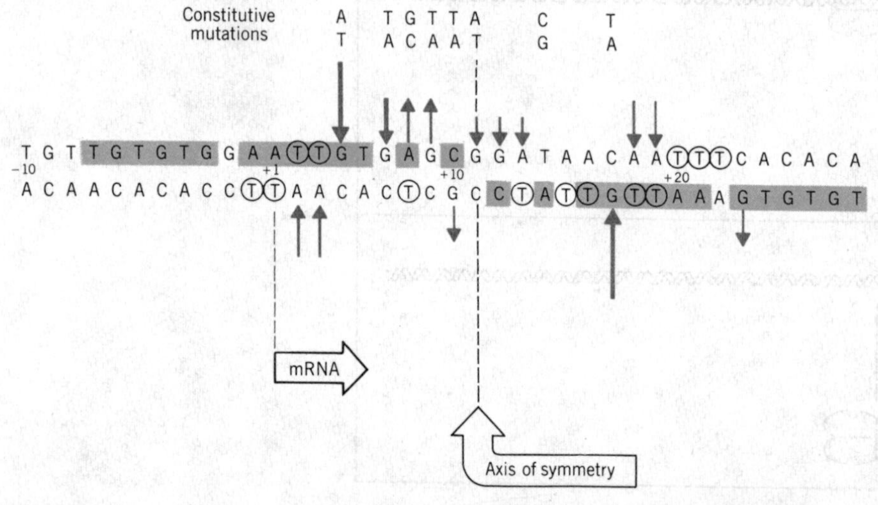

We have mentioned that it comprises the 26 bp region from −5 to +21. The area identified by constitutive mutations is even smaller. Within a central region extending over the 13 bp from +5 to +17, there are eight sites at which single base-pair substitutions cause constitutivity. This emphasizes the same point made by the promoter mutations summarized earlier in Figure 12.5. *A small number of essential specific contacts within a larger region can be responsible for sequence-specific association of DNA with protein.*

The pattern of enhancement and protection of bases shows some features of symmetry within a general region of close contacts. Figure 14.7 shows that all but two of the thymine residues within the region from +1 to +22 can be cross-linked to repressor. Methylation experiments show that a large proportion of the purines (A and G) between +3 and +19 are protected quite strongly by repressor binding. A few display increased susceptibility, presumably due to the creation of "hydrophobic pockets" in the repressor-operator complex.

The contacts between +1 and +6 are essentially symmetrical with those made between +21 and +16; contacts closer to the axis of symmetry are not symmetrical. Thus the repressor binds to the operator in such a way as to sit symmetrically about the outlying points of contact, while not lying symmetrically in the immediate vicinity of the axis of symmetry.

The inverted repeats of the *lac* operator are not quite identical. There are three differences between them (shown by the breaks in the shaded blocks). The distribution of the sites of mutation suggests that the left side of the operator may be more susceptible to damage: it contains six mutations, compared with the two on the right. Also, mutations that occur at equivalent positions on the left side and right side have greater effect on the left. Thus symmetry is clearly relevant to repressor-operator contacts, but nonetheless the repressor seems to bind to the left side more intimately.

The idea that the left half of the operator has greater affinity for the repressor is supported by the properties of a synthetic operator, consisting of a perfect inverted repeat of the left half (the sequence from −7 to +10 in Figure 14.7). This DNA sequence has an affinity for repressor about 10 times greater than the natural operator. The increase is probably due largely to the conversion of the right half to an exact mirror image of the left half; also, the elimination of the central base pair (+11), bringing the two halves closer together, may help.

Does the symmetry of the DNA sequence reflect a symmetry in the protein? This is likely to be the case, because the repressor is a tetramer of identical subunits, each of which must therefore have the same DNA-binding site. Each half of the operator is probably contacted in the same way by a repressor dimer.

The central region that is crucial for repressor binding occupies roughly the first 20 bp downstream from the startpoint for transcription. But if we recall the points contacted by RNA polymerase at the promoter, Figure 12.5 shows that these tend to lie upstream from the startpoint. The major points of contact for the two proteins are therefore adjacent rather than overlapping.

THE INTERACTION OF REPRESSOR SUBUNITS

The repressor protein has two types of binding site that (as we have seen) interact to control gene expression in response to the environment. The *DNA-binding site* recognizes the sequence of the operator. The *inducer-binding site* binds the small-molecule inducer; and as a result of this interaction, the DNA-binding site *loses* its ability to hold the operator DNA.

The two types of binding site can be identified within the repressor subunit by mutations in *lacI* that inactivate them. Their proper relationship *in vivo* depends on the multimeric structure of the repressor.

Repressor subunits associate at random in the cell to form the active protein tetramer. When two different alleles of the *lacI* gene are present, the subunits made by each can associate to form a heterotetramer, whose properties may differ from those of either homotetramer. This type of interaction between subunits is a characteristic feature of multimeric proteins and is described as **interallelic complementation** (see Chapter 2).

A particular type of interaction, called **negative complementation,** occurs with some repressor mutants. This is seen in the combination of $lacI^{-d}$ with $lacI^{+}$ genes. The $lacI^{-d}$ mutation alone results in the production of a repressor that cannot bind the operator, and is therefore of the constitutive type representative of all *lacI*⁻ alleles. Because the *lacI*⁻ type of mutation inactivates the repressor, it is recessive to the wild type. However, the −*d* notation indicates that this variant of the negative type is dominant when paired with a wild-type allele.

The reason for the dominance is that the *lacI*$^{-d}$ allele produces a "bad" subunit, which is not only itself unable to bind to operator DNA, but is also able as part of a tetramer to prevent any "good" subunits from binding. This demonstrates that the repressor tetramer as a whole, rather than the individual monomer, is needed to achieve repression. The poisoning effect also can be produced *in vitro* by mixing appropriate "good" and "bad" subunits.

The *lacI*$^{-d}$ mutations identify the DNA-binding site of the repressor subunit. This explains their ability to prevent mixed tetramers from binding to the operator; a reduction in the number of binding sites must reduce the specific affinity too much. The map of the *lacI* gene shown in **Figure 14.8** shows that the *lacI*$^{-d}$ mutations are clustered at the extreme left end of the gene. This identifies the immediate N-terminal region of the protein as the DNA-binding site. Mutations of the recessive *lacI*⁻ type also occur elsewhere in the molecule, but could exert their effects on DNA binding indirectly.

The role of the N-terminal region in specifically binding DNA is shown also by its location as the site of occurrence of "tight binding" mutations. These increase the affinity of the repressor for the operator, sometimes so much that it cannot be released by inducer. They are rare.

Uninducible mutations of the *lacI*^{s} type render the repressor unresponsive to the inducer. This could happen either because the protein has lost its inducer-binding site, or because it has become unable to transmit its effect to the DNA-binding site. As can be seen from Figure 14.8, the *lacI*^{s} mutations occur in clusters that are rather regularly spaced along the gene. The spacing may represent turns in the polypeptide chain. (A systematic study of the effects of amino acid replacements shows that about 60% of the positions are neutral, with no effect on repression.)

REPRESSOR AS A DNA-BINDING PROTEIN

The behavior of the isolated repressor protein *in vitro* shows directly that it possesses a DNA-binding site whose ability to remain attached to operator DNA is influenced by the structure of the rest of the molecule. When the repressor is treated with trypsin, it is cleaved preferentially at amino acid 59. The C-terminal fragment of the protein, containing residues 60–360, is known as the **trypsin-resistant core.** It retains the ability to aggregate into a tetramer and to bind in-

Figure 14.8
Mutations in the *lacI* gene identify domains for different functions.
Distance along the gene is numbered in codons. The upper map shows that *lacI*⁻ mutations occur throughout the gene, but the *lacI*$^{-d}$ variety occurs only at the N-terminal end. Red indicates mutations that are *lacI*⁻ because the monomers cannot aggregate into tetramers. The lower map shows that *lacI*^{s} mutations do not start until residue 62, and tend to occur in clusters roughly between positions 150 and 300, generally separated by 26 residues or so.

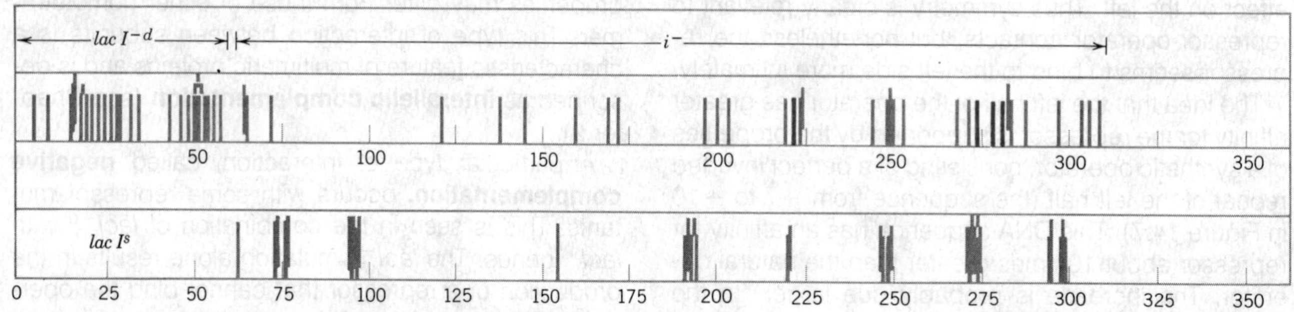

ducer; but it cannot bind the operator. The amino-terminal fragment, consisting of residues 1–59, is known as the **long headpiece.** It may be cleaved again by trypsin at amino acid 51, generating a fragment of residues 1-51 called the **short headpiece.**

The short and long headpieces retain the ability to bind to DNA; and when presented with operator DNA, they make the same pattern of contacts achieved by intact repressor (although they bind more weakly than the intact protein). The ability of the headpieces to bind to the operator suggests that their structure per se is independent of the rest of the protein.

This accords with models in which the DNA-binding site lies as an **arm** or **protrusion** of the N-terminal 50 amino acids from the body of the protein. The arm may be connected to the core by a **hinge** region, constructed from amino acids 50–80. The remainder of the protein, amino acids 81–360, is responsible for aggregating into the tetrameric structure and for binding inducer.

If the headpieces are protrusions that have the DNA-binding ability, what is the function of the tetrameric structure, and how does it contact DNA? One model suggests that the repressor has the shape of an elongated dumbbell or cylinder, about 115Å long, whose axis may lie across the DNA at a shallow angle of 15–20°. The cores of all four subunits are in contact in a central region; the headpiece binding sites lie at the extremities of the molecule, arranged in pairs. It is possible that actually only two of the four subunits contact DNA, leaving the other two DNA-binding sites unoccupied.

Although the core of the subunit does not recognize the operator, it binds to DNA in a nonsequence-specific manner, so it is possible that it too makes contacts (although these are not pronounced enough to influence the pattern of protection of DNA relative to the headpiece alone). This model therefore pictures the repressor-DNA complex as a tetrameric protein symmetrically disposed on the DNA, and with only a small proportion (about 20%) of the polypeptide chain intimately contacting DNA. This last feature is shared by another quite different model, in which the axis of the repressor is perpendicular to the axis of DNA, so that only the two headpieces at one end are bound to the operator.

GETTING OFF DNA

A repressor tetramer is bound tightly to the operator. An inducer comes along and binds to the repressor. How does the protein get off the DNA?

Various inducers cause characteristic reductions in the affinity of the repressor for the operator *in vitro*. These changes correlate with the effectiveness of the inducers *in vivo*. This means that the mechanism of induction relies on a reduction in the attraction between operator and repressor. How is this accomplished? The rate at which the repressor dissociates from the operator is rather slow, typically displaying a half-life that varies *in vitro* from 10–20 minutes, depending on the conditions. But when IPTG is added, there is an immediate reduction in the stability of the complex, as seen by a drastic reduction in its half-life.

This result distinguishes between the two models for repressor action illustrated in **Figure 14.9.** The equilibrium model (left) calls for repressor bound to DNA to be in rapid equilibrium with free repressor; inducer would bind to the free form of repressor, and thus unbalance the equilibrium by preventing reassociation with DNA. But the rate of dissociation of the repressor from the operator (as measured in the absence of inducer) is much too slow to be compatible

Figure 14.9
Does the inducer bind to free repressor to upset an equilibrium (left) or directly to repressor bound at the operator (right)?

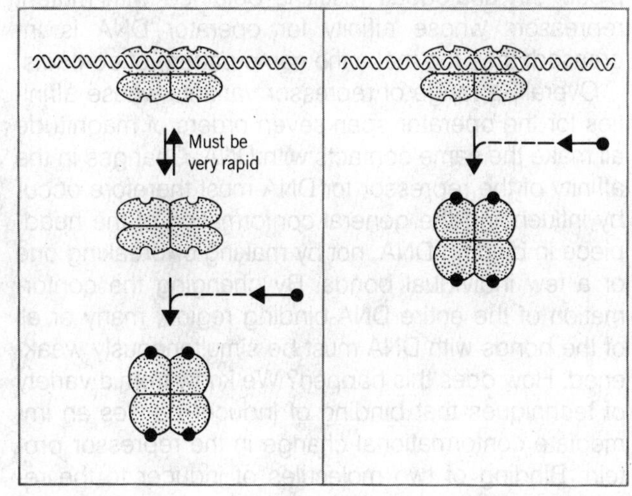

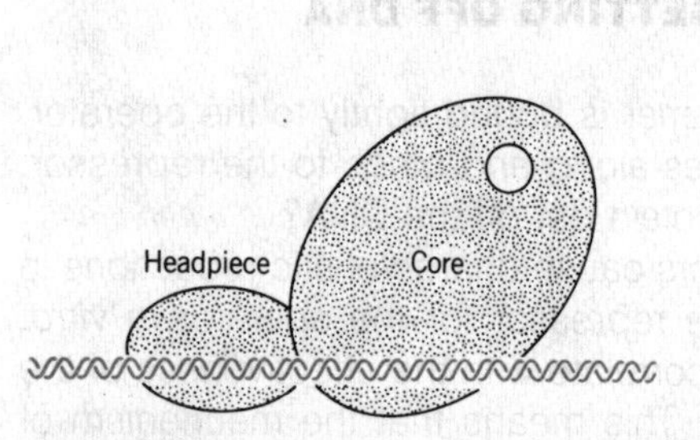

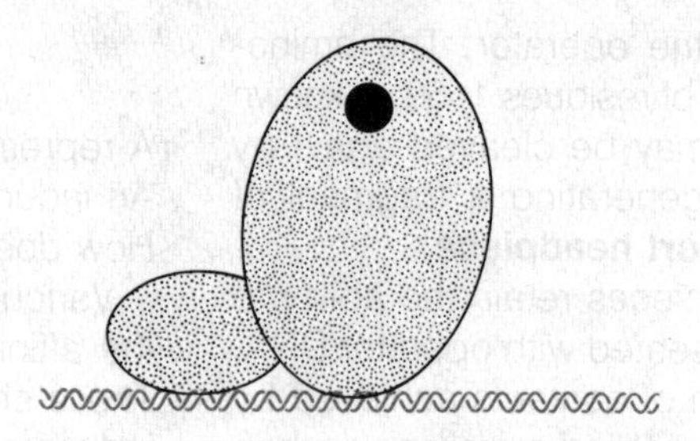

Figure 14.10
Repressor displacement involves a general change in structure that weakens all its contacts with the operator.
Left. The headpiece is specifically bound to the operator, lying along the DNA to make contacts with the double helix.
Right. Inducer changes the conformation (illustrated by showing a change in the angle between headpiece and core), so that the headpiece can no longer contact the DNA tightly.

with this model. This means that instead the *inducer must bind directly to repressor protein complexed with the operator.* As indicated in the model on the right, inducer binding must produce a change in the properties of the repressor that makes it let go of the operator.

Binding of the repressor-IPTG complex to the operator can be studied by using greater concentrations of the protein in the methylation protection/enhancement assay. The large amount compensates for the low affinity of the repressor-IPTG complex for the operator. The complex makes exactly the same pattern of contacts with DNA as does free repressor or headpiece. An analogous result is obtained with mutant repressors whose affinity for operator DNA is increased; they too make the same pattern of contacts.

Overall, a range of repressor variants whose affinities for the operator span seven orders of magnitude all make the same contacts with DNA. Changes in the affinity of the repressor for DNA must therefore occur by influencing the general conformation of the headpiece in binding DNA, not by making or breaking one or a few individual bonds. By changing the conformation of the entire DNA-binding region, many or all of the bonds with DNA must be simultaneously weakened. How does this happen? We know from a variety of techniques that binding of inducer causes an immediate conformational change in the repressor protein. Binding of two molecules of inducer to the repressor tetramer is adequate to release repression. But we do not yet know how the overall conformational change is related to the affinity of repressor for the operator.

The most likely model is that a change in conformation is transmitted from the core via the hinge to the headpiece. **Figure 14.10** represents such a model diagrammatically. The conformation of the DNA-binding site is altered from a state in which it exactly fits the DNA sequence to a state in which the headpiece is held in a slightly different register that cannot contact the DNA tightly enough.

STORING SURPLUS REPRESSOR

Probably all proteins that have a great affinity for a specific sequence also possess a low affinity for any (random) DNA sequence. A large number of low affinity sites will compete just as well for a repressor tetramer as a small number of high-affinity sites. There is only one high-affinity site in the *E. coli* genome: the operator. The remainder of the DNA can be considered to provide low-affinity binding sites. How is the repressor partitioned between the operator and the rest of DNA? What happens to the repressor when it has bound inducer and dissociated from the operator?

To achieve specificity for the operator within the context of the genome, the repressor's specific binding affinity must be very much greater than its random affinity. In fact, the repressor binds about 4×10^6 times better to operator DNA than to any random DNA sequence of the same length. (This is the ratio of the association constants of the repressor for *lacO* versus other DNA.) Now the operator represents 26 bp in a genome of 4.2×10^6 bp, so that the excess of non-operator DNA over operator DNA is about 2×10^5. Put another way, there are 2×10^5 low-affinity binding sites for the repressor, compared with the one high-affinity binding site.

The relative preference of the repressor for the operator DNA within the bacterial genome is given by dividing the preference per se for the high-affinity site (4×10^6) by the number of low-affinity sites (2×10^5). The answer indicates that the repressor has about a twentyfold better probability of binding the operator than any other site.

This explains why mutations that reduce the affinity of the operator for the repressor by as little as twenty- or thirtyfold have sufficient effect to be constitutive. Within the genome, the mutant operators can be overwhelmed by the preponderance of random sites. Their specific affinity for repressor relative to these sites is simply no longer great enough to offset their minority status.

The consequence of these affinities is that in an uninduced cell, one tetramer of repressor will usually be bound to the operator. All or almost all of the remaining tetramers will be bound at random to other regions of DNA, as illustrated in **Figure 14.11.** There are likely to be very few or no repressor tetramers free within the cell.

When inducer binds to the repressor, the affinity for

Figure 14.11
Repressor loves DNA.
Virtually all the repressor in the cell is bound to DNA.
(*1*) A tetramer is present at the operator.
(*2*) When the tetramer is induced to leave, it binds at another, random site on DNA. The other tetramers also bind inducer, but this does not affect their general affinity for DNA.
(*3*) Repressor may move from one site to another on DNA by direct displacement.
(*4*) When inducer is removed, the operator is contacted by a repressor tetramer that moves from a random site to restore the uninduced state.

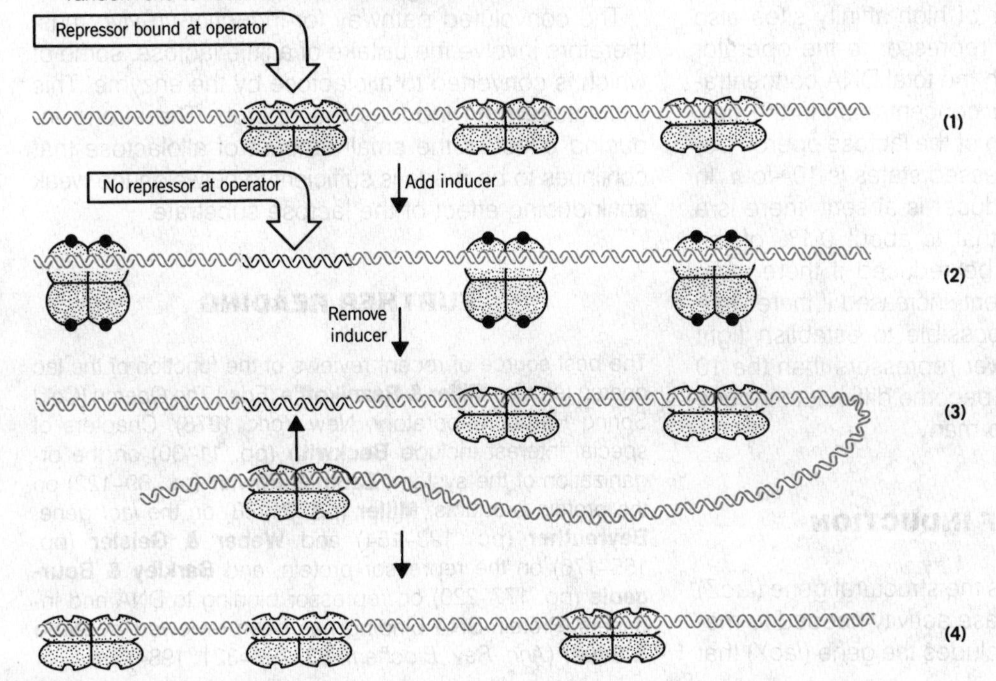

the operator is reduced by about 10^3-fold. The affinity for general DNA sequences remains unaltered. Thus in an induced cell, probably all or virtually all the repressor tetramers will be "stored" on random DNA sites. The effect of induction is therefore not to generate free repressor, but to change its distribution on DNA.

Several important biological conclusions follow from the storage of repressor on DNA. Most directly, the ability to bind to the operator very rapidly is not consistent with the time that would be required for multiple dissociations from, and reassociations with, DNA. The discrepancy excludes random-hit mechanisms for finding the operator, suggesting that the repressor may be able to move directly from a random site on DNA to the operator. Movement could be accomplished either by sliding along the DNA very rapidly (which seems improbable because there would be obstacles in the form of other bound proteins) or by direct displacement from site to site. The latter is the same mechanism that we have already considered for RNA polymerase (see Chapter 11). A displacement reaction might be aided by the presence of more binding sites per tetramer (four) than are actually used to contact DNA at any one time (two).

The ability of a large number of nonspecific sites to overpower a small number of high-affinity sites also implies that the binding of repressor to the operator will be very sensitive to both the total DNA concentration and the total repressor concentration in the cell. The difference in expression of the lactose operon between its induced and repressed states is 10^3-fold. In other words, even when inducer is absent, there is a basal level of expression that is about 0.1% of the induced level. This would be reduced if there were more repressor protein present, increased if there were less. Thus it could be impossible to establish tight repression if there were fewer repressors than the 10 found per cell; and it might become difficult to induce the operon if there were too many.

A PARADOX OF INDUCTION

The lactose operon contains the structural gene (*lacZ*) coding for the β-galactosidase activity needed to metabolize the sugar; it also includes the gene (*lacY*) that codes for the protein needed to transport the substrate into the cell. But if the operon is repressed, how does the inducer enter the cell to start the process of induction by binding to repressor protein?

The basal level of expression of the operon may be crucial in ensuring that there is always a minimal amount of the protein present in the cell, enough to start the process off. This turns out to be necessary not just for uptake of the inducer (some of which could enter anyway via another uptake system), but also to start the metabolic pathway.

Although lactose is the substrate for β-galactosidase, it is not an inducer of the operon. In fact, it has a slight effect as an antiinducer (it binds to the repressor to increase its affinity for the operator). So how is induction to be accomplished?

The natural inducer of the operon has been obtained by isolating repressor from induced cells. The protein is bound to **allolactose.** The predominant reaction of β-galactosidase when presented with lactose is to break it to glucose and galactose. But a side reaction transfers the galactose moiety to certain acceptor molecules, producing (principally) allolactose and galactobiose. Since the acceptor used to form allolactose is in fact glucose, the reaction ends up as a molecular rearrangement.

The convoluted pathway for induction *in vivo* must therefore involve the uptake of a little lactose, some of which is converted to allolactose by the enzyme. This allolactose then induces the operon. The strong inducing effect of the small amount of allolactose that continues to be made is sufficient to outweigh the weak antiinducing effect of the lactose substrate.

FURTHER READING

The best source of recent reviews of the function of the lac operon remains **Miller & Reznikoff's** (Eds.) *The Operon* (Cold Spring Harbor Laboratory, New York, 1978). Chapters of special interest include **Beckwith** (pp. 11–30) on the organization of the system, **Zabin & Fowler** (pp. 89–122) on the protein products, **Miller** (pp. 31–88) on the *lacI* gene, **Beyreuther** (pp. 123–154) and **Weber & Geisler** (pp. 155–176) on the repressor protein, and **Barkley & Bourgeois** (pp. 177–220) on repressor binding to DNA and inducers. Protein-DNA binding has been reviewed by **Pabo & Sauer** (*Ann. Rev. Biochem.* **53**, 293–321, 1984).

CHAPTER 15
CONTROL CIRCUITS: A PANOPLY OF OPERONS

Sometimes it seems that every conceivable mechanism is used for controlling gene expression in one situation or another. But through this variety of control mechanisms, there runs a common thread. Regulation results from the interaction of a regulatory macromolecule with a sequence of nucleic acid, often one that has dyad symmetry. Most commonly the macromolecule is a protein, but in some cases it is an RNA. It may even be a macromolecular assembly that functions elsewhere in the cell in another capacity; the paradigm for this type of interaction is the ability of translation by the ribosome to influence transcription by RNA polymerase.

In this chapter, we shall consider examples of these interactions from the perspective of their linkage into circuits that control individual operons. Similar mechanisms occur in slightly different guises in many systems, and the intricate pattern of connections varies from system to system.

The operon represents an arrangement that permits *coordinate* regulation of a group of structural genes. But, superimposed on this, further controls may create *differences* in the extent to which individual genes are expressed.

DISTINGUISHING POSITIVE AND NEGATIVE CONTROL

Positive and negative control systems are defined by the response of the operon when no regulator protein is present. The characteristics of the two types of control system are mirror images.

Genes under negative control are expressed *unless they are switched off by a repressor protein*. Any action that interferes with gene expression can provide a negative control, but there is a uniformity in these mechanisms: a repressor protein either binds to DNA to prevent RNA polymerase from initiating transcription, or binds to mRNA to prevent a ribosome from initiating translation.

Negative control provides a fail-safe mechanism: if the regulator protein is inactivated, the system functions and so the cell is not deprived of these enzymes. It is easy to see how this might evolve. Originally a system functions constitutively, but then cells able to interfere specifically with its expression acquire a selective advantage by virtue of their increased efficiency.

For genes under positive control, expression is pos-

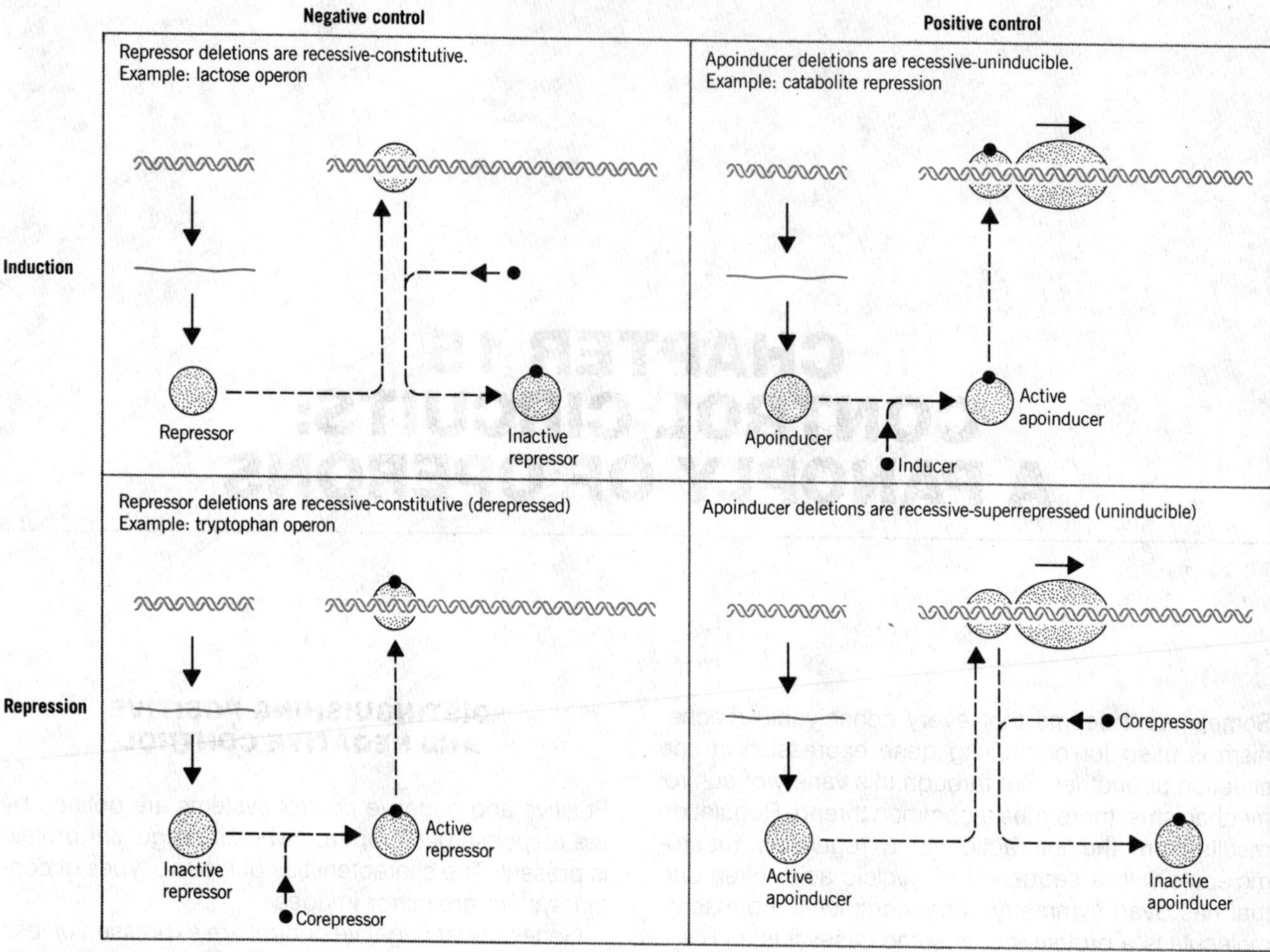

Figure 15.1
Control circuits are versatile and can be designed to allow positive or negative control of induction or repression.

sible *only when an active regulator protein is present*. How is this to be achieved? Positive controls that act at the initiation of transcription in particular operons are exact counterparts of negative control systems. But instead of *interfering* with initiation, the regulator protein is *essential* for it. It interacts with DNA and with RNA polymerase to *assist the initiation event*. Such a protein may be called an **apoinducer.** Other positive controls provide for the global substitution of RNA polymerase subunits involved in setting initiation specificity (Chapter 12), or act as antitermination factors in transcription (Chapter 13).

It is more difficult to see how positive control evolved, since the cell must have had the ability to express the regulated genes even before any control existed. Presumably some component of the control system must have changed its role. Perhaps originally it was used as a regular part of the apparatus for gene expression; then later it became restricted to act only in a particular system or systems.

Operons are defined as **inducible** or **repressible** by the nature of their response to the small molecule that regulates their expression. Just as it may be advantageous for a bacterium to induce a set of enzymes only after addition of the inducer substrate that they metabolize, so also may it be useful to repress the enzymes that synthesize some compound if it is provided in adequate amounts by the medium. Thus inducible operons function only in the *presence* of the small-molecule inducer. Repressible operons function only in the *absence* of the small-molecule **corepressor** (so called to distinguish it from the repressor protein).

The terminology used for repressible systems describes the active state of the operon as **derepressed;** this has the same meaning as *induced*. The condition in which a (mutant) operon cannot be derepressed is sometimes called **superrepressed;** this is the exact counterpart of *uninducible*.

Either positive or negative control could be used to achieve either induction or repression by utilizing appropriate interactions between the regulator protein and the small-molecule inducer or corepressor. **Figure 15.1** summarizes four simple types of control circuit. Induction is achieved when an inducer inactivates a repressor protein or activates an apoinducer protein. Repression is accomplished when a corepressor activates a repressor protein or inactivates an apoinducer protein.

The genetic consequences of inactivating the regulator protein can be used to discriminate between negative and positive control systems. Inactivation of a repressor protein creates the recessive constitutivity (or derepression) typical of negative control systems. On the other hand, a mutation that inactivates an apoinducer causes the recessive uninducibility or superrepression that typifies positive control systems.

THE TRYPTOPHAN OPERON IS REPRESSIBLE

The five structural genes of the tryptophan operon are arranged in a contiguous series, coding for the three enzymes that convert chorismic acid to tryptophan by the pathway given in **Figure 15.2.** Transcription starts

Figure 15.2
The *trp* operon consists of five contiguous structural genes preceded by a control region.

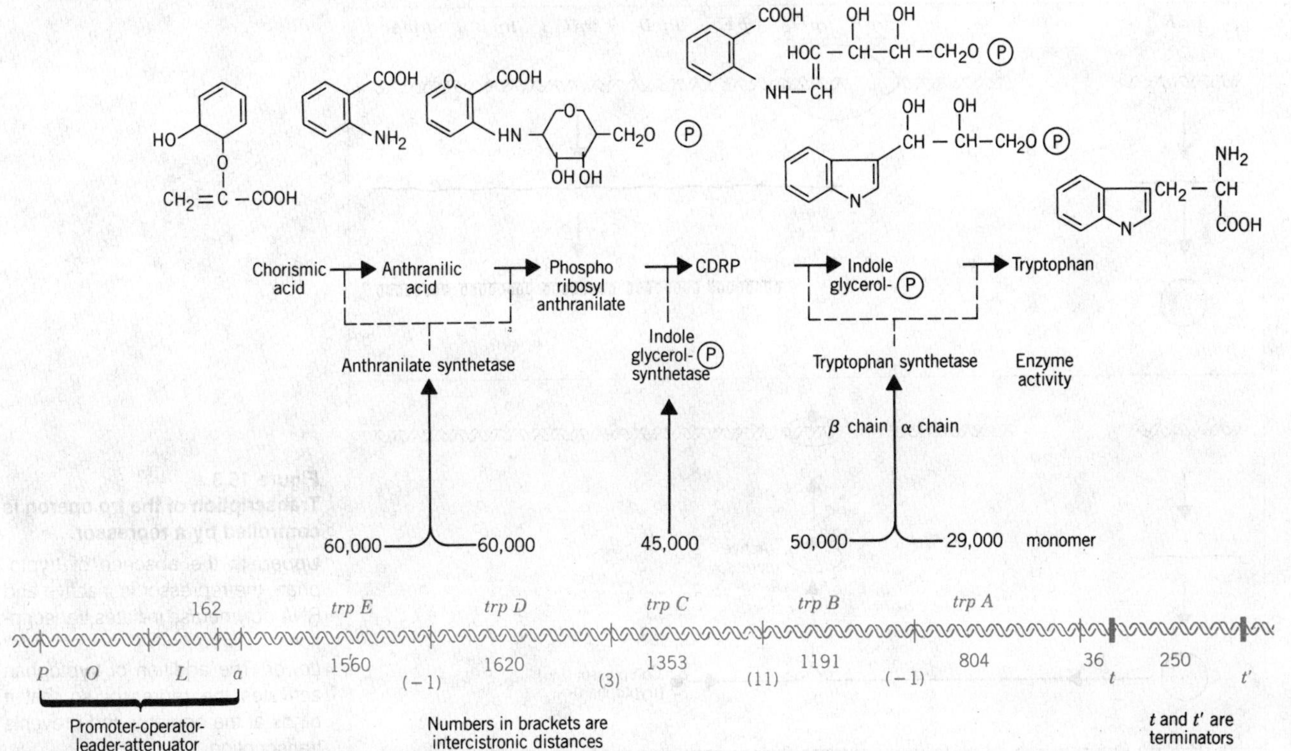

at a promoter at the left end of the cluster. Adjacent to it is the operator that binds the repressor protein coded by the unlinked gene *trpR*. Between the operator and the coding region of the first gene there is a leader sequence. Transcription of the structural genes is terminated at a rho-independent site, *trpt*, 36 bp beyond the end of the last coding region. About 250 bp later there is a rho-dependent terminator, *trpt'*. Essentially the same operon is present in *E. coli* and in *S. typhimurium*.

Both the activity and the synthesis of the tryptophan enzymes are controlled by the level of tryptophan in the cell. The classic feedback loop of **end-product inhibition** applies to the enzymes: the catalytic activities of the first enzyme of the pathway are inhibited by tryptophan, the ultimate product. This means that when the cell has sufficient tryptophan, it is able to cut off the synthesis of further molecules of the amino acid by inhibiting the beginning of the pathway.

Tryptophan also functions as the corepressor that activates the repressor protein. This is the classic mechanism for repression, one of the examples given in Figure 15.1, and the circuit is illustrated in more detail in **Figure 15.3.** In conditions when the supply of tryptophan is plentiful, the operon is repressed because the repressor protein•corepressor complex is bound at the operator. When tryptophan is in short supply, the corepressor is inactive, therefore has reduced specificity for the operator, and is stored elsewhere on DNA.

The *trp* repressor functions as a tetramer of four identical subunits of about 12,500 daltons each. There are about 20 molecules of tetramer per bacterium. The repressor can bind to the operator *in vitro* only in the presence of tryptophan. The only significant difference from the interactions in the *lac* operon is that the presence of tryptophan *increases* the affinity of the repressor protein for the operator.

The sequence of the *trp* promoter-operator region is given in **Figure 15.4.** RNA polymerase binds to a region extending from about −40 to +18. The promoter has the usual consensus-like sequences at −35 and −10.

The operator lies entirely within the promoter. The points contacted by the repressor mostly lie symmetrically on the two DNA strands and occupy the region

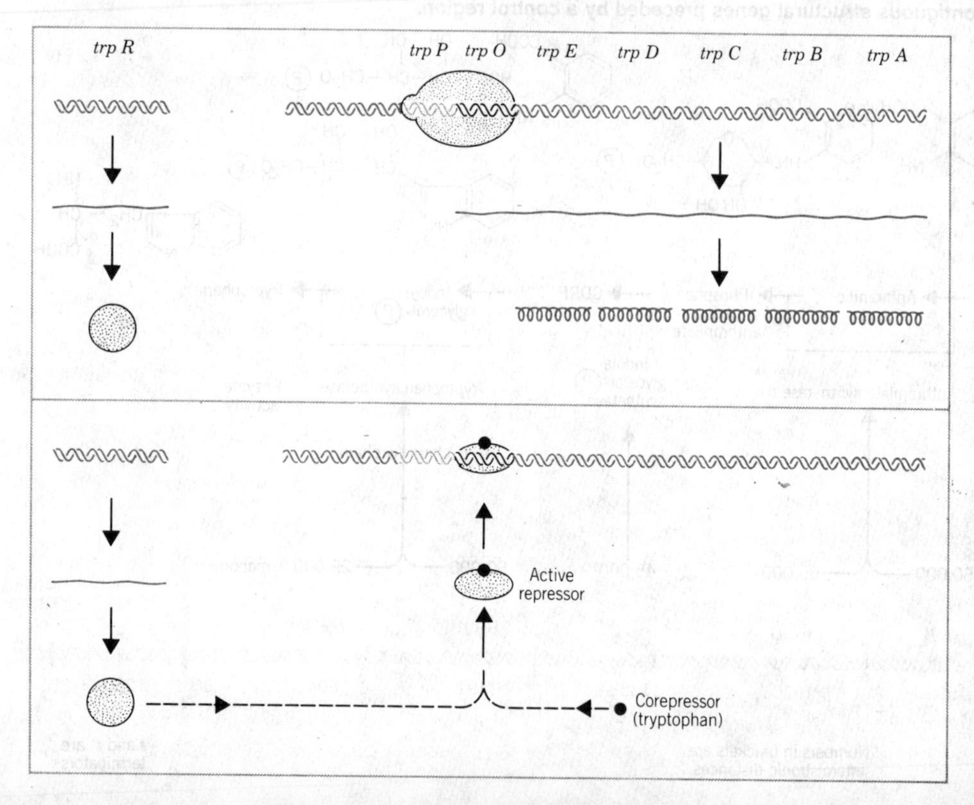

Figure 15.3
Transcription of the *trp* operon is controlled by a repressor.
Upper. In the absence of tryptophan, the repressor is inactive and RNA polymerase initiates transcription.
Lower. The addition of tryptophan activates the repressor so that it binds at the operator and prevents transcription.

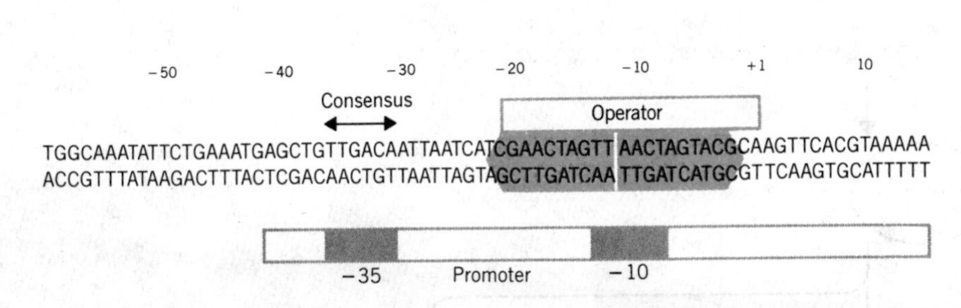

Figure 15.4
The operator for the *trp* genes lies within the promoter.
Transcription initiates at +1. The usual hexamers related to the consensus sequences are present around −10 and −35. There is a 10 bp repeat within the operator (with one nonmatching position). All the o^c mutations lie within the dyad symmetry, but outside the −10 consensus sequence with which it overlaps.

from positions −23 to −3. The operator has a region of dyad symmetry, which includes the consensus sequence of the promoter at −10.

Although the exact relationship of the operator and promoter sequences is different from that of the *lac* operon, the general mechanism of repression is the same: binding of RNA polymerase and binding of repressor are mutually exclusive.

Deprivation of repressor causes roughly a seventy-fold increase in the frequency of initiation events at the *trp* promoter. Even under repressing conditions, the structural genes continue to be expressed at a low **basal** or **repressed level.** This means that the efficiency of repression at the operator is much lower than that seen in the *lac* operon (where the basal level is only about 1/1000 of the induced level).

THE TRYPTOPHAN OPERON IS CONTROLLED BY ATTENUATION

It has been known almost since studies of the tryptophan operon began that, in addition to the promoter-operator complex, another site is involved in regulation. Its existence was first revealed by the observation that deleting a sequence between the operator and the *trpE* coding region can increase the expression of the structural genes. This effect is independent of repression: both the basal and derepressed levels of transcription are increased. So this site influences events that occur *after* RNA polymerase has set out from the promoter (irrespective of the conditions prevailing at initiation).

The regulator site is called the **attenuator.** It lies within the transcribed leader sequence of 162 nucleotides that precedes the initiation codon for the *trpE* gene. The attenuator is a barrier to transcription. It consists of a rho-independent termination site like those we have discussed in Chapter 13. A short G-C-rich palindrome is followed by eight successive U residues. RNA polymerase terminates there, either *in vivo* or *in vitro,* to produce a 140-base transcript.

The termination event at this site responds to the level of tryptophan, as illustrated in **Figure 15.5.** In the presence of adequate amounts of tryptophan, termination is efficient. But in the absence of tryptophan, RNA polymerases can continue into the structural genes. Such regulation of termination of transcription is called **attenuation.**

We see that attenuation controls the operon in the same direction as repression. When tryptophan is present, the operon is repressed; and most of the RNA polymerases that escape from the promoter then terminate at the attenuator. When tryptophan is removed, RNA polymerase has free access to the promoter, and also is no longer compelled to terminate prematurely.

Attenuation has an eight- to tenfold effect on transcription. Under conditions of efficient termination, about 10% of the RNA polymerases that encounter the terminator can get past. Under conditions of attenuation, probably almost all of the RNA polymerases can continue transcription. Together with the seventyfold increase in initiation of transcription that results from the release of repression, this allows roughly a 600-fold range of regulation of the operon.

ALTERNATIVE SECONDARY STRUCTURES CONTROL ATTENUATION

How can termination of transcription at the attenuator respond to the level of tryptophan? The sequence of the leader region suggests a mechanism.

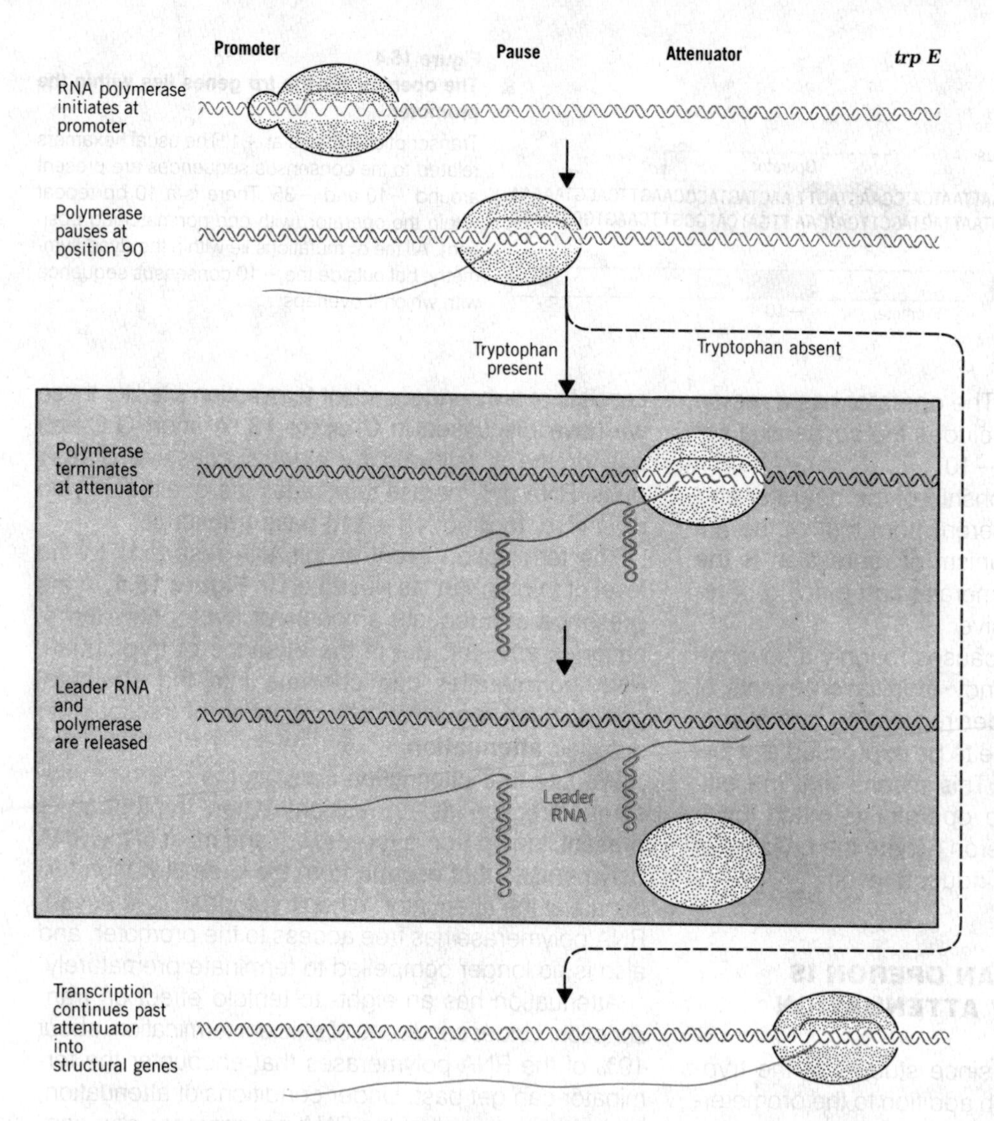

Figure 15.5
An attenuator controls the progression of RNA polymerase into the *trp* genes.
After initiating at the promoter (whether under basal expression or derepression), RNA polymerase proceeds to position 90, where it pauses. When it reaches the attenuator, in the presence of tryptophan there is a (roughly) 90% probability of termination to release the 140 base leader RNA. In the absence of tryptophan, the polymerase continues into the structural genes (*trpE* starts at +163).

Figure 15.6 shows that it contains a ribosome binding site whose AUG codon is followed by a coding region of 13 codons. Is this sequence translated into a **leader peptide?** Although no product has been detected *in vivo*, probably this is because it is unstable. We know that the ribosome binding site is functional (because when it is fused to a structural gene it sponsors effective translation).

What is the function of the leader peptide? It contains two tryptophan residues in immediate succession. Tryptophan is a rare amino acid in *E. coli* proteins, so this is unlikely to be mere coincidence. When the cell runs out of tryptophan, ribosomes initiate translation of the leader peptide, but stop when they reach the Trp codons. The sequence of the mRNA suggests, that this **ribosome stalling** may in turn influence termination at the attenuator.

The leader sequence can be written in alternative base-paired structures. The ability of the ribosome to proceed through the leader region may control transitions between these structures. The structure determines whether the mRNA can provide the features needed for termination.

Figure 15.7 draws these structures. In the first, re-

Leader peptide

Met Lys Ala Ile Phe Val Leu Lys Gly Trp Trp Arg Thr Ser

pppAAGUUCACGUAAAAAGGGUAUCGACAAUGAAAGCAAUUUUCGUACUGAAAGGUUGGUGGCGCACUUCCUGA

140

AACGGGCAGUGUAUUCACCAUGCGUAAAGCAAUCAGAUACCCAGCCCGCCUAAUGAGCGGGCUUUUUUUU

162

GAACAAAAUUAGAGAAUAACAAUGCAAACACAAAAACCG ---

Met Gln Thr Gln Lys Pro . . .

trpE

Figure 15.6
The leader sequence of the *trp* genes codes for a short peptide.

gion **1** pairs with region **2;** and region **3** pairs with region **4.** The pairing of regions **3** and **4** generates the hairpin that precedes the U_8 sequence: this is the essential signal for termination. Probably the RNA would take up this structure in lieu of any outside intervention.

A different structure is formed if region **1** is prevented from pairing with region **2.** In this case, region **2** is free to pair with region **3.** Then region **4** has no available pairing partner; so it is compelled to remain single-stranded. Thus the terminator hairpin cannot be formed.

Figure 15.8 shows that the position of the ribosome can determine which structure is formed, in such a way that termination is attenuated only in the absence of tryptophan.

Figure 15.7
The *trp* leader region can exist in alternative base-paired conformations.
In the center is a diagrammatic representation showing the four regions that can base pair. Region **1** is complementary to region **2,** which is complementary to region **3,** which is complementary to region **4.** On the left is the conformation produced when region **1** pairs with region **2,** and region **3** pairs with region **4.** On the right is the conformation when region **2** pairs with region **3,** leaving regions **1** and **4** unpaired.

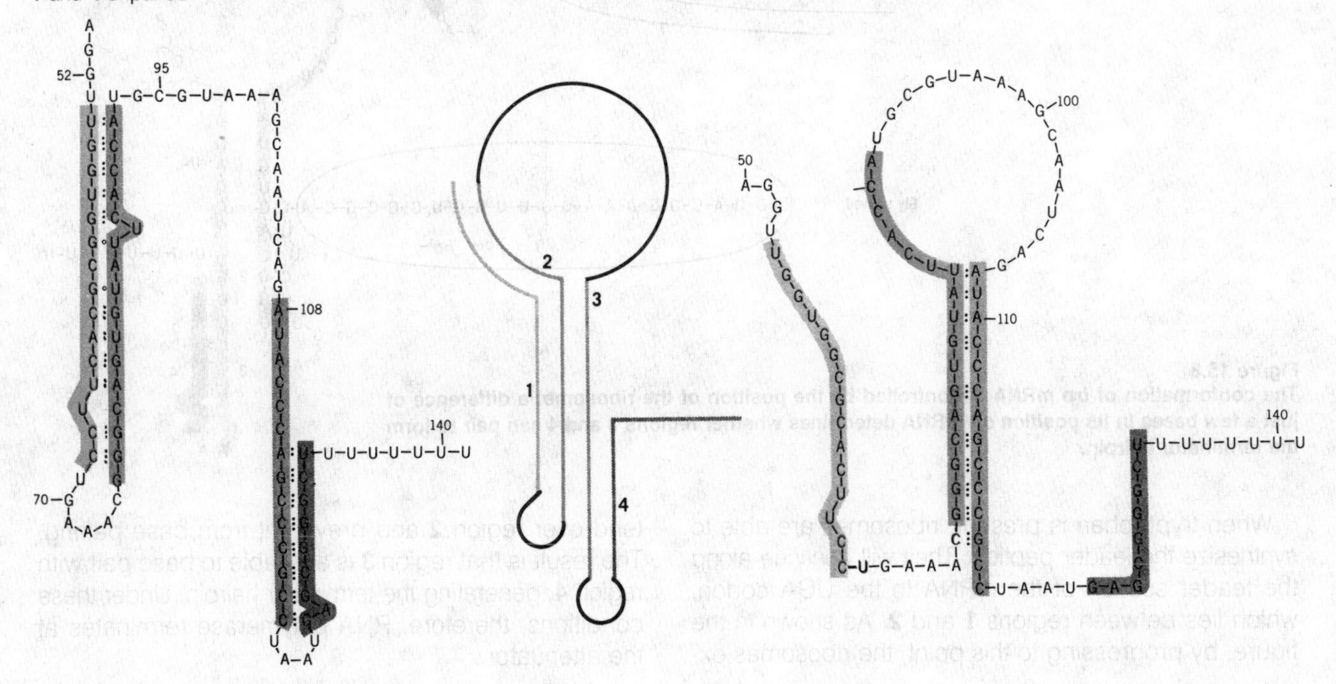

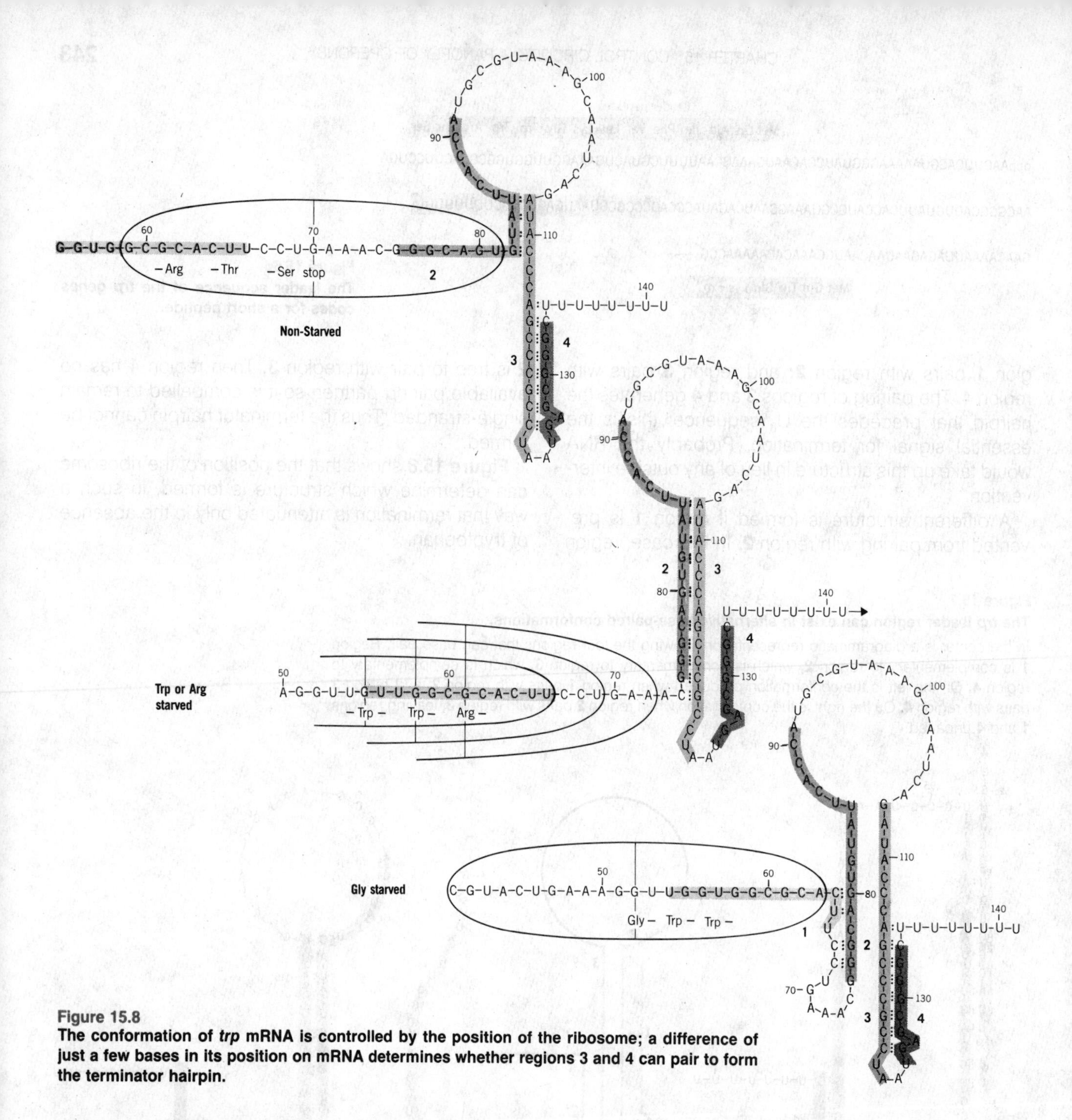

Figure 15.8

The conformation of *trp* mRNA is controlled by the position of the ribosome; a difference of just a few bases in its position on mRNA determines whether regions 3 and 4 can pair to form the terminator hairpin.

When tryptophan is present, ribosomes are able to synthesize the leader peptide. They will continue along the leader section of the mRNA to the UGA codon, which lies between regions **1** and **2.** As shown in the figure, by progressing to this point, the ribosomes extend over region **2** and prevent it from base pairing. The result is that region **3** is available to base pair with region **4,** generating the terminator hairpin. Under these conditions, therefore, RNA polymerase terminates at the attenuator.

When there is no tryptophan, ribosomes stall at the Trp codons, which are part of region **1.** Thus region **1** is sequestered within the ribosome and cannot base pair with region **2.** If this happens even while the mRNA itself is being synthesized, regions **2** and **3** will be base paired before region **4** has been transcribed. This compels region **4** to remain in a single-stranded form. In the absence of the terminator hairpin, RNA polymerase continues transcription past the attenuator.

Starvation for other amino acids does not have this result, because the positions at which the ribosome stalls leave regions **1** and **2** able to base pair, so that regions **3** and **4** in turn can base pair to form the terminator hairpin. The example of glycine starvation is shown in the figure.

This model is supported by the properties of mutations that influence events at the attenuator. The most common are mutations that cause a deficiency in termination, increasing expression of the structural genes. These mutations are similar to those found in other terminators. One class eliminates base pairs in the double-stranded region **3:4,** thus reducing the stability of the terminator hairpin. Another type lies in the stretch of U residues, with similar effects.

Some mutations in the leader region increase termination at the attenuator. This may prevent tryptophan starvation from relieving termination. One of these mutations destabilizes the pairing between regions **2** and **3.** Thus region **3** remains available to pair with region **4** to form the terminator hairpin, even when cells are starved for tryptophan. Another mutation changes the AUG initiation codon of the leader peptide, so that translation is prevented. This shows that transcription through the attenuator depends on the ability to translate the leader region.

Attenuation therefore provides a mechanism to sense the inadequacy of the supply of Trp-tRNA. Thus it responds directly to the need of the cell for tryptophan in protein synthesis. **Figure 15.9** summarizes the status quo when either termination or readthrough is occurring at the attenuator.

The critical aspect of the sensing mechanism is the position of the Trp codons in the leader peptide relative to the location of the attenuator. A feature that may ensure that all the events occur with the right timing is the presence of a site that causes the RNA

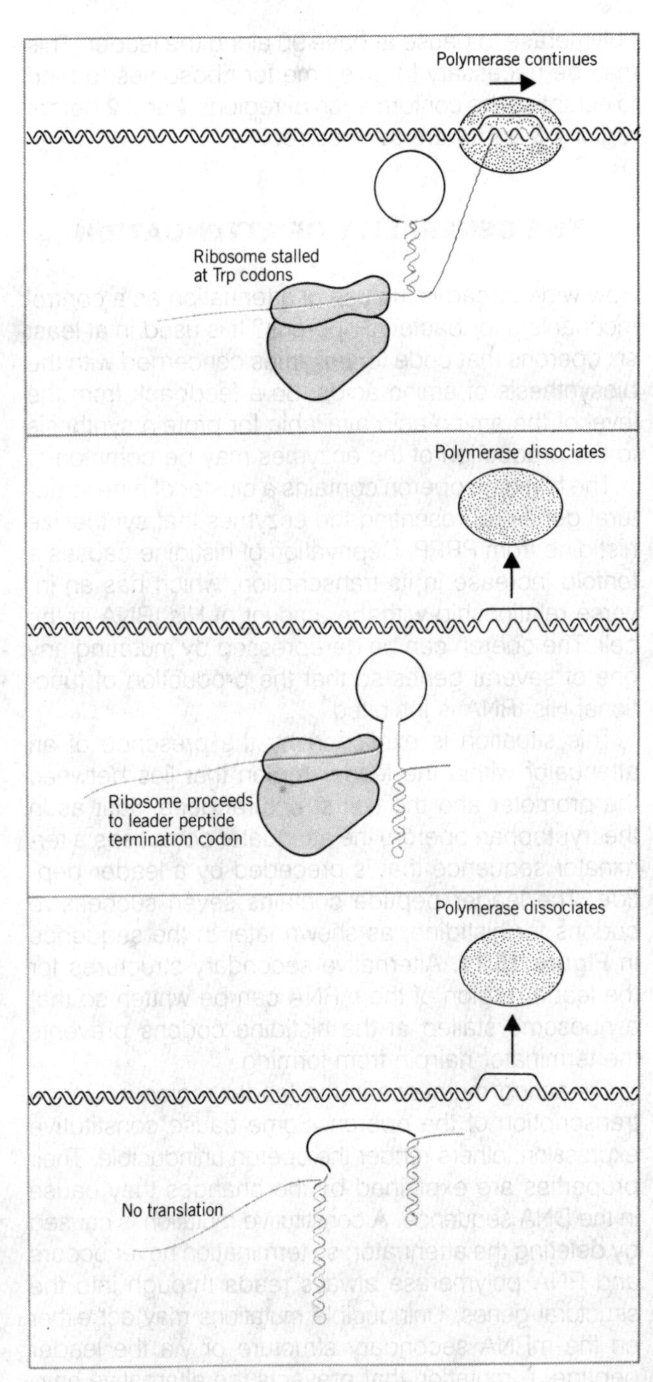

Figure 15.9
The alternatives for RNA polymerase at the attenuator depend on the location of the ribosome.

polymerase to pause at base 90 along the leader. This may be necessary to give time for ribosomes to bind to establish the conformation of regions **1** and **2** before regions **3** and **4** are synthesized.

THE GENERALITY OF ATTENUATION

How widespread is the use of attenuation as a control mechanism for bacterial operons? It is used in at least six operons that code for enzymes concerned with the biosynthesis of amino acids. So a feedback from the level of the amino acid available for protein synthesis to the production of the enzymes may be common.

The histidine operon contains a cluster of nine structural genes representing the enzymes that synthesize histidine from PRPP. Deprivation of histidine causes a tenfold increase in its transcription, which has an inverse relationship with the amount of His-tRNA in the cell. The operon can be derepressed by mutating any one of several genes so that the production of functional His-tRNA is inhibited.

This situation is explained by the presence of an attenuator within the leader region that lies between the promoter and the first structural gene. Just as in the tryptophan operon, the attenuator comprises a terminator sequence that is preceded by a leader peptide. The leader peptide contains seven successive codons for histidine, as shown later in the sequence in **Figure 15.11.** Alternative secondary structures for the leader region of the mRNA can be written so that a ribosome stalled at the histidine codons prevents the terminator hairpin from forming.

Cis-acting mutations in the leader region influence transcription of the operon. Some cause constitutive expression; others render the operon uninducible. Their properties are explained by the changes they cause in the DNA sequence. A constitutive mutation is caused by deleting the attenuator; so termination never occurs and RNA polymerase always reads through into the structural genes. Uninducible mutations may act either on the mRNA secondary structure or via the leader peptide. A mutation that prevents the alternative hairpin from forming ensures that the terminator hairpin *always* forms, so that the polymerase is unable to transcribe the structural genes. A mutation causing premature termination of leader peptide synthesis effectively creates a "no translation" situation similar to that shown in Figure 15.9, in which the terminator hairpin always forms.

The similarities between the tryptophan and histidine attenuation mechanisms are pronounced. In each case, it is deprivation of the aminoacyl-tRNA that directly prevents termination of transcription and so causes the structural genes to be transcribed. A difference is that the tryptophan operon also has a repressor-operator interaction; whereas in the histidine operon, attenuation provides the sole control. (The mutations in the *his* leader region were originally named *hisO,* because they were thought to identify an operator; this makes the point that the nature of such mutations cannot be determined until the molecular mechanism of control is investigated biochemically.)

An additional feature that may represent a new type of control mechanism is found in the histidine operon. In the attenuating conformation, the *his* leader region has a remarkable homology with the secondary structure of $tRNA^{His}$. **Figure 15.10** shows that they share a clover-leaf structure in which more than half of the nucleotides are identical.

This relationship could be involved in several ways in regulation. The tRNA is a target for several modifying enzymes, which might also act on the homologous sites in the leader. Such modifications might influence the choice between alternate conformations of the leader or might directly influence the initiation of translation or the stability of the mRNA. Some modifying activities respond to metabolic stress; a particular stress prevents certain modifications. Thus expression of the *his* operon could be related to metabolic conditions by this means.

Attenuation occurs in several operons. Their leader peptide sequences are summarized in **Figure 15.11;** in each case stalling of the ribosome at the codons representing the regulator amino acid(s) can cause the mRNA to take up a secondary structure in which a terminator hairpin cannot form. The *thr* and *ilv* operons are subject to **multivalent repression.** Each is derepressed by starvation for more than one amino acid, two in the case of *thr,* three in the case of *ilv.* The sequence of the leader peptide shows how this is accomplished: codons for the various amino acids

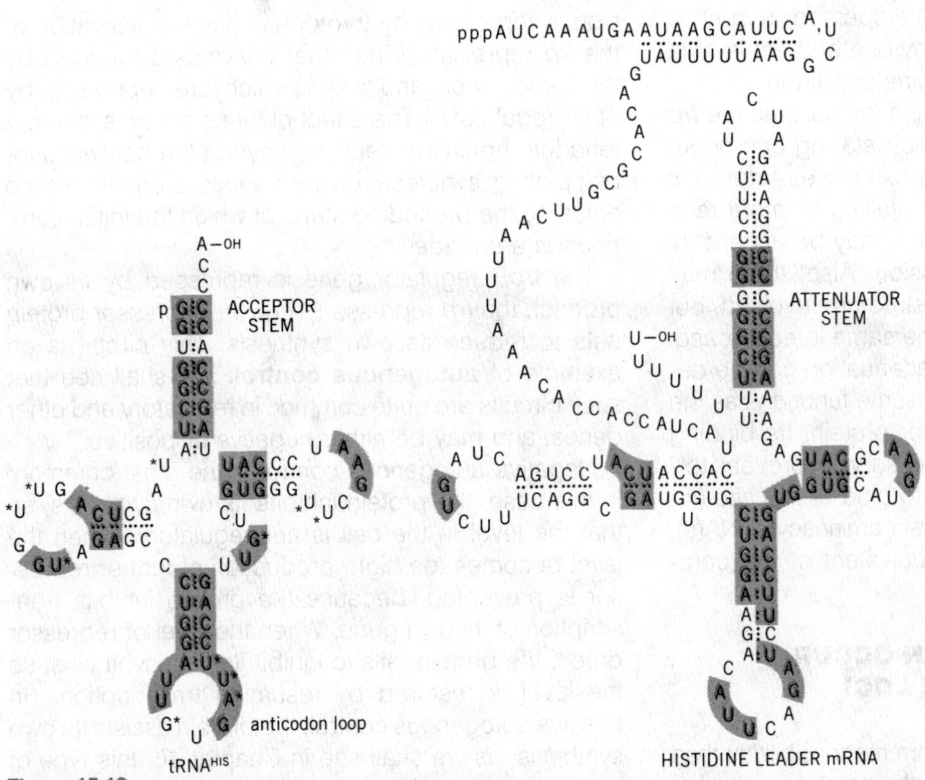

Figure 15.10
The leader sequence of *his* mRNA is 60% homologous to tRNA^{His} and can form a similar secondary structure.

Figure 15.11
Leader peptide sequences for amino acid biosynthetic operons contain multiple codons for the amino acid(s) that regulate the operon.

Operon	Leader peptide sequence	Regulatory amino acids
his	Met-Thr-Arg-Val-Gln-Phe-Lys-His-His-His-His-His-His-His-Pro-Asp	His
pheA	Met-Lys-His-Ile-Pro-Phe-Phe-Phe-Ala-Phe-Phe-Phe-Thr-Phe-Pro	Phe
leu	Met-Ser-His-Ile-Val-Arg-Phe-Thr-Gly-Leu-Leu-Leu-Leu-Asn-Ala-Phe-Ile-Val-Arg-Gly-Arg-Pro-Val-Gly-Gly-Ile-Gln-His	Leu
thr	Met-Lys-Arg-Ile-Ser-Thr-Thr-Ile-Thr-Thr-Thr-Ile-Thr-Ile-Thr-Thr-Gly-Asn-Gly-Ala-Gly	Thr Ile
ilv	Met-Thr-Ala-Leu-Leu-Arg-Val-Ile-Ser-Leu-Val-Val-Ile-Ser-Val-Val-Val-Ile-Ile -Ile-Pro-Pro-Cys-Gly-Ala-Ala-Leu-Gly-Arg-Gly-Lys-Ala	Leu, Val, Ile

that regulate the operon are interspersed in such a way that ribosome stalling at any one of them is able to prevent formation of the terminator hairpin.

Several of the leaders are much longer than the *trp* leader, and the positions at which stalling can occur to cause derepression are more extensive. Actually, it is possible that in these cases stalling of more than one ribosome in the leader region may be necessary to achieve maximum derepression. Also, there may be several loops of secondary structure involved, but the general principle remains the same in each case.

Formally, the mechanism of attenuation can be described as one in which the ribosome functions as the equivalent of a positive regulator protein. Its binding at the site(s) of stalling is necessary to prevent termination. It is rendered unable to bind at this site (inactivated) by the binding of certain aminoacyl-tRNA(s), which therefore provide the equivalent of the corepressor.

REPRESSION CAN OCCUR AT MULTIPLE LOCI

The *trp* repressor protein has more responsibilities than just to act on the operator at the *trp* locus. It also represses transcription at two further loci.

The *aroH* gene codes for one of the three enzymes that catalyze the initial reaction in the common pathway of aromatic amino acid biosynthesis. Its expression is repressed by tryptophan, via the activation of the *trp* repressor. (The other enzymes are coded by the genes *aroF* and *aroG*, which are repressed by other regulators.) The effect of these controls is to extend the regulatory network, beyond the pathways for completing synthesis of the various aromatic amino acids, to the preceding steps at which the initial compounds are made.

The *trpR* regulator gene is repressed by its own product, the *trp* repressor. Thus the repressor protein acts to reduce its own synthesis. This circuit is an example of **autogenous control.** We shall see that such circuits are quite common in regulatory and other genes, and may be either negative or positive.

Negative autogenous control is the most common; in this case, the protein inhibits its own synthesis, so that the level in the cell is autoregulatory. When the level becomes too high, production of further repressor is prevented, because the protein inhibits transcription of its own gene. When the level of repressor drops, the protein fails to inhibit its own synthesis, so the level is restored by resuming transcription. (In positive autogenous control, the protein assists its own synthesis; as we shall see in Chapter 16, this type of amplification provides an on/off switch.)

A related operator sequence, extending over 21 bp, is present at each of the three loci at which the *trp* repressor acts. The conservation of sequence is indicated in **Figure 15.12.** Each operator contains ap-

Figure 15.12
The *trp* repressor recognizes operators at three loci.
Vertical lines show the conserved bases in the operators. Arrows indicate the regions of dyad symmetry. Boxes indicate promoter consensus sequences (with good fit in *aroH* and *trp*, poor fit in *trpR*). There is no −35 sequence in *aroH*.

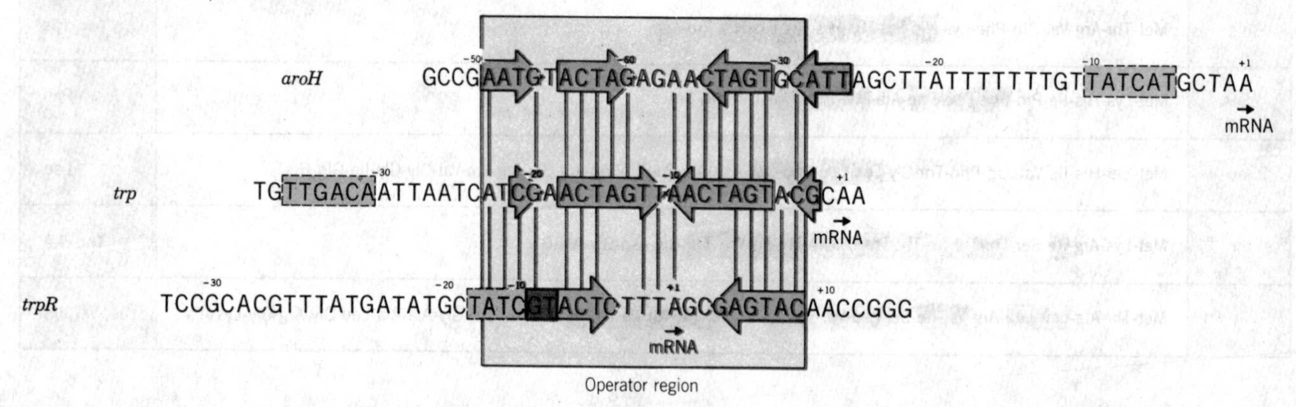

preciable (but not identical) dyad symmetry. Presumably the features present at all three operators include the important points of contact for *trp* repressor. This explains how one repressor protein acts on several loci: each locus has a copy of a specific DNA-binding sequence recognized by the repressor (just as each promoter shares consensus sequences with other promoters).

A notable feature of the dispersed operators is their presence at different locations relative to the startpoint in each locus. Figure 15.12 shows that in the *trp* operon the operator occupies positions −23 to −3, while in *trpR* it lies between positions −12 and +9, but in the *aroH* locus it lies farther upstream, between −49 and −29. We have already seen that different repressors bind at sites located at various positions within or adjacent to the promoters: by revealing that the *same* repressor is effective at differently positioned operators, the *trp* situation reinforces the conclusion that repression is a matter of blocking access to the promoter.

Figure 15.13 summarizes the variety of relationships between operators and promoters, in which the operator may lie downstream from the promoter (as in *lac*), within the promoter (as in the various loci responding to Trp repressor), or apparently just upstream of the promoter (as in *gal*, where the nature of the repressive effect is not quite clear).

Figure 15.13
Operators may lie at various positions relative to the promoter.

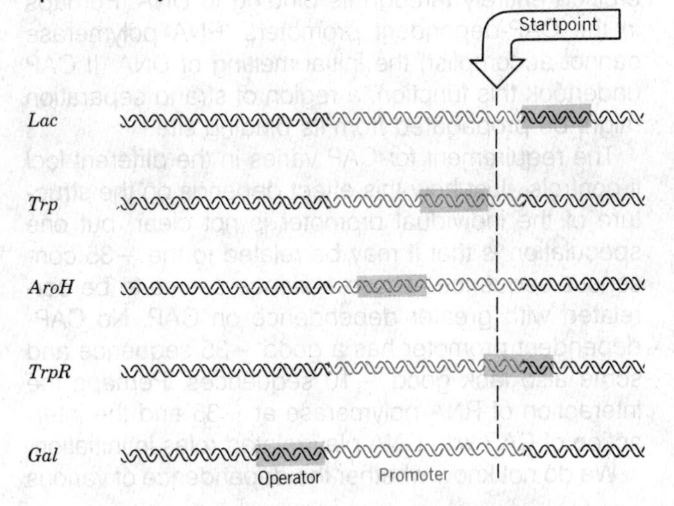

CATABOLITE REPRESSION INVOLVES POSITIVE REGULATION AT THE PROMOTER

So far we have dealt with the promoter as a DNA sequence that is competent to bind RNA polymerase and then to initiate transcription. But there are some promoters at which RNA polymerase cannot initiate transcription without assistance from an ancillary protein. Such proteins constitute **positive regulators,** because their presence is necessary to switch on the transcription unit. Several positive regulators are known, some coded by phages and some present within the host cell. At the outset, it has to be said that we do not properly understand the nature of the difference between promoters that can function per se and those that need a positive regulator.

The best characterized positive regulator is a protein that controls the activity of a large set of operons in *E. coli*. When glucose is available as an energy source, it is used in preference to other sugars. Thus when *E. coli* finds (for example) both glucose and lactose in the medium, it metabolizes the glucose and represses the use of lactose. This is accomplished by preventing expression of the genes of the *lactose* operon, an effect called **catabolite repression.** The same effect is found with other operons, including galactose and arabinose. Thus catabolite repression represents a general coordinating system that exercises a preference for glucose by inhibiting the expression of the operons that code for the enzymes of alternative metabolic pathways.

Catabolite repression is set in train by the ability of glucose to reduce the level of cyclic AMP (cAMP) in the cell. We do not know exactly how this is accomplished; it may involve the rate of synthesis. At all events, expression of the catabolite-regulated operons shows an inverse relationship with the level of cyclic AMP.

Two types of mutation abolish catabolite repression. The first, *cya*$^-$, introduces a defect in the enzyme **adenylate cyclase** that synthesizes cyclic AMP. (The reaction uses ATP as substrate and consists of making a 3′–5′ link via phosphodiester bonds; the result is a monophosphate joined to *both* of the positions at which phosphates usually can be found.) The second type of mutation, known alternatively as *cap*$^-$ or *crp*$^-$, identifies the regulator protein that acts directly on the target operons. The protein is known as CAP (for catabolite activator protein) or CRP (for cyclic AMP receptor protein).

CAP is a positive control factor whose presence is necessary to initiate transcription at dependent promoters. The protein is active *only in the presence of cyclic AMP*, which behaves as the classic small-molecule inducer (see Figure 15.1). Reducing the level of cyclic AMP renders the protein unable to bind to the control region, which in turn prevents RNA polymerase from initiating transcription. So the effect of glucose in reducing cyclic AMP levels is to deprive the relevant operons of a control factor necessary for their expression.

The CAP factor acts by binding directly to DNA, and complexes of cyclic AMP•CAP•DNA can be isolated at each promoter at which it functions. The factor is a dimer of two identical subunits of 22,500 daltons. One dimer seems to bind to each binding site on DNA, covering about 25 bp.

At each promoter, the CAP binding site is upstream from the RNA polymerase-binding site. The binding sites include variations of a rather short consensus sequence, TGTGA, which is probably bound by one of the factor's subunits. Mutations preventing CAP action usually are located within this sequence. Some binding sites contain an inverted repeat of the consensus, probably bound by the other dimer. However, most binding sites are not symmetrical, so the second subunit must bind a different sequence.

The action of CAP is puzzling, because in each target operon it binds to a site at a different location relative to the startpoint. This has made it difficult to construct a uniform model for CAP action. The three examples summarized in **Figure 15.14** encompass the range of locations.

Figure 15.14
The CAP protein can bind at different sites relative to RNA polymerase.

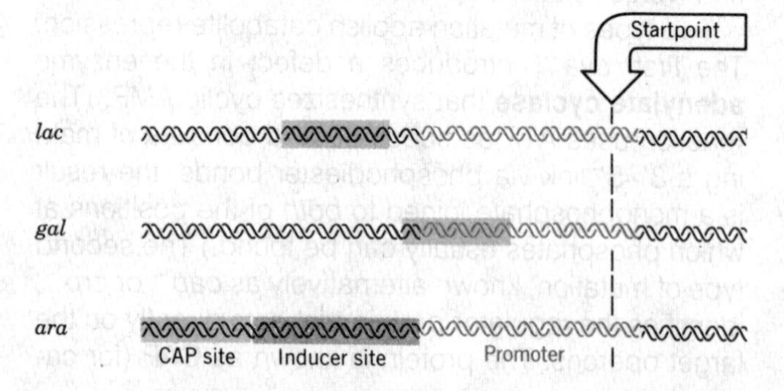

In the *lac* operon, the region of DNA protected by CAP extends from about −72 to −52. It is possible that two dimers of CAP are bound. The binding pattern is consistent with the presence of CAP largely on one face of DNA, the same face that is bound by RNA polymerase. This location would place the two proteins just about in reach of each other.

At the *gal* locus, however, the CAP binding site lies between −50 and −23. It is likely that only a single CAP dimer is bound, probably in quite intimate contact with RNA polymerase, since the CAP binding site extends well into the region generally protected by the RNA polymerase.

In the *ara* region, the binding site for a single CAP is the farthest from the startpoint, at −107 to −78. Here the CAP cannot be in contact with RNA polymerase, because *another* regulatory protein binds in the region between the CAP and RNA polymerase sites.

One model for CAP action supposes that the main effect is mediated by protein-protein interactions when two (or more) proteins are bound to DNA. Quite a modest interaction between two proteins would be sufficient to achieve a substantial increase in the ability of RNA polymerase to bind a promoter. The geometry of the CAP-RNA polymerase interaction would have to be different in *lac* and in *gal*, while in *ara* there would need to be a triple-protein interaction.

Another possibility is that the effect of CAP is exercised entirely through its binding to DNA. Perhaps in the CAP-dependent promoters, RNA polymerase cannot accomplish the initial melting of DNA. If CAP undertook this function, a region of strand separation might be propagated from its binding site.

The requirement for CAP varies in the different loci it controls. Just how this effect depends on the structure of the individual promoter is not clear, but one speculation is that it may be related to the −35 consensus sequence, whose absence seems to be correlated with greater dependence on CAP. No CAP-dependent promoter has a good −35 sequence and some also lack good −10 sequences. Perhaps the interaction of RNA polymerase at −35 and the interaction of CAP with DNA play related roles in initiation.

We do not know whether the dependence of various

operons on CAP necessarily reflects a common mode of regulation at the level of individual protein-protein or protein-DNA interactions. However, it accomplishes the same purpose: to turn off alternative metabolic pathways when they become unnecessary because the cell has an adequate supply of glucose. Again, this makes the point that coordinate control, of either negative or positive type, can extend over dispersed loci by repetition of binding sites for the regulator protein.

AUTOGENOUS CONTROL OF RIBOSOMAL PROTEIN TRANSLATION

About 70 or so proteins constitute the apparatus for gene expression. The ribosomal proteins are the major component, together with the ancillary proteins involved in protein synthesis. The subunits of RNA polymerase and its accessory factors make up the remainder. Coordinate controls ensure that these proteins are synthesized in amounts appropriate for the growth conditions: when bacteria grow more rapidly, they devote a greater proportion of their efforts to the production of the apparatus for gene expression. An array of mechanisms is used to control the expression of the genes coding for this apparatus; and probably we have not yet identified all of the regulatory systems that are employed.

Almost every one of these proteins is represented only by a single gene in *E. coli*. The genes coding for ribosomal proteins, protein-synthesis factors, and RNA polymerase subunits all are intermingled and organized into a small number of operons.

The organization of the operons characterized so far is summarized in **Table 15.1.** About half of the genes for ribosomal proteins (often abbreviated as **r-proteins**) map in four operons that lie close together. These are known as *str*, *spc*, *S10*, and α (each named simply for the first one of its functions to have been identified). The *rif* and *L11* operons lie together at a different location.

Table 15.1
Genes for ribosomal proteins, protein synthesis factors, and RNA polymerase subunits are interspersed in a few operons.

Operon	Genes and Proteins (in order from promoter)	Regulator
str	*rpsL-rpsG-fusA-tufA* S12 S7 EF-G EF-Tu	S7
spc	*rplN-rplX-rplE-rpsN-rpsH-rplF-rplR-rpsE-rplO-rpmD* L14 L24 L5 S14 S8 L6 L18 S5 L15 L30	S8
S10	*rpsJ-rplC-rplB-rplD-rplW-rplS-rplV-rpsC-rpsQ-rplP-rpmC* S10 L3 L2 L4 L23 S19 L22 S3 S17 L16 L29	L4
α	*rpsM-rpsK-rpsD-rpoA-rplQ* S13 S11 S4 α L17	S4
L11	*rplK-rplA* L11 L1	L1
rif	*rplJ-rplL-rpoB-rpoC* L10 L7/L12 β β′	L10

The overall organization of these operons identifies two major gene clusters. One contains the operons *str*-14,000 bp-*S10-spc*-α; the other contains the adjacent operons *L11-rif*. Each operon is written so that it has its promoter at the left end. The regulator protein is indicated at the right, and the proteins subject to regulation are underlined (in red). Genes whose inclusion in the regulation is uncertain are indicated by broken lines.

Each operon contains a melange of functions. The *str* operon has genes for small subunit ribosomal proteins as well as for EF-Tu and EF-G. The *spc* and *S10* operons both have genes interspersed for both small and large ribosomal subunit proteins. The α operon has genes for proteins of both ribosomal subunits as well as for the α subunit of RNA polymerase. The *rif* locus has genes for large subunit ribosomal proteins and for the β and β' subunits of RNA polymerase.

In most cases, there is no evident correlation with the functions of the proteins. Certainly the 30S subunit proteins are not coded in any pattern related to their assembly into the subunit. The group of 50S subunit proteins coded in the *rif/L11* cluster does show a correlation with the assembly map, in which L7/L12, L10, and L11 may be related. So in this case the proteins could assemble into some subparticle as they are synthesized.

All except one of the ribosomal proteins are needed in equimolar amounts, which must be coordinated with the level of rRNA. (There could also be variations for ribosomal proteins that have extracurricular activities; for example, S10 is implicated in the antitermination function of pN.) The dispersion of genes whose products must be equimolar, and their intermingling with genes whose products are needed in different amounts, pose some interesting problems for coordinate regulation.

The exceptional ribosomal protein is L7/L12, present in four copies per ribosome. Another exception is EF-Tu, which is present in amounts roughly equimolar with aminoacyl-tRNA—that is, about 10 times greater than the ribosomes. (This is the one case in which there is more than one gene, so the need for extra synthesis is divided between the two genes *tufA* and *tufB*.) Another difference occurs between ribosomes and RNA polymerase, which is present in somewhat smaller amounts. So some mechanism must increase the synthesis of L7/L12 and EF-Tu, and decrease the synthesis of RNA polymerase subunits, relative to the level of ribosomal proteins.

A feature common to all of the operons described in Table 15.1 is autogenous regulation of some of the genes by one of the products. Usually the regulatory protein inhibits expression of a contiguous set of genes within the operon, always including its own gene.

The exact pattern of inhibition varies from operon to operon. Thus S7 inhibits synthesis of itself and EF-G, but not of S12 or EF-Tu. Similarly, S8 inhibits expression of its entire operon except for the first two genes. On the other hand, L4 and L1 probably inhibit the expression of all genes in their respective operons. In the operons containing RNA polymerase subunit genes, S4 inhibits the expression of all the other small ribosomal subunit genes; and similarly L10 inhibits its own production, but not that of L7/L12 or the RNA polymerase subunits.

In each case, accumulation of the protein inhibits further synthesis of itself and of whatever other genes are involved. The effect often is caused at the level of translation of the polycistronic mRNA, and in several cases can be reproduced *in vitro*. Thus an excess of free ribosomal protein triggers a repression of translation.

Each of the regulators is a ribosomal protein that binds directly to rRNA. Its effect on translation is a result of its ability also to bind to its own mRNA. In some cases, the binding site has been characterized.

In the example of the L11 operon, protein L1 binds to a site in the vicinity of the initiation codon for the first gene. This action probably inhibits ribosome binding. The inhibition affects translation of both genes of the operon, probably because there are only three nucleotides between the L11 and L1 genes. Thus it is plausible to suppose that the genes can only be translated sequentially.

In the case of the *rif* operon, mutations in autogenous control map in a region 70–195 bp upstream from the initiation site of *rplJ*; this region probably contains a site at which L10 binds, perhaps to alter the conformation of the mRNA so that initiation is prevented downstream.

The use of r-proteins that bind rRNA to establish autogenous regulation immediately suggests that this may serve as a mechanism to link r-protein synthesis to rRNA synthesis. A generalized model is depicted in **Figure 15.15.** Suppose that the binding sites for the autogenous regulator r-proteins on rRNA are much stronger than those on the mRNAs. Then so long as any free rRNA is available, the newly synthesized r-proteins will associate with it to start ribosome assembly. There will be no free r-protein available to bind to the mRNA, so its translation will continue. But as soon as the synthesis of rRNA slows or stops, free

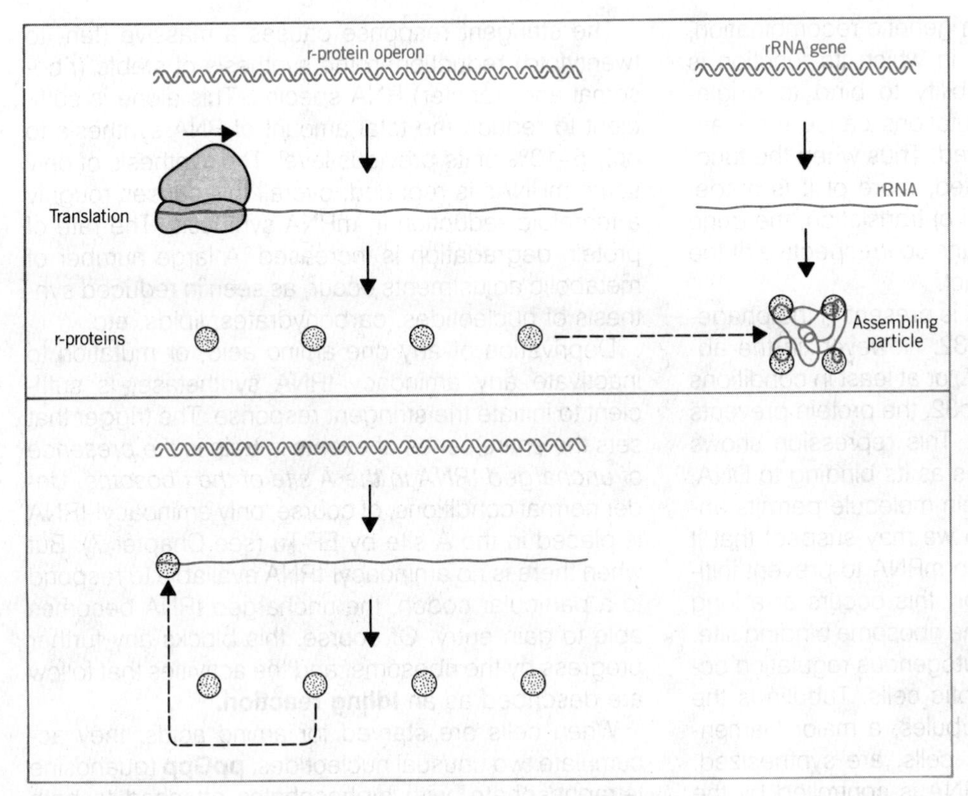

Figure 15.15
Translation of the r-protein operons is autogenously controlled and responds to the level of rRNA.
Upper. When rRNA is being produced, ribosomal proteins associate with it, so there is no free pool, and translation of the r-protein mRNA continues.
Lower. When all rRNA has been assembled into ribosomes and none is free, the r-proteins accumulate. One of them binds to the r-protein mRNA and prevents further translation.

r-proteins begin to accumulate. Then they are available to bind their mRNAs, repressing further translation. This circuit ensures that each r-protein operon responds in the same way to the level of rRNA.

Two objectives are accomplished by this regulation of the r-protein operons. First, the level of r-proteins corresponds with the growth conditions of the cell. By controlling the level of rRNA, the cell controls production of all ribosomal components. Second, the other proteins coded by these operons are synthesized at their own rates, divorced from the translation of the r-protein genes. The β subunits of RNA polymerase may be subject to their own autogenous regulation. Both EF-Tu and L7/L12 are translated with increased efficiency. So within the control circuits of these operons, there is provision for allowing disparate rates of synthesis for coordinately regulated proteins.

Autogenous control of translation is not the only mechanism employed to allow variation to occur within the operon. Some operons contain additional, internal promoters; attenuation may be employed; processing of mRNA can be influenced.

AUTOGENOUS CONTROL AND MACROMOLECULAR ASSEMBLIES

Autogenous regulation may turn out to be a common type of control among proteins that are incorporated into macromolecular assemblies. The advantage is evident. The assembled particle itself may be unsuitable as a regulator, because it is too large, too numerous, or too restricted in its location. But the need for synthesis of its components may be reflected in the pool of free precursor subunits. If the assembly pathway is blocked for any reason, free subunits accumulate and shut off the unnecessary synthesis of further components.

Actually, the first example of autogenous regulation was provided by the gene *32* protein of phage T4. The protein plays a central role in genetic recombination, DNA repair, and replication, in which its function is exercised by virtue of its ability to bind to single-stranded DNA. Nonsense mutations cause the inactive protein to be overproduced. Thus when the function of the protein is prevented, more of it is made. This effect occurs at the level of translation; the gene *32* mRNA is stable, and remains so irrespective of the behavior of the protein product.

When single-stranded DNA is present in the phage-infected cell, it sequesters p32. However, in the absence of single-stranded DNA, or at least in conditions in which there is a surplus of p32, the protein prevents translation of its own mRNA. This repression shows the same cooperative features as its binding to DNA, in which binding of one protein molecule permits another to bind more easily. So we may suspect that it is mediated by p32 binding to mRNA to prevent initiation of transcription. Probably this occurs at a long single-stranded region near the ribosome binding site.

Another system in which autogenous regulation occurs is presented by eukaryotic cells. Tubulin is the monomer from which microtubules, a major filamentous system of all eukaryotic cells, are synthesized. The production of tubulin mRNA is controlled by the free tubulin pool. When this pool reaches a certain concentration, the production of further tubulin mRNA is prevented. Again, the principle is the same: tubulin sequestered into its macromolecular assembly plays no part in regulation, but the level of the free precursor pool determines whether further monomers are added to it.

HARD TIMES PROVOKE THE STRINGENT RESPONSE

When bacteria find themselves in such poor growth conditions that they lack a sufficient supply of amino acids to sustain protein synthesis, they shut down a wide range of activities. This is called the **stringent response.** We can view it as a mechanism for surviving hard times: the bacterium husbands its resources by engaging in only the minimum of activities until nutrient conditions improve, when it reverses the response and again engages its full range of metabolic activities.

The stringent response causes a massive (ten- to twentyfold) reduction in the synthesis of stable (ribosomal and transfer) RNA species. This alone is sufficient to reduce the total amount of RNA synthesis to only 5–10% of its previous level. The synthesis of only some mRNAs is reduced; overall this causes roughly a threefold reduction in mRNA synthesis. The rate of protein degradation is increased. A large number of metabolic adjustments occur, as seen in reduced synthesis of nucleotides, carbohydrates, lipids, etc.

Deprivation of any one amino acid, or mutation to inactivate any aminoacyl-tRNA synthetase, is sufficient to initiate the stringent response. The trigger that sets the entire series of events in train is *the presence of uncharged tRNA in the A site of the ribosome.* Under normal conditions, of course, only aminoacyl-tRNA is placed in the A site by EF-Tu (see Chapter 7). But when there is no aminoacyl-tRNA available to respond to a particular codon, the uncharged tRNA becomes able to gain entry. Of course, this blocks any further progress by the ribosome, and the activities that follow are described as an **idling reaction.**

When cells are starved for amino acids, they accumulate two unusual nucleotides, **ppGpp** (guanosine tetraphosphate, with diphosphates attached to both 5′ and 3′ positions) and **pppGpp** (guanosine pentaphosphate, with a 5′ triphosphate group and a 3′ diphosphate). In all strains of *E. coli,* the stringent response is accompanied by the accumulation of ppGpp; the pppGpp may or may not be produced. These nucleotides are typical small-molecule effectors whose activity is mediated by the ability to bind to protein(s) to alter the conformation.

The components involved in producing ppGpp and pppGpp have been identified by mutations that eliminate the stringent response. In **relaxed mutants,** starvation for amino acids does not cause any reduction in stable RNA synthesis or in any of the other reactions comprising the stringent response.

The most common site of relaxed mutation lies in the gene *relA,* which codes for a protein called the **stringent factor.** This factor is associated with the ribosomes, although the amount is rather low—say, less than 1 molecule for every 200 ribosomes. So perhaps only a minority of the ribosomes are able to produce the stringent response.

Ribosomes obtained from stringent bacteria can

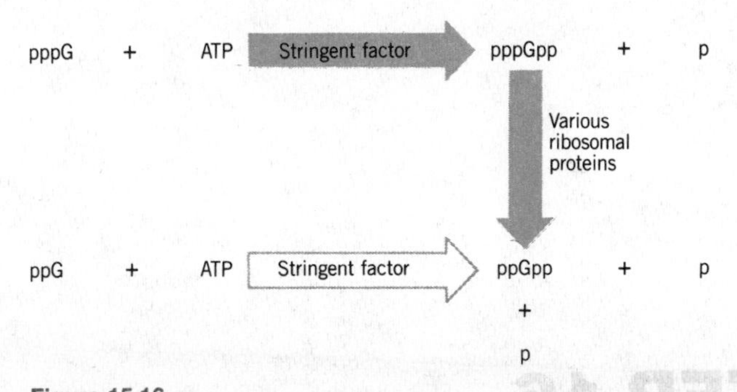

Figure 15.16
Stringent factor catalyzes the synthesis of pppGpp and ppGpp; ribosomal proteins can dephosphorylate pppGpp to ppGpp.

synthesize ppGpp and pppGpp *in vitro,* provided that the A site is occupied by an uncharged tRNA *specifically responding to the codon.* Ribosomes extracted from relaxed mutants cannot perform this reaction; but they are able to do so if the stringent factor is added.

Figure 15.16 shows the pathways for synthesis of the unusual guanine nucleotides. The stringent factor is an enzyme that catalyzes the synthetic reaction in which ATP is used to donate a pyrophosphate group to the 3′ position of either 5′ GTP or GDP. The former is used as substrate more frequently. However, pppGpp may be converted to ppGpp by several enzymes; among those able to perform this dephosphorylation are the translation factors EF-Tu and EF-G. The production of ppGpp via pppGpp is the most common route.

What is involved in the idling reaction? Mutations in another locus able to cause the relaxed type turn out to lie in the 50S subunit protein L11, which is located in the vicinity of the A and P sites. The stringent response could involve the TψC region of tRNA, since mutations in tRNA, especially involving this region, can influence the reaction. It is possible that this region of tRNA (located at the hinge of the L tertiary structure) could be involved in transmitting a signal that the right tRNA is paired with the codon, thus allowing the ribosome to proceed with peptide transfer. The same pathway could be part of the route for establishing the idling reaction. The presence of a properly paired uncharged tRNA in the A site may trigger a conformational change in the ribosome; but because no amino acid is present on the tRNA, the idling reaction occurs instead of polypeptide transfer from the peptidyl-tRNA.

What does ppGpp do? It may be an effector for controlling several reactions, including the inhibition of transcription. Many conflicting effects have been reported, among which two stand out. There is a specific inhibition of *initiation* of transcription at the promoters of operons coding for rRNA. And the addition of ppGpp slows the *elongation* phase of transcription of many or most templates *in vitro.* The cause is increased pausing by the enzyme. This effect is responsible for the general reduction in transcription efficiency when ppGpp is added *in vitro.* We do not yet know the specificity of such inhibition, but it will not be surprising if there is variation in magnitude from operon to operon in such a way that particular operons are more inhibited.

It is interesting that unusual nucleotides are used in at least two control systems with a general coordinating function. Both appear to be specific to bacteria. Although cyclic AMP is the classic "second messenger" in eukaryotic cells, there is no evidence that it plays any role in the specific control of initiation of transcription. Similarly, an extensive search for ppGpp in eukaryotes has failed to detect any.

FURTHER READING

The circuitry of several operons has been reviewed in **Miller & Reznikoff's** (Eds.) *The Operon* (Cold Spring Harbor Laboratory, New York, 1978). This includes **Platt** on the *trp* operon (pp. 263–302), **Lee** on the *ara* operon (pp. 389–410), and **De Crombrugghe & Pastan** on cyclic AMP and the *gal* operon (pp. 303–324). Work on attenuation of the *trp* operon has been brought further up to date by **Yanofsky** (*Nature* **289,** 751–758, 1981) and **Platt** (*Cell* **24,** 10–23, 1981). The functions of this operon have been summarized in terms of its sequence by **Yanofsky et al.** (*Nuc. Acids Res.* **9,** 6647–6668, 1981). Their theory that modified nucleotides have a role in regulating gene expression was developed by **Ames et al.** (*Proc. Nat. Acad. Sci.* **80,** 5240–5242, 1983). The control of ribosome synthesis and feedback mechanisms in relation to growth control has been reviewed by **Nomura et al.** (*Ann. Rev. Biochem.* **53,** 75–117, 1984). Stringent control has been generally reviewed by **Gallant** (*Ann. Rev. Genet.* **13,** 393–415, 1979).

CHAPTER 16
LYTIC CASCADES AND LYSOGENIC REPRESSION

Some phages have only a single strategy for survival. On infecting a susceptible host, they subvert its functions to the purpose of producing a large number of progeny phage particles. As the result of this **lytic infection,** the host bacterium dies. In the typical lytic cycle, the phage DNA (or RNA) enters the host bacterium, its genes are transcribed in a set order, the phage genetic material is replicated, and the protein components of the phage particle are produced. Finally, the host bacterium is broken open **(lysed)** to release the assembled progeny particles.

Other phages have a dual existence. They are able to perpetuate themselves via the same sort of lytic cycle in what amounts to an open strategy for producing as many copies of the phage as rapidly as possible. But they also have an alternative, closet existence, in which the genome is present in the bacterium in a latent form known as **prophage.** This form of propagation is called **lysogeny.** In lysogenic bacteria, the prophage is **integrated** into the bacterial genome, and is inherited in the same way as bacterial genes. By virtue of its possession of a prophage, a lysogenic bacterium has **immunity** against infection by further phage particles of the same type, so usually there is only one copy of a prophage of any particular type in a bacterial genome.

Transitions between the lysogenic and lytic modes of existence may occur in either direction. When a phage produced by a lytic cycle enters a new bacterial host cell, it may either repeat the lytic cycle or enter the lysogenic state. The outcome depends on the conditions of infection and the genotypes of phage and bacterium. A prophage may be freed from the restrictions of lysogeny by the process called **induction,** in which it is **excised** from the bacterial genome, to generate a free phage DNA that then proceeds through the lytic pathway.

Another type of existence within bacteria is represented by **plasmids.** These are autonomous elements whose genomes exist in the cell as **extrachromosomal** units. Plasmids are self-replicating circular molecules of DNA that are maintained in the cell in a stable and characteristic number of copies; that is, the number remains constant from generation to generation.

Some of these elements also have alternative lifestyles. They can exist either in the autonomous extrachromosomal state; or they can be inserted into the bacterial chromosome, and then be carried as part of it like any other sequence. These are called **episomes.** (As a note of caution, since their original definition, the terms "plasmid" and "episome" have sometimes been used loosely as though interchangeable, so they do not always carry the precise connotations described here.)

Like lysogenic phages, plasmids and episomes maintain a selfish possession of their bacterium and often make it impossible for another element of the same type to become established. This effect also is called **immunity,** although the basis for plasmid immunity is different from lysogenic immunity (see Chapter 31).

LYTIC DEVELOPMENT IS CONTROLLED BY A CASCADE

Phage genomes of necessity are small. Indeed, as with all viruses, a principal restriction is the need to package the nucleic acid within its protein coat. This limitation dictates many of the viral strategies for reproduction. Typically a virus takes over the apparatus of the host cell, which, instead of replicating and expressing bacterial genes, replicates and expresses phage genes.

Usually the phage genes include functions that ensure preferential replication of phage DNA. These functions may be concerned with the initiation of replication or may even provide a new DNA polymerase. Some change is always made in the capacity of the host cell to engage in transcription. This may involve replacing the RNA polymerase or modifying its capacity for initiation or termination. The result is always the same: phage mRNAs are preferentially transcribed. So far as protein synthesis is concerned, usually the phage is content to use the host apparatus, redirecting its activities principally by replacing bacterial mRNA with phage mRNA.

Lytic development is accomplished by a pathway in which the phage functions are expressed in a particular order. This ensures that the right amount of each component is present at the appropriate time. The cycle can be divided into the two general parts illustrated in **Figure 16.1. Early infection** describes the period from entry of the DNA to the start of its replication. **Late infection** defines the period from the start of replication to the final step of lysing the bacterial cell to release progeny phage particles.

In the usual order of battle, the early phase is devoted to the production of enzymes involved in the reproduction of DNA. These include the enzymes concerned with DNA synthesis, recombination, and sometimes modification. Their activities cause a **pool** of phage genomes to accumulate. In this pool, genomes are continually replicating and recombining, so that the events even during one lytic cycle concern a population of phage genomes.

When the late phase begins, the protein components of the phage particle are synthesized. Often many different proteins are needed to make up head and tail structures, so the largest part of the phage genome consists of late functions. In addition to the structural proteins, there may also be assembly proteins, whose presence is necessary to construct the particle, but which are not themselves part of it. By the time these components are assembling into heads and tails, replication of DNA has reached its maximum rate. The genomes then are inserted into the empty protein heads, tails are added, and the cell is lysed.

Many phages display an organization in which the genetic map closely reflects the sequence of lytic development. The concept of the operon is taken to somewhat of an extreme, in which the genes coding for proteins with related functions are clustered to allow their control with the maximum economy. This allows the pathway of lytic development to be controlled with a small number of regulatory switches.

To arrange for the expression of phage genes in a particular order, the lytic cycle is under positive control. Each group of phage genes can be expressed only when an appropriate signal is given.

Usually only a few phage genes can be expressed by the transcription apparatus of the host cell. Their promoters are indistinguishable from those of host genes. The name of this class of genes depends on the phage. In most cases, they are known as the **early genes.** In phage lambda, they are given the evocative description of **immediate early.** Irrespective of the name, they constitute only a preliminary, representing just the initial part of the early period. Sometimes they are exclusively occupied with the transition to the next period. At all events, one of these genes always codes for a protein that is necessary for transcription of the next class of genes.

This second class of genes is known variously as the **delayed early** or **middle** group. As the name sug-

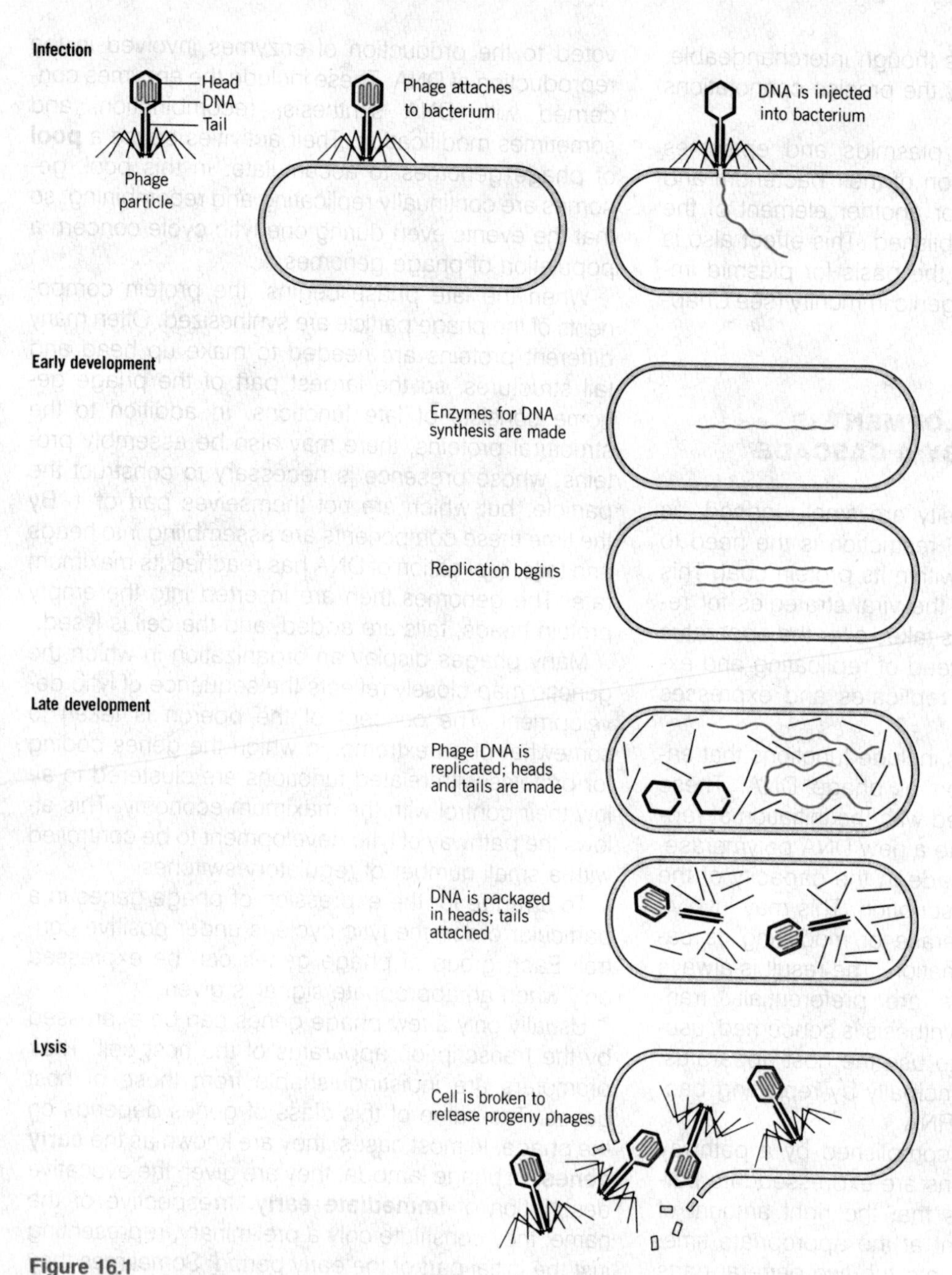

Figure 16.1
Lytic development takes place by producing phage genomes and protein particles that are assembled into progeny phages.

gests, its expression starts in the early period, typically as soon as the regulator protein is available. Depending on the nature of the control circuit, the initial set of early genes may or may not continue to be expressed at this stage (see Figure 13.4). Often the expression of host genes is reduced. Together the two sets of early genes account for all the phage functions except those needed to assemble the particle coat itself and to lyse the cell.

When the replication of phage DNA begins as the result of early gene expression, it is time for the **late genes** to be expressed. Their transcription at this stage usually is provided for by embedding a further regulator gene within the previous (delayed early or middle) set of genes. This regulator may be another antitermination factor (as in lambda) or it may be another sigma factor (as in SPO1).

The use of these successive controls, in which each set of genes contains a regulator that is necessary for expression of the next set, creates a cascade in which groups of genes are turned on (and sometimes off) at particular times. The detailed construction of each phage cascade is different, but the results are similar, as the following sections show.

FUNCTIONAL CLUSTERING IN PHAGES T7 AND T4

The genome of phage T7 has three classes of genes, each constituting a group of adjacent loci. As **Figure 16.2** shows, the class I genes are the immediate early type, expressed by host RNA polymerase as soon as the phage DNA enters the cell. Among the products of these genes are enzymes that interfere with host gene expression (and therefore with their own expression) and a phage RNA polymerase. The phage enzyme is responsible for expressing the class II genes (concerned principally with DNA synthesis functions) and the class III genes (concerned with assembling the mature phage particle).

Phage T4 has one of the larger genomes, organized with extensive functional grouping of genes. **Figure 16.3** presents the genetic map. Genes that are numbered are **essential:** a mutation in any one of these loci prevents successful completion of the lytic cycle. Genes indicated by three-letter abbreviations are **nonessential,** at least under the usual conditions of infection. We do not really understand the inclusion of so many nonessential genes, but presumably they confer a selective advantage. (In most phage genomes, most or all genes are essential.)

There are three phases of gene expression, and a

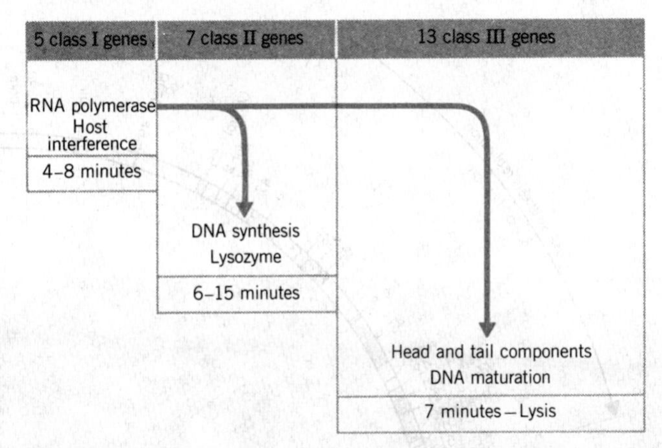

Figure 16.2
Phage T7 contains three classes of genes that are expressed sequentially. The genome is about 38,000 bp.

summary of the functions of the genes expressed in each is given in **Figure 16.4.** The early genes are transcribed by host RNA polymerase. At a slightly later point, the quasi-late genes are transcribed, but we do not yet understand how this is controlled (their name reflects a mode of expression that does not fit with the categories we have described for other phages). Together the early and quasi-late genes account for virtually all of the phage functions concerned with the synthesis of DNA, modifying cell structure, and transcribing and translating phage genes.

The two essential genes in the "transcription" category fulfill a regulatory function: their products are necessary for late gene expression. Phage T4 infection depends on a mechanical link between replication and late gene expression. Only actively replicating DNA can be used as template for late gene transcription. The connection is generated by modifying the host RNA polymerase so that it requires some feature present only in replicating DNA (probably nicks). This establishes a correlation between the synthesis of phage protein components and the number of genomes available for packaging.

THE LAMBDA LYTIC CASCADE RELIES ON ANTITERMINATION

One of the most intricate cascade circuits is provided by phage lambda. Actually, the cascade for lytic de-

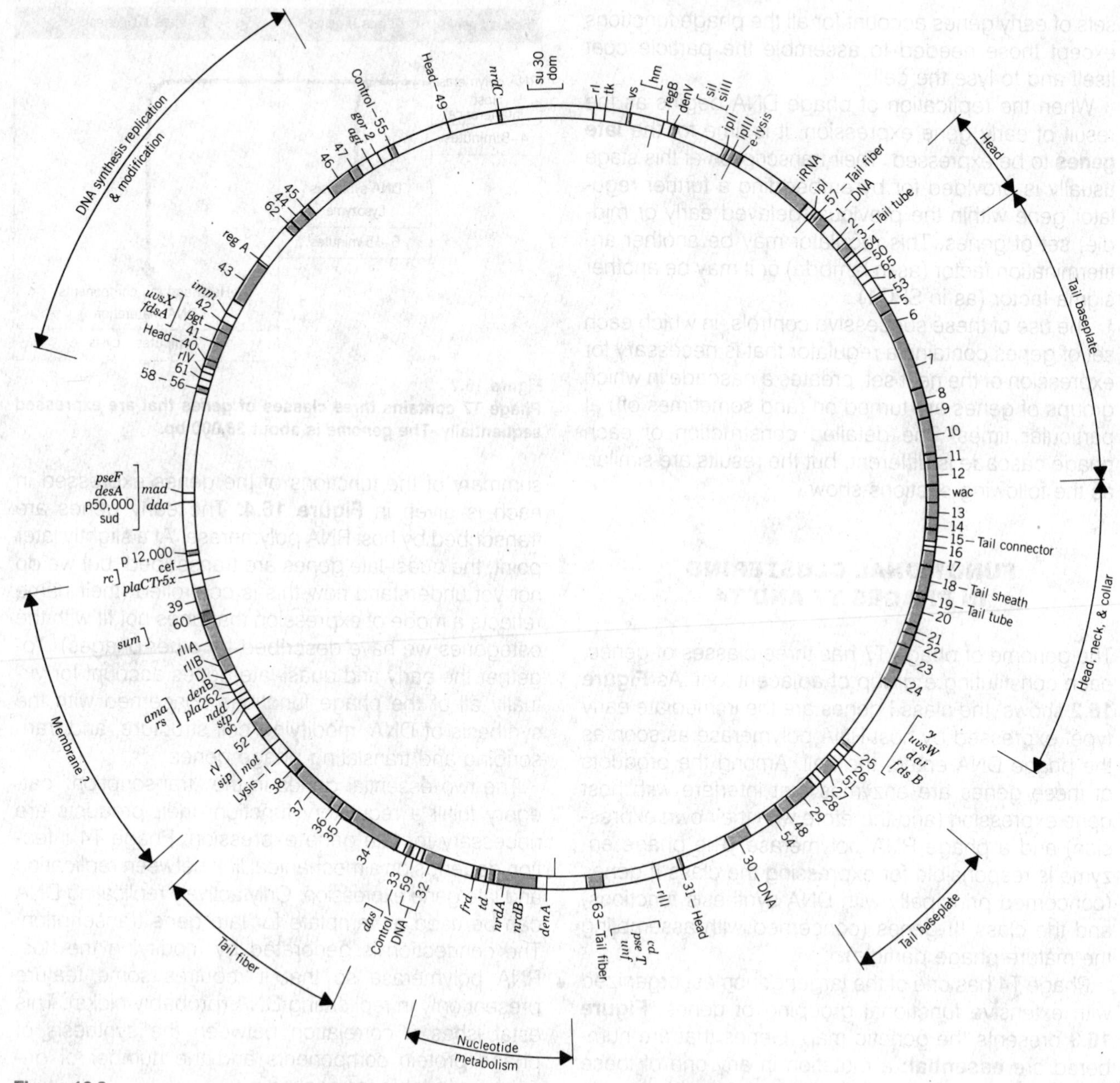

Figure 16.3
The map of phage T4 is circular and shows extensive clustering of related functions in the 165,000 bp genome.

velopment itself is straightforward, with two regulators controlling the successive stages of development. But the circuit for the lytic cycle is interlocked with the circuit for establishing lysogeny, as summarized in

Figure 16.5. When lambda DNA is introduced into a new host cell, the lytic and lysogenic pathways start off the same way. Both require expression of the immediate early and delayed early genes. But then they

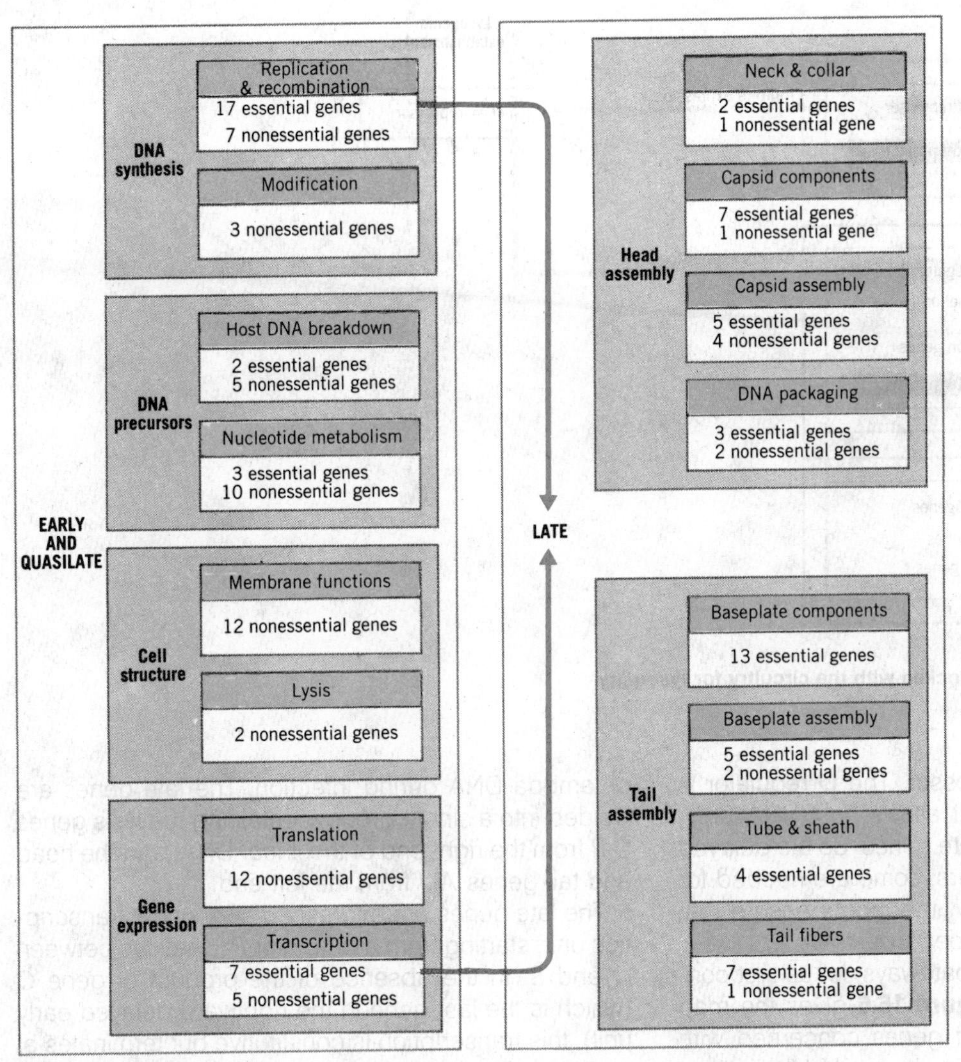

Figure 16.4
The phage T4 lytic cascade falls into two parts: early and quasi-late functions are concerned with DNA synthesis and gene expression; late functions are concerned with particle assembly.

diverge, lytic development following if the late genes are expressed, lysogeny ensuing if the synthesis of the repressor is established.

Lambda has only two immediate early genes, transcribed independently by host RNA polymerase. One of these is *cro,* which has dual functions: it prevents synthesis of the lysogenic repressor (a necessary action if the lytic cycle is to proceed); and it turns off expression of the immediate early genes (which are not needed later in the lytic cycle). The other immediate early gene is *N*, which, as we have seen in Chapter 13, codes for an antitermination factor whose action at the *nut* sites allows transcription to proceed into the delayed early genes.

The delayed early genes include two replication genes (needed for lytic infection) and seven recombination genes (some involved in recombination during lytic infection, two necessary to integrate lambda DNA into the bacterial chromosome for lysogeny). The *cII-cIII* pair of regulators is needed to start up the syn-

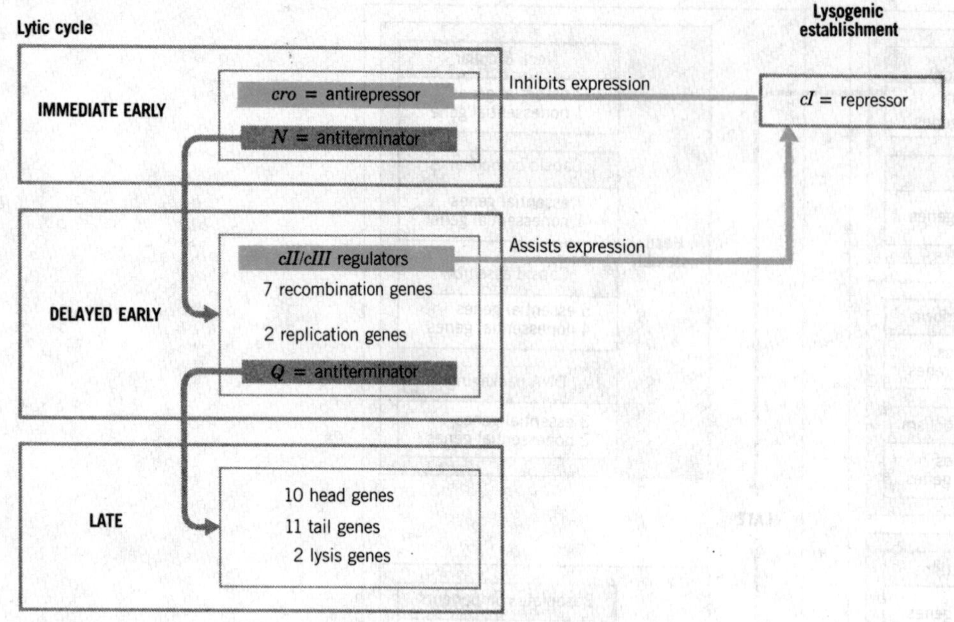

Figure 16.5
The lambda lytic cascade is interlocked with the circuitry for lysogeny.

thesis of the lysogenic repressor. The *Q* regulator is an antitermination factor that allows host RNA polymerase to proceed into the late genes. So the delayed early genes serve two masters: some are needed for the phage to enter lysogeny, the others ensure that the lytic cycle follows its proper order.

To disentangle these two pathways, let us first consider just the lytic cycle. **Figure 16.6** gives the map of lambda DNA. A group of genes concerned with regulation is surrounded by genes needed for recombination and replication. Within the regulatory group are the immediate early genes, *N* and *cro*. They are transcribed from different strands of DNA—*N* toward the left, and *cro* toward the right. In the presence of the pN antitermination factor, transcription continues to the left of *N* into the recombination genes, and to the right of *cro* into the replication genes (see Figure 13.7).

The map shown in Figure 16.6 gives the organization of the lambda DNA as it exists in the phage particle. But shortly after infection, the ends of the DNA join to form a circle. **Figure 16.7** shows the true state of lambda DNA during infection. The late genes are welded into a single group, containing the lysis genes *S-R* from the right end of the linear DNA, and the head and tail genes *A-J* from the left end.

The late genes are expressed as a single transcription unit, starting from a promoter $P_{R'}$ that lies between *Q* and *S*. In the absence of the product of gene *Q* (which is the last gene in the rightward delayed early unit), this transcription is constitutive but terminates at a site t_{R3}. The transcript is known as 6S RNA and is 194 bases long. When pQ becomes available, however, it suppresses termination at t_{R3} (probably by acting at a site called *qut*, separate from the promoter and terminator) and the 6S RNA is extended, with the result that the late genes are heavily expressed.

Late gene transcription does not seem to terminate at any specific point, but continues through all the late genes into the region beyond. A similar event happens with the leftward delayed early transcription, which continues past the recombination functions. Transcription in each direction is probably terminated before the polymerases could crash into each other.

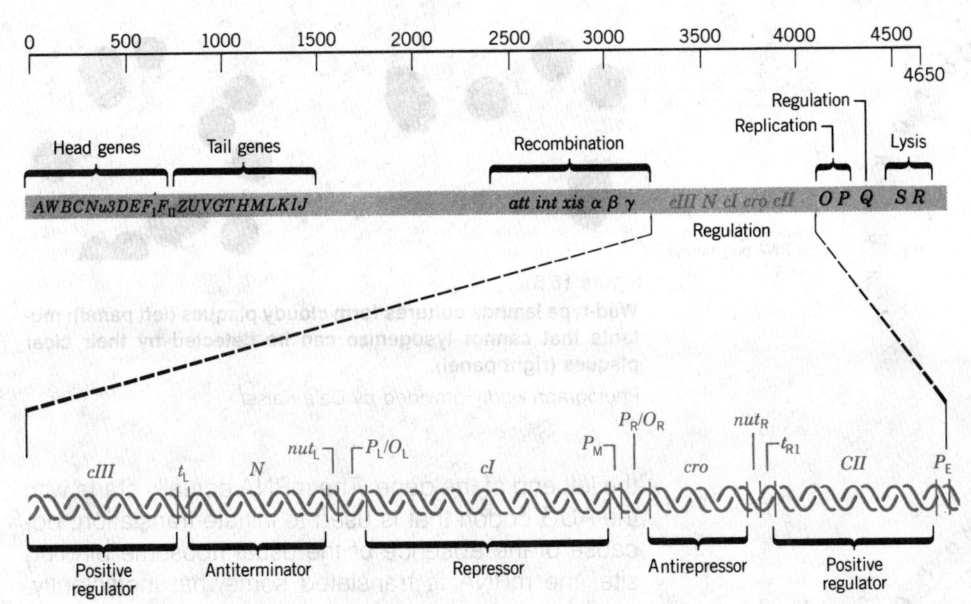

Figure 16.6
The lambda map shows clustering of related functions. The genome is about 46,500 bp.

LYSOGENY IS MAINTAINED BY AN AUTOGENOUS CIRCUIT

Looking at the lambda lytic cascade, we can see that the entire program is set in train by initiating transcription at the two promoters P_L and P_R for the immediate early genes *N* and *cro*. Because lambda uses antitermination to proceed to the next stage of (delayed early) expression, the same two promoters continue to be used right through the early period.

The expanded map of the regulatory region in **Figure 16.6** shows that the promoters P_L and P_R lie on either side of the *cI* gene. Associated with each promoter is an operator (O_L, O_R) at which a repressor protein binds to prevent RNA polymerase from initiating transcription. The sequence of each operator overlaps with the promoter that it controls; so often these are described as the P_L/O_L and P_R/O_R control regions.

Because of the sequential nature of the lytic cascade, the control regions provide a pressure point at which entry to the entire cycle can be controlled. By denying RNA polymerase access to these promoters, the repressor prevents the phage genome from initiating the lytic cycle.

The operators are revealed as the targets for repressor action by **virulent** mutations. These mutations prevent the repressor from binding at O_L or O_R, with the result that the phage inevitably proceeds into the lytic pathway. Virulent mutations in phages are the equivalent of operator-constitutive mutations in bacterial operons.

The repressor protein is coded by the *cI* gene. Mutants in this gene cannot maintain lysogeny, but are fated always to enter the lytic cycle. This gives rise to the name of the gene, which reflects the **clear plaque** phenotype of the resulting infection. When a bacterial culture is infected with a phage, the cells are lysed to generate regions that can be seen on a culture plate as small areas of clearing called **plaques.** With wild-type phages, the plaques are cloudy, because they contain some cells that have established lysogeny instead of being lysed. The effect of a cI^- mutation is to prevent lysogeny, so that the plaques contain only lysed cells, and are therefore clear in appearance. **Figure 16.8** shows both types of plaque.

The *cI* gene is transcribed from a promoter P_M, sometimes also called P_{RM}, that lies at its right end (see Figure 16.6). (The subscript "M" indicates maintenance of repressor.) Transcription is terminated at

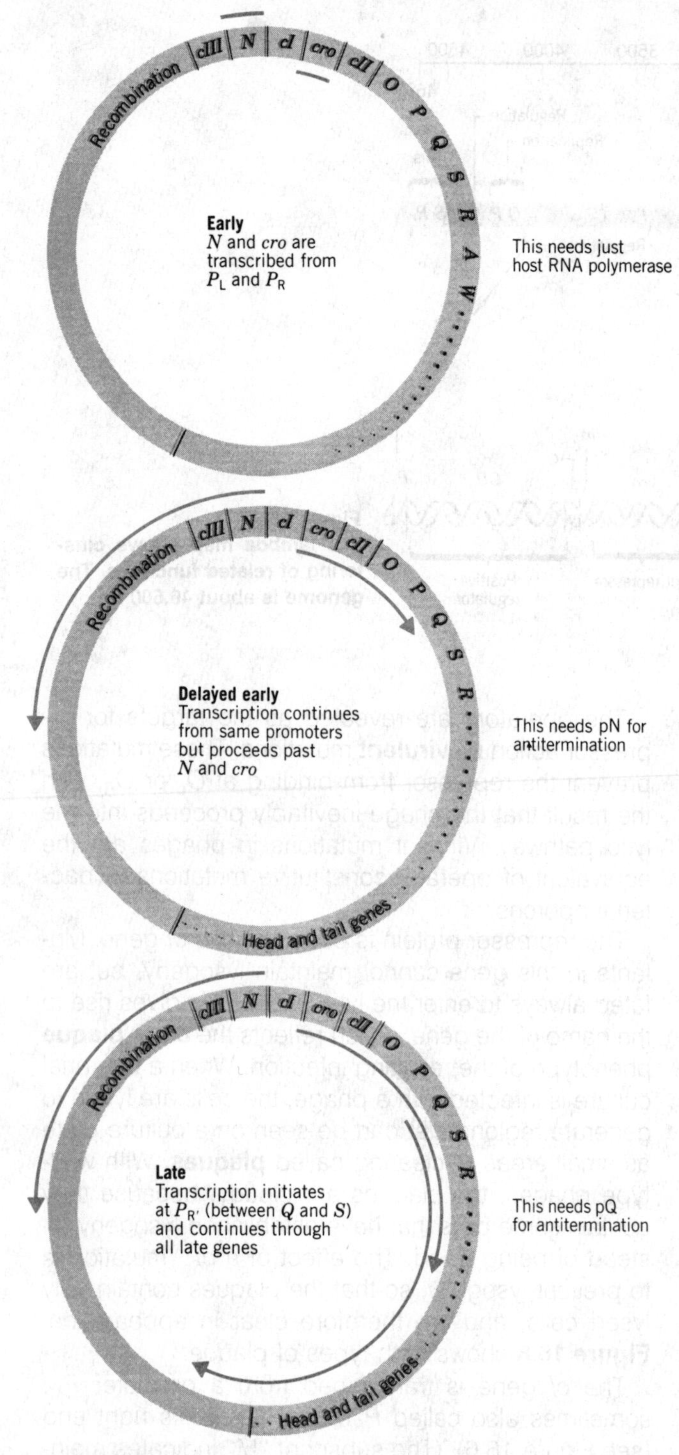

Figure 16.7

Lambda DNA circularizes during infection, so that the late gene cluster is intact in one transcription unit.

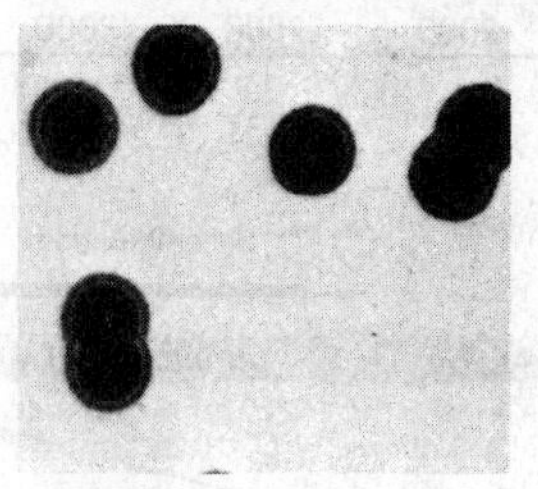
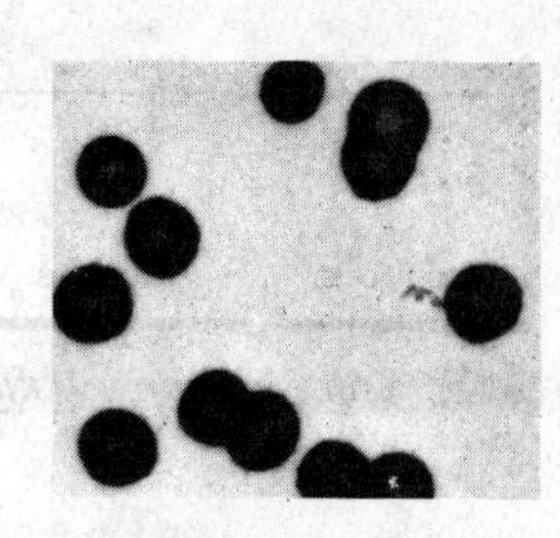

Figure 16.8

Wild-type lambda cultures form cloudy plaques (left panel); mutants that cannot lysogenize can be detected by their clear plaques (right panel).

Photograph kindly provided by Dale Kaiser.

the left end of the gene. The mRNA actually starts with the AUG codon that is used to initiate translation; because of the absence of the usual ribosome binding site, the mRNA is translated somewhat inefficiently, producing only a low level of repressor protein.

The repressor binds independently to the two operators. It has a single function at O_L, but has dual functions at O_R. These are illustrated in **Figure 16.9.**

At O_L the repressor has the same sort of effect that we have already discussed for several other systems: it prevents RNA polymerase from initiating transcription at P_L. This stops the expression of gene *N*. Since P_L is used for all leftward early gene transcription, this action prevents expression of the entire leftward early transcription unit. Thus the lytic cycle is stymied before it can begin.

At O_R, repressor binding prevents the use of P_R. Thus *cro* and the other rightward early genes cannot be expressed. (We shall see later why it is important to prevent the expression of *cro* when lysogeny is being maintained.)

But the presence of repressor at O_R also has another effect. The promoter for repressor synthesis, P_M, is adjacent to the rightward operator O_R. It turns out that RNA polymerase can initiate at P_M *only when repressor is bound at* O_R. In this action, the repressor is acting as a positive regulator protein that is necessary for transcription of the *cI* gene. Since it is in fact the product of this gene, this interaction creates an autogenous circuit, in which the presence of repressor is necessary to support its own continued synthesis.

The nature of this control circuit explains the biolog-

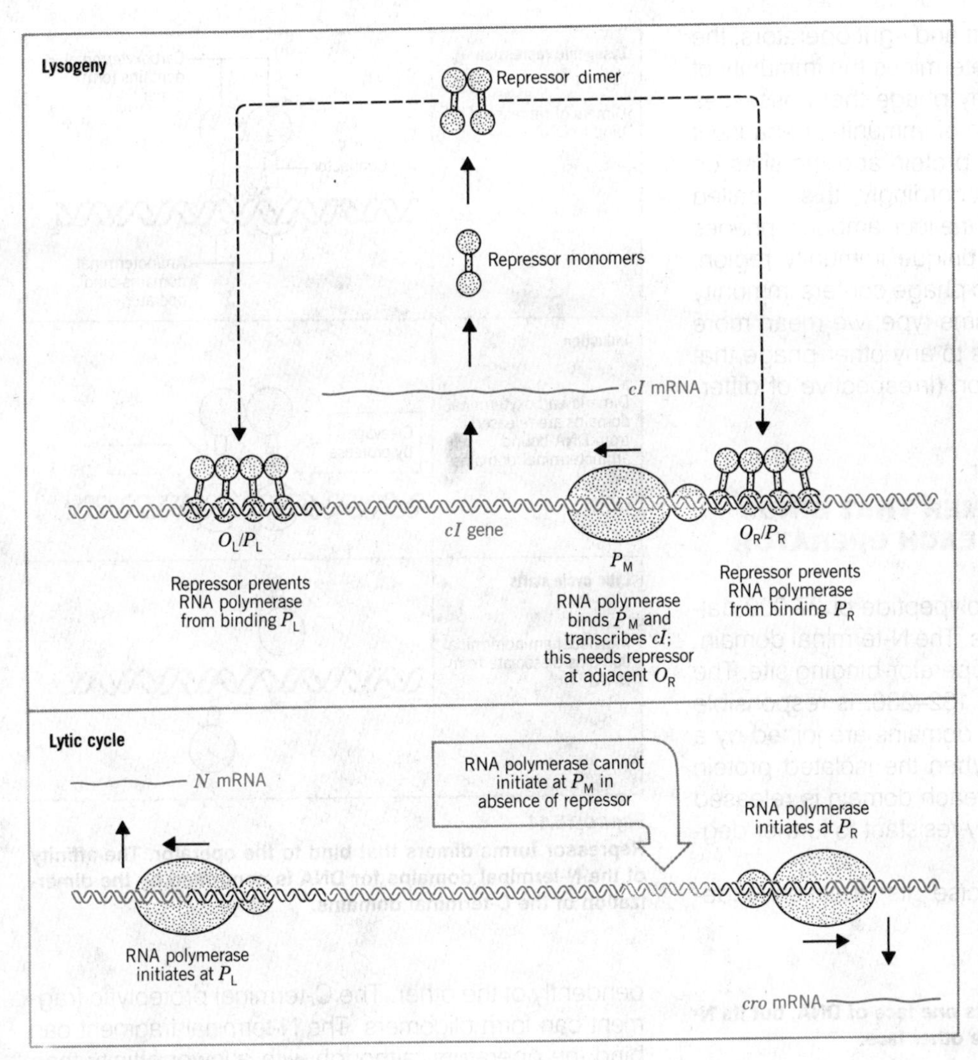

Figure 16.9
Lysogeny is maintained by an autogenous circuit (upper). If this circuit is interrupted, the lytic cycle starts (lower).

ical features of lysogenic existence. Lysogeny is stable because the control circuit ensures that, so long as the level of repressor is adequate, there is continued expression of the *cI* gene. The result is that O_L and O_R remain occupied indefinitely. By repressing the entire lytic cascade, this action maintains the prophage in its inert form.

The presence of repressor explains the phenomenon of immunity. If a second lambda phage DNA enters a lysogenic cell, repressor protein synthesized from the resident prophage genome will immediately bind to O_L and O_R in the new genome. This prevents the second phage from entering the lytic cycle.

Prophage is induced to enter the lytic cycle when the lysogenic circuit is broken. This happens when the repressor is inactivated. The absence of repressor allows RNA polymerase to bind at P_L and P_R, starting the lytic cycle as shown in Figure 16.9. The autogenous nature of the repressor-maintenance circuit creates a very sensitive response. Because the presence of repressor is necessary for its own synthesis, expression of the *cI* gene stops as soon as the existing repressor is destroyed. Thus no repressor is synthesized to replace the molecules that have been damaged. So the lytic cycle can start without any undue interference from the circuit that maintains lysogeny.

The region including the left and right operators, the *cI* gene, and the *cro* gene determines the immunity of the phage. That is to say, any phage that possesses this region has the same type of immunity, because it specifies both the repressor protein and the sites on which the repressor acts. Accordingly, this is called the **immunity region.** Each of the four lambdoid phages φ80, *21*, *434*, and λ has a unique immunity region. When we say that a lysogenic phage confers immunity to any other phage of the same type, we mean more precisely that the immunity is to any other phage that has the same immunity region (irrespective of differences in other regions).

REPRESSOR IS A DIMER THAT BINDS COOPERATIVELY AT EACH OPERATOR

The repressor subunit is a polypeptide of 27,000 daltons with two distinct domains. The N-terminal domain, residues 1–92, provides the operator-binding site. The C-terminal domain, residues 132–236, is responsible for forming a dimer. The two domains are joined by a connector of 40 residues. When the isolated protein is digested with a protease, each domain is released as a fragment that is relatively resistant to further degradation.

Each domain can exercise its function independently of the other. The C-terminal proteolytic fragment can form oligomers. The N-terminal fragment can bind the operators, although with a lower affinity than the intact repressor. Thus the information for specifically contacting DNA is contained within the N-terminal domain, but (as also with the *lac* repressor) the efficiency of the process is enhanced by the attachment of the C-terminal domain.

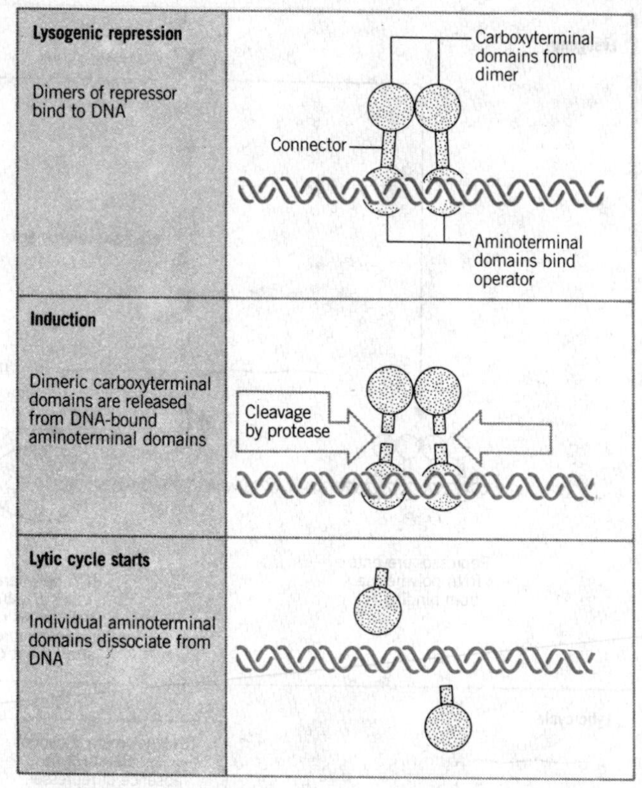

Figure 16.11
Repressor forms dimers that bind to the operator. The affinity of the N-terminal domains for DNA is controlled by the dimerization of the C-terminal domains.

Figure 16.10
The bulk of the repressor contacts one face of DNA, but its N-terminal arms reach around to the other face.

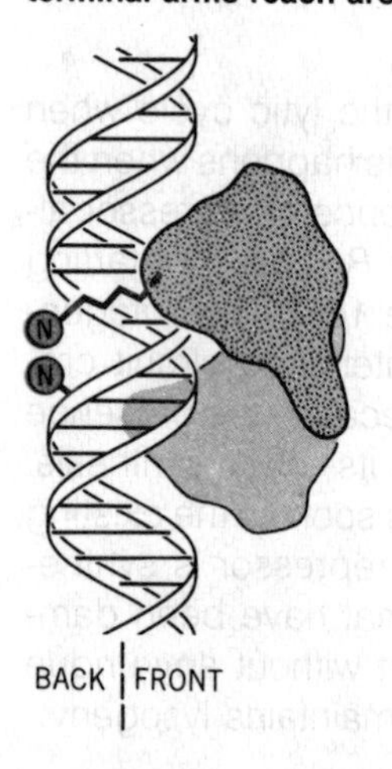

The isolated N-terminal domain makes the same contacts with DNA as the intact repressor. However, removing the last three N-terminal amino acids eliminates some of the contacts. This observation provided the basis for the model illustrated in **Figure 16.10,** in which the bulk of the N-terminal domain contacts one face of DNA, while the last three N-terminal amino acids form an arm extending around the back.

The dimeric structure of the repressor is crucial in

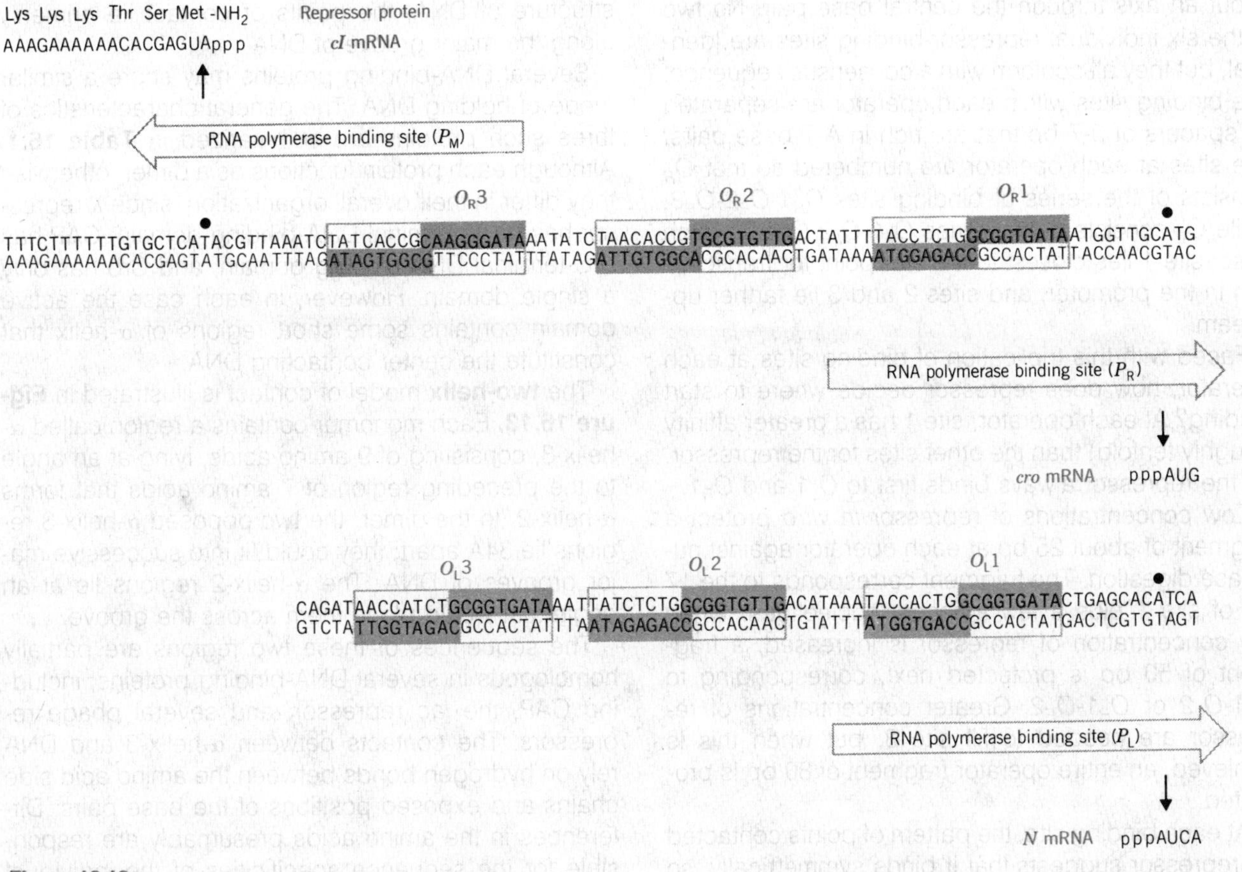

Figure 16.12

Each operator contains three repressor-binding sites, and overlaps with the RNA polymerase binding site.

The orientation of O_L has been reversed from usual to facilitate comparison with O_R.

maintaining lysogeny. The induction of a lysogenic prophage to enter the lytic cycle is caused by cleavage of the repressor subunit in the connector region, between residues 111 and 112. (In a sense, this is the counterpart to the allosteric change in conformation that results when a small-molecule inducer inactivates the repressor of a bacterial operon, a capacity not possessed by the lysogenic repressor.)

In the intact state, dimerization of the C-terminal domains ensures that when the repressor binds to DNA its two N-terminal domains each contact DNA simultaneously. But cleavage releases the dimeric C-terminal domains from the N-terminal domains. As illustrated in **Figure 16.11,** the N-terminal domains then no longer have sufficient affinity to remain on the operator. So they dissociate from DNA, allowing lytic infection to start.

The balance between lysogeny and the lytic cycle depends on the concentration of repressor. Intact repressor is present in the cell at a concentration sufficient to ensure that the operators are occupied. But if the repressor is cleaved, this concentration is inadequate, because of the lower affinity of the separate N-terminal domain for the operator. Too high a concentration of repressor would make it impossible to induce the lytic cycle in this way; too low a level, of course, would make it impossible to maintain lysogeny.

Each operator contains three repressor-binding sites. As can be seen from **Figure 16.12,** each binding site is a sequence of 17 bp displaying partial symmetry about an axis through the central base pair. No two of the six individual repressor-binding sites are identical, but they all conform with a consensus sequence. The binding sites within each operator are separated by spacers of 3-7 bp that are rich in A-T base pairs. The sites at each operator are numbered so that O_R consists of the series of binding sites O_R1-O_R2-O_R3, while O_L consists of the series O_L1-O_L2-O_L3. In each case, site 1 lies closest to the startpoint for transcription in the promoter, and sites 2 and 3 lie farther upstream.

Faced with this triplication of binding sites at each operator, how does repressor decide where to start binding? At each operator, site 1 has a greater affinity (roughly tenfold) than the other sites for the repressor. So the repressor always binds first to O_L1 and O_R1.

Low concentrations of repressor *in vitro* protect a fragment of about 25 bp at each operator against nuclease digestion. The fragment corresponds to the 17 bp of site 1 plus a few adjacent nucleotides. When the concentration of repressor is increased, a fragment of 50 bp is protected next, corresponding to O_L1-O_L2 or O_R1-O_R2. Greater concentrations of repressor are needed to fill site 3, but when this is achieved, an entire operator fragment of 80 bp is protected.

At each binding site, the pattern of points contacted by repressor suggests that it binds symmetrically, so that each N-terminal domain of the dimer contacts a similar set of bases. Thus the individual N-terminal region contacts a half-binding site. The side of the binding site that the repressor contacts is indicated by the shading in Figure 16.12. In the double-helical structure of DNA, the points of contact lie primarily along the major groove of DNA.

Several DNA-binding proteins may share a similar mode of holding DNA. The general characteristics of three such proteins are summarized in **Table 16.1.** Although each protein functions as a dimer, otherwise they differ in their overall organization, since λ repressor has an N-terminal DNA-binding domain, CAP has a C-terminal DNA-binding domain, and Cro has only a single domain. However, in each case the active domain contains some short regions of α-helix that constitute the center contacting DNA.

The **two-helix** model of contact is illustrated in **Figure 16.13.** Each monomer contains a region called α-helix-3, consisting of 9 amino acids, lying at an angle to the preceding region of 7 amino acids that forms α-helix-2. In the dimer, the two opposed α-helix-3 regions lie 34Å apart; they could fit into successive major grooves of DNA. The α-helix-2 regions lie at an angle that would place them across the groove.

The sequences of these two regions are partially homologous in several DNA-binding proteins, including CAP, the *lac* repressor, and several phage repressors. The contacts between α-helix-3 and DNA rely on hydrogen bonds between the amino acid side chains and exposed positions of the base pairs. Differences in the amino acids presumably are responsible for the sequence specificities of the individual proteins. Contacts from α-helix-2 to the DNA take the form of hydrogen bonds connecting with the phosphate backbone. In addition to these contacts, a large part of the overall energy of interaction with DNA is provided by ionic interactions with the phosphate

Table 16.1
DNA-binding domains may occupy different parts of regulator proteins.

Protein	Monomer Length (residues)	Active Form	DNA-Binding Domain	Other Domain
λ repressor	236	Dimer	N-terminal 92 residues	C-terminal (aggregates)
Cro	66	Dimer	Whole protein	None
CAP activator	209	Dimer	C-terminal 74 residues	N-terminal (binds cAMP)

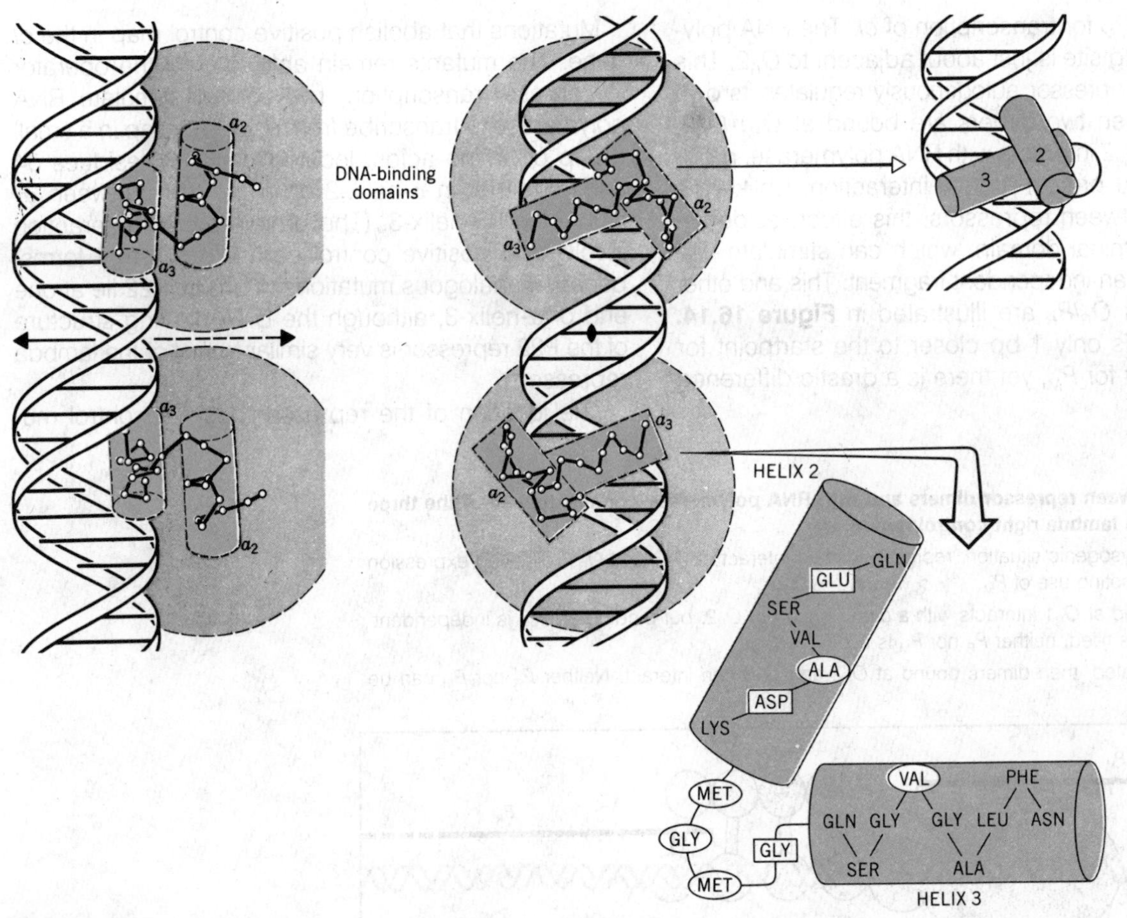

Figure 16.13
In the two-helix model for DNA binding, α-helix-3 of each monomer lies in the wide groove on the same face of DNA, and α-helix-2 lies across the groove.

backbone, which do not depend on the sequence of DNA.

Lambda repressor binds to each operator in a cooperative manner. First, a repressor dimer binds to site 1, because of its greater affinity. However, the presence of a dimer at site 1 greatly increases the affinity with which a second dimer can bind to site 2. When both sites 1 and 2 are occupied, this interaction does *not* extend farther, to site 3. At the concentrations of repressor usually found in a lysogen, both sites 1 and 2 are filled at each operator, but site 3 is not occupied.

If site 1 is inactive (because of mutation), then repressor binds cooperatively to sites 2 and 3. That is, binding at site 2 assists another dimer to bind at site 3. This interaction occurs directly between repressor dimers and not via conformational change in DNA. Probably the connector region of the first repressor orients the C-terminal regions of the dimer in such a way that they contact the C-terminal regions of just the second dimer.

From the sequences shown in Figure 16.12, we can see that O_L1 and O_R1 lie more or less in the center of the RNA polymerase binding sites of P_L and P_R, respectively. Occupancy of O_L1-O_L2 and O_R1-O_R2 thus physically blocks access of RNA polymerase to the corresponding promoter.

A different relationship is shown between O_R and

the promoter P_M for transcription of *cI*. The RNA polymerase binding site is just about adjacent to O_R2. This explains how repressor autogenously regulates its own synthesis. When two dimers are bound at O_R1-O_R2, the dimer at O_R2 interacts with RNA polymerase, probably through a protein-protein interaction. Unlike the interaction between repressors, this effect resides in the amino terminal domain, which can stimulate use of P_M even as an independent fragment. This and other interactions at O_R/P_R are illustrated in **Figure 16.14.** The O_R2 site is only 1 bp closer to the startpoint for P_R than to that for P_M, yet there is a drastic difference in effect!

Mutations that abolish positive control map in the *cI* gene. The mutants remain able to bind the operator to repress transcription, but cannot stimulate RNA polymerase to transcribe from P_M. They map in a small group of amino acids, located between residues 34 and 43 either in α-helix-2 or in the turn between α-helix-2 and α-helix-3. (The protein-protein interaction involved in positive control can take different forms, because analogous mutations in phage P22 lie at one end of α-helix-3, although the DNA-binding structure of the P22 repressor is very similar to that of the lambda repressor.)

The location of the repressor positive control mu-

Figure 16.14

Interactions between repressor dimers and with RNA polymerase control the use of the three promoters in the lambda right control region.

(*a*) In the usual lysogenic situation, repressor dimers interact at O_R2 and O_R1, blocking expression from P_R, but promoting use of P_M.

(*b*) A dimer bound at O_R1 interacts with a dimer bound at O_R2, but binding at O_R3 is independent. With all three sites filled, neither P_R nor P_M is used.

(*c*) If O_R1 is mutated, then dimers bound at O_R2 and O_R3 can interact. Neither P_R nor P_M can be used in this case.

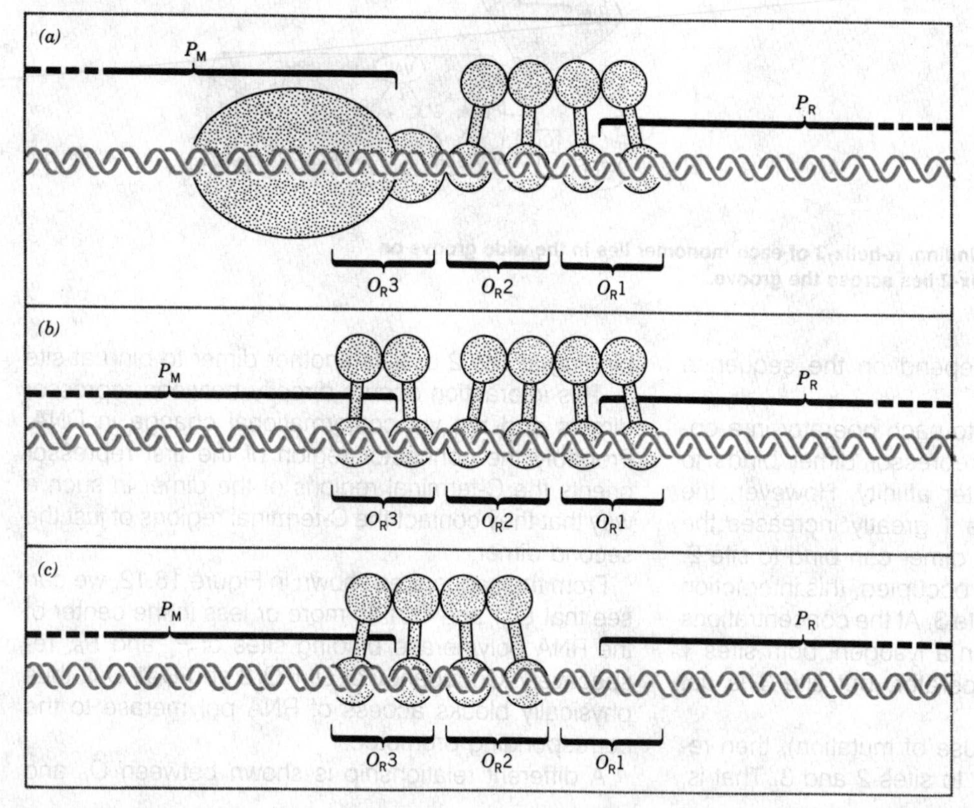

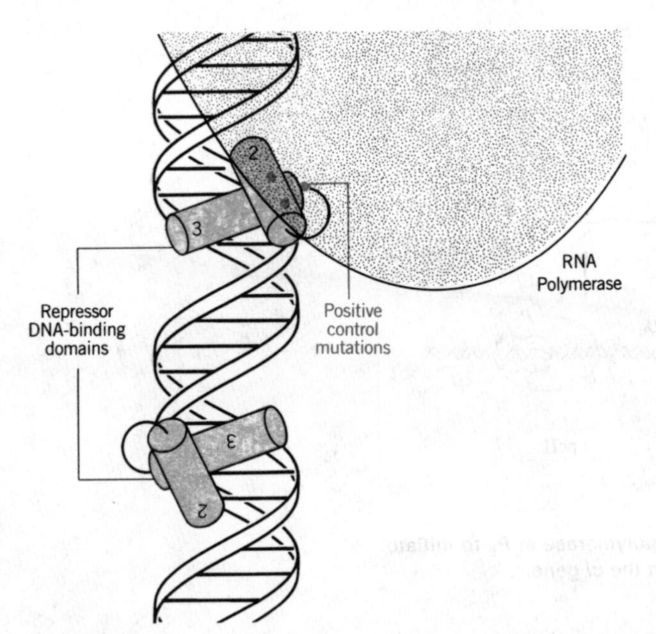

Figure 16.15
Positive control mutations identify a small region at α-helix-2 that interacts directly with RNA polymerase.

tations is indicated on **Figure 16.15.** The site lies close to a phosphate group on DNA that is "shared" by both repressor and RNA polymerase. Thus the group of amino acids is in an appropriate position on the repressor to contact the polymerase. The interaction between repressor and polymerase is needed for the polymerase to make the transition from a closed complex to an open complex.

What happens if a repressor dimer binds to O_R3? This site overlaps with the RNA polymerase binding site at P_M. Thus if the repressor concentration becomes great enough to cause occupancy of O_R3, the transcription of *cI* is prevented. This leads in due course to a reduction in repressor concentration; O_R3 then becomes empty, and the autogenous loop can start up again because O_R2 remains occupied. This provides a mechanism for preventing the concentration of repressor from becoming too great. (However, repressor concentration in lysogens usually is not high enough to trigger this effect, whose importance *in vivo* is therefore unclear.) In the formal sense, the repressor is an autogenous regulator of its own expression that functions positively at low concentrations and negatively at high concentrations.

Virulent mutations have been found in sites 1 and 2 of both O_L and O_R. Every one of these changes the base composition. The mutations vary in their degree of virulence, according to the extent to which they reduce the affinity of the binding site for repressor, and also depending on the relationship of the afflicted site to the promoter. Consistent with the conclusion that O_R3 and O_L3 usually are not occupied, virulent mutations are not found in these sites.

HOW IS REPRESSOR SYNTHESIS ESTABLISHED?

The control circuit for maintaining lysogeny presents a paradox. The presence of repressor protein is necessary for its own synthesis. This explains how the lysogenic condition is perpetuated. But how is the synthesis of repressor established in the first place?

When a lambda DNA enters a new host cell, the bacterial RNA polymerase cannot transcribe *cI*, because there is no repressor present to aid its binding at P_M. But this same absence of repressor means that P_R and P_L are available. So the first event when lambda DNA infects a bacterium is for genes *N* and *cro* to be transcribed. Then pN allows transcription to be extended farther. This allows *cIII* (and other genes) to be transcribed on the left, while *cII* (and other genes) are transcribed on the right (see Figure 16.7).

The *cII* and *cIII* genes share with *cI* the property that mutations in them cause clear plaques. But there is a difference. Mutants that are cI^- can neither establish nor maintain lysogeny. Mutants that are cII^- or $cIII^-$ have some difficulty in establishing lysogeny, but once established, they are able to maintain it by the same autogenous circuit we have already discussed. This implicates the *cII* and *cIII* genes as positive regulators whose products are needed for an alternative system for repressor synthesis. The system is needed only to *initiate* the expression of *cI* in order to circumvent the inability of the autogenous circuit to engage in de novo synthesis.

The pcII protein acts directly on gene expression, where pcIII acts indirectly. Between the *cro* and *cII* genes is another promoter, called P_E, sometimes also known as P_{RE}. (The "E" indicates its use for repressor-

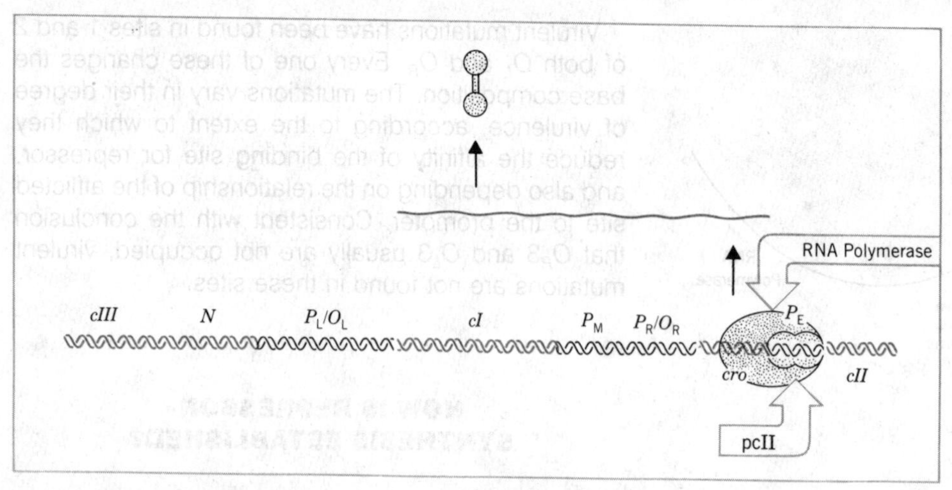

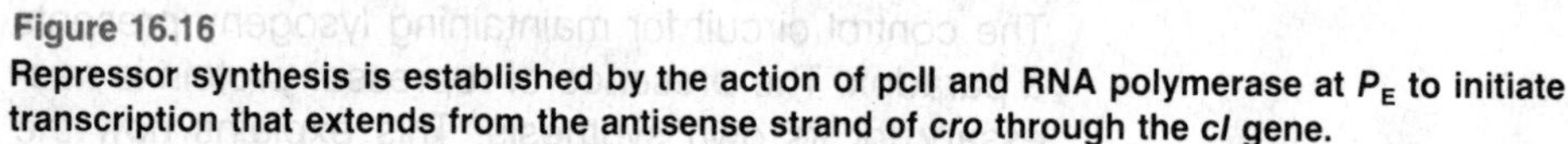

Figure 16.16
Repressor synthesis is established by the action of pcII and RNA polymerase at P_E to initiate transcription that extends from the antisense strand of *cro* through the *cI* gene.

establishment.) This promoter can be recognized by RNA polymerase only in the presence of pcII, whose action is illustrated in **Figure 16.16.** The pcII protein is extremely unstable *in vivo*, because it is degraded as the result of the activity of a host protein called HflA. The role of pcIII is to protect pcII against this degradation.

The P_E promoter has an unusual sequence that lacks both the normal consensus sequences. The deficiency may explain its dependence on *cII*. The development of a system for *in vitro* transcription at P_E has made it possible to define the action of pcII. The promoter is completely inactive with RNA polymerase alone, but can be transcribed when pcII is added. The regulator binds to a region extending from about −25 to −45. When RNA polymerase is added, an additional region is protected, extending from −12 to +11. As summarized in **Figure 16.17,** binding of the two proteins is adjacent or overlapping.

The importance of the −35 and −10 regions for promoter function, in spite of their lack of resemblance with the consensus, is indicated by the existence of *cy* mutations. These have effects similar to those of cII^- and $cIII^-$ in preventing the establishment of lysogeny; but they are *cis*-acting instead of *trans*-acting. They fall into two groups, *cyL* and *cyR*, located around the −10 and −35 positions of P_E, respectively.

The *cyL* mutations around −10 probably prevent RNA polymerase from recognizing the promoter. The *cyR* mutations at −35 fall into two types. In the region corresponding to the −35 box, mutations do not affect pcII binding; presumably they prevent RNA polymerase binding. Although lacking homology with the consensus, this region is exactly comparable in size and location. On either side of this region, mutations prevent pcII from binding. These mutations fall into the short tetrameric repeats in the sequence

TTGCNNNNNNTTGC
AACGNNNNNNAACG

Figure 16.17
RNA polymerase binds to P_E only in the presence of pcII, which contacts the region around −35.
Note that the orientation of the gene is shown in the usual 5′-3′ direction, that is, in the opposite orientation from Figure 16.16.

Bound by pcII and RNA polymerase together

Bound by pcII alone

−40 −30 −20 −10 +10

Startpoint

Each base in the tetramer is 10 bp (one helical turn) separated from its homologue in the other tetramer, so that when pcII recognizes the two tetramers, it lies on one face of the double helix.

The P_E transcript contains the **antisense** strand of the *cro* gene, which usually is transcribed in the opposite direction (i.e., from P_R). The antisense *cro* sequence is not translated on the P_E transcript, but the *cI* coding region is very efficiently translated (in contrast with the weak translation of the P_M transcript mentioned earlier). In fact, repressor is synthesized about seven or eight times more effectively via expression from P_E than from P_M.

For a long time it seemed puzzling that the *cro* gene could simultaneously be transcribed in opposite directions. We really don't know what happens if two RNA polymerases meet when traveling toward each other. The sensible presumption is that a tangle would develop, blocking transcription, but it is possible that the polymerases are able to pass in some way we do not understand. The explanation may be that, because the P_E promoter is stronger, its use suppresses the use of P_R.

Now we can see how lysogeny is established on a new infection. **Figure 16.18** recapitulates the initial stages of transcribing *N* and *cro* and extending this transcription via pN action into *cIII* and *cII*. The presence of pcII allows P_E to be used for transcription extending through *cI*. Repressor protein is synthesized in high amounts from this transcript. Immediately it binds to O_L and O_R.

By directly inhibiting any further transcription from P_L and P_R, the binding turns off the expression of all phage genes. This halts the synthesis of pcII and pcIII, which are unstable, so they decay rapidly, with the result that P_E can no longer be used. Thus the synthesis of repressor via the establishment circuit is brought to a halt.

But repressor now is present at O_R. It switches on the maintenance circuit for expression from P_M. Repressor continues to be synthesized, although now at the lower level typical of P_M function. So the establishment circuit starts off repressor synthesis at a high level; then repressor turns off all other functions, while at the same time turning on the maintenance circuit, which functions at the low level adequate to sustain lysogeny.

We shall not now deal in detail with the other functions needed to establish lysogeny, but we can just briefly remark that the infecting lambda DNA must be inserted into the bacterial genome (see Chapter 34). The insertion requires the product of gene *int*, which is expressed from its own promoter P_I, at which pcII also is necessary. The sequence of P_I shows homology with P_E in the pcII binding site (although not in the −10 region). The functions necessary for establishing the lysogenic control circuit are therefore under the same control as the function needed physically to manipulate the DNA. Thus the establishment of lysogeny is under a control that ensures all the necessary events occur with the same timing.

ANTIREPRESSOR IS NEEDED FOR LYTIC INFECTION

We started this chapter by saying that lambda has the alternatives of becoming lysogenic or starting a lytic infection. We have seen that lysogeny is initiated by establishing an autogenous maintenance circuit that inhibits the entire lytic cascade through applying pressure at two points. The program for establishing lysogeny actually proceeds through some of the same events that we described earlier in terms of the lytic cascade (expression of delayed early genes via expression of *N* is needed). We now face a problem. How does the phage enter the lytic cycle?

What we have left out of this account so far is the role of gene *cro*, which codes for another regulator protein, sometimes known as the **antirepressor.** This regulator is responsible for preventing the synthesis of the repressor protein; this action shuts off the possibility of establishing lysogeny. Mutants of the cro^- type usually establish lysogeny rather than entering the lytic pathway, because they lack the ability to switch events away from the expression of repressor.

The Cro product is a small protein (only 9000 daltons) that forms a dimer that acts within the immunity region. It has two effects. It prevents the synthesis of repressor via the maintenance circuit; that is, it prevents transcription via P_M. It also inhibits the expression of early genes from both P_L and P_R. This means that, when a phage enters the lytic pathway, Cro has responsibility both for preventing the synthesis of re-

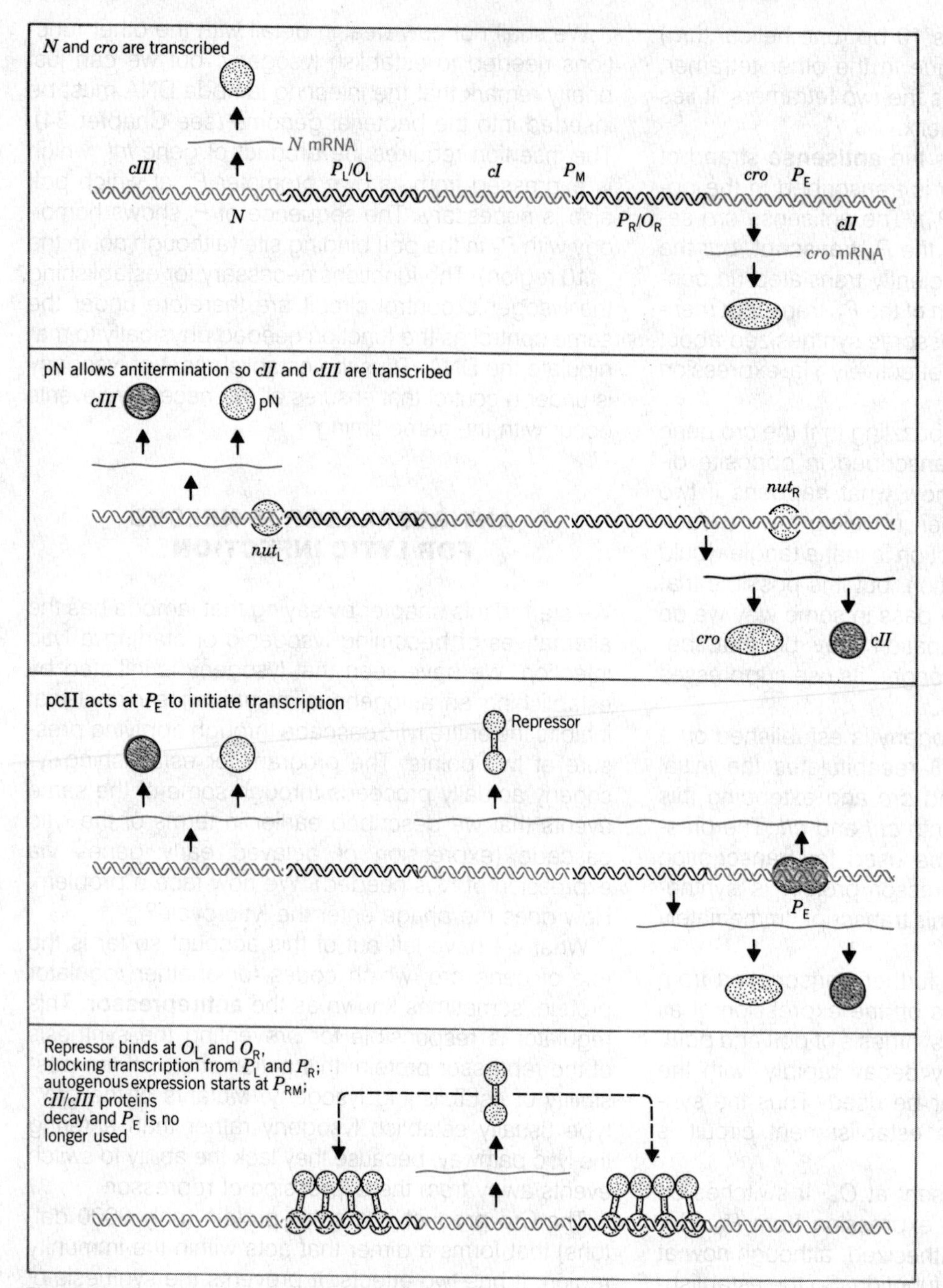

Figure 16.18

A cascade is needed to establish lysogeny, but then this circuit is switched off and replaced by the autogenous repressor-maintenance circuit.

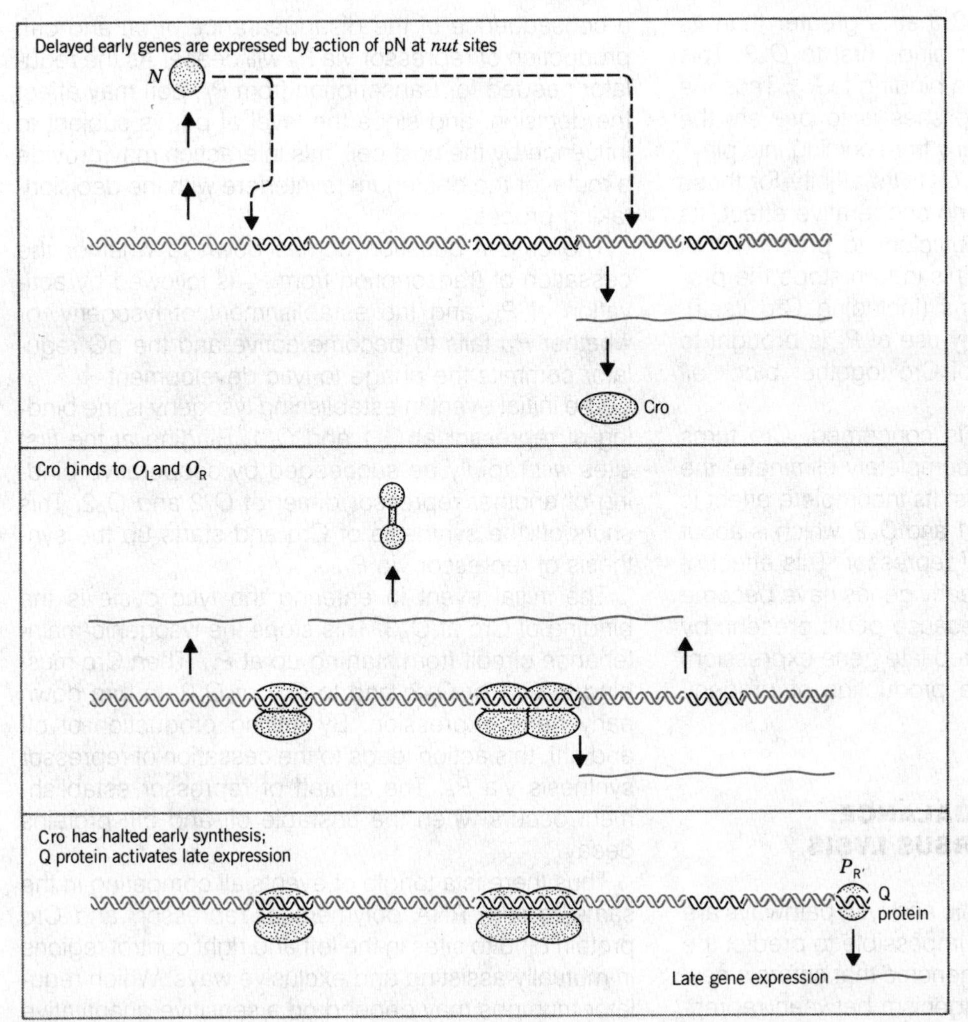

Figure 16.19
The lytic cascade requires Cro protein, which directly prevents repressor maintenance via P_M, as well as turning off delayed early gene expression, indirectly preventing repressor establishment via P_E.

pressor and for turning down the expression of the early genes.

Cro achieves its function by binding to the operators. It binds symmetrically at these sites, making contacts similar to those made by repressor. In fact, Cro includes a region with the same general structure as the repressor; an α-helix-2 is offset at an angle from α-helix-3. The sequences of Cro and repressor in this region are partially homologous, explaining their ability to contact the same DNA sequences. However, Cro contacts only one face of DNA; it lacks the N-terminal arms by which repressor reaches around to the other side.

How can two proteins have the same sites of action, yet have such opposite effects? The answer lies in the different affinities that each protein has for the individual binding sites within the operators. Let us just consider O_R, where more is known, and where Cro exerts both its effects. The series of events is illustrated in **Figure 16.19.**

Cro has an affinity for O_R3 that is greater than its affinity for O_R2 or O_R1. So it binds first to O_R3. This inhibits RNA polymerase from binding to P_M. Thus the first action that Cro accomplishes is to prevent the maintenance circuit for lysogeny from coming into play.

Then Cro binds to O_R2 or O_R1. Its affinity for these sites is similar, and there is no cooperative effect. Its presence at either site is sufficient to prevent RNA polymerase from using P_R. This in turn stops the production of the early functions (including Cro itself). Because pcII is unstable, any use of P_E is brought to a halt. So the two actions of Cro together block *all* production of repressor.

So far as the lytic cycle is concerned, Cro turns down (although it does not completely eliminate) the expression of the early genes. Its incomplete effect is explained by its affinity for O_R1 and O_R2, which is about eight times lower than that of repressor. This effect of Cro does not occur until the early genes have become more or less superfluous, because pQ is present; by this time, the phage has started late gene expression, and is concentrating on the production of progeny phage particles.

A DELICATE BALANCE: LYSOGENY VERSUS LYSIS

The programs for the lysogenic and lytic pathways are so intimately related that it is impossible to predict the fate of any individual phage genome that enters a new host bacterium. Will the antagonism between repressor and Cro be resolved by establishing the autogenous maintenance circuit shown in Figure 16.18, or by turning off repressor synthesis and entering the late stage of development shown in Figure 16.19? The same pathway is followed in both cases right up to the brink of decision. Both involve the expression of the immediate early genes and extension into the delayed early genes. The difference between them comes down to the question of whether repressor or Cro will obtain occupancy of the two operators.

The early phase during which the decision is taken is limited in duration in either case. No matter which pathway the phage follows, expression of all early genes will be prevented as P_L and P_R are repressed; and, as a consequence of the disappearance of cII and cIII, production of repressor via P_E will cease. As the regulator needed for transcription from P_E, pcII may affect the decision; and since the level of pcII is subject to influence by the host cell, this interaction may provide a route for the bacterium to interfere with the decision-taking process.

The critical question comes down to whether the cessation of transcription from P_E is followed by activation of P_M and the establishment of lysogeny, or whether P_M fails to become active and the pQ regulator commits the phage to lytic development.

The initial event in establishing lysogeny is the binding of repressor at O_L1 and O_R1. Binding at the first sites will rapidly be succeeded by cooperative binding of another repressor dimer at O_L2 and O_R2. This shuts off the synthesis of Cro and starts up the synthesis of repressor via P_M.

The initial event in entering the lytic cycle is the binding of Cro at O_R3. This stops the lysogenic-maintenance circuit from starting up at P_M. Then Cro must bind to O_R1 or O_R2, and to O_L1 or O_L2, to turn down early gene expression. By halting production of cII and cIII, this action leads to the cessation of repressor synthesis via P_E. The shutoff of repressor establishment occurs when the unstable cII and cIII proteins decay.

Thus there is a tangle of events all competing in the same region. RNA polymerase, repressor, and Cro protein bind to sites in the left and right control regions in mutually assisting and exclusive ways. Which regulator triumphs may depend on a sensitive quantitative and temporal balance. Fluctuations in the timing and frequencies of transcription and translation in individual cells may determine which pathway is followed in each case.

FURTHER READING

The basis for this view of lambda is analyzed in detail in **Lewin's** *Gene Expression*, **3,** *Plasmids and Phages* (John Wiley, New York, 1978, pp. 274–535), which also deals with phage T4 (pp. 536–681), phages T3 and T7 (pp. 682–723), and other phages. An early view of the rationale underlying lambda repressor function was laid out by **Ptashne** (pp. 325–343) in *The Operon* (Miller & Reznikoff, Eds., Cold Spring Harbor Laboratory, New York, 1978), and the interplay between repressor and Cro has been reviewed by **Ptashne et al.** (*Cell* **19,** 1–11, 1980). Current progress on lambda has been summarized in *Lambda II*, edited by Hendrix, Roberts, Stahl & Weisberg (Cold Spring Harbor Laboratory, New York, 1983). Chapters dealing with the life cycle include **Friedman & Gottesman** (pp. 43–51) on lytic development, **Wulff & Rosenberg** (pp. 53–73) on repressor establishment, and **Gussin et al.** (pp. 93–121) on repressor and antirepressor structure and function. The two-helix model for DNA-binding proteins has been reviewed by **Takeda et al.** (*Science* **221,** 1020–1026. 1983).

PART 5
CONSTITUTION OF THE EUKARYOTIC GENOME

The present day genetics concept visualizes the appearance of an organism as a result of an interaction of the whole set of genes the organism possesses and the environment in which it develops. A change in any of the genes, called a mutation, is liable to upset the balance of that system and show up on the organism as a character, usually as an abnormality, in some respect poorer than the wild type. . . Studies with deficiencies, *viz.*, material where one or several genes are missing, show that the majority of deficiencies are lethal to the organism when present in a homozygous condition. This suggests that the presence of at least the majority of genes is essential in order that an organism may live. Moreover, the work with *D. melanogaster* deficiencies indicates that many of them are cell-lethal, *viz.*, that even a small group of cells located among normal tissues but containing a homozygous deficiency cannot exist. This suggests that genes are active in every cell and that, probably, the majority of them perform there a function highly important in the vital processes of the cell.

Milislav Demerec, 1935

CHAPTER 17
THE EXTRAORDINARY POWER OF DNA TECHNOLOGY

The technology for dealing with DNA has advanced dramatically in the past few years. We can subject DNA to a remarkable range of manipulations, amongst which are procedures for reconstructing DNA molecules by joining sequences from quite different sources. The product is often described as **recombinant DNA** and the techniques (more colloquially) as **genetic engineering.** They are equally applicable to prokaryotes and eukaryotes, although the power of this approach is especially evident with eukaryotic genomes. With these techniques, we can isolate and characterize genes that otherwise would not be accessible.

At the heart of these techniques lies the ability of restriction enzymes to cleave DNA at particular, rather short nucleotide sequences. We have already discussed in Chapter 5 the use of restriction enzymes to make physical maps of DNA and to generate fragments whose nucleotide sequences can be determined directly. Thus when the DNA corresponding to a particular gene is available, it is now a relatively routine, not to say mundane, task to determine its sequence. Using such information, together with several chemical and biochemical techniques, we can break and rejoin DNA molecules at virtually any desired site.

In this chapter, we shall be concerned with the first steps in characterizing particular genes. How do we isolate the RNA (from the cytoplasm) or the DNA (from the genome) that represents a particular gene? Having identified this nucleic acid, how do we proceed to obtain a sufficient amount of material to characterize? Can we direct the isolated coding sequence to synthesize its product *in vitro* or *in vivo*? Can we introduce changes in this sequence that will influence its expression and help reveal the nature of regulatory signals?

ANY DNA SEQUENCE CAN BE CLONED IN BACTERIA

The **cloning** of a fragment of DNA allows indefinite amounts to be produced from even a single original molecule. (A **clone** is defined as a large number of cells or molecules all identical with an original ancestral cell or molecule.) Cloning of DNA is made possible by the ability of bacterial plasmids and phages to continue their usual life-style after additional sequences of DNA have been incorporated into their genomes.

An insertion generates a **hybrid** or **chimeric** plasmid or phage, consisting in part of the authentic DNA of the original genome and in part of the additional "foreign" sequences. These chimeric elements replicate in bacteria just like the original plasmid or phage and so can be obtained in large amounts. Copies of the original foreign fragment then can be retrieved from the progeny. Since (with some rare exceptions) the properties of the chimeric species are unaffected by the particular foreign sequences that are involved, virtually any sequence of DNA can be cloned in this way. Because the phage or plasmid is used to "carry" the foreign DNA as an inert part of the genome, it is often referred to as the **cloning vector.**

A critical feature of any cloning vector is that it should possess a site at which foreign DNA can be inserted without disrupting any essential function. The simplest approach is to use a restriction enzyme that makes only a single break in the vector DNA. The insertion procedure generates only a small proportion of chimeric genomes from the starting material, so it is important to have some means of *selecting* the chimeric genome from the original vector.

Plasmids are small autonomous elements that replicate in bacteria independently of the bacterial chromosome (see Chapter 31). A plasmid genome always consists of a circular duplex of DNA, with a replication control system that maintains it in the bacterium at a characteristic level.

There are two general types of control system. **Single-copy** plasmids are maintained at 1 plasmid DNA per host chromosome. **Multicopy** plasmids are present in greater amounts, typically about 10–20 genomes per cell. Some plasmids are under **relaxed replication control,** which means that they accumulate in very large amounts (~1000 genomes per cell) when the bacteria stop growing. They are often used to provide cloning vectors, because of the high yield of material.

A single cleavage in a circular plasmid DNA converts the genome into a linear molecule, as depicted in **Figure 17.1.** Then the two ends can be joined to the ends of a linear foreign DNA, regenerating a circular chimeric plasmid. There is no formal limit on the length of foreign DNA that can be inserted. The chimeric plasmid can be perpetuated indefinitely in bacteria. It may be isolated by virtue of its size or circularity—for example, by gel electrophoresis.

Many plasmids carry genes that specify resistance to antibiotics. This feature is useful in designing cloning systems. A common procedure is to use a plasmid that has two resistance genes, each specifying resistance to a different antibiotic. One of the genes is used just to identify bacteria that carry the plasmid.

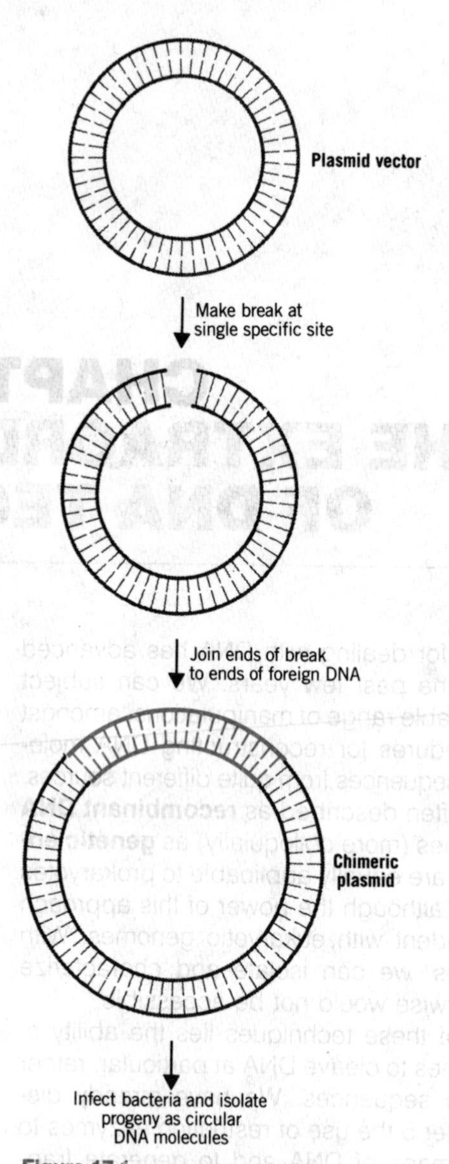

Figure 17.1
Plasmid vectors can be used to clone any fragment of DNA that is inserted at an appropriate site.

The other is used to distinguish chimeric plasmids from parental vectors. If the site used to insert foreign DNA lies within this second gene, the chimeric plasmid *loses* the antibiotic resistance. Thus a parental vector can be selected by its resistance to both antibiotics; whereas a chimeric plasmid can be selected by its retention of resistance to one antibiotic, but sensitivity to the other.

Often natural plasmids have some, but not all, of the desired properties of a cloning vector. Several series of developments have been undertaken to engineer better vectors from the natural starting material. This may involve the introduction of changes in the replication control system or the addition of genes determining resistance to particular antibiotics. One of the standard cloning vectors most widely used at present is **pBR322,** which was derived by several sequential alterations of earlier cloning vectors. It is a multicopy plasmid that has genes for resistance to tetracycline and ampicillin; several restriction enzymes have unique cleavage sites at useful locations.

Phages provide another type of vector system. Usually the phage is a linear DNA molecule, so that a single restriction break generates two fragments. These are joined together with the foreign DNA to generate a chimeric phage as shown in **Figure 17.2.** Chimeric phage genomes can be conveniently isolated by allowing the phage to proceed through the lytic cycle to produce particles. However, this procedure imposes a limit on the length of foreign DNA that can be cloned, because the capacity of the phage head prevents genomes that are too long from being packaged into progeny particles.

Figure 17.2
Phage vectors can be used to clone a foreign DNA that is inserted into a nonessential region of the genome.

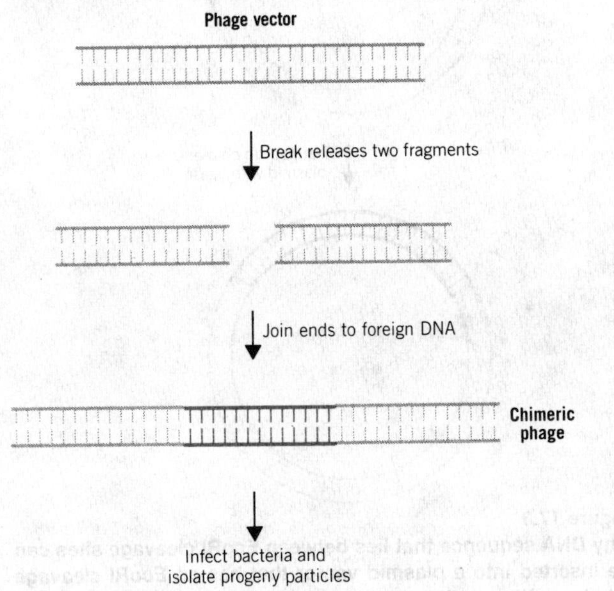

To ameliorate this problem, a fragment of the vector that does not carry any essential phage genes can be *replaced* by the foreign DNA. This approach has been taken to a fine art with phage lambda, where a new vector has been created by manipulating the DNA to produce a shorter genome (lacking nonessential genes) that has only a single restriction site for the enzyme EcoRI. This cloning vector actually is *too short to be packaged into the phage head,* which has a minimum as well as maximum length requirement. Thus it is *necessary* for a foreign DNA fragment to be inserted into the cleaved parental vector in order to generate a phage that can be perpetuated as progeny particles. This demand creates an automatic selective system for obtaining chimeric phage genomes.

The utility of this type of vector has been increased by the development of systems for packaging the DNA into the phage particle *in vitro.* An attempt to combine some of the advantages of plasmids and phages led to the construction of **cosmids.** These are plasmids into which have been inserted the particular DNA sequences (*cos* sites) needed to package lambda DNA into its particle. These vectors still can be perpetuated in bacteria in the plasmid form, but can be purified by packaging *in vitro* into phages. They are still subject to the length limitation imposed by the particle head, but it is not necessary for the genome to carry the genes for lytic development.

We have dealt with cloning vectors in the context of using bacterial hosts. Sometimes it is useful to use a eukaryotic host. The only known authentic eukaryotic plasmids reside in yeast. By reconstruction of DNA, some "dual-purpose" or "shuttle" plasmids have been obtained that have the necessary sequences for surviving in either *E. coli* or *S. cerevisiae.* Thus the one vector can be used with either host.

CONSTRUCTING THE CHIMERIC DNA

To join a foreign DNA fragment to a cloning vector requires a reaction between the ends of the fragment and the vector. This can be accomplished by generating complementary sequences on the fragment and vector, so that they can recombine into a chimeric DNA when mixed together.

The most common method is to use restriction enzymes that make **staggered cuts** to generate short, complementary single-stranded **sticky ends.** The classic example is the enzyme EcoRI, which cleaves each of the two strands of duplex DNA at a different point. These sites lie on either side of a short palindromic sequence that is part of the site recognized by the enzyme.

In a sequence of DNA in which *N* indicates bases not involved in recognition, the EcoRI enzyme cuts its recognition site (highlighted in color) at the bonds indicated by the vertical arrows.

↓
5′...pNpNpNp GpApApTpTpCpNpNpN ...3′
3′... NpNpNpCpTpTpApApGpNpNpNp ...5′
↑

The DNA fragments on either side of the target site fall apart, and because of the stagger of individual cutting sites, they have protruding single-stranded regions that are complementary—the sticky ends indicated by the color.

pApApTpTpCpNpNpN ...3′
GpNpNpNp ...5′

and

5′ ... pNpNpNpG
3′... NpNpNpCpTpTpApAp

Their complementarity allows the sticky ends to anneal by base pairing. When two different molecules are cleaved with EcoRI, the *same* sticky ends are generated on both. This enables one to anneal with the other, as illustrated in the protocol of **Figure 17.3.** The procedure generates a chimeric plasmid that is intact except for the lack of covalent bonds between the vector and the foreign DNA. The missing bonds are replaced by the enzyme DNA ligase *in vitro*. This technique for recombining two DNA molecules has both pros and cons.

An advantage is that the chimeric plasmid possesses regenerated EcoRI sites at either end of the inserted DNA. Thus the foreign DNA fragment can be retrieved rather easily from the cloned copies of the chimeric vector, just by cleaving with EcoRI.

A disadvantage is that any EcoRI sticky end can anneal with any other EcoRI sticky end. Thus some vectors reform by direct reaction between their ends, without gaining an insert, while others may gain an

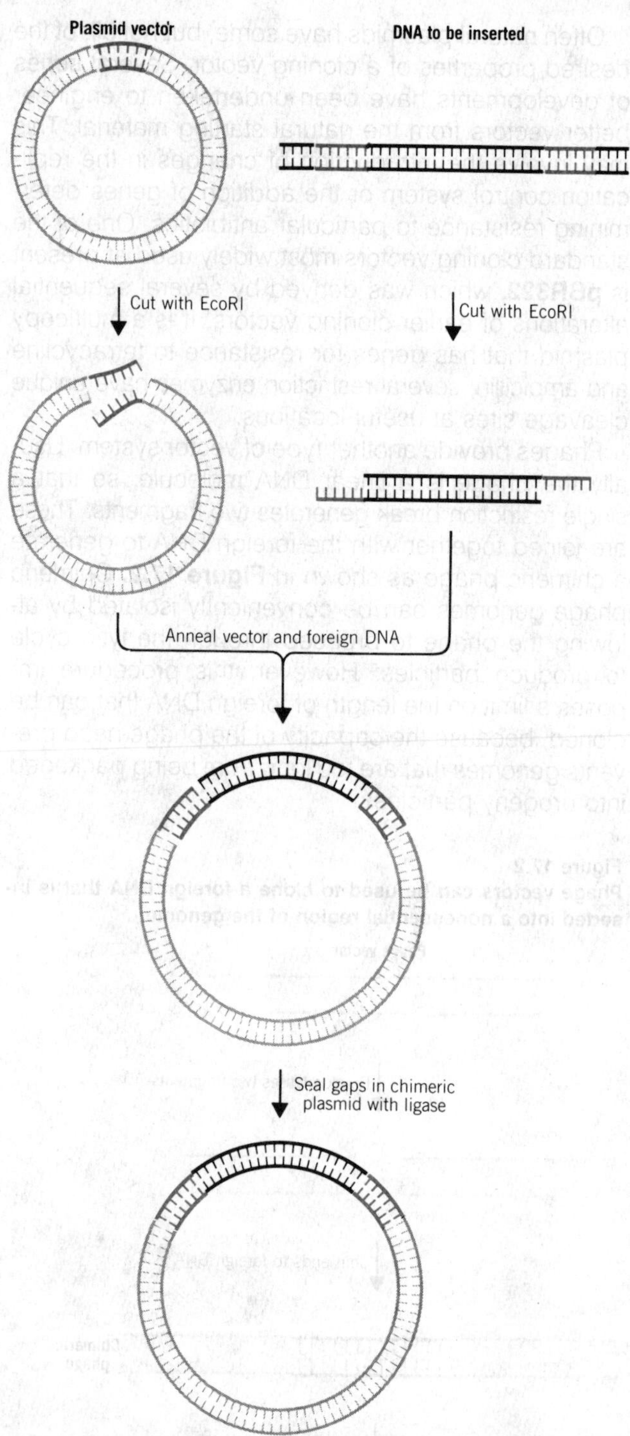

Figure 17.3
Any DNA sequence that lies between EcoRI cleavage sites can be inserted into a plasmid vector that has an EcoRI cleavage site by cutting and annealing the two DNA molecules.

insert of several foreign fragments joined end to end. It is therefore necessary to select chimeric plasmids that have gained only a single insert.

A problem in relying exclusively on restriction enzymes to generate ends for the joining reaction is that their recognition sites may not lie at convenient points in the foreign DNA sequence. Another method allows any DNA end to be used for recombination with a plasmid.

The plasmid is cleaved as before with an enzyme that recognizes a single site in a suitable location, but this need not necessarily generate staggered ends. To both the 3′ ends of the plasmid DNA, poly(dA) is added, by using the enzyme **terminal transferase** with the precursor dATP. In the same way, poly(dT) is added to the 3′ ends of the foreign DNA molecule that is to be inserted; these ends also can have been generated in any convenient manner. As **Figure 17.4** demonstrates, the poly(dA) on the plasmid can anneal *only* with the poly(dT) on the insert fragment. Thus only one reaction is possible: the insertion of a single foreign fragment into the vector.

A drawback of this technique is that it is not so easy to retrieve the inserted fragment from the cloned chimeric plasmid, because the recognition site for the restriction enzyme in the original vector has been abolished by the insertion of the foreign DNA. The inserted sequence is flanked on either side by the poly(dA): poly(dT) paired region. (The technique also can be used with poly(dC) and poly(dG), which has the advantage that, if the plasmid has been cleaved with PstI, the poly(dC):poly(dG) joining regenerates the PstI cleavage site on either side of the insert.)

Another useful technique is **blunt-end ligation,** which relies on the ability of the T4 DNA ligase to join together two DNA molecules that have blunt ends, that is, they lack any protruding single strands. (This reaction is in addition to the usual activity of joining broken bonds within a duplex). When DNA has been cleaved with restriction enzymes that cut across both strands at the same position, blunt-end ligation can be used to join the fragments directly together.

The great advantage of this technique is that any pair of ends may be joined together, irrespective of sequence. This is especially useful when we want to join two defined sequences without introducing any additional material between them. A problem inherent in this technique is that there is no control over which

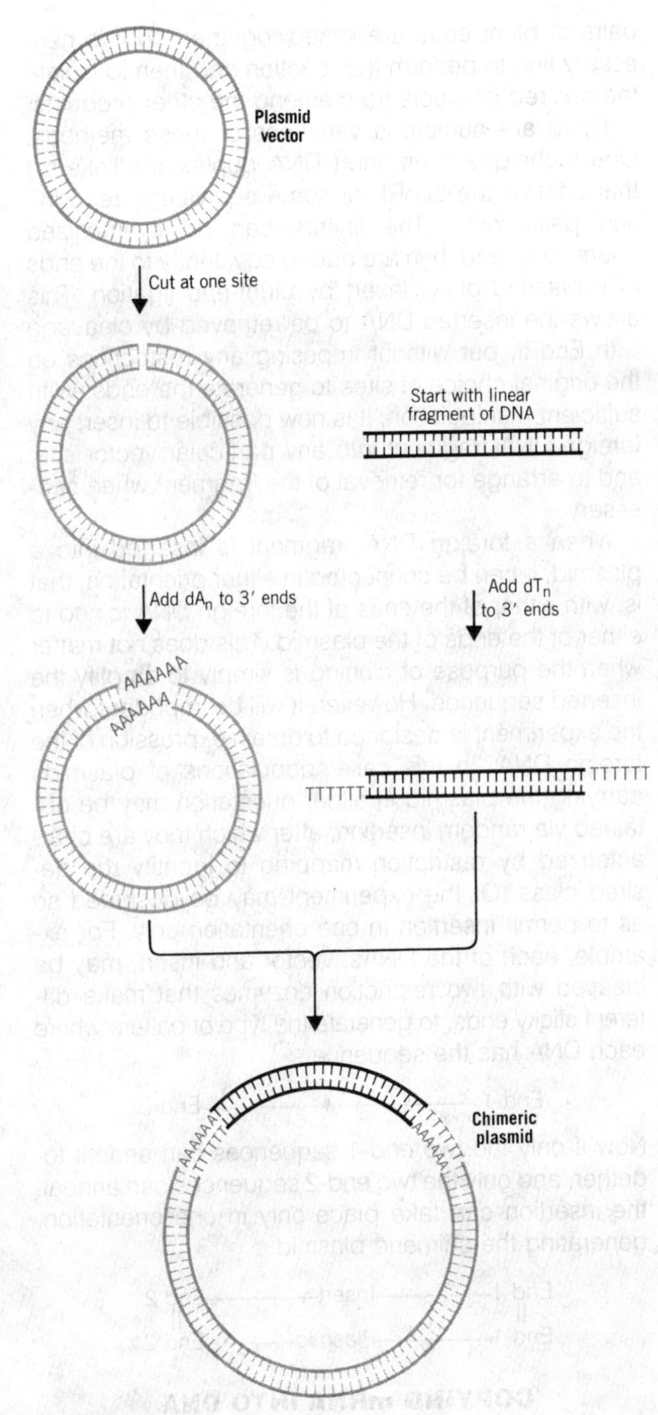

Figure 17.4
The poly(dA:dT) tailing technique allows any two DNA molecules to be joined by adding poly(dA) to the 3′ ends of one and adding poly(dT) to the 3′ ends of the other.

pairs of blunt ends are joined together, so it is necessary first to perform the reaction and then to isolate the desired products from among the other products.

There are numerous variations of these methods. One technique uses short DNA duplexes ("linkers") that contain the EcoRI (or some equivalent) recognition palindrome. The linkers can be synthesized chemically, and then are added covalently to the ends of a plasmid or an insert by blunt-end ligation. This allows the inserted DNA to be retrieved by cleavage with EcoRI, but without imposing any restrictions on the original choice of sites to generate the ends. With sufficient manipulation, it is now possible to insert any foreign DNA fragment into any particular vector site, and to arrange for retrieval of the fragment when necessary.

When a foreign DNA fragment is inserted into a plasmid, it can be connected in either orientation, that is, with either of the ends of the foreign DNA joined to either of the ends of the plasmid. This does not matter when the purpose of cloning is simply to amplify the inserted sequence. However, it will be important when the experiment is designed to obtain expression of the foreign DNA. In this case, populations of plasmids carrying the plasmid in either orientation may be obtained via random insertion, after which they are characterized by restriction mapping to identify the desired class. Or the experiment may be designed so as to permit insertion in one orientation only. For example, each of the DNAs, vector and insert, may be cleaved with two restriction enzymes that make different sticky ends, to generate the type of pattern where each DNA has the sequence

End 1 ———————— End 2

Now if only the two end-1 sequences can anneal together, and only the two end-2 sequences can anneal, the insertion can take place only in one orientation, generating the chimeric plasmid

End 1 ——— Insert ——— End 2
|| ||
End 1 ——— Plasmid ——— End 2

COPYING mRNA INTO DNA

For the purpose of obtaining a DNA sequence that represents a particular protein, the ideal approach is to start with mRNA, which, after all, is the template used to produce the protein *in vivo*. The existence of reverse transcription makes it possible to synthesize a duplex DNA from an mRNA. This is especially easy for mRNAs that carry a poly(A) tail at the 3′ end, as illustrated in **Figure 17.5.**

First, a **primer** is annealed to the poly(dA). It is a short sequence of oligo(dT), whose purpose is to provide a free 3′ end that can be used for extension by the enzyme reverse transcriptase. (Like other enzymes that synthesize DNA, reverse transcriptase cannot initiate formation of a polynucleotide chain without a priming end.) The enzyme engages in the usual 5′–3′ synthesis, adding deoxynucleotides one at a time, as directed by complementary base pairing with the mRNA template.

The product of the reaction is a hybrid molecule, consisting of a template RNA strand base paired with the complementary DNA strand. The only practical problem is the propensity *in vitro* of reverse transcriptase to stop at various points before it has reached the 5′ end of the mRNA. In this case, the resulting reverse transcript falls short of representing the entire

Figure 17.5
mRNA can be copied into double-stranded DNA.

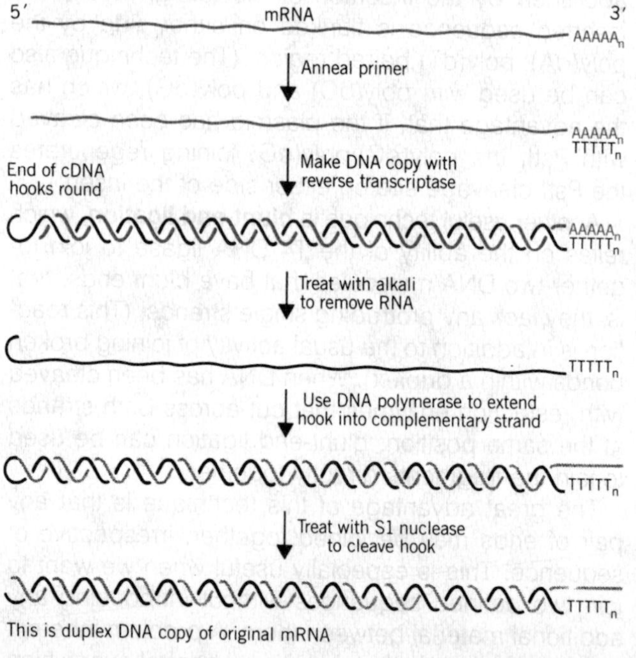

This is duplex DNA copy of original mRNA

mRNA, because it lacks some of the sequences complementary to the 5′ end. However, by judicious adjustment of the experimental conditions, usually it is possible to persuade reverse transcriptase to proceed all the way.

A useful reaction tends to occur at the end of reverse transcription. Upon reaching the end of the mRNA, the enzyme may cause the reverse transcript to "loop back" on itself, by using the last few bases of the reverse transcript as a template for synthesis of a complement. That is, the end of the complementary DNA is used to direct synthesis of a short sequence that is identical with the mRNA, and which displaces it. This creates a short hairpin, usually 10–20 bp long, at this end of the reverse transcript.

At this juncture, the original mRNA is degraded by treatment with alkali (a procedure that does not affect DNA). This produces a single-stranded DNA that is complementary to the mRNA; it is called **cDNA.** The hairpin at its 3′ end provides a natural primer for the next step, the use of the enzyme DNA polymerase I of *E. coli* to convert the single-stranded cDNA into a duplex DNA via synthesis of the complementary strand. In this reaction, the enzyme uses the cDNA as template for synthesis of a sequence identical with the original mRNA. The product is a duplex molecule with a hairpin at one end. The hairpin is cut by the enzyme S1 nuclease (which specifically degrades single-stranded DNA) to generate a conventional DNA duplex.

The duplex DNA can be cloned to generate large amounts of a synthetic gene that represents the mRNA sequence in the form of duplex DNA. This is called a **cDNA clone.** (From this terminology, a somewhat looser use of the term "cDNA" has emerged, being taken to describe the duplex insert and not just the original single-stranded reverse transcript.)

ISOLATING SPECIFIC GENES FROM THE GENOME

One of the principal uses of cloning technology is to isolate individual genes directly from the genome. Any particular gene represents only a very small part of a eukaryotic genome. For example, in a typical mammal, the size of the genome is about 10^9 bp, so that a single gene of (say) 5000 bp represents only 0.00005% of the total nuclear DNA. To identify such a tiny proportion, we need to have a very specific **probe** that reacts *only* with the particular sequence in which we are interested, to pick it out from the vast excess of other sequences. The usual technique is to use a highly labeled radioactive probe of RNA or DNA, whose hybridization with the gene is assayed by autoradiography.

The crux of the matter is to obtain the mRNA that represents a particular protein. This can be quite difficult when the product is rare. There are several techniques for isolating an mRNA via the properties of its product. One is to rely on highly sensitive detection of newly-synthesized proteins after injecting the RNA into a *Xenopus* oocyte. A powerful technique that isolates cDNA complementary to an mRNA is **hybrid-arrested translation.** This identifies a cDNA that can *prevent* translation of an mRNA by hybridizing to it, with the result that the product disappears from an *in vitro* translation system.

Originally, mRNA itself was used as a probe for gene isolation. This suffers from several practical disadvantages. The mRNA is never completely pure and may be hard to obtain in sufficient amounts. In fact, the technique works well enough for genes whose mRNAs are abundant, but cannot be used for genes that are expressed at low levels. Another difficulty is that often the mRNA cannot be obtained with a sufficiently intense radioactive label.

The use of mRNA as a probe has therefore been superseded by the use of a cloned cDNA copy of the mRNA. Large amounts of purified material can be obtained by the techniques we have already described. This material can be labeled *in vitro* to achieve a very high specific radioactivity.

The first step toward identifying the gene corresponding to a particular probe is to break the DNA of the genome into fragments of a manageable size. It is desirable to obtain the gene in as few fragments as possible (ideally only one). Usually the maximum genome lengths that can be handled are in the range of 15–20 kb. Sometimes it is not possible to obtain a gene in the form of a single fragment, and then its structure must be determined by piecing together the information gained from its various fragments.

Two techniques can be used to fragment a genome. One is to make a restriction digest. Then every fragment ends in a site that was recognized by that particular enzyme. An alternative is to shear the DNA mechanically, by introducing breaks in it at a frequency that is controlled by the conditions of shearing. Then the sites of breakage are entirely random.

There are technical difficulties in cloning the fragments generated by mechanical shear; it is easier to insert restriction fragments into the vector. However, one problem is that restriction sites may occur at inconvenient locations—for example, in the middle of a gene that is to be cloned. One way to avoid this is to use more than one restriction enzyme, that is, to repeat the experiment with different enzymes whose recognition sites lie at different locations. But this is time consuming, and when a long sequence is involved, it may be difficult to find an enzyme that does not cleave within it.

When the DNA of an entire genome is digested with a restriction enzyme, the frequency of breakage is controlled by the length of the sequence recognized by the enzyme. The longer the sequence, the less often it occurs by chance. For example, the probability that a particular 4 bp sequence will occur is $0.25^4 = 1/256$, so that an enzyme with this short recognition sequence will cleave DNA rather frequently. The frequency declines to 1/1000 for a 5 bp sequence and to 1/4000 for a 6 bp sequence. (This calculation assumes that each base is equally well represented in DNA, which usually is not the case. The frequency of cutting can be decreased by using an enzyme whose target sequence contains base pairs that are less predominant in the base composition of the DNA, and vice versa.)

To make a useful set of fragments from a restriction digest, a trick is employed to reduce the frequency of cutting. An enzyme with a short (4 bp) recognition sequence is used under conditions that generate a *partial* digest. Any particular target site is cleaved only occasionally, so the presence of a target site in some sequence does not prevent it from being obtained intact. The infrequent cleavage at each site, together with the frequent distribution of sites, means that the fragment distribution approaches an entirely random cleavage of the genome. But each fragment ends in the same sequence, which can contain a sticky end and is therefore useful for cloning.

The distribution of sites recognized by an enzyme becomes a matter of chance taken over the genome as a whole. So a restriction digest of eukaryotic DNA generates a continuum of fragments. When electrophoresed on a gel, these fragments form a smear in which no distinct bands are evident (with the exception of some repeated sequences, which are discussed in Chapter 24). However, when a specific probe is available, it is possible to detect the corresponding sequences in the restriction smear.

DNA fragments cannot be handled directly on an agarose gel. The key feature in the protocol for identifying fragments is the ability to transfer the DNA to a medium on which hybridization reactions can occur. So the DNA is denatured to give single-stranded fragments that are transferred from the agarose gel to a nitrocellulose filter on which they become immobilized. The procedure used to transfer the DNA is somewhat akin to blotting, and this is used colloquially as a description of the procedure. When performed with DNA, it is known as **Southern blotting** (named for the inventor of the procedure).

Figure 17.6 illustrates the protocol for transferring the DNA fragments. The agarose gel is placed on a filter paper that has been soaked in a concentrated salt solution. Then the nitrocellulose filter is placed on the gel, and some dry filter paper is placed in contact with the nitrocellulose. The salt solution is attracted to the dry filter paper. To get there, it must pass through the agarose gel and then through the nitrocellulose

Figure 17.6
Southern blotting transfers DNA fragments from an agarose gel to a nitrocellulose filter.

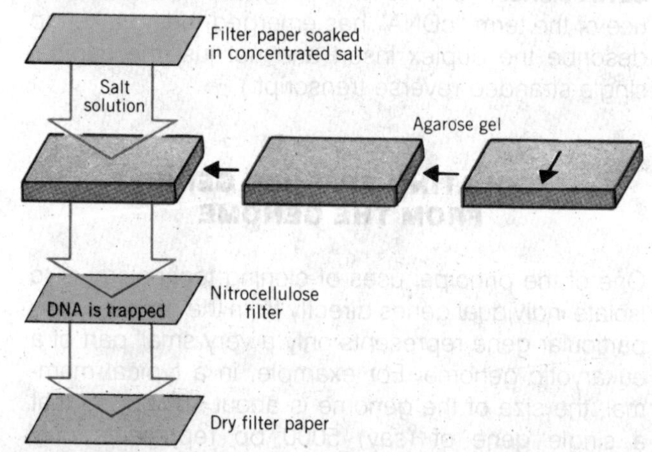

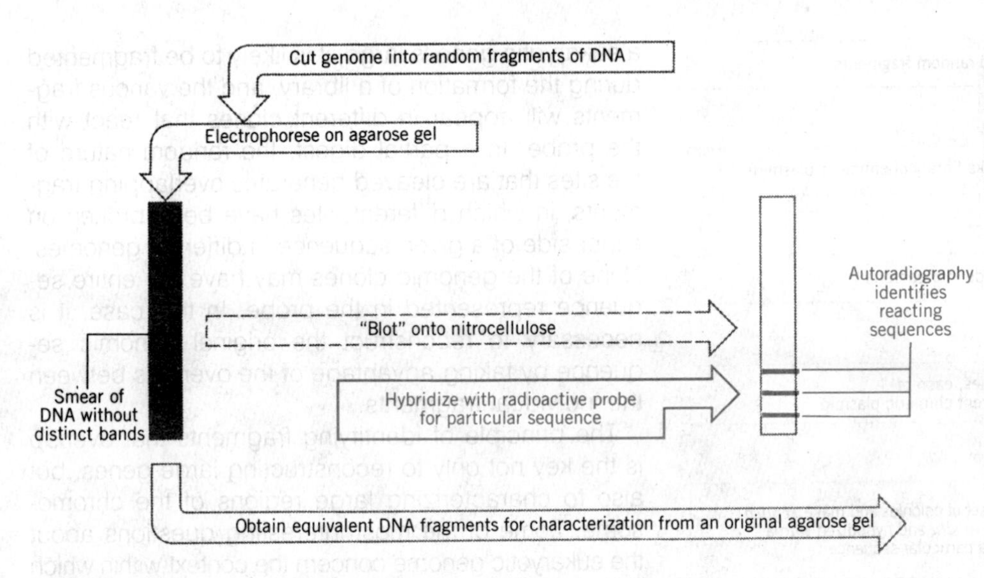

Figure 17.7
Southern blotting allows DNA fragments corresponding to a particular probe to be isolated directly from a digest of the DNA of the genome.

filter. The DNA is carried along with it, but becomes trapped in the nitrocellulose.

Figure 17.7 shows that when the DNA has been immobilized on the nitrocellulose, it can be hybridized *in situ* with a radioactive probe. Only those fragments complementary to a particular probe will hybridize with it. Because the probe is radioactive, the hybridization can be visualized by autoradiography. Each complementary sequence gives rise to a labeled band at a position determined by the size of the DNA fragment.

The technique can also be performed with RNA. To blot RNA from agarose onto a medium suitable for hybridization, some changes in the technique are necessary. The procedure then is known as **Northern blotting.**

WALKING ALONG THE CHROMOSOME

There are practical difficulties in directly isolating genomic fragments corresponding to a probe, and a powerful technique for isolating a particular gene is to reverse the order of events by cloning the genome first. Then clones containing a particular sequence are selected. Vectors carrying DNA derived from the genome itself are called **genomic** or **chromosomal DNA clones** (in distinction to the cDNA clones that are representations of the mRNA).

Cloning an entire genome (as opposed to specific fragments) is often called a **shotgun experiment.** It is done by breaking the whole genome into fragments of a manageable size. Then the fragments are inserted into a cloning vector to generate a population of chimeric vectors. A set of cloned fragments of this sort is called a **genome library.** Once a library has been obtained with either a phage or plasmid vector (more often a phage, because it is easier to store the necessary large numbers of chimeric DNAs), it can be perpetuated indefinitely, and is readily retrieved whenever a new probe is available to seek out some particular fragment.

The number of random fragments that must be cloned to ensure a high probability that *every* sequence of the genome is represented in at least one chimeric plasmid decreases with the fragment size and increases with the genome size and the desired probability. For a probability level of 99%, 1500 cloned fragments are needed with *E. coli,* but the necessary library size increases to 4600 with yeast, 48,000 with *D. melanogaster,* and 800,000 with mammals. Libraries of cloned fragments reaching this probability level have been established in all these cases.

How is a particular genomic clone to be selected from the library? The technique of **colony hybridization** is illustrated in **Figure 17.8.** Bacterial colonies carrying chimeric vectors are lysed on nitrocellulose filters. Then their DNA is denatured *in situ* and fixed on the filter. The filter is hybridized with a radioactively

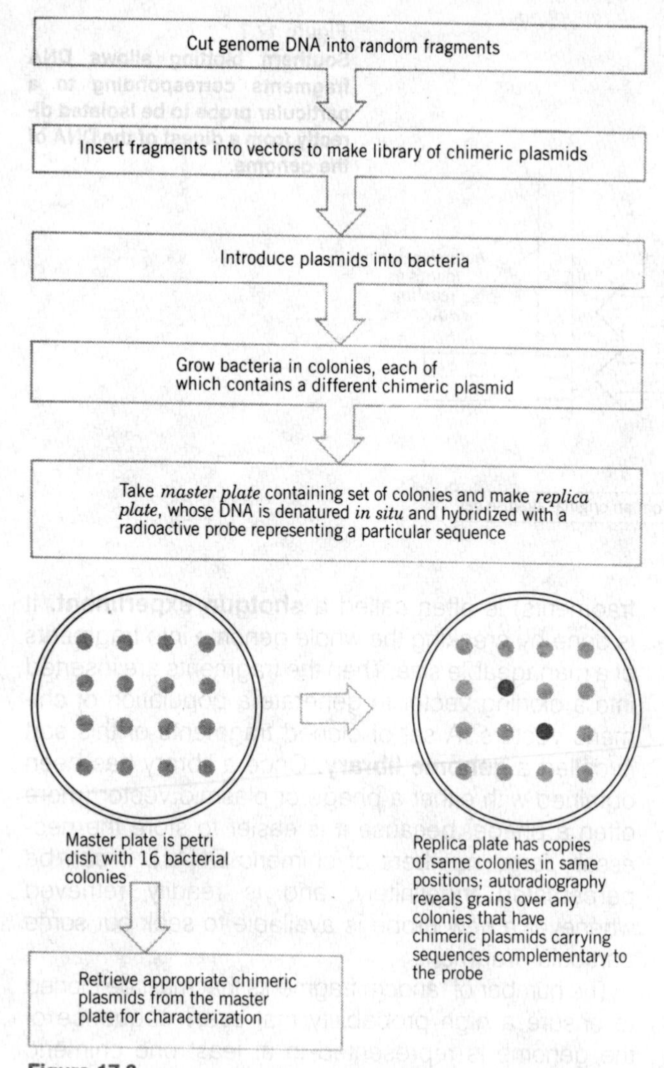

Figure 17.8

Colony hybridization allows a chimeric plasmid carrying a particular sequence to be selected by its complementarity with a radioactively labeled probe.

labeled probe that represents the desired sequence (usually a cloned cDNA). Any colonies in which it occurs are visualized as dark spots by autoradiography. Then the corresponding chimeric vectors can be recovered from the original reference library.

To react with a probe, a genomic clone needs only to carry some of the probe's sequence (usually the minimum is reckoned to be about 50 bp). Indeed, when a eukaryotic gene is large, it is likely to be fragmented during the formation of a library, and the various fragments will appear in different clones that react with the probe. In a partial digest, the random nature of the sites that are cleaved generates overlapping fragments, in which different sites have been broken on either side of a given sequence in different genomes. None of the genomic clones may have the entire sequence represented in the probe. In this case, it is necessary to reconstruct the original genomic sequence by taking advantage of the overlaps between the individual fragments.

The principle of identifying fragments that overlap is the key not only to reconstructing large genes, but also to characterizing large regions of the chromosome. Some of the most interesting questions about the eukaryotic genome concern the context within which a gene resides and its relationship to adjacent genes. In characterizing a region beyond a gene itself, we are venturing into unknown territory in the sense that there are no probes derived from gene products. Each successive fragment of the genome is isolated purely by virtue of its relationship with another, partially overlapping fragment.

The technique of **chromosome walking** is illustrated in **Figure 17.9.** We start with a clone that may have been isolated because it contains a known gene or because we know (from genetic mapping) that it lies near some region of interest. Because the fragment may be quite large, the process is made more controllable by using a subfragment from one end of the first clone to isolate the next set. Then we identify other chimeric vectors in our library that overlap with this clone. These new vectors will extend on one side or the other of the fragment carried in the first clone. We can determine the direction of extension by making a restriction map of each fragment. The process can be repeated indefinitely.

It is possible in principle to walk for hundreds of kb, and regions of the genome approaching 1000 kb have been mapped in this way. Some technical problems are involved whenever a clone contains sequences that are not unique, but which are also present elsewhere in the genome, because then the next set of overlapping clones may react with this other region as well as the region in which we are interested. How-

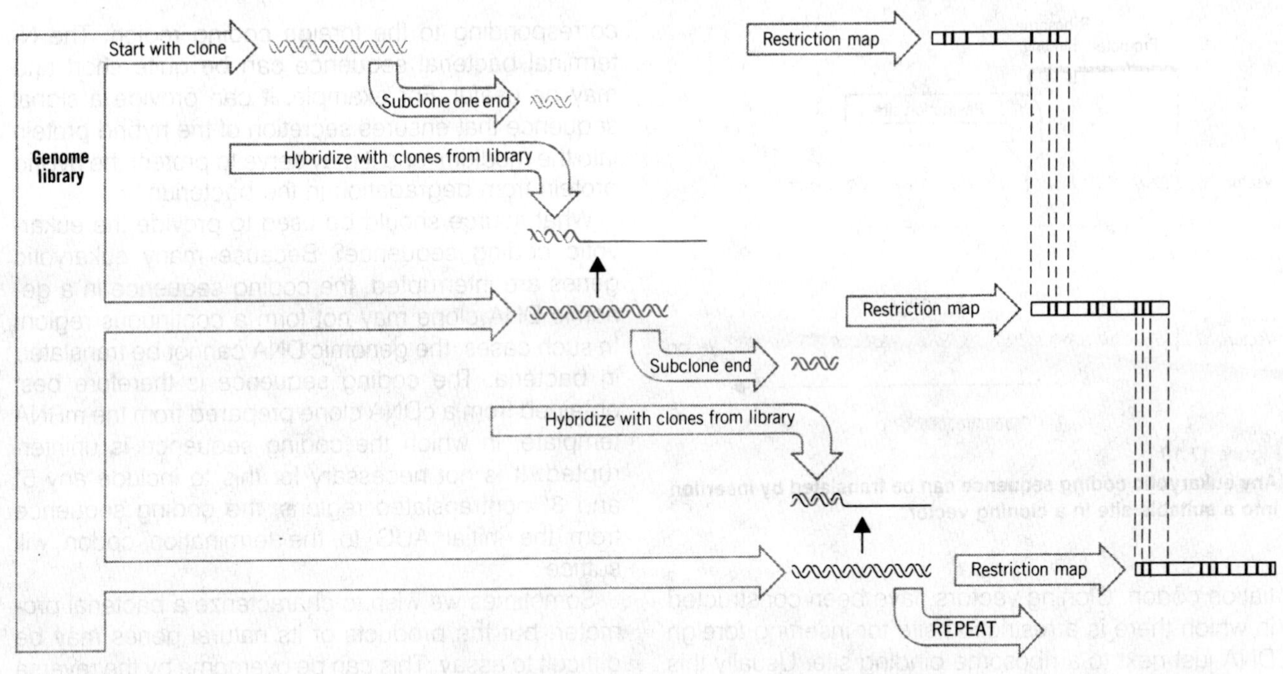

Figure 17.9
Chromosome walking is accomplished by successive hybridizations between overlapping genomic clones.

ever, it is usually possible to walk of the order of 100 kb a month, making it practical to characterize quite large regions.

A critical assumption underlying the entire cloning approach is that a cloned eukaryotic sequence will be perpetuated with complete fidelity in bacteria. This is borne out by the data available. There are some exceptional cases in which deletions have occurred during passage through the bacterial host.

EUKARYOTIC GENES CAN BE TRANSLATED IN BACTERIA

Because the genetic code is universal, a given coding sequence should have the same meaning in all situations. (The sole exception lies with mitochondrial genes, where there are differences in the genetic code, as described in Chapter 6.) Thus when an intact coding sequence for a eukaryotic protein is carried on a chimeric vector, it should be possible to transcribe the sequence into an mRNA that can be translated within the bacterial host. The only caveats are that modifications made to the protein in its natural habitat may not occur, and, of course, there is always the risk that the polypeptide chain may be unstable in the bacterial environment. However, in an appropriate environment, it should be possible to translate any eukaryotic coding sequence into the corresponding polypeptide.

What conditions must be fulfilled to obtain effective expression of eukaryotic genes in bacteria? Because eukaryotic and prokaryotic promoters differ in sequence, the eukaryotic gene must be placed under the control of a bacterial promoter. For efficient expression, a eukaryotic insert is placed adjacent to a promoter from which it is transcribed as a monocistronic mRNA. A promoter in the plasmid may be used or a bona fide bacterial promoter may be inserted into the vector for the purpose of expressing the foreign gene.

To be translated, the mRNA must have a proper ribosome binding site in juxtaposition to the AUG ini-

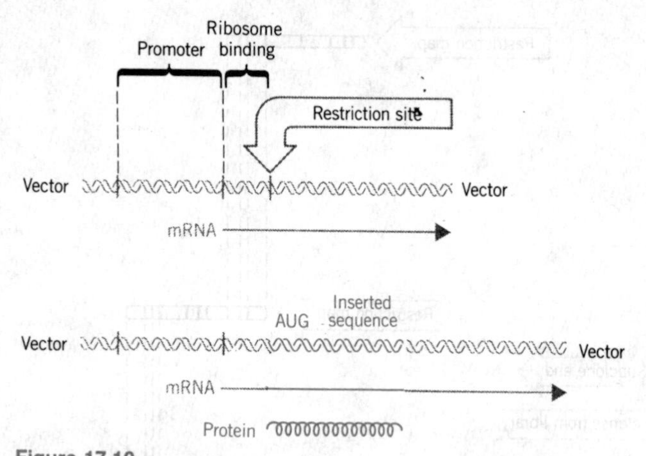

Figure 17.10

Any eukaryotic coding sequence can be translated by insertion into a suitable site in a cloning vector.

tiation codon. Cloning vectors have been constructed in which there is a restriction site for inserting foreign DNA just next to a ribosome binding site. Usually this is arranged so that the inserted DNA contributes the AUG codon. **Figure 17.10** shows that any foreign DNA sequence (prokaryotic or eukaryotic) starting with an AUG codon can be placed at this site and will be transcribed and translated in bacteria. This produces a protein that corresponds precisely to the inserted coding region.

Another technique is to insert a coding sequence shortly after the start of a bona fide protein. This produces a hybrid protein, with an N-terminal region consisting of the bacterial protein whose sequence has been interrupted, and the remainder of the protein corresponding to the foreign coding region. The N-terminal bacterial sequence can be quite short and may be useful. For example, it can provide a signal sequence that ensures secretion of the hybrid protein into the medium; and it may serve to protect the hybrid protein from degradation in the bacterium.

What source should be used to provide the eukaryotic coding sequence? Because many eukaryotic genes are interrupted, the coding sequence in a genomic DNA clone may not form a continuous region. In such cases, the genomic DNA cannot be translated in bacteria. The coding sequence is therefore best obtained from a cDNA clone prepared from the mRNA template, in which the coding sequence is uninterrupted. It is not necessary for this to include any 5′ and 3′ nontranslated regions; the coding sequence from the initial AUG to the termination codon will suffice.

Sometimes we wish to characterize a bacterial promoter, but the products of its natural genes may be difficult to assay. This can be overcome by the reverse type of manipulation, in which the readily assayable *lac* genes are connected to the promoter. Then its activity can be followed by the standard assays for β-galactosidase. A vector that accomplishes this is a variant of phage Mu, called the **Mud phage.**

FURTHER READING

The topics of this chapter are dealt with in detail in **Lewin's** *Gene Expression,* **2,** *Eucaryotic Chromosomes* (Wiley, New York, 1980, pp. 761–785).

CHAPTER 18
EUKARYOTIC GENOMES: A CONTINUUM OF SEQUENCES

The "eukaryotic genome" should be a somewhat nebulous concept. The essence of being eukaryotic is that the major part of the genome is sequestered in the nucleus, where it is safeguarded by the nuclear membrane from exposure to the cytoplasm. At a gross level, the total amount of nuclear DNA varies enormously, the number of chromosomes into which it is segregated is extremely variable, the relatively minuscule part of the genome contained in organelles also shows wide characteristic variations, and there can be major differences in the types of sequences that constitute the nuclear genome.

Yet certain features of eukaryotic genomes are absent from prokaryotic genomes. The integrity of individual genes can be interrupted, there can be multiple and (sometimes) identical copies of particular sequences, and there may be large blocks of DNA that do not code for protein. Because of the division between nucleus and cytoplasm, the arrangements for gene expression must necessarily be different in eukaryotes from those in prokaryotes.

Do these features represent any underlying uniformity in the organization of eukaryotic DNA? We should remember that the eukaryotic kingdom is extremely broad, and at present we have detailed information about the genetic organization of only a few types of species. In describing the characteristics of eukaryotic DNA, we are dealing with features that may be represented to widely varying degrees in different individual genomes. We are therefore concerned with establishing the options available for the organization of eukaryotic DNA, rather than with defining a hypothetical "typical" pattern.

THE C-VALUE PARADOX DESCRIBES VARIATIONS IN GENOME SIZE

The total amount of DNA in the (haploid) genome is a characteristic of each living species known as its **C value.** It may be measured chemically in picograms (pg) of DNA or can be determined by the technique of DNA reassociation (see later), when it is generally given in base pairs or daltons. The equivalence is 1 pg = 0.965×10^9 bp = 6.1×10^{11} daltons. There is enormous variation in the range of C values, from as little as a mere 10^4 bp for a mycoplasma to as much as 10^{11} bp for some plants and amphibians.

Figure 18.1 summarizes the range of C values found in different evolutionary phyla. There is some increase in the minimum genome size found in each phylum as the complexity increases.

Unicellular eukaryotes (whose life-styles can somewhat resemble the prokaryotic) get by with genomes

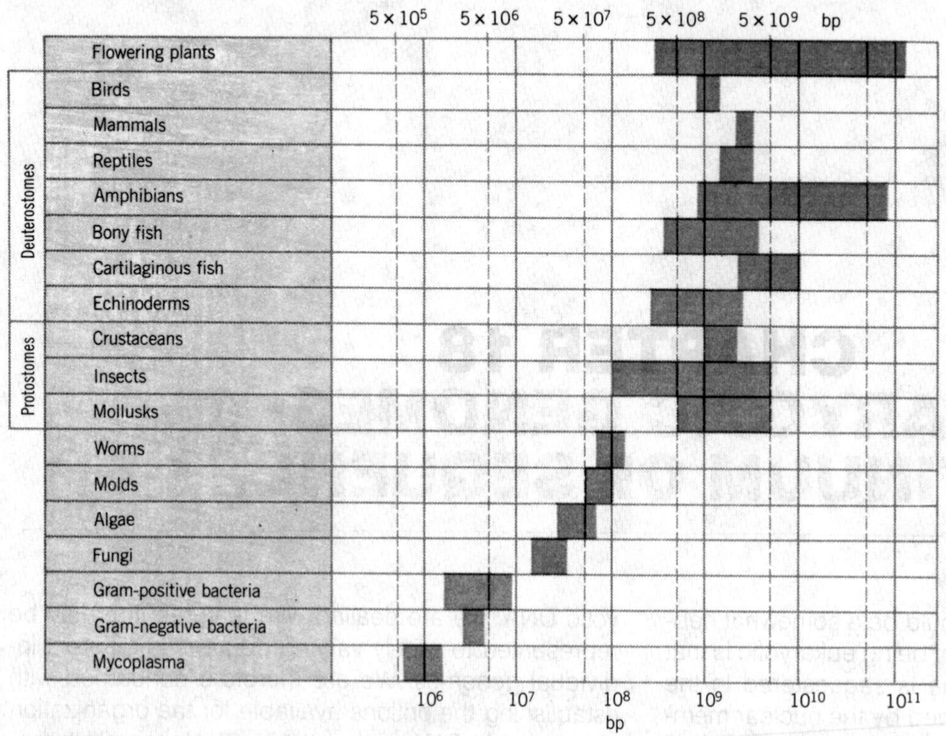

Figure 18.1
DNA content of the haploid genome is not closely related to the morphological complexity of the species. Genome size does increase through the prokaryotes and the protostomes, but extends over a wide range for the deuterostomes. The range of DNA values within a phylum is indicated by the shaded area.

hat are small, although larger than those of the bacteria. Being eukaryotic per se does not imply a vast increase in genome size; for example, the yeast *S. cerevisiae* has a genome size of about 2.3×10^7 bp, only five times greater than that of the bacterium *E. coli*.

A further modest increase in genome size (just over twofold) is adequate to support the slime mold *D. discoideum,* able to live in either unicellular or multicellular modes. Another increase in complexity is necessary to produce the first fully multicellular organisms. For example, the nematode worm *C. elegans* has a DNA content of 8×10^7 bp. Then any close relationship between complexity of the organism and content of DNA becomes obscure, although it is necessary to have a genome of $>10^8$ bp to make an insect, $>4 \times 10^8$ bp to assemble an echinoderm, $>8 \times 10^8$ bp to produce a bird or amphibian, and $>2 \times 10^9$ bp to develop a mammal.

In some cases, the spread of genome sizes is quite small. For example, birds, reptiles, and mammals all show little variation within the phylum, with a range of genome sizes in each case about twofold. But in other cases there is quite a wide range of values, often more than tenfold.

The **C value paradox** takes its name from our inability to account for the content of the genome in terms of known functions.

One puzzling feature is the existence of huge variations in C values between certain species whose apparent complexity does not vary correspondingly. An extraordinary range of C values is found in amphibians, where the smallest genomes are just below 10^9 bp while the largest are almost 10^{11} bp. It is hard to believe that this could reflect a 100-fold variation in the number of genes needed to specify different amphibians.

To reinforce this skepticism, in some cases rather closely related species show surprising variations in total genome size. For example, two amphibian species whose overall morphologies are very similar may have a difference of, say, 10 in their relative amounts of DNA. Again, it seems unlikely that there could be a tenfold difference in the gene number. Yet if the gene number is roughly similar, most of the DNA in the species with the larger genome cannot be concerned with coding for protein: what can be its function?

This leads us to another aspect of the C-value paradox: the apparent absolute excess of DNA compared with the amount that could be expected to code for proteins. Indeed, this is often referred to as the problem of excess eukaryotic DNA. We now know that some of the excess is accounted for because genes are much larger than the sequences needed to code for proteins (principally because of the intervening sequences that may break up a coding region into different segments). However, this does not seem sufficient to resolve the problem. Let us consider a couple of examples.

Even if we suppose that the average mammalian gene is as large as 5000–8000 bp (which actually is longer than most known interrupted genes), the number of genes in the mammalian genome would be between 400,000 and 600,000. Is this plausible?

Probably not: for although gene numbers are not known with much precision, it seems likely that they are to be counted in tens of thousands rather than in hundreds of thousands.

Some direct evidence is gained by counting the number of genes via estimates of the number of different mRNAs (see Chapter 19). While this cannot be done for every cell type (to generate a sum total for the organism), it seems that the number of genes expressed in a given cell type is roughly about 10,000. Most of these are common to most or all cells of the organism. So this value is probably within a factor of (say) 2–4 of the total expressed gene number. Given some uncertainties about estimating the numbers of genes present in multiple copies, we might say that the mammalian genome looks to be of the order of 30,000–40,000 gene functions.

A less direct line of evidence is provided by attempts to estimate the number of essential genes by identifying all the loci that can be mutated. Much of this work has been performed with *D. melanogaster*, where we can infer that the total essential gene number is likely to be about 5000. With a reasonable estimate for the size of the insect gene as 2000 bp, this corresponds to a total length of 10^7 bp, just 10 times less than the amount of available DNA.

Of course, the genes identified by mutation are those in which damage has visible or lethal effects. Perhaps only some genes fall into this class. This would imply that at least a large proportion, possibly even the majority, of genes are concerned with specifying proteins that are not essential for the survival of the organism (at least in the sense that mutational damage does not cause any detectable effect).

So we are left with several key questions. What proportion of the DNA of the genome actually is concerned with representing proteins in the sense that it lies in a gene, either in the coding region itself or in an intervening or transcribed flanking sequence? Of the number of genes, how many are essential and how many dispensable? What is the function (if any) of DNA that does not reside in genes? What effect does a large change in total size have on the operation of the genome, as in the case of the related amphibians?

REASSOCIATION KINETICS DEPEND ON SEQUENCE COMPLEXITY

The reassociation of two complementary sequences of DNA occurs by base pairing in a reversal of the process of denaturation by which they were separated (see Figure 4.9). We have already discussed many experiments that rely on the extension of this technique to isolate individual DNA or RNA sequences by their ability to hybridize with a particular probe. Now we shall see that the kinetics of the reassociation reaction reflect the variety of sequences that are present; and the reaction can be used to quantitate genes and their RNA products. When performed in solution, such reactions are described as **liquid hybridization.**

Renaturation of DNA depends on random collision of the complementary strands; it therefore follows second-order kinetics. This means that the rate of reaction is governed by the equation

$$\frac{dC}{dt} = -kC^2 \quad (1)$$

where C is the concentration of DNA that is single-stranded at time *t*, and *k* is a reassociation rate constant.

By integrating this equation between the limits of the initial concentration of DNA, C_0 at time $t = 0$, and the concentration C that remains single-stranded after time *t*, we can describe the progress of the reaction as

$$\frac{C}{C_0} = \frac{1}{1 + k.C_0t_{1/2}} \quad (2)$$

Thus when the reaction is half complete, at time $t_{1/2}$,

$$\frac{C}{C_0} = \frac{1}{2} = \frac{1}{1 + k.C_0t_{1/2}} \quad (3)$$

so that

$$C_0t_{1/2} = \frac{1}{k} \quad (4)$$

The reassociation of any particular DNA can be described by the rate constant *k* (in units of liter-nucleotide-moles^{-1}-sec^{-1}), or in the form of its reciprocal $C_0t_{1/2}$ (given in nucleotide-moles x sec/liter).

This equation shows that the parameter controlling the reassociation reaction is the product of DNA concentration (C_0) and time of incubation (*t*). This is usually described simply as the **Cot.** Similarly, the value required for half-reassociation is called the **Cot$_{1/2}$**. Since the Cot$_{1/2}$ is the product of the concentration and time required to proceed halfway, a greater Cot$_{1/2}$ implies a slower reaction.

The reassociation of DNA usually is followed in the form of a **Cot curve,** which plots the fraction of DNA that has reassociated ($1 - C/C_0$) against the log of the Cot. **Figure 18.2** gives Cot curves for several genomes. The form of each curve is similar, with renaturation occurring over about a 100-fold range of Cot values between the points of 10% reaction and 90% reaction. But the Cot required in each case is very different. It is described by the Cot$_{1/2}$ (the Cot value at 50% reassociation).

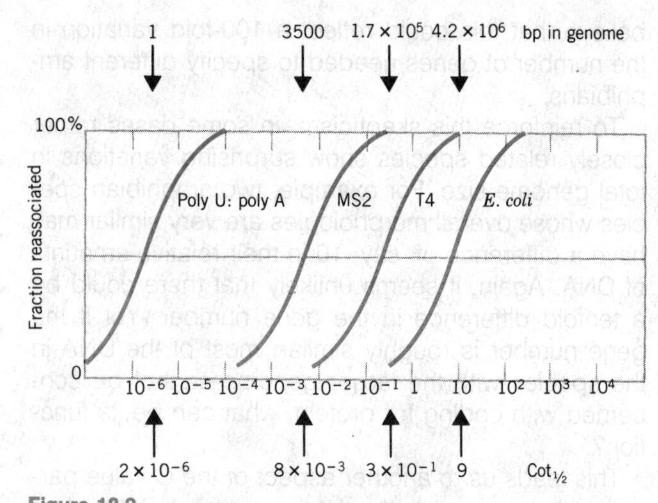

Figure 18.2
Rate of reassociation is inversely proportional to the length of the reassociating DNA.

The Cot$_{1/2}$ is directly related to the amount of DNA in the genome. This reflects a situation in which, as the genome becomes more complex, there are fewer copies of any particular sequence within a given mass of DNA. For example, if the C_0 of DNA is 12 pg, it will contain 3000 copies of each sequence in a bacterial genome whose size is 0.004 pg, but will contain only 4 copies of each sequence present in a eukaryotic genome of size 3 pg. Thus the same *absolute* concentration of DNA measured in moles of nucleotides per liter (that is, the C_0) will provide a concentration of each eukaryotic sequence that is 750 times (3000/4) lower than that of each bacterial sequence.

Since the rate of reassociation depends on the concentration of complementary sequences, for the eukaryotic sequences to be present at the same *relative* concentration as the bacterial, it is necessary to have 750 times more DNA (or to incubate the same amount of DNA for 750 times longer). Thus the Cot$_{1/2}$ of the eukaryotic reaction is 750 times the Cot$_{1/2}$ of the bacterial reaction.

The Cot$_{1/2}$ of a reaction therefore indicates the *total length of different sequences* that are present. This is described as the **complexity.** It is usually given in base pairs, but can be expressed in daltons or any other mass unit.

The renaturation of the DNA of any genome (or part of a genome) should display a Cot$_{1/2}$ that is proportional to its complexity. Thus the complexity of any DNA can be determined by comparing its $Cot_{1/2}$ with that of a standard DNA of known complexity. Usually *E. coli* DNA is used as a standard. Its complexity is taken to be identical with the length of the genome (implying that every sequence in the *E. coli* genome of 4.2×10^6 bp is unique). Thus we can write

$$\frac{Cot_{1/2}\ \text{(DNA of any genome)}}{Cot_{1/2}\ (\textit{E. coli}\ \text{DNA})} = \frac{\text{Complexity of any genome}}{4.2 \times 10^6\ \text{bp}} \qquad (5)$$

EUKARYOTIC GENOMES CONTAIN SEVERAL SEQUENCE COMPONENTS

When the DNA of a eukaryotic genome is characterized by reassociation kinetics, usually the reaction occurs over a range of Cot values spanning up to eight orders of magnitude. This is much broader than the 100-fold range expected from equation **2** and shown for the examples of Figure 18.2. The reason is that the equation applies to a single **kinetically pure** reassociating component. A genome actually may have several of these components, each reassociating with its own characteristic kinetics.

Figure 18.3 shows the reassociation of a (hypothetical) eukaryotic genome, starting at a Cot of 10^{-4} and terminating at a Cot of 10^4. The reaction falls into three distinct phases, outlined by the shaded boxes. A plateau separates the first two phases, but the second and third overlap slightly. Each of these phases represents a different kinetic component of the genome.

The first fraction to reassociate is called the **fast component.** In this case, it represents 25% of the total DNA, renaturing between Cot values of 10^{-4} and about 2×10^{-2}, with a $Cot_{1/2}$ value of 0.0013.

The next fraction is called the **intermediate component.** This represents 30% of the DNA. It renatures between Cot values of about 0.2 and 100, with a $Cot_{1/2}$ value of 1.9.

The last fraction to renature is called the **slow component.** This is 45% of the total DNA; it extends over a Cot range from <100 to about 10,000, with a $Cot_{1/2}$ of 630.

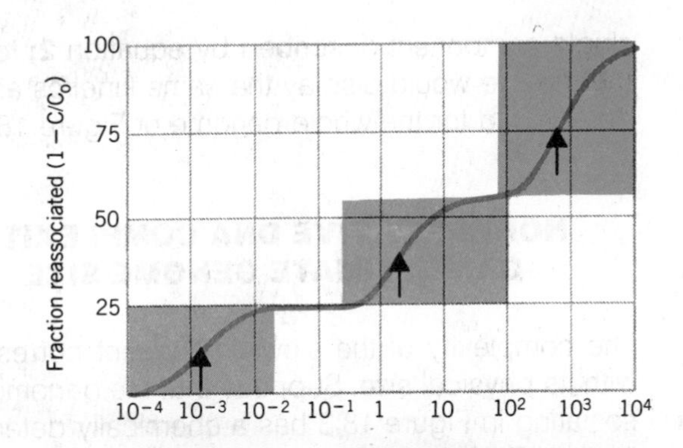

	Fast Component	Intermediate Component	Slow Component
Percent of genome	25	30	45
$Cot_{1/2}$	0.0013	1.9	630
Complexity, bp	340	6.0×10^5	3.0×10^8
Repetition frequency	500,000	350	1

Figure 18.3
The reassociation kinetics of eukaryotic DNA show three types of component (indicated by the shaded areas).

To calculate the complexities of these fractions, each must be treated as an independent kinetic component whose reassociation is compared with a standard DNA. The slow component represents 45% of the total DNA, so its concentration in the reassociation reaction is 0.45 of the measured C_0 (which refers to the total amount of DNA included). Thus the $Cot_{1/2}$ applying to the slow fraction alone is $0.45 \times 630 = 283$.

So if the slow DNA were isolated as a pure component free of the other fractions, it would renature with a $Cot_{1/2}$ of 283. Suppose that under these conditions, *E. coli* DNA reassociates with a $Cot_{1/2}$ of 4.0. Comparing these two values, we see from equation **5** that the complexity of this fraction is 3.0×10^8 bp. Treating the other components in a similar way shows that the intermediate component has a complexity of 6×10^5 bp, and the fast component has a complexity of only 340 bp. This provides a quantitative basis for our statement that the faster a component reassociates, the lower its complexity.

Reversing the argument, if we took three DNA preparations, each containing a unique sequence of the appropriate length (340 bp, 6×10^5 bp, and 3×10^8 bp, respectively) and mixed them in the proportions 25:30:45, each would renature as though it were a single component described by equation **2;** together the mixture would display the same kinetics as those determined for the whole genome of Figure 18.3.

NONREPETITIVE DNA COMPLEXITY CAN ESTIMATE GENOME SIZE

The complexity of the slow component corresponds with its physical size. Suppose that the genome reassociating in Figure 18.3 has a chemically determined haploid DNA content of 7.0×10^8 bp. Then 45% of it is 3.15×10^8 bp, which is only marginally greater than the value of 3.0×10^8 bp measured by the kinetics of reassociation. In fact, given the errors of measurement inherent in both techniques, we can say that the complexity of the slow component is the same whether measured chemically or kinetically. The two values are referred to as the **chemical complexity** and **kinetic complexity.**

The coincidence of these values means that the slow component comprises sequences that are unique in the genome: on denaturation, *each single-stranded sequence is able to renature only with the corresponding complementary sequence.* This part of the genome is the sole component of prokaryotic DNA (as witnessed by the examples of Figure 18.2 and implied by the use of *E. coli* DNA as a standard); and it is usually a major component in eukaryotes. It is called **nonrepetitive DNA.**

We can use the kinetic complexity of nonrepetitive DNA to estimate the complexity of the genome. This just reverses the calculation that we performed earlier. For the example of Figure 18.3, the complexity of nonrepetitive DNA is 3.0×10^8 bp. If this fraction is unique and represents 45% of the genome, the whole genome should have a size of $3.0 \times 10^8 \div 0.45 = 6.6 \times 10^8$ bp. This provides an independent assessment for genome size that we can compare with the chemical complexity, which in this case is 7.0×10^8 bp.

Virtually every eukaryotic genome has a nonrepetitive DNA component. The sole exceptions are some plants that were generated by a recent polyploidization, where as a result there is more than one copy of every sequence.

Figure 18.4 plots the relationship between genome size, as determined by reassociation kinetics of nonrepetitive DNA, and the haploid DNA content, as determined by chemical analysis. The good agreement verifies the underlying assumption that the nonrepetitive DNA component consists of individual sequences present in only one copy per genome.

This is an important point in light of the C value paradox. The presence of nonrepetitive DNA implies that larger genomes are not generated simply by increasing the number of copies of the same sequences present in smaller genomes. If this were the case, larger genomes would behave as though polyploid, so there would be no nonrepetitive DNA. The reassociation data therefore exclude models that would explain the C value paradox in terms of simple increases in the number of copies of each gene present in the haploid genome. This means that we must account for differences in genome size on the basis that larger genomes genuinely contain greater diversity of sequences.

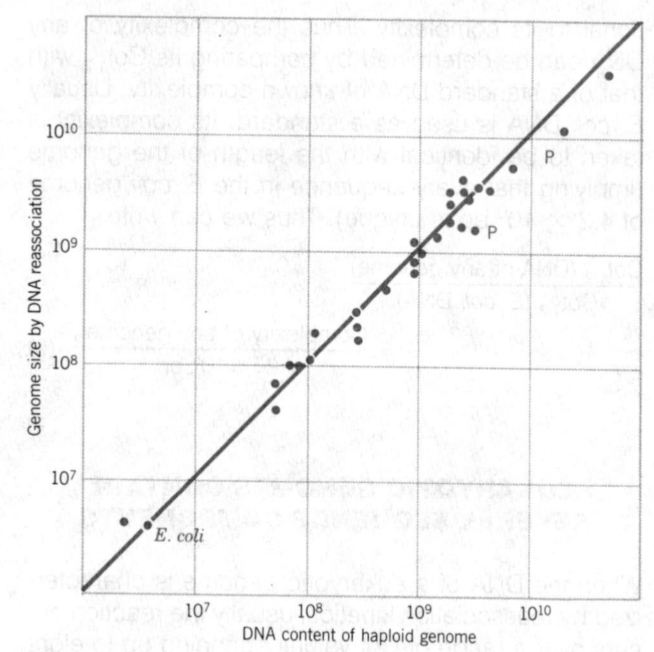

Figure 18.4
There is a good correlation between the kinetic complexity and chemical complexity of eukaryotic genomes, with the exception of polyploid genomes (indicated by *P*).

EUKARYOTIC GENOMES CONTAIN REPETITIVE SEQUENCES

What is the nature of the components that renature more rapidly than the nonrepetitive (slow) DNA? In the example of Figure 18.3, the intermediate component occupies 30% of the genome, which, according to the chemical complexity, should correspond to a total amount of material of $0.3 \times 7 \times 10^8 = 2.1 \times 10^8$ bp. But the kinetic complexity of this component is only 6×10^5 bp.

The unique length of DNA that corresponds to the $Cot_{1/2}$ for reassociation is much shorter than the total length of the DNA chemically occupied by this component in the genome. In other words, the intermediate component behaves as though consisting of a sequence of 6×10^5 bp that is present in 350 copies in every genome (because $350 \times 6 \times 10^5 = 2.1 \times 10^8$). Following denaturation, *the single strands generated from any one of these copies are able to renature with their complements from any one of the 350 copies*. This effectively raises the concentration of reacting sequences in the reassociation reaction, explaining why the component renatures at a lower $Cot_{1/2}$.

Sequences that are present in more than one copy in each genome are described as **repetitive DNA.** The number of copies present per genome is called the **repetition frequency (*f*).** Formally, the repetition frequency is defined by the ratio

$$f = \frac{\text{Chemical complexity}}{\text{Kinetic complexity.}} \qquad (6)$$

Together, the complexity and repetition frequency describe the properties of the sequence components of the genome. For example, if a given genome consists of 1 sequence that is *A* bp long, 10 copies of a sequence that is *B* bp long, and 100 copies of a sequence that is *C* bp long, the complexity is $A + B + C$. The repetition frequencies of sequences *A, B,* and *C* are 1, 10, and 100, respectively. Sequence *A* is nonrepetitive; sequences *B* and *C* are repetitive.

The repetitive DNA fraction includes all components whose repetition frequency is significantly greater than 1. For practical purposes, repetition frequencies between 1 and 2 are taken to indicate sequences that belong in the nonrepetitive component. (It is impossible for a sequence to be represented a nonintegral number of times in the haploid genome!) In the example of Figure 18.3, the application of equation **6** to the nonrepetitive DNA gives a value of $3.15 \times 10^8 / 3.0 \times 10^8 = 1.05$. The usual range of values for nonrepetitive DNA in real genomes is from about 0.8 to 1.6.

There may be several components in the repetitive DNA of any genome, but two general types are distinguished by the range of Cot values within which reassociation occurs. These correspond approximately to the intermediate and fast components of Figure 18.3. The intermediate fraction, reassociating usually in a range between a Cot of 10^{-2} and that of nonrepetitive DNA, is said to consist of **moderately repetitive DNA.** The fast fraction, reassociating before a Cot of 10^{-2} is reached, is said to consist of **highly repetitive DNA.**

Highly repetitive DNA takes its name from the very large number of copies of the basic reassociating sequence present. Generally this sequence is rather short. In the example of Figure 18.3, the fast component consists of a sequence only 340 bp long, present in 500,000 copies per genome. Because of the short length of the reassociating unit, sometimes this is also called **simple sequence DNA.**

Rather than proceed through the series of calculations we have described, the repetition frequency of any component in a reassociation curve can be determined directly by comparison with the nonrepetitive DNA component. The inverse ratio of their $Cot_{1/2}$ values gives the relationship between the repetition frequencies of any two components in the same reassociation curve. So if we assume that nonrepetitive DNA really is unique (its *f* is defined to be 1.0), we can write an alternative equation for the frequency of any repetitive component.

$$f = \frac{Cot_{1/2} \text{ of nonrepetitive DNA component}}{Cot_{1/2} \text{ of repetitive DNA component}} \qquad (7)$$

MODERATELY REPETITIVE DNA CONSISTS OF MANY DIFFERENT SEQUENCES

The behavior of a repetitive DNA component represents only an average that is useful for describing its sequences. The relevant parameters do not necessarily represent the properties of any particular sequence. The moderately repetitive component of Figure 18.3 includes a total length of 6×10^5 bp of DNA, repeated about 350 times per genome. But this does not correspond to a single, identifiable, continuous length of DNA. Instead, it is made up of a variety of individual sequences, each much shorter, whose total length together comes to 6×10^5 bp. These individual sequences may be dispersed about the genome. Their average repetition is 350, but some will be present in more copies than this and some in less.

When a eukaryotic genome is analyzed by reassociation kinetics, the individual sequence components are rarely so well separated as shown in Figure 18.3. In fact, they often overlap too extensively to be distinguished by eye. An example is drawn in **Figure 18.5.** The genome of the sea urchin *S. purpuratus* renatures over eight orders of magnitude, and points of inflection can be seen that distinguish the various components, but the proportion and $Cot_{1/2}$ of each is not immediately evident.

To identify individual components, a computer program is used to fit individual curves (for hypothetical kinetically pure components) to each region of the reassociation curve. Together the individual curves overlap to give a sum that corresponds to the behavior actually observed for the genome. As we noted earlier, if DNA preparations of the appropriate complexities were mixed in the specified proportions, they would give the same curve as that obtained experimentally.

An example of such a resolution is shown in Figure 18.5, where the three individual curves represent components occupying 19%, 27%, and 50% of the total DNA (in order of reassociation). In this case, the first two components both would be classified as moderately repetitive DNA, with $Cot_{1/2}$ values of 0.53 and 8.3; they correspond to components of complexity 1.0×10^6 bp repeated 160 times, and of complexity 2.3×10^7 repeated 10 times. The third component in the curve is nonrepetitive DNA.

The usual approach to making computer fits is to seek the minimum number of individual kinetic components whose sum gives a good enough approximation to the total curve. But this need not necessarily reflect the biological construction of the genome. Often it is possible to achieve an equally good resolution with a greater number of components. For example, the data shown in Figure 18.5 can be interpreted just as well with three moderately repetitive components, occupying 10%, 27%, and 25% of the genome, with repetition frequencies of 8000, 240, and 19. This analysis also reduces the size of the nonrepetitive DNA component.

The real situation may therefore comprise a continuum of repetitive components reassociating over a range from about 10 times to more than 20,000 times that of the nonrepetitive component. The kinetic components introduced to solve the curve do not correspond to discrete repetitive fractions of the genome, but are just useful approximations for descriptive purposes.

Figure 18.5

Several components can be discerned in the reassociation kinetics of *S. purpuratus* DNA.

The experimental curve is drawn through the actual data points; the curves without data points show the three minimum ideal components that could account for the experimental curve.

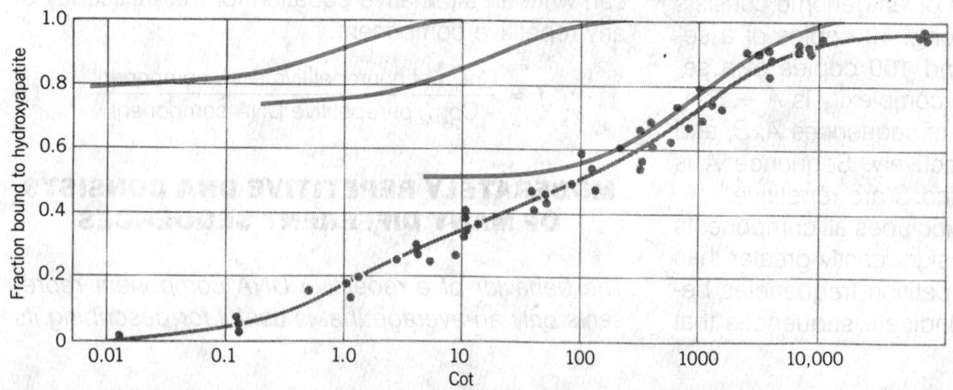

The proportion of the genome occupied by repetitive DNA varies widely. For lower eukaryotes, most of the DNA is nonrepetitive, with only 10–20% falling into one or more moderately repetitive components. In animal cells, up to half of the DNA often is occupied by moderately and highly repetitive components. In plants and amphibians, the repetitive proportion may be even greater, sometimes representing the majority of the genome, with the moderately and highly repetitive components accounting for up to 80%. The length of the nonrepetitive DNA component tends to increase with overall genome size, but is not tightly related to it.

This offers a modest relief from the C value paradox, in the sense that the range of complexities of nonrepetitive DNA is less than the range of genome sizes. If we assume that genes are contained within nonrepetitive DNA, we still have to explain rather a substantial variation in complexities that does not seem to accord with the morphological features of the organism, but the range is reduced. For example, a fruit fly has a nonrepetitive DNA complexity of 1×10^8 bp and a housefly has one of 2.8×10^8 bp, so the difference is only threefold, compared with a discrepancy in genome sizes of sixfold. Seen in these terms, the mammalian nonrepetitive DNA complexity tends to be about 9×10^8 bp, the frog *X. laevis* is 1.6×10^9 bp, and plants and amphibians extend to about 9×10^9 bp.

Finally, we should emphasize that no single feature characterizes all eukaryotic genomes. We have mentioned that plants that have become polyploid very recently may have no nonrepetitive DNA, so that the slowest reassociating fraction has a repetition frequency of 2–3 or more. (True, this is probably a temporary stage in evolution, which usually may lead to divergence to regenerate nonrepetitive DNA.) In crab genomes, there may be no moderately repetitive DNA at all—just highly repetitive and nonrepetitive DNA. In lower eukaryotes, there may be no highly repetitive DNA.

MEMBERS OF REPETITIVE SEQUENCE FAMILIES ARE RELATED BUT NOT IDENTICAL

The different components of eukaryotic DNA can be isolated by separating the duplex molecules according to the Cot needed for their renaturation. **Figure 18.6** illustrates the procedure. The renaturation reaction is allowed to proceed only to a particular Cot value. At this point, the DNA that has renatured is separated by virtue of its double-stranded structure; the remainder remains single-stranded. Usually the duplex and single-stranded molecules are separated by using hydroxyapatite columns, which preferentially retain duplex DNA. The double-stranded preparation represents more rapidly reassociating material (repetitive sequences); the single-stranded preparation represents more slowly reassociating material (nonrepetitive DNA and, depending on the chosen Cot value, some of the repetitive sequences).

When nonrepetitive DNA is renatured, it forms duplex material that behaves very much like the original preparation of DNA before its denaturation. When subjected to a denaturation procedure, the duplex molecules melt sharply at a T_m only very slightly below that of the original native DNA. This shows that strand

Figure 18.6
Different kinetic components of the genome can be isolated by reassociation to an intermediate Cot level.

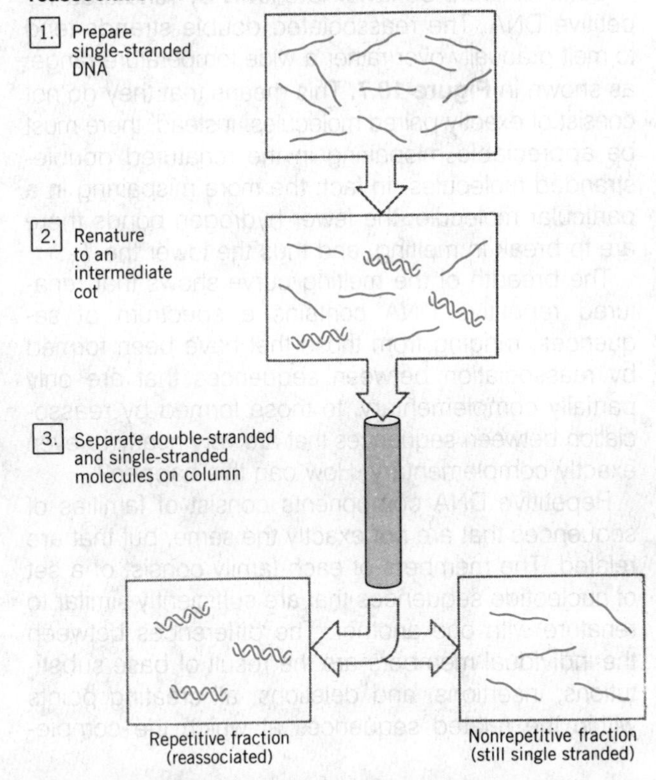

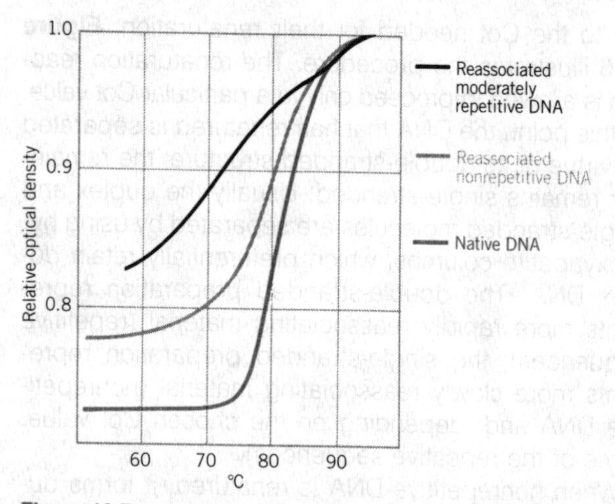

Figure 18.7
The denaturation of reassociated nonrepetitive DNA takes place over a narrow temperature range close to that of native DNA, but reassociated moderately repetitive DNA melts over a wide temperature range.

reassociation has been accurate: each unique sequence has annealed with its exact complement.

Quite different behavior is shown by renatured repetitive DNA. The reassociated double strands tend to melt gradually over rather a wide temperature range, as shown in **Figure 18.7.** This means that they do not consist of exactly paired molecules. Instead, there must be appreciable mispairing in the renatured double-stranded molecules. In fact, the more mispairing in a particular molecule, the fewer hydrogen bonds there are to break in melting, and thus the lower the T_m.

The breadth of the melting curve shows that renatured repetitive DNA contains a spectrum of sequences, ranging from those that have been formed by reassociation between sequences that are only partially complementary, to those formed by reassociation between sequences that are very nearly or even exactly complementary. How can this happen?

Repetitive DNA components consist of families of sequences that are not exactly the same, but that are related. The members of each family consist of a set of nucleotide sequences that are sufficiently similar to renature with one another. The differences between the individual members are the result of base substitutions, insertions, and deletions, all creating points within the related sequences at which the complementary strands cannot base pair. The proportion of these changes establishes the relationship between any two sequences. When two closely related members of the family renature, they form a duplex with high T_m; when two more distantly related members associate, they form a duplex with a lower T_m.

The ability of related but not identical complementary sequences to recognize each other can be controlled by the **stringency** of the conditions imposed for reassociation. A higher stringency is imposed by (for example) an increase in temperature, which requires the formation of a greater number of base pairs in a given length of DNA. So by performing the hybridization reaction at high temperatures, reassociation may be restricted to rather closely related members of a family; at lower temperatures, more distantly related members may anneal. Note also that, as the stringency of hybridization increases, the proportion of the genome characterized as nonrepetitive also increases.

Analysis of the members of repetitive families shows that there is wide variety in their construction. In some cases, well-defined families can be identified, whose members are quite closely related. Such families remain intact even when the stringency of hybridization is increased. In other cases, there is a continuum of variously related sequences, so that the size of the family decreases continuously as the stringency is increased. The measured size of such repetitive families is arbitrary, since it is determined by the particular hybridization conditions that are used. These results tell us that eukaryotic genomes contain sequences related to one another, most likely derived by divergence from some common ancestral sequence that was multiplied, but the extent of such relationships varies enormously.

The reassociation of two sequences that are related but not identical occurs more slowly than the reaction between identical sequences. This means that a greater Cot is required for the reassociation of related sequences. So the $Cot_{1/2}$ values observed for repetitive fractions may be higher than really corresponds to the complexity. Thus most repetitive components have lower complexities and greater repetition frequencies than implied by their reassociation kinetics. We shall see that the effect is especially marked for highly repetitive DNA (Chapter 24).

Where does moderately repetitive DNA reside in the genome? In most genomes, neither moderately repetitive nor nonrepetitive DNA takes the form of a long continuous sequence uninterrupted by the sequences of other components. Instead, both components are present in the form of individual sequences interspersed with one another. In many cases, there is some degree of regularity in the type of interspersion.

The most common type is **short period interspersion,** in which the moderately repetitive DNA is present as individual sequences of average length about 300 bp. They alternate with nonrepetitive sequences whose average length varies from 800 to 1500 bp in various genomes. This type of pattern may occupy about half of the total genome, and usually it accounts for most of both the moderately repetitive and nonrepetitive DNA.

Some genomes display a **long period interspersion,** in which there still is an alternation of moderately repetitive and nonrepetitive DNA components, but each type of sequence is very much longer. In neither type of interspersion is there any clear significance to attribute to the pattern.

FURTHER READING

A principal source for the information discussed in this chapter is **Lewin's** *Gene Expression,* **2,** *Eucaryotic Chromosomes* (Wiley, New York, 1980, pp. 479–530). The reassociation technique was introduced by **Britten & Kohne** (*Science* **161,** 529–540, 1968) and subsequent results were reviewed by **Britten & Davidson** (*Quart. Rev. Biol.* **48,** 565–613, 1973).

CHAPTER 19
STRUCTURAL GENES: AS REPRESENTED IN mRNA

Mendelian genetics for simple traits imply that there is only one copy of each determining factor in the haploid genome. The factor can be mapped to a particular locus; and the simplest assumption is that each such locus is occupied by a DNA sequence representing a single protein, as exemplified by the definition of the gene in Chapter 2. This is the classic view of a structural gene: a unique component of the genome, the only sequence coding for its protein product, and therefore identifiable by mutations that impede the protein function. The sequence of a unique structural gene, unrelated to any other sequences in the genome, should form part of the nonrepetitive DNA component.

Another class of genes takes the form of multiple sequences all coding for the same protein (or sometimes for closely related proteins). These genes may be difficult or even impossible to identify by point mutation, because inactivation of any one copy does not impede the activity of the remainder. Genetic data are therefore biased toward characterizing the first type of gene, so we must turn to direct analysis of the DNA of the genome to determine the numbers and proportions of unique and repeated genes.

Genes that are not unique could appear in either nonrepetitive or moderately repetitive DNA, depending on the number of copies and the relationship between them. A structural gene can be unique in the sense that it is indeed the only DNA sequence coding for its exact protein product, but other sequences in the genome may be related to it because they code for related proteins. Usually a family of this sort consists of only a few members; and, given the effects of mismatching on reassociation, the relevant genes are likely to appear in nonrepetitive DNA or to be very slightly repeated.

Some structural genes are repeated: there may be more than one copy of a sequence coding for a particular protein. Because of the degeneracy of the genetic code, DNA sequences coding for the same polypeptide need not actually be identical. However, any sequence present in a reasonable number of copies (more than two or three) should behave as part of the moderately repetitive sequence component. (Of course, such sequences constitute only a part of this component, because much of the moderately repetitive DNA consists of individual sequences that are too short to code for proteins, interspersed with nonrepetitive DNA.)

For the purpose of identifying and characterizing structural genes, mRNA provides the ideal intermediate. The protein to which it corresponds can be determined by translating the mRNA. The gene from which it is derived can be obtained by hybridizing the mRNA with the genomic DNA. An individual mRNA provides a handle, as it were, for proceeding back from its protein to the gene.

A population of mRNAs, manifested as the spectrum of sequences found in the polysomes, defines the entire set of genes that is expressed in a cell or tissue. Thus the constitution of the mRNA reveals both the nature and number of structural genes. By means of nucleic acid hybridization, we can come to grips with some central questions. How many copies of each gene are there? How many genes are expressed in a particular cell type? How much overlap is there between the sets of genes whose expression defines different cell types?

The kinetics of hybridization can be used to determine the number of copies of each gene corresponding either to members of the mRNA population or to individual mRNAs. More precisely, the question asked by such analysis is: which sequence component of the genome—nonrepetitive or repetitive—is represented in mRNA?

Given the limitations of these experiments, structural genes that are found to be part of nonrepetitive DNA need not necessarily be unique. They could be present in a small number of copies, but this should certainly be less than, say, three or four. To determine the precise number, it is necessary to isolate the DNA corresponding to individual mRNAs.

For structural genes that lie in repetitive DNA, the repetition frequency of the DNA sequence gives only a crude estimate of the gene number. To determine the number of genes and the relationships between them, again it is necessary to isolate the individual members of the family. The properties of repetitive DNA as a whole correspond to families that consist of related but not identical sequences, but this does not exclude the presence within this component of some families all of whose members are identical or very closely related.

MOST STRUCTURAL GENES LIE IN NONREPETITIVE DNA

The genome sequence components represented in mRNA can be determined by using the RNA as a **tracer** in a reassociation experiment. A very small amount of radioactively labeled RNA (or DNA) is included together with a much larger amount of cellular DNA. Because the cellular DNA is present in great excess, the amount withdrawn into a hybrid with the tracer RNA does not change the concentration of single-stranded DNA sequences.

The reaction is governed by the reassociation of the complementary cellular DNA strands (as though the tracer RNA were not present). This is described as a **DNA-driven** reaction. The reassociation of the whole DNA is followed by the usual means (change in optical density or retention on hydroxyapatite). The reassociation of the tracer is followed by the entry of its radioactivity into duplex form.

The tracer RNA (or DNA) participates in the reaction as though it were just another member of the sequence component from which it was transcribed. The component is identified by comparing the radioactive tracer curve with the reassociation curve for whole DNA. Thus the Cot values at which the labeled RNA hybridizes can be taken to identify the repetition frequencies of the corresponding genomic sequences.

When an individual mRNA molecule is used, it hybridizes with a single $Cot_{1/2}$ value determined by the repetition frequency of the gene or genes representing it. For a unique gene, this will be the same as nonrepetitive DNA; while for a repeated gene, there will be a corresponding decrease in the $Cot_{1/2}$. When a population of mRNA molecules is used, each mRNA hybridizes with a characteristic $Cot_{1/2}$, so that overall the curve is the sum of the individual components. It can be resolved in the same way as the reassociation curve of the genomic DNA itself.

With a population of mRNAs, a typical result resembles that shown in **Figure 19.1.** A small proportion of the RNA, generally 10% or less, hybridizes with a $Cot_{1/2}$ corresponding to moderately repetitive sequences. The major component hybridizes with a $Cot_{1/2}$ identical with or very close to that of nonrepetitive DNA. Usually this represents up to 50% of the total RNA.

What is the relationship between the mRNA sequences that hybridize with nonrepetitive DNA and those that hybridize with repetitive DNA? They can be separated into different classes (by retrieving the RNA that hybridizes first). This shows that independent molecules are involved: one class represents genes

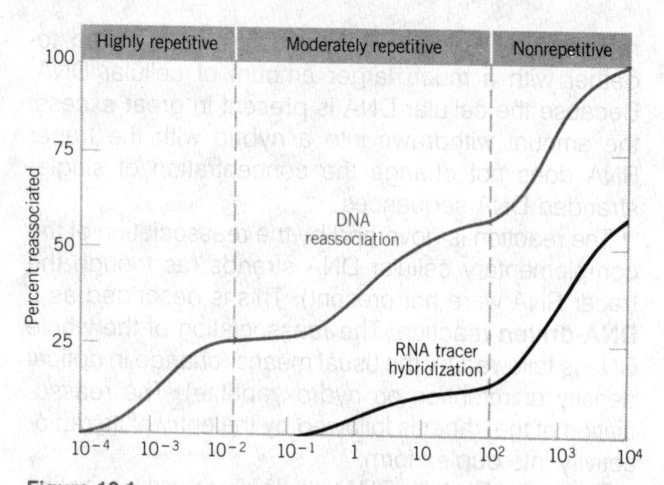

Figure 19.1
The hybridization of an mRNA tracer preparation in a reassociation curve shows that most mRNA sequences are derived from nonrepetitive DNA, the remainder from moderately repetitive DNA, and none from highly repetitive DNA.

that lie in nonrepetitive DNA, and the other corresponds to genes that lie in repetitive DNA.

The main problem in interpreting these results is to decide on the nature of the sequences that do not hybridize at all, which often correspond to up to half of the total RNA. Probably most of this material is derived from the nonrepetitive sequences. One reason why they fail to hybridize may be that they are mRNAs present in very large amounts, large enough to interfere with the assumption that their quantity in the tracer is small relative to the DNA. In other words, if the DNA excess is not sufficient in a DNA-driven reaction, some of the RNA may fail to hybridize.

From these results, it is clear that most of the mRNA, perhaps up to 80%, is derived from sequences that reassociate with the nonrepetitive DNA component. Because of the difficulties in detecting very low degrees of repetition (especially when the repeated copies are related rather than identical), these genes are not necessarily unique, but at least should be present in fewer than three or four copies per genome. Only a small proportion of genes are openly present in repetitive DNA; whether these multiple copies are identical or related is not revealed by the hybridization kinetics.

Note that because these experiments are conducted in terms of the *total mass of mRNA,* and different genes are expressed as different amounts of mRNA, this technique does not prove that 80% of the *genes* lie in the nonrepetitive DNA component, although this class does appear to comprise a majority.

The hybridization kinetics of individual mRNAs have been determined for several genes. Usually the results suggest that there are one or two copies of each gene. Actually, this technique underestimates the number of related sequences in the genome, since experiments to isolate the genes directly often identify further sequences related to them (see Chaper 21).

HOW MANY NONREPETITIVE GENES ARE EXPRESSED?

The number of nonrepetitive DNA sequences represented in RNA can be determined directly in terms of the proportion of the DNA that is able to hybridize with RNA. When a small amount of single-stranded DNA is hybridized with a large amount of RNA, all the sequences in the DNA that are complementary to the RNA should react to form an RNA-DNA hybrid. This is called a **saturation experiment,** because the critical feature is that the excess of RNA should be great enough to ensure that every available complementary DNA sequence actually is hybridized. In this **RNA-driven** reaction, the controlling parameter is the product of RNA concentration and time, known as the **Rot.** This is exactly analogous to the use of Cot values to describe DNA-driven reactions.

The usual way to follow a saturation hybridization is to plot the percent of hybridizing nonrepetitive DNA against the Rot value. A curve of this sort is shown in **Figure 19.2.** In this example, the reaction is complete by a Rot value of about 300, but it is necessary to extend it further to be sure that a plateau has been reached. No further DNA is hybridized as the Rot is increased up to 1200, which demonstrates that all the available DNA has indeed been hybridized.

From the proportion of hybridized DNA, we can calculate the number of genes represented in the mRNA population used to drive the reaction. At saturation, 1.35% of the available nonrepetitive DNA is hybridized. Because only one strand of DNA is transcribed into RNA, only half of the DNA in principle is potentially able to hybridize with RNA (the other half is identical

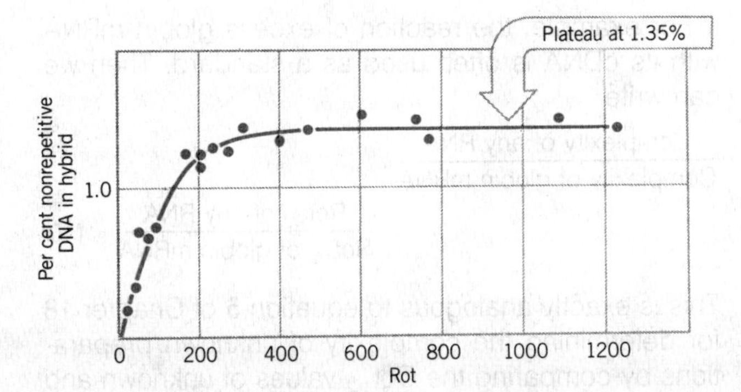

Figure 19.2
Hybridizing an excess of mRNA with nonrepetitive DNA until saturation is reached shows that only a small proportion of the DNA is represented in the mRNA.
The experiment is performed by using purified nonrepetitive DNA that carries a radioactive label; so the proportion of radioactivity entering the hybrid is measured. The data in this figure show the reaction between nonrepetitive DNA of the sea urchin *S. purpuratus* and the mRNA extracted from polysomes of the gastrula embryo.

in sequence with it). Thus 2.70% of the total sequences of nonrepetitive DNA are represented in the mRNA.

The nonrepetitive DNA itself represents 75% of a genome of 8.1×10^8 bp. Thus the complexity of DNA represented in the RNA population is $0.027 \times 0.75 \times 8.1 \times 10^8 = 1.7 \times 10^7$ bp. The mRNA population had an average length of 2000 bases. Thus the total number of different messengers is $1.7 \times 10^7/2000 = 8500$. This corresponds to the total number of genes expressed in the tissue from which the mRNA was taken (the gastrula embryo of the sea urchin).

Similar experiments have been performed for quite a large number of cases. For a lower eukaryote—for example, yeast—the total number of expressed genes is relatively low, roughly 4000. For somatic tissues of higher eukaryotes, the number usually is between 10,000 and 15,000. The value is similar for plants and for vertebrates. The total amount of DNA represented in mRNA typically is therefore a very small proportion of the genome, of the order of 1–2%. The only consistent exception to this type of value is presented by mammalian brain, where much larger numbers of genes appear to be expressed, although the exact quantitation is not certain.

This type of experiment can be performed only for nonrepetitive DNA, in which each DNA sequence can be hybridized only by the mRNA originally derived from it. Because of the presence of multiple copies of identical or related sequences in moderately repetitive DNA, an RNA derived from this component might be able to hybridize with other genomic sequences in addition to the particular sequence from which it was originally transcribed. At saturation, therefore, a large number of additional DNA sequences could be contained in the hybrid, thus overestimating the number of expressed genes.

In fact, to be sure that this has not happened, the DNA that has been saturated with RNA usually is recovered (by degrading the RNA) and then used in a reassociation experiment to show that its $Cot_{1/2}$ does indeed correspond to that expected of nonrepetitive DNA. This is called a **playback experiment.** So the number of expressed genes estimated by this technique refers strictly to the majority whose sequences lie in nonrepetitive DNA.

ESTIMATING GENE NUMBERS BY THE KINETICS OF RNA-DRIVEN REACTIONS

Because RNA is in excess over the DNA, in an RNA-driven reaction the RNA concentration remains essentially unchanged by the small amount drawn into hybrid form during the reaction. This means that a saturation analysis follows first-order kinetics, as described by the equation

$$\frac{D}{D_0} = e^{-k.Rot} \quad (8)$$

so that when the reaction is half complete, and $D/D_0 = 0.5$,

$$K = \frac{ln2}{Rot_{1/2}} \quad (9)$$

However, the $Rot_{1/2}$ displayed by an RNA-driven saturation reaction cannot be used to determine the complexity of the RNA population. The reason is that different RNA sequences are present at different concentrations, depending on the characteristic levels at which their genes are expressed. This means that the *Rot* as measured by the mass of RNA is not applicable for any particular sequence.

In fact, usually quite a large proportion of the mass of mRNA is provided by just a few sequences. They very quickly saturate their complements in DNA, leaving the reaction to be driven by the remaining sequences. Thus the real mass of RNA driving the reaction is much less than the measured R_0. The measured $Rot_{1/2}$ is therefore much too high; it includes the large mass of RNA that does not participate in the reaction because its complementary sequences are very quickly saturated.

Another technique has been developed to allow the kinetics of an RNA-driven hybridization reaction to be used to determine complexity. The protocol is illustrated in **Figure 19.3.** It makes use of a labeled cDNA prepared by reverse transcription of the mRNA. Remember that the cDNA consists of single-stranded DNA that is complementary to the mRNA.

The hybridization consists of a reaction between excess mRNA and the labeled cDNA previously prepared from it. Each mRNA sequence should be represented in the cDNA population with a frequency corresponding to its proportion in the RNA. Since all of the cDNA sequences have been derived from the mRNA, all of the labeled cDNA should be driven into the hybrid form. For a single component, the $Rot_{1/2}$ of reaction is proportional to complexity in the same way that $Cot_{1/2}$ is determined by complexity in a reassociation reaction. Thus the complexity of an unknown RNA population may be determined by comparing its $Rot_{1/2}$ with the $Rot_{1/2}$ of a standard reaction.

Figure 19.3

The hybridization of excess mRNA x cDNA is used to determine mRNA complexity.

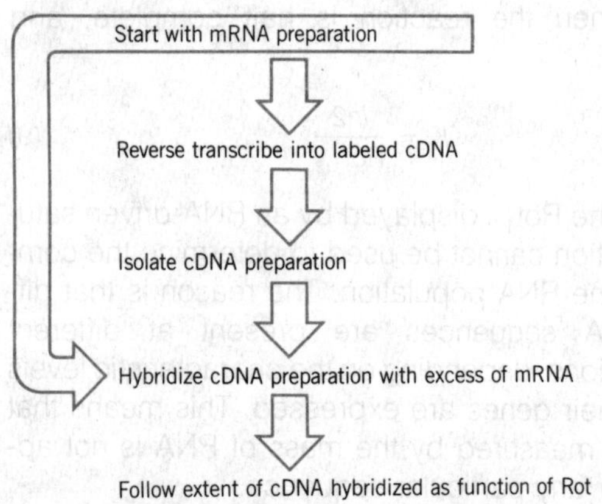

For example, the reaction of excess globin mRNA with its cDNA is often used as a standard. Then we can write

$$\frac{\text{Complexity of any RNA}}{\text{Complexity of globin mRNA}} = \frac{Rot_{1/2} \text{ of any RNA}}{Rot_{1/2} \text{ of globin mRNA}} \quad (10)$$

This is exactly analogous to equation **5** of Chapter 18 for determining the complexity of unknown preparations by comparing the $Cot_{1/2}$ values of unknown and standard DNAs. In the Rot analysis also, the calculated complexity gives the total length of different sequences without making any implication about organization in terms of individual sequences.

Just as with a DNA reassociation curve, a single component hybridizes over about two decades of Rot values, and a reaction extending over a greater range must be resolved by computer curve-fitting into individual components. Again this represents what may really be a continuous spectrum of sequences. To determine the complexity of each component, it is treated as though it were the only one. The $Rot_{1/2}$ value measured for a component is multiplied by the proportion of the reaction occupied by that component. Essentially this corrects the R_0 so that it includes only the mass of RNA present in this component. The corrected $Rot_{1/2}$ value is used to calculate the complexity of the component.

An example of an excess mRNA × cDNA reaction that generates three fairly clear components is given in **Figure 19.4.** The control for this reaction was the hybridization between purified ovalbumin mRNA (2000 bases long) and its cDNA, which had a $Rot_{1/2}$ of 0.0008. This means that a $Rot_{1/2}$ of 0.0004 is demanded for each 1000 bases of complexity.

The first component has an observed $Rot_{1/2}$ of 0.0015 and represents 50% of the total reaction. Thus the $Rot_{1/2}$ applicable to the purified component should be $0.5 \times 0.0015 = 0.00075$, almost exactly the same as the control $Rot_{1/2}$ for ovalbumin. This suggests that the first component is in fact just ovalbumin mRNA! This fits well with the known properties of the oviduct tissue, where ovalbumin is indeed the predominant product of gene expression, with its RNA corresponding to about half of the messenger mass.

The next component has an observed $Rot_{1/2}$ of 0.04,

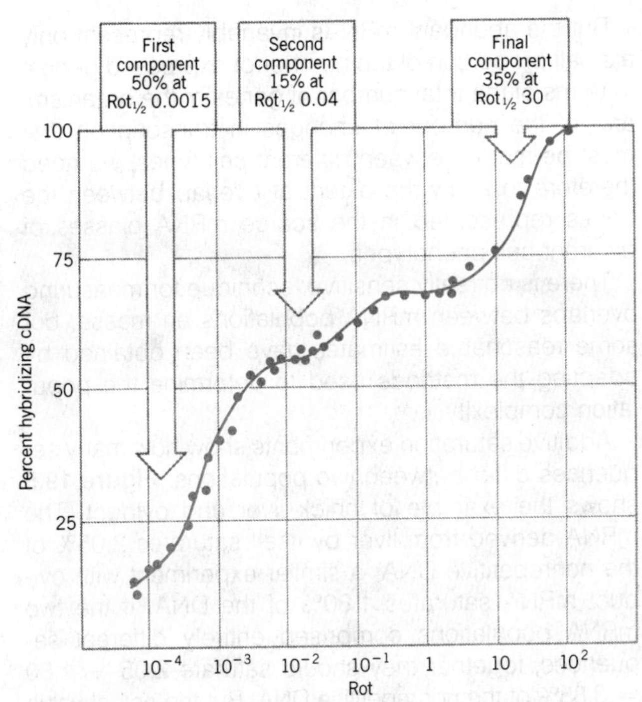

Figure 19.4

Hybridization between excess mRNA and cDNA identifies several components, each characterized by the $Rot_{1/2}$ of reaction. The reaction is expressed relative to the final amount of cDNA hybridized, but usually this is close to 100%, almost always more than 90%. These data show the reaction using mRNA derived from the chick oviduct.

and provides 15% of the reaction. Thus its purified $Rot_{1/2}$ should be $0.15 \times 0.04 = 0.006$. Its complexity is therefore $0.006/0.0004 \times 1000 = 15{,}000$ bases. This would correspond to 7–8 mRNA species of average length 2000 bases.

The last component has a $Rot_{1/2}$ of 30 and provides 35% of the reaction; so its purified $Rot_{1/2}$ of $0.35 \times 30 = 10.5$ corresponds to a complexity of $10.5/0.0004 \times 1000 = 26{,}000{,}000$ bases. This corresponds to about 13,000 mRNA species of average length 2000 bases.

From this analysis, we can see that about half of the mass of mRNA in the cell represents a single mRNA, about 15% of the mass is provided by a mere 7–8 mRNAs, and about 35% of the mass is divided into the large number of 13,000 mRNA species. It is therefore obvious that the mRNAs present in the various components must be present in very different amounts.

GENES ARE EXPRESSED AT WIDELY VARYING LEVELS

The average number of molecules of each mRNA per cell is called its **abundance** or **representation.** It can be calculated quite simply if the total mass of RNA in the cell is known. For each component, total mass = abundance × complexity, so that as a general equation,

$$\text{Abundance} = \frac{\text{gm of mRNA in cell} \times \text{fraction in component} \times 6 \times 10^{23}}{\text{Complexity of component in daltons}} \quad (11)$$

The equation is usually expressed in this form, since total mRNA is measured (chemically) in picograms, while complexity is determined (by hybridization) in bases or daltons.

In the example shown in Figure 19.4, there are 0.275 pg mRNA per cell. This corresponds to 100,000 copies of the first component (ovalbumin mRNA), 4000 copies of each of the 7–8 mRNAs in the second component, but only about 5 copies of each of the 13,000 mRNAs that constitute the last component.

The oviduct is a somewhat extreme case, with so much of the mRNA represented in only one mRNA species, but most cells do contain a small number of RNAs present in many copies each. This **abundant** component typically consists of fewer than 100 different mRNAs present in more than 1000 or even 10,000 copies per cell. It often corresponds to a major part of the mass, approaching 50% of the total mRNA. The abundant component may be resolved into two components, as seen in the example of the oviduct, or it may seem to constitute one component; but for mRNAs that are frequently expressed, probably each gene has a characteristic and different level.

In all cases, around half of the mass of the mRNA consists of a large number of sequences, of the order of 10,000, each represented by only a small number of copies in the mRNA—say, less than 10. This is the **scarce** or **complex** mRNA class. (It is this class that drives a saturation reaction.)

The total numbers of expressed genes as estimated

by the saturation technique or the kinetic technique usually are in fairly good agreement. The kinetic technique provides a lower estimate (because some sequences that are very scarce may fail to react). The saturation technique provides a higher estimate (because there are always likely to be some sequences included that are present in more than one copy per genome). Thus for chick oviduct the kinetic technique identifies about 13,000 mRNAs, while the saturation technique corresponds to 15,000.

OVERLAPS BETWEEN mRNA POPULATIONS

Many somatic tissues of higher eukaryotes have an expressed gene number in the range of 10,000 to 20,000. How much overlap is there between the genes expressed in different tissues? For example, the expressed gene number of chick liver is in the range of 11,000 to 17,000, compared with the value we have just quoted for oviduct of 13,000 to 15,000. How many of these two sets of genes are identical? How many are specific for each tissue?

We see immediately that there are likely to be substantial differences among the genes expressed in the abundant class. Ovalbumin, for example, is synthesized only in the oviduct, not at all in the liver. This means that 50% of the mass of mRNA in the oviduct is specific to that tissue. Taking a more general view, in many cases the abundant mRNAs represent genes whose proteins are a major product of the cell type. In terms of the mass of mRNA and of protein synthesized, there may therefore be appreciable differences between cell types.

But the abundant mRNAs invariably represent only a small proportion of the number of expressed genes. In terms of the total number of genes of the organism, and of the number of changes in transcription that must be made between different cell types, we need therefore to know the extent of overlap between the genes represented in the scarce mRNA classes of different cell phenotypes.

There is no really sensitive technique for measuring overlaps between mRNA populations en masse, but some reasonable estimates have been obtained by adapting the methods used to determine the population complexity.

Additive saturation experiments show how many sequences differ between two populations. **Figure 19.5** shows the example of chick liver and oviduct. The mRNA derived from liver by itself saturates 2.05% of the nonrepetitive DNA; a similar experiment with oviduct mRNA saturates 1.80% of the DNA. If the two mRNA populations comprised entirely different sequences, together they should saturate 2.05 + 1.80 = 3.85% of the nonrepetitive DNA. But the actual value is 2.4%, only slightly greater than that of liver alone.

Thus about 75% of the sequences expressed in liver and oviduct are the same (though since this is a saturation experiment, the data do not show whether they are present in the same or different abundances in the two tissues). In other words, about 12,000 genes are expressed in both liver and oviduct, about 5000 additional genes are expressed only in liver, and about 3000 additional genes are expressed only in oviduct.

Another way to estimate the extent of overlap is to hybridize the mRNA of a tissue with nonrepetitive DNA. Then the DNA that reacts is isolated to constitute the

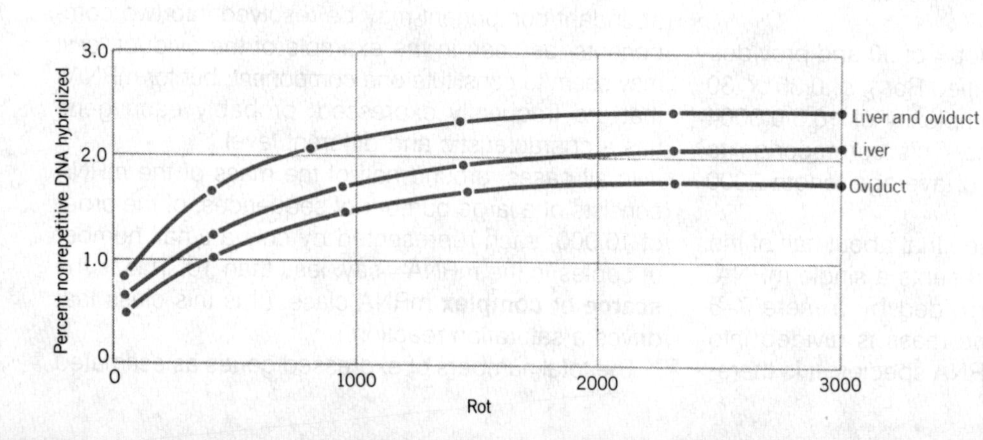

Figure 19.5
Additive saturation hybridization shows how many different sequences are present in two mRNA preparations.

mDNA preparation, while the DNA that does not react constitutes the **null DNA** preparation. These DNA preparations are then hybridized separately with an excess of mRNA from some other tissue. The proportion of the mDNA preparation that reacts identifies the proportion of genes expressed in the second as well as in the first tissue. The amount of the null DNA preparation that reacts identifies the number of genes expressed in the second, but not in the first, preparation.

Heterologous experiments also can be performed by using excess mRNA of one source to drive a reaction with cDNA representing the mRNA of another source. These experiments show that the sequences that are abundant in one tissue usually are not abundant in another tissue, although sometimes they are present (at much lower levels, in the scarce mRNA class).

The scarce mRNAs overlap extensively. Between mouse liver and kidney, about 90% of the scarce mRNAs are identical, leaving a difference between the tissues of only 1000–2000 in terms of the number of expressed genes. The general result obtained in several comparisons of this sort is that only about 10% of the mRNA sequences of a cell are unique to it. The majority of sequences are common to many, perhaps even all, cell types.

This suggests that the common set of expressed gene functions, numbering perhaps about 10,000 in a mammal, may comprise functions that are needed in all cell types. Sometimes this type of function is referred to as a **housekeeping** or **constitutive** activity. It contrasts with the activities represented by specialized functions (such as ovalbumin or globin) needed only for particular cell phenotypes. These are sometimes called **luxury** genes.

If we take into account all the various cell phenotypes of the organism, certainly there may be as many luxury functions as housekeeping functions. Still, the total number of types of luxury function (at least as represented in nonrepetitive DNA) is unlikely to be more than, say, 2–3 times the number of housekeeping functions—within the range of 20,000 to 40,000.

Some specialized cell types can function with a relatively small number of expressed genes. A detailed examination of expressed genes in the sea urchin *S. purpuratus* has revealed that the oocyte contains about 18,000 mRNA sequences. During embryogenesis, the number of expressed genes falls to about 13,000 at blastula, 8500 at gastrula, and 7000 at pluteus. But in some adult tissues—tubefoot, intestine, coelomocyte—the number of expressed genes falls in the range of only 2500 to 3000. Thus these cells can be maintained with much less than the housekeeping number of 10,000 that has been implied for mammalian cells.

An especially significant feature of this series of cells is that there is very extensive overlap in the expressed sequences, so that genetic development occurs via a progressive reduction in the expressed gene number. Experiments with the mDNA/null DNA assay have shown that most of the genes expressed at a later stage have also been expressed at an earlier, embryonic stage. It is even possible that all the structural genes of the organism are expressed in the oocyte, which would imply that the sea urchin genome consists of fewer than 20,000 genes, whose coding sequences correspond to only 3% of the total haploid DNA. Even if we allow the possibility that each gene is not unique, but is repeated (say) 3 times in the genome, we still have accounted for only a very small part of the DNA in coding for proteins.

FURTHER READING

A detailed review of the estimates of gene number was presented in **Lewin's** *Gene Expression*, **2,** *Eucaryotic Genomes* (Wiley, New York, 1980, pp. 694–727). An earlier view of these experiments was summarized by **Lewin** (*Cell* **4,** 77–93, 1975).

CHAPTER 20
THE ORGANIZATION OF INTERRUPTED GENES

For a long time there were somewhat uneasy suspicions that there might be something unusual about eukaryotic genes, that they might differ in some fundamental way from bacterial genes. The roots of this idea lay in the apparent discrepancies of DNA content described by the C value paradox. The problem was reinforced by measurements of structural gene complexity, which implied that only a small part of the genome can be accounted for by its representation in the mRNA template.

One escape from this situation supposes that a large proportion (even a majority) of DNA sequences are not part of the structural genes. The amount of such DNA, and what functions it may (or may not) have, remains to be established.

Another idea is that the structural gene—or at least the transcription unit—is *much larger than the sequence represented in mRNA.* This view is supported indirectly by observations that the size of the nuclear RNA is much larger than that of mRNA. A large part of the additional length must be removed when the RNA is **processed** for transport to the cytoplasm.

An increase in gene size originally was taken to imply that the transcription unit might include extensive sequences on one or the other side of the region represented in mRNA (most likely upstream from it), perhaps involved in regulation. There could also be extensive nontranscribed regions needed for regulation, so the unit of gene expression might be rather large relative to the size of the mRNA.

Indeed it turns out that many eukaryotic genes are much longer than their mRNAs. But the cause of the discrepancy is the existence of *interruptions that separate different parts of the coding region in DNA.*

The existence of interrupted genes was revealed by experiments designed to identify the DNA corresponding to a particular mRNA. When this work began, its purpose was not to compare the two nucleic acid sequences per se, but to locate the genomic DNA sequence corresponding to the mRNA. This in turn allows the flanking sequences—including putative regulatory regions—to be identified and characterized.

The intention of this approach was to proceed back from the mRNA to the gene, to allow the recovery of intact transcription units, including promoters and other elements not necessarily represented in the mRNA. The aim of characterizing the context within which a structural gene resides still remains an object of these experiments, one that has been achieved in several cases, as we shall see in Chapter 21.

Originally it was assumed that an mRNA would have the same sequence as the DNA from which it is transcribed. As a control to show that an isolated genomic DNA sequence did indeed correspond with the mRNA used to isolate it, their sequences were compared, either by electron microscopy or by restriction mapping. But what this revealed was a discrepancy between the sequences in the form of additional regions present in the genomic DNA but absent from mRNA.

The presence of additional sequences explains at least part of the size discrepancy between mRNA and nuclear RNA; these **intervening sequences** are included in the primary transcript, but are removed from it during maturation into mRNA. Although comprising part of the structural gene, the intervening sequences do not have any coding function.

At this point, we should recapitulate the terminology mentioned in Chapter 3 for describing the relationship between a gene and its RNA product. An interrupted gene consists of an alternating series of **exons** and **introns.** The exons are the sequences represented in the RNA (which can be mRNA, rRNA, or tRNA, so the exons may or may not have a protein-coding function). The introns are the intervening sequences that are removed from the primary transcript and are therefore absent from the mature RNA. A gene naturally must start and end with exons (corresponding to the 5′ and 3′ ends of the RNA), but there may be any number of introns within it. Often the exons and introns are numbered or assigned letters in order along the gene.

The interruption of coding regions is unique to the eukaryotes, but is not found with all eukaryotic genes. Indeed, an insufficient number of genes has been characterized yet to determine any average relationship between the size of the gene and that of the mRNA. So we do not yet know to what extent the presence of introns helps to resolve the C value paradox.

Proceeding from the structure of the gene itself, the next question is to determine its context. How much material on either side of a gene is involved in its function? How far is it to the next gene, and how much of the genome comprises sequences that lie between transcription units? Do adjacent genes tend to be related and is there any mechanism for regulating regions of the chromosome longer than single genes?

VISUALIZING INTERRUPTED GENES BY ELECTRON MICROSCOPY

Electron microscopy can be used to visualize RNA-DNA hybrids in two ways. In both the object of the exercise is to distinguish RNA-DNA hybrids from unpaired DNA regions. The length and relative position of the hybrid region identify the location of the DNA sequence representing the mRNA.

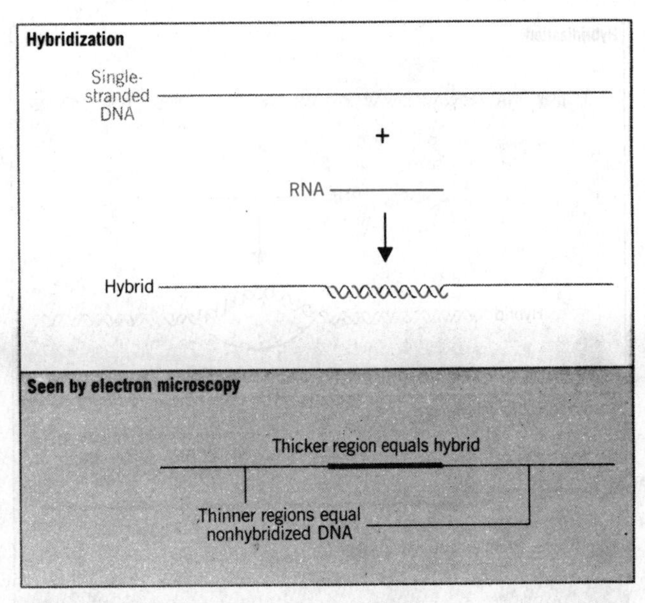

Figure 20.1
Single-stranded DNA is hybridized with RNA to form a structure in which the hybrid region can be visualized by electron microscopy. The exact length of duplex nucleic acid varies with the conditions, but roughly 1 μm = 3250 bp.

In this and the subsequent figures, the upper part of the figure shows the events that are involved in terms of the individual strands of DNA and RNA, and the lower part shows what is actually seen by electron microscopy, where thicker and thinner regions have to be interpreted in terms of double- and single-stranded regions.

When an RNA is hybridized with a single-stranded DNA, the complementary regions form a double-stranded hybrid while the other regions remain single-stranded. The double-stranded region is thicker than the single-stranded, although sometimes it is difficult to distinguish the exact junctions between the two regions. **Figure 20.1** shows that this technique visualizes a continuous DNA sequence representing an RNA as a thicker region within the thinner single strand of DNA.

In the alternative technique of **R-loop mapping,** an RNA is hybridized with a double-stranded DNA under conditions in which an RNA-DNA hybrid is more stable than the original DNA duplex. This allows the RNA to displace one DNA strand from the duplex in the region in which it can pair with its complement. **Figure 20.2**

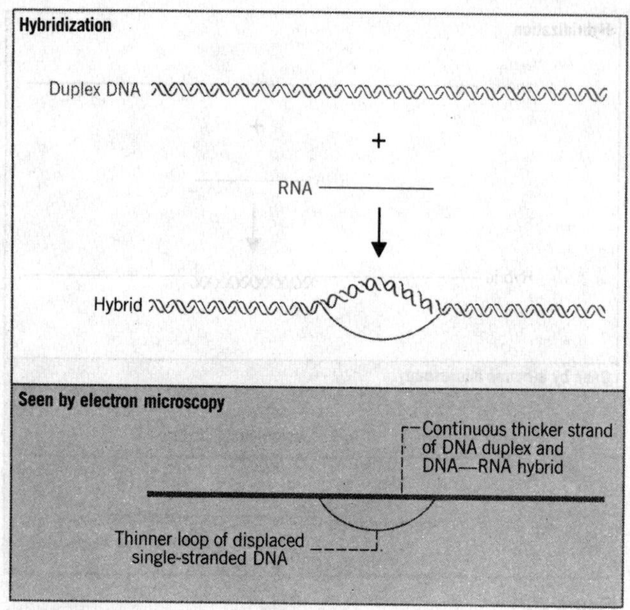

Figure 20.2
Under the conditions for R-loop formation, RNA hybridizes with DNA by displacing one strand of the double helix.

shows that the displaced strand forms a loop flanked by the undisturbed double-stranded DNA. This gives rise to the name of the technique.

The critical feature to be noted in both Figures 20.1 and 20.2 is that when an RNA corresponds to a continuous sequence of DNA, the two molecules are colinear. There is a single, continuous hybrid region within the DNA.

But when a gene possesses an internal sequence that is *not* represented in mRNA, this part of the DNA cannot pair with the RNA. Yet the RNA can pair to form a hybrid with the sequences on either side of the interruption. This reaction results in the formation of a characteristic type of structure in each method.

When RNA is hybridized with single-stranded DNA, the intervening sequence is extruded as a single-stranded loop of DNA from within the region of the RNA-DNA hybrid. The structure is illustrated in **Figure 20.3.** The length and relative position of the loop within the hybrid define the size and location of the intron within the transcription unit.

When the technique of R-loop mapping is used, the RNA displaces one strand of the DNA to form an RNA-DNA hybrid with the regions on either side of the intervening sequence. But the intervening sequence itself remains inviolate in its original double-stranded form. This produces the structure illustrated in **Figure 20.4,** in which the two RNA-coding regions are brought together in the hybrid, as seen by the two displaced single-stranded DNA loops. At the junction of these loops, a double-stranded loop of DNA is extruded, corresponding to the intervening sequence.

An example in which a single intron is visualized in the mouse minor β-globin gene is shown in **Figure 20.5.** (The gene also contains a second intron that is too small to be visualized by electron microscopy; see later.)

The first results of electron microscopic visualization were obtained with the viral genomes of adenovirus and SV40. In each case, when late mRNAs were hybridized with the viral DNA, the main part of the mRNA formed a hybrid with one region of the viral DNA, but the 5′ terminal part of the mRNA hybridized with a sequence or sequences far upstream in the viral DNA.

Figure 20.3
When RNA hybridizes with two regions of single-stranded DNA that are separated by an intervening sequence, the interruption protrudes as a nonhybridized, single-stranded loop of DNA from the RNA-DNA hybrid.

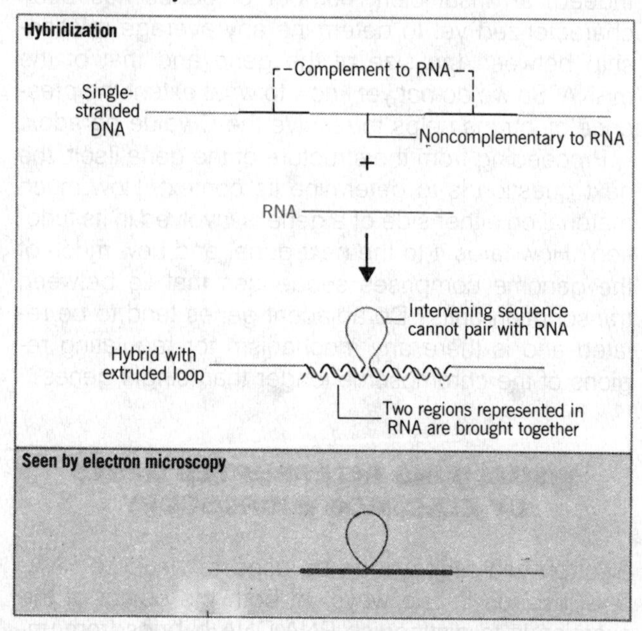

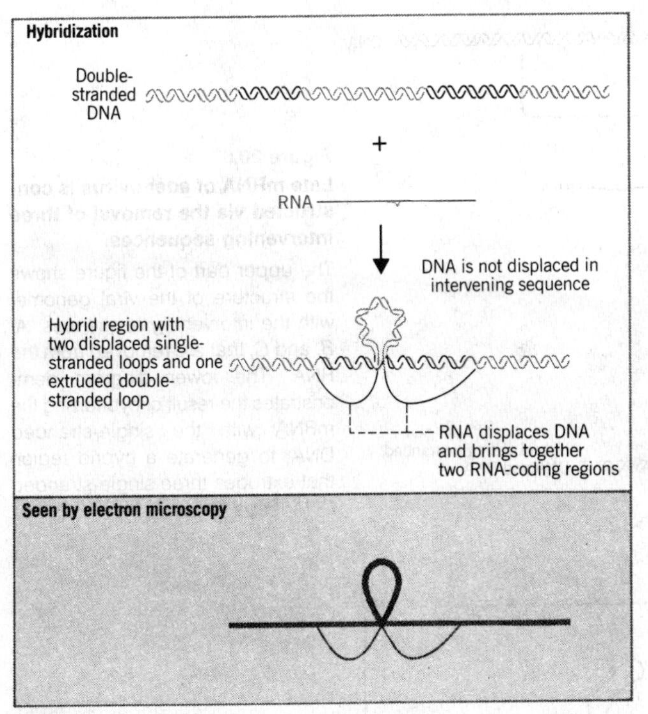

Figure 20.4
When RNA is hybridized with an interrupted gene by R-loop mapping, the intervening sequence cannot hybridize and remains as double-stranded DNA. It is extruded as a thick duplex loop from the RNA-DNA hybrid region. The DNA sequences displaced by RNA form single-stranded loops that identify the individual RNA-coding segments.

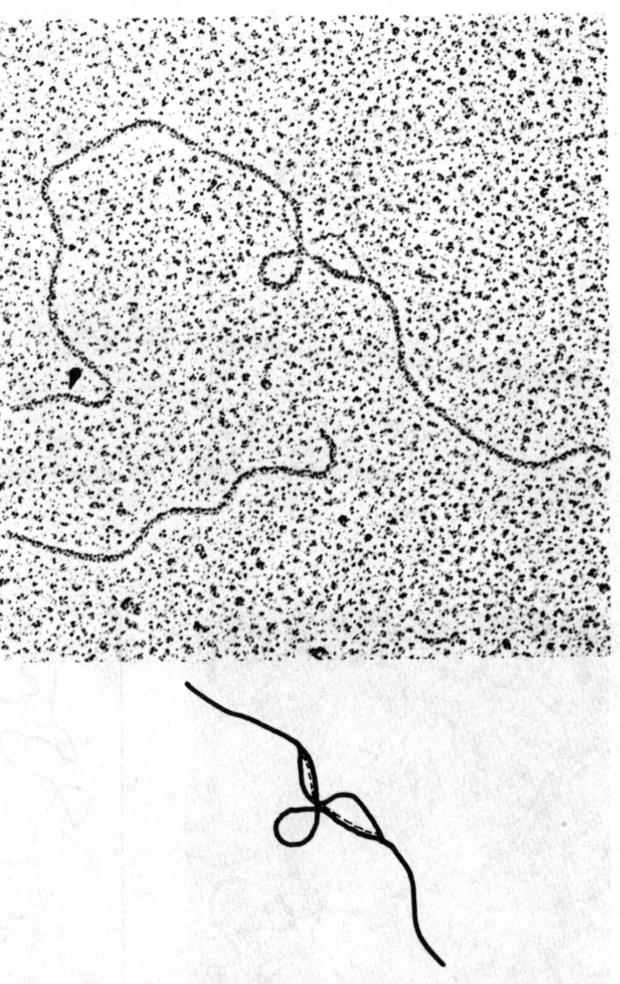

Figure 20.5
R-loop mapping of the mouse minor β-globin gene identifies a long intervening sequence. The electron micrograph is shown on top and an interpretive drawing below it.
Photograph kindly provided by Philip Leder.

The situation in adenovirus is depicted in **Figure 20.6,** which demonstrates that the 5′ end of the late mRNA actually is an amalgam of three very short RNA sequences that represent widely separated sites in the viral genome. In this instance, the tripartite 5′ region corresponds to a nontranslated leader sequence, while the coding region is provided by the main body of the mRNA. **Figure 20.7** visualizes this organization by electron microscopy. As evident from this hybrid, when there is a series of intervening sequences, the structures can become tangled and difficult to interpret.

The resolution of electron microscopic mapping is good to within about 50–100 bp, depending on the conditions. Segments smaller than this can be missed on an electron microscopic map.

RESTRICTION MAPPING OF INTERRUPTED GENES

When a gene is uninterrupted, the restriction map of its DNA corresponds exactly with the map of its mRNA (obtained by characterizing a cDNA reverse transcript). When a gene contains a sequence that is not present within the corresponding mRNA (cDNA), the discrepancy may appear in either of two ways.

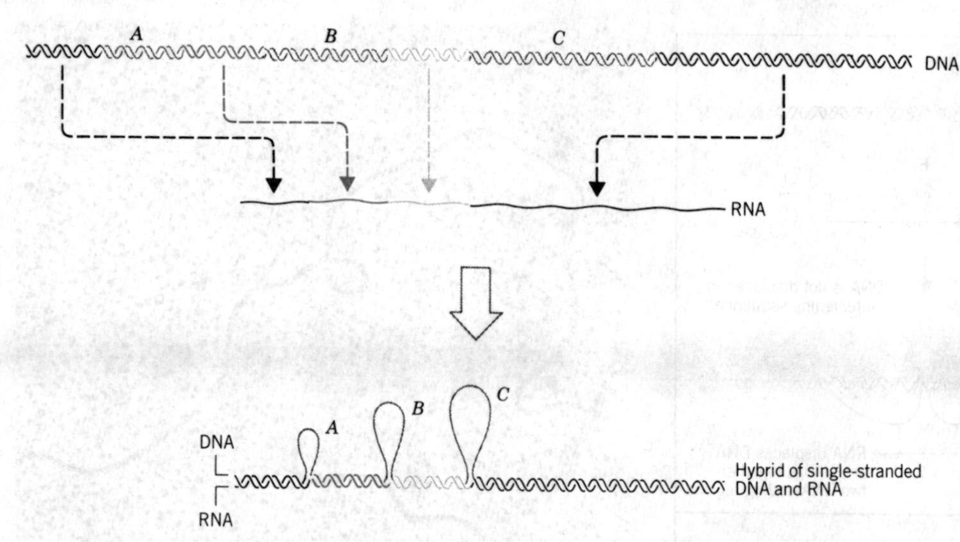

Figure 20.6
Late mRNA of adenovirus is constructed via the removal of three intervening sequences.
The upper part of the figure shows the structure of the viral genome, with the intervening sequences, *A*, *B*, and *C*, that are removed from the RNA. The lower diagram demonstrates the result of hybridizing the mRNA with the single-stranded DNA, to generate a hybrid region that extrudes three single-stranded DNA loops.

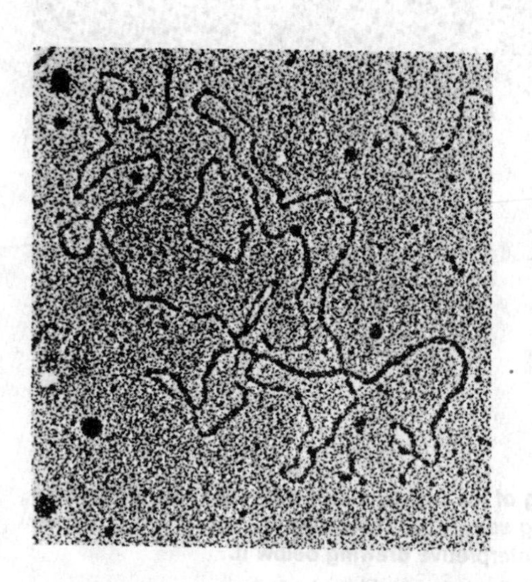

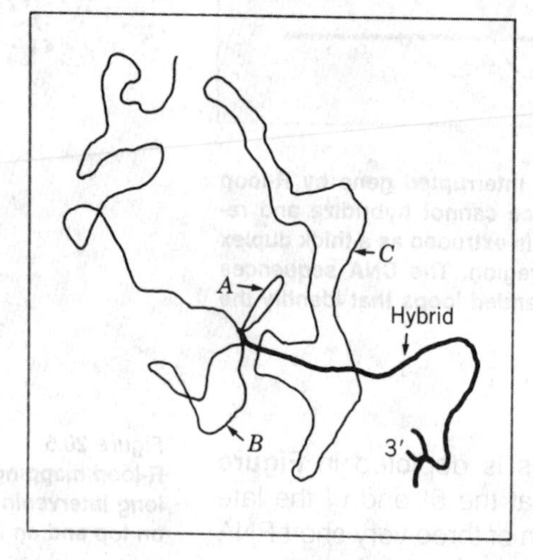

Figure 20.7
One of the original electron micrographs of an RNA hybridization with single-stranded adenovirus DNA identifying the mosaic structure of the late transcription unit.
Photograph kindly provided by Phillip Sharp.

Figure 20.8 shows that if the intervening sequence does not have a recognition site for whatever restriction enzyme is being employed, there is a change in the length of the corresponding fragment. The presence of the intron means that a greater distance separates the two restriction sites on either side of it in the gene than in the mRNA. Thus the length of this fragment is increased in the digest of the gene.

On the other hand, if the intervening sequence possesses a recognition site for the enzyme, an extra cut is made within it. Since this cut occurs only in the gene and not in the mRNA sequence, cleavage of the gene generates two fragments corresponding to the one generated by the message. This situation is illustrated in **Figure 20.9.**

When a map is constructed by using several restriction enzymes, a mixture of these two results is obtained. Some enzymes will not have target sites within an intron, so the lengths of the corresponding fragments are increased. Other enzymes will cleave the intron, generating additional fragments.

Putting the results together, when an entire restric-

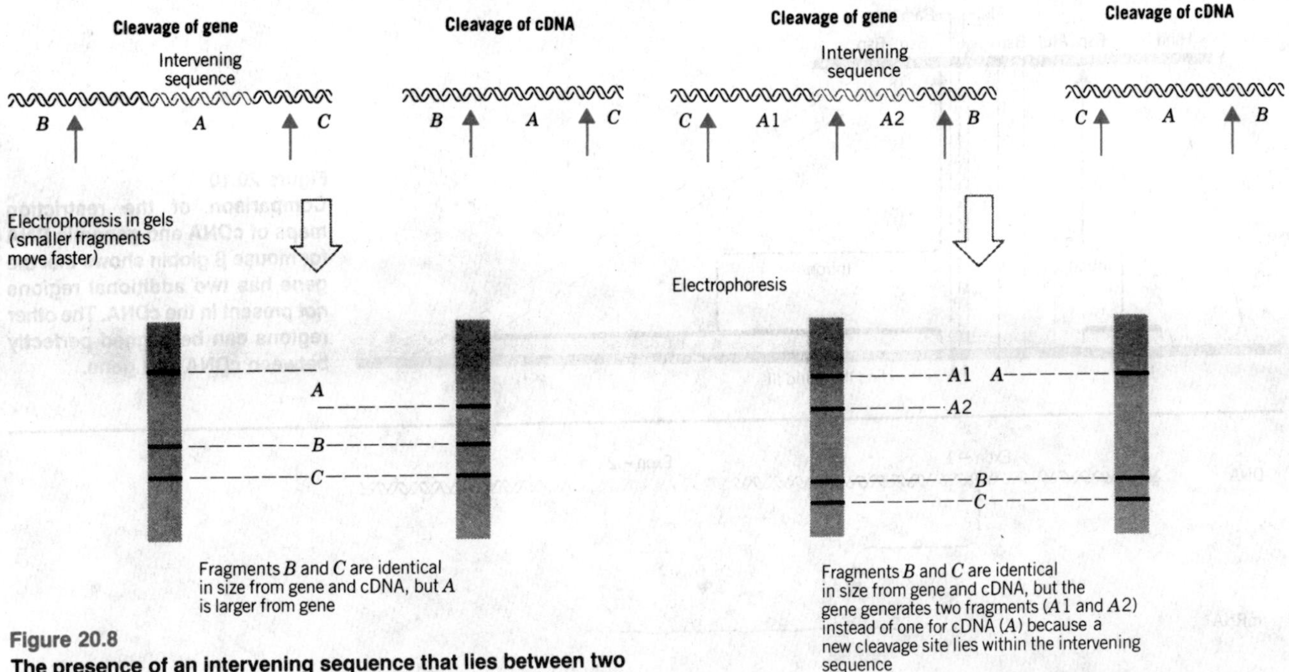

Figure 20.8

The presence of an intervening sequence that lies between two cleavage sites increases the size of the corresponding restriction fragment.

The figure compares a gene and cDNA that have two cleavage sites for a restriction enzyme. This generates three fragments, labeled *A*, *B*, and *C* in order of decreasing size. The intervening sequence lies between the two cleavage sites that generate *A*, so just the size of this fragment is increased in the gene.

Figure 20.9

The presence of an intervening sequence creates new restriction fragments when it contains a cleavage site for a particular enzyme.

An additional cleavage site in the gene increases the number of cuts to three from the two found with cDNA. Thus fragment *A* from cDNA is replaced by *A1* and *A2* from genomic DNA. The combined size of *A1* + *A2* exceeds that of *A* by the length of the intervening sequence.

tion map is constructed for a gene possessing an intron, we find that the map at each end of the gene corresponds with the map at each end of the message sequence. But at some point within the gene, the maps diverge, and an additional region is present within the gene, containing a series of recognition sites not represented in the message. The resolution of restriction mapping is adequate to detect gene segments down to about 20–30 bp.

Figure 20.10 compares the restriction maps of a β-globin gene and mRNA. Two introns can be recognized. The larger is the same as the intron seen by electron microscopy in Figure 20.5. The smaller shows up only on the restriction map.

A useful procedure for analyzing an interrupted gene is to characterize the hybrid formed between the mRNA and DNA, as summarized in **Figure 20.11.** In addition to the RNA-DNA hybrid regions, the molecule contains single-stranded flanking DNA and the extruded single-stranded intron. The molecule can be treated in two ways.

The flanking DNA can be removed by the enzyme exonuclease VII. Then the single-stranded DNA corresponding to the length of the gene itself can be isolated by degrading the RNA with alkali.

The individual regions of the gene can be isolated by treatment with S1 nuclease, an enzyme that specifically cleaves any single-stranded DNA. This treatment removes the flanking regions and the intron. The product is an RNA-DNA hybrid, of the length of the mRNA, with a gap in the covalent integrity of the DNA strand. Treatment with alkali removes the RNA, releasing the individual fragments of DNA. These identify the individual exons of the gene.

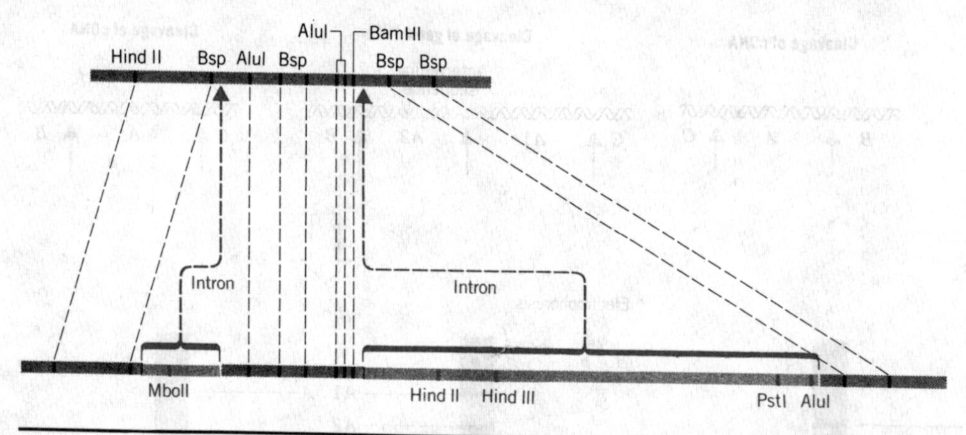

Figure 20.10
Comparison of the restriction maps of cDNA and genomic DNA for mouse β globin shows that the gene has two additional regions not present in the cDNA. The other regions can be aligned perfectly between cDNA and gene.

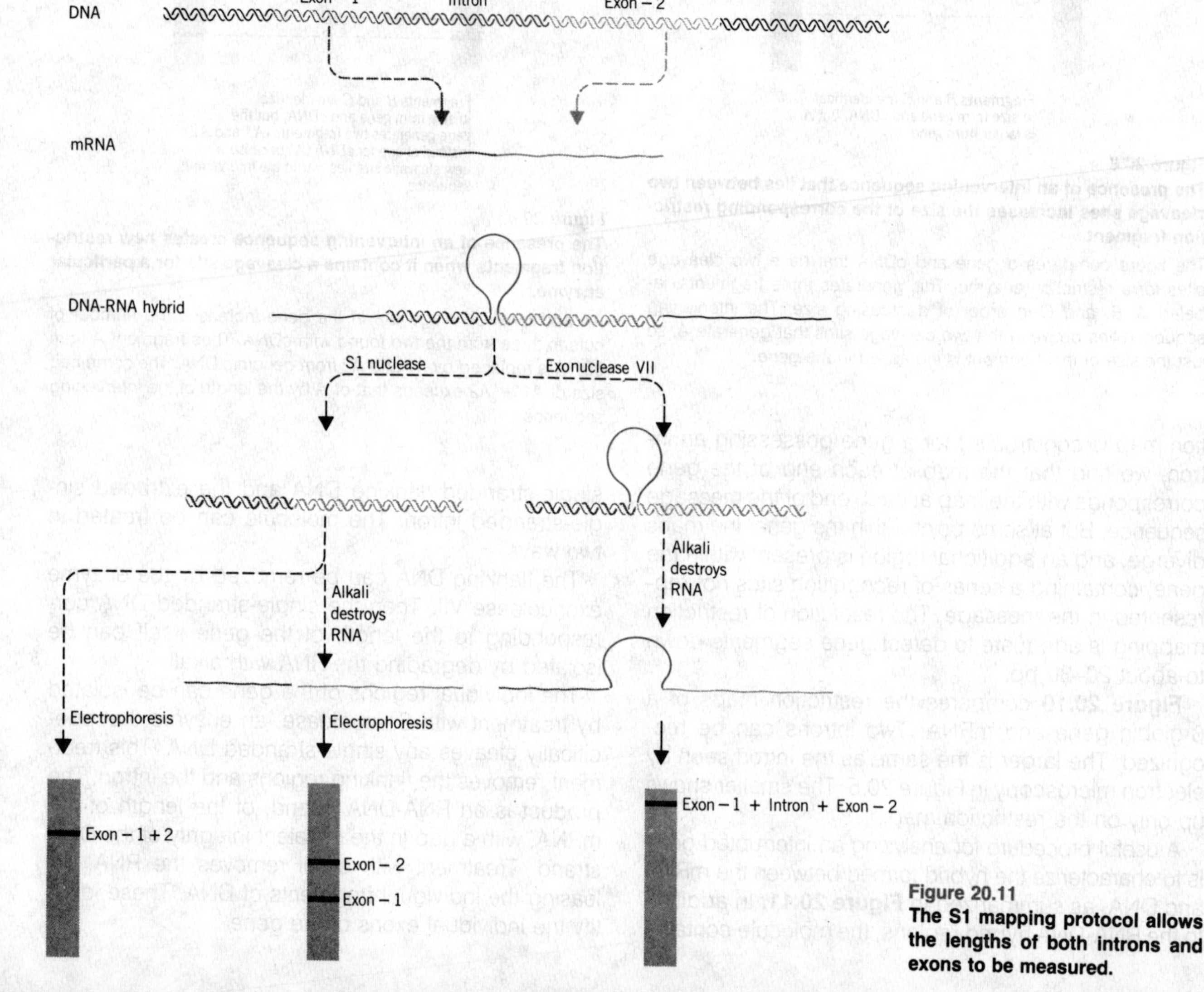

Figure 20.11
The S1 mapping protocol allows the lengths of both introns and exons to be measured.

In the form illustrated in the figure, the technique is often known as Berk-Sharp mapping, named for its authors. In more generalized form, it is sometimes called S1 nuclease mapping. The property of this enzyme for breaking all bonds in unpaired DNA is very useful. Often it is used to locate the end of an RNA sequence on DNA. For example, DNA is hybridized with RNA and then S1 nuclease is used to remove the flanking regions. The DNA then is terminated *exactly* at the ends of the RNA molecule (as we have mentioned previously in Figure 12.2). By comparing the sequence of this DNA fragment with the sequence of the original DNA, the exact startpoint for the mRNA can be located.

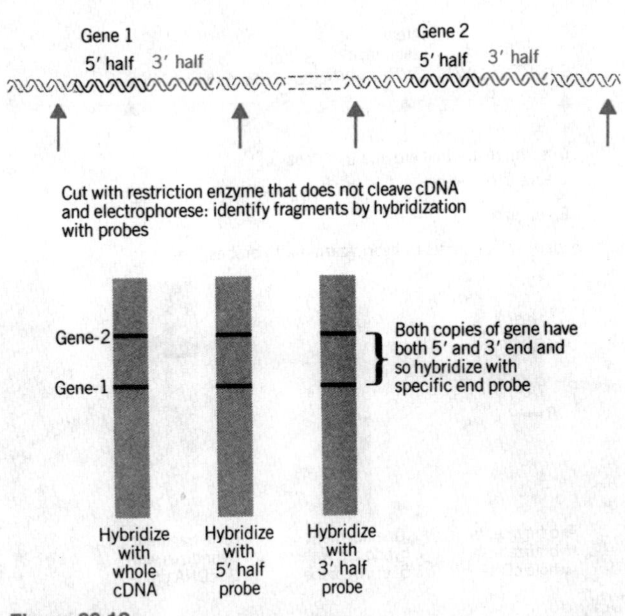

Figure 20.12
Multiple genes produce intact copies when the genome is cleaved with a restriction enzyme that does not cut within the cDNA. Two identical genes, each consisting of a 5′ half and a 3′ half, give different fragments because the restriction sites lie at different relative positions in the flanking DNA. The fragments are identified by blotting a gel of the electrophoresed DNA and hybridizing with a labeled cDNA probe. Since each fragment has both ends of the gene, each hybridizes with both the specific 5′ and 3′ cDNA probes.

CHARACTERIZING GENOMIC DNA FRAGMENTS

When genomic DNA is cleaved with a restriction enzyme that does not cut within a particular message sequence, cuts will be made on either side of the gene. These cuts per se should generate a single fragment. What are we to conclude when two (or more) fragments that hybridize with the mRNA actually are released in such an experiment?

One possibility is that there is more than one copy of the gene. (The copies need not necessarily be identical, just closely related enough to hybridize with one another.) In this case, the sequences on either side of each copy could be different, so that the restriction sites lie at different chance locations with regard to the gene. Then each copy will be released in a restriction fragment of different size.

The alternative is that the gene contains an intervening sequence (or sequences) with a recognition site for the enzyme. In fact, an additional fragment will be generated for every intron in which there is such a site. (The reverse result, that a single fragment is obtained, does not prove the absence of an intervening sequence, because the intron need not possess a site for the enzyme. Nor does it exclude the possibility that there is a small number of tandem copies.)

How are the alternatives of multiple gene copies and interrupted gene structures to be resolved? The usual approach is to use probes derived from **subclones** that represent only specific regions of the mRNA. These are obtained by cleaving the cloned cDNA into 5′ and 3′ regions with a suitable restriction enzyme (that is, one that cuts at an appropriate place within the message sequence). Then these end-specific probes are recloned separately into new plasmids.

Figure 20.12 shows that if each of two fragments represents a different copy of the gene, each will have its complete sequence; so both fragments will hybridize with both 5′ and 3′ probes.

Figure 20.13 shows that if two genomic fragments represent different parts of the gene that have been separated by cleavage within an intervening sequence, one will hybridize with the 5′ probe of cDNA; the other will hybridize with the 3′ probe.

When an interrupted gene is obtained in fragments, how are we to be certain of identifying every intron and every exon? The structure is deduced by obtaining *overlapping* fragments that correspond to different

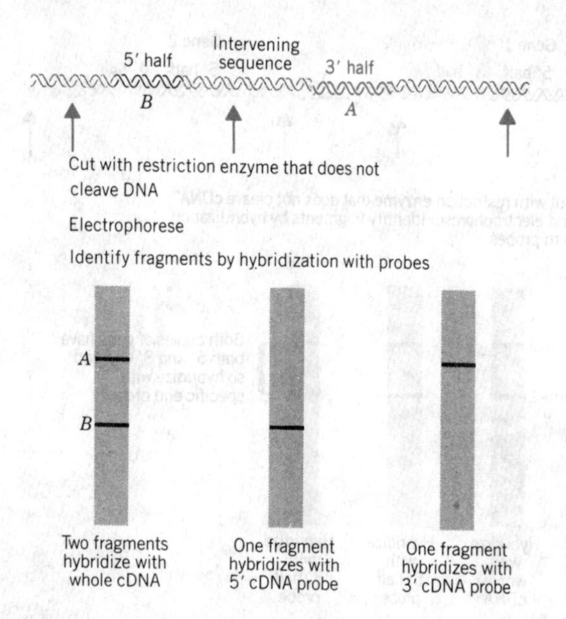

Figure 20.13
A single gene with an intron(s) can produce multiple fragments when the genome is cleaved with a restriction enzyme that does not cut within the cDNA.
If the enzyme cuts within the intron, two fragments are released that can hybridize with a labeled cDNA probe. But when 5′- and 3′-specific probes are used, each fragment hybridizes with only one probe.

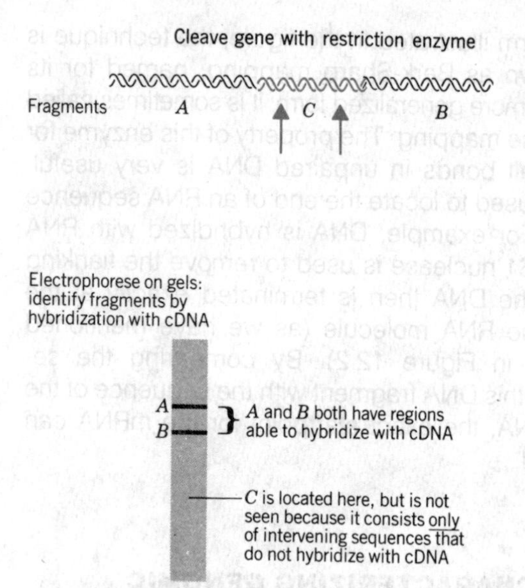

Figure 20.14
When genomic fragments are identified by hybridization with cDNA or mRNA, regions of introns that lie between two restriction sites may be missed.

regions. This is a general principle of restriction mapping and DNA sequencing (as described in Chapter 5), and is necessary to exclude the possibility that part of the gene has been missed.

Suppose, for example, that within the gene there is an intron carrying two restriction sites for some enzyme, as illustrated in **Figure 20.14.** When these are cleaved, the fragments to the left (*A*) and to the right (*B*) each consist in part of the sequence represented in mRNA and in part of the intron. The presence of the first component allows each to hybridize with the message probe. But the central fragment (*C*) consists *only* of intron sequences; because it lacks any message sequence, it does not hybridize with a probe representing the message. So the sequence will not be detected in the restriction digest. This type of occurrence is avoided by using several enzymes, extending the map in such a way that the *junction between each pair of adjacent fragments also is represented intact in some other fragment.* This may involve using regions of the intervening sequence as probes.

Ultimately a comparison of the nucleotide sequences of the gene and cDNA clone shows exactly where and how large the intervening sequences are. Resolution at the sequence level is necessary before we can be sure that all the segments of the gene have been identified. Indeed, there have been several cases where the genomic sequence presented a surprise in the form of an additional short coding segment. When a coding segment is shorter than about 50 bp, it may fail to hybridize with the cDNA probe, and can therefore pass unnoticed within the intervening sequences on either side.

GENES COME IN ALL SHAPES AND SIZES

No particular rhyme or reason yet has been discerned in the extremely varied structures of eukaryotic genes. Some genes are uninterrupted, so that the sequence in the genome is colinear with that of the mRNA. More genes at present appear to have interruptions, but the intervening sequences vary enormously in both number and size.

Some important features are common to all interrupted genes. The order of the parts of an interrupted gene is the same in the genome as in its mature RNA product. Genes are thus split rather than dispersed. Introns of nuclear genes generally have termination codons in all reading frames, and so cannot code for any polypeptides. An interrupted gene retains the same structure in all tissues, including the germ line and somatic tissues in which it is or is not expressed. Thus the presence of an intervening sequence is an invariant feature.

All classes of genes may be interrupted: nuclear genes coding for proteins, nucleolar genes coding for rRNA, and genes coding for tRNA. Interruptions also are found in mitochondrial genes in yeast and in chloroplast genes. Interrupted genes do not appear to be excluded from any class of eukaryotes, and have even been found in an archebacterium. They appear to be entirely absent only from the eubacteria.

Intervening sequences must be removed before a gene can be translated (or before an interrupted rRNA or tRNA can fulfill its function). They are present in the primary transcript, but absent from the mature RNA product. The process of their removal is called **RNA splicing.** Essentially it involves a controlled deletion of an intervening sequence from the primary transcript; the ends of the RNA on either side are joined together to form a covalently intact RNA. The general reaction is illustrated in **Figure 20.15;** mechanisms of splicing are discussed in Chapter 26.

Some interrupted genes possess only one or a few introns. The globin genes provide an extensively studied example (see Chapter 21). The two general types of globin gene, α and β, share a common type of

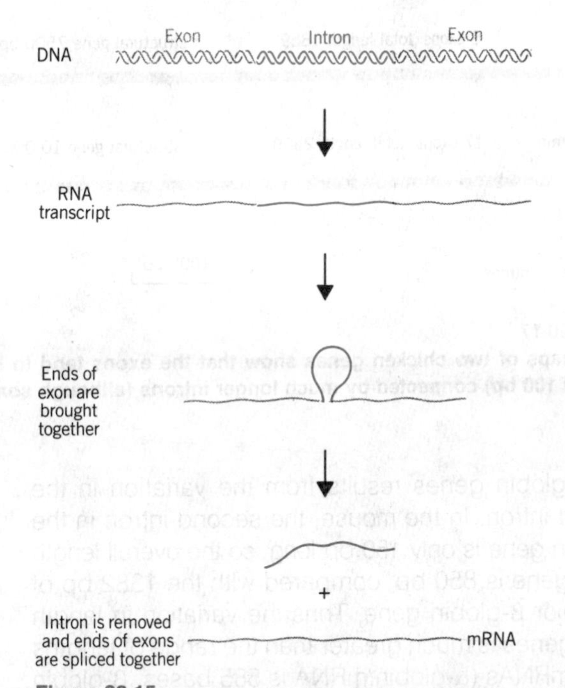

Figure 20.15
An interrupted gene consists of alternating exons and introns; the introns are removed by splicing of the RNA transcript, generating an mRNA consisting only of exon sequences.

structure in which two introns occur at constant positions relative to the coding sequence. The consistency of the organization of mammalian β-globin genes is evident from the summary in **Figure 20.16.** Interruptions occur at homologous positions in all known active globin genes, including those of several mammals, birds, and a frog. The first intron is always fairly short, and the second is usually longer, but the actual lengths can vary.

Most of the variation in overall lengths between dif-

Figure 20.16
All functional globin genes have an interrupted structure with three exons. The lengths indicated in the figure apply to the mammalian β-globin genes.

	Exon 1	Intron 1	Exon 2		Exon 3
Length in bp	142-145	116-130	222	573-904	216-255
Represents	5′ nontranslated + amino acids 1-30		Amino acids 31-104		Amino acids 105-end + 3′ nontranslated

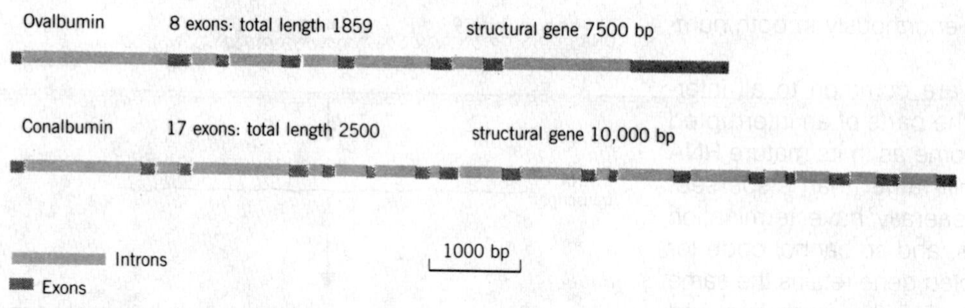

Figure 20.17

Scale maps of two chicken genes show that the exons tend to be rather small units (of the order of 100 bp) connected by much longer introns (although some exons are also long).

ferent globin genes results from the variation in the second intron. In the mouse, the second intron in the α-globin gene is only 150 bp long, so the overall length of the gene is 850 bp, compared with the 1382 bp of the major β-globin gene. Thus the variation in length of the genes is much greater than the range of lengths of the mRNAs (α-globin mRNA is 585 bases, β-globin mRNA is 620 bases).

Other genes may have more introns. Two examples are shown in **Figure 20.17.** The ovalbumin and conalbumin genes of the chicken have 8 exons and 17 exons, respectively. This makes the total lengths of the genes 7500 and 10,000 bp, compared with their respective mRNAs of 1859 and 2500 bases.

(In this and other cases, we define the length of the structural gene as the distance in the genome between points corresponding to the 5′ and 3′ terminal bases of mRNA. We know that transcription starts at the 5′ end of the mRNA; but it is not certain whether it terminates at the 3′ end or whether this end is generated by cleavage from a longer precursor. In the latter case, the length of the gene would be greater, including additional sequences downstream.)

Some genes are much longer than these examples and others are more highly mosaic. The present extremes are presented by mouse DHFR and chicken α collagen.

The gene for mouse DHFR is organized into 6 exons that correspond to the 2000 base mRNA. But these extend over more than 31,000 bp of DNA. Here the introns are exceedingly long.

The gene for chicken proα2 collagen is split into more than 50 exons, each rather short. The lengths of characterized exons vary from 45 to 249 bp. Some of the introns are short, comparable in length to the exons; but some are much longer, so that the total length of exons—about 5000 bp—is spread out over 40,000 bp of the genome.

In some eukaryotic structural genes, therefore, the exons actually comprise only a small proportion of the total length of the gene. Together with their often rather short lengths, this gives a picture of the gene as comprising a series of brief exons spread out over a large area of the genome. In such cases, the map visualizes the exons as interruptions in the introns, rather than vice versa. This is clear from Figure 20.17. Many genes must be painstakingly pieced together, as it were, by joining many short exons that reside considerable distances apart.

It seems possible that introns will actually fill a greater proportion of the higher eukaryotic genome than is provided by the exons, although present data are too sketchy to reach a firm conclusion. In lower eukaryotes, introns may be relatively less frequent, and genes are more likely to be uninterrupted, or at least possess few introns. Certainly the more startling examples of intricate gene structures all have been provided by higher eukaryotes. The mosaic form of the gene goes some way toward explaining the minor proportion of DNA that is represented in mRNA. Just how far remains to be seen.

INTRONS IN GENES CODING FOR rRNA and tRNA

Introns occur in the gene for large rRNA in several cases. Actually, interrupted genes were first discovered in *D. melanogaster*, where about two thirds of the genes for 28S rRNA have an interruption of up to 5000 bp in length. (Genes for rRNA are always present in multiple, usually identical, copies; see Chapter 23.) But since the other third are not interrupted, and since the interrupted genes do not seem to be used to synthesize rRNA, their structure left open the possibility that they constitute damaged, nonfunctional genes.

In some lower eukaryotes, however, *all* of the copies of the large rRNA gene are interrupted. Examples include *Tetrahymena pigmentosa* (a ciliate) and *Physarum polycephalum* (a slime mold). Their situation is therefore analogous to that of the nonrepetitive nuclear structural genes: the intron must be removed by splicing to make an active gene product. Yet in other lower eukaryotes, such as *S. cerevisiae* (yeast) and *D. discoideum* (a slime mold), the rRNA genes all are uninterrupted. In the mitochondrion of *S. cerevisiae* and *N. crassa* (a fungus), and in the chloroplast of *C. reinhardii* (an alga), again the gene for large rRNA is interrupted; but it is uninterrupted in the mitochondria and chloroplasts of other species.

No systematic relationship is discernible in the distribution of interrupted and uninterrupted rRNA genes. However, in all cases but one, there is only a single intron, always located at a position about two thirds of the way along the gene. (The rRNA molecules vary in length from about 3000 to 4000 bases.) Could this be a homologous position in these very different organisms? There are no interruptions in the genes for small rRNAs.

The presence of the intron is polymorphic in two instances: some strains of the organism possess an interrupted rRNA type of gene; other strains possess the uninterrupted type. This occurs in the nucleus of *T. pigmentosa* and in the mitochondrion of *S. cerevisiae*. In neither case is there any difference in the ability to utilize the gene. The coexistence of the two forms of the gene implies that the intron is neither essential nor deleterious.

Introns have been found in tRNA genes of the yeast nucleus. An interesting feature is that the intron is always located at the same relative location, at the start of the anticodon loop. Its presence allows the interrupted tRNA precursor to take up a conformation in which the anticodon is base paired with part of the intron. This may be relevant to the mechanism of splicing, and is illustrated later in Figure 26.1.

The length of the intron varies quite widely, being as short as 14 or as long as 46 bp in different tRNA genes. Yet again it cannot be an essential feature of the yeast nuclear tRNA gene per se, because some tRNA genes are not interrupted. In families of tRNA genes, some members of a family may be interrupted, while other, related genes are continuous.

EXON-INTRON JUNCTIONS HAVE A CONSENSUS SEQUENCE

By comparing the nucleotide sequence of mRNA with that of the structural gene, the junctions between exons and introns can be assigned. Two features (or lack thereof) are significant. First, there is no extensive homology or complementarity between the two ends of an intron. This excludes the possibility that an extensive secondary structure could form to link them directly together as a preliminary step in splicing. Second, the junctions do prove to have a well-conserved, though rather short, consensus sequence; its ubiquity argues that it may be involved in splicing in the nucleus (see Chapter 26).

In many cases, it is not possible to assign a unique site for the exon-intron junction solely on the basis of the mRNA/gene comparison. The difficulty is that a short sequence, usually between 1 and 4 bp, is repeated at each end of the intron. **Figure 20.18** shows an example from mouse (or rabbit) β globin, in which any one of four pairs of sites could mark the ends of the intron.

In spite of this difficulty, it is possible to assign a specific end to each intron by relying on the homology of exon-intron junctions. They can all be aligned in

Figure 20.18
Repetition of the trinucleotide AGG at both ends of the second intron in mouse β globin allows several possible frames for splicing.

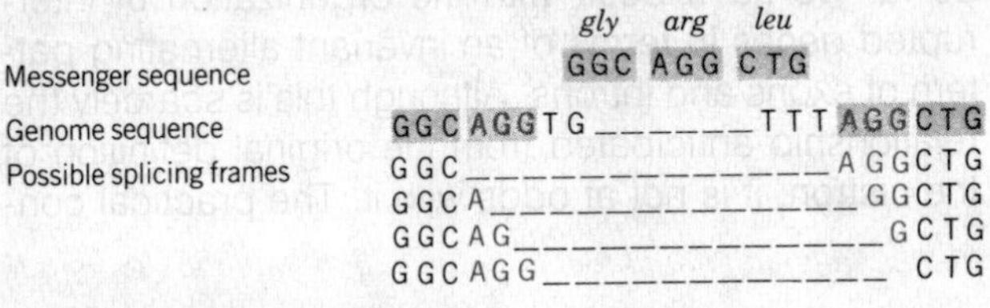

```
                            gly arg leu
Messenger sequence          GGC AGG CTG
Genome sequence             GGC AGG TG_______TTT AGG CTG
Possible splicing frames    GGC_________________AGGCTG
                            GGC A________________GGCTG
                            GGC AG________________GCTG
                            GGC AGG________________CTG
```

such a way as to relate closely to the consensus sequence

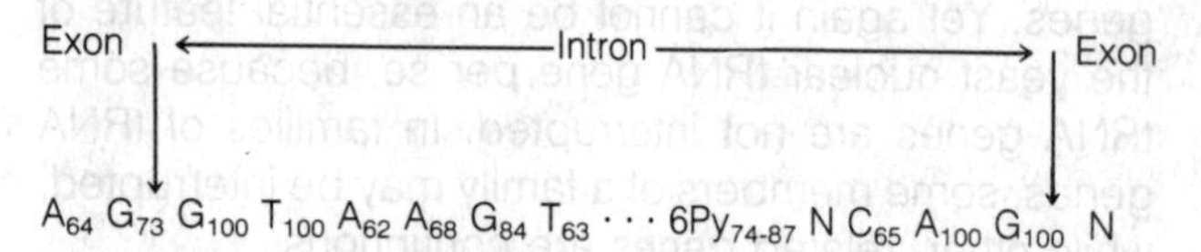

where the arrows mark the putative ends of the intron. (In this as in other cases, we write just the sequence of the DNA strand that is identical with the RNA product.)

The subscripts indicate the percent of junctions at which the specified base (or type of base) is found. The really high conservation is found only immediately within the intron at the presumed junctions. This identifies the essential ends

GT · · · · · · AG

Because the intron defined in this way starts with the dinucleotide GT and ends with the dinucleotide AG, the junctions are often described as conforming to the **GT–AG** rule.

Note that the two junctions have different sequences and so define the ends of the intron directionally. They are named proceeding from left to right along the intron, that is, as the **left** and **right** splicing sites. Sometimes they are called the **donor** and **acceptor** sites.

The consensus sequence is found in nuclear structural genes of many eukaryotes. So far as we know, it is common to all higher eukaryotes; this implies that there may be a common mechanism for splicing the introns out of RNA. However, the consensus does not apply to the introns of mitochondria and chloroplasts, nor to the yeast tRNA genes; at least in these cases, a fundamental difference in splicing mechanism may apply (see Chapter 26).

ONE GENE'S INTRON CAN BE ANOTHER GENE'S EXON

So far we have dealt with the organization of interrupted genes in terms of an invariant alternating pattern of exons and introns. Although this is scarcely the relationship anticipated from the original definition of the cistron, it is not at odds with it. The practical consequence of the mosaic structure is that mutations in a protein should map in clusters that correspond to the exons, rather than in a continuous segment of DNA. None of the mutations in a cluster will complement either with one another or with the mutations in other clusters. Any mutations in an intron that affect splicing of the RNA will behave in the same way as the exonic mutations, exercising a *cis*-acting control over expression. In other words, a mosaic gene is expressed *only* via the splicing together of exons carried by one molecule of RNA, thus conforming to the original concept of the *cis/trans* test (see Chapter 2).

In contrast with the common situation in which a gene is mosaic but otherwise conforms to the traditional view, there are some cases in which the fixed relationship between gene and protein does not hold. Here more than one mRNA sequence can be derived from a single stretch of DNA. The meaning of a particular region of DNA therefore is not invariant, but depends on the pathway selected for its expression. In this case, the pattern of complementation between mutations cannot be expected to conform entirely to the traditional interactions.

A tissue-specific change in the expression of mouse amylase mRNA is summarized in **Figure 20.19.** The same amylase protein is synthesized in the liver and the salivary gland of the mouse, and is coded by a single gene. The mRNAs present in each tissue are identical in their coding region, but differ in the first part of the 5′ nontranslated leader. The leader is longer in liver and different in sequence from that of salivary gland.

The amylase coding sequence starts about 50 bp within exon 2, and is formed by joining exon 2 to exon 3 and the subsequent exons. This sequence may be preceded by either of two exons. In liver, the first 161 bases of the mRNA are coded by exon L, which lies about 4500 bp upstream. In salivary gland, the first 50 bases of the mRNA are coded by exon S, which lies about 7300 bp upstream. Thus exons S and L provide alternative initial sequences for the amylase mRNA. The sequence of exon L is in fact part of the long intron that is removed by splicing in the salivary gland.

Another change in a processing pathway for mRNA maturation occurs during the expression of immunoglobulin genes. It involves the substitution of one exon by another at the 3′ end of the transcription unit. This

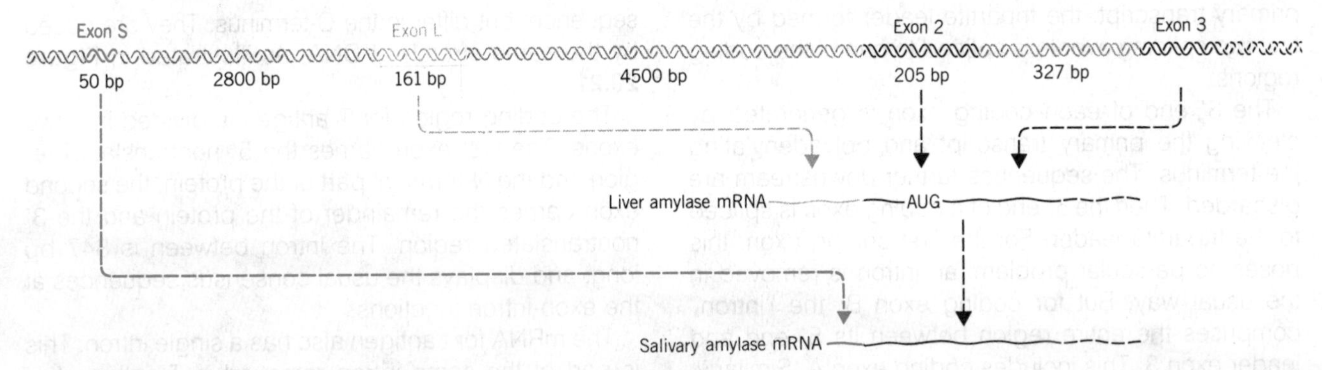

Figure 20.19
Alternative splicing generates two amylase mRNAs from one mouse gene.

changes the C-terminal amino acid sequence of the protein, with the result that a region responsible for membrane attachment is replaced by a region that allows secretion. The mechanism is discussed in Chapter 37.

In late adenovirus expression, the same leader sequence can be spliced to any one of several different coding sequences. **Figure 20.20** presents a diagrammatic version of these events. There is a single transcription unit whose expression is initiated at a single point. The first part of the transcription unit carries three sequences that are spliced together to form a nontranslated leader (as depicted in Figures 20.6 and 20.7). The components of the leader are quite short and are indicated as leader exons 1, 2, and 3. Downstream from these sequences, there are several coding regions, each representing a late viral protein. They are indicated as coding exons A, B, C, etc. In each

Figure 20.20
The same leader can be joined to different messenger bodies during late adenovirus expression. Only three coding exons are shown, but actually there are five groups of messenger bodies defined by common 3′ ends, and these include at least nine individual different coding regions.

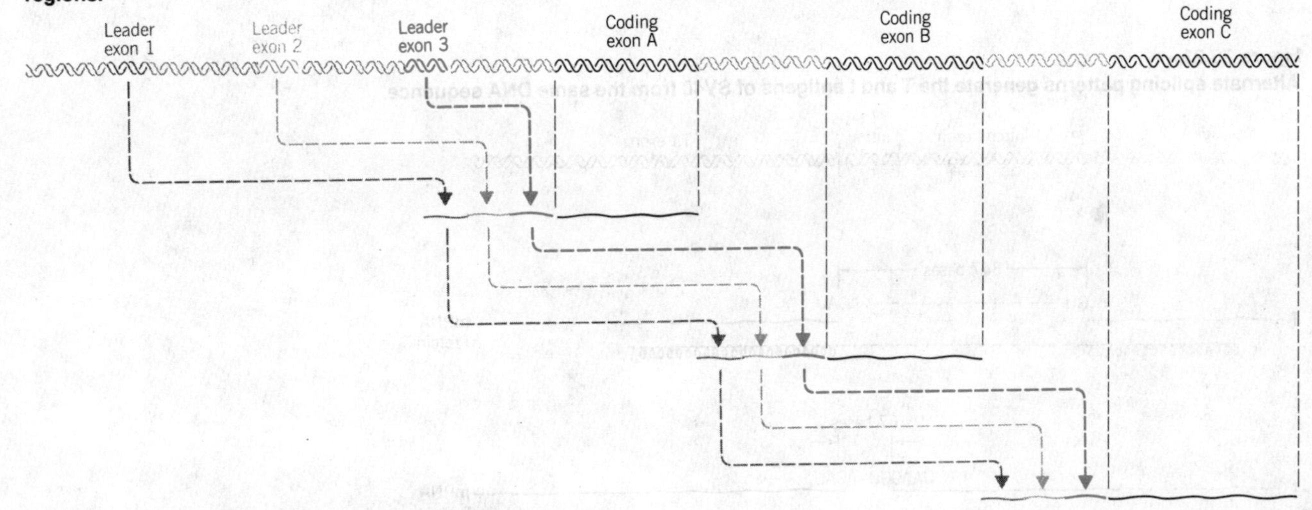

primary transcript, the tripartite leader formed by the first two splicing events is spliced to *one* of the coding regions.

The 3′ end of each coding exon is generated by cleaving the primary transcript and polyadenylating the terminus. The sequences farther downstream are discarded. Then the 5′ end of a coding exon is spliced to the tripartite leader. For the first coding exon, this poses no particular problem: an intron is removed in the usual way. But for coding exon B, the "intron" comprises the entire region between its 5′ end and leader exon 3. This includes coding exon A. Similarly, when coding exon C is spliced to the leader, both coding exons A and B are part of the "intron" that is removed.

So none of these coding exons is inevitably used to direct protein synthesis in a particular transcript molecule: in fact, only one can be utilized per RNA molecule. The others are discarded if they lie downstream or spliced out if they are upstream from the chosen exon. The generation of the 3′ end of the nuclear RNA is probably responsible for determining which exon is spliced to the tripartite leader.

Another situation occurs in the small DNA viruses SV40 and polyoma, where alternative splicing patterns allow the simultaneous production of proteins whose amino acid sequences overlap. This happens during both early and late transcription.

The early transcription of SV40 leads to the synthesis of two proteins, **T** (called "big T") and **t** (called "little t"). They have the identical N-terminal amino acid sequence, but differ in the C-terminus. They are coded by the same stretch of DNA, as illustrated in **Figure 20.21.**

The coding region for T antigen is divided into two exons. The first exon carries the 5′ nontranslated region and the N-terminal part of the protein; the second exon carries the remainder of the protein and the 3′ nontranslated region. The intron between is 347 bp long, and displays the usual consensus sequences at the exon-intron junctions.

The mRNA for t antigen also has a single intron. This is part of the same intron removed in T-antigen formation, but consists of only its last 67 bases. Thus the right splicing junction is identical in the two cases, and the difference lies in the choice of the left junction that is spliced to it. The left junction used in t-mRNA formation also follows the consensus rules.

Translation of the T and t antigens starts at the same AUG codon in their common initial sequence. The amino acid sequences diverge at the point where the T sequence is spliced to its second exon. The result is to produce two proteins, sharing the same N-terminal half, but having a different C-terminal half.

(The t antigen actually is coded entirely by the first exon of the t mRNA. The termination codon lies immediately before the end of the exon. Although the second T exon is present in t mRNA, it is not translated because protein synthesis terminates at the t-UAA codon; this codon does not terminate T synthesis, because it is removed as part of the T intron.)

In these circumstances, the definition of introns and

Figure 20.21

Alternate splicing patterns generate the T and t antigens of SV40 from the same DNA sequence.

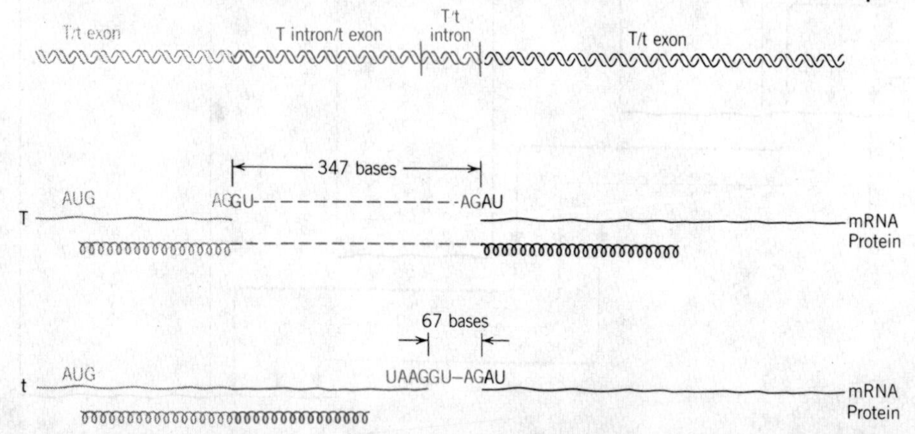

exons does not have the clear significance attributable in cellular genes.

MUTATIONS MAP MOSTLY IN EXONS

Will mutations in introns be detected by impairment of gene function? We might expect most changes in an intron to pass unnoticed, since, after all, these sequences are removed during expression of the gene. Mutations are detected in introns when they affect the ability of the RNA to be recognized by the splicing apparatus. Usually they change the consensus sequences, preventing proper removal of the introns (see Chapter 26).

In very few cases has it been possible to make any systematic study of mutations in a higher eukaryotic gene. We are too limited by the need to identify mutants on the basis of phenotypic change. One of the most detailed studies concerns the *white* locus of *D. melanogaster*, and its general outcome supports the view that exons are much more important than introns in gene function.

The *w (white)* locus was identified by the very first mutation isolated in the fruit fly, in the form of a male whose eye appeared bleached or translucent instead of displaying the usual wild-type red color. The wild-type eye possesses both red and brown pigments, unrelated in structure, and synthesized by different pathways. Several genes are identified by their ability to interfere with one or the other pathway (*vermilion* mutants lack brown pigment, *brown* mutants lack red pigment, and so on).

Mutations of the *w* locus fall over a region of about 0.04 map units and identify only a single complementation group. The products of *both* pathways are absent from *w* mutant eyes, which suggests that w^+ does not itself code for an enzyme involved in pigment production, but regulates, either directly or indirectly, the production or deposition of the pigments.

The *w* mutants are classified phenotypically by the magnitude of their effect on pigment formation. Null or bleached alleles—such as the original isolate w^1—entirely lack both red and brown pigments. Other alleles have intermediate phenotypes in which reduced amounts of one or both pigments are produced. The alleles usually are identified by superscripts whose names reflect the phenotype. For example, w^a is the apricot allele, w^i is the ivory allele, w^e is the eosin allele, and so on. Most of the alleles affect the quality of pigment, but the w^{sp} group also affect its distribution within the eye. This is consistent with the view that *w* governs the deposition of pigment.

Figure 20.22 shows a map of a region of about 15 kb that spans all known sites of *w* mutations. Analysis of the RNA corresponding to this region identifies a long stretch that represents either a single exon or (more likely) several exons separated by rather short introns. This segment corresponds to the 3′ end of the unit. At the 5′ end is a much shorter exon, separated from the 3′ segment by an intron of 3 kb.

Potential point mutations are identified by *w* loci in which no change can be detected in the restriction map. Genetic mapping locates all of these mutations to the left of the w^a insertion, in the region of the map corresponding to the major exon(s). Many of these mutations entirely abolish the function of the locus (the eye is bleached); others retain some residual pigment. Insertions in this segment usually cause the bleached null-phenotype that would be expected if a protein-coding region were disrupted.

Many mutations result from insertions of material (see Chapter 36). The only detectable deletions are the alleles w^{sp3} and w^{sp4}, which occur at or close to the

Figure 20.22
The *white* locus may be mutated by insertions, deletions or (potential) point mutations.
The region in which potential point mutations occur is white. (They are identified as such by the lack of any alteration in the restriction map.) Some of these mutations have a null phenotype. Insertions are indicated by arrows; those that cause a null phenotype are indicated by open arrows; those allowing residual pigment formation are shown by filled arrows.

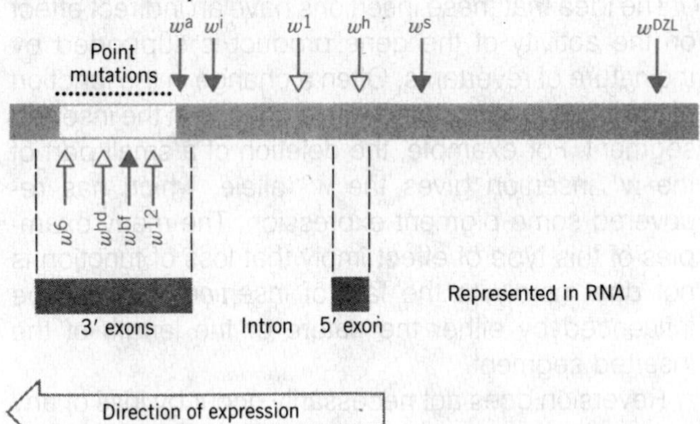

site of the w^{sp1} insertion. Most insertions map in exonic regions, usually in the large 3′ segment (as would be expected statistically). Not all of them have the null phenotype expected for disruption of a protein-coding region; but we must wait to see whether the leaky insertions perhaps lie in small introns.

Mutations mapping within the major intron and upstream from the 5′ exon often are leaky, displaying some residual pigment. No potential point mutations are found in this region; all mutants have changes in the restriction map. The great majority are insertions. The implication is that loss of function by mutation outside the coding region is more likely to result from a major disruption of sequence. Presumably point mutations have occurred just as frequently in the introns as in the exons, but have not been detected, because they have no effect on function. (Ironically, the atypical allele in this region is w^1, the original isolate with the bleached phenotype.)

The leaky insertions extend well to the right of the coding region. Their leakiness suggests that they do not interrupt a coding region, but interfere with some other function. For example, an insertion lying in an intron could prevent splicing, perhaps by interfering with the structure of the RNA.

At the w^{sp} site are several insertions, deletions, and combinations of them. The w^{DZL} insertion is the farthest on the right, possibly identifying the limits of the locus. It is possible that these sites of insertion identify individual regulatory regions; insertions at other sites in the right segment might not produce detectable mutant phenotypes. All mutations that affect potential regulatory functions, such as timing or pattern of deposition of pigment, or dosage compensation, map in the right segment.

The idea that these insertions have an indirect effect on the activity of the gene product is supported by the nature of revertants. Often a change in the function of the locus is associated with a change in the inserted segment. For example, the deletion of a small part of the w^1 insertion gives the w^e allele, which has recovered some pigment expression. The many examples of this type of effect imply that loss of function is not due merely to the fact of insertion, but can be influenced by either the nature or the length of the inserted segment.

Reversion does not necessarily occur by loss of any part of the inserted material, as highlighted by the derivatives of the w^i mutation. Unlike other changes that increase the size of the locus, w^i is not due to insertion of extraneous material, but represents a tandem duplication of a region of 3 kb from within *w*. The mutation reverts by precisely losing the duplication at the relatively high rate of $\sim 10^{-5}$, probably by intrachromosomal recombination.

An exceptional revertant has been found in the form of the w^c mutant, which has a crimson eye; it results from the insertion of an 11 kb sequence into the w^i duplication. The w^c allele is unstable, reverting at a frequency of $\sim 10^{-3}$ to wild type, back to w^i, or to other forms, including stable derivatives with deletions ending in *w;* it seems likely that these various rearrangements are sponsored by the inserted sequence. It seems extraordinary that an insertion as large as this sequence can restore the phenotype that had been abolished by the internal duplication. Perhaps the insert substitutes a new promoter for a damaged site.

An intriguing regulatory effect at the *w* locus is identified by the phenomenon of **transvection.** This term describes the ability of chromosomal rearrangements that prevent synapsis between homologous regions (as seen in polytene chromosomes) to change the phenotype, although the overall genotype is unaltered. The basis for the effect lies in the ability of a mutation in one allele to influence the activity of the *other* allele *only so long as the two alleles are synapsed.* Thus transvection may depend on some structural effect mediated by physical contact. Its existence raises the possibility that chromosome pairing (or some related effect) occurs in normal diploid cells.

An opportunity for insight into the basis for transvection is presented by the w^{DZL} mutation (DZL is an abbreviation for dominant zeste-like). So far as the *w* region to its left is concerned, the DZL insertion exerts a *cis*-acting inhibition of function. The unusual feature of this effect is illustrated in **Figure 20.23;** the w^{DZL} allele also prevents expression of a w^+ allele synapsed with it. This is not a conventional dominant *trans*-acting effect, because the same w^+ allele is expressed normally if it is translocated to another location where it no longer comes under the influence of a synapsed w^{DZL}.

The w^{DZL} allele has an insertion of a 13 kb sequence consisting of two nonidentical but related sequences

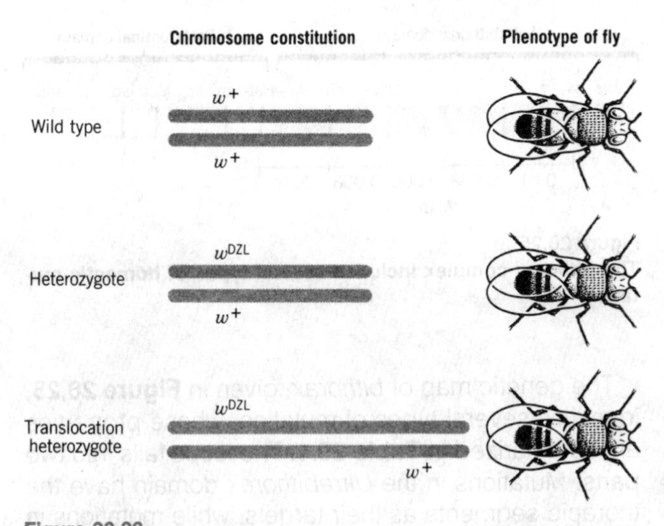

Figure 20.23
The w^{DZL} mutation prevents expression of a w^+ allele synapsed with it, but cannot act on a w^+ allele that has been translocated to a different location.

that flank a nonhomologous 6 kb segment of DNA (a nonrepetitive sequence from chromosome II). Like the similar insertion (in w^c), it is unstable, reverting at a frequency of 10^{-2} to 10^{-3}. Revertants to wild type have an excision of part of the DZL insertion, always including the central 6 kb region. This partial excision probably occurs by recombination between the sequences at the ends of the insertion; the revertants remain unstable because of the presence of the ends of these sequences. However, since the presence of an insertion at this site does not cause the DZL phenotype, the central 6 kb sequence rather than the insertion per se may be responsible for the transvection effect. Could this region code for a *cis*-acting product that also recognizes contact between DNAs?

Transvection can be exercised over long distances. The original example is the effect of the *z* mutation, 0.75 map units to the left of *w*, a distance probably corresponding to 300–400 kb. The *z* mutation is recessive. It has no effect in males (which have only one X chromosome); but in females that are homozygous for *z*, it represses a pair of synapsed w^+ alleles, giving a pale-colored (zeste) eye. The effect is abolished if one of the w^+ alleles is translocated elsewhere, or even if it is deleted. It may be altered by certain alleles of *w* created by mutations in the right segment of the locus.

The nature of the interaction between *z* and *w* suggests not only that it extends over synapsed alleles, but further that synapsis is necessary for the effect to occur at all. The propagation of this effect over such a long distance implies also that it may be a function of different loci (*z* and *w*); the corollary is that the DZL mutation need not necessarily be considered to comprise part of the *w* locus itself, but may be adjacent to it.

The basis for transvection is mysterious. Because it depends on contact between the alleles, the obvious explanation is that it is mediated through some effect on their expression. But it does not appear to result from inhibition of transcription, because the same amount of *white* RNA is found in *zeste* as in wild-type cells. We are therefore faced with the paradox that the effect must act before the product RNA is released from the locus, but it does not seem to affect the amount of product.

COMPLEX LOCI ARE EXTREMELY LARGE AND INVOLVED IN REGULATION

Several loci in *D. melanogaster* behave in a manner that is difficult to reconcile with the notion that each contains a single gene coding for a unique product. They are called **complex loci.**

A conventional gene—even an interrupted one—is identified at the level of the genetic map by a tightly linked cluster of noncomplementing mutations. A hallmark of complex loci is the presence of rather well-spaced groups of mutations, extending over a relatively large map distance, and often displaying a complex pattern of complementation. For example, the mutations may fall into several overlapping complementation groups. The individual mutations may have different and complex morphological effects on the phenotype.

At least some of the complex loci appear to be involved in regulating development of the adult insect body. The bodies of adult insects are divided into a series of **segments. Figure 20.24** locates the three thoracic and eight abdominal segments of *D. melanogaster*. Each segment consists of an anterior and a posterior compartment, each compartment derived from a relatively small group of "founder" cells.

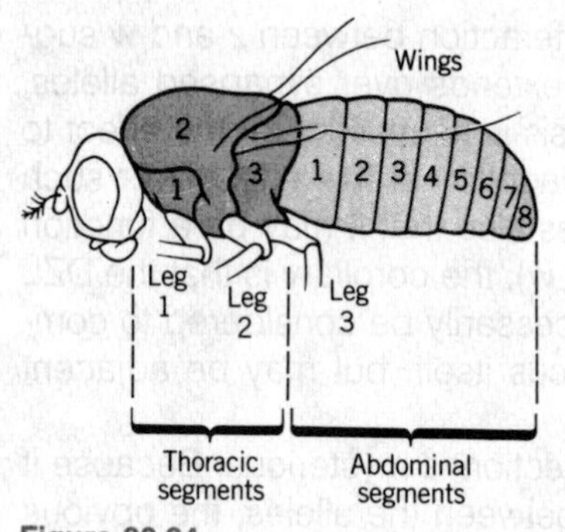

Figure 20.24
***D. melanogaster* has three thoracic segments and eight abdominal segments.**

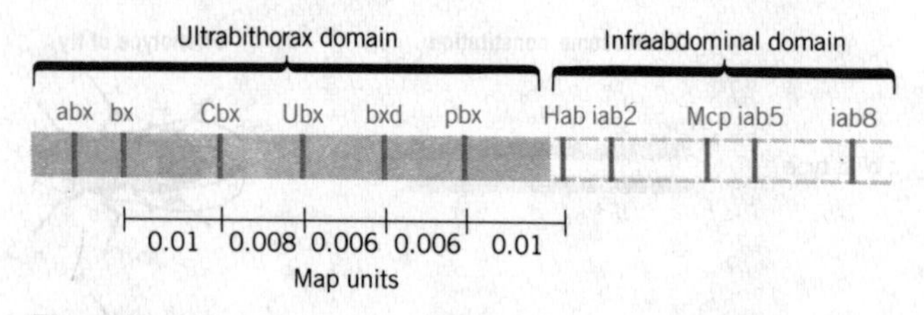

Figure 20.25
The *bithorax* complex includes several types of homeotic mutation.

Mutants affecting pattern formation fall into two general classes. The **segmentation mutants** alter the number or polarity of body segments. The **homeotic** mutations actually are able to transform part of a segment or an entire segment into another type of segment. These mutations identify loci that are particularly interesting because they may be concerned with determining the nature of body parts.

The classic complex locus is *bithorax*, characterized by several groups of homeotic mutations that affect sequential stages of development of the thorax, causing major morphological changes in the abdomen. When the whole complex is deleted, the animal dies late in embryonic development. Within the complex, however, are mutations that are viable, but change the phenotype of certain segments.

The genetic map of *bithorax*, given in **Figure 20.25**, identifies several types of mutation, whose properties are summarized in **Table 20.1**. The locus falls into two parts. Mutations in the *Ultrabithorax* domain have the thoracic segments as their targets, while mutations in the *infraabdominal* domain have the abdominal segments as targets.

Most is known about the *Ultrabithorax* domain, in which there are several different complementation groups. The mutations in each group form a set of mutant alleles that show a similar phenotype, varying in strength, but displaying the same types of tissue transformation. Each set identifies a gene that acts on a particular compartment. *Within the* bithorax *complex as a whole, therefore, are several individual loci whose products have related functions in the development of the fly.*

The individual types of mutations are given names related to *bithorax*. The original mutation, simply called *bithorax*—so the same name describes both one of the types of mutation within the complex as well as

Table 20.1
Each type of mutation within *bithorax* has a characteristic phenotype.

Mutation	Affected Segment	Phenotype
abx (anterobithorax)	T3	similar (but not identical) to *bx*
bx (bithorax)	T3	converts anterior T3 to anterior T2
Cbx (Contrabithorax)	T2	converts T2 toward T3
Ubx (Ultrabithorax)	T3	sum of *bx, bxd,* and *pbx*
bxd (bithoraxoid)	A1	converts anterior A1 to T3
pbx (postbithorax)	T3	converts posterior T3 to T2
Hab (Hyperabdominal)	Abdominal	details unknown
Mcp (Miscadestral pigmentation)	Abdominal	details unknown
iab2 (infraabdominal)	A2	converted to resemble A1

Figure 20.26
The *bithorax* mutation produces a four-winged fly.
Photograph kindly provided by Ed Lewis.

the complex as a whole—has the dramatic effect of converting the third thoracic segment (which carries the halteres [truncated wings]) into the tissue type of the second segment (which carries the wings). This type of mutation creates the fly shown in **Figure 20.26,** with four wings instead of the usual two.

The original series of *bithorax* and related mutations affect the thoracic segments or convert the first abdominal segment into a thoracic segment. These mutations identify the ultrabithorax domain that occupies the left half of the genetic map of Figure 20.25. Adjacent to them are further mutations that affect the abdominal segments per se and define the infraabdominal domain.

Chromosome walking has been used to construct a map of >100 kb covering the left half of the complex. The locations of the types of mutation are shown in **Figure 20.27.**

Mutations of the *bx* and *bxd* types each span a large, but apparently discrete region. Each region could be involved in coding for a protein needed to specify a particular segment (or part of segment).

The *pbx* and *Cbx* mutations are complementary in their phenotypic effects and turn out to be related at the molecular level. The *pbx*[1] deletion has lost a region of 17 kb, which is inserted elsewhere (in reverse orientation) to generate the *Cbx*[1] mutation. If this region codes for the information to specify the afflicted region (the posterior third thoracic segment), its loss in *pbx* mutants may cause that segment instead to resemble the preceding segment (the second thoracic). How-

Figure 20.27
Most mutations in the bithorax complex are due to insertions or deletions, which map in groups according to the affected function.

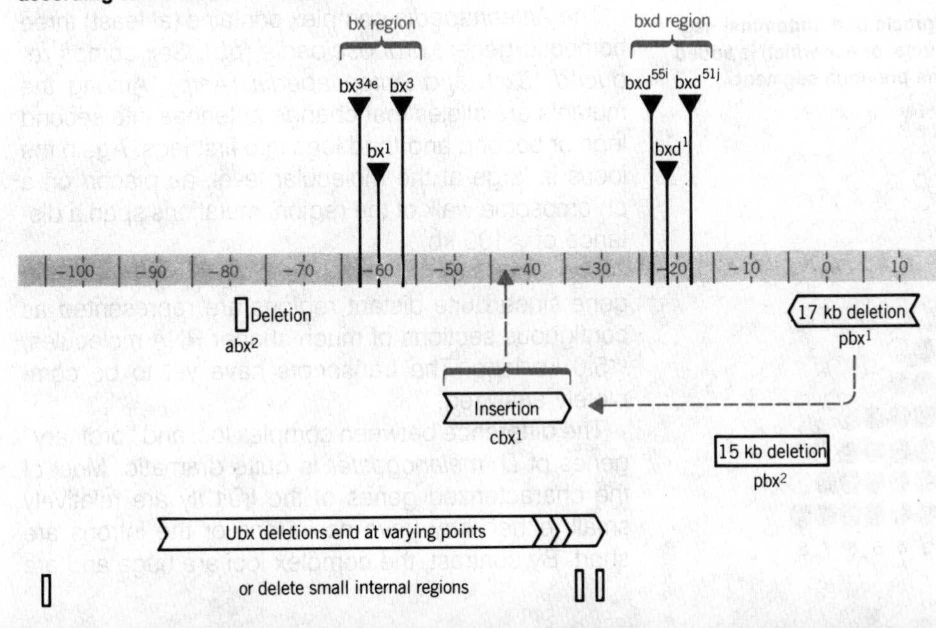

ever, when it is inserted at the -44 kb site, it is expressed in the second thoracic segment, thus causing the production instead of third thoracic structures.

The *Ubx* type behaves as though possessing mutations in the three functions of *bx, bxd,* and *pbx.* Most *Ubx* alleles have cytologically visible rearrangements of the chromosome, which at the molecular level turn out to represent deletions eliminating varying lengths of the left half of the cluster. Some other *Ubx* mutations have small deletions at points within the cluster. We do not yet understand the nature of the *Ubx* mutations, but they seem to imply that Ubx^+ functions require the entire left half of the complex.

The positions of mutations on the genetic map tend to correspond with their effects on the phenotype. Moving along the gene complex from left to right, mutations usually affect segments moving down the fly. **Figure 20.28** illustrates a model that relates the genetic constitution in diagrammatic form to the fly morphology. Each segment is distinguished from its predecessor by the synthesis of an additional protein, coded by the next gene in line along the genetic map. In assessing this model, it is a pity that we have most information about the region concerned with the three thoracic loci, rather than what should be a larger set of functions concerned with the abdominal regions.

The complex nature of the *bithorax* locus as a whole can be explained by supposing that there are interactions between the synthesis or function of the individual genes within it. We should like to know whether these genes are transcribed individually or in some coordinate manner. Is there some sort of regulation of the locus as a whole? How are the products of the locus deployed geographically within the embryo?

Figure 20.28

The *bithorax* region could define thoracic and abdominal segments if it codes for a series of products, one of which is added to distinguish each segment from the previous segment.

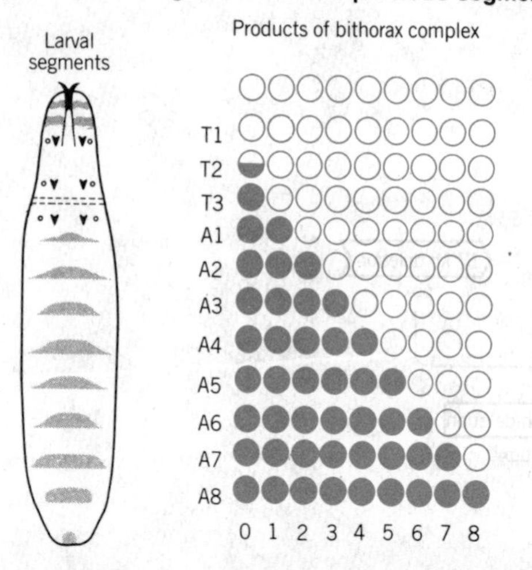

Notch (N) is another complex locus, at which mutations have several apparently unrelated effects on the phenotype. *N* mutants produce a dominant notched wing phenotype in *N*/+ heterozygotes, and are recessive lethals as seen by the death of *N/N* homozygotes. *N* mutations do not complement one another.

Within the locus are several groups of recessive mutations with different effects on the phenotype. Some alter wing formation; others affect the eyes or bristles. These mutations all are defined as contributing to *Notch* by their failure to complement *N* mutations, but tested among themselves they fall into (at least) four complementation groups.

The *Notch* locus contains a transcription unit that extends over 37 kb and includes 9 exons, ranging in length from 130 bp to 7250 bp.The corresponding RNA is about 11.7 kb long. Most of the *N* mutations fall in the exons; some of the recessive mutations represent insertions in introns. When we can characterize the mutations by determining their effects on the DNA sequence of the locus, we may be able to understand how the complex nature of *Notch* mutations is caused.

The *Antennapedia* complex contains (at least) three homeotic genes, *proboscipedia (pb), Sex combs reduced (Sxr),* and *Antennapedia (Antp).* Among the mutants are alleles that change antennae into second legs or second and third legs into first legs. Again the locus is large at the molecular level; as placed on a chromosome walk of the region, mutations span a distance of >100 kb.

The locus has the characteristics of an interrupted gene since quite distant regions are represented as contiguous sections of much shorter RNA molecules, <5.0 kb long. The transcripts have yet to be completely analyzed.

The difference between complex loci and "ordinary" genes of *D. melanogaster* is quite dramatic. Most of the characterized genes of the fruit fly are relatively small; either they have no introns or the introns are short. By contrast, the complex loci are huge and are spliced into RNA molecules an order of magnitude shorter. Why are loci involved in regulating development of the adult insect from the embryonic larva different from genes coding for the everyday proteins of the organism?

One possibility is that alternate patterns of gene expression generate various products from the complex. The alternate products could be generated by DNA rearrangement (which at present seems unlikely because no genomic changes have been detected), or alternate splicing patterns. Such models could explain the overlapping but related complementation patterns, in which mutations sometimes appear to belong to independent genes within the locus and sometimes appear to be part of a single complex locus.

The expression of these loci is difficult to analyze. It is extremely hard to identify RNA molecules ~100 kb in length; and in any case, if splicing occurred before transcription is completed, intact primary transcripts may not exist. Differently spliced RNA molecules might be present at rather low concentrations, making it difficult to identify the individual products.

A potential relationship between the functions of homeotic genes has emerged with the discovery of sequence homologies between the homeotic loci *Ultrabithorax, Antennapedia,* and the segmentation locus *fushi tarazu (ftz)* (in which mutations halve the number of segments). Each of these genes is expressed in a spatially restricted way during embryonic development. For example, *ftz* is required in alternating segment primordia.

A 180 bp region located near the 3′ end of the transcription unit of each of these loci is common to all of them. In pairwise comparisons of the corresponding sequence from each locus, the homology is ~80% by nucleotide sequence. This **homeo box** is part of an open reading frame, in which many of the base changes are located at third-base positions in the codons.

Even more striking, a DNA probe representing the homeo box hybridizes with frog, mouse, and human DNA. A gene from *X. laevis* has been cloned by virtue of its possession of the homeo box; it shares 55 of the 60 amino acids coded by the *Antp* homeo box. The gene is expressed during early development, and produces three transcripts. The genes in mouse and man also possess highly conserved homeo boxes.

What common function could there be between genes engaged in developmental regulation in fruit flies, frogs, and mammals? Perhaps the gene products give universal types of instructions, telling the cells to stop dividing, to divide faster, or to freeze the state of development.

The homeotic genes have been regarded as regulators of developmental switching, operating within specific compartments, perhaps coding for proteins that bind to particular batteries of genes whose products give the cells of each compartment their unique identity. Could similar mechanisms operate in invertebrates and vertebrates in spite of their very different superficial organization? Have we discovered a common mechanism of regulation of embryonic development?

HOW DID INTERRUPTED GENES EVOLVE?

By using restriction fragments that correspond to specific regions of a structural gene, it is possible to test whether they are related to other sequences in the genome. A cloned restriction fragment can either be used as a tracer in hybridization with whole genome DNA to determine its repetition frequency, or it can be used as a probe to detect corresponding sequences in a restriction digest. By obtaining appropriate restriction fragments, regions of either exons or introns can be investigated.

Often it turns out that the exons of a gene are related to those of another gene; when an exonic probe is used, it detects fragments that are part of another gene or genes. This implies that the two genes originated in a duplication of some common ancestral gene, after which differences accumulated between the copies.

Usually the introns are not related to other sequences. And when two genes have related exons, the relationship between their introns is more distant than the relationship between the exons. Because the exons are under the constraint of needing to code for a particular sequence of amino acids, they are limited in their potential to change sequence. Free of such restriction, the introns can accumulate more mutations.

Our original description considered nonrepetitive DNA to represent sequences that are unique in the genome. Introns come closer than exons to fitting this definition. In asking whether structural genes are nonrepetitive, we see therefore that the answer is ambiguous: the entire length of the gene may be unique as such, but its exons often are related to those of other genes. At least for some genes, the exons constitute slightly repetitive sequences embedded in a unique context of introns.

The overall degree of divergence between two exons is related to the differences between the proteins. It is caused mostly by base substitutions. However, many of the changes do not affect codon meanings, either because they lie in third-base positions or because they are in nontranslated regions.

In corresponding introns, the pattern of divergence involves both changes in size (due to deletions and insertions) and base substitutions. In some pairwise combinations of β-globin genes, there has been so much variation in the second intron that corresponding sequences cannot even be recognized. In other words, introns evolve much more rapidly than exons.

These changes imply that the intron does not have a sequence-specific function (or at least not one that is the same in different species). Whether its presence is at all necessary for gene function is impossible to say, except to note that no functional β-globin genes have been found in which it is absent. However, an intron may be able quite freely to accumulate point mutations and other changes. Sequence comparisons imply that changes occur at the same rate in both exons and introns, but are not removed from the introns by adverse selection. The example of the *white* locus confirms that mutations are rarely detected in introns, and the few exceptions turn out to involve major changes of sequence.

What was the original form of genes that today are interrupted? Did the ancestral protein-coding units consist of uninterrupted sequences of DNA, into which introns were subsequently inserted during their evolution to the present form? Or did these genes initially arise as interrupted structures, which since have been maintained in this form? Another form of this question is to ask whether the difference between eukaryotic and prokaryotic genes is to be accounted for by the acquisition of introns in the eukaryotes or by the loss of introns from the prokaryotes.

Could the mosaic structure be a remnant of an ancient approach to the reconstruction of genes to make novel proteins? Suppose that an early cell had a number of separate protein-coding sequences. One aspect of its evolution is likely to have been the reorganization and juxtaposition of different polypeptide units to build up new proteins.

If the protein-coding unit must be a continuous series of codons, every such reconstruction would require a precise recombination of DNA to place the two protein-coding units in register, end to end in the same reading frame. Furthermore, if this combination should not be successful, the cell has been damaged, because it has lost the original protein-coding units.

But if an approximate recombination of DNA could place the two protein-coding units within the same transcription unit, splicing patterns could be tried out to combine the two proteins into a single polypeptide chain. And if these combinations are not successful, the original protein-coding units remain available for further trials. Such an approach essentially allows the cell to try out controlled deletions in RNA without suffering the damaging instability that could occur from applying this procedure to DNA.

If current proteins evolved by the combination of originally separate ancestral proteins, the accretion of units is likely to have occurred sequentially over some period of time, with one exon added at a time to existing proteins. If this model is realistic, we can ask whether the different functions from which these genes were pieced together are discernible in their present structure. In other words, can we identify particular functions of current proteins with individual exons?

In some cases, there is indeed a clear relationship between the structures of the gene and protein. The example par excellence is provided by the immunoglobulin proteins, which are coded by genes in which every exon corresponds exactly with a known functional domain of the protein (see Chapter 37).

In other instances some of the exons can be identified with particular functions. Often the first exon, coding for the N-terminal region of the polypeptide, specifies the region that provides the signal sequence involved in membrane secretion. An example is insulin.

Sometimes the evolution of a gene may involve the duplication of exons, creating an internally repetitious sequence in the protein. For example, in chicken collagen a 54 bp exon appears to have been multiplied

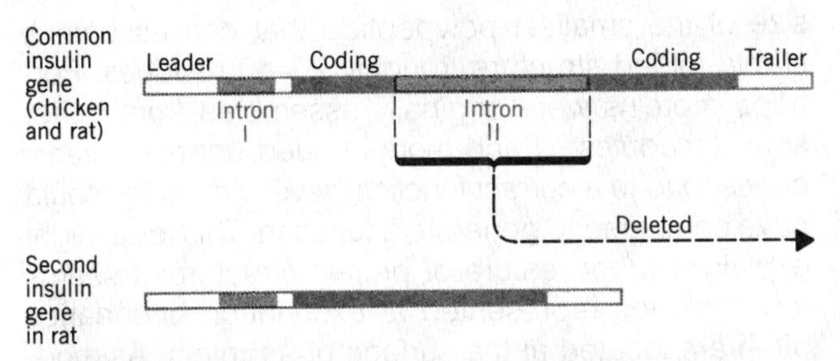

Figure 20.29
The rat insulin gene with one intron evolved by losing an intron from an ancestor with two interruptions.

many times, generating a series of exons that are either 54 bp or multiples of 54 bp in length.

In some instances, exons can be identified that code for segments that clearly extend across the junction between two different domains of the protein. Examples are found in collagen. In the ADH gene of *D. melanogaster,* there is a curious duality, with some exons corresponding exactly to protein domains, while others divide a domain into two parts.

There are some interesting cases in which homologous genes differ in structure. This may provide information about their evolution. An example is insulin. Mammals and birds have only one gene for insulin, except for the rodents, which have two genes. **Figure 20.29** illustrates the structures of these genes.

In chicken, the single insulin gene has two introns; one of the two rat genes has the same structure. The common structure implies that the ancestral insulin gene had two introns. However, the second rat gene has only one intron. It must have evolved from the first by a process in which a gene duplication in rodents was followed by the precise removal of one intron from one of the copies.

A fascinating case of evolutionary conservation is presented by the globins, all of whose genes have three exons. The two introns are located at constant positions relative to the coding sequence. The central exon appears to represent the heme-binding domain of the globin chain. The active protein is a tetramer containing two globin chains of the α type and two of the β type.

Another perspective on this structure is provided by the existence of two other types of protein that are related to globin. Myoglobin is a monomeric oxygen-binding protein of animals, whose amino acid sequence suggests a common (though ancient) origin with the globin subunits. Leghemoglobins are oxygen-binding proteins present in the legume class of plants; like myoglobin, they are monomeric. They too share a common origin with the other heme-binding proteins. Together, the globins, myoglobin, and leghemoglobin constitute the globin "super-family," a set of gene families all descended from some common ancestor.

Myoglobin is represented by a single gene in the human genome, whose structure is essentially the same as that of the globin genes. The three exon structure therefore predates the evolution of separate myoglobin and globin functions.

Leghemoglobin genes contain three introns, the first and last of which occur at points in the coding sequence that are homologous to the locations of the two introns in the globin genes. This remarkable similarity suggests an exceedingly ancient origin for the heme-binding proteins in the form of a split gene, as illustrated in **Figure 20.30.**

The central intron of leghemoglobin separates two exons that together code for the sequence corresponding to the single central exon in globin. Could the central exon of the globin gene have been derived by a fusion of two central exons in the ancestral gene, bringing together the sequences coding for two parts of the protein chain that together form the heme-binding structure?

Figure 20.30
The exon structure of globin genes corresponds with protein function, but leghemoglobin has an extra intron in the central domain.

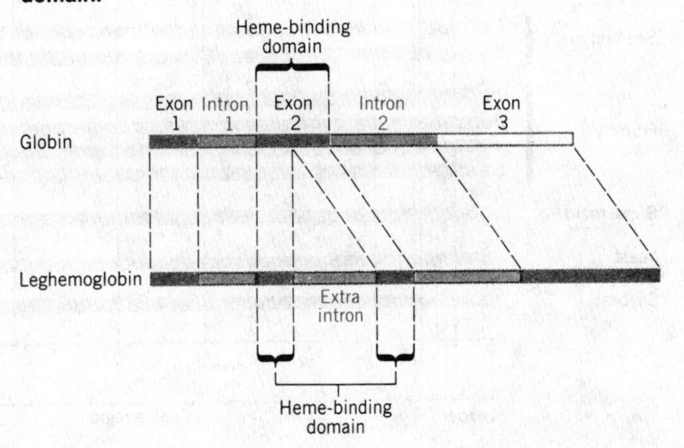

Considering all the genes whose organization is known, the relationship between exons and protein domains is somewhat erratic. In some cases there is a clear 1:1 relationship; in others no pattern is to be discerned. One possibility is that introns are sometimes removed, fusing the adjacent exons. A difficulty in this idea is the need to suppose that the intron removal was precise, not changing the integrity of the coding region. An alternative idea is that some introns arose by insertion into a coherent domain; here the difficulty is that we must suppose that the intron carried with it the ability to be spliced out.

The equation of at least some exons with protein domains supports the idea that this has been a fundamental relationship in the evolution of genes. It is clear that the duplication and juxtaposition of exons may have played an important role in evolution. We cannot trace the actual events involved in the evolution of every gene; many relationships between exons and protein domains do not conform to a simple equation, but they could be accounted for if events such as exon fusions have modified the ancestral structure during the evolution of nuclear genes.

Exons tend to be fairly small, on average coding for between 20 and 80 amino acids. This is around the size of the smallest polypeptide that can assume a stable folded structure, roughly 20–40 residues. Perhaps proteins were originally assembled from rather small "modules." Each module need not necessarily correspond to a current function; several modules could have combined to generate a function. This idea might explain another feature of protein structure: it seems that the sites represented at exon-intron boundaries often are located at the surface of a protein. As modules are added to a protein, the connections, at least of the most recently added modules, could tend to lie at the surface.

There are some situations in which we are faced with rather widespread discrepancies in gene structures. In these cases, there must have been extensive removal or insertion of introns during evolution.

The best characterized case is represented by the actin genes. The typical actin gene has a nontranslated leader of less than 100 bases, a coding region of just under 1200 bases, and a trailer of about 200 bases. Most actin genes are interrupted; the positions of the introns can be aligned with regard to the coding sequence (except for a single intron sometimes found in the leader).

Figure 20.31 shows that almost every actin gene is

Figure 20.31
Actin genes vary widely in their organization.
The sites of introns are indicated by red triangles; the number identifies the codon that the intron interrupts.

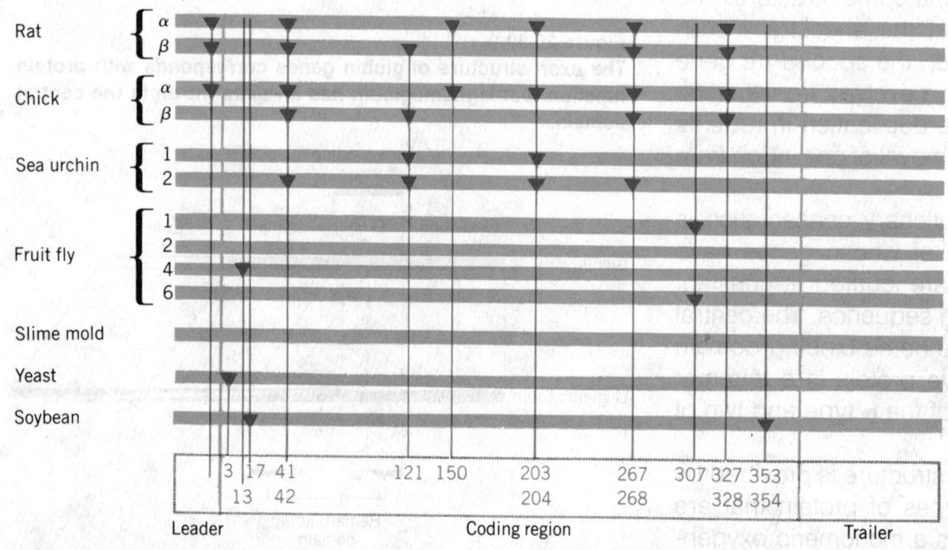

different in its pattern of interruptions. Taking all the genes together, introns occur at 12 different sites. However, no individual gene has more than 6 introns; some genes have only one intron, and one is uninterrupted altogether. Some relationships are found between different species; for example, 4 of the intron locations are found in rat, chick, and sea urchin. One site is common to plant and mammal.

How did this situation arise? If we suppose that the primordial actin gene was interrupted, and all current actin genes are related to it by loss of introns, different introns have been lost in each evolutionary branch. Probably some introns have been lost entirely, so the primordial gene could well have had 20 or more.

A similar situation may apply to fibrinogen, which is even more dramatic, since three genes, linked in the rat genome, each have different patterns of interruption. Out of a total of 14 introns in all three genes, 11 occur at a unique site in one of the genes.

The mitochondrial genome represents another striking case. The genes of yeast and mammalian mitochondria code for virtually identical mitochondrial proteins, in spite of a striking difference in gene organization. Vertebrate mitochondrial genomes are very small, with an extremely compact organization of continuous genes (see Chapter 22), whereas yeast mitochondrial genomes are larger and have some complex interrupted genes. Which is the ancestral form?

Is the presence of introns in the yeast mitochondrion at odds with the view that mitochondria are the result of an ancient endosymbiotic event in which a bacterial prototype was inserted into eukaryotic cytoplasm? This is not a problem if we assume that the insertion predated the loss of introns from the prokaryotic line, although then we have to assume that the ability to splice out introns was lost independently during the evolution of bacteria and vertebrate mitochondria.

Polymorphisms seem common in genes for rRNA and tRNA, where alternative forms can often be found, with and without introns. In the case of the tRNAs, where all the molecules conform to the same general structure, it seems unlikely that evolution brought together the two regions of the gene. After all, the different regions are involved in the base pairing that gives significance to the structure. So here it may be that the introns were inserted into continuous genes.

FURTHER READING

An extensive discussion of structures of interrupted genes was given in **Lewin's** *Gene Expression,* **2,** *Eucaryotic Chromosomes* (Wiley, New York, 1980, pp. 790–847). Some further thoughts were given in *Cell* (**22,** 645–646, 1980). The first major review of gene structure since the discovery of interrupted genes was provided by **Breathnach & Chambon** (*Ann. Rev. Biochem.* **50,** 349–383, 1981). The evolution of interrupted genes has been analyzed by **Perler et al.** (*Cell* **20,** 555-566, 1980). The biology of *bithorax* has been reviewed by **Lawrence** (*Cell* **35,** 595–601, 1983) and a start on its molecular characterization was made by **Bender et al.** (*Science* **221,** 23–29, 1983). The homeo box was discovered by **McGinnis et al.** (*Cell,* **37,** 403–408, 1984), and **Carrasco et al.** (*Cell* **37,** 409–414, 1984).

PART 6
CLUSTERS OF RELATED SEQUENCES

In the genetic programme, therefore, is written the result of all past reproductions, the collection of successes, since all traces of failures have disappeared. The genetic message, the programme of the present-day organism, therefore, resembles a text without an author, that a proof-reader has been correcting for more than two billion years, continually improving, refining and completing it, gradually eliminating all imperfections.

François Jacob, 1973

CHAPTER 21 STRUCTURAL GENES BELONG TO FAMILIES

Few genes appear to be alone in the eukaryotic genome. Although most genes behave as members of nonrepetitive DNA in liquid hybridization analysis, the isolation of genomic fragments usually reveals the presence of other sequences related to their exons. Sometimes these other sequences are functional genes, less or more distantly separated in evolution; sometimes they are nonfunctional **pseudogenes,** relics of evolution, descended from genes that once must have been functional, but that now have become inactive and accumulated many mutational changes.

Considering the interrupted, sometimes very extensively spread out structure of eukaryotic genes, we can picture the eukaryotic genome as a sea of introns (mostly but not exclusively unique in sequence), in which islands of exons (sometimes very short) are strung out in individual archipelagoes that constitute genes.

Comparing the corresponding exons in related genes, we see a relationship that emphasizes the importance of gene duplication as a mechanism for generating new genes. One copy can evolve via mutation, while the other retains its original function.

The history of a gene is likely to encompass a series of events in which first its component exons were brought together to form a coding region; later the entire series of exons and introns constituting the gene may have been duplicated. The duplication will have been followed by lesser divergence in exon sequences and greater divergence in intron sequences. From the sequences of related genes, we can therefore attempt to reconstruct the history of their evolution.

A set of genes descended by duplication and variation from some ancestral gene is called a **gene family.** Its members may be clustered together or dispersed on different chromosomes (or a combination of both). They may be identical or may be different, but related to one another.

Clustering is a prerequisite for maintaining identity between genes, although clustered genes are not necessarily identical. **Gene clusters** range from extremes where a duplication has generated two adjacent related genes to cases where hundreds of identical genes lie in a tandem array.

Situations where related genes are dispersed at different locations must have arisen by **translocation** of one gene at some time after a duplication event. After their separation, the genes usually diverge in sequence.

The members of a structural gene family have related or identical functions, although they may be expressed at different times or in different cell types. In cases where there is an extensive tandem repetition of a gene or genes, the product is needed in unusually large amounts. Examples are the genes for rRNA or histone proteins (see Chapter 23). A lesser degree of repetition may be used to provide slightly different proteins for particular circumstances. For example, different globin proteins are provided for use in embryonic and adult red blood cells, while different actins are utilized in muscle and nonmuscle cells. Sometimes no significance is discernible in a repetition; for example, no difference is apparent in the expression or function of the duplicate insulin genes of rodents (and we know that a single insulin gene indeed is adequate in other mammals).

GLOBIN GENES ARE ORGANIZED IN TWO CLUSTERS

The transport of oxygen through the bloodstream is a function central to the animal kingdom, coded by an ancient gene family. The major constituent of the red blood cell is the globin tetramer, associated with its heme (iron-binding) group in the form of hemoglobin. Genes coding for the globin polypeptide chains have been analyzed in several species. All functional globin genes have the same general structure, divided into three exons as shown previously in Figure 20.16. We have concluded that all globin genes are derived from a single ancestral gene (see Chapter 20). We may therefore learn a great deal about the general mechanisms involved in the evolution of gene systems by tracing the development of individual globin genes within and between species.

In adult cells, the globin tetramer consists of two identical α chains and two identical β chains. The α- and β-globin genes are coded by independent genetic loci whose expression must be coordinated to ensure equivalent production of both polypeptides. This system therefore provides an example of the need for simultaneous control of dispersed genes to generate a particular cell phenotype.

Embryonic blood cells contain hemoglobin tetramers that are different from the adult form. Each tetramer contains two identical α-like chains and two identical β-like chains, each of which is related to the adult polypeptide and is later replaced by it. This is an example of developmental control, in which different genes are successively switched on and off to provide alternative products that fulfill the same function at different times.

The details of the relationship between embryonic and adult hemoglobins vary with the organism. The human pathway consists of three stages (which still are not completely delineated). The hemoglobins found at different stages of development are

Embryonic (up to 8 weeks)	$\xi_2\epsilon_2$ and $\xi_2\gamma_2$ and $\alpha_2\epsilon_2$
Fetal	$\alpha_2\gamma_2$
Adult	$\alpha_2\delta_2$ and $\alpha_2\beta_2$

Zeta and alpha are the two α-like chains. Epsilon, gamma, delta, and beta are the β-like chains. In terms of developmental events, the simplest pathway is to suppose that zeta is the first α-like chain to be expressed, but is soon replaced by alpha itself. While in the β-pathway, epsilon and gamma are expressed first, with delta and beta replacing them later. In adults, the $\alpha_2\beta_2$ form provides 97% of the hemoglobin, $\alpha_2\delta_2$ is about 2%, and about 1% is provided by persistence of the fetal form $\alpha_2\gamma_2$.

The division of globin chains into α-like and β-like reflects the organization of the genes. Each type of globin is coded by genes organized into a single cluster. In man, the α cluster lies on chromosome 16, and the β cluster lies on chromosome 11.

Each cluster was discovered by the ability of genomic fragments to hybridize with probes representing adult globin mRNA. Extension of this analysis along the DNA led to the discovery of further genes, some representing known or unknown embryonic globins, others comprising nonfunctional pseudogenes.

The structures of the two clusters in the human genome are illustrated in **Figure 21.1.** The organization of these genes in other higher primates is very similar. For example, the β clusters of gorilla and baboon are virtually indistinguishable from that of man. All of the genes are transcribed from left to right, and they lie in the order of developmental expression: 5′—embryonic—fetal (in β cluster)—adult—3′.

Stretching over 50 kb, the β cluster contains five functional genes—epsilon, two gamma, delta, and beta—and one pseudogene. The two gamma genes differ in their coding sequence in only one amino acid; the G variant has glycine at position 136, where the A variant has alanine.

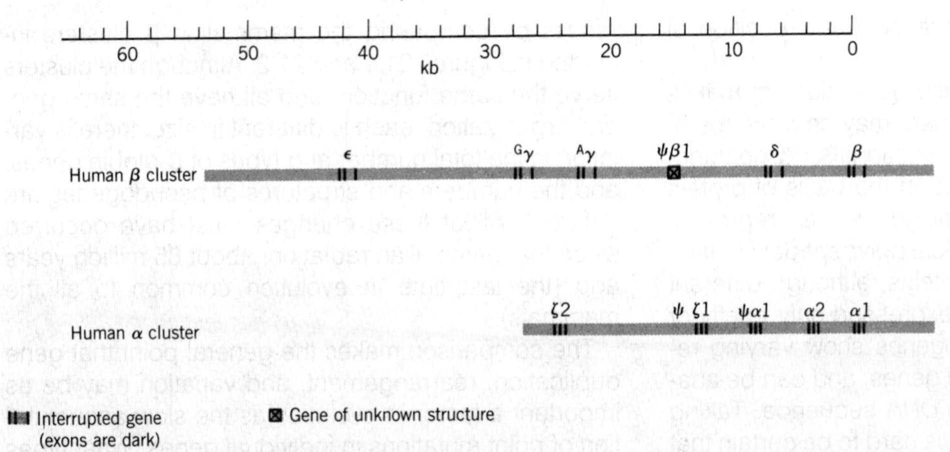

Figure 21.1
The human α-like and β-like globin gene families each are organized into a single cluster that includes functional genes and pseudogenes (ψ).

The more compact α cluster extends over 28 kb and includes one active zeta gene, one zeta pseudogene, two alpha genes, and one alpha pseudogene. The two α genes code for the same protein. Two (or more) identical genes present on the same chromosome are described as **nonallelic** copies.

A similar general organization is found in other vertebrate globin gene clusters, but details of the types and numbers of genes may vary. Some examples of β-globin clusters are illustrated in **Figure 21.2.**

Four β-like genes have been identified in the rabbit: two embryonic, one pseudo, and one adult, again lying in order of expression. Seven β-like genes have been found in the goat. Seven mouse genes include two early embryonic, one late embryonic, two adult genes, and two pseudogenes. They are not arranged in order of expression.

Clusters can be smaller and need not include pseudogenes. In the chicken, the β-globin cluster is less than 14 kb in length, and seems to include only four functional β-like genes. The outside two are embryonic and the inside two are adult, so there is no cor-

Figure 21.2
Clusters of β-globin genes and pseudogenes are found in other vertebrates.

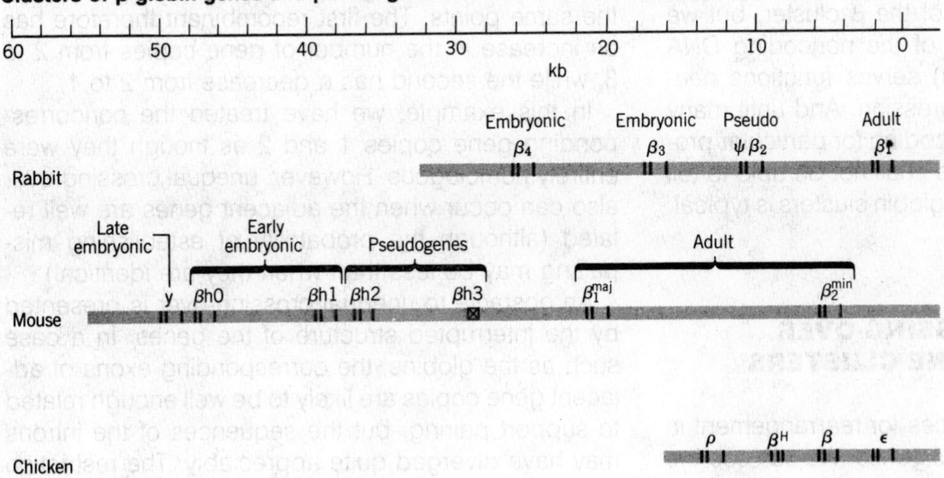

relation between time of expression and position of the gene.

The characterization of these gene clusters makes an important general point. There may be more members of a gene family, both functional and nonfunctional, than we would suspect on the basis of protein analysis. The extra functional genes may represent duplicates that code for identical polypeptides; or they may be related to known proteins, although different from them, and presumably expressed only briefly or in low amounts. The pseudogenes show varying relationships with the functional genes, and can be analyzed only at the level of the DNA sequence. Taking these features into account, it is hard to be certain that all the members of a cluster have been identified until flanking regions have been analyzed well beyond the terminal members.

Are *all* globin genes confined to the clusters? In each genome, all the *active* globin genes lie in the α and β clusters, sometimes together with pseudogenes. In the mouse genome, some additional pseudogenes have been found away from the clusters, located on different chromosomes. They must have originated by translocation events that moved a gene away from the cluster.

With regard to the question of how much DNA is needed to code for a particular function, we see that coding for the β-like globins requires a range of from about 14 kb to 50 kb in different mammals and a bird. Needless to say, this is much greater than we would expect just from scrutinizing the known β-globin proteins. It does not seem likely that any other proteins are coded within the region of the β cluster, but we do not yet know how much of the noncoding DNA (both flanking and in introns) serves functions necessary for β-globin gene expression. And until many more genes or gene clusters coding for particular proteins have been identified, we shall not be able to tell whether the complexity of the globin clusters is typical.

UNEQUAL CROSSING-OVER REARRANGES GENE CLUSTERS

There are frequent opportunities for rearrangement in a cluster of related or identical genes. We can see the results by comparing the mammalian β clusters included in Figures 21.1 and 21.2. Although the clusters serve the same function, and all have the same general organization, each is different in size, there is variation in the total number and types of β-globin genes, and the numbers and structures of pseudogenes are different. All of these changes must have occurred since the mammalian radiation, about 85 million years ago (the last time in evolution common to all the mammals).

The comparison makes the general point that gene duplication, rearrangement, and variation may be as important a factor in evolution as the slow accumulation of point mutations in individual genes. What types of mechanism are responsible for reorganization of the genes?

A gene cluster can expand or contract by **unequal crossing-over,** when recombination occurs between nonallelic genes, as illustrated in **Figure 21.3.** Usually, recombination involves corresponding sequences of DNA held in exact alignment between the two homologous chromosomes. However, when there are two copies of a gene on each chromosome, occasionally a misalignment may occur to allow pairing between them. (This requires some of the adjacent regions to go unpaired.)

When a recombination event occurs between the mispaired gene copies, it generates **nonreciprocal recombinant chromosomes,** one of which has a duplication of the gene and one a deletion. The duplication includes all the material between the points of unequal recombination; the deletion extends between the same points. The first recombinant therefore has an increase in the number of gene copies from 2 to 3, while the second has a decrease from 2 to 1.

In this example, we have treated the noncorresponding gene copies 1 and 2 as though they were entirely homologous. However, unequal crossing-over also can occur when the adjacent genes are well related (although the probability of establishing mispairing may be less than when they are identical).

An obstacle to unequal crossing-over is presented by the interrupted structure of the genes. In a case such as the globins, the corresponding exons of adjacent gene copies are likely to be well enough related to support pairing; but the sequences of the introns may have diverged quite appreciably. The restriction

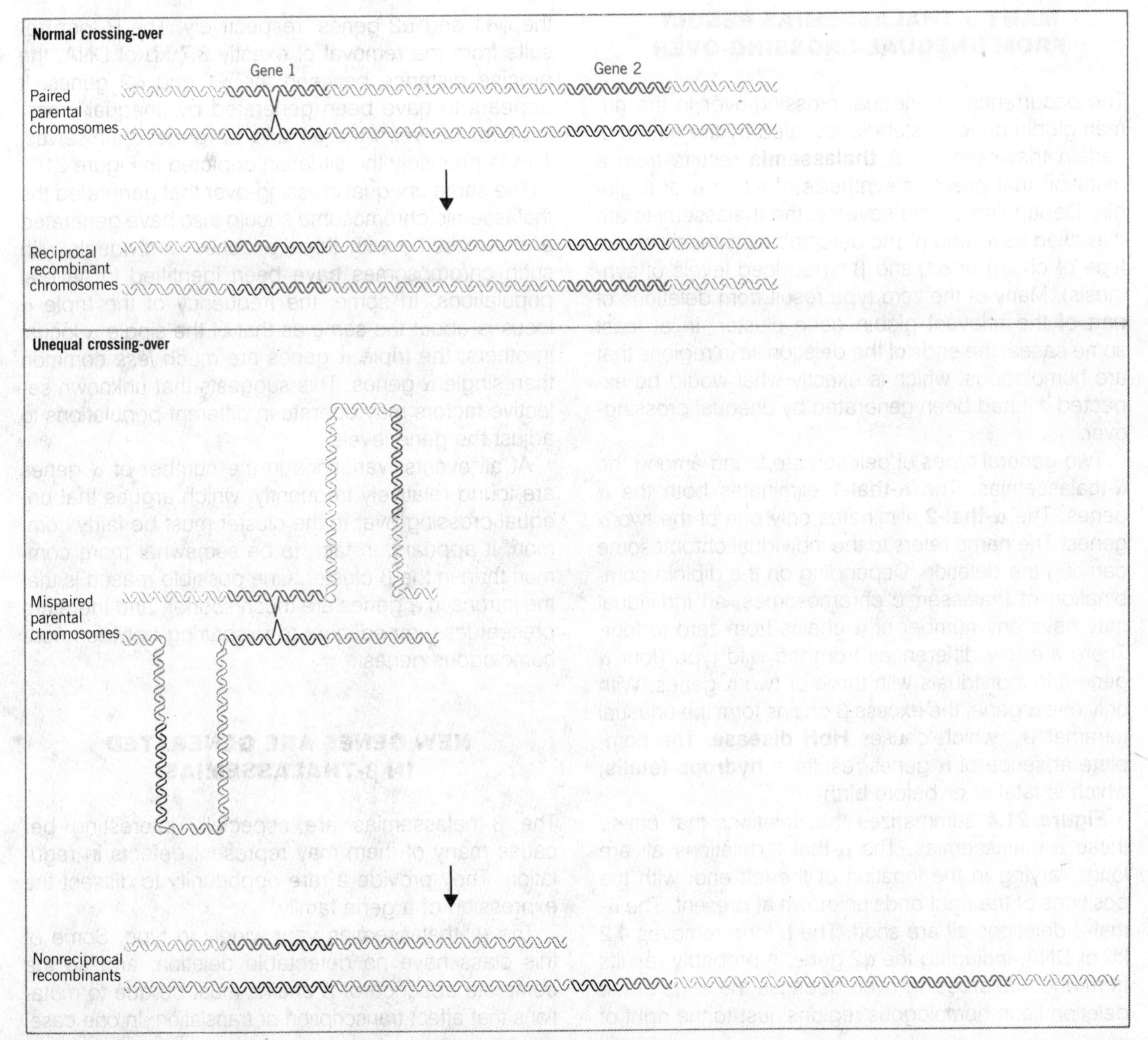

Figure 21.3
Gene number can be changed by unequal crossing-over.
If gene 1 of one chromosome pairs with gene 2 of the other chromosome, the other gene copies are excluded from pairing, as indicated by the extruded loops. Recombination between the mispaired genes produces one chromosome with a single (recombinant) copy of the gene and one chromosome with three copies of the gene (one from each parent and one recombinant).

of pairing to the exons considerably reduces the continuous length of DNA that can be involved and may correspondingly lower the chance of unequal crossing-over. Indeed, one idea is that divergence between introns may enhance the stability of gene clusters by hindering the occurrence of unequal crossing-over.

MANY α-THALASSEMIAS RESULT FROM UNEQUAL CROSSING-OVER

The occurrence of unequal crossing-over in the human globin gene clusters is revealed by the nature of certain thalassemias. A **thalassemia** results from a mutation that prevents synthesis of either α or β globin. Depending on the severity, the thalassemias are classified as α^0 and β^0 (no detectable synthesis of one type of chain) or α^+ and β^+ (reduced levels of synthesis). Many of the zero type result from deletions of part of the relevant globin gene cluster. In at least some cases, the ends of the deletion lie in regions that are homologous, which is exactly what would be expected if it had been generated by unequal crossing-over.

Two general types of deletion are found among the α-thalassemias. The **α-thal-1** eliminates both the α genes. The **α-thal-2** eliminates only one of the two α genes. The name refers to the individual chromosome carrying the deletion. Depending on the diploid combination of thalassemic chromosomes, an individual may have any number of α chains from zero to four. There are few differences from the wild type (four α genes) in individuals with three or two α genes. With only one α gene, the excess β chains form the unusual tetramer β_4, which causes **HbH disease.** The complete absence of α genes results in **hydrops fetalis,** which is fatal at or before birth.

Figure 21.4 summarizes the deletions that cause these α-thalassemias. The α-thal-1 deletions all are long, varying in the location of the left end, with the positions of the right ends unknown at present. The α-thal-2 deletions all are short. The L form removes 4.2 kb of DNA, including the α2 gene. It probably results from unequal crossing-over, because the ends of the deletion lie in homologous regions, just to the right of the ψα1 and α2 genes, respectively. The R form results from the removal of exactly 3.7 kb of DNA, the precise distance between the α1 and α2 genes. It appears to have been generated by unequal crossing-over between the α1 and α2 genes themselves. This is precisely the situation depicted in Figure 21.3.

The same unequal crossing-over that generated the thalassemic chromosome should also have generated a chromosome with three α genes. Individuals with such chromosomes have been identified in several populations. In some, the frequency of the triple α locus is about the same as that of the single α locus; in others, the triple α genes are much *less* common than single α genes. This suggests that unknown selective factors may operate in different populations to adjust the gene levels.

At all events, variations in the number of α genes are found relatively frequently, which argues that unequal crossing-over in the cluster must be fairly common. It appears, in fact, to be somewhat more common than in the β cluster. One possible reason is that the introns in α genes are much shorter, and therefore present less impediment to mispairing between nonhomologous genes.

NEW GENES ARE GENERATED IN β-THALASSEMIAS

The β-thalassemias are especially interesting because many of them may represent defects in regulation. They provide a rare opportunity to dissect the expression of a gene family.

The β^0-thalassemias vary widely in type. Some of this class have no detectable deletion, and so the complete absence of β chains must be due to mutations that affect transcription or translation. In one case,

Figure 21.4
Thalassemias result from various deletions in the α-globin gene cluster.

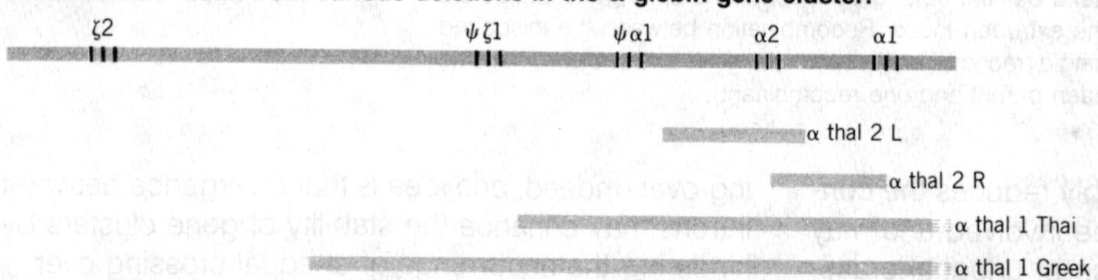

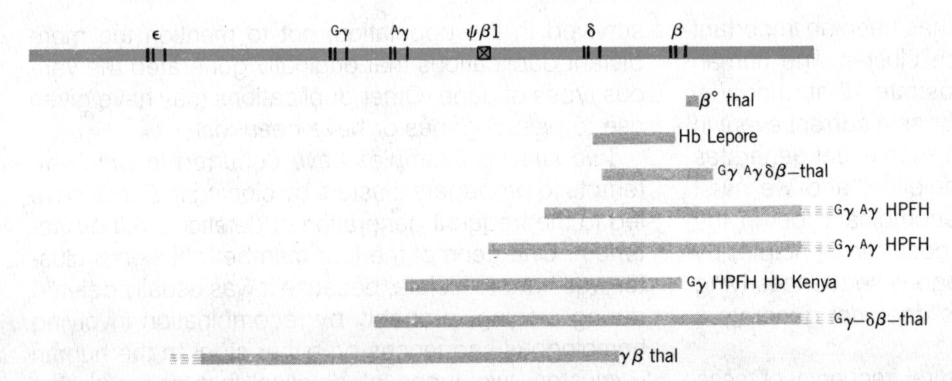

Figure 21.5
Deletions in the β-globin gene cluster cause several types of thalassemia.

mutation has generated a nonsense codon at position 17, so the defect is in translation. In another, the defect lies at an earlier stage of gene expression; nuclear RNA is produced, but there is no mRNA in the cytoplasm. The cause is a mutation that interferes with the splicing of mRNA (see Chapter 26).

The β^0 deletions are summarized in **Figure 21.5.** In some (rare) cases, only the β gene is affected. These have a deletion of 600 bp, extending from the second intron through the 3′ flanking regions. In the other cases, more than one gene of the cluster is affected. Many of the deletions are very long, extending from the 5′ end indicated on the map for 50 kb or more toward the right.

The **Hb Lepore** type provided the classic evidence that deletion can result from unequal crossing-over between linked genes. The β and δ genes differ only about 7% in sequence. Unequal recombination deletes the material between the genes, thus fusing them together (see Figure 21.3). The fused gene produces a single β-like chain that consists of the N-terminal sequence of δ joined to the C-terminal sequence of β.

Several types of Hb Lepore now are known, the difference between them lying in the point of transition from the δ to the β sequences. Thus when the δ and β genes pair for unequal crossing-over, the exact point of recombination determines the position at which the switch from δ to β sequence occurs in the amino acid chain.

The reciprocal of this event has been found in the form of **Hb anti-Lepore,** which is produced by a gene that has the N-terminal part of β and the C-terminal part of δ. This fusion gene should lie between normal δ and β genes.

Evidence that unequal crossing-over can occur between more distantly related genes is provided by the identification of **Hb Kenya,** another fused hemoglobin. This contains the N-terminal sequence of the $^{A}\gamma$ gene and the C-terminal sequence of the β gene. The fusion must have resulted from unequal crossing-over between $^{A}\gamma$ and β, which differ about 20% in sequence.

The absence of both δ and β can result in either of two phenotypes. In **HPFH** (hereditary persistence of fetal hemoglobin), there are no clinical symptoms; the disease is ameliorated because the synthesis of fetal hemoglobin ($\alpha_2\gamma_2$) continues after the time in development at which it would usually have been turned off. In **δβ thalassemia,** there are anemic symptoms, because, although γ gene expression may continue in adult life, it is less effective than in HPFH.

A variety of deletions prevent synthesis of both δ and β, and they are named in Figure 21.5 according to the class of phenotype. We do not understand the difference between the HPFH and δβ thalassemias. Perhaps it depends on the nature of the sequences on the 3′ side that the deletion brings into juxtaposition with the globin genes.

GENE CLUSTERS SUFFER CONTINUAL REORGANIZATION

From the differences between the globin gene clusters of various mammals, we may infer that duplication followed (sometimes) by variation has been an important feature in the evolution of each cluster. The human thalassemic deletions demonstrate that unequal crossing-over continues to occur as a current event in both globin gene clusters. Each such event generates a duplication as well as the deletion, and we must account for the fate of both recombinant loci in the population. Deletions can also occur (in principle) by recombination between homologous sequences lying on the *same* chromosome. This does not generate a corresponding duplication.

It is difficult to estimate the natural frequency of these events, because selective forces rapidly adjust the levels of the variant clusters in the population. There may be a rough correlation between the likelihood of an unequal crossing-over and the relationship of the genes—the more closely related (including both exons and introns), the greater the chance of mispairing. (However, some unequal recombinations do not involve the genes themselves, but rely on repetitive sequences nearby.)

Generally a contraction in gene number is likely to be deleterious and selected against. However, in some populations, there may be a balancing advantage that maintains the deleted form at a low frequency.

What is the result of an expansion? The only examples that have been characterized are the triple α locus and the anti-Lepore. Individuals who possess 5 α-globin genes (one normal locus and one triple) do not display any change in hemoglobin synthesis. However, it is possible that a further increase (to the triple homozygote) could be deleterious, by unbalancing globin synthesis through the production of excess α chains. Individuals with anti-Lepore have the fusion βδ gene as an addition to the normal β and δ genes. It is possible that the additional chain is deleterious because it interferes with the assembly of normal hemoglobin.

These particular changes in gene number are unlikely to have a selective advantage that will cause them to spread through the population. However, the structures of the present human clusters show several duplications that attest to the importance of such mechanisms. The *functional* sequences include two α genes coding the same protein, fairly well-related β and δ genes, and two almost identical γ genes. These comparatively recent independent duplications have survived in the population, not to mention the more distant duplications that originally generated the various types of gene. Other duplications may have given rise to pseudogenes or have been lost.

Two striking examples have occurred in which attempts to propagate clusters by cloning in *E. coli* have led to the frequent generation of deletions in the bacterium. One gene of the four-member chicken β cluster was hard to isolate, because it was usually deleted during cloning, probably by recombination involving homologous sequences on either side. In the human α cluster, two types of deletion that occur during propagation in *E. coli* correspond exactly with the two types of α-thal-2 deletion shown in Figure 21.4.

The occurrence of exchanges in bacteria with the same consequences as those observed naturally in the human population suggests that these events are an inevitable side effect of the mechanisms involved in DNA recombination. Thus we may expect continual duplication and deletion to be a feature of all gene clusters.

From the organization of globin genes in a variety of species, we should be able eventually to trace the evolution of present globin genes from a single ancestral globin gene. Our present view of the evolutionary descent is pictured in **Figure 21.6.** Let us follow it backward from the present.

Since there are separate clusters for α and β globins in both birds and mammals, the α and β genes must have been physically separated before the mammals and birds diverged from their common ancestor, an event that occurred probably about 270 million years ago.

The preceding stage of evolution is represented by the state of the globin genes in the frog *X. laevis,* which has two globin clusters. However, each cluster contains both α and β genes, of both larval and adult types. The cluster must therefore have evolved by internal duplications of a linked α-β pair, followed by divergence between the individual copies, and duplication of the entire cluster. The amphibians separated from the mammalian/avian line about 350 million years ago, so the separation of the α- and β-globin genes must have resulted from a transposition in the mammalian/avian forerunner somewhere around the period of early vertebrate evolution.

For the preceding stage, we have evidence of pro-

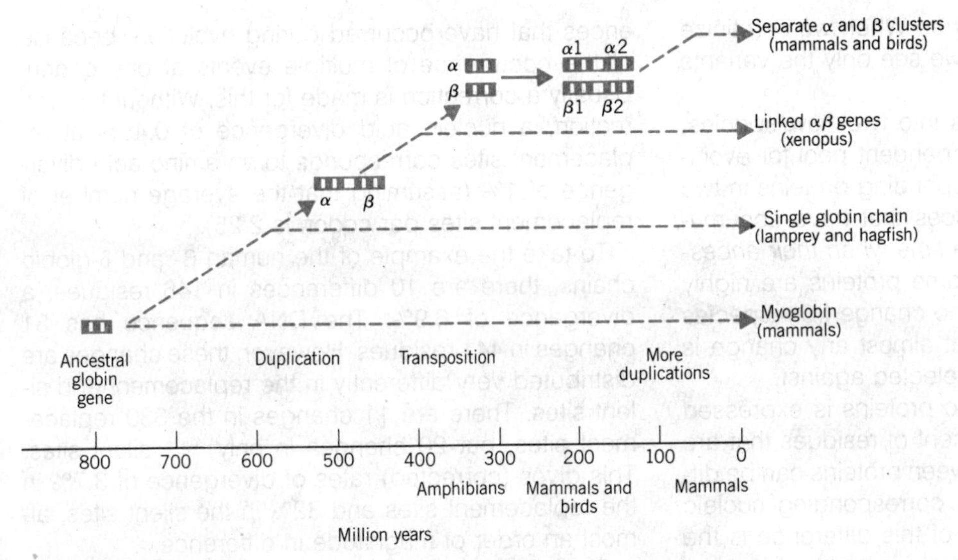

Figure 21.6
All globin genes have evolved by a series of duplications, transpositions, and mutations from a single ancestral gene.

tein chains, but not yet of genes. Some "primitive" fish have only a single type of globin chain, so they must have diverged from the line of evolution before the ancestral globin gene was duplicated to give rise to the α and β variants. This appears to have occurred roughly 500 million years ago, during the evolution of the bony fish.

We can trace the globin gene even further back, since the amino acid sequence of the single chain of mammalian myoglobin argues that it diverged from the globin line of descent about 800 million years ago. The myoglobin gene has the same organization as globin genes, so we may take the three exon structure to represent their common ancestor. Representing an even earlier separation, the leghemoglobin gene of plants is related to the globin gene, but has an extra intron (see Chapter 20).

SEQUENCE DIVERGENCE DISTINGUISHES TWO TYPES OF SITES IN DNA

Most changes in protein sequences occur by small mutations that accumulate slowly with time. Point mutations and small insertions and deletions occur by chance, probably with more or less equal probability in all regions of the genome, except for hotspots at which the frequency is much increased. Most mutations that change the amino acid sequence are *deleterious* and will be eliminated fairly rapidly by natural selection (the rate of removal depending on the extent of the damaging effect).

Few mutations will be advantageous, but these may spread through the population, eventually replacing the former sequence. When a new variant replaces the previous version of the gene, it is said to have been **fixed** in the population. A contentious issue is what proportion of mutational changes in an amino acid sequence may be **neutral,** that is, without any effect on the function of the protein, and able therefore to accrue as the result of **random drift and fixation.**

The rate at which mutational changes accumulate is a characteristic of each protein, presumably depending at least in part on its flexibility with regard to change. Within a species, a protein evolves by mutational substitution, followed by elimination or fixation within the single breeding pool. The presence in the population of two (or more) allelic variants is called a **polymorphism.**

A polymorphism can be stable, in which case neither form has any relative advantage. Or the polymorphism may be transient, as must be the case while one variant is replacing another. When we scrutinize the gene pool of a species, we see only the variants that have survived.

When a species separates into two new species, each now constitutes an independent pool for evolution. By comparing the corresponding proteins in two species, we see the differences that have accumulated between them *since the time when their ancestors ceased to interbreed.* Some proteins are highly **conserved,** showing little or no change from species to species. This indicates that almost any change is deleterious and therefore is selected against.

The difference between two proteins is expressed as their **divergence,** the percent of residues that are different. The divergence between proteins can be different from that between the corresponding nucleic acid sequences. The source of this difference is the representation of each amino acid in a three-base codon, in which often the third base has no effect on the meaning.

We can divide the nucleotide sequence into potential **replacement sites** and **silent sites.** At replacement sites, a mutation alters the amino acid that is coded. The effect of the mutation (deleterious, neutral, or advantageous) depends on the result of the amino acid replacement. At silent sites, mutation only substitutes one synonym codon for another, so there is no change in the protein. Usually the replacement sites account for 75% of a coding sequence and the silent sites provide 25%.

Although silent mutations are neutral with regard to the protein, they could affect gene expression via the sequence change in RNA. Potential effects are changes in the secondary structure of the RNA, which might influence transcription, processing, or translation. Another possibility is that the change in synonym codons calls for a different tRNA to respond; this could influence the efficiency of translation. And, of course, in addition to the coding sequences of a gene, there are nontranslated regions. Here again, mutations are potentially neutral, apart from their effects on either secondary structure or (usually rather short) regulatory signals.

The mutations in replacement sites should correspond with the amino acid divergence, which essentially is a count of the percent of changes. Actually, the measured divergence underestimates the differences that have occurred during evolution, because of the occurrence of multiple events at one codon. Usually a correction is made for this. Without the correction, a nucleic acid divergence of 0.45% at replacement sites corresponds to an amino acid divergence of 1% (assuming that the average number of replacement sites per codon is 2.25).

To take the example of the human β- and δ-globin chains, there are 10 differences in 146 residues, a divergence of 6.9%. The DNA sequence has 31 changes in 441 residues. However, these changes are distributed very differently in the replacement and silent sites. There are 11 changes in the 330 replacement sites, but 20 changes in only 111 silent sites. This gives (corrected) rates of divergence of 3.7% in the replacement sites and 32% in the silent sites, almost an order of magnitude in difference.

The striking difference in the divergence of replacement and silent sites demonstrates the existence of much greater constraints on nucleotide positions that influence protein constitution relative to those that do not. So it is likely that very few of the amino acid changes are neutral.

If we take the rate of mutation at silent sites to indicate the underlying rate of mutational fixation (this is to assume that there is no selection at all at the silent sites), then over the period since the β and δ genes diverged, there should have been changes at 32% of the 330 replacement sites, a total of 105. All but 11 of these have been eliminated, which means that roughly 90% of the mutations did not survive.

THE EVOLUTIONARY CLOCK TRACES THE DEVELOPMENT OF GLOBIN GENES

When a particular protein is examined in a range of species, the divergence between the sequences in each pairwise comparison is (more or less) proportional to the time since they separated. This provides an **evolutionary clock** that measures the accumulation of mutations at an apparently even rate during the evolution of a given protein.

The rate at which divergence has developed can be measured as the percent difference per million years, or as its reciprocal, the unit evolutionary period (UEP), the time in millions of years that it takes for 1% divergence to develop. Once the clock has been established by pairwise comparisons between species (remembering the practical difficulties in establishing the actual time of speciation), it can be applied to related genes *within* a species. From their divergence, we can therefore calculate the time since the duplication in which they originated.

By comparing the sequences of homologous genes in different species, the rate of divergence at both replacement and silent sites can be determined.

In pairwise comparisons, there is an average divergence of 10% in the replacement sites of either the α- or β-globin genes of mammals that have been separated since the mammalian radiation occurred about 85 million years ago. This corresponds to a replacement divergence rate of 0.12% per million years.

The rate is steady when the comparison is extended to genes that diverged in the more distant past. For example, the average replacement divergence between corresponding mammalian and chicken globin genes is 23%. Relative to a separation roughly 270 million years ago, this gives a rate of 0.09% per million years.

Going further back, we can compare the α with the β-globin genes within a species. These have been diverging since they separated ≥500 million years ago (see Figure 21.6). They have an average replacement divergence of about 50%, which gives a rate of 0.1% per million years.

These data are plotted in **Figure 21.7,** which shows that replacement divergence in the globin genes has a clock with an average rate of about 0.096% per million years (or a UEP of 10.4). Considering the uncertainties in estimating the times at which the species diverged, the results lend good support to the idea that there is a linear clock.

The data on silent site divergence are much less clear. In every case, it is evident that the silent site divergence is much greater than the replacement site divergence, by a factor that varies from 2 to 10. But the spread of silent site divergences in pairwise comparisons is too great to show whether a clock is applicable (so we must base temporal comparisons on the replacement sites).

From Figure 21.7, it is clear that the rate at silent sites is not linear with regard to time. *If we assume that there must be zero divergence at zero years of separation,* we see that the rate of silent site divergence is much greater for about the first 100 million years of separation. One interpretation is that a fraction of roughly half of the silent sites is rapidly (within 100 million years) saturated by mutations; this fraction behaves as neutral sites. The other fraction accumulates mutations more slowly, at a rate approximately the same as that of the replacement sites; this fraction identifies sites that are silent with regard to the protein, but that come under selective pressure for some other reason.

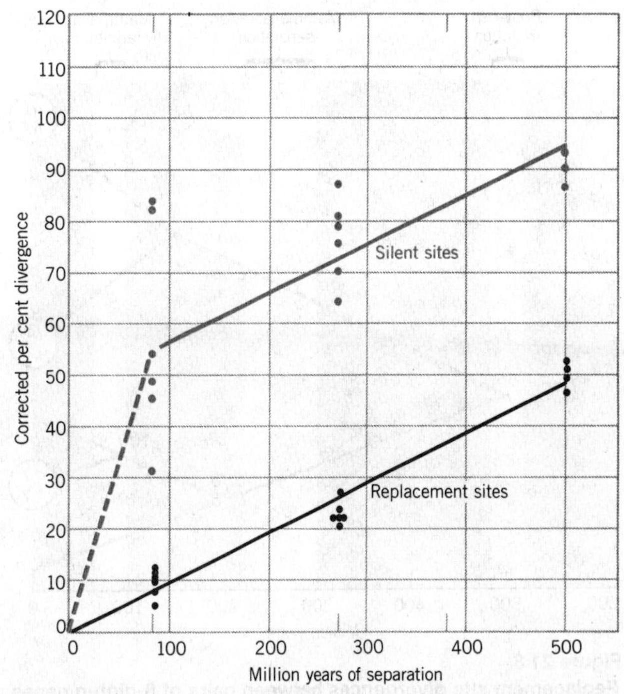

Figure 21.7
Divergence of DNA sequences depends on evolutionary separation. Each point on the graph represents a pairwise comparison.

Now we can reverse the calculation of divergence rates to estimate the times since genes within a species have been apart. The difference between the human β and δ genes is 3.7% for replacement sites. At a UEP of 10.4, these genes must have diverged 10.4 × 3.7 ≈ 40 million years ago—about the time of the separation of the lines leading to New World monkeys, Old World monkeys, great apes, and humans. All of these higher primates have both β and δ genes,

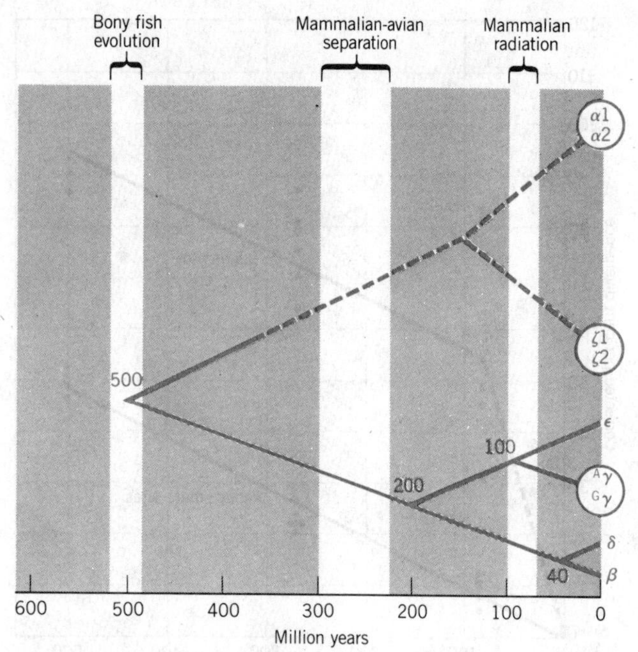

Figure 21.8

Replacement site divergences between pairs of β-globin genes allow the history of the human cluster to be reconstructed. This tree accounts for the separation of *classes* of globin genes. Duplications of individual genes are of unknown origin, as indicated by the circles. The time of the α-ξ divergence is not known.

which suggests that the gene divergence commenced just before this point in evolution.

Proceeding further back, the divergence between the replacement sites of γ and ε genes is 10%, which corresponds to a time of separation about 100 million years ago. The separation between embryonic and fetal globin genes may therefore have just preceded or accompanied the mammalian radiation.

An evolutionary tree for the human globin genes is constructed in **Figure 21.8.** Features that evolved before the mammalian radiation—such as the separation of adult from pre-adult β-like genes—should be found in all mammals. Features that evolved afterward—such as the separation of β- and δ-globin genes—must have evolved independently in individual mammalian lines.

In each species, there have been comparatively recent changes in the structures of the clusters, since we see differences in gene number (one β-globin gene in man, two in mouse) or in type (we are not yet sure whether there are separate embryonic and fetal β-like globins in rabbit and mouse).

MECHANISMS FOR MAINTAINING ACTIVE SEQUENCES

The duplication of a gene is likely to result in an immediate relaxation of the evolutionary pressure on its sequence. Now that there are two identical copies, a change in the sequence of either one will not deprive the organism of a functional protein, since the original amino acid sequence continues to be coded by the other copy. Thus the selective pressure on the two genes is diffused, until one of them mutates sufficiently away from its original function to refocus all the selective pressure on the other.

Immediately following a gene duplication, changes might accumulate more rapidly in one of the copies, leading eventually to a new function (or to its disuse in the form of a pseudogene). If a new function develops, the gene then may evolve at the same, slower rate characteristic of the original function. Probably this is the sort of mechanism responsible for the separation of functions between embryonic and adult globin genes.

Yet there are some instances where duplicated genes retain the same function, coding for the identical or nearly identical proteins. Identical proteins are coded by the two human α-globin genes, and there is only a single amino acid difference between the two γ-globin proteins. Is this not paradoxical in view of the expectation that duplication should be followed by variation? In these cases, we can probably exclude the only simple explanation, that both genes are needed and remain under selective pressure because of the necessity to make enough protein. So how is selective pressure maintained on both genes?

Two general types of mechanism have been proposed. They share the principle that nonallelic genes are not independently inherited, but that both must be continually regenerated from *one* of the copies of a preceding generation. When a mutation occurs in one copy, either it is by chance eliminated (because the sequence of the other copy takes over), or it is spread to both duplicates (because the mutant copy becomes the dominant version). Spreading exposes a mutation to selection. The result is that the two genes evolve together as though only a single locus existed. This is called **coincidental** or **concerted evolution** (occasionally **coevolution**).

One mechanism supposes that the sequences of the nonallelic genes are directly compared with one another and homogenized by enzymes that recognize any differences. This can be done by exchanging single strands between them, to form genes one of whose strands derives from one copy, one from the other copy. Any differences show as improperly paired bases, which may be the subject of attention from enzymes able to excise and replace a base, so that only A-T and G-C pairs survive. This type of event is called **gene conversion** and is associated with genetic recombination as described in Chapter 34.

Another mechanism is for both copies of the gene physically to be regenerated from one copy. This could occur by unequal crossing-over. Following the sort of event depicted in Figure 21.3, for example, the chromosome carrying a triple locus could suffer deletion of one of the genes. Of the two remaining genes, 1½ represent the sequence of one of the original copies; only ½ of the sequence of the other original copy has survived. Any mutation in the first region now exists in both genes and is subject to selective pressure. This type of event is called **crossover fixation** and is discussed in the context of satellite DNA evolution in Chapter 24.

We should be able to ascertain the scope of these mechanisms by comparing the sequences of duplicate genes. If they are subject to concerted evolution, we should not see the accumulation of silent site substitutions between them (because the homogenization process applies to these as well as to the replacement sites). We know that the extent of the maintenance mechanism need not extend beyond the gene itself, since there are cases of duplicate genes whose flanking sequences on either side are entirely different. Indeed, we may see abrupt boundaries that mark the ends of the sequences that were homogenized.

We must remember that the existence of such mechanisms can invalidate the determination of the history of such genes via their divergence, *because the divergence reflects only the time since the last homogenization/regeneration event, not the original duplication.*

PSEUDOGENES ARE DEAD ENDS OF EVOLUTION

Pseudogenes are defined as such by their possession of sequences that are related to those of the functional genes, but that cannot be translated into a functional protein.

Some pseudogenes have the same general structure as functional genes, with sequences corresponding to exons and introns in the usual locations. They are rendered inactive by mutations that prevent any or all of the stages of gene expression. The changes can take the form of abolishing the signals for initiating transcription, preventing splicing at the exon-intron junctions, or terminating translation prematurely.

Usually a pseudogene has several deleterious mutations, presumably because once it ceased to be active, there was no impediment to the accumulation of further mutations. Pseudogenes that represent inactive versions of currently active genes have been found in many systems, including globin, immunoglobulins, and histocompatibility antigens, where they are located in the vicinity of the gene cluster, often interspersed with the active genes.

A typical example is the rabbit pseudogene, $\psi\beta2$, which has the usual organization of exons and introns, and is related most closely to the functional gene $\beta1$. But the deletion of a base pair at codon 20 of $\psi\beta2$ has caused a frameshift that would lead to termination shortly after. Several point mutations have changed later codons representing amino acids that are highly conserved in the β globins. Neither of the two introns any longer possesses boundary sequences that conform adequately to the GT–AG rule, so probably the introns could not be spliced out even if the gene were transcribed. However, there are no transcripts corresponding to the gene, possibly because there have been changes in the 5′ flanking region.

Since this list of defects includes mutations potentially preventing each stage of gene expression, we have no means of telling which event originally inactivated this gene. However, from the divergence between the pseudogene and the functional gene sequences, we can estimate when the pseudogene originated and when its mutations started to accumulate.

If the pseudogene had become inactive as soon as

it was generated by duplication from β1, we should expect both replacement site and silent site divergence rates to be the same. (There is no reason for them to be different if the gene is not translated.) But actually there are fewer replacement site substitutions than silent site substitutions. This suggests that at first there was selection against replacement site substitution. From the relative extents of substitution in the two types of site, we can calculate that ψβ2 diverged from β1 about 55 million years ago, remained a functional gene for 22 million years, but has been a pseudogene for the last 33 million years.

Similar calculations can be made for other pseudogenes. Some also appear to have been active for some time before becoming pseudogenes. Some appear to have been inactive from the very time of their original generation. The general point made by the structures of these pseudogenes is that each has evolved independently during the development of the globin gene cluster in each species. This reinforces the conclusion that the creation of new genes, followed by their acceptance as functional duplicates, variation to become new functional genes, or inactivation as pseudogenes, is a continuing process in the gene cluster.

The mouse ψα3 gene has an interesting property: it precisely lacks both introns. Its sequence can be aligned (allowing for accumulated mutations) with the α-globin mRNA. The apparent time of inactivation coincides with the original duplication, which suggests that the original inactivating event may have been associated with the loss of introns. Without introns, the gene may have been unable to function, and thus will have begun immediately to collect mutations.

The main question is how the introns could have been lost so precisely. There is no reason why the systems for reconstructing DNA should recognize the exon-intron boundaries, which argues for the involvement at some level of the mRNA itself. The most likely model is that a reverse transcript of the mRNA was inserted into the genome (perhaps carried by a retrovirus for which such action is normal). Another possible mechanism is that pairing might occur between the mRNA and one strand of DNA at some stage of a recombination process.

Pseudogenes that resemble the RNA transcript are called **processed pseudogenes.** Supporting the idea that they originated by insertion at some random location of a product derived from the RNA, they may be located anywhere in the genome, not necessarily even on the same chromosome as the active gene.

How common are pseudogenes? Most gene families have members that are pseudogenes. Usually the pseudogenes represent a small minority of the total gene number. In an exceptional case, however, there is one active gene coding for a mouse ribosomal protein; and it has about 15 processed pseudogene relatives. This type of effect must be taken into account when we try to calculate the number of genes from hybridization data.

If pseudogenes are evolutionary dead ends, simply an unwanted accompaniment to the rearrangement of functional genes, why are they still present in the genome? Do they fulfill any function or are they entirely without purpose, in which case there should be no selective pressure for their retention? It is necessary to remember that we see those genes that have survived in present populations. In past times, any number of other pseudogenes may have been eliminated. This elimination could occur by deletion of the sequence as a sudden event or by the accretion of mutations to the point where the pseudogene can no longer be recognized as a member of its original sequence family (probably the ultimate fate of any pseudogene that is not suddenly eliminated).

Even relics of evolution can be duplicated. In the β-globin genes of the goat, there are two adult species, β^A and β^C. Each of these has a pseudogene a few kilobases upstream from it (called $\psi\beta^Z$ and $\psi\beta^X$, respectively). The two pseudogenes are better related to each other than to the adult β-globin genes; in particular, they share several inactivating mutations. Also, the two adult β-globin genes are better related to each other than to the pseudogenes. This implies that an original ψβ-β structure was itself duplicated, giving two functional β genes (which diverged further into the β^A and β^C genes) and two nonfunctional genes (which diverged into the current pseudogenes).

The general implication to read into this is that the mechanisms responsible for gene duplication, deletion, and rearrangement act on all sequences that are recognized as members of the cluster, whether or not they are functional. It is left to selection to discriminate among the products.

GENE FAMILIES ARE COMMON FOR ABUNDANT PROTEINS

From experiments to count the number of genes by isolating the genomic fragments corresponding to a particular mRNA or cDNA probe, we know that in several cases there are multiple sequences in the genome that correspond to the probe. Of course, until the nucleotide sequence of each of these is determined, we do not know whether all of them are active or some are pseudogenes. Usually, it is assumed that most of the genes are likely to be active.

By this criterion, there are multiple genes for many, perhaps most, structural proteins (for example, actin, and tubulin). Within the gene families, there may be more repetition of some subtypes of gene than others. For example, there seem to be at least eight genes relating to the α (leucocyte) type of interferon in man, but only one for β (fibroblast) interferon.

The multiple genes may be clustered or dispersed. In the case of interferon, there is evidence for clustering. In the example of the actin and tubulin genes of *D. melanogaster,* the copies seem to lie at different chromosomal locations, so they are effectively unlinked.

Is the preservation of structure a usual feature of multiple gene copies? Globin genes are well conserved. The interferon genes seem to have a common structure, entirely lacking introns. But the actin genes have interrupted structures that vary widely, implying a more frequent rate of change than we have seen in the globin genes. In actin genes, the protein-coding regions are highly homologous, but there is little or no relationship between the flanking or even the nontranslated regions within a species (see Figure 20.31).

These data are somewhat anecdotal, showing that the preservation of gene organization may vary considerably in scope, but not telling us what we should take as the norm. The only general conclusions we can draw from these somewhat restricted results are that functional genes may be interrupted or uninterrupted and that changes in the pattern of interruptions need not affect gene activity. Genes coding for the same or related proteins need not be located in tandem, but may have become dispersed as individual genes or perhaps as small clusters.

What can we conclude about the total number of genes? When we look for the gene representing a particular protein, we tend to find other related active genes and (sometimes) pseudogenes. We do not have enough data to extrapolate to a total gene number, but we may ask how far this type of repetition may go toward resolving the apparently excessive amount of DNA.

The repetition of sequences increases the amount of DNA concerned with a particular function. There may be several active genes, either coding for the same protein or coding for unexpected variants. When the copies are clustered, the regions between the genes may be quite considerable, thus increasing the total amount of DNA devoted to specifying the function. Pseudogenes may account for further sequences connected with the function, but no longer contributing to it.

It would not be surprising if the amount of DNA concerned with a given function were 3–4 times greater than expected if a single gene coded for each function. In analyzing the number of genes in Chapters 18 and 19, we predicted a number of roughly 30,000–40,000 gene functions for a mammal. If we now argue that this might require 150,000 genes, each gene being several times longer than the coding region, and occupying (say) 5000 bp, we have accounted for $<10^9$ bp, about a third of the size of the genome.

The primary assumption in accounting for this part of the genome is that most functions are represented by a few genes instead of being unique as they had previously appeared. A secondary assumption is that a good deal of DNA is wasted either within the genes (in the form of introns) or between them (because gene clusters can be rather spread out and it is likely to be quite a long way from one to another). This extremely large amount of DNA is only peripherally concerned with gene functions.

We have made a tangible advance in accounting for DNA content with the notion that there are simply more genes than expected, because we are often dealing with small families rather than individuals when we come to characterize a particular function. By this argument, we can account for $\sim 2 \times 10^8$ bp of a mammalian genome (about 10% of the DNA) in directly coding for proteins.

Including extra DNA within these genes, or surrounding them with intergenic sequences that have no other function, accounts for about 3 times as much DNA, another $\sim 6 \times 10^8$ bp of DNA. The assignment of DNA to functions could be an underestimate if the genes are larger or the distance between them is greater than we have supposed. Could our 150,000 notional genes account for the entire mammalian genome of 3×10^9 bp?

We have disposed of part of the genome by saying that it is associated with the coding regions; although it does not itself code for protein, it has no other apparent function. Does this leave us with a nagging suspicion that something is awry with a system that is so wasteful; have we missed some important feature?

Even if we have successfully explained the content of the mammalian genome, we have not accounted for the C-value paradox that total DNA content can vary far more widely than we would expect the number of genes to vary. Does this mean that the size of the gene itself, or the distance between genes, is relatively unimportant to the organism and is subject to willful adjustment?

FURTHER READING

The structure and function of human hemoglobins have been reviewed by **Maniatis et al.** (*Ann. Rev. Genet.* **14,** 145–178, 1980). The general characteristics of gene and protein evolution were reviewed by **Wilson et al.** (*Ann. Rev. Biochem.* **46,** 573–639, 1977). An original report of a pseudogene lacking introns was made by **Nishioka et al.** (*Proc. Nat. Acad. Sci. USA* **77,** 2806–2809, 1980). Pseudogenes have been encapsulated in review by **Little** (*Cell* **28,** 683–684, 1982). The basis for thalassemic defects was last brought up to date by **Weatherall & Clegg** (*Cell* **29,** 7–9, 1982).

CHAPTER 22
GENOMES SEQUESTERED IN ORGANELLES

No counterpart to the operon has been found in higher eukaryotic genomes, but in some situations a defined region of DNA contains a set of genes that share the task of contributing to the same aspect of the phenotype. Mitochondria and chloroplasts both possess DNA genomes that code for all of the RNA species and for some of the proteins involved in the functions of the organelle.

The organization of an organelle genome may be more akin to the construction of a viral genome than to a cellular genome, since it codes for some, but not all, of the functions needed to perpetuate the organelle. The others are coded in the nucleus, expressed via the cytoplasmic protein synthetic apparatus, and imported into the organelle. Of course, the organelle is a permanent component of the cell and not a virus merely passing through. In effect, the organelle genome comprises a length of DNA that has been physically sequestered in a defined part of the cell, and is accordingly subject to its own form of expression and regulation.

The term **cytoplasmic inheritance** has often been used to describe the behavior of genes in organelles. However, we shall not use this description, since it is important to be able to distinguish between events in the general cytoplasm and those in specific organelles. A better general term for genes not residing within the nucleus is **extranuclear,** which leaves us able to use the term **cytoplasmic protein synthesis** to describe the final stage of expression of *nuclear* genes; while organelle genes are transcribed and translated in the *same* organelle compartment in which they reside.

ORGANELLE GENES SHOW NONMENDELIAN INHERITANCE

The first evidence for the presence of genes outside the nucleus was provided by nonMendelian inheritance in plants (observed in the early years of this century, just after the rediscovery of Mendelian inheritance).

NonMendelian inheritance is revealed by abnormal segregation ratios when a cross is made between mutant and wild type. The extreme form is uniparental inheritance, when the genotype of only one parent is inherited and that of the other parent is permanently lost. In less extreme examples, the progeny of one parental genotype exceed the other.

Usually, it is the mother whose genotype is preferentially inherited. This effect is sometimes described as **maternal inheritance.** The important point is that the genotype contributed by the parent of one particular sex predominates, in contrast with the behavior of Mendelian genetics when reciprocal crosses show the contributions of both parents to be equally inherited (see Chapter 1).

The representation of organelle genetic information in a zygote depends on the relative contributions of each parent or on differential ability of the information to survive (or on both). This contrasts with nuclear genetic information, which derives equally from each parent.

Maternal inheritance in higher animals can be predicted by supposing that the mitochondria are contributed probably entirely by the ovum and not at all by the sperm. Thus the mitochondrial genes are derived exclusively from the mother; and in males they are discarded each generation.

A different cause for uniparental inheritance is found in the alga *Chlamydomonas reinhardii*, in which gametes of equal size fuse to form the zygote. Both parents contribute chloroplast DNA to a zygote. But the chloroplast DNA provided by one parent is degraded shortly after zygote formation, leaving the DNA of the other parent in exclusive possession of the organelle.

The existence of these two mechanisms makes it impossible to predict the reason for the predominance of one parental genotype in other situations (for example, in higher plants). Even in those cases where the parents contribute different amounts of cytoplasm to the zygote, so that one is likely to provide more copies of the organelle genome than the other, selective mechanisms could operate to alter the balance further.

Another feature of nonMendelian inheritance is that, in those cases where both genotypes survive, the wild and mutant phenotypes segregate during somatic growth of the plant. Thus in a single heterozygous plant, some tissues may have one parental phenotype, while other tissues display the alternate parental phenotype. This **somatic segregation** contrasts with the stability of the entire nuclear genotype, as defined by Mendelian genetics.

The association of somatic segregation with nonMendelian inheritance suggests that there are multiple copies of the organelle genes in the zygote, and that they may be distributed into different cells during somatic division.

The question of which organelle carries a particular genetic marker is not trivial. Since the only organelle shown to possess DNA in higher animal cells is the mitochondrion, probably it is the sole extranuclear residence of genetic material. But in plants and in some unicellular eukaryotes, both chloroplasts and mitochondria are present. They are the only organelles shown to possess DNA; and it is likely (although not proven) that any marker displaying nonMendelian inheritance resides in one or the other compartment. In some cases, the location is clear from the nature of the mutant phenotype; but in other instances, the defect takes a more general nature that does not reveal the compartment in which the relevant gene resides.

ORGANELLE GENOMES ARE CIRCULAR DNA MOLECULES

Almost all organelle genomes so far characterized take the form of a single molecule of DNA of unique sequence. It can usually be recovered as a circular duplex, although sometimes breakage during isolation is so frequent that most of the material is recovered in the form of linear fragments. Ciliate protozoans provide some exceptions where mitochondrial DNA is a linear molecule. Usually, there are several copies of the genome in the individual organelle. Since there are multiple organelles per cell, there may be a large number of organelle genomes per cell. Although the organelle genome itself is unique, it constitutes a repetitive sequence relative to any nonrepetitive nuclear sequence.

Chloroplast genomes are relatively large, usually about 140 kb in higher plants, and up to 200 kb in lower eukaryotes. This is comparable to the size of a large bacteriophage, for example, T4 at about 110 kb. **Table 22.1** shows that in two lower eukaryotes, many of the circular DNA molecules are present per chloroplast. The total amount of DNA in the chloroplasts forms a proportion of total cellular DNA as high as several percent. Unfortunately, we do not have comparable figures for higher plants, but here the amount of nuclear DNA is very much greater, and so the proportion of mass present in the chloroplast must be rather small.

Mitochondrial genomes vary in total size by almost an order of magnitude, as can be seen from **Table 22.2.** Mammals have small mitochondrial genomes,

Table 22.1
Chloroplast DNA shows a narrow range of complexities.

Species	Complexity of DNA	Genomes /Organelle	Organelles /Cell	ctDNA /total DNA
C. reinhardii	195 kb	70–100	1	14%
E. gracilis	135 kb	40	15	3%
Higher plants	140 kb	Unknown	Unknown	Unknown

Table 22.2
Mitochondrial DNA varies widely in complexity.

Species	Complexity of DNA	Genomes /Organelle	Organelles /Cell	mtDNA /total DNA
Mouse (L cell)	16.2 kb	2	500	0.2%
Human (HeLa)	16.6 kb	10	800	1%
D. melanogaster	18.4 kb	Unknown	Unknown	Unknown
X. laevis	18.4 kb	Unknown	Unknown	Unknown
S. cerevisiae	84 kb	4	22	18%
P. sativum	110 kb	Unknown	Unknown	Unknown

small enough to have been completely sequenced in man, mouse, and cow; all are about 16.5 kb. Data on the number of organelles per cell are available only for cultured cell lines, but the number is large (several hundred). The total amount of mitochondrial DNA relative to nuclear DNA is small, less than 1%. There are between 1 and 10 genomes per mitochondrion. The size of the mitochondrial genome is slightly larger in the fruit fly and frog, where we lack information about the number of genomes and organelles.

In yeast, the mitochondrial genome is much larger. In *S. cerevisiae,* the actual size varies quite widely among different strains, but is about 84 kb. There are about 22 mitochondria per cell, which corresponds to about 4 genomes per organelle. In growing cells, the proportion of mitochondrial DNA can be as high as 18%. (Other yeasts may have much larger mitochondrial genomes.)

Plants show an extremely wide range of variation in mitochondrial DNA size, with a minimum of about 100 kb. We have no idea at all about the significance of the extra DNA in the larger mitochondrial genomes. There is some evidence that the plant mitochondrial genome does not conform to the pattern of a single molecule of DNA, but may consist of several molecules, varying in size. Together the plant mitochondrial and chloroplast genomes may carry quite a substantial amount of information.

ORGANELLES EXPRESS THEIR OWN GENES

Both of the organelles concerned with energy conversion run a mixed economy. Most of their constituent proteins are imported from the surrounding cytoplasm, where their synthesis represents the final stage of expression of nuclear genes. But each organelle also engages in its own protein synthesis. As illustrated in **Figure 22.1** for the example of mitochondria, this effort is devoted to the production of a small number of proteins, each of which is a component of an oligomeric aggregate that includes some protein subunits imported from the cytoplasm.

The apparatus for organelle protein synthesis is itself of mixed origin. Most or all of its protein components are transported into the organelle from the sur-

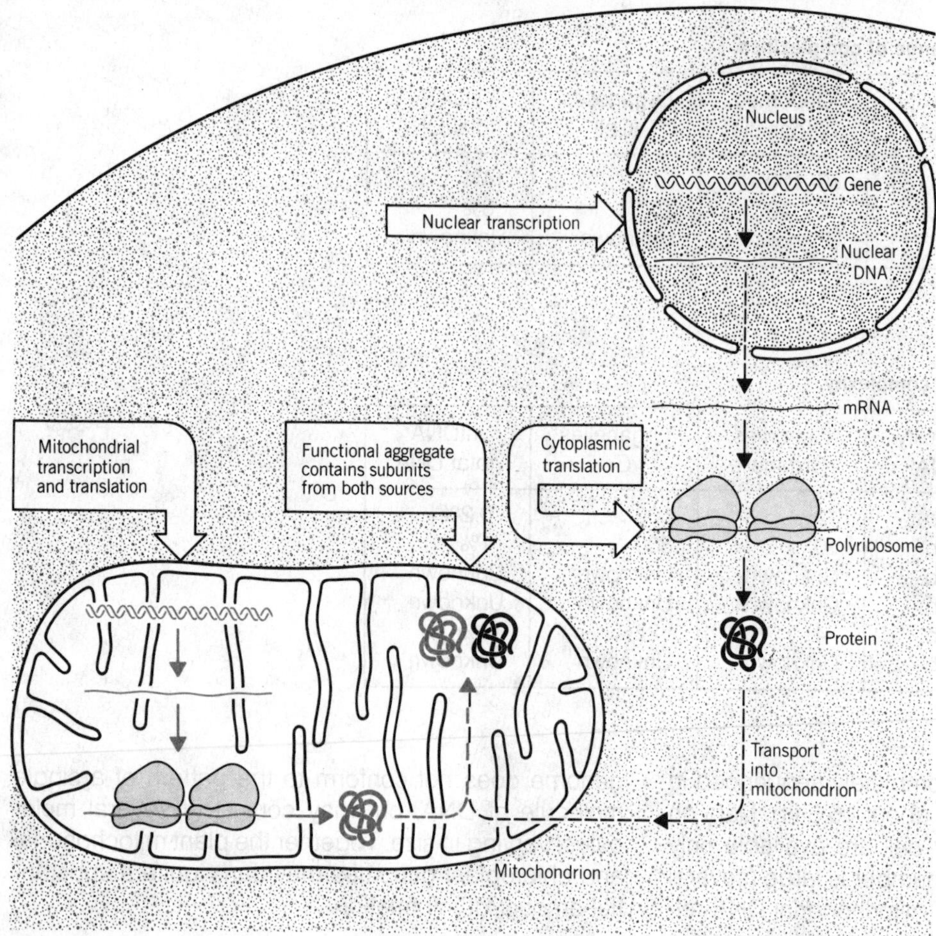

Figure 22.1
Mitochondrial protein aggregates are assembled from the products of expression of nuclear genes and mitochondrial genes.

rounding cytoplasm. But in both mitochondria and chloroplasts, no passage is permitted for nucleic acids in either direction through the organelle membrane. Thus among the products of the organelle itself are all of the RNA components of the protein synthetic apparatus.

The origin of the apparatus for gene expression in yeast mitochondria is summarized in **Table 22.3.** It is generally similar in the mitochondria of other species and in chloroplasts. In the yeast mitochondrion, eight mRNAs can be identified; they are transcribed by an RNA polymerase that (presumably) is coded within the nucleus. Only the mRNAs that have been transcribed from mitochondrial genes can be translated in the organelle; and conversely, this is the only compartment in which they can be expressed.

The mitochondrial ribosome consists of the usual larger and smaller subunits, each with a single rRNA coded within the mitochondrion, and with a large number of proteins imported from the cytoplasm. However, the ribosome structures are diverse in different species (see Chapter 9).The set of tRNAs is smaller than needed in the cytoplasm, as described in Chapter 8. The aminoacyl-tRNA synthetases responsible for charging the tRNAs are brought in from the cytoplasm. All of the protein components of the synthetic appa-

Table 22.3
The mitochondrial apparatus for protein synthesis is assembled from RNA synthesized in the mitochondrion and proteins imported from the cytoplasm.

Component	RNAs Synthesized in Mitochondrion	Proteins Synthesized in Cytoplasm
Template	~ 8 mRNAs	RNA Polymerase
Ribosome	2 rRNAs	~75 ribosomal proteins
Adaptor	22 tRNAs	22 aminoacyl-tRNA synthetases

ratus appear to be unique to the mitochondrion; although coded by nuclear genes, they are different from the proteins of the protein synthetic apparatus in the surrounding cytoplasm.

Corresponding with the eight mitochondrial mRNAs, eight proteins can be identified as the products of mitochondrial synthesis. The synthetic origin of a protein can be determined *in vivo* by virtue of its susceptibility to drugs that preferentially inhibit either cytoplasmic or organelle protein synthesis. Chloroplast and mitochondrial protein synthesis usually is sensitive to antibiotics that inhibit bacterial protein synthesis—for example, erythromycin and chloramphenicol. Cytoplasmic protein synthesis is susceptible to cycloheximide. The origins of organelle proteins can be determined *in vitro* by characterizing the products made in isolated organelle preparations.

The complexes containing the proteins synthesized in the yeast mitochondrion are summarized in **Table 22.4.** The ATPase consists of two units, a membrane factor of 3 subunits coded by the mitochondrial genome, and the soluble F1 ATPase of 5 subunits synthesized in the cytoplasm. The cytochrome *c* oxidase similarly consists of subunits from both sources. The cytochrome bc_1 complex consists of a single protein from the mitochondrion associated with six subunits from the cytoplasm. The ribosome small subunit contains one protein (Var1) coded in the mitochondrion. Mutations identifying almost all of the mitochondrial genes have been isolated.

Nuclear mutations that abolish each of these complexes also have been found. In both organelle and nuclear mutants, the protein complex fails to assemble properly, so that several subunits appear to be absent. This suggests that the assembly of each of these complexes into its usual membrane-associated form is an intricate process in which the absence of one component may prevent the assembly of the others. This relationship has the practical consequence of making it difficult to equate nuclear mutations with individual subunits. Also, there are some nuclear mutations that prevent complex assembly, but seem to

Table 22.4
Mitochondrial (yeast) protein complexes are assembled from proteins synthesized in different cell compartments.

Protein Complex	Mass (daltons)	Subunits Synthesized in Mitochondrion	Subunits Synthesized in Cytoplasm
Oligomycin-sensitive ATPase	340,000	ATPase 6, 8, 9 (F0 membrane factor)	ATPase 1,2,3,4,7 (F1 ATPase)
Cytochrome c oxidase	137,000	CO 1, 2, 3	CO 4, 5, 6, 7
Cytochrome bc1 complex	160,000	Cytochrome b apoprotein	6 subunits

leave the individual subunits unaffected. This may mean that some nuclear-coded products are necessary for assembly, but do not actually comprise part of the final aggregate.

If the nuclear mutations reside in unique structural genes, there must be a numerical discrepancy between the representation of organelle-coded and nuclear-coded components, since there are many more copies of the mitochondrial genome (see Table 22.2). Presumably the nuclear-coded genes are expressed more efficiently. The discrepancy has been confirmed directly in the case of a chloroplast function, the RuBP carboxylase of maize, in which the large subunit is coded by an organelle gene present in many copies per cell, while the small subunit is coded by a nuclear gene whose DNA is nonrepetitive.

Is the particular division of labors displayed by the yeast mitochondrion and nucleus typical of other species? The overall constitution of the mixed protein aggregates is similar at least in mammalian and other fungal mitochondria. Genes coding for the same functions are present in yeast and mammalian mitochondrial genomes. Our general expectation is therefore that there will prove to have been a substantial conservation of the coding functions found in the mitochondrial genomes of different species.

One exception to this rule has been found. The smallest subunit of the ATPase is coded by the mitochondrial genome in yeast and in *Aspergillus nidulans*. But in another fungus, *Neurospora crassa,* the corresponding protein appears to be synthesized in the cytoplasm. Nor is any gene corresponding to it present in mammalian mitochondrial DNA. If this subunit is coded by a nuclear gene in these cases, at some time there must have been a transfer of the gene between mitochondrial and nuclear genomes.

Are mitochondrial and chloroplast genomes unique or are some sequences shared between organelles or by an organelle and the nucleus? Most of an organelle genome is unique to the organelle, but a 12 kb homology has been found between mitochondrial and chloroplast DNAs of maize. Some mitochondrial sequences in the sea urchin *S. purpuratus* have homologous regions in the nucleus. Exchanges of DNA between organelles or with the nucleus undoubtedly are rare, but do seem to occur occasionally. We do not yet know what functions are exercised by the sequences involved in these homologies.

THE LARGE MITOCHONDRIAL GENOME OF YEAST

The fivefold discrepancy in size between the *S. cerevisiae* (84 kb) and mammalian (16 kb) mitochondrial genomes alone alerts us to the fact that there must be a great difference in their genetic organization in spite of their common function. We have already mentioned the absence of any substantial increase in yeast in the number of endogenously synthesized products concerned with mitochondrial enzymatic functions. Does the additional genetic material in yeast mitochondria represent other proteins, perhaps concerned with regulation, or is it unexpressed?

The existence of genetic recombination in yeast mitochondria (see later) has made it possible to construct a genetic map, which has been correlated with the physical map by the use of naturally occurring deletions in the mitochondrial genome (called petites). The extent of each gene has been determined by mapping the corresponding mRNA.

The map shown in **Figure 22.2** accounts for the major RNA and protein products of the yeast mitochondrion (excluding the tRNA molecules, of which there are about 22, and which have not yet been completely mapped). The most notable feature is the dispersion of loci on the map.

This genome has the distinction of providing the extremely rare situation where the genes for rRNA are separated. The gene for 15S rRNA is uninterrupted and lies about 25,000 bp away from the gene for 21S rRNA. In some strains of *S. cerevisiae* the latter gene has a single intron (as shown on the map); in other strains it is uninterrupted.

The two most prominent loci are the mosaic genes *box* (coding for cytochrome *b*) and *oxi3* (coding for subunit 1 of cytochrome oxidase). Together these two genes are almost as long as the entire mitochondrial genome in mammals! Many of the long introns in these genes have open reading frames in register with the preceding exon; at least in some cases, there is evidence for translation of the intron (see Chapter 27).

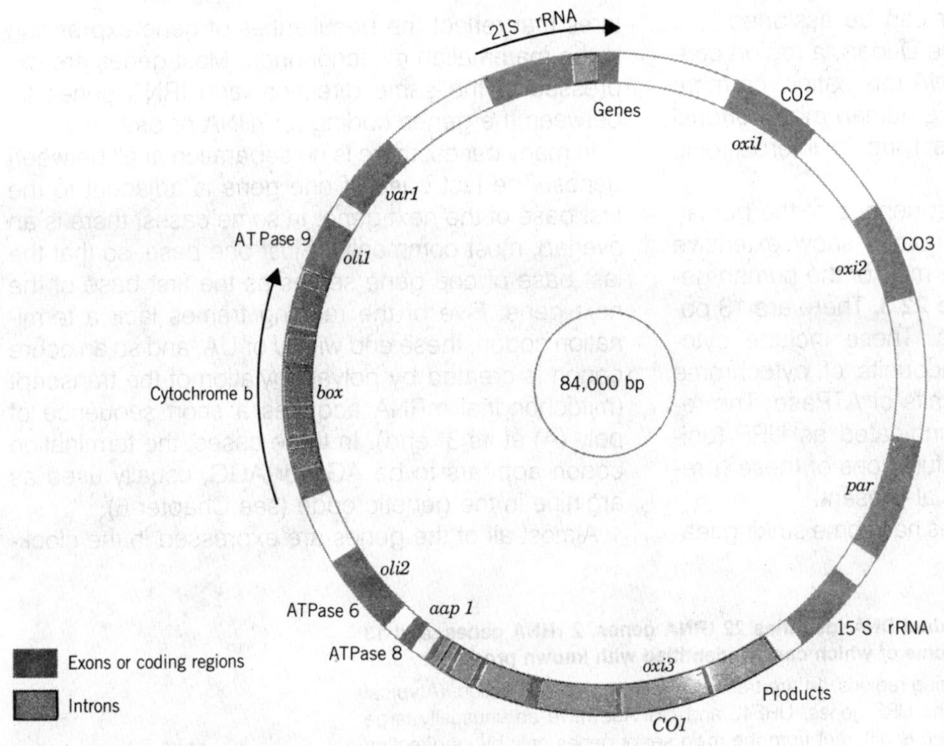

Figure 22.2
The mitochondrial genome of *S. cerevisiae* contains both interrupted and uninterrupted protein-coding genes, rRNA genes, and tRNA genes (positions not indicated). Arrows indicate direction of transcription.

This adds several proteins, presumably all synthesized in low amounts, to the complement of the yeast mitochondrion.

The remaining genes appear to be uninterrupted. They correspond to the other two mitochondrially coded subunits of cytochrome oxidase, to the subunit(s) of the ATPase, and (in the case of *var1*) to a mitochondrial ribosomal protein.

Quite a large proportion of the yeast mitochondrial genome, about 25%, consists of short stretches rich in A-T base pairs, which probably lack any coding function. However, this still leaves a considerable amount of genetic material to be accounted for, and it will be surprising if some further genes are not discovered in the regions that have been left open on the map. Even given this likelihood, though, we can conclude that the total number of yeast mitochondrial genes is unlikely to exceed about 20.

THE COMPACT MITOCHONDRIAL GENOME OF MAMMALS

A very different picture is seen in mammalian mitochondrial DNA. Its organization is extremely compact; there are no introns, some genes actually overlap, and almost every single base pair can be assigned to a gene. With the exception of the D-loop, a region concerned with the initiation of DNA replication, no more than 87 of the 16,569 bp of the human mitochondrial genome can be regarded as lying in intercistronic regions.

The complete nucleotide sequences of the human and murine mitochondrial genomes show extensive homology in organization. The map of the human genome is summarized in **Figure 22.3.** There are 13 potential protein-coding regions. These include cytochrome *b*, the usual three subunits of cytochrome oxidase, and one of the subunits of ATPase. The remaining reading frames are indicated as URF (unidentified reading frame). The functions of these (presumed) proteins are unknown at present.

The organization of the genes has some striking features that reflect the peculiarities of gene expression in the mammalian mitochondrion. Most genes are expressed in the same direction; and tRNA genes lie between the genes coding for rRNA or protein.

In many cases, there is no separation at all between genes. The last base of one gene is adjacent to the first base of the next gene. In some cases, there is an overlap, most commonly of just one base, so that the last base of one gene serves as the first base of the next gene. Five of the reading frames lack a termination codon; these end with U or UA, and so an ochre codon is created by polyadenylation of the transcript (mitochondrial mRNA acquires a short sequence of poly (A) at its 3′ end). In three cases, the termination codon appears to be AGA or AGG, usually used as arginine in the genetic code (see Chapter 6).

Almost all of the genes are expressed in the clock-

Figure 22.3

The sequence of human mitochondrial DNA identifies 22 tRNA genes, 2 rRNA genes, and 13 potential protein-coding regions, some of which can be identified with known products.

Of the 15 protein-coding or rRNA-coding regions, 14 are transcribed in the same direction. (Atypical features are presented by three of the URF genes. URF4L and URFA6L have an unusually large overlap with the adjacent genes. URF6 is different from the main set of genes only by its direction of expression.) Of the 22 tRNA genes, 14 are expressed in the clockwise direction and 8 are read counterclockwise.

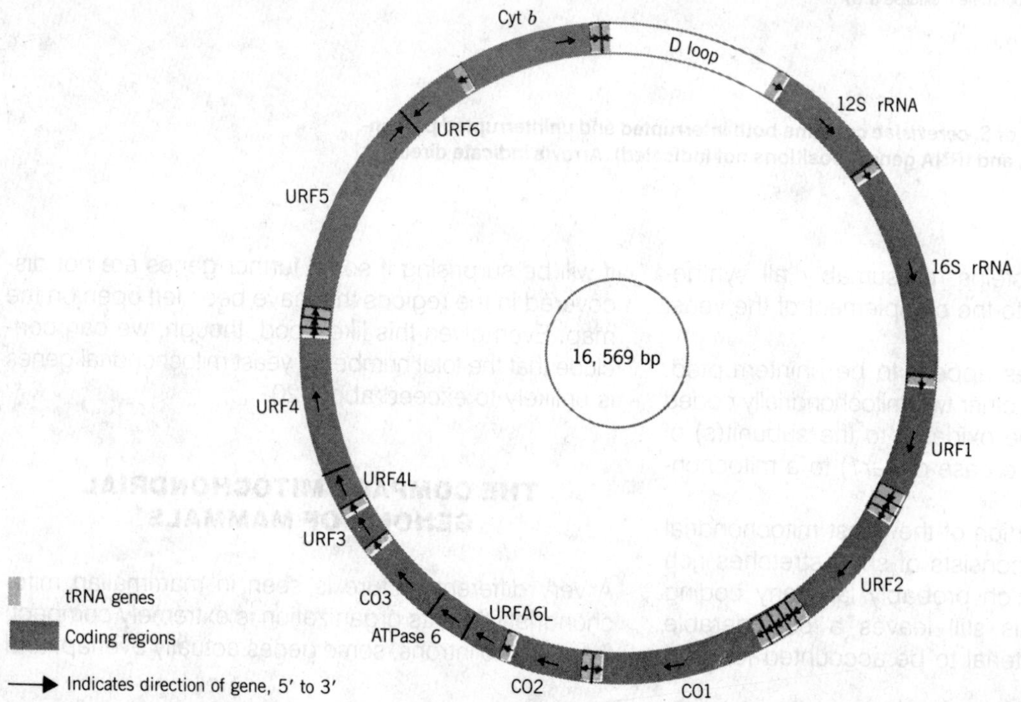

wise direction, as indicated by the arrows on the map. In only one case are two clockwise coding regions found in the form of contiguous genes (ATPase 6 and CO3). In every other case, at least one tRNA gene separates adjacent coding regions. The mRNA species corresponding to all the protein-coding regions have been identified; and in each case, an initiation codon lies within six bases of the start. The codon used for initiation apparently can be AUG, AUA, or AUU. The mRNAs terminate with nonsense codons or incomplete nonsense codons. So nontranslated 5′ and 3′ regions are virtually absent from human mitochondrial mRNAs.

The punctuation of the rRNA- and protein-coding regions by the tRNA genes does not leave room for promoters comparable to those found in eukaryotic nuclei or in bacteria. A single promoter for clockwise transcription is located in the region of the D-loop. This suggests the model illustrated in **Figure 22.4.** Transcription starts just before the tRNA gene in front of the 12S rRNA gene, and continues almost all the way around the circle, to terminate in the D-loop. The transcribed strand is called the **H strand** (because mitochondrial DNA can be separated into heavy and light single strands on the basis of their density).

Figure 22.4
Human mitochondrial DNA is transcribed into a single transcript, from which tRNAs are cleaved to release the rRNAs and mRNAs.

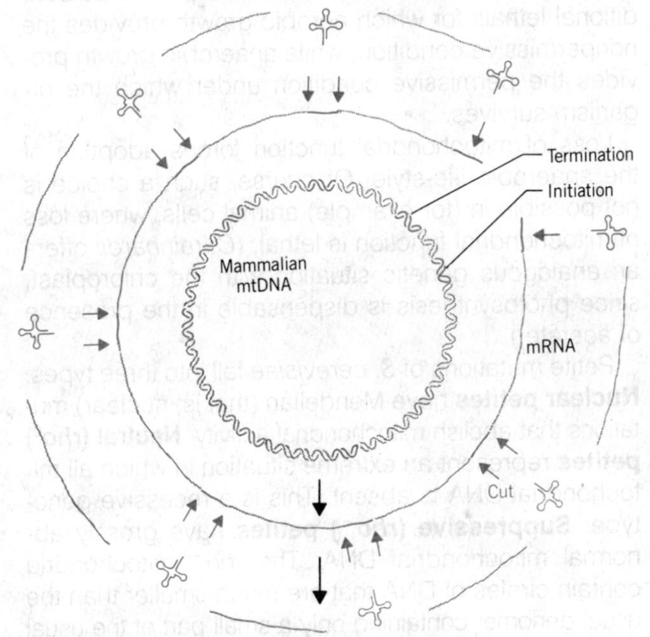

The significance of the alternation of tRNA genes with rRNA- and protein-coding genes is that the tRNAs indicate sites of cleavage. By cleaving the primary transcript on either side of each tRNA gene, all except two of the genes (ATPase 6 and CO3) give rise to monocistronic products. The rRNA molecules appear to be synthesized in greater amounts than the mRNAs. This could be caused by premature termination of some proportion of the transcripts at some point after the two rRNA genes have been transcribed.

This mitochondrial DNA therefore presents the closest analogy to a bacterial operon in the eukaryotes. It is transcribed from a single promoter region, but individual tRNAs, rRNAs, and mRNAs are released from the transcript. Processing of the transcript therefore becomes the central event in gene expression. (There are also bacterial operons in which the separation of tRNA from rRNA or protein-coding regions must be accomplished after transcription; see Chapter 23.)

How are the tRNA genes coded by the L strand expressed? Probably a similar mechanism applies, with a giant transcript produced for the L strand, from which the tRNA sequences and perhaps some mRNA sequences are conserved, while the rest is degraded. The origin for L-strand (counterclockwise) transcription also lies in the region of the D-loop. Very large L-strand transcripts can be isolated, so that their processing probably takes place some time subsequently to transcription. Large H-strand transcripts cannot be found, which suggests that they are processed more or less immediately after the transcription of each tRNA junction.

RECOMBINATION OCCURS IN (SOME) ORGANELLE DNAs

The prerequisite for recombination is the coexistence of genomes of both parents, a condition that is prevented in situations where absolute uniparental inheritance prevails. But the existence of instances where both organelle genomes survive immediately raises the question of whether they can interact. Is complementation possible between different mutations; can recombination occur between genomes?

Complementation requires intermingling of the products of gene expression; recombination requires physical exposure of the two genomes to each other. Both demands are somewhat at odds with the static view of the individual organelle that might be inferred from the description of its endogenous gene expression as an event strictly confined to its own compartment. Yet in at least one species, recombination can occur between chloroplast DNAs; and in another, both complementation and recombination can occur with mitochondrial genomes.

In the alga *C. reinhardii,* the chloroplast DNA usually is inherited uniparentally. However, under the unusual circumstance when the genes of the usually predominant parent are irradiated with ultraviolet, the chloroplast genes of the other parent also survive. In the resulting biparental zygotes, recombination occurs between the chloroplast genomes, and this can be used to construct a genetic map.

In the yeast *S. cerevisiae,* recombination between mitochondrial markers is the norm. The nature of the recombination event is unknown, but a cross between two parents whose mitochondrial DNAs have gross differences generates progeny possessing physically recombinant DNAs. This demonstrates directly that the mitochondrial DNA of one parent comes into contact with the mitochondrial DNA provided by the other parent.

How do the parental organelle DNAs reach each other? The most likely explanation is that organelles can undergo fusion, to generate a single compartment containing genomes from both parents. This concept has been taken further in the case of yeast, where it is possible that instead of many individual mitochondria, a large branched structure (or structures) corresponds to many mitochondria. Perhaps there is a continual state of flux, with making and breaking of membranes. This raises the possibility that the presence of a genome is related to some unit smaller than the organelle itself; then the total number of genomes per organelle would be determined by the number of these constituent subunits.

In yeast it takes some hours after the production of a zygote before recombination occurs. But in the interim period, complementation can be detected between different mutations. This provides a unique opportunity to define the genetic organization of the organelle genome, since mutations can be both mapped and assigned to complementation groups. (A similar approach also has proved successful with the *C. reinhardii* chloroplast.)

REARRANGEMENTS OF YEAST MITOCHONDRIAL DNA

The yeast mitochondrial genome has a surprising fluidity of structure. We have already mentioned that different forms of some genes are found in different strains of *S. cerevisiae.* Indeed, the total size of the mitochondrial genome can vary substantially, by as much as 10,000 bp between strains. Extensive deletions of mitochondrial DNA can occur, as found in the existence of **petite** strains of *S. cerevisiae* (and certain other yeasts). They may arise spontaneously or may be induced by certain treatments, such as addition of ethidium bromide.

All petite mutations lack mitochondrial function. Their existence is possible because the alternative life-styles of yeast allow it to survive aerobically (when respiration is essential) or anaerobically (when it is dispensable). Thus mitochondrial mutations behave as conditional lethals for which aerobic growth provides the nonpermissive condition, while anaerobic growth provides the permissive condition under which the organism survives.

Loss of mitochondrial function forces adoption of the anaerobic life-style. Of course, such a choice is not possible in (for example) animal cells, where loss of mitochondrial function is lethal. (*C. reinhardii* offers an analogous genetic situation with the chloroplast, since photosynthesis is dispensable in the presence of acetate.)

Petite mutations of *S. cerevisiae* fall into three types. **Nuclear petites** have Mendelian (that is, nuclear) mutations that abolish mitochondrial activity. **Neutral (*rho*^{0}) petites** represent an extreme situation in which all mitochondrial DNA is absent. This is a recessive genotype. **Suppressive (*rho*$^-$) petites** have grossly abnormal mitochondrial DNA. The *rho*$^-$ mitochondria contain circles of DNA that are much smaller than the usual genome, containing only a small part of the usual complexity, ranging from 0.2% to 36% in different cases. Thus more of the mitochondrial genome is deleted than is retained.

The sequence retained in a *rho*$^-$ petite often is amplified, to generate a large number of copies. Amplification occurs either by increase of the ploidy (the number of DNA molecules) or by formation of multimeric DNA molecules, each containing many copies of the sequence. The multiple copies can be arranged tandemly as direct or inverted repeats. The sequence of the petite need not necessarily represent a formerly contiguous region of the mitochondrial genome, but can have different regions juxtaposed. Some petites are stable; others are unstable, and further rearrangements of the sequence may occur.

The DNA retained in a petite strain can recombine with the DNA of another petite or of a wild-type strain (called *grande,* by comparison). This allows the genetic markers present in each petite to be correlated with the sequence of DNA that has been retained. In fact, it is the use of such petites for this deletion mapping that allows construction of the mitochondrial genetic map shown in Figure 22.2.

Any sequence of mitochondrial DNA can be retained in a petite. This implies that all segments of the DNA can be independently replicated. We do not know how this facility is provided and whether it relies on any feature equivalent to an origin for replication (see Chapter 31).

When a rho$^-$ petite strain of yeast is crossed with a wild-type strain, in a certain proportion of the progeny, only the petite genotype is found. The wild-type mitochondrial DNA has disappeared. This effect is called **suppression;** and the characteristic proportion of progeny in which it occurs gives the degree of **suppressiveness.**

The cause of suppression in highly suppressive petites is preferential replication of the petite mitochondrial DNA. This probably happens because the petite DNA possesses an increased concentration of origin-like sequences, called *rep* regions, relative to wild-type DNA. So we see that yeast mitochondrial DNA does contain a specific sequence(s) that can be used to initiate replication. But, unlike its counterpart in other genomes, this sequence is dispensable, because petite DNA lacking it can survive (although it is less suppressive). The cause of suppression therefore is not necessarily the same in less suppressive petites.

The general point made by these studies of petite and wild-type mitochondrial DNA is that there is extensive flexibility in the organization of the genome, with regard both to its expression and replication. Perhaps the difference from other mitochondrial DNAs lies in the ability of yeast altogether to dispense with mitochondrial functions. It will be interesting to characterize the enzyme systems that act on this DNA.

FURTHER READING

A general review of chloroplast and mitochondrial inheritance was provided by **Gillham** in *Organelle Heredity* (Raven Press, 1978). The topics of this chapter are reviewed in **Lewin's** *Gene Expression,* **2,** *Eucaryotic Chromosomes* (Wiley, New York, 1980, pp. 583–637). The sequence of the human mitochondrial genome was analyzed by **Anderson et al.** (*Nature* **290,** 457–465, 1981). Transcription of mammalian mitochondrial DNA has been reviewed by **Clayton** (Ann. Rev. Biochem. **53,** 573–594, 1984).

CHAPTER 23
IDENTITY AND VARIATION IN TANDEM GENE CLUSTERS

From the perspective of the individual gene, all extremes are found in genome organization. A gene may exist in a location where it is the only sequence of its kind (although it is likely to have relatives elsewhere). It may be a member of a small cluster of related genes descended from a common ancestor and retaining similar functions (as in the globin system).

Other gene clusters may be larger, displaying extensive repetition of genes that are either identical or very closely related. Genes (or other sequences) that exist as adjacent multiple copies are said to be **tandemly repeated.** One type of tandem gene cluster codes for histone proteins, another for rRNAs. In both cases, the extensive repetition may reflect the need to produce large amounts of the product.

Tandem clustering of genes into a defined region whose only purpose is to code for a particular type of function suggests a way to describe the organization of sequences. Each tandem cluster can be defined in terms of a **repeating unit.** This is the basic entity whose end-to-end repetitions make up the cluster. The copies of the repeating unit may be identical, or they may show variations.

Usually this repeating unit is longer than the sequence that is transcribed; and so it may be divided into two regions. The **transcription unit** occupies the distance between the first and last bases transcribed into RNA. The **nontranscribed spacer** is the remaining part of the repeating unit; it comprises the region lying between adjacent transcription units. This form of organization is illustrated in **Figure 23.1.**

The restriction map of a tandem gene cluster has a characteristic feature. It is circular. When the repeating units are identical, each generates exactly the same pattern of restriction cuts. Adjacent repeats share their terminal fragments. In the example shown in the figure, fragment *A* is adjacent to fragment *B*, which is adjacent to fragment *C*, which in turn is adjacent to fragment *A*, thus generating a circular map.

This pattern is broken only at the ends of the cluster, shown in the figure by fragments *X* and *Y*. But usually these are not detected, because each is unique, compared with the multiple repetition of internal fragments. For example, if there were 100 copies of the repeating unit, fragments *X* and *Y* would constitute only 1% of the material present in fragments *A*, *B*, and *C*.

The nontranscribed spacer is usually shorter than the transcription unit. We can define the character of the spacer because of the clarity with which regions can be delineated within a tandem gene cluster. The spacer is simply a part of the repeating unit that is not expressed in RNA, but whose functions (if any) remain concerned with the sole purpose of the cluster; production of the RNA coded by the transcription unit.

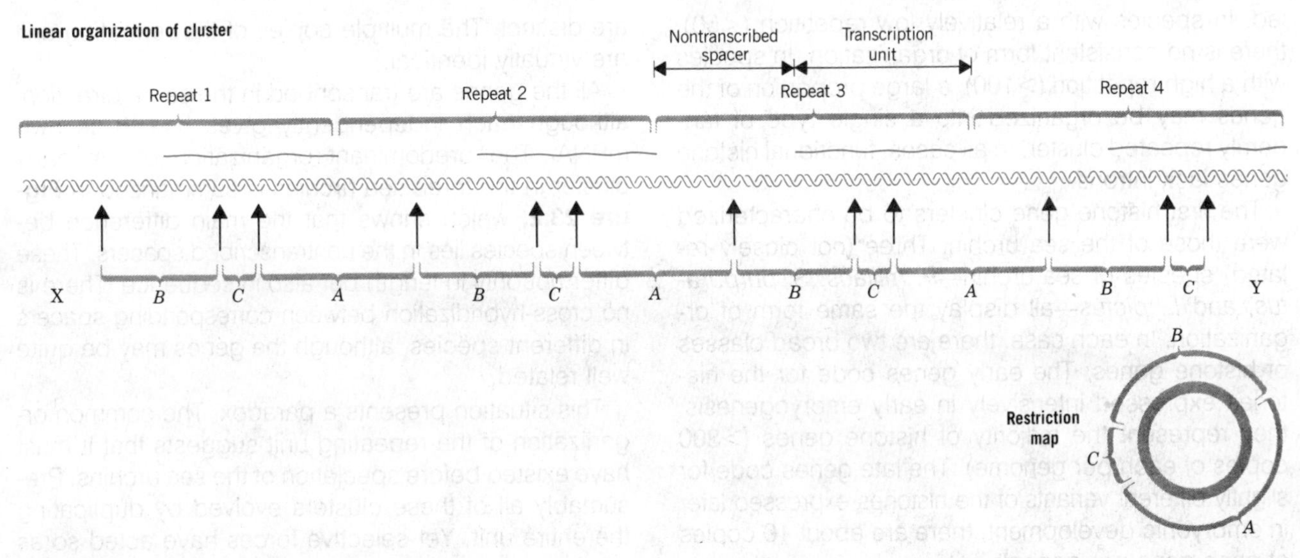

Figure 23.1
A tandem gene cluster has an alternation of transcription unit and nontranscribed spacer and generates a circular restriction map.

This view may prompt us to ask whether the somewhat longer sequences between individual genes in (for example) the globin gene cluster should be regarded in the same light. However, there is an important difference in the two situations. Every intergenic region in a globin gene cluster is different, whereas the nontranscribed spacers of a tandem gene cluster are repeated along with the genes.

A VARIETY OF TANDEM GENE CLUSTERS CODE FOR HISTONES

Histones are nuclear structural proteins par excellence. In (probably) all eukaryotes, the chromosomes contain five types of histone protein: H1 present in half-molar; and H2A, H2B, H3, and H4 present in equimolar amounts (see Chapter 29). The total mass of histones is roughly equivalent to the mass of DNA itself.

In dividing somatic cells, histone synthesis occurs pari passu with DNA replication; as soon as a stretch of DNA is replicated, histones are available to associate with it. This means that a large amount of histone protein must be synthesized during the short period while DNA is synthesized. The need for the somatic cell to synthesize a genome's worth of histones in a relatively short time may be the basic cause of histone gene repetition.

There is no particular relationship between genome size and histone gene number (so the efficiency of histone gene expression must vary in each species). But the repetition of histone genes is a common aspect of the arrangements for producing histones. Usually there is the same number of copies of each histone gene. In the chicken the repetition frequency is about 10, in mammals it is about 20. It increases to roughly 40 in *X. laevis* and to about 100 in *D. melanogaster*. Several sea urchin species have a range of 300–600 copies of each histone gene.

In the sea urchins, the initial nuclear divisions of embryogenesis occur very rapidly indeed. The large gene number in sea urchins may be needed to allow de novo histone gene expression to keep up with DNA synthesis. In amphibians, histone gene expression precedes the start of embryogenesis, and a large amount of histone mRNA accumulates in the oocyte. Also, some histone proteins may be stored. This provides an exception to the rule that histone genes are expressed only during the phase of DNA synthesis.

The arrangement of histone genes is somewhat varied. In species with a relatively low repetition (<50), there is no consistent form of organization. In species with a high repetition (>100), a large proportion of the genes may be organized into a single type of tandemly repeated cluster. In all cases, functional histone genes lack introns.

The first histone gene clusters to be characterized were those of the sea urchin. Three (not closely related) species of sea urchin—*P. miliaris, S. purpuratus,* and *L. pictus*—all display the same form of organization. In each case, there are two broad classes of histone genes. The early genes code for the histones expressed intensively in early embryogenesis; they represent the majority of histone genes (>300 copies of each per genome). The late genes code for slightly different variants of the histones expressed later in embryonic development; there are about 10 copies of each gene per genome.

All five early histone genes are part of a single repeating unit that constitutes a single tandem gene cluster in the genome. Each gene is separated from the next by a nontranscribed spacer; the five spacers are distinct. The multiple copies of the repeating unit are virtually identical.

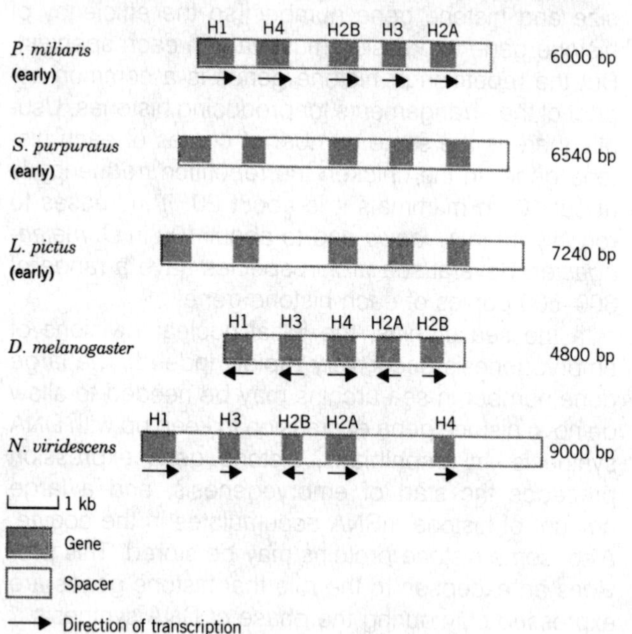

Figure 23.2
Repeating units of histone gene clusters vary in organization.

All the genes are transcribed in the same direction, although each independently gives rise to its own mRNA. The predominant organization of the early cluster in the three sea urchins is summarized in **Figure 23.2,** which shows that the main difference between species lies in the nontranscribed spacers. These differ not only in length but also in sequence. There is no cross-hybridization between corresponding spacers in different species, although the genes may be quite well related.

This situation presents a paradox. The common organization of the repeating unit suggests that it must have existed before speciation of the sea urchins. Presumably all of these clusters evolved by duplicating the entire unit. Yet selective forces have acted so as to preserve the functions of the genes, while allowing the intermingled spacers to diverge entirely *between* species, although remaining constant *within* each species. The situation implies that some corrective mechanism must act on the cluster within each species.

Actually, each of the five spacers is not perfectly preserved. Some variation is seen in the lengths of the corresponding spacers of individual repeating units, as detected by (slight) heterogeneity in the lengths of the appropriate restriction fragments. These variations between repeating units are referred to as **microheterogeneity.**

In one sea urchin, two types of repeating unit specifying the early embryonic genes have been characterized. They differ in their spacers. Each type of unit appears to be tandemly repeated; the two types of cluster are not intermingled. This implies that whatever mechanisms are responsible for maintaining homogeneity must be able to act on subregions of the cluster and need not extend throughout it.

The late histone genes are organized in a different way. The genes may be found on individual fragments of DNA; sometimes they occur as a pair of genes loosely linked (>10 kb apart). There is no consistent form of organization.

Within the same organism, the histone gene family may therefore be organized in two quite different ways: early genes fall into a large discrete cluster, while late genes are dispersed. The dispersed genes probably arose by translocation of individual members away from an ancestral cluster, freeing them from the selective constraints that maintain homogeneity of sequence, and thus allowing them to diverge to acquire the functions now associated with late genes.

All five histone genes are organized into a single repeating unit in *D. melanogaster,* although here the organization of the repeating unit is different. The genes lie in a different order and are transcribed in different directions. Two translocations would be needed for conversion between the sea urchin and fruit fly types of cluster. Spacer variants also are present in the *D. melanogaster* cluster.

An intriguing variation on this form of organization is found in the newt *N. viridescens,* in which there is a single type of repeating unit (different from both sea urchin and fruit fly). However, the individual repeating units do not lie adjacent to one another, but are separated by up to 10–50 kb of highly repetitive DNA.

A more variable form of organization is found in some cases. In *X. laevis,* the histone genes are organized in a cluster, but there is heterogeneity in their organization, since the neighbors of a given type of gene may vary. If there are tandem repeats, there must be more than one class, each containing genes in a different order. In chickens, there is a cluster of histone genes, but the genes lie in a variety of orders, and there are no tandem repeats. Proceeding to mammals, there is again evidence that the genes may not be systematically organized into repeating units, but may lie in smaller groups or even as individual genes. These patterns therefore seem to be somewhat intermediate between the small gene cluster (such as globin) and the tandem gene cluster.

There is an interesting contrast between the structures of individual histone genes and their overall organization. All histone genes share the same general uninterrupted structure; and corresponding genes in different species code for well-related proteins. Yet there is appreciable flexibility in the relationships between the various classes of genes, which vary from rigorous tandem clustering to an apparently erratic ordering. All of this reinforces the general conclusions suggested by the example of the globin gene cluster. Individual genes may be tailored to their function by particular mutations in an otherwise conserved sequence, but there is continual generation of new copies and subsequent rearrangement of the overall organization of the cluster.

GENES FOR rRNA AND tRNA ARE REPEATED

Each of the RNA components of the protein synthetic apparatus is coded by many genes. In a sense, the structural genes coding for mRNA form the largest class, although since there are so many different mRNAs, each gene is represented in only a relatively small number of messenger molecules. To match the large number of mRNAs that need translation, the cell manufactures many ribosomes and tRNA molecules.

Ribosomal RNA is by far the predominant product of transcription, constituting some 80–90% of the total mass of cellular RNA in both eukaryotes and prokaryotes. There are several tRNA molecules per ribosome, but of course they are much smaller than the rRNAs. Both rRNA and tRNA are represented by multiple genes. **Table 23.1** summarizes the numbers present in some haploid genomes.

The number of major rRNA genes varies from 7 in *E. coli,* to between 100 and 200 in lower eukaryotes, to several hundred in higher eukaryotes. In virtually every case, the genes for the large and small rRNA form a tandem pair. (The sole exception is the yeast mitochondrion.) In bacteria and some lower eukaryotes, the gene for 5S RNA is part of the same unit,

Table 23.1
There are multiple rRNA and tRNA genes in all genomes.

Species	Ratio rDNA/Total	Number of Genes 18S/28S	5S	tRNA
E. coli	1.0%	7	7	60
S. cerevisiae	5.5%	140	140	250
D. discoideum	17.0%	180	180	?
D. melanogaster	1.3%		165	850
(X)		250		
(Y)		150		
Man	0.4%	280	2000	1300
X. laevis	0.18%	450	24000	1150

so the total number of 5S genes is the same as that of the major rRNAs. In bacteria, the 5S gene is cotranscribed with the major rRNA genes; in eukaryotes it is transcribed independently. In higher eukaryotes, the genes for 5S RNA are separately organized into their own clusters, and their number exceeds that of the major rRNA genes.

The lack of any detectable variation in the sequences of the rRNA molecules implies that all the copies of each gene must be identical, or at least must have differences below the level of detection in rRNA (about 1%). In bacteria, the multiple 16S-23S rRNA gene pairs are dispersed. In most eukaryotic nuclei, the rRNA genes are contained in a tandem cluster or clusters. These regions are sometimes known as **rDNA.** (In some cases, the proportion of rDNA in the total DNA, together with its atypical base composition, is great enough to allow its isolation as a separate fraction directly from sheared genomic DNA.)

In the case of tRNA, the exact number of genes is hard to determine, because the extensive secondary structure of the molecule creates technical difficulties in the hybridization reaction. Probably we have underestimated the actual number.

Figure 23.3
A single transcription unit contains the sequence for both the small and large rRNA molecules.

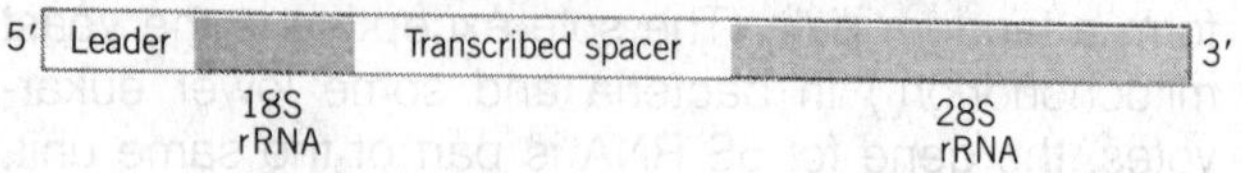

5′ Leader | 18S rRNA | Transcribed spacer | 28S rRNA 3′

A TANDEM REPEATING UNIT CONTAINS BOTH rRNA GENES

The pair of major rRNAs is transcribed as a single precursor in both bacteria and eukaryotic nuclei. Following transcription, the precursor is cleaved to release the individual rRNA molecules by the pathways discussed in more detail in Chapter 25. The salient features of the gene organization are described in **Figure 23.3.** The transcript starts with a 5′ leader, followed by the sequence of the small rRNA, then a region called the **transcribed spacer,** and finally the sequence of the large rRNA toward the 3′ end of the molecule.

(The *transcribed spacer* is not to be confused with the *nontranscribed spacer* that separates transcription units. Its name reflects the fact that it is part of the transcription unit, but is not represented in the mature RNA products.)

The transcription unit is therefore longer than the combined length of the mature rRNAs. **Table 23.2** summarizes some examples of the relationship between the primary transcript and the rRNAs.

The transcription unit is shortest in bacteria, where the rRNA sequences constitute 80% of its total length of 6000 bases. No particularly systematic pattern is seen in a range of eukaryotes in which the transcript length varies from about 7000 to 8000 bases, with 70–80% of the sequence representing the rRNAs. The precursor is longest in mammals (where it is known as 45S RNA, according to its rate of sedimentation).

Table 23.2
Each rRNA precursor is longer than the two mature rRNAs.

Species	Length of Transcript	Length of rRNA		rRNA Percent of Transcript
		Large	Small	
E. coli	5600	2904	1542	80
S. cerevisiae	7200	3750	2000	80
D. melanogaster	7750	4100	2000	78
X. laevis	7875	4475	1925	79
N. tabacum	7900	3700	1900	71
G. domesticus	11250	4625	1800	57
M. musculus	13400	5000	1950	52

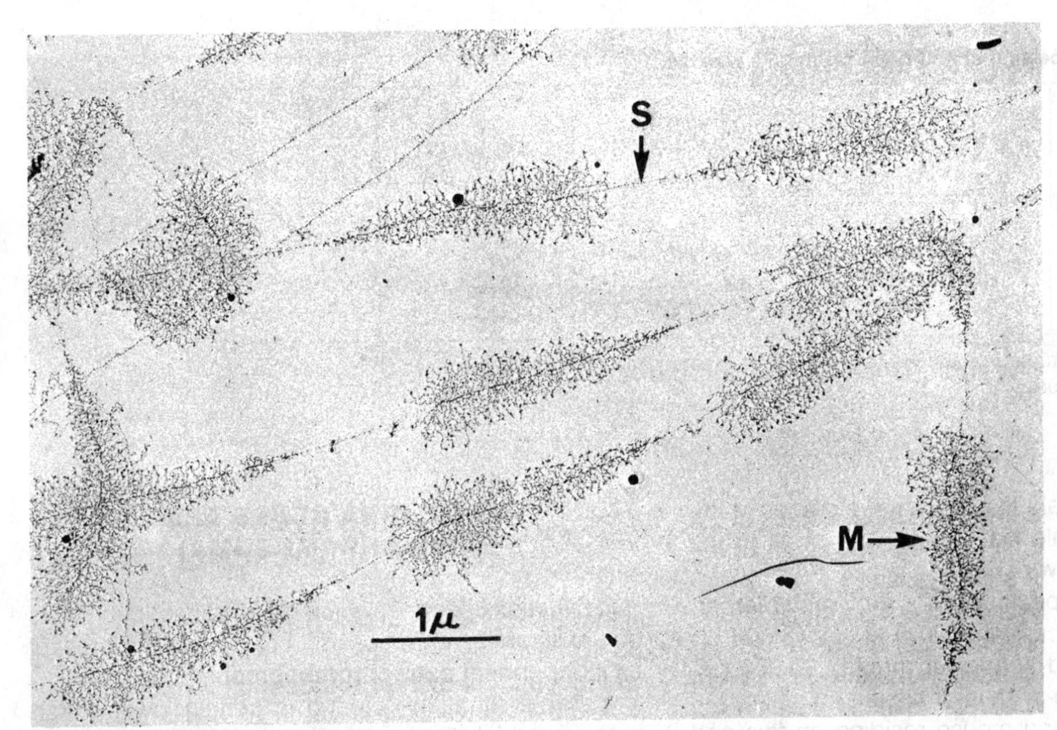

Figure 23.4
Transcription of rDNA clusters generates a series of matrices, each corresponding to one transcription unit and separated from the next by the nontranscribed spacer.
Photograph kindly provided by Oscar Miller.

Here the mature rRNA sequences occupy only just over 50% of the length of the complete transcript.

What happens to the nonribosomal parts of the RNA precursor? The leader and the transcribed spacer regions are discarded during the maturation of rRNA. The transcribed spacer contains some short sequences that are released by cleavage. In mammals and amphibians, one short sequence forms the 5.8S RNA, a small molecule that hydrogen bonds with the 28S rRNA in the ribosome. In bacteria, as we shall see later, tRNA sequences may lie in the transcribed spacer and (sometimes) also at the 3′ end of the primary transcript. The rest of the transcribed spacer, and all of the leader sequences are presumably degraded to nucleotides.

An rDNA cluster contains many transcription units, each separated from the next by a nontranscribed spacer. The alternation of transcription unit and nontranscribed spacer can be seen directly in electron micrographs. The example shown in **Figure 23.4** is taken from the newt *N. viridescens,* in which each transcription unit is intensively expressed, so that many RNA polymerases are simultaneously engaged in transcription on one repeating unit. The polymerases are so closely packed that the increase in the length of their products forms a characteristic matrix moving along the transcription unit.

The nontranscribed spacer varies widely in length between and (sometimes) within species. **Table 23.3** summarizes the situation.

In yeast there is a short nontranscribed spacer, relatively constant in length. In *D. melanogaster,* there is almost a twofold variation in the length of the nontranscribed spacer between different copies of the repeating unit. A similar situation is seen in *X. laevis.* In each of these cases, all of the repeating units are present as a single tandem cluster on one particular chromosome. (In the example of *D. melanogaster,* this happens to be the sex chromosome. The cluster on the X chromosome is larger than that on the Y chro-

Table 23.3
Length of the tandem repeating unit of rDNA clusters is variable.

Species	Length of Repeating Unit	Length of Nontranscribed Spacer	Length of Transcript
S. cerevisiae	8950	1750	7200
D. melanogaster	11500–14200	3750–6450	7750
X. laevis	10500–13500	2300–5300	7875
M. musculus	44000	30000	13400

These data refer to uninterrupted genes. The range would be greater if the interrupted 28S genes of *D. melanogaster* were included.

mosome, so that female flies have more copies of the rRNA genes than male flies. However, not all of the genes are active, as we shall see later.)

In mammals the repeating unit is very much larger, comprising the transcription unit of about 13,000 bp and a nontranscribed spacer of roughly 30,000 bp. Usually, the genes lie in several dispersed clusters—in the case of man and mouse residing on five and six chromosomes, respectively. One interesting question is how the corrective mechanisms that presumably function within a single cluster to ensure constancy of rRNA sequence are able to work when there are several clusters.

The region of the nucleus where rRNA synthesis occurs has a characteristic appearance, with a core of fibrillar nature surrounded by a granular cortex. The fibrillar core is the region at which the rRNA is transcribed from the DNA template; and the granular cortex is formed by the ribonucleoprotein particles into which the rRNA is assembled. The whole area is called the **nucleolus.**

Particular chromosomal regions can be seen to be associated with a nucleolus. These are called **nucleolar organizers.** Each nucleolar organizer corresponds to a cluster of tandemly repeated rRNA genes, so in a diploid nucleus there are twice as many nucleolar organizers as there are tandem clusters in the haploid genome. The concentration of the tandemly repeated rRNA genes, together with their very intensive transcription, is responsible for creating the characteristic morphology of the nucleoli.

SOME rRNA GENES ARE EXTRACHROMOSOMAL

Two situations occur in which rRNA transcription units do not lie on the chromosome but instead are present as independent **extrachromosomal** molecules.

A stable situation is found in certain lower eukaryotes, in which the genes responsible for producing rRNA are carried by extrachromosomal DNA molecules, present in many copies per nucleus. The example of *D. discoideum* is shown in **Figure 23.5.** The extrachromosomal molecule is a large palindromic dimer, containing two transcription units in opposite orientation, separated by about 20,000 bp. Each extrachromosomal dimer also contains two 5S genes, although these are transcribed independently from the major transcription unit. We know very little about the arrangements for inheritance of this molecule.

A transient state occurs during the development of some animals. During oogenesis in many species, a large number of *additional* rRNA genes are suddenly generated. The process has been characterized most

Figure 23.5
The rRNA genes in *D. discoideum* lie on an extrachromosomal palindrome.

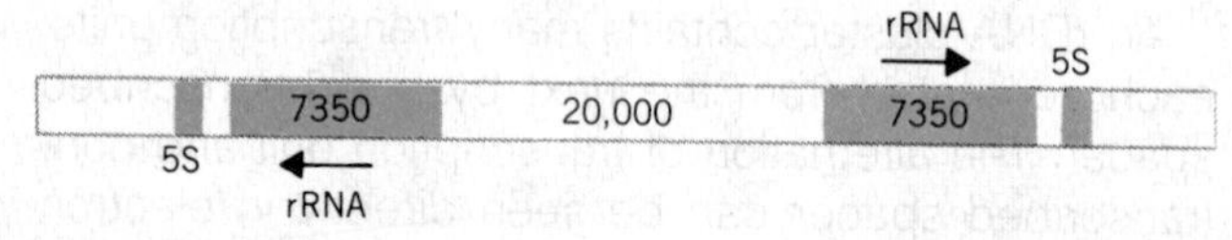

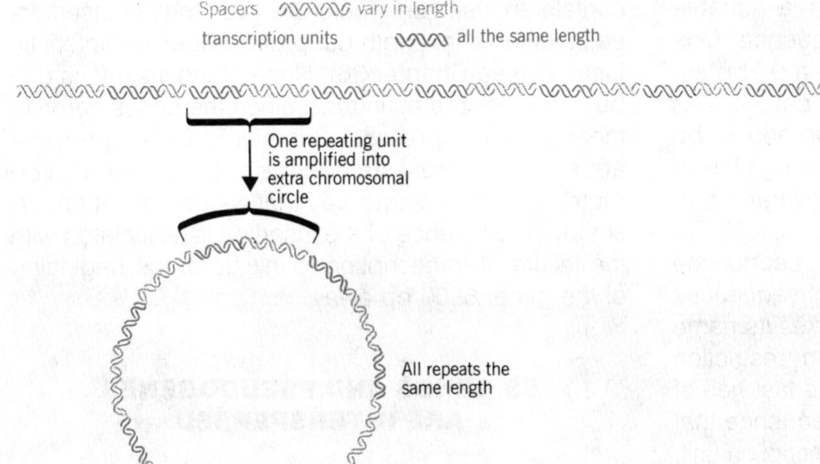

Figure 23.6
Each extrachromosomal rDNA in the *X. laevis* oocyte is a circle generated by end-to-end amplification of a single repeating unit of chromosomal rDNA.

fully in *X. laevis*. Circular extrachromosomal molecules of DNA each containing many repeating units are produced by **amplification** of the chromosomal repeating units. Unlike the chromosomal genes, in which adjacent repeating units can have spacers of different length, on the extrachromosomal circles the adjacent units all have spacers of the *same* length. This suggests that each amplified circle is produced by making multiple end-to-end copies from a single chromosomal repeating unit, as illustrated in **Figure 23.6.**

The additional genes are needed to achieve sufficient production of rRNA in the oocyte, and are not a feature of somatic cells, which rely on the chromosomal genes. (The mechanism of this amplification is a form of replication called the rolling circle, which is discussed in Chapter 31 and illustrated in Figure 31.11.)

ABOUT NONTRANSCRIBED SPACERS AND PROMOTERS

The variation in length of the nontranscribed spacer in a single gene cluster contrasts with the conservation of sequence of the transcription unit. In spite of this variation, the sequences of longer nontranscribed spacers remain homologous with those of the shorter nontranscribed spacers. This implies that each nontranscribed spacer is *internally repetitious,* so that the variation in length results from changes in the number of repeats of some subunit.

The general nature of the nontranscribed spacer is illustrated by the example of **Figure 23.7,** which shows its organization in *X. laevis*. Regions that are fixed in length alternate with regions that may vary. Each of

Figure 23.7
The nontranscribed spacer of *X. laevis* rDNA has an internally repetitious structure that is responsible for its variation in length.

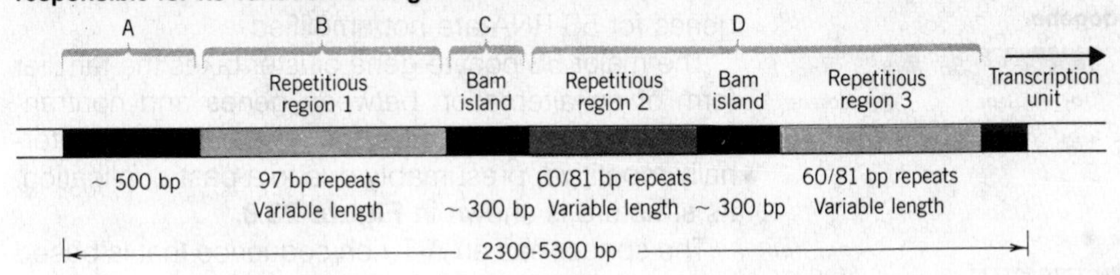

the three repetitious regions comprises a variable number of repeats of a rather short sequence. One type of repetitious region has repeats of a 97 bp sequence; the other, which occurs in two locations, has a repeating unit found in two forms, 60 bp and 81 bp long. The variation in the number of repeating units in the repetitious regions accounts for the overall variation in spacer length.

One of the fixed regions (*A*) is unique in sequence and length. The others are short constant sequences called **Bam islands.** (This description takes its name from their isolation via the use of the Bam restriction enzyme.) They are puzzling, because the last half of each Bam island is very similar to the sequence that immediately precedes the start of the transcription unit, which contains the promoter. The same type of sequence is also found in the 60/81 bp units of repetitious regions 2 and 3. Its occurrence raises the question of why transcription does not start at these sequences as well as at the proper promoter.

The nontranscribed spacer does have a function in transcription. The 60/81 bp repeats play a role in initiation, probably comparable to the role that enhancers play for genes transcribed by RNA polymerase I (see Chapter 12). The presence of a spacer increases the frequency of initiation, apparently by causing several RNA polymerase II molecules to bind in succession at the promoter. We do not yet understand the exact nature of this relationship, and how it depends on the similarities of sequence between the spacer and the promoter.

Another situation in which there are seemingly identical promoters, some of which initiate transcription while others do not, is provided in *D. melanogaster.* Here about two thirds of the genes for the large rRNA contain an insertion. There are two types of insertion, each variable in length but present at a characteristic location (see Chapter 36). None of the insertions can be characterized as introns, since the genes carrying them are not expressed. But their 5′ flanking regions appear no different from those of the active uninterrupted genes. In some way that we do not yet understand, the presence of the insertion is associated with the failure of transcription to initiate at the beginning of the gene, 6800 bp away.

Figure 23.8
The repeating unit of the major oocyte 5S gene cluster of *X. laevis* contains a gene and a pseudogene.

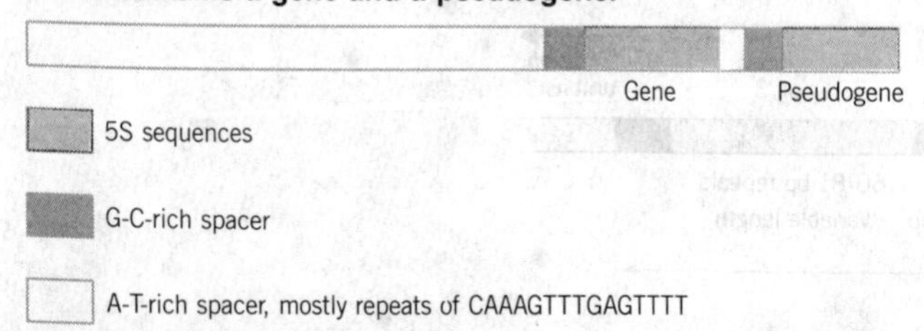

5S GENES AND PSEUDOGENES ARE INTERSPERSED

In most eukaryotes, the organization of 5S genes is distinct from that of the major rRNA genes. The total number of 5S genes may be greater, their location(s) different, and they are transcribed by RNA polymerase III instead of by RNA polymerase I (the nucleolar enzyme) We really have little idea whether and how the transcription of the 5S genes and the major rRNA genes is coordinated.

The 5S genes usually are organized as a cluster or clusters of tandem repeats. They have been analyzed in detail in *X. laevis,* in which most or all of the chromosomes have a cluster of 5S genes located at or very close to the telomere (the end).

Several types of 5S RNA sequence are found in *X. laevis,* falling into two principal types, differing in sequence at six positions. One is found exclusively in oocytes; the other is synthesized in somatic cells. The same division of types is found in the related frog *X. borealis.*

About 20,000 genes code for 5S RNA in *X. laevis,* but the vast majority represent the oocyte type. The reason for this large number may be to provide sufficient 5S genes to keep up with the expression of the amplified major rRNA genes during oogenesis. The genes for 5S RNA are not amplified.

The major 5S oocyte gene cluster takes the familiar form of an alternation between genes and nontranscribed spacers. However, the repeating unit is internally repetitive, presumably due to a past duplication. Its structure is shown in **Figure 23.8.**

The spacer has an A-T-rich sequence that is based on varying numbers of repeats of a 15 bp sequence and some variants closely related to it. The length of each nontranscribed spacer therefore varies according to the number of tandem copies of this repeat. The average length of the A-T-rich sequence is about 400 bp.

The remaining 300 bp of the repeating unit is G-C-rich. It starts with a 49 bp sequence, which is followed by the 120 bp sequence of the 5S gene itself. Next come about 1½ of the A-T-rich repeating units, succeeded by another copy of the 49 bp G-C-rich spacer sequence. Then there is a repeat of the first 101 bp of the gene; as an incomplete sequence that is not represented in RNA, it constitutes a pseudogene. The region including the pseudogene represents a repeat of the sequence around the active gene from −73 to +101. A surprising feature is that the cluster as a whole must contain an equal number of genes and pseudogenes.

The promoter for the transcription of 5S RNA lies entirely within the gene (see Chapter 12). The sequence of the pseudogene differs from the active gene by 9 point mutations; presumably these changes account for the failure of the pseudogene to give rise to an RNA product *in vivo*.

Not all the oocyte 5S genes are identical. A minor component of oocyte 5S RNA, called trace 5S RNA, displays some difference in sequence. It also is coded by a tandem array of genes and spacers; but the repeat length is only 350 bp, and the spacer is unrelated to that of the major cluster. Again, we see that a gene may undergo rather small changes in sequence, while the spacer changes completely. In a similar way, a 5S gene cluster in *X. borealis* has a spacer sequence unrelated to either of those characterized in *X. laevis*.

AN EVOLUTIONARY DILEMMA

The same problem is encountered by all clusters of identical genes. How is constancy of sequence maintained when there are so many copies of each gene? In more practical terms, the question becomes how selection can be imposed to prevent the accumulation of deleterious mutations.

When there are many copies of a gene, the immediate effects of mutation in any one copy must be very slight. The consequences of an individual mutation are diluted by the large number of copies of the gene that retain the wild-type sequence. This suggests that an appreciable proportion of mutant copies could accumulate before the effect becomes strong enough to be eliminated by evolution.

Lethality becomes quantitative, a conclusion reinforced by the observation that half of the units of the rDNA cluster of *X. laevis* or *D. melanogaster* can be deleted without ill effect. So how are these units prevented from gradually accumulating deleterious mutations? And what chance is there for the rare favorable mutation to display its advantages in the cluster?

The inevitable conclusion seems to be that some mechanism must allow particular variants to be scrutinized by evolution. Two types of mechanism have been proposed.

The **sudden correction** model supposes that every so often the entire gene cluster is replaced by a new set of copies that is derived from one or at least from only a very few of the copies present in the previous generation. To impose sufficient selective forces, "every so often" need be only every few generations; but in practical terms, any mechanism must be constructed on a regularly recurring basis—that is, every generation. The small number of copies that give rise to the new cluster form a "master" set on which selection acts; the discarded copies are irrelevant to selection. The master copies could constitute a particular set or might be chosen at random.

The difficulty with this model is that it predicts regeneration of the cluster from a few *repeating units*. This means that both the transcription units and spacers should be regenerated, so that the entire repeating unit should be homogeneous in sequence (or at least, should show no more variation than the small number of master copies). But in fact the spacers show a continual range of variation, while only the transcription units are constant.

The **crossover fixation** model supposes that the entire cluster is subject to continual rearrangement by the mechanism of unequal crossing-over. Essentially this applies to the tandem cluster on a grand scale the mechanisms we have already discussed for globin genes on a more restricted and occasional basis. We shall explore their consequences in more detail in Chapter 24, but it is immediately evident that tandem clustering provides frequent opportunities for "mispairing" of genes whose sequences are actually the same, but that lie in different positions in their clusters.

By continually expanding and contracting the number of units via unequal crossing-over, it is possible for all the units in one cluster to be derived from rather a small number of those in an ancestral cluster. The variable lengths of the spacers are consistent with the idea that unequal crossing-over events take place in spacers that are internally mispaired. This can explain the homogeneity of the genes compared with the variability of the spacers. For the genes are subject to selection when individual repeating units are amplified within the cluster; but the spacers are irrelevant and can accumulate changes.

BACTERIAL rRNA GENES FORM MIXED OPERONS WITH tRNA GENES

The seven operons coding for rRNA in *E. coli* are named *rrnA–G*. They are not closely linked on the chromosome; and we have no information on how sequence homogeneity is maintained. Because the number of genes is small, it should be possible to determine their exact sequences, and thus to see directly whether all the rRNA genes are identical or have minor variations not detectable in the rRNA population.

Each of these operons is transcribed into a single RNA precursor from which the mature rRNA molecules are released by cleavage. In bacteria in which the cleavage reaction has been blocked by mutation of the responsible enzyme, the precursor accumulates as an RNA sedimenting at about 30S. The overall organization of all the *rrn* operons is the same, containing all three rRNA molecules in the order 16S—23S—5S, as illustrated in **Figure 23.9.** The two major rRNAs lie in the same order that is found in the eukaryotic transcription unit; the presence of the 5S RNA in the same transcription unit is unique to the prokaryotes.

All the *rrn* operons have a dual promoter structure. The first promoter, *P1*, lies about 300 bp upstream from the start of the 16S rRNA sequence. Probably this is the principal promoter. Up to the first 150 bp of the transcription unit may be different in the various *rrn* operons. Within this region, about 110 bp from *P1*, there is a second promoter, *P2*. The 5′ end of the individual precursor to 16S rRNA (which is slightly longer than the mature rRNA) lies after the start of the sequence that is common for all *rrn* operons.

Between the sequences for 16S rRNA and 23S rRNA, there is a transcribed spacer region of 400–500 bp. Within this region, however, lies a sequence or sequences coding for tRNA. As summarized in the generalized diagram of Figure 23.9, in four of the *rrn* operons the transcribed spacer contains a single tRNA sequence, that of $tRNA_2^{Glu}$. In the other three *rrn* operons, the transcribed spacer contains the sequences of two tRNAs ($tRNA_1^{Ile}$ and $tRNA_{1B}^{Ala}$). Thus on the basis of the "spacer tRNA" sequences, we can distinguish

Figure 23.9
The *rrn* operons contain genes for both rRNA and tRNA.
Each functional sequence is separated from the next by a transcribed spacer region; the lengths of the leader and trailer depend on which promoters and terminators are used. Each RNA product must be released from the transcript by cuts on either side.

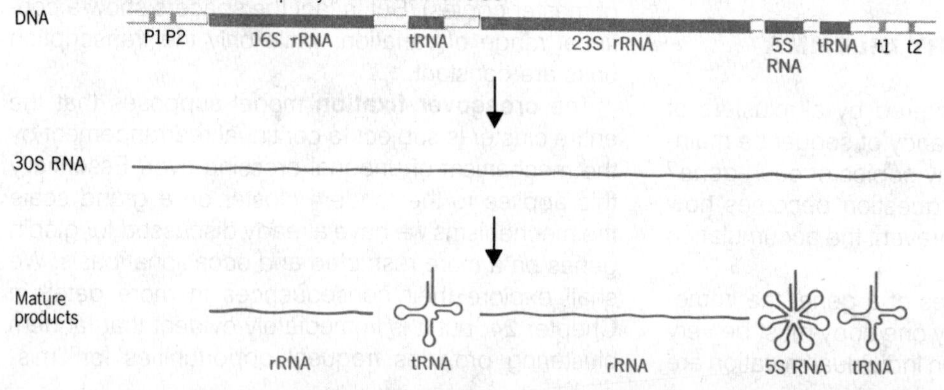

two types of rRNA operon. The tRNAs are released from the precursor RNA by cleavage; the remaining regions of the spacer presumably are degraded.

Yet another variation is found in the *rrn* operons. In some, the last sequence coded in the precursor is the 5S RNA. In others, there may be some additional tRNA sequences. For example, in *rrnC* there are two genes, $tRNA^{Trp}$ and $tRNA_1^{Asp}$, located at the 3′ end. (The presence of the $tRNA^{Trp}$ gene makes this particular *rrn* operon indispensable for *E. coli,* since this is the only copy of the gene, and thus provides sole capacity for the utilization of tryptophan in protein synthesis.) There is no correlation between spacer tRNA sequences and distal tRNAs.

tRNA GENES MAY LIE IN CLUSTERS

We have only an incomplete picture of the organization of tRNA genes in either bacteria or eukaryotes, but there seems to be a tendency for tRNA genes to lie in clusters.

In eukaryotes, the scanty information available is provided by cases where genomic sequences corresponding to a particular tRNA probe have been cloned. Often there turn out to be sequences close by that code for other tRNAs. In contrast to tandem gene clusters, the tRNAs coded in the same region all may be different. We do not know whether the clustered tRNA genes are transcribed independently or form a common unit of gene expression.

A similar situation is seen in *E. coli,* in which tRNA genes form clusters, each of which is concerned with several amino acid acceptors. At least some of these form operons, transcribed from a single promoter into one long precursor RNA. Phage T4 contains a cluster of eight tRNA genes expressed in a similar manner. This contrasts with the situation in mammalian mitochondria, in which tRNA genes punctuate the sequence of structural genes (see Chapter 22).

In every known case, a common principle applies. Each tRNA is formed as part of a longer precursor RNA from which it is released by cleavage and other processing reactions. These are discussed in Chapter 25.

Two of the tRNA gene clusters in *E. coli* have been characterized in detail, and both prove to contain other sequences as well as the tRNAs. These two clusters are named for their possession of tyrosine tRNA genes. The *tyrU* cluster includes the genes for four tRNAs, $tRNA_4^{Thr}$, $tRNA_2^{Tyr}$, $tRNA_2^{Gly}$, and $tRNA_3^{Thr}$. The *tyrT* cluster contains two identical genes coding for $tRNA_1^{Tyr}$.

In each cluster, the single promoter is located before the first tRNA gene, as shown in the example of **Figure 23.10.** Beyond the last tRNA gene, there is a sequence coding for protein. In the *tyrU* cluster this is

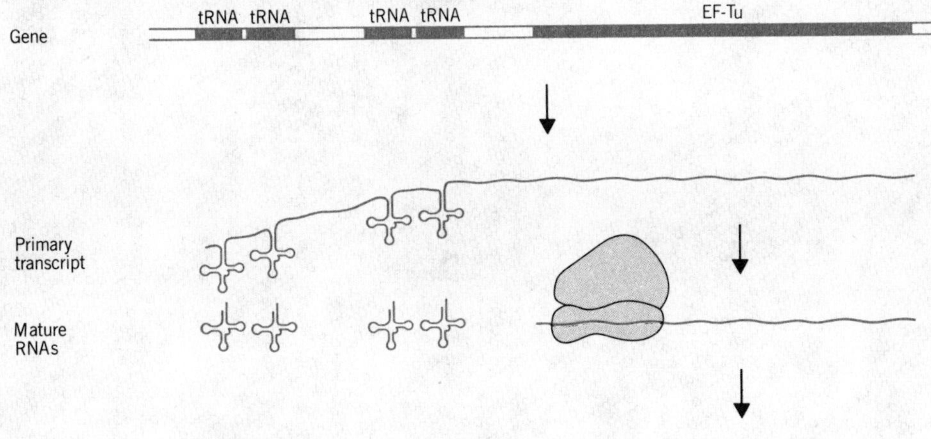

Figure 23.10
The *tyrU* operon codes for 4 tRNAs and the protein EF-Tu.

the gene *tufB* (one of two coding for the protein synthesis factor EF-Tu). In the *tyrT* cluster, the coding sequence represents protein P, a protamine-like polypeptide (protamines are very basic proteins associated with DNA in some spermatozoa).

Presumably the mRNA for each of these proteins is generated by cleavage of the primary transcript between the last tRNA sequence and the start of the structural gene. There is a short common sequence in the two operons that might be used for this purpose. Unlike other bacterial operons, in which ribosomes probably attach to the mRNA almost as soon as the RNA polymerase has left the promoter, in this case the mRNA is likely first to be generated by cleavage from the preceding tRNA sequences.

FURTHER READING

The elucidation of histone gene structure, and changing views on its degree of conservation, can be traced back through a series of reviews, from **Hentschel & Birnstiel** (*Cell* **25,** 301–313, 1981) to **Kedes** (*Ann. Rev. Biochem.* **48,** 837–870, 1979). The structure of eukaryotic rDNA clusters is dealt with in **Lewin's** *Gene Expression*, **2,** *Eucaryotic Genomes* (Wiley, New York, 1980, pp. 875–906). Spacers in rDNA have been analyzed by **Reeder** (*Cell* **38,** 349–351, 1984).

CHAPTER 24
ORGANIZATION OF SIMPLE SEQUENCE DNA

Within the eukaryotic genome are many sequences that do not code for protein (or RNA). Very likely constituting a majority of the DNA, these sequences are under different evolutionary constraints from those imposed by the need to represent a series of amino acids. Some of them, of course, are part of transcriptional units—for example, the nontranslated flanking regions in the mRNA or the introns that are removed during maturation of mRNA.

Others may provide signals that are recognized by proteins, including elements such as promoters for transcription, origins for DNA replication, sites for folding the chromosome, points for attaching the kinetochore, and other cellular functions. Very few of these have been characterized to a point where a particular function can be associated with a particular sequence. Virtually the only elements whose sequences have been delineated are the promoters and enhancers of transcription. If we include viral genomes, origins for DNA replication also are known. From the little information that is available, control signals of this type are likely to be relatively short overall, and the conserved sequences may be extremely short.

Examining the problem of noncoding DNA from the reverse perspective, we can enquire what we know about short sequences that are repeated in identical or related copies in the genome. Of course, this amounts to a definition of repetitive DNA as characterized by the DNA reassociation kinetics described in Chapter 18. To recapitulate, there are two classes of repetitive DNA.

The *moderately repetitive* component consists of a variety of sequence families, present in varying degrees of repetition and with differing internal relationships. The members of these families often are interspersed in a more or less regular way with longer stretches of nonrepetitive DNA. A large part of the moderately repetitive component of mammalian genomes turns out to comprise members of a single family, discussed in Chapter 36.

By contrast, the *highly repetitive* component generally consists of very short sequences repeated many times in tandem in large clusters. Because of its short repeating unit, this is sometimes described as "simple sequence DNA." This type of component is present in almost all higher eukaryotic genomes, but its overall amount is extremely variable. We do have some information about its evolution, which is generally informative about the mechanisms involved in manipulating DNA sequences over long periods of time, but still we have no good ideas about the functions of such sequences (or even whether they have any).

HIGHLY REPETITIVE DNA FORMS SATELLITES

Highly repetitive DNA takes the form of very short sequences repeated many times in tandem. This sequence component therefore forms blocks of material, each block consisting of very long arrays of multiple copies of the highly repetitive sequence. In some cases, the multiple copies are identical; in others, they are related.

The tandem repetition of a short sequence often creates a fraction with distinctive physical properties that may be used to isolate it. In some cases, the repetitive sequence has a base composition that is different from the genome average, which may allow it to form a separate fraction by virtue of its buoyant density.

The buoyant density of a duplex DNA depends on its G-C content according to the empirical formula

$$\rho = 1.660 + 0.00098\ (\%\text{G-C})\ \text{g-cm}^{-1} \qquad (12)$$

Buoyant density usually is determined by centrifuging DNA through a **density gradient** of CsCl. The DNA forms a band at the position corresponding to its own density. Fractions of DNA differing in buoyant density by more than about 0.005 g-cm^{-3} can be separated on a gradient. The discrimination corresponds to a difference in G-C content of about 5%.

When a eukaryotic DNA is centrifuged on a density gradient, most of the genome forms a continuum of fragments that appears as a rather broad peak centered on the buoyant density corresponding to the average G-C content of the genome. This is called the **main band.** Sometimes an additional, smaller peak (or peaks) is seen at a different value. This material is called **satellite DNA.**

A good example is provided by mouse DNA, shown in **Figure 24.1.** The graph is a quantitative scan of the bands formed when mouse DNA is centrifuged through a CsCl density gradient. The main band contains 92% of the genome and is centered on a buoyant density of 1.701 g-cm^{-3} (corresponding to its average G-C of 42%, typical for a mammal). The smaller peak represents 8% of the genome and has a distinct buoyant density of 1.690 g-cm^{-3}. It contains the mouse satellite DNA, whose G-C content is much lower than any other part of the genome (it is roughly 30%).

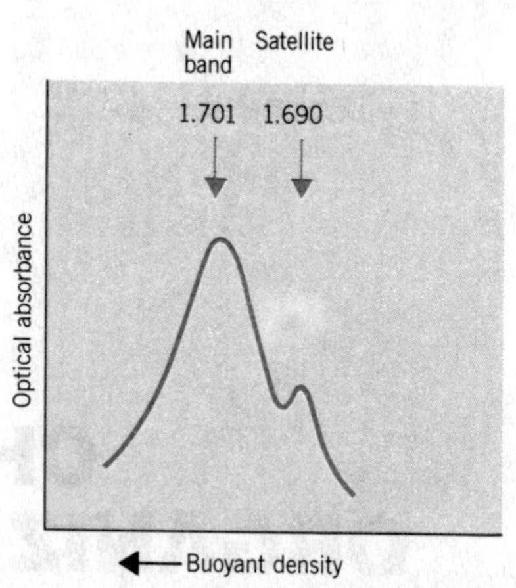

Figure 24.1
Mouse DNA is separated into a main band and a satellite by centrifugation through a density gradient of CsCl.

Satellites are present in many eukaryotic genomes. They may be either heavier or lighter than the main band; but it is uncommon for them to represent more than 5% of the total DNA. The resolution of satellite DNA is often much improved by using centrifugation through gradients of Cs_2O_4 containing silver ions, or certain other reagents, including various dyes. Sometimes a single satellite is more clearly separated from the main band; sometimes multiple satellite sequences can be resolved.

The behavior of satellite DNA on density gradients is anomalous more often than not. Thus when the actual base composition of a satellite is determined, it is often different from what had been predicted from its buoyant density. The reason is that ρ is a function not just of base composition, but of the constitution in terms of nearest neighbor pairs. For simple sequences, these are likely to deviate from the random pairwise relationships needed to obey equation **12.** Also, satellite DNA may be methylated, which changes its density.

Often most of the highly repetitive DNA of a genome can be isolated in the form of satellites. When a highly repetitive DNA component does not separate as a satellite, on isolation its properties often prove to be similar to those of satellite DNA. That is to say that it consists of multiple tandem repeats with anomalous centrifugation. Material isolated in this manner is sometimes referred to as a **cryptic satellite.** Together the cryptic and apparent satellites usually account for all the tandemly repeated blocks of highly repetitive DNA. When a genome has more than one type of highly repetitive DNA, each exists in its own satellite block (although sometimes different blocks are adjacent).

SATELLITE DNA OFTEN LIES IN HETEROCHROMATIN

Where are the blocks of highly repetitive DNA located? An extension of nucleic acid hybridization techniques allows the location of satellite sequences to be determined directly in the chromosome complement. In the technique of **in situ** or **cytological hybridization,** the chromosomal DNA is denatured by treating cells that have been squashed on a cover slip. Then a solution containing a radioactively labeled DNA or RNA probe is added. The probe hybridizes with its complements in the denatured genome. The location of the sites of hybridization can be determined by autoradiography.

Labeled satellite DNAs often are confined to the heterochromatin present around the centromeres of mitotic chromosomes. **Heterochromatin** is the term used to describe regions of chromosomes that are permanently tightly coiled up and inert, in contrast with the **euchromatin** that represents most of the genome (see Chapter 28).

An example of the localization of satellite DNA for the mouse chromosomal complement is shown in **Figure 24.2.** In this case, the ends of the chromosomes are labeled, because this is where the centromeres are located in the *M. musculus* chromosomes.

The identification of satellite DNA with heterochromatin implies that highly repetitive DNA is not expressed in the form of RNA or protein (the last is evident also from its simple sequence). Attempts to detect RNA products corresponding to satellite DNA have confirmed that the highly repetitive sequences usually are not transcribed.

The centromeric location of satellite DNA suggests

Figure 24.2
Cytological hybridization shows that mouse satellite DNA is located at the centromeres.
Photograph kindly provided by Mary Lou Pardue and Joe Gall.

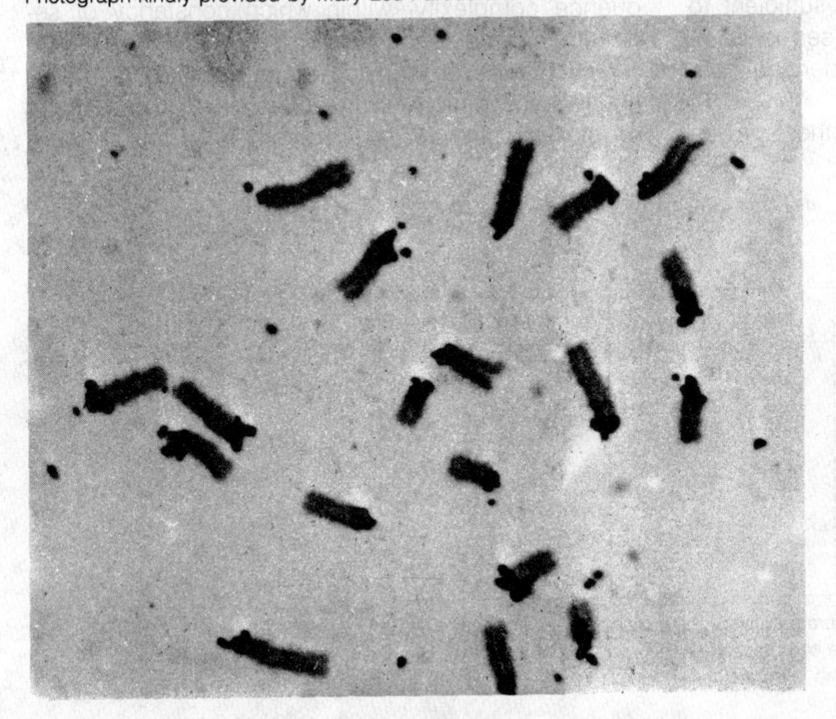

that it may have some structural function in the chromosome. Since the centromeres are the regions where the kinetochores are formed at mitosis and meiosis for controlling chromosome movement, this function could be connected with the process of chromosome segregation. But that is all we know about its role, which apart from this general suggestion remains quite mysterious.

ARTHROPOD SATELLITES HAVE VERY SHORT IDENTICAL REPEATS

In the arthropods, as typified by insects and crabs, each satellite DNA appears to be rather homogeneous in sequence constitution. Usually, a single very short repeating unit accounts for more than 90% of the satellite. This makes it relatively straightforward to determine the sequence, which has been done in several cases.

In *Drosophila virilis*, there are three major satellites and also a cryptic satellite, together representing a large proportion of the genome, about 40%. The sequences of the satellites are summarized in **Table 24.1.** The three major satellites have closely related sequences. A single base substitution is sufficient to generate either satellite II or III from the sequence of satellite I. The cryptic satellite can be derived from satellite II by 2 base changes.

The satellite I sequence is present in other species of *Drosophila* related to *virilis*, and so may have preceded speciation. The sequences of satellites II and III seem to be specific to *D. virilis*, and so may have evolved from satellite I after speciation.

The main feature of these satellites is their very short repeating unit: only 7 bp. Similar satellites are found in other species. For example, *D. melanogaster* has a variety of satellites, several of which have very short repeating units (5, 7, 10, or 12 bp). Comparable satellites are found in the crabs.

The close sequence relationship found among the *D. virilis* satellites is not necessarily a feature of other genomes, where the satellites may have unrelated sequences. It is clear that each satellite has arisen by a lateral amplification of a very short sequence. This sequence may represent a variant of a previously existing satellite (as in *D. virilis*), or could have some other origin.

At all events, it seems likely that satellites may continually be generated and lost from genomes. This makes it difficult to ascertain evolutionary relationships, since a current satellite could have evolved from some previous satellite that has since been lost. The important feature of these satellites is that they represent very long stretches of DNA of very low sequence complexity, within which constancy of sequence can be maintained.

One feature of many of these satellites is a pronounced asymmetry in the orientation of base pairs on the two strands. In the example of the *D. virilis*

Table 24.1
Satellite DNAs of *D. virilis* are related.

Satellite	Predominant Sequence	No. of Copies	Percent of DNA
I	A C A A A C T T G T T T G A	1.1×10^7	25
II	A T A A A C T T A T T T G A	3.6×10^6	8
III	A C A A A T T T G T T T A A	3.6×10^6	8
Cryptic	A A T A T A G T T A T A T C		

More than 95% of each satellite consists of a tandem repetition of the predominant sequence. Note that satellites II and III have exactly the same base composition (1 G–C pair out of 7 = 14% G–C), but have buoyant densities of 1.688 and 1.671 g-cm^{-1}, respectively.

satellites shown in Table 24.1, in each of the major satellites one of the strands is much richer in T and G bases. This increases its buoyant density, so that upon denaturation this **heavy strand** (**H**) can be separated from the complementary **light strand** (**L**). This can be useful in sequencing the satellite.

MAMMALIAN SATELLITES CONSIST OF HIERARCHICAL REPEATS

In the mammals, as typified by various rodents, the sequences comprising each satellite show appreciable divergence between tandem repeats. Common short sequences can be recognized by their preponderance among the oligonucleotide fragments released by chemical or enzymatic treatment. However, the predominant short sequence usually accounts for only a small minority of the copies. The other short sequences are related to the predominant sequence by a variety of substitutions, deletions, and insertions.

But a series of these variants of the short unit can constitute a longer repeating unit that is itself repeated in tandem with some variation. Thus mammalian satellite DNAs are constructed from a *hierarchy* of repeating units. These longer repeating units constitute the sequences that renature in reassociation analysis. They can also be recognized by digestion with restriction enzymes.

Uncertainty in evaluating the complexity of a repetitive DNA is caused by the effect of mismatching on the reassociation reaction. The effect is particularly pronounced with satellite DNA; it means that the length of the reassociating unit can be assessed only within a factor of two or so at best.

When any satellite DNA is digested with an enzyme that has a recognition site in its repeating unit, one fragment will be obtained for every repeating unit in which the site occurs. In fact, when the DNA of a eukaryotic genome is digested with a restriction enzyme, most of it gives a general smear, due to the random distribution of cleavage sites. But satellite DNA generates sharp bands, because a large number of fragments of identical or almost identical size are created by cleavage at restriction sites that lie a regular distance apart.

Determining the sequence of satellite DNA can be difficult. Using the discrete bands generated by restriction cleavage, we can attempt to obtain a sequence directly. However, if there is appreciable divergence between individual repeating units, there will be different nucleotides present at the same position in different repeats, so the sequencing gels will be obscure. If the divergence is not too great—say, within about 2%—it may be possible to determine an average repeating sequence.

Individual segments of the satellite can be inserted into plasmids for cloning. A difficulty is that the satellite sequences sometimes tend to be excised from the chimeric plasmid by recombination in the bacterial host. However, when the cloning succeeds, it is possible to determine the sequence of the cloned segment unambiguously. While this gives the actual sequence of a repeating unit or units, we should need to have many individual such sequences to reconstruct the type of divergence typical of the satellite as a whole.

By either sequencing approach, the information that can be gained is limited to the distance that can be analyzed on one set of sequence gels. The repetition of divergent tandem copies makes it impossible to reconstruct longer sequences by obtaining overlaps between individual restriction fragments.

The satellite DNA of the mouse *M. musculus* is cleaved by the enzyme EcoRII into a series of bands, including a predominant monomeric fragment whose electrophoretic mobility corresponds to a length of 230–240 bp. Using this fragment, direct sequencing identifies a single nucleotide sequence of 234 bp. This sequence must be repeated with few variations throughout the 60–70% of the satellite that is cleaved into the monomeric band. We shall now analyze this sequence in terms of its successively smaller constituent repeating units.

Figure 24.3 depicts the sequence in terms of two half-repeats. By writing the 234 bp sequence so that the first 117 bp are aligned with the second 117 bp, we see that the two halves are quite well related. They differ at 22 positions, corresponding to 19% divergence. This means that the current 234 bp repeating unit must have been generated at some time in the past by duplicating a 117 bp repeating unit, after which differences accumulated between the duplicates.

Within the 117 bp unit, we can recognize two further subunits. Each of these is a quarter-repeat relative to

10 20 30 40 50 60 70 80 90 100 110

GGACCTGGAATATGGCGAGAAAACTGAAAATCACGGAAAATGAGAAATACACACTTTAGGACGTGAAATATGGCGAG^{G}AAACTGAAAAAGGTGGAAAATT^{T}AGAAATGTCCACTGTA

GGACGTGGAATATGGCAAGAAAACTGAAAATCATGGAAAATGAGAAACATCCACTTGACGACTTGAAAAATGACGAAATCACTAAAAACGTGAAAAATGAGAAATGCACACTGAA

120 130 140 150 160 170 180 190 200 210 220 230

Figure 24.3
The repeating unit of mouse satellite DNA contains two half-repeats, whose homologies are indicated in color.

the whole satellite. The four quarter-repeats are aligned in **Figure 24.4.** The upper two lines represent the first half-repeat of Figure 24.3; the lower two lines represent the second half-repeat. We see that the divergence between the four quarter-repeats has increased to 23 out of 58 positions, or 40%. Actually, the first three quarter-repeats are somewhat better related, and a large proportion of the divergence is due to changes in the fourth quarter-repeat.

Looking within the quarter-repeats, we find that each consists of two related subunits (one-eighth-repeats), shown as the α and β sequences in **Figure 24.5.** The α sequences all have an insertion of a *C* and the β sequences all have an insertion of a trinucleotide, relative to a common consensus sequence. This suggests that the quarter-repeat originated by the duplication of a sequence like the consensus sequence, after which changes occurred to generate the components we now see as α and β. Further changes then took place between tandemly repeated αβ sequences to generate the individual quarter- and half-repeats that exist today. Among the one-eighth-repeats, the present divergence is 19/31 = 61%.

Within the consensus sequence, we can recognize three related 9 bp sequences. If in one of the repeats we take the next most frequent base at two positions instead of the most frequent, we obtain three well-related 9 bp sequences.

G A A A A A C G T
G A A A A A T G A
G A A A A A A C T

This is as good a guess as we can make for the original constitution of the satellite.

The consensus sequence is analyzed directly in **Figure 24.6,** which demonstrates that the current satellite sequence can be treated as derivatives of a 9 bp sequence. The overall consensus sequence is $\mathrm{GAAAA}^{\mathrm{ATGA}}_{\mathrm{TACT}}$, which is effectively an amalgam of the three 9 bp repeats of Figure 24.5.

RECONSTRUCTING THE STAGES OF MOUSE SATELLITE DNA EVOLUTION

From the sequence of the satellite DNA, we can both reconstruct its evolution and account for its present properties. A model showing the possible steps involved in its evolution is given in **Figure 24.7.**

Figure 24.4
The alignment of quarter-repeats shows that there are homologies between the first and second half of each half-repeat.

10 20 30 40 50
GGACCTGGAATATGGCGA AAAACTGAAAATCACGGAAAATGAGAAATACACACTTTA

60 70 80 90 100 110
GGACGTG· AATATGGCGAG^{G}AAAACTGAAAAAGGTGGAAAATT^{T}AGAAATGTCCACTGTA

120 130 140 150 160 170
GGACGTGGAATATGGCAAGAAAACTGAAAATCATGGAAAATGAGAAACATCCACTTGA

180 190 200 210 220 230
CGACTTGAAAAATGACGAAATCACTAAAAACGTGAAAAATGAGAAATGCACACTGAA

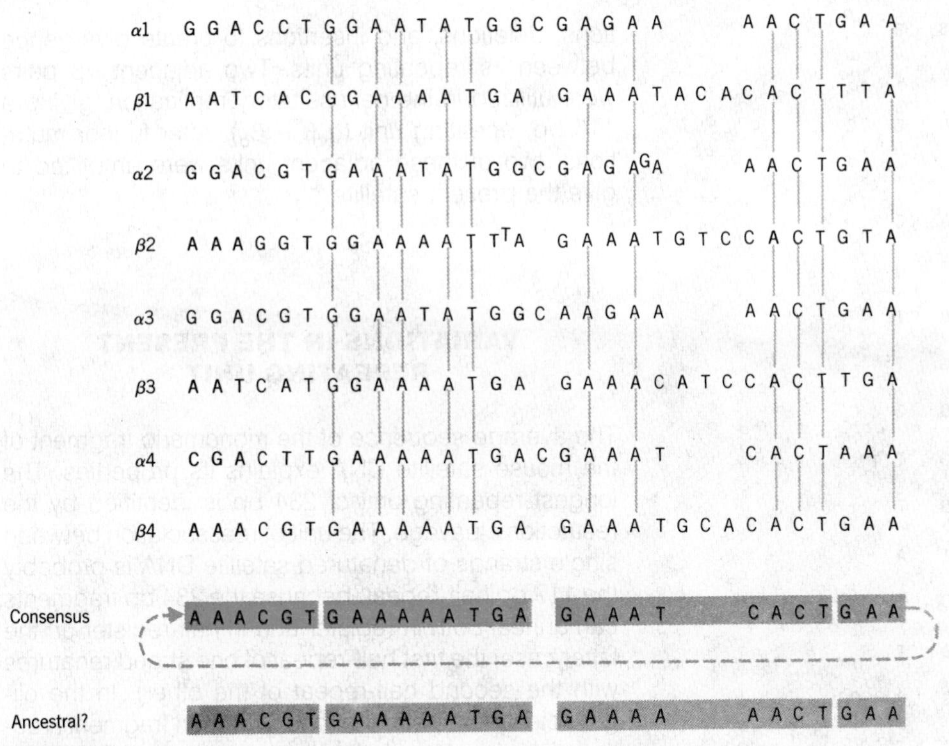

Figure 24.5

The alignment of eighth-repeats shows that each quarter-repeat consists of an α and a β half. The consensus sequence gives the most common base at each position. The ancestral sequence shows three 9 bp units, very closely related to the consensus sequence, that could be predecessor to the α and β units.

(Remember that the satellite sequence is continuous, so that for the purpose of deducing the consensus sequence, we can treat it as a circular permutation, as indicated by joining the last GAA triplet to the first 6 bp.)

The general principle of this model is that at various times a group of repeating units may be suddenly amplified laterally to generate a large number of identical tandem copies. An event of this sort is called a **saltatory replication.** Then the copies diverge in sequence as mutations accumulate in them. At some subsequent time, a group of these copies again may be subject to a saltatory replication. The extent of divergence among the copies that are amplified at each time will depend on the period that has passed since the last saltatory replication imposed identity on the satellite. The satellite can evolve by a series of these saltatory replications, alternating with accumulation of mutations.

Suppose that the present satellite originated with the tandem repetition of a sequence such as GAAAAATGT or something closely related to it. (This sequence might have been part of a satellite or could have some quite different origin: its existence is as far back as we can pursue the satellite.) All the original 9 bp units were identical, but with time mutations created differences between them. Then three adjacent units with the ancestral sequence hypothesized in Figure 24.5 were amplified, giving a tandem repeat of 27 bp.

Mutations occurred in this unit, including cases in which one unit gained an additional C while its neighbor gained a triplet insertion (of unknown sequence). This pair of repeats, now together 58 bp long, was subject to saltatory replication and gave a satellite that we can describe as $(\alpha\beta)_n$.

Once again, the satellite accumulated point muta-

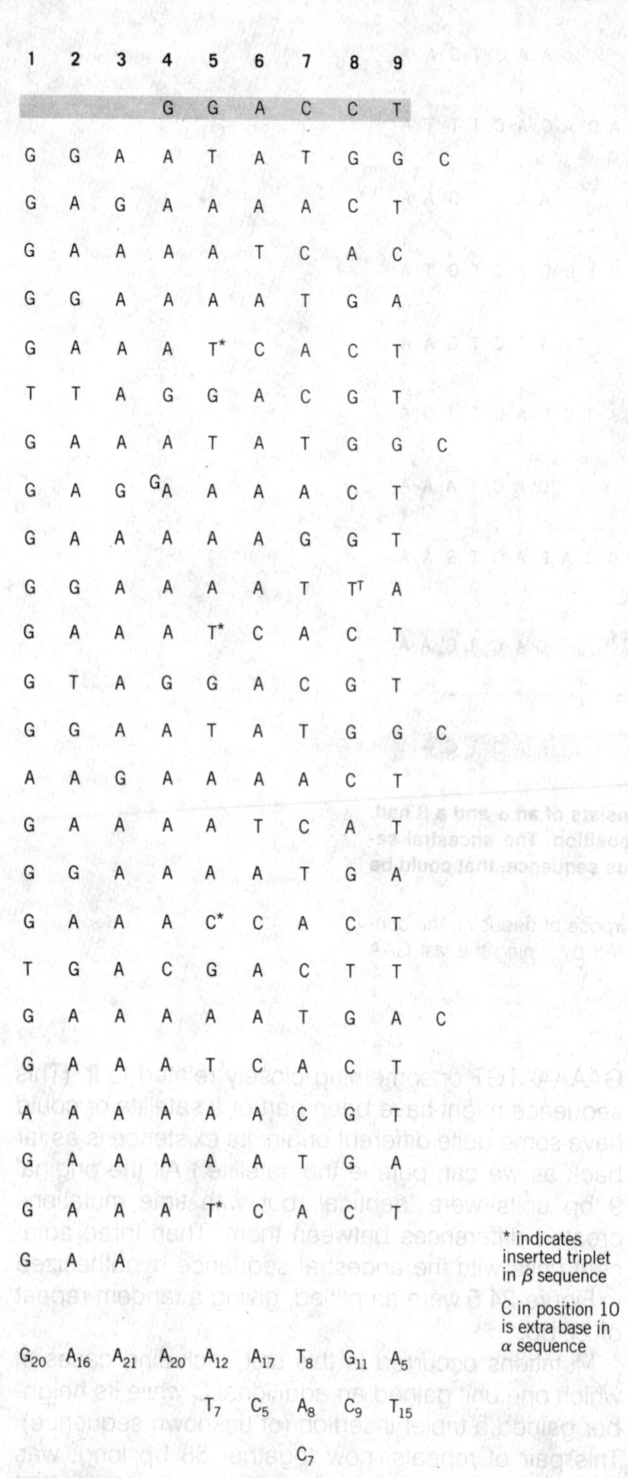

Figure 24.6
The existence of an overall consensus sequence is shown by writing the satellite sequence in terms of a 9 bp repeat.

tions, deletions, and insertions to create divergence between its repeating units. Two adjacent αβ pairs were utilized in the next saltatory replication, giving a 116 bp repeating unit ($\alpha_c\beta_c\alpha_d\beta_d$). After further mutations, two of these adjacent units were amplified to give the present satellite.

VARIATIONS IN THE PRESENT REPEATING UNIT

The average sequence of the monomeric fragment of the mouse satellite DNA explains its properties. The longest repeating unit of 234 bp is identified by the restriction cleavage. The unit of reassociation between single strands of denatured satellite DNA is probably the 117 bp half-repeat, because the 234 bp fragments can anneal both in register and in half-register (in the latter case, the first half-repeat of one strand renatures with the second half-repeat of the other). In the oligonucleotide digest, the most common fragments, accounting for 4% of the total amount of DNA, are GA_5TGA, GA_4TGA, and GA_4CTGA, all of which can be found in the 234 bp unit and are related to the proposed ancestral units.

So far, we have treated the present satellite as though it consisted of identical copies of the 234 bp repeating unit. Although this unit accounts for the majority of the satellite, variants of it also are present. Some of these are scattered at random throughout the satellite; others are clustered.

The existence of variants is implied by our description of the starting material for the sequence analysis as the "monomeric" fragment. When the satellite is digested by an enzyme that has one cleavage site in the 234 bp sequence, it also generates dimers, trimers, and tetramers relative to the 234 bp length. These arise when a repeating unit has lost the enzyme cleavage site as the result of mutation.

The monomeric 234 bp unit is generated when two adjacent repeats each have the recognition site. A dimer occurs when one unit has lost the site, a trimer is generated when two adjacent units have lost the site, and so on. With the enzymes EcoRII or Sau96I, most of the satellite is cleaved into a member of this repeating series, as shown in the example of **Figure**

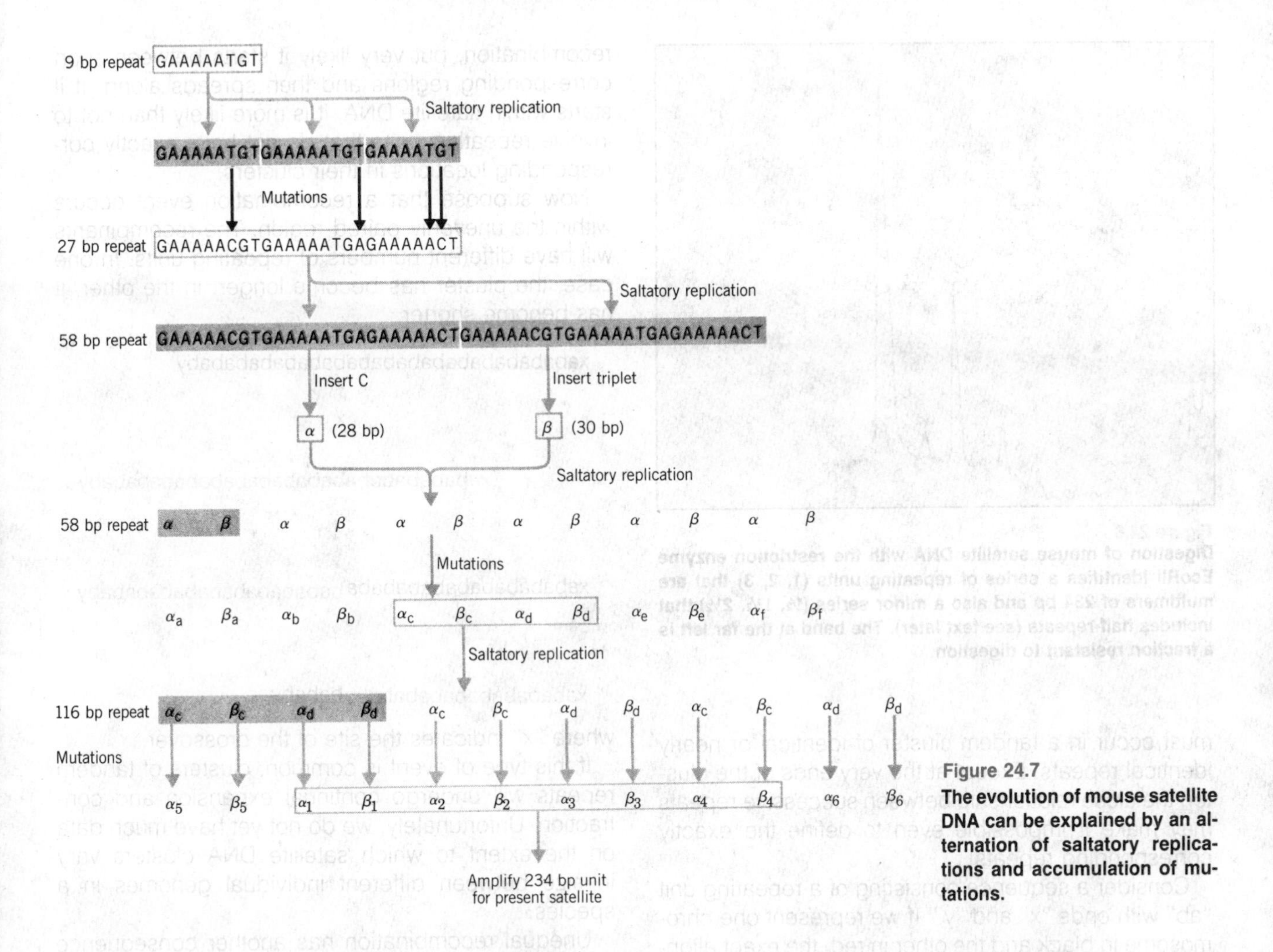

Figure 24.7
The evolution of mouse satellite DNA can be explained by an alternation of saltatory replications and accumulation of mutations.

24.8. The declining number of dimers, trimers, etc., shows that there is a random distribution of the repeats in which the enzyme's recognition site has been eliminated by mutation.

Some restriction enzymes show a different type of behavior with the satellite DNA. They continue to generate the same series of bands. But they cleave only a small proportion of the DNA, say 5–10%. This implies that a particular region of the satellite contains a concentration of the repeating units that have this particular restriction site. Presumably the series of repeats in this domain all are derived from an ancestral variant that possessed this recognition site (although in the usual way, some members since have lost it by mutation).

THE CONSEQUENCES OF UNEQUAL CROSSING OVER

In a region of nonrepetitive DNA, recombination occurs between precisely matching points on the two homologous chromosomes, generating reciprocal recombinants. The basis for this precision is the ability of two duplex DNA sequences to align exactly. We know that unequal recombination can occur when there are multiple copies of genes whose exons are related, even though their flanking and intervening sequences may differ (see Chapter 21). This happens because of the mispairing between corresponding exons in *nonallelic* genes.

Imagine how much more frequently misalignment

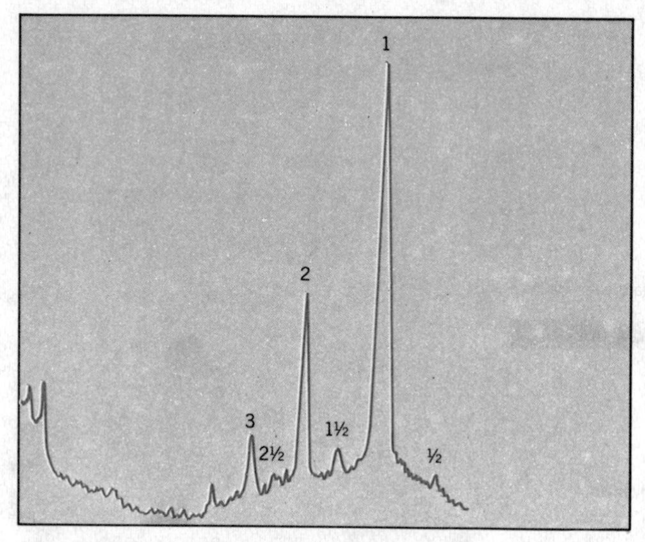

Figure 24.8
Digestion of mouse satellite DNA with the restriction enzyme EcoRII identifies a series of repeating units (1, 2, 3) that are multimers of 234 bp and also a minor series (½, 1½, 2½) that includes half-repeats (see text later). The band at the far left is a fraction resistant to digestion.

must occur in a tandem cluster of identical or nearly identical repeats. Except at the very ends of the cluster, the close relationship between successive repeats may make it impossible even to define the exactly corresponding repeats!

Consider a sequence consisting of a repeating unit "ab" with ends "x" and "y." If we represent one chromosome in black and the other in red, the exact alignment between "allelic" sequences would be

xababababababababababababababababy

xababababababababababababababababy

But probably *any* sequence ab in one chromosome could pair with *any* sequence ab in the other chromosome. In a misalignment such as

xababababababababababababababababy

xababababababababababababababababy

the region of pairing is no less stable than in the perfectly aligned pair, although it is shorter. We do not know very much about how pairing is initiated prior to recombination, but very likely it starts between short corresponding regions and then spreads along. If it starts within satellite DNA, it is more likely than not to involve repeating units that do not have exactly corresponding locations in their clusters.

Now suppose that a recombination event occurs within the unevenly paired region. The recombinants will have different numbers of repeating units. In one case, the cluster has become longer; in the other, it has become shorter,

xababababababababababababababababy

x

xababababababababababababababababy

↓

xabababababababababa bababababababababababy

+

xababababababababababababy

where "x" indicates the site of the crossover.

If this type of event is common, clusters of tandem repeats will undergo continual expansion and contraction. Unfortunately, we do not yet have much data on the extent to which satellite DNA clusters vary in size between different individual genomes in a species.

Unequal recombination has another consequence when there is internal repetition in the repeating unit. In the example just given, the two clusters are misaligned with respect to the positions of the repeating units within each cluster, but they are aligned **in register,** as seen by the correspondence between individual ab repeats and ab repeats.

But suppose that the "a" and "b" components of the repeating unit are themselves sufficiently well related to pair. Then the two clusters can align in **half-register,** with the "a" sequence of one aligned with the "b" sequence of the other. How frequently this occurs will depend on the closeness of the relationship between the two halves of the repeating unit. (We have already mentioned that reassociation between the denatured satellite DNA strands *in vitro* commonly occurs in the half-register.)

When a recombination event occurs, it changes the length of the repeating units that are involved in the reaction.

xababababababababababababababy

x

xababababababababababababababy

↓

xabababababababababaababababaaby

\+

xababababababababbabababaaby

In the upper recombinant cluster, an "ab" unit has been replaced by an "aab" unit. In the lower cluster, the "ab" unit has been replaced by a "b" unit.

This type of event explains a feature of the restriction digest of mouse satellite DNA. In addition to the integral repeating series in Figure 24.8, there is a fainter series of bands at lengths of ½, 1½, 2½, and 3½ repeating units. Suppose that in the preceding example, "ab" represents the 234 bp repeat of mouse satellite DNA, generated by cleavage at a site in the "b" segment. The "a" and "b" segments correspond to the 117 bp half-repeats.

Then in the upper recombinant cluster, the "aab" unit generates a fragment of 1½ times the usual repeating length. And in the lower recombinant cluster, the "b" unit generates a fragment of half of the usual length. (The multiple fragments in the half-repeat series are generated in the same way as longer fragments in the integral series, when some repeating units have lost the restriction site by mutation.)

Turning the argument the other way around, the identification of the half-repeat series on the gel shows us that the 234 bp repeating unit consists of two half-repeats well enough related to pair sometimes for recombination. Also visible in Figure 24.8 are some rather faint bands corresponding to ¼- and ¾-spacings. These will be generated in the same way as the ½-spacings, when recombination occurs between clusters aligned in a quarter-register. The decreased relationship between quarter-repeats compared with half-repeats explains the reduction in frequency of the ¼- and ¾-bands compared with the ½-bands.

CROSSOVER FIXATION COULD MAINTAIN IDENTICAL REPEATS

A general presumption about satellite DNA is that its sequence is not under high selective pressure (if indeed under any at all). Unlike a sequence that codes for protein, where mutation may inactivate the product, satellite DNA seems likely to serve any functions it has by its presence instead of by virtue of its exact sequence. This idea fits perfectly well with the structure of satellites whose repeating units are related rather than identical. Mutations have accumulated since the satellite was last rendered uniform in sequence. But how are we to explain the presence in arthropods of satellites most of whose repeating units remain identical? Even if the sequence were important, there still

Figure 24.9
Unequal recombination allows one particular repeating unit to occupy the entire cluster.
The numbers indicate the length of the repeating unit at each stage

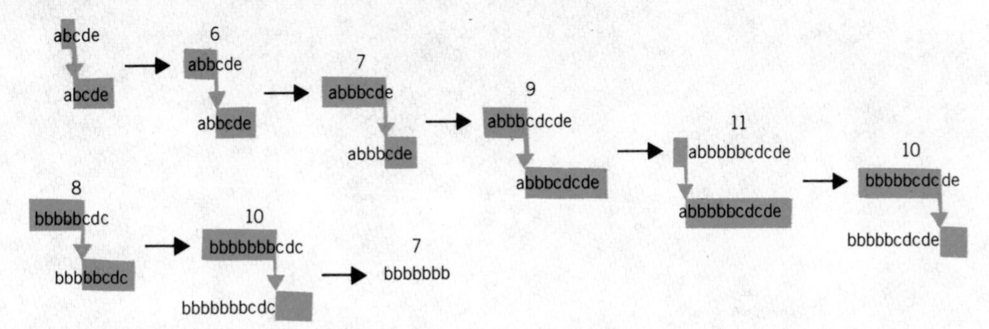

remains the difficulty of how selection could be imposed on so many copies.

Unequal recombination has been proposed as an alternative model to saltatory replication to explain the evolution of satellite DNA. The basic idea is that unequal crossing-over occurs frequently at random sites. A series of random events cause one repeating unit to take over the entire satellite. This process is called **crossover fixation.**

The spread of a particular repeating unit through the satellite is illustrated in **Figure 24.9.** Suppose that a satellite consists initially of a sequence *abcde,* where each letter represents a repeating unit. The different repeating units are closely enough related to each other to mispair for recombination. Then by a series of unequal recombinations, the size of the repetitive region may increase or decrease, and also one unit may spread through to replace all the others.

This model actually predicts that *any* sequence of DNA that is not under selective pressure will be taken over by a series of identical tandem repeats generated in this way. The critical assumption is that the process of crossover fixation is fairly rapid relative to mutation, so that new mutations either are eliminated (their repeats are lost) or come to take over the entire cluster (as shown for the spread of variant "b" in Figure 24.9).

We may wonder whether the existence of domains that contain particular repeating units (as in mouse satellite DNA) could represent a partial spreading of the unit from its origin. For example, we can see that in the intermediate stages of the spreading illustrated in Figure 24.9, there are clusters that have a domain consisting of a series of "b" variants.

FURTHER READING

The idea of saltatory replication was introduced by **Southern** (*J. Mol. Biol.* **94,** 51–70, 1975). The implications of unequal crossing-over were developed by **Smith** (*Science* **191,** 528–535, 1976). The organization of inverted and tandem repeats has been reviewed in **Lewin's** *Gene Expression,* **2,** *Eucaryotic Chromosomes* (Wiley, New York, 1980, pp. 531–569).

PART 7
REACHING MATURITY: RNA PROCESSING

The code-script must itself be the operative factor bringing about the development [of the organism]. But . . . with the molecular picture of the gene it is no longer inconceivable that the miniature code should precisely correspond with a highly complicated and specified plan of development and should somehow contain the means to put it into operation.

Erwin Schrödinger, 1945

CHAPTER 25
CUTTING AND TRIMMING STABLE RNA

The "processing" of a mature molecule from its precursor is a crucial stage in the production of many forms of RNA. All stable RNAs in bacteria are synthesized as primary transcripts from which the mature rRNA or tRNA must be released. A similar pathway is followed for eukaryotic rRNA and tRNA. By contrast, most bacterial mRNAs are primary transcripts, so only in the exceptional case is processing necessary for translation. In eukaryotes, however, producting an mRNA from an interrupted gene is the most labor-intensive of all RNA processing, sometimes involving a very large number of splicing reactions in which the exons are connected via removal of the introns (see Chapter 26).

Processing reactions are highly specific, producing mature RNA molecules with unique 5′ and 3′ ends. The reactions are the responsibility of particular ribonucleases: for they depend on the enzymatic ability to cleave phosphodiester bonds in RNA. As with all nucleases, the enzymes fall into the general classes of **exonucleases** and **endonucleases.**

The endonucleases cut individual bonds *within* RNA molecules, generating discrete fragments. They are involved in cutting reactions, when the mature sequence is separated from the flanking sequence.

Exonucleases remove residues one at a time from the end of the molecule, generating mononucleotides. Their attitude toward the substrate may be **random** or **processive.** A randomly acting enzyme removes a base from one RNA molecule and then dissociates; for its next catalytic event, the enzyme may attack a different RNA substrate molecule. Processive action means that the enzyme stays with one substrate molecule, removing further bases until its mission is accomplished.

Exonucleases are involved in trimming reactions. From the perspective of the RNA substrate, the extra residues are whittled away, base by base. All the characterized exonucleases proceed along the nucleic acid chain from the 3′ end. Thus additional material at the 3′ end can be trimmed, but additional material at the 5′ end can be released only by cutting.

As well as participating in specific maturation reactions, ribonucleases accomplish the degradation of superfluous RNA. "Superfluous" molecules include mature species that turn over, as well as the extraneous material that is discarded during the cutting of precursors. In particular, mRNA whose time has come must be degraded (its time comes rapidly in bacteria, more slowly in eukaryotes, but in very few cases is mRNA stable enough to survive for a protraoted period).

We might expect that the ribonucleases engaged in RNA turnover should be less specific than those involved in particular maturation pathways, since their role is to degrade the RNA completely rather than to generate specific termini. Both endoribonucleases and exoribonucleases are found in this class. In a sense, of course, the degradation of an RNA to mononucleotides is its final maturation, returning its components for reutilization via the appropriate metabolic pathways.

Ribonucleases in fact vary widely in their specificities. None is known that has any multibase sequence requirement in its substrate. Thus there is no counterpart to the restriction nucleases that attack specific sequences in DNA. Many ribonucleases have some small specificity in the sense that they cleave phosphodiester bonds at particular individual bases. But any greater specificity seems to be accomplished by relying on the conformation of the RNA substrate.

Substrate recognition may take the form of seeking a specific feature of secondary structure (such as a hairpin of certain size), or examining the overall secondary/tertiary structure (as seems to apply in tRNA processing).

The only situation in which there is any hope of gaining sufficient information to define the entire cellular set of ribonucleases is in *E. coli*. Here not only can the enzymes be defined biochemically, but their effects *in vivo* can also be determined by selecting bacteria in which a particular enzyme has been mutated. Indeed, both lines of evidence often are needed to distinguish enzymes or to show that two apparently different activities reside in the same enzyme. (In time, yeast also may become approachable in these terms.)

We do not yet know precisely how many ribonucleases there are in *E. coli*, but the current count is less than 10. It seems likely that combinations of a small number of ribonuclease activities provide the necessary range of reactions to process rRNA and tRNA. Although mRNA is actively degraded, the process has been intractable to analysis so far, and we do not yet have any detailed information about the relevant enzymatic activities.

PHOSPHODIESTER BONDS CAN BE CLEAVED ON EITHER SIDE

Nucleases vary in their enzymatic activities as well as in their substrate specificities. When the phosphodiester bond linking two nucleotides is cleaved, it can in principle be cut on either side of the phosphate group. The consequences are illustrated in **Figure 25.1.** Cleavage on one side generates 3′-hydroxyl and 5′-phosphate termini. Cleavage on the other side generates 3′-phosphate and 5′-hydroxyl termini.

Two ribonucleases of practical importance, although of uncharacterized physiological function, are ribonuclease T1 and pancreatic ribonuclease. Both are endoribonucleases with rather simple sequence requirements. Ribonuclease T1 is obtained from *Aspergillus oryzae* (but is common in other fungi and in bacteria). It hydrolyzes the bond on the 3′ side of a G residue. Thus it cleaves a sequence such as

5′. . .pNpNpNpNpNpNpNpNpNpNpGp XpXpX. . .3′

to generate the products

5′. . .pNpNpNpNpNpNpNpNpNpNpGp3′

5′ HO- XpXpX. . .3′

Note that the new ends have 3′-phosphate and 5′-hydroxyl termini. The same types of ends are generated by pancreatic ribonuclease, which cleaves on the 3′ side of either pyrimidine.

Before the advent of the current methods for nucleic acid sequencing, these enzymes were used to digest RNA into oligonucleotides from which the sequence of the molecule could be pieced together. Given their low specificity, presumably the enzymes are concerned with the degradation of RNA *in vivo*. This may explain the type of ends that they produce.

In the natural state, all RNA molecules have a polarity in which the 3′ end bears a hydroxyl group and the 5′ end is phosphorylated (it may have a triphosphate or monophosphate or even a cap, depending on whether and what changes have been introduced at the original 5′ terminus). Enzymes concerned with RNA processing generate the same type of termini. All the cutting and trimming enzymes that we discuss in this chapter generate 5′-phosphate and 3′-hydroxyl ends.

RNAase III RELEASES THE PHAGE T7 EARLY mRNAs

The processing of mRNA is a relatively rare event, not well characterized in either eukaryotes or prokaryotes.

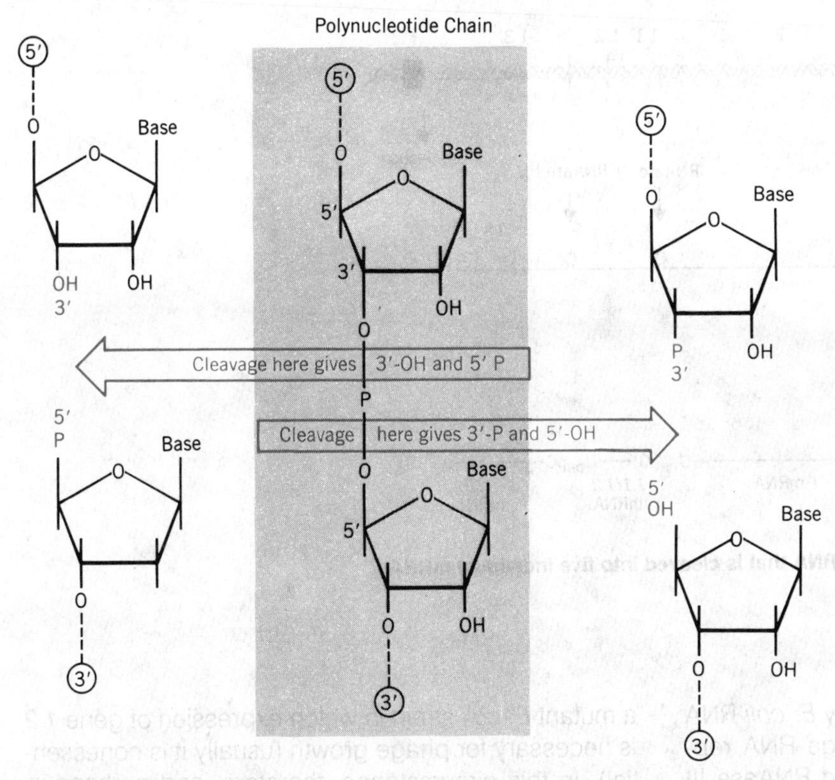

Figure 25.1
The nature of the new termini is determined by which side of the phosphodiester bond is cleaved.

We have mentioned in Chapter 10 that cleavage is used to generate individual mRNAs from certain polycistronic RNAs that are transcribed from eukaryotic viruses. In this case, we understand the purpose, which is to generate monocistronic mRNAs suitable for translation, but we do not know anything about the mechanism.

Bacterial mRNAs almost all are translated in the same form in which they are transcribed. Most are polycistronic, some are monocistronic. An exceptional situation, however, is presented by the early genes of phage T7. These are transcribed into a polycistronic mRNA. But then the molecule is cleaved at several sites to release mRNAs for the individual genes. (A similar cleavage may also occur with late RNA, but is less well characterized.)

This form of expression is found in phage T3 as well as in T7. Since the two phage DNA sequences are quite divergent, the retention of a common pathway suggests that it offers some distinct selective advantage. But we do not know why this unique mRNA processing pathway should be favored by these two phages.

The transcription of the early region of phage T7 is illustrated in **Figure 25.2.** Initiation occurs at one of a group of three promoters (*A1, A2, A3*) and continues to a terminator (*t*) about 7000 bp later. There are six genes within this region, and cuts are made at intercistronic boundaries to release five individual mRNAs (four monocistronic and one bicistronic).

The enzyme responsible for making the cuts is **RNAase III,** which was originally discovered as an endoribonuclease (then of unknown function) that is active with double-stranded RNA. The duplex structure is an important feature of the sites recognized by the enzyme.

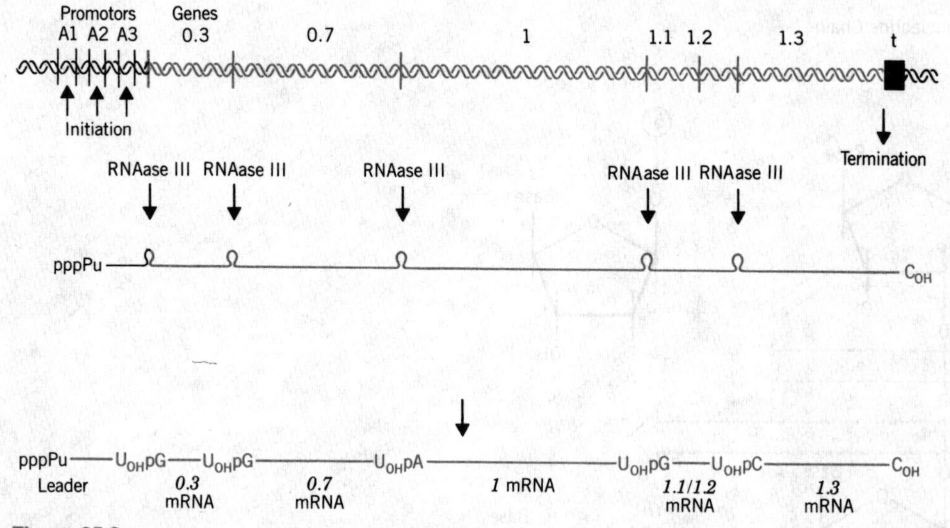

Figure 25.2
The T7 early region is transcribed into a single RNA that is cleaved into five individual mRNAs by RNAase III.

When T7 DNA is transcribed *in vitro* by *E. coli* RNA polymerase, the product is a single large RNA representing the entire early region. Purified RNAase III can cleave this precursor into the individual mRNAs found *in vivo*. Bacterial mutations (called *rnc*$^-$) that inactivate RNAase III block the processing reaction *in vivo*. Thus in the *rnc*$^-$ mutant cells, T7 DNA is transcribed into the same large precursor that is generated *in vitro*.

Although the precursor is not cleaved in a mutant bacterial host, it can be translated *in vivo* as a polycistronic mRNA, giving the usual early proteins and thus sustaining infection. The successful infection of T7 in *rnc*$^-$ mutants makes it seem that the cleavage reaction is unnecessary for early expression. However it turns out that cleavage is necessary for translation of the bicistronic *1.1/1.2* mRNA.

Some phage mutants have an altered pattern of cleavage that extends the *1.1/1.2* mRNA for an extra 29 bases. These mutants fail to translate either coding region. The reason is that the longer mRNA takes up a secondary structure that prevents initiation at the *1.1* coding sequence (and failure to translate the *1.1* coding sequence in turn prevents translation of *1.2*).

This effect was revealed by attempts to grow T7 on a mutant *E. coli* strain in which expression of gene *1.2* is necessary for phage growth (usually it is nonessential). In this circumstance, therefore, and perhaps in others in the natural habitat, cleavage to generate the normal end may be a necessary reaction.

Another instance in which RNAase III may affect translation is presented by an mRNA of phage lambda that carries the sequence of gene *int*. This mRNA usually is not translated (Int protein is synthesized from another mRNA). Mutations in a site called *sib*, located downstream from the *int* coding sequence, allow Int protein to be translated from the usually inactive mRNA.

The *sib* mutation relieves an inhibitory effect that is produced by the wild-type sequence. This effect is called **retroregulation.** Retroregulation does not occur in *rnc*$^-$ hosts, in which the mRNA is translated. This suggests that the wild-type *sib* site may be a target site for RNAase III; when the mRNA is cleaved here, either it acquires a conformation that precludes initiation, or it is degraded. Mutations in either *sib* or *rnc* have the same effect; they prevent cleavage and thus allow translation to occur.

In both the T7 *1.1/1.2* genes and the lambda *int* system, there is a connection between the ability of the RNA to be translated and its cleavage by RNAase

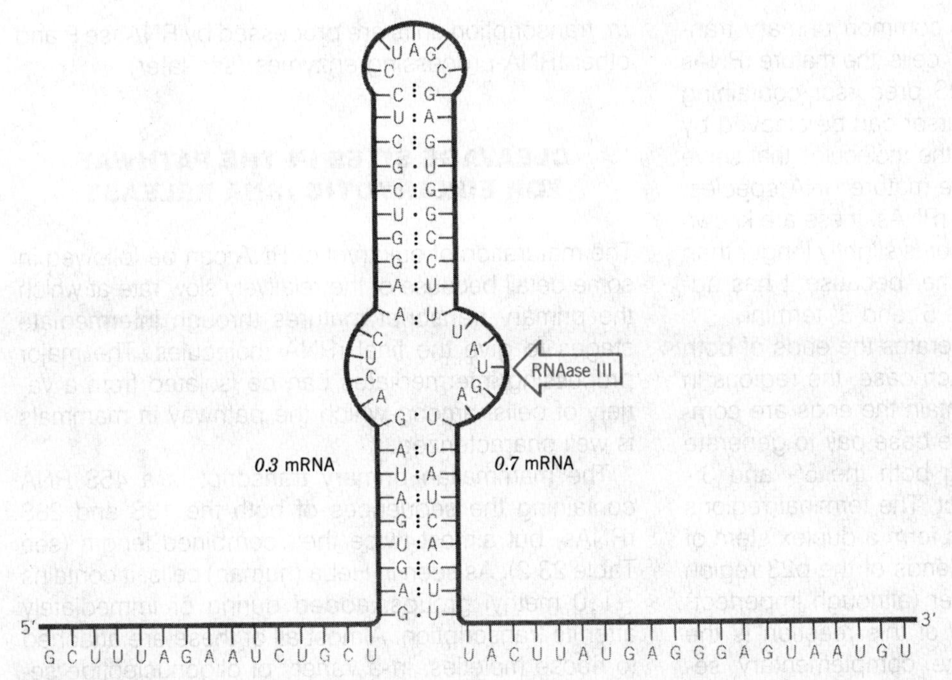

Figure 25.3
A typical cleavage site for RNAase III in T7 RNA lies within a short hairpin that contains an unpaired bubble.

III at a particular site. In both cases, the probable mechanism of the relationship is the effect that cleavage has on the secondary structure of the RNA. By releasing part of the molecule, the cleavage event changes the structure of other regions, including sites that are important in the RNA's functions.

Affected functions can involve initiation of translation, susceptibility to degradation, ability to be spliced (in eukaryotes; see next chapter), and no doubt other events. We saw in Chapter 15 that alternate secondary structures mediate termination of transcription, resulting in the regulatory network of attenuation. The connection between secondary (and perhaps also tertiary) structure and RNA function is a theme becoming of increasing importance as we expand our realization of the range of activities of which RNA is capable (see next chapter).

What is the nature of the sites recognized by RNAase III? The targets have been identified by determining the sequences at the 5′ and 3′ ends of the mRNA molecules generated by cleavage. These sequences can be aligned with the known DNA sequence of phage T7 to reveal the complete environment of the cleavage site.

Three of the T7 early cleavage sites show quite strong sequence homology, but the other two are different. The longest homology between all five is a triplet sequence close to the site of cleavage. However, all the sites conform to the same general structure, a duplex hairpin containing an unpaired bubble. An example is drawn in **Figure 25.3.** RNAase III cleaves at a single point at four of the sites, and at two points in one site, within or near to the bubble. The only information we have about the dependence of cleavage on the RNA structure is that mutations that reduce the stability of the hairpin may decrease or eliminate cutting by the enzyme.

RNAase III SEPARATES rRNAs FROM THEIR PRECURSOR

In the host bacterium, RNAase III is involved in the processing of rRNAs. The 16S, 23S, and 5S rRNAs all are synthesized as part of a common primary transcript (see Figure 23.9). In *rnc*$^-$ cells, the mature rRNAs do not appear; instead, a 30S precursor containing them accumulates. This precursor can be cleaved by RNAase III *in vitro* to generate the molecules that serve as individual precursors to the mature RNA species. In the case of the 16S and 23S rRNAs, these are known as p16 and p23. Each precursor is slightly longer than the rRNA found in the ribosome, because it has additional sequences at both the 5′ and 3′ termini.

A common mechanism generates the ends of both the p16 and p23 RNAs. In each case, the regions in the primary transcript that contain the ends are complementary. They can therefore base pair to generate a duplex structure containing both the 5′- and 3′-terminal regions of the product. The terminal regions around the p16 sequence can form a duplex stem of 26 bp. The sequences at the ends of the p23 region can base pair to form a longer (although imperfect) duplex. A remarkable feature of this reaction is the large distance separating the complementary sequences, about 1600 nucleotides for the p16 and about 2900 nucleotides for the p23.

Figure 25.4 shows that in each case RNAase III cleaves on both sides of the double-stranded stem, thus simultaneously generating the 5′ and 3′ ends of the p16 or p23 molecules. This action of RNAase III differs in two respects from that most commonly seen with the T7 early RNA. The sequence that is recognized is a continuous stem, without any prominent bubble. And two cuts are made rather than one (this also happens in an atypical T7 site).

There is no extensive sequence homology between either the p16 and p23 RNAase III sites or between either of these and any of the T7 sites. The features required by RNAase III for specific cutting of RNA therefore are mysterious, except for the common demand for a double-stranded structure either surrounding or including the actual site of cleavage. The enzyme could depend on a combination of primary sequence and secondary structure.

Since the cleavages accomplished by RNAase III do not release the actual 5′ or 3′ termini of either 16S or 23S rRNA, further processing reactions are necessary. Not very much is known about them, except that their occurrence is contingent on the prior action of RNAase III. The tRNA sequences included in the *rrn* transcription units are processed by RNAase P and other tRNA-processing enzymes (see later).

CLEAVAGE SITES IN THE PATHWAY FOR EUKARYOTIC rRNA RELEASE

The maturation of eukaryotic rRNA can be followed in some detail because of the relatively slow rate at which the primary transcript matures through intermediate stages to give the final rRNA molecules. The major processing intermediates can be isolated from a variety of cells, among which the pathway in mammals is well characterized.

The mammalian primary transcript is a 45S RNA containing the sequences of both the 18S and 28S rRNAs, but almost twice their combined length (see Table 23.2). As seen in HeLa (human) cells, it contains ~110 methyl groups, added during or immediately after its transcription. Almost all of these are attached to ribose moieties, in a variety of oligonucleotide sequences.

The methyl groups are *conserved* during processing of the precursor into mature rRNAs. There are ~39 of the original methyl groups on mature 18S rRNA, plus another 4 that are added later in the cytoplasm. There are ~74 methyl groups (all original) on the 28S rRNA. This suggests that methylation is used to distinguish the regions of the primary transcript that mature into rRNA.

More than one pathway has been found for maturation, as seen from the sizes of the intermediate RNA molecules. But all the pathways can be reconciled by supposing that a small number of cleavage sites can be utilized in varying orders. Two mammalian pathways, originally characterized in HeLa (human) and L (mouse) cells, are illustrated in **Figure 25.5.** In each case, there are cleavage sites on the 5′ side of the 18S gene, within the spacer between the 18S and 5.8S gene, and in the spacer between the 5.8S and 28S sequences. (The 5.8S sequence becomes a small RNA that associates with the 28S rRNA by base pairing.) The difference between the two pathways shown in the figure lies just in the order with which the cleavage sites are utilized. Other variations also are possible. In some cases, more than one pathway is found in a given cell type.

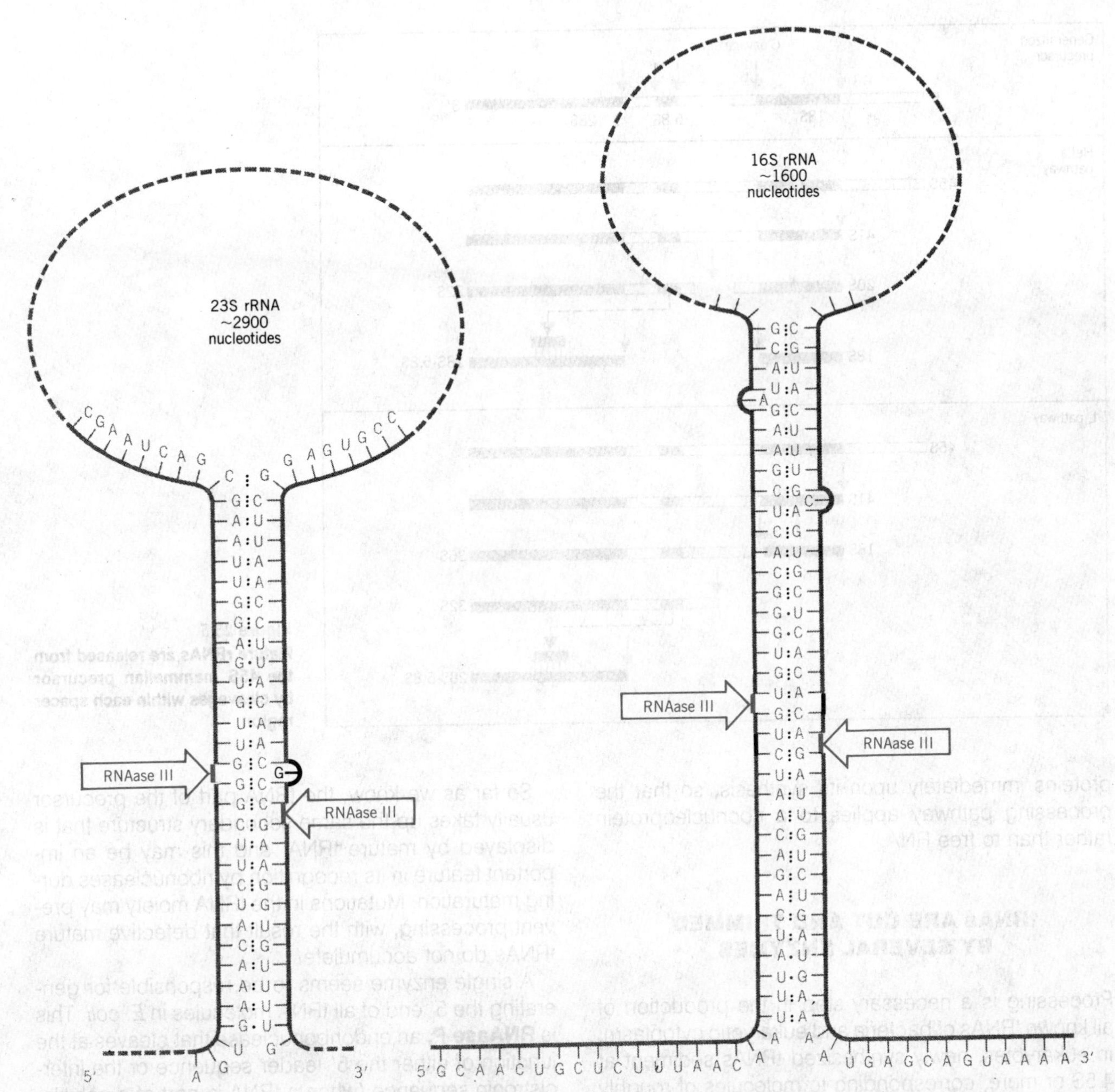

Figure 25.4
Regions containing the 5′ and 3′ ends of the p16 and p23 RNAs base pair to form duplex stems, each of which is cleaved at two points by RNAase III.

The figure shows the minimum number of likely cleavage sites used to generate all the 5′ and 3′ ends. We do not know whether these cleavages actually generate the mature ends, or whether cleavage releases individual precursors that then are cut or trimmed further. Nor do we have any information about the ribonuclease activities responsible for the process. We do know, however, that the 45S RNA associates with

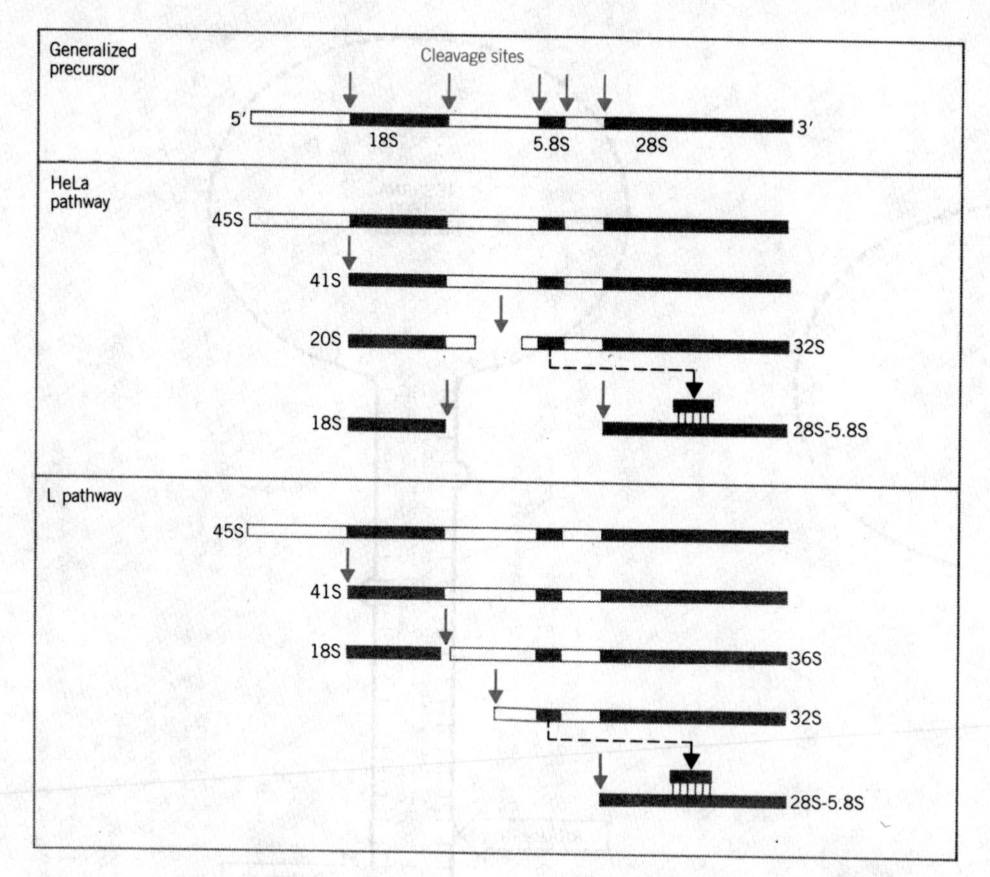

Figure 25.5
Mature rRNAs are released from the 45S mammalian precursor by cleavages within each spacer region.

proteins immediately upon its synthesis, so that the processing pathway applies to a ribonucleoprotein rather than to free RNA.

tRNAs ARE CUT AND TRIMMED BY SEVERAL ENZYMES

Processing is a necessary step in the production of all known tRNAs of bacteria and eukaryotic cytoplasm. In eukaryotes, newly synthesized tRNAs sediment at 4.5S or more, corresponding to molecules of roughly 100 nucleotides or so, compared with the mature size of 4S tRNA at 70–80 nucleotides. In bacteria, all tRNAs are synthesized in the form of precursors with additional sequences at both the 5′ and 3′ ends. In some cases, a single primary transcript may contain several tRNA sequences, up to about six, which must be separated.

So far as we know, the tRNA part of the precursor usually takes up the same secondary structure that is displayed by mature tRNA, and this may be an important feature in its recognition by ribonucleases during maturation. Mutations in the tRNA moiety may prevent processing, with the result that defective mature tRNAs do not accumulate.

A single enzyme seems to be responsible for generating the 5′ end of all tRNA molecules in *E. coli*. This is **RNAase P,** an endoribonuclease that cleaves at the junction of either the 5′ leader sequence or the intercistronic sequence (when a tRNA is part of a polycistronic transcript). The primary sequences in which the cleavage occurs are different in various precursors. Mutations in tRNAs that prevent processing by RNAase P occur in several regions of the molecule. This suggests that RNAase P must recognize a common feature of tRNA tertiary structure maintained in the precursor, and relying on different parts of the tRNA.

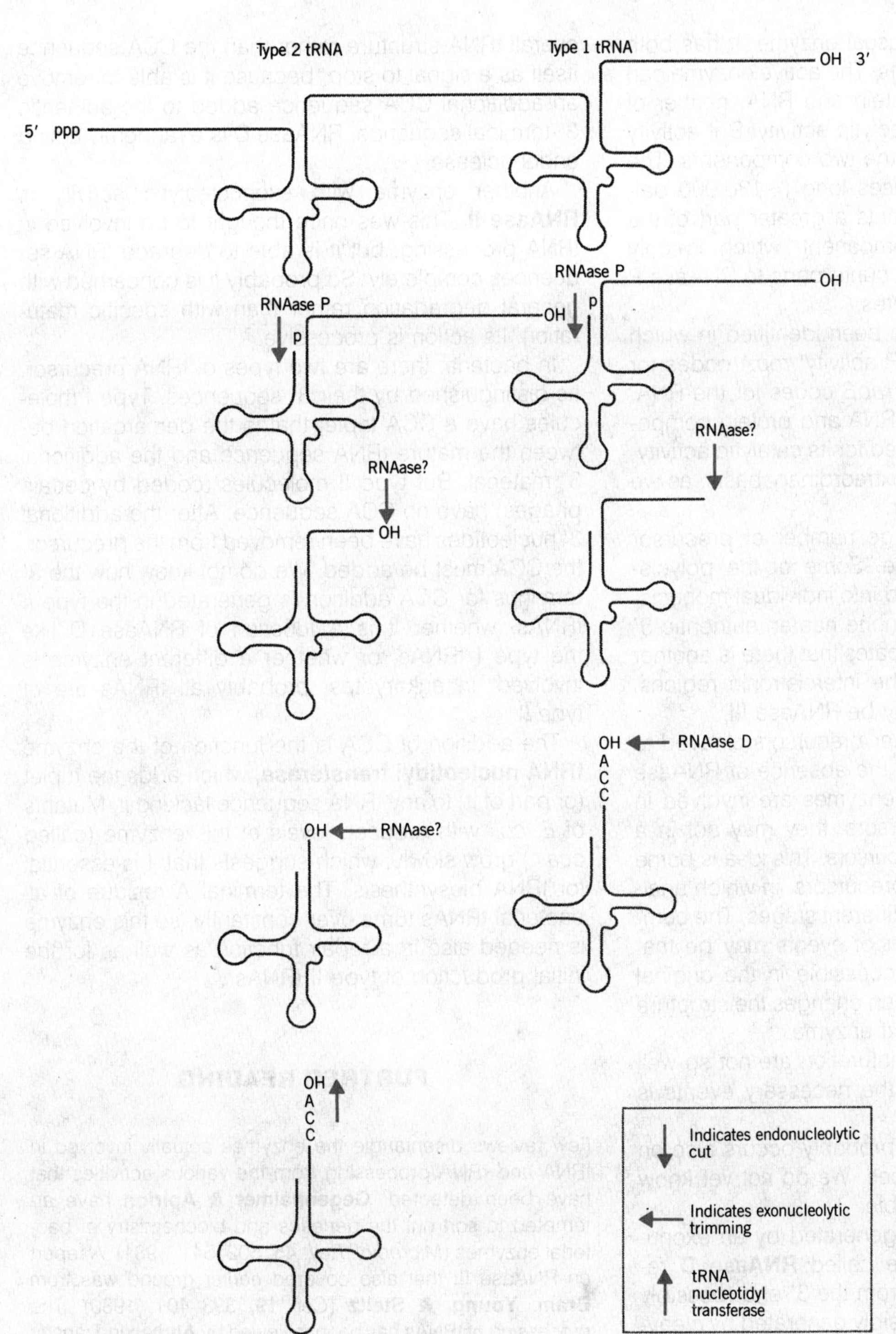

Figure 25.6
Several enzymes are needed to process a (hypothetical) dimeric precursor containing a type I and a type II tRNA.

Ribonuclease P is an unusual enzyme. It has both protein and RNA components. The active enzyme can be dissociated into the protein and RNA, neither of which retains the endonucleolytic activity. But activity can be restored by mixing the two components. The RNA component is 375 bases long (~130,000 daltons), and therefore represents a greater part of the mass than the protein component, which is only ~20,000 daltons. Potential counterparts to RNAase P have been found in eukaryotes.

Two genes in *E. coli* have been identified in which mutations abolish RNAase P activity: *rnpA* codes for the protein component and *rnpB* codes for the RNA. They lie far apart. Both the RNA and protein components of RNAase P are needed for its catalytic activity, which turns out to have an extraordinary basis, as we shall see in the next chapter.

In the *rnp* mutants, a large number of precursor tRNA molecules accumulate. Some of the polycistronic precursors are cleaved into individual monocistronic precursors, although none has an authentic 5′ terminus. This cleavage indicates that there is another enzyme able to cleave in the intercistronic regions. The enzyme responsible may be RNAase III.

Why are some but not other precursors cleaved to monocistronic precursors in the absence of RNAase P? Even if the same set of enzymes are involved in cleavage of all tRNA precursors, they may act in a different order in different precursors. This idea is borne out by studies of individual precursors, in which analogous events may occur at different stages. The compulsion for a particular order of events may be that only one cleavage site is accessible in the original transcript, but its cleavage then changes the structure to expose the site for the next enzyme.

The other steps in tRNA maturation are not so well established. A summary of the necessary events is given in **Figure 25.6.**

Endonucleolytic cleavage probably occurs also on the 3′ side of tRNA sequences. We do not yet know which enzyme(s) is responsible.

The actual 3′ terminus is generated by an exonucleolytic activity. An enzyme called **RNAase D** removes bases one at a time from the 3′ end (possibly of the original precursor, possibly generated by cleavage) until it arrives at the -CCA terminus common to all mature tRNAs. The enzyme seems to recognize the overall tRNA structure rather than the CCA sequence itself as a signal to stop, because it is able to remove an additional CCA sequence added to the authentic 3′-terminal sequence. RNAase D is a randomly acting endonuclease.

Another enzyme with exonucleolytic activity is **RNAase II.** This was once thought to be involved in tRNA processing, but it is able to degrade tRNA sequences completely. So probably it is concerned with general degradation rather than with specific maturation. Its action is processive.

In bacteria, there are two types of tRNA precursor, as distinguished by their 3′ sequences. Type I molecules have a CCA triplet that is the demarcation between the mature tRNA sequence and the additional 3′ material. But type II molecules (coded by certain phages) have no CCA sequence. After the additional 3′ nucleotides have been removed from the precursor, the CCA must be added. We do not know how the 3′ terminus for CCA addition is generated in the type II tRNAs, whether it is a function of RNAase D like the type I tRNAs, or whether a different enzyme is involved. In eukaryotes, probably all tRNAs are of type II.

The addition of CCA is the function of the enzyme **tRNA nucleotidyl transferase,** which adds the triplet (or part of it) to any tRNA sequence lacking it. Mutants of *E. coli* with reduced levels of this enzyme (called *cca*$^-$) grow slowly, which suggests that it is essential for tRNA biosynthesis. The terminal A residue of all bacterial tRNAs turns over constantly, so this enzyme is needed also in a repair function as well as for the initial production of type II tRNAs.

FURTHER READING

Few reviews disentangle the enzymes actually involved in tRNA and rRNA processing from the various activities that have been detected. **Gegenheimer & Apirion** have attempted to sort out the genetics and biochemistry of bacterial enzymes (*Microbiol. Rev.* **45,** 502–541, 1981). A report on RNAase III that also covered earlier ground was from **Bram, Young, & Steitz** (*Cell* **19,** 393–401, 1980). The processing of tRNAs has been reviewed by **Altman** in *Transfer RNA* (MIT Press, Cambridge, Mass. 1978, pp. 48–77) and updated in *Cell* (**23,** 3–4, 1981).

CHAPTER 26
RNA AS CATALYST: MECHANISMS OF RNA SPLICING

The discovery that interrupted genes are transcribed into precursor RNAs from which the exons are recovered via splicing raises a series of questions about gene expression. Viewing RNA splicing simply as one of the mechanisms involved in gene expression, we want to know at what point it occurs vis à vis the other modifications of RNA (capping and polyadenylation). Are introns excised from a precursor in a particular order? What is responsible for the remarkable precision of the process? Does it occur at a particular location in the nucleus; is it connected with other events, for example, nucleocytoplasmic transport? Do such connections create differences in the pathways for expression of interrupted and uninterrupted genes? Is the maturation of RNA (via splicing or other mechanisms) used to *regulate* gene expression by discriminating among the available precursors?

RNA splicing involves breakage of the phosphodiester bonds at exon-intron boundaries and the formation of a bond between the ends of the exon. This could occur either as two successive, independent reactions, or as one coordinated transfer reaction. Breakage of phosphodiester bonds is common to other RNA processing reactions, but their creation in RNA is unique to the splicing reaction. Whatever the enzymatic mechanism(s), a critical feature is that the reacting groups must be brought into juxtaposition for bond formation.

There is no unique mechanism for RNA splicing. We can identify at least four distinct types of splicing system. The removal of introns from yeast nuclear tRNA precursors involves enzymatic activities whose dealings with the substrate seem to resemble those of the tRNA processing enzymes, since a critical feature is the conformation of the tRNA precursor. Excision of the intron from a ciliate rRNA precursor occurs as an autonomous property of the RNA. The removal of introns from yeast mitochondrial RNAs probably lacks a single mechanism, but at least in some cases depends on a protein coded by the intron that is to be removed. Introns are removed from the nuclear RNAs of higher eukaryotes by a system that apparently recognizes nothing more specific than the short consensus sequences conserved at exon-intron boundaries in a wide range of species. So if there was a single evolutionary origin for the interrupted gene, considerable divergence since has occurred in the means of its reconstitution in RNA.

YEAST tRNA SPLICING INVOLVES CUTTING AND REJOINING

About 40 of the ~400 nuclear tRNA genes in yeast are interrupted. Each has a single intron, located just one nucleotide beyond the 3′ side of the anticodon. The introns vary in length from 14 to 46 bp. Those in related tRNA genes are related in sequence, but the introns in tRNA genes representing different amino acids are unrelated. *There is no consensus sequence that could be recognized by the splicing enzymes.*

All the introns include a sequence that is complementary to the anticodon of the tRNA. This creates an alternative conformation for the anticodon arm in which the anticodon is base paired to form an extension of the usual arm. An example is drawn in **Figure 26.1.** Only the anticodon arm is affected—the rest of the molecule retains its usual structure. *Splicing may therefore depend on recognition of a common secondary structure rather than a common sequence.*

The features of the precursor that are recognized in splicing can be identified by introducing changes into the RNA molecule, either via manipulation *in vitro* of the cloned gene, or by mutation *in vivo.* The irrelevance of the exact sequence and size of the intron is indicated by the natural variation between genes. It is confirmed by an experiment in which an extra 21 bp segment was inserted into the intron; this extends the

Figure 26.1
The intron in yeast tRNA^{Phe} base pairs with the anticodon to change the structure of the anticodon arm.

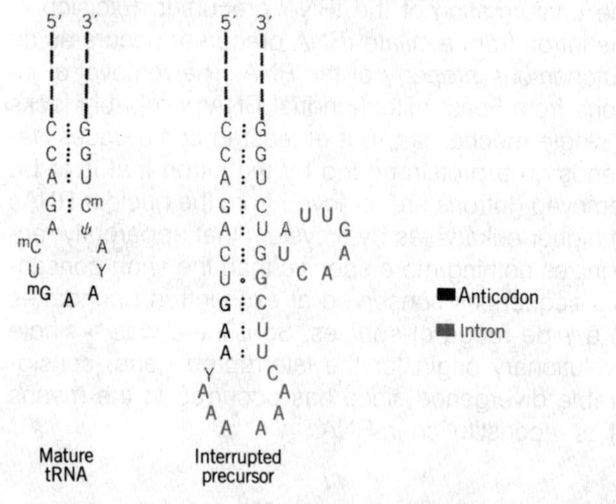

length of the additional arm present in the precursor, but has no effect on splicing.

An advantage of the tRNA system is that mutants can be obtained rather readily. For example, there is a yeast variant in which an interrupted tRNA^{Tyr} gene has been mutated to an ochre suppressor. From these mutant cells, we can select further variants that are *ineffective* in suppression. Some of these lack the suppressor tRNA because mutations have prevented processing of the precursor. The series of three tRNA structures drawn in **Figure 26.2** consists of the wild-type precursor, the suppressor tRNA precursor (which is spliced normally), and a mutant that is spliced seven times less efficiently.

The secondary structure of the intronic region is different in each case. The suppressor has a shorter base-paired stem, and one end of the intron is base paired instead of single-stranded. The *a122* mutation actually lies at the splicing junction and creates a large unpaired bubble in the duplex stem; the other end of the intron is base paired instead of lying in the unpaired loop. So the splicing sites are recognized whether they lie in base-paired or in single-stranded regions, and even a change at the boundary reduces only the efficiency and does not alter the sites used in splicing.

This seems reminiscent of the conclusions we advanced in Chapters 8 and 25 about tRNA structure and expression. Specific bases do not seem consistently to be implicated in recognition reactions, which instead rely on the *conformation of the molecule as a whole.* In particular, it may be important to recognize in the precursor those structural features that are conserved in the mature tRNA. This is consistent with the observation that mature tRNA competitively inhibits the splicing of precursors, presumably by being recognized by the enzyme(s).

An opportunity to study the splicing reaction is provided by the existence of a temperature-sensitive mutant of yeast that fails to remove the introns and accumulates the interrupted precursors in the nucleus. Flanking sequences can be processed as usual. We do not know the molecular basis for the defect; but it is specific to interrupted genes, because the tRNAs coded by uninterrupted genes can mature and be transported to the cytoplasm.

This is very useful, because it allows the tRNA pre-

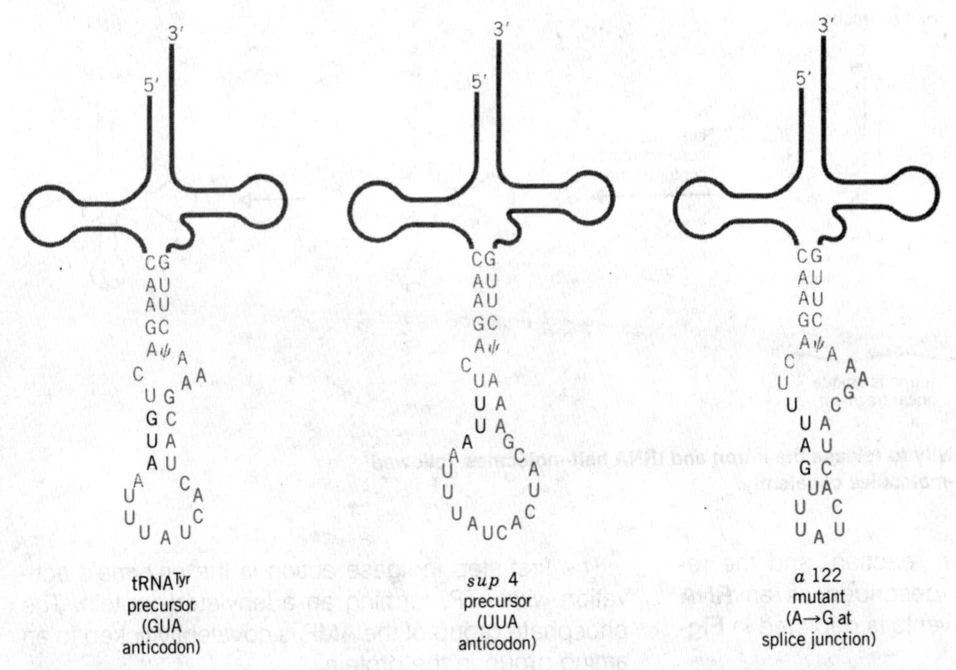

Figure 26.2
Three variants of yeast tRNA^{Tyr} have different conformations in the anticodon region of the interrupted precursor.

cursors to be isolated in a form in which their only difference in sequence from the mature tRNA is the presence of the intron. (The precursors may also lack some base modifications.) These molecules can be used as substrates for a cell-free system extracted from wild-type cells. The splicing of the precursor can be followed quite simply by virtue of the resulting size reduction. This is seen by the change in position of the band on gel electrophoresis, as illustrated in **Figure 26.3.** The reduction in size can be accounted for by the appearance of a band representing the intron.

The cell-free extract can be purified by assaying the ability to splice the tRNA. The *in vitro* reaction requires ATP and can be split into two steps by omitting the nucleotide. Characterizing the reactions that occur with and without ATP shows that the *two separate stages of the reaction are catalyzed by different enzymes.*

The first step does not require ATP. It involves phosphodiester bond cleavage, taking the form of an atypical nuclease reaction. It is catalyzed by an endonuclease that behaves like a membrane-bound protein. The second step requires ATP and involves bond for-

Figure 26.3
Splicing of yeast tRNA *in vitro* can be followed by assaying the RNA precursor and products by gel electrophoresis.

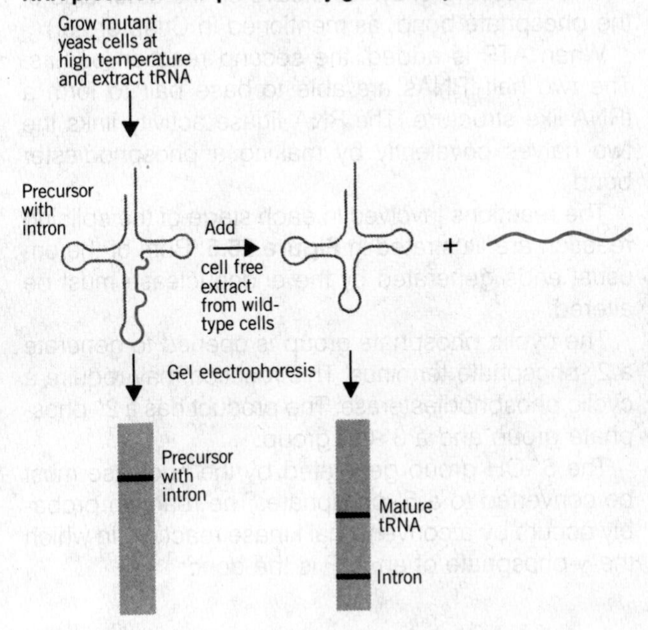

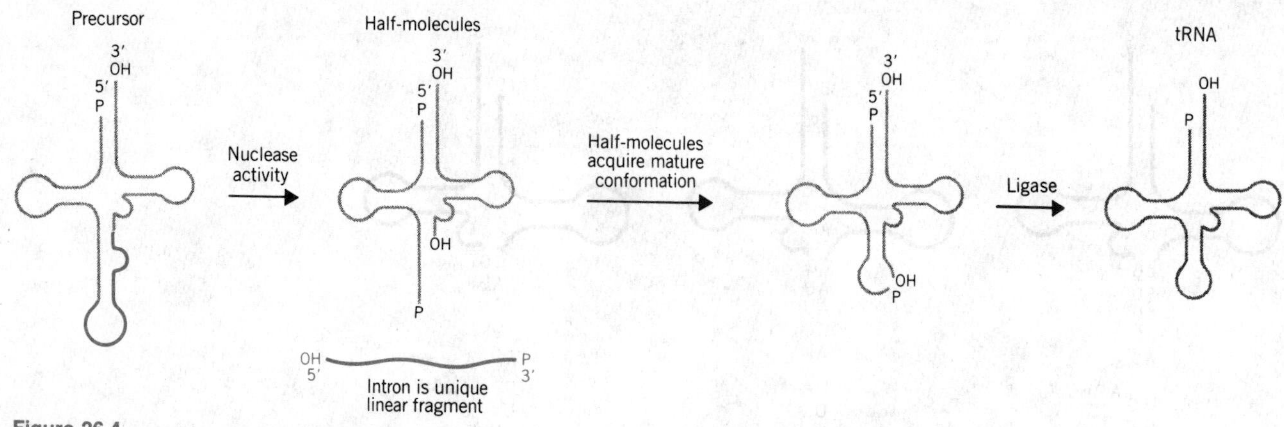

Figure 26.4

Splicing occurs via a nuclease activity to release the intron and tRNA half-molecules, followed by a ligase activity to link the half-molecules covalently.

mation; it is called a **ligation** reaction, and the responsible enzyme activity is described as an **RNA ligase.** The overall series of events is depicted in **Figure 26.4.**

In the absence of ATP, the endonuclease cleaves the precursor at both ends of the intron, releasing two half-tRNA molecules and a linear intervening sequence. These intermediates have unique ends. Each 5′ terminus ends in a hydroxyl group; each 3′ terminus ends in a 2′,3′-cyclic phosphate group. (All other known RNA processing enzymes cleave on the other side of the phosphate bond, as mentioned in Chapter 25.)

When ATP is added, the second reaction occurs. The two half-tRNAs are able to base pair to form a tRNA-like structure. The RNA ligase activity links the two halves covalently by making a phosphodiester bond.

The reactions involved in each stage of the splicing reaction are illustrated in **Figure 26.5.** Both of the unusual ends generated by the endonuclease must be altered.

The cyclic phosphate group is opened to generate a 2′-phosphate terminus. This reaction may require a cyclic phosphodiesterase. The product has a 2′-phosphate group and a 3′-OH group.

The 5′-OH group generated by the nuclease must be converted to a 5′-phosphate. The reaction probably occurs by a conventional kinase reaction, in which the γ-phosphate of an ATP is the donor.

The first step in ligase action is the enzyme's activation with ATP, forming an adenylated protein. The phosphate group of the AMP is covalently linked to an amino group in the protein.

Release of the AMP from the enzyme is coupled with the ligation reaction itself. If the reaction follows the ligation pathway previously characterized for the RNA ligase of phage T4, it will occur in two steps.

First the AMP will be transferred from the enzyme to the 5′-phosphate group, forming a 5′-5′ phosphate-phosphate linkage. Then the AMP is displaced by an attack from the 3′-OH group of the other half of the tRNA. Note that the phosphate group linking the two exons is not an original part of the RNA, but has been provided by ATP during the reaction.

The spliced molecule is now covalently intact, with a 5′-3′ phosphate linkage at the site of splicing, but it also has a 2′-phosphate group marking the event. The surplus group must be removed by a phosphatase. (In the interim it could be useful in marking the site where the ligation occurred.)

The occurrence of a 2′,3′-cyclic phosphate is not unique to yeast. The cyclic group is generated during the tRNA-splicing reaction in plants and mammals. The reaction in plants seems to be the same as in yeast.

However, the reaction is different in mammals. The HeLa (human) ligase directly joins an RNA end with a 2′,3′-cyclic phosphate group to an RNA end with a

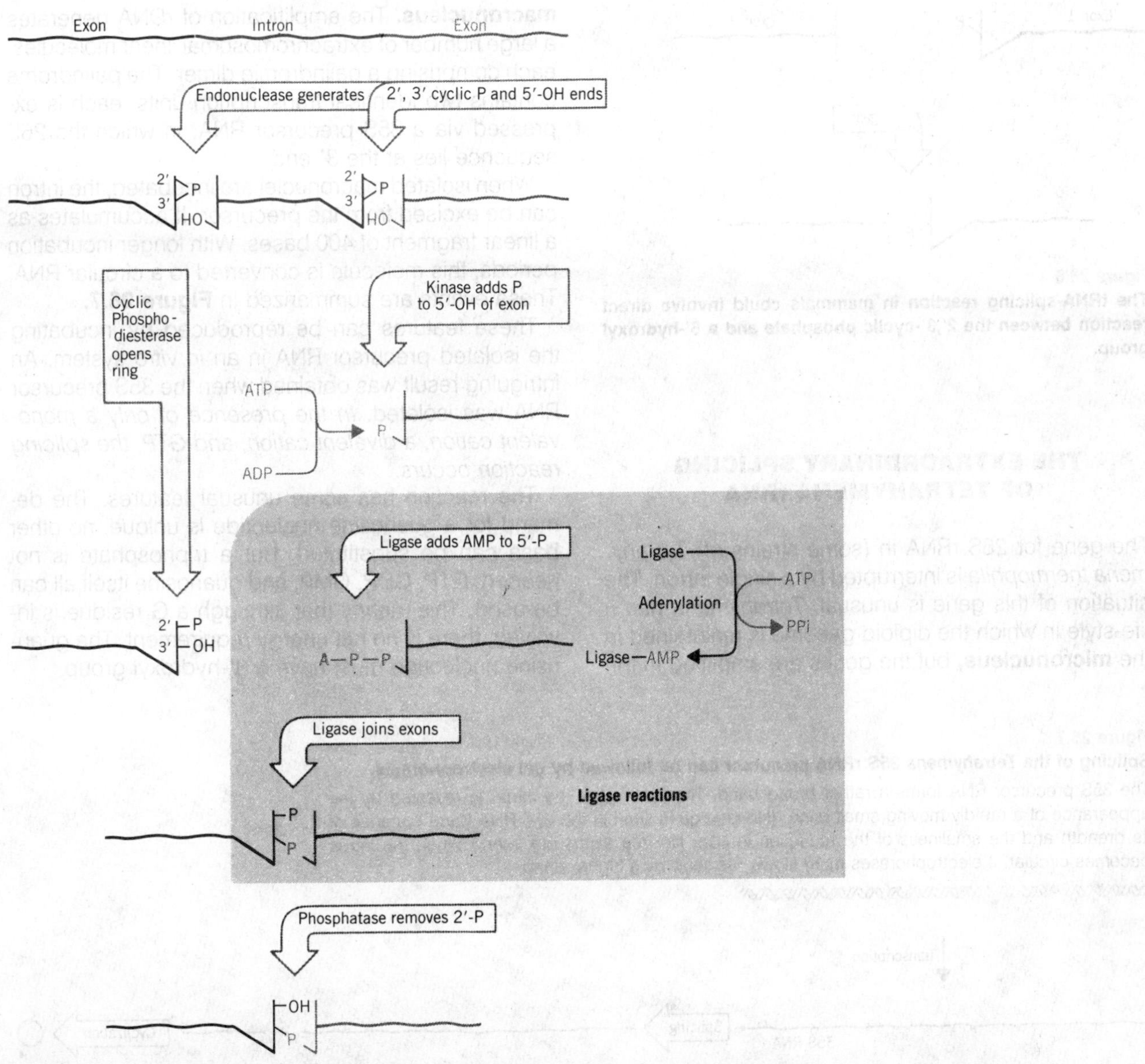

Figure 26.5
Splicing of yeast and plant tRNAs involves a series of reactions that rely on unusual 5′ and 3′ termini in the RNA.

5′-hydroxyl group. Thus the phosphate group that links the two exons is the phosphate originally present at the end of the left exon, as illustrated in **Figure 26.6.**

The yeast tRNA precursors also can be spliced in an extract obtained from the germinal vesicle (nucleus) of *Xenopus* oocytes. This shows that the reaction is not species-specific. *Xenopus* must have enzymes able to recognize the introns in the yeast tRNAs.

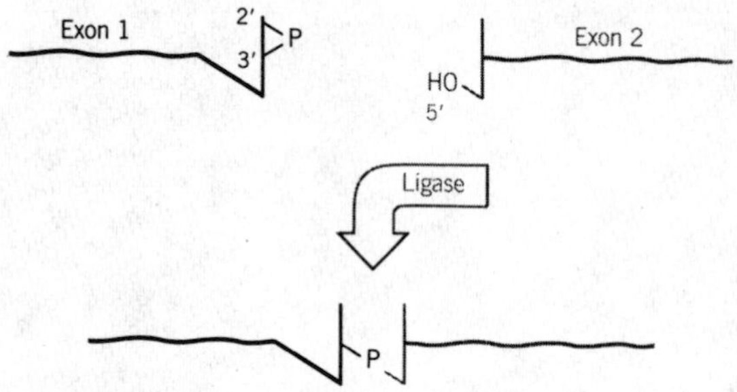

Figure 26.6
The tRNA-splicing reaction in mammals could involve direct reaction between the 2′,3′-cyclic phosphate and a 5′-hydroxyl group.

THE EXTRAORDINARY SPLICING OF *TETRAHYMENA* rRNA

The gene for 26S rRNA in (some strains of) *Tetrahymena thermophila* is interrupted by a single intron. The situation of this gene is unusual. *Tetrahymena* has a life-style in which the diploid genome is maintained in the **micronucleus,** but the genes are amplified in the **macronucleus.** The amplification of rDNA generates a large number of extrachromosomal linear molecules, each comprising a palindromic dimer. The palindrome contains two identical transcription units; each is expressed via a 35S precursor RNA, in which the 26S sequence lies at the 3′ end.

When isolated macronuclei are incubated, the intron can be excised from the precursor. It accumulates as a linear fragment of 400 bases. With longer incubation periods, this molecule is converted to a circular RNA. These events are summarized in **Figure 26.7.**

These features can be reproduced by incubating the isolated precursor RNA in an *in vitro* system. An intriguing result was obtained when the 35S precursor RNA was isolated. *In the presence of only a monovalent cation, a divalent cation, and GTP, the splicing reaction occurs.*

The reaction has some unusual features. The demand for a guanosine nucleotide is unique; no other base can be substituted. But a triphosphate is not needed; GTP, GDP, GMP, and guanosine itself all can be used. This means that although a G residue is involved, there is no net energy requirement. The guanosine nucleotide must have a 3′-hydroxyl group.

Figure 26.7
Splicing of the *Tetrahymena* 35S rRNA precursor can be followed by gel electrophoresis.
The 35S precursor RNA forms a rather broad band. The removal of the intron is revealed by the appearance of a rapidly moving small band. (No change is seen in the 35S RNA band because of its breadth and the smallness of the reduction in size. No free exons are seen.) When the intron becomes circular, it electrophoreses more slowly, as seen by a higher band.

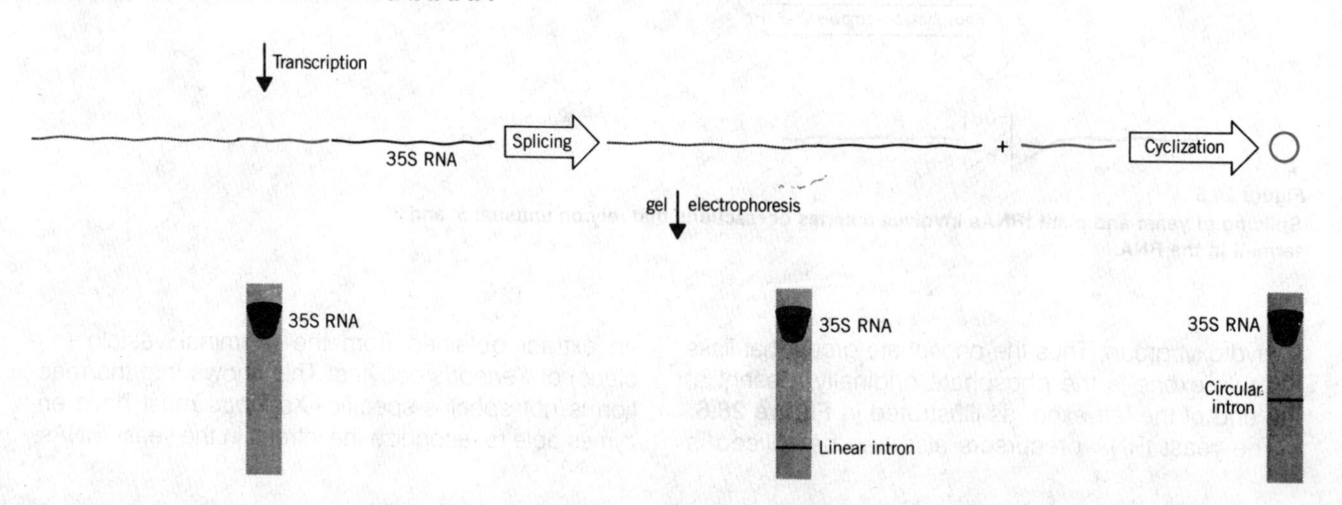

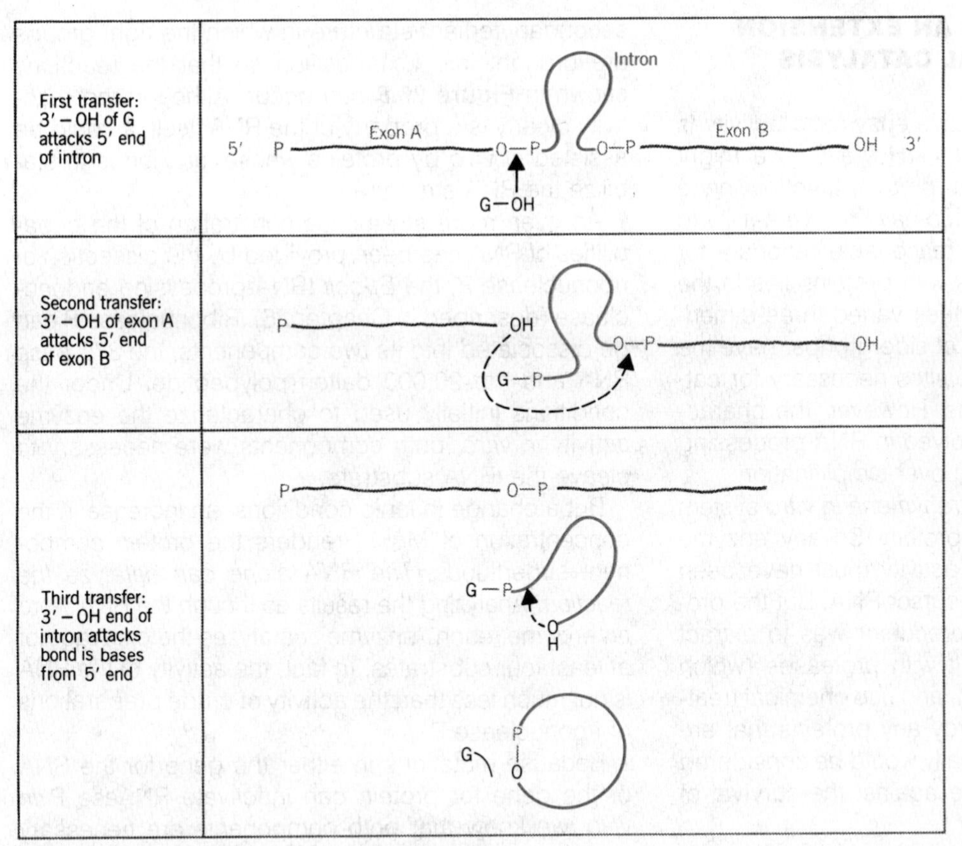

Figure 26.8
The molecules formed by the *Tetrahymena* splicing reaction can be explained by phosphoester transfer reactions.

When the guanosine nucleotide carries a radioactive label, the radioactivity enters the excised linear intron fragment. It turns out that the G is linked to the 5′ end of the intron by a normal phosphodiester bond. When the splicing reaction is allowed to continue *in vitro,* the linear intron becomes a circle. The reaction involves joining of the 3′ end to a site 15 bases from the 5′ end (the terminal 15 base segment, including the additional G, is released).

The splicing reaction is equivalent to a phosphoester transfer, as illustrated in **Figure 26.8.** The guanine nucleotide is a cofactor that provides a free 3′-hydroxyl group to which the 5′ end of the intron is transferred. This reaction could be followed by a second, similar reaction, in which the 3′-hydroxyl created at the end of the first exon is used to attack the second exon. Because one phosphate ester is converted directly into another, without any intermediary hydrolysis, the energy of the phosphoester bond is conserved; this explains why the reaction does not require input of energy from hydrolysis of ATP or GTP.

The two transfers seem to be connected. No free exons have been observed, so their ligation may occur as part of the same reaction that releases the intron. The cyclization of the linear intron fragment could be explained by another phosphoester transfer, in which the bond 15 bases from the end is broken and the released 5′ end is transferred to the 3′-hydroxyl group of the intron molecule. (There is a parallel for this type of enzymatic activity, involved in transferring 3′-OH to 5′-P without net input of energy, in the DNA nicking-closing enzymes discussed in Chapter 34.)

RNA AS CATALYST: AN EXTENSION OF BIOCHEMICAL CATALYSIS

The idea that only proteins have enzymatic activity is deeply rooted in biochemistry. (However, we might remember that devotees of protein function once thought that only proteins could have the versatility to be the genetic material!) A reasonable rationale for the identification of enzymes with proteins lies in the view that only proteins, with their varied three-dimensional structures and variety of side groups, have the flexibility to create the active sites necessary for catalyzing biochemical reactions. However, the characterization of two systems involved in RNA processing has shown this view to be an over-simplification.

The components of the *Tetrahymena in vitro* system do not include any added protein. So any enzyme responsible for the splicing activity must have been purified together with the precursor RNA. But the procedure used to isolate the precursor was to extract the RNA with phenol, treat it with proteases (which should degrade any proteins), and use chemical treatments that also should destroy any proteins that are present. Such treatments usually would be considered to constitute strong evidence against the survival of any protein.

This result leaves two possibilities (either of which a priori would be thought rather unlikely). The splicing activity could reside in an enzyme that is very tightly bound indeed to the RNA (perhaps through covalent bonds) and that retains its enzymatic activity through this series of severe treatments. Or there might be no splicing enzyme, and the reaction is an intrinsic property of the RNA molecule itself. This would imply that the RNA acquires some particular conformation that brings the relevant bonds into juxtaposition for an autonomous bond breakage and reunion reaction.

The situation can be resolved by analyzing a precursor RNA transcribed from a copy of the gene cloned in *E. coli*. This RNA has never been exposed to the proteins of *Tetrahymena*. But it retains the ability to splice itself. This demonstrates dramatically that *the property is intrinsic to the RNA.*

The activity has been called **auto-catalysis,** because the RNA is able to sponsor its own rearrangement. It does not catalyze the reaction of other molecules. We assume that the RNA is able to form a specific secondary/tertiary structure in which the right groups are brought into juxtaposition so that the reactions shown in **Figure 26.8** can occur. Although the reactivity clearly is a property of the RNA itself, it could be assisted *in vivo* by proteins whose function is to stabilize the RNA structure.

An even more striking demonstration of the capabilities of RNA has been provided by the dissection of ribonuclease P, the *E. coli* tRNA-processing endonuclease (described in Chapter 25). Ribonuclease P can be dissociated into its two components, the 375 base RNA and the 20,000 dalton polypeptide. Under the conditions initially used to characterize the enzyme activity *in vitro,* both components were necessary to cleave the tRNA substrate.

But a change in ionic conditions, an increase in the concentration of Mg^{2+}, renders the protein component superfluous. *The RNA alone can catalyze the reaction!* Analyzing the results as though the RNA were an enzyme, each "enzyme" catalyzes the cleavage of at least four substrates. In fact, the activity of the RNA is not much less than the activity of crude preparations of ribonuclease P.

Because mutations in either the gene for the RNA or the gene for protein can inactivate RNAase P *in vivo,* we know that both components are necessary for natural enzyme activity. Naturally it had been assumed that the protein provided the catalytic activity, while the RNA filled some subsidiary role, for example, assisting in the binding of substrate (it has some short sequences complementary to exposed regions of tRNA). Now it turns out that these roles actually should be reversed!

How can RNA provide a catalytic center? Its ability seems reasonable if we think of an active center as a surface that exposes a series of active groups in a fixed relationship. In a protein, the active groups are provided by the side-chains of the amino acids, which have appreciable variety, including positive and negative ionic groups and hydrophobic groups (see Figure 1.2). In an RNA, the available moieties are more restricted, consisting primarily of the exposed groups of bases. We might suppose that short regions are held in a particular structure by the secondary/tertiary conformation of the molecule, and they provide a surface of active groups able to maintain an environment in which bonds can be broken and made in another molecule. It seems inevitable that the interaction between the RNA catalyst and the RNA substrate will rely on base pairing to create the environment.

The evolutionary implications of these discoveries are obvious. The split personality of the genetic apparatus, in which RNA is present in all components, but proteins undertake the catalytic reactions, has always been puzzling. It seems unlikely that the very first replicating systems could have contained both nucleic acid and protein.

But suppose that the first systems contained only a self-replicating nucleic acid with primitive catalytic activities, just those needed to make and break phosphodiester bonds. If we suppose that the involvement of 2′ bonds in current splicing reactions is derived from these primitive catalytic activities, we may argue that the original nucleic acid was RNA, since DNA lacks the 2′-OH group and therefore could not undertake such reactions.

We can see that proteins could have been added for their ability to stabilize the RNA structure, which was likely to have been precariously maintained. Then the greater versatility of proteins could have allowed them to take over catalytic reactions, leading eventually to the complex and sophisticated apparatus of modern gene expression.

Remnants of the original system are still to be found, most clearly in the examples of RNAase P and *Tetrahymena* rRNA, more speculatively in organelles such as the ribosome. Think of the ribosome as originally consisting of a catalytic RNA, with proteins slowly accreting to the structure, taking over catalytic functions, and eventually relegating the RNA to a structural role—an exact reversal of the original roles of the two components.

AN INTRON THAT MAY CODE FOR A REGULATOR PROTEIN

Two interrupted genes in the yeast mitochondrion display unprecedented genetic properties that are so much at odds with the traditional view of the gene that they could be explained only by virtue of the molecular analysis of DNA. The organization of these genes is illustrated in **Figure 26.9.**

The locus *oxi3* codes for subunit 1 of cytochrome oxidase. The gene has a coding region of 1530 bp, extending over about 10,000 bp of DNA. It is organized into eight exons, two of which are very short.

The gene coding for cytochrome-b is known variously as *box* and *cob*. It exists in two versions. Some strains of yeast have a "long" gene, in which the coding region of 1155 bp extends over about 6400 bp, organized into six exons (numbered B1 to B6), some of which are very short. Other strains have a "short" gene, about half the length, in which the first four exons of the "long" gene all form a single, continuous exon. In other words, the first three introns of the gene all may be present or all may be absent. Both forms of the gene are expressed equally well. The genetic analysis that we shall discuss has been performed with the long gene. The introns are identified according to their numbering in the long gene, I1 to I5.

When RNA molecules corresponding to the *box* gene are isolated, they fall into a discrete series of sizes. The shortest species is about 3000 bases long and provides the mRNA. This is longer than the coding region and includes 5′ and 3′ nontranslated regions (not shown on the map in Figure 26.9). Among the larger RNA molecules, each size possesses one or more of the introns. The largest molecule, of 8500 bases, is probably the intact primary transcript. The others are intermediates in which some introns have been spliced out while others remain.

Mutations preventing the synthesis of cytochrome-b can be mapped into a series of clusters. Each cluster covers a short genetic distance, of the order of 1% recombination. The clusters are separated from each other by greater distances, in the range of 2–9% recombination. Each cluster is indicated as *box* followed by a number. The clusters fall into three classes.

Mutations in four of the clusters affect the cytochrome-b protein directly. All of these mutants synthesize normal mRNA. The effects of the mutations are mediated at the level of translation, by causing missense or nonsense readings of the message. None of these mutations complements any other, in either the same or a different cluster. By this criterion, they all lie in the same gene.

These clusters identify some of the exons; the correspondence is *box4* = B1, *box8* = B3, *box1* = B4, *box6* = B6 (see Figure 26.9). Mutations have not been found in the other two exons, perhaps because of their

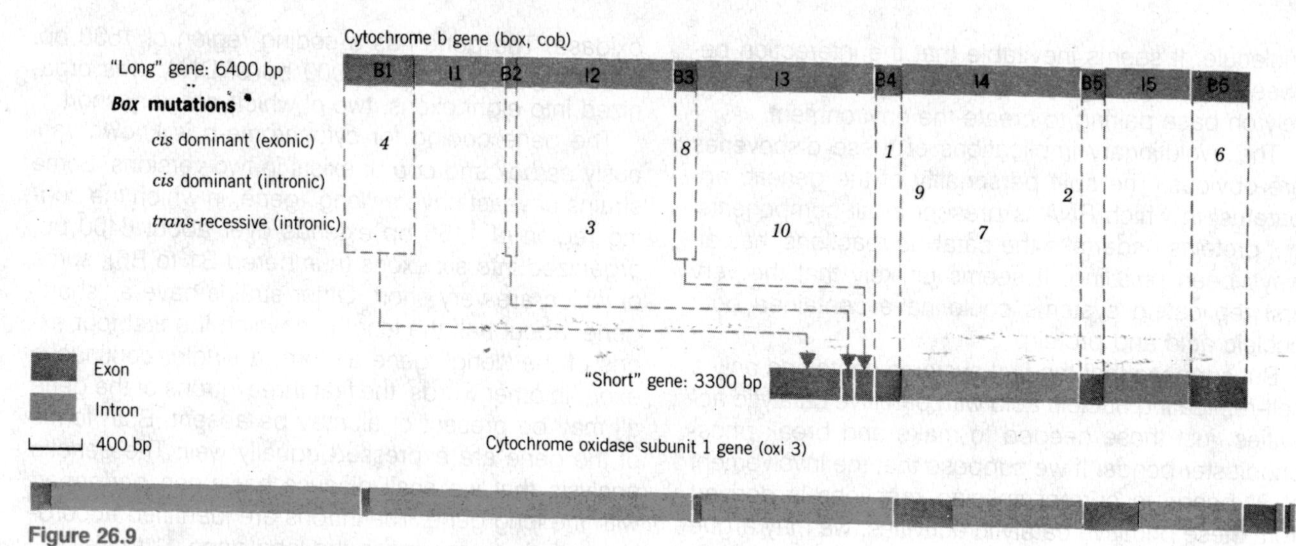

Figure 26.9

Two genes of the yeast mitochondrion have mosaic structures.

The gene *oxi3* has 8 exons that code for subunit 1 of cytochrome oxidase, (probably) 510 amino acids long. The "short" and "long" versions of the *box* gene code for the same cytochrome-b protein of 385 amino acids. The clusters of mutations are indicated as *box* followed by a number that reflects order of isolation (not map position). Their locations on the map indicate the exons or introns to which they are assigned. Both genes as depicted correspond only to the protein-coding regions of the mRNA and do not include 5′ or 3′ nontranslated regions.

small sizes (B2 = 14 bp, B5 = 51 bp). The properties of these groups of mutations correspond exactly with those predicted for an interrupted gene. In fact, this is the only case known so far in which the genetics of a mosaic gene can be studied.

Mutations in the *box9* and *box2* clusters also fail to complement with mutations of the other groups. By this genetic criterion, they are therefore indistinguishable from the exonic mutations. But their biochemical properties are different, as indicated by a failure to synthesize the normal mRNA. Sequence analysis of the DNA shows that both these clusters lie in I4. The *box9* mutations lie in a sequence of 8 bp about 350 bp downstream from the junction with B4. The *box2* mutations lie toward the other end of the intron, about 25 bp upstream from the junction with B5. Both groups of mutations prevent the splicing together of B4 and B5 by removal of I4.

The existence of this class of mutations reveals two important general points. First, mutations in specific sites can prevent recognition of particular splicing junctions; and these sites may be quite distant from the actual junctions themselves. Second, such mutations cannot be distinguished genetically from mutations in the protein itself. (The same lack of discrimination applies to mutations in promoters or operators and their structural genes, the classic *cis*-acting type; see Chapter 14.)

Three other groups, *box3*, *box10*, and *box7*, each represent a small cluster of mutations located within an intron. Their behavior in complementation tests reveals that these mutations are of a novel type. The mutations within each of these three clusters behave as a complementation group. But although they cannot complement other mutations in the cluster, they *can* complement mutations in any other cluster, either of this type or of the exon type.

Formally, this means that each of these three clusters codes for a *trans*-acting, diffusible product that is distinct from the cytochrome-b protein, but nonetheless is necessary for its synthesis. In practical terms, this means that each of these three introns must contain sequences whose function is to code for some product that has an independent existence and plays some regulatory role in cytochrome-b production.

The nature of this role is suggested by another

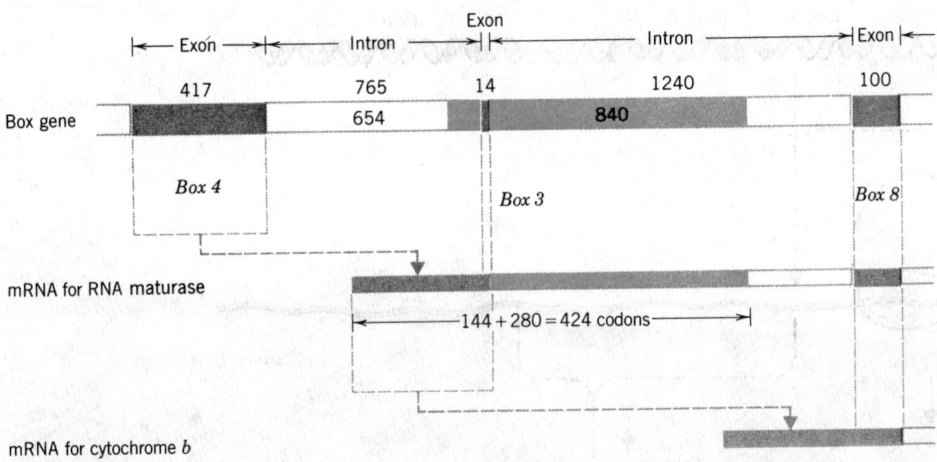

Figure 26.10

Successive splices in the *box* gene could generate mRNAs coding for overlapping proteins. Removal of only the first intron generates the mRNA for RNA maturase; removal of the second intron then gives the start of the coding sequence for cytochrome b.

property of the mutations. They all block the production of cytochrome-b mRNA by causing precursor RNAs to accumulate. These precursors are: 7500 bases for *box3* mutants, 7100 bases for *box10,* 3500 for *box7.* This suggests that each cluster of mutations may block the maturation of the RNA at a specific stage, by inactivating a diffusible product that probably is needed to remove a particular intron.

The nature of this function is suggested by the sequence of the gene. The salient features are shown in diagrammatic form in **Figure 26.10.** The first exon (B1) carries the N-terminal 139 codons (417 bp) of cytochrome-b. An intron of 765 bp (I1), blocked in all reading frames, separates the first exon from the second, very short exon (B2, 5 codons). This exon is followed by the long second intron (I2). The significant feature about this intron is that *the first 840 bp represent an open reading frame in exact register with the reading frame of the preceding exon.*

The *box3* mutations all lie in this region; they create nonsense codons in the open reading frame. This part of the intron may therefore have a coding function; and the protein may provide the diffusible function involved in splicing. It has been called the **RNA maturase.** (The remainder of the intron is blocked in all reading frames.)

What is the structure of the RNA maturase? Indirect evidence suggests that its translation is not initiated within the intron, but occurs via readthrough from the second exon. This could occur if the first splicing intermediate shown in Figure 26.10 is translated. The RNA is generated by removing just the first intron. This connects the first and second exons, creating a reading frame that continues into the second intron. Its total length is 424 codons; and its translation would produce a protein with the 144 N-terminal amino acids of cytochrome-b and 279 amino acids coded by the intron.

If the RNA maturase coded by this sequence is needed specifically to splice out the second intron, the enzyme action creates the exquisitely sensitive negative feedback loop illustrated in **Figure 26.11.** The removal of the second intron to join the first two exons to the third exon disrupts the sequence coding for the maturase. Thus the splicing activity of RNA maturase leads to loss of the capacity to synthesize the enzyme. So an equilibrium will be achieved between the levels of the two forms of the RNA and the RNA maturase protein (which is indeed present in exceedingly small amounts and therefore has been very difficult to characterize).

Is this a general model for other introns in the mitochondrion? We lack detailed evidence, but in both the *box* and *oxi3* genes the predominant form taken by introns is to display an open reading frame in exact register with the preceding exon. This cannot be a coincidence. It makes it seem likely that these reading frames are translated.

Unlike nuclear structural genes of yeast, mitochon-

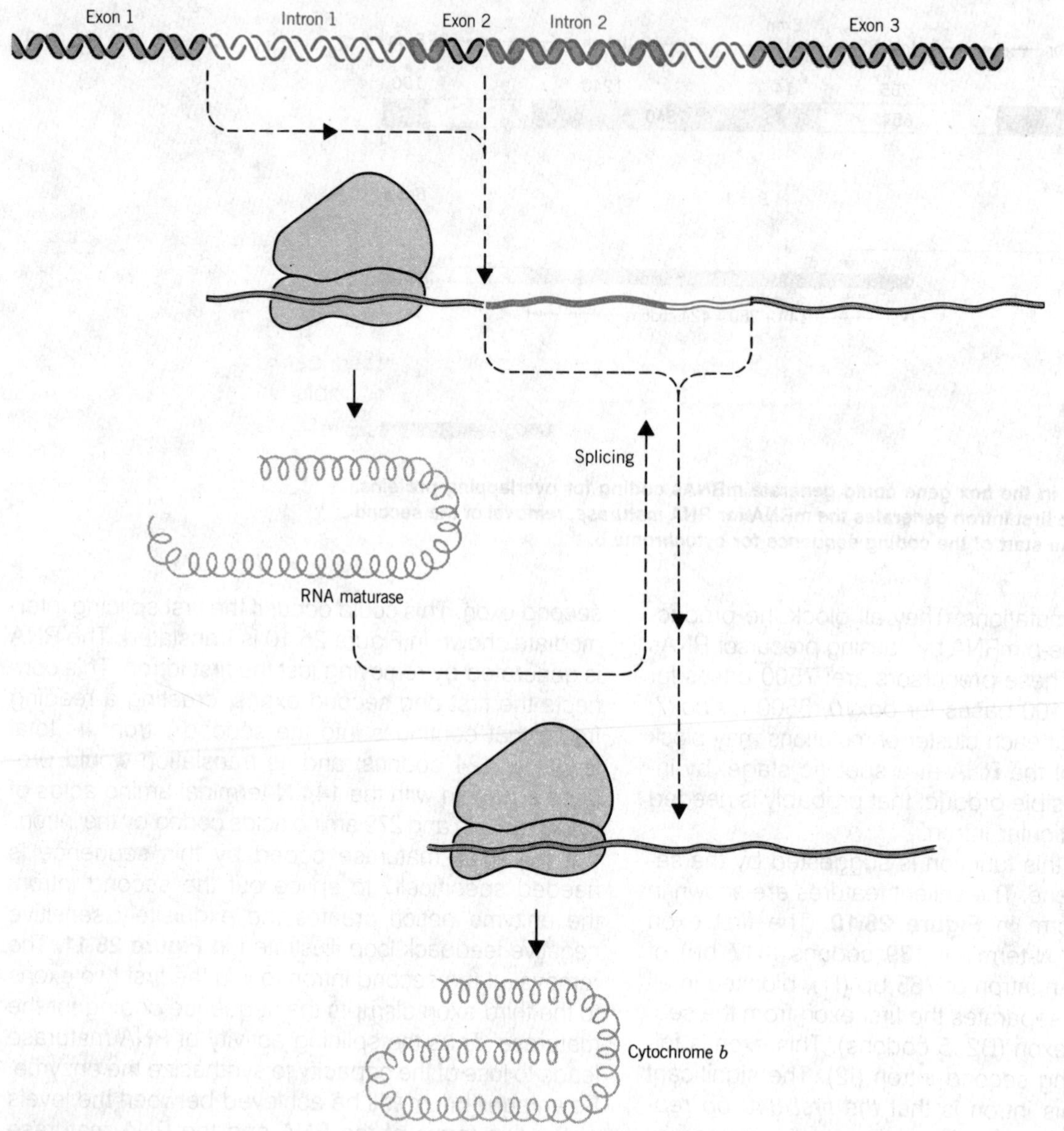

Figure 26.11

If the RNA maturase specifically splices out the intron that codes for it, a negative feedback loop will control the amounts of the protein and the two mRNAs.

Shaded regions on the DNA and RNA indicate open reading frames; regions in outline are blocked in all reading frames.

drial exon-intron junctions do not follow the GT–AG rule; and they do not display any other consensus sequence. So an RNA maturase could be specific for a particular intron or introns. The probable function of the *box3* RNA maturase is to recognize only the ends of the second intron, so that the flanking exons can be spliced together.

In another case, there is an indication that an intron-coded function may be involved in two splicing events; since *box7* mutants (in the fourth intron of the *box* gene) are defective not only in the splicing of the *box* mRNA, but also of *oxi3* mRNA. The *box7* mutations create nonsense codons in the open reading frame of the intron, so they may disrupt synthesis of another maturase, one needed to remove I4 of the *box* gene and an intron of *oxi3*.

What could be the function of the maturase? The idea that each intron reading frame codes for a maturase specific for that intron (or at least for very few introns) suggests that its role is more likely to be concerned with particular recognition events than with the catalytic activities per se. For example, the maturase could provide a "specificity subunit" of a complex in which some other component is responsible for the actual bond breaking and making events.

An intriguing connection has been noticed between the yeast mitochondrial introns and some nuclear rRNA-coding genes. They contain several rather short (~12 base) common sequences. The conserved sequences are located at a distance from the splicing junctions (which do not themselves show conservation of sequence). Some of these sequences are implicated in the splicing event at least in the *box* gene, because they provide the sites of *cis*-acting mutations that block splicing, such as *box9* and *box2* (see Figure 26.9).

In at least several of the genes that have these conserved sequences, a secondary structure can be drawn in which they are responsible for creating base paired regions. **Figure 26.12** shows in diagrammatic form that this base pairing creates a secondary structure in which the ends of the intron are brought near one another.

Figure 26.12
Base pairing between short conserved sequences could provide a framework for a secondary/tertiary structure in the intron that brings together the exons for splicing.

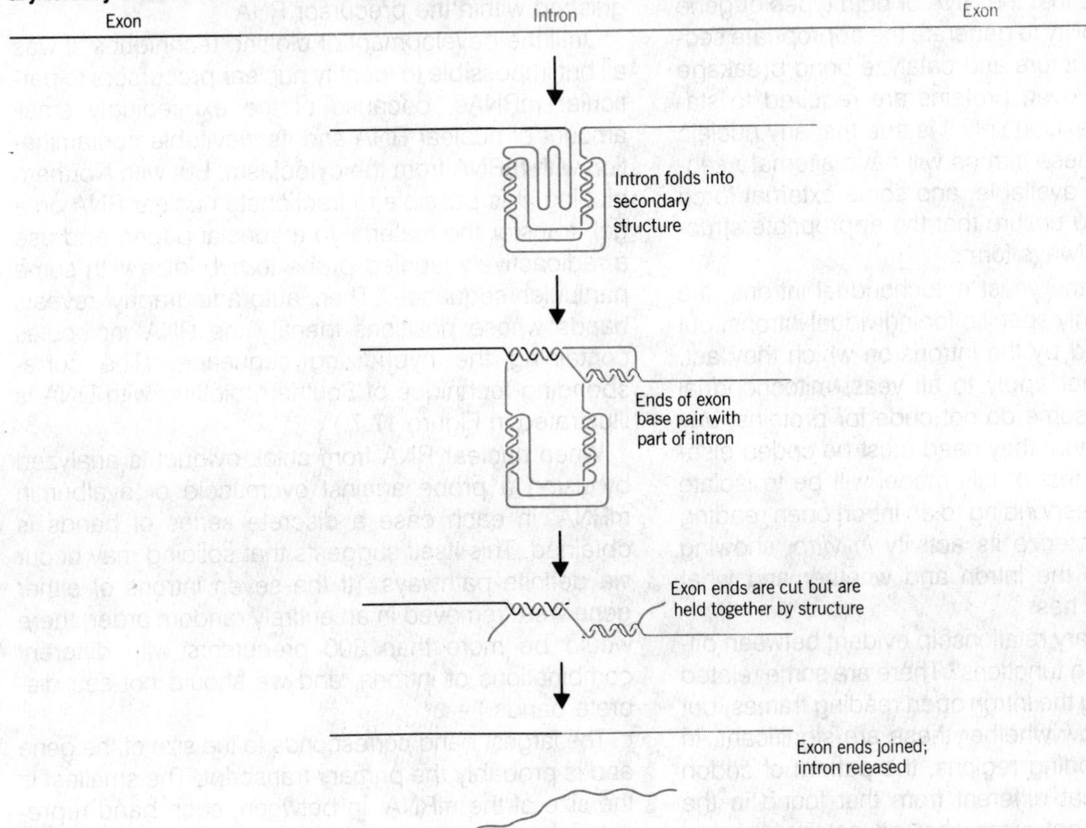

One model proposes that this proximity then enables another sequence (an "internal guide sequence") actually to base pair with the ends of the exons, holding them together while the enzymatic cutting and joining occurs.

Probably all of the yeast mitochondrial introns have these consensus sequences. Thus they are found in introns that have open coding regions and in the other introns in which all possible reading frames are blocked. Among the nuclear genes that display them is the *Tetrahymena* gene that codes for the self-splicing rRNA precursor. This situation gives a remarkable display of an unexpected evolutionary relationship, and implies that we may expect splicing of the RNAs to share a common mechanism. This conclusion poses an interesting paradox about splicing mechanisms.

On the one hand, a consensus sequence occurs in introns that code for a protein needed for splicing; on the other hand, they occur in a gene whose RNA is able to splice itself. One model to unify the mechanism of splicing argues that the RNA of both types of gene has an intrinsic ability to generate the appropriate secondary/tertiary structure and catalyze bond breakage and reunion. However, proteins are required to stabilize the structure. Certainly it is true that any nucleic acid as long as these introns will have alternate secondary structures available, and some external force may be needed to ensure that the appropriate structure for splicing always forms.

In the case of the yeast mitochondrial introns, the proteins are not only specific for individual introns, but actually are coded by the introns on which they act. (This model cannot apply to all yeast mitochondrial introns, because some do not code for proteins; any ancillary proteins that they need must be coded elsewhere.) The acid test of this model will be to isolate the maturase corresponding to an intron open reading frame and characterize its activity *in vitro*, showing where it binds to the intron and whether and what catalytic activity it has.

Is any evolutionary relationship evident between different intron coding functions? There are some related sequences among the intron open reading frames, but we do not yet know whether these are significant. In all of the intron coding regions, the pattern of codon usage is somewhat different from that found in the exons. While it is not clear what advantage accrues to the mitochondrion from maintaining such an intricate system for expressing its interrupted genes, we could speculate that at some time there was an independent protein involved in splicing the second *box* intron. Its gene might later have been translocated into the intron, giving rise to the present arrangement. Of course, it is still true that the sole purpose of the intron coding function appears to be to remove the sequence that codes for it from the mRNA.

NUCLEAR RNA SPLICING FOLLOWS PREFERRED PATHWAYS

Highly mosaic genes present some interesting problems for gene expression. A large number of introns must be removed; and the exons must be connected in the correct sequence. The reaction does not proceed sequentially along the precursor, which means that individual pairs of intron ends must be distinguished within the precursor RNA.

Until the development of blotting techniques, it was all but impossible to identify nuclear precursors to particular mRNAs, because of the exceedingly small amount of nuclear RNA and its inevitable contamination with mRNA from the cytoplasm. But with Northern blotting, it is possible to fractionate nuclear RNA on a gel, transfer the material to a special paper, and use a radioactively labeled probe to hybridize with some particular sequence. Then autoradiography reveals bands whose positions identify the RNA molecules containing the hybridizing sequence. (The corresponding technique of Southern blotting with DNA is illustrated in Figure 17.7.)

When nuclear RNA from chick oviduct is analyzed by using a probe against ovomucoid or ovalbumin mRNA, in each case a discrete series of bands is obtained. This itself suggests that splicing may occur via definite pathways. (If the seven introns of either gene were removed in an entirely random order, there would be more than 300 precursors with different combinations of introns, and we should not see discrete bands.)

The largest band corresponds to the size of the gene and is probably the primary transcript. The smallest is the size of the mRNA. In between, each band represents a particular precursor(s) from which some but

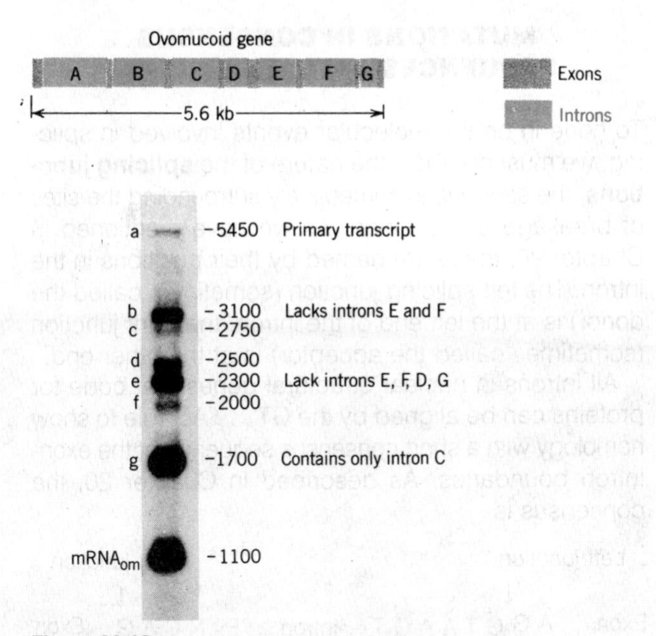

Figure 26.13

Northern blotting of nuclear RNA with an ovomucoid probe identifies discrete precursors to mRNA.

not other of the introns have been removed. **Figure 26.13** shows a Northern blot analysis of the precursors to ovomucoid mRNA, in which the most prominent intermediates are *b* (lacks introns E and F), *d–e* (lack introns E, F, D, and G), and *g* (contains only intron C).

Further information about the pathway can be gained by electron microscopic analysis of individual nuclear RNA molecules. Each molecule is hybridized with the DNA of the intact gene and then examined to see which introns have been removed. Any intron that has been lost shows up as a loop in the DNA. The results of such an analysis are summarized in **Table 26.1.** They show that there does not seem to be an *obligatory* pathway, since intermediates can be found in which different combinations of introns have been removed. However, there is evidence for a *preferred* pathway or pathways.

When only one intron has been lost, it is virtually always E or F. But either can be lost first. When two introns have been lost, E and F are again the most frequent, but there are other combinations. Intron C is never or very rarely lost at one of the first three splicing steps. From this pattern, we can see that there is a preferred pathway in which F and E are lost first, G and D are lost next, B tends to be removed next, and A and C are excised last. But clearly there are other pathways, since (for example) there are some molecules in which D or G is lost last. A caveat in interpreting these results is that we do not have proof that all these intermediates actually lead to mature mRNA, but this is a reasonable assumption.

The general conclusion suggested by this analysis is that the conformation of the RNA may influence the accessibility of the splicing junctions. As particular introns are removed, the conformation changes, and new pairs of splicing sites become available. But the ability of the precursor to remove its introns in more than one order suggests that either the molecule may take up alternative conformations, or more than one pair of splicing junctions might be available in any particular conformation.

In effect, we must deal with the situation depicted for one intron in Figure 26.12 in a molecule in which

Table 26.1

Introns are removed from ovomucoid nuclear RNA in a preferred order.

Total No. of Introns Removed	Frequency of Loss of Individual Introns						
	A	B	C	D	E	F	G
1	5	0	0	0	30	60	5
2	20	20	0	2	60	60	25
3	5	5	5	30	100	95	60
4	10	25	35	95	90	90	55
5	40	75	65	85	100	75	60
6	55	100	80	90	100	100	80

there may well be 10 or more comparable introns, each demanding a particular secondary structure. The ability of one intron to acquire its structure may depend on events occurring in other introns. The longer the molecule, the more structural options become available; and when we consider larger genes, it is difficult to see how provision is made for the imposition of specific secondary structure.

One point emerging from this type of analysis is that all the intermediates are polyadenylated. This implies that usually transcription is completed, and poly(A) is added to at least an appreciable proportion of the molecules, before any splicing occurs. In this case, splicing succeeds transcription and is not concomitant with it. However, this sequence of events is not obligatory, because splicing can continue when polyadenylation is inhibited (see Chapter 27).

We have assumed so far that each intron is excised in a single step. But this need not necessarily be so. In the mouse β-globin gene, a 15S precursor can be identified that retains only the longer intron. This may mean that the shorter intron always is removed first. But then the longer intron may be removed in two discrete stages, for an intermediate can be recovered that retains only part of it. In seeking to identify the sites involved in splicing, therefore, we must remember that although they must include the exon-intron junctions, other sites between also could be utilized in intermediate "hops."

What happens to the excised intron fragments? Hybridization experiments with the nuclear RNA population show that the concentration of ovalbumin intron sequences is about tenfold lower than the concentration of exon sequences. Probably this means that the released intron fragments are degraded rather rapidly. This also makes the point that the precursors identified on Northern blots or in the electron microscopic analysis represent the *steady-state* population of the nucleus.

Any intermediates that are more rapidly processed will be present in relatively reduced amounts. So the precursors we identify could be biased by an increased concentration of any intermediates that are processed slowly. To elucidate the splicing pathway in detail, we will need to obtain *pulse-chase* data, where a radioactive label is briefly incorporated into the primary precursor and then followed through the stages of maturation.

MUTATIONS IN CONSENSUS SEQUENCES AFFECT SPLICING

To hone in on the molecular events involved in splicing, we must consider the nature of the **splicing junctions,** the sequences immediately surrounding the sites of breakage and reunion. As we have mentioned in Chapter 20, these are named by their positions in the intron. The left splicing junction (sometimes called the donor) is at the left end of the intron; the right junction (sometimes called the acceptor) is at the other end.

All introns in nuclear structural genes that code for proteins can be aligned by the GT . . . AG rule to show homology with a short consensus sequence at the exon-intron boundaries. As described in Chapter 20, the consensus is

```
 Left junction                              Right junction
          ↓                                        ↓
Exon. . . A G G T A A G T . . Intron . . Py N C A G . . Exon
```

The arrows identify bonds whose splicing would convert the precursor to the mature RNA. To confirm their use it is necessary to determine the actual sites of splicing by examining the ends of the intermediates. (In many cases, a sequence repetition means that the splicing sites cannot be identified unequivocally on the basis of sequence alone; in the example of the consensus itself, the arrows could be moved one or two bases to the left at each junction.)

In scrutinizing the sequences of splicing junctions, these short consensus sequences are the only common features that are apparent. They describe the exon-intron boundaries of an extremely wide range of species, from animals to yeast. But there is no more extensive conservation to be discerned even when we consider particular groups of introns (such as those in a species, those in a gene family, etc.).

Can such short sequences provide for the specificity of splicing? We know that at least in some cases, no extensive sequence is needed at the boundary. For example, a deletion in SV40 that removes almost all sequences upstream from a left consensus sequence does not prevent its use in splicing. We need further experiments of this sort to see whether "splicing junction" is synonymous with "consensus sequence." We do know that the junctions do not always function autonomously. A short deletion in SV40 some 200 bp distant from a left splicing junction may prevent its use (presumably via the introduction of conformational changes in the RNA). Nuclear introns may therefore have *cis*-acting sites involved in splicing, analogous to those of the yeast mitochondrial introns.

Our implication of the consensus sequences as the sites that are recognized in splicing is really faute de mieux. But their importance is shown directly by the existence of some point mutations that affect consensus sequences and influence splicing. These have been found in the form of human thalassemias, in which the production of either α- or β-globin protein is prevented. (There are many causes of thalassemia, some of which we have reviewed in Chapter 21.)

In an α-thalassemia, the production of mRNA from the α2 gene is prevented by a mutation that deletes a pentanucleotide sequence from the left junction of intron *1*. No stable RNA product is found.

A β^0-thalassemia, in which there is no production of β globin, is produced by a point mutation in a consensus sequence. This changes the GT at the left junction of intron *2* into an AT. Presumably this prevents the production of mRNA.

In β^+-thalassemia, there is a large reduction in the *amount* of β-globin mRNA and protein synthesis. But the mRNA and protein that are made appear to be normal. Some β^+-thalassemic genes have been cloned to identify the basis for the defect.

In one case, there is a single point mutation, located 21 bp before the right end of the small (first) intron. This changes the wild-type sequence from a G into an A, creating the mutant sequence

Wild type G (↓) — Intron/exon (↓)

C C T A *T T A G T C T A T T T T C C C A C C C T T A G* G C T

The sequence surrounding the mutant site is almost identical with the sequence that surrounds the splicing junctions—the mutation *creates* homology at 6 out of 7 positions. (And the base that is different is at a position whose member varies at authentic splicing junctions in other β-like globin genes.) Note that the 7 base sequence extends to the right of the splicing site, a region not included in the consensus sequence at the right junctions.

The result of this resemblance is that splicing takes place at the site of mutation instead of at the proper junction. This happens both in human patients with thalassemia, and when a cloned copy of the gene is expressed in either monkey or human (tissue culture) cells. Up to 90% of the precursor RNA molecules are spliced at the mutant instead of at the authentic splicing site. The small amount of splicing at the proper site generates the active β-globin mRNA that is present in the patients. We do not know why the mutant site is used more efficiently; it could be because it lies closer to the left junction, or because of some feature of the conformation of the RNA.

The overall efficiency of splicing is reduced with the mutant RNA, and there is also an accumulation in the nucleus of entirely unspliced precursor RNA. This implies that failure to splice intron *1* may cause a retardation in the splicing of intron *2*. This type of effect could be mediated by secondary structure.

What happens to the misspliced RNA? This is not entirely clear and may depend on circumstance. In the thalassemic individual, it does not accumulate in the cytoplasm of reticulocytes. This implies that either the precursor or the misspliced product is unstable, or there is a deficiency in transporting it from nucleus to cytoplasm. But in tissue culture cells, the misspliced RNA makes its way to the cytoplasm as effectively as the normal mRNA. We do not know whether it enters the polysomes, but if it does it would be translated into an aberrant and prematurely terminated protein. (The use of the mutant splicing junction introduces a nonsense codon into the reading frame.)

NUCLEAR SPLICING JUNCTIONS MAY BE INTERCHANGEABLE

Does recognition of the correct pair of left and right splicing junctions depend on the individual sequences or is it imposed by some external feature of the interaction with the splicing apparatus? In other words, is each correct pair in some way marked by a unique feature of sequence or structure, or are all left junctions functionally equivalent and all right junctions similarly indistinguishable?

From the sequences at and around the splicing junctions, we know that there is no complementarity between corresponding left and right junctions, which excludes the idea that they might be brought together directly by base pairing. This leaves the possibility that individual pairs of junction sequences might be recognized specifically by proteins or by RNA sequences. Models of this sort may apply in the yeast mitochondrion or nuclear rRNA genes, where only a few pairs of junctions are involved, but seem harder to visualize for the much larger numbers in the nucleus.

An associated question is whether there is any tissue specificity in RNA splicing. Is the mechanism of splicing the same for all genes, irrespective of circumstance, so that recognition of the proper junctions depends only on the RNA and the common splicing apparatus? Or is the capacity to splice a particular nuclear RNA present only in the cells in which it is usually expressed? These issues can be explored by studying the splicing of authentic RNA precursors in novel cellular situations, and by constructing synthetic genes with new combinations of splicing junctions.

The same techniques that we described in Chapter 17 for cloning foreign sequences in bacterial phages or plasmids can be used to insert genes into eukaryotic viruses. By making appropriate restriction cleavages and then rejoining the fragments, globin, insulin, and other genes from a variety of sources have been inserted into eukaryotic viral vectors. One of the most common vectors is the monkey virus SV40, which can be perpetuated in the cells of several mammals. The new genes can be inserted into either the early or late

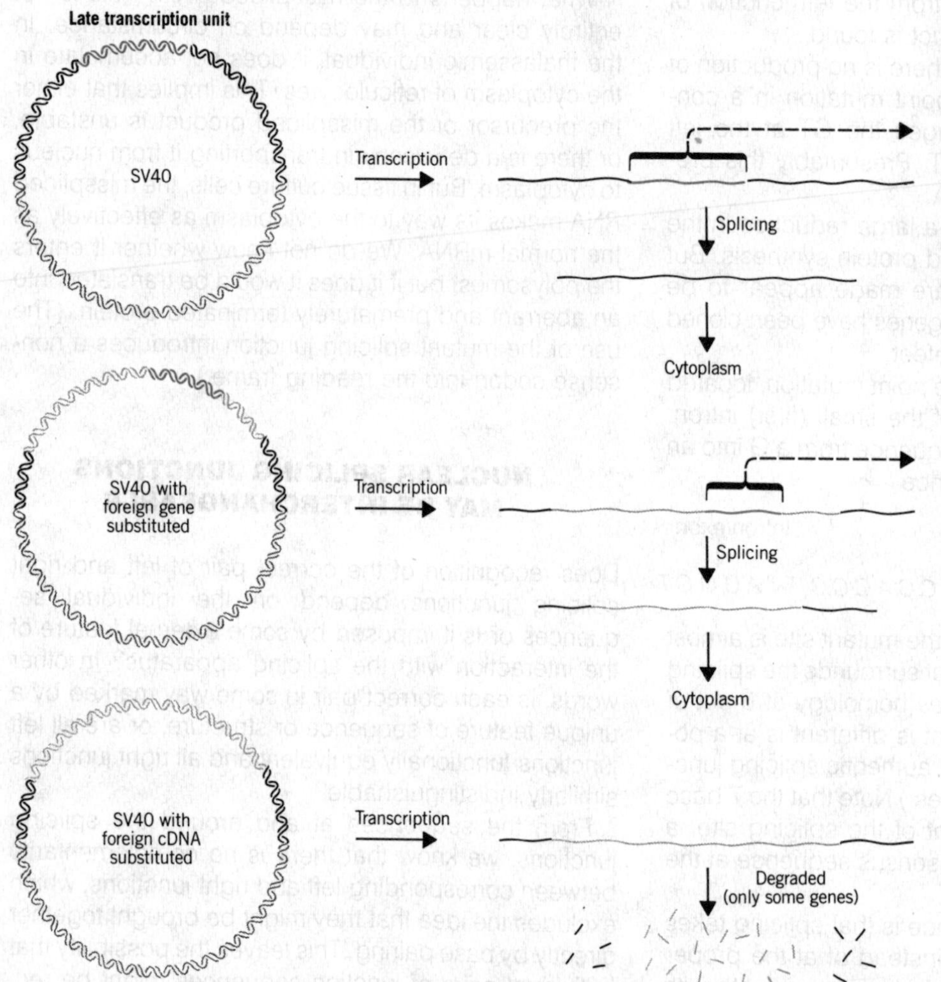

Figure 26.14
SV40 vectors carrying foreign interrupted genes in place of the late transcription unit are transcribed, spliced normally, and the product exported to the cytoplasm. When the foreign insert is an uninterrupted cDNA representing the gene, it is transcribed but in some cases the product is degraded in the nucleus.

transcription unit of the virus. They are expressed via the formation of a precursor RNA that usually contains viral sequences at the 5′ terminus followed by the foreign gene. Such RNAs can be spliced normally, as indicated in the example of **Figure 26.14.**

This provides two important pieces of information. First, correct splicing does not depend on the integrity of the natural precursor RNA, because a foreign gene (or sometimes just part of it) can be properly processed in the context of viral sequences. Second, the information carried in the sequence is adequate to support splicing in cells derived from a different tissue and/or species from the usual place of expression. The signals involved in splicing must therefore be highly conserved.

Is splicing *necessary* for the production of mRNA from interrupted genes? (We know that any requirement can be bypassed by mRNA from naturally uninterrupted genes.) The use of SV40 vectors allows direct comparison of the expression of interrupted and uninterrupted versions of the same gene. An interrupted gene is provided by a genomic fragment; an uninterrupted gene is created by the cDNA that represents the mRNA.

In these cases, when a transcription unit lacks any splicing junctions, it is transcribed normally. But in some cases (for example, globin), the resulting RNA is unstable, as illustrated in Figure 26.14. This suggests that the transcripts of certain interrupted nuclear genes may have either of two fates: to be spliced and then transported to the cytoplasm; or to be degraded in the nucleus. This effect is not seen for other genes, which can be expressed equally well in either interrupted or uninterrupted form.

The reconstruction approach has been taken a step further by producing synthetic genes in which an exon from one authentic gene is linked to an exon from a different gene. This is illustrated diagrammatically in **Figure 26.15.** In an actual experiment, the first exon of the early SV40 transcription unit was linked to the third exon of mouse β globin. The hybrid intron was spliced out perfectly. Thus the left junction of an SV40 intron (*l1* in the figure) can be spliced to the right junction of a mouse β-globin intron (*r2* in the figure). This implies that in principle any left splicing junction may be able to react with any right splicing junction.

The paradox between this conclusion and the usual circumstances in which splicing occurs only between

Figure 26.15
Splicing occurs normally in a transcript of a unit constructed by linking the exons of two different genes.

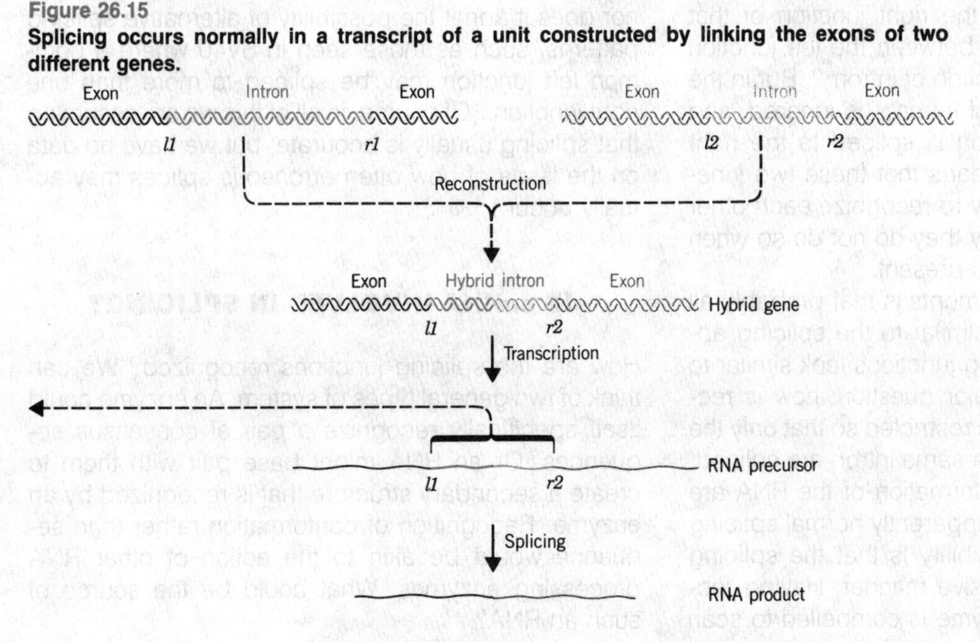

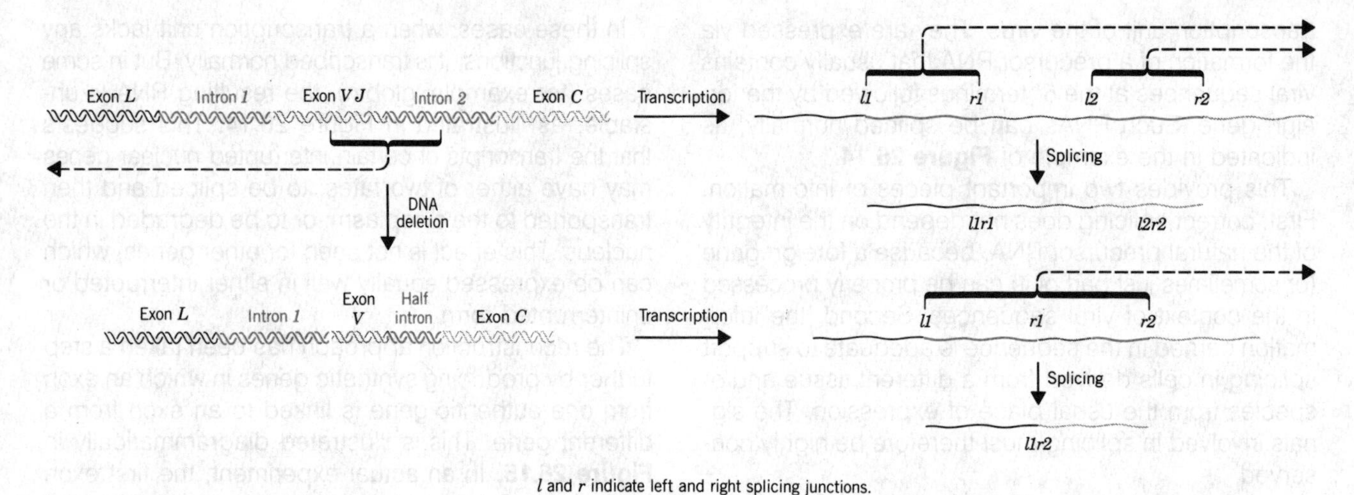

Figure 26.16

Exons *L* and *C* are spliced together when the left junction of intron 2 is deleted but not when it is intact.

the left and right junctions of the *same* intron is highlighted by the situation summarized in **Figure 26.16.** This compares the expression of a wild-type immunoglobulin gene with that of a mutant in which a deletion has removed the region including the junction between exon *VJ* and intron *2*.

In the wild type, the left junction of each intron is spliced together only with the right junction of that intron. There is no reaction between the left junction of intron *1* and the right junction of intron *2*. But in the deletion, the right junction of intron *1* is ignored, and the left junction of this intron is spliced to the right junction of intron *2*. This means that these two junctions have an inherent ability to recognize each other for splicing, although usually they do not do so when the left junction of intron *2* is present.

The moral of these experiments is that probably all left splicing junctions look similar to the splicing apparatus, and all right splicing junctions look similar to it. This leaves us with a major question: how is recognition of splicing junctions restricted so that only the left and right junctions of the same intron are spliced. Models that invoke the conformation of the RNA are rendered less likely by the apparently normal splicing of hybrid genes. One possibility is that the splicing apparatus acts in a processive manner. Having recognized a junction, the enzyme is compelled to scan the RNA in the appropriate direction until it meets a junction of the other type. This would restrict splicing to adjacent junctions.

This model could explain those cases in which the junctions are uniquely defined and interact directly (although some introns are very long indeed). It does not explain the use of multiple steps in splicing an intron, nor does it admit the possibility of alternative splicing patterns, such as those seen in SV40 where a common left junction may be spliced to more than one right junction. (Of course, in all of this we are assuming that splicing usually is accurate; but we have no data on the issue of how often erroneous splices may actually occur.)

IS snRNA INVOLVED IN SPLICING?

How are the splicing junctions recognized? We can think of two general types of system. An enzyme could itself specifically recognize a pair of consensus sequences. Or an RNA might base pair with them to create a secondary structure that is recognized by an enzyme. Recognition of conformation rather than sequence would be akin to the action of other RNA processing enzymes. What could be the source of such an RNA?

It could be a sequence located elsewhere in the same transcript, involving a mechanism analogous to that suggested for yeast mitochondrial introns and nuclear rRNA genes. An example is provided by yeast nuclear genes. All their known introns contain a short conserved sequence, located not far upstream of the right splicing junction. Mutations or deletions of this sequence prevent splicing; the introduction of a synthetic sequence within an intron introduces aberrations in splicing. The conserved consensus sequence is complementary to the left splicing consensus at 5 bases:

left splicing junction	5′	A G G T A A G T 3′
internal conserved consensus	3′	A C A A T C A T 5′

Read in the usual direction, the internal sequence is sometimes called the **TACTAAC box.** It is tempting to speculate that it pairs with a left splicing junction, thereby ensuring that the adjacent right splicing junction is utilized.

An alternative idea is that an independent RNA might base pair with the unspliced precursor to bring the intron ends together. A ready candidate for this role is available. Both the nucleus and cytoplasm of eukaryotic cells contain many discrete small RNA species. They range in size from 100 to 300 bases, and vary in abundance from about 10^5 to 10^6 molecules per cell. Some are synthesized by RNA polymerase III, others by RNA polymerase II, among which some of the products are capped like mRNAs.

Those restricted to the nucleus are called **small nuclear RNAs** (**snRNA**); those found in the cytoplasm are called **small cytoplasmic RNAs** (**scRNA**). In their natural state, they exist as ribonucleoprotein particles (snRNP and scRNP). Colloquially, these are sometimes known as *snurps* and *scyrps*. Some sequence relationships are found among the various snRNAs; and the reactions of the snRNPs with antibodies suggest that some of them may share common protein subunits.

The 5′-terminal sequence of one of these species, U1 snRNA, is complementary to the consensus sequences at the splicing junctions. The U1 snRNA is present, with only modest variations of sequence, in animal, bird, and insect cells. The human U1 snRNP contains 8 proteins as well as the RNA. The probable secondary structure of the human U1 snRNA is drawn in **Figure 26.17.** Note that the 5′-terminal 11 nucleotides are single-stranded and therefore in principle available for reaction.

A model for the base pairing of this region with the splicing junctions is illustrated in **Figure 26.18.** The upper part depicts the overall structure of the intron vis à vis the U1 snRNA. The lower part shows the pattern of base pairing between the U1 snRNA sequence and the sequences at the ends of a typical intron. The exact pattern of pairing is difficult to predict, because the sequence UCC in the U1 snRNA could pair with the AGG found at either of the exon-intron junctions.

The base pairing is not extensive, and we do not know whether it would by itself be adequate to bring the splicing junctions together. It is shown at its best in the figure. In actual junctions, there seem always to be 4–6 bp able to form at the left junction site, but there may be only 2 instead of 3 at the right junction site.

The intact U1 snRNP particle can bind *in vitro* to a left splicing junction. Binding is a property of the entire particle; purified U1 RNA cannot bind. There is no evidence yet on whether U1 snRNP can bind to right splicing junctions.

In considering the role that U1 binding might play in splicing, we must remember that the specificity of the base-pairing reaction may depend on the environment. Thus we would not a priori predict that the triplet codon-anticodon reaction would bring sufficient specificity to translation, but it does so within the ribosomal environment. Could the specificity of the U1 snRNA•nuclear RNA interaction similarly rest on the environment provided by protein components?

Some indirect evidence that U1 snRNA is involved in splicing is provided by antibodies obtained from patients with the autoimmune disease **lupus erythematosus.** Such patients have a variety of reactions in which their antibodies react against normal components of the human cell; among them it is common to find antibodies directed against nucleic acids. In some cases, the target is one or several of the snRNP types. The reaction is directed against the ribonucleoprotein particle, and probably recognizes a protein subunit or subunits.

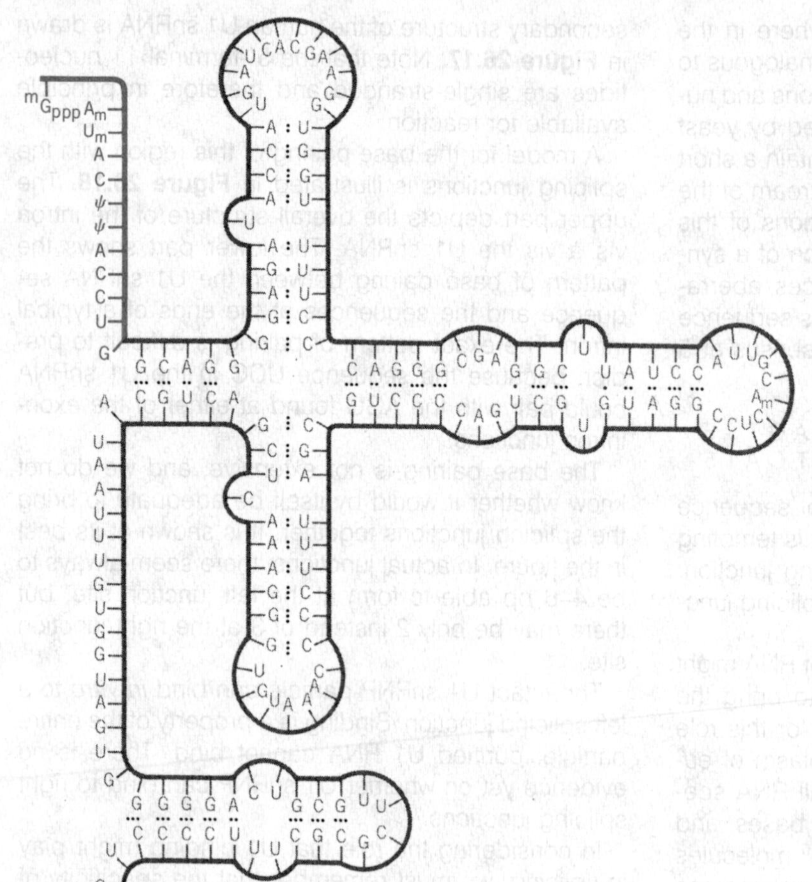

Figure 26.17
Human U1 snRNA has a secondary structure in which the 5′-terminal end is single-stranded. Other species have similar sequences and structures.

Nuclei isolated from HeLa cells infected with adenovirus are able to continue splicing the viral RNAs. When these nuclei are incubated with the anti-snRNP antibodies, the splicing reaction is halted. The inhibition seems to be specific to splicing, and does not represent a general interference with gene expression, because other stages (such as transcription or polyadenylation) continue as normal. Only antibodies that react with U1 snRNP are effective; other snRNPs are not involved.

Some other evidence for a possible connection between U1 snRNA and nuclear RNA is provided by experiments with psoralens. These are cross-linking reagents that act specifically on base-paired regions of nucleic acids. The cross-links can be reversed by irradiation with ultraviolet light. When nuclear RNP preparations are treated with psoralen, and then the links are reversed with ultraviolet, some U1 snRNA is released. This implies that it was trapped with the nuclear RNA because it was base paired to it. Only a very small proportion of the U1 snRNA in the nucleus participates in this reaction.

Is U1 snRNA the only possible RNA assistant in splicing or are there others? Some viruses code for small RNA molecules. One of these is adenovirus, which produces the VAI RNA. It is complementary to the splicing junctions of some adenovirus genes; it could be involved in their recognition. If small, diffusible RNAs are involved in creating structures that are recognized for splicing, the production of specific RNA assistants could be needed to accomplish particular splicing events.

A decisive account of the components involved in splicing will not be possible until we have available *in*

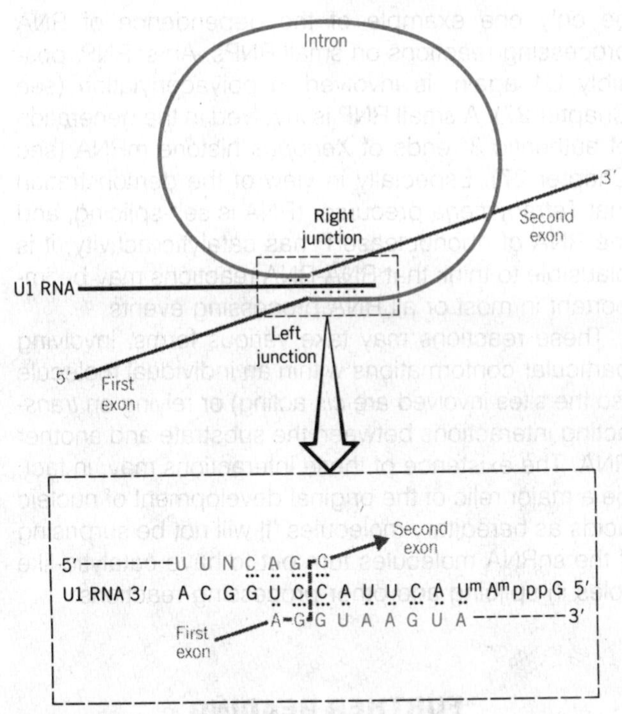

Figure 26.18

The 5′ end of U1 snRNA could base pair with the left splicing junction and (perhaps) also with the right splicing junction.

vitro systems in which introns can be removed from RNA precursors. Substantial progress has been made recently in developing such systems. The early results suggest that there may be some mechanistic similarities with other splicing systems, and they tend to confirm our suspicions about the involvement of U1 RNA.

Nuclear extracts from HeLa cells have been tested with adenovirus precursor RNAs. One such system requires Mg^{2+} ions, ATP, and is stimulated by a crude protein fraction, whose active component purifies with small ribonucleoprotein particles. The ability of this system to function with a purified RNA precursor shows that the action of splicing is not linked to the process of transcription.

Nor is splicing obligatorily linked to modification of RNA: Globin RNAs that lack poly(A) tails are spliced perfectly well; although capped RNAs are spliced rather more efficiently than noncapped RNAs, there is no other difference in their reaction. Precursor RNAs derived from thalassemic patients use the same abnormal splicing sites *in vitro* that are recognized *in vivo*, which opens a route towards investigating the detailed sequence requirements of the reaction. This result suggests that it may be possible to introduce other mutations in the precursor RNA to determine the roles of particular sites.

Splicing *in vitro* involves distinct cutting and ligation reactions. An intermediate with a curious structure has been identified using adenovirus or globin substrates; it may represent the common pathway of nuclear splicing.

This splicing pathway is illustrated in **Figure 26.19.** A cut is made at the left end of an intron, releasing as separate RNA molecules the left exon and the right intron-exon molecule. Presumably the two RNA species are held together by proteins involved in splicing.

Figure 26.19

Splicing *in vitro* involves cutting at the left junction, followed by formation of a lariat in which the left end of the intron is joined by a 5′-2′ bond to a site near to the right end of the intron. Subsequent stages involve cutting at the right junction and covalent linkage between the exons. The excised intron is "debranched."

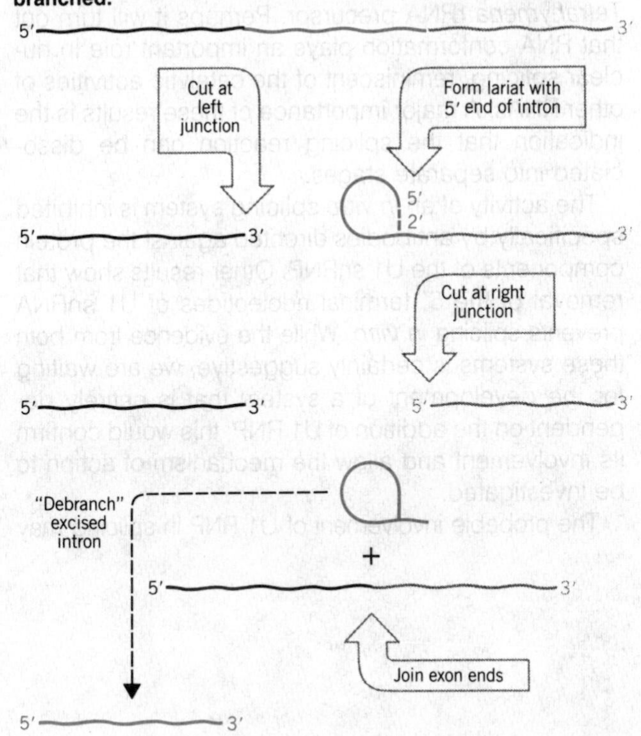

The left exon is released as a linear molecule, but the intron-exon molecule is not linear. The 5′ terminus generated at the left end of the intron becomes linked by a 5′–2′ bond to the A of the sequence CTG AC, located about 30 bases upstream of the right end of the intron. This target sequence is related to the last five bases of the TACTAAC box of yeast nuclear introns. Its involvement in splicing may be a central part of the mechanism that removes the intron. The reaction generates a "lariat" or "panhandle" molecule.

Cutting at the right splicing junction then releases the intron as a lariat and the right exon as a linear molecule. The lariat is "debranched" to give a linear excised intron. The left and right exons are ligated together.

Why should splicing proceed through a structure involving the end of an intron that is to be discarded and a site within the intron? Perhaps it is necessary to sequester the free 5′ end of the intron so that it does not impede the reaction. Perhaps it is a "marker" indicating that splicing is occurring in this intron.

The reaction is interesting because of the evident parallels with involvement of the 2′ position in tRNA splicing and the sequestration of the 5′ end of the *Tetrahymena* rRNA precursor. Perhaps it will turn out that RNA conformation plays an important role in nuclear splicing, reminiscent of the catalytic activities of other RNAs. A major importance of these results is the indication that the splicing reaction can be dissociated into separate stages.

The activity of an *in vitro* splicing system is inhibited specifically by antibodies directed against the protein components of the U1 snRNP. Other results show that removal of the 5′ terminal nucleotides of U1 snRNA prevents splicing *in vitro.* While the evidence from both these systems is certainly suggestive, we are waiting for the development of a system that is entirely dependent on the addition of U1 RNP; this would confirm its involvement and allow the mechanism of action to be investigated.

The probable involvement of U1 RNP in splicing may be only one example of the dependence of RNA processing reactions on small RNPs. An snRNP, possibly U1 again, is involved in polyadenylation (see Chapter 27). A small RNP is involved in the generation of authentic 3′ ends of *Xenopus* histone mRNA (see Chapter 27). Especially in view of the demonstration that *Tetrahymena* precursor rRNA is self-splicing, and the RNA of ribonuclease P has catalytic activity, it is plausible to think that RNA-RNA reactions may be important in most or all RNA processing events.

These reactions may take various forms, involving particular conformations within an individual molecule (so the sites involved are *cis*-acting) or relying on *trans*-acting interactions between the substrate and another RNA. The existence of these interactions may, in fact, be a major relic of the original development of nucleic acids as hereditary molecules. It will not be surprising if the snRNA molecules turn out to have catalytic-like roles in splicing and other processing reactions.

FURTHER READING

Analyses of splicing have perhaps produced more surprises and revealed more fundamental biochemical features than any other topic in recent memory. The flavor of these discoveries is by far best obtained from the original research reports. The *in vitro* autocatalytic system from *Tetrahymena* was discovered by **Cech et al.** (*Cell* **27,** 487–496, 1981). Catalytic activity of RNA was discovered by **Guerrier-Takada et al.** (*Cell* **35,** 849–857, 1983). A Lariat structure was reported by **Ruskin et al.** (*Cell* **38,** 317–331, 1984). There have been several short reviews in *Cell* devoted to the issue of splicing. Enzymatic mechanisms were reviewed by **Cech** (**34,** 713–716, 1983); the view of RNA as catalyst was extended by **Altman** (**36,** 237–239, 1984). The unique features of organelle splicing were emphasized by **Lewin** (**22,** 645–646, 1980). General mechanisms involving consensus sequences have been addressed by **Lewin** (**22,** 324–326, 1980) and **Sharp** (**23,** 643–646, 1981). The structures and functions of snRNAs have been addressed by **Zieve** (**25,** 296–297, 1981) and by **Lerner & Steitz** (**25,** 298–300, 1981).

CHAPTER 27
CONTROL OF RNA PROCESSING

Protein-coding structural genes in the nucleus of a higher eukaryotic cell are transcribed in the nucleoplasm. But the nucleoplasmic RNA is not like mRNA. Its average size is much larger, it is very unstable, and it has a much greater sequence complexity. Taking its name from its very broad size distribution, it is called **heterogeneous nuclear RNA (hnRNA)**.

Before the discovery of interrupted genes, there was a heated debate as to whether mRNA is derived from hnRNA via a size reduction, or whether the hnRNA might serve some other purpose altogether. Now some of the discrepancies between hnRNA and mRNA can be reconciled by invoking the reduction in size and complexity that occurs via RNA splicing. Many protein-coding structural genes are much longer than their mRNAs, and the primary transcript must be at least as long as the known gene (it could be longer if it included additional flanking sequences).

But we shall see that its inclusion of introns as well as exons does not entirely explain the high instability of the hnRNA. On average, only a small proportion of the transcripts of each gene actually yield an mRNA product. And in some cases, there are transcription units whose products are found in the nucleus, but that do not seem at all to give rise to cytoplasmic mRNA. This prompts the question of whether the expression of some genes might be controlled at the level of hnRNA, by virtue of deciding whether or not the processing pathway should produce mRNA from a particular nuclear RNA.

hnRNA IS LARGE AND UNSTABLE

Transcription of genes in the nucleoplasm represents only a small part of the total RNA-synthesizing activity of the cell. The intensive transcription of rRNA genes in the nucleolus is responsible for most (up to 90%) of the synthesis of RNA. In fact, this makes it difficult to follow the synthesis of hnRNA en masse, because its small proportion is hard to distinguish from the rRNA. However, this can be accomplished by scrutinizing only part of the hnRNA population or by using special circumstances.

One option is to examine the very largest hnRNA fractions (which are distinct from rRNA precursors because they sediment more rapidly than 45S). Or it is possible to inhibit rRNA synthesis preferentially with actinomycin. Another technique is to follow just for a short time the fate of a pulse (i.e., very brief) radioactive label, which enters hnRNA more rapidly than rRNA. None of these conditions is entirely satisfactory, but they are adequate to outline the general features of nucleoplasmic transcription.

Two prominent features are the size and instability of hnRNA. Most of a radioactive label incorporated into hnRNA actually turns over in the nucleus. It never even enters the cytoplasm. The lifetime of hnRNA varies somewhat in different cell types. There may be only one, or there may be two, components with regard to the decay rate; but the half-lives generally lie in the range of from a few minutes to about an hour or so.

The nuclear turnover demonstrates that *most of the mass of hnRNA is synthesized and degraded entirely within the nucleus*. This does not make any implication about the fate of *individual* molecules or parts of molecules, so from these data per se we do not know whether the nuclear-degraded material represents only introns or includes entire molecules of hnRNA.

One of the difficulties in determining the true size of hnRNA is its tendency to aggregate. There is some secondary structure within the hnRNA, a large part of which represents reactions between inverted repeats of a single family, the Alu family or its equivalent (see Chapter 36). But there are also *inter*molecular reactions; probably these are responsible for forming the very rapidly sedimenting material.

The size distribution of hnRNA is compared with the average size of mRNA in **Figure 27.1.** The histogram measures the lengths of hnRNA molecules after spreading for electron microscopy under denaturing conditions (which should prevent aggregation between molecules). The distribution probably underestimates the lengths of the primary transcripts, for two reasons. Any breakage in a long molecule during its isolation will generate two shorter molecules. And some of the molecules could be partially processed intermediates (remember the series of ovomucoid precursors in Figure 26.13, all of which are present in the steady-state hnRNA population).

But even with these caveats, it is clear that most hnRNA is much longer than mRNA. Based on the analysis of pulse labels incorporated into nascent transcripts, it seems now that the size distribution of

Figure 27.1
hnRNA has a broad size distribution extending from about the size of mRNA to more than 14,000 bases.
The very short (<1500 base) hnRNA molecules are probably mostly the result of breakage during preparation.

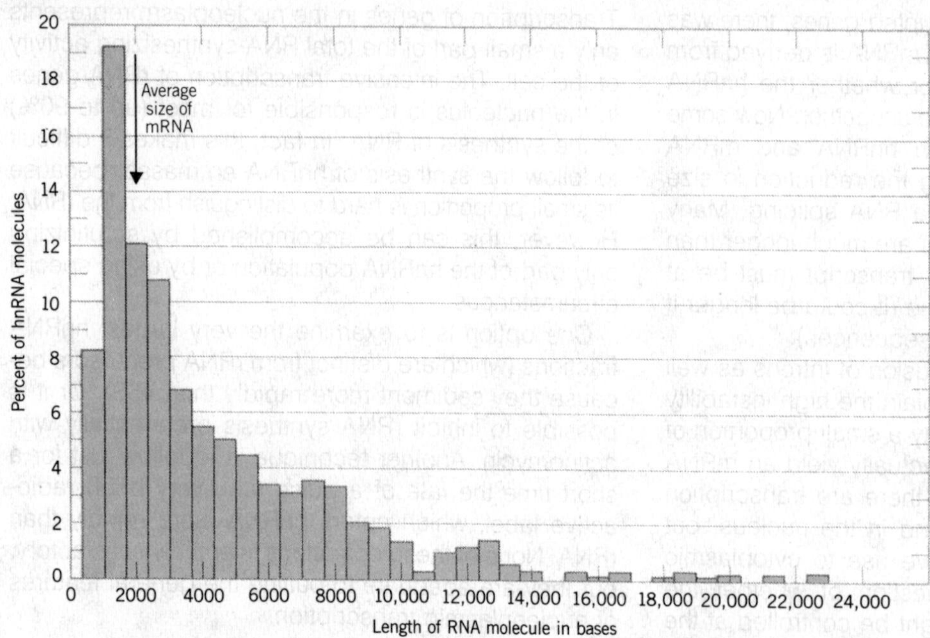

(mammalian) transcription units has an average in the range of 8000–10,000 bp. The overall distribution extends from ~2000 to ~14,000 bp. So the average hnRNA molecule is 4–5 times larger than the average mRNA of 1800–2000 bases.

The range of lengths suggests at once that hnRNA is rather long to exist in the form of an extended RNA chain within the nucleus. In fact, it is found in the form of a ribonucleoprotein (hnRNP), in which the hnRNA is associated with several proteins. They appear to comprise a more complex set than those associated with mRNA. The mass of protein in the hnRNP is about four times the mass of the hnRNA.

The structure of the hnRNP is unclear. The ribonucleoprotein particle begins to form as soon as the RNA is transcribed—there is probably little free RNA even at the beginning of transcription. The hnRNP may fold into a fiber which in turns folds itself into a thicker fiber; there may be several levels of such packing.

The hnRNP is associated with the **nuclear matrix.** This is a dense fibrillar network that lies on the nuclear side of the membrane. It forms a sort of inner shell around the nucleus and also extends into the interior. **Figure 27.2** is an electron micrograph showing the nuclear matrix within the remnants of the surrounding cytoplasm. The structure of the fibrils is not yet understood, although we do know that they are composed of a large number of proteins.

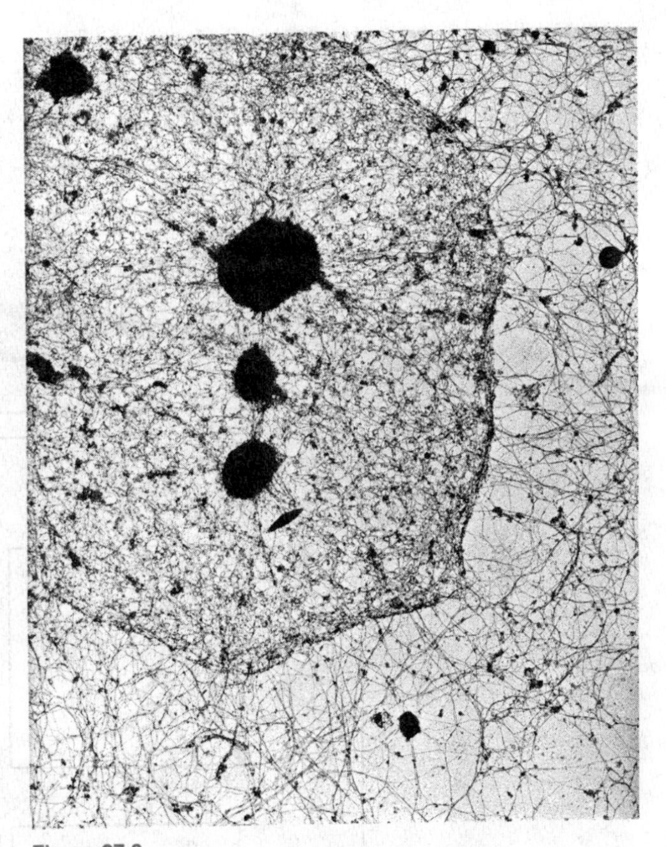

Figure 27.2
The nuclear matrix consists of a fibrillar network.
A HeLa cell was first subjected to gentle extraction with a nonionic detergent to remove membranous material, and then treatment with DNAase and high salt concentration was used to remove chromatin and some of the cytoskeleton (the fibers of the surrounding cytoplasm). The small ring structures at the surface of the nuclear matrix may be the nuclear pores. The large dark objects are the nucleoli. Photograph kindly provided by David Capco and Sheldon Penman.

The chromatin itself is intermittently attached to the nuclear matrix, and it is likely that the primary transcripts individually become attached soon after or even during their transcription. In fact, their processing may take place on the matrix. We do not know whether this location is essential in the sense that processing can occur *only* here. But in thinking about the events involved in RNA maturation, we should remember that the process may be complex topographically and is not accomplished simply by a set of "soluble" enzymes floating around the interior of the nucleus.

We have virtually no information on how mature mRNA is exported to the cytoplasm. It is possible that its emigration could be linked spatially to processing. As for the route of export, around the nucleus is a series of "pores" that could provide a route for RNA transport, but we do not know whether in fact they are used for this purpose.

mRNA IS DERIVED FROM hnRNA

The ends of some hnRNA molecules carry the same modifications that are seen in mRNA. The 5′ end gains a (partially) methylated cap; the 3′ end is polyadenylated. The similarities of these end modifications suggest that hnRNA and mRNA may have a precursor-product relationship. [Also, among the proteins bound to the hnRNP, the major 74,000-dalton species asso-

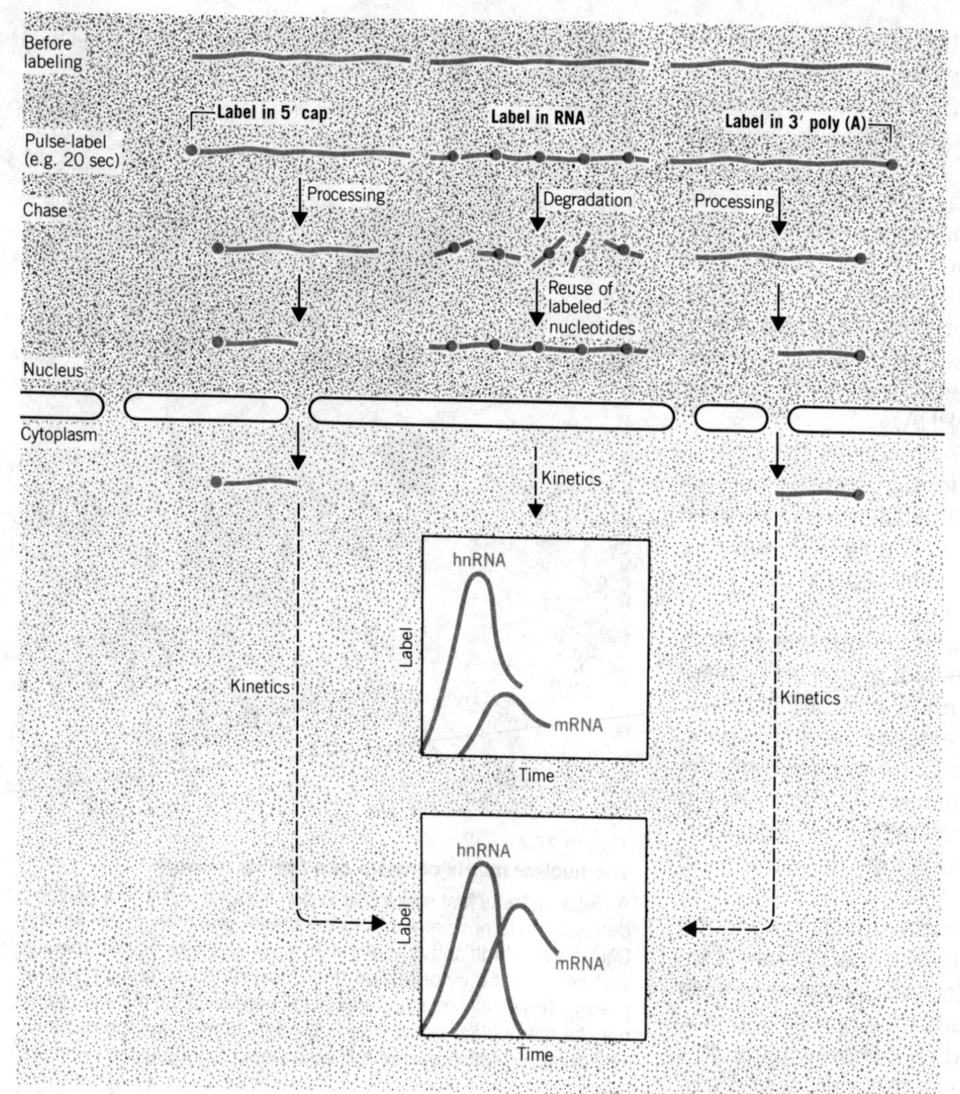

Figure 27.3
Radioactive labels in either 5′ caps or 3′ poly(A) first appear in hnRNA and then are chased into mRNA, but most of a label in the body of hnRNA is released by degradation and then reused.

ciated with poly(A) may be the same as that present in mRNA.] The paradox originally presented by the apparent identity of the extremities of hnRNA and mRNA was that the hnRNA is so much larger. It was not apparent how both ends of the molecule could find their way into mRNA. Now we see that the discrepancy is resolved by removing the introns.

Pulse-chase experiments can be used to follow the fate of the modified ends. The results are illustrated diagrammatically in **Figure 27.3.** A radioactively labeled precursor is added for a very brief period of time. This is the pulse label. Then it is removed and replaced by an unlabeled precursor. This is the chase. The amount of the label in hnRNA and in mRNA is measured at intervals of time.

For a short time after the pulse, the radioactive label

accumulates in the hnRNA. Because the hnRNA molecules are unstable, it then declines rapidly. Different results are obtained depending on whether the label was incorporated into one of the ends or into the body of the RNA.

Because the end-modifications are stable, they can be chased from hnRNA into mRNA. In fact, the principal route for the labeled ends to leave the hnRNA population is via the processing of the RNA, followed by its departure for the cytoplasm. So the decline of the curve representing the amount of label in hnRNA is accompanied by an increase in the curve representing the amount of label in mRNA. The coincidence of the decline in hnRNA and rise in mRNA demonstrates that the ends of the mRNA are derived from the ends of the hnRNA. For these parts of the molecule there is a conventional precursor-product relationship between hnRNA and mRNA.

Probably about 20% of the hnRNA is capped following its transcription. At least a large proportion of the caps can be chased from hnRNA into mRNA. Although the data do not prove that *all* the 5′ caps in hnRNA give rise to mRNA, this relationship seems plausible. Capping occurs very rapidly after the start of transcription, as witnessed by our inability to characterize original 5′-triphosphate ends on primary transcripts.

At the other end of the molecule, the length of the nuclear poly(A) is slightly greater than that found in mRNA. A few bases are removed at the stage of nucleocytoplasmic passage. There has been some discussion about whether *all* of a label in poly(A) can be chased from the nucleus to the cytoplasm. The answer is not unequivocal—it may vary in different cells and in different circumstances—but it is clear that at least the great majority of poly(A) tails added to hnRNA do find their way into mRNA.

When labeled nucleotides are incorporated into the internal regions of hnRNA, their main fate is to be released from the molecule when it is degraded. Then they are reutilized in the synthesis of further RNA, entering both rRNA and hnRNA. The result is a generalized decline in the label in hnRNA. A small proportion of the label does enter mRNA in the cytoplasm, but because of its minority status, the appearance in mRNA does not coincide with the decline in hnRNA in a way that demonstrates a conventional precursor-product relationship.

Early attempts to pulse-chase nucleotides from the body of hnRNA into mRNA thus were clouded by the nuclear turnover of the majority of the label. But improvements in the technique later made it possible to show that mRNA indeed is derived from hnRNA. From such experiments, it is possible to calculate the proportion of the *mass* of the hnRNA that is converted to mRNA. This is about 5% in mammalian cells. Given the fivefold discrepancy in average size, it corresponds to a conversion of (roughly) 25% of the hnRNA molecules.

This means that reduction in size of the transcript via loss of introns can account for the fate of only about a quarter of the hnRNA population. The rest consists of molecules that turn over *entirely* within the nucleus: no part of these molecules is used to provide cytoplasmic RNA. [The only caveat is that this analysis relies on the isolation of the poly(A)$^+$ mRNA; so some further part of the hnRNA population might give rise to poly(A)$^-$ mRNA. But although this would increase the proportion that consists of precursors, it would not change the basic conclusion.]

POLYADENYLATION AND THE GENERATION OF 3′ ENDS

Poly(A) is added to the 3′ ends of hnRNA molecules after transcription, by an enzyme, poly(A) polymerase, that uses ATP as substrate. The addition of poly(A) to hnRNA can be prevented by the analog **3′-deoxyadenosine,** also known as **cordycepin.** This is a specific inhibitor of polyadenylation; it does not stop the transcription of hnRNA. But its addition prevents the appearance of mRNA in the cytoplasm. This shows that polyadenylation is *necessary* for the maturation of mRNA from hnRNA. [This is true, of course, of the poly(A)$^+$ mRNA; the poly(A)$^-$ mRNA can bypass this requirement.]

Only some of the hnRNA is polyadenylated, about 30% in mammalian cells, compared with a value of about 70% for mRNA. There is often a rough correlation between the proportion of hnRNA molecules that is polyadenylated and the proportion that gives rise to cytoplasmic poly(A)$^+$ mRNA molecules. This suggests that polyadenylation could provide a signal that a particular hnRNA molecule is to be processed. However, although necessary, polyadenylation is unlikely itself to be sufficient; because, in this case, there would always be strict conservation of the poly(A) between nucleus and cytoplasm.

One of the important unresolved questions is the origin of the 3′ ends to which the poly(A) is added. We have virtually no information about the mechanism of termination by RNA polymerase II, so we do not know what an authentic 3′ end should look like. However, we do know that in the case of some viruses, the ends are generated by cleavage of longer precursors, as illustrated in **Figure 27.4.**

A common feature of both cellular and viral mRNAs in higher eukaryotes (but not in yeast) is the presence of a sequence AAUAAA in the region from 11 to 30 nucleotides upstream from the site of poly(A) addition. When this sequence is deleted from a viral transcription unit, cleavage and polyadenylation no longer occur at the usual site. The AAUAAA sequence is necessary, but not by itself sufficient, for the reaction. It probably provides a signal for cleavage. When the U is mutated to a G, the efficiency of cleavage is much reduced; but those molecules that are cleaved can then be polyadenylated.

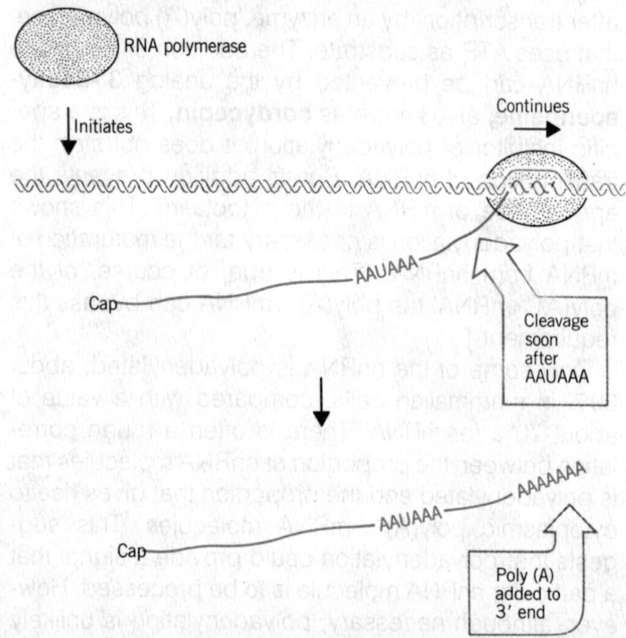

Figure 27.4

The sequence AAUAAA is necessary for cleavage and polyadenylation of SV40 mRNA.

The ubiquity of this signal provides a hint that the same mechanism—cleavage and polyadenylation—may be employed in cellular transcription units.

Polyadenylation usually precedes RNA splicing. In the case of cellular transcripts, both the full length and the various intermediate RNA precursor molecules are polyadenylated. In the case of the processing of mRNAs from the late transcript of adenovirus, kinetic studies show that the cleavage to generate a 3′ end for polyadenylation occurs *before* the tripartite leader is spliced to a messenger body. (This is the system illustrated in Figure 20.20.)

The relationship between polyadenylation and splicing is not causal mechanistically [although in those cases where alternative splicing junctions are used, cleavage of the 3′ end may control the *choice* of splicing junction (see, for example, Chapter 37).]

The development of a system in which polyadenylation occurs *in vitro* opens the route to analyzing these events. The system works only with RNA synthesized in the extract, although there is a substantial delay between the completion of transcription and the polyadenylation event. This could mean that the RNA can form an appropriate substrate structure, susceptible to processing and polyadenylation, *only* during the act of transcription, even though the actual addition of poly(A) may not occur until later.

The *in vitro* reaction is inhibited by antibodies directed against snRNPs, in particular against U1 snRNP. Is U1 snRNP involved in polyadenylation as well as splicing? Perhaps the U1 snRNP is not itself directly involved, but is part of some more complex structure that is necessary for the reaction.

An *in vitro* system has been developed that generates authentic 3′ ends in *X. laevis* histone mRNAs (which are not polyadenylated). When a cloned histone gene (the H3 gene of a sea urchin) is injected into the *Xenopus* oocyte, it is faithfully initiated and transcribed, but termination occurs at variable sites. However, when a nuclear extract from the sea urchin is simultaneously injected with the gene, the transcribed mRNA has the proper 3′ end.

The active component of the nuclear extract is a 12S factor, with a molecular mass of about 2.5×10^5. It behaves like a protein complex when fractionated. However, a small RNA from the sea urchin also can generate authentic 3′ ends; and this RNA can be isolated from the 12S fraction.

The RNA is about 60 bases long (20,000 daltons); it is not related to any known snRNA. Probably it works in the *Xenopus* oocyte because of the presence of a pool of snurp proteins, from which it is able to reassemble a ribonucleoprotein particle.

We do not yet know the stage at which this snurp acts: it could be termination or cleavage. However, the involvement of an snRNP in this process greatly strengthens the view that many—perhaps all—hnRNA processing events will prove to depend on interactions with snRNP.

hnRNA IS MORE COMPLEX THAN mRNA

Given the difference in size between hnRNA and mRNA, we expect hnRNA to have a greater sequence complexity. For those transcription units in which a primary transcript is spliced to yield a smaller mRNA, the sequence complexity of hnRNA should include both exons and introns, while the sequence complexity of mRNA should include only the introns. Unfortunately, it is difficult to measure the sequence complexity of hnRNA with as much precision as that of mRNA. First, we lack a functional test to isolate hnRNA free of other RNAs. Second, the use of hybridization kinetics is restricted by the presence in hnRNA of repetitive sequences that are interspersed with the nonrepetitive sequences forming the majority of mRNAs.

From measurements of the ability of hnRNA to saturate nonrepetitive DNA, its complexity falls in a range that varies from as little as four to as much as ten times greater than that of mRNA. Hybridization experiments confirm that the hnRNA population does indeed include all the sequences that are present in cytoplasmic mRNA. What is the nature of the additional sequences?

This question can be asked in another form. Do the hnRNA molecules that turn over entirely within the nucleus comprise transcripts of the *same* or of *different* genes as those that give rise to mRNA. The consequences are illustrated in **Figure 27.5.**

If the *same* genes are involved, in the typical mammalian cell the average gene is transcribed into hnRNA molecules, 75% of which are entirely degraded, and 25% of which donate 20% of their length (the exons) to become mRNA. Thus processing is extremely wasteful. In this case, the difference in complexities between hnRNA and mRNA is accounted for entirely by the loss of introns. In the example of mammalian cells, the difference between hnRNA and mRNA complexities is about fourfold, very similar to the size discrepancy. The data are not precise enough to say whether *all* the extra sequences in hnRNA can be ac-

Figure 27.5
Is processing of all hnRNAs wasteful as shown on the left, or are some transcripts efficiently processed while others do not give rise at all to mRNA as shown on the right?

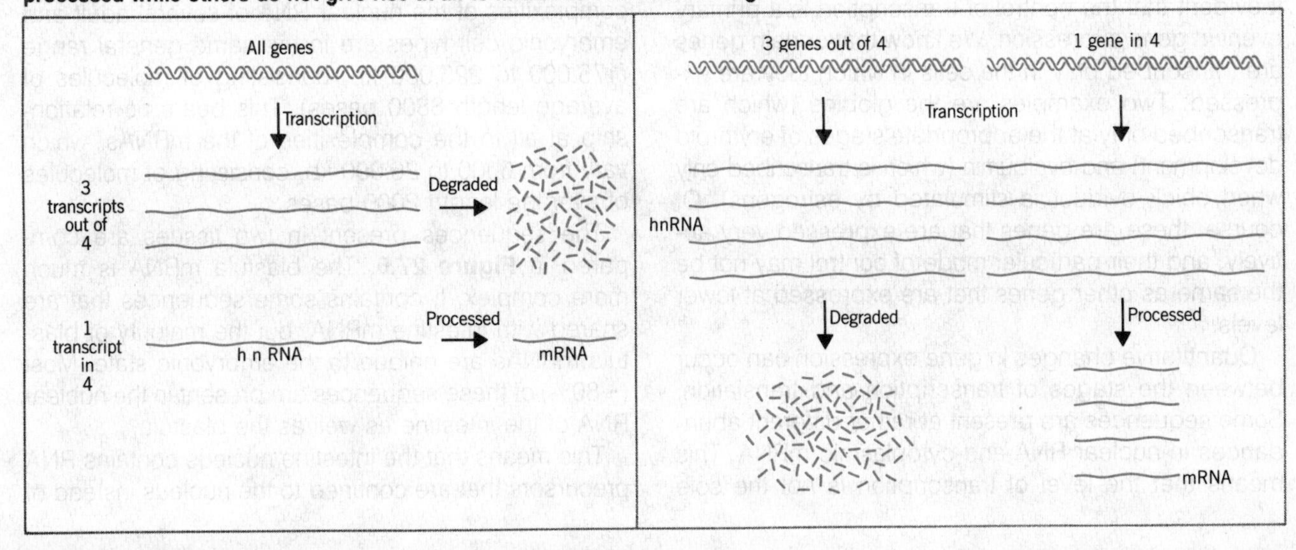

counted for by introns, but certainly it seems likely that this is the major cause.

If *different* genes give rise to hnRNA molecules that are conserved or that are degraded entirely, 75% of the transcription units must represent hnRNA molecules all copies of which are degraded: none of these donates mRNA to the cytoplasm. (The strict conclusion is that none gives rise to mRNA in this particular cell type; it could do so in another.) The other 25% of the genes give transcripts all copies of which mature to mRNA (conserving 20% of their length). In this case, the complexity of hnRNA should be much greater than that of mRNA, not only because the molecules are longer, but also because there is an entirely independent set of transcripts from those that yield mRNA. In some sea urchin tissues, where the difference in nuclear and cytoplasmic RNA complexities is tenfold, the discrepancy seems too great to be accounted for just in terms of loss of introns.

Of course, these two models are extremes, and no doubt the situation in actual nuclei lies somewhere between. It is clear, however, that either the processing of expressed genes is inefficient or some transcription units are expressed only in nuclear RNA.

IS THERE CONTROL AFTER TRANSCRIPTION?

The small proportion of the genome represented in hnRNA (say, less than 5% in mammalian cells) makes it evident that the control of transcription is a primary event in gene expression. We know that certain genes are transcribed only in the cells in which they are expressed. Two examples are the globins (which are transcribed only at the appropriate stages of erythroid development) and ovalbumin (which is transcribed only when chick oviduct is stimulated by estrogens). Of course, these are genes that are expressed very actively, and their particular mode of control may not be the same as other genes that are expressed at lower levels.

Quantitative changes in gene expression can occur between the stages of transcription and translation. Some sequences are present at rather different abundances in nuclear RNA and cytoplasmic mRNA. This means that the level of transcription is not the sole factor responsible for establishing the level of mRNA. Changes in stability or in the efficiency of processing could intervene to alter the relative abundance of an RNA sequence between nucleus and cytoplasm.

A situation in which general quantitative changes occur is provided by the transition that cells undergo in culture from the resting (that is, nondividing) to the growing state (when they are actively proceeding through the cell cycle).

Growing cells have more RNA than resting cells, and in particular they have an increased proportion of mRNA relative to rRNA. But there are no changes in the rate of transcription of hnRNA, in the proportion of molecules that are polyadenylated, or in the stability of the mRNA. What happens is that a greater proportion of the poly(A)$^+$ hnRNA is converted into poly(A)$^+$ mRNA in growing cells.

The major part of the change is quantitative rather than qualitative. Some new sequences are expressed, but for the most part, the change represents an increase in the efficiency with which a constant set of sequences in hnRNA is converted into mRNA. Put the other way, in resting cells there is a decrease in the efficiency of conversion. In the light of our previous question about the nature of the nuclear-restricted sequences, this makes it clear that, at least in resting cells, some of them do represent surplus copies of expressed sequences.

A situation in which control over the selection of sequences is exercised at the level of nuclear RNA is found in the sea urchin. In *S. purpuratus,* the total complexities of the nuclear RNA of several adult and embryonic cell types are in the same general range (175,000 to 225,000 kb, consisting of molecules of average length 8800 bases). This bears no relationship at all to the complexities of the mRNAs, which vary from 6000 to 26,000 kb, consisting of molecules of average length 2000 bases.

The sequences present in two tissues are compared in **Figure 27.6.** The blastula mRNA is much more complex. It contains some sequences that are shared with intestine mRNA, but the majority of blastula mRNAs are unique to the embryonic state. Most (~80%) of these sequences are present in the nuclear RNA of the intestine as well as the blastula.

This means that the intestine nucleus contains RNA precursors that are confined to the nucleus instead of

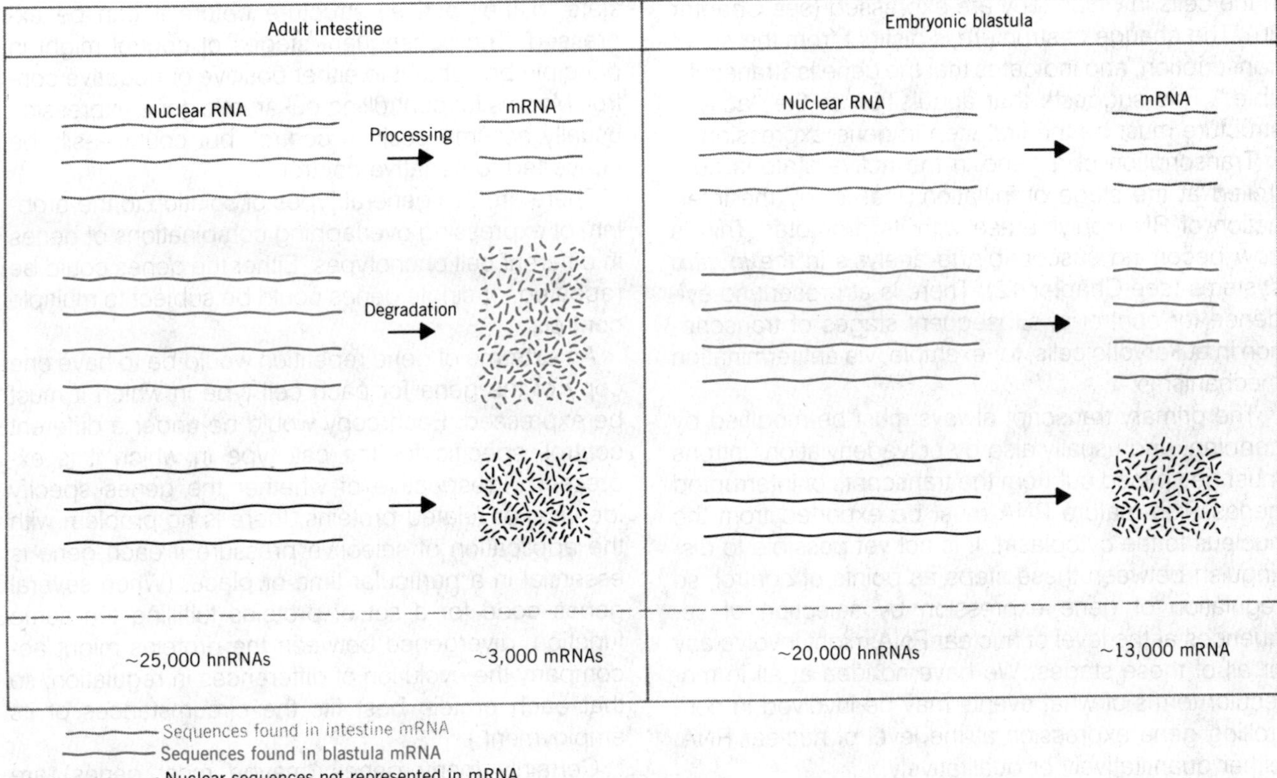

Figure 27.6
The same nuclear transcripts are degraded in the adult intestine but used to give mRNA in the embryonic blastula of the sea urchin.

being used to produce mRNA. But in the blastula, *these same sequences* are used to give rise to mRNAs. So in this case, many genes are transcribed in tissues irrespective of whether they are finally expressed. The transcripts are selected for processing and nucleo-cytoplasmic transport only in the appropriate tissues. Only a minority of genes seem to be controlled at the level of transcription.

MODELS FOR CONTROLLING GENE EXPRESSION

It is too early to form any general view of the frequency with which different levels of control are used to provide the stage at which gene expression is regulated. The concept of the "level of control" implies that gene expression is not necessarily an automatic process once it has begun. It could in principle be regulated in a gene-specific way at any one of several sequential steps. We can distinguish (at least) four potential control points, forming the series

Activation of gene structure
↓
Initiation of transcription
↓
Processing transcript and transport to cytoplasm
↓
Translation of mRNA

The existence of the first step is implied by the discovery that genes themselves may exist in either of two structural conditions. Relative to the state of most of the genome, genes are found in an "active" state in the cells in which they are expressed (see Chapter 30). The change of structure is distinct from the act of transcription, and indicates that the gene is "transcribable." This suggests that acquisition of the "active" structure must be the first step in gene expression.

Transcription of a gene in the active state is controlled at the stage of initiation, that is, by the interaction of RNA polymerase with its promoter. This is now becoming susceptible to analysis in the *in vitro* systems (see Chapter 12). There is at present no evidence for control at subsequent stages of transcription in eukaryotic cells, for example, via antitermination mechanisms.

The primary transcript always must be modified by capping, and usually also by polyadenylation. Introns must be spliced out from the transcripts of interrupted genes. The mature RNA must be exported from the nucleus to the cytoplasm. It is not yet possible to distinguish between these steps as points of control, so regulation of gene expression by selection of sequences at the level of nuclear RNA might involve any or all of these stages. We have no idea at all in molecular terms of what events may be involved in controlling gene expression at the level of nuclear RNA, either quantitatively or qualitatively.

Finally, the translation of the mRNA in the cytoplasm can be specifically controlled. There is little evidence for the employment of this mechanism in adult somatic cells, but it does occur in some embryonic situations, as described in Chapter 10. The mechanism is presumed to involve the blocking of initiation of translation of some mRNAs by specific protein factors.

We must look beyond the question of the use and type of mechanism employed at each potential level of control to the general issue of how gene expression is coordinated between different loci. The genes that are expressed to generate any particular cell phenotype may lie in a large number of different locations. Each gene function probably is required to contribute to a characteristic spectrum of cell phenotypes. How does each cell activate the right combination of genes to produce its phenotype?

Eukaryotic gene expression generally is regarded as being under positive control: genes are inactive unless they are turned on. This is consistent with the need for a gene to change from the usual inactive state to the "active" structure before it can be expressed. The subsequent stages of control might in principle be subject to either positive or negative control. Models for controlling eukaryotic gene expression usually assume positive control, but could easily be converted to negative control.

There are two general types of solution to the problem of expressing overlapping combinations of genes in different cell phenotypes. Either the genes could be repeated, or single genes could be subject to multiple controls.

An extreme of gene repetition would be to have one copy of the gene for each cell type in which it must be expressed. Each copy would be under a different control, specific for the cell type in which it is expressed. Irrespective of whether the genes specify identical or related proteins, there is no problem with the application of selective pressure if each gene is essential in a particular time or place. (When several genes code for a set of proteins fulfilling the same function, divergence between the proteins might accompany the evolution of differences in regulation, so that each protein best fits the circumstances of its employment.)

Certainly many genes (maybe most genes) are present in more than one copy per genome. It is beginning to seem that the copies are usually at least slightly different. From the limited data at present available, it appears that different copies tend to be expressed in different cellular situations.

Rather than repeat the entire gene, an alternative way to generate variation in its product is to provide alternate exons. Changes in the splicing pattern then may generate proteins differing in just a particular part of their sequence. The same mechanism may also be used to generate mRNAs whose nontranslated regions differ, although the protein is unaltered.

When there is only one copy of a gene, if its expression is required in more than one set of circumstances, one option is to provide it with multiple regulator elements, each able independently to sponsor its expression; such arrangements are found in bacterial genomes and may be postulated for eukaryotes.

A cell phenotype is the result of the coordinated expression of a particular set of genes. When dispersed genes must be coordinately controlled, the

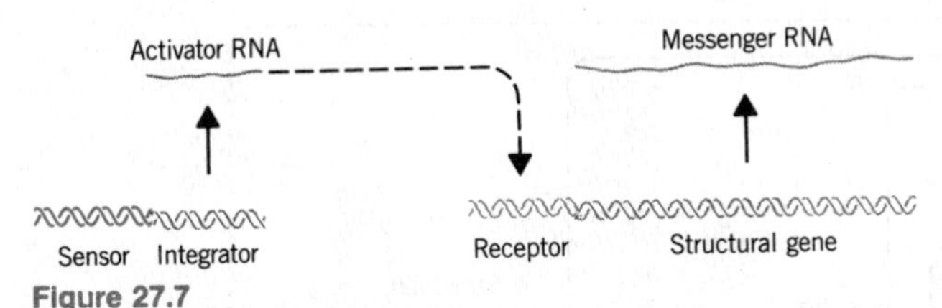

Figure 27.7
The basic components of eukaryotic gene control could comprise a sensor site-integrator gene that regulates a receptor site-structural gene.

same regulator element may be repeated at each locus.

Relatively short sequences of DNA may be used as recognition sites to control the activity of contiguous structural genes. We need not specify *how* they control gene activity: their action could involve a structural change prior to transcription, it could concern initiation at the promoter, or it could be exercised in the nuclear RNA transcript.

A formal network to provide multiple controls of individual genes was promulgated by Britten and Davidson from the basic model illustrated in **Figure 27.7.** A structural gene is positively controlled by an adjacent **receptor site.** This responds to an **activator** molecule, considered in the model to be RNA, but it could alternatively be protein. The activator is synthesized by an **integrator gene,** which (of course) is the direct equivalent of a bacterial regulator gene. The only difference is that the integrator gene is itself subject to external control. The controlling element, adjacent to it, is called a **sensor site.** It is supposed to respond to functions that control gene expression, for example, hormones. The model is shown for control prior to transcription, but easily can be adapted to other levels.

A single activator can control many genes if each of its target loci possesses a copy of the appropriate receptor. There is ample precedent for such repetition in dispersed bacterial loci that are under common control (see Chapter 15). All the genes possessing the same receptor comprise a **set** that is expressed in response to the activator. The set is somewhat equivalent to an operon since all its genes always are expressed together. It is therefore likely to include genes specifying proteins with related functions, such as the enzymes of a metabolic pathway.

A structural gene can be expressed in different circumstances if it possesses more than one receptor. Each receptor is able to activate the gene by virtue of its reaction with the activator molecule. By placing a series of receptors adjacent to a structural gene, it can be included in several different sets, as depicted in **Figure 27.8.** By appropriate repetition of receptors, each of the three structural genes (*x*, *y*, *z*) belongs to two of the three partially overlapping sets (*a*, *b*, *c*).

Different sets of genes can be controlled together by supposing that a sensor may control several integrators. This allows a single stimulus to one sensor to activate several sets of genes. The genes all controlled by one sensor are called a **battery.** A set can be included in more than one battery if the integrator elements are repeated as illustrated in **Figure 27.9.**

Figure 27.8
Each structural gene can be activated by more than one receptor and therefore belongs to more than one set.

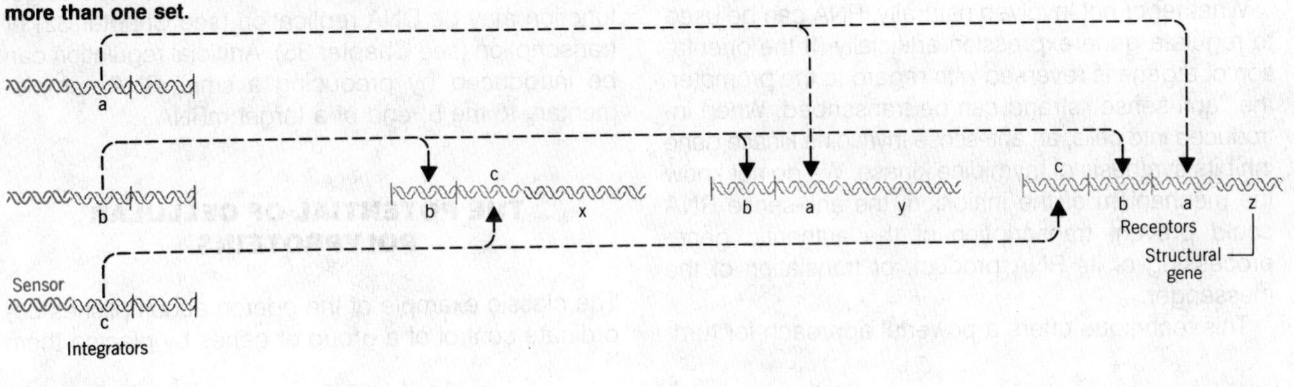

a
b
c
b
x
b
a
y
c
b
a
z
Receptors
Structural gene
Sensor
c
Integrators

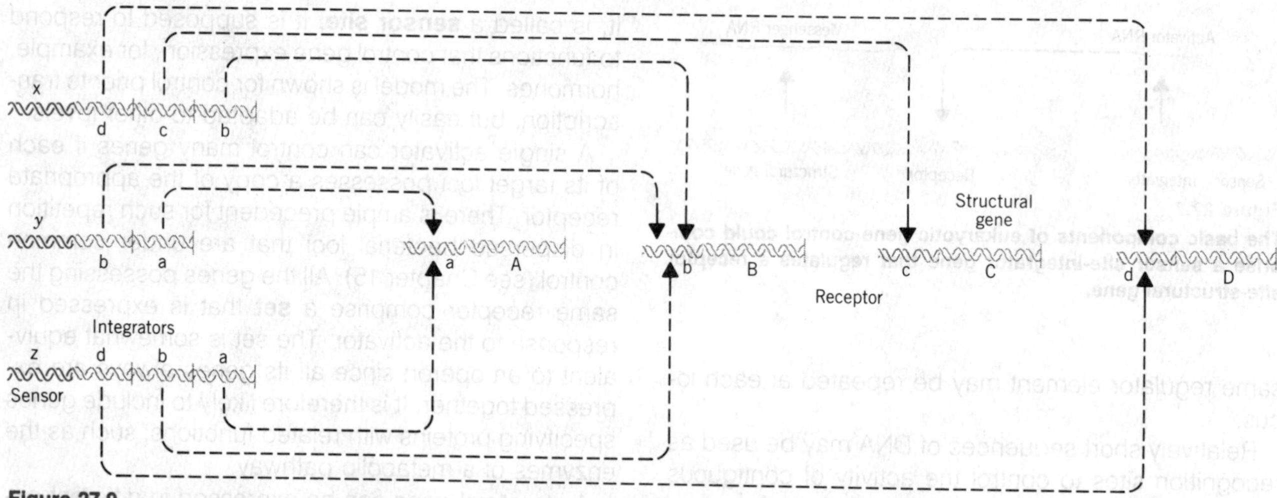

Figure 27.9
Integrators can be repeated and each sensor may control several integrators, so that a signal to one sensor may activate several sets of genes, and each set can be included in more than one battery.

This hierarchy of elements allows any particular gene to be combined for expression with any other gene with economy of control. The model requires the integrator and receptor sequences to be repeated, which therefore defines them as members of the repetitive DNA component. Indeed, an impetus for the construction of the model was to find a role for repetitive sequences in the genome.

Ironically, however, it turns out that repetitive sequences, at least as defined by hybridization kinetics probably have nothing to do with regulation. If there are repetitive regulatory sequences, they are likely to take the form of very short consensus sequences, much too short to be detected in hybridization analysis.

Whether or not involved naturally, RNA can be used to regulate gene expression artificially. If the orientation of a gene is reversed with regard to the promoter, the "anti-sense" strand can be transcribed. When introduced into cells, an anti-sense thymidine kinase gene inhibits synthesis of thymidine kinase. We do not know the mechanism of the inhibition; the anti-sense RNA could prevent transcription of the authentic gene, processing of its RNA product, or translation of the messenger.

This technique offers a powerful approach for turning off genes at will; for example, the function of a regulatory gene could be investigated by introducing an anti-sense version. If the anti-sense gene were placed under control of a promoter itself subject to regulation, the target gene could be turned off and on at will by regulating the production of anti-sense RNA. This technique could allow investigation of the importance of the timing of expression of the target gene.

The approach is effective in both eukaryotic cells and bacteria. In the eukaryotic example, the anti-sense gene represented a normal gene reversed in orientation. In some authentic cases of bacterial regulation, RNA provides a regulatory function by virtue of complementarity with a target RNA sequence. The target function may be DNA replication (see Chapter 32) or transcription (see Chapter 35). Artificial regulation can be introduced by producing a small RNA complementary to the 5′ end of a target mRNA.

THE POTENTIAL OF CELLULAR POLYPROTEINS

The classic example of the operon accomplishes coordinate control of a group of genes by placing them in a single unit of transcription. Few cases are known in eukaryotic genomes in which related genes are expressed as part of a common unit. Some small proteins, however, are coded by sequences that are repeated, each copy existing as part of a unit that includes other, different functions. The concept that genes may be repeated with differing companions is in a sense the antithesis of the Britten-Davidson model for repeating the regulatory elements.

Some of the options for handling repeated sequences are displayed dramatically by the synthesis of **polyproteins.** The name was first used to describe the multifunctional products of retroviral RNA genomes. Several viral proteins are synthesized in the form of a common precursor, which is cleaved at specific points to release the individual functional proteins. This situation formally complies with the apparent rule that eukaryotic translation systems only handle monocistronic messengers, but allows several proteins to be synthesized as the result of a single transcription event.

Several forms of cellular polyproteins since have been discovered. In some cases, the polyprotein contains a series of identical or related polypeptides. Thus an enkephalin precursor contains six copies of Met-enkephalin (the peptide Tyr-Gly-Gly-Phe-Met) and also one copy of Leu-enkephalin (Tyr-Gly-Gly-Phe-Leu). The polyprotein here functions principally as an amplification mechanism, so that each cycle of transcription and translation produces several copies of the polypeptide product.

The more complex example of the gene for pro-opiomelanocortin (POMC) is summarized in **Figure 27.10.** Genes from several mammals have a similar structure, in which a single intron is present in the signal part of the coding region, and several protein functions are coded by a single exon. In each species, a single gene has been characterized, although other (not necessarily identical) copies may also be present.

The single polyprotein product of 32,000 daltons carries several known polypeptide hormones and also has other regions whose sequences suggest that they too may be of biological importance. All of the actual or potential individual products have dipeptide borders consisting of pairs of basic amino acids that may provide trypsin-like cleavage points. The use of these

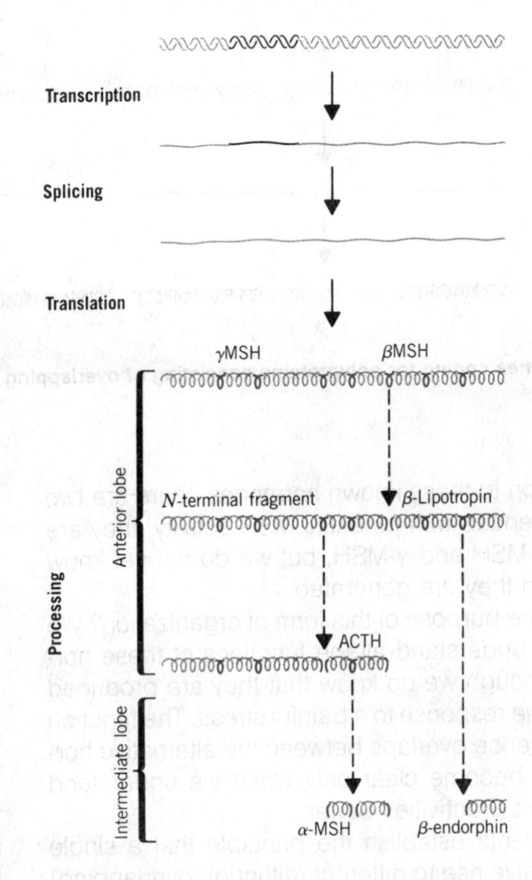

Figure 27.10
The POMC gene is translated into a polyprotein whose products depend on tissue-specific processing.

cleavage sites is different in two tissues in which the gene is expressed.

In the anterior lobe of the pituitary, the POMC protein first is cleaved once, releasing the N-terminal fragment and β-lipotropin (the C-terminal 70–90 amino acids). Then the N-terminal fragment is cleaved to release the ACTH hormone (39 amino acids long). In this tissue, the reaction proceeds no further.

In the intermediate lobe, ACTH is cleaved again, releasing α-melanotropin (α-MSH), which consists of the 13 N-terminal amino acids. The β-lipotropin also is cleaved once, releasing the analgesic β-endorphin, consisting of the 31 C-terminal amino acids.

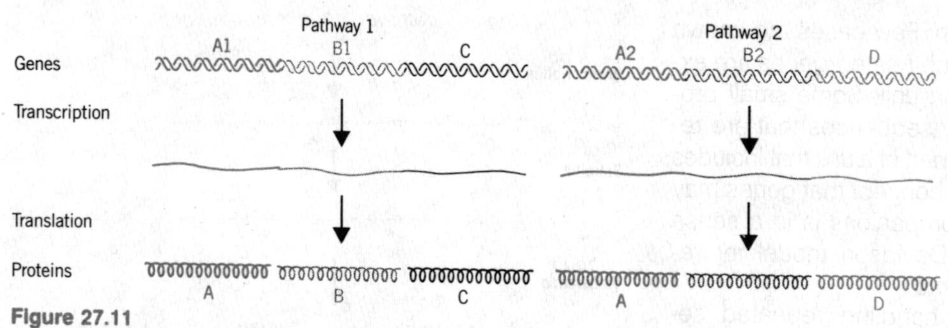

Figure 27.11
Are there genes coding for polyproteins consisting of overlapping products?

In addition to these known hormones, there are two other sequences with potential MSH activity; they are denoted β-MSH and γ-MSH, but we do not yet know if and when they are generated.

What is the purpose of this form of organization? We still do not understand all the functions of these hormones, although we do know that they are produced as part of the response to a painful stress. The function of the sequence overlaps between the alternative hormones will become clear only when we understand their biological activities better.

These events establish the principle that a single gene may give rise to different (although overlapping) products by virtue of variation in the pathway for protein processing (presumably regulated by the processing enzymes of the relevant tissues). A further possibility is that repetitive copies of genes for polyproteins could allow the *same protein* to be synthesized in the company of alternative companions. The principle is illustrated in **Figure 27.11.**

Two polyprotein genes respectively contain the sequences of the individual proteins A, B, C and A, B, D. Thus activation of either gene leads to synthesis of A and B; but in one case this is accompanied by production of C, in the other case by production of D. This situation is formally equivalent to repetition of structural genes in polycistronic transcription units, although the products become separated at the stage of protein processing instead of at translation.

A family of genes coding for ELH, a neuropeptide of the marine mollusc *Aplysia,* may fall into this category. *Aplysia* DNA contains about 5 genes for the ELH hormone (a polypeptide of 36 amino acids). Although the full set of genes has yet to be characterized, several different mRNA species coding for ELH have been identified. Each of these is translated into a different polyprotein. The polyproteins all have an ELH sequence; but the other sequences represent different products, which may include other polypeptide hormones. Different mRNAs may be synthesized in different tissues, allowing the release in each case of a specific combination of hormones. This mechanism therefore allows overlapping combinations of hormones to be synthesized by activating appropriate members of a repetitive gene family.

It is notable that these examples of cellular polyproteins involve the generation of small products (generally less than 40 amino acids each). Does this reflect some difficulty in synthesizing small proteins de novo, which has led to the necessity for cleavage from longer precursors (and thus opened the way for evolution of alternative products)? It will be interesting to characterize the genes that code for multiple, related polyproteins. Will they prove to have been assembled by exon shuffling of the sort described in Chapter 20, or has the entire process evolved at some subsequent stage of expression of repeated genes whose ancestors did not have these multiple functions?

FURTHER READING

The properties of nuclear RNA have been discussed in detail in **Lewin's** *Gene Expression,* **2,** *Eucaryotic Genomes*

(Wiley, New York, 1980, pp. 728–760). The developing relationship between hnRNA and mRNA was previously reviewed in *Cell* (**4,** 11–20 & 77–94, 1975). More recent information on the signals for polyadenylation and its relationship with splicing is to be found in the original papers from **Fitzgerald & Shenk** (*Cell* **24,** 251–260, 1981) and **Zeevi, Nevins & Darnell** (*Cell* **26,** 39–46, 1981). The involvement of snurps in generating 3′ ends was discovered by **Galli et al.** (*Cell* **34,** 823-828, 1983). Their (now rather dated) model for gene control was published by **Britten & Davidson** in *Science* (**165,** 349–357, 1969). Mammalian polyproteins have been reviewed by **Douglas, Civelli & Herbert** (*Ann. Rev. Biochem.* **53,** 665–715, 1984). The case of *Aplysia* was discovered by **Scheller et al.** (*Cell* **28,** 707–719, 1982).

PART 8
THE PACKAGING OF DNA

Do bacteria contain a nucleus? This question has agitated many workers; but at the present moment it appears no more meaningful than the everlasting controversies about the authorship of Shakespeare's plays which was finally resolved by the recognition that another fellow with the same name must have been the author. The important property of the nucleus of animal and plant cells is not so much its appearance or shape as the orderly changes, connected with definite biological tasks, which it can be seen to undergo.

Erwin Chargaff, 1947

CHAPTER 28
ABOUT GENOMES AND CHROMOSOMES

A common problem is presented by the packaging of DNA into phages and viruses, into bacterial cells and eukaryotic nuclei. The length of the DNA as an extended molecule would vastly exceed the dimensions of the compartment that contains it. The DNA (or in the case of some viruses, the RNA) must therefore be compressed exceedingly tightly to fit into the space available.

The magnitude of the discrepancy between the length of the nucleic acid and the size of its compartment is evident from the examples summarized in **Table 28.1.** For bacteriophages and for eukaryotic viruses, whether long and thin (filamentous) or approximately spherical (icosahedral), a large amount of nucleic acid, whether DNA or RNA, whether single-stranded or double-stranded, effectively fills the container.

For bacteria or for eukaryotic cell compartments, the discrepancy is even greater. The DNA is contained in a compact area that occupies only part of the compartment. This is seen as the **nucleoid** in bacteria and as the mass of **chromatin** in eukaryotic nuclei at interphase (between divisions). Among the examples given in the table, the problem culminates in the need to fit 1.8 m of human DNA into a nucleus whose diameter may be as little as 6 μm (6×10^{-6} m).

The packaging of chromatin is flexible; it changes during the eukaryotic cell cycle. At the time of division (mitosis or meiosis), the genetic material becomes even more tightly packaged, and individual mitotic **chromosomes** become recognizable.

The overall compression of the DNA can be described by the **packing ratio,** the ratio of the length of DNA to the length of the unit that contains it. For example, the smallest human chromosome contains about 4.6×10^7 bp of DNA (about 10 times the genome size of the bacterium *E. coli*). This is equivalent to 14,000 μm (= 1.4 cm) of extended DNA. At the most condensed moment of mitosis, the chromosome is about 2 μm long. Thus the packing ratio of DNA in the chromosome can be as great as 7000.

Packing ratios cannot be established with this certitude for the more amorphous overall structures of the bacterial nucleoid or eukaryotic chromatin. However, the usual reckoning is that mitotic chromosomes are likely to be 5–10 times more tightly packaged than interphase chromatin, which therefore has a typical packing ratio of 1000–2000.

Condensation of nucleic acid depends on the presence of specific proteins to which it binds. Usually these proteins are basic, and their positive charges neutralize the negative charges of the nucleic acid. The relationship of the proteins with the DNA (or RNA) determines its structure.

A general principle is evident in the organization of

Table 28.1
The length of nucleic acid is much greater than the dimensions of the surrounding compartment.

Compartment	Shape	Dimensions	Type of Nucleic Acid	Length
TMV	filament	0.008 × 0.3 μm	1 single-stranded RNA	2 μm (6.4 kb)
Phage fd	filament	0.006 × 0.85 μm	1 single-stranded DNA	2 μm (6 kb)
Adenovirus	icosahedron	Diameter 0.07 μm	1 double-stranded DNA	11 μm (35 kb)
Phage T4	icosahedron	0.065 × 0.10 μm	1 double-stranded DNA	55 μm (170 kb)
E. coli	cylinder	1.7 × 0.65 μm	1 double-stranded DNA	1300 μm (1.3 mm = 4.2 × 10^3 kb
Mitochondrion (human)	oblate spheroid	3.0 × 0.5 μm	~10 identical double-stranded DNAs	50 μm (each 5 μm = 16 kb)
Nucleus (human)	spheroid	Diameter 0.06 μm	46 chromosomes of double-stranded DNA	1.8 m (6 × 10^6 kb)

all cellular genetic material. It exists as a compact mass, in a delineated area; and its various activities, such as replication and transcription, must be accomplished within these confines.

One of the major unanswered questions concerns the *specificity* of this packaging. Is the DNA folded into a *particular* pattern, or is it different in each individual copy of the genome? How does the pattern of packaging change when a segment of DNA is replicated or transcribed?

CONDENSING VIRAL GENOMES INTO THEIR COATS

From the perspective of packaging the *individual* sequence, there is an important difference between a cellular genome and a virus. The cellular genome is essentially indefinite in size; the number and location of individual sequences can be changed by duplication, deletion, and rearrangement. Thus it requires a *generalized* method for packaging its DNA, insensitive to the total content or distribution of sequences. By contrast, two restrictions define the needs of a virus. The amount of nucleic acid to be packaged is *predetermined* by the size of the genome. And it must all fit within a coat assembled from a protein or proteins coded by the viral genes.

A virus particle is deceptively simple in its superficial appearance. The nucleic acid genome is contained within a **capsid.** This is a symmetrical or quasi-symmetrical structure assembled from one or only a few proteins. Attached to the capsid, or incorporated into it, may be other structures, assembled from distinct proteins, and necessary for infection of the host cell.

Viewed in these terms, the tolerances for constructing the virus particle are rather tight. The internal volume of the capsid is rarely much greater than the volume of the nucleic acid it must hold. Simple calculations suggest that the difference is usually less than twofold, and often the internal volume is barely larger than the nucleic acid.

In its most extreme form, the restriction that the capsid must be assembled from virally coded proteins means that the entire shell is constructed from a single type of subunit. The rules for assembly of identical subunits into closed structures restrict the capsid to one of two types. The protein subunits may stack sequentially in a helical array to form a **filamentous** or rodlike shape. Or they may form a pseudospherical shell; this type of structure actually is a polyhedron with **icosahedral symmetry.** It now turns out that some viral capsids are assembled from more than a single type of protein subunit, but although this extends the exact type of structure that can be formed, viral capsids still all conform to the general classes of quasi-crystalline filaments or icosahedrons.

There are two types of solution to the problem of how to construct a capsid that contains nucleic acid. The protein shell can be assembled around the nucleic acid, condensing the DNA or RNA by protein-

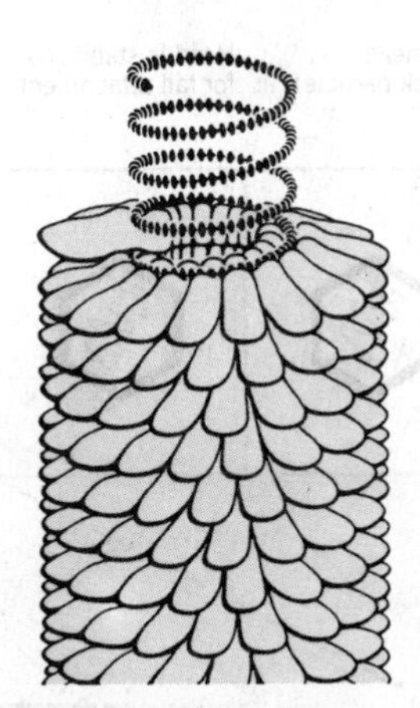

Figure 28.1
A helical path for TMV RNA is created by the stacking of protein subunits in the virion.

nucleic acid interactions during the process of assembly. Or the capsid can be constructed from its component(s) in the form of an empty shell, after which the nucleic acid must be inserted, being condensed as it enters.

Assembly of the capsid around the genome occurs in the case of single-stranded RNA viruses. The best characterized example is TMV (tobacco mosaic virus). Assembly starts at a duplex hairpin that lies within the RNA sequence. From this **nucleation center,** it proceeds bidirectionally along the RNA, until reaching the ends. The unit of the capsid is a two-layer disk, each layer containing 17 identical protein subunits. The disk forms a circular structure, which is converted to the helical form by interaction with the RNA. The RNA is coiled in a helical array on the inside of the protein shell, as illustrated in **Figure 28.1.**

The entire length of this RNA genome exists in a structure determined by its interaction with the protein shell. This arrangement is not unique to the filamentous viruses. The same general type of responsibility for internal organization is presented by the single-stranded RNA genome of TYMV (turnip yellow mosaic virus), which has a spherical capsid.

TMV and TYMV virions are therefore organized according to the same principle. *The position of the RNA within the capsid is determined directly by its binding to the proteins of the shell.*

The spherical capsids of DNA viruses are assembled in a different way, as best characterized for the phages lambda and T4. In both cases, an empty head shell is assembled from a small set of proteins. *Then the duplex genome is inserted into the head,* a process that is accompanied by a structural change in the capsid.

Figure 28.2 summarizes the assembly of lambda. It starts with a small head shell that contains a protein "core." This is converted to an empty headshell of more distinct shape. Then DNA packaging begins, the head shell expands in size though remaining the same shape, and finally the full head is sealed by the addition of the tail.

For both lambda and T4, the DNA that is to be inserted into the empty head takes the form of **concatemeric** molecules. These are multiple genomes joined end to end, as depicted in **Figure 28.3.** Each phage has its own mechanism for recognizing the proper amount of DNA to insert.

The ends of the lambda genome are marked by sequences called *cos* sites. Cleavage occurs at the left *cos* site (as defined on the usual map) to generate a free end that is inserted into the capsid. The insertion of DNA continues until the right *cos* site is encountered, when it is cleaved to generate the other end. The end that goes into the capsid last during assembly comes out first when a new host cell is infected.

Any DNA contained between two *cos* sites can be packaged. This is the basis of the "cosmid" cloning technique described in Chapter 17. However, the distance between the *cos* sites can be varied only slightly from the usual length of lambda DNA. Packaging does not occur if the distance is either too great or too small. This demonstrates that there must be *enough* DNA to complete the packaging reaction, as well as showing that there is room in the head only for a very little extra DNA (about 15%).

With phage T4, insertion starts at a *random* point in the concatemeric precursor. It continues until a genome's worth of DNA (a "headful") has been inserted. This implies the existence of some mechanism for measuring the amount of DNA. [Actually, the amount that is inserted is slightly greater than the length of the unit genome, with the result that there is a **terminal redundancy** corresponding to the additional length. In the terms of Figure 28.3, the first virion might contain the DNA from *A* to *A*, the next from *B* to *B*, and so on,

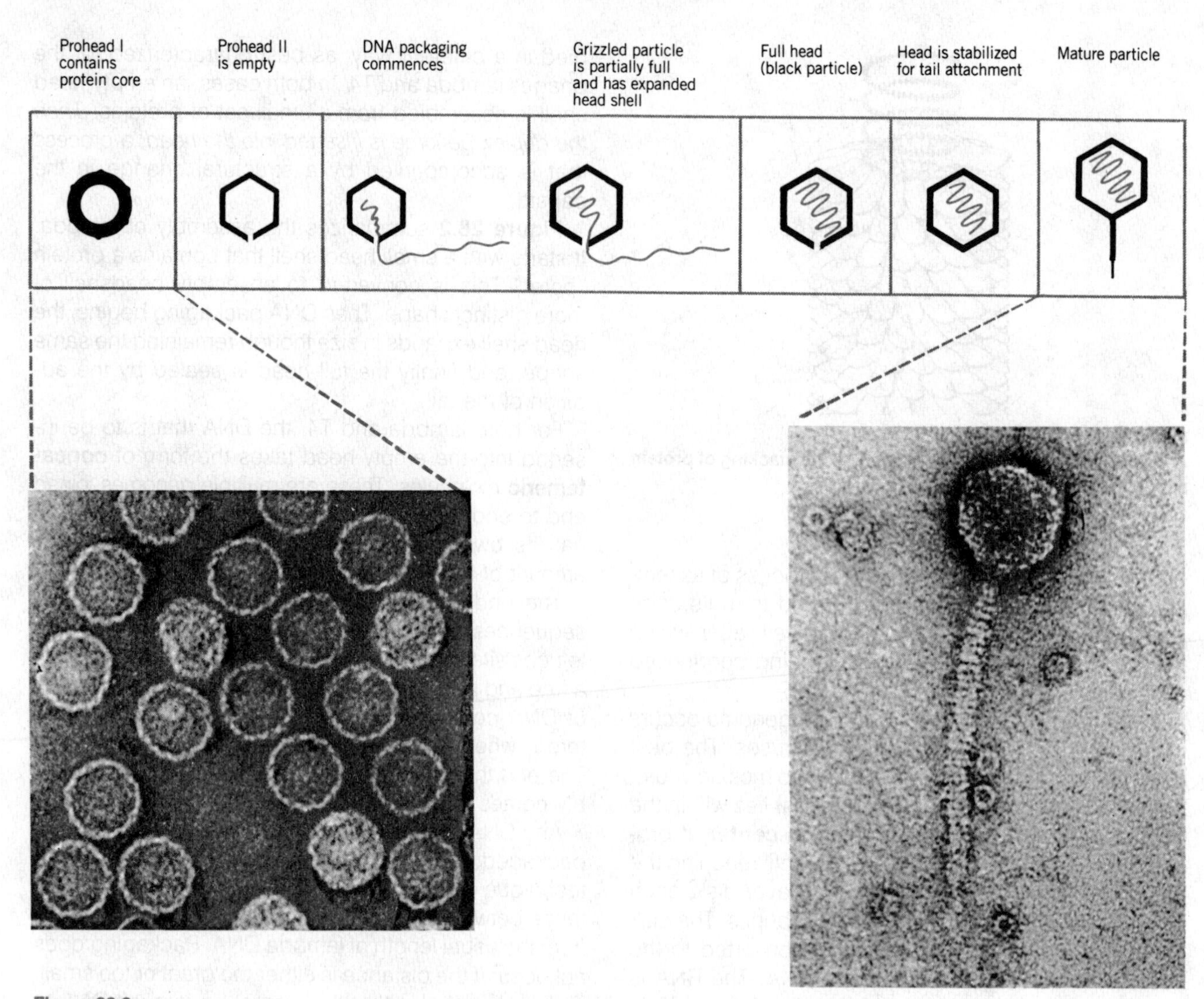

Figure 28.2
The empty head of phage lambda changes shape and expands when it becomes filled with DNA.
The electron micrographs show the particles at the start and end of the maturation pathway. Photographs kindly provided by A. F. Howatson.

Figure 28.3
Concatemeric DNA consists of a tandem series of phage genomes.

Concatemer

ABC– – – – – XYZABC – – – – – XYZABC – – – – – XYZABC – – – – – XYZ

Individual genome

ABC – – – – – XYZ

so that each genome has a (different) letter repeated at each end.]

Now a double-stranded DNA considered over short distances is a fairly rigid rod. Yet it must be folded into a very compact structure to fit within the capsid. Little is known about how this is accomplished, except that there are "internal proteins" in the capsid, together with the DNA. One possibility is that they provide some sort of "scaffolding" onto which the DNA condenses. (This would be a counterpart to the use of the proteins of the shell in the plant viruses mentioned earlier.)

How specific is this packaging? It cannot depend on particular sequences, because deletions, insertions, and substitutions all fail to interfere with the assembly process. The relationship between DNA and the headshell has been investigated directly by determining which regions of the DNA can be chemically crosslinked to the proteins of the capsid. The surprising answer is that all regions of the DNA are more or less equally susceptible to crosslinking. This probably means that when DNA is inserted into the head, it follows a general rule for condensing, but the pattern is not determined by particular sequences.

Probably the DNA condenses by a "spooling" mechanism, in which it winds around a central axis (like a spool of cotton). If any region of the DNA can be in contact with the protein shell, however, the spool cannot take a form in which first the outside is filled and then the inside, or vice versa. The model illustrated in **Figure 28.4** supposes that the end of the DNA that enters first is fixed to a point near the site of entry. DNA continues to enter along a central axis, coiling around at the "top" of the head. The turns of the spool are spaced about equally, so that as more DNA enters, the coil is forced to tighten. The response of the DNA to the tightening varies from head to head, so that the points of contact with the protein shell fluctuate in individual phages.

These varying mechanisms of virus assembly all accomplish the same end: packaging a single DNA or RNA molecule into the capsid. However, there are some viruses whose genomes consist of multiple nucleic acid molecules. Reovirus contains ten double-stranded RNA segments, all of which must be packaged into the capsid. Nothing is known about how the assembly process selects one copy of each different molecule in order to collect a complete set of genetic information.

Some plant viruses are multipartite: their genomes consist of segments each of which is packaged into a *different* capsid. An example is alfalfa mosaic virus, which has four different single-stranded RNAs, each packaged independently into a coat comprising the same protein subunit. A successful infection depends on the entry of one of each type into the cell.

The four components of the virus exist as particles of different sizes. This means that the same capsid protein can package each RNA into its own characteristic particle. This is a departure from the packaging of a unique length of nucleic acid into a capsid of fixed shape. However, in the case of viruses whose capsids have only one authentic form, the assembly pathway often displays a side effect. This is the formation of some aberrant (although inactive) **monster** particles in which the head is longer than usual.

A capsid protein(s) therefore has an intrinsic ability to assemble into a particular type of structure, but the exact size and shape need not be invariant. Often there are other, **assembly proteins,** needed for head formation, but not themselves part of the head shell; they may limit the options of the capsid protein so that it assembles only along the desired pathway. Cellular genomes also may employ proteins whose functions are to direct the assembly of others (see Chapter 30).

Figure 28.4
Lambda DNA may spool around a central axis as it enters the headshell, gradually tightening; the exact arrangement is subject to individual fluctuations.

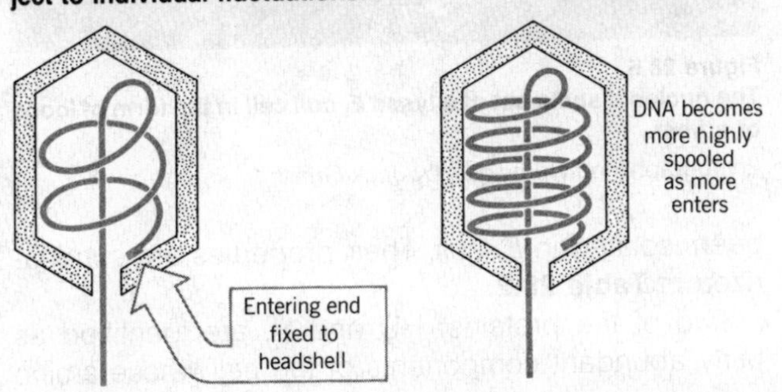

THE BACTERIAL GENOME IS A NUCLEOID WITH MANY SUPERCOILED LOOPS

Although bacteria do not display structures with the distinct morphological features of eukaryotic chromosomes, their genomes nonetheless are organized into definite bodies. The genetic material can be seen as a fairly compact clump or series of clumps that occupies (very roughly) about one third of the volume of the cell. **Figure 28.5** displays a thin section through a bacterium in which this **nucleoid** is evident.

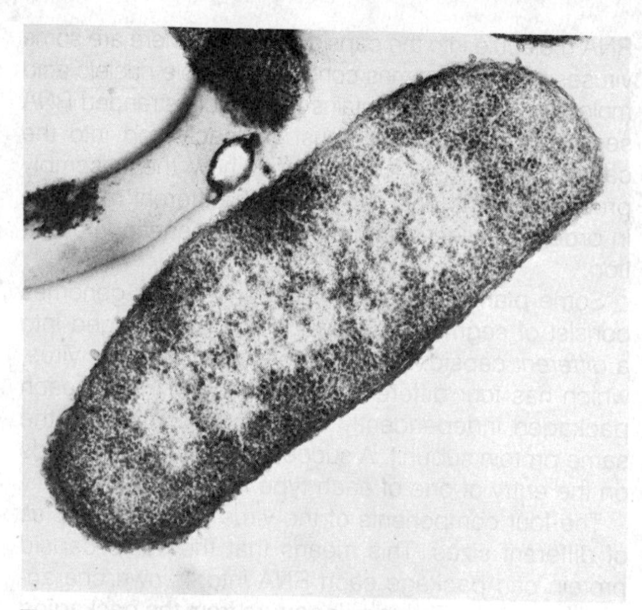

Figure 28.5
A thin section shows the bacterial nucleoid as a compact mass in the center of the cell.
Photograph kindly provided by Jack Griffith.

In bacteria that have partially replicated their DNA, the nucleoid may contain more than one genome's worth of DNA. By the time of cell division, the material has separated into two nucleoids that are partitioned into the daughter cells. The segregation mechanism probably involves the attachment of the bacterial genome to the membrane. This mechanism provides a counterpart to the mitotic segregation of eukaryotic chromosomes. Instead of being pulled apart on a spindle, the sequences of a specific locus on each bacterial genome may be connected to a site on the membrane. The two membrane sites move apart as the cell grows; and when division occurs, each lies in a different compartment, taking its nucleoid with it (see Chapter 31).

When *E. coli* cells are lysed, fibers are released in the form of loops attached to the broken envelope of the cell. As can be seen from **Figure 28.6,** the DNA of these loops is not found in the extended form of a free duplex, but is folded into a more compact shape, presumably by virtue of its association with proteins.

Several DNA-binding proteins with a superficial resemblance to eukaryotic chromosomal proteins have

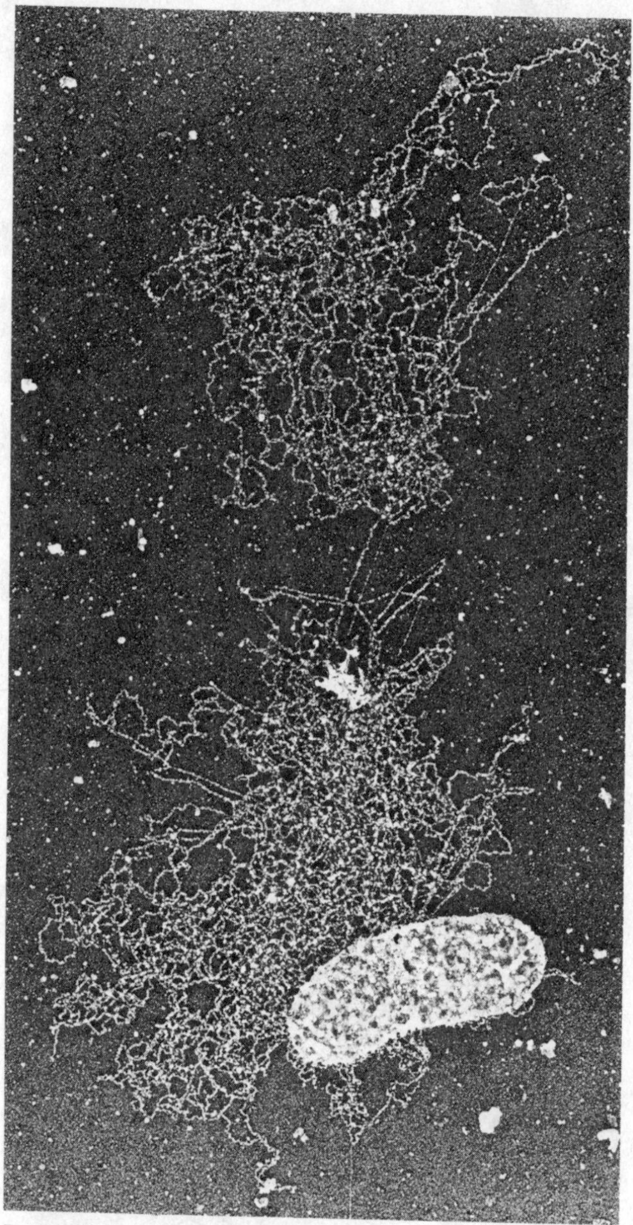

Figure 28.6
The nucleoid spills out of a lysed *E. coli* cell in the form of loops of a fiber.
Photograph kindly provided by Jack Griffith.

been isolated in *E. coli*. Their properties are summarized in **Table 28.2.**

Two of the proteins, HU and H, are identified as fairly abundant components of the cell whose amino

Table 28.2
***E. coli* contains several DNA-binding proteins.**

Protein	Composition	Content/Cell	Eukaryotic Relatives	Locus
HU	2 nonidentical subunits of 9000 daltons	40,000 dimers	Histone H2B	Not known
H	2 identical subunits of 28,000 daltons	30,000 dimers	Histone H2A	Not known
HLP1	Monomer of 17,000 daltons	20,000 copies	Not known	*firA*
P	Subunit of 3000 daltons	Not known	Protamines	Not known

acid compositions resemble the histones that bind to eukaryotic DNA (see Chapter 29). The proteins are small, highly basic, and bind strongly to DNA. Protein HU stimulates DNA replication. Nothing is known about the function of protein H.

A potential DNA-binding protein has been identified by virtue of the presence of its gene in a tRNA operon (see Chapter 23). This is protein P. It has an amino acid composition resembling the protamines that bind to DNA in certain sperm. Its quantity and functions are not known.

The one protein firmly implicated in interacting with the genetic apparatus is HLP1. This is coded by the locus *firA*, in which mutations can abolish the resistance to rifampicin conferred by some mutations in the β subunit of RNA polymerase. The properties of the mutants do not show whether the protein interacts with RNA polymerase or with DNA; HLP1 is small, basic, and can bind DNA like the other proteins.

The nucleoid can be isolated directly in the form of a very rapidly sedimenting complex, consisting of about 80% DNA by mass. (The analogous complexes in eukaryotes have about 50% DNA by mass, as described in Chapter 29.) In bacteria that contain more than a single copy of the genome (because DNA replication has occurred), the complex is proportionately increased in size.

The complex can be unfolded to a less compact state by treatment with reagents that act on RNA or protein. This suggests that its structure is stabilized by these components. The possible role of proteins is evident. The role that RNA might have has been quite refractory to analysis. Attempts to isolate any specific RNA involved in a structural capacity all have failed. So have attempts to implicate nascent RNA chains (that is, the incomplete products of currently proceeding transcription).

The DNA of the compact body isolated *in vitro* behaves as a closed duplex structure, as judged by its response to ethidium bromide. This small molecule intercalates between base pairs to generate *positive* superhelical turns in "closed" circular DNA molecules, that is, molecules in which both strands have covalent integrity. In "open" circular molecules, which contain a nick in one strand, or with linear molecules, the DNA can rotate freely in response to the intercalation, thus relieving the tension.

In a natural closed DNA that is *negatively* supercoiled, the intercalation of ethidium bromide first removes the negative supercoils and then introduces positive supercoils. The amount of ethidium bromide needed to achieve zero supercoiling is a measure of the original density of negative supercoils.

Some nicks occur in the compact nucleoid during its isolation; they can also be generated by limited treatment with DNAase. But this does not abolish the ability of ethidium bromide to introduce positive supercoils. This capacity of the genome to retain its response to ethidium bromide in the face of nicking means that it must have many independent **domains;** the supercoiling in each domain is not affected by events in the other domains.

This autonomy suggests that the structure of the bacterial chromosome may have the general organization depicted diagrammatically in **Figure 28.7.** Each domain consists of a loop of DNA, the ends of which are secured in some (unknown) way that does not allow rotational events to propagate from one domain to another. There are about 100 such domains per genome, so that each consists of about 40 kb (13 μm)

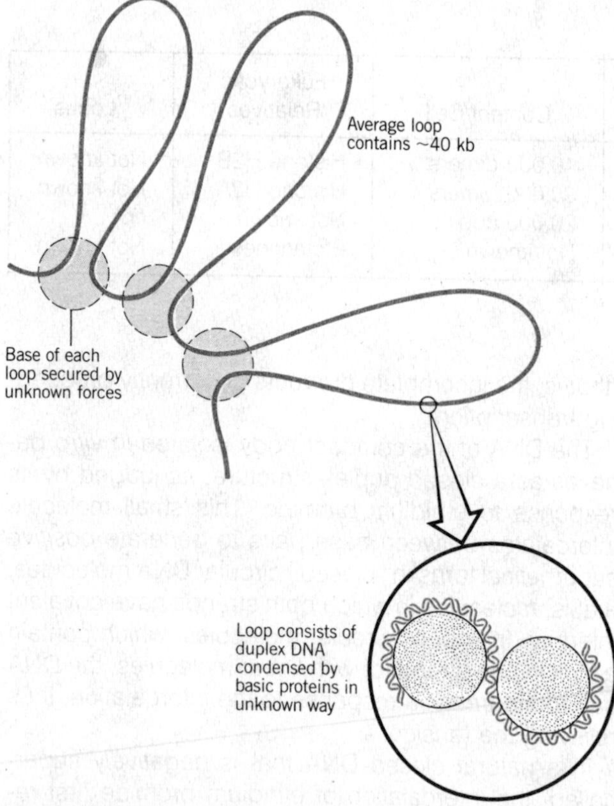

Figure 28.7
The bacterial genome consists of a large number of loops of duplex DNA (in the form of a fiber), each secured at the base to form an independent structural domain.

of DNA, organized into some more compact fiber whose structure has yet to be characterized.

The existence of separate domains could permit different degrees of supercoiling to be maintained in different regions of the genome. This is a pertinent factor in considering the different susceptibilities of particular bacterial promoters in response to supercoiling (see Chapter 12).

A supercoiled DNA can be organized in two ways. If the DNA remains free, its path is **unrestrained,** and negative supercoils generate a state of torsional tension that can be relieved by unwinding the double helix, as described in Chapter 4. The DNA may be in a dynamic equilibrium between the states of tension and unwinding. However, the supercoiling can be **restrained** if proteins are bound to the DNA to hold it in a particular three-dimensional configuration. In this case, the supercoils are represented by the path the DNA follows in its fixed association with the proteins. The energy of interaction between the proteins and the supercoiled DNA influences the stability of the double helix, which could be stabilized in duplex form or single-stranded form, so that no tension is transmitted along the molecule.

Are the supercoils in *E. coli* DNA mostly restrained *in vivo* or is the double helix subject to the torsional tension characteristic of free DNA? Measurements of supercoiling *in vitro* encounter the difficulty that restraining proteins might have been lost during isolation. However, the situation *in vivo* can be investigated by use of the cross-linking reagent psoralen, which binds more readily to DNA when it is under torsional tension. The reaction of psoralen with *E. coli* DNA *in vivo* corresponds to an average density of one negative superhelical turn for every 200 bp. This result demonstrates that supercoils *do* create torsional tension *in vivo*. Most of the natural supercoils in *E. coli* DNA may be unrestrained.

Many of the important features of the structure of the compact nucleoid remain to be established. What is the specificity with which domains are constructed—do the same sequences always lie at the same relative locations, or can the contents of individual domains shift? How is the integrity of the domain maintained? What structural roles are played by the DNA-binding proteins? Biochemical analysis by itself may be unable to answer these questions fully, but if it is possible to devise suitable selective techniques, the properties of structural mutants should lead to a molecular analysis of nucleoid construction.

THE CONTRAST BETWEEN INTERPHASE CHROMATIN AND MITOTIC CHROMOSOMES

Individual eukaryotic **chromosomes** come into the limelight for only a brief period, during the act of cell division. Only then can each be seen as a compact unit, with a packing ratio of up to 10,000. **Figure 28.8** is an electron micrograph of a sister chromatid pair, captured at metaphase. (The sister chromatids are daughter chromosomes produced by the previous replication event, still joined together at this stage of

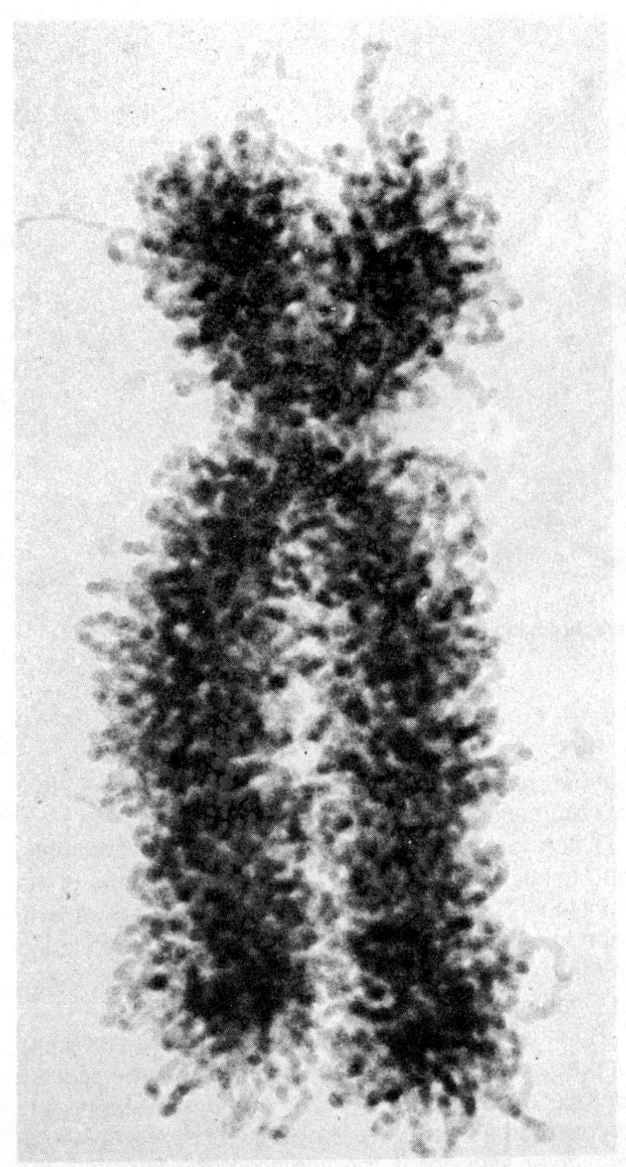

Figure 28.8
The sister chromatids of a mitotic pair each consist of a fiber (~ 30 nm in diameter) compactly folded into the chromosome.
Photograph kindly provided by E. J. DuPraw.

mitosis, as described in Chapter 2.) Each consists of a fiber with a diameter of about 30 nm and a nubbly appearance, folded into the chromosomal structure.

During most of the life cycle of the eukaryotic cell, however, its genetic material occupies an area of the nucleus in which individual chromosomes cannot be distinguished. The structure of the interphase **chromatin** does not change visibly between divisions. For example, usually no disruption is evident during the period of replication, when the amount of chromatin doubles. Chromatin is fibrillar, although the overall configuration of the fiber in space is hard to discern in detail. The fiber itself, however, is similar to (and could be identical with) that found in the mitotic chromosomes.

Chromatin can be divided into two types of material. These are visible in **Figure 28.9.** In most regions, the fibers are much less densely packed than they are in the chromosome. This material is called **euchromatin.** Its overall packing ratio is probably about 1000–2000 during interphase.

Some regions of chromatin are very densely packed with fibers, displaying a condition comparable to that of the chromosome at mitosis. This material is called **heterochromatin.** It passes through the cell cycle with relatively little change in its degree of condensation. Often the various heterochromatic regions aggregate into a densely staining **chromocenter.**

The same fibers run continuously between euchromatin and heterochromatin, which implies that these states represent different degrees of condensation of the genetic material. In the same way, euchromatic regions exist in different states of condensation during interphase and during mitosis. The important conclusion to be drawn from these features is that the genetic material is organized in a manner that both permits alternative states to be maintained side by side in chromatin, and allows cyclical changes to occur in the packaging of euchromatin between interphase and division.

There is a correlation between the structural condition of the genetic material and its transcriptional activity. Mitotic chromosomes are transcriptionally inert; cells virtually cease transcription during the process of division. There are two classes of heterochromatin, each containing a different type of sequence; and in neither type is the DNA transcribed.

Constitutive heterochromatin consists of particular regions that usually are *never* expressed. This includes satellite DNA sequences. **Facultative heterochromatin** takes the form of entire chromosomes that are inactive in one cell lineage, although they can be

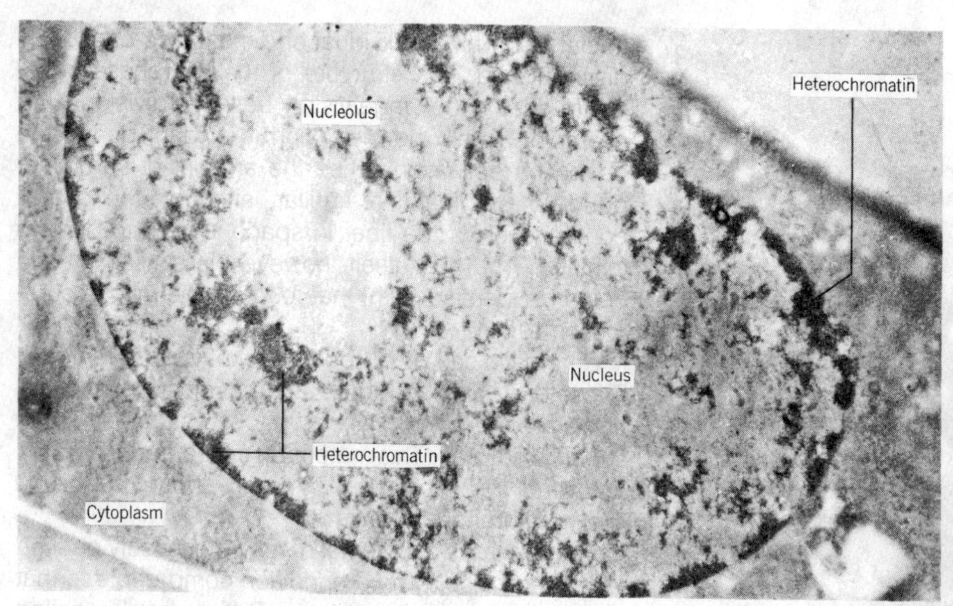

Figure 28.9
A thin section through a nucleus stained with Feulgen shows heterochromatin as compact regions clustered near the nucleolus and nuclear membrane.
Photograph kindly provided by Edmund Puvion.

expressed in other circumstances. The example par excellence is the mammalian X chromosome, one copy of which (selected at random) is entirely inactive in a given female. (This compensates for the presence of two X chromosomes, compared with the one present in males.) The inactive X chromosome is perpetuated in a heterochromatic state; the active X chromosome is part of the euchromatin. Here it is possible to see a correlation between transcriptional activity and structural organization when the *identical DNA sequences* are involved in both states.

A very tight condensation of the genetic material is thus associated with (perhaps is responsible for) its inactivity. Note, however, that the reverse is not true. Active genes are indeed contained within euchromatin; but only a small minority of the sequences in euchromatin are transcribed at any time. Thus location in euchromatin is *necessary* for gene expression, but is not *sufficient* for it. We may wonder whether the gross changes seen between euchromatin and heterochromatin are mimicked in a lesser manner by changes in the structure of euchromatin, to give transcribed regions a less condensed structure than that of nontranscribed regions.

When isolated *in vitro,* both interphase chromatin and mitotic chromosomes possess large loops of the constituent fiber, apparently secured at the base to form independent domains analogous to those found in the bacterial nucleoid. Their structure is discussed in Chapter 29.

Because of the diffuse state of chromatin, we cannot at present determine the specificity of its organization. But we can ask how ordered is the structure of the chromosome? Do particular sequences always lie at particular sites, or is the folding of the fiber into the overall structure a more random event? Some years ago, the development of the **G-banding** technique showed that each chromosome has a different and reproducible ultrastructure. When subjected to certain treatments and then stained with the chemical dye Giemsa, chromosomes generate a series of bands. An example of the human set is presented in **Figure 28.10.**

Until the development of this technique, chromo-

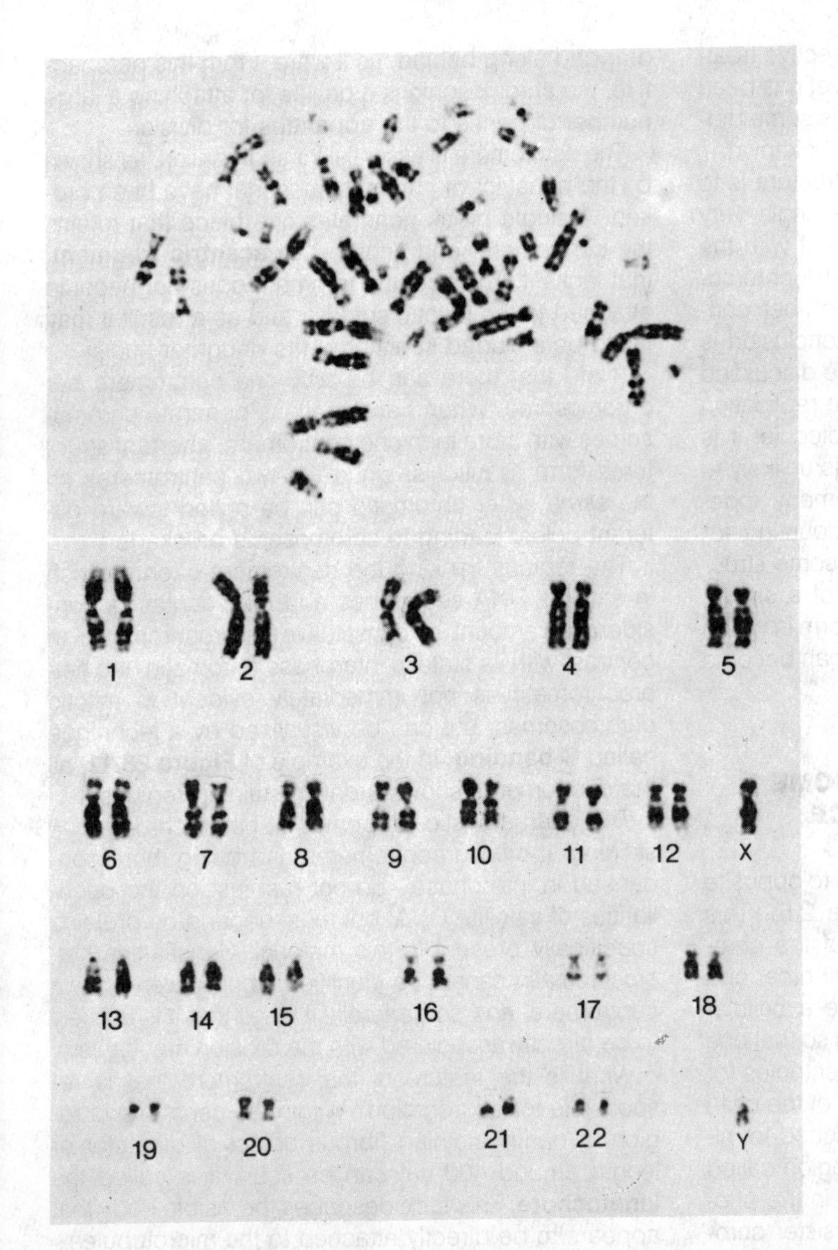

Figure 28.10
G-banding generates a characteristic lateral series of bands in each member of the chromosome set.
This example of the human complement was provided by N. Davidson.

somes could be distinguished only by their overall size and the relative location of the centromere (see later). Now each chromosome can be identified by its characteristic banding pattern. This pattern is reproducible enough to allow translocations from one chromosome to another to be identified by comparison with the original diploid set.

The banding technique is of enormous practical use, but the mechanism of banding remains a mystery. All that is certain is that the dye stains untreated chromosomes more or less uniformly. So the generation of bands depends on a variety of treatments that change the response of the chromosome (presumably by extracting the component that binds the stain from the nonbanded regions). But the variety of effective treatments is so great that no common cause yet has been discerned. These results imply that there is some definite long-range structure, but its basis is unknown.

The simplest model for chromosome structure is to suppose that each chromosome contains a single, very long duplex of DNA. This view is consistent with the continuity of the 30 nm fiber throughout the chromosome. Early experiments showed that the fiber contains a single duplex of DNA, and this conclusion is confirmed in detail by experiments that are discussed in Chapter 29. The pattern of chromosome replication resembles that of the individual DNA molecule: it is semiconservative (see Figure 1.19). This is unlikely to be the case if a chromosome carries many independent molecules of DNA. Thus in accounting for interphase chromatin and mitotic chromosome structure, we have to explain the packaging of a single, exceedingly long molecule of DNA into a form in which it can be transcribed and replicated, and can become cyclically more and less compressed.

THE EUKARYOTIC CHROMOSOME AS A SEGREGATION DEVICE

During mitosis, the sister chromatids move to opposite poles of the cell (as illustrated in Figure 2.5). This movement depends on the attachment of the chromosome to microtubules connected at their other end to the poles. [The microtubules comprise a cellular filamentous system, reorganized at mitosis so that one end of each microtubule is joined to the centrioles (or a point close to them) that lie at each pole of the cell.]

The site at which a chromosome is attached to microtubules is identified by a constricted region called the **centromere.** This constriction is clear in the photograph of Figure 28.8, which shows the sister chromatids at the metaphase stage of mitosis, when they are still attached to one another in the centromeric region.

The term "centromere" historically has been used in both the functional and structural sense to describe the feature of the chromosome responsible for its movement. The centromere is pulled toward the pole during mitosis, and the attached chromosome is dragged along behind, as it were. From this perspective, the chromosome is a device for attaching a large number of genes to the apparatus for division.

The centromere is essential for segregation, as shown by the behavior of chromosomes that have been broken. A single break generates one piece that retains the centromere, and another, an **acentric fragment,** that lacks it. The acentric fragment does not become attached to the mitotic spindle; and as a result it may fail to be included in either of the daughter nuclei.

(Note that there can be *only* one centromere per chromosome. When translocations generate chromosomes with more than one centromere, aberrant structures form at mitosis, since the two centromeres on the *same* sister chromatid can be pulled toward different poles, leading to chromosome breakage.)

The regions flanking the centromere often are rich in satellite DNA sequences and may contain a considerable amount of constitutive heterochromatin. In contrast with its state in interphase chromatin, the heterochromatin is not immediately evident in mitotic chromosomes, but can be visualized by a technique called **C-banding.** In the example of **Figure 28.11,** all the centromeres show as darkly staining regions.

The characteristic properties of heterochromatin—staining at mitotic centromeres and being more condensed in interphase—do not rest just on the peculiarities of satellite DNA, but must depend on proteins specifically present in this material. Constitutive heterochromatin cannot be identified around *every* known centromere, and so generally it is felt that it is unlikely to be directly associated with the division mechanism.

What is the feature of the centromere that is responsible for segregation? Within the centromeric region, a darkly staining fibrous object of diameter or length around 400 nm can be seen. It is called the **kinetochore.** This term describes the visible body that appears to be directly attached to the microtubules.

Usually it is assumed that a specific sequence of DNA in some way defines the site at which the kinetochore should be established. The structure of the kinetochore varies, and has been difficult to analyze in relation to its function. However, there have been claims that it contains DNA, which would imply that it may form *directly* on the responsible chromosomal sequence.

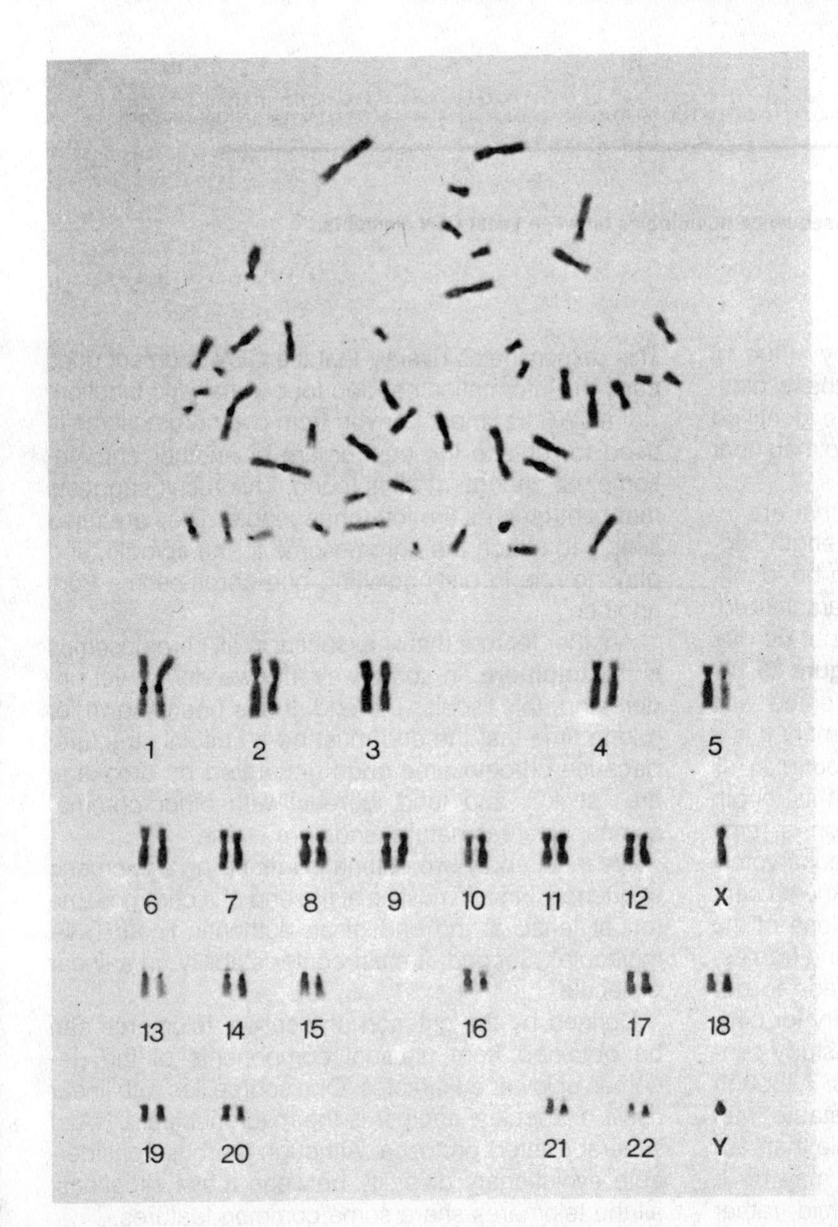

Figure 28.11
C-banding generates intense staining at the centromeres of all chromosomes.
Photograph kindly provided by N. Davidson.

If a centromeric sequence of DNA is responsible for segregation, any molecule of DNA possessing this sequence should move properly at cell division, while any DNA lacking it will fail to segregate. This prediction has been used to isolate centromeric DNA in the yeast, *S. cerevisiae*. Yeast chromosomes do not display visible kinetochores comparable to those of higher eukaryotes, but otherwise divide at mitosis and segregate at meiosis by the same mechanisms.

Genetic engineering has produced plasmids of yeast that are replicated like chromosomal sequences (see Chapter 31). However, they are unstable at mitosis and meiosis, disappearing from a majority of the cells because they segregate erratically. Fragments of

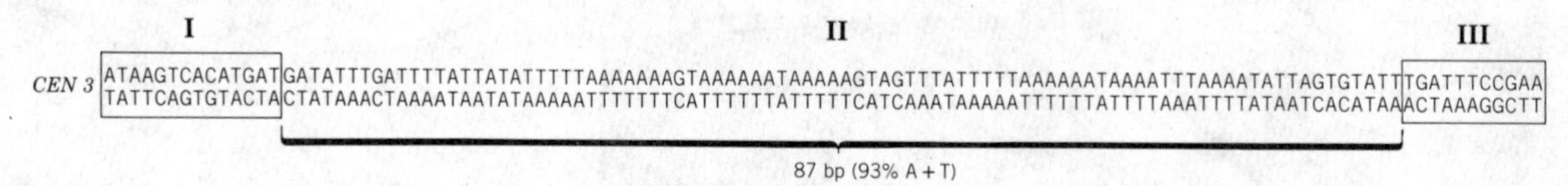

Figure 28.12
Three conserved regions can be identified by the sequence homologies between yeast *CEN* elements.

chromosomal DNA have been isolated by virtue of their ability to confer mitotic stability on these plasmids. Their chromosomal derivations can be identified when they contain genetic markers known to map near a centromere.

By reducing the sizes of the fragments that are incorporated into the plasmid, the minimum length necessary for mitotic centromeric function can be identified. Several *CEN* fragments have been characterized. They lack extensive regions of homology, but do display the common features summarized in **Figure 28.12.**

Element I is a 14 bp sequence represented with some variations in all the centromeres. Element II is a >90% A-T-rich sequence of 82–89 bp found in all centromeres; its function could depend on its length rather than exact sequence. Its constitution is reminiscent of some satellite DNAs in higher eukaryotes. Element III is an 11 bp region found in all known centromeres with almost perfect homology. None of the centromeric sequences has any open reading frames.

Deletions and other changes can be made to determine directly what features are necessary for centromeric function. However, it is difficult to study centromeric functions directly in the plasmids. Although plasmids with *CEN* sequences are more stable than those without, still they are much less stable than authentic centromeres. The lack of stability may be a consequence of the small size of the plasmid, rather than due to any deficiency in the *CEN* region.

Another way to use the availability of the centromeric sequences is to modify them *in vitro* and then reintroduce them into the yeast cell, where they may replace the corresponding centromere on the chromosome. If the conserved region is removed from the *CEN* fragment, when the fragment replaces the chromosomal centromere, the chromosome behaves like an acentric fragment in mitosis: it cannot segregate. This demonstrates directly that the *CEN* fragment does carry the information needed for centromeric function.

If a *CEN* fragment derived from one chromosome is used to replace the centromere of another chromosome, no change at all is found. This result suggests that centromeres are interchangeable. They are used simply to attach the chromosome to the spindle, and play no role in distinguishing one chromosome from another.

Another feature that is essential in all chromosomes is the **telomere.** In some way that we do not yet understand, this "seals" the end. It has been known for a long time that the end must be a special structure, because chromosome ends generated by breakage are "sticky" and tend to react with other chromosomes, whereas natural ends are stable.

We must apply two criteria in identifying a telomeric sequence. First, it must lie at the end of a chromosome (or, at least, at the end of an authentic linear DNA molecule). Second, it must confer stability on a linear molecule.

Defined by the criterion of location, telomeres can be obtained from unusual components of the genomes of lower eukaryotes. One source lies with linear rDNA molecules; another is the macronuclear DNA of several ciliated protozoa. Although there is considerable evolutionary diversity between these situations, all the telomeres share some common features.

The telomere consists of a long series of short, tandemly repeated sequences. **Table 28.3** lists the repeating units that have been identified at the ends of linear DNA molecules. They can all be written in the general form $C_n(A/T)_m$, where n>1 and m is 1–4.

Within the telomeric region is a specific array of discontinuities, taking the form of single-strand breaks whose structure prevents them from being sealed by the ligase enzyme that normally acts upon nicks in

Table 28.3
Telomeres have a common type of short tandem repeat.

Organism	Source of DNA	Repeating Unit (5′ – 3′)
Holotrichous ciliates (e.g., *Tetrahymena, Paramecium*	macronucleus	CCCCAA
Hypotrichous ciliates (e.g., *Stylonchia, Oxytricha*	macronucleus	CCCCAAAA
Flagellates (e.g., *Trypansoma, Leishmania*	mini-chromosome	CCCTA
Slime molds: *Physarum*	rDNA	$CCCTA_n$
Dictyostelium	rDNA	$C_{1-8}T$
Yeast (*Saccharomyces*)	chromosome	$C_{2-3}A(CA)_{1-3}$

The repeating unit gives the sequence of one strand, going in the direction from the telomere toward the interior.

one DNA strand. The very terminal bases are blocked in some way—they may be organized in a hairpin—so that they are not recognized by nucleases.

The problem of finding a system that offers an assay for function again has been brought to the molecular level by using yeast. The plasmids that survive in yeast all are circular DNA molecules. Linear plasmids are unstable (because they are degraded). Could an authentic telomeric DNA sequence confer stability on a linear plasmid?

A region from the end of a known natural linear DNA molecule—the extrachromosomal rDNA of *Tetrahymena*—is able to render the yeast plasmid stable in linear form. The nicks in the telomeric sequence are perpetuated at the same sites in yeast, a remarkable interspecies conservation.

The *Tetrahymena* sequence can be replaced by fragments from yeast DNA that have related sequences. They are likely to prove to be the natural telomeres of yeast chromosomes. The same type of analysis performed with the centromeric DNA, using size reduction and inducing mutations, should reveal how they function.

Some indications about how a telomere functions are given by some unusual properties of the ends of linear DNA molecules. In a trypanosome population, the ends are variable in length. When an individual cell clone is followed, the telomere grows longer by 7-10 bp per generation. Even more revealing is the fate of ciliate telomeres introduced into yeast. After replication in yeast, yeast telomeric repeats *are added onto the ends of the Tetrahymena repeats*.

It seems likely that addition of telomeric repeats to the end of the chromosome occurs in every replication cycle (probably providing a solution to the problem of replicating linear DNA molecules discussed in Chapter 32). Some mechanism must prevent the ends from growing too long, possibly by removing some of the repeats.

If telomeres are continually being lengthened, their exact sequence may be irrelevant. All that is required is for the present end to be recognized as a suitable substrate for addition. This explains how the ciliate telomere functions in yeast. However, we do not yet understand how the telomeric sequence is related to the resistance of chromosome ends to being damaged.

The minimum features required for existence as a chromosome are the possession of proper telomeres to ensure survival, a centromere to support segregation, and an origin where replication is initiated (see Chapter 31). All of these elements have been put together to construct a synthetic yeast chromosome. It turns out that the synthetic chromosome is stable only if it is longer than 20–50 kb. We do not know the basis for this effect, but the ability to construct a synthetic chromosome offers the potential to investigate the nature of the segregation device in a controlled environment.

THE EXTENDED STATE OF LAMPBRUSH CHROMOSOMES

It would be extremely useful to visualize gene expression in its natural state, to see what structural changes are associated with transcription. But the nature of the material restricts such analysis to only some unusual circumstances.

The compression of DNA in chromatin, coupled with the difficulty of identifying particular genes within it, makes it impossible to visualize the transcription of individual active genes. (However, they do display some distinctive biochemical properties that can be analyzed *in vitro*, as described in Chapter 30.)

Mitotic chromosomes are inert in gene expression, and, in any case, are so compact as to preclude the identification of individual loci. The distinct regions that are rendered discernible by the G-banding technique, as in the example of Figure 28.10, each contain of the order of 10^7 bp of DNA, which could include many hundreds of genes.

Lateral differentiation of structure is evident in many chromosomes when they first appear for meiosis. At this stage, the chromosomes resemble a series of beads on a string. The beads are densely staining granules, properly known as **chromomeres.** However, usually there is little gene expression at meiosis, and it is not practical to use this material to identify individual genes.

Figure 28.13
A lampbrush chromosome is a meiotic bivalent in which the two pairs of sister chromatids are held together at chiasmata (indicated by arrows).
Photograph kindly provided by Joe Gall.

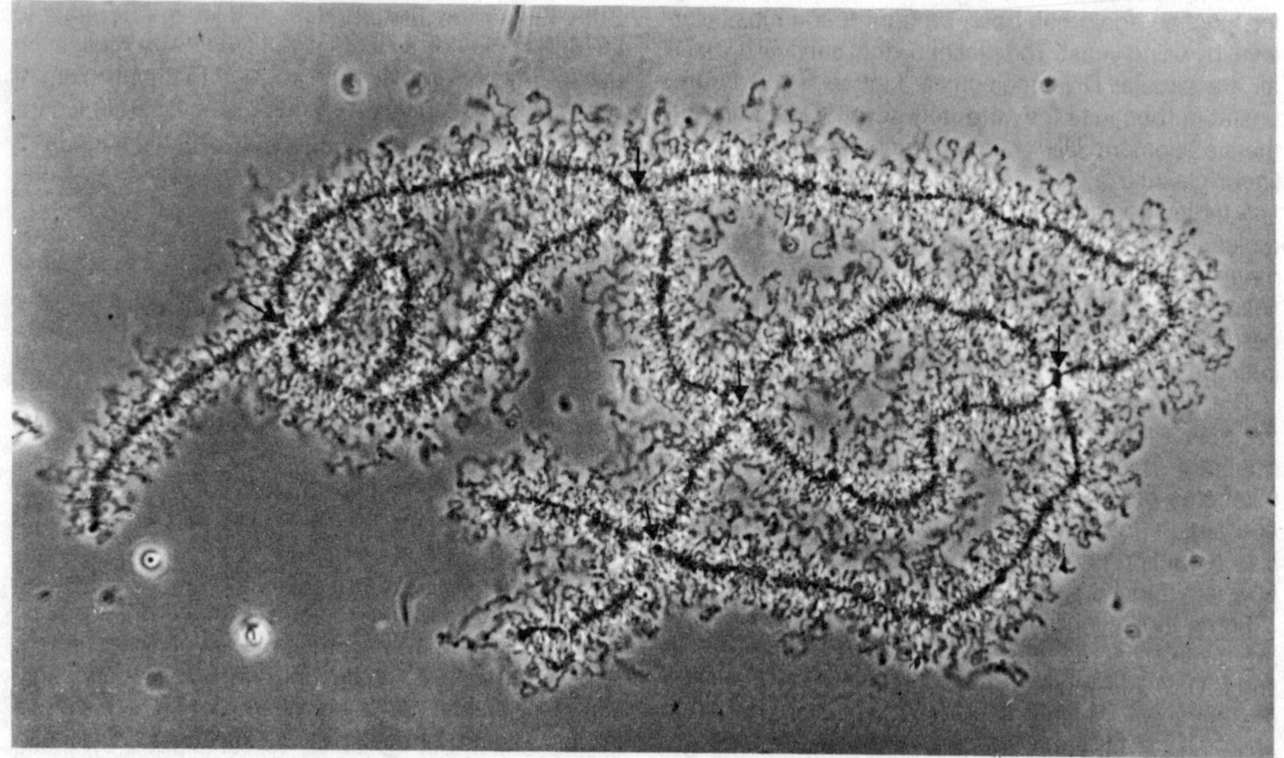

Gene expression can be visualized directly in two situations, in which the chromosomes are found in a highly extended form that allows individual loci (or groups of loci) to be distinguished. These are the lampbrush chromosomes (best characterized in certain amphibians), and the polytene chromosomes (best characterized in some insects).

The **lampbrush chromosomes** are formed during an unusually extended meiosis, which can last up to several months! During this period, the chromosomes are held in a stretched-out form in which they can be visualized in the light microscope. Later during meiosis, the chromosomes revert to their usual compact size. So the extended state essentially proffers an unfolded version of the normal condition of the chromosome.

The lampbrush chromosomes are meiotic bivalents, so each consists of two pairs of sister chromatids. **Figure 28.13** shows an example in which the sister chromatid pairs have mostly separated so that they are held together only by chiasmata (the sites of crossing-over). Each sister chromatid pair forms a series of ellipsoidal chromomeres, about 1–2 μm in diameter, which are connected by a very fine thread. This thread contains the two sister duplexes of DNA and runs continuously along the chromosome, through the chromomeres.

The lengths of the individual lampbrush chromosomes in the newt *Notophthalmus viridescens* range from 400 to 800 μm, compared with the range of 15–20 μm seen later in meiosis. So the lampbrush chromosomes are about 30 times less tightly packed. The total length of the entire lampbrush chromosome set may be 5–6 mm, organized into about 5000 chromomeres.

The lampbrush chromosomes take their name from the lateral loops that extrude from the chromomeres at certain positions. (These resemble a lampbrush, a now-extinct object.) The loops extend in pairs, one from each sister chromatid. The loops are continuous with the axial thread, which suggests that they represent chromosomal material that has been extruded from its more compact organization in the chromomere.

The loops are surrounded by a matrix of ribonucleoproteins. These contain nascent RNA chains. Often a transcription unit can be defined by the increase in the length of the RNP moving around the loop. An example is shown in **Figure 28.14.**

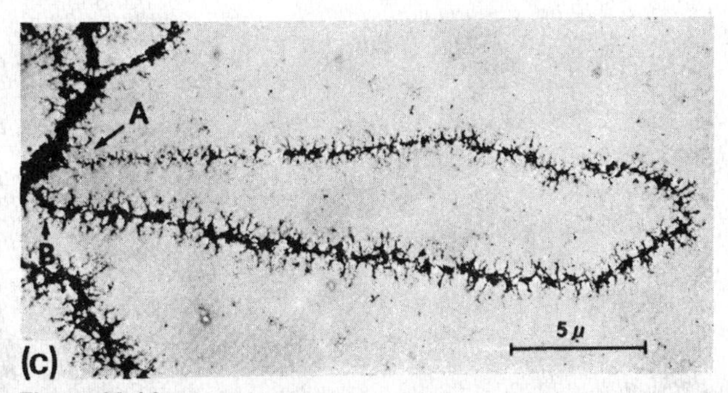

Figure 28.14
A lampbrush chromosome loop is surrounded by a matrix of ribonucleoprotein.
Photograph kindly provided by Oscar Miller.

So the loop is an extruded segment of DNA that is being actively transcribed. In some cases, loops corresponding to particular genes have been identified. Then the structure of the transcribed gene, and the nature of the product, can be scrutinized *in situ*.

POLYTENY FORMS GIANT CHROMOSOMES

The interphase nuclei of some tissues of the larvae of Dipteran flies contain chromosomes that are greatly enlarged relative to their usual condition. They possess both increased diameter and greater length. **Figure 28.15** shows an example of a chromosome set from the salivary gland of *D. melanogaster*. They are called **polytene chromosomes.**

Each chromosome consists of a visible series of **bands** (more properly, but rarely, described as chromomeres). There is an approximately tenfold range in the sizes of the bands, the largest having a breadth of ~0.5 μm and the smallest ~0.05 μm. (The smallest can be distinguished only under the electron microscope.) The bands contain most of the mass of DNA and stain intensely with appropriate reagents. The regions between them stain more lightly and are called **interbands.** There are about 5000 bands in the *D. melanogaster* set.

The centromeres of all four chromosomes of *D. melanogaster* aggregate to form a chromocenter that

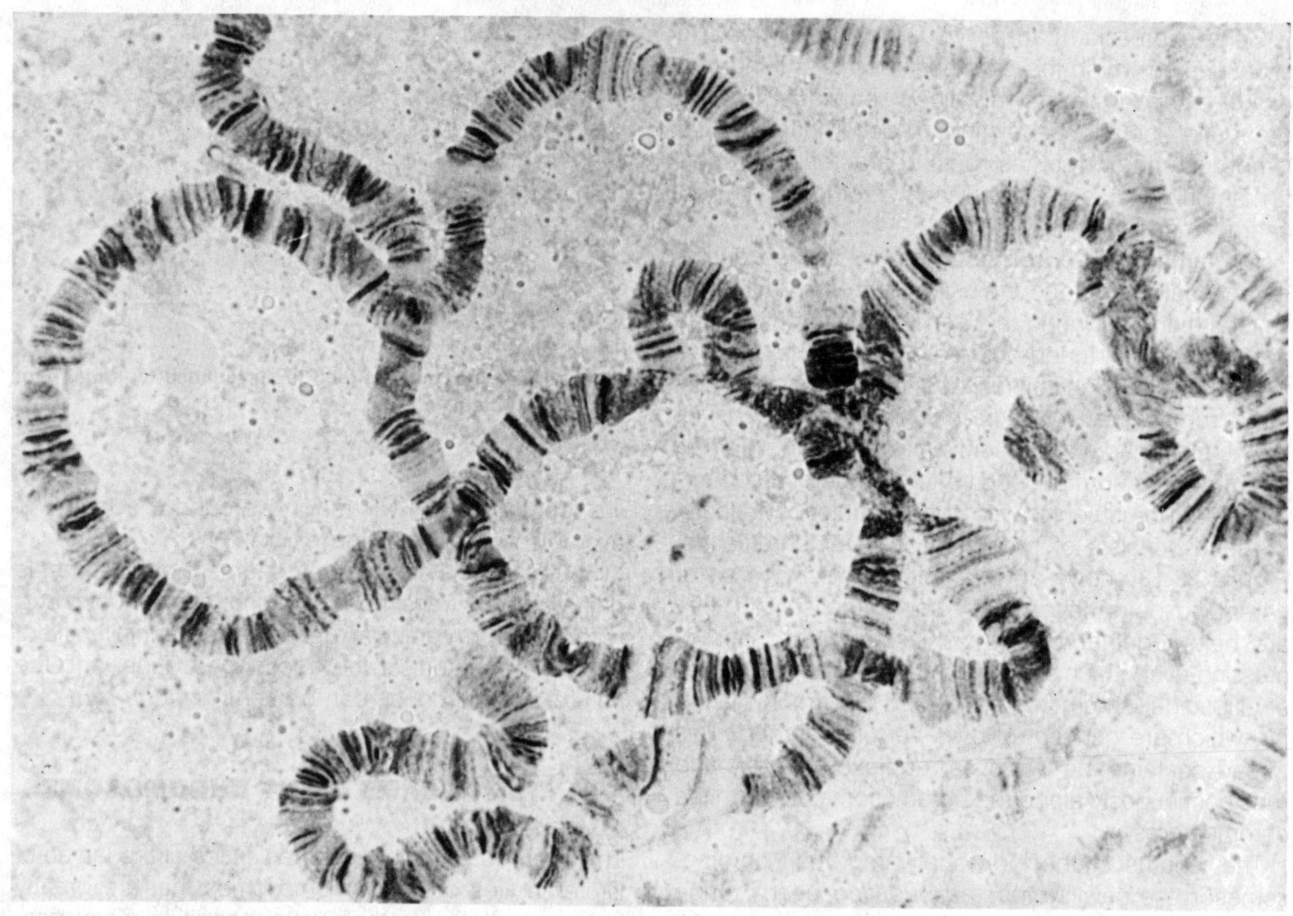

Figure 28.15
The polytene chromosomes of *D. melanogaster* form an alternating series of bands and interbands.
Photograph kindly provided by Jose Bonner.

consists largely of heterochromatin (and in the male includes the entire Y chromosome). Allowing for this, about 75% of the haploid DNA set is organized into the band-interband alternation. The length of the chromosome set is about 2000 μm; 75% of the DNA is 1.3 x 10^8 bp, which would extend for about 40,000 μm, so the average packing ratio is ~20. This demonstrates vividly the extension of the genetic material relative to the usual states of interphase chromatin or mitotic chromosomes.

What is the structure of these giant chromosomes? Each is produced by the successive replications of a synapsed diploid pair. The replicas do not separate, but remain attached to each other in their extended state. At the start of the process, each synapsed pair has a DNA content of 2C (where C represents the DNA content of the individual chromosome). Then this doubles up to 9 times, at its maximum giving a content of 1024C. The number of doublings is different in the various tissues of the *D. melanogaster* larva and, of course, in other Dipteran flies.

Each chromosome can be visualized as a large number of parallel fibers running longitudinally, tightly condensed in the bands, less condensed in the interbands. Probably each fiber represents a single (C) haploid chromosome content. This gives rise to the name "polyteny." The degree of polyteny is the number of haploid chromosomes contained in the giant chromosome.

The banding pattern is characteristic for each strain of *Drosophila*. The constant number and linear arrangement of the bands was first noted in the 1930s, when it was realized that they form a **cytological map** of the chromosomes. Rearrangements—such as deletions, inversions, or duplications—result in alterations of the order of bands.

The linear array of bands can be equated with the linear array of genes. Thus genetic rearrangements, as seen in a linkage map, can be correlated with structural rearrangements of the cytological map. Ultimately, a particular mutation can be located in a particular band.

The positions of particular genes on the cytological map can be determined directly by the technique of ***in situ*** or **cytological hybridization,** described in Chapter 24 in the context of highly repetitive DNA. The protocol is recapitulated in **Figure 28.16.**

A radioactive probe representing a gene (most often a labeled cDNA clone derived from the mRNA) is hybridized with the denatured DNA of the polytene chromosomes *in situ*. By the superimposition of grains at a particular band or bands, autoradiography identifies the position or positions of the corresponding genes. An example is shown in **Figure 28.17.** With this type of technique at hand, it is therefore possible to determine directly the band within which a particular sequence lies.

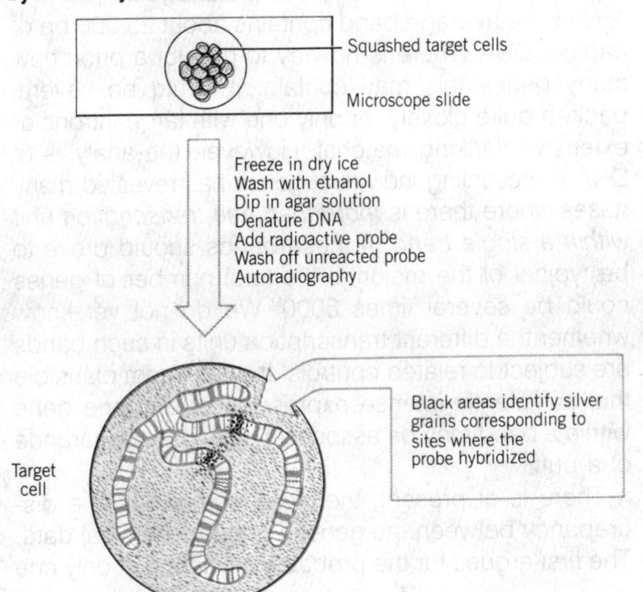

Figure 28.16
Individual bands containing particular genes can be identified by *in situ* hybridization.

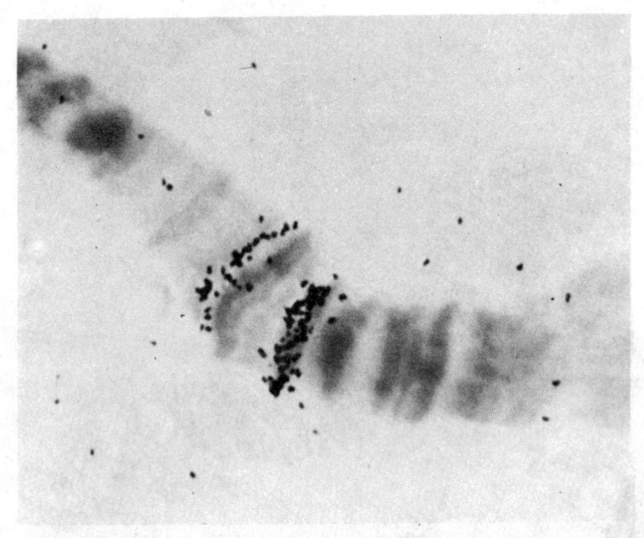

Figure 28.17
A magnified view of bands 87A and 87C shows their *in situ* hybridization with labeled RNA extracted from heat-shocked cells.
Photograph kindly provided by Jose Bonner.

TRANSCRIPTION DISRUPTS THE CHROMOSOME STRUCTURE

One of the intriguing features of the polytene chromosomes is that active sites can be visualized. Some of the bands are found at times in an expanded or **puffed** state, in which chromosomal material is extruded from the axis. An example of some very large puffs (called Balbiani rings) is shown in **Figure 28.18.**

What is the nature of the puff? It consists of a region in which the chromosome fibers unwind from their usual state of packing in the band. The fibers remain continuous with those in the chromosome axis. Puffs usually emanate from single bands, although when they are very large, as typified by the Balbiani rings, the swelling may be so extensive as to obscure the underlying array of bands.

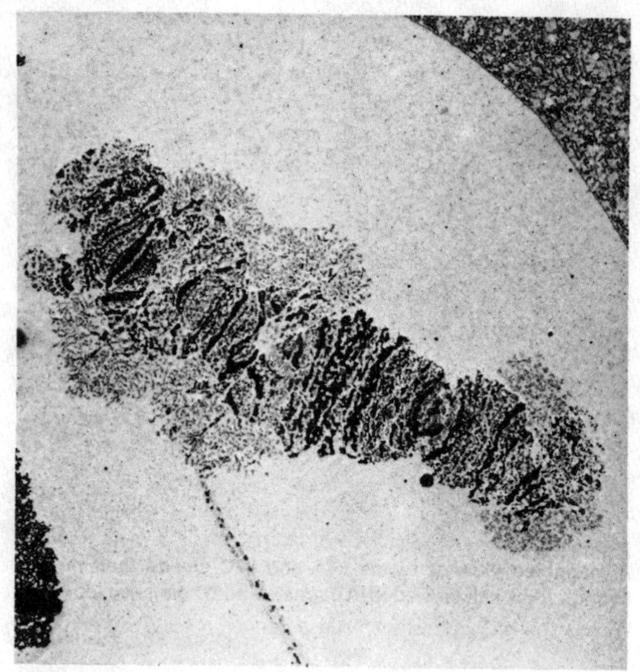

Figure 28.18
Chromosome IV of *C. tentans* has three Balbiani rings in the salivary gland.
Photograph kindly provided by Bertil Daneholt.

The pattern of puffs is related to gene expression. During larval development, puffs appear and regress in a definite, tissue-specific pattern. A characteristic pattern of puffs is found in each tissue at any given time. Puffs are induced by the hormone ecdysone that controls *Drosophila* development. Some puffs are induced directly by the hormone; others are induced indirectly by the products of earlier puffs.

The puffs are *sites where RNA is being synthesized*. The accepted view of puffing has been that expansion of the band is a consequence of the need to relax its structure in order to synthesize RNA. Puffing has therefore been viewed as a consequence of transcription.

A hint that puffing instead might be a requirement for transcription is provided by a mutant that has a puff at a site that usually synthesizes glue proteins, even though the mutation prevents transcription of the genes located at the band. Perhaps the mutation has stopped gene expression at some stage after puff formation but before transcription. The idea that puffing can be dissociated from the act of transcription is novel, because this mutant is the only exception to an apparent rule that the puff directly reflects the extent of RNA synthesis.

The sites of puffing differ from the ordinary bands in accumulating additional proteins. Characterization of these proteins at present is only rather primitive, but in due course it should be possible to establish whether they are a cause or a consequence of the puffing. Then it should be possible to determine the nature of the molecular events that are responsible for the expansion of material.

So a puff represents a band containing a gene that is being transcribed actively. Is the band a unit of gene expression? To this somewhat vexed question, there is at present no unequivocal answer.

Genetic experiments have concentrated on correlating the number of lethal loci with the bands. A **lethal locus** is defined by mutations that prevent the organism from surviving. Thus it identifies an essential gene. The number of lethal loci appears to be close to the number of bands; and in defined areas of the chromosomes, individual bands can be equated with lethal loci on virtually a 1:1 basis. This would argue that there are about 5000 lethal loci in *D. melanogaster*.

Now the average band contains about 25,000 bp of haploid DNA. There is no way to predict a priori how many genes this may contain: it could be several packed quite closely, or only one with large introns or extensive flanking material. However, the analysis of DNA representing individual genes has revealed many cases where there is *more than one transcription unit within a single band*. If these bands should prove to be typical of the majority, the total number of genes could be several times 5000. We do not yet know whether the different transcription units in such bands are subject to related controls. It does seem plausible that sufficiently intense expression of any one gene within a band may be associated with the appearance of a puff.

There is at present, therefore, somewhat of a discrepancy between the genetic and biochemical data. The first argues for the probable existence of only one essential gene per band, the second argues for the presence of several genes per band. A possible reconciliation argues that many genes may be nonessential, at least as defined by the criterion of the lethal locus. This leaves the puzzle of why lethal loci should be distributed regularly at approximately one per band.

The features displayed by lampbrush and polytene chromosomes suggest a general conclusion. In order to be transcribed, the genetic material is dispersed from its usual more tightly packed state. The question to keep in mind is whether this dispersion at the gross level of the chromosome mimics the events that occur at the molecular level within the mass of ordinary interphase euchromatin.

FURTHER READING

Phage morphogenesis has been dealt with in detail in **Lewin's *Gene Expression,*** 3, *Plasmids and Phages* (Wiley, New York, 1978, pp. 496–535, 642–673). The unusual chromosomes discussed in this chapter are the subject of pp. 455–475 of *Gene Expression*, 2, *Eucaryotic Chromosomes*. Centromeres have been reviewed by **Carbon** (*Cell* **37,** 351–353, 1984). Telomeres have been reviewed by **Blackburn** (*Cell* **37,** 7–8, 1984). A review of both chromosome regions has been provided by **Blackburn & Szostak** (*Ann. Rev. Biochem.* **53,** 163–194, 1984). The functional elements needed by a chromosome were put together to make a synthetic chromosome by **Murray & Szostak** (*Nature* **305,** 189–193, 1983).

CHAPTER 29
CHROMATIN STRUCTURE: THE NUCLEOSOME

Chromatin contains DNA, RNA, and protein in a compact form of organization in which the majority of DNA sequences are structurally inaccessible and functionally inactive. Within this mass are the minority of active sequences. What is the general structure of chromatin and how does it differ between active and inactive sequences?

Enormous progress has been made since the critical insight in 1974 that chromatin is constructed from a subunit that has the *same type of design in all eukaryotes.* It contains about 200 bp of DNA, organized by an octamer of small, basic proteins (histones) into a beadlike structure. It turns out that DNA lies on the surface of this particle, while the protein components constitute the interior.

At the level of the fundamental DNA-protein interaction, we can define the structure of the particle in terms of the path of the DNA and the contacts between the nucleic acid and the proteins. We can investigate the binding between the individual proteins.

We can ask whether the conformation of the particle can vary, and how such variations might be related to the functions of chromatin. What happens to the particle when chromatin is transcribed or replicated? These processes may require additional proteins to modulate the behavior of the particle.

When chromatin is replicated, the series of particles must be reproduced on both daughter duplex molecules. As well as asking how the particle is assembled, we must inquire what happens to other proteins present in chromatin. Since replication is likely to disrupt the structure of chromatin, it both poses a problem for maintaining regions with specific structure and offers an opportunity to change the structure.

The high overall packing ratio of the genetic material immediately suggests that DNA cannot be directly packaged into the final structure of chromatin. There must be *hierarchies* of organization.

The first level is the winding of the DNA into the beadlike particles, with a packing ratio of about 6. These particles are an invariant component of euchromatin, heterochromatin, and chromosomes.

The second level of organization is the coiling of the series of beads into a helical array to constitute the ~30 nm fiber that is found in both interphase chromatin and mitotic chromosomes (see Figure 28.8). In chromatin this brings the packing ratio of DNA to about 40. It is not certain whether the fiber is unaffected or has an increased packing ratio in mitotic chromosomes.

The final packing ratio is determined by the third level of organization, the packaging of the fiber itself.

This gives an overall packing ratio of ≥1000 in euchromatin, cyclically interchangeable with the packing into mitotic chromosomes to achieve an overall ratio of ≤10,000. This too is likely to be a function modulated by accessory proteins (and also possibly involving chemical modification of histones), as is the difference between euchromatin and heterochromatin.

THE PROTEIN COMPONENTS OF CHROMATIN

The mass of chromatin contains roughly twice as much protein as DNA. The proteins are divided into two types: histones and nonhistones. The mass of RNA is less than 10% of the mass of DNA. Much of the RNA consists of nascent chains that are still associated with the template DNA.

The **histones** are the most basic proteins in chromatin and account for just about the same mass as the DNA. Five classes of histones were originally characterized by the relative proportions of each type of basic amino acid, as expressed in the lysine/arginine ratio. A relic of this description still lingers in the names of the histones. The same classes can be recognized in virtually all eukaryotes. Their properties are summarized in **Table 29.1.**

Histones H2A, H2B, H3, and H4 interact directly with DNA to form the first-level series of particles in chromatin. Histone H1 is concerned with the interactions between particles. The constancy of the histones suggests that the histone-DNA interactions, histone-histone interactions, and histone-nonhistone interactions may be generally similar in different species; so we may be able to deduce general mechanisms for the formation of both the primary particle and the folding of a series of particles into a higher-order structure.

The histones involved in particle formation have an appreciable content of acidic amino acids, as well as the basic amino acids. These proteins are therefore highly charged. The basic/acidic ratios lie in the range of 1.4 to 2.5. These histones fall into two groups.

The **arginine-rich** histones are the two species **H3** and **H4.** They are among the most conserved proteins known in evolution. They even have identical sequences in species as far distant as the cow and the pea. Only a few amino acid replacements have been found in the H3 and H4 sequences of other species. The conservation of the entire sequence shows that all of it must be important in the function of the protein; and by implication, this function must be identical in greatly different species, which lends support to the concept of a common basis for chromatin structure.

The **slightly lysine-rich** histones comprise two proteins. They are called **H2A** and **H2B** (but, in contrast with the implication of the nomenclature, they are independent proteins, not related variants). The same two types can be recognized in various eukaryotes, but show appreciable interspecific variation in sequence.

The other type of histone is different. The **very lysine-rich** class consists of several rather closely related proteins, with overlapping amino acid sequences. These are the **H1** histones (also there is a variant in avian red blood cells called **H5**). They show appreciable variation between tissues and between species

Table 29.1
Histones are highly basic proteins.

Histone	Type	Basic Amino Acids		Acidic Amino Acids	Basic/Acidic Ratio	Molecular Weight
		Lys	Arg			
H1	(Very) Lysine-rich	29%	1%	5%	5.4	23,000
H2A	Slightly Lysine-rich	11%	9%	15%	1.4	13,960
H2B	Slightly Lysine-rich	16%	6%	13%	1.7	13,774
H3	Arginine-rich	10%	13%	13%	1.8	15,342
H4	Arginine-rich	11%	14%	10%	2.5	11,282

(and this class apparently is absent from yeast). Although these are the most basic histones, they can be extracted readily from chromatin with quite dilute (0.5 M) salt solution.

As their disphonious name suggests, the **nonhistones** include all the other proteins of chromatin. They are therefore presumed to be more variable between tissues and species—although good evidence still is lacking on the extent of the variability—and they comprise a relatively smaller proportion of the mass than the histones. They also comprise a much larger number of proteins, so that any individual protein is present in amounts much smaller than any histone.

The nonhistone proteins include functions concerned with gene expression and with higher-order structure. Thus RNA polymerase may be considered to be a prominent nonhistone. The HMG (high-mobility group) proteins comprise a discrete and well-defined subclass of nonhistones. A major problem in working with other nonhistones is they tend to be contaminated with other nuclear proteins.

CHROMATIN CONTAINS DISCRETE PARTICLES

When interphase nuclei are suspended in a solution of low ionic strength, they swell and rupture to release fibers of chromatin. **Figure 29.1** shows a lysed nucleus in which fibers are streaming out. In some regions, the fibers consist of tightly packed material, but in regions that have become stretched, they can be seen to consist of discrete particles. These are called **nucleosomes.** In especially extended regions, individual nucleosomes are connected by a fine thread; this is a free duplex of DNA. So a continuous duplex thread of DNA runs through the series of particles.

Individual nucleosomes can be obtained by treating chromatin with the enzyme **micrococcal nuclease,** an endonuclease that cuts the DNA thread at the junction between nucleosomes. First, it releases groups of particles; finally, it releases single nucleosomes. Monomeric nucleosomes can be seen clearly in **Figure 29.2** as compact particles. They sediment at a rate of about 11S, corresponding to a total mass in the range of 250,000–300,000 daltons. The protein/DNA ratio is about 1.25.

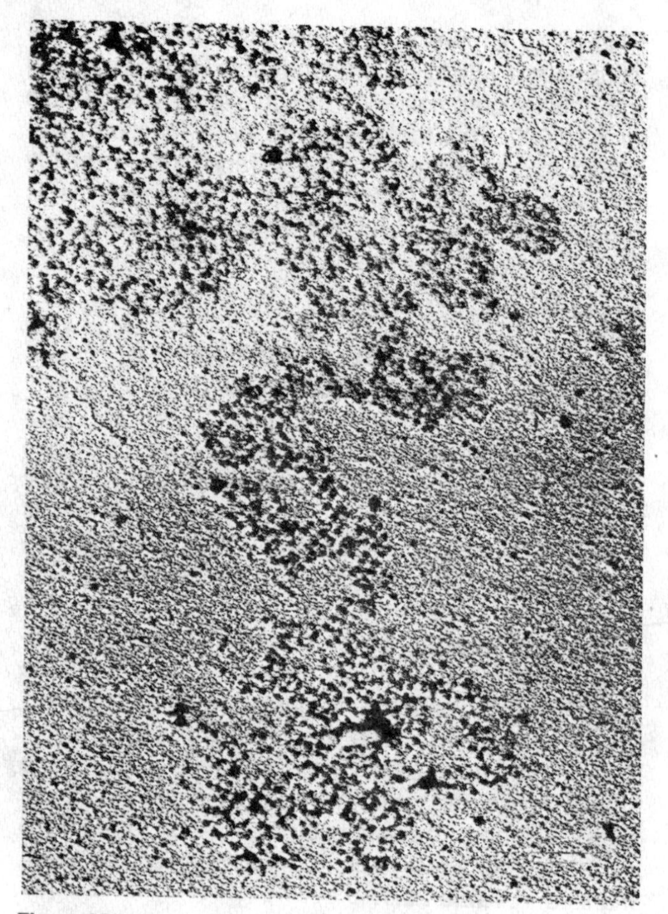

Figure 29.1
Chromatin spilling out of lysed nuclei consists of a compactly organized series of particles. The bar is 100 nm.
Photograph kindly provided by Pierre Chambon.

The monomeric nucleosome contains ~200 bp of DNA associated with a histone octamer. This octamer consists of two copies each of H2A, H2B, H3, and H4. Sometimes these are known as the **core histones.** Their association is illustrated diagrammatically in **Figure 29.3.** This model explains the stoichiometry of the core histones in chromatin, in which H2A, H2B, H3, and H4 are present in equimolar amounts, with 1 molecule of each per ~100 bp of DNA.

Associated with each nucleosome is a single molecule of H1. It is present in chromatin overall with a stoichiometry of half that of the other histones. Usually it is present in (up to) molar amounts when multimeric

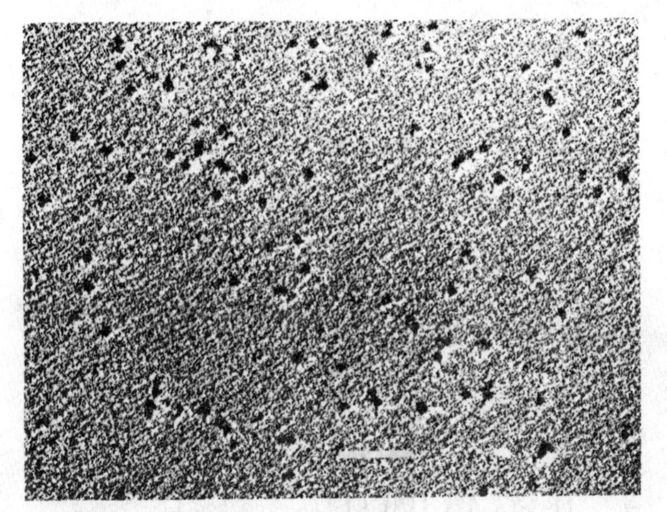

Figure 29.2
Individual nucleosomes are released by digestion of chromatin with micrococcal nuclease. The bar is 100 nm.
Photograph kindly provided by Pierre Chambon.

nucleosomes are isolated, but the amount is reduced in preparations of monomers. All of the H1 can be removed without affecting the structure of the nucleosome, which suggests that it has an external location.

Including the H1, the combined mass of protein and DNA should be about 262,000 daltons per nucleosome, with roughly equal amounts of DNA and protein. This is within the range of experimentally determined values. Any additional protein content probably represents small amounts of nonhistone proteins that are associated with the nucleosomes.

Histone H1 may lie in the region joining adjacent nucleosomes. This would be consistent with old results that H1 is removed the most readily from chromatin, and that H1-depleted chromatin is more readily "solubilized;" and also with more recent results that show it is easier to obtain a stretched-out fiber of beads on a string when the H1 has been removed (see later).

THE NUCLEOSOME IS THE BASIC SUBUNIT OF ALL CHROMATIN

When isolated nuclei or chromatin are digested with the enzyme micrococcal nuclease, the DNA is cleaved into integral multiples of a unit length. Upon fractionation by gel electrophoresis, this forms the "ladder" presented in **Figure 29.4.** Such ladders extend for 10 or more steps, and the unit length, determined by the increments between successive steps, is ~200 bp.

This observation can be correlated with the prop-

Figure 29.3
The nucleosome consists of approximately equal masses of DNA and histones (if HI is included).

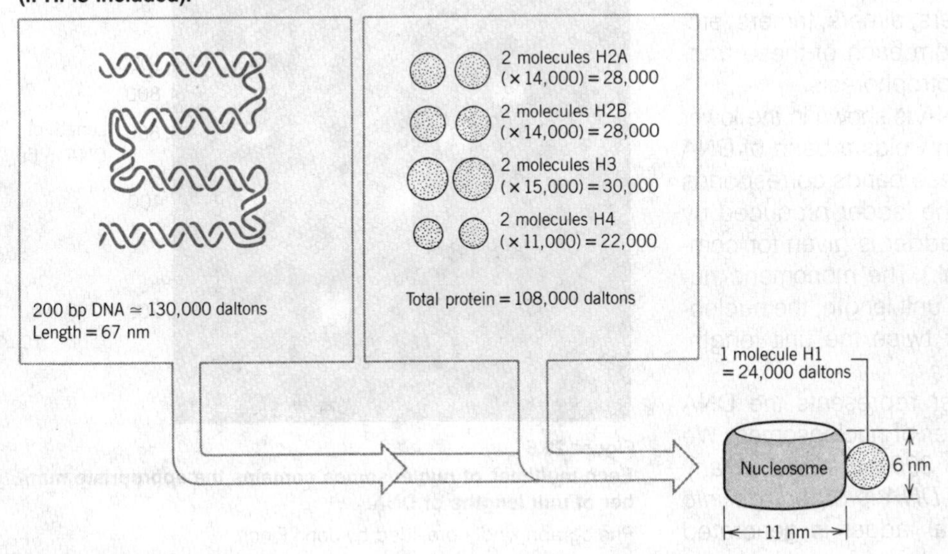

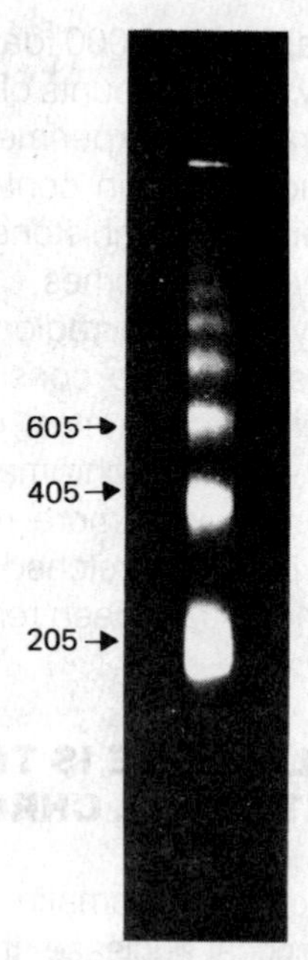

Figure 29.4
Micrococcal nuclease digests chromatin in nuclei into a multimeric series of DNA bands that can be separated by gel electrophoresis.
Photograph kindly provided by Markus Noll.

erties of individual nucleosomes, as shown in **Figure 29.5.** The analysis of isolated particles is given in the upper part of the figure. The preparation of nucleosomes was fractionated by sedimentation on a sucrose gradient to give monomers, dimers, trimers, etc. Then the DNA was purified from each of these fractions and analyzed by gel electrophoresis.

The electrophoresis of the DNA is shown in the lower part of the figure. Each fraction yields a band of DNA of a particular size. Each of these bands corresponds with the appropriate step on the ladder produced by digestion of chromatin. (The ladder is given for comparison in the rightmost panel.) The monomeric nucleosome contains DNA of the unit length, the nucleosome dimer contains DNA of twice the unit length, and so on.

So each step on the ladder represents the DNA derived from a discrete number of nucleosomes. *We can take the existence of the 200 bp ladder in any chromatin to indicate that the DNA is organized into nucleosomes.* The micrococcal ladder is generated

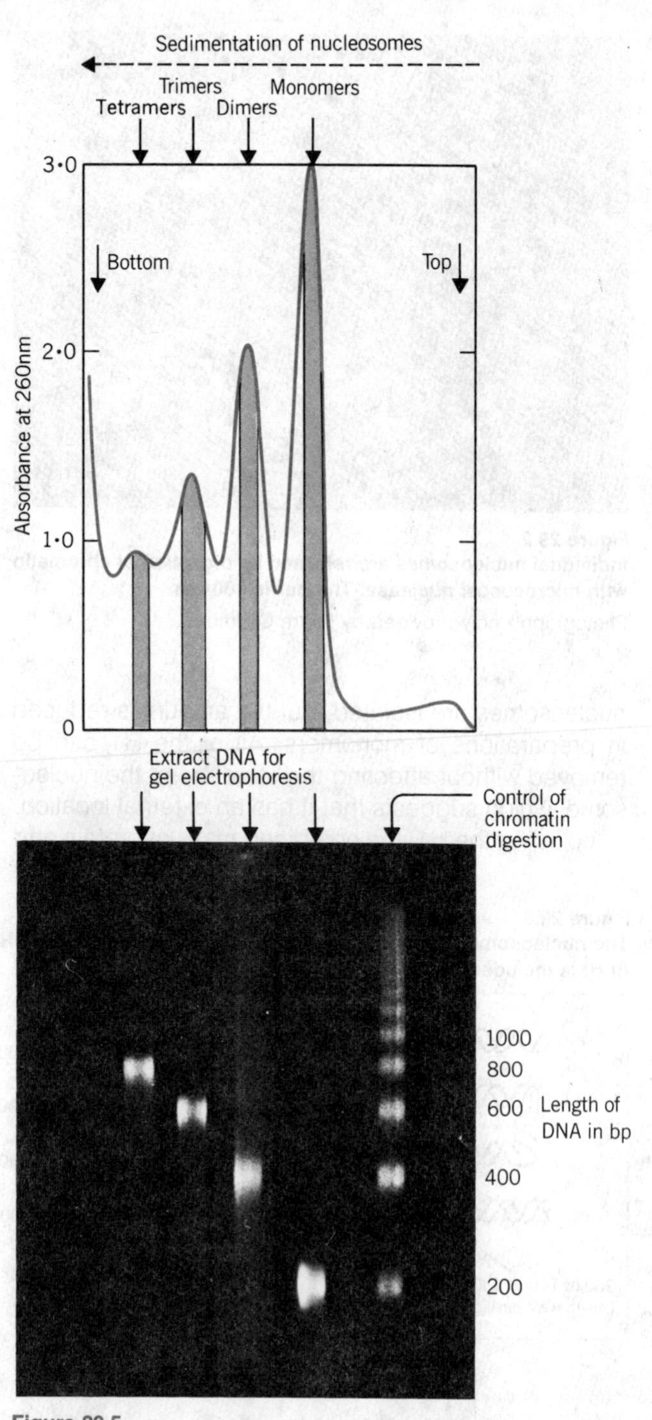

Figure 29.5
Each multimer of nucleosomes contains the appropriate number of unit lengths of DNA.
Photograph kindly provided by John Finch.

when only about 2% of the DNA in the nucleus is rendered acid-soluble (degraded to small fragments) by the enzyme. Thus a small proportion of the DNA is specifically attacked; this must represent especially susceptible regions.

When chromatin is spilled out of nuclei, we often see a series of nucleosomes connected by a thread of free DNA (the beads on a string). Is this the natural situation, with nucleosomes separated by free DNA, or is it an artifact of conditions *in vitro*? The need for tight packaging of DNA *in vivo* suggests *a priori* that probably there is usually little (if any) free DNA.

This is confirmed by the fact that *more than 90% of the DNA of chromatin can be recovered in the form of the 200 bp ladder*. DNA must therefore be organized in, rather than between, nucleosomes. In their natural state, nucleosomes are likely to be closely packed, with DNA passing directly from one to the next. Free DNA is probably generated by the loss of some histone octamers during isolation.

The length of DNA present in the nucleosome may vary somewhat from the value of 200 bp characterized in the first experiments. When entire genomes are characterized from particular cells, each has a fairly well-defined average value (±5 bp). The average most often is between 180 and 200, but there are extremes as low as 154 bp (in a fungus) or as high as 260 bp (in a sea urchin sperm).

The value is not necessarily fixed. It may change during embryonic development. In the case of the sea urchin, it is reduced to a more typical level during the early embryonic cell divisions. The average value may be different in individual tissues of the adult organism. And there can be differences between different parts of the genome in a single cell type; for example, known cases of variation from the genome average include satellite DNAs and 5S RNA genes.

THE CORE PARTICLE IS HIGHLY CONSERVED

The histone octamer forms the protein component of the nucleosome (probably) in all eukaryotes. This constancy of function and structure explains the striking conservation of histone sequences. The extreme conservation of H3 and H4 implies that their role may be central and unchanging, while H2A and H2B provide species-specific variation.

All nucleosomes consist of a histone octamer associated with a particular length of DNA. What is responsible for the variation in the length of the DNA in nucleosomes from different sources? It is easiest to answer by defining what is *not* responsible. The variation does not represent a change in the association of DNA with the histone octamer. This always forms a **core particle** containing 146 bp of DNA, irrespective of the total length of DNA in the nucleosome. So the variation in length of DNA present on the nucleosome is superimposed on this basic core structure.

The core particle was identified by the effects of micrococcal nuclease on the nucleosome monomer. The initial reaction of the enzyme is to cut between nucleosomes, but if it is allowed to continue after monomers have been generated, then it proceeds to digest some of the DNA of the individual nucleosome. This occurs by a reaction in which DNA is "trimmed" from the ends of the nucleosome.

The length of the DNA is reduced in discrete steps, as shown in **Figure 29.6.** With rat liver nuclei, for example, the nucleosome monomers initially have 205 bp of DNA. Then some monomers are found in which the length of DNA has been reduced to 160–170 bp. Then this is reduced to the length of the DNA of the core particle, 146 bp. (This is reasonably stable, but if digestion is continued further, cuts can be made within the core DNA to generate a **limit digest,** in which the longest fragments are the 146 bp DNA of the core, while the shortest are as small as 20 bp.)

This analysis suggests that the nucleosomal DNA can be divided into two regions. The **core DNA** has an invariant length of 146 bp, and is relatively resistant to digestion by nucleases. (Other nucleases as well as the micrococcal enzyme also stop at this length.) The **linker DNA** comprises the rest of the repeating unit. It can vary in length *in vivo* from as little as 8 bp to as much as 114 bp per nucleosome.

The well-defined nature of the band of DNA generated by the initial cleavage with micrococcal nuclease suggests that the region immediately available to the enzyme is restricted. It represents only part of each linker. (If the entire linker DNA were susceptible, the band would range from 146 bp to more than the repeating length.) But once a cut has been made in

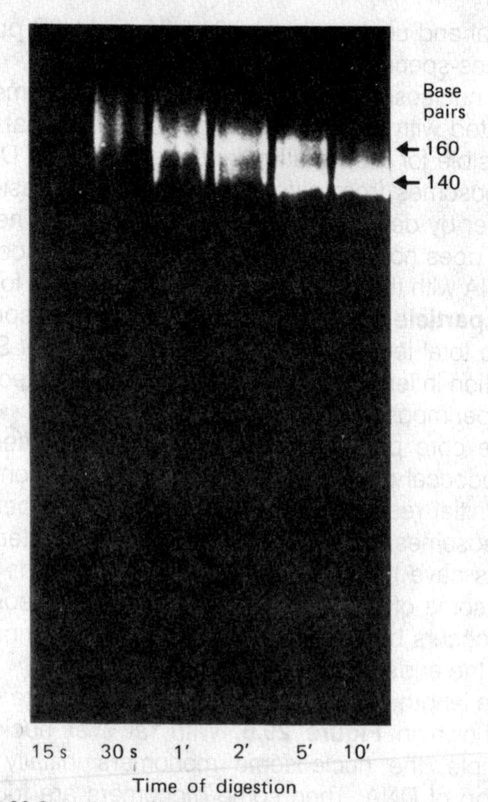

Figure 29.6

Micrococcal nuclease reduces the length of DNA of nucleosome monomers in discrete steps.

Photograph kindly provided by Roger Kornberg.

the linker DNA, the rest of this region becomes susceptible, and can be removed relatively rapidly by further enzyme action.

The core particles have properties similar to those of the nucleosomes themselves, although they are smaller. Their shape and size are similar to nucleosomes, which suggests that the essential geometry of the particle is established by the interactions between DNA and the protein octamer in the core particle. Because core particles are more readily obtained as a homogeneous population, they have been used for many of the structural studies in preference to nucleosome preparations. (Nucleosomes tend to vary more because it is difficult to obtain a preparation in which there has been no end-trimming of the DNA.)

What is the physical nature of the core and the linker regions? These terms are *operational definitions that describe the regions relatively less and relatively more susceptible to nuclease treatment*. This does not make any implication about their actual structure; in particular, it does not imply that the linker DNA has a more extended conformation. On the one hand, the path of DNA in the nucleosome could be continuous, with no distinction evident between these regions of the monomer. Indeed, this is a convenient working assumption often made in attempts to extrapolate from core particle structure to nucleosome structure. On the other hand, it is possible that the path of the linker DNA does differ from the path of the core particle DNA, especially given the pronounced variations that occur in its length.

The existence of linker DNA depends on factors other than the four core histones. Reconstitution experiments *in vitro* show that histones have an intrinsic ability to organize DNA into core particles, but do not form nucleosomes with the unit length of DNA characteristic of the *in vivo* state. The degree of supercoiling of the DNA is an important factor. Histone H1 and/or nonhistone proteins may influence the length of DNA that is associated with the histone octamer in a natural series of nucleosomes. "Assembly proteins" that are not part of the nucleosome structure may be involved *in vivo* in constructing nucleosomes from histones and DNA (see Chapter 30).

We have mentioned that H1 is lost during the degradation of monomeric nucleosomes. It can be retained on monomers that still have 160–170 bp of DNA; but it is always lost with the final reduction to the 146 bp core particle. This suggests that H1 could be located in the region of the linker DNA immediately adjacent to the core DNA.

DNA IS COILED AROUND THE HISTONE OCTAMER

Two types of data suggest independently that DNA lies on the surface of the nucleosome, following a path around the outside of the histone octamer. Biophysical studies suggest that the diameter of the protein component of the nucleosome is smaller than that of the DNA component. Biochemical studies show that the DNA is accessible to nucleases at rather regular intervals (see later).

The shape of the nucleosome is generally ellipsoidal. Its axial ratio of 0.5 corresponds to dimensions of about 11 × 11 × 6 nm. This is usually represented as a flat cylinder, diameter 11 nm and height 6 nm. Even just from the dimensions, it seems likely that the DNA lies on the exterior. The circumference of the particle is about 34 nm, compared with the length of DNA of 67 nm (=200 bp). It is difficult to see how the DNA could be squashed into the interior of such a particle.

The properties of the individual components can be measured by neutron scattering, a technique that allows the scatter due to DNA to be distinguished from that due to protein. This shows that the protein component (the histone octamer) has a radius of gyration of about 3.2 nm. But the radius of gyration of the DNA component is about 5.2 nm. The difference corresponds to the 2 nm diameter of the DNA double helix. This suggests that the protein forms a compact body, around which the DNA is wrapped.

Many models have been suggested for the path that DNA follows on the nucleosome. The common features are that the structure should be symmetrical and that the DNA is likely to pass around the octamer twice. **Figure 29.7** is a diagrammatic illustration of a path in which the DNA lies in a helical coil that makes two turns. One consequence of this particular path is that the DNA "enters" and "leaves" the nucleosome at points that are close to each other. A question that has not received much attention is how this path might be modified to account for the variation in length of DNA on the nucleosome.

Considering this model in terms of a cross-section through the nucleosome, as depicted in **Figure 29.8,** we see that the two circumferences made by the DNA lie close to each other. This has a possible functional

Figure 29.7
The nucleosome may be a cylinder with up to two turns of DNA around the surface.

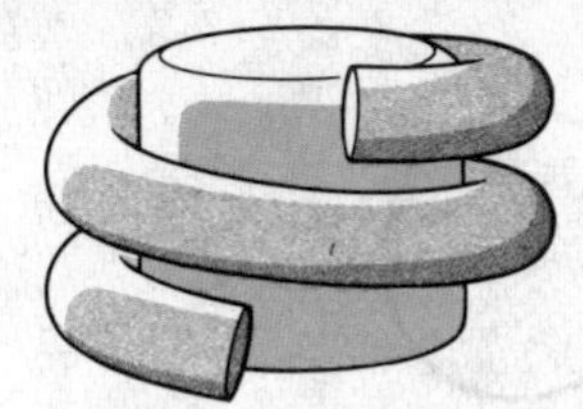

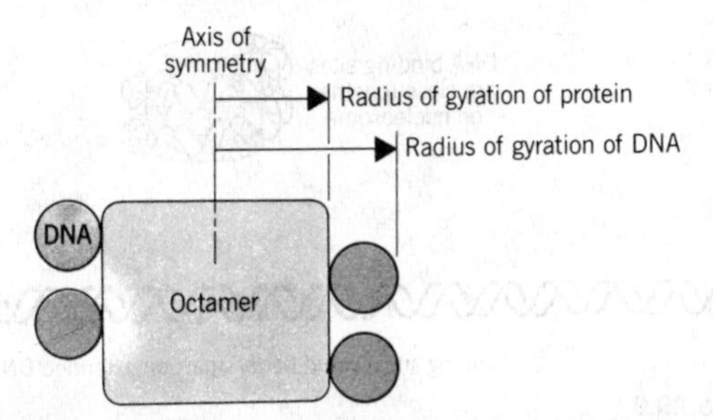

Figure 29.8
The two turns of DNA on the nucleosome must lie quite close together.

consequence. One turn around the nucleosome is about 80 bp of DNA, so that points this distance apart in the free double helix may actually be rather close to each other on the nucleosome. Thus if a DNA-binding protein simultaneously contacted both turns of DNA as illustrated in **Figure 29.9,** the sequences it recognizes could lie farther apart in the duplex DNA than the apparent span of the polypeptide. One possibility that is often considered is whether the startpoint for transcription might be close in space to the −80 position, so that both are contacted simultaneously by RNA polymerase. (This would provide a particular example of the less detailed model illustrated in Figure 12.11.)

The packing ratio of the individual nucleosome is ~6 (since 67 nm of DNA is packaged into a particle of length about 11 nm). It is possible to pack a series of nucleosomes tightly enough to maintain this ratio.

Much work on the structure of sets of nucleosomes has been carried out with the virus SV40. The DNA of SV40 is a circular molecule of 5200 bp, that is, of contour length about 1500 nm. Both in the virion and in the infected nucleus, it is packaged into a series of nucleosomes, called a **minichromosome.** As usually isolated, the contour length of the minichromosome is about 210 nm, corresponding to a packing ratio of ~7. Changes in the salt concentration can convert it to a flexible string of beads with a much lower overall packing ratio. This emphasizes the point that nucleosome strings can take more than one form *in vitro,* depending on the conditions.

An important parameter in describing the structure of chromatin is the degree of supercoiling, which can be created at several levels. First, there may be su-

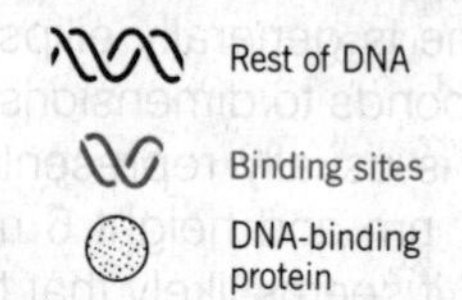

Figure 29.9
A protein could contact sequences on the DNA that lie on different turns around the nucleosome.

percoiling as a result of the path that DNA follows on the nucleosome. Second, there may be supercoiling as a result of the path that the nucleosomes follow in higher-level structures. Some of this supercoiling may be restrained by proteins. Direct measurements of the supercoiling density therefore will show what average torsional tension applies to DNA as the result of free supercoils, but will not reveal any restrained supercoiling in the DNA path.

The degree of supercoiling on the nucleosome itself can be measured directly by using the SV40 minichromosome. The procedure to measure supercoiling due just to nucleosome structure is illustrated in **Figure 29.10.** First, the free supercoils of the minichromosome itself are relaxed, so that the nucleosomes form a circular string with a superhelical density of 0. Then the histone octamers are extracted. This releases the DNA to follow a free path. Thus every su-

Figure 29.10
The supercoils of the minichromosome can be relaxed to generate a circular structure, whose loss of histones then generates supercoils in the free DNA.

percoil that was present but restrained in the minichromosome will appear in the deproteinized DNA as −1 turn. So now the total number of supercoils in the SV40 DNA is measured.

The value actually observed is close to the number of nucleosomes. The reverse result is seen when nucleosomes are assembled *in vitro* on to a supercoiled SV40 DNA: the formation of each nucleosome removes ~1 negative supercoil. Thus the DNA follows a path on the nucleosomal surface that generates ~1 negative supercoiled turn when the restraining protein is removed. We note that there is a potential discrepancy between the measured value of −1 turn per nucleosome and the model that shows DNA in a path equivalent to −2 superhelical turns (see later).

DNA IS SYMMETRICALLY EXPOSED TO NUCLEASES

The enzymes DNAase I and DNAase II make single-strand nicks in DNA. This means that they cleave a bond in one strand, but the other strand remains intact at this point. Thus no effect is visible in the double-stranded DNA. But upon denaturation, shorter fragments are released instead of full-length single strands. This effect is illustrated in **Figure 29.11.**

When DNA is free in solution, it is nicked at random. The DNA on nucleosomes also can be nicked by the enzymes, *but only at certain points*. Cutting occurs at regular intervals, so that when DNA is denatured and electrophoresed, a ladder is obtained. In this type of experiment, fragments of denatured DNA can represent the distance from the end of the nucleosome to the site of cutting, or can be generated by two internal nicks.

The actual points of cutting can be determined by using radioactively end-labeled DNA, and identifying the fragments by autoradiography. *Only* those fragments with a labeled end are detected, as indicated in **Figure 29.12.** (This is exactly analogous to the restriction mapping technique shown in Figure 5.4.)

A ladder obtained by DNAase I treatment is displayed in **Figure 29.13.** Core particles usually are used for such experiments *in vitro* because of the difficulty in obtaining nucleosome preparations with a homogeneous size distribution of DNA. The interval between successive steps on the ladder is approximately 10 bases. The ladder extends for the full distance of core DNA, and the cleavage sites are numbered as S1 through S13 (where S1 is ~10 bases from the labeled 5′ end, S2 is ~20 bases from it, and so on).

This result makes two important points. First, DNA is exposed on the nucleosome so that it is accessible to DNAase I; it is not covered by proteins. Second, there is a periodicity in its exposure.

Not all sites are cut with equal frequency: some are cut rather effectively, others are cut scarcely at all. The enzymes DNAase I and DNAase II generate the same ladder, although with some differences in the intensities of the bands. This shows that the pattern of cutting represents a unique series of targets in DNA, determined by its organization, with only some slight preference for particular sites imposed by the individual enzyme.

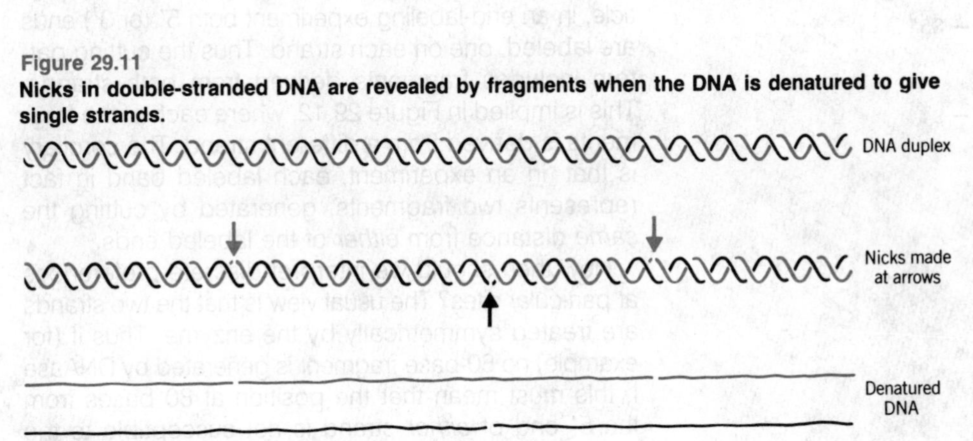

Figure 29.11
Nicks in double-stranded DNA are revealed by fragments when the DNA is denatured to give single strands.

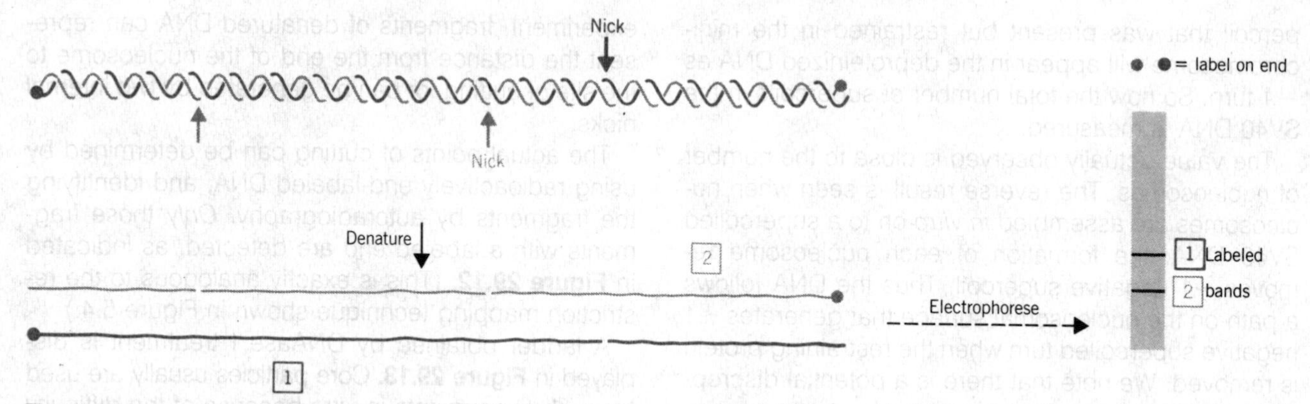

Figure 29.12
The end-labeling technique identifies the distance of nicks from the labeled ends.

Figure 29.13
Sites for nicking lie at regular intervals along core DNA, as seen in a DNAase I digest of nuclei.
Photograph kindly provided by Leonard Lutter.

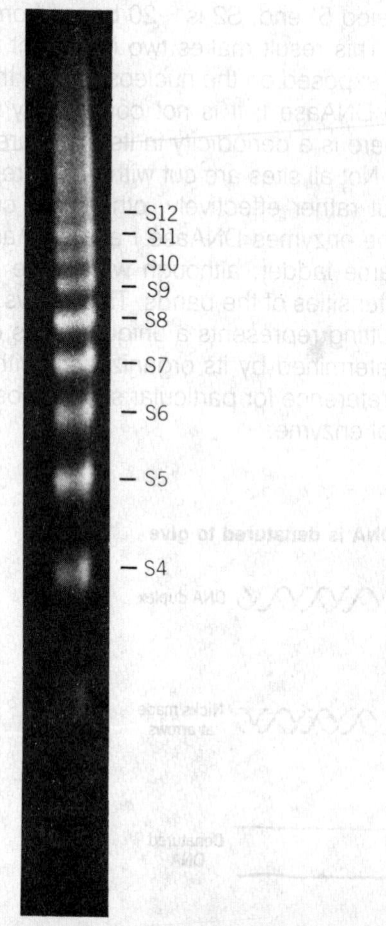

What is the nature of the target sites? **Figure 29.14** displays a high-resolution gel of the sites cut by DNAase I in end-labeled core particles. At each site, there are actually 3-4 positions at which cutting may occur; that is, the cutting site is defined ± 2 bp. So a cutting site represents a short stretch of bonds on both strands, exposed to nuclease action over 3-4 base pairs. The relative intensities indicate that some sites are preferred to others.

From this pattern, we can calculate the "average" point that is cut. This shows that pairs of sites from S1 to S4 lie apart a distance of 10.0 bases each, the separation from sites S4 to S10 is 10.7 bases, and for sites S10 to S13 the distance returns to 10.0 bases. (Because this analysis deals with *average* positions, sites need not lie an integral number of bases apart.)

Since there are two strands of DNA in the core particle, in an end-labeling experiment both 5′ (or 3′) ends are labeled, one on each strand. Thus the cutting pattern includes fragments derived from both strands. This is implied in Figure 29.12, where each of the fragments is derived from a different strand. The corollary is that in an experiment, each labeled band in fact represents two fragments, generated by cutting the *same* distance from *either* of the labeled ends.

How then should we interpret discrete preferences at particular sites? The usual view is that the two strands are treated symmetrically by the enzyme. Thus if (for example) no 80-base fragment is generated by DNAase I, this must mean that the position at 80 bases from the 5′ end of *either* strand is not susceptible to the enzyme.

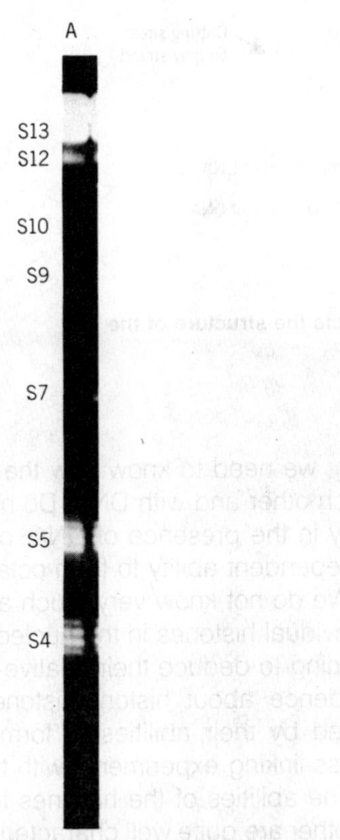

Figure 29.14
High-resolution analysis shows that each site for DNAase I consists of several adjacent susceptible phosphodiester bonds as seen in this example of sites S4 and S5 analyzed in end-labeled core particles.
Photograph kindly provided by Leonard Lutter.

THE UNRESOLVED QUESTION OF THE PERIODICITY OF DNA

Experiments in which DNA on nucleosomes is analyzed for its sensitivity to nucleases are analogous to footprinting experiments. Thus we can assign the lack of reaction at particular target sites to the structure of the nucleosome, in which certain positions on DNA are rendered inaccessible. But what is the reason for the 10.0-10.7 periodicity of cutting?

This periodicity is close to the helical periodicity of B-form DNA. Does the recurrence of the exposed site simply reflect the immobilization of the DNA on the histone octamer? This could happen as shown in **Figure 29.15,** where a peak of exposed sites occurs on each strand with a periodicity determined by the number of base pairs per turn of the double helix.

Some light on this question is cast by comparable studies with DNA immobilized on a flat surface (the situation envisaged in Figure 29.15). In this case, sites are cut evenly at a separation of 10.6 bases. This probably means that free DNA (that is, as obtained in solution) has a double-helical B-type structure in which the repeat length is 10.6 bp. Other measurements by independent techniques of the number of base pairs per turn have produced values close to this, notably 10.4.

We assume that when the DNA is immobilized on a flat surface, the **cutting periodicity** (the spacing between cleavage points) coincides with, indeed, is a reflection of, the **structural periodicity** (the number of base pairs per turn of the double helix). Because of the variation in cutting periodicity along the core DNA (10.0 at the ends, 10.7 in the middle), we are immediately faced with a problem in applying this interpretation.

One possibility is that the structural periodicity of core DNA actually varies, being close to its solution value in the middle, but being less tightly screwed at the ends. Another possibility is that the DNA has a constant structural periodicity, but that the geometry of the nucleosome causes a change in the average separation of cuts. If the approach of the enzyme is restricted at some positions, there could be a shift in its selection between the 3-4 available bonds. Then the cutting periodicity is no longer identical with the structural periodicity.

If this is true, what is the structural periodicity? In the absence of data, it is necessary to resort to model building to see which assumptions can be reconciled with the needs of the enzyme for access to DNA. Changing the structural periodicity, or fixing a periodicity of 10.6 bp, both fit less well than supposing there is a constant structural periodicity on the nucleosome of 10.0 bp per turn.

Some interesting implications follow from the possibility that the structural periodicity of DNA could be different on the nucleosome (10.0) from its value in solution (10.6). When DNA is released from the nucleosome, its structure would become more tightly wound, since there are more base pairs per turn. This change would reduce the apparent degree of supercoiling. Suppose that on the nucleosome the DNA fol-

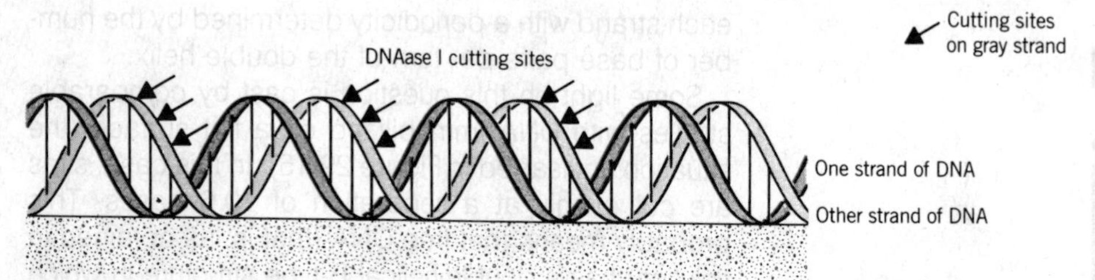

Figure 29.15
The most exposed positions on DNA recur with a periodicity that reflects the structure of the double helix.
(For clarity, sites are shown for only one strand.)

lows a path equivalent to 2 superhelical turns. Then the histone octamer is removed. Some of the torsional strain goes into increasing the winding of DNA; only the rest is left to be measured as a supercoil.

This provides a possible way to reconcile models for −2 superhelical turns per nucleosome with data that identify only −1 superhelical turn, since the difference in periodicity (0.6 bp per turn of the helix) multiplied by the number of helical turns per nucleosome (>15) is roughly 10 bp, corresponding to 1 turn of the double helix, and thus potentially absorbing 1 negative superhelical turn.

One extremely important point must be made about the material used in these experiments. All of these data have been obtained with core particles. The models for reconciling discrepancies in the number of superhelical turns assume that linker DNA follows the same path as core DNA; but it could be different. An indication that the same pattern actually may extend beyond the core is given by some DNAase I ladders that extend for more than 200 bp. This implies that the same periodic exposure is maintained not just on the individual nucleosomes, but even between nucleosomes. In other words, DNA passes from one nucleosome to the next without any disruption in its path, at least as seen by the cutting periodicity.

THE ARRANGEMENT OF HISTONES AND DNA

So far we have considered the construction of the nucleosome from the perspective of how the DNA is organized on the surface. From the perspective of protein, we need to know how the histones interact with each other and with DNA. Do histones react properly only in the presence of DNA, or do they possess an independent ability to form octamers?

We do not know very much about the structures of individual histones in the nucleosome, but we are beginning to deduce their relative locations. Most of the evidence about histone-histone interactions is provided by their abilities to form aggregates, and by cross-linking experiments with the nucleosome.

The abilities of the histones to aggregate with one another are quite well characterized. The arginine-rich histones can be obtained as a well-defined tetramer ($H3_2H4_2$). The slightly lysine-rich histones form a product that has been less clearly characterized, but that seems to be a dimer (H2A·H2B), with a tendency to aggregate further. One of the forms of aggregation could be the tetramer ($H2A_2H2B_2$).

Intact histone octamers can be obtained either by extraction from chromatin or (with more difficulty) by association of the histones *in vitro* under conditions of high-salt and high-protein concentration. The octamer can dissociate to generate a hexamer of histones that has lost an H2A·H2B dimer. Then the other H2A·H2B dimer is lost separately, leaving the $H3_2H4_2$ tetramer. These forms all can be extracted from chromatin. This argues for a form of organization in which the nucleosome may have a central "kernel" consisting of the $H3_2H4_2$ tetramer, associated with two independent H2A·H2B dimers.

Cross-linking studies extend these relationships to show which pairs of histones lie near each other in the nucleosome. (A difficulty with such data is that usually only a small proportion of the histones become cross-

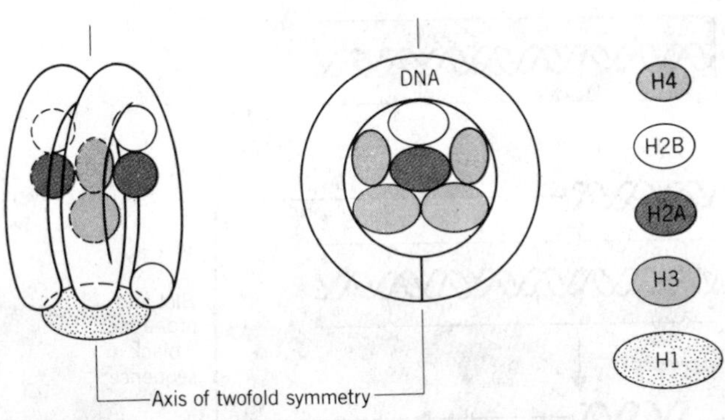

Figure 29.16
A model for the nucleosome as seen from the "front" or "top" is symmetrical, with the $H3_2H4_2$ tetramer providing a kernel for the shape.

linked, so it is necessary to be cautious in deciding whether the results typify the major interactions.) From these data, a model has been constructed for the organization of the nucleosome. It is shown in diagrammatic form in **Figure 29.16.**

Structural studies show that the overall shape of the isolated histone octamer is similar to that of the core particle. This suggests that it is the histone-histone interactions that establish the general structure. The positions of the individual histones have been assigned to regions of the octameric structure on the basis of their aggregation behavior and response to cross-linking.

The $H3_2H4_2$ tetramer accounts for the diameter of the octamer. This explains the ability of the tetramer to organize DNA *in vitro* into particles that display some of the properties of the core particle (see later). The H2A·H2B pairs fit in as two dimers. DNA is wound twice around the octamer. Histone H1 could "seal" the DNA in the nucleosome by binding at the point where it enters and leaves. The model displays twofold symmetry.

Although the overall shape of the octamer is now quite well defined, the individual histones are shown as amorphous blobs, because we lack information on their shapes. The core histones all share a common pattern for the distribution of amino acids, with an N-terminal region that is highly charged; the rest of the molecule contains the hydrophobic amino acids that are likely to generate a globular structure and be involved in protein-protein interactions.

For this reason, histones sometimes have been visualized as globular proteins with charged N-terminal "tails." One would think that the tails are likely to possess the predominant DNA-binding activity, while the globular regions constitute the interior of the particle. But arguing against this model is the observation that the N-terminal regions can be cleaved from the histones of core particles (by treatment with trypsin) without greatly disrupting the structure. And the N-tailless histones can participate in nucleosome assembly *in vitro.* At present, we are left without any assignment of individual functions to particular regions of the histone molecules.

ARE NUCLEOSOMES ARRANGED IN PHASE?

We know that nucleosomes can be reconstituted *in vitro* without regard to DNA sequence, but this does not exclude the possibility that their formation *in vivo* is controlled in a sequence-dependent manner. Does a particular DNA sequence always lie in a certain position *in vivo* with regard to the topography of the nucleosome? Or are nucleosomes arranged randomly on DNA, so that a particular sequence may occur at any location, for example, in the core region in one copy of the genome and in the linker region in another?

To investigate this question, it is necessary to use a defined sequence of DNA; or, more precisely, we need to determine the position relative to the nucleosome of a defined point in the DNA. **Figure 29.17** illustrates the principle of a procedure that is used to achieve this.

Suppose that the DNA sequence is organized into nucleosomes in only one particular configuration, so that each site on the DNA always is located at a particular position on the nucleosome. This is called **nucleosome phasing.** In a series of phased nucleosomes, the linker regions of DNA comprise unique sites.

Consider the consequences for just a single nucleosome. Cleavage with micrococcal nuclease generates a monomeric fragment that constitutes a *specific sequence.* If the DNA is isolated and cleaved with a restriction enzyme that has only one target site in

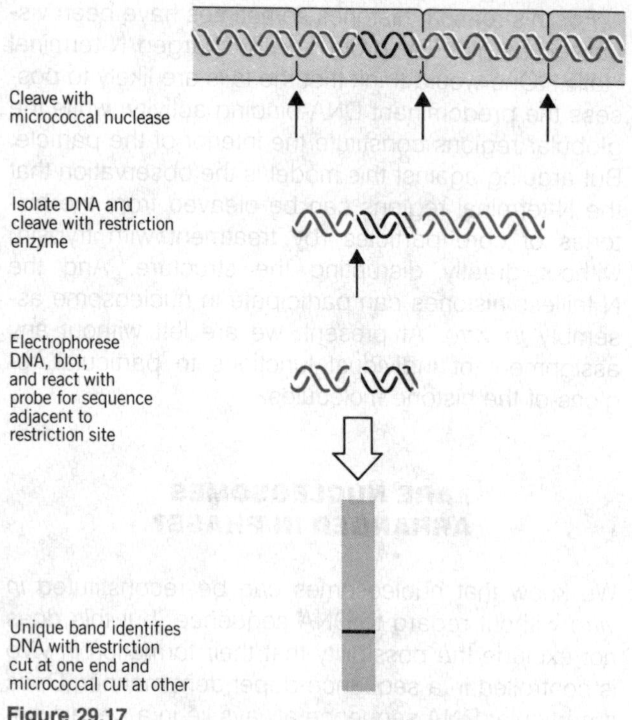

Figure 29.17

Nucleosome phasing places restriction sites at unique positions relative to the linker sites cleaved by micrococcal nuclease.

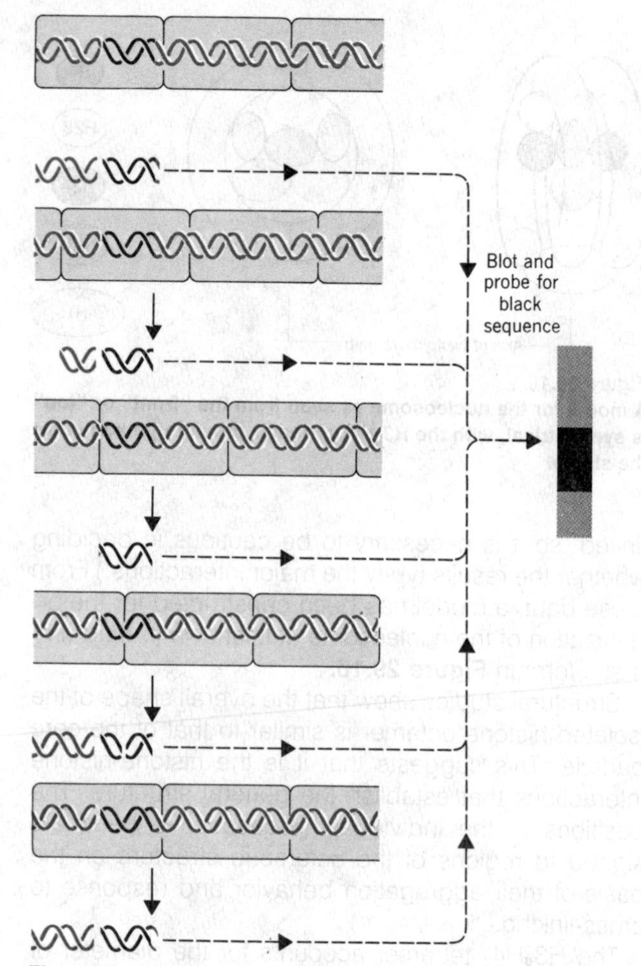

Figure 29.18

In the absence of nucleosome phasing, a restriction site lies at all possible locations in different copies of the genome.

The figure shows the result of making cleavages at the right end of the black sequence and at the junctions between nucleosomes.

this fragment, it should be cut at a unique point. This produces two fragments, each of unique size.

The products of the micrococcal/restriction double digest are separated by gel electrophoresis. A probe representing the sequence immediately adjacent to (one side of) the restriction site is used to identify the corresponding fragment in the double digest. This technique is called **indirect end labeling** (not altogether an appropriate name).

Reversing the argument we have just made, the identification of a single sharp band demonstrates that the position of the restriction site is uniquely defined with respect to the end of the nucleosomal DNA (as defined by the micrococcal cut). Thus the nucleosome has a unique sequence of DNA.

What happens if the nucleosomes do *not* lie at a single position? Now the linkers consist of *different* DNA sequences in each copy of the genome. Thus the restriction site lies at a different position each time; in fact, it lies at all possible locations relative to the ends of the monomeric nucleosomal DNA. **Figure 29.18** shows that the double cleavage will therefore generate a broad smear, ranging from the smallest detectable fragment (about 20 bases) to the length of the monomeric DNA.

In discussing these experiments, we have treated micrococcal nuclease as an enzyme that cleaves DNA at the exposed linker regions without any sort of sequence specificity. This indeed originally was thought to be the case. However, it has since turned out that the enzyme actually does have some sequence specificity (biased toward selection of certain A-T-rich sequences). So we cannot assume that the existence of a specific band in the indirect end-labeling technique represents the distance from a restriction cut to the linker region. It could instead represent the distance from the restriction cut to a preferred micrococcal nuclease site!

This possibility is dealt with by performing a control experiment in which the naked DNA is treated in exactly the same way as the chromatin. If there are preferred sites for micrococcal nuclease in the particular region, specific bands are found. Then this pattern of bands can be compared with the pattern generated from chromatin.

If there is a random arrangement of nucleosomes, all the preferred sites for micrococcal nuclease will at one time or another be available in a linker region. The pattern of cutting in chromatin and in DNA will be identical. But if the nucleosomes lie in a particular phase, some of the preferred micrococcal cleavage sites will be unavailable, because they lie in core regions. Some bands that are present in the DNA pattern will disappear from the chromatin pattern. Other bands may appear in the chromatin pattern if the distribution of nucleosomes makes new sites preferentially accessible.

So a *difference* between the control DNA band pattern and the chromatin pattern provides evidence for nucleosome phasing. This is how experiments on nucleosome phasing actually have been performed.

These results also cast some light on the nature of the site cut by micrococcal nuclease between nucleosomes. Probably the enzyme does not cut exactly at the most exposed point in the linker DNA, but cuts the preferred site nearest to that position (if there is one). Since the relationship will vary, this increases the breadth of the bands. So the ability of preferred sites to seduce the enzyme may have increased the range of positions in linker DNA that appear to be most exposed to micrococcal nuclease.

The analysis illustrated in Figures 29.17 and 29.18 applies to a single short sequence. However, with large genomes, it may be impractical to identify a single unique fragment with sufficient precision. And we want to know whether nucleosomes are phased over some substantial region, not just over one particular short sequence. So attempts to investigate nucleosome phasing often have made use of tandemly repeated sequences. This approach effectively amplifies the fragment that is investigated: a single probe will detect whether the corresponding sequence is phased in every one of its genomic repeating units.

Thus probes have been used to investigate phasing on satellite DNA, tDNA, 5S DNA, and histone genes. In such cases, nucleosome phasing can occur only if there is a simple relationship between the nucleosomal repeat length and the repeating unit of the tandem array. This means that there must be an integral number of nucleosomes per tandem repeat.

In most cases, experimental data have not generated the clear type of result illustrated in Figure 29.17 for a unique disposition of nucleosomes; but sometimes they fall short of the wide bands or smears expected of random location. What we actually see is a confinement of the micrococcal cutting sites to a small number of positions, relative to the defined restriction site. This suggests that nucleosomes may be limited so that they lie in only a few (2–4) alternative phases.

An intriguing situation has been noted in some tRNA genes of *X. laevis*. In erythrocyte chromatin, in which the genes are not expressed, they are organized into nucleosomes with a DNA repeat length of 198 bp. The nucleosomes lie mostly in a single phase over a distance of at least 3 kb (within which there is a stretch of 250 bp that may be nonnucleosomal). A different situation occurs in the chromatin of liver and kidney cells, in which the tRNA genes are expressed. The repeating length of the nucleosomes is altered to 185 bp. And now there is no phasing. This suggests that both the repeat length and the location of nucleosomes can be influenced by circumstance.

Nucleosome phasing might be accomplished in either of two ways. One is for *every* nucleosome to be deposited specifically at a particular DNA sequence. This is somewhat at odds with the view of the nucleosome as a subunit able to form between any sequence of DNA and a histone octamer. An alternative is for some particular sequence preferentially to assemble the first nucleosome in a region; then the adjacent nucleosomes may be assembled sequentially, with a defined repeat length. (If there is some variability in the construction of the nucleosomes—for example, if the length of the linker can vary by, say, 10 bp—the specificity of location would decline continuously, proceeding away from the first, defined nucleosome.) A starting point for nucleosome formation could be determined by binding of nonhistone proteins at a specific site of DNA.

The occurrence of a fixed series of nucleosomes does not necessarily mean that they are deliberately phased. Apparent phasing could result from the presence of a region from which nucleosomes happen to be excluded. Such regions may be created by complexes concerned with controlling gene expression or generating chromatin higher-order structure (see Chapter 30). The excluded region may provide a *boundary* that restricts the positions available to the adjacent nucleosomes. The statistical distribution of nucleosomes then may give the same results as phasing, although the individual positions are not necessarily fixed in any particular copy of the genome.

Although phasing adjacent to fixed sites may be a consequence of boundary effects, phased arrays of nucleosomes may also exist. An example is presented by the α-satellite chromatin of African green monkey. The DNA consists of a 172 bp tandemly repeated sequence. The repeating length of the nucleosomes is the same, so there is one nucleosome per satellite repeat.

An abundant nonhistone, called α-protein, binds preferentially to the α DNA, recognizing two types of short A-T-rich sequence. The arrangement of these sequences in the satellite DNA repeating unit is illustrated in **Figure 29.19.** Each DNA repeating unit contains one copy of the first sequence (site I) and two copies of the second sequence (sites II and III), present in inverted orientation. The inverted sequences are 145 bp apart, the distance around the core DNA. The other site is midway between them.

Nucleosomes on the α satellite are phased in such a way that the positions of sites II and III define the ends of the core DNA. As the DNA winds around the histone octamer, site I should come into close proximity with them. *So a cluster of sites bound by α-protein lie in the same area of the nucleosome surface.*

The location of these binding sites suggests that α-protein could be responsible for phasing the nucleosomes. By binding to a regularly spaced series of sites on DNA, α-protein could ensure that histones form octamers whose DNA is organized in phase. This

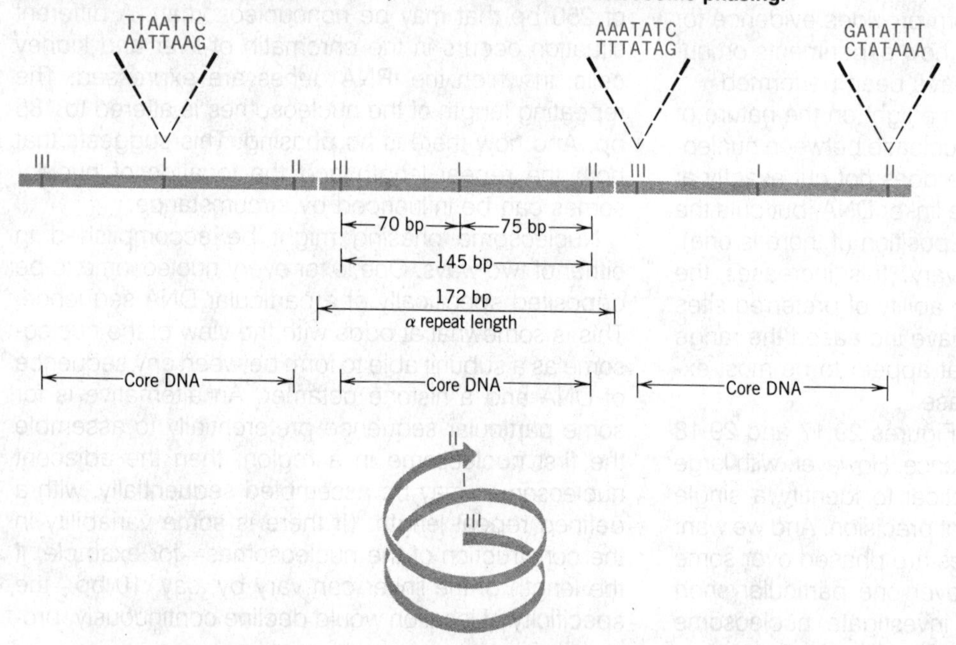

Figure 29.19
Binding sites for α-protein are located in the same region of the nucleosome when α-satellite DNA winds around a histone octamer, and could control nucleosome phasing.

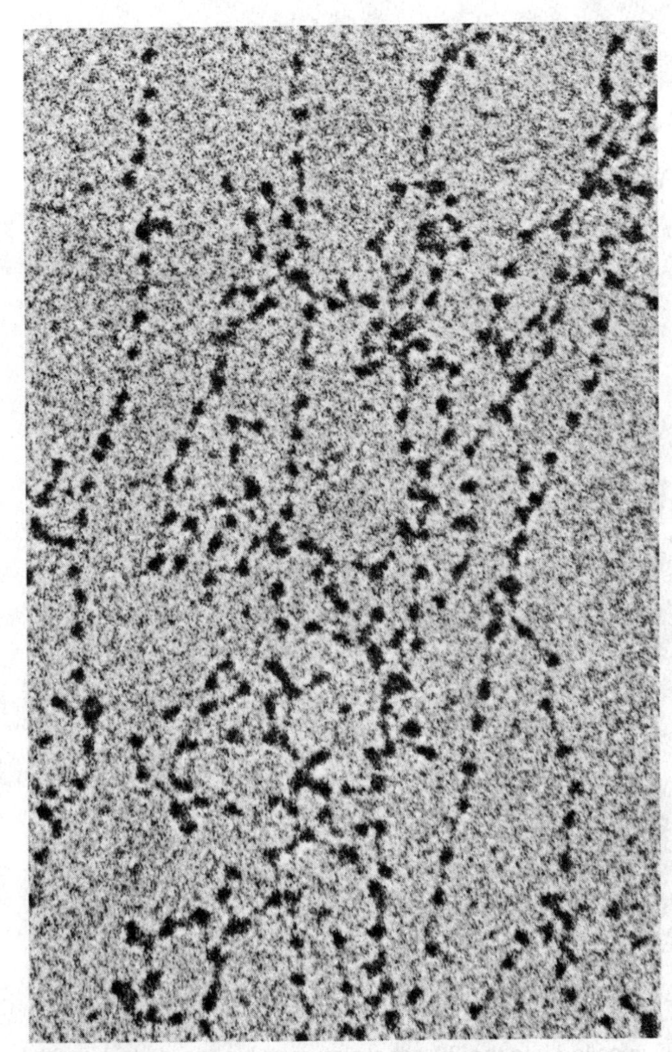

Figure 29.20
The 10-nm fiber in partially unwound state can be seen to consist of a string of nucleosomes.
Photograph kindly provided by Barbara Hamkalo.

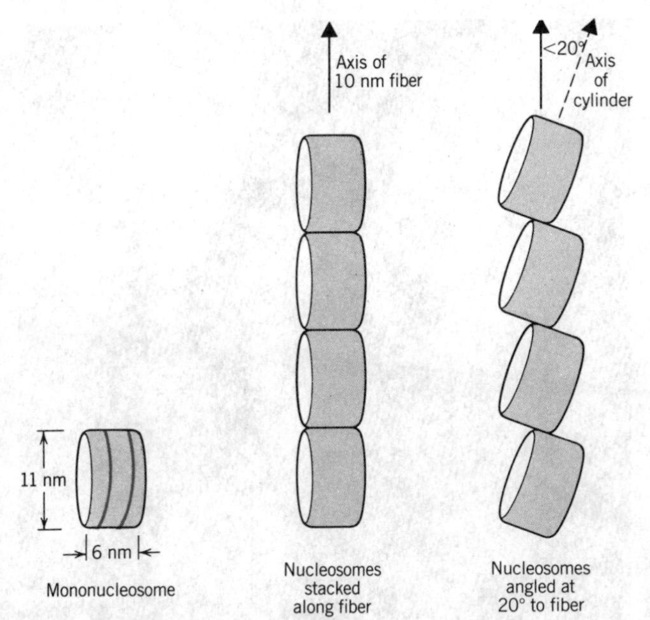

Figure 29.21
The 10-nm fiber consists of a series of nucleosomes organized edge to edge, stacked along the fiber or tilted less than 20° to the axis.

could be a common mechanism for phasing nucleosomes over a protracted distance or, indeed, for starting off a series of nucleosomes at a fixed position.

THE PATH OF NUCLEOSOMES IN THE CHROMATIN FIBER

When chromatin is examined in the electron microscope, two types of fiber are seen: the 10-nm fiber and 30-nm fiber. They are described by the approximate diameter of the thread (that of the 30-nm fiber actually varies from about 25 nm to 30 nm).

The **10-nm fiber** is essentially a continuous string of nucleosomes. Sometimes, indeed, it runs continuously into a more stretched-out region in which nucleosomes are seen as a string of beads, as indicated in the example of **Figure 29.20.** The 10-nm fibril structure is obtained under conditions of low ionic strength and in the absence of histone H1. This means that it is a function strictly of the nucleosomes themselves.

How are the nucleosomes arranged in the 10-nm fiber? Viewing the particle itself as a somewhat flat cylinder, adjacent cylinders might be arranged either edge to edge or with their faces touching. These arrangements can be distinguished by biophysical techniques, such as neutron scattering or electric dichroism, that depend on the orientation of the individual subunit relative to the axis of the fiber. The results suggest the type of model illustrated in **Figure 29.21,** in which the cylinders are oriented edge to edge, with their faces parallel (or at least, not much inclined)

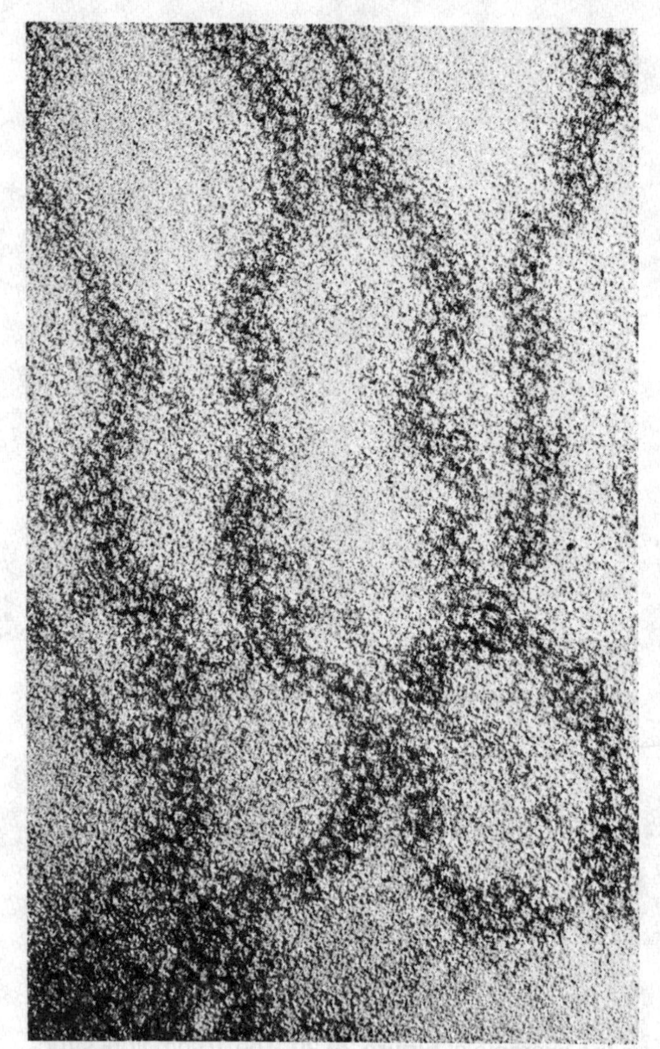

Figure 29.22
The 30-nm fiber has a coiled structure.
This fiber is shown at the same magnification as the 10-nm fiber of Figure 29.20.
Photograph kindly provided by Barbara Hamkalo.

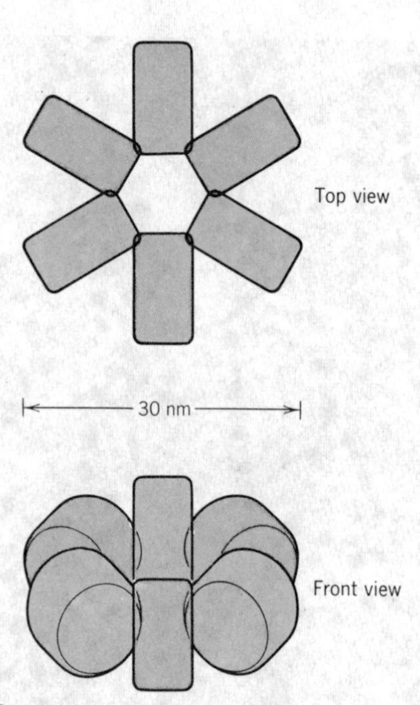

Figure 29.23
The 30-nm fiber may have a helical coil of 6 nucleosomes per turn, organized radially.

to the axis of the fiber. The data imply that the maximum angle between the faces and the axis can be about 20°.

When chromatin is visualized in conditions of greater ionic strength, and H1 remains, the **30-nm fiber** is obtained. An example is given in **Figure 29.22.** The fiber can be seen to have an underlying coiled structure. It has about 6 nucleosomes for every turn, which corresponds to a packing ratio of 40 (that is, each μm along the axis of the fiber contains 40 μm of DNA). This fiber is the basic constituent of both interphase chromatin and mitotic chromosomes (see, for example, Figure 28.7).

The 30-nm and 10-nm fibers can be reversibly converted by changing the ionic strength. This suggests that the linear array of nucleosomes in the 10-nm fiber is coiled into the 30-nm structure at higher ionic strength and in the presence of H1.

Again there are several possibilities for the packing of nucleosomes into the fiber. The most likely at the present seems to be a radial model, illustrated in **Figure 29.23,** in which the nucleosomes turn in a helical array, with an angle of about 60° between the faces of adjacent nucleosomes.

It seems likely that the parameters of the 30-nm fiber are not rigidly fixed, but can vary. This would accommodate the variation in the length of DNA per nucleosome, as well as allowing for other changes in the density of packing. It is not certain whether the fiber has the identical structure in both interphase chromatin and mitotic chromosomes.

Although the presence of H1 is necessary for the formation of the 30-nm fiber, information about its location is conflicting. Its relative ease of extraction from chromatin seems to argue that it may be present on the outside of the superhelical fiber axis; but other data on its accessibility suggest it is harder to find in 30-nm fibers than in 10-nm fibers that retain it, which would argue for an inside location.

How do we get from the 30-nm fiber to the specific structures displayed in mitotic chromosomes? And is there any further specificity in the arrangement of interphase chromatin; do particular regions of 30-nm

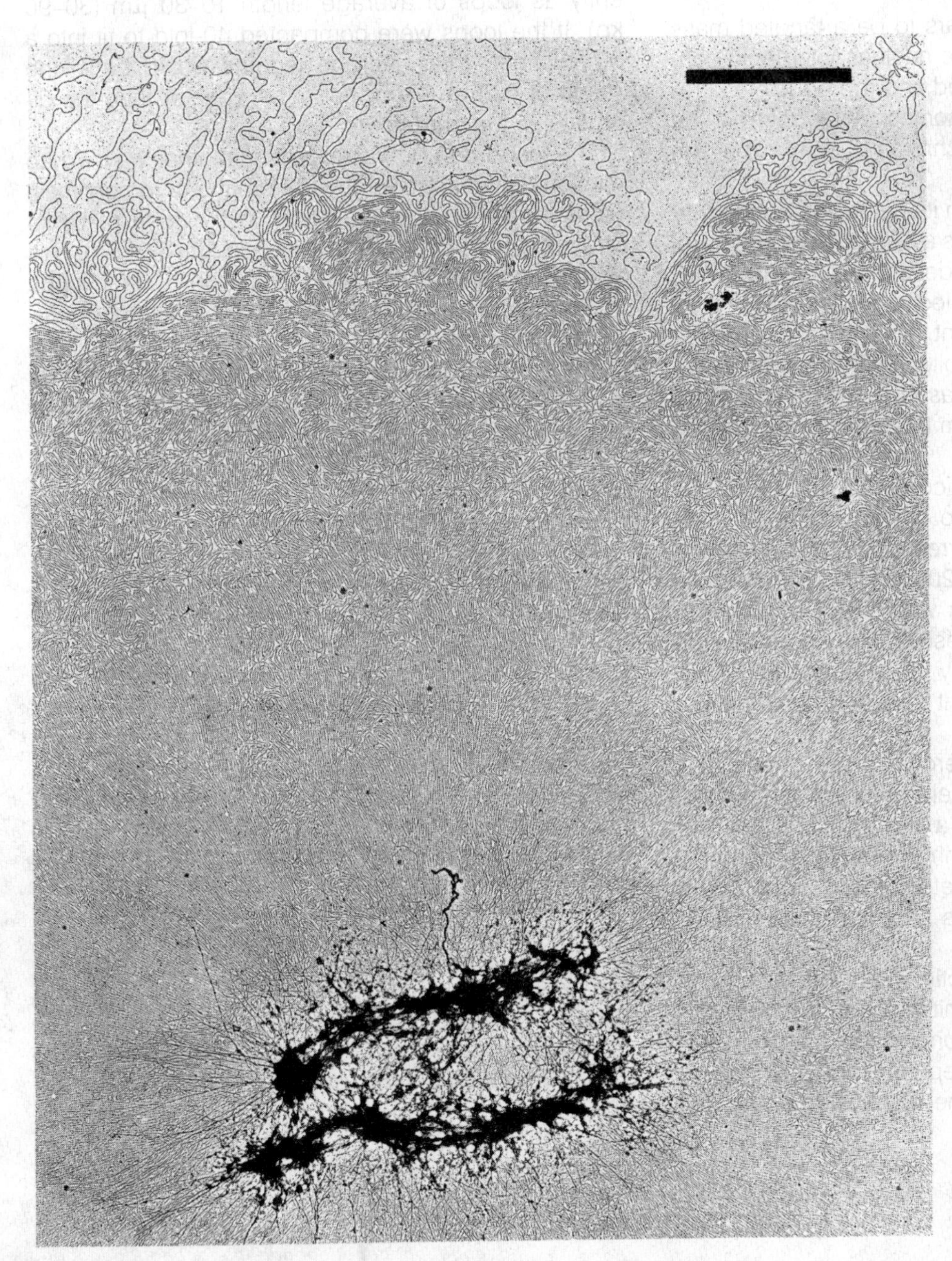

Figure 29.24
Histone-depleted chromosomes consist of a protein scaffold to which loops of DNA are anchored.
Photograph kindly provided by Ulrich K. Laemmli.

fibers bear a fixed relationship to one another or is their arrangement random? To such questions we have no answers at present.

LOOPS, DOMAINS, AND SCAFFOLDS

Interphase chromatin appears to be a tangled mass occupying a large part of the nuclear volume, in contrast with the highly organized and reproducible ultrastructure of mitotic chromosomes. What controls the distribution of interphase chromatin within the nucleus?

Some indirect evidence on its nature is provided by the isolation of the genome as a single, compact body. Using the same technique described in Chapter 28 for isolating the bacterial nucleoid, nuclei can be lysed on top of a sucrose gradient. This releases the genome in a form that can be collected by centrifugation. As isolated from *D. melanogaster*, it can be visualized as a compactly folded 10-nm fiber consisting of DNA and the four core histones.

The supercoiling of the compact body can be measured by its response to ethidium bromide (see Chapter 28). The level corresponds to about one negative supercoil for every 200 bp. These supercoils can be removed by nicking with DNAase, although the nucleosomes remain present. This suggests that the supercoiling is caused by the arrangement of nucleosomes. It must represent torsion across the nucleosome junctions.

Full relaxation of the supercoils requires one nick for every 85 kb. Thus the average length of "closed" DNA is ~85 kb. It could be a loop or domain similar in nature to those identified in the bacterial genome. We should like to know whether these loops correspond to specific sequences and whether they have functional significance.

Loops can be seen directly when the majority of proteins are extracted from mitotic chromosomes. The histones are removed by competition with the polyanions dextran sulfate and heparin. This also removes a large part of the nonhistone proteins. The resulting complex consists of the DNA associated with about 8% of the original protein content. As seen in **Figure 29.24,** the histone-depleted chromosomes take the form of a central **scaffold** surrounded by a halo of DNA.

The scaffold consists of a dense network of fibers. Threads of DNA emanate from the scaffold, apparently as loops of average length 10–30 μm (30–90 kb). If the loops were compacted 40-fold to fit into a 30-nm fiber, their average lengths would be in the range of 0.25–1.0 μm, not much greater than the diameter of the chromosome.

The loops can be visualized in another way. When divalent cations are removed from the chromosomes, cross sections show the loops in the form of radial arrays of the 10-nm fiber, average length 3–4 μm. This again is consistent with an organization in which loops of DNA of about 60 kb are anchored in a central proteinaceous scaffold.

The DNA can be digested without affecting the integrity of the scaffold. The scaffold looks like a mitotic pair of sister chromatids. The sister scaffolds usually are tightly connected, but sometimes are separate, joined only by a few fibers. Could this be the structure responsible for maintaining the shape of the mitotic chromosomes? Could it be generated by bringing together the protein components that usually secure the bases of the loops in interphase chromatin?

FURTHER READING

The development of the nucleosome can be traced from the somewhat flimsy evidence on which the model was originally propounded by **Kornberg** (*Science* **184,** 868–871, 1974) to the massive weight of evidence now assembled in reviews such as those of **Kornberg** (*Ann. Rev. Biochem.* **46,** 931–954, 1977), **McGhee & Felsenfeld** (*Ann. Rev. Biochem.* **49,** 1115–1156, 1980), and **Lewin** in *Gene Expression,* **2,** *Eucaryotic Chromosomes* (Wiley, New York, 1980, pp. 332–393). The path of DNA on the nucleosome has been succinctly analyzed by **Wang** (*Cell* **29,** 724–726, 1982). A model for the histone octamer was developed by **Klug et al.** (*Nature* **287,** 509–516, 1980), who also cite earlier structural analyses.

CHAPTER 30
THE NATURE OF ACTIVE CHROMATIN

The description of chromatin as a thread of duplex DNA coiled around a series of nucleosomes is the crucial first step toward visualizing the state of the genetic material in the nucleus. This somewhat static view accounts for the structure of the individual subunit and (to some degree) for its relationship with the next subunit. However, the organization of nucleosomes must be *flexible* enough to satisfy the various structural and functional demands that are made on chromatin.

The mass of euchromatin exists in an intermediate state of packaging. On the one hand, during cell division, euchromatic regions must become even more tightly packaged in mitotic chromosomes. This cyclical change affects all euchromatin more or less simultaneously, and most likely is controlled by changes in proteins that are widely distributed through chromatin. In the opposite direction of change, two types of local event seem likely to require some dispersion of structure because of their topological demands. These are replication and transcription.

During the interphase of a growing cell, all the chromatin must be reproduced. Replication occurs as a series of individual events in local regions (replicons), generating duplicate double-stranded DNA regions each associated with a set of histone octamers. The events involved in reproducing the nucleosome particle have yet to be defined. We should like to know what happens to the nucleosome during replication, and how new nucleosomes are assembled.

At all events, it seems inevitable that the separation of parental DNA strands must disrupt the structure at least of the 30-nm fiber and probably also of the 10-nm fiber. We should like to know the extent of this disruption. Is it confined to the immediate vicinity of the point where DNA is being synthesized, or does it extend farther? Are there discernible structural differences between regions that have replicated and those that have yet to do so? The transience of the replication event is a major difficulty in analyzing the structure of a particular region while it is being replicated.

Transcription also involves the unwinding of DNA, and presumably therefore requires unfolding of the fiber in restricted regions of chromatin. A simple-minded view suggests that some "elbow-room" must be needed for the process. The features of polytene and lampbrush chromosomes described in Chapter 28 offer hints that a more expansive structural organization may be associated with gene expression.

We should like to know what structural changes occur when a gene is being transcribed. Does the overall structure of the region change? Does the transcribed sequence remain in the form of nucleosomes; and if so, what happens to them when RNA polymerase transcribes the DNA? What ensures that the promoter is initially accessible to the enzyme?

Can we identify sets of nucleosomes whose different properties explain the structure or function of particular regions? And are nucleosomes the sole type of protein-DNA structure in the duplex thread, or are other structures present to delineate particular sites?

Important though these questions are, they are really a prolegomenon to the major issue in thinking about gene expression. *What changes the state of a gene to enable it to be transcribed at the right time and place?* The obverse question is whether and how genes can be turned off.

NUCLEOSOME ASSEMBLY VERSUS CHROMATIN REPRODUCTION

Core histones possess an inherent ability to wrap DNA around the protein octamer. But there are several routes by which DNA might associate with the histones.

The reconstitution reaction at first could be performed *in vitro* only under strikingly nonphysiological conditions, by mixing the components in high salt and urea, which were then dialyzed away. But now the reaction can be partially controlled by adjusting the order of addition and the proportions of histones and DNA. This allows self-assembly to occur under conditions that approach the physiological. It is a slow process, still limited by the tendency of the assembling particles to precipitate. Two possibilities for de novo nucleosome assembly pathways are summarized in **Figure 30.1.**

One pathway is revealed by the ability of the $H3_2H4_2$ tetramer to organize DNA into particles that partially resemble cores (as judged by their susceptibility to micrococcal nuclease). These bodies can be converted into core particles by the addition of H2A·H2B. This was the origin of the idea that there is a "kernel" of arginine-rich histones in the nucleosome structure. The possible use of this pathway in assembly *in vivo* is consistent with observations that H3 and H4 are incorporated into replicating chromatin before H2A and H2B enter the product.

Another result makes use of cross-linked histone octamers that cannot dissociate into individual proteins. Still it is possible to bind DNA to form a particle. This shows that in principle it is possible for DNA to be wrapped around a previously intact octamer.

Attempts to produce nucleosomes *in vitro* essen-

Figure 30.1

In certain conditions *in vitro*, DNA can assemble with the $H3_2H4_2$ tetramer, to which two H2A·H2B dimers are added. In other conditions, DNA can interact directly with an intact (cross-linked) histone octamer.

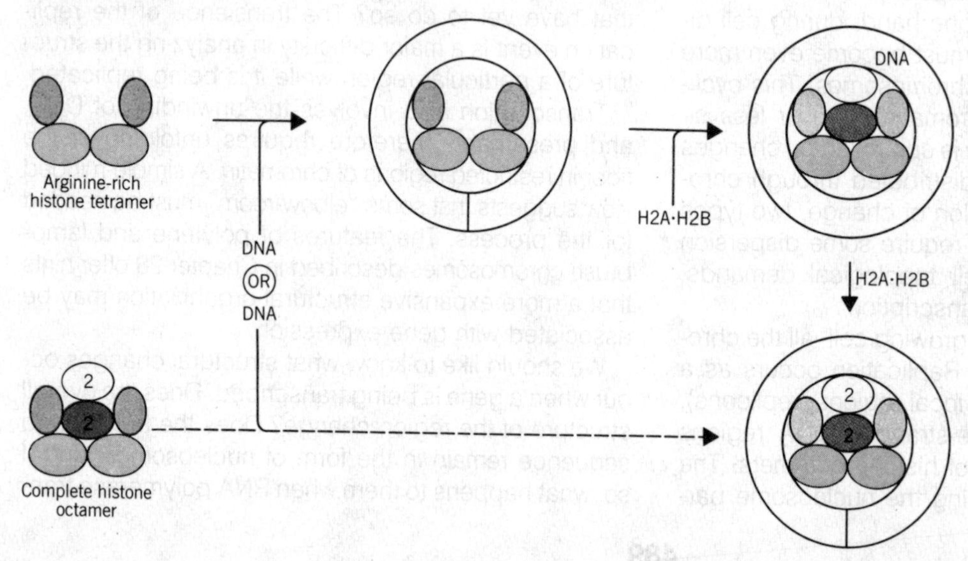

tially consider the process of assembly as one involving free DNA and histones. But what actually happens *in vivo* is that *chromatin* is reproduced. A stretch of DNA *already associated with nucleosomes* is replicated, giving rise to two daughter duplexes. What happens to the preexisting nucleosomes at this point? Are the histone octamers dissociated into free histones for reuse, or do they remain assembled? Some possibilities are illustrated in **Figure 30.2.**

Cross-linking experiments suggest that the histone octamer is **conserved,** surviving as such through cycles of replication. In this case, do the "old" octamers associate in any particular pattern with the duplicate DNAs? For example, all the old octamers might stay with one daughter duplex, while the "new" octamers assemble on the other. However, when the assembly of new histones into nucleosomes is prevented (by treatment with cycloheximide to inhibit protein synthesis), the "old" octamers are dispersed to both of the daughter duplexes.

Nucleosomes may therefore originate in two ways. When chromatin is replicated, the existing histone octamers are displaced from the DNA to allow it to replicate. The octamers are conserved and may reassociate with either daughter duplex. However, an equal number of octamers must be formed from newly synthesized histones. Are these octamers also assembled before they associate with DNA, or do they follow an alternative pathway, assembling on the DNA?

Reproduction of chromatin does not involve any protracted period during which the DNA is free of nucleosomes. Once DNA has been replicated, nucleosomes are quickly generated on both the duplicates. This point is illustrated by the electron micrograph of **Figure 30.3,** which shows a recently replicated stretch of DNA, already covered with nucleosomes on both daughter duplex segments.

The structure of the point at which DNA is actually being replicated does appear to differ from other regions. It is more resistant to micrococcal nuclease and is digested into bands that differ in size from nucleosomal DNA. This suggests that quite a large protein complex is engaged in replicating the DNA, but the nucleosomes reform more or less immediately behind as it moves along.

Chromatin contains nonhistone proteins as well as the nucleosomal histones, and these additional components also must be reproduced. Since the nonhistone complement is likely to vary with the phenotype of the cell, its reproduction involves the maintenance of cell-specific features. Thus the possible existence of segregation patterns for proteins during DNA replication has a significance extending beyond nucleosome assembly. One of the principal questions that we should like to answer is how alternate states of chromatin structure are perpetuated through cell division.

Consider a gene that is activated (or repressed) by the binding to DNA of some specific regulator protein and/or by some change in chromatin structure. How is this particular state to be inherited by the duplicate chromosomes?

If all the proteins dissociate from DNA during replication, the specific state must be reestablished in every cell cycle. However, a definite pattern of segregation could be used to propagate information about the state of gene expression. One possibility is that a specific structure might be *perpetuated* by segregation and duplication during DNA replication. Thus a complex of nonhistone proteins might be established on the DNA, split into half-complexes at replication, and rebuild complete complexes on each daughter duplex.

The other side of this issue is how a *change* is made in the pattern of gene expression. One possibility often discussed is that changes might be connected with DNA replication, taking advantage of the transient disruption of chromatin structure. Once made, a change could be inherited via a segregation mechanism, unless and until another change is made.

NUCLEOSOME ASSEMBLY REQUIRES NONHISTONE PROTEINS

Procedures for reconstituting histones with DNA generate a structure more akin to the core particle than to the complete nucleosome. At low concentrations, when individual reconstituted particles are connected by a thread of DNA, more than 146 bp are associated with the protein octamer. But this is not a firm characteristic, since as soon as the concentration is increased so that the particles become more tightly packed, each octamer is found to organize only the

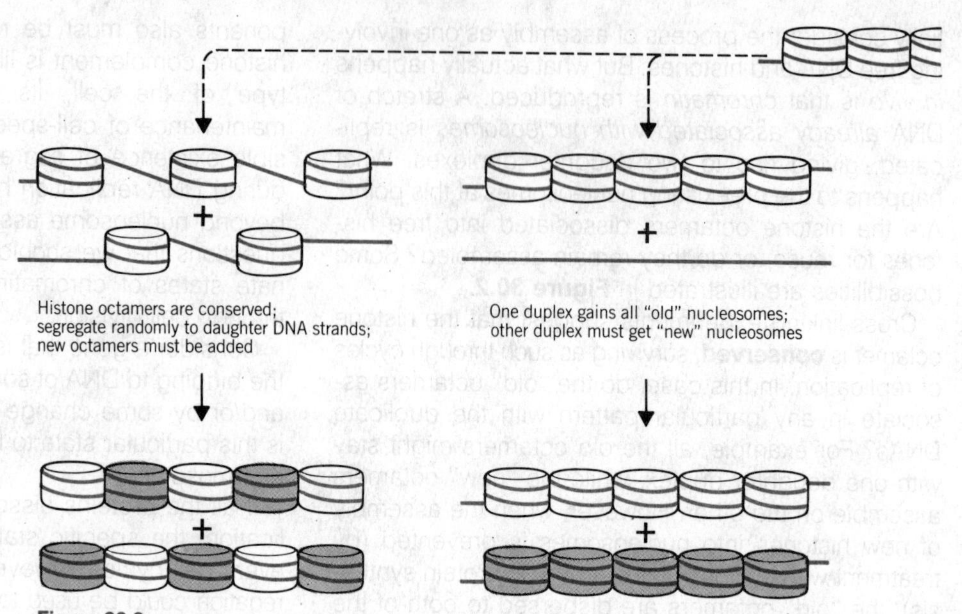

Figure 30.2
Many possibilities have been considered for the fate of nucleosomes when DNA is replicated.

core length of 146 bp of DNA. It is as though, in the absence of an adjacent core particle, a histone octamer will opportunistically mobilize more DNA, but gives up all but the core length as soon as an immediate neighbor is formed.

What is responsible for modulating this reaction *in vivo* so that nucleosomes are formed with the spacing that is characteristic of their genome or even of their particular location within it?

The only system in which an authentic spacing of reconstituted nucleosomes has been obtained is provided by the *Xenopus* oocyte. When SV40 DNA is injected into the oocyte, the circular molecules can be assembled into minichromosomes. When a sufficient excess of DNA is used, the pool of endogenous histones becomes exhausted, and assembly depends on the injection of additional histones. The features of the system are retained in a cell-free extract, which assembles nucleosomes onto free DNA with a spacing of 195 bp.

This is important information because it shows that proper assembly of nucleosomes can occur de novo with free DNA. *Assembly is not necessarily associated with the act of replication; and it is independent of the sequence of the added DNA.*

These features have not yet been reproduced entirely in a fractionated system, but some progress has been made toward identifying the components involved. An assembly protein has been purified from the *Xenopus* oocyte. It is a pentamer containing identical subunits of 29,000 daltons. This protein is the most abundant in the oocyte, and is located in the nucleoplasm. An antibody prepared against it reacts with proteins in the nucleoplasms of a wide range of eukaryotes, which suggests that it could represent a universal function that has been conserved in evolution. It is called **nucleoplasmin.**

In the presence of nucleoplasmin, the histones can associate with DNA to generate particles under physiological (low-salt) conditions. When digested with micrococcal nuclease, these particles generate bands of DNA at 146 bp and 165 bp. This suggests that core particles are formed, some with an additional length of DNA, but not yet extending to the construction of the full nucleosome. Nucleoplasmin is therefore involved in controlling the reaction of histones with DNA in such a way that it becomes productive, and forms nucleosomes instead of random aggregates. However, it does not establish the spacing characteristic of nucleosomes.

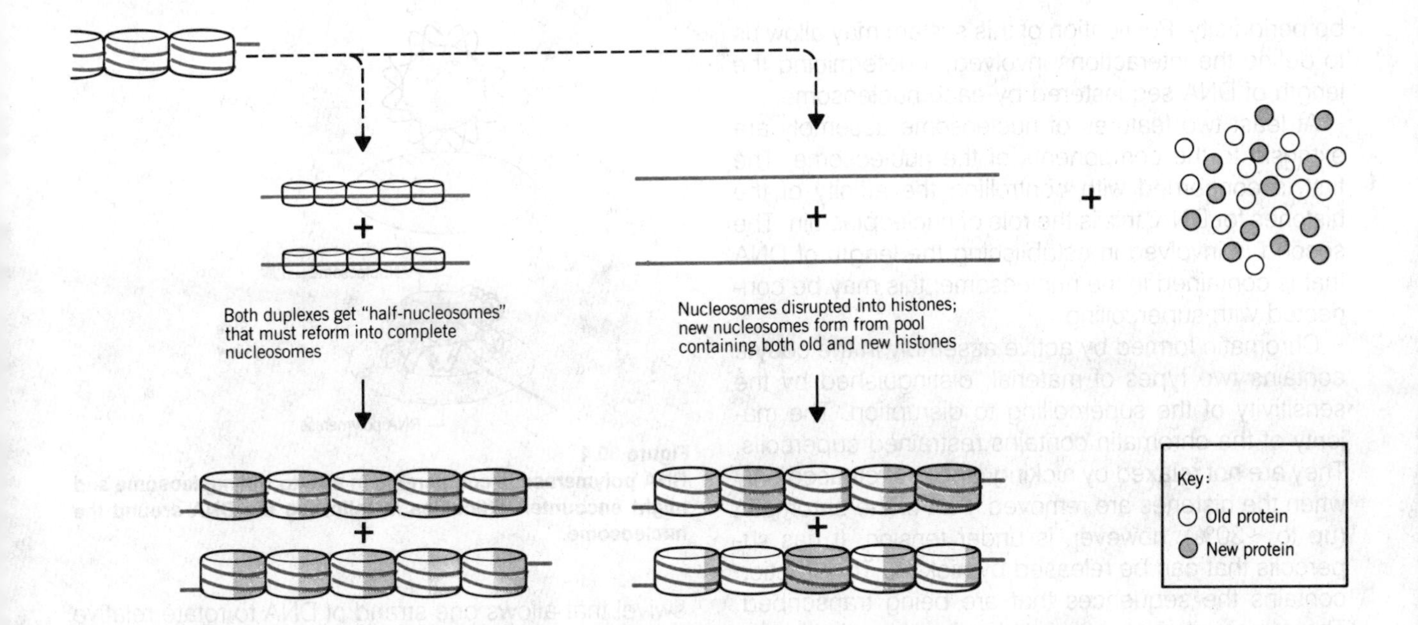

What is the function of the nucleoplasmin? It is an acidic protein that does not bind to free DNA or to intact nucleosomes, but it does bind to all of the individual histones. The reaction saturates at a level equivalent to one pentamer of nucleoplasmin per octamer of histones. Nucleoplasmin may be a "molecular chaperone," binding to the histones and releasing them to DNA in a more controlled manner than is possible when the DNA has no competitor for their attention.

The basis for its effect may lie in the ability of nucleoplasmin to bind to histones to reduce the net positive charge. The use of high salt concentration for histone octamer assembly *in vitro* may mimic this situation. In this context, we should also mention an early idea that modification of charged groups on histones might be involved in regulating the affinity of the protein for DNA (see later). *Such interactions may allow histones to form the thermodynamically most stable aggregate (the nucleosome) without becoming trapped in kinetic intermediates (that is, other complexes resulting from the high affinity of histones for DNA).*

The overall structure of the chromatin thread may be an important factor in the assembly of chromatin. When a circular DNA is injected into the oocyte, first nucleosomes assemble, and then the thread becomes supercoiled. These features can be reproduced in an *in vitro* system containing ATP, which provides the energy needed to drive the supercoiling reaction. This "active assembly" generates nucleosomes with a 200

Figure 30.3
Replicated regions of chromatin contain nucleosomes on both daughter DNA duplexes.
Photograph kindly provided by Steven L. McKnight.

Nonreplicated regions
Replicated region

bp periodicity. Purification of this system may allow us to define the interactions involved in determining the length of DNA sequestered by each nucleosome.

At least two features of nucleosome assembly are extrinsic to the components of the nucleosome. The first is concerned with controlling the affinity of the histones for DNA; this is the role of nucleoplasmin. The second is involved in establishing the length of DNA that is contained in the nucleosome; this may be connected with supercoiling.

Chromatin formed by active assembly in the oocyte contains two types of material, distinguished by the sensitivity of the supercoiling to disruption. The majority of the chromatin contains restrained supercoils. They are not relaxed by nicking, and are released only when the histones are removed. Part of the chromatin (up to ~30%), however, is under tension. It has supercoils that can be released by nicking. This fraction contains the sequences that are being transcribed. The supercoiled region could coincide with the domains previously seen in whole genomes (see Chapter 28). We do not yet know what produces the difference between the two types of material.

ARE TRANSCRIBED GENES ORGANIZED IN NUCLEOSOMES?

Transcription involves the movement along DNA of a region of local unwinding that is generated by RNA polymerase (as described in Chapter 11). Usually both strands of DNA are envisaged as being intimately associated with, or even surrounded by, the enzyme, as imagined in Figure 11.4. About 50 bp or so of DNA are actually bound to the enzyme tightly enough to be protected against nucleases.

The need to unwind the DNA makes it seem unlikely that the segment engaged by RNA polymerase could remain on the surface of the nucleosome. On the other hand, little is known about the stereochemistry of RNA polymerase action (especially in eukaryotes), so it is possible that the enzyme might be able to use one strand of DNA as template while the other strand remains bound to histones.

The mechanism by which the DNA unwinds to generate a free template strand for RNA synthesis is not clear. Usually it is assumed to involve some sort of swivel that allows one strand of DNA to rotate relative to the other. The problem of unwinding would be exacerbated if DNA were immobilized on the nucleosome during transcription.

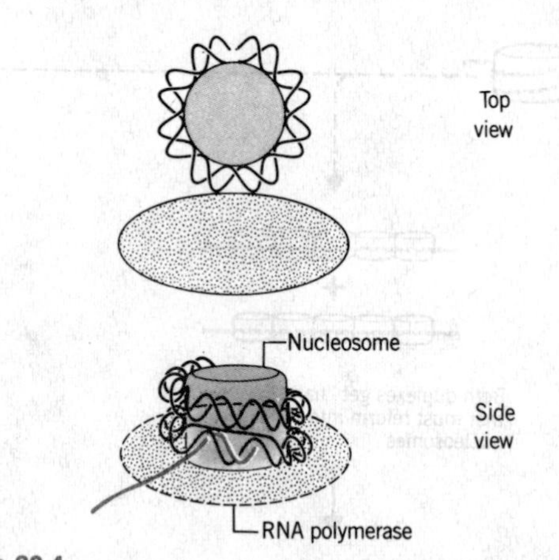

Figure 30.4
RNA polymerase is comparable in size to the nucleosome and might encounter difficulties in following the DNA around the nucleosome.

In considering these issues, we must bear in mind the relative sizes of RNA polymerase and the nucleosome. The eukaryotic enzymes are large proteins, typically >500,000 daltons. Compare this with the 262,000 daltons of the nucleosome. **Figure 30.4** illustrates the approach of RNA polymerase to nucleosomal DNA. Even without detailed knowledge of the interaction, it is evident that what is involved would be the approach of two comparable bodies. It seems inevitable that this must involve a structural change.

Nucleosomal DNA is relatively inaccessible to micrococcal nuclease; and even on isolated core particles, only some of the bonds in DNA are accessible to nicking by DNAase I (see Chapter 29). Remember that the nucleosome is not an isolated object but is adjacent to others, and consider the two turns that DNA makes around it. All of this highlights the question of whether RNA polymerase would have sufficient access to DNA if the nucleic acid were confined to its customary path on the nucleosome. It is hard to imagine that during transcription the polymerase could follow the DNA around the nucleosome.

So the first question to ask about the structure of active genes is whether DNA that is being transcribed remains organized in nucleosomes.

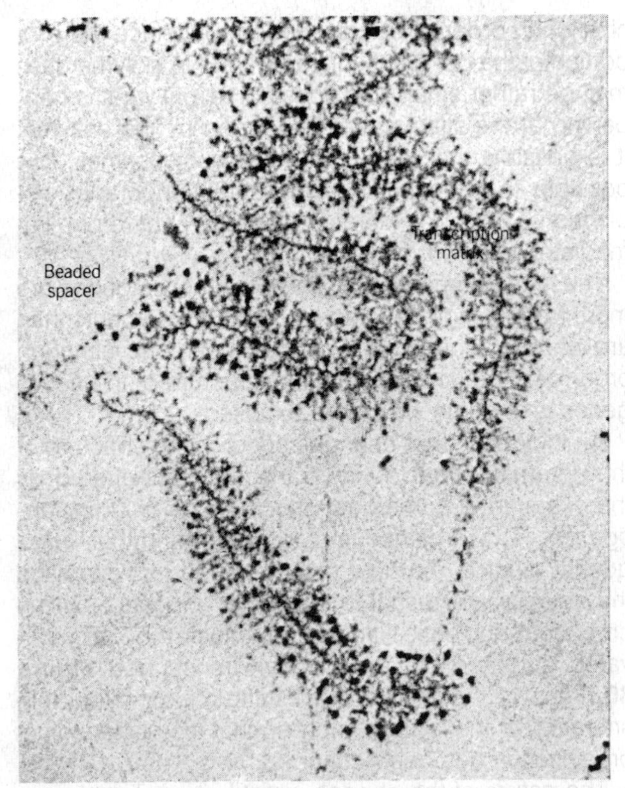

Figure 30.5
The extended axis of an rDNA transcription unit alternates with the only slightly less extended nontranscribed spacer.
Photograph kindly provided by Charles Laird.

The genes coding for rRNA can be examined by electron microscopy. The structure of the DNA is hard to see, because of the intense packing of RNA polymerases. This is evident in Figure 23.4 and is shown in another example in **Figure 30.5.** The packing ratio of the DNA can be calculated by dividing the known length of the transcription unit by the measured length of the axis of the transcription matrix. The ratio is about 1.2. Thus the DNA is almost completely extended; it cannot be organized in nucleosomes.

The nontranscribed spacers between the transcription matrices also appear to be quite extended, with a packing ratio of about 1.4. Under some conditions of preparation, the spacers contain beads that could be nucleosomes; but if this is what they are, there must be free spacer DNA connecting them. Under other conditions, the spacer appears to consist of free DNA.

All of this argues that the state of DNA in the very intensively transcribed rRNA tandem genes is a far cry from the compact organization that would be seen even of a simple string of adjacent nucleosomes in a 10-nm fiber (which would have a packing ratio of ~6).

On the other hand, transcription complexes of SV40 minichromosomes can be extracted from infected cells. They contain the usual complement of histones and display a beaded structure. Chains of RNA can be seen to extend from the minichromosome, as in the example of **Figure 30.6.** This argues that transcription can proceed while the SV40 DNA is organized into nucleosomes. Of course, the SV40 minichromosome is transcribed less intensely than the rRNA genes.

Another approach is to digest chromatin with micrococcal nuclease, and then to use a probe to some specific gene or genes to determine whether the corresponding fragments are present in the usual 200 bp ladder. The problem with this procedure is that we cannot be sure whether all the copies of the gene are being transcribed. A nucleosomal ladder could be produced by a minority of inactive copies. To deal with this difficulty, the concentration of sequences that hybridize with the probe is compared using monomeric fragments and whole DNA.

The conclusions that we can draw from these experiments are limited but important. *Genes that are being transcribed contain nucleosomes at the same frequency as nontranscribed sequences.* Thus genes do not necessarily enter an alternative form of orga-

Figure 30.6
An SV40 minichromosome can be transcribed.

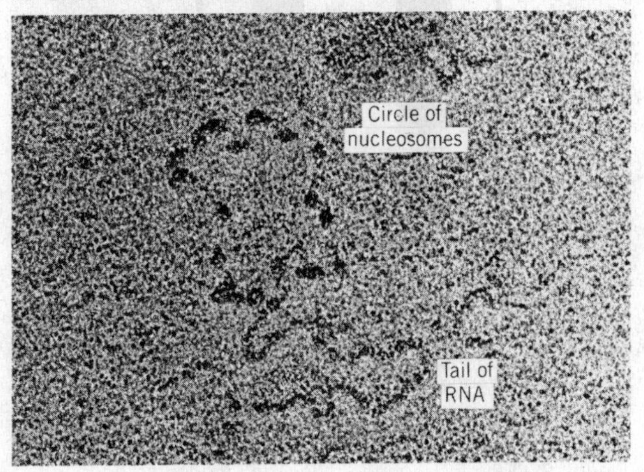

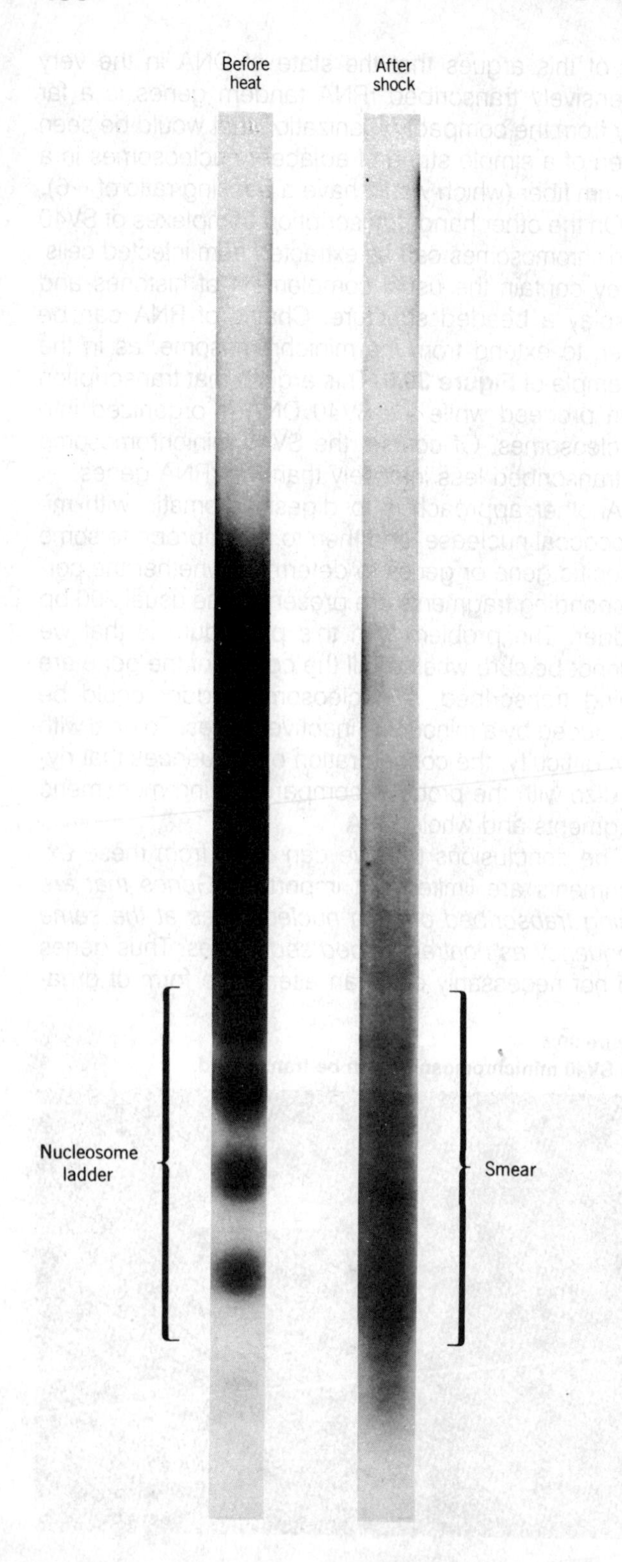

nization in order to be transcribed. But since the proportion of the gene associated with RNA polymerases may be rather small, this does not reveal what is happening at the sites actually engaged by the enzyme. It is possible that they retain their nucleosomes; it is possible that the nucleosomes are temporarily displaced as RNA polymerase passes by, but reform immediately afterward.

These experiments also show that active genes are more rapidly digested into monomeric fragments than are inactive sequences. This may mean that, although organized in nucleosomes, the structure of the active genes is in some way more exposed.

An indication that changes occur in the structure of the chromatin fiber during transcription is offered by the example of some heat-shock genes of *D. melanogaster*. These genes are transcribed rather infrequently prior to the heat shock, and in this condition they display the usual ladder when digested by micrococcal nuclease. Upon heat shock they are activated; and then they are intensely transcribed. **Figure 30.7** shows that activation results in a considerable smearing of the ladder, implying that the nucleosomal organization has been changed.

The nature of the change is not known. Treatments such as shearing of chromatin can abolish the characteristic response to micrococcal nuclease, even though nucleosomes remain present. Thus we cannot tell whether the ladder disappears because it is obscured by the presence of RNA polymerase (or other proteins), because there is a disruption of higher-order structure, or because the nucleosomes have been modified or even displaced altogether.

It is clear that an important structural change occurs when a gene is intensely transcribed. In the case of the rRNA genes, it looks as though the nucleosomes are displaced. But this could be an exceptional case. It might be reconciled with the presence of nucleosomes in less heavily transcribed genes by supposing that RNA polymerase displaces the nucleosome at the

Figure 30.7

The micrococcal ladder is evident before activation of heat shock genes, but becomes smeared soon after intense transcription begins.

Photograph kindly provided by Sarah Elgin.

point of transcription, but that the histone octamer immediately recaptures its position unless another RNA polymerase is present to prevent it from doing so.

THE DNAase-SENSITIVE DOMAINS OF TRANSCRIBABLE CHROMATIN

Since only a characteristic minority of DNA sequences are transcribed in any particular cell, some feature must distinguish at least the promoters (and possibly the entire lengths) of these transcription units. Something must indicate to the transcription apparatus that a locus is available for expression.

Some perturbation of structure must occur in a gene when it is being transcribed, if only as a result of the movement of RNA polymerase along the DNA. Thus structural changes could be a consequence of the act of transcription, rather than a cause of it. So in assessing the properties of transcribed loci, we need to determine which of their particular features occur prior to transcription as a prerequisite for it, and which are induced subsequently as a result of the events involved in the synthesis of RNA.

When chromatin is digested with DNAase I, it is eventually degraded into acid-soluble material (very small fragments of DNA). The progress of the overall reaction can be followed in terms of the proportion of DNA that is rendered acid soluble. *When only 10% of the total DNA has become acid soluble, more than 50% of the DNA of an active gene has been lost.* This suggests that active genes are preferentially degraded.

The fate of individual genes can be followed by quantitating the amount of DNA that survives to react with a specific probe (using the usual techniques of restriction cleavage, blotting and hybridization). The protocol is outlined in **Figure 30.8.** The principle is that the loss of a particular band indicates that the corresponding region of DNA has been degraded by the enzyme.

Figure 30.9 shows what happens to β-globin genes and an ovalbumin gene in chromatin extracted from chicken red blood cells (in which globin genes are expressed and the ovalbumin gene is inactive). The restriction fragments representing the β-globin genes are rapidly lost, while those representing the ovalbumin gene show little degradation. (The ovalbumin gene in fact is digested at the same rate as the bulk of DNA.)

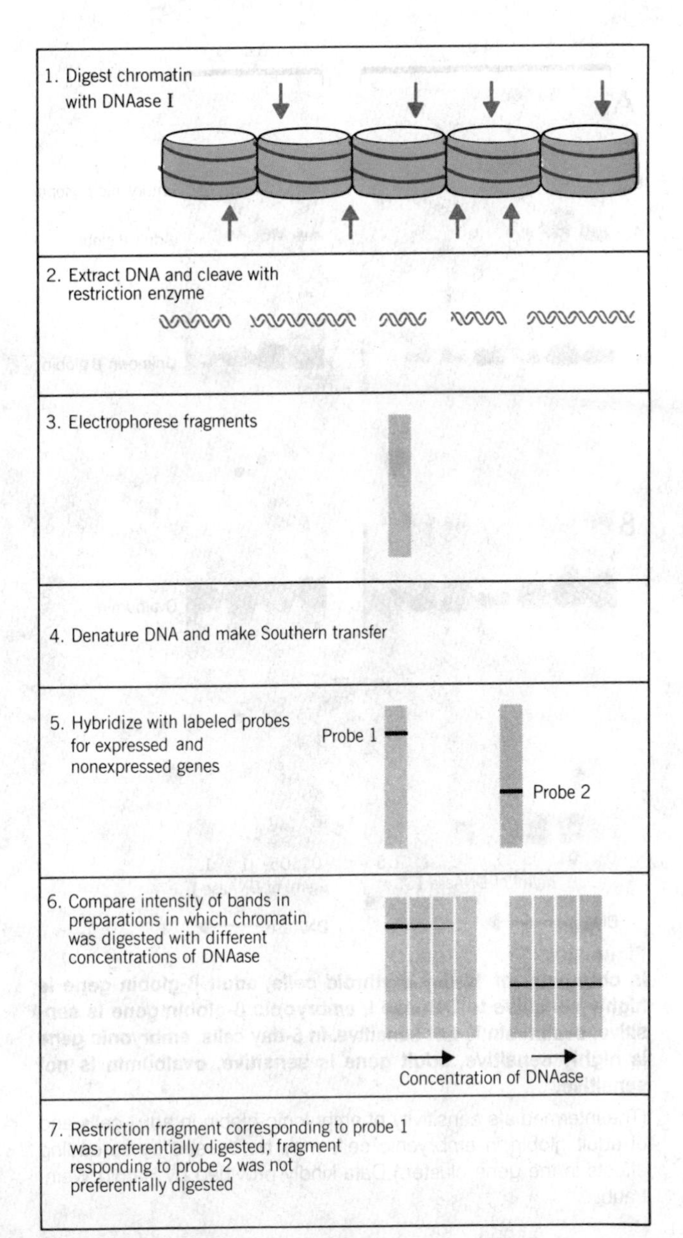

Figure 30.8
Sensitivity to DNAase I can be measured by determining the rate of disappearance of the material hybridizing with a particular probe.

So the bulk of chromatin is relatively resistant to DNAase I and contains nonexpressed genes (as well

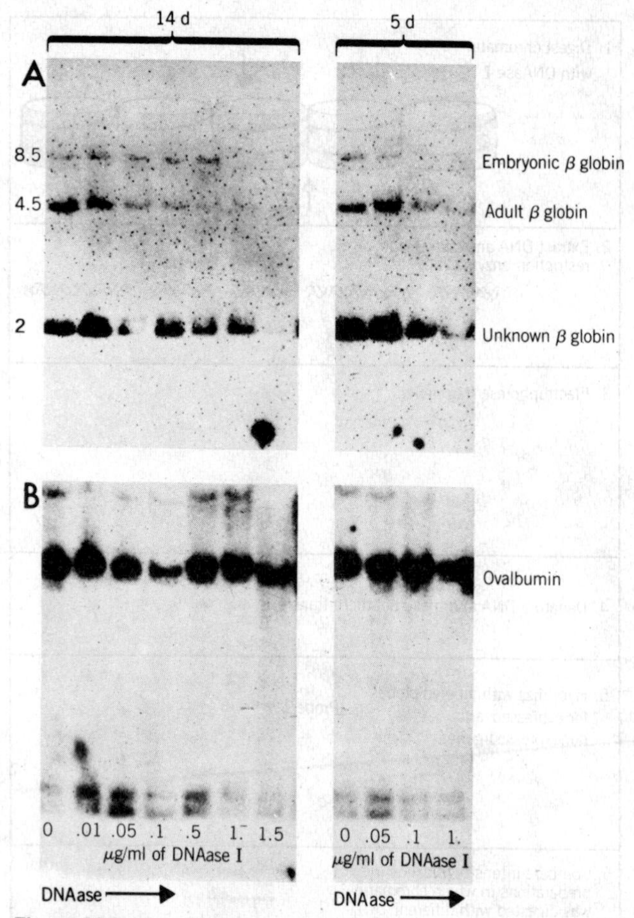

Figure 30.9
In chromatin of 14-day erythroid cells, adult β-globin gene is highly sensitive to DNAase I, embryonic β-globin gene is sensitive, ovalbumin is not sensitive. In 5-day cells, embryonic gene is highly sensitive, adult gene is sensitive, ovalbumin is not sensitive.

(The intermediate sensitivity of embryonic globin in adult cells and of adult globin in embryonic cells may be caused by spreading effects in the gene cluster.) Data kindly provided by Harold Weintraub.

as other sequences). A gene becomes relatively susceptible to the enzyme specifically in the tissue(s) in which it is expressed.

Is preferential susceptibility a characteristic only of rather actively expressed genes, such as globin, or of all active genes? Experiments using cDNA probes representing the entire cellular mRNA population suggest that all active genes, whether coding for abundant or for rare mRNAs, are preferentially susceptible to DNAase I. (However, there may be differences in the degree of susceptibility.) Since the rarely expressed genes are likely to have very few RNA polymerase molecules actually engaged in transcription at any moment, this implies that the sensitivity to DNAase I does not result from the act of transcription, but is a feature of *genes that are able to be transcribed.*

How extensive is the preferentially sensitive region? Its extent can be determined by using a series of probes representing the flanking regions as well as a transcription unit itself. The sensitive region always extends for a considerable distance on either side of the transcribed region. This has been taken to suggest that DNAase sensitivity defines a chromosomal **domain,** a region including at least one active transcription unit, but whose structural change extends well beyond it. (Note that the use of the term "domain" does not imply any necessary connection with the structural domains identified by the loops of chromatin or chromosomes.)

It is not clear whether the entire domain is uniformly sensitive. In the β-globin gene cluster of the chicken, the sensitive domain extends for 6–7 kb on the 5′ side of the cluster and for at least 8 kb on the 3′ side. Within the gene cluster itself, regions that are being transcribed appear to be more sensitive to DNAase I than the flanking regions that are not actually expressed (and which are said to have "intermediate" sensitivity). Similarly, the ovalbumin gene and the related X and Y genes may lie in a DNAase-sensitive domain that extends for about 100 kb. But here there do not seem to be any differences in sensitivity between coding and noncoding DNA within a domain.

The critical concept implicit in the description of the domain is that a region of high sensitivity to DNAase I extends over at least the length of the transcription unit. Often we think of regulation as residing in events that occur at a discrete site in DNA—for example, in the ability to initiate transcription at the promoter. Even if this is true, such regulation must determine, or must be accompanied by, a more wide-ranging change in structure. This may be a difference between eukaryotes and prokaryotes.

NONHISTONE PROTEINS CONFER DNAase SENSITIVITY

What is responsible for the DNAase-sensitive state? We must look to the nonhistone proteins for differences that may be associated with changes in chromatin structure. Sensitive chromatin displays a difference in the behavior of some HMG proteins. HMG is an abbreviation for high-mobility group, a set of small nonhistone proteins. These proteins are extracted from chromatin with 0.35 M NaCl, which means that they are relatively loosely bound.

Two nonhistone proteins, HMG14 and HMG17, are preferentially released from chromatin on digestion with DNAase I. After the extraction of these proteins, a globin gene in red blood cell chromatin does not show preferential sensitivity to DNAase I. But sensitivity is restored when the proteins are added back to the salt-extracted chromatin.

The proteins do *not* confer preferential susceptibility on globin genes if brain chromatin is used instead of red blood cell chromatin. So the HMG proteins cannot be solely responsible for the effect. Some other factor, resident in the NaCl-extracted chromatin, must differ between the globin genes of the red blood cell and brain. This difference allows *only the red blood cell chromatin* to bind the HMG14 and HMG17 proteins into a structure that makes the globin gene sensitive to DNAase.

Either HMG14 or HMG17 is effective by itself. About 1 mole of added HMG is sufficient to confer full sensitivity for every 10–20 nucleosomes extracted from bulk chromatin. The ratio is the same for genes that are transcribed actively, such as globin, or for genes that are transcribed only rarely.

Is DNAase sensitivity conferred at the level of the individual nucleosome? Monomers isolated by treatment with micrococcal nuclease show exactly the same sensitivity to DNAase I as whole nuclei or chromatin. The effect is lost when HMG14 and HMG17 are removed; it is regained when they are restored. *The same result is obtained with core particles.* Attempts to identify the feature that allows core particles to bind HMG14 and HMG17 have not yet succeeded. Note the clear implication in this work that the globin gene remains in the form of nucleosomes when it is transcribed.

The intermediate level of DNAase sensitivity displayed by the regions flanking the transcribed chicken globin genes is lost upon treatments that tend to destroy higher-order structure in chromatin, and is not sustained in individual nucleosomes. It cannot be restored by addition of HMG14 and HMG17.

Sensitivity to DNAase I may therefore be created by a highly sensitive region very closely surrounding the transcribable sequence. Apparently this extends less than 1 kb on either side. The adjacent regions could show intermediate sensitivity because of the spreading of a structural effect (although the distance of spreading is quite large relative to the length of the transcribed region itself).

What is the timing of the acquisition of the DNAase-sensitive state in relation to the occurrence of transcription? Does the sensitive state represent an activation event—perhaps a dispersion of structure—that occurs in preparation for transcription? We cannot investigate this question with the globin genes, because sensitivity may spread from embryonic to adult genes, or vice versa. To determine whether DNAase sensitivity appears before transcription starts, we need to investigate a gene that occurs in isolation, where there is no spreading effect from adjacent loci.

It is clear, however, that the sensitive state may persist after transcription has ceased. Chick erythrocytes are mature red blood cells that continue to translate globin mRNA, but do not transcribe the genes. However, the globin genes remain in the sensitive state. And withdrawal of estrogen from chickens brings transcription of the ovalbumin gene to a halt, but the coding region remains in the active state. We do not yet have data from a situation that reveals at what point a previously active gene returns to an inactive (DNAase-resistant) structure, for example, in the course of embryonic development.

The continuation of the sensitive state after transcription has ceased shows that the gene structure can be active without necessarily possessing active RNA polymerase molecules. This result, together with the sensitivity of rarely transcribed genes, suggests that the change in structure is independent of the act of transcription. DNAase sensitivity is therefore necessary, but not sufficient, for transcription.

We should like to know more about the nature of the

sensitive domain, in particular how the borders are defined. And we must ask why it is necessary to change the structure of every nucleosome in the transcription unit. A possible answer is that the change is needed to allow RNA polymerase to proceed along the template—for example, perhaps the histones can be displaced from the DNA only if the nucleosomes are modified.

HISTONES SUFFER TRANSIENT MODIFICATIONS

Each histone has an invariant sequence in the somatic cells of an adult organism. However, all of the histones are subject to modifications in which additional moieties are linked covalently to the free groups of certain amino acids. These modifications reduce the overall positive charge (that is, the basicity) of the protein molecule. This allows the unmodified and the various modified forms to be separated, for example, by gel electrophoresis.

Acetylation and methylation occur on the free (ϵ) amino group of lysine. As seen in **Figure 30.10,** this removes the positive charge that resides on the NH_3^+ form of the group. Methylation also occurs on arginine and on histidine.

Phosphorylation of histones occurs on the hydroxyl group of serine and also on histidine. This introduces a negative charge in the form of the phosphate group (see Figure 30.10).

All of these modifications affect internal residues and are transient. A contrast is provided by the stable acetylation of the N-termini of some histones that occurs at the time of synthesis.

The transient modifications occur at one point in the cell cycle and (usually) are reversed at another point. Because they reduce the positive charge of the protein molecule, they have been viewed as potentially able to change the functional properties of the histones. At present there is no evidence that these changes are related to chromatin functions, although there are some quite provocative correlations.

A cycle of histone acetylation and deacetylation is well characterized in the trout testis (where the histones are displaced by other basic proteins called protamines). Here H3 and H4 are the principal targets, with up to four acetyl groups added to each; H2A and H2B are affected also, but to a lesser degree. The positions that can be acetylated (or methylated) have been identified in this and some other situations. They are not identical in different organisms, although they do constitute only a small number of overlapping sites.

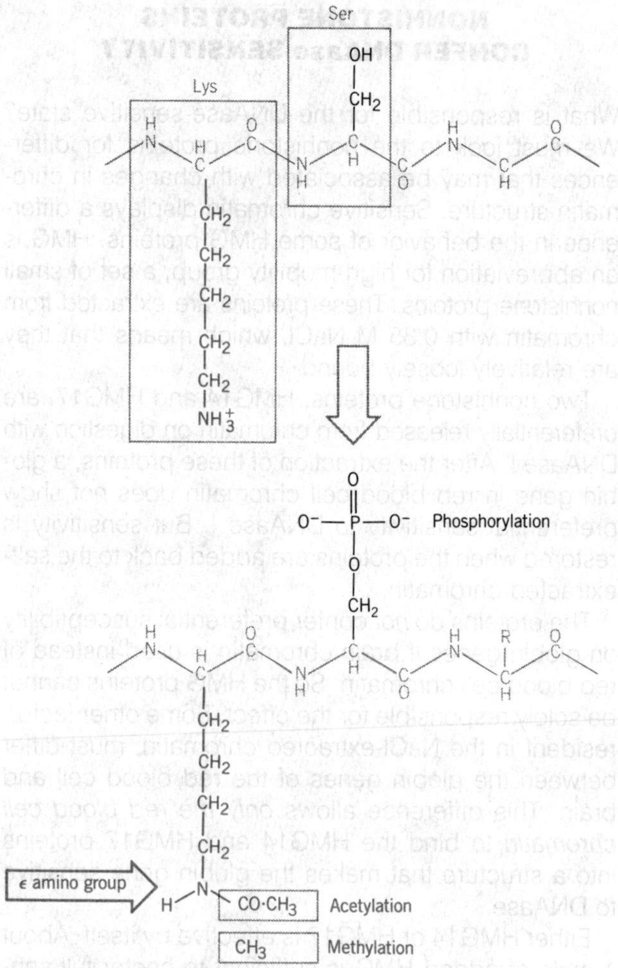

In synchronized cells in culture, both the preexisting and the newly synthesized core histones appear to be acetylated and methylated during S phase. (This is the part of the cell cycle during which DNA is replicated and when the histones also are synthesized.)

Figure 30.10
Acetylation of lysine or phosphorylation of serine reduces the overall positive charge of a protein.

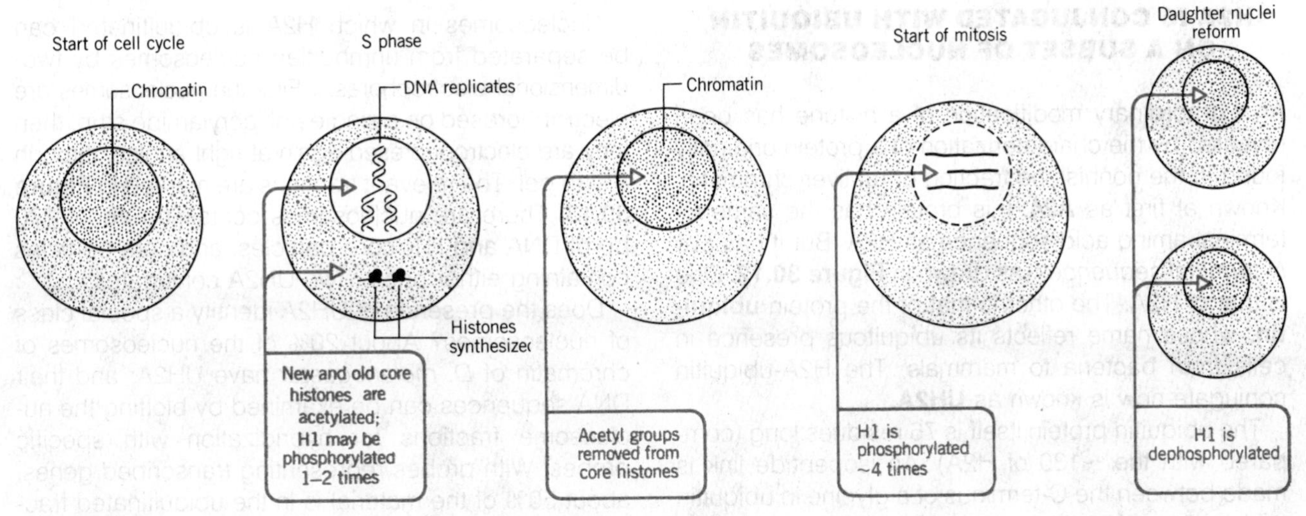

Figure 30.11
Core histone acetylation occurs during DNA synthesis and then is reversed; the major phosphorylation of H1 occurs at the start of mitosis and is reversed at the end.

During the cell cycle, the modifying groups are later removed. These events are repeated in each cell cycle, and their relative timing is summarized in **Figure 30.11.**

The coincidence of modification and replication suggests that acetylation (and methylation) could be connected with nucleosome assembly. One speculation has been that the reduction of positive charges on histones might be useful in reducing their affinity for DNA. This could allow the reaction to be better controlled. The idea has lost some ground in view of the observation that nucleosomes can be reconstituted, at least *in vitro,* with unmodified histones (but see earlier).

The transience of the acetylation event has been an obstacle to its study. The difficulty can be overcome by adding butyric acid to cells growing in culture. The treatment inhibits the enzyme histone deacetylase, so that acetylated nucleosomes accumulate. All the core histones are acetylated.

Acetylation is associated with changes in chromatin similar to those found on gene activation. The chromatin is more sensitive to DNAase I and (possibly) to micrococcal nuclease. However, it has not been possible to demonstrate any decisive relationship; and there is no evidence for preferential acetylation of active genes. This result therefore tells us that acetylation can indeed affect the structure of chromatin, but the significance of the change remains to be seen.

A cycle of phosphorylation and dephosphorylation occurs with H1, but its timing is different from the modification cycle of the other histones. With cultured mammalian cells, one or two phosphate groups may be introduced at S phase. But the major phosphorylation event is the later addition of more groups, to bring the total number up to as many as six. This occurs at mitosis, as indicated in Figure 30.11. All the phosphate groups are removed at the end of the process of division. The introduction of some of the phosphate groups is catalyzed by a phosphokinase enzyme whose activity increases sharply at the very start of mitosis. Not much is known about the phosphatase that removes the groups later.

The timing of the major H1 phosphorylation has prompted speculation that it may be involved in mitotic condensation. Certainly this is consistent with the need for H1 to generate the 30-nm chromatin fiber (see Chapter 29). Some temperature-sensitive mutants that are defective in H1 phosphorylation are unable to complete DNA replication and cannot pass through cell division. But this correlation does not reveal whether the phosphorylation is cause or effect of the mitotic condensation.

H2A IS CONJUGATED WITH UBIQUITIN ON A SUBSET OF NUCLEOSOMES

An extraordinary modification of a histone has been revealed by the characterization of a protein originally found in the nonhistone fraction of rat liver chromatin. Known at first as **A24,** this protein has the same C-terminal amino acid sequence as H2A. But it has *two N-terminal sequences,* as drawn in **Figure 30.12.** One is that of H2A. The other is that of the protein **ubiquitin,** whose name reflects its ubiquitous presence in cells from bacteria to mammals. The H2A-ubiquitin conjugate now is known as **UH2A.**

The ubiquitin protein itself is 76 residues long (compared with the ~130 of H2A). An isopeptide link is made between the C-terminus of a glycine in ubiquitin and the free ϵ-NH_2 of the lysine at position 119 of H2A. (It is called an *isopeptide* bond to indicate that the NH_2 group is not the usual amino group involved in peptide bond formation.) Ubiquitin is an acidic protein, whose content of glutamic and aspartic acids reduces the basic/acidic ratio of the conjugated protein.

Some 5–15% of the H2A may be in the form of UH2A. Usually only one of the two H2A molecules in a histone octamer carries ubiquitin, so that about 10–30% of the nucleosomes may be ubiquitinated. The ubiquitin probably lies at the surface of the nucleosome. It does not have any discernible effect on nucleosome structure. A rather small proportion of the H2B also can be conjugated with ubiquitin.

Figure 30.12
UH2A consists of ubiquitin linked to H2A.

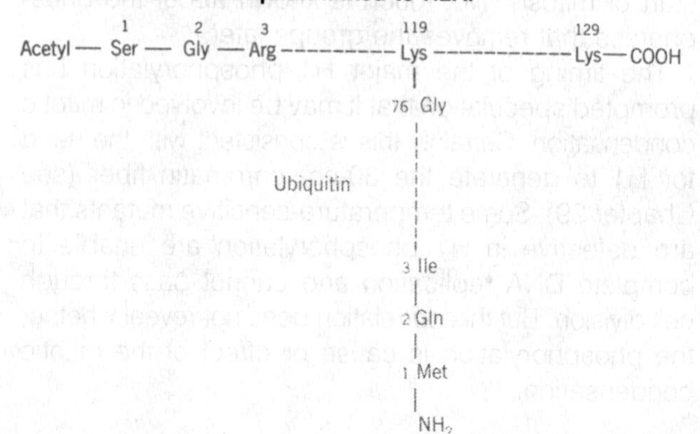

Nucleosomes in which H2A is ubiquitinated can be separated from unmodified nucleosomes by two-dimensional electrophoresis. First the nucleosomes are electrophoresed on a single polyacrylamide strip; then they are electrophoresed again at right angles, through a slab gel. The relevant fractions are marked in **Figure 30.13.** There are nucleosomes containing about 170 bp of DNA and H1, core particles, and core particles containing either one or two UH2A conjugates.

Does the presence of UH2A identify a special class of nucleosomes? About 20% of the nucleosomes of chromatin of *D. melanogaster* have UH2A; and their DNA sequences can be examined by blotting the nucleosome fractions for hybridization with specific probes. With probes representing transcribed genes, about 50% of the material is in the ubiquitinated fraction. With a satellite DNA probe, less than 4% of the material is ubiquitinated. UH2A therefore seems to be excluded from heterochromatin. There is a tendency for the UH2A to be concentrated in the nucleosomes of transcribed sequences.

Very little is known about the function of the ubiquitin. It appears to be released from chromatin at mitosis. This is disconcerting, because it raises the question of how the ubiquitinated state of particular regions might be perpetuated through cell division.

A possible function has been reported for ubiquitin in the cytoplasm, where it is a participant in a system for protein degradation. One or more ubiquitin moieties are covalently linked to the "target" protein in a reaction that uses ATP. Then the target protein is degraded. So the ubiquitin provides a marker to identify the substrate for the degradation system. We do not know whether this is relevant to nuclear events.

Figure 30.13
Nucleosomes, core particles, and ubiquitinated particles can be separated by two-dimensional gel electrophoresis.
Photograph kindly provided by Alex Varshavsky.

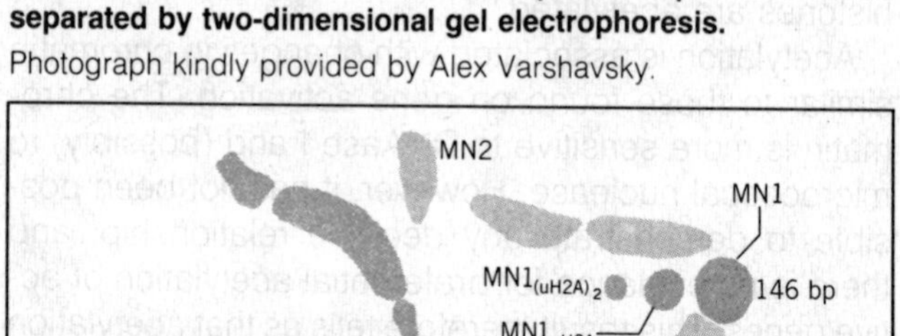

MN2
MN1
$MN1_{(uH2A)_2}$
146 bp
$MN1_{uH2A}$

GENE EXPRESSION IS ASSOCIATED WITH DEMETHYLATION

Between 2% and 7% of the C residues of animal cell DNA are methylated (the value varies with the species). Often satellite DNA is extensively methylated. The remaining methyl groups are present throughout the genome. Most of the methyl groups are found in CG "doublets," and, in fact, the majority of the CG sequences may be methylated. This short sequence is palindromic and usually the C residues on both strands are methylated, giving the structure

$$\begin{array}{ccc} 5' & {}^{m}\mathrm{CpG} & 3' \\ 3' & \mathrm{GpC}^{m} & 5' \end{array}$$

A doublet that instead is methylated on only one of the two strands is said to be **hemimethylated.**

The distribution of methyl groups can be examined by taking advantage of restriction enzymes that cleave target sites containing the CG doublet. Two types of restriction activity are compared in **Figure 30.14.** These **isoschizomers** are enzymes that cleave the same target sequence in DNA, but have a different response to its state of methylation.

In this pair, the enzyme HpaII cleaves the sequence CCGG (writing the sequence of only one strand of DNA). But if the second C is methylated, the enzyme can no longer recognize the site. However, the enzyme MspI cleaves the same target site *irrespective* of the state of methylation at this C. So MspI can be used to identify all the CCGG sequences; and HpaII can be used to determine whether or not they are methylated.

With a substrate of nonmethylated DNA, the two enzymes would generate the same restriction bands. But in methylated DNA, the modified positions are not cleaved by HpaII. For every such position, one larger HpaII fragment replaces two MspI fragments. An example is given in **Figure 30.15.**

Is the pattern of methylation invariant or is it subject to alteration by circumstance? Individual sites have been analyzed in several cases, including cellular protein-coding genes, the tandem rDNA cluster, and sequences of several integrated or free viral genomes. Those sites identified by the use of restriction enzymes comprise only some of the methylated sequences, but we assume that their behavior is typical of the entire set.

A similar result is obtained in all these cases. Some sites are methylated in all tissues examined; some sites are unmethylated in all tissues. *A minority of sites are methylated in tissues in which the gene is not expressed, but are not methylated in tissues in which the gene is active.* Thus an active gene may be described as "undermethylated."

In the example of the human β-globin gene cluster, the correlation tends to be best in the embryonic genes.

Figure 30.14

The restriction enzyme MspI cleaves all CCGG sequences whether or not they are methylated, but HpaII cleaves only nonmethylated CCGG tetramers.

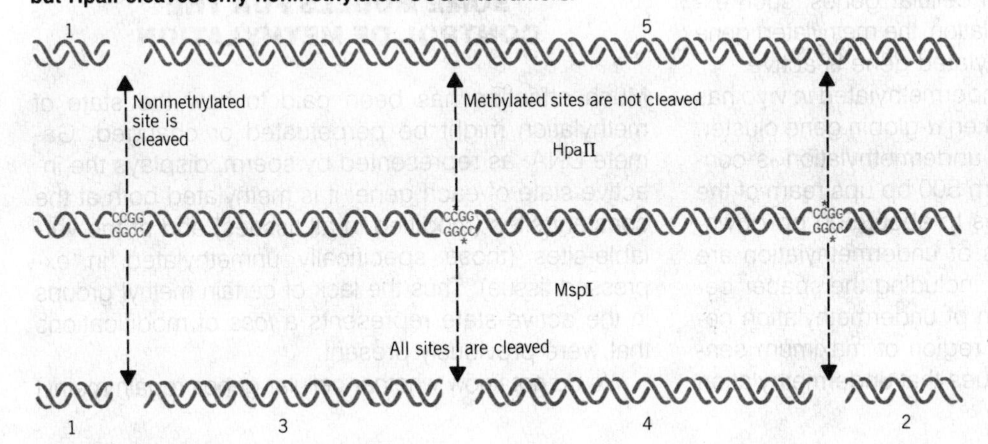

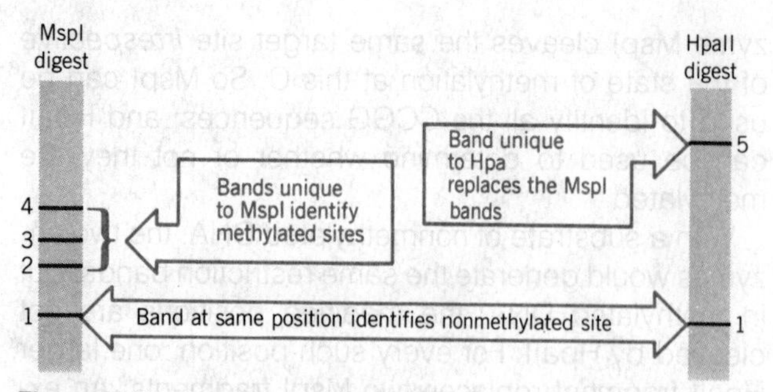

Figure 30.15
The results of MspI and HpaII cleavage are compared by gel electrophoresis of the fragments.
The numbers of the bands refer to the fragments identified in Figure 30.14.

The majority (~15) of identified sites are unmethylated in expressed erythroid tissue, are partially methylated in adult erythroid tissue, and elsewhere are methylated. With the adult genes, there is only a partial reduction in methylation at some sites in the expressed tissue. ("Partial" methylation implies variability, with a given site methylated in some cells, but not methylated in others apparently of the same phenotype.)

In the rRNA genes of *X. laevis*, most of the CG sites remain methylated in all circumstances. Two sites (located near the promoter) become unmethylated at the time in embryonic development when transcription of rRNA starts.

As well as examining the state of methylation of resident genes, we can compare the results of introducing methylated or nonmethylated DNA into new host cells. First performed with viral genes, and since extended to several purified cellular genes, such experiments show a clear correlation: the methylated gene is inactive, but the nonmethylated gene is active.

The extent of the region undermethylated *in vivo* has been determined in the chicken α-globin gene cluster. In adult erythroid cells, the undermethylation is confined to sites that extend from 500 bp upstream of the first of the two adult α genes to about 500 bp downstream of the second. Sites of undermethylation are present in the entire region, including the spacer between the genes. The region of undermethylation coincides rather well with the region of maximum sensitivity to DNAase I. This argues that undermethylation may be a feature of a domain that contains a transcribed gene or genes.

Our problem in interpreting the general association between undermethylation and gene activation is that only a minority (sometimes a small minority) of the methylated sites are involved. One possibility is that the state of methylation is critical at specific sites; for example, demethylation at the promoter might be involved in making this region available for the initiation of transcription. Another possibility is that a reduction in the level of methylation is part of some structural change needed to permit transcription to proceed.

In some genes, it is indeed methylation at the 5′ end that is related to expression. Several genes are not methylated at the 5′ end of the gene, but remain methylated at the 3′ end, when they are expressed. As with other changes in chromatin, it seems likely that the absence of methyl groups is associated with the *ability to be transcribed* rather than with the act of transcription itself. The importance of demethylation at the promoter should be relatively easy to test directly, since it should be possible to construct genes *in vitro* that differ only in their state of 5′ methylation.

There are exceptions to the general relationship we have described. Some genes can be expressed even when they are extensively methylated. A striking exception is the case of the H2 histocompatibility locus of an early embryonic cell line), which seems to be activated by methylation. The connection between methylation and expression thus is not universal.

SOME MODELS FOR THE CONTROL OF METHYLATION

Much attention has been paid to how the state of methylation might be perpetuated or changed. Gamete DNA, as represented by sperm, displays the inactive state of each gene: it is methylated both at the constant sites (modified in all tissues) and at the variable-sites (those specifically unmethylated in expressed tissue). Thus the lack of certain methyl groups in the active state represents a *loss* of modifications that were previously present.

We do not know whether cellular genes regain methyl

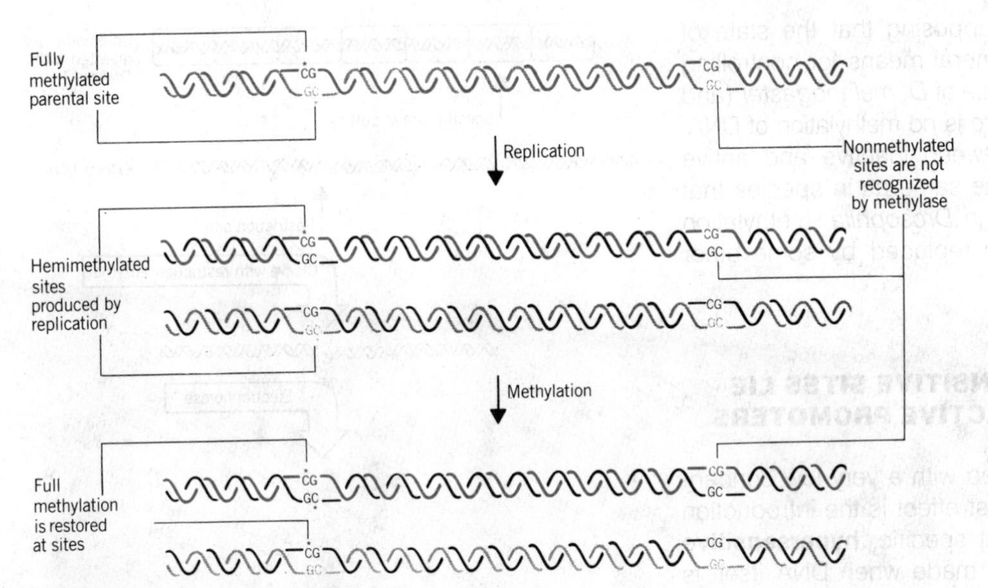

Figure 30.16
The state of methylated sites could be perpetuated by an enzyme that recognizes only hemimethylated sites as substrates.

groups if and when they cease to be expressed. A critical question we should like to answer is how sequences are selected as the targets for tissue-specific changes in the state of methylation.

A simple model for the perpetuation of methylated sites is to suppose that the DNA methylase acts on hemimethylated DNA. As can be seen from **Figure 30.16,** replication of a fully methylated CG doublet produces two hemimethylated daughter duplexes. Recognizing each of the hemimethylated sites, the enzyme could convert it to the normal fully methylated state.

Such a model accords with the observation that, when methylated DNA is introduced into a cell, it continues to be methylated through an indefinite number of replication cycles, with a fidelity approaching 95% per site. If nonmethylated DNA is introduced, it is not methylated de novo. This implies that the enzyme recognizes *only* the hemimethylated sites. Its action allows the condition of a $\frac{\text{CG}}{\text{GC}}$ doublet—methylated or nonmethylated—to be perpetuated.

If this model is correct, an entirely different enzyme activity must be involved in any creation of new sites of methylation (for example, if the methylated condition of a gene is restored when transcription ceases).

The methylated condition might be lost in two ways. Methyl groups might be actively removed by a demethylase enzyme. Or methylation might simply fail to occur at a hemimethylated site generated by replication. One of the DNA duplexes produced by the next replication then would lack the methylated site.

Experiments with the drug 5-azacytidine produce indirect evidence that demethylation can result in gene expression. The drug is incorporated into DNA in place of cytidine, and cannot be methylated, because the 5′ position is blocked. This leads to the appearance of demethylated sites in DNA.

The phenotypic effects of 5-azacytidine include the induction of changes in the state of cellular differentiation; for example, muscle cells are induced to develop from nonmuscle cell precursors. The drug also activates genes on a silent X chromosome, which raises the possibility that inactivation of the chromosome could be due to its state of methylation.

Although we do not fully understand the effects of methylation, it is clear that the absence of methyl groups is associated with gene expression. However, there are some difficulties in supposing that the state of methylation provides a general means for controlling gene expression. In the case of *D. melanogaster* (and other Dipteran insects), there is no methylation of DNA. The other differences between inactive and active chromatin appear to be the same as in species that display methylation. Thus in *Drosophila,* methylation either is superfluous or is replaced by some other mechanism.

DNAase HYPERSENSITIVE SITES LIE UPSTREAM FROM ACTIVE PROMOTERS

When chromatin is digested with a very low concentration of DNAase I, the first effect is the introduction of breaks in the duplex at specific, **hypersensitive sites.** The breaks are not made when DNA itself is digested; they result from the (tissue-specific) structure of chromatin.

The locations of the hypersensitive sites can be determined by the technique of indirect end labeling that we introduced earlier in the context of nucleosome phasing. The protocol is recapitulated in **Figure 30.17.** Chromatin is digested with DNAase I, the DNA is isolated, and then it is cleaved with a restriction enzyme. The material produced by the double digest is electrophoresed and hybridized with a probe that represents a region adjacent to the restriction site. The existence of a particular cutting site for DNAase I generates a discrete fragment, identified as a band whose size indicates the distance of the DNAase I hypersensitive site from the restriction site.

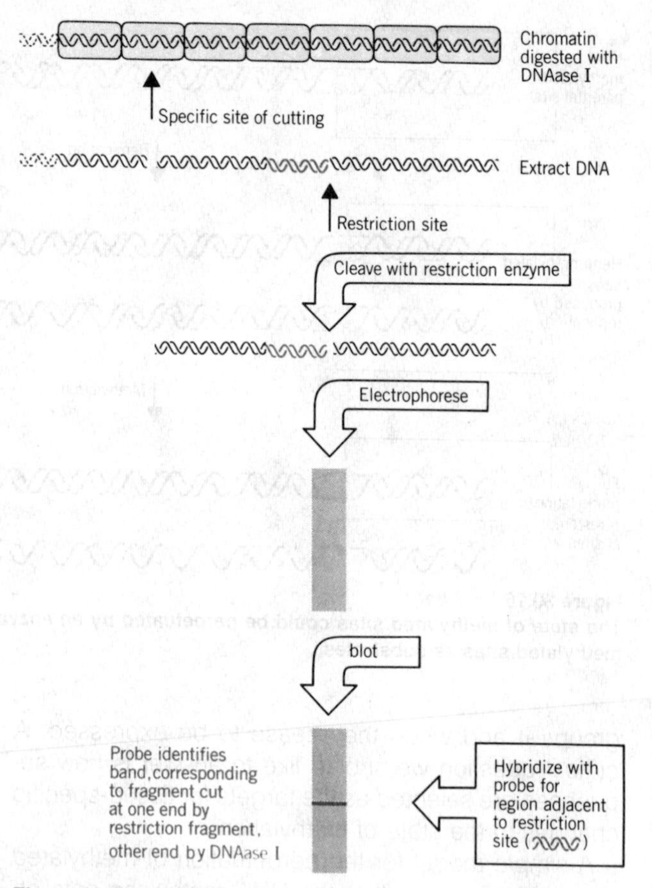

Figure 30.17
Indirect end-labeling identifies the distance of a DNAase hypersensitive site from a restriction cleavage site.

Many of the hypersensitive sites are related to gene expression. Every active gene has a site of cutting, or sometimes more than one site, in the region immediately upstream from the promoter. Some examples of the locations of hypersensitive sites are summarized in **Figure 30.18.**

A 5′ cutting site is found *only in chromatin of cells in which the gene is being expressed;* it does not occur when the gene is inactive. In the globin genes, for example, hypersensitive sites are found upstream from embryonic genes in embryonic cells but not in adult cells, and vice versa. The sites may vary somewhat in sensitivity (as judged from the intensity of the corresponding band). When multiple sites lie close together, often some are appreciably more sensitive than others. In addition to these sites, there may be others located elsewhere, within a transcribed gene or downstream from it. The significance of the sites in these other regions is unclear.

The correlation between transcription and upstream hypersensitive sites is greatly strengthened by data on the *Sgs4* gene of *D. melanogaster.* This gene codes for the glue protein of salivary glands. In this tissue, the locus has two prominent hypersensitive sites, lying about 330 and 405 bp upstream from the 5′ end of the gene. A mutant strain of the fly has a deletion of about 100 bp that removes both the hypersensitive sites. In spite of the retention of the remaining region, which includes more than 250 bp of the sequence

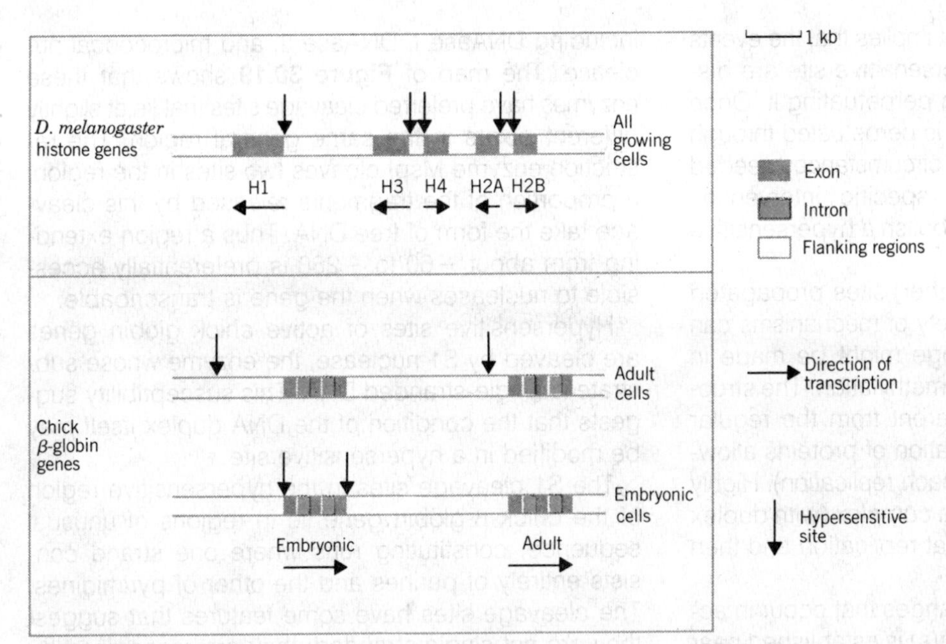

Figure 30.18
Hypersensitive sites lie near the 5′ ends of active genes and (sometimes) also in other places.

upstream from the startpoint, this mutant does not produce glue protein. This suggests that removal of the hypersensitive site prevents activation of the gene.

Although necessary, the presence of a hypersensitive site is not sufficient to ensure transcription. The timing of the appearance of the hypersensitive site relative to the onset of transcription is hard to establish in authentic situations, where it is difficult to obtain cells all at the same stage of development. All the indications, however, are that the 5′ hypersensitive site(s) appear before transcription starts, very likely as a prerequisite for initiation.

How are hypersensitive sites established? Analysis of their structure requires an *in vitro* system. The hypersensitive region of the adult chick β-globin gene has been reconstructed on a plasmid. When a plasmid carrying the gene is recombined with histones in the presence of an extract from red blood cell nuclei, the relevant region becomes hypersensitive. Purification of the active factor should be fairly straightforward, simply using as an assay the ability to generate a hypersensitive region.

The extract cannot confer hypersensitivity if it is added *after* the histones, which suggests that it must recognize DNA directly and in some way change the organization of the region as nucleosomes are deposited. At least under these conditions, the relevant component cannot displace nucleosomes after they have formed, which takes us back to the issue of how the structure of chromatin is changed when a gene is to be activated.

The stability of hypersensitive sites is revealed by the properties of chick fibroblasts transformed with temperature-sensitive tumor viruses. These experiments take advantage of an unusual property: although fibroblasts do not belong to the erythroid lineage, transformation of the cells at the normal temperature leads to activation of the globin genes. The activated genes have hypersensitive sites. If transformation is performed at the higher (nonpermissive) temperature, the globin genes are not activated; and hypersensitive sites do not appear. When the globin genes have been activated by transformation at low temperature, they can be inactivated by raising the temperature. But the hypersensitive sites are retained through at least the next 20 cell doublings.

This result again demonstrates that acquisition of a hypersensitive site is only one of the features necessary to initiate transcription; and it implies that the events involved in establishing a hypersensitive site are distinct from those concerned with perpetuating it. Once the site has been established, it is perpetuated through replication in the absence of the circumstances needed for induction. (Presumably a specific intervention therefore would be needed to abolish a hypersensitive site.)

How are hypersensitive (or other) sites propagated through DNA replication? A variety of mechanisms can be imagined. A heritable change might be made in the DNA itself—for example, demethylation. The structure of the DNA might be different from the regular double helix (through the mediation of proteins allowing the structure to reform after each replication). Highly cooperative proteins may form a complex (with duplex DNA) that is able to segregate at replication and then rebuild.

We have described three changes that occur in active genes. A hypersensitive site(s) is established near the promoter. The nucleosomes of a domain including the transcribed region become more sensitive to DNAase I. The DNA of the same region is undermethylated. All of these changes may be necessary for transcription.

HYPERSENSITIVE SITES EXCLUDE NUCLEOSOMES

What is the nature of the hypersensitive sites? Each seems to represent a region of 100–200 bp to which access is not restricted in the manner typical of nucleosomes. Some of these sites have been investigated in detail.

The site at the 5′ end of the chicken adult β-globin gene is preferentially digested by several enzymes, including DNAase I, DNAase II, and micrococcal nuclease. The map of **Figure 30.19** shows that these enzymes have preferred cleavage sites that lie at slightly different points in the same general region. The restriction enzyme MspI cleaves two sites in the region; a proportion of the fragments released by this cleavage take the form of free DNA. Thus a region extending from about −60 to −260 is preferentially accessible to nucleases when the gene is transcribable.

Hypersensitive sites of active chick globin genes are cleaved by S1 nuclease, the enzyme whose substrate is single-stranded DNA. This susceptibility suggests that the condition of the DNA duplex itself may be modified in a hypersensitive site.

The S1 cleavage sites in the hypersensitive region of the chick β-globin gene lie in regions of unusual sequence, constituting runs where one strand consists entirely of purines and the other of pyrimidines. The cleavage sites have some features that suggest they are not single-stranded: they are recognized by restriction enzymes that act only on duplex DNA, and they are not recognized by another enzyme specific for single-stranded DNA. These features raise the possibility that these sites have some unusual structure, one that allows them to be recognized by S1 nuclease without actually forming single strands.

A hypersensitive site represents a structural change in chromatin. The hypersensitive sequences probably do not form the usual nucleosomal structure, perhaps because other proteins are complexed with the DNA and prevent the histones from binding. If nucleosomes do intrude into the hypersensitive region, their structure must be different from usual.

A significant feature of the hypersensitive sites is their location at varying distances from the startpoint for transcription. In the *Sgs* gene of *D. melanogaster* the site is more than 250 bp upstream, in the chicken

Figure 30.19
The hypersensitive site of a chicken β-globin gene is in an extended region that is susceptible to several nucleases.

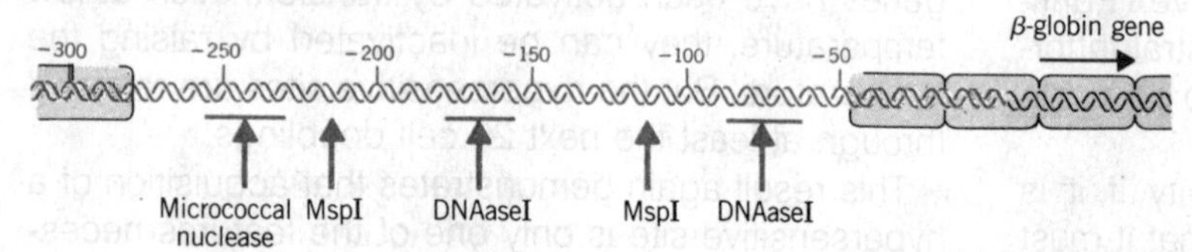

β-globin gene it is just upstream, and in the thymidine kinase gene it includes all of the sequences necessary for promoter function.

The hypersensitive sites could be involved directly in promoter function, most likely in providing upstream components needed for recognition by RNA polymerase (see Chapter 12). Alternatively, the sites may play a more general role in gene activation—for example, triggering a change in local structure that is necessary for RNA polymerase binding or even for transcriptional activity.

A particularly well-characterized nuclease-sensitive region lies on the SV40 minichromosome. A short segment near the origin of replication, just upstream from the promoter for the late transcription unit, is cleaved preferentially by DNAase I, micrococcal nuclease, and other nucleases (including restriction enzymes). The stretch over which the preferential digestion occurs is about 400 bp long, and the segment can be released as a fragment of free DNA.

The state of the SV40 minichromosome can be visualized by electron microscopy. In up to 20% of the samples, a "gap" is visible in the nucleosomal organization. This is evident in **Figure 30.20.** The gap is a region of about 120 nm in length (about 350 bp), surrounded on either side by the nucleosomes that occupy the rest of the genome.

The location of the gap can be determined by cleaving the circular minichromosome with a restriction enzyme that has a single known target. The visible gap corresponds with the nuclease-sensitive region. This shows directly that increased sensitivity to nucleases is associated with the exclusion of nucleosomes. It does not imply that the DNA is free *in vivo,* because some other proteinaceous structure may have been lost in isolating the minichromosome.

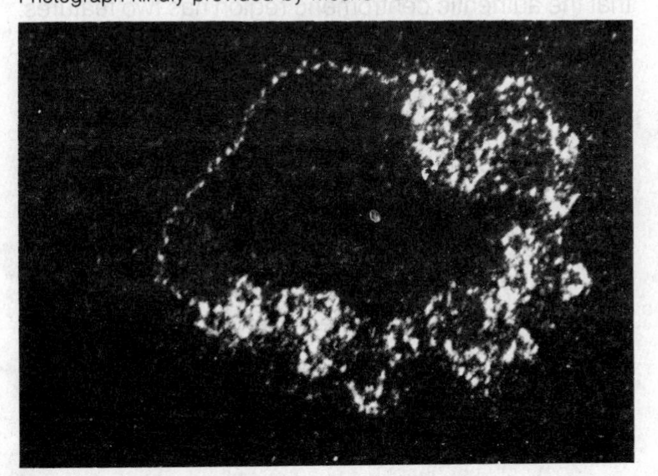

Figure 30.20
The SV40 minichromosome may have a nucleosome-free gap.
Photograph kindly provided by Moshe Yaniv.

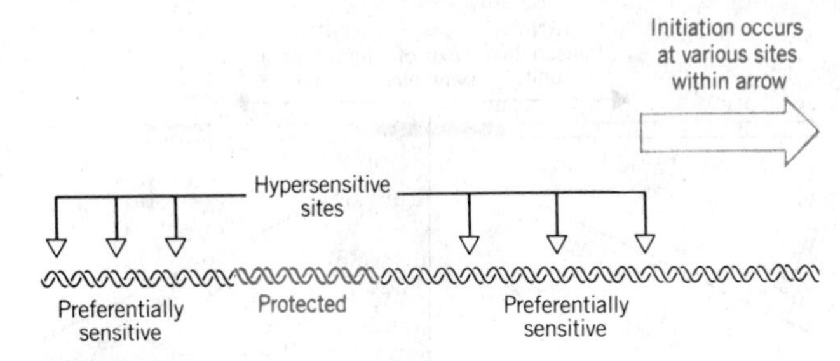

Figure 30.21
The SV40 gap includes hypersensitive sites, sensitive regions, and a protected region of DNA.

A similar nuclease-sensitive gap is found in the related polyoma virus, and here it has been shown that the entire region is not uniformly sensitive to nucleases. Within a sensitive region of about 260 bp, there are two hypersensitive DNAase I sites and a "protected" region. The map is given in **Figure 30.21.** The protected region presumably reflects the association of protein with the DNA. Indeed, such an association might well be the cause of the exclusion of the nucleosomes from this region.

In a mutant of polyoma in which the sequences present in this region are substantially rearranged, *the same result is obtained.* This implies that the presence of the gap is not determined by the entire sequence of the sensitive region itself, but must be "measured" in some way from a fixed point that is either part of the region or is adjacent to it.

The region of SV40 or polyoma surrounding the nuclease-sensitive gap has several potential functions, including the initiation of replication and transcription, so the function of the gap cannot simply be equated with a particular activity. The presence of the gap may be associated with the sequences of the enhancer elements, which lie in this region and are necessary for promoter function (see Chapter 12).

An interesting structure has been found in the ex-

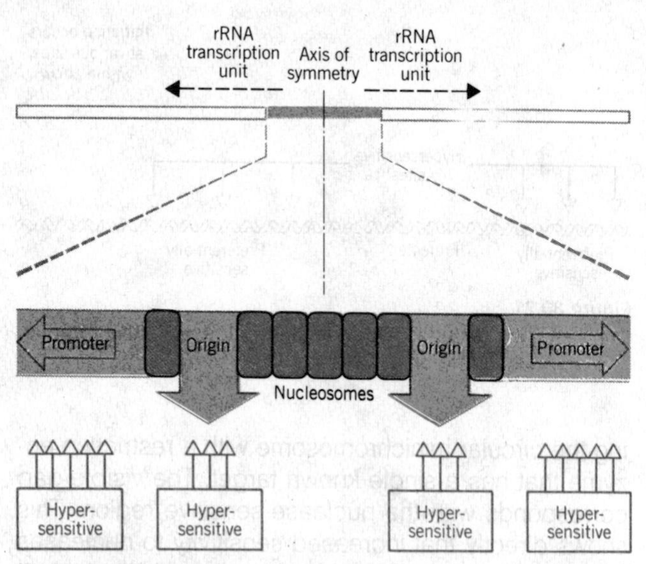

Figure 30.22
The central region of the *Tetrahymena* rDNA has a closely defined chromatin structure, with hypersensitive sites at the promoters and origins, and a phased series of nucleosomes between the origins.

trachromosomal rDNA of the macronucleus of *Tetrahymena pyriformis*. Recall that the rDNA molecule is a linear palindrome, in which the two rRNA transcription units lie some distance apart and are transcribed in opposite directions. The organization of the rDNA is shown in **Figure 30.22**. Close to the center of the palindrome lie two origins of replication, short sequences at which the replication of DNA is initiated. The structure of the chromatin of this region is shown in the expanded portion of the map.

Several hypersensitive sites are located in the area. Each promoter has one, and so does each origin. Between the two origins is a series of exactly 5 nucleosomes. The origins lie 1000 bp apart, and the spacing of the nucleosomes is precisely 200 bp, so every nucleosome has a defined position. In the related organism, *T. pyriformis*, the central region between the origins is 1400 bp long, and is filled by 7 nucleosomes.

Tetrahymena rDNA provides an example of nucleosome phasing par excellence. The fact that the two versions of the rDNA differ by multiples of the nucleosome repeat suggests that there might be selective pressure for the ability to package the central region into an exact number of nucleosomes. The locations of the histone octamers are probably defined by the boundaries of the hypersensitive sites on either side, rather than by any intrinsic property of the sequence that is phased.

Comparable structural alterations in chromatin may exist also in the region of the yeast centromere. In lieu of a visible kinetochore in yeast, it seems likely that the chromatin thread itself might be involved in centromeric function. The structure of the centromeric region has been investigated by using a short fragment of the DNA as a probe in the indirect end-labeling technique. The probe identifies two DNAase I hypersensitive sites, located on either side of regions I and III, two of the short highly conserved sequences in the centromeric DNA, whose sequences are summarized in Figure 28.12. The region of about 220–250 bp between the two hypersensitive sites is protected against nuclease digestion. Thus the role of the conserved region could be to bind specific centromeric proteins that form a nonnucleosomal structure.

Another effect is seen when the same experiment is performed using micrococcal nuclease to digest the chromatin. The same protected region is evident; and, in addition, a phased series of nucleosomes extends on either side. The phasing lasts for more than 12 nucleosomes, each containing 160 bp of DNA.

Does the nucleosome phasing propagate from the centromeric sequence itself, or is it a property of the phased sequences themselves? In a plasmid that contains these sequences but lacks the centromeric region, the phasing is retained. This result suggests that the authentic centromeric region has two features. A conserved sequence establishes some nonnucleosomal structure. And the flanking sequences appear to possess an intrinsic ability to phase their nucleosomes.

What happens to the array of nucleosomes at the end of a chromosome? When cleaved with micrococcal nuclease, the telomeres of the ciliate *Oxytricha* show a series of repeating bands at distances of 100, 300, 500, 700, 900 bp from the terminus. This separation suggests that the very last 100 bp at the telomere are protected by a nonnucleosomal protein complex, adjacent to which is a series of phased nucleosomes.

SPECULATIONS ABOUT THE NATURE OF GENE ACTIVATION

Sites detectable by hypersensitivity thus occur in various situations: upstream from transcribable genes, within the nuclease-sensitive gap in SV40 and polyoma, at the origins of replication in *Tetrahymena* rDNA, and at the yeast centromere; they may also be associated with transposition of yeast mating type genes (see Chapter 36). Sometimes a series of nucleosomes reside at fixed positions adjacent to the site, either because the site provides a boundary or because they are intrinsically phased.

The variety of situations in which hypersensitive sites occur suggests that their existence reflects a general principle. *Sites at which the double helix initiates an activity are kept free of nucleosomes.* In each case, some (unknown) nonhistone proteins, concerned with the particular function of the site, modify the properties of a short region of DNA so that nucleosomes are excluded. The structures formed in each situation need not necessarily be similar (except that each, by definition, creates a site hypersensitive to DNAase I).

We may speculate whether the converse of this principle is true: *genes whose control regions are organized in nucleosomes cannot be expressed.* Suppose that the formation of nucleosomes occurs in a manner independent of sequence to any region of DNA from which histones are not specifically excluded. Then in the absence of specific regulatory proteins, promoters, enhancers, and any other regulatory regions will be organized by histone octamers into a state in which very possibly *they cannot be activated.* (There is no evidence for any protein able to *displace* histones from DNA.)

There are hints that disruption of chromatin structure is necessary for gene activation. The 5S transcription factor cannot activate genes *in vitro* if they are complexed with histones. However, the factor can form the necessary complex with free DNA, after which the addition of histones does not prevent the gene from remaining active. Once the factor has bound, it remains at the site, allowing a succession of RNA polymerase molecules to initiate transcription. Whether the factor or histones get to the control site first may be the critical factor.

Replication provides an obvious opportunity for disrupting chromatin structure, since nucleosomes must be transiently absent from at least one of the duplicate chromosomes. This may provide an opportunity for other proteins to bind to control sites, preventing nucleosomes from forming. The excluded region may be recognized as a hypersensitive site, and could provide a boundary from which nucleosomes are phased.

How might we test the idea that replication is often involved in changing the state of expression of chromatin? The induction of hypersensitive sites in genes that respond (for example) to hormones may be the sort of system we need. A clear answer on the question of how active regions form, however, will depend on the development of a system in which chromatin structure can be activated *in vitro* (which may be difficult if DNA replication indeed is necessary).

Once a hypersensitive site (or some other activating structure) has formed, it may be propagated through replication in a stable manner. This could happen if the factor has the ability to segregate at replication. **Figure 30.23** illustrates a hypothetical case in which a protein that binds cooperatively to DNA forms a complex that splits at replication, each of the half-complexes then reassembling a full complex on the daughter chromosomes.

Since the reconstitution of the complex is necessary every generation, a corollary is that the gene could be turned off in a dividing cell simply by restricting the amount of factor. When the factor becomes diluted out by division, the complex will become unable to reassemble, and the histones will form nucleosomes at the control sites. This could explain how genes are turned off during embryonic development. It poses the question of how an active gene might be turned off in a nondividing cell. Indeed, can active genes be turned off in terminally differentiated cells (that is, cells that have reached their final state of phenotypic expression)?

Exact mechanisms for chromatin activation (or inactivation) are too speculative to consider now, but the general thread of these views is that active chromatin and inactive chromatin cannot be considered in a dynamic equilibrium. Sudden, disruptive events are

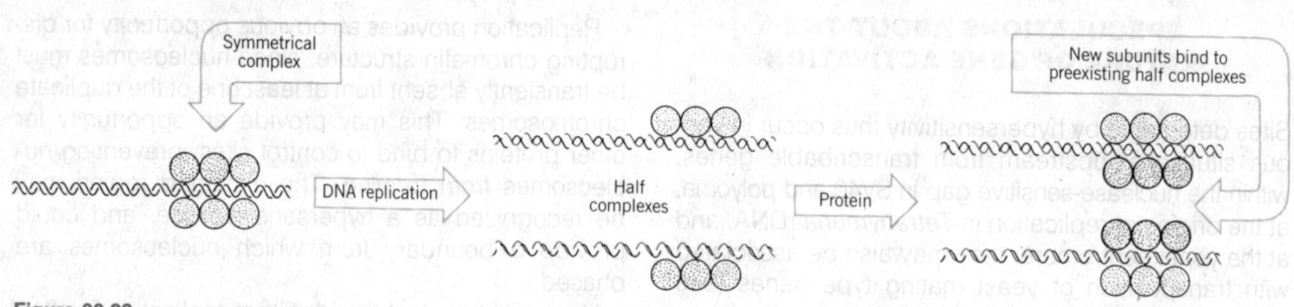

Figure 30.23
A protein complex could perpetuate itself on DNA by segregating into half-complexes at replication and cooperatively recruiting additional protein molecules to restore the full complex.

needed to convert one to the other. One implication of the types of model we have discussed is that a much greater concentration of a regulatory protein may be needed to initiate the activation event than to maintain the state of active chromatin.

The events involved in activating a gene need not coincide with the initiation of transcription. Formation of the appropriate complex may render the gene *transcribable,* but further action may be needed actually to transcribe it. In the language of developmental biology, complex formation may be a *determination* event, preceding by some time the *differentiation* event, when the gene is actually switched on. A possible example is that genes whose expression is hormone-dependent may first enter the determined state in the appropriate cell; and then may actually start transcription when the hormone arrives. (In these terms, prokaryotic genes are controlled by differentiation-like mechanisms, since the activation event always involves the actual start of transcription, rather than creating a state of readiness.)

FURTHER READING

The topics of this chapter were reviewed in **Lewin's** *Gene Expression,* **2,** *Eucaryotic Chromosomes* (Wiley, New York, 1980, pp. 394–427). Functional aspects of chromatin were brought up to date by **Igo-Kemenes, Horz, & Zachau** (*Ann. Rev. Biochem.* **51,** 89–121, 1982). Nucleosome assembly was reviewed by **Laskey & Earnshaw** (*Nature* **286,** 765–767, 1980). The existence of hypersensitive sites was summarized briefly by **Elgin** (*Cell* **27,** 413–415, 1981). Their propagation has been followed by **Groudine & Weintraub** (*Cell* **30,** 131–139, 1982). A series of experiments to distinguish the changes in transcribable genes culminated in the report of **Weintraub, Larsen & Groudine** (*Cell* **24,** 333–344, 1981). A splendid tour d'horizon of the issues involved in establishing and maintaining active gene structures has been given by **Brown** (*Cell* **37,** 359–365, 1984). Speculations on methylation were the subject of a review from **Razin & Riggs** (*Science* **210,** 604–610, 1980). Results on the viral gap were brought together in the report by **Herbomel et al.** (*Cell* **25,** 651–658, 1981). The structure of the centromere was revealed by **Bloom & Carbon** (*Cell* **29,** 305–317, 1982).

PART 9
PERPETUATION OF DNA

If it be true that the essence of life is the accumulation of experience through the generations, then one may perhaps suspect that the key problem of biology, from the physicist's point of view, is how living matter manages to record and perpetuate its experiences. Look at a single bacterium in a large volume of fluid of suitable chemical composition. It assimilates substance, grows in length, divides in two. The two daughters do the same, like the broomstick of the Sorcerer's apprentice. Occasionally the replica will be slightly faulty and an individual arises with somewhat different properties, and it perpetuates itself in this modified form. It is quite easy to believe that the game of evolution is on once the trick of reproduction, covariant on mutation, has been discovered, and that the variety of types will be multiplied indefinitely.

Max Delbruck, 1949

CHAPTER 31
THE REPLICON: UNIT OF REPLICATION

Whether a cell has only one chromosome (as in prokaryotes) or has many chromosomes (as in eukaryotes), the entire genome must be replicated precisely once for every cell division. How is the act of replication linked to the cell cycle?

There are two points at which the state of replication must be compared with the condition of the cell.

The *initiation* of replication *commits the cell (prokaryotic or eukaryotic) to a further division.* From this standpoint, the number of descendants that a cell generates is determined by a series of decisions on whether or not to initiate DNA replication.

If replication does occur, the consequent division cannot be permitted to occur *until the replication event has been completed.* Indeed, the completion of replication could provide a trigger for cell division. Then the duplicate genomes are segregated one to each daughter cell, via mitosis in eukaryotes, and by some other mechanism in prokaryotes. The unit of segregation is the chromosome.

The unit in which the cell controls individual acts of replication is called the **replicon.** Each replicon "fires" only once in each cell cycle. The replicon is defined by its possession of the control elements needed for replication. It has an **origin** at which replication is initiated. It may have a **terminus** at which replication stops.

Any sequence attached to an origin—or, more precisely, not separated from an origin by a terminus—is replicated as part of that replicon. In genetic terms, the origin is a *cis*-acting site, able to affect only that molecule of DNA of which it is physically part.

Replication is controlled at the stage of initiation. *Once replication has started, it continues until the entire replicon has been duplicated.* The frequency of initiation is controlled by the interaction of regulator protein(s) with the origin. Thus the original formulation of the replicon (in prokaryotes) viewed it as the possessor of both the origin *and* the gene coding for the regulator protein. Now, however, "replicon" is often applied to eukaryotic chromosomes to describe a unit of replication that contains an origin; any necessary regulator protein(s) can be provided by another unit.

The bacterial chromosome constitutes a single replicon: so the units of replication and segregation coincide. Initiation at a single origin sponsors replication of the entire genome.

In addition to the chromosome, bacteria may contain plasmids; a plasmid is an autonomous circular DNA genome that constitutes a separate replicon. Some plasmids share the stringency of replication of the bacterial chromosome: these **single-copy** plasmids are present as one copy for every copy of the bacterial chromosome. Other plasmids are **multicopy** and exist in a characteristic number >1 per bacterial chromosome. Each phage or virus DNA also constitutes a replicon, able to initiate many times during an infectious cycle.

Perhaps a better way to view the prokaryotic replicon is to reverse the definition: any DNA molecule that contains an origin can be replicated autonomously in the cell. The number of replication events will depend on the interaction of the origin with the appropriate regulator proteins.

A eukaryotic chromosome contains a large number of replicons. So the unit of segregation includes many units of replication. This adds another dimension to the problem of control. All the replicons on a chromosome must be fired during one cell cycle; however, they are not active simultaneously, but are fired over a fairly protracted period. *Yet each of these replicons must be activated only once in the cell cycle.* Some signal must distinguish replicated from nonreplicated replicons and indicate when the entire process has been completed.

We should like to know the set of sequences that function as origins and how they are recognized by the appropriate proteins of the apparatus for replication. From the perspective of the DNA, we want to know whether a duplex sequence of base pairs is recognized, or whether some alternative secondary structure is formed. From the perspective of the regulator proteins, we want to know how they manage to initiate just one cycle of replication for every cycle of cell division.

SEQUENTIAL REPLICATION FORMS EYES

Consider a molecule of DNA engaged in replication. As illustrated in **Figure 31.1,** its nonreplicated region consists of the parental duplex, opening into the replicated region where the two daughter duplexes have formed. The point at which replication is occurring is called the **replication fork** (sometimes also known as the **growing point**). The replication fork moves *sequentially* along the DNA, from its starting point at the origin.

Replication may occur in either a **unidirectional** or **bidirectional** manner. The type of event is determined by whether one or two replication forks set out from the origin. In unidirectional replication, one replication fork leaves the origin and proceeds along the DNA. In bidirectional replication, two replication forks are formed, and they proceed away from the origin in opposite directions.

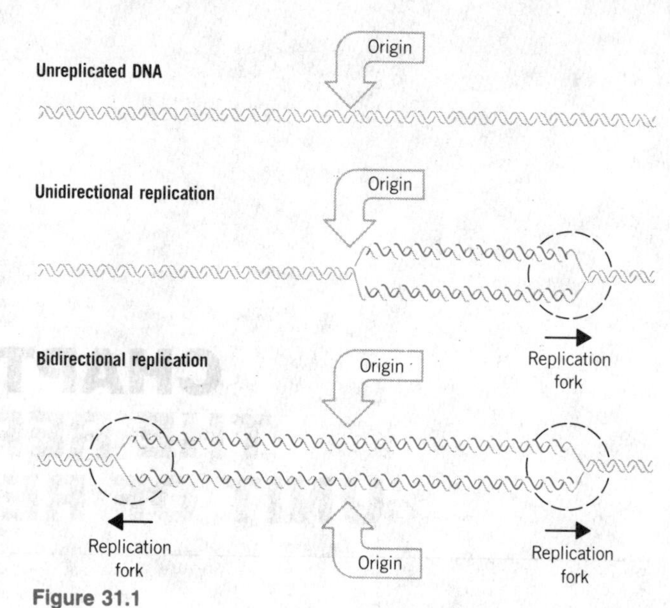

Figure 31.1

Replicons may be unidirectional or bidirectional, depending on whether one or two replication forks are formed at the origin.

When replicating DNA is viewed by electron microscopy, the region that has been replicated appears as an **eye** within the nonreplicated DNA. However, its appearance does not distinguish between unidirectional and bidirectional replication. As depicted in **Figure 31.2,** the eye can represent either of two structures. If generated by unidirectional replication, the eye represents a fixed origin and a moving replication fork. If generated by bidirectional replication, the eye represents a pair of replication forks. In either case, the progress of replication expands the eye until ultimately it encompasses the whole replicon.

When the replicon is circular, the presence of an eye forms the **θ-structure** drawn in **Figure 31.3.** The successive stages of replication of the circular DNA of polyoma virus are visualized by electron microscopy in **Figure 31.4.** Note that the "eye" becomes larger than the nonreplicated segment. (The two sides of the eye can be defined because they are both the same length.)

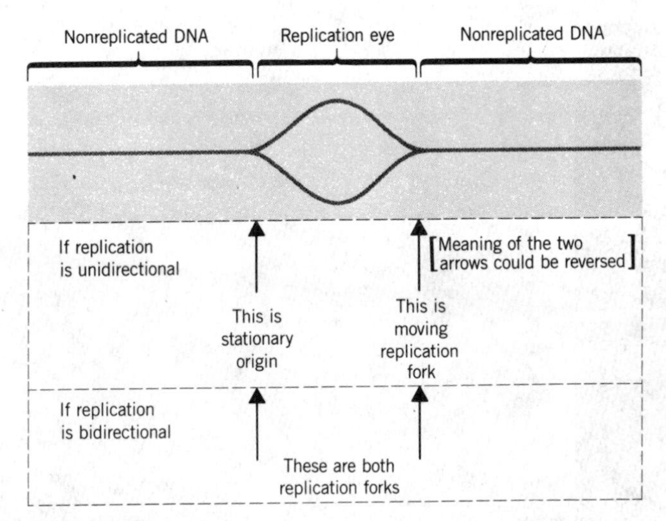

Figure 31.2
A replication eye can represent either a unidirectional or bidirectional replicon.

Whether a replicating eye has one or two replication forks can be determined in two ways. The choice of method depends on whether the DNA is a defined molecule or an unidentified region of a cellular genome.

With a defined linear molecule, we can use electron microscopy to measure the distance of each end of the eye from the end of the DNA. Then the positions of the ends of the eyes can be compared in molecules that have eyes of different sizes. If replication is unidirectional, only one of the ends will move; the other is the fixed origin. If replication is bidirectional, both will move; the origin is the point midway between them. With a defined circular molecule, the DNA is converted before electron microscopy into a linear form (by treatment with a restriction enzyme that cleaves at a unique site).

With undefined regions of large genomes, two successive pulses of radioactivity can be used to label the movement of the replication forks. One pulse has a more intense label than the other, so the pulses can be distinguished by the relative intensities of labeling, which are visualized by autoradiography. **Figure 31.5** shows that unidirectional replication produces the appearance of one type of label followed by the other at *one* end of the eye. Bidirectional replication produces a (symmetrical) pattern at *both* ends of the eye.

Figure 31.3
A replication eye forms a theta structure in circular DNA.

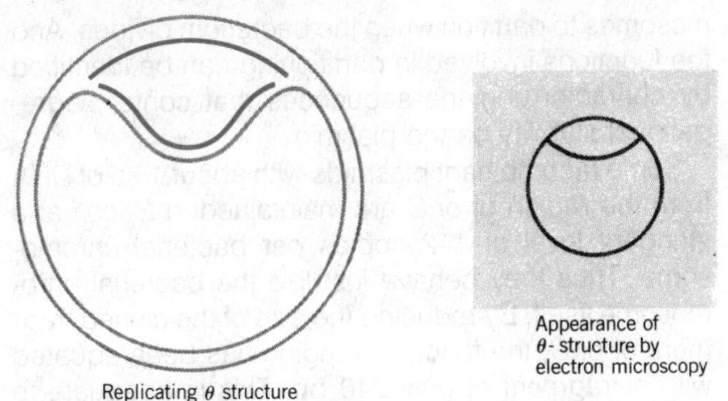

THE BACTERIAL GENOME IS A SINGLE REPLICON

The genome of *E. coli* is replicated bidirectionally from a single origin, identified as the genetic locus *oriC*. The DNA of the origin can be isolated by virtue of its ability to support replication of any DNA sequence to which it is joined. The principle of this approach is to clone DNA from the region of the origin into a molecule (carrying suitable genetic markers) that lacks an origin. This reconstruction will create a plasmid capable of autonomous replication *only if the DNA from the origin region contains all the sequences needed to identify itself as an authentic origin for replication.* (A comparable approach has been used to identify centromeric or telomeric DNA in yeast, by virtue of their characteristic effects on plasmid survival; see Chapter 28.)

A problem in analyzing origins is that plasmids carrying *oriC* tend to be rather unstable. However, they can be perpetuated in some bacterial strains. (The presence of additional origins could cause some problems for the cell; if it is not possible to distinguish among the various origins, the plasmid and the authentic chromosome will divide the number of initiation events, with the result that the number of chromosomes may be reduced. The obvious corollary is that genuine plasmid origins may be expected to differ from the chromosomal origin.)

The isolated origin fragment can be used for inves-

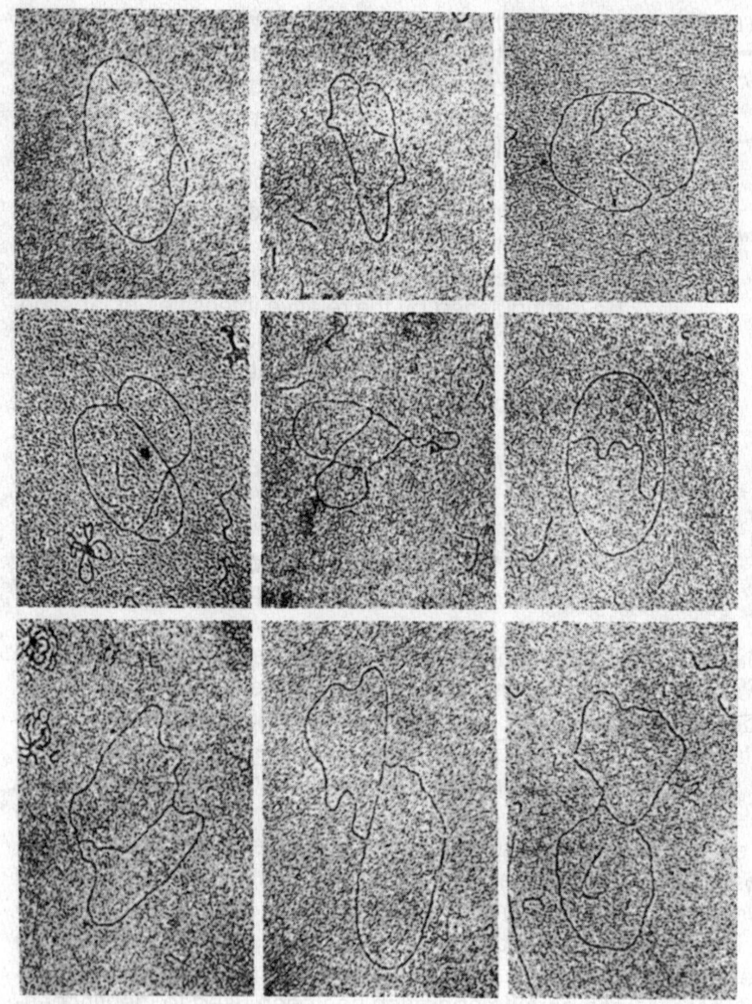

Figure 31.4
The replication eye becomes larger as the replication forks proceed along the replicon.
Photograph kindly provided by Bernard Hirt.

tigating several functions involved in replication, including the act of initiating a replication cycle, control of the frequency of initiation events, and segregation of replicated chromosomes to duplicate cells. We may look for mutants that are deficient in any one of these functions, in this way identifying sequences involved in each activity.

Plasmids carrying *oriC* segregate irregularly, but can be stabilized by introducing additional sequences. This result makes two points. The origin itself does not carry the necessary information to enable duplicate chromosomes to partition when the bacterium divides. And the functions involved in partitioning can be identified by characterizing the sequences that confer segregational stability on the plasmid.

Some recombinant plasmids with about 1 kb of DNA from the region of *oriC* are maintained in *E. coli* at a stringent level of 1–2 copies per bacterial chromosome. Thus they behave just like the bacterial chromosome itself. By reducing the size of the cloned fragment of *oriC*, the functional region has been equated with a fragment of only 240 bp. This is adequate to

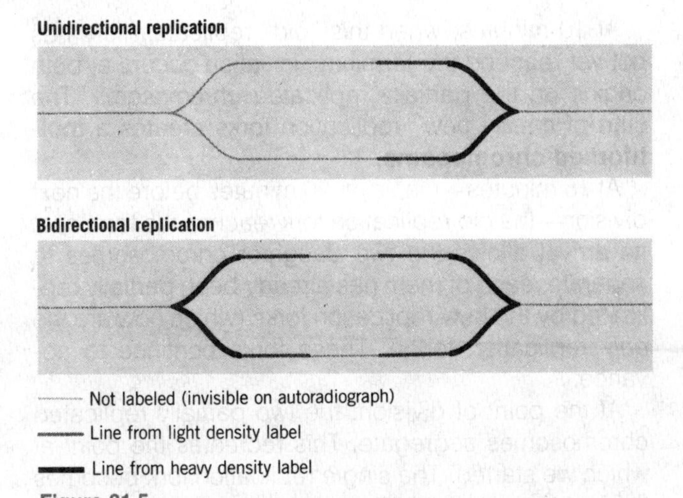

Figure 31.5
Different densities of radioactive labeling can be used to distinguish unidirectional and bidirectional replication.

ensure survival of a plasmid that carries it. However, such plasmids exist in the cell at a copy number of about 20 per bacterial chromosome. Thus they have lost some feature that restricts the frequency of initiation events. The determination of copy number and the act of initiating replication may therefore depend on different sequences.

The origin of the bacterium *Salmonella typhimurium* has been located in a 296 bp fragment of DNA. Comparison with *E. coli* shows that 86% of the bases are conserved between the origins. Origins from other bacteria can function in *E. coli,* although some are only distantly related. Comparison of several origins identifies clusters of short conserved sequences, but we do not yet know whether and how these are involved in origin function.

The sequences of the origins of some lambdoid phages (of which lambda is one example) also have been determined. These origins are related to one another, but differ in sequence from the bacterial origin.

To identify the features that enable sequences to act as origins, we will need to perform *in vitro* mutagenesis, in which individual changes are introduced into the putative origins, after which they are reintroduced into bacteria to determine what effect the change has on the ability to function as an origin.

So far we have dealt with the bacterial chromosome as though it were linear. Because it is really circular, the two replication forks each move around the genome to a meeting point. We do not know what happens at this encounter. Do the forks crash into each other, or is there a specific terminus at which they stop? How do the enzymes involved in replication disengage from the chromosome, remembering that the DNA must be replicated right across the region where the forks meet? A mutation of *E. coli* called *dnaT* may lie in a gene that codes for a protein involved in terminating the round of replication, but the product has not been characterized.

The *E. coli* chromosome may have a terminus at which the replication forks cease to move. It is not located exactly halfway around the chromosome; the disparity implies that the two forks must move different distances. In the plasmid R6K, a terminus has been identified as a region that stops replication from proceeding; this region remains effective when it is inserted into another DNA molecule. We do not know how it functions.

CONNECTING REPLICATION TO THE CELL CYCLE

Bacteria have two links between replication and cell growth. The frequency with which cycles of replication are initiated is adjusted to fit the rate at which the cell is growing. And the completion of a replication cycle is connected with division of the cell.

Cells of *E. coli* can grow at rates ranging from doubling times as quick as 18 minutes to slower than 180 minutes. Because the bacterial chromosome is a single replicon, the frequency of replication cycles is controlled by the number of initiation events at the single origin. The rate of DNA synthesis is more or less invariant at a constant temperature; it proceeds at the same speed unless and until the supply of precursors becomes limiting.

The replication cycle can be defined in terms of two constants. **C** is the fixed time of about 40 minutes required to replicate the entire bacterial chromosome. Its duration corresponds to a rate of movement by the individual replication fork of about 50,000 bp per minute. **D** is the fixed time of about 20 minutes that elapses between the completion of a round of replication and the cell division with which it is connected. This period may represent the time required to assemble the components needed for division.

(The constants C and D can be viewed as representing the maximum speed of which the bacterium is capable; they apply for all growth rates between doubling times of 18 and 60 minutes, but both constant phases become longer when the cell cycle occupies more than 60 minutes.)

A cycle of chromosome replication must be initiated a fixed time, C + D = 60 minutes, before a cell division. For bacteria dividing more frequently than at 60-minute intervals, a cycle of replication must therefore be initiated before the end of the *preceding* division cycle. Consider the example of cells dividing every 35 minutes. The cycle of replication connected with a division must have been initiated 25 minutes before the preceding division. This situation is illustrated in **Figure 31.6,** which shows the chromosomal complement of a bacterial cell at 5-minute intervals throughout the cycle.

At division (35/0 minutes), the cell receives a partially replicated chromosome. The replication fork continues to advance.

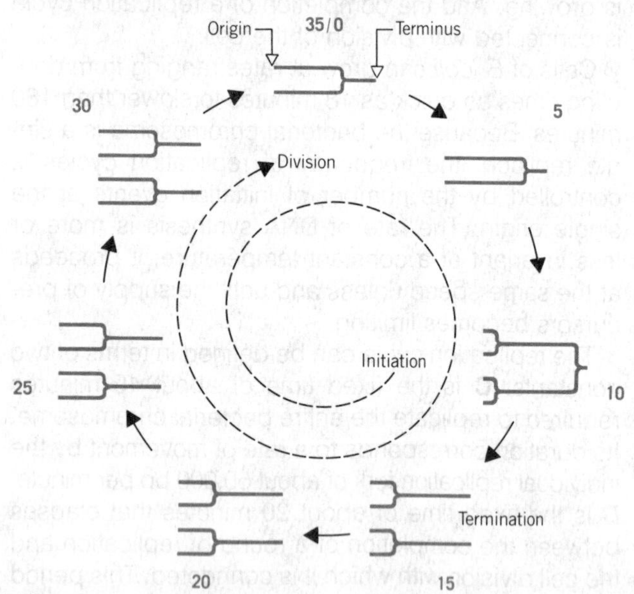

Figure 31.6
The fixed interval of 60 minutes between initiation of replication and cell division produces multiforked chromosomes in rapidly growing cells.

At 10 minutes, when this "old" replication fork has not yet reached the terminus, initiation occurs at both origins on the partially replicated chromosome. The start of these "new" replication forks creates a **multiforked chromosome.**

At 15 minutes—that is, at 20 minutes before the next division—the old replication fork reaches the terminus. Its arrival allows the two daughter chromosomes to separate; each of them has already been partially replicated by the new replication forks (which now are the *only* replication forks). These forks continue to advance.

At the point of division, the two partially replicated chromosomes segregate. This recreates the point at which we started. The single replication fork becomes "old," it terminates at 15 minutes, and 20 minutes later there is a division. We see that the initiation event occurs $1\frac{25}{35}$ cell cycles before the division event with which it is associated.

The general principle of the link between initiation and the cell cycle is that as cells grow more rapidly (the cycle is shorter), the initiation event occurs an increasing number of cycles before the related division. There are correspondingly more chromosomes in the individual bacterium. This relationship can be viewed as the cell's response to its inability to reduce the periods of C and D to keep pace with the shorter cycle.

How does the cell know when to initiate the replication cycle? The initiation event occurs at a constant ratio of cell mass to the number of chromosome origins. Cells growing more rapidly are larger, have greater mass, and therefore possess a greater number of origins. In terms of Figure 31.6, it is at the point 10 minutes after division that the cell mass has increased sufficiently to support an initiation event at both available origins.

Two types of model have been proposed for titrating cell mass. An initiator protein could be synthesized continuously throughout the cell cycle; accumulation of a critical amount would trigger initiation. Or an inhibitor protein might be synthesized at a fixed point, and diluted below an effective level by the increase in cell volume. There is evidence to suggest that a titration model actually does regulate initiation, but the data do not distinguish between accumulation of an initiator and dilution of an inhibitor. Current thinking favors an initiator, which is consistent with evidence that protein synthesis is needed for the initiation event.

Mutations in several *E. coli* genes affect the initiation of replication cycles; and there are many genes whose products are necessary for cell division. Analysis of these mutants, however, has not yet identified any proteins responsible for controlling the frequency of initiation or for linking replication to the division cycle.

Whichever type of titration model applies, the growth of the bacterium can be described in terms of the **unit cell,** an entity 1.7 μm long. A bacterium contains one origin per unit cell; a rapidly growing cell with two origins will be <3.4 μm long. A topological link between the initiation event and the structure of the cell could take the form of a growth site, a physical entity in the cell that provides the only place at which initiation can occur. There should be one growth site per unit cell.

There have been suspicions for years that a physical link may exist between bacterial DNA and the membrane. Bacterial DNA can be found in membrane fractions, which tend to be enriched in genetic markers near the origin, the replication fork, and the terminus. The growth site could be a structure on the membrane to which the origin must be attached for initiation. Mammalian forks may be bound to the nuclear matrix.

A link between DNA and the membrane also could account for segregation. If daughter chromosomes are attached to the membrane, they could be physically separated as the membrane grows between them. **Figure 31.7** shows that then a septum might form to segregate each chromosome into a different daughter cell. If we recall the segregation event depicted in Figure 31.6, we would expect the terminus of the chromosome to be involved.

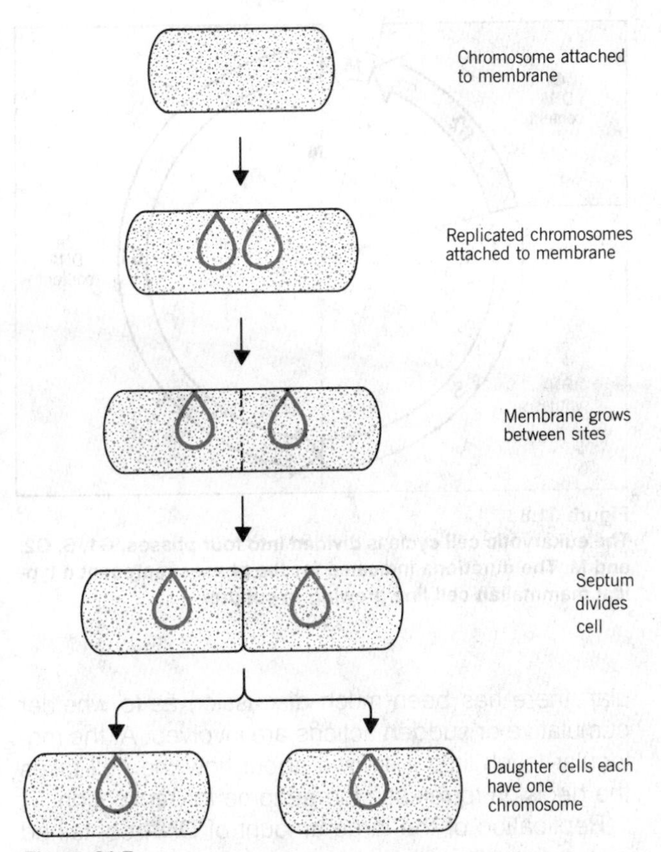

Figure 31.7
Attachment of bacterial DNA to the membrane could provide a mechanism for segregation.

EACH EUKARYOTIC CHROMOSOME CONTAINS MANY REPLICONS

In eukaryotic cells, the replication of DNA is confined to part of the cell cycle. The period between mitoses (interphase) can be divided into the phases indicated in **Figure 31.8.**

The typical cell starts its cycle in the diploid condition. It remains in this state through the **G1** period, which often comprises the major part of the cycle (and is the most variable in length between cells of different phenotype).

The synthesis of DNA defines the period of **S phase,** which often lasts a few hours in a higher eukaryotic cell.

At the end of S phase, the cell is in a tetraploid condition, in which it remains for the **G2** period. Then mitosis (M) reduces the chromosome complement to the diploid number in each daughter cell. (The periods preceding and succeeding S phase are called G1 and G2 to indicate that they represent "gaps" in DNA synthesis.)

The events responsible for the initiation of S phase occur during G1. Protein synthesis is needed, but the exact nature of the events remains unclear; in partic-

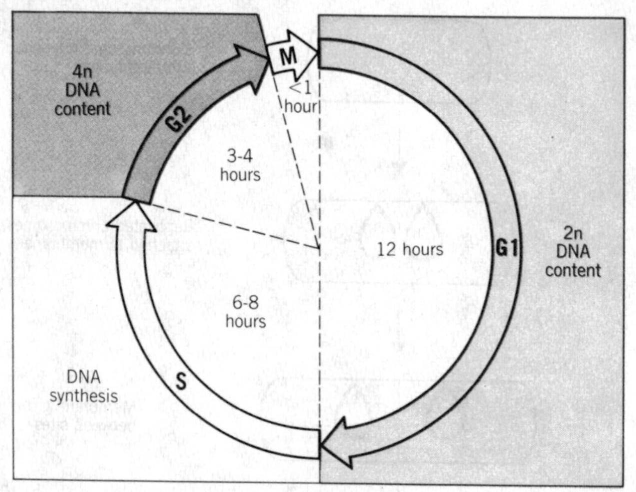

Figure 31.8
The eukaryotic cell cycle is divided into four phases, G1, S, G2, and M. The durations indicated for the phases represent a typical mammalian cell line growing in culture.

ular, there has been much discussion as to whether cumulative or sudden actions are involved. At the molecular level, little is known about how the cell takes the decision to release the genome for replication.

Replication of the large amount of DNA contained in a eukaryotic chromosome is accomplished by dividing it into many individual replicons. These replicons are not all active simultaneously; at any point during S phase, only some are engaged in replication. Presumably each replicon is activated at a specific time during S phase, although the evidence on this issue is not decisive.

The important point is that the start of S phase is signaled by the activation of the first replicons. Then, over the next few hours, many further initiation events occur at other replicons. The control of S phase therefore involves two processes: release of the cell from G1; and initiation of replication at individual replicons in an ordered manner.

Much of our knowledge about the properties of the individual replicons is derived from autoradiographic studies, generally using the type of protocol illustrated in Figure 31.5. Chromosomal replicons usually display bidirectional replication, as seen by the existence of matched pairs of tracks.

How large is the average replicon, and how many are there in the genome? A difficulty in characterizing the individual unit is that adjacent replicons may fuse to give large replicated eyes, as illustrated in **Figure 31.9.** The approach usually used to distinguish individual replicons from fused eyes is to rely on measurements of stretches of DNA in which several replicons can be seen to be active, presumably captured at a stage when all have initiated around the same time, but before the forks of adjacent units have met.

(There is some evidence that "regional" controls might produce this sort of activation pattern, in which groups of replicons are initiated more or less coordinately, as opposed to a mechanism in which individual replicons are activated one by one in dispersed areas of the genome.)

In groups of active replicons, the average size of the unit is measured by the distance between the origins (that is, between the midpoints of adjacent replicons). For many of the higher eukaryotes, this lies in the range of 100–200 kb. Thus in a haploid mammalian genome, there should be 20,000–30,000 replicons. In *D. melanogaster* or in *S. cerevisiae,* the replicons are smaller, averaging about 40 kb. This corresponds to about 3500 replicons in the haploid set of the fruit fly, and to about 500 replicons in the yeast. The range of individual lengths is high, more than tenfold, so the average gives only an approximate value for the number of replicons.

The rate at which the replication fork moves can be estimated from the maximum distance that the autoradiographic tracks travel during a given time. This is in the range of 1000–3000 bp/minute for mammals; it is of the order of 1000 bp/minute for plants, possibly because of their lower temperatures. These values are very much slower than the rate of bacterial replication fork movement (~50,000 bp/minute).

We should like to know what constitutes the origin of each replicon. Is it a specific DNA sequence, possibly a member of a repetitive sequence family? Is it a feature of the higher order structure of the chromatin fiber, not necessarily associated with a single type of sequence?

Then we must ask how origins are selected for initiation at different times during S phase. Does the structure of the origin carry some information that establishes its time of use, or is this a consequence of some ordered rearrangement of structure? An especially intriguing question is how the replication appa-

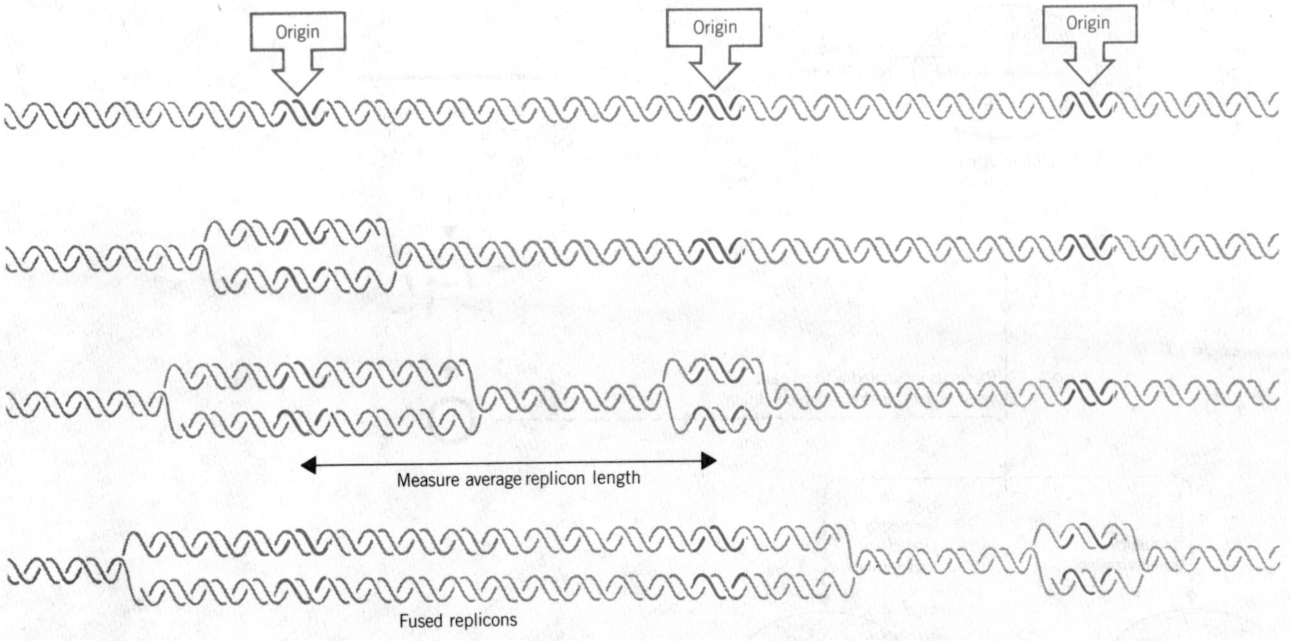

Figure 31.9
Measuring the size of the replicon requires a stretch of DNA in which adjacent replicons are active.

ratus distinguishes origins that have already been replicated from those that have yet to be replicated. Is the utilization of each origin once and only once during S phase a property of the DNA (for example, as seen by its state of methylation) or of proteins associated with it?

We know of one case in which origins are not fixed, but depend on the cell type. The replicons of early embryonic divisions in *D. melanogaster* are much smaller than those of somatic cells. They function simultaneously, thus reducing the time required to complete chromosome replication, a need presumably resulting from the rapidity of cell divisions at this time. A possible explanation is that the genome contains a set of tissue-specific origins, activated only in embryonic divisions, and not recognized as such in somatic cell divisions. We do not know whether tissue-specific replication patterns are common.

It is not clear whether chromosomal replicons have discrete termini at which the replication forks cease movement and (presumably) dissociate from the DNA. The alternative is that a replication fork continues from its origin until it meets a fork proceeding toward it from the adjacent replicon. We have already mentioned the potential topological problem of joining the newly synthesized DNA at the junction of the replication forks.

There is a difference in the timing of the replication of euchromatin and heterochromatin. Both constitutive and facultative heterochromatin are **late replicating**, as seen by their later start and finish to replication. This may be a consequence of their more condensed structure. Usually (although not always) the last material in the cell to synthesize DNA during S phase is heterochromatic.

ISOLATING THE ORIGINS OF YEAST REPLICONS

Any segment of DNA that has an origin should be able to replicate. So although plasmids are rare in eukaryotes, it may be possible to construct such molecules by suitable manipulation *in vitro*. Potential plasmids can be tested in suitable recipient cells.

One system in which extrachromosomal DNA molecules can replicate is the *Xenopus* egg. An injected

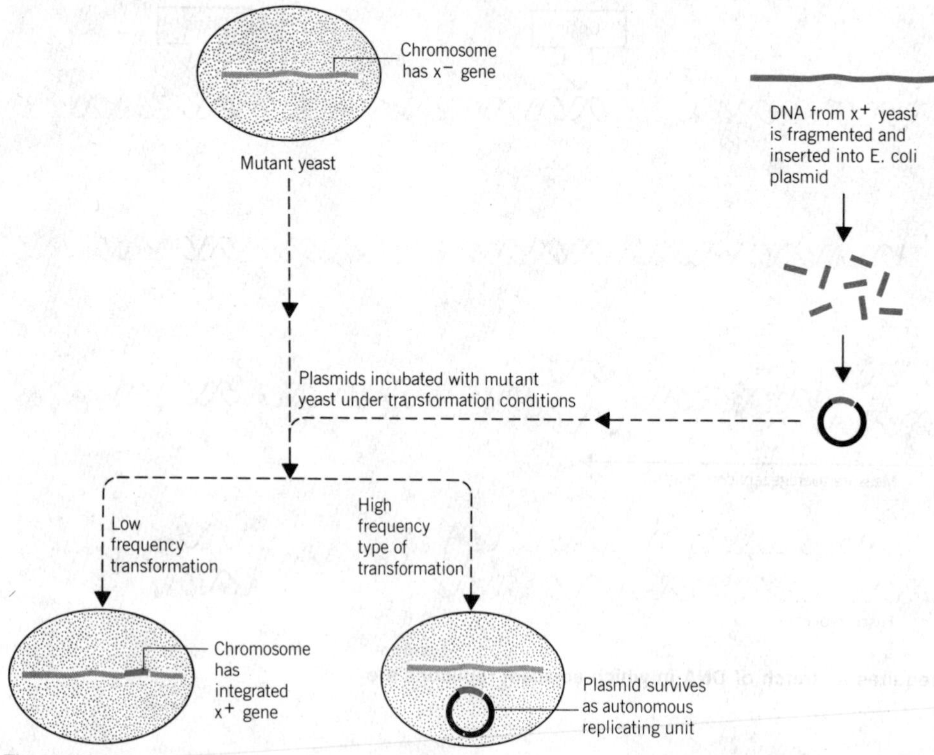

Figure 31.10
Transformation frequency in yeast is influenced by whether a fragment integrates into chromosomal DNA or survives as a plasmid.

DNA is replicated once in the cell cycle, which suggests that it is under proper control. However, *any* injected (circular) DNA is able to replicate, which argues against the need for a specific origin of replication in this system.

Another system offers the opportunity to isolate discrete origins. Cells of the yeast *S. cerevisiae* that are mutant in some function can be "transformed" by addition of DNA that carries a wild-type copy of the gene. The protocol is illustrated in **Figure 31.10.** The DNA of wild-type cells is extracted, cleaved into fragments, and cloned in *E. coli* plasmids. The chimeric plasmids are incubated with the mutant yeast cells, which are placed in conditions in which they can survive *only if they express the function of the wild-type gene.* Two types of transformation can be distinguished by the frequency with which transformed cells arise.

Any gene can be transformed at a *low frequency*. The surviving cells possess a copy of the wild-type donor gene integrated into the chromosome. Probably it has replaced the original mutant gene by a homologous recombination. The frequency of this type of transformation is low because integration is a rare event.

Some yeast DNA fragments are able to transform defective cells at much *higher* (~1000-fold) *frequencies*. These fragments can survive in the cell in the unintegrated (autonomous) state, that is, in the form of the originally added circular DNA molecule. The lack of any need for an (inefficient) integration event allows transformation to occur at a high frequency.

A high-frequency transforming fragment possesses a sequence that confers on a chimeric plasmid the ability to replicate efficiently in yeast. This segment is called an *ars* (for autonomously replicating sequence). We should like to know whether the *ars* sequences are authentic origins of replication.

Sequences with *ars* function occur at about the same average frequency as origins of replication (every 30–40 kb). The only homology between known *ars* elements consists of an 11 bp consensus sequence, represented in each element with up to 3 substitutions. The consensus is usually surrounded by A-T-rich DNA.

Point mutations of an *ars* element implicate a 14 bp "core" region in origin function. The core includes a consensus sequence. It is possible that the consensus is the recognition site for a protein involved in origin function; if other sequences are also involved, they must be different in each of the known *ars* elements.

Yeast origins of replication cannot simply be members of a repetitive sequence family. It is possible that many different sequences (containing in common only the very short consensus sequence) may be able to provide origin function in yeast.

REPLICATION CAN PROCEED THROUGH EYES, ROLLING CIRCLES, OR D LOOPS

The structures generated by replication depend on the relationship between the template and the replication fork. The critical features are whether the template is circular or linear, and whether the replication fork is engaged in synthesizing both strands of DNA or only one.

So far we have treated the replication fork as a site at which both new strands of DNA are synthesized. In a linear molecule, whether replicated unidirectionally or bidirectionally, the movement of the fork(s) generates an eye, as seen in Figure 31.2. If the template is a circular DNA, the replicating molecule takes the form of a θ structure, as drawn in Figure 31.3.

Another outcome is possible with a circular molecule. Suppose that replication of *one* strand of a duplex molecule starts at the origin. A nick opens one strand, and then the free 3′-OH end generated by the nick is extended by the DNA polymerase. The newly synthesized strand displaces the original parental strand. The ensuing events are depicted in **Figure 31.11.**

This type of structure is called a **rolling circle,** because the growing point can be envisaged as rolling around the circular template strand. It could in principle continue to do so indefinitely. As it moves, the replication fork extends the outer strand and displaces the previous partner.

Because the newly synthesized material is covalently linked to the original material, the displaced strand has the original unit genome at its 5′ end. The original unit is followed by any number of unit genomes, synthesized by continuing revolutions of the template. Each revolution displaces the material synthesized in the previous cycle.

An example is shown in the electron micrograph of **Figure 31.12.** The rolling circle is put to several uses *in vivo.*

In certain phages the genome consists of a single-stranded circle of DNA. This circle is first converted to a duplex form, which is then replicated by a rolling circle mechanism. The displaced tail generates a series of unit genomes; these can be cleaved and inserted into phage particles, or they can be used for further replication cycles. (These replication systems are discussed in Chapter 32.)

The displaced single strand of a rolling circle can be converted to duplex DNA by synthesis of a complementary strand. This action can be used to generate the concatemeric duplex molecules needed for maturation of certain phage DNAs (see Chapter 28). This provides the principal pathway through which phage lambda DNA matures. (It is not the only way that concatemeric molecules can be generated; recombination is another pathway, and appears to predominate with phage T4.)

Rolling circles also are used to generate amplified rDNA in the *Xenopus* oocyte. This explains why the amplified material consists of a large number of identical repeating units (see Figure 23.6). In this case, a single genomic repeating unit is converted into a rolling circle. The displaced tail is converted into duplex DNA; and at some point this is cleaved from the circle so that the two ends can be joined together to generate the amplified circle of rDNA.

Another type of action for a replication fork has been identified in some mitochondria. Replication starts at a specific origin in the circular duplex DNA. But initially only one of the two parental strands (the H strand in mammalian mitochondrial DNA) is used as a template for synthesis of a new strand. Synthesis proceeds for only a short distance, displacing the original partner (L) strand, which remains single-stranded, as illus-

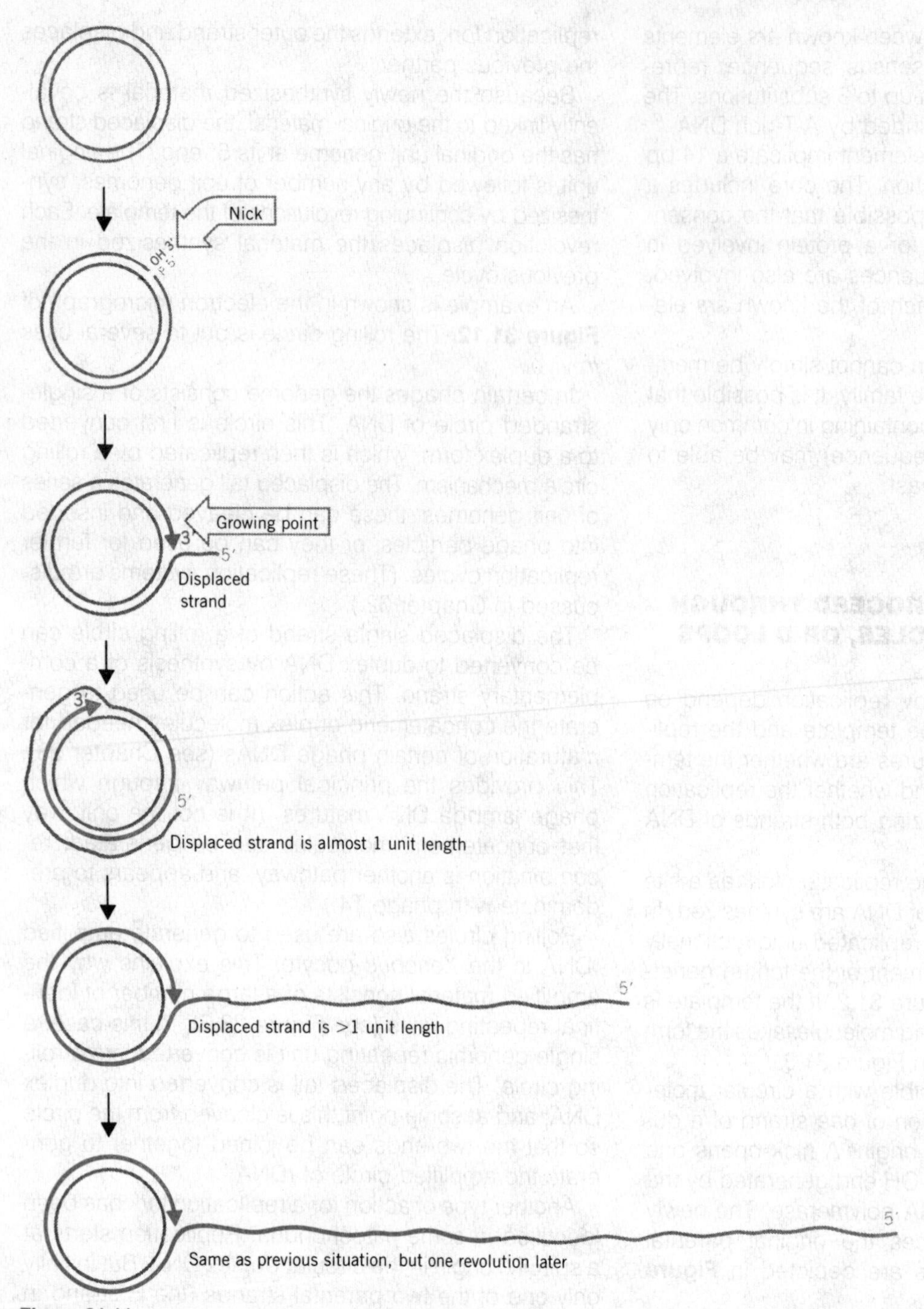

Figure 31.11
The rolling circle generates a multimeric single-stranded tail (which can be converted to a duplex form by synthesis of a complement; not shown).

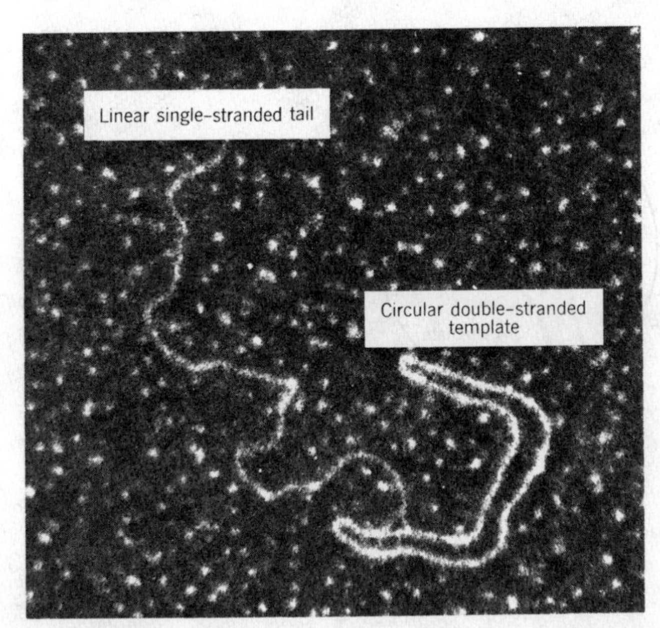

Figure 31.12
A rolling circle appears as a circular molecule with a linear tail by electron microscopy.
Photograph kindly provided by David Dressler.

trated in **Figure 31.13.** The condition of this region gives rise to its name as the **displacement** or **D loop.**

A single D loop is found as a stable structure of 500–600 bases in mammalian mitochondria. The short strand has a unique 5′ end and a variable 3′ end, probably as the result of termination of synthesis at any one of 3–4 discrete sites. This short strand is unstable and turns over; it is frequently degraded and resynthesized to maintain the opening of the duplex at this site.

Some mitochondrial DNAs, such as *X. laevis,* possess a single but longer D loop. Others may possess several D loops; for example, there may be as many as six in the linear mitochondrial DNA of *Tetrahymena.* The same mechanism is employed in chloroplast DNA, where (in higher plants) there are two D loops.

To replicate mammalian mitochondrial DNA, the short strand in the D loop is extended. The displaced region of the original L strand becomes longer, expanding the D loop. This expansion continues until it reaches a point about 67% of the way around the circle. Replication of this region exposes an origin in the displaced L strand. Synthesis of an H strand initiates at this site, proceeding around the displaced single-stranded L template in the opposite direction from L-strand synthesis. Because of the lag in its start, H-strand synthesis has proceeded only 30-40% of the way around the circle when L-strand synthesis finishes. This releases one completed duplex circle and one gapped circle, which remains partially single-stranded until synthesis of the H strand is completed. Finally, the new strands are sealed to become covalently intact.

The existence of rolling circles and D loops exposes a general principle. *An origin can be a sequence of DNA that serves to initiate DNA synthesis using one strand as template.* The opening of the duplex does not necessarily lead to the initiation of replication on the other strand. This might need its own origin for replication. A D loop results when the origins of the two strands are separated by some distance. The corollary is that an origin that serves to initiate replication of *both* strands may do so specifically because of its features; that is, because the origins of both strands are located at the same point. These features also will determine whether replication forks set out unidirectionally or bidirectionally.

PLASMID INCOMPATIBILITY IS CONNECTED WITH COPY NUMBER

Plasmids are selfish. Having obtained possession of a bacterium, the resident plasmid will seek to prevent any other plasmid of the same type from establishing residence. Two independent devices are used to establish these territorial rights: **surface exclusion** and **plasmid incompatibility.**

Surface exclusion describes the inability of a plasmid to enter cells already carrying another plasmid of the same type. The effect is mediated at the surface of the bacterium and is a feature of the system used to transfer the plasmid between bacteria. In the case of the sex (F) factor of *E. coli,* surface exclusion is achieved by inhibiting the DNA from leaving its present bacterial host in order to enter another. Thus it is an emigration rather than immigration control.

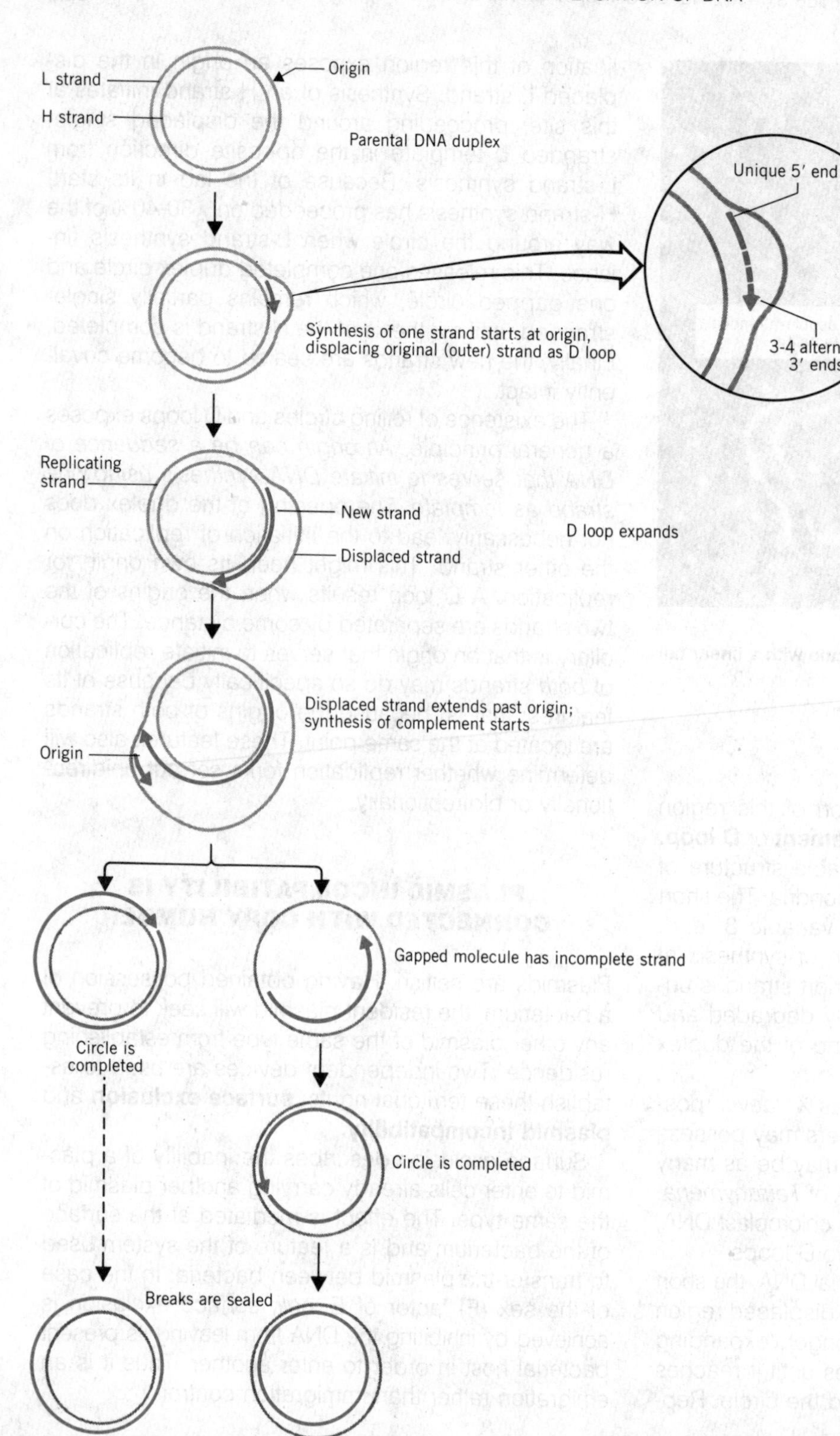

Figure 31.13
The D loop is a stable structure in mammalian mitochondrial DNA, which has separate origins for the replication of each strand.

Incompatibility is related to the regulation of plasmid copy number. A **compatibility group** is defined as a set of plasmids whose members are unable to coexist in the same bacterial cell. The presence of a member of one compatibility group does not directly affect the survival of a plasmid belonging to a different group. Thus only one replicon of a given compatibility group can be maintained in the bacterium, but it does not interact with replicons of other compatibility groups (although in limiting conditions they may compete for lebensraum).

The negative control model for plasmid incompatibility follows the idea that copy number control is achieved by synthesizing a repressor that measures the number of origins. (Formally this is the same as the titration model for regulating bacterial chromosomal replication.) The introduction of a new origin in the form of a second plasmid of the same compatibility group mimics the result of replication of the resident plasmid; two origins now are present. Thus any further replication is prevented until after the two plasmids have been segregated to different cells to create the correct prereplication copy number.

Mutations may either increase or decrease the copy number. Such mutants are important because they define the genetic locus responsible for controlling copy number; and their analysis opens the way to defining the mechanism involved. These mutations also influence incompatibility, demonstrating its connection with copy number.

The best characterized copy number and incompatibility system is that of the plasmid ColE1 of *E. coli*. Before considering it, we must digress briefly into the mechanism for initiating replication at the ColE1 origin. The relevant events are illustrated in **Figure 31.14.** Replication starts with the transcription of an RNA that initiates 555 bp upstream from the origin. Transcription continues through the origin. The enzyme RNAase H (whose name reflects its specificity for a substrate of RNA hybridized with DNA) cleaves the transcript at the origin. This generates a 3′-OH end that is used as the "primer" for synthesis of DNA.

Two regulatory systems exert their effects on the RNA primer. One involves a protein coded by a nearby locus; the other involves synthesis of an RNA complementary to the primer.

A small (63 amino acid) protein is coded by the *rop* gene, located downstream from the origin. The Rop protein inhibits formation of the primer; its mechanism of action is unknown at present, but seems to involve

Figure 31.14
Replication of ColE1 DNA is initiated by cleaving the primer RNA to generate a 3′-OH end.

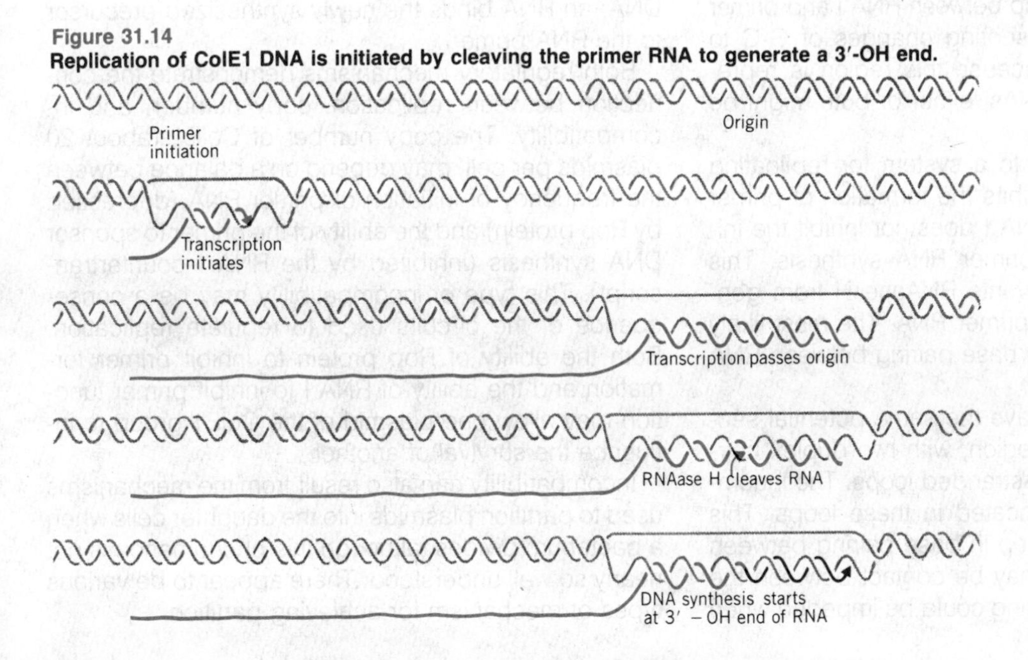

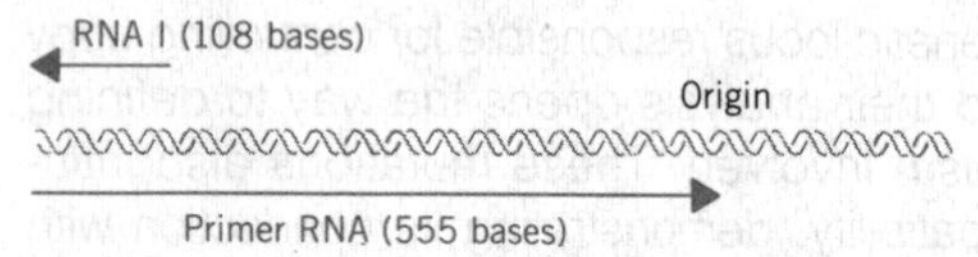

Figure 31.15
The sequence of RNA I is complementary to the 5′ region of primer RNA.

sequences just downstream from the primer start. Whatever its molecular action, the result is to reduce the copy number.

A second RNA species is synthesized from ColE1 DNA in the same region as the primer. This is RNA I, a molecule of about 108 bases, coded by the opposite strand from that specifying primer RNA. The relationship between the primer RNA and RNA I is illustrated in **Figure 31.15.** The RNA I molecule terminates very close to the site where the primer RNA initiates. Thus the RNA I species is complementary to the 5′-terminal region of the primer RNA. An RNA molecule that exercises some function by virtue of its complementarity with another RNA coded in the same region may be called a **countertranscript.**

Mutants that reduce or eliminate incompatibility can be obtained by selecting plasmids of the same group for their ability to coexist. Mutants in the ColE1 group map in the region of overlap between RNA I and primer RNA, four of them representing changes of G-C to A-T, one the reverse. Because this region is represented in two different RNAs, either or both might be involved in the effect.

When RNA I is added to a system for replicating ColE1 DNA *in vitro,* it inhibits the formation of primer RNA. The presence of RNA I does not inhibit the initiation or elongation of primer RNA synthesis. This suggests that RNA I prevents RNAase H from generating the 3′ end of the primer RNA. The most likely basis for this effect lies in base pairing between RNA I and primer RNA.

Both RNA molecules have the same potential secondary structure in this region, with two duplex hairpins terminating in single-stranded loops. The incompatibility mutations are located in these loops. This suggests that an initial step in base pairing between RNA I and primer RNA may be contact between the unpaired loops. Their pairing could be impeded when the RNA I and primer RNA have different sequences because of a mutation in one of the parent plasmids. Each RNA I would continue to pair with the primer RNA coded by the same plasmid, but might be unable to pair with the primer RNA coded by the other plasmid.

How does pairing with RNA I prevent cleavage to form primer RNA? A model is illustrated in **Figure 31.16.** In the absence of RNA I, the primer RNA has a secondary structure in which its 5′ terminal region is paired with some other part of the molecule. The intervention of RNA I sequesters the 5′ end of the potential primer RNA, altering its secondary structure. This new secondary structure in some way prevents the cleavage reaction, although we do not understand the basis for this effect. (The action of RNAase H has been thought to depend on recognition of RNA-DNA hybrid regions, not on RNA structure.)

The model is somewhat reminiscent of the mechanism involved in attenuation of transcription, in which the alternative pairings of an RNA sequence permit or prevent formation of the secondary structure needed for termination by RNA polymerase (see Chapter 15). Formally, the model is equivalent to postulating the presence in the cell of a repressor that prevents newly introduced DNA from functioning, analogous to the role of the lambda lysogenic repressor (see Chapter 16). Instead of a repressor protein that binds the new DNA, an RNA binds the newly synthesized precursor to the RNA primer.

Both regulatory mechanisms demonstrate the connection between replication, copy number, and incompatibility. The copy number of ColE1 (about 20 plasmids per cell) may depend on a balance between the frequency of initiation of primer RNA (depressed by Rop protein) and the ability of the primer to sponsor DNA synthesis (inhibited by the RNA I countertranscript). This type of incompatibility may be a consequence of the circuits used to regulate replication. Both the ability of Rop protein to inhibit primer formation and the ability of RNA I to inhibit primer function may allow one plasmid of the ColE1 group to influence the survival of another.

Incompatibility can also result from the mechanisms used to partition plasmids into the daughter cells when a bacterium divides, although this phenomenon is not nearly so well understood. There appear to be various types of mechanism for achieving partition.

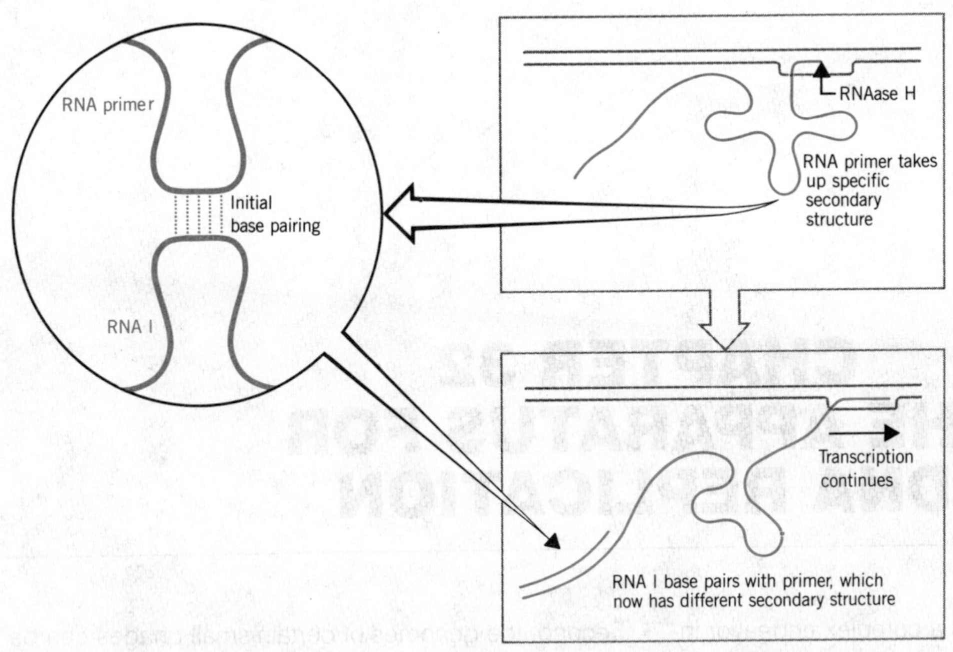

Figure 31.16
Base pairing with RNA I may change the secondary structure of the primer RNA sequence and thus prevent cleavage from generating a 3′-OH end.

In the plasmid pSC101, a 370 bp region called *par* is necessary for partition. It is a *cis*-acting locus and does not code for protein. The parts of this region involved in partition consist of three discrete segments; deletion of any one segment impedes partition. The three sequence elements are related in such a way that (considering each strand of DNA separately) the central sequence could form a hairpin by base pairing with either of the flanking sequences on that strand. Deletion of any one of the segments impedes partition.

By contrast, the prophage form of phage P1 is a plasmid whose *par* function codes for a protein that apparently recognizes a *cis*-acting site called *incB;* incompatibility mutations map in *incB*.

Cis-acting partition sites may be equivalent to eukaryotic centromeres in providing target sequences involved in attachment to cellular structures. Competition for these sites may be involved in incompatibility. The proteins may provide the missing link between partition and cell division.

FURTHER READING

General reviews of bacterial and eukaryotic replicons are to be found in **Lewin's** *Gene Expression*, [pp. 550–576 of **1,** *Bacterial Genomes*, and pp. 570–582 of **2,** *Eucaryotic Chromosomes* (Wiley, New York, 1974 & 1980)]. An origin was isolated by **Zyskind & Smith** (*Proc. Nat. Acad. Sci. USA* **77,** 2460–2464, 1980). **Laskey & Harland** (*Cell* **24,** 283–284, 1981) have speculated about eukaryotic replication origins. Mitochondrial DNA replication has been reviewed by **Clayton** (*Cell* **28,** 693–705, 1982). The control of plasmid regulation has been reviewed by **Scott** (*Microbiol. Rev.* **48,** 1–23, 1984). The ColE1 system was elucidated by **Tomizawa & Itoh** (*Proc. Nat. Acad. Sci. USA* **78,** 6096–6100, 1981).

CHAPTER 32
THE APPARATUS FOR DNA REPLICATION

Replication of duplex DNA is a complex endeavor involving a conglomerate of enzyme activities. The parental strands must be separated and (temporarily) stabilized in the single-stranded state. Synthesis of daughter strands must be initiated; then they must be elongated. At the end of the replicon, joining and/or termination reactions are necessary. All of these individual activities must be undertaken within the constraints imposed by the complex topology of DNA.

Although the various enzymatic activities can be equated with individual proteins or groups of proteins, they do not function independently; all may be contained in a discrete multiprotein structure, sometimes called the **replisome.** This is not evident as a free cellular body (for example, analogous to the ribosome), but may be assembled from its components at the onset of replication. Thus the replisome may exist *only as a protein complex associated with the particular structure that DNA takes at the replication fork.*

With eukaryotes, we have at present made little more than a start toward defining the proteins involved in replication. With prokaryotes, two approaches converge in defining the components of the replication apparatus. First, mutants defective in replication identify genes whose products either are components of the apparatus or are involved in peripheral functions. Second, the genomes of certain small phages can be used to provide templates that are replicated *in vitro* by a system of defined host (or phage-coded) proteins. There has therefore been appreciable progress toward defining a replication apparatus in which each component is available for study *in vitro* as a biochemically pure product, and is implicated *in vivo* by the effect of mutations in its gene.

Enzymes that are able to synthesize new DNA strands on a template strand are called **DNA polymerases.** Both prokaryotic and eukaryotic cells contain more than one DNA polymerase activity. But only one particular enzyme provides the replicase function in either the bacterium or eukaryotic nucleus. The others are involved in subsidiary roles in replication and/or participate in "repair" synthesis of DNA to replace damaged sequences (see Chapter 33).

It is convenient to think of DNA-synthesizing activities in terms of the DNA polymerase enzymes; but it is a moot point whether a DNA replicase enzyme exists as a discrete entity. (We have already considered a similar issue for RNA polymerase.) In bacteria, a protein complex with DNA replicase activity can be identified, but it is recovered in the form of several (overlapping) aggregates of various subunits. It is not clear which of these should properly be considered a component of the replicase per se and which consti-

Table 32.1
There are three mammalian DNA polymerases.

	DNA Polymerase α	DNA Polymerase β	DNA Polymerase γ
Location	Nuclear	Nuclear	Nuclear & mitochondrial
Proportion of Activity	~80%	10–15%	2–15%
Daltons	110–220,000	45,000	60,000
Subunits	Several	1	1

tute accessory factors needed for proper replicase function. In the normal process of replication, the DNA-synthesizing activity is only one of several functions associated in the replisome; and perhaps we should not be unduly influenced to think of it as a separate function because of our ability to assay its particular step in the overall process.

No DNA polymerase activity yet has been discovered that can initiate a deoxyribonucleotide chain. All the enzymes extend previously started chains. So an essential feature of the replication apparatus is the provision for a "priming" activity to start off the DNA chain. This usually involves the synthesis of a short ribonucleotide sequence that is later removed.

EUKARYOTIC DNA POLYMERASES

Eukaryotic DNA polymerases have been investigated the most thoroughly in mammals, where the first such enzyme was discovered as long ago as 1960 (in calf thymus). Most eukaryotic enzymes possess only the DNA-synthesizing activity, a contrast with the wide capacities of bacterial enzymes, most of which also possess exonucleolytic activity (see later).

Three classes of eukaryotic DNA polymerase have been identified. Their properties are summarized in **Table 32.1.** Can one be identified as the authentic replicase?

DNA polymerase α is the only one whose levels increase during S phase. It is resistant to 2′3′-dideoxythymidine triphosphate; this analog does not inhibit *in vitro* replication systems, but does inhibit polymerases β and γ. A fungal antibiotic, aphidicolin, inhibits eukaryotic DNA replication *in vivo;* the only enzyme that it inhibits *in vitro* is DNA polymerase α. Mutant cells resistant to aphidicolin contain a DNA polymerase α that is resistant to the antibiotic *in vitro*.

These results identify DNA polymerase α as the nuclear DNA replicase. In some systems, preparations of DNA polymerase α have an RNA priming activity, probably because a priming enzyme is closely associated with the DNA polymerase.

Activities other than replication in which DNA polymerases are involved include "repair" functions, in which segments of damaged DNA are replaced. Probably the DNA polymerase β has a repair function.

The role of DNA polymerase γ presumably is to undertake the replication of mitochondrial DNA. Its role in the nucleus (if any) is unknown.

DNA replication requires many activities in addition to the DNA polymerase. A difficulty in analyzing the role of the eukaryotic DNA polymerases is caused by the lack of systems containing all the components needed for replication. However, there is progress toward obtaining suitable *in vitro* systems that can replicate defined templates. Some nuclear systems have been obtained that contain all of the components necessary to continue replication fork movement. Initiation of synthesis at the origins of new replicons has not yet been obtained *in vitro*.

PROKARYOTIC DNA POLYMERASES HAVE SEVERAL ENZYMATIC ACTIVITIES

Three DNA polymerase enzymes have been characterized in *E. coli*. The first to be discovered, DNA polymerase I, is involved in the repair of damaged DNA and, in a subsidiary role, in semiconservative replication. Nothing is known about the function *in vivo* of the second enzyme, DNA polymerase II. The third

Table 32.2
***E. coli* has three DNA polymerase enzymes.**

	DNA Polymerase I	DNA Polymerase II	DNA Polymerase III
Daltons	109,000	120,000	>250,000
Constitution	Monomer	Not known	Heteromultimeric
Number/cell	400	Not known	10–20
Enzymatic activities:			
5′–3′ elongation from 3′-OH primer	Yes	Yes	Yes
3′–5′ exonuclease	Yes	Yes	Yes
5′–3′ exonuclease	Yes	No	Yes
Mutant loci	*polA*	*polB*	*polC, dnaN, dnaX, dnaZ, dnaQ*
Mutant phenotype	Defective in repair	None	Prevents replication
Lethality	Viability reduced only when 5′–3′ exonuclease affected	No effect	Conditional lethal

enzyme, DNA polymerase III, is a multisubunit protein; it is the replicase responsible for de novo synthesis of new strands of DNA. The enzymatic activities of the three enzymes are summarized in **Table 32.2.**

All prokaryotic and eukaryotic DNA polymerases share the same fundamental type of synthetic activity. Each can extend a DNA chain by adding nucleotides one at a time to a 3′-OH end. Thus the new chain is synthesized in the usual direction, 5′-3′. The reaction is illustrated diagrammatically in **Figure 32.1.** The precursor for DNA synthesis is a nucleoside triphosphate, which loses the terminal two phosphate groups in the reaction. The choice of which nucleotide is added to the chain is dictated by base pairing with the template strand.

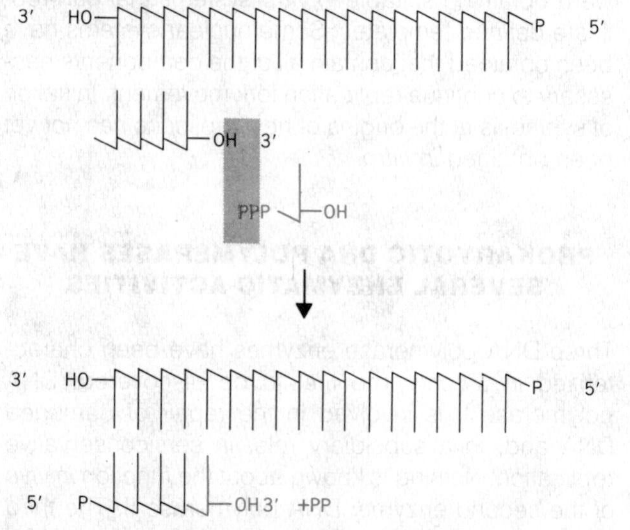

Figure 32.1
DNA synthesis occurs by addition of nucleotides to the 3′-OH end of the growing chain.

The fidelity of replication poses the same sort of problem we have already encountered in considering (for example) the fidelity of translation. It relies on the specificity of base pairing. Yet when we consider the interactions involved in base pairing, we would expect errors to occur with a frequency of 10^{-4} to 10^{-5} per base pair replicated. The actual rate in *E. coli* cells or in phage T4 replication seems to be in the range of 10^{-8} to 10^{-10}. (This corresponds to approximately 1 error per 1000 bacterial replication cycles, or, in other terms, to a rate of $\sim 10^{-6}$ per gene per generation.)

The enzyme might improve the specificity of complementary base selection at either (or both) of two stages. It could scrutinize the incoming base for the proper complementarity with the template base; for example, by specifically recognizing matching chemical features. This would be a *presynthetic* error control. Or it could scrutinize the base pair *after* the new base has been added to the chain, and, in those cases in which a mistake has been made, remove the most recently added base. All of the bacterial enzymes possess a 3′–5′ exonucleolytic activity that appears to exercise just such a **proofreading** function. The absence of this activity from the eukaryotic enzymes implies that they must have some other mechanism for controlling the error rate.

The 3′-5′ exonucleolytic activity proceeds in the reverse direction from DNA synthesis. The proofreading action is illustrated diagrammatically in **Figure 32.2.** In the chain elongation step, a precursor nucleotide enters the position at the end of the growing chain. A bond is formed. The enzyme moves one base pair farther, ready for the next precursor nucleotide to enter. If a mistake has been made, the enzyme moves backward, excising the last base that was added, creating a site for a replacement precursor nucleotide to enter. This activity has been the most fully characterized with DNA polymerase I, but is thought to function in an analogous manner with other prokaryotic DNA polymerases.

DNA polymerase I is a talented enzyme with several activities. Coded by the locus *polA,* it is a single polypeptide of 109,000 daltons. The chain can be cleaved into two regions by proteolytic treatment. The large fragment (76,000 daltons) contains the 5′–3′ polymerase and the 3′–5′ exonuclease activities. The small fragment (36,000 daltons) possesses another activity of the enzyme, a 5′–3′ exonucleolytic activity. This activity is therefore independent of the synthetic/proofreading activities. The 5′–3′ exonuclease action excises small groups of nucleotides, up to ~10.

DNA polymerase I has a unique ability to start replication *in vitro* at a nick in DNA. At a point where a phosphodiester bond has been broken in a double-

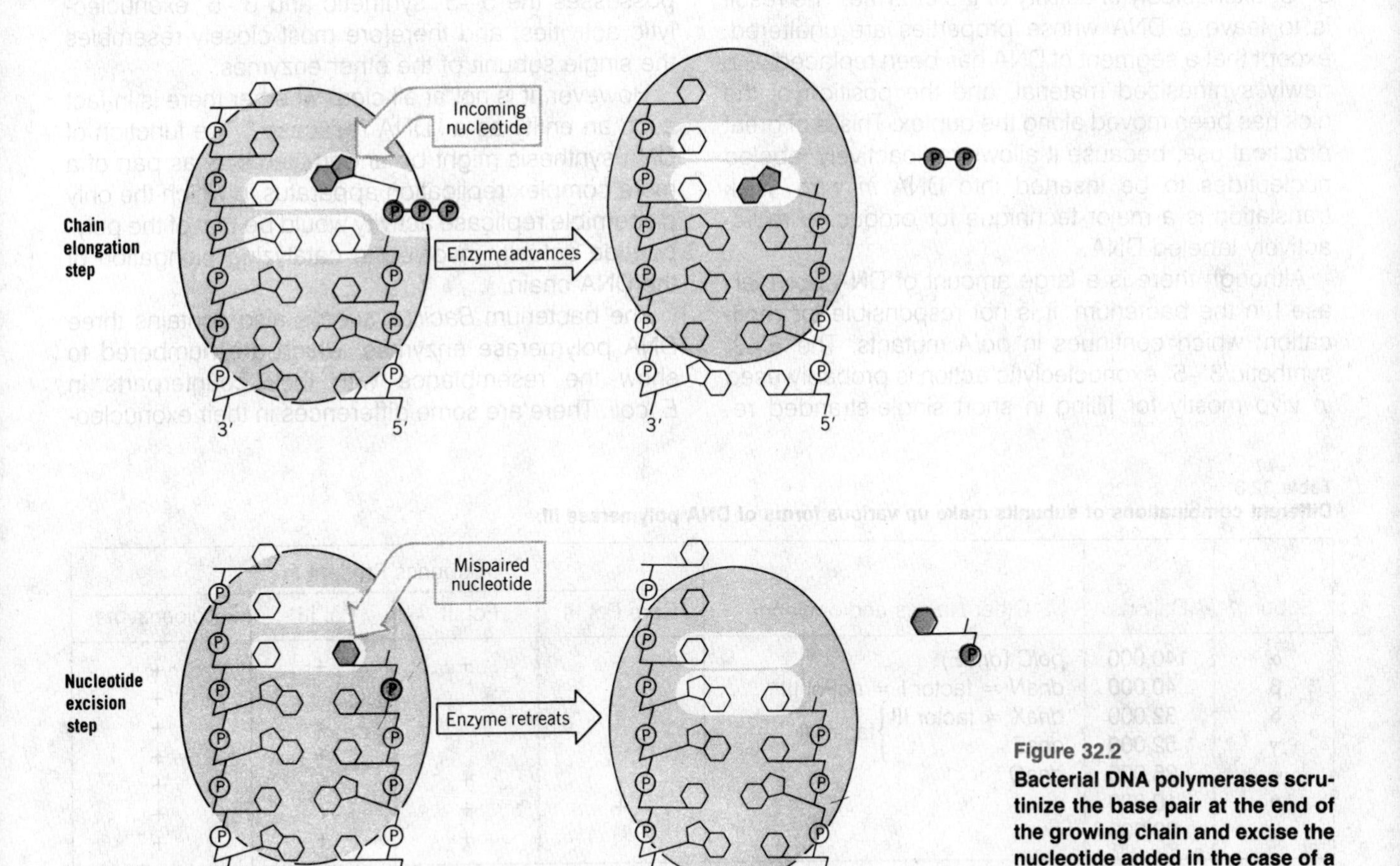

Figure 32.2
Bacterial DNA polymerases scrutinize the base pair at the end of the growing chain and excise the nucleotide added in the case of a misfit.

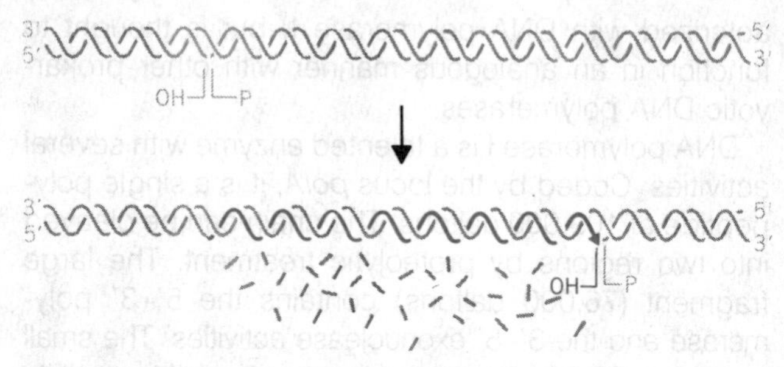

Figure 32.3
Nick translation replaces part of a preexisting strand of duplex DNA with newly synthesized material.

stranded DNA, the enzyme extends the 3′-OH end. As the new segment of DNA is synthesized, it *displaces the existing homologous strand in the duplex.*

This process of **nick translation** is illustrated in **Figure 32.3.** The displaced strand is degraded by the 5′–3′ exonucleolytic activity of the enzyme. The result is to leave a DNA whose properties are unaltered, except that a segment of DNA has been replaced with newly synthesized material, and the position of the nick has been moved along the duplex. This is of great practical use, because it allows radioactively labeled nucleotides to be inserted into DNA *in vitro*. Nick translation is a major technique for producing radioactively labeled DNA.

Although there is a large amount of DNA polymerase I in the bacterium, it is not responsible for replication, which continues in *polA* mutants. The 5′–3′ synthetic/3′–5′ exonucleolytic action is probably used *in vivo* mostly for filling in short single-stranded regions in double-stranded DNA. These regions arise during replication, and when bases that have been damaged are removed from DNA (see Chapter 33).

The enzyme DNA polymerase III was originally discovered by the conditional lethality of a mutation in the *polC* locus, which codes for the 140,000 α subunit that is a component of all the enzyme preparations. Purified DNA polymerase III activities have several subunits. We use the plural "activities" because the enzyme has been obtained in several forms, and it is unclear which (if any) represents the unit involved in replication *in vivo*. These forms are summarized in **Table 32.3.**

All the subunits listed in the table are needed for full activity in *in vitro* replication systems. There have been attempts to divide the subunits into components of the "DNA replicase" and "accessory proteins" needed to assist the replicase. The α subunit, coded by *polC*, possesses the 5′–3′ synthetic and 3′–5′ exonucleolytic activities, and therefore most closely resembles the single subunit of the other enzymes.

However, it is not at all clear whether there is in fact such an entity as a "DNA replicase." The function of DNA synthesis might be undertaken only as part of a more complex replication apparatus in which the only discernible replicase activity would be that of the polypeptide actually involved in catalyzing elongation of the DNA chain.

The bacterium *Bacillus subtilis* also contains three DNA polymerase enzymes, which are numbered to show the resemblance with their counterparts in *E. coli*. There are some differences in their exonucleo-

Table 32.3
Different combinations of subunits make up various forms of DNA polymerase III.

Subunit	Daltons	Other Names and/or Gene	Subunits Present in			
			Core Pol III	Pol III′	Pol III*	Holoenzyme
α	140,000	*polC* (*dnaE*)	+	+	+	+
β	40,000	*dnaN* = factor I = coPol III*				+
δ	32,000	*dnaX* = factor III } factor II			+	+
γ	52,000	*dnaZ* } factor II			+	+
ϵ	25,000	*dnaQ*	+	+	+	+
θ	10,000		+	+	+	+
τ	83,000			+	+	+

The stoichiometry of the various complexes is not certain, although in most cases there is only one copy of each subunit.

lytic activities relative to the *E. coli* enzymes. The subunit identified with the DNA synthetic activity of *B. subtilis* DNA polymerase III is stimulated by factors I and II of *E. coli* (subunits βδγ), which suggests that there may be a related division of labors in the replication apparatus of the two bacteria. The level of *B. subtilis* DNA polymerase II is related to physiological events; it is absent from dormant spores, but appears about 30 minutes after the start of germination.

Some phages also code for DNA polymerases. They include T4, T5, T7, and SPO1. The enzymes all possess 5′–3′ synthetic activities and 3′–5′ exonuclease proofreading activities. In each case, a mutation in the gene that codes for a single phage polypeptide prevents phage development. Each phage polymerase polypeptide associates with other proteins, of either phage or host origin. The examples of phages T4 and T7 are discussed later.

DNA SYNTHESIS IS SEMIDISCONTINUOUS

The antiparallel structure of the two strands of duplex DNA poses a problem for replication. As the replication fork advances, daughter strands must be synthesized on both of the exposed parental single strands. The fork is moving in the direction from 5′ to 3′ on one strand, and in the direction from 3′ to 5′ on the other strand. Yet nucleic acids are synthesized only from a 5′ end toward a 3′ end. The problem is solved by synthesizing the strand that grows overall from 3′ to 5′ in a series of short fragments, each actually synthesized with the customary 5′ to 3′ polarity.

Consider the region immediately behind the replication fork, as illustrated in **Figure 32.4.** On one strand, DNA synthesis can proceed continuously in the 5′ to 3′ direction as the parental duplex is unwound. This is called the **leading strand.** On the other strand, however, a stretch of single-stranded parental DNA must be exposed, and then a segment is synthesized in the reverse direction (relative to fork movement). This is called the **lagging strand.** A series of these fragments are synthesized, each 5′ to 3′, and then they are joined together to create an intact lagging strand. This behavior is called **discontinuous replication.**

Discontinuous replication is studied in terms of the fate of a very brief label of radioactivity. The label enters newly synthesized DNA in the form of short fragments, sedimenting in the range of 7–11S, corresponding (approximately) to sequences of 1000–2000 bases in length. These are called **Okazaki fragments.** They are found in all replicating DNA, both prokaryotic and eukaryotic. After longer periods of incubation, the label enters larger segments of DNA. This transition represents the covalent linkage of Okazaki fragments.

The lagging strand *must* be synthesized in the form of Okazaki fragments. There has been a protracted controversy as to whether the leading strand is syn-

Figure 32.4
The leading strand is synthesized continuously while the lagging strand is synthesized discontinuously.

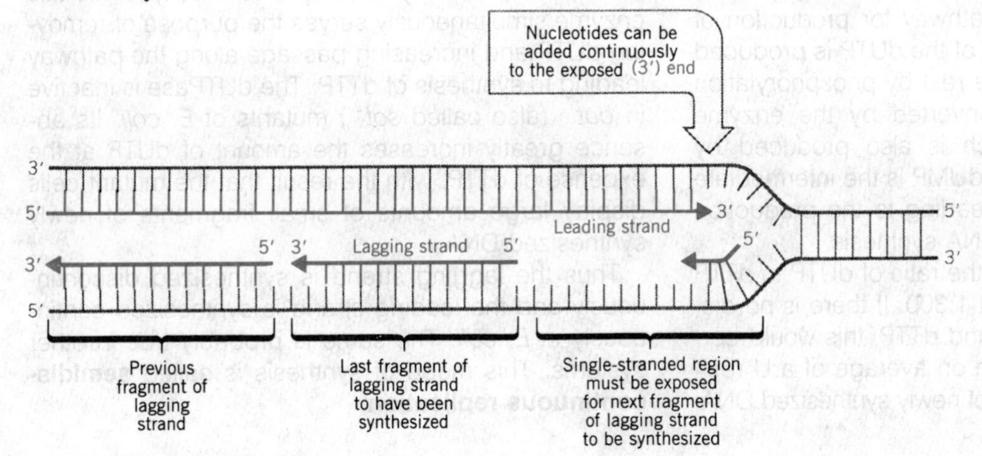

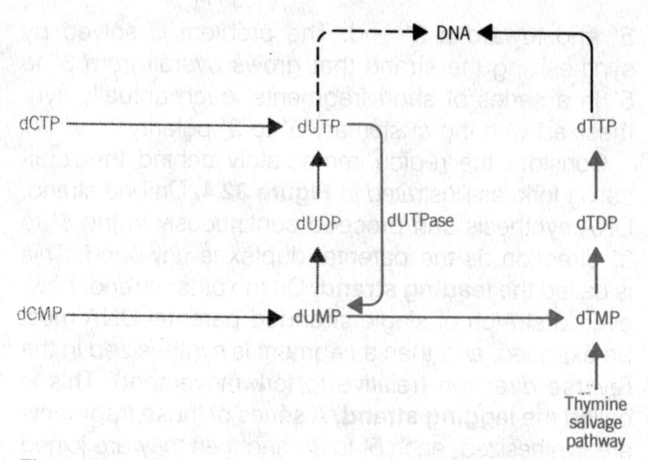

Figure 32.5
The synthesis of dUTP and dTTP is connected.

thesized in the same way or is synthesized continuously.

All of the newly synthesized DNA is found in the form of short fragments in *E. coli*. Superficially, this suggests that both strands are synthesized discontinuously. However, it turns out that not all of the fragment population represents bona fide Okazaki fragments; some are pseudofragments, generated by breakage in a DNA strand that actually was synthesized as a continuous chain.

The source of this breakage is the incorporation of some uracil into DNA in place of thymine. At one time, it was thought that such misincorporation is a rare event. But now it seems that *E. coli* DNA polymerase III does not discriminate well between dUTP and dTTP *in vitro* and (presumably) *in vivo*.

Figure 32.5 shows the pathway for production of dUTP and dTTP. Some 75% of the dUTP is produced by deamination of dCTP, the rest by phosphorylation of dUDP. The dUTP is converted by the enzyme **dUTPase** into dUMP (which is also produced by deamination of dCMP). The dUMP is the intermediate used to synthesize dTMP, leading to the production of dTTP, the precursor for DNA synthesis.

Under normal conditions, the ratio of dUTP to dTTP in *E. coli* seems to be about 1:300. If there is no discrimination between dUTP and dTTP, this would correspond to the incorporation on average of a U residue once every 1200 bases of newly synthesized DNA.

What happens to this uracil residue? It is removed by an enzyme, **uracil-DNA-glycosidase,** which is important as a defense against mutation as well as against replication errors. (The mutation pathway produces uracil by deaminating cytosine in DNA, as described in Chapter 3.) The uracil-DNA-glycosidase activity has been identified in both prokaryotes and eukaryotes. It is likely to be a ubiquitous system that protects DNA against the presence of uracil.

The removal of the uracil is accomplished by cleaving the base from the sugar. This reaction is followed by the introduction of a break in the phosphodiester backbone. The break may be widened into a gap by the removal of further nucleotides. These gaps are filled by synthesizing DNA to replace the lost material. This series of events is illustrated in **Figure 32.6.** During the period between the creation and filling of the gap, the breaks in the newly synthesized DNA generate fragments that are comparable in size to the Okazaki fragments. (The same events also reduce the size of the Okazaki fragments of the lagging strand.)

The importance of two enzymes, the uracil-DNA-glycosidase and the dUTPase, is shown by the effects of mutations in their genes.

The excision of uracil is prevented in *ung*$^-$ mutants of *E. coli,* which lack the glycosidase enzyme. In these mutants, about half of a radioactive label appears in newly synthesized DNA in the form of (genuine) Okazaki fragments; the other half directly enters much larger fragments. In this situation, we can see that the leading strand is synthesized continuously.

The cellular balance between dUTP and dTTP is greatly influenced by the dUTPase activity, since this enzyme simultaneously serves the purpose of removing dUTP and increasing passage along the pathway leading to synthesis of dTTP. The dUTPase is inactive in *dut*$^-$ (also called *sof*$^-$) mutants of *E. coli*. Its absence greatly increases the amount of dUTP at the expense of dTTP, with the result that the mutant cells display large amounts of small fragments of newly synthesized DNA.

Thus the lagging strand is synthesized discontinuously and the leading strand is synthesized continuously in *E. coli*. The same is probably true in other systems. This mode of synthesis is called **semidiscontinuous replication.**

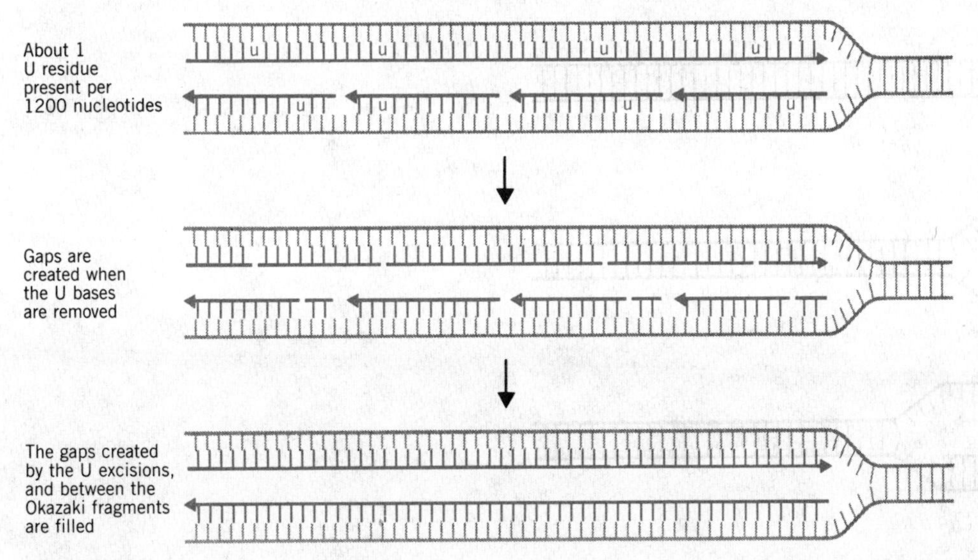

Figure 32.6

Excision of uracil from newly synthesized DNA creates gaps in both the continuously synthesized leading strand and the Okazaki fragments of the lagging strand. The gaps must be filled and the Okazaki fragments must be linked to generate continuous DNA strands.

OKAZAKI FRAGMENTS ARE PRIMED BY RNA

We have already mentioned that all of the known DNA polymerases of both prokaryotes and eukaryotes can extend a deoxyribonucleotide chain from a free 3′-OH end; but none can initiate synthesis. Yet there must be a series of initiation events, since each Okazaki fragment of the lagging strand requires its own start de novo. Each Okazaki fragment starts with a **primer,** a (short) sequence that provides the 3′-OH end for extension by DNA polymerase. The primer is RNA, about 10 bases long.

We can now expand our consideration of the actions involved in joining Okazaki fragments, as illustrated in **Figure 32.7.** The order of events is uncertain, but must involve removal of the RNA primer, its replacement by a stretch of DNA, and the covalent linking of adjacent Okazaki fragments.

The figure suggests that synthesis of an Okazaki fragment terminates just before the start of the RNA primer of the preceding fragment. When the primer is removed, there will be a gap. DNA polymerase I is likely to be involved in filling the gap, since *polA* mutants may fail to join their Okazaki fragments properly. Indeed, this enzyme might use its 5′–3′ exonuclease activity to remove the RNA primer while simultaneously replacing it with a DNA sequence extended from the 3′-OH end of the next Okazaki fragment. This is equivalent to the activity of nick translation, except that the new DNA replaces a stretch of RNA rather than a segment of DNA.

Once the RNA has been removed and replaced, the adjacent Okazaki fragments must be linked together. The 3′-OH end of one fragment will be adjacent to the 5′-phosphate end of the previous fragment. The responsibility for sealing this nick lies with the enzyme **DNA ligase** (which we have mentioned previously in connection with its use in the *in vitro* reconstruction of DNA). Ligases are present in both prokaryotes and eukaryotes; the two most intensively studied are those of *E. coli* and phage T4. The enzymes are implicated in joining Okazaki fragments *in vivo* by the persistence of unconnected fragments in *lig*$^-$ mutants.

The *E. coli* and T4 enzymes share the property of sealing nicks that have 3′-OH and 5′-phosphate termini, as illustrated in **Figure 32.8.** Both enzymes undertake a two-step reaction, involving an enzyme-AMP

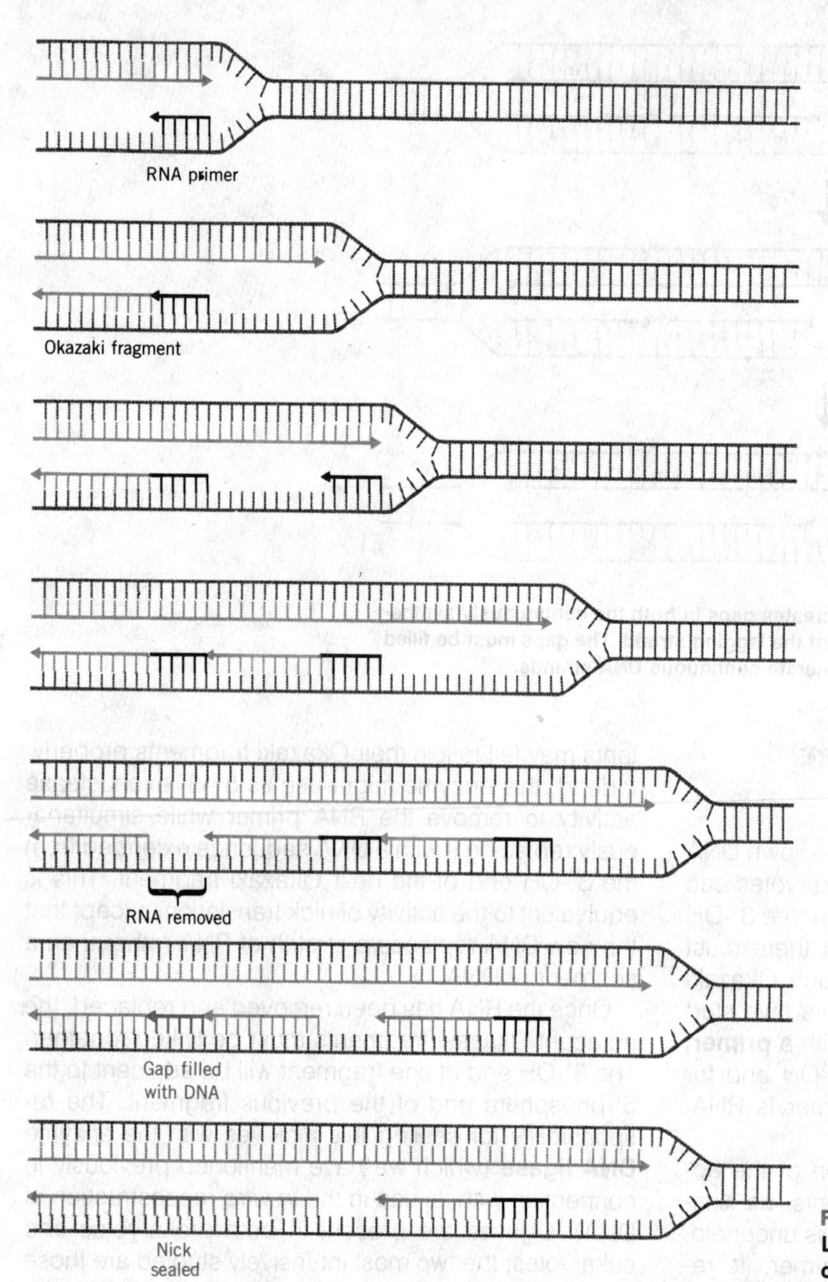

Figure 32.7
Linkage of Okazaki fragments requires removal of RNA, gap filling, and nick ligation.

complex. The AMP of the enzyme complex becomes attached to the 5′-phosphate of the nick; and then a phosphodiester bond is formed with the 3′-OH terminus of the nick, releasing the enzyme and the AMP.

The discovery of these events raises questions about the specificity of synthesis of the Okazaki fragments. What enzyme is responsible for synthesizing the RNA primer? Does it start and stop at specific sequences, or is its action determined by the organization of the replication fork? Does the synthesis of an Okazaki fragment continue until it reaches the adjacent fragment, or are there specific termination points?

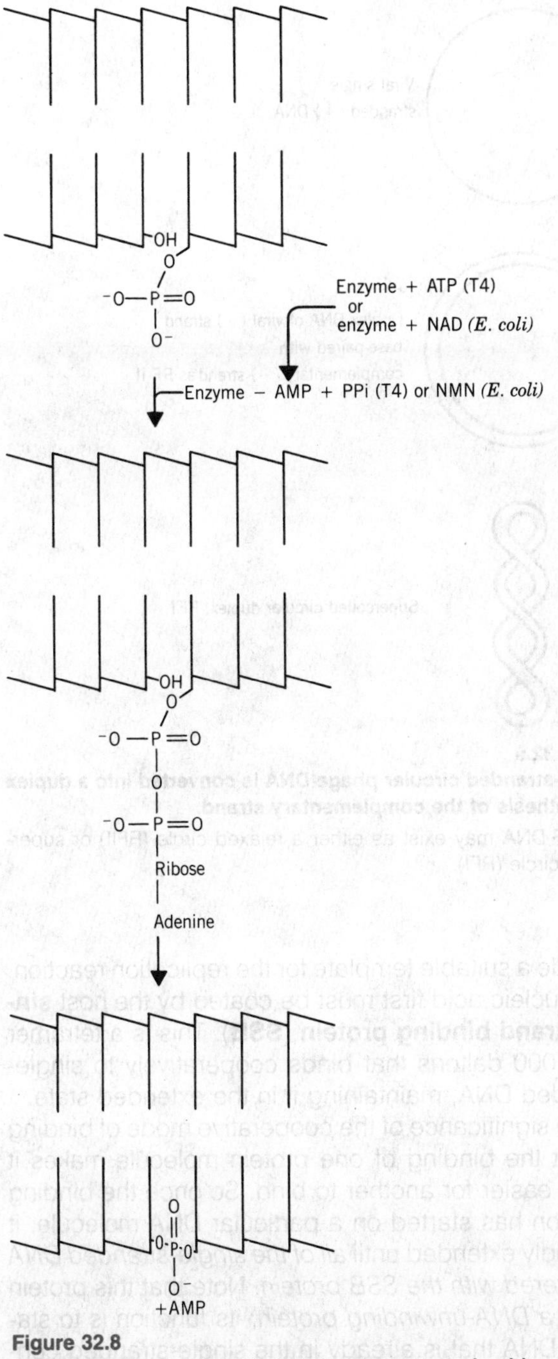

Figure 32.8
DNA ligase seals nicks between adjacent nucleotides by employing an enzyme-AMP intermediate.
The *E. coli* and T4 enzymes use different cofactors. The *E. coli* enzyme uses NAD (nicotinamide adenine dinucleotide) as a cofactor, while the T4 enzyme uses ATP.

THE COMPLEXITY OF THE BACTERIAL REPLICATION APPARATUS

The inability to replicate DNA is fatal for any cell. Mutants in replication must therefore be obtained as conditional lethals, able to accomplish replication under permissive conditions (provided by the normal temperature of incubation), and displaying their defect only under nonpermissive conditions (provided by the higher temperature of 42°C). These temperature-sensitive mutants identify a series of loci described as **dna** genes. The *dna* mutants are divided into two general classes on the basis of their behavior when the temperature is raised.

The **slow-stop mutants** complete the current round of replication, but cannot start another. They are therefore defective in the events involved in initiating a cycle of replication at the origin. Not a great deal is known about them (see Table 32.6). The main point to be made at present is that several different proteins are needed to initiate a cycle of replication.

The **quick-stop mutants** cease replication *immediately* on a temperature rise. Those concerned directly with replication are defective in the components of the replisome (see Table 32.4). Mutants assigned to the quick-stop class could also be defective in initiating cycles of replication, although this would be obscured by the quick-stop phenotype.

Some quick-stop mutants are defective in the production of the precursors needed for DNA synthesis. For example, mutations in *nrdA* were originally called *dnaF;* this gene codes for a subunit of ribonucleotide reductase, which is essential for providing precursors. Similarly, the dut^- mutations that eliminate dUTPase activity were originally called *dnaS*. These mutants illustrate the importance of the precursor supply chain. Thus the isolation of a mutation with the Dna phenotype does not imply that its gene product is directly involved in replication.

Although efforts to isolate *dna* mutants have been intensive, the mutations do not (yet) identify all the functions needed for replication. Several additional proteins, coded by unknown genes, are needed for the function of *in vitro* replication systems (see Table 32.5 later).

When extracts of *E. coli* are assayed simply for their ability to synthesize DNA, the predominant enzyme activity is that of DNA polymerase I. In fact, its activity is so great as to make it impossible to detect the activities of the enzymes actually responsible for DNA replication. To develop *in vitro* systems in which replication can be followed, extracts are therefore prepared from *polA* mutant cells.

The characterization of *in vitro* replication systems in terms of their individual components requires a template that allows the nature of the DNA synthesis to be assayed. We must be sure that synthesis can be equated with an authentic *in vivo* replication process, not an alternative activity, such as the repair of damaged DNA. Different templates can be used to establish *in vitro* systems whose characteristics reflect the synthesis of either lagging strands or leading strands.

The roles of the individual components can be distinguished by two approaches. First, the complete system can be fractionated into components that are purified and reconstituted. Second, the system can be prepared from a *dna* mutant and operated under conditions in which the mutant gene product is inactive. Then extracts from wild-type cells can be tested for their ability to restore activity. This **in vitro complementation assay** can be used to purify the protein coded by the *dna* locus.

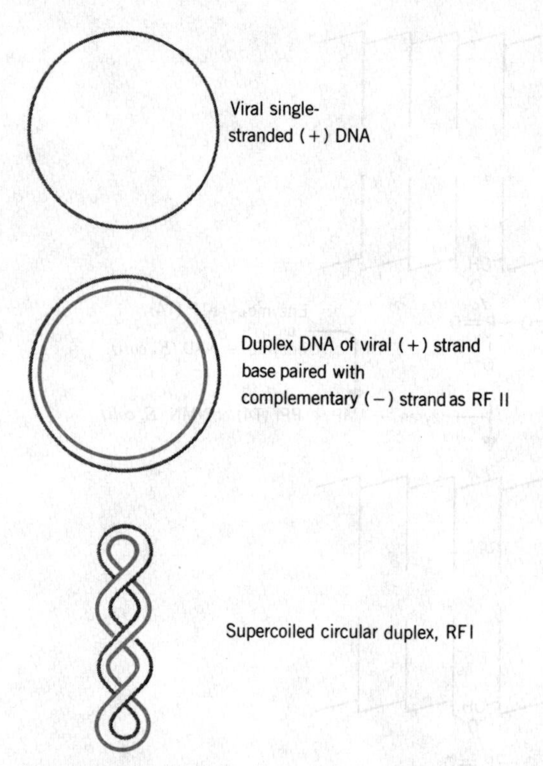

Figure 32.9
Single-stranded circular phage DNA is converted into a duplex by synthesis of the complementary strand.
The RF DNA may exist as either a relaxed circle (RFII) or supercoiled circle (RFI).

INITIATING SYNTHESIS OF A SINGLE DNA STRAND

The production of an Okazaki fragment involves synthesis of a DNA complement to the single-stranded region exposed by the movement of the replication fork. Three single-stranded DNA phages have been used as templates to provide a paradigm for the events involved in initiating and then elongating the lagging strand. Each of these phage genomes consists of a circular single strand of DNA, called the **viral** or (+) strand. **Figure 32.9** shows that synthesis of a **complementary** or (−) strand converts the genome into a duplex circular **replicative form** (**RF**) DNA. Synthesis of the complementary strand is regarded as an analog for synthesis of the lagging strand of duplex DNA.

Although each of these phage genomes shows little intrastrand base pairing and therefore remains largely in the single-stranded state, the DNA itself does not provide a suitable template for the replication reaction. The nucleic acid first must be coated by the host **single-strand binding protein** (**SSB**). This is a tetramer of 74,000 daltons that binds cooperatively to single-stranded DNA, maintaining it in the extended state.

The significance of the cooperative mode of binding is that the binding of one protein molecule makes it much easier for another to bind. So once the binding reaction has started on a particular DNA molecule, it is rapidly extended until *all of the single-stranded DNA is covered with the SSB protein.* Note that this protein is *not a DNA-unwinding protein;* its function is to stabilize DNA that is already in the single-stranded condition.

The simplest of the replication systems is presented by M13. The relevant stages are illustrated in **Figure 32.10.** An exception to the generally single-stranded

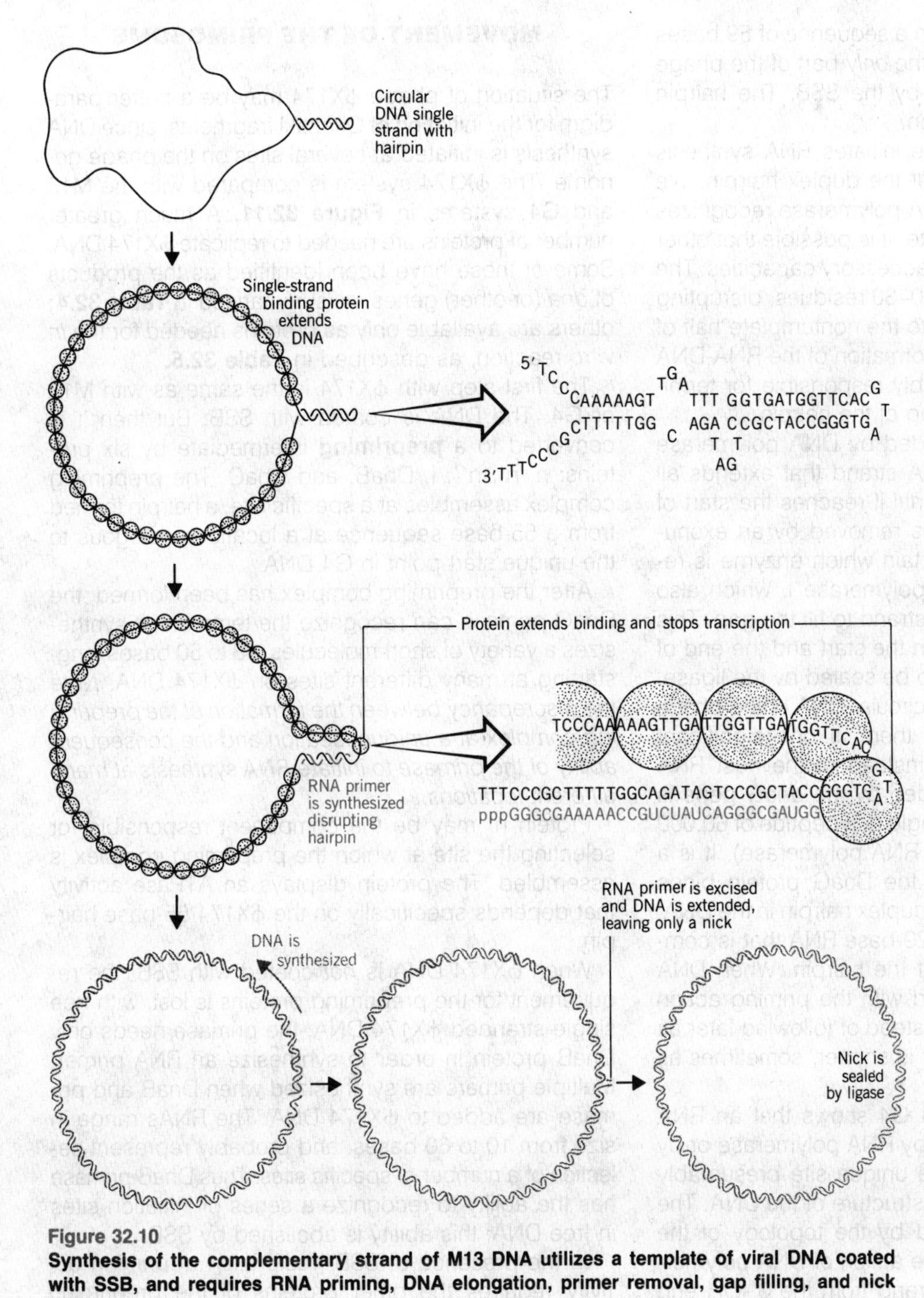

Figure 32.10
Synthesis of the complementary strand of M13 DNA utilizes a template of viral DNA coated with SSB, and requires RNA priming, DNA elongation, primer removal, gap filling, and nick sealing.

structure of the DNA exists in a sequence of 59 bases that forms a hairpin. This is the *only* part of the phage genome that is not bound by the SSB. The hairpin provides a signal for initiation.

The host RNA polymerase initiates RNA synthesis six bases before the start of the duplex hairpin. We do not yet know how the RNA polymerase recognizes the hairpin as an initiation site; it is possible that other proteins may be involved in accessory capacities. The RNA chain is extended for 20–30 residues, disrupting the hairpin. The SSB binds to the nontemplate half of the hairpin released by the formation of the RNA-DNA hybrid. This action is probably responsible for terminating transcription at the top of the hairpin.

The RNA primer is extended by DNA polymerase III, which synthesizes a DNA strand that extends all the way around the circle until it reaches the start of the RNA primer. The RNA is removed by an exonucleolytic action; it is not certain which enzyme is responsible. It may be DNA polymerase I, which also could extend the new DNA strand to fill the gap. This leaves only the nick between the start and the end of the complementary strand to be sealed by the ligase.

When the single-stranded circular DNA of the phage G4 provides the template, there is one significant difference in the reaction. Instead of the host RNA polymerase, the protein coded by the *dnaG* gene is needed. This enzyme is a single polypeptide of 60,000 daltons (much smaller than RNA polymerase). It is a **primase.** One molecule of the DnaG protein binds directly to the SSB-resistant duplex hairpin in the DNA. The primase synthesizes a 29-base RNA that is complementary to one strand of the hairpin. When DNA synthesis is directly coupled with the priming action (occurring simultaneously instead of following later as a separate step), the primer is shorter, sometimes as short as six bases.

The situation of M13 and G4 shows that an RNA primer may be synthesized by RNA polymerase or by primase, in either case at a unique site presumably indicated by the secondary structure of the DNA. The reaction may be terminated by the topology of the initiation region and/or by the action of DNA polymerase. Extension of a DNA strand from the 3′-OH end of the primer involves the same synthetic events for either phage.

MOVEMENT OF THE PRIMOSOME

The situation of phage ϕX174 may be a better paradigm for the initiation of Okazaki fragments, since DNA synthesis is initiated at several sites on the phage genome. The ϕX174 system is compared with the M13 and G4 systems in **Figure 32.11.** A much greater number of proteins are needed to replicate ϕX174 DNA. Some of these have been identified as the products of *dna* (or other) genes, as summarized in **Table 32.4;** others are available only as proteins needed for the *in vitro* reaction, as described in **Table 32.5.**

The first step with ϕX174 is the same as with M13 or G4. The DNA is coated with SSB. But then it is converted to a **prepriming** intermediate by six proteins: n, n′, n″, i, DnaB, and DnaC. The prepriming complex assembles at a specific site, a hairpin formed from a 55-base sequence at a location analogous to the unique start point in G4 DNA.

After the prepriming complex has been formed, the DnaG primase can recognize the template; it synthesizes a variety of short molecules, 15 to 50 bases long, starting at many different sites on ϕX174 DNA. *Note the discrepancy between the formation of the prepriming complex at a unique location and the consequent ability of the primase to initiate RNA synthesis at many different locations.*

Protein n′ may be the component responsible for selecting the site at which the prepriming complex is assembled. The protein displays an ATPase activity that depends specifically on the ϕX174 55-base hairpin.

When ϕX174 DNA is *not* coated with SSB, the requirement for the prepriming proteins is lost; with free single-stranded ϕX174 DNA, the primase needs only DnaB protein in order to synthesize an RNA primer. Multiple primers are synthesized when DnaB and primase are added to ϕX174 DNA. The RNAs range in size from 10 to 60 bases, and probably represent selection of a number of specific sites. Thus DnaB•primase has the ability to recognize a series of initiation sites in free DNA; this ability is abolished by SSB.

In the presence of SSB, restoration of priming activity requires the other proteins of the prepriming complex: n, n′, n″, i, and DnaC. We know that this situation reflects the *in vivo* reality, because SSB is

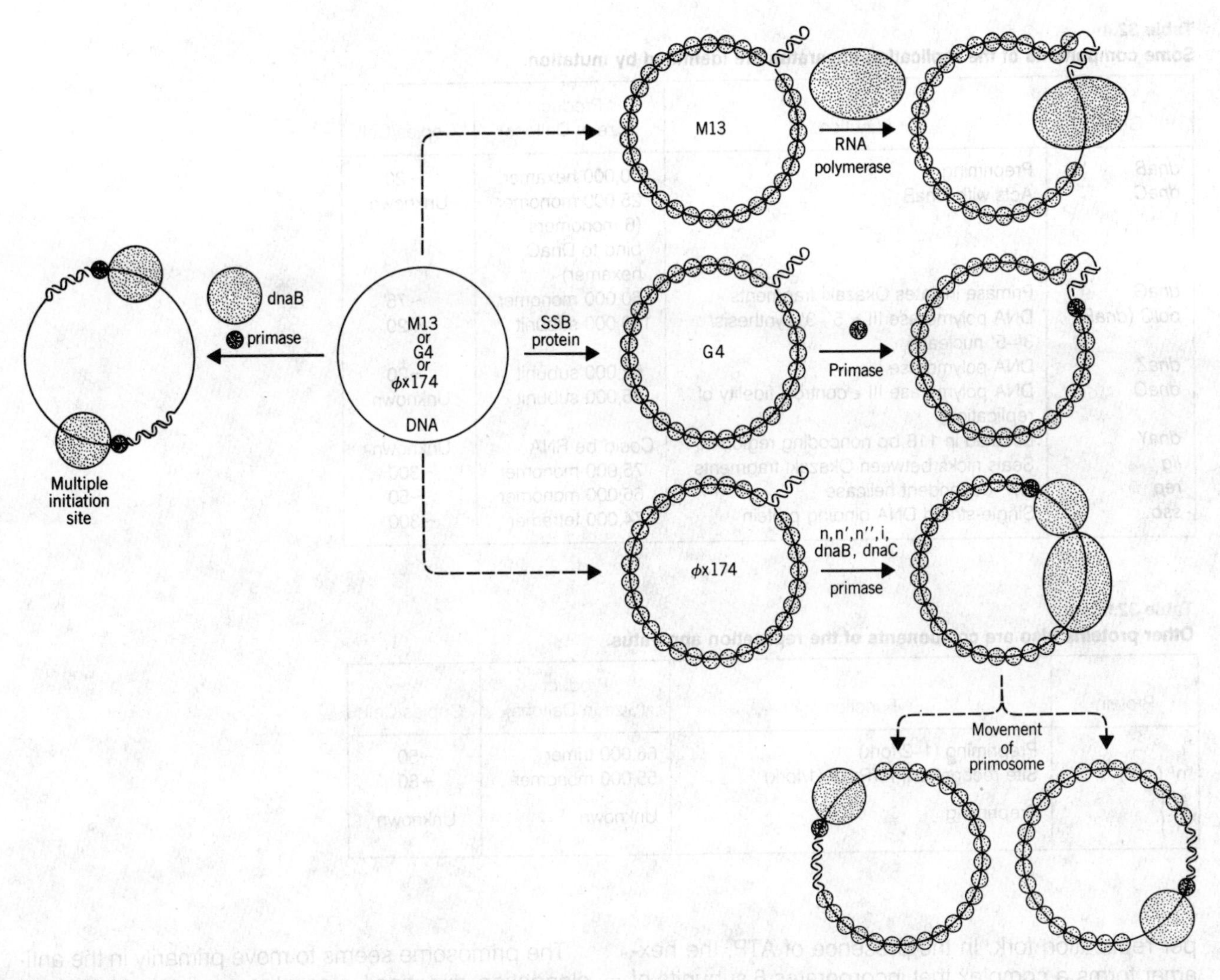

Figure 32.11
Free DNA can be primed at multiple sites by DnaB·primase. SSB-coated M13 or G4 DNA is primed at only one site. SSB-coated ϕX174 DNA requires several additional proteins to form a prepriming complex at a unique site; then priming occurs at several sites.

needed for replication in bacteria, as indicated by the quick-stop phenotype produced by mutation in its gene.

The complex of proteins involved in the priming reaction has been called the **primosome.** One interpretation of the discrepancy between the unique prepriming site and the multiple priming sites is to suppose that the primosome initially can assemble only at one location; but then some or all of the proteins can move along the single-stranded DNA to the sites at which priming actually occurs. The moving complex must include at least the DnaB and primase proteins; the n′ protein is also likely to be a component, because it has the ability to displace SSB from DNA, probably a necessary function for movement.

The behavior of DnaB also may be important in movement. There is probably one hexamer of DnaB

Table 32.4
Some components of the replication apparatus are identified by mutation.

Gene	Function	Product (Size in Daltons)	Copies/Cell
dnaB	Prepriming	300,000 hexamer	~20
dnaC	Acts with DnaB	25,000 monomer (6 monomers bind to DnaC hexamer)	Unknown
dnaG	Primase initiates Okazaki fragments	60,000 monomer	~75
polC (*dnaE*)	DNA polymerase III α 5′–3′ synthesis/ 3′–5′ nuclease	140,000 subunit	~20
dnaZ	DNA polymerase III	52,000 subunit	~20
dnaQ	DNA polymerase III ε controls fidelity of replication	25,000 subunit	Unknown
dnaY	Located in 118 bp noncoding region	Could be RNA	Unknown
lig	Seals nicks between Okazaki fragments	75,000 monomer	~300
rep	ATP-dependent helicase	66,000 monomer	~50
ssb	Single-strand DNA binding protein	74,000 tetramer	~300

Table 32.5
Other proteins also are components of the replication apparatus.

Protein	Function	Product (Size in Daltons)	Copies/Cell
i	Prepriming (1–2/fork)	66,000 trimer	~50
n′ (=Y)	Site recognition, ATPase (1/fork)	55,000 monomer	~80
n″ } n	Prepriming	Unknown	Unknown

per replication fork. In the presence of ATP, the hexamer forms a complex that incorporates 6 subunits of DnaC; this may be involved in initial binding to DNA.

DnaB binds to single-stranded DNA with high affinity in the presence of ATP. Replacement of the ATP by ADP effects an allosteric change that reduces the affinity of DnaB for the DNA. Thus the hydrolysis of ATP after the priming reaction could be necessary to allow DnaB to dissociate from DNA to move on to another priming site (where the ADP is replaced by ATP). The primase stabilizes the complex of DnaB•ATP with single-stranded DNA.

Sites in several plasmids are now known at which a primosome may form. By comparing and mutating them, we should know fairly soon what features of DNA sequence are important.

The primosome seems to move primarily in the antielongation direction—along the parental strand from 5′ toward 3′. **Figure 32.12** shows that this would be suitable for initiating the Okazaki fragments of the lagging strand of a duplex DNA. As the replication fork advances, it creates a single-stranded region ahead of the primosome (on its 3′ side). After each initiation event, the primosome moves along the single-stranded stretch to the site for starting the next Okazaki fragment. Thus the primosome moves in the same direction as the replication fork, but in the opposite direction from DNA synthesis of the lagging strand.

To bring the reaction characterized with φX174 DNA further into the context of the replication of duplex DNA, we must consider how the single-stranded template is prepared for synthesis of the lagging strand. The

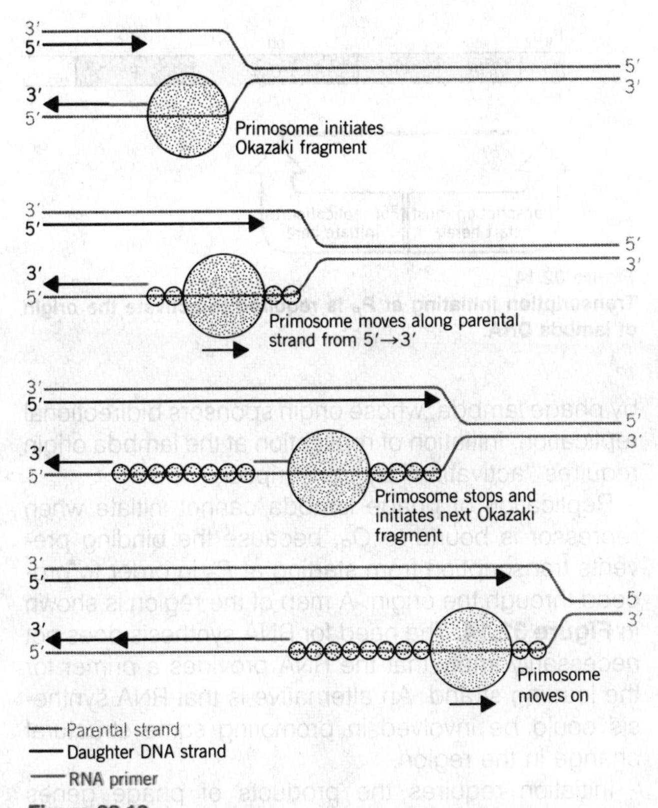

Figure 32.12
The primosome could follow the replication fork, moving 5′-3′ along the template for lagging strand synthesis, periodically initiating synthesis of Okazaki fragments.

SSB presumably binds to DNA as the replication fork advances, keeping the two parental strands separate so that they (or at least the lagging strand) are in the appropriate condition to act as template. The SSB is needed in stoichiometric amounts at the replication fork.

What is responsible for initially unwinding the two strands of the parental duplex? The probable function is identified by *rep* mutations, which slow the movement of the replication fork. The Rep protein is a single strand-dependent ATPase. In the presence of ATP, the combined action of Rep protein and SSB can separate a nicked duplex of ϕX174 RF DNA into its constituent single strands. Probably the Rep protein unwinds the strands, which are then trapped in the single-stranded state by the SSB.

A model for Rep action is illustrated in **Figure 32.13.**

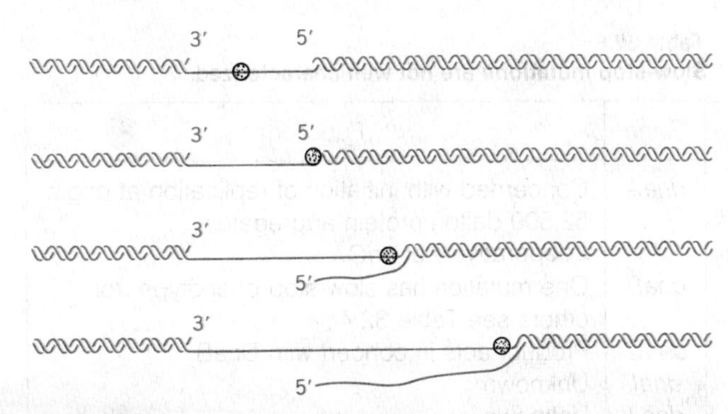

Figure 32.13
Rep protein may move along one strand of DNA, displacing the partner strand.

The protein binds to a single-stranded region within a duplex DNA, migrating along in a fixed direction. When it reaches the duplex segment, it displaces the 5′ end. Every base pair unwound requires the hydrolysis of two ATP molecules.

A problem in assigning Rep the role of unwinding at the replication fork is that *rep* mutants continue to replicate bacterial DNA, even though they cannot separate the strands of ϕX174 DNA. Survival of M13 DNA also depends on Rep; in rep^- mutant hosts, progeny (single-stranded) phage DNA cannot be generated from the circular RF DNA. However, if an *E. coli* origin is inserted into M13 DNA, replication can occur. Thus two phage DNAs require Rep for replication, but bacterial DNA can manage without it. The *E. coli* origin probably can initiate and then sustain strand separation by some other (unknown) mechanism.

INITIATING REPLICATION AT DUPLEX ORIGINS

Starting a cycle of replication of duplex DNA requires two types of activity in addition to the functions needed to synthesize Okazaki fragments. First, the two strands of DNA must suffer their initial separation. Second, synthesis of the leading strand must be initiated. This may or may not occur by the same or a similar mechanism as that involved in initiating Okazaki fragment synthesis.

The slow-stop mutations of *E. coli* identify several loci whose products may participate in the events at

Table 32.6
Slow-stop mutations are not well characterized.

Gene	Function
dnaA	Concerned with initiation of replication at origin 52,500 dalton protein aggregates cooperatively at *oriC*
dnaB	One mutation has slow-stop phenotype; for others see Table 32.4
dnaC	Product acts in concert with DnaB
dnaI	Unknown
dnaJ	Unknown
dnaK	Related to 70,000 dalton *Drosophila* heat shock protein
dnaP	Resistance to phenethyl alcohol; concerned with membrane?

the origin, but very little is known about their molecular activities. Of the loci summarized in **Table 32.6,** information is available about the functions of only three. Mutations of either the slow-stop or quick-stop phenotype may occur in *dnaB* and *dnaC,* and we have some information about the function of the proteins in the continuation of replication (see earlier). We assume that their role at the origin is related. Some mutations in *dnaA* can be suppressed by mutations in RNA polymerase, which suggests that DnaA could be involved in an initiation step that requires RNA synthesis.

The other loci are identified only by mutation, except for *dnaK,* which is one of a group of several loci whose expression is stimulated by an increase in temperature. A remarkable relationship of sequence has been noted between the DnaK protein and the major "heat shock" protein of *D. melanogaster,* a product whose production becomes predominant when insect cells are subjected to a high temperature. The basis for this relationship between prokaryotic and eukaryotic proteins is quite mysterious.

An involvement of RNA polymerase in initiation at the origin also is implied by the observation that rifampicin can inhibit the start of bacterial replication at a time when inhibitors of protein synthesis are ineffective. This means that the synthesis of RNA per se is important (rather than its translation into protein). A suitable system for studying this function is provided

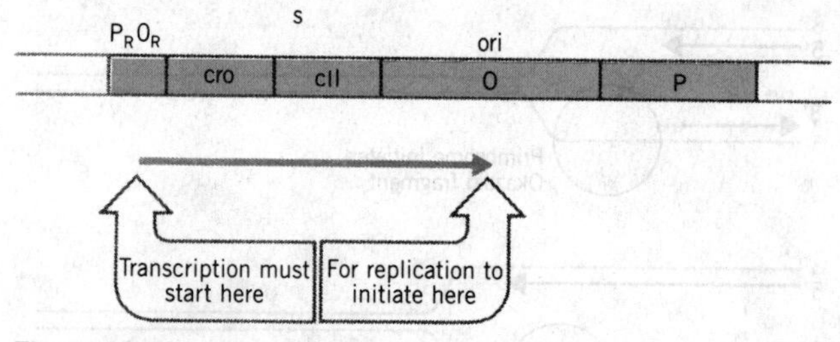

Figure 32.14
Transcription initiating at P_R is required to activate the origin of lambda DNA.

by phage lambda, whose origin sponsors bidirectional replication. Initiation of replication at the lambda origin requires "activation" by transcription.

Replication of phage lambda cannot initiate when repressor is bound to O_R, because the binding prevents transcription from starting at P_R in order to proceed through the origin. A map of the region is shown in **Figure 32.14.** The need for RNA synthesis does not necessarily imply that the RNA provides a primer for the leading strand. An alternative is that RNA synthesis could be involved in promoting some structural change in the region.

Initiation requires the products of phage genes *O* and *P,* as well as several host functions, including the products of slow-stop genes. The phage O protein binds to the lambda origin; the phage P protein interacts with the O protein and with the bacterial proteins. The origin lies within gene *O,* so the protein acts close to its site of synthesis.

Variants of the phage called λ*dv* consist of shorter genomes that carry all the information needed to replicate, but lack infective functions; λ*dv* DNA survives in the bacterium in the form of plasmids. The λ*dv* DNA can be replicated *in vitro* by a system consisting of the phage-coded proteins O and P as well as bacterial replication functions. With this system, it should become possible to analyze the role of transcriptional activation.

The availability of plasmids carrying the *E. coli oriC* sequence makes it possible to develop a cell-free system for replication from this origin. Initiation of replication of the plasmid is inhibited by rifampicin and also by drugs that prevent DNA gyrase action. Antibodies against DnaB or SSB proteins inhibit the reaction. An extract prepared from *dnaA* mutant cells is

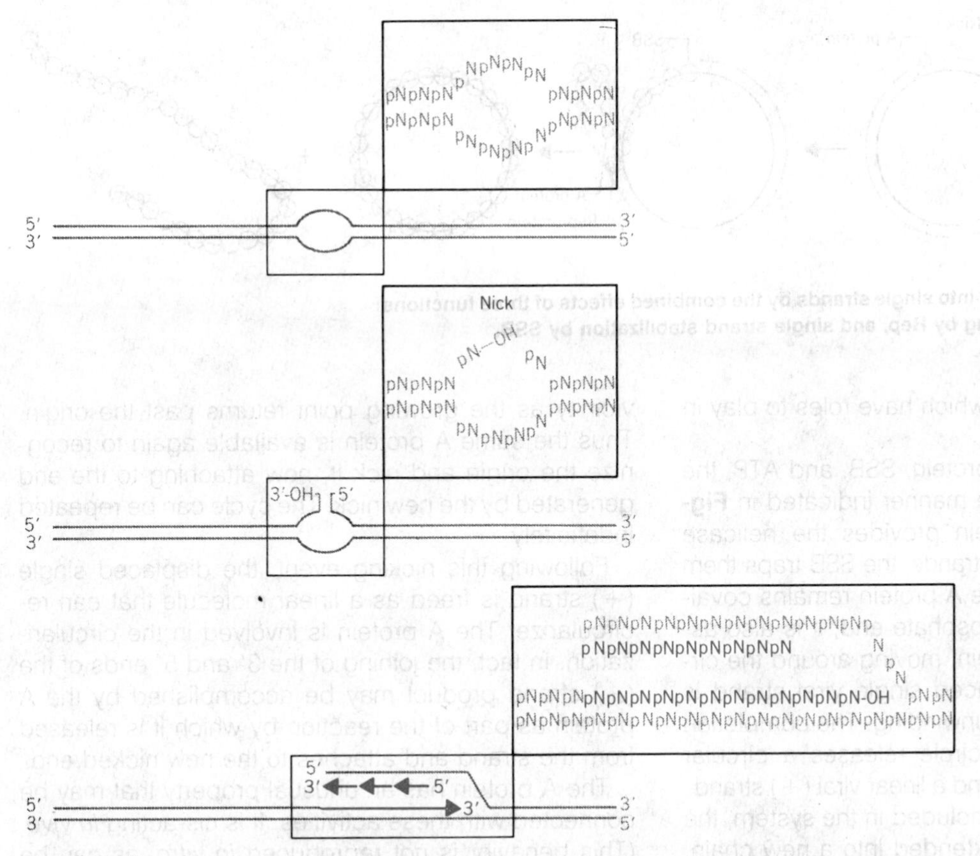

Figure 32.15
Leading strand synthesis could be initiated by nicking to generate a 3′-OH end that primes synthesis of a new DNA strand to displace the original strand.

inactive. The reaction is stimulated by protein HU, one of the histone-like proteins of *E. coli* (see Table 28.2).

One model for the initiation of leading strand synthesis relies on a primer RNA, just like the initiation of Okazaki fragments on the lagging strand, except that the priming event is followed by continuous elongation.

Another model for initiation at an origin requires that a nick is made in the parental strand, whose 3′-OH end is used as a primer, as illustrated in **Figure 32.15.** Eventually the covalent junction between the daughter strand and parental strand must be broken.

Nicking followed by extension of a parental strand is a well-established mechanism in the replication of some DNA phages, which proceed through the rolling circle illustrated previously in Figure 31.11. This has been investigated with the RF DNA of ϕX174. Replication via the rolling circle requires fewer enzymatic functions than are involved in producing RF DNA from the original viral single strand. Supercoiled circles of ϕX174 RF DNA can be replicated *in vitro* by a system containing the product of phage gene *A*, the host Rep protein, SSB, and DNA polymerase III.

The A protein nicks the viral (+) strand of the duplex DNA at a specific site that defines the origin for RF replication. The DNA can be nicked *only when it is supercoiled.* The A protein is able to bind to a single-stranded decamer fragment of DNA that surrounds the site of the nick. This suggests that the supercoiling may be needed to assist the formation of a single-stranded region that provides the A protein with its binding site. The nick generates a 3′-OH end and a

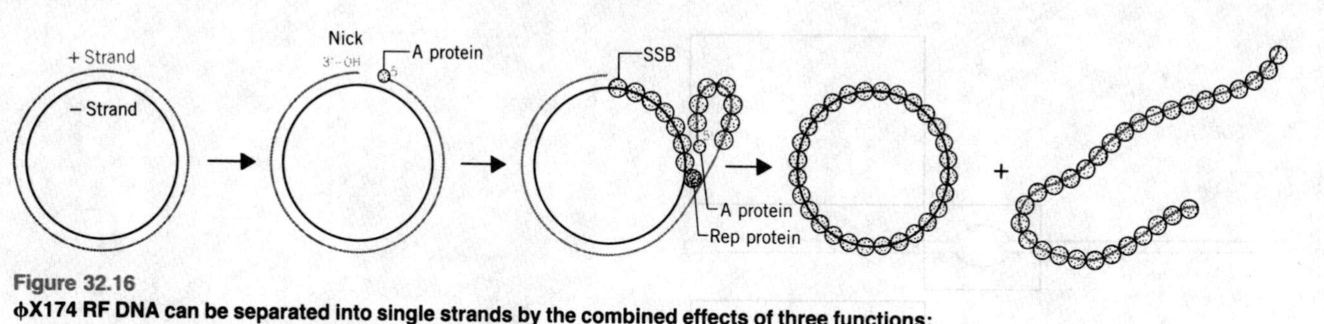

Figure 32.16
ϕX174 RF DNA can be separated into single strands by the combined effects of three functions: nicking with A protein, unwinding by Rep, and single strand stabilization by SSB.

5′-phosphate end, both of which have roles to play in ϕX174 replication.

In the presence of Rep protein, SSB, and ATP, the nicked DNA unwinds in the manner indicated in **Figure 32.16.** The Rep protein provides the helicase function that separates the strands; the SSB traps them in single-stranded form. The A protein remains covalently attached to the 5′-phosphate end; it is also associated with the Rep protein, moving around the circle with it. Thus the displaced single viral strand is looped out from the site of unwinding. The completion of movement around the circle releases a circular complementary (–) strand and a linear viral (+) strand.

If DNA polymerase III is included in the system, the 3′-OH end of the nick is extended into a new chain. The chain is elongated around the circular (–) strand template, until it reaches the starting point and displaces the origin, as indicated in **Figure 32.17.** Since the A protein has remained associated with Rep protein during the traverse of the circle, it is now in the vicinity as the growing point returns past the origin. Thus the same A protein is available again to recognize the origin and nick it, now attaching to the end generated by the new nick. The cycle can be repeated indefinitely.

Following this nicking event, the displaced single (+) strand is freed as a linear molecule that can recircularize. The A protein is involved in the circularization. In fact, the joining of the 3′ and 5′ ends of the (+) strand product may be accomplished by the A protein as part of the reaction by which it is released from the strand and attaches to the new nicked end.

The A protein has an unusual property that may be connected with these activities. It is *cis*-acting *in vivo.* (This behavior is not reproduced *in vitro,* as can be seen from its utilization by any DNA template when it is added to a cell-free system.) The implication is that *in vivo* the A protein synthesized by a particular genome can attach only to the DNA of that genome. We do not know how this is accomplished. However, its

Figure 32.17
ϕX174 RF DNA can be used as a template for synthesizing single-stranded viral circles. The A protein remains attached to the same genome through indefinite revolutions, each time nicking the origin on the viral (+) strand and transferring to the new 5′ end. At the same time, the released viral strand is circularized.

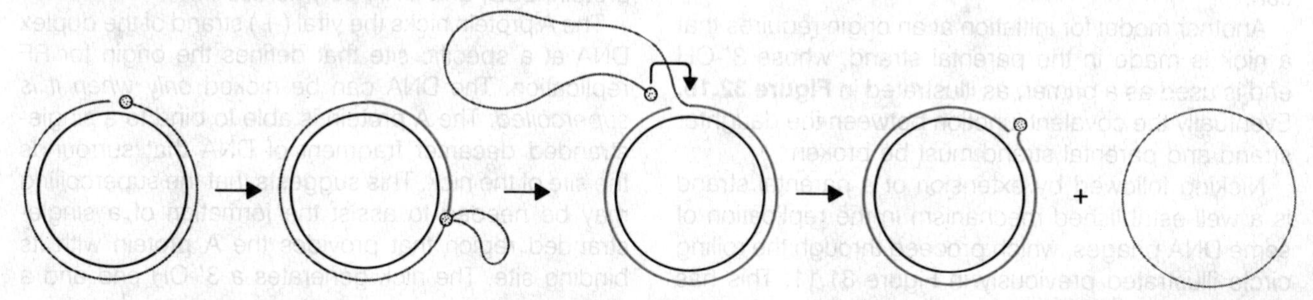

activity *in vitro* shows how it may remain associated with the same parental (−) strand template.

For phage morphogenesis, a modification is made in the cycle shown in Figure 32.17. Instead of the displaced (+) strand being covered with SSB (and reentering the replication cycle), it enters the phage virion.

THE REPLICATION APPARATUS OF PHAGE T4

When phage T4 takes over an *E. coli* cell, it provides a large number of functions of its own that either replace or augment the host functions. Because the host DNA is degraded, the phage places little reliance on expression of host functions, except for utilizing enzymes that are present in large amounts in the bacterium. The degradation of host DNA is important in releasing nucleotides that are reused in the synthesis of phage DNA. (The phage DNA differs in base composition from cellular DNA in using hydroxymethylcytosine instead of the customary cytosine.)

The phage-coded functions concerned with DNA synthesis in the infected cell can be identified by mutations that prevent or impede the production of mature phages. Essential phage functions are identified by conditional lethal mutations, which fall into three phenotypic classes. Those in which there is *no DNA synthesis* at all identify genes whose products either are components of the replication apparatus or are involved in the provision of precursors (especially the hydroxymethylcytosine). Those in which the onset of DNA synthesis is *delayed* are concerned with the initiation of replication. Those in which DNA synthesis starts but then is *arrested* include regulatory functions, the DNA ligase, and some of the enzymes concerned with host DNA degradation. There are also nonessential genes concerned with replication; for example, including those involved in glucosylating the hydroxymethylcytosine in the DNA.

Synthesis of T4 DNA is catalyzed by a multienzyme aggregate assembled from the products of seven essential genes. Each of the components has been purified by an *in vitro* complementation assay. The system is active with several DNA templates *in vitro*, it proceeds at a velocity close to the natural rate, and its frequency of errors is about 10^{-5} to 10^{-6} (compared with an *in vivo* error rate of $\sim 2 \times 10^{-8}$). *In vitro*, CTP is used as efficiently as hydroxymethyl-CTP.

The proteins of the T4 replication complex are described in **Table 32.7.**

The gene *32* protein is a highly cooperative single-strand binding protein, needed in stoichiometric amounts. In fact, it was the first example of its type to be characterized. The geometry of the T4 replication fork may specifically require the phage-coded protein, since the *E. coli* SSB cannot substitute. The gene *32* protein forms a complex with the T4 DNA polymerase; this interaction could be important in constructing the replication fork.

The T4 system depends on RNA priming. The nature of the primer, however, is different from that pro-

Table 32.7
The T4 replication apparatus consists of defined components.

Gene	Function	Product (Size in Daltons)		Relative Number
43	DNA polymerase	110,000		1
32	Single-strand binding protein	35,000		150
44	ATPase, increases rate 3–4 fold	4 x 34,000	180,000	5
62		2 x 20,000		
41	Required for RNA priming single-strand dependent GTPase helicase	58,000		20
61	Required for RNA priming			1
45	Assists DNA polymerase	2 x 27,000		10

vided in *E. coli*. With single-stranded T4 DNA as template, the gene *41* and *61* products act together to synthesize short primers. These are all pentaribonucleotides with the general sequence pppApCpNpNpNp. Thus the dinucleotide starting sequence is specific, and the length is limited to five bases, but the last three positions can be occupied by any of the bases.

If the complete replication apparatus is present, these primers are extended into DNA chains. The primers may be formed at specific sites, although these sites are likely to occur more frequently than the average 2000-base spacing between the starts of Okazaki fragments. Possibly the replication fork passes a series of potential primer sites, only some of which are used, as dictated by the distance moved since the last initiation event.

The gene *41* protein is a helicase (able to unwind a double helix). And it has a GTPase activity stimulated by single-stranded DNA. When the protein binds to single-stranded DNA, it may be able to move from its initial binding site, migrating at a rate of about 400 nucleotides per second. Its role could be analogous to the host DnaB protein, finding periodic sites at which to initiate primer synthesis. The hydrolysis of GTP is presumed to provide the energy for movement.

The gene *61* protein is needed in much smaller amounts than most of the T4 replication proteins. This has impeded its characterization. (It is required in such small amounts that originally it was missed as a necessary component, because enough was present as a contaminant of the gene *32* protein preparation.) It is possible that the gene *61* protein is the primase, analogous to DnaG of *E. coli*. There may be as few as 10 copies of *61* protein per cell.

The gene *43* DNA polymerase has the usual 5′–3′ synthetic activity associated with a 3′–5′ exonuclease proofreading activity. The remaining three proteins are referred to as "polymerase accessory proteins." The gene *45* product is a dimer. The products of genes *44* and *62* form a tight complex, which has ATPase activity and increases the rate of movement of the DNA polymerase by 3–4-fold, from 250 nucleotides/second to about 800 nucleotides/second, close to the *in vivo* rate of about 1000 nucleotides/second. A similar increase is seen whether the template is single-stranded or double-stranded.

When T4 DNA polymerase uses a single-stranded DNA as template, its rate of progress is uneven. The enzyme moves rapidly through single-stranded regions, but proceeds much more slowly through regions that have a base-paired intrastrand secondary structure. The accessory proteins assist the DNA polymerase in passing these roadblocks. Thus their function in increasing the rate of replication may be to keep the DNA polymerase up to the same speed in "difficult" regions.

The presence of the proteins increases the affinity of the DNA polymerase for the DNA, and also its processivity—its ability to stay on the same template molecule without dissociating. Possibly the proteins act as a "clamp," holding the DNA polymerase subunit more tightly on the template. The combined action of all three proteins is needed for this effect. As we have mentioned before, this type of intimate relationship makes it a moot point what is a component of DNA polymerase and what is an accessory factor.

We have dealt with DNA replication so far solely in terms of the progression of the replication fork. The need for other functions is shown by the DNA-delay and DNA-arrest mutants. The four genes of the DNA-delay mutants are *39*, *52*, *58*, and *60*. Using the *in vitro* complementation assay, the gene *39* and *52* products have been shown to be associated with another protein (of unknown origin) in a complex of three subunits (of 51,000, 64,000, and 110,000 daltons). This constitutes the T4 topoisomerase II, an activity needed for removing supercoils in the template (see Chapter 34). The essential role of this enzyme suggests that T4 DNA does not remain in a linear form, but lacks free ends. The topoisomerase could be needed to unwind the DNA ahead of the replication fork.

THE REPLICATION APPARATUS OF PHAGE T7

As further replication systems are characterized, we see that the same types of activity are present in each, although they may be provided in different ways. Phage T7 has a replication apparatus whose activities are analogous to those of *E. coli* or phage T4, although apparently less complex.

The phage codes for one subunit of the DNA

polymerase. A subunit of the host protein thioredoxin is sequestered to provide the second subunit of the T7 DNA polymerase. The host protein is a 12,000-dalton polypeptide that is used as a redox coenzyme in ribonucleotide reduction. Its function as a partner for T7 DNA polymerase is a mystery, but presumably represents a different activity from its role in the host bacterium.

The T7 DNA polymerase employs an auxiliary protein, the product of phage gene *4*. It combines the roles of unwinding the duplex template and synthesizing the RNA primer. The DNA polymerase and gene *4* protein together can accomplish replication *in vitro*, although *in vivo* the exonuclease coded by phage gene *6* may also be involved, in removing the RNA primer.

The replication fork probably contains ten molecules of gene *4* protein with one DNA polymerase per growing strand. The gene *4* product hydrolyzes NTPs over and above the extent of their use as substrates for RNA synthesis. Up to four additional hydrolytic events may occur for every base incorporated into the DNA. Probably this hydrolysis provides the energy for strand separation.

The gene *4* primase activity synthesizes tetraribonucleotide primers with the almost fixed sequence $pppApCpCp^{C}_{A}p$. The sequence recognized for the priming reaction includes a preceding G residue. The primer sequence should occur in T7 DNA about once every 500 bases, compared with the Okazaki fragment length of 1000 to 6000 bases. So probably only some primer sites are used. The point of switching from the primer to DNA synthesis is not absolutely specified, because some primers extend into pentaribonucleotides.

THE PROBLEM OF LINEAR REPLICONS

The ability of all known nucleic acid polymerases, DNA or RNA, to proceed only in the 5′–3′ direction poses a problem for synthesizing DNA at the end of a linear replicon. Consider the two parental strands depicted in **Figure 32.18.** The lower strand presents no problem: it can act as template to synthesize a daughter strand that runs right up to the end, where presumably the polymerase falls off. But to synthesize a comple-

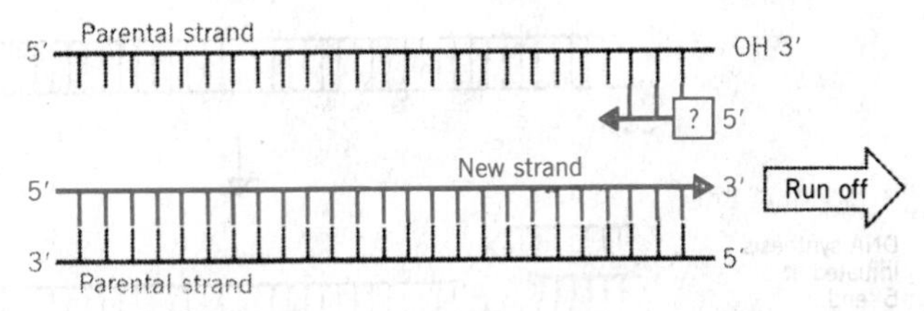

Figure 32.18
Replication could run off the 3′ end of a newly synthesized linear strand, but could it initiate a 5′ end?

ment at the end of the upper strand, a priming event must occur right at the very last base. (This applies whether it is a lagging strand or a leading strand.)

We do not know whether terminal priming is feasible. We usually think of a polymerase as binding at a site *surrounding* the position at which a base is to be incorporated. Four types of solution may be imagined to accommodate the need to copy a terminus.

The problem may be circumvented by converting a linear replicon into a circular or concatemeric mole-

Figure 32.19
The 5′ terminal phosphate at each end of adenovirus DNA is covalently linked to serine in the 55,000-dalton Ad-binding protein.

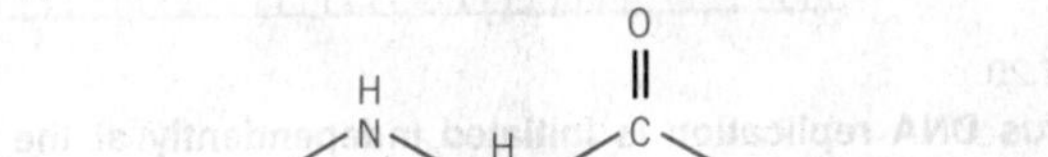

Polypeptide chain
Polypeptide chain
Cytosine
Adenovirus DNA

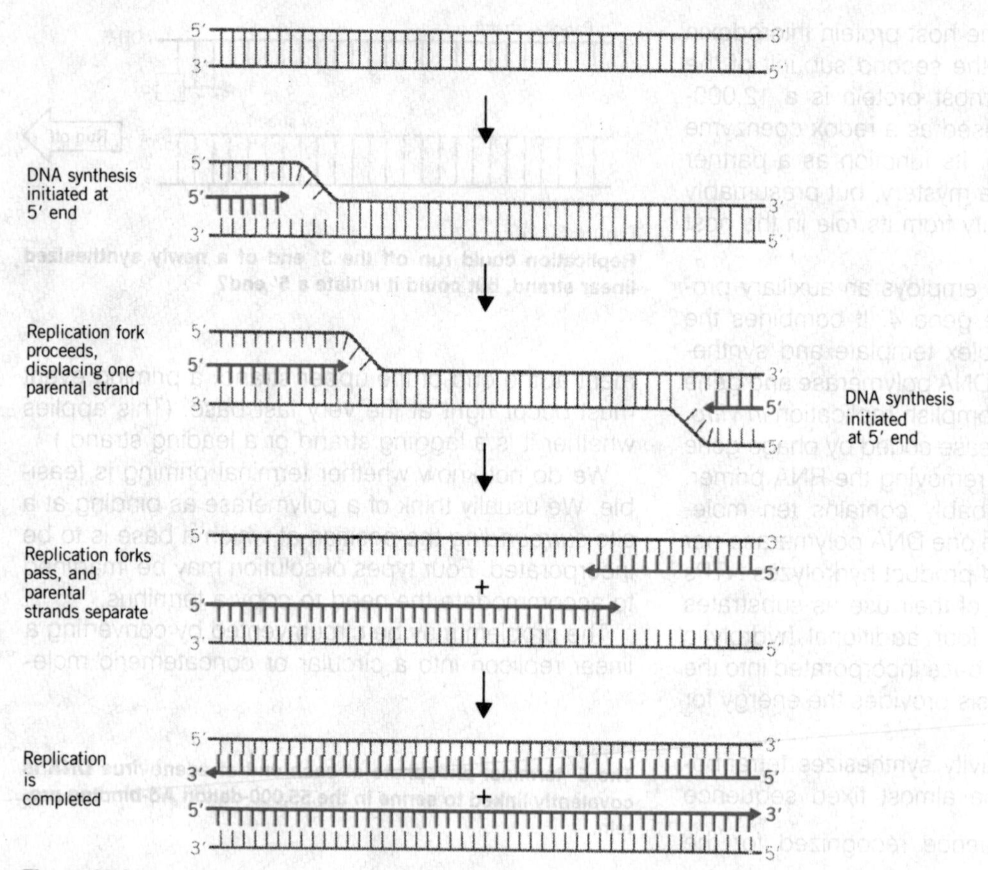

Figure 32.20
Adenovirus DNA replication is initiated independently at the two ends of the molecule and proceeds by strand displacement.

cule. Phages such as T4 or lambda provide obvious examples.

The DNA may form an unusual structure—for example, by creating a hairpin at the terminus, so that there is in fact no free end. For example, formation of a cross-link is involved in replication of the linear mitochondrial DNA of *Paramecium*.

Instead of being precisely determined, the end may be variable. The telomeres of chromosomes of lower eukaryotes may adopt this solution, in which the number of copies of a short repeating unit may change (see Chapter 28). If there is a mechanism to add or remove units, it becomes unnecessary to replicate right up to the very end.

The most direct solution is for a protein to intervene to make a priming reaction possible. Several linear viral nucleic acids have proteins that are *covalently linked to the 5′ terminal base*. The best characterized examples are adenovirus DNA, phage φ29 DNA, and poliovirus RNA.

Adenovirus DNA is a large linear duplex molecule; both 5′ ends are covalently attached to a 55,000-dalton protein. The linkage involves a phosphodiester bond to serine, as indicated in **Figure 32.19.** The same type of arrangement is found in φ29, where a 27,000-dalton protein is affixed to each 5′ end. In the single-stranded RNA poliovirus, the VPg protein of only 22 amino acids is linked via the hydroxyl group of tyrosine to the 5′ terminal base. In each case, the attached protein is coded by the virus and is implicated in replication.

The DNA of adenovirus or φ29 replicates from its ends by the mechanism of **strand displacement** il-

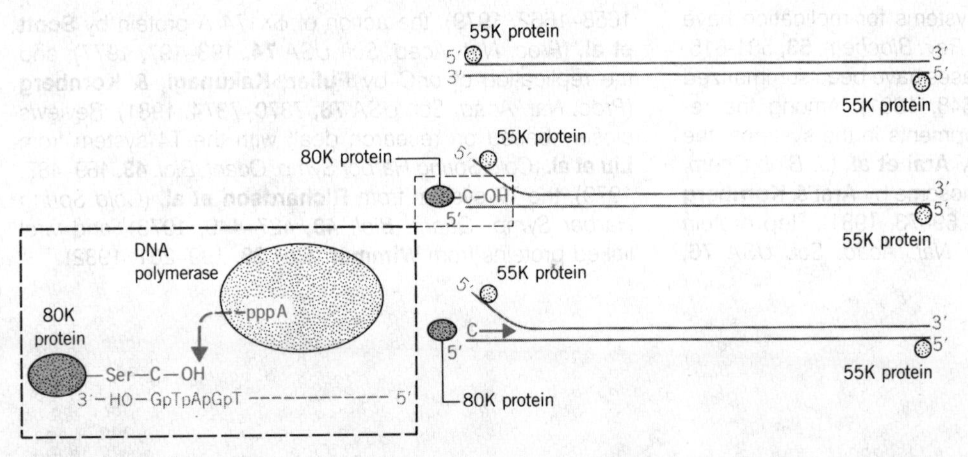

Figure 32.21
The 80,000-dalton adenovirus protein displaces the 5′ end of DNA (which is bound to the 55,000 dalton protein), and provides CTP to prime synthesis of a new DNA strand.

lustrated in **Figure 32.20.** The same events occur independently—that is, not simultaneously—at either end. Synthesis of a new strand starts, displacing the homologous strand that was previously paired in the duplex. Eventually, either the replication fork reaches the other end of the molecule, or two replication forks proceeding from opposite ends meet somewhere in the middle. Either event releases the parental strands from each other.

The protein present on replicating strands of adenovirus DNA is 80,000 daltons, larger than the protein found on mature DNA isolated from the virion. The larger protein is related to the smaller species. This suggests that the 80,000-dalton protein is involved in initiating replication, but, at some point during the maturation of the virus, is cleaved to the smaller size.

How could the attachment of a protein overcome the priming problem? The role of the 80,000-dalton adenovirus protein is suggested by its ability to become covalently linked to dCTP in extracts of adenovirus-infected cells. It is intimately connected with another protein, a 140,000 dalton DNA polymerase. This suggests the model illustrated in **Figure 32.21.**

The 80,000 dalton protein forms a complex with the DNA polymerase. The complex of polymerase and terminal protein, bearing a cytidine that has a free 3′-OH group, approaches the end of the adenovirus DNA. It displaces the current 5′ end, which may be recognized by its possession of the 55,000-dalton version of the protein. Then the free 3′-OH end is proffered to an incoming nucleotide, which pairs with the template strand under direction from DNA polymerase. A similar model probably applies to ϕ29 replication.

The entire series of reactions may be coordinated, because the 80,000-dalton protein can become linked to dCTP only in the presence of the polymerase and adenovirus DNA. This suggests that the linkage reaction may be catalyzed in a similar way to the synthesis of nucleotide-nucleotide bonds.

The RNA polymerase responsible for replicating poliovirus is strictly primer-dependent, like DNA polymerases. Here again, a larger precursor to the covalently linked VPg protein has been found. It is 12,000 daltons in size, bound to the membrane; the precursor could undertake the priming reaction and then be cleaved to the vestigial VPg. The VPg protein is removed from poliovirus RNA that is destined to be used as mRNA and is found only on those genomes intended for replication.

FURTHER READING

An extensive account of the replication apparatus has been given by **Kornberg's** *DNA Replication* (Freeman & Co., San

Francisco, 1980). Prokaryotic systems for replication have been surveyed by **Nossal** (*Ann. Rev. Biochem.* **53,** 581-615, 1983). Eukaryotic DNA polymerases have been summarized by **Huberman** (*Cell* **23,** 647–648, 1981). Among the research papers reporting developments in the systems, the *E. coli* system was dealt with by **Arai et al.** (*J. Biol. Chem.* **256,** 5239–5246, 1981), the primosome by **Arai & Kornberg** (*Proc. Nat. Acad. Sci. USA* **78,** 69–73, 1981), Rep protein by **Yarranton & Gefter** (*Proc. Nat. Acad. Sci. USA* **76,** 1658–1662, 1979), the action of ϕX174 A protein by **Scott et al.** (*Proc. Nat. Acad. Sci. USA* **74,** 193–197, 1977), and the replication of *oriC* by **Fuller, Kakunagi, & Kornberg** (*Proc. Nat. Acad. Sci. USA* **78,** 7370–7374, 1981). Reviews closely based on research dealt with the T4 system from **Liu et al.** (*Cold Spring Harbor Symp. Quant. Biol.* **43,** 469–487, 1979), the T7 system from **Richardson et al.** (*Cold Spring Harbor Symp. Quant. Biol.* **43,** 427–440, 1978), and viral linked proteins from **Wimmer** (*Cell* **28,** 199–201, 1982).

CHAPTER 33
SYSTEMS THAT SAFEGUARD DNA

DNA fulfills its hereditary functions via replication and transcription. Each process involves several enzymatic or regulatory activities in addition to the appropriate polymerase. In addition to these systems, a variety of enzymes interact with DNA to modify its structure or repair damage that has occurred to it. These activities are important in understanding how DNA is perpetuated through indefinite numbers of generations; the act of replication itself is insufficient to safeguard its role in evolution.

Although the sequence of base pairs in DNA is sacrosanct, some of the bases may be chemically modified after their incorporation into DNA. Prokaryotes and eukaryotes both contain enzymes that methylate DNA, although the function of the methylation appears different.

The particular enzyme(s) present in a bacterial strain determine the pattern of its **modification** of DNA; viewed in reverse, the pattern identifies the origin of the DNA. Thus the bacterium can distinguish between its own DNA and any invading "foreign" DNA, which lacks the characteristic modification pattern of the host. This difference renders the foreign DNA susceptible to attack by **restriction enzymes** that recognize the absence of methyl groups at the appropriate sites.

Such **restriction and modification** systems are widespread in bacteria; in the context of safeguarding DNA, their object is xenophobic—to protect the resident DNA against contamination by sequences of foreign origin (although their presence is not obligatory; some bacterial strains lack any restriction system).

Methylation in eukaryotes has another purpose: distinguishing genes in different functional conditions as described in Chapter 30; this is not associated with restriction of unmodified sites.

A bacterium contains several systems that protect its DNA against the consequences of damage by external agents or faults in replication. Any event that causes the structure of DNA to deviate irreversibly from its regular double helix is recognized as inadmissible. The change can be a point mutation that converts one base into another, thus creating a pair of bases not related by the Watson-Crick rules. Or it can be a structural change that adds a bulky adduct to DNA or links two bases together.

The sites of mutation are recognized by special nucleases that excise the damaged region from DNA; then further enzymes synthesize a replacement sequence. Together, these activities form a **repair system.** As well as the direct repair of damage by excision and replacement, there are systems to cope with the adverse consequences of replicating damaged DNA. These **retrieval systems** are related to those involved in genetic recombination.

The various systems include enzymes with some remarkable abilities to recognize sequences or structures in DNA. Restriction enzymes bind to DNA at specific target sequences, offering another perspective on the nature of protein-nucleic acid interactions. Some of these enzymes bind to DNA at one site, but then cleave at another site far away, offering insights into the nature of the ability of proteins to move along duplex DNA. Some repair enzymes recognize damaged sites in DNA apparently by the distortion in the regular structure of the double helix. An enzyme involved in recombination can bind two molecules of DNA to promote base pairing between them.

From these various activities, we gain two types of information. We learn about the interactions of proteins with DNA in a sequence- or structure-specific manner, an issue at the crux of the biology of the gene. And we begin to obtain a general view of the multifarious nature of the systems needed to preserve and protect DNA against insults from the environment or the errors of other cellular systems.

THE OPERATION OF RESTRICTION AND MODIFICATION

Bacterial restriction endonucleases all recognize specific, rather short sequences of DNA as binding sites. Binding is followed by cleavage, at the recognition site itself or elsewhere, depending on the enzyme. We have already discussed the cleavage reaction in some detail from the perspective of its use in mapping and reconstructing DNA *in vitro* (see Chapters 5 and 17). Now we must ask how the natural function of these enzymes is exercised.

In addition to its restriction activity (or activities), a

Figure 33.1
Methylated sites are perpetuated indefinitely and are safe from restriction; unmethylated sites are cleaved.

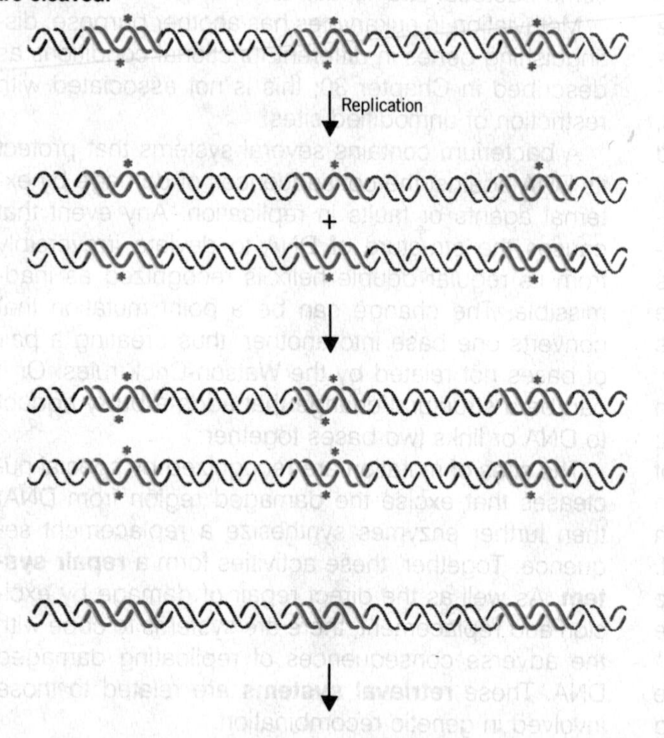

bacterial strain possesses a DNA methylase activity with the *same sequence specificity* as the restriction activity. The methylase adds methyl groups (to adenine or cytosine residues) in the same target sequence that constitutes the restriction enzyme binding site. The methylation renders the target site resistant to restriction. Thus methylation protects the DNA against cleavage.

In the bacterium, the resident DNA is methylated at the appropriate target sites, and so is immune from attack by the restriction activity, as indicated in **Figure 33.1.** This is clearly necessary to prevent the restriction enzyme from degrading the DNA of the cell in which it resides. But the protection does not extend to any foreign DNA that gains entry to the cell. Such DNA has unmodified target sites, and therefore is attacked by the restriction enzyme. The combination of modification and restriction allows the cell to distinguish foreign DNA from its own, protected sequences.

In this context, "foreign" DNA means any DNA derived from a bacterial strain *lacking the same modification and restriction activity*. There is no distinction among *types* of DNA, that is, between bacterial, plasmid, or phage sequences. The same modification pattern is possessed by *all* DNA that is resident in, or has passed through, a particular strain of bacteria. The restriction and modification system offers no hindrance to the passage of DNA between bacteria of the same strain.

Indeed, these systems were originally discovered by their effects on infecting phage DNAs. The phage DNA released from a bacterium of one strain can successfully prosecute an infection in another bacterium of the same strain, because it has the same modification pattern as the host DNA. But phage DNA that proceeds from one bacterial strain to another is attacked by the restriction activity. The phage is "restricted" to one strain of bacteria, hence the origin of the term.

The restriction is not absolute. Some infecting phages escape restriction, because of mutations in the target sites or negligence by the host system. In the latter case, they acquire the modification type of the new host.

Plasmids may contribute to the strain specificity of bacteria. Some plasmids (and phages) possess genes for restriction and modification systems, so a bacterium carrying one of these elements is defined as belonging to a different strain from a bacterium that lacks the element. At all events, the existence of these systems serves to counter the mobility of bacterial sequences conferred by the passage of plasmids and phages between bacteria.

The restriction enzymes fall into two general classes, which can be further subdivided. Their properties are summarized in **Table 33.1.**

The first class consists of the **type II** activities, which include the restriction enzymes we have discussed previously. A type II system is found in about one in three bacterial strains. Each of these enzymes is re-

Table 33.1
Restriction and methylation activities may be associated or may be separate.

Class of Enzyme	Type II	Type III	Type I
Protein Structure	Separate endonuclease and methylase	Bifunctional enzyme of 2 subunits	Bifunctional enzyme of 3 subunits
Recognition Site	Short sequence (4–6 bp), often palindromic	Asymmetrical sequence of 5–7 bp	Bipartite and asymmetrical (e.g., $TGAN_8TGCT$)
Cleavage Site	Same as or close to recognition site	24–26 bp downstream of recognition site	Nonspecific >1000 bp from recognition site
Restriction & methylation	Separate reactions	Simultaneous	Mutually exclusive
ATP needed for restriction?	No	Yes	Yes

sponsible only for the act of restriction; a separate enzyme is responsible for methylating the same target sequence.

The protein structures tend to be uncomplicated. The best characterized examples are the components of the EcoRI system, where the restriction enzyme is a dimer of identical subunits and the methylase is a monomer. (The recognition of a common target could have arisen either by duplication of the part of the protein that recognizes DNA or by convergent evolution.)

The target sites are often palindromes of 4–6 bp. Because of the symmetry, the bases to be methylated occur on both strands of DNA. Thus a target site may be *fully methylated* (both strands are modified), *hemimethylated* (only one strand is methylated), or *nonmethylated.*

A fully methylated site is a target for neither restriction nor modification.

A hemimethylated site is not recognized by the restriction enzyme, but may be converted by the methylase into the fully modified condition.

In the bacterium, most methylation events are concerned with perpetuating the current state of modification. Replication of fully methylated DNA produces hemimethylated DNA (see Figure 33.1). Recognition of the hemimethylated sites is likely to be the usual mode of action of the methylase *in vivo.*

A nonmethylated target site may be a substrate for either restriction or modification *in vitro.* In the cell, unmodified DNA is more likely to be restricted.

The methylase adds only one methyl group at a time. With a nonmethylated substrate, a single methylation is followed by dissociation from DNA. A separate binding and methylation event occurs to introduce the second group.

Most of the type II restriction enzymes cleave the DNA *at* an unmethylated target site. One bond is cleaved in each strand of DNA; the cleavages may occur sequentially. Some enzymes introduce staggered cuts, others generate blunt ends (see Chapter 17). A subclass consists of enzymes that cleave the DNA a few bases to one side of the recognition site.

The second class of restriction and modification activities consists of the type I and type III enzymes, both of which comprise multimers that undertake *both* the endonuclease and methylation functions. Their mechanisms of action are somewhat different from each other and from the type II activities. It is uncertain what proportion of bacteria have type I or type III systems, but they are less common than type II.

THE ALTERNATE ACTIVITIES OF TYPE I ENZYMES

The **type I** enzymes, represented by the EcoK and EcoB activities of *E. coli* strains K and B, were the first to be discovered. A type I restriction and modification enzyme consists of three types of subunit, as indicated in **Figure 33.2.** The R subunit is responsible for restriction, the M subunit is responsible for methylation, and the S subunit is responsible for recognizing the target site on DNA. When a target site has been recognized by the S subunit, the enzyme's binding to DNA can be followed by *either* restriction *or* modification; the activities of the R and M subunits are mutually exclusive.

Each subunit is coded by a single gene, denoted *hsdR, hsdM,* and *hsdS.* Mutants in *hsdR* are phenotypically r^-m^+; they cannot restrict DNA, but can modify it. Mutants in *hsdS* are r^-m^-; they can neither modify nor restrict DNA. Mutants in *hsdM* turn out to prevent

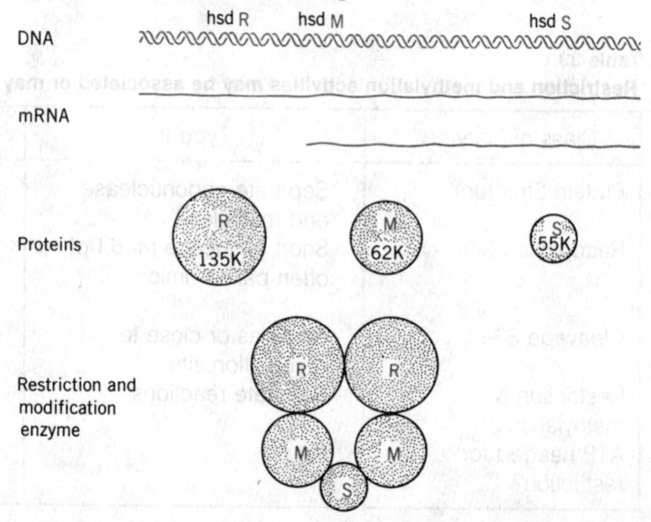

Figure 33.2
Multifunctional type I enzymes have different subunits for restriction, modification, and recognition.

restriction as well as modification. Thus the M subunit is involved in R subunit function. This may be a fail-safe mechanism, ensuring that *hsdM* mutants are r^-m^-, rather than the r^+m^- phenotype they would otherwise have, which presumably would be lethal (because the restriction activity would degrade the cell's own unmodified DNA).

The EcoB and EcoK enzymes are allelic. Their target sequences are different, but in diploids, the S subunit of one strain can direct the activities of the R and M subunits of the other bacterial strain. This confirms that the recognition step is independent of the succeeding cleavage or methylation events.

The EcoK enzyme has a molecular weight of about 400,000 daltons. It contains two copies of the 135,000-dalton R subunit, two copies of the 62,000-dalton M subunit, and one copy of the 55,000-dalton S subunit.

The EcoB enzyme has subunits of similar size, but they may be present in different molar ratios. The M and S subunits of strain B can form a 1:1 complex that exercises the methylation function independently of the restriction function. This may be a natural occurrence, because the three genes lie in an operon, in the order: *P1-hsdR-P2-hsdM-hsdS,* where *P1* and *P2* are independent promoters (see Figure 33.2). Thus *hsdM* and *hsdS* can be expressed separately from *hsdR*.

The recognition sites for EcoB and EcoK are bipartite structures, each consisting of a specific sequence of 3 bp separated by a certain distance (8 bp for EcoB, 6 bp for EcoK) from a specific sequence of 4 bp. The distance between the two parts of the recognition site means that both lie on one face of the DNA; the sequence of the intervening region does not seem to be important. Note that neither of the recognition sites is symmetrical. However, each of them possesses an adenine for methylation on each strand, as indicated by the asterisks in the EcoB recognition sequence.

```
  *
TGANNNNNNNNTGCT
ACTNNNNNNNNACGA
             *
```

Whether a given DNA is to be cleaved or modified is determined by the state of the target site, as indicated in **Figure 33.3.** If the target site is fully methylated, the enzyme may bind to it, but is released without any further action. If the target is hemimethylated, the enzyme methylates the unmethylated strand. This action perpetuates the state of methylation in host DNA, as noted previously. If the target site is unmethylated, its recognition triggers a cleavage reaction.

The groups for methylation are provided by the cofactor S-adenosyl-methionine (SAM), which is converted to S-adenosyl-homocysteine (SAH) in the reaction. The SAM binds to the M subunit. In the initial stage of the reaction, the SAM acts as an allosteric effector that changes the conformation of S subunit to allow it to bind to DNA.

After binding to DNA, the next step is a reaction with ATP. If the enzyme is bound at a completely methylated site, the arrival of ATP releases it from the DNA. At an unmethylated site, the ATP converts the enzyme into a state ready to sponsor cleavage. This depends on the R subunit. Hydrolysis of ATP is needed to cleave DNA. The SAM is released from the enzyme before the restriction step occurs.

The type I enzymes display an extraordinary relationship between the sites of recognition and cleavage. *The cleavage site is located more than 1000 bp away from the recognition site.* Cleavage does not occur at a specific sequence, but the selection of a site for cutting does not seem to be entirely random, because some regions of DNA are preferentially cleaved. (The discrepancy between the sites of recognition and cleavage means that recognition sites cannot be defined by characterizing the broken ends of DNA; the target sites are identified by their modification in the methylation event, and by the locations of mutations that abolish recognition.)

The cleavage reaction itself involves two steps. First, one strand of DNA is cut; then the other strand is cut nearby. There may be some exonucleolytic degradation in the regions on either side of the site of cleavage. Extensive hydrolysis of ATP occurs; its function is not yet known.

How does the enzyme recognize one site and cleave another so far away? An important feature is that *the protein never lets go of the DNA molecule to which it initially binds.* If the enzyme is incubated with a mixture of modified and unmodified DNA, it preferentially cleaves the unmodified DNA. This implies that, having recognized a binding site, it does not dissociate from

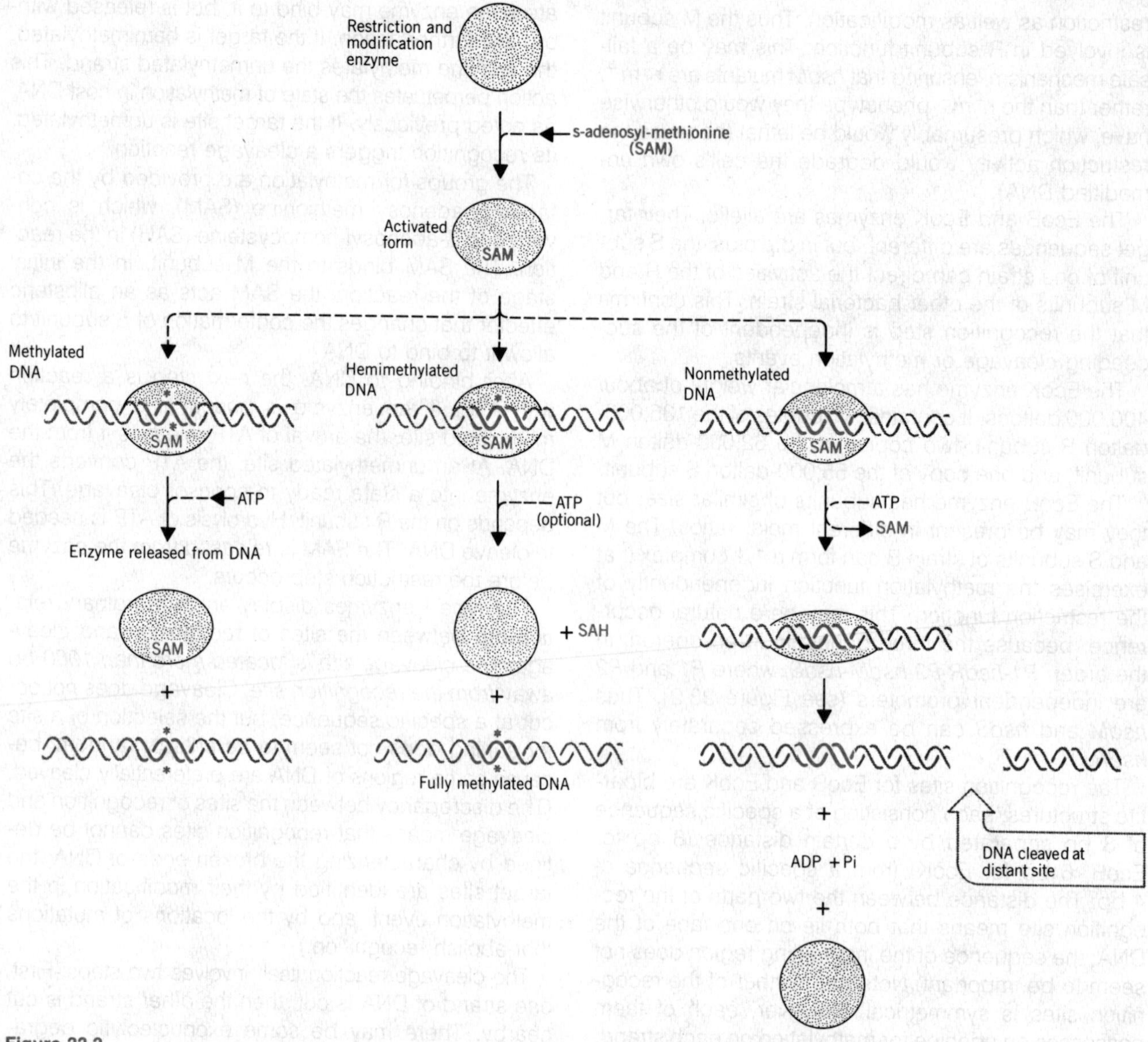

Figure 33.3
Type I enzymes bind to target sites, after which they are released from fully methylated sites, complete the methylation of hemimethylated sites, and move along DNA from nonmethylated sites to cleave the molecule elsewhere.

the unmethylated DNA in order to find its cleavage site.

Two types of model could explain the relationship between the recognition and cleavage sites: the enzyme moves, or the DNA moves. They are illustrated in **Figure 33.4.** If the enzyme moves, it could translocate along the DNA until (for some unknown reason) it exercises the cleavage option. If the DNA moves, the enzyme could remain attached at the recognition site, while it pulls the DNA through a second binding

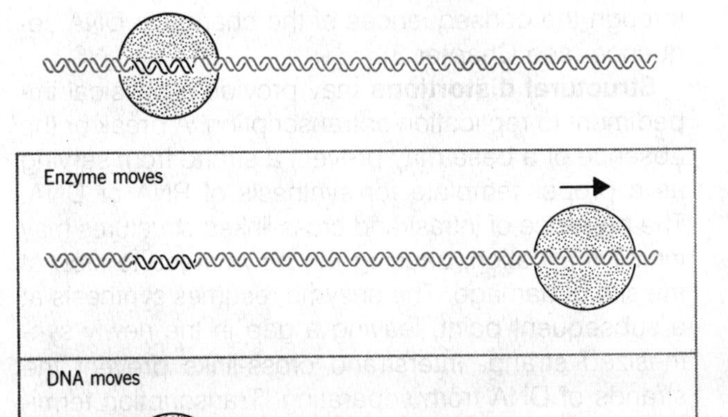

Figure 33.4
Does a type I enzyme move along DNA or does it remain at its target site, simultaneously pulling the DNA through the protein?

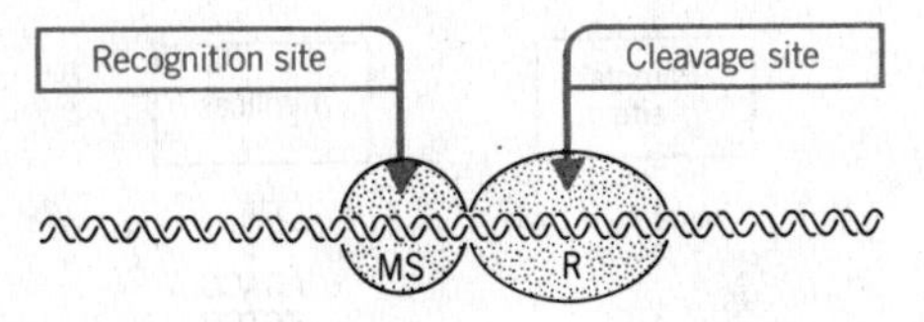

Figure 33.5
Type III enzymes have two subunits: recognition and methylation by the MS subunit occur at the target site; the restriction event may occur at a nearby site contacted by the R subunit.

site on the enzyme, winding along until it reaches a cleavage region (again undefined in nature). There is electron microscopic evidence that the enzyme generates loops in DNA, and it seems to remain attached to its recognition site after cleavage; this supports the second model.

THE DUAL ACTIVITIES OF TYPE III ENZYMES

Three type III restriction and modification enzymes have been investigated; EcoP1 and EcoP15 are coded by plasmids P1 and P15 in *E. coli*, and Hinf is found in *H. influenzae* of serotype R_f. Each enzyme consists of two types of subunit. The R subunit is responsible for restriction; *hsdR* mutants have the r^-m^+ phenotype. The MS subunit is responsible for *both* modification and recognition. Mutations in the *hsdMS* gene generate the r^-m^- phenotype if they are located in the recognition part of the sequence; they are reported to generate the r^+m^- phenotype if located in the modification part (presumably this is lethal). The recognition part of the gene probably lies in the middle of its sequence, as judged by comparisons of *hsdMS* genes from the P1 and P15 systems. The R subunit is 106,000–110,000 daltons; the MS polypeptide is 73,000–80,000 daltons.

In the type III enzymes, the restriction and modification activities are expressed *simultaneously*. The enzyme first binds to its site on DNA, an action that requires ATP. (The subsequent dependence on ATP varies among the enzymes.) Then the methylation and restriction activities *compete* for reaction with the DNA. The methylation event takes place at the binding site, consistent with the combination of methylation and recognition in the MS subunit. The restriction cleavage occurs 24–26 bases on one side, probably because the enzyme is large enough for the restriction subunit to contact DNA at this point, as depicted rather schematically in **Figure 33.5.** Restriction involves staggered cuts, 2–4 bases apart.

The enzymes methylate adenine residues, but the target sites of P1 and P15 have an intriguing feature: they can be methylated *only on one strand*. The sequences of the sites are

sP1	AGACC TCTGG
sP15	CAGCAG GTCGTC

How is the state of methylation perpetuated? **Figure 33.6** demonstrates that two types of site are generated by replication. One replica has the original methylated strand; in fact it is indistinguishable from the parental site. The other replica is entirely unmethylated. It is therefore in principle a target for either restriction or modification. What determines that it is methylated rather than restricted? We do not know the answer, but one idea is that the modification reaction is linked to the act of replication.

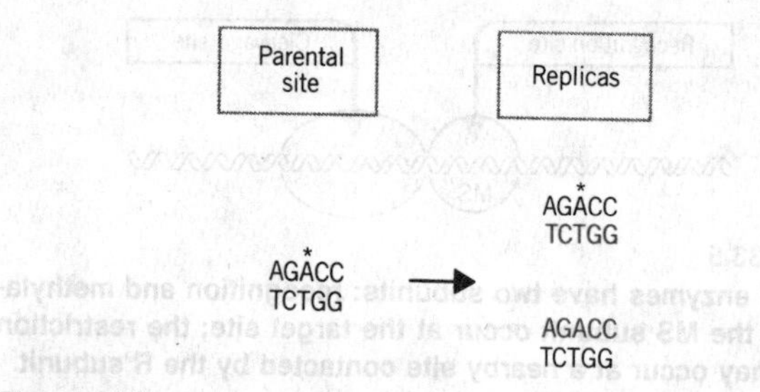

Figure 33.6
Replication of a methylated sP1 sequence generates one methylated and one unmethylated replica.

DEALING WITH INJURIES IN DNA

Injury to DNA is minimized by damage-containment systems that recognize the occurrence of a change and then rectify it. The measured mutation rate—spontaneous or induced—reflects a balance between the number of damaging events occurring in DNA and the number that have been corrected (or miscorrected). These repair systems are likely to be as complex as the replication apparatus itself, an indication of their importance for the survival of the cell.

"Damage" to DNA is defined as any change introducing a deviation from the usual double-helical structure. Events in this category include:

Introduction of single strand breaks.

Removal of a base to leave its former partner unpaired.

Covalent addition of new groups to a base to modify its properties.

Conversion of one base into another that is not properly paired with the partner base.

Introduction of covalent links between bases on one strand of DNA or between bases on opposite strands.

Some repair systems are not specific for particular types of damage, but recognize a range of distortions in DNA as signals indicating the need for action. We can divide the changes listed above into two general classes.

Point mutations may be recognized because they disrupt proper pairing of the partner bases. This does not affect transcription or replication, when the strands of the DNA duplex are separated. Thus these changes exert their damaging effects on future generations through the consequences of the change in DNA sequence (see Chapter 3).

Structural distortions may provide a physical impediment to replication or transcription. A break or the absence of a base may prevent a strand from serving as a proper template for synthesis of RNA or DNA. The presence of intrastrand cross-linked structures may inhibit replication, causing the polymerase to stop at the site of damage. The enzyme resumes synthesis at a subsequent point, leaving a gap in the newly synthesized strand. Interstrand cross-links prevent the strands of DNA from separating. Transcription terminates prematurely at either type of cross-link.

The most studied example is the effect of ultraviolet irradiation, which introduces covalent bonds between two adjacent thymine bases, giving the intrastrand **thymine dimer** drawn in **Figure 33.7.**

Three types of event may cope with the damage. The links in thymine dimers may be directly reversed by a light-dependent enzyme. This process is called **photoreactivation;** in *E. coli* it depends on the product of a single gene (*phr*). Alternatively, the damaged material may be removed and replaced by a **repair system.** Finally, the damage may remain at the original site, but the daughter strands generated by replication may be repaired by a **retrieval system.** A given cell may have several systems engaging in repair and retrieval.

Excision-repair systems remove the mispaired or damaged bases from DNA and then synthesize a new stretch of DNA to replace them. The main type of pathway for excision-repair is illustrated in **Figure 33.8.** In

Figure 33.7
A thymine dimer is generated by covalent links between adjacent bases.

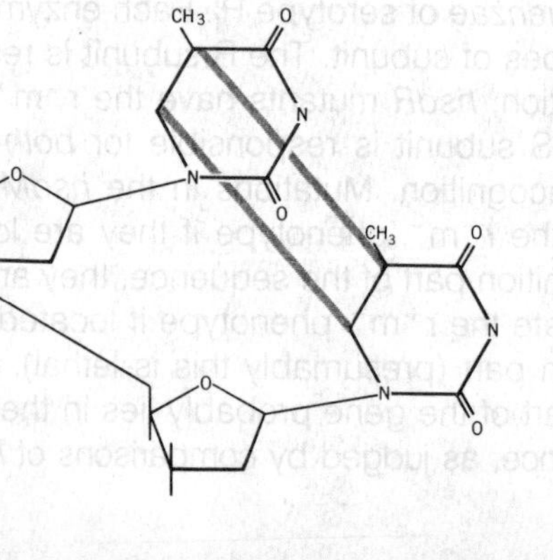

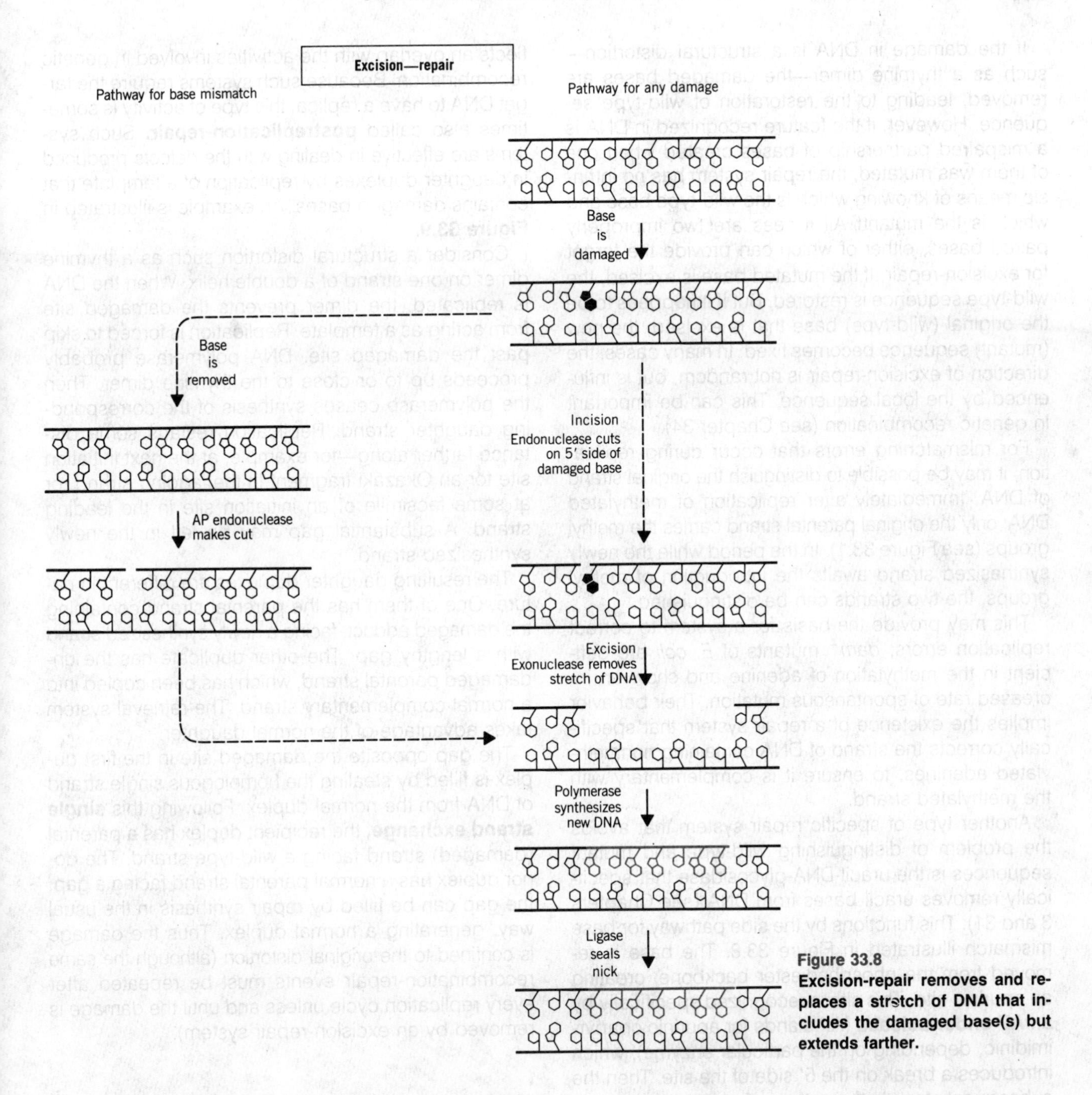

Figure 33.8
Excision-repair removes and replaces a stretch of DNA that includes the damaged base(s) but extends farther.

the **incision** step, the damaged structure is recognized by an endonuclease that cleaves the DNA strand on the 5′ side of the damage. In the **excision** step, a 5′–3′ exonuclease removes a stretch of this strand, extending past the site of damage. In the **repair** step, the resulting single-stranded region serves as a template for a DNA polymerase to synthesize a replacement for the excised sequence. Finally, DNA ligase covalently links the 3′ end of the new material to the old material.

If the damage in DNA is a structural distortion—such as a thymine dimer—the damaged bases are removed, leading to the restoration of wild-type sequence. However, if the feature recognized in DNA is a mispaired partnership of bases created when one of them was mutated, the repair system has no intrinsic means of knowing which is the wild-type base and which is the mutant! All it sees are two improperly paired bases, either of which can provide the target for excision-repair. If the mutated base is excised, the wild-type sequence is restored. But if it happens to be the original (wild-type) base that is excised, the new (mutant) sequence becomes fixed. In many cases, the direction of excision-repair is not random, but is influenced by the local sequence. This can be important in genetic recombination (see Chapter 34).

For mismatching errors that occur during replication, it may be possible to distinguish the original strand of DNA. Immediately after replication of methylated DNA, only the original parental strand carries the methyl groups (see Figure 33.1). In the period while the newly synthesized strand awaits the introduction of methyl groups, the two strands can be distinguished.

This may provide the basis for a system to correct replication errors; *dam*$^-$ mutants of *E. coli* are deficient in the methylation of adenine and show an increased rate of spontaneous mutation. Their behavior implies the existence of a repair system that specifically corrects the strand of DNA containing nonmethylated adenines, to ensure it is complementary with the methylated strand.

Another type of specific repair system that avoids the problem of distinguishing wild-type and mutant sequences is the uracil-DNA-glycosidase that specifically removes uracil bases from DNA (see Chapters 3 and 31). This functions by the side pathway for base mismatch illustrated in Figure 33.8. The base is removed from the phosphodiester backbone, creating an apurinic site. This site is recognized specifically by an **AP endonuclease** (AP stands for apurinic or apyrimidinic, depending on the particular enzyme), which introduces a break on the 5′ side of the site. Then the subsequent steps in the pathway for general excision-repair can be followed.

Retrieval systems are provided by **recombination-repair** enzymes that function by retrieving material from one duplex of DNA to repair another. The name reflects an overlap with the activities involved in genetic recombination. Because such systems require the target DNA to have a replica, this type of activity is sometimes also called **postreplication-repair.** Such systems are effective in dealing with the defects produced in daughter duplexes by replication of a template that contains damaged bases. An example is illustrated in **Figure 33.9.**

Consider a structural distortion such as a thymine dimer on one strand of a double helix. When the DNA is replicated, the dimer prevents the damaged site from acting as a template. Replication is forced to skip past the damaged site. DNA polymerase probably proceeds up to or close to the thymine dimer. Then the polymerase ceases synthesis of the corresponding daughter strand. Replication restarts some distance farther along—for example, at the next initiation site for an Okazaki fragment in the lagging strand, or at some facsimile of an initiation site in the leading strand. A substantial gap may be left in the newly synthesized strand.

The resulting daughter duplexes are different in nature. One of them has the parental strand containing the damaged adduct, facing a newly synthesized strand with a lengthy gap. The other duplicate has the undamaged parental strand, which has been copied into a normal complementary strand. The retrieval system takes advantage of the normal daughter.

The gap opposite the damaged site in the first duplex is filled by stealing the homologous single strand of DNA from the normal duplex. Following this **single strand exchange,** the recipient duplex has a parental (damaged) strand facing a wild-type strand. The donor duplex has a normal parental strand facing a gap; the gap can be filled by repair synthesis in the usual way, generating a normal duplex. Thus the damage is confined to the original distortion (although the same recombination-repair events must be repeated after every replication cycle unless and until the damage is removed by an excision-repair system).

EXCISION-REPAIR SYSTEMS IN *E. COLI*

Mutations in many loci of *E. coli* affect the response to agents that damage DNA. Many of these genes are likely to code for enzymes that participate in DNA re-

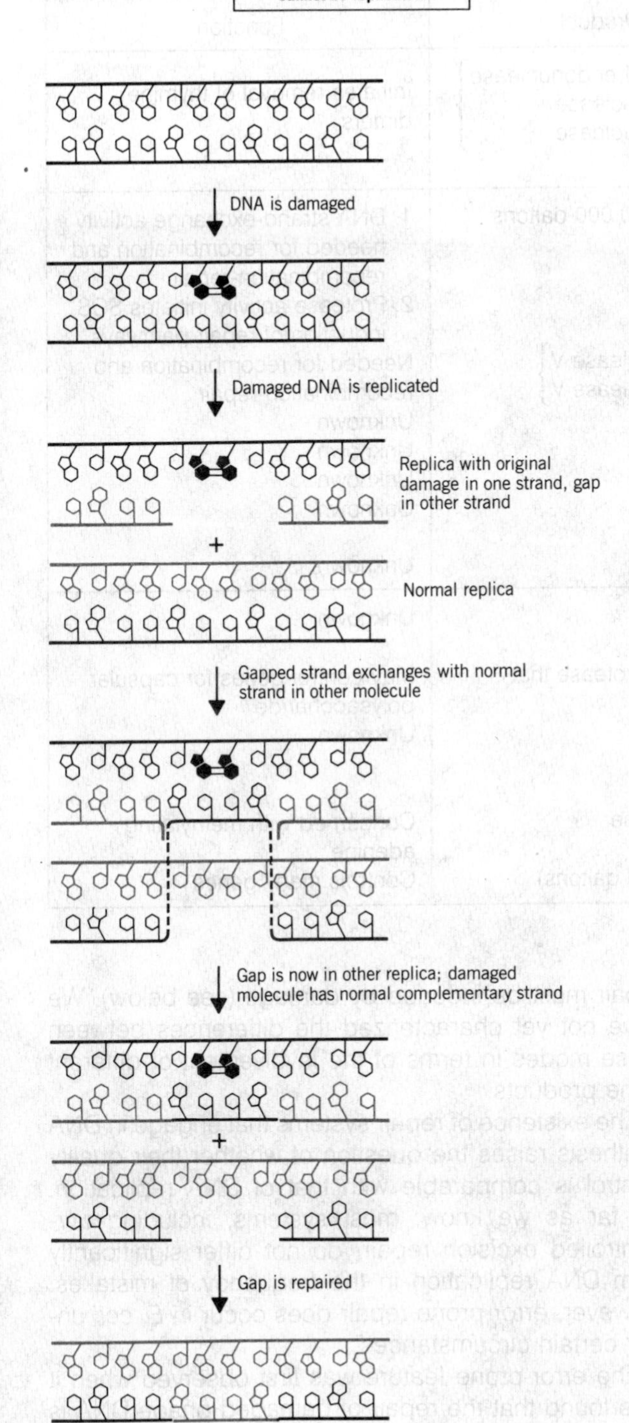

pair systems, although in the majority of cases the gene product and its function remain to be identified. When the repair systems are eliminated, cells become *exceedingly* sensitive to ultraviolet irradiation. The introduction of UV-induced damage has been a major test for repair systems, and so in assessing their activities and relative efficiencies, we should remember that the emphasis might be different if another damaged adduct were studied.

Table 33.2 summarizes the properties of some of the mutations that affect the ability of *E. coli* cells to engage in DNA repair. We know of at least three repair pathways. Two different excision-repair modes are identified by the heterogeneity of the lengths of the segments of repaired DNA. These pathways are described as **short-patch repair** and **long-patch repair.** A recombination-repair pathway (or pathways) overlaps extensively with the enzymes involved in recombination itself (see Chapter 34).

The *uvr* system is involved in both modes of excision-repair and includes three genes, *uvrA,B,C,* that code for the components of a repair endonuclease. *In vitro*, the endonuclease recognizes the thymine dimer (or other distortion) and makes an incision on each side, one just on the 5′ side of the damaged site, and the other on the 3′ side about 12 nucleotides away.

Several *E. coli* enzymes can excise thymine dimers *in vitro* from DNA in which such incisions have been made. The enzymes include the 5′-3′ exonuclease activities of DNA polymerase I and DNA polymerase II (see Table 32.2) and the single-strand specific exonuclease VII. However, mutation of any one of these putative excision nucleases does not result in a measurable diminution in the ability of *E. coli* cells to remove thymine dimers. Based on the behavior of various double mutants, current opinion is that DNA polymerase I is likely to perform most of the excision *in vivo*.

The average length of excised DNA is about 20 nucleotides, which gives rise to the description of this mode as the short-patch repair. The enzyme involved in the repair synthesis probably also is DNA polymer-

Figure 33.9
Recombination-repair retrieves a normal strand of DNA to replace the gap left in a newly synthesized strand opposite a site of unrepaired damage.

Table 33.2
Many gene products are involved in repairing DNA damage in *E. coli*.

Gene	Effect of Mutation *in vivo*	Gene Product	Function
uvrA	Sensitivity to UV	ATPase subunit of endonuclease	Initiates removal of thymine dimers
uvrB	"	Subunit of endonuclease	
uvrC	"	Subunit of endonuclease	
uvrE	"	Unknown	
recA	1. Deficiency in recombination 2. Does not induce repair pathways after damage	RecA protein of 40,000 daltons	1. DNA strand-exchange activity needed for recombination and recombination-repair 2. Protease activity initiates SOS induction of repair pathways
recB	Deficiency in recombination	Subunit of exonuclease V	Needed for recombination and recombination-repair
recC	"	Subunit of exonuclease V	
sbcB	Suppresses *recBC* mutation	Exonuclease I	Unknown
recF	Deficiency in recombination-repair; and deficiency in recombination in *recBCsbcB* mutants	Unknown	Unknown
recJ		Unknown	Unknown
recK		Unknown	Unknown
recE	Deficiency in recombination	Exonuclease VIII	Unknown
umuC	Abolishes mutagenic effects of error-prone repair	Unknown	Unknown
lon	UV-sensitive, septum-deficient	ATP-dependent protease that binds to DNA	May control genes for capsular polysaccharide
mut	At least 6 genes whose mutation increases the rate of mutation	Unknown	Unknown
dam	UV-sensitive, increased mutation rate	Could be methylase	Concerned with methylating adenine
lexA	Abolishes SOS response	Repressor (22,000 daltons)	Controls many genes

ase I (although polymerases II and III can substitute for it).

Other potential components of the *uvr* system have been identified by mutations that impede short-patch repair, but none has yet been identified with a product or function in the pathway.

Short-patch repair accounts for 99% of the excision-repair events. The remaining 1% involve the replacement of stretches of DNA mostly ~1500 nucleotides; a minority of the long patches extend for >9000 nucleotides. This mode also requires the *uvr* genes and involves DNA polymerase I. A difference between the two modes of repair is that short-patch repair is a constitutive function of the bacterial cell, but long-patch repair must be induced by damage (see below). We have not yet characterized the differences between these modes in terms of the involvement of different gene products.

The existence of repair systems that engage in DNA synthesis raises the question of whether their quality control is comparable with that of DNA replication. So far as we know, most systems, including *uvr*-controlled excision repair, do not differ significantly from DNA replication in the frequency of mistakes. However, error-prone repair does occur in *E. coli* under certain circumstances.

The error-prone feature was first observed when it was found that the repair of damaged phage DNA is accompanied by the induction of mutations if the phage is introduced into cells that had previously been irradiated with UV. This suggests that the UV irradiation of the host has activated some protein(s) that do not function in the unirradiated cell, and whose activity in some repair pathway generates mutations. The mutagenic response also operates on the bacterial host DNA.

What is the actual error-prone activity? Current thinking focuses on the idea that the relevant component of the repair pathway may be responsible for permitting or compelling replication to proceed past the site of damage. When the DNA polymerase passes the thymine dimer, it does not insert the correct bases; the bases it does insert represent mutations. The error-prone activity requires DNA polymerase III, the usual replicase, which is consistent with the idea that the relevant function acts in concert with the normal replication apparatus.

Mutations in the genes *umuD* and *umuC* abolish UV-induced mutagenesis, but do not interfere with any known enzymatic functions. The genes constitute the *umuDC* operon, whose expression is induced by DNA damage (see later). Some plasmids carry genes called *mucA* and *mucB*, which are homologs of *umuD* and *umuC*, and whose introduction into a bacterium increases resistance to UV killing and susceptibility to mutagenesis.

RECOMBINATION-REPAIR SYSTEMS IN *E. COLI*

In cells deficient in excision-repair, mutation in the *recA* gene essentially abolishes all the remaining repair and recovery facilities. Thus attempts to replicate DNA in uvr^- $recA^-$ cells produce fragments of DNA whose size corresponds with the expected distance between thymine dimers. This result implies that the unremoved dimers provide a lethal obstacle to replication in the absence of RecA function. It explains why the double mutant cannot tolerate more than 1-2 dimers in its genome (compared with the ability of a wild-type bacterium to shrug off as many as 50).

The dramatic effects of mutation in *recA* bring us to a major feature of the response of *E. coli* to damage in its DNA. The first interpretation of this result was that the *recA* gene must be part of another pathway. Since *recA* mutants are almost completely deficient in genetic recombination as well as in the recovery response, this pathway seemed likely to be a recombination-repair system. This view was supported by the isolation of other *rec* mutants with similar (although lesser) deficiencies in both repair and recombination (see Table 33.2). The components involved in recombination-repair overlap with those involved in recombination, but are not entirely identical with them, since some mutations affect one but not the other activity.

The RecA protein has the function of exchanging strands between DNA molecules, a central activity in recombination (see Chapter 34), related to the proposed single-strand exchange involved in recombination-repair. The properties of double mutants suggest that *recA* may participate in two Rec pathways.

To test whether two genes with related functions are part of the same or different pathways, the phenotypes of the single mutants are compared with the phenotype of the double mutant. If the genes are in the same pathway, the phenotype of the double mutant will be no different from that of the individual mutants. If the genes are in different pathways, the double mutant will lack both instead of one, and so has a more severe phenotype. By this criterion, one *rec* pathway involves the *recBC* genes; the other involves *recF*.

The *recBC* genes code for the two subunits of exonuclease V, whose action is limited by some other component of the pathway. (In *recA* mutants, the enzyme is unlimited, and degrades an excessive amount of DNA. This is the "reckless" phenotype that gave rise to the name of these loci.) The function of *recF* is unknown. Other *rec* loci have been identified by the effects of mutations on recombination-repair or recombination. Several are identified by their ability to suppress other *rec* mutants. As with the additional *uvr* loci, they have not been equated with products or functions.

The designations of these genes are based on the phenotypes of the mutants; but sometimes a mutation isolated in one set of conditions and named as a *uvr* locus turns out to have been isolated in another set of conditions as a *rec* locus. This uncertainty makes an important point. We cannot yet define how many functions belong to each pathway or how the pathways interact. The *uvr* and *rec* pathways may not be entirely independent, because *uvr* mutants show reduced efficiency in recombination-repair. We must expect to find a network of nuclease, polymerase, and other activities, constituting repair systems that may be partially overlapping (or in which one enzyme usually used to provide some function can be substituted by another from a different pathway).

The direct involvement of RecA protein in recombination-repair is only one of its activities. This extraordinary protein also has another, quite distinct function. It is a protease activated by ultraviolet irradiation (or by other treatments that block replication). The protease activity is responsible for inducing the expression of many genes, whose products include repair functions. Among them, for example, is the long-patch repair mode; so RecA is needed to induce this excision-repair system as well as to participate in recombination-repair.

These dual activities of the RecA protein make it difficult to know whether a deficiency in repair in *recA* mutant cells is due to loss of the DNA strand-exchange function of RecA or to some other function whose induction depends on the protease activity.

AN SOS SYSTEM OF MANY GENES

Many treatments that damage DNA or inhibit replication in *E. coli* induce a complex series of phenotypic changes called the **SOS response.** This is set in train by the interaction of the RecA protein with the LexA repressor.

The damage can take the form of ultraviolet irradiation (the most studied case) or can be caused by cross-linking or alkylating agents. Inhibition of replication by any of several means, including deprivation of thymine, addition of drugs, or mutations in several of the *dna* genes, has the same effect.

The response takes the form of increased capacity to repair damaged DNA, achieved by inducing synthesis of the components of both the long-patch excision-repair system and the Rec recombination-repair pathways. In addition, cell division is inhibited. Lysogenic prophages may be induced.

The initial event in the response is the activation of the RecA protease activity by the damaging treatment. We do not know very much about the relationship between the damaging event and the sudden change in RecA activity. Because a variety of damaging events can induce the SOS response, current work focuses on the idea that RecA may be stimulated by some common intermediate in DNA metabolism.

The inducing signal could consist of a small molecule released from DNA; or it might be some structure formed in the DNA itself. *In vitro,* the protease activity of RecA requires the presence of single-stranded DNA and ATP. Thus the activating signal could be the presence of a single-stranded region at a site of damage.

Whatever form the signal takes, its interaction with RecA is rapid: the SOS response occurs within a few minutes of the damaging treatment. On activation, the RecA protease cleaves the 22,000-dalton repressor protein coded by the *lexA* gene. The LexA protein is relatively stable in untreated cells, where it functions as a repressor at many operons. Proteolytic cleavage of the repressor coordinately induces all these operons.

The target genes for LexA repression include many repair functions, only some of which are identified at present. A systematic screen for LexA-responsive genes has been constructed by fusing random operons to the *lacZ* gene, and then assaying for increased levels of the product β-galactosidase when cells are treated with mitomycin C. At least five responsive genes have been identified; they are called *din* (damage inducible). Genes previously known to be part of the SOS response by the activation of their products include *recA, lexA, uvrA, uvrB, umuC,* and *himA.*

Some of the SOS genes may be active only in treated cells; others are active in untreated cells, but the level of expression is increased by cleavage of LexA. In the case of *uvrB,* which is a component of the excision-repair system, the gene has two promoters; one functions independently of LexA, the other is subject to its control. Thus after cleavage of LexA, the gene can be expressed from the second promoter as well as from the first.

Purification of the LexA repressor allows its action to be characterized *in vitro.* It represses its target genes by binding to a stretch of DNA of about 20 bp, called an **SOS box;** this sequence displays symmetry, and a copy is present at each target locus. The SOS boxes at different loci are not identical, but conform to a consensus sequence with 8 absolutely conserved posi-

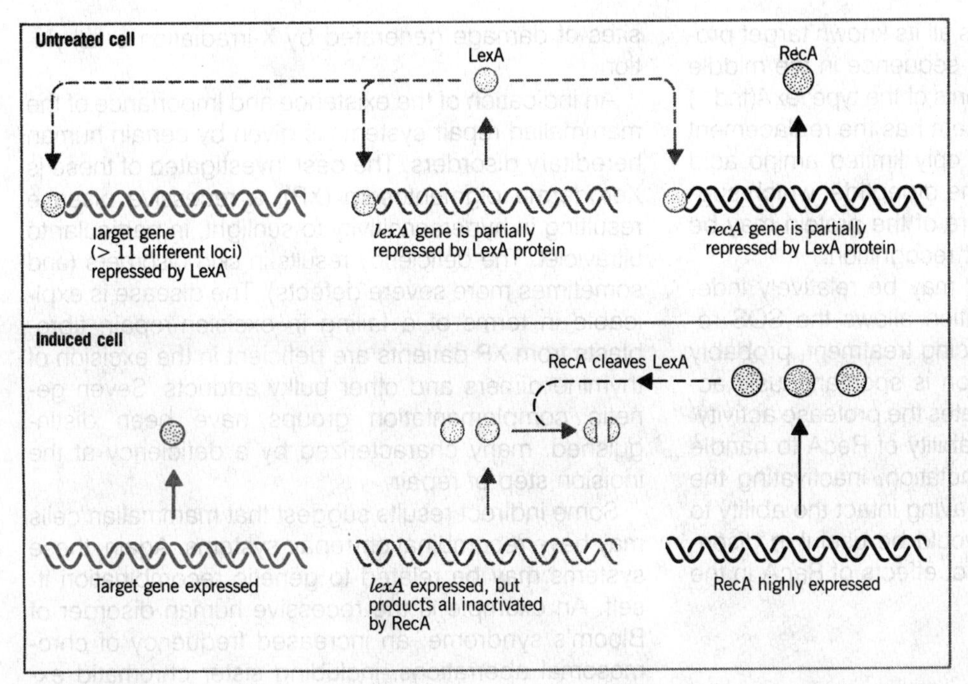

Figure 33.10
The LexA protein represses many genes, including repair functions, *recA* and *lexA*. Activation of the RecA protease cleaves LexA and induces all of these genes.

tions. Like other operators, the SOS boxes overlap with the respective promoters. At the *lexA* locus, the subject of autogenous repression, there are two adjacent SOS boxes.

RecA and LexA are mutual targets in the SOS circuit: RecA cleaves LexA, which represses *recA*. The SOS response therefore causes amplification of both the RecA protein and the LexA repressor. The results are illustrated in **Figure 33.10;** they are not so contradictory as might at first appear.

The increase in expression of RecA protein is necessary (presumably) for its direct role in the recombination-repair pathways. On induction, the level of RecA is increased from its basal level of about 1200 molecules/cell by up to 50-fold. The high level in induced cells means that there is sufficient protease to ensure that all the additional LexA protein is cleaved. This should prevent LexA from reestablishing repression of the target genes.

But the main importance of this circuit for the cell may lie in the ability to return rapidly to normalcy. When the inducing signal is removed, the RecA protein loses its protease activity. At this moment, the *lexA* gene is being expressed at a high level; in the absence of RecA protease, the LexA protein rapidly accumulates in the uncleaved form and turns off the SOS genes. This may explain why the SOS response is freely reversible.

The RecA protease also acts on some other repressor proteins, those of several prophages. Among these is the lambda repressor (with which the protease activity in fact was discovered). This explains why lambda is induced by ultraviolet irradiation; RecA cleaves the lysogenic repressor, releasing the phage to enter the lytic cycle.

This reaction is not a cellular SOS response, but instead represents a recognition by the prophage that the cell is in trouble, so that survival is best assured by entering the lytic cycle to generate progeny phages. In this sense, the prophage induction is piggybacking onto the cellular system by responding to the same indicator (activation of RecA).

The RecA protease cleaves all its known target proteins at an Ala-Gly dipeptide sequence in the middle of the polypeptide chain. Mutants of the type *lexA*(Ind$^-$) cannot be induced; one of them has the replacement sequence Ala-Asp. There is only limited amino acid homology on either side of the dipeptide, which suggests that the tertiary structure of the protein may be an important feature in target recognition.

The two activities of RecA may be relatively independent. The *recA441* mutation allows the SOS response to occur without inducing treatment, probably because the protease function is spontaneously active. Another mutation inactivates the protease activity. Neither mutation affects the ability of RecA to handle DNA. The reverse type of mutation, inactivating the recombination function but leaving intact the ability to induce the SOS response, would be useful in disentangling the direct and indirect effects of RecA in the repair pathways.

MAMMALIAN REPAIR SYSTEMS

Biochemical characterization of repair systems in eukaryotic cells is only primitive, for the most part confined to the isolation of the occasional enzyme preparation whose properties *in vitro* suggest that it could be part of a repair system. The existence of excision-repair pathways can be established in cultured cells by following the replacement of DNA segments in response to damaging treatments. One technique is to follow events in synchronized cells outside S phase, when the occurrence of **unscheduled DNA synthesis** can be entirely attributed to the operation of repair systems.

Mammalian cells show heterogeneity in the amount of DNA incorporated after damage. However, the longest patches seen in mammalian cells are comparable to those of the short-patch bacterial repair system. This pathway operates on damage caused by ultraviolet irradiation or treatments with related effects. Another pathway introduces only 3-4 repair bases at sites of damage generated by X-irradiation or alkylation.

An indication of the existence and importance of the mammalian repair systems is given by certain human hereditary disorders. The best investigated of these is Xeroderma pigmentosum (XP), a recessive disease resulting in hypersensitivity to sunlight, in particular to ultraviolet. The deficiency results in skin disorders (and sometimes more severe defects). The disease is explicable in terms of a failing in excision-repair; fibroblasts from XP patients are deficient in the excision of thymine dimers and other bulky adducts. Seven genetic complementation groups have been distinguished, many characterized by a deficiency at the incision step of repair.

Some indirect results suggest that mammalian cells may have recombination-repair systems. Again, these systems may be related to genetic recombination itself. An example is the recessive human disorder of Bloom's syndrome; an increased frequency of chromosomal aberrations, including sister chromatid exchanges, could be related to the operation of recombination systems.

FURTHER READING

The separate (type II) restriction and modification activities have been reviewed by **Modrich** (*Quart. Rev. Biophys.* **3,** 315-369, 1979). A valuable tour of multifunctional (type I and type III) enzymes, emphasizing the enzymatic activities, was provided by **Yuan** (*Ann. Rev. Biochem.* **50,** 285-315, 1981), who developed a model for type I enzyme translocation (*Cell* **20,** 237-244, 1980). The functions of AP endonucleases and associated systems have been reviewed by **Lindahl** (*Prog. Nucleic Acid Res.* **22,** 135-192, 1979). An integrated view of the systems and mechanisms for DNA repair in prokaryotes and eukaryotes was presented by **Hanawalt et al.** (*Ann. Rev. Biochem.* **48,** 783-836, 1979). The SOS control network of *E. coli* has been reviewed extensively by **Little & Mount** (*Cell* **29,** 11-22, 1982) and **Walker** (*Microbiol. Rev.* **48,** 60–93, 1984).

CHAPTER 34
RECOMBINATION AND OTHER TOPOLOGICAL MANIPULATIONS OF DNA

Without genetic recombination, the content of each individual chromosome would be irretrievably fixed in its particular alleles, subject only to the changes caused by mutations. The length of the target for mutation damage would be increased from the gene to the chromosome. Deleterious mutations would accumulate, eliminating each chromosome (and therefore removing any favorable mutations that may have occurred).

By shuffling the genes, recombination allows favorable and unfavorable mutations to be separated and tested as individual units in new assortments. It provides a means of escape and spreading for favorable alleles, and a means to eliminate an unfavorable allele without bringing down all the other genes with which this allele may have been associated in the past. From the long perspective of evolution, a chromosome is a bird of passage, a temporary association of particular alleles. Recombination is responsible for this flighty behavior.

This type of recombination, involving reaction between homologous sequences of DNA, is called **general recombination.** Its critical feature is that the enzymes responsible can use *any* pair of homologous sequences as substrates (although there is always the possibility that some types of sequence are favored over others). In the genome as a whole, the frequency of recombination is not constant, but may be influenced by chromosome structure; for example, crossing-over may be suppressed in the vicinity of heterochromatin.

Another type of event sponsors recombination between *specific* pairs of sequences. **Site-specific recombination** is responsible for the integration of phage genomes into the bacterial chromosome. The recombination event involves specific sequences of the phage DNA and bacterial DNA. Within these sequences, there is only a short stretch of homology, necessary for the recombination event, but not sufficient for it. The enzymes involved in this event cannot recombine other pairs of (homologous or nonhomologous) sequences, but act only on the particular pair of phage and bacterial sequences. This type of recombination is also called **conservative recombination,** because it relies entirely on the reaction of the two preexisting DNA molecules.

Another type of recombination event allows one DNA sequence to be inserted into another without reliance on sequence homology. **Replicative recombination** is responsible for the ability of certain elements to move from one location to another. The movable element of DNA is copied, and then the replica is transferred elsewhere. This is the subject of Chapter 35

TOPOLOGICAL MANIPULATION OF DNA

The manipulations involved in recombination may be related to the reactions by which the higher order (supercoiled) structure of a single molecule of DNA is altered. In fact, the need for topological manipulation of DNA is a central aspect of *all* its functional activities—recombination, replication, and (perhaps) transcription—as well as of the organization of higher order structure. In considering all of these processes, we must remember that the duplex structure of DNA is, as it were, an obstacle to be overcome by any reaction involving strand separation. In a broad sense, all recombination events represent different types of topological reconstruction.

All synthetic activities involving double-stranded DNA require the strands to separate, a reaction that can be accomplished by input of sufficient energy to disrupt the noncovalent forces that stabilize the double helix. However, the strands do not simply lie side by side; they are intertwined. Their separation therefore requires the strands to rotate about each other in space. Some possibilities for the unwinding reaction are illustrated in **Figure 34.1.**

We might envisage the structure of DNA in terms of a free end that would allow the strands to rotate about the axis of the double helix for unwinding (part *A*). Given the length of the double helix, however, this would involve the separating strands in a considerable amount of flailing about, which seems unlikely in the confines of the cell.

A similar result is achieved by placing an apparatus to control the rotation at the free end (part *B*). However, the effect must be transmitted over a considerable distance, again involving the rotation of an unreasonable length of material. Remember also that replication converts each of the rotating single strands into the more cumbersome structure of a daughter duplex.

The demonstration that DNA behaves as a closed structure lacking free ends (see Chapter 28) excludes these models as a matter of principle and brings home

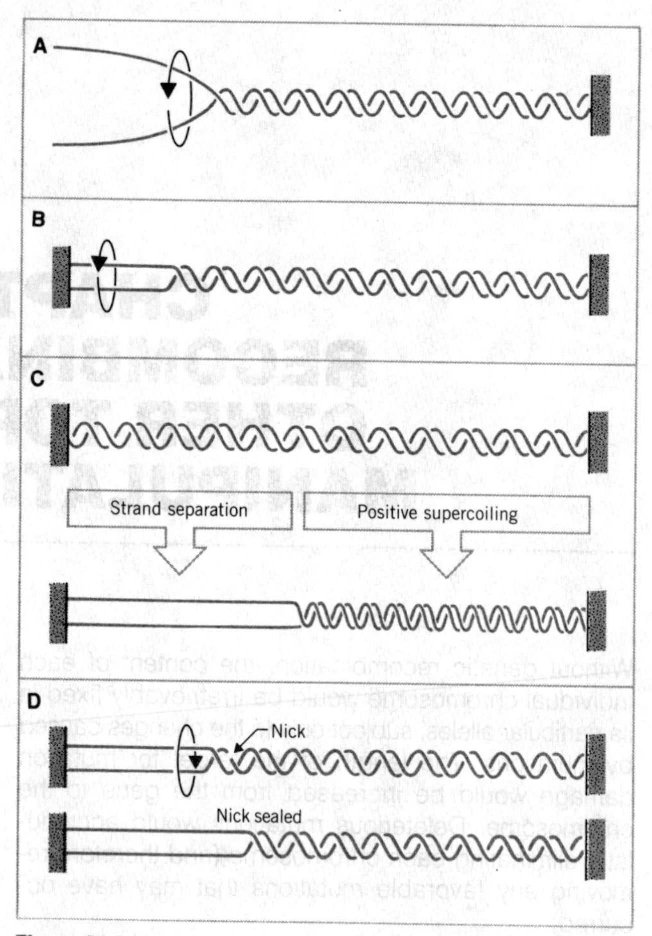

Figure 34.1
Separation of the strands of a DNA double helix could be achieved by one of
A rotation about a free end.
B an apparatus that holds the strands while it rotates.
C compensating positive supercoiling elsewhere.
D nicking, rotation, and ligation.

the severity of the topological problem. Consider the effects of separating the two strands in a molecule whose ends are not free to rotate (part *C* of the figure). When two intertwined strands are pulled apart from one end, the result is to *increase their winding about each other farther along the molecule*. Thus movement of a replication fork per se would generate increasing positive supercoiling ahead of it, rapidly generating insuperable resistance to further movement.

The problem can be overcome by introducing a transient nick in one strand. Part *D* of the figure shows that the internal free end allows the nicked strand to rotate about the intact strand, after which the nick can be sealed. The nicking and sealing reaction can be repeated as the replication fork advances. Such reactions occur with substrates of closed DNA; they are mediated by enzymes called topoisomerases, whose actions are crucial in replication.

Recall from Chapter 4 that a closed molecule of DNA can be characterized by its linking number, the number of times one strand crosses over the other in space. Closed DNA molecules of identical sequence may have different linking numbers, reflecting different degrees of supercoiling. Molecules of DNA that are the same except for their linking numbers are called **topological isomers.**

Any change in the linking number requires at least one strand to be broken, so that one strand can be rotated about the other, after which the break is made good. Such a reaction converts one topological isomer into another. Enzymes with this ability are called **DNA topoisomerases.** Some topoisomerases can relax (remove) only negative supercoils from DNA; others can relax both negative and positive supercoils. Some can introduce negative supercoils.

Topoisomerases are divided into two classes, according to the nature of the mechanisms they employ. **Type I** enzymes act by making a transient break in one strand of DNA. **Type II** enzymes act by introducing a transient double-strand break.

The best characterized type I topoisomerase is the enzyme of *E. coli*. This enzyme relaxes highly negatively supercoiled DNA with high efficiency; the rate of relaxation becomes slower as the linking number approaches that of completely relaxed DNA. The enzyme does not act on positively supercoiled DNA. It is coded by the gene *top;* mutations cause an increase in the level of supercoiling in the nucleoid (and may affect transcription, as described in Chapter 12).

In addition to the relaxation of negative supercoils in duplex DNA, the enzyme interacts with single-stranded DNA. This suggests that it likes negative supercoils because they tend to stabilize single-stranded regions, which could provide the substrate bound by the enzyme.

When *E. coli* topoisomerase I binds to DNA, it forms a stable complex in which one strand of the DNA has been nicked and its 5′-phosphate end is covalently linked to a tyrosine residue in the enzyme. This suggests a mechanism for the action of the enzyme; it transfers a phosphodiester bond in DNA to the protein, manipulates the structure of the two DNA strands, and then rejoins the original strand.

The transfer of bonds from nucleic acid to protein explains how the enzyme can function without requiring any input of energy. There has been no irreversible hydrolysis of bonds; their energy has been conserved through the transfer reactions.

A model for the action of topoisomerase I is illustrated in **Figure 34.2.** The enzyme binds to a region in which duplex DNA becomes separated into its single strands; then it breaks one strand, pulls the other strand through the gap, and finally seals the gap.

The reaction changes the linking number in steps of 1. Each time one strand is passed through the break in the other, there is a ΔL of $+1$. The figure illustrates the enzyme activity in terms of moving the individual

Figure 34.2
Bacterial type I topoisomerases recognize partially unwound segments of DNA and pass one strand through a break that is made in the other.

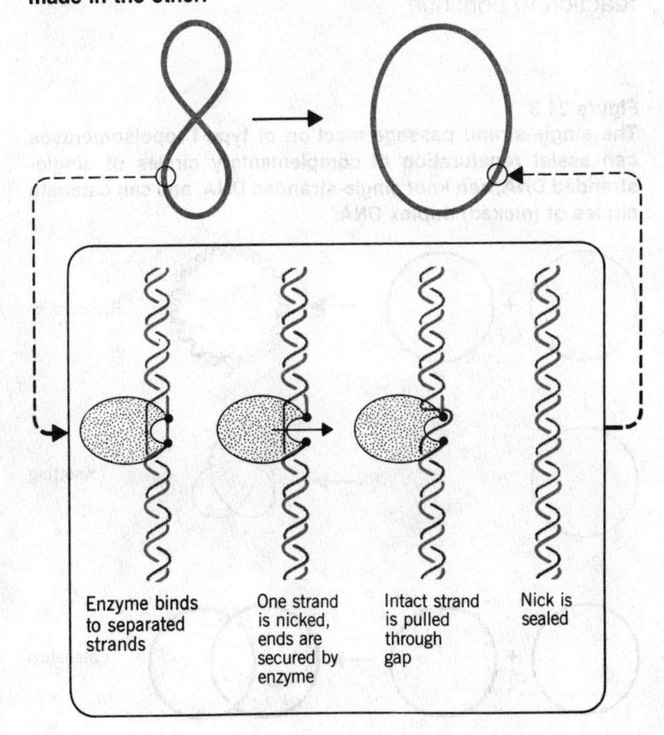

strands; in a highly supercoiled molecule, the free interchangeability of W and T should let the change in linking number be taken up by a change of $\Delta W = +1$, that is, by one less turn of negative supercoiling (see Chapter 4).

The reaction is equivalent to the rotation illustrated in part *D* of Figure 34.1, with the restriction that the enzyme limits the reaction to a single strand passage per event. (By contrast, the introduction of a nick in a supercoiled molecule allows free strand rotation to relieve all the tension by multiple rotations per event.)

The type I topoisomerase also can pass one segment of a single-stranded DNA through another. Some consequences of this **single-strand passage** reaction are illustrated in **Figure 34.3.**

Two single-stranded complementary circles may be permitted to renature into a duplex circle. Formation of a duplex region connecting the two circles must generate an equivalent underwinding, that is, negative T or W, elsewhere in each circle. In the absence of the enzyme, this prevents continuation of the pairing. Topoisomerase I relieves the tension by passing the two strands about each other in the opposite direction from the rotation involved in pairing. This allows the reaction to continue.

Figure 34.3
The single-strand passage reaction of type I topoisomerases can assist renaturation of complementary circles of single-stranded DNA, can knot single-stranded DNA, and can catenate circles of (nicked) duplex DNA.

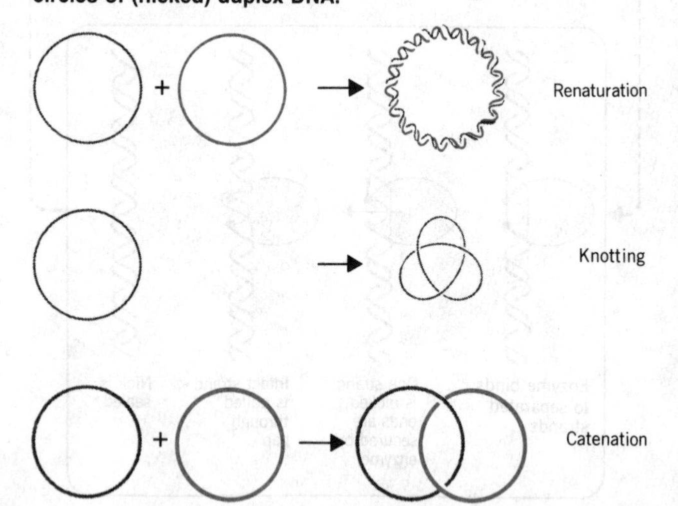

If the single-strand passage reaction is performed with two parts of the *same* circular molecule, a **knot** is produced.

These actions of topoisomerase I are extended by another activity discovered more recently. The enzyme can act on two circular *duplex* molecules of DNA, provided that one of them has a nick in one strand. When one circle is passed through another, they become interlocked. The structure is called a **catenane;** the activity is described as **catenating** the DNA. (Formally, catenation and knotting are topologically similar, since the first involves interlocking of different circles, while the second involves interlocking within a single circle.)

The catenation of double-stranded circles suggests the model illustrated in **Figure 34.4,** in which the enzyme binds opposite the nick, introduces a break in the other strand, passes a duplex of DNA through the break, and reseals the break. The only difference from the procedure involved in knotting single-stranded DNAs is that the passage reaction involves transfer of a duplex DNA. Since double-stranded and single-stranded DNA molecules have different structures, it seems surprising that the enzyme is able to handle passage of either. This could be a matter of appearance, however, if the enzyme actually passes one strand of the duplex through the break at a time (the predicted intermediates would have a complex topology).

The formal properties of eukaryotic type I topoisomerases are similar, but they can relax positive as well as negative supercoils.

Type II topoisomerases have been characterized from both prokaryotes and eukaryotes. Most relax both negative and positive supercoils; the reaction is ATP-dependent. Probably one unit of ATP is hydrolyzed for each catalytic event. The reaction is mediated by making a double-stranded break in one DNA duplex, and passing another duplex region through it. This is illustrated in **Figure 34.5.** A formal consequence of two-strand transfer is that the linking number is always changed in multiples of two. The topoisomerase II activity can be used also to introduce or resolve catenated duplex circles and knotted molecules.

The reaction probably represents a nonspecific recognition of duplex DNA in which the enzyme binds any two double-stranded segments that cross each

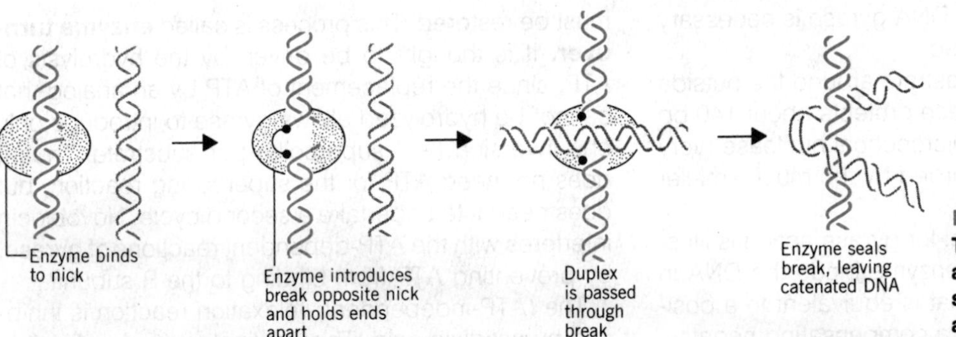

Figure 34.4
Type I topoisomerases can pass a duplex DNA through the intact single strand opposite a nick in another duplex DNA.

other. The hydrolysis of ATP may be used to drive the enzyme through conformational changes that provide the force needed to push one DNA duplex through the break made in the other. Because of the topology of supercoiled DNA, the relationship of the crossing segments will be such that the supercoils will be removed from either positively or negatively supercoiled circles. The best characterized example of this type of enzyme may be the T4 topoisomerase II, which is the product of three genes, and whose activity is essential for replication of phage DNA.

GYRASE INTRODUCES NEGATIVE SUPERCOILS IN DNA

Bacterial DNA gyrase (best characterized in *E. coli*) is a topoisomerase of type II that is able to *introduce* negative supercoils into a relaxed closed circular molecule. DNA gyrase binds to a circular DNA duplex and supercoils it processively and catalytically: it continues to introduce supercoils into the same DNA molecule. One molecule of DNA gyrase can introduce about 100 supercoils per minute.

The supercoiled form of DNA has a higher free energy than the relaxed form, and the energy needed to accomplish the conversion is supplied by the hydrolysis of ATP. In the absence of ATP, the gyrase can *relax* negative but not positive supercoils, although the rate is more than 10 times slower than the rate of introducing supercoils.

The *E. coli* DNA gyrase consists of two types of subunit. The active enzyme has the structure A_2B_2, with an overall size of about 400,000 daltons. The GyrA subunit is ~105,000 daltons; the GyrB subunit is ~95,000 daltons. Gyrase is inhibited by two types of antibiotic, each of which acts on one of the subunits. Mutations that confer resistance to the antibiotics identify the loci that code for the subunits. The *gyrA* locus (formerly called *nalA*) is the site of resistance to nalidixic acid and oxilinic acid. The *gyrB* locus (formerly called *cou*) is the site of resistance to coumermycin A1 and novobiocin. These drugs inhibit repli-

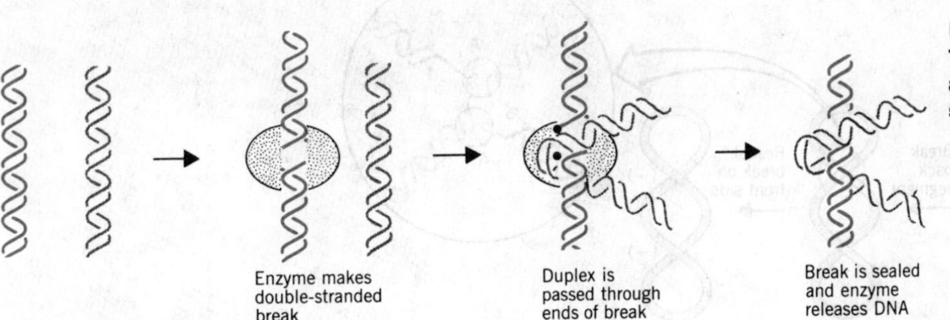

Figure 34.5
Type II topoisomerases can pass a duplex DNA through a double strand break in another duplex.

cation, which suggests that DNA gyrase is necessary for DNA synthesis to proceed.

Gyrase binds its DNA substrate around the outside of the protein tetramer. Gyrase protects about 140 bp of DNA from digestion by micrococcal nuclease (very similar to the protection afforded by the much smaller histone octamer).

The **sign inversion** model for gyrase action is illustrated in **Figure 34.6.** The enzyme binds the DNA in a crossover configuration that is equivalent to a positive supercoil. This induces a compensating negative supercoil in the unbound DNA. Then the enzyme breaks the double strand at the crossover of the positive supercoil, passes the other duplex through, and reseals the break.

The reaction directly inverts the sign of the supercoil: it has been converted from a +1 turn to a −1 turn. Thus the linking number has changed by $\Delta L = -2$, conforming with the demand that all events involving double-strand passage must change the linking number by a multiple of two. The gyrase then releases one of the crossing segments of the now negative bound supercoil, the negative turns redistribute along DNA (as change in either T or W or both), and the cycle begins again. The same type of topological manipulation is responsible for catenation and knotting, although the roles of these rearrangements *in vivo* are unknown.

When the gyrase releases the inverted supercoil, its conformation changes. For the enzyme to undertake another cycle of supercoiling, its original conformation must be restored. This process is called **enzyme turnover.** It is thought to be driven by the hydrolysis of ATP, since the replacement of ATP by an analog that cannot be hydrolyzed allows gyrase to introduce only one inversion (−2 supercoils) per substrate. Thus it does not need ATP for the supercoiling reaction, but does need it to undertake a second cycle. Novobiocin interferes with the ATP-dependent reactions of gyrase, by preventing ATP from binding to the B subunit.

The (ATP-independent) relaxation reaction is inhibited by nalidixic acid. This implicates the A subunit in the breakage and reunion reaction. Treating gyrase with nalidixic acid allows DNA to be recovered in the form of fragments generated by a staggered cleavage across the duplex. The termini all possess a free 3′-OH group and a 4-base 5′ single-strand extension covalently linked to the A subunit (see Figure 34.6). The covalent linkage retains the energy of the phosphate bond; this can be used to drive the resealing reaction, explaining why gyrase can undertake relaxation without ATP. The sites of cleavage are fairly specific, occurring about once every 100 bp.

RECOMBINATION REQUIRES SYNAPSIS OF HOMOLOGOUS DUPLEX DNAs

Recombination between chromosomes involves a physical exchange of parts (see Figure 2.10). The structure created by this exchange is visible at meiosis in the form of a chiasma (see Figure 2.9). The chiasma

Figure 34.6
DNA gyrase may introduce negative supercoils in duplex DNA by inverting a positive supercoil.

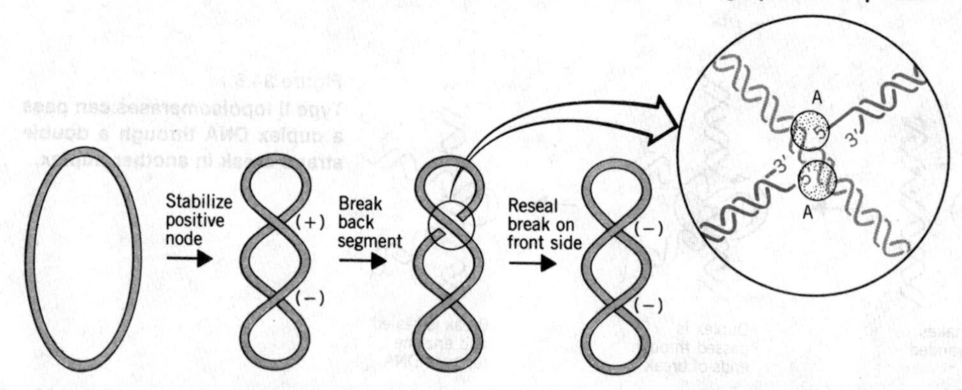

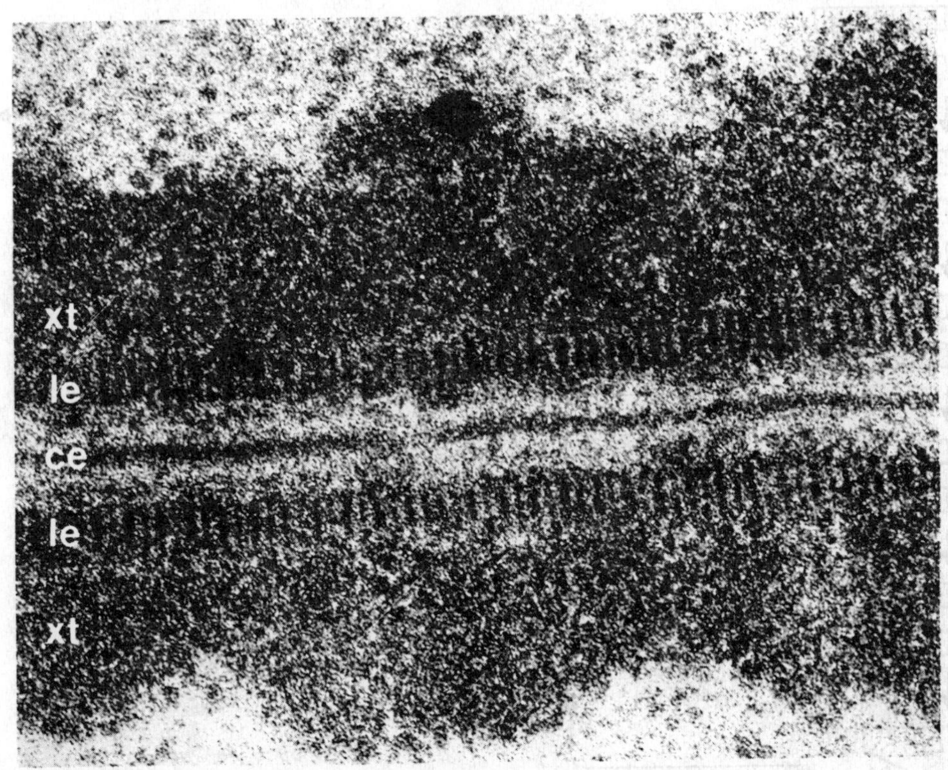

Figure 34.7
The synaptonemal complex of *Neotellia* shows that pairing of chromosomes does not bring the homologous DNAs into juxtaposition.
xt indicates chromatin; *le* indicates the lateral elements, each about 50 nm in diameter; *ce* indicates the central element, about 18 nm in diameter. The distance between the lateral elements is about 120 nm.
Photograph kindly provided by M. Westergaard and D. Von Wettstein.

represents the results of a **breakage and reunion,** in which two nonsister chromatids (each containing a duplex of DNA) have been broken and then linked each with the other. Recombination occurs between precisely corresponding sequences, so that not a single base pair is added to or lost from the recombinant chromosomes.

Yet this is the act of recombination itself. Its description begs the issue of the first, crucial step: two homologous duplex molecules of DNA must be brought into close contact so that the corresponding sequences can be exchanged. The nature of the problem may be somewhat different in eukaryotes and prokaryotes.

In prokaryotic systems, we are dealing largely with the DNA itself (albeit in the nucleoid). There is no visible guide to the events responsible for recognition of homology, but probably they involve only those regions actually participating in the recombination crossover. The recognition may be part and parcel of the recombination mechanism, which may rely on the comparison of single strands of DNA to identify complementary sequences (see later).

In eukaryotic cells, the DNA is tightly packaged into the discrete structures of the chromosomes. How do the DNA molecules come into juxtaposition? Contact between a pair of parental *chromosomes* occurs early in meiosis. The process is called **synapsis** or **chromosome pairing.** Homologous chromosomes (each of which actually consists of the two sister chromatids produced by the prior replication) approach one another. They become laterally associated in the form of a **synaptonemal complex,** which has a characteristic structure in each species, although there is wide variation in the details between species.

An example of a synaptonemal complex is shown in **Figure 34.7.** Each chromosome at this stage appears as a mass of chromatin bounded by a **lateral element** (which in this case has a striated structure). The two lateral elements are separated from each other by a **central element.** The triplet of parallel dense strands lies in a single plane that curves and twists

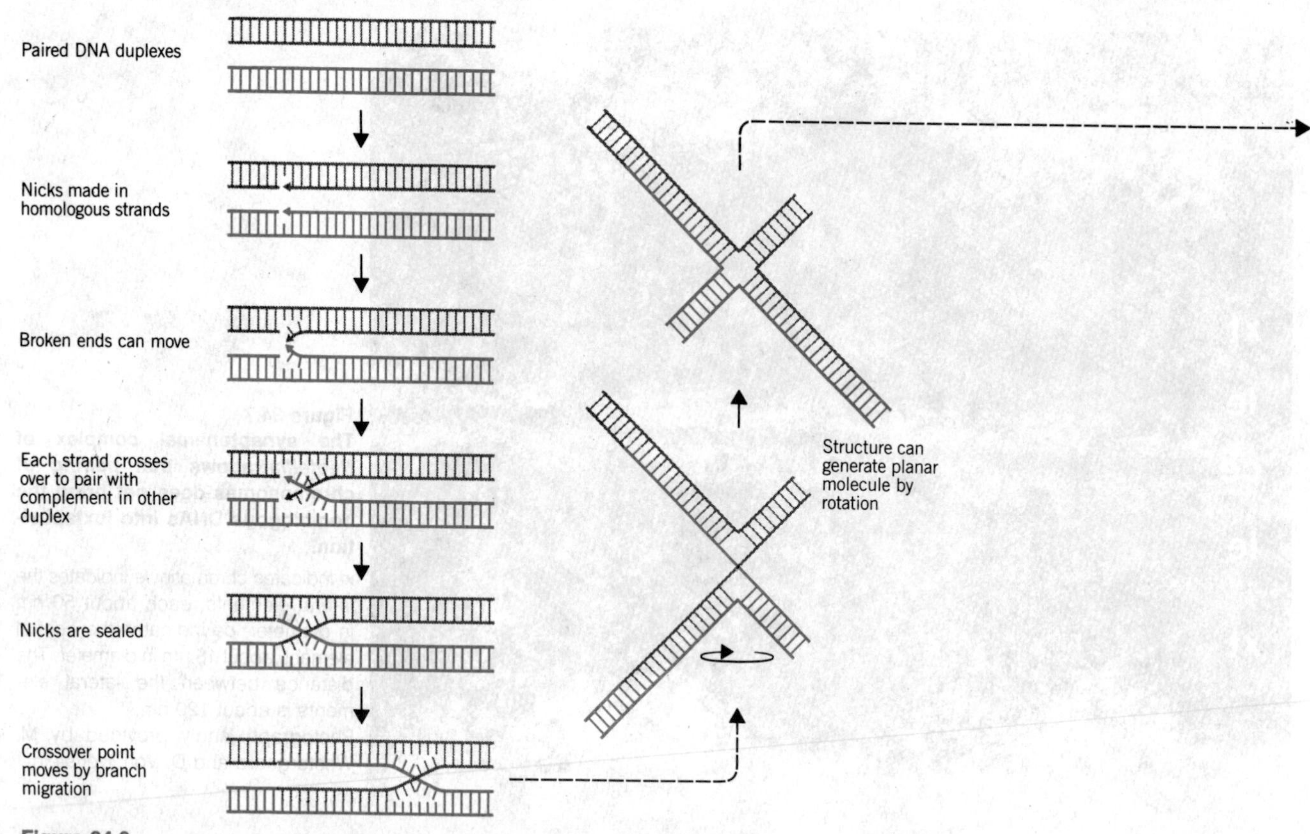

Figure 34.8
Recombination between two paired duplex DNAs could be initiated by reciprocal single-strand exchange, extended by branch migration, and resolved by nicking.

along its axis. The distance between the homologous chromosomes is considerable in molecular terms, more than 200 nm (the diameter of DNA is 2 nm).

The generation of the synaptonemal complex coincides with the presumed time of crossing-over, although there is no direct evidence that recombination occurs at the stage of synapsis. A major problem in understanding the role of the complex is that, although it brings homologous chromosomes into contact, it is far from bringing homologous DNA molecules into contact.

The only visible link between the two sides of the synaptonemal complex is provided by spherical or cylindrical structures observed in fungi and insects. They lie across the complex and are called **nodes** or **recombination nodules;** they occur with the same frequency and distribution as the chiasmata. Their name reflects the hope that they may prove to be the sites of recombination.

At the next stage of meiosis, the chromosomes shed the synaptonemal complex; then the chiasmata become visible as points at which the chromosomes are connected. This is presumed to indicate the occurrence of a genetic exchange. Later in meiosis, the chiasmata may move toward the ends of the chromosomes. This flexibility suggests that they may represent some remnant of the recombination event, rather than providing the actual intermediate.

BREAKAGE AND REUNION INVOLVES HETERODUPLEX DNA

In contrast with the unknown basis for the ability of homologous duplex DNA sequences to recognize one

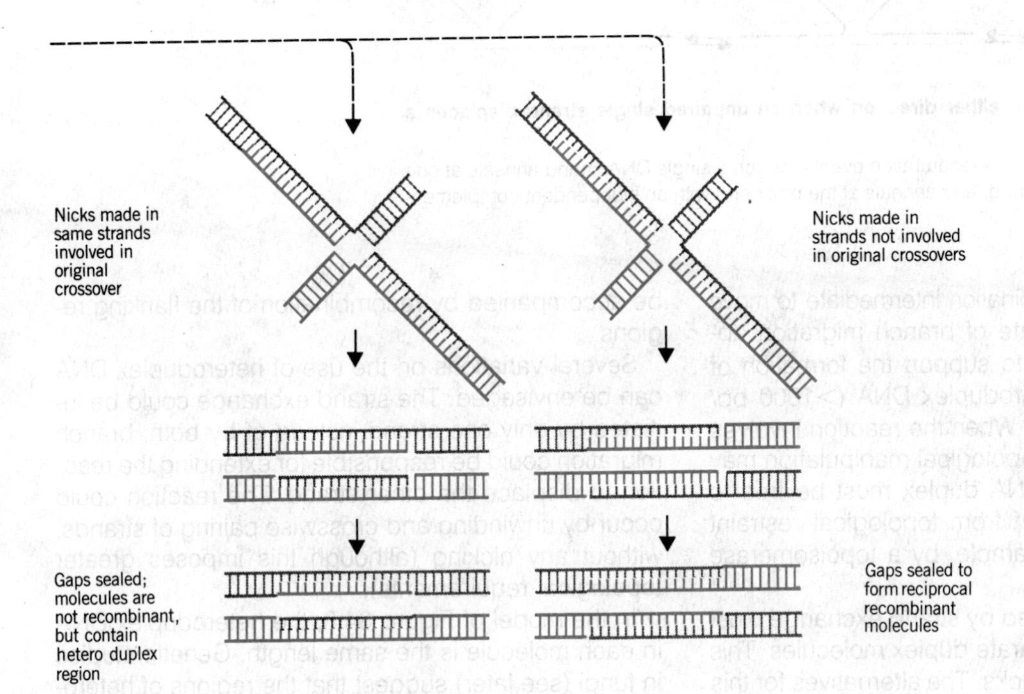

another, a mechanism of considerable precision—base pairing—exists for recognition between complementary single strands. This is used in the formation of recombination intermediates. One of the several models for the reaction is illustrated in **Figure 34.8.**

The process starts with breakage at the corresponding points of the homologous strands of two paired DNA duplexes. The breakage allows movement of the free ends created by the nicks. Each strand leaves its partner and crosses over to pair with its complement in the other duplex. This reciprocal exchange creates a connection between the two DNA duplexes. Initially, this is sustained only by hydrogen bonding; at some point, it is made covalent by sealing the nicks at the sites of exchange (the sealing could occur at a later stage than shown in the figure). The connected pair of duplexes is called a **joint molecule.**

At the site of recombination, each duplex has a region consisting of one strand from each of the parental DNA molecules. This region is called **hybrid DNA** or **heteroduplex DNA.** Model building shows that there is (surprisingly) little steric hindrance of the formation of the reciprocal heteroduplex regions between the paired DNA molecules; thus virtually all of the bases can remain base paired (that is, in either the parental duplex or heteroduplex regions). It is not necessary to invoke the existence of extensive single-stranded regions.

Once the strand exchange has been initiated, it can move along the duplex. This ability is called **branch migration.** Branches in a DNA structure can involve either single or double strands; either type can migrate. **Figure 34.9** shows an example of single-strand branch migration. The point of branching can migrate in either direction as one strand is displaced by the other.

Branch migration is important for both theoretical and practical reasons. As a matter of principle, it confers a dynamic property on recombinant structures. As a practical feature, its existence means that the point of branching cannot be established by examining a molecule *in vitro* (because the branch may have migrated since the molecule was isolated).

The same type of movement could allow the point

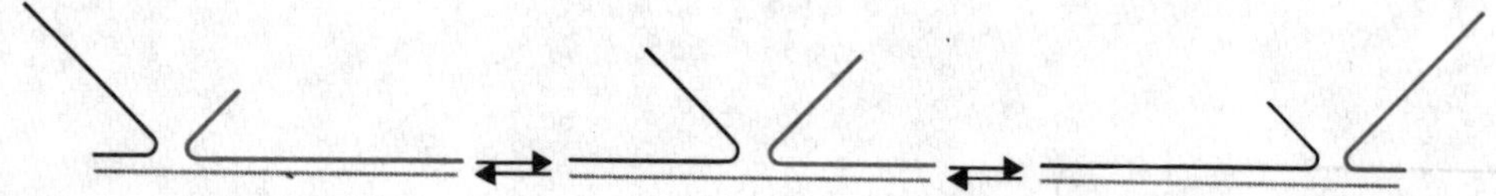

Figure 34.9
Branch migration can occur in either direction when an unpaired single strand displaces a paired strand.
This structure could be created by a renaturation event in which a single DNA strand anneals at one end with one complementary strand, and anneals at the other end with an independent complementary strand.

of crossover in the recombination intermediate to move in either direction. The rate of branch migration appears to be fast enough to support the formation of extensive regions of heteroduplex DNA (>1000 bp/sec) in natural conditions. When the reaction involves duplex DNA molecules, topological manipulation may be required; either the DNA duplex must be free to rotate, or equivalent relief from topological restraint must be provided, for example, by a topoisomerase activity (see later).

The joint molecule formed by strand exchange must be **resolved** into two separate duplex molecules. This requires a further pair of nicks. The alternatives for this reaction are visualized in the right side of Figure 34.8 in terms of the planar molecule generated by rotating one of the duplexes of the recombination intermediate. The consequences depend on which pair of strands is nicked.

If the nicks are made in the pair of strands that were *not* originally nicked (the pair that did not initiate the strand exchange), all four of the original strands have been nicked. This releases recombinant DNA molecules. The duplex of one DNA parent is covalently linked to the duplex of the other DNA parent, via a stretch of heteroduplex DNA. There has been a conventional recombination event between markers located on either side of the heteroduplex region.

If the *same* two strands involved in the original nicking are nicked again, the other two strands remain intact. The nicking releases the original parental duplexes, intact except that each has a residuum of the event in the form of a length of heteroduplex DNA.

These alternative resolutions of the joint molecule establish the principle that a strand exchange between duplex DNAs always leaves behind a region of heteroduplex DNA, but the exchange may or may not be accompanied by recombination of the flanking regions.

Several variations on the use of heteroduplex DNA can be envisaged. The strand exchange could be initiated by only one strand instead of by both; branch migration could be responsible for extending the reaction to displace the other strand. The reaction could occur by unwinding and crosswise pairing of strands, without any nicking (although this imposes greater topological requirements).

In the model of Figure 34.8, the heteroduplex DNA in each molecule is the same length. Genetic studies in fungi (see later) suggest that the regions of heteroduplex DNA in recombinant chromosomes can vary in length. This can be accommodated by a modification of the model in which a new stretch of DNA is synthesized to replace part of one of the exchanging strands.

All of these factors affect the details of the mechanism and the order of events, but all models rely on the same general principle: the formation of an intermediate involving heteroduplex DNA that can be extended by branch migration.

DO DOUBLE-STRAND BREAKS INITIATE RECOMBINATION?

Another model for recombination supposes that *genetic exchange is initiated by a double-strand break*. Then one of the single strands migrates to the other duplex, so that the duplex molecules become connected by a stretch of heteroduplex DNA. The model is illustrated in **Figure 34.10.**

Recombination is initiated when an endonuclease

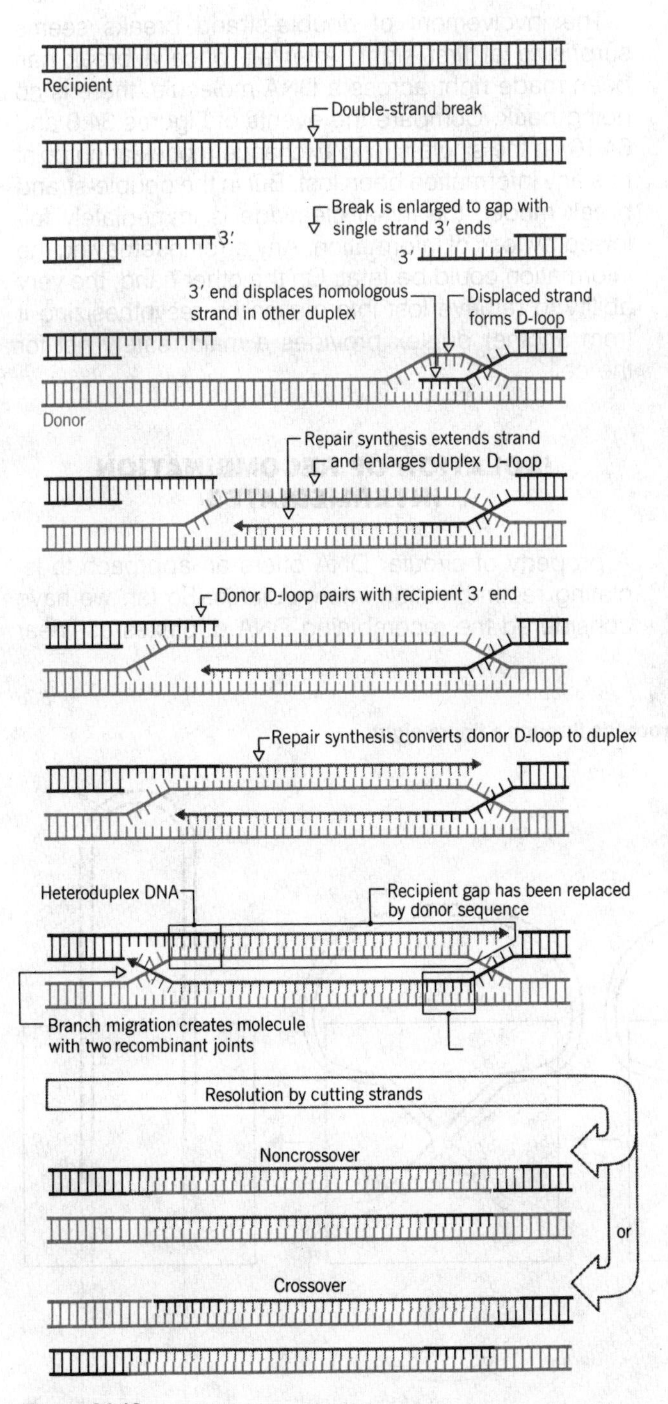

Figure 34.10

Recombination could be initiated by a double-strand break, followed by formation of single-stranded 3′ ends, one of which migrates to a homologous duplex.

makes a double-strand break in one chromatid, the "recipient." The cut is enlarged to a gap, probably by exonuclease action. The exonuclease(s) nibble away one strand on either side of the break, generating 3′ single-stranded termini. One of the free 3′ ends then invades a homologous region in the other, "donor" duplex. The formation of heteroduplex DNA generates a D-loop, in which one strand of the donor duplex is displaced. The D-loop is extended by repair synthesis that uses the free 3′ end as a primer.

Eventually the D-loop becomes large enough to correspond to the entire length of the gap on the recipient chromatid. When the extruded single strand reaches the far side of the gap, the complementary single-stranded sequences anneal. Now there is heteroduplex DNA on either side of the gap, and the gap itself is represented by the single-stranded D-loop.

The duplex integrity of the gapped region can be restored by repair synthesis using the 3′ end on the left side of the gap as a primer. Overall, the gap has been repaired by two individual rounds of single-strand DNA synthesis.

Branch migration converts this structure into a molecule with two recombinant joints. If both junctions are resolved in the same way, for example, the inner strands are cut at each joint, the original noncrossover molecules will be released, each with a region of altered genetic information that is a footprint of the exchange event. If the two junctions are resolved in opposite ways—one is cut on the inner strand and the other on the outer strand—a genetic crossover results.

The structure of the two-jointed molecule before it is resolved illustrates some of the differences between the double-strand break model and models that invoke only single strand actions. Heteroduplex DNA has been formed at each end of the region involved in the exchange. Between the two heteroduplex segments is the region corresponding to the gap, which now has the sequence of the donor DNA in both molecules.

So the arrangement of heteroduplex sequences is asymmetric, and part of one molecule has been converted to the sequence of the other (which is why the initiating chromatid is called the recipient). Single-strand exchange models make the opposite prediction, that the initiating chromatid will be the donor of genetic information. Data in yeast are consistent with the demand of the double-strand break model that initiation is involved with receiving genetic information.

This model makes predictions about the relationships between the sites involved in the genetic exchange event which fit well with results obtained by analyzing fungal systems (see later). One of its main strengths is that double-strand breaks are known to be involved in certain recombination-like events in yeast (see Chapter 36). When a plasmid with a double-strand break is introduced into yeast, it stimulates recombination, and, furthermore, can give rise to products in which a gap has been repaired.

The double-strand break model does not reduce the importance of the formation of heteroduplex DNA, which remains the only plausible means by which two duplex molecules can interact. However, by shifting the responsibility for initiating recombination from single-strand to double-strand breaks, it requires a change in our perspective about the ability of the cell to manipulate DNA.

The involvement of double-strand breaks seems surprising at first sight. After all, once a break has been made right across a DNA molecule, there is no going back. Compare the events of Figures 34.8 and 34.10. In the single-strand exchange model, at no point has any information been lost. But in the double-strand break model, the initial cleavage is immediately followed by loss of information. Any error in retrieving the information could be fatal. On the other hand, the very ability to retrieve lost information by resynthesizing it from another duplex provides a major safety net for the cell.

ISOLATION OF RECOMBINATION INTERMEDIATES

A property of circular DNA offers an approach to isolating recombination intermediates. So far, we have considered the recombining DNA duplexes as linear

Figure 34.11
Reciprocal recombination between homologous duplex circles proceeds through a figure-eight intermediate to generate a dimeric circle.

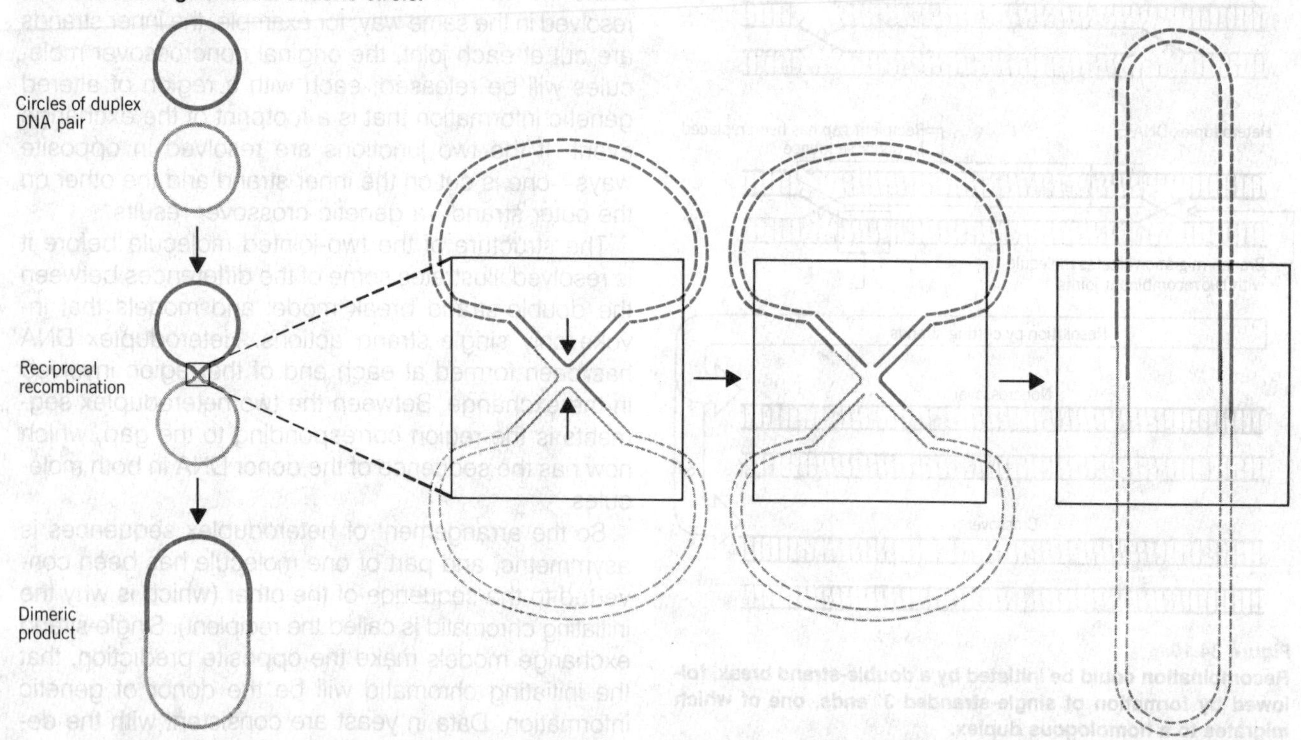

molecules, but many genomes (especially in viruses and plasmids) are circular.

Figure 34.11 illustrates the consequences of a reciprocal recombination between homologous sites on two circular DNA molecules. The expanded part of the figure shows that the structure at the recombinant joint is the same as that shown in Figure 34.8; the only difference is that the ends of each duplex parental DNA are joined together.

This structure has an important topological consequence. Resolution of the structure by the recombination route generates a dimeric circle, with the two original parental sequences joined head to tail.

What does this mean for the life-style of circular DNA elements? If they indulge in recombination, the original (monomeric) circular genome can be maintained only by *pairs* of recombination events. A single reciprocal recombination always generates the dimer shown in Figure 34.11. This dimer can return to the monomeric condition by a second recombination involving any pair of homologous sequences, as illustrated in **Figure 34.12.** In any population of circular genomes, we may therefore expect to find some multimeric circles if recombination is occurring.

A practical consequence of recombination between circular DNAs is that the recombination intermediate should take the form of a figure-eight, as depicted in the expanded structure of Figure 34.11. Can we isolate such molecules? Dimeric DNA molecules with this appearance have been detected among the DNAs of several phages in infected cells and also in a yeast plasmid population. However, the mere outline of these structures does not unequivocally identify them as recombination intermediates; for example, a figure-eight could represent two interlocked (catenated) monomeric circles.

The nature of the figure-eight can be distinguished by cleaving the isolated molecule with a restriction enzyme that cuts each monomeric circle only once. The two parts of the figure-eight then will fall apart *unless they are covalently connected.* If the figure-eight is a recombination intermediate, cleavage will generate a structure in which the parental duplex molecules are held together by a region of heteroduplex DNA at the point of fusion. The cleaved molecule has four arms, each pair of the arms corresponding to each of the original monomeric circles.

In the cleaved form, this molecule is called a **chi structure,** because of its resemblance to the Greek letter χ. An example is shown in **Figure 34.13** together with an interpretive drawing. The four arms of duplex DNA are connected by a region in which the heteroduplex segments have been pulled apart into their constituent single strands (probably by the conditions of preparation). This structure corresponds precisely with the predicted intermediate of Figures 34.8 and 34.11; its isolation directly demonstrates the existence of heteroduplex DNA *in vivo*.

Chi structures are found only in wild-type bacteria, not in $recA^-$ mutants. Thus their presence is correlated with the ability of the cell to undertake genetic recombination. This strengthens the case that they are authentic intermediates in the process and not some irrelevant oddity.

These observations offer an approach to purifying the recombination enzymes. Cell-free extracts can be assayed for their ability to generate dimeric circles from monomeric circles. Extracts from wild-type cells can accomplish this reaction, generating figure-eight molecules that can be cleaved into chi structures. Extracts from $recA^-$ cells are inactive. In principle, it should be possible to fractionate this system into its components and to use *in vitro* complementation as-

Figure 34.12
A dimeric circle can generate two monomeric circles by a reciprocal recombination between homologous sequences.

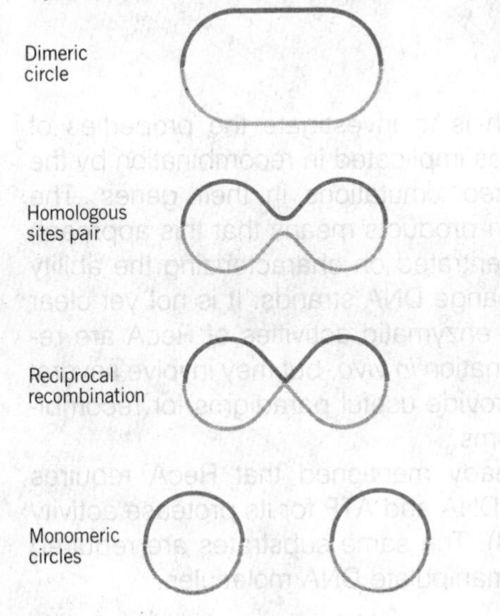

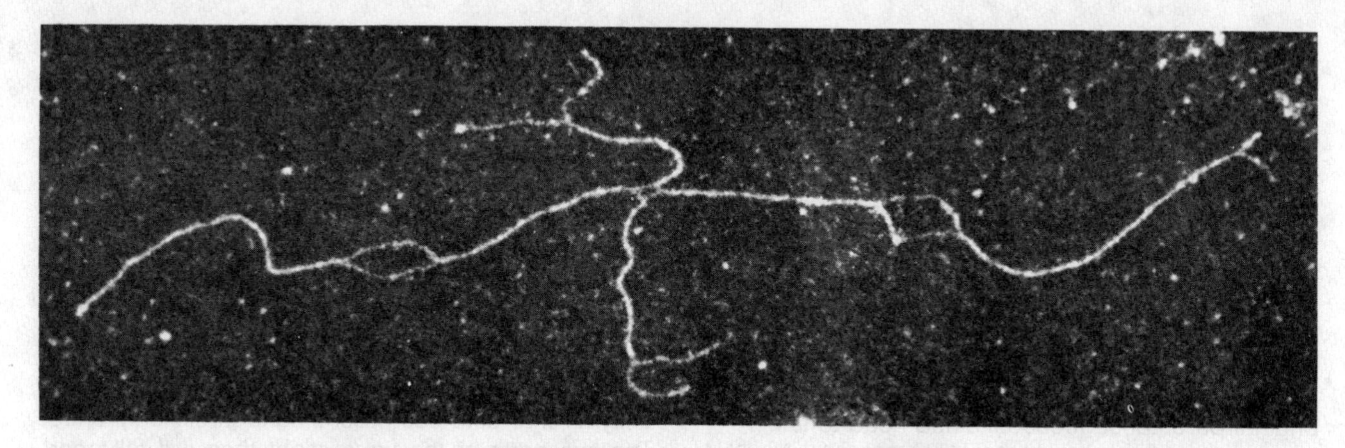

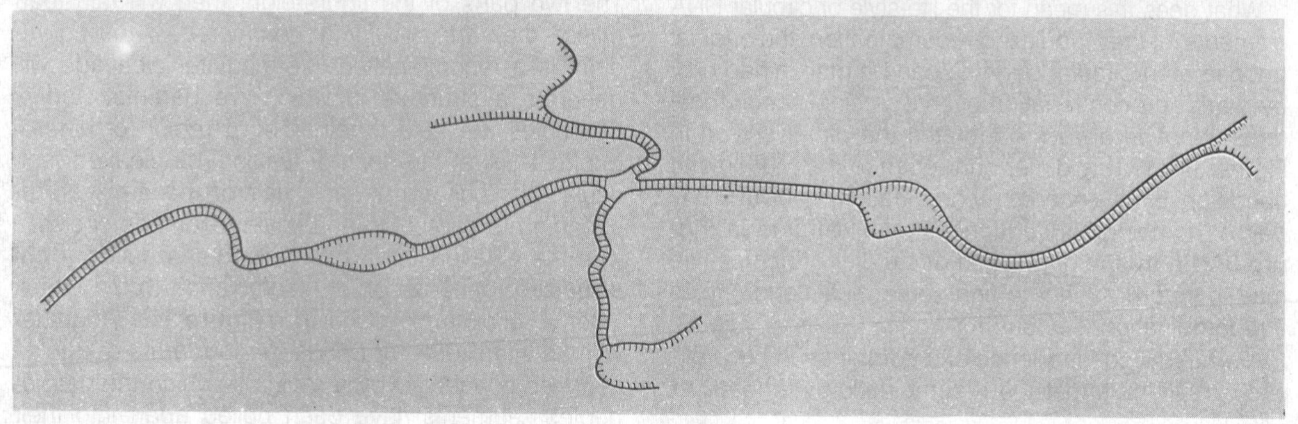

Figure 34.13
Cleavage of a figure-eight at homologous sites on each circle generates a chi structure in which the four duplex arms are held together at a site of strand exchange.
Photograph kindly provided by David Dressler.

says, along lines similar to the approach developed previously for DNA replication (see Chapter 32).

THE STRAND-EXCHANGE FACILITY OF RecA

Very little is known about the apparatus responsible for general recombination. Enzymes involved in some specialized recombination events have been characterized, and some related topological activities have been described, but it has proved very difficult to identify the enzymes that undertake homologous recombination.

One approach is to investigate the properties of bacterial enzymes implicated in recombination by the occurrence of *rec*$^-$ mutations in their genes. The paucity of known products means that this approach so far has concentrated on characterizing the ability of RecA to exchange DNA strands. It is not yet clear exactly how the enzymatic activities of RecA are related to recombination *in vivo*, but they involve several reactions that provide useful paradigms for recombination mechanisms.

We have already mentioned that RecA requires single-stranded DNA and ATP for its protease activity (see Chapter 33). The same substrates are required for its ability to manipulate DNA molecules.

RecA promotes base pairing between a single strand of DNA and its complement in a supercoiled duplex circle. The single strand invades the circle to displace the original partner to its complement, forming a D-loop, as illustrated in the upper row of **Figure 34.14.** This reaction is called **single-strand uptake** or **single-strand assimilation.** Single-strand assimilation is potentially related to the model for heteroduplex formation in Figure 34.8. The difference is that the recombination model invokes two invasions, with each single strand reciprocally displacing its homologue in the partner duplex.

The RecA protein can bind both single-stranded DNA (an action stimulated by the presence of ATP) and double-stranded DNA (an action requiring the presence of ATP). The assimilation reaction requires hydrolysis of the ATP. Assimilation works at an optimum rate with 1 monomer of ATP for every 5–10 nucleotides of single-stranded DNA. Thus the reaction is driven by the binding of RecA to single-stranded DNA.

We do not yet understand the mechanism of the preliminary stages that precede D-loop formation, but probably the RecA protein first binds to single-stranded DNA, and then binds duplex DNA to search for a complement to the single-stranded region. The main query about this procedure is how the duplex DNA is melted to allow examination of its potential complementarity.

In the D-loop reaction, the displacement of the homologous circular strand is only partial; the structure of the covalently closed duplex circle may impose topological constraints on the progress of the reaction. A more extensive reaction can occur if we reverse the structures of the double- and single-stranded molecules. A circular single-stranded molecule can react with a homologous duplex molecule to displace one strand in its entirety. The reaction is shown in the center row of Figure 34.14. It starts at one end of the duplex and continues to the other end, forming a heteroduplex up to several thousand base pairs long.

This pair of substrates has been used to ask some questions about the mechanisms of the reaction. Does it proceed in a specific direction? **Figure 34.15** shows that this can be tested by using a linear duplex DNA that is complementary with the single strand only at

Figure 34.14
RecA promotes the assimilation of invading single strands into duplex DNA so long as one of the reacting strands has a free end.

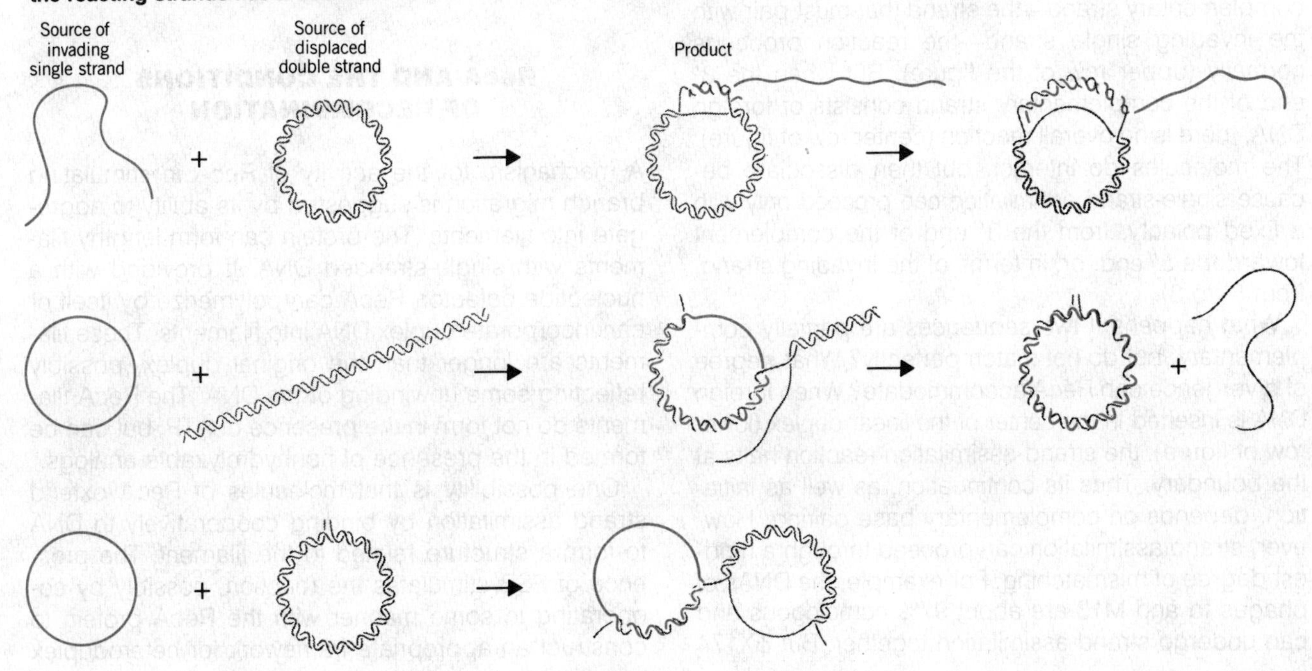

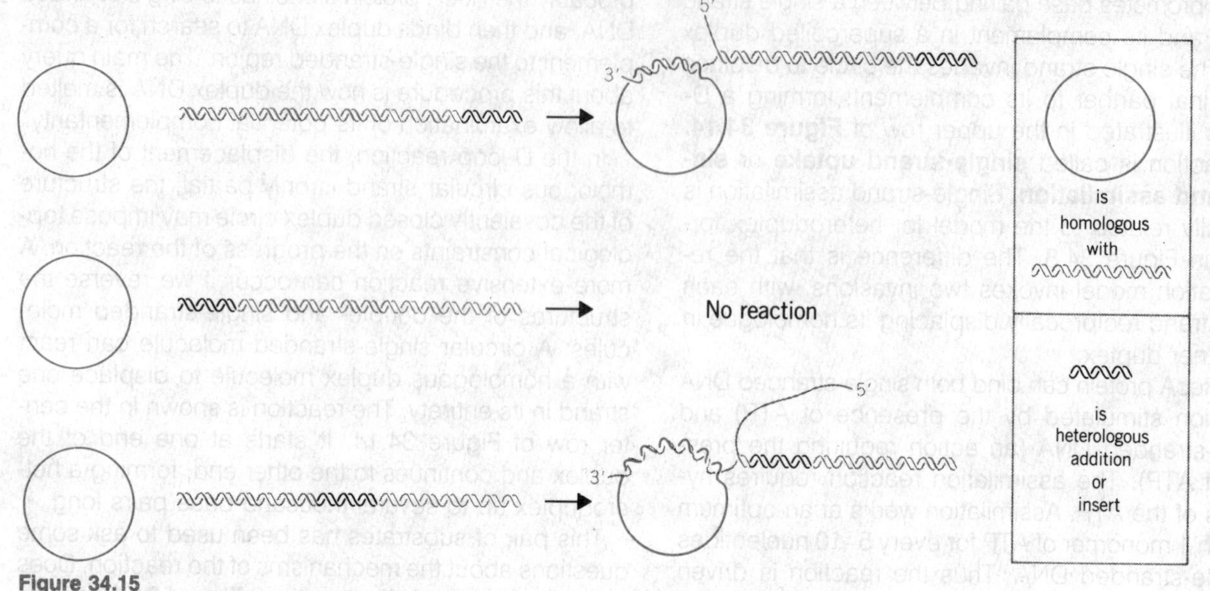

Figure 34.15
RecA can assimilate a circular single strand into a duplex only if there is a complementary 3′ end with which the single strand can base pair. If foreign DNA is present within the duplex, the assimilation reaction stops at the end of the complementary sequence.

one end. This substrate is generated by adding a foreign DNA fragment to one end of the linear molecule.

When the fragment is located at the 5′ end of the complementary strand—the strand that must pair with the invading single strand—the reaction proceeds normally (upper row of the figure). But when the 3′ end of the complementary strand consists of foreign DNA, there is no overall reaction (center row of figure). The molecules do interact, but then dissociate because single-strand assimilation can proceed only with a fixed polarity, from the 3′ end of the complement toward the 5′ end, or, in terms of the invading strand, from 5′ to 3′.

What happens if two sequences are partially complementary, but do not match perfectly? What degree of divergence can RecA accommodate? When foreign DNA is inserted in the center of the linear duplex (lower row of figure), the strand-assimilation reaction halts at the boundary. Thus its continuation, as well as initiation, depends on complementary base pairing. However, strand assimilation can proceed through a modest degree of mismatching. For example, the DNAs of phages fd and M13 are about 97% homologous and can undergo strand assimilation together. But ϕX174 DNA and G4 DNA, which are 70% homologous, cannot do so. The maximum tolerable divergence is not yet known.

RecA AND THE CONDITIONS OF RECOMBINATION

A mechanism for the activity of RecA in stimulating branch migration is suggested by its ability to aggregate into filaments. The protein can form lengthy filaments with single-stranded DNA. If provided with a nucleotide cofactor, RecA can polymerize by itself or can incorporate duplex DNA into filaments. These filaments are longer than the original duplex, possibly reflecting some unwinding of the DNA. The RecA filaments do not form in the presence of ATP, but can be formed in the presence of nonhydrolyzable analogs.

One possibility is that molecules of RecA extend strand assimilation by binding cooperatively to DNA to form a structure related to the filament. The presence of SSB stimulates the reaction, possibly by cooperating in some manner with the RecA protein to construct an appropriate framework for heteroduplex formation. Such a model explains the requirement for stoichiometry between RecA and single-stranded DNA (which would not be found if individual molecules of RecA moved along the nucleic acid). The extension reaction requires hydrolysis of ATP.

Single-strand assimilation requires a free end of DNA. As **Figure 34.14** shows, the free end can be on the invading strand (upper row) or on the duplex molecule (center row). If both molecules are circular, the duplex circle must have a break in the strand that is to be displaced (lower row). The need for a free end could reflect an inherent property of the RecA protein, or it could result from topological constraints. A single strand with a free end can wrap around its complement to form a duplex, but if there is no free end, the molecules are rather quickly constrained from further pairing.

When a single-stranded molecule reacts with a duplex DNA, the duplex molecule becomes unwound in the region of the recombinant joint. The initial region of heteroduplex DNA may not even lie in the conventional double helical form, but could consist of the two strands associated side by side. A region of this type is called a **paranemic joint** (compared with the classical intertwined **plectonemic** relationship of strands in a double helix).

A paranemic joint is unstable; further progress of the reaction requires its conversion to the double-helical form. This reaction is equivalent to removing negative supercoils and may require topoisomerase I. The topoisomerase presumably solves the unwinding/rewinding problem by making transient breaks that allow the strands to rotate about each other. An implication of its involvement is that recombination could occur by unpairing and cross-pairing without the initial strand breakages shown in Figure 34.8.

All of the reactions we have discussed so far represent only a part of the potential recombination event: the invasion of one duplex by a single strand. Two duplex molecules can interact with each other under the sponsorship of RecA, provided that one of them has a single-stranded region of at least 50 bases. The single-stranded region can take the form of a tail on a linear molecule or of a gap in a circular molecule.

The reaction between a partially duplex molecule and an entirely duplex molecule leads to the exchange of strands. An example is illustrated in **Figure 34.16.** Assimilation starts at one end of the linear molecule, where the invading single strand displaces its homologue in the duplex in the customary way. But when the reaction reaches the region that is duplex in both molecules, the invading strand unpairs from its partner, which then pairs with the other displaced strand.

At this stage, the molecule has a structure indistinguishable from the recombinant joint in Figure 34.8. When the reacting molecules are circular, the product can be visualized as a chi structure. The reaction sponsored *in vitro* by RecA can generate chi structures, which suggests that the enzyme can mediate reciprocal strand transfer.

RecA requires a single-stranded region in one of the reacting DNA molecules. How is this provided *in vivo?* One possibility is that there is a limited degradation of one strand of DNA. Another is that some recombination activity unwinds the DNA. It is possible that the RecBC product may be involved. Originally identified as a potent nuclease, in the presence of SSB the enzyme also is able to unwind DNA to generate single-stranded regions.

A protein with abilities to manipulate DNA in a manner similar to RecA has been purified from the fungus *Ustilago maydis*. The Rec1 protein can synapse homologous duplex DNA molecules in the same single-strand-dependent manner as RecA. In the presence of type I topoisomerase, synapsed DNA molecules can be linked.

We have dealt with the actions of RecA in a context independent of DNA sequence. However, certain hotspots stimulate the RecA recombination system. They were discovered in phage lambda in the form of mutants, called *chi,* that have single base-pair changes creating sites that stimulate recombination.

These sites share the same nonsymmetrical sequence of 8 bp. Any change in this sequence abolishes its ability to act as a hotspot. The *chi* sequence occurs naturally in *E. coli* DNA about once every 10 kb. Its absence from wild-type lambda DNA, and also from other genetic elements, shows that it is not essential for RecA-mediated recombination.

The *chi* sequence is able to stimulate recombination in its general vicinity, say, within a distance of up to 10 kb from the site. A chi site can be activated by a double-strand break made several kilobases away *on*

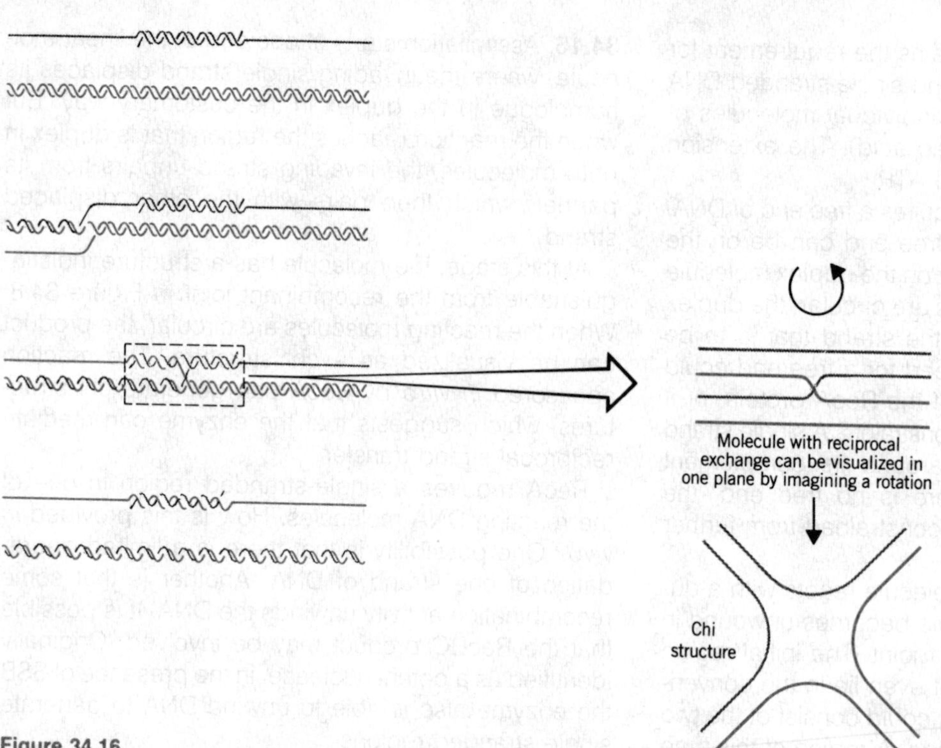

Figure 34.16
RecA-mediated strand exchange between partially duplex and entirely duplex DNA generates a joint molecule with the same structure as a recombination intermediate.

one particular side. This dependence on orientation suggests that the recombination apparatus must associate with DNA at a broken end, and then can move along the duplex in only one direction. We do not know what role *chi* plays when the recombination machinery reaches it; perhaps the *chi* sequence is needed to initiate or resolve recombination intermediates.

GENE CONVERSION ACCOUNTS FOR INTERALLELIC RECOMBINATION

The involvement of heteroduplex DNA in recombination explains the characteristics of recombination between alleles; indeed, these data provided the impetus for the development of the heteroduplex model. When recombination between alleles was discovered, the natural assumption was that it takes place by the same mechanism of reciprocal recombination that applies to more distant loci. That is to say that an individual breakage and reunion event occurs between the loci to generate a reciprocal pair of recombinant chromosomes. However, in the close quarters of a single gene, the formation of heteroduplex DNA itself is usually responsible for the recombination event.

Individual recombination events can be studied in the *Ascomycetes* fungi, because the products of a single meiosis are held together in a large cell, the ascus. Even better, the four haploid nuclei produced by meiosis are arranged in a linear order. Actually, a mitosis occurs after the production of these four nuclei, giving a linear series of eight haploid nuclei. **Figure 34.17** shows that each of these nuclei effectively represents the genetic character of one of the eight strands of the four chromosomes produced by the meiosis.

Meiosis in a heterozygote should generate four copies of each allele. This is seen in the majority of spores.

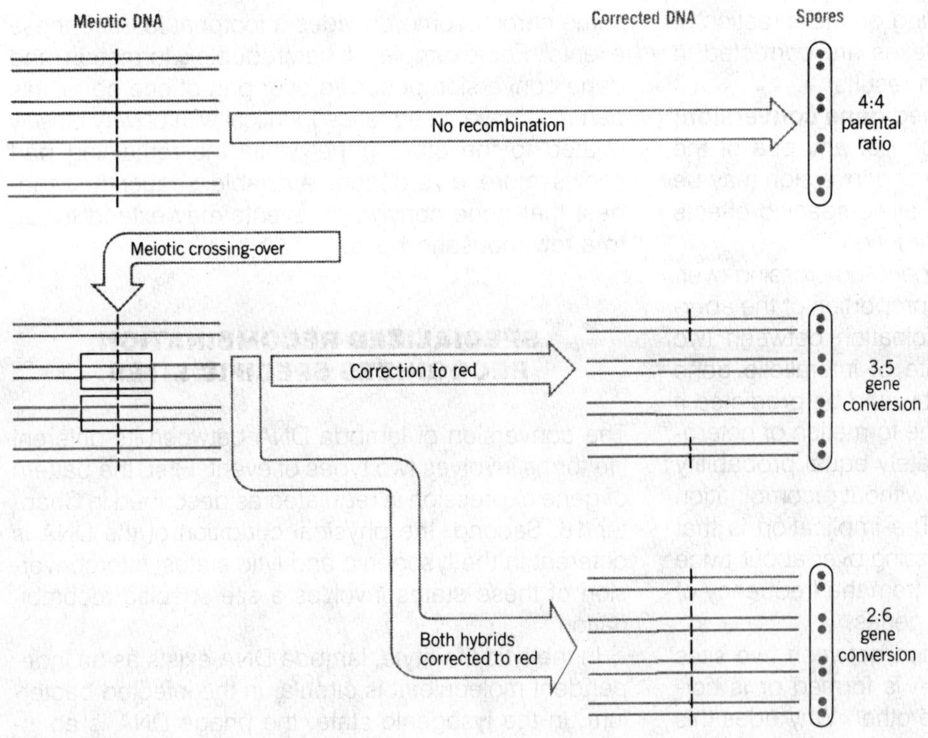

Figure 34.17
Spore formation in the *Ascomycetes* allows determination of the genetic constitution of each of the DNA strands involved in meiosis.
A recombination event forms hybrid DNA in the region indicated by the shaded box. Follow the genetic fate of a locus in this region, indicated by the vertical line. If one of the hybrid DNA molecules is corrected, the spores form a 3:5 segregation. If both are corrected the same way, the ratio is 2:6. Each spore represents one DNA strand, because a further replication occurs between meiosis and spore formation.

But there are some spores with abnormal ratios. They are explained by the formation and correction of heteroduplex DNA in the region in which the alleles differ.

Suppose that two alleles differ by a single point mutation. When a strand exchange occurs to generate heteroduplex DNA, the two strands of the heteroduplex will be mispaired at the site of mutation. In effect, each strand of DNA carries different genetic information. If no change is made in the sequence, the strands separate at the ensuing replication, each giving rise to a duplex that perpetuates its information. The result is the abnormal 4:4 ratio, in which the *order* of the spores is altered, because of the single-strand exchanges. This ratio is called **postmeiotic segregation,** because its occurrence depends on the separation of the DNA strands at the replication that succeeds meiosis.

Some asci display ratios of 3:5 or 2:6, in which one or two spores, respectively, that should have been of one allelic type actually are of the other type. This effect is caused by the correction of mismatches in heteroduplex DNA.

When a repair system recognizes mispaired bases in heteroduplex DNA, it may excise and replace one of the strands to restore complementarity. Such an event changes the strand of DNA representing one allele into the sequence of the other allele. If this happens at only one of the reciprocal heteroduplexes, a

3:5 or 5:3 ratio results (depending on the direction of conversion). If both heteroduplexes are corrected in the same way, a 2:6 or 6:2 ratio results.

The correction process is called **gene conversion;** it may be recognized in the form of any one of the aberrant ratios. (Either direction of conversion may be equally likely, or there may be allele-specific effects causing a preference for one direction.)

Gene conversion does not depend on crossing-over, but is correlated with it. A large proportion of the aberrant asci show genetic recombination between two markers on either side of a site of interallelic gene conversion. This is exactly what would be predicted if the aberrant ratios result from the formation of heteroduplex DNA, with an approximately equal probability of resolving the structure with or without recombination (as indicated in Figure 34.8). The implication is that fungal chromosomes initiate crossing-over about twice as often as would be expected from the frequency of recombination between distant genes.

Within a gene, a recombination between two sites occurs when heteroduplex DNA is formed or is corrected at one site but not at the other. Why does this process yield a linear genetic map for recombination within a gene? The closer two sites lie together, the greater the probability they will suffer **coconversion** in the same stretch of heteroduplex DNA. In this case, both sites are corrected in the same direction, upsetting the ratios of genotypes emerging from the cross, but not resulting in a recombination between the markers.

Heteroduplex DNA may extend for appreciable distances. Some information about the extent of gene conversion is provided by the sequences of members of gene clusters. Usually, the products of a recombination event will separate and become unavailable for analysis at the level of DNA sequence. However, if an unequal exchange takes place between two members of a gene cluster, as illustrated previously in Figure 21.3, a heteroduplex may be formed between the two nonallelic genes. This heteroduplex may suffer gene conversion to the sequence of one of its strands. In effect, this converts one of the nonallelic genes to the sequence of the other. A similar heteroduplex could occur by a direct reaction between the two copies of the gene on the same chromosome.

The presence of more than one gene copy on the same chromosome provides a footprint to trace these events. For example, if heteroduplex formation and gene conversion occurred over part of one gene, this part may have a sequence identical with or very closely related to the other gene, while the remaining part shows more divergence. Available sequences suggest that gene conversion events may extend for up to a few thousand bases.

SPECIALIZED RECOMBINATION RECOGNIZES SPECIFIC SITES

The conversion of lambda DNA between its different life forms involves two types of event. First, the pattern of gene expression is regulated as described in Chapter 16. Second, the physical condition of the DNA is different in the lysogenic and lytic states; interconversion of these states involves a site-specific recombination.

In the lytic life-style, lambda DNA exists as an independent molecule; it is circular in the infected bacterium. In the lysogenic state, the phage DNA is an integral part of the bacterial chromosome (called prophage). To enter the lysogenic condition, free lambda DNA must be **integrated** into the host DNA. To be released from lysogeny into the lytic cycle, prophage DNA must be **excised** from the chromosome.

Integration and excision occur by recombination at specific loci on the bacterial and phage DNAs called **attachment (att) sites.** The attachment site on the bacterial chromosome is called *att*$^{\lambda}$ in bacterial genetics; the locus is defined by its possession of prophage λ in lysogenic strains and by mutations that prevent integration of lambda.

For analyzing the recombination reaction illustrated in **Figure 34.18,** the bacterial attachment site is called *attB*, consisting of the sequence components *BOB'*. The attachment site on the phage, *attP*, consists of the components *POP'*.

The sequence *O* is common to *attB* and *attP*. It is called the **core** sequence; and the recombination event occurs within it. The flanking regions *B*, *B'* and *P*, *P'* are referred to as the **arms,** and each is distinct in sequence.

Because the phage DNA is circular, the recombination event inserts it into the bacterial chromosome

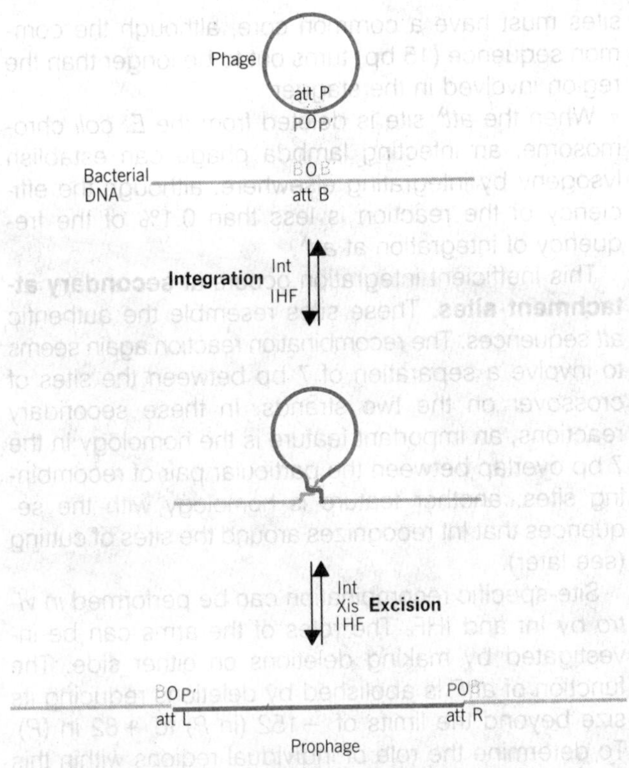

Figure 34.18
Circular phage DNA is converted to an integrated prophage by a reciprocal recombination between *attP* and *attB*; the prophage is excised by reciprocal recombination between *attL* and *attR*.

as a linear sequence. The prophage is bounded by two new *att* sites, the products of the recombination. As the map is usually written, at the left is *attL* consisting of *BOP'*, and at the right is *attR* consisting of *POB'*.

An important consequence of the constitution of the *att* sites is that the integration and excision reactions do not involve the same pair of reacting sequences. Integration requires recognition between *attP* and *attB*; while excision requires recognition between *attL* and *attR*. Thus although the recombination event is reversible, different conditions prevail for each direction of the reaction. This may be important in the life of the phage, since it offers a means to ensure that an integration event is not immediately reversed by an excision, and vice versa.

The difference in the pairs of sites reacting at integration and excision is reflected by a difference in the proteins that mediate the two reactions. Integration requires the product of the phage gene *int* and a bacterial protein called integration host factor (IHF). The IHF consists of two subunits, one of which is coded by the SOS-regulated locus *himA*. In addition to Int and IHF, the excision reaction requires the product of phage gene *xis*.

All four types of *att* site are available for study. They are provided in a convenient form by "illegitimate recombination," in which the lambda genome is released by reciprocal recombination between a sequence in the prophage and a sequence in the flanking bacterial DNA. The reaction occurs by the generalized (RecA-dependent) pathway, probably reflecting a spurious homology between the recombining sites. The consequences are illustrated in **Figure 34.19.**

Illegitimate recombination generates a defective phage in which some of the authentic phage DNA has been replaced by some of the flanking bacterial DNA. The excised genome carries *attL* if the recombination involves a bacterial site on the left; it carries *attR* if the event occurs on the other side.

Figure 34.19
Illegitimate recombination between a site in prophage and a site in bacterial DNA excises a transducing phage in which bacterial genes have replaced some phage genes.

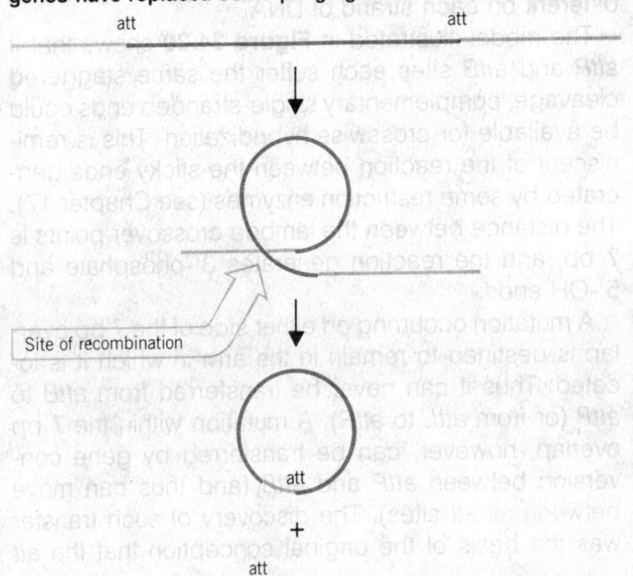

The presence of some bacterial DNA in the defective phage is a very useful feature. The transport of bacterial genes via such a phage is called **transduction;** before the advent of recombinant DNA technology, this provided an important technique for obtaining the bacterial genes residing on either side of the att^{λ} site.

Using phages of this sort, it is possible to obtain similar DNA molecules carrying each type of *att* site. These substrates can be used to test the abilities of all pairwise combinations of *att* sites in site-specific recombination. The Int protein is needed for *all* combinations; it is aided by Xis most dramatically in the *attL* × *attR* reaction, but Xis has a stimulatory effect on some other combinations. We do not yet understand the differences in the various reactions that control the need for Xis.

STAGGERED BREAKAGE AND REUNION IN THE CORE

The core region lies within an A-T-rich sequence of 15 bp that is common to all *att* sites. Does recombination occur at any position within the core sequence or at a specific point? The site-specific recombination involves a precise breakage and reunion in the absence of any synthesis of DNA. The points of exchange are different on each strand of DNA.

The model illustrated in **Figure 34.20** shows that if *attP* and *attB* sites each suffer the same staggered cleavage, complementary single-stranded ends could be available for crosswise hybridization. This is reminiscent of the reaction between the sticky ends generated by some restriction enzymes (see Chapter 17). The distance between the lambda crossover points is 7 bp, and the reaction generates 3′-phosphate and 5′-OH ends.

A mutation occurring on either side of the 7 bp overlap is destined to remain in the arm in which it is located. Thus it can never be transferred from *attB* to *attP* (or from *attL* to *attR*). A mutation within the 7 bp overlap, however, can be transferred by gene conversion between *attP* and *attB* (and thus can move between all *att* sites). The discovery of such transfer was the basis of the original conception that the *att* sites must have a common core, although the common sequence (15 bp) turns out to be longer than the region involved in the stagger.

When the att^{λ} site is deleted from the *E. coli* chromosome, an infecting lambda phage can establish lysogeny by integrating elsewhere, although the efficiency of the reaction is less than 0.1% of the frequency of integration at att^{λ}.

This inefficient integration occurs at **secondary attachment sites.** These sites resemble the authentic *att* sequences. The recombination reaction again seems to involve a separation of 7 bp between the sites of crossover on the two strands. In these secondary reactions, an important feature is the homology in the 7 bp overlap between the particular pair of recombining sites; another feature is homology with the sequences that Int recognizes around the sites of cutting (see later).

Site-specific recombination can be performed *in vitro* by Int and IHF. The roles of the arms can be investigated by making deletions on either side. The function of *attP* is abolished by deletions reducing its size beyond the limits of −152 (in *P*) to +82 in (*P*). To determine the role of individual regions within this stretch of 240 bp, it will be necessary to introduce changes within each segment. The *attB* site is smaller; its function can be exercised by the 23 bp fragment extending from −11 to +11, in which there are only 4 bp on either side of the core. A fragment of *attP* of similar size behaves like an *attB* site in this assay. The disparity in their sizes suggests that *attP* and *attB* play different roles in the recombination, with *attP* providing additional information necessary to distinguish it from *attB*.

When the reaction is performed *in vitro* between two supercoiled DNA molecules, almost all of the supercoiling is retained by the products. Thus there cannot be any free intermediates in which strand rotation could occur. Either the free ends generated by breakage are held in a fixed configuration by the recombination enzyme(s), or one end is transferred directly to the other end via some four-stranded intermediate structure. The breakage and reunion could be accomplished by an activity resembling topoisomerase I, except that the nicked strand is sealed together with a nicked strand of a different instead of the same duplex. Int protein

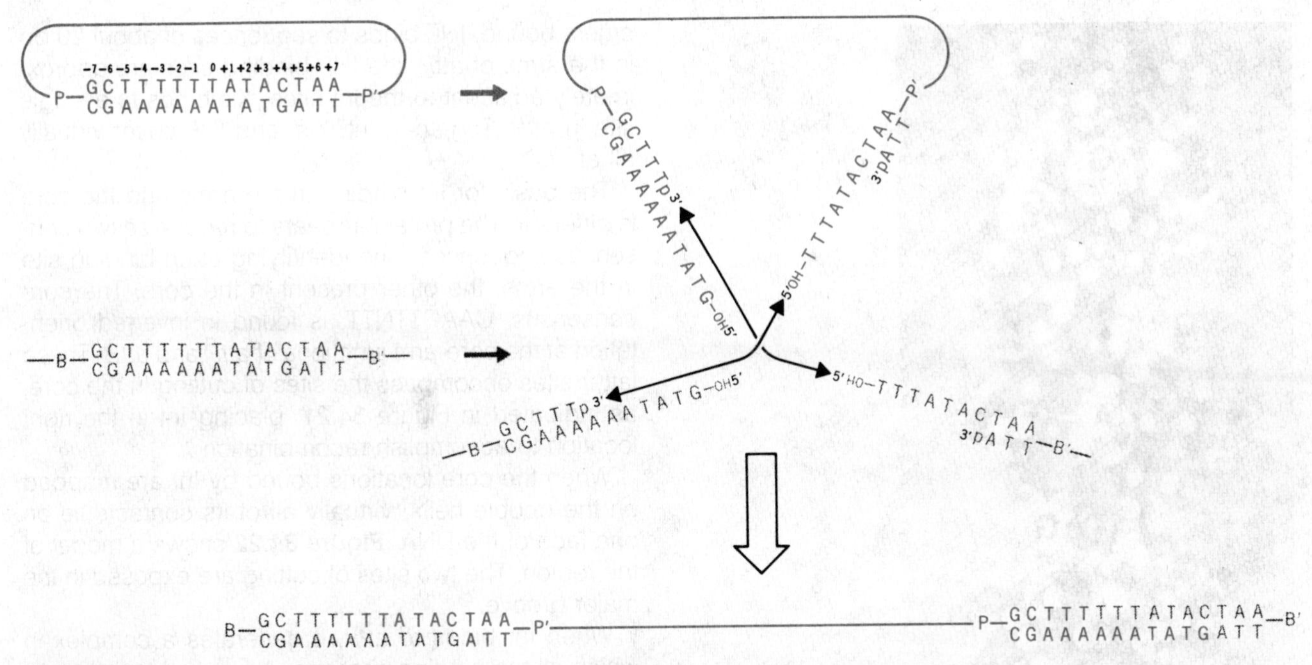

Figure 34.20

Staggered cleavages in the common core sequence of *attP* and *attB* allow crosswise reunion to generate reciprocal recombinant junctions.

The sequence is numbered relative to the center of the core.

may be involved in such a function, since it has a (rather ineffectual) topoisomerase I ability to relax negatively supercoiled DNA.

Both the Int and IHF proteins bind to specific sites in *att* DNA *in vitro*. For recombination *in vitro*, about 20–40 molecules of the 40,000-dalton Int protein are needed per recombinant DNA. Even more of the IHF protein is needed, about 70 molecules of the 20,000-dalton protein per recombinant. (IHF consists of subunits of 11,000 and 9500 daltons.) The high stoichiometry suggests that the proteins do not function catalytically, but form some structure that supports only a single recombination event.

Both the Int and IHF proteins bind to specific sites in the *att* region. The binding sites in *attP* are summarized in **Figure 34.21.** Int binds to a segment of 30 bp encompassing the core region, to a sequence of similar length in *P′*, and to two separate shorter 15 bp sequences in *P*. The differences in the lengths of the individual sites probably reflect the number of Int mon-

Figure 34.21

Int and IHF bind to different sites in *attP*. The Int recognition sequences in the core region include the sites of cutting.

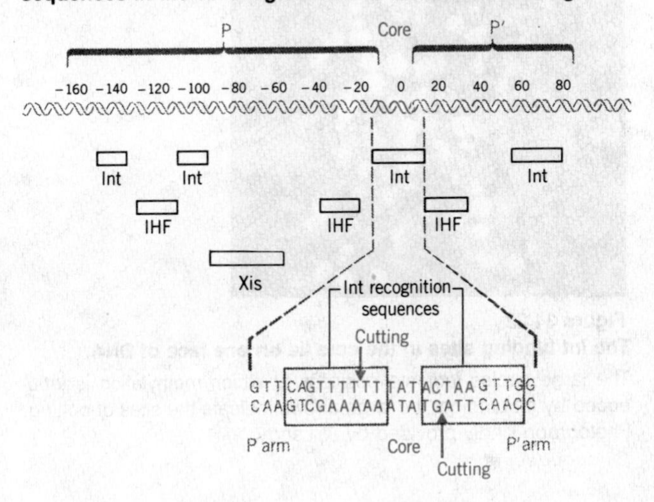

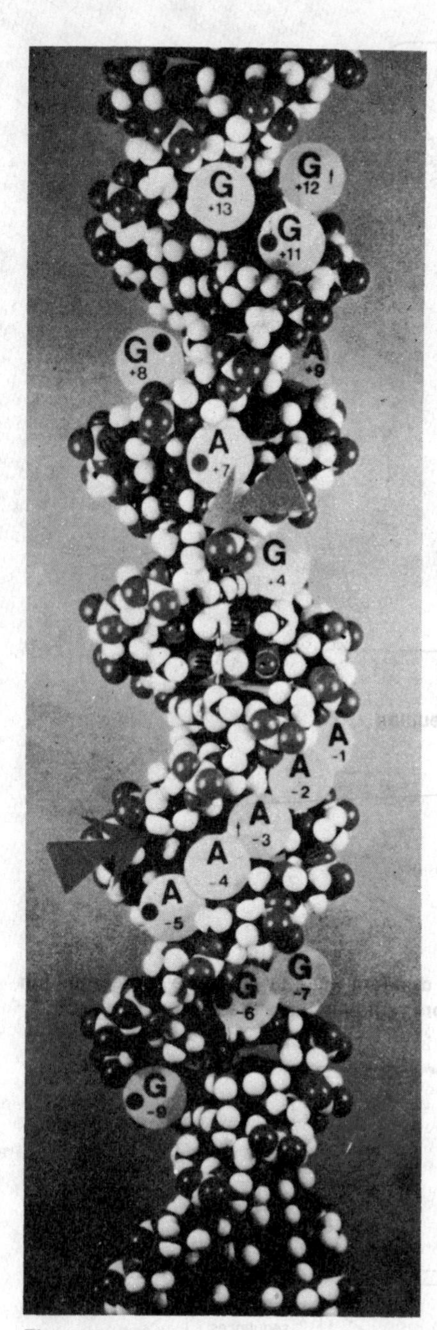

Figure 34.22
The Int binding sites in the core lie on one face of DNA.
The large circles indicate positions at which methylation is influenced by Int binding; the large arrows indicate the sites of cutting. Photograph kindly provided by A. Landy.

omers bound. IHF binds to sequences of about 20 bp in the arms of *attP;* the IHF binding sites are approximately adjacent to the Int sites. Xis binds to a single site in *attP*. Together, Int, Xis, and IHF cover virtually all of *attP*.

The basis for Int binding in the arms and the core is different. The protein appears to recognize two consensus sequences, one identifying each binding site in the arms, the other present in the core. The core consensus, CAACTTNTT, is found in inverted orientation at the core-arm junctions of *attP* and *attB*. These latter sites encompass the sites of cutting in the core, as indicated in Figure 34.21, placing Int in the right location to accomplish recombination.

When the core locations bound by Int are mapped on the double helix, virtually all of its contacts lie on one face of the DNA. **Figure 34.22** shows a model of the region. The two sites of cutting are exposed in the major groove.

When Int binds to *attP,* it generates a complex in which all the binding sites may be pulled together on the surface of a protein oligomer. The reaction involves wrapping about 230 bp of DNA around 4–8 Int monomers. This complex could be the intermediate that reacts with *attB,* whose core sequence is the only region recognized by Int.

The asymmetry of the integration and excision reactions is shown by the fact that Int can form a similar complex with *attR* only if Xis is added. This complex can pair with a condensed complex that Int forms at *attL*. IHF is not needed for this reaction.

Much of the complexity of site-specific recombination may be caused by the need to regulate the reaction so that integration occurs preferentially when the virus is entering the lysogenic state, while excision is preferred when the prophage is entering the lytic cycle. By controlling the available amount of Int and Xis, the appropriate reaction will occur and will not be recklessly reversed.

FURTHER READING

Recombination has been a focus for many reviews over the past few years. A splendid view of mechanisms, delving into the role of RecA, was given by **Dressler & Potter** (*Ann. Rev. Biochem.* **51,** 727–761, 1982). A review relating mechanisms to earlier observations in fungi, and also considering the role of hotspots, was by **Stahl** (*Ann. Rev. Genet.* **13,** 7–24, 1979). The influential paper introducing the idea of the double-strand break was written by **Szostak et al.** (*Cell* **33,** 25–35, 1983). The mechanism of lambda site-specific recombination was brought up to date by **Weisberg & Landy** (pp. 211–250 in *Lambda II,* Eds. Hendrix et al., Cold Spring Harbor Laboratory, New York, 1983). The distinction between topoisomerase types has been considered by **Liu, Liu, & Alberts** (*Cell* **19,** 697–707, 1980) and by **Cozzarelli** (*Cell* **22,** 327–328, 1980). Gyrase actions have been reviewed by **Cozzarelli** (*Science,* **207,** 953–960, 1981). A general review of topoisomerases has been accomplished by **Gellert** (*Ann. Rev. Biochem.* **50,** 879–910, 1981).

PART 10
THE DYNAMIC GENOME: DNA IN FLUX

Elements carried in the maize chromosomes. . . serve to control gene action and to induce, at the site of the gene, heritable modifications affecting this action. These elements were initially discovered because they do not remain at one position in the chromosome complement. They can appear at new locations and disappear from previously determined locations. The presence of one such element at or near the locus of a known gene may affect the action of this gene. In so doing, it need not alter the action potential of the genic substances at the locus. Therefore, these elements were called controlling elements. . . It might be considered that a controlling element represents some kind of extrachromosomal substance that can attach itself or impress its influence in some manner at various positions in the chromosome complement and so affect the action of the genic substances at these positions. The modes of operation of controlling elements do not suggest this, however. Rather they suggest that controlling elements are integral components of the chromosomes themselves, and that they have specific activities and modes of accomplishing them, much as the genes are presumed to have. . . Transpositions of controlling elements either arise from some yet-unknown mechanism or occur during the chromosome reduplication process itself and are a consequence of it.

Barbara McClintock, 1956

CHAPTER 35
TRANSPOSABLE ELEMENTS IN BACTERIA

The genomes of both prokaryotes and eukaryotes are usually regarded as somewhat static, changing only on the leisurely time scale of evolution. We are accustomed to the idea that the construction of a genetic map identifies the loci at which known genes reside; by implication, other (unidentified) sequences also may be expected to remain at constant positions in the population of genomes.

The stability of genetic organization is indicated by the retention of linkage relationships even after speciation—for example, between man and the apes. The difference of generation times between prokaryotes and eukaryotes suggests that their evolutionary scales might be different in terms of real time; but even in the prokaryotes, the overall organization of the genome changes only relatively slowly. For example, a very similar genetic map describes the different bacterial species *E. coli* and *S. typhimurium*.

Genomes evolve both by acquiring new sequences and by rearranging existing sequences. New sequences may be introduced by vectors or may arise by mutation of existing sequences. Rearrangements may create new sequences and may change the functions of existing sequences by placing them in new regulatory situations.

In prokaryotes, extrachromosomal elements are responsible for mediating genetic exchanges. Plasmids sponsor conjugation (mating) between bacteria, while phages are released by infection. Both may occasionally transfer host genes along with the autonomous replicon.

In eukaryotes, reciprocal recombination occurs each generation between corresponding sites on homologous chromosomes; very infrequently, it is accompanied by duplication or rearrangement of loci. Such reorganization is essentially a side effect of the usual mechanisms involved in genetic recombination and/or DNA synthesis (such as unequal crossing-over or gene conversion). The mechanisms responsible for translocations between nonhomologous chromosomes remain unknown.

A potent force for change within both prokaryotic and eukaryotic genomes is provided by the ability of certain sequences to move from one site to another. These sequences are called **transposable elements** or **transposons.**

Bacterial transposition involves duplication of the element; one copy remains at the original "donor" site, while the other copy appears at a new "recipient" or "target" site. *Transposition does not rely on any relationship between the sequences at the donor and recipient sites,* an important difference from all other mechanisms involved in reconstruction of DNA.

Each transposon carries the genes required for its

own transposition, although it may also require functions of the genome in which it resides (such as DNA polymerase and DNA gyrase). The occurrence of comparable events in eukaryotes may be inferred from the general structural similarity and variety in location of analogous elements (see Chapter 36).

What consequences follow from the presence of transposable sequences able to reproduce and move to new locations? A possible advantage for the organism lies in the ability of transposition to sponsor rearrangements, directly or indirectly. The transposition event itself may cause deletions or inversions or may lead to the transport of a host sequence to a new location. Transposons could provoke cellular systems by functioning as "portable regions of homology;" two copies of a transposon at different locations (even on different chromosomes) may provide sites for reciprocal recombination. Such exchanges could result in deletions, insertions, inversions, or translocations.

Transposable elements have indeed played an active role in the evolution of some bacterial plasmids. The components of plasmids may be brought together by reactions involving recognition of transposon sequences. The integration of the F (sex) plasmid of *E. coli* into the bacterial chromosome often occurs by a host (RecA)-mediated reciprocal recombination between a transposon in the plasmid DNA and a homologous transposon in the bacterial DNA.

On some occasions, a cellular or phage genome may have imprisoned a transposable function for its own purposes. The element becomes static and plays a role in regulation, engaging in specific transposition-like events. Examples in *Salmonella* and in phage Mu show that regulation may involve inversion of a specific segment of DNA, apparently by a recombination mechanism related to transposition functions. Thus transposition mechanisms may run the gamut from the ability to link nonhomologous DNA sequences to the sponsorship of specific recombination events.

Considering the transposon itself, the advantages conferred by its intermittent activities provide somewhat nebulous targets for natural selection. This view has prompted suggestions that (at least some) transposable elements may confer neither advantage nor disadvantage on the phenotype, but could constitute "selfish DNA," concerned only with their own propagation.

According to this concept, the relationship of the transposon to the genome resembles that of a parasite with its host. Presumably the spreading of the elements by duplication and transposition is balanced by the harm done if a transposition event inactivates a necessary gene, or if the number of transposons becomes a burden on cellular systems. Yet we must remember that any transposition event conferring a selective advantage—for example, a genetic rearrangement—will lead to preferential survival of the genome carrying the active transposon.

THE DISCOVERY OF TRANSPOSITION IN BACTERIA

Transposable elements were discovered in the form of spontaneous insertions in bacterial operons. Located within a gene, such an insertion prevents transcription and/or translation of the gene, and may also reduce the expression of subsequent genes in the operon. Initially detected simply as a negative mutant, the insertion is distinguished from point mutations by the property that the only form of reversion is the deletion of the inserted material. By comparing the sequences present in various insertion mutants, it is possible to classify several discrete elements.

Transposons can be followed through multiple cycles of transposition. **Figure 35.1** summarizes a protocol that allows both the donor and the recipient to be examined. A plasmid carrying the transposon is introduced into a cell that has another plasmid lacking the transposon. Recipients are selected for transposition (for example, by the inactivation of a drug resistance marker on the target plasmid). Then the recipient and donor plasmid DNA molecules are isolated. The recipient proves to have gained an insertion of the transposon; but the donor has not lost its copy. *Thus transposition is associated with the generation of an additional copy of the transposon at the recipient site.*

Transposition is often studied by isolating many independent insertions of an element in a selected target region. Fine-structure analysis then reveals whether there is any pattern in the selection of sites for insertion. Most transposons have multiple sites of insertions; some show (varying degrees of) preference for hotspots. At the molecular level, selection of

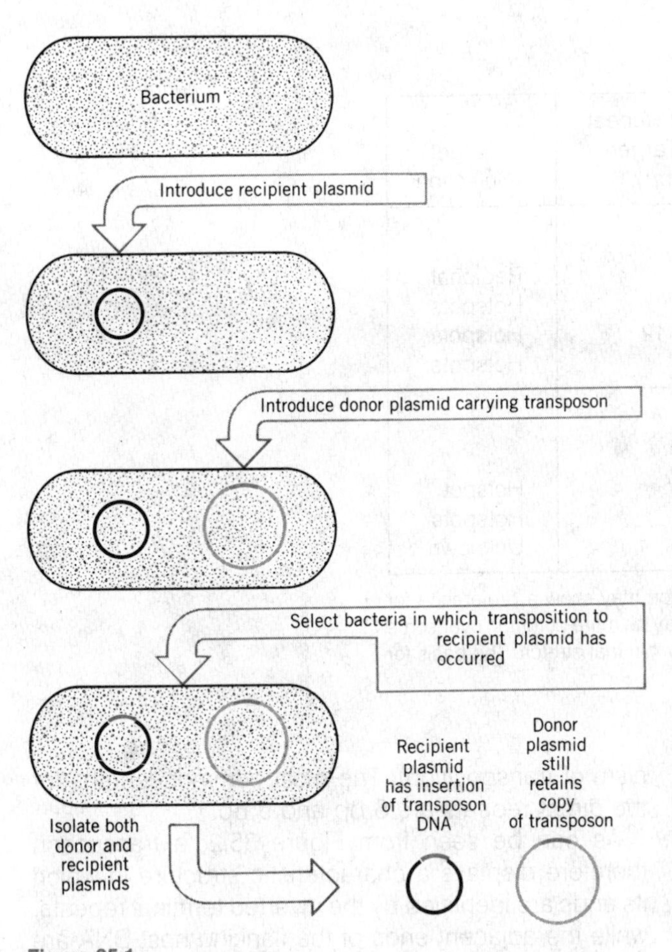

Figure 35.1
Transposition generates a new copy of the transposon at the target site while the original copy remains at the donor site.

a target is independent of nucleotide sequence (with one prominent exception).

The frequency of transposition varies among different elements, but typically is in a range comparable to the spontaneous mutation rate, usually within 10^{-5} to 10^{-7} per generation. Reversion (by deletion) is infrequent, with a range of rates of 10^{-6} to 10^{-10} per generation, up to some 10^3 times less frequent than insertion.

Although transposons vary in type, certain features are common to all transposable elements. Each element is an autonomous unit that codes for the proteins needed to sponsor its transposition. The reaction involves recognition of the ends of the transposable element. The smallest transposons code only for the genes involved in transposition. (They fit the definition of selfish DNA.) Larger transposons carry additional genetic markers.

INSERTION SEQUENCES ARE BASIC TRANSPOSONS

The first set of transposons to be isolated were called **insertion sequences** by virtue of the way in which they were detected. Each type is given the prefix **IS,** followed by a number that identifies the type. (The original classes were numbered IS1-4; later classes have numbers reflecting the history of their isolation, but not corresponding to the total number of elements so far isolated!)

The IS elements are normal constituents of bacterial chromosomes and plasmids. For example, one standard strain of *E. coli* has eight copies of IS1 and five copies of IS2. To describe an insertion into a particular site, a double colon is used; thus lambda::IS1 describes an IS1 element inserted into phage lambda.

The IS elements identify the simplest class of transposons, since their genetic functions are concerned only with the ability to transpose. Although each IS element is different in sequence, all share a common form of organization. The parameters of some common IS elements are summarized in the upper part of **Table 35.1.**

Each element possesses short **inverted terminal repeats.** Usually the repeats are between 15 and 25 bp long and the two copies are closely related rather than identical. As illustrated in **Figure 35.2,** the presence of the inverted terminal repeats means that the same sequence is encountered proceeding toward the element from the flanking DNA on either side of it.

When an IS element transposes, a sequence of host DNA at the site of insertion is duplicated. The nature of the duplication is revealed by comparing the sequence of the target site before and after an insertion has occurred.

In a gene carrying an insert, the IS DNA is always flanked by very short **direct repeats.** (In this context, "direct" indicates that two copies of a sequence are repeated in the same orientation, not that the repeats are adjacent.) But in the original (uninserted) gene,

Table 35.1
Some transposons consist of or contain individual modules.

	Length (bp)	Inverted Terminal Repeat (bp)	Direct Repeat at Target (bp)	Target Selection
IS elements				
IS1	768	23	9	Regional
IS2	1327	41	5	Hotspots
IS4	1428	18	11 or 12	Hotspots
IS5	1195	16	4	Hotspots
IS-like elements				
IS10R	1329	22	9	Hotspot
IS50R	1531	9	9	Hotspots
IS903	1057	18	9	Unknown

Transposons whose sites of insertion are random within a small region may show a preference for one general region compared with another. For example, insertions may be more common over some particular stretch of 3 kb or so, but sites may be chosen at random within that stretch. The basis for this selection is unknown; it is described as a "regional" pattern.

the target site has the sequence of only *one* of these repeats. In the hypothetical example of Figure 35.2, the target site consists of the sequence ATGCA/TACGT. After transposition, one copy of this sequence is present on either side of the transposon.

For most transposons, the *sequence* of the direct repeat is different in each individual transposition event, but the *length* is constant (a reflection of the mechanism of transposition). The most common lengths for the direct repeats are 5 bp and 9 bp.

As can be seen from Figure 35.2, a transposon therefore displays a characteristic structure in which its ends are identified by the inverted terminal repeats, while the adjacent ends of the flanking host DNA are identified by the short direct repeats. When observed in a sequence of DNA, this type of organization is taken to be diagnostic of a transposon, and makes a

Figure 35.2
Transposons have inverted terminal repeats and generate direct repeats of flanking DNA at the target site.
In this example, the target is a 5 bp sequence. The ends of the transposon consist of invertéd repeats of 9 bp, where the numbers 1 through 9 indicate a sequence of base pairs.

```
                         Target
                          site
NNNNNNNNNNNNNNNNNNATGCAMMMMMMMMMMMMMMMMMMMM
NNNNNNNNNNNNNNNNNNTACGTMMMMMMMMMMMMMMMMMMMM

NNNNNNNNNNNNNNNNNATGCA123456789————————987654321ATGCAMMMMMMMMMMMMMMM
NNNNNNNNNNNNNNNNNTACGT123456789—Transposon—987654321TACGTMMMMMMMMMMMMMMM
                      Inverted                  Inverted
                       repeat                    repeat
```

prima facie case that the sequence originated in a transposition event.

A transposon consists of a constant linear sequence. Thus all copies of a given transposon have the same inverted terminal repeats at the junction with host DNA. The importance of the ends of the element is confirmed by the terminal location of mutations that prevent an element from transposing. The *cis*-acting nature of such mutations shows that the ends are recognized by the protein(s) responsible for transposition. The protein is called a **transposase** (and see later).

The shortest transposon, IS1, contains two open reading frames in the same direction; they are translated into the proteins InsA and InsB (whose functions have yet to be delineated). All the other IS elements contain a single long coding region, starting just inside the inverted repeat at one end, and terminating just before or within the inverted repeat at the other end. Usually there are one or more shorter potential coding regions on the other strand of DNA—that is, reading the same sequence in the opposite direction. In some cases, the overlapping protein has been identified. Every transposon therefore codes for 2 (and possibly more) proteins.

COMPOSITE TRANSPOSONS HAVE *IS* MODULES

Some transposons carry drug resistance (or other) markers in addition to the functions concerned with transposition. These transposons are named **Tn** followed by a number. One class of larger transposons comprises **composite elements** that consist of a central region carrying the drug marker(s) flanked on either side by **arms.**

The arms consist of related sequences that may be in either the same or (more commonly) an inverted orientation. Thus a composite transposon with arms that are direct repeats has the structure

Arm L | Central region | Arm R

If the arms are inverted repeats, the structure is

Arm L | Central region | Arm R

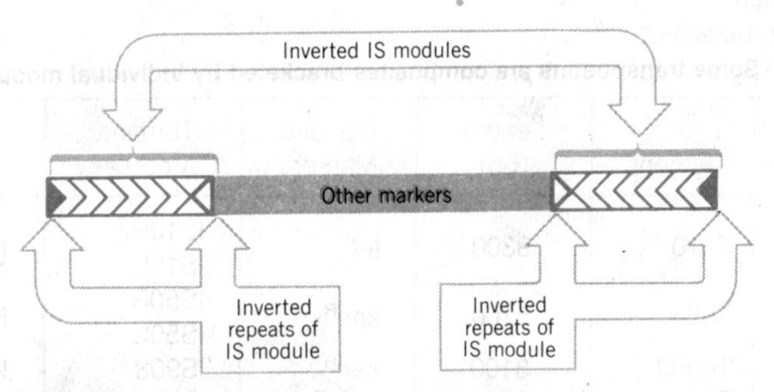

Figure 35.3
A composite transposon has a central region carrying markers unconnected with transposition (such as drug resistance) flanked by IS modules. The modules have short inverted terminal repeats. If the modules themselves are in inverted orientation (as drawn), the short inverted terminal repeats at the ends of the transposon are identical.

The arrows indicate the orientation of the arms, which are identified as L and R according to an (arbitrary) orientation of the genetic map of the transposon from left to right. The structure of a composite transposon is illustrated in more detail in **Figure 35.3.**

The arms may consist of identical IS elements. In the case of Tn9, a direct repeat of an IS1 element is present at each end of the transposon.

In other cases, the arms *resemble* IS elements. Each consists of a single continuous open reading frame, flanked by the usual short inverted terminal repeats. However, because these elements are not found as individual components in bacterial DNA, they are described as **IS-like** elements. They are named in the same way as IS elements, and usually are given a number related to the transposon in which they are found. The properties of IS-like elements are compared with those of the IS elements in the lower part of Table 35.1.

The properties of some composite transposons are summarized in **Table 35.2.** Since the arms on the one hand are part of the composite transposon, but on the other resemble IS elements or even are IS elements, they are sometimes described as IS or IS-like **modules.** Since each module has the usual structure ending in inverted repeats, the composite transposon also ends in the same short inverted repeats.

In some cases, the modules of a composite transposon are identical, such as Tn9 (direct repeats of IS1) or Tn903 (inverted repeats of IS903). In other cases,

Table 35.2
Some transposons are composites bracketed by individual modules.

Element	Length (bp)	Genetic Markers	Terminal Modules	Module Orientation	Module Relationship	Module Functions
Tn10	9300	*tet*^{R}	IS10R IS10L	Inverted	2.5% divergence	Fully functional Reduced function
Tn5	5700	*kan*^{R}	IS50R IS50L	Inverted	1 bp change	Functional Nonfunctional
Tn903	3100	*kan*^{R}	IS903	Inverted	Identical	Both functional
Tn9	2500	*cam*^{R}	IS1	Direct	Presumed identical	Presumed functional

the modules are closely related, but not identical. Thus we can distinguish the modules such as IS10L and IS10R. (In this case, the L and R terminology identifies *different modules;* when the modules are identical, the terminology merely indicates their location relative to the central region of the transposon.)

Although the IS-like modules are not usually found as independent elements in bacterial DNA (hence their name), a functional module may be able to transpose itself, as shown by the examples of IS10R of Tn10 or IS50R of Tn5. When the modules of a composite transposon are identical, presumably either module can sponsor movement of the transposon. When the modules are different, they may differ in functional ability, so transposition can depend entirely or principally on one of the modules.

We assume that composite transposons evolved by an association of two originally independent modules with the central region. Such a situation could arise when an IS element transposes to a recipient site close to the donor site. Originally identical, therefore, the modules may have remained identical or may have diverged. The ability of a single module to transpose the entire composite element may explain the lack of selective pressure for both modules to remain active.

We should like to know how two modules cooperate to transpose a composite transposon instead of just the individual module. This question is especially pressing in cases where both the modules *are* functional. In the example of Tn9, where the modules are IS1 elements, presumably each is active in its own right as well as on behalf of the composite transposon. Of course, since transposition involves the generation of a new copy at the target site, the existence of a composite transposon is not imperiled if one of its modules sponsors an independent transposition of itself (although this does not further the aim of spreading the entire transposon).

Two IS elements in fact are able to transpose any sequence residing between them, as well as themselves. An experiment that demonstrates this ability is summarized in **Figure 35.4.** If Tn10 resides on a circular replicon, its two modules can be considered to flank *either* the *tet*^{R} gene of the original Tn10 *or* the sequence in the other part of the circle. Thus a transposition event can involve either the original Tn10 transposon or the "inside-out" transposon with the alternate central region.

Note that both transposons have inverted modules, but these modules evidently can function in either orientation relative to the central region. In events of this nature, the frequency of transposition declines with the distance between the modules. The length dependence probably reflects problems that occur during the process of transposition. It may have helped determine the sizes of the common composite transposons.

To summarize the features that are necessary for transposition of IS or IS-like elements, all except IS1 have a single major gene product, corresponding to a reading frame that runs the length of the element. This single product is probably responsible for creating a target site and for recognizing the ends of the transposon. *Only the ends are needed for a transposon to serve as a substrate for transposition.* Assistance from host functions may be needed to replicate

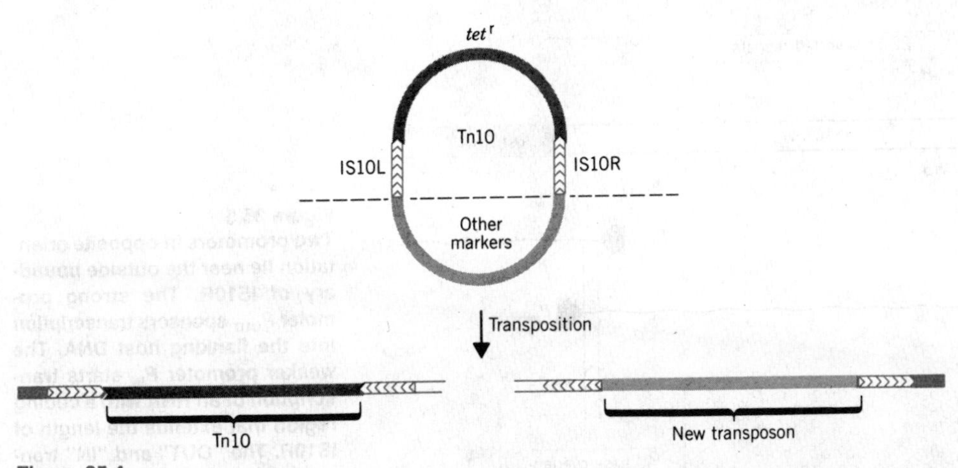

Figure 35.4
The IS10 modules can cooperate to transpose any region of DNA that lies between them.
When Tn10 is part of a small circular molecule, the IS10 repeats can transpose either side of the circle. Transposition of *tet*^{R} corresponds to the movement of Tn10. Transposition of the markers on the other side creates a new "inside-out" transposon.

the transposon. We know little about the detailed processes and order of events involved in these reactions, but we surmise that they may be similar for all the IS elements and composite transposons (perhaps excepting IS1).

ONLY ONE MODULE OF Tn10 IS FUNCTIONAL

The relationship between modules that are no longer identical may offer insights into the evolution of the composite transposon. The examples of Tn10 and Tn5 are the best characterized.

Only a few bases at each end of Tn10 are needed to recognize the element as a substrate for transposition. The inverted terminal repeats are 22 bp long; probably only the last 13 are needed for the transposition reaction. Mutations in this region prevent transposition and are *cis*-acting.

Tn10 is an exceptional element that has a specific target sequence. The 9 bp direct repeats of flanking DNA generated by transposition display a consensus of a 6 bp sequence symmetrically disposed within the target. Thus the repeats on either side of Tn10 often take the form $\frac{\text{NGCTNAGCN}}{\text{NCGANTCGN}}$, where N identifies any base pair. The stronger the hotspot, the more closely it conforms to the consensus. Probably the same transposon-coded function recognizes the target sequence and the ends of the transposon.

The element of IS10R provides the active module of Tn10. The IS10L module is functionally defective and provides only 1–10% of the transposition activity of IS10R. The accumulation of mutations in IS10L relative to IS10R corresponds to about 2.5% divergence between the modules.

The organization of IS10R is summarized in **Figure 35.5.** It has a continuous reading frame that codes for a transposase of 47,000 daltons; the significance of another, overlapping reading frame is unknown. The level of the transposase is enhanced by "high-hopper" mutations that increase expression of the IS10R gene. Thus the amount of this protein may limit the rate of transposition.

Mutants in this gene can be complemented in *trans* by another, wild-type element only with some difficulty. This reflects a strong preference of the transposase for *cis*-action; the enzyme functions efficiently only with the DNA template from which it was transcribed and translated. *Cis*-preference is a common feature of proteins involved in IS-mediated transposition. (We have described the *cis*-preference of another function, the A protein involved in φX174 replication, in Chapter 32.)

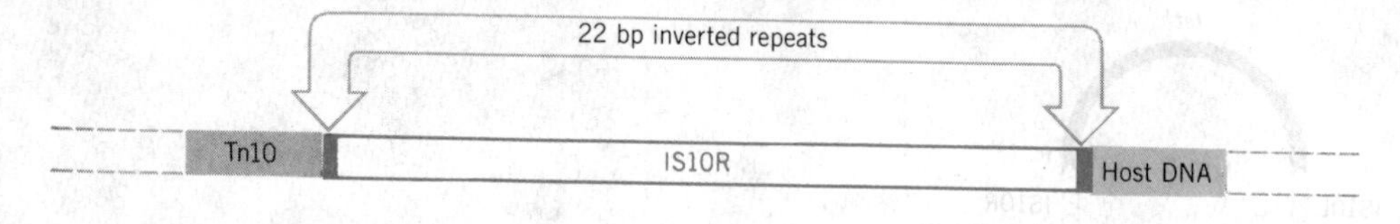

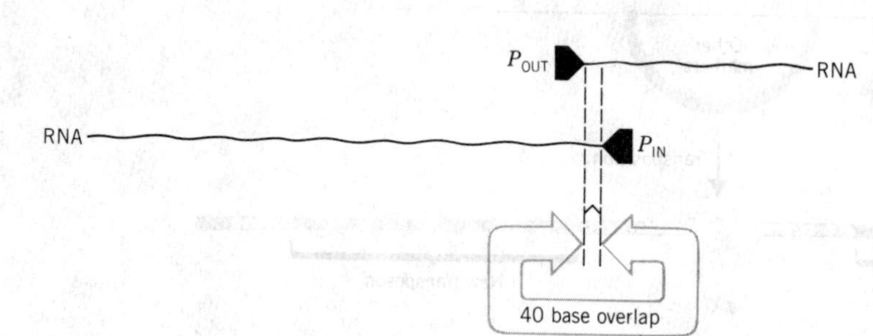

Figure 35.5
Two promoters in opposite orientation lie near the outside boundary of IS10R. The strong promoter P_{OUT} sponsors transcription into the flanking host DNA. The weaker promoter P_{IN} starts transcription of an RNA with a coding region that extends the length of IS10R. The "OUT" and "IN" transcripts have a 40-base overlap.

When the distance of the IS10 transposase gene from the transposon ends is increased, the efficiency of the transposase declines. The reason for its *cis*-preference is therefore that the transposase recognizes a DNA target sequence near the site where it is synthesized more efficiently than a target sequence farther away. Either the transposase binds to DNA so rapidly after (or even during) protein synthesis that it has a very low probability of diffusing elsewhere, or it may be so unstable that protein molecules failing to bind quickly (and therefore nearby) never have a chance to become active.

Two promoters are found close to the outside boundary of IS10R. The promoter P_{IN} is responsible for transcription of IS10R. The promoter P_{OUT} causes transcription to proceed into the adjacent flanking DNA, an effect that could be responsible for activating bacterial genes adjacent to Tn10.

The phenomenon of "multicopy inhibition" reveals that expression of the IS10R gene is regulated. Transposition of a Tn10 element on the bacterial chromosome is reduced when additional copies of IS10R are introduced via a multicopy plasmid. The inhibition requires the P_{OUT} promoter; and it is effective only for an IS10R gene expressed from the P_{IN} promoter (the effect is lost if the gene is placed under control of a different promoter). However, multicopy inhibition is exercised at the level of translation, not transcription. The basis for the effect may lie with the 40 bp overlap in the 5′ terminal regions of the transcripts from P_{IN} and P_{OUT}. Because P_{OUT} is a much stronger promoter, a large amount of the OUT RNA may be present; it could base pair with the IN RNA to prevent translation of the protein needed for transposition.

THE MODULES OF Tn5 ARE ALMOST IDENTICAL BUT VERY DIFFERENT

The inverted modules of Tn5 provide a striking example of the ability of small mutational changes to produce major functional effects. The module IS50R is functional; module IS50L is nonfunctional with regard to transposition. The sole difference between the two modules lies in the substitution of a single base pair. The effects of this substitution are illustrated in **Figure 35.6.**

Two proteins are produced from the same reading frame in IS50R. Their only difference is that protein **1** has an additional N-terminal ~40 amino acids that are absent from protein **2.** The precise events involved in transcription and translation of these proteins have yet to be defined; all we know for certain is that more is produced of protein **2** than of protein **1.**

The single base-pair change in IS50L simultaneously affects translation of these proteins and controls transcription of the central region. The substitution creates an ochre codon that prematurely terminates translation of both proteins **1** and **2.** The truncated proteins (sometimes called **3** and **4**) lack transposition

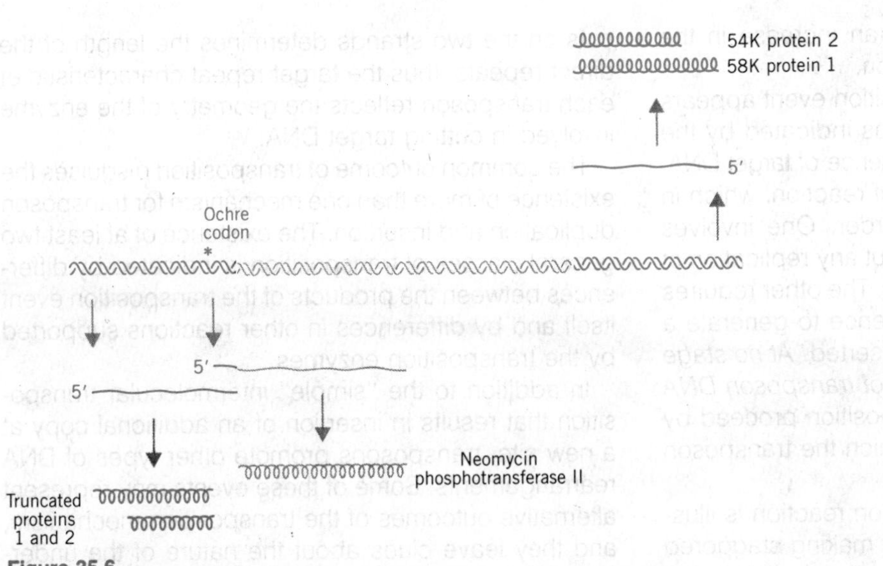

Figure 35.6

A single base-pair change in IS50L prematurely terminates synthesis of proteins 1 and 2 and creates a promoter for transcription of the gene for neomycin (and kanamycin) resistance.

activity. The same substitution also creates a promoter for transcription of the gene of the central region that codes for neomycin phosphotransferase II, the enzyme responsible for resistance to antibiotics such as neomycin and kanamycin. The change is therefore necessary for this function of the transposon.

The functions of proteins **1** and **2** are related but different. Protein **1** is essential for transposition of either the IS50 module or the intact Tn5 transposon. The wild-type protein complements a defective module in *trans* only exceedingly poorly. Thus protein **1** may be a typical *cis*-acting transposition function.

Protein **2** is an inhibitor of transposition. Its action is exercised at some currently unknown stage of the transposition process itself (that is, rather than by regulating gene expression). Protein **2** is *trans*-acting; one possibility for its action is that it preempts the transposition function of protein **1** by binding to the same sites that protein **1** must recognize in order to sponsor transposition. An alternative is that proteins **1** and **2** form some oligomeric complex in which the activity of protein **1** is inhibited.

The behavior of protein **2** may explain a property of Tn5. On entering a new host bacterium, Tn5 transposes at a high frequency. Once the element has become established, the frequency decreases. An established element also can inhibit transposition by an incoming element. This *trans*-acting function may reflect the ability of protein **2** to prevent transposition; when a new element enters the cell, sufficient amounts of protein **2** already are present to inhibit the *cis*-acting function of the protein **1** of the new as well as the old Tn5 element. The truncated proteins of IS50L also may be involved in this effect.

The relationship between protein **2** and protein **1** of Tn5 is naturally different from the relationship between the P_{OUT} and P_{IN} transcripts of Tn10. Yet these interactions share a common feature: they limit the frequency of transposition. Mechanisms to restrict transposons from multiplying ad infinitum may be important to prevent them from interfering too much with cellular functions.

TRANSPOSITION INVOLVES REPLICATIVE RECOMBINATION

"Transposition" is a misnomer, in the sense that a transposon does not leave its former site in the course of transposition. Since the process involves duplication of the transposon, one copy remains at the original site, while the other is found at the new site. Thus transposition is accompanied by an increase in the number of copies of the transposon.

The basic nature of the transposition event appears to be similar for all transposons, as indicated by the diagnostic repetition of a short sequence of target DNA. Transposition involves two types of reaction, which in principle could occur in either order. One involves replication of the transposon without any replication of adjacent chromosomal sequences. The other requires breakage of the target DNA sequence to generate a site into which the transposon is inserted. *At no stage in this process is a free molecule of transposon DNA generated;* all the events of transposition proceed by manipulation of the DNA within which the transposon resides.

The general nature of the insertion reaction is illustrated in **Figure 35.7.** It consists of making staggered breaks in the target DNA, joining the transposon to the protruding single-stranded ends, and filling in the gaps. The generation and filling of the staggered ends explain the occurrence of the direct repeats of target DNA at the site of insertion. The stagger between the cuts on the two strands determines the length of the direct repeats; thus the target repeat characteristic of each transposon reflects the geometry of the enzyme involved in cutting target DNA.

The common outcome of transposition disguises the existence of more than one mechanism for transposon duplication and insertion. The existence of at least two general classes of transposition is indicated by differences between the products of the transposition event itself and by differences in other reactions supported by the transposition enzymes.

In addition to the "simple" intermolecular transposition that results in insertion of an additional copy at a new site, transposons promote other types of DNA rearrangements. Some of these events may represent alternative outcomes of the transposition mechanism, and they leave clues about the nature of the underlying events. Others are consequences of the relationship between the multiple copies of the transposon.

Rearrangements of host DNA may result when a transposon inserts a copy at a second site near its original location. Host systems may undertake recip-

Figure 35.7
The direct repeats of target DNA flanking a transposon are generated by the introduction of staggered cuts whose protruding ends are linked to the transposon.

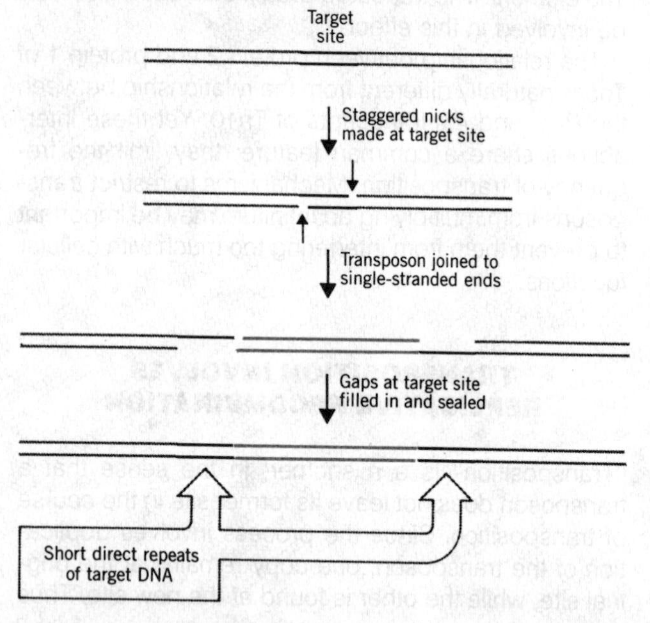

Figure 35.8
Reciprocal recombination between direct repeats excises the material between them; each product of recombination has one copy of the direct repeat.

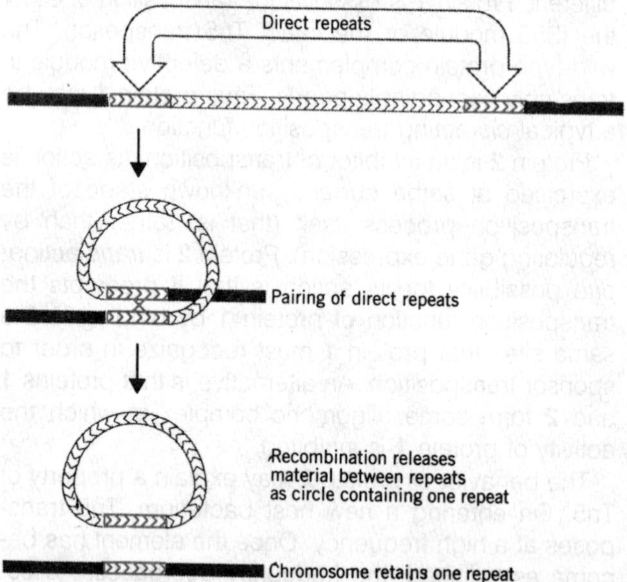

rocal recombination between the two copies of the transposon; the consequences are determined by whether the repeats are the same or in inverted orientation.

Figure 35.8 illustrates the general rule that recombination between any pair of direct repeats will delete the material between them. The intervening region is excised as a circle of DNA (which is lost from the cell); the chromosome retains a single copy of the direct repeat. Note that a recombination between the directly repeated IS1 modules of the composite transposon Tn9 would therefore replace the transposon with a single IS1 module.

Deletion of sequences adjacent to a transposon could therefore result from a two-stage process; transposition generates a direct repeat of a transposon, and recombination occurs between the repeats. However, the majority of deletions that arise in the vicinity of transposons probably result from a variation in the pathway followed in the transposition event itself.

Figure 35.9 depicts the consequences of a reciprocal recombination between a pair of inverted repeats. The region between the repeats becomes inverted; the repeats themselves remain available to sponsor further inversions.

Duplicative inversions are associated with transposition, and are identified by transposons that lie in inverted orientation on either side of a central region that has been inverted from its usual orientation. Note that a composite transposon whose modules are inverted is a stable component of the genome, although the direction of the central regic with regard to the modules could be inverted by recombination.

A significant omission from the list of events sponsored by transposons is **excision**. The loss of a transposon can be detected by restoration of function at the site of insertion. This must involve **precise excision;** it occurs at a frequency of $\sim 10^{-6}$ for Tn5. Loss can also be detected by relief of polarity in adjacent genes; this can be accomplished by **imprecise excision** (which leaves a remnant of the transposon, sufficient to prevent reactivation of the target gene, but insufficient to cause polar effects). Imprecise excision occurs at a frequency of $\sim 10^{-5}$ for Tn5.

Neither type of excision is associated with transposition to new sites. The process does not rely on transposon-coded functions; it is RecA-independent and could occur by some cellular mechanism that generates spontaneous deletions between closely spaced repeated sequences.

The most revealing of the transposon-mediated reactions is **replicon fusion** to form a **cointegrate** structure. A replicon containing a transposon may become fused with a replicon lacking the element, as illustrated in **Figure 35.10.** The resulting cointegrate has two copies of the transposon, one at each junction between the original replicons, oriented as direct repeats. The cointegrates generated by composite transposons may have a duplication of either the whole element (Tn cointegrates) or the IS module (IS cointegrates).

The relative frequencies with which cointegrates are generated by different transposons indicate that there may be different pathways for transposition. For example, Tn903 and IS1 generate "simple" transpositions and cointegrate structures at about the same frequency. But Tn5 and Tn10 generate cointegrates infrequently or not at all, and are likely to transpose by some mechanism not involving a cointegrate structure.

A model for a transposition pathway in which a cointegrate is an obligatory intermediate is illustrated in **Figure 35.11.** Four single-strand cleavages initiate the

Figure 35.9
Reciprocal recombination between inverted repeats inverts the region between them.

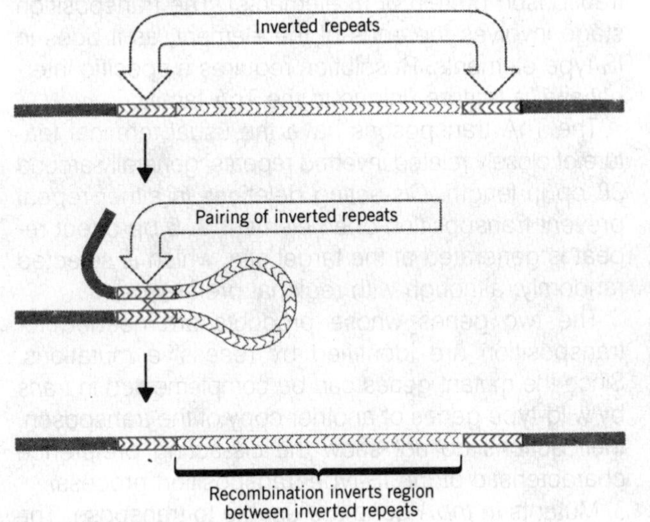

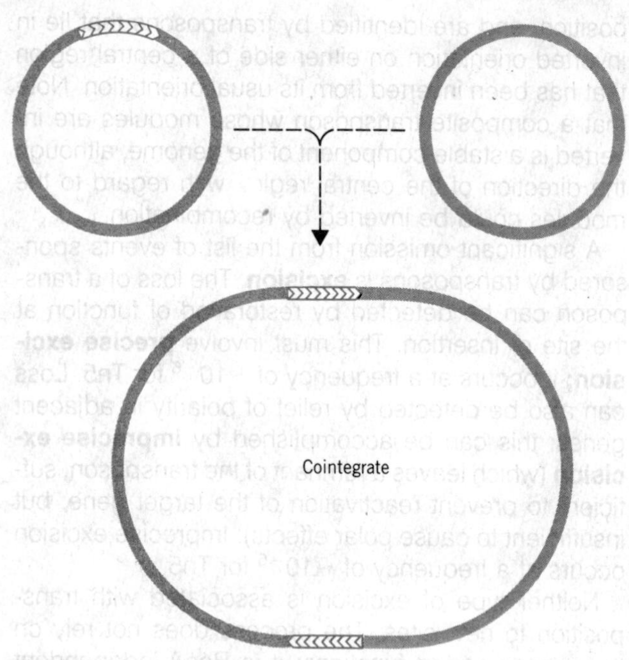

Figure 35.10
Transposition may fuse a donor and recipient replicon into a cointegrate.

process. The donor molecule is cleaved at either end of the transposon by a site-specific enzyme that recognizes the termini. The target molecule is cleaved at sites staggered (most commonly) by 5 or 9 bases. The order and time at which the cleavages are made during the transposition reaction could be changed.

The donor and target strands are ligated at the nicks. Each end of the transposon sequence is joined to one of the protruding single strands generated at the target site. The linkage generates a chi-shaped structure held together at the duplex transposon. The formation of this structure is responsible for the movement of the transposon; the enzyme activity responsible for it is called the **transposase.**

The chi structure contains a single-stranded region at each of the staggered ends. These regions are pseudoreplication forks that provide a template for DNA synthesis. (Use of the ends as primers for replication implies that the strand breakage must occur with a polarity that generates a 3′-OH terminus at this point.)

If replication continues from both the pseudoreplication forks, it will proceed through the transposon, separating its strands, and terminating at its ends. Replication is probably accomplished by host-coded functions. At this juncture, the structure has become a cointegrate, possessing direct repeats of the transposon at the junctions between the replicons (as can be seen by tracing the path around the cointegrate).

A site-specific recombination between the two copies of the transposon can regenerate the original donor replicon, releasing a target replicon that has gained a transposon flanked by short direct repeats of the host target sequence. This reaction is called **resolution;** the enzyme activity responsible is called the **resolvase.**

TRANSPOSITION OF Tn3 PROCEEDS BY COINTEGRATE RESOLUTION

Some large (~5 kb) transposons are not composites relying on IS-type transposition modules, but comprise independent units carrying genes for transposition as well as for features such as drug resistance. The TnA family is the best characterized. Originally thought to be a single transposon, it is now known to include many related transposons, some of which are described in **Table 35.3.**

Transposition of these elements proceeds in two stages: first a cointegrate is formed; then it is resolved. Each stage is accomplished by a different product of the transposon (in contrast with the putative single-transposon protein of IS elements). The transposition stage involves the ends of the element, as it does in IS-type elements. Resolution requires a specific internal site, a feature unique to the TnA family.

The TnA transposons have the usual terminal feature of closely related inverted repeats, generally around 38 bp in length. *Cis*-acting deletions in either repeat prevent transposition of an element. A 5 bp direct repeat is generated at the target site, which is selected randomly, although with regional preference.

The two genes whose products are needed for transposition are identified by recessive mutations. Since the mutant genes can be complemented in *trans* by wild-type genes of another copy of the transposon, their actions do not show the *cis*-acting preference characteristic of the IS-type transposition process.

Mutants in *tnpA* gene are unable to transpose. The

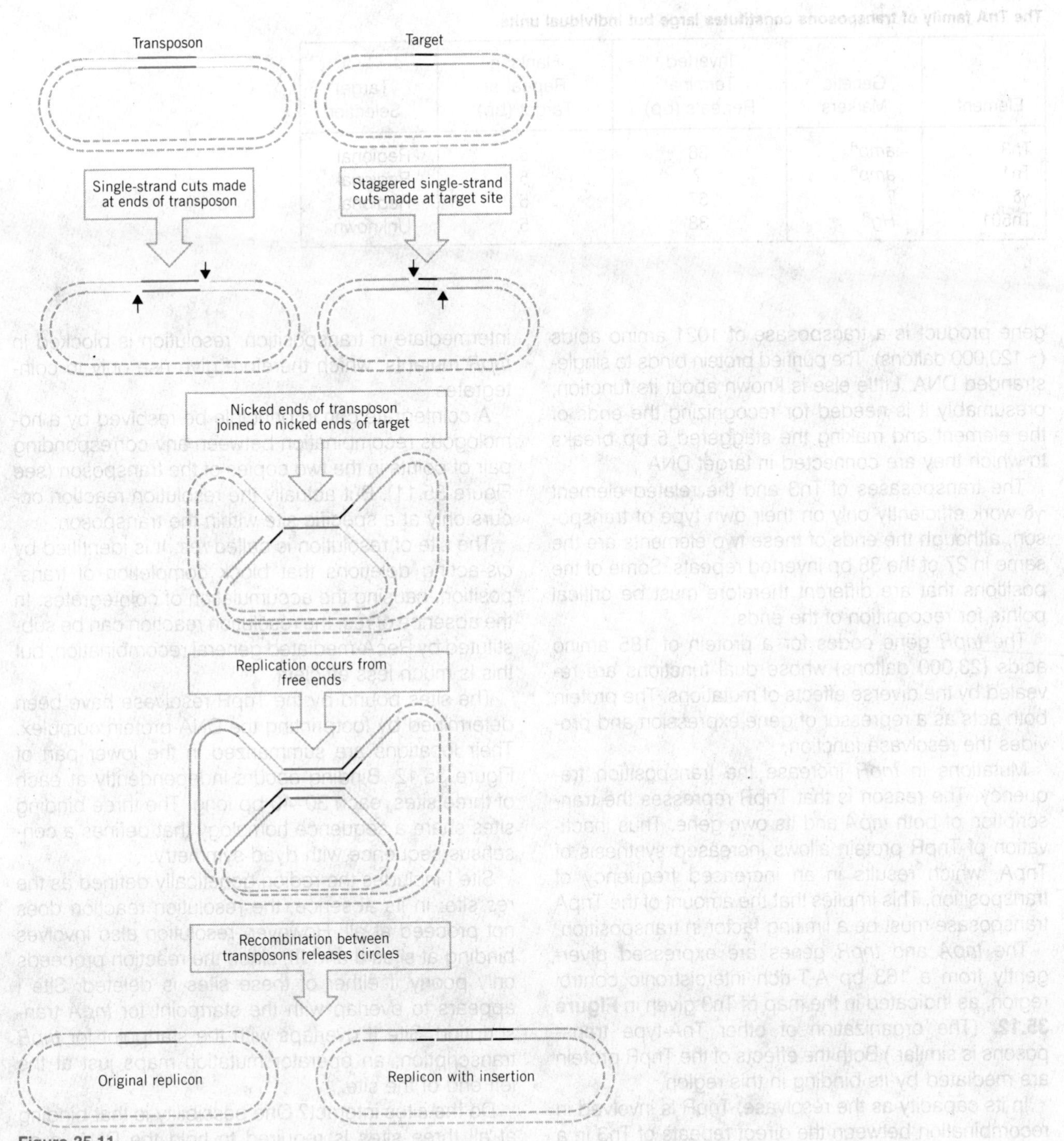

Figure 35.11
One pathway for transposition proceeds through the formation of a cointegrate that is resolved by recombination.

Table 35.3
The TnA family of transposons constitutes large but individual units.

Element	Genetic Markers	Inverted Terminal Repeats (bp)	Flanking Repeat at Target (bp)	Target Selection
Tn3	*amp*^{R}	38	5	Regional
Tn1	*amp*^{R}	?	5	Regional
γδ	?	37	5	Regional
Tn501	*Hg*^{R}	38	5	Unknown

gene product is a transposase of 1021 amino acids (~120,000 daltons). The purified protein binds to single-stranded DNA. Little else is known about its function; presumably it is needed for recognizing the ends of the element and making the staggered 5 bp breaks to which they are connected in target DNA.

The transposases of Tn3 and the related element γδ work efficiently only on their own type of transposon, although the ends of these two elements are the same in 27 of the 38 bp inverted repeats. Some of the positions that are different therefore must be critical points for recognition of the ends.

The *tnpR* gene codes for a protein of 185 amino acids (23,000 daltons) whose dual functions are revealed by the diverse effects of mutations. The protein both acts as a repressor of gene expression and provides the resolvase function.

Mutations in *tnpR* increase the transposition frequency. The reason is that TnpR represses the transcription of both *tnpA* and its own gene. Thus inactivation of TnpR protein allows increased synthesis of TnpA, which results in an increased frequency of transposition. This implies that the amount of the TnpA transposase must be a limiting factor in transposition.

The *tnpA* and *tnpR* genes are expressed divergently from a 163 bp A-T-rich intercistronic control region, as indicated in the map of Tn3 given in **Figure 35.12.** (The organization of other TnA-type transposons is similar.) Both the effects of the TnpR protein are mediated by its binding in this region.

In its capacity as the resolvase, TnpR is involved in recombination between the direct repeats of Tn3 in a cointegrate structure. The cointegrate is an obligatory intermediate in transposition; resolution is blocked in *tnpR* mutants, which therefore give rise only to cointegrates.

A cointegrate can in principle be resolved by a homologous recombination between any corresponding pair of points in the two copies of the transposon (see Figure 35.11). But actually the resolution reaction occurs only at a specific site within the transposon.

The site of resolution is called *res*. It is identified by *cis*-acting deletions that block completion of transposition, causing the accumulation of cointegrates. In the absence of *res*, the resolution reaction can be substituted by RecA-mediated general recombination, but this is much less efficient.

The sites bound by the TnpR resolvase have been determined by footprinting the DNA-protein complex. Their locations are summarized in the lower part of Figure 35.12. Binding occurs independently at each of three sites, each 30–40 bp long. The three binding sites share a sequence homology that defines a consensus sequence with dyad symmetry.

Site I includes the region genetically defined as the *res* site; in its absence, the resolution reaction does not proceed at all. However, resolution also involves binding at sites II and III, since the reaction proceeds only poorly if either of these sites is deleted. Site I appears to overlap with the startpoint for *tnpA* transcription. Site II overlaps with the startpoint for *tnpR* transcription; an operator mutation maps just at the left end of the site.

Do the sites interact? One possibility is that binding at all three sites is required to hold the DNA in an appropriate topology; the resolution reaction could be

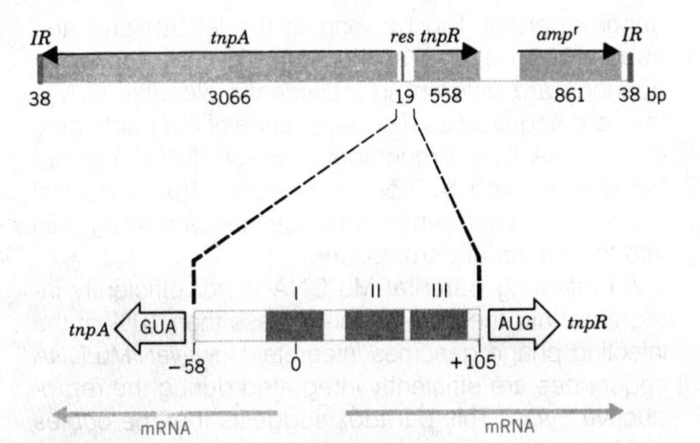

Figure 35.12

Transposons like Tn3 have inverted terminal repeats, an internal *res* site, and three known genes.

Arrows indicate the direction of expression of the coding regions. Numbers in the upper diagram indicate the length in base pairs of each region. Numbers in the lower diagram give the distance in base pairs from the crossover point (0).

triggered when the protein-DNA complex of one transposon interacts with that of another transposon. Binding at a single set of sites may repress *tnpA* and *tnpR* transcription without introducing any change in the DNA.

The region in which the site-specific recombination occurs can be identified by comparing the reactions of transposons Tn3 and γδ. Their *tnpR* genes are highly homologous and can substitute for each other in complementation assays. The sequences of recombinants show that the resolution reaction occurs within a 19 bp part of site I that is identical in both plasmids.

More information about the reaction has been gained from an *in vitro* resolution assay that uses a cointegrate-like DNA molecule as substrate. (The substrate was constructed by cloning *res* to form direct repeats at two locations in the plasmid pBR322.) The substrate must be supercoiled; its resolution produces two catenated circles, each containing one *res* site. The reaction requires large amounts of the TnpR resolvase; resolution works best at enzyme to DNA ratios of ~20:1. No host factors are needed (a contrast with lambda site-specific recombination).

Resolution is a conservative reaction; bonds are broken and rejoined without demand for input of energy. The products identify an intermediate stage in cointegrate resolution; they consist of resolvase covalently attached to both 5′ ends of double-stranded cuts made at the *res* site. The cleavage occurs symmetrically at a short palindromic region to generate two base extensions. Expanding the view of the crossover region located in site I, we can describe the cutting reaction as

```
5′ TTATAA 3′
3′ AATATT 5′
      ↓
5′ TTAT
3′ AA—protein
      +
   protein—AA 3′
           TATT 5′
```

A similarity with the lambda site-specific recombination system is implied by the relationship of the *res* site with the lambda *att* core described in Chapter 34. Immediately surrounding the point of crossover, there is homology between the two sites at 10 of the 15 bp. The sequences are

```
                     ↓
res  G A T A A T T T A T A A T A T
att  G C T T T T T T A T A C T A A
             ↑
```

The arrows identify the sites cleaved in *res* by TnpR and in *att* by Int. The reactions themselves are analogous, although resolution occurs only between intramolecular sites, whereas the recombination between *att* sites is intermolecular and directional (as seen by the differences in *att* sites). The resemblance to the cutting and rejoining reactions of topoisomerases is evident in both cases.

Finally, we should mention a remarkable effect of Tn3 called **transposition immunity.** A plasmid carrying Tn3 is immune to further insertions from any Tn3 element carried on a different DNA molecule (although the Tn3 on the plasmid appears able itself to transpose within the plasmid). There is a lag between the insertion of Tn3 into a new plasmid and the establishment of immunity. The effect is specific; it applies only to Tn3 and does not prevent other transposons from inserting into the plasmid. The basis for this *cis*-acting long-range effect is mysterious.

SOME UNUSUAL FEATURES OF TRANSPOSING PHAGE MU

Mu is a phage with the alternate life-styles of lytic cycle and lysogeny. It was discovered by its ability to cause mutations in *E. coli*. The mutations result from insertion of the 37 kb phage genome into random sites (with regional specificity) in the bacterial DNA. The insertions do not revert at a detectable frequency.

The map of phage Mu is shown in **Figure 35.13.** The integrated Mu genome has the same gene order as the free phage DNA, which is a linear molecule. (Note the contrast with phage lambda, whose linear free DNA circularizes on infection, so that integration generates a linear prophage whose gene order is a permutation of that of the free λ DNA.)

The linear Mu genomes have neither sticky ends nor repeats, which raises the question of how the ends act in concert during integration. Possibly the reaction involves homologies that occur up to ~100 bp from each end. The existence of a mechanism to recognize the specific ends of an integrated Mu sequence was revealed by the discovery of Mu excision, a reaction that occurs only when Mu prophage itself has gained an insertion of IS1. The mechanism is unknown, but involves Mu-coded functions as well as IS1-coded functions.

The ends of free Mu DNA possess additional material that is not part of the phage genome. The actual termini consist of sequences obtained from the host genome, about 100 bp long at the left (*c*) end and about 1500 bp long at the right (*S*) end. These sequences are different in individual molecules of Mu; they are acquired as a consequence of the packaging of Mu DNA from precursors in which the phage genome is flanked by host sequences. The additional regions are shed when a phage genome integrates into the bacterial chromosome.

An infecting parental Mu DNA is *not* efficiently integrated into the host genome. Less than 10% of the infecting phage genomes integrate. However, Mu DNA sequences are efficiently integrated during the reproductive cycle. This paradox suggests that the *copies* of Mu produced by replication have the ability to integrate. In a related reaction, induction of a Mu prophage to enter the lytic cycle is not accompanied by excision of phage DNA from the host chromosome. It is (again) the replication product that is important.

A connection between integration and DNA replication is revealed by the nature of mutants of Mu that are defective in integration. These mutants are unable to replicate; they lie in either of the two genes *A* and *B*, needed for phage replication. Thus integration seems to depend on the replication functions rather than on separate recombination enzymes (a contrast with lambda).

There may be two pathways for Mu insertion. First, lysogens contain prophages generated by simple insertions of Mu. The similarity of this event to transposition is indicated by the presence of direct repeats

Figure 35.13
The order of Mu genes is the same in the prophage and free DNA, but the ends are different. Prophage DNA is flanked by 5 bp direct repeats of a target host sequence. Free phage DNA has different additional sequences of ~100 bp at the left end and ~1500 bp at the right end.

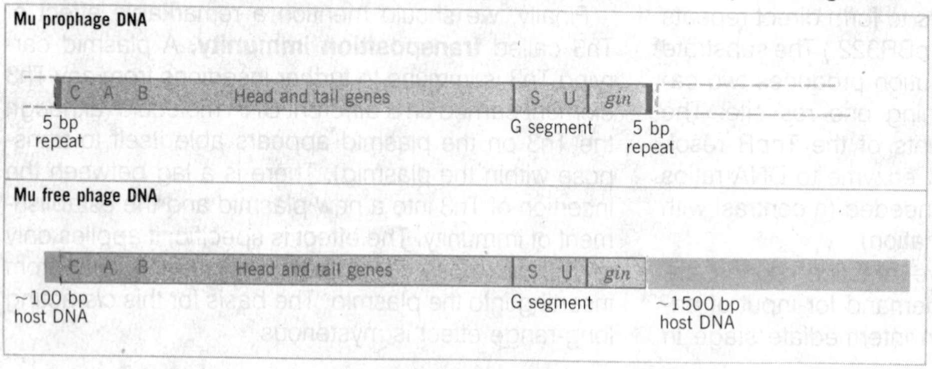

Mu prophage DNA
C A B Head and tail genes S U *gin*
5 bp repeat
G segment
5 bp repeat
Mu free phage DNA
C A B Head and tail genes S U *gin*
~100 bp host DNA
G segment
~1500 bp host DNA

of 5 bp of target DNA on either side of the prophage. Second, during lytic growth, almost all of the products of Mu-mediated transposition are cointegrate structures. We do not know whether the simple insertions and cointegrates represent alternative interactions of Mu DNA with its target or are alternative fates of a common intermediate. The basic mechanisms of Mu transposition have yet to be worked out.

Another unusual feature of phage Mu is found near the right end of the genome. The **G** or **invertible segment** comprises a 3 kb segment found in different orientations in different molecules of the Mu DNA. In phage grown by lytic infection of *E. coli* strain K12, the G segment is always in the orientation called G(+). However, the DNA of phages generated by induction may contain the G segment in either this or the opposite G(−) orientation. A Mu gene called *gin,* located just beyond the G segment, is required for the inversion reaction.

The G segment contains the genes coding for proteins involved in phage adsorption. The orientation of the segment controls the expression of these genes, with the result that G(+) and G(−) phages have different specificities for bacterial strains. The dependence of strain specificity on the G segment is illustrated diagrammatically in **Figure 35.14.**

The G segment carries alternate sets of genes. In the G(+) orientation, genes *S* and *U* are expressed. Their products allow the phage to adsorb to *E. coli* K12, but not to *E. coli* C. In the G(−) orientation, genes *S′* and *U′* are expressed. Their products allow adsorption to *E. coli* C but not K.

Figure 35.14
Inversion of the G segment determines whether the S and U proteins or the S′ and U′ proteins are synthesized. One model supposes that their expression is initiated to the left of the invertible segment. Proteins S and S′ share an N-terminal sequence (S_c) coded at the left of the invertible segment. S_v and U are coded by one strand of DNA while S'_v and U′ are coded by the complementary strand.

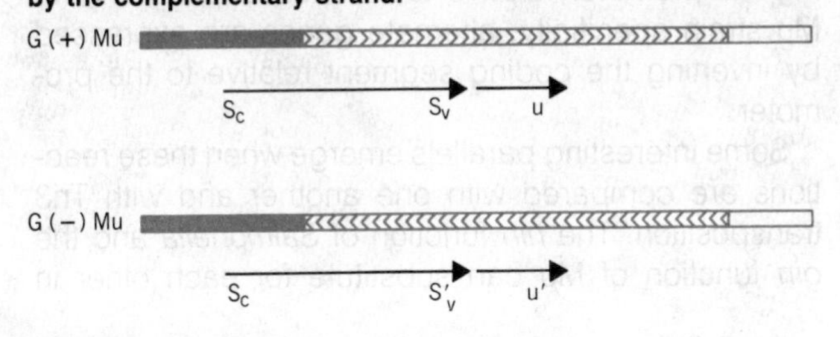

The proteins have been tentatively identified: S is 56,000 daltons, U is 21,000 daltons, S′ is 48,000 daltons, and U′ is 26,000 daltons. A significant feature is that their combined lengths would require a coding sequence of about 4 kb, substantially longer than the G segment itself.

A current model for the organization of the two gene sets is displayed in **Figure 35.14.** The two sets of genes are coded on opposite strands of DNA and transcribed from a promoter located to the left of the inverted region. The size discrepancy is explained because the S proteins actually start with a common N-terminal sequence, S_c, which is coded outside the inverted region. Depending on the orientation of the region, the variable C-terminal sequences S_v or S'_v may be connected to S_c. The two U genes are independently coded.

SALMONELLA PHASE VARIATION OCCURS BY INVERSION

The movement of bacteria is accomplished by the waving of their flagella. Many *Salmonella* species are **diphasic** because they possess two nonallelic genes for flagellin (the protein subunit of the flagellum). A given clone of bacteria may express either the **H1** type (it is said to be in **phase 1**) or the **H2** type (the bacteria are in **phase 2**). Transition from one phase to the other occurs about once in every 1000 bacterial divisions and is called **phase variation.**

The genes for the two types of flagellin reside at different chromosomal locations. The circuit for control of flagellin synthesis is illustrated in **Figure 35.15.** The H2 gene is closely linked to another gene (*rh1*) that codes for a repressor of H1 synthesis. These two genes are coordinately expressed. Thus when H2 is expressed, the repressor also is expressed and prevents any synthesis of H1. When H2 is not expressed, neither is the repressor, so synthesis of H1 occurs. The phase of the bacterium is determined by whether the *H2-rh1* transcription unit is active.

Expression of this transcription unit is controlled by the orientation of a segment of DNA adjacent to it. The

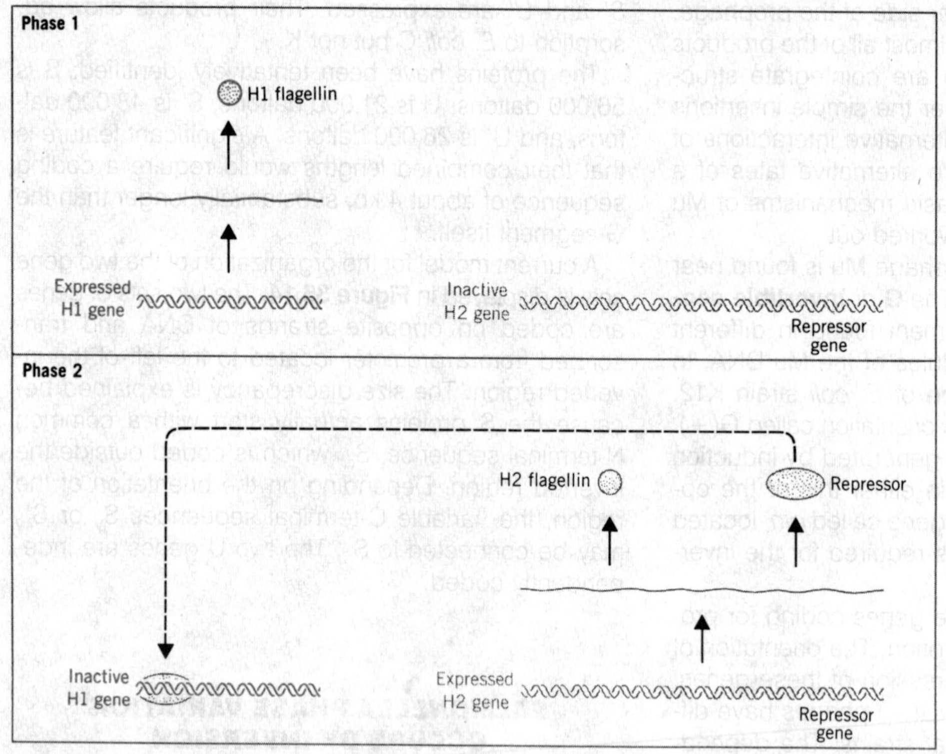

Figure 35.15
***Salmonella* phase is determined by the activity of the H2 transcription unit.**
In phase 1, the H2 unit is inactive, so the H1 flagellin is synthesized. In phase 2, the H2 unit is active, and synthesizes H2 flagellin and a repressor that prevents expression of the H1 flagellin gene.

segment is 995 bp long and is bounded by 14 bp repeats (*IRL* and *IRR*). The initiation codon for H2 lies 16 bp to the right of the adjacent inverted repeat.

The segment of DNA between *IRL* and *IRR* contains the *hin* gene, whose product is a *trans*-acting protein that mediates the inversion of the entire segment by a reciprocal recombination between the inverted repeats (see Figure 35.9 previously). Mutations in the *hin* gene reduce the frequency of inversion by 10^4 times.

The consequences of the inversion reaction are illustrated in **Figure 35.16.** The promoter for the *H2-rh1* transcription unit lies within the invertible segment. In one orientation, transcription initiates at the promoter and continues through *H2-rh1*, resulting in phase 2 expression. In the other orientation, the promoter faces in the other direction, and the transcription unit is not expressed (although transcription probably occurs in the other direction, with unknown consequences). The absence of promotion of *H2-rh1* results in phase 1 expression.

Phase variation and Mu strain specificity therefore both use inversion to regulate gene expression. In phase variation, an on/off switch is provided by moving the promoter relative to the transcription unit. In Mu strain specificity, alternate genes are expressed by inverting the coding segment relative to the promoter.

Some interesting parallels emerge when these reactions are compared with one another and with Tn3 transposition. The *hin* function of *Salmonella* and the *gin* function of Mu can substitute for each other in

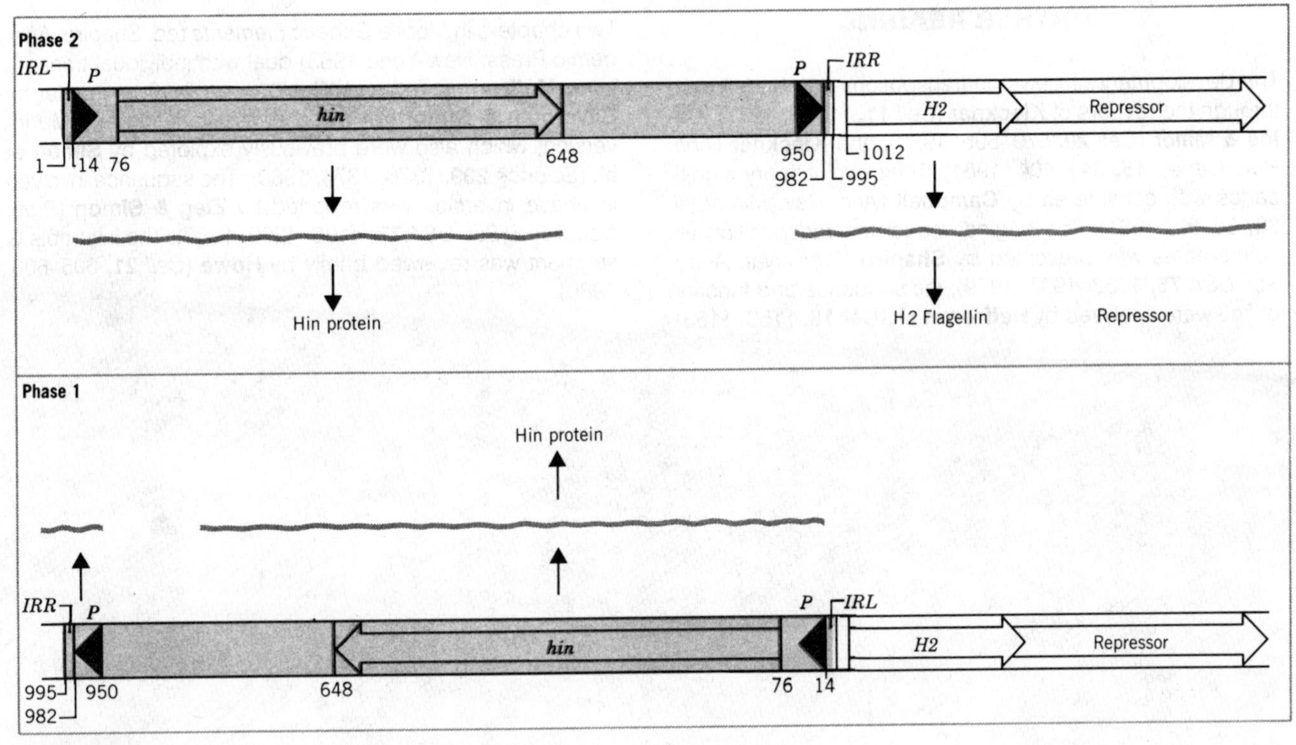

Figure 35.16
Expression of H2 flagellin and the phase 1 repressor is controlled by the orientation of a 995 bp invertible region.
Phase 2 occurs when the promoter at position 950 sponsors transcription through *IRR* into the H2 unit. Phase 1 occurs when the promoter faces in the opposite direction. In both phases, *hin* is expressed from its own promoter (whose location is not determined). Numbers indicate base pairs within the invertible segment.

complementation assays. (Another function interchangeable with these is the *cin* gene responsible for inversion of the *C* segment of phage P1.) Although the Tn3 *tnpR* gene cannot replace either of these sequences, the amino acid sequence of the resolvase shows considerable homology with that of the Hin protein.

There are sequence homologies between *IRL*/*IRR*, the inverted repeats at either end of the G segment, and the *res* target for TnpR. This suggests that inversion of the *hin* segment of *Salmonella* and the G segment of Mu occurs by a site-specific recombination between the inverted repeats, related in mechanism to the reaction involved in resolution of Tn3 cointegrates. Thus these static inversion events could have evolved by interchanges between transposons and bacterial loci, possibly even by some ancestral immobilization of a transposable element, whose functions became limited.

A related gene is also found in *E. coli*. Called *pin*, it is able to complement *gin* mutations in phage Mu. It appears to catalyze the inversion of an adjacent segment of 1800 bp, which is surrounded by 29 bp inverted repeats. We have no idea what function this reaction serves.

There are good homologies (60–70%) between all three inversion functions (Gin, Hin, and Pin). The sequence relationships among the genes suggest that *pin* may be an intermediate stage in evolution between *gin* and *hin*.

FURTHER READING

The development of views on transposons can be followed through the reviews of **Kleckner** (*Cell* **11,** 11–23, 1977), **Calos & Miller** (*Cell* **20,** 579–595, 1980), and **Kleckner** (*Ann. Rev. Genet.* **15,** 341–404, 1981). Their evolutionary significance was considered by **Campbell** (*Ann. Rev. Microbiol.* **35,** 55–83, 1981). The original model for transposition via cointegrates was presented by **Shapiro** (*Proc. Nat. Acad. Sci. USA* **76,** 1933–1937, 1979); the sequence and function of Tn3 were analyzed by **Heffron et al.** (*Cell* **18,** 1153–1163).

Two chapters in *Mobile Genetic Elements* (ed. Shapiro, Academic Press, New York, 1983) deal with individual transposons: **Heffron** on Tn3 and **Kleckner** on Tn10; a chapter by **Silverman & Simon** reviewed phase variation and Mu inversion, which also were previously explored by **Simon et al.** (*Science* **209,** 1370–1374, 1980). The sequence involved in phase inversion was reported by **Zieg & Simon** (*Proc. Nat. Acad. Sci. USA* **77,** 4196–4200, 1980). The invertible G segment was reviewed briefly by **Howe** (*Cell* **21,** 605–606, 1980).

CHAPTER 36
MOBILE ELEMENTS IN EUKARYOTES

The genetic background of eukaryotes includes elements comparable to bacterial transposons, with the ability to move to randomly selected new locations in the genome. Little is known about the mechanism of transposition within eukaryotic DNA, but in general it appears that transposable elements can propagate in eukaryotes as successfully as in prokaryotes.

The existence of transposition was in fact first deduced many years ago in McClintock's genetic studies of maize. Although her results remained in an intellectual void until the more recent discovery of transposition in bacteria, now these events can be studied at the molecular level in eukaryotes also.

Elements directly comparable to bacterial transposons have been characterized in *S. cerevisiae* and *D. melanogaster*, as well as maize. They are small, have terminal repeats, and are found at different locations in different genomes. We assume that they carry genes needed to code for transposition; we do not know whether they carry accessory functions unconnected with the transposition event itself.

All of these elements are analogous to bacterial transposons in comprising sequences able to transport themselves around the genome in which they reside. Both prokaryotic and eukaryotic transposons apparently have no life outside the genome. Whether valuable components playing a role in cellular survival or selfish parasites concerned only with their own survival, they have no independent existence and do not generate free molecules of DNA.

The functions involved in eukaryotic transposition are unknown; we have yet to identify the relevant gene products of a eukaryotic transposon. However, the presence of the diagnostic feature of the short direct repeats of target DNA suggests an underlying similarity of mechanism in which this "generalized" transposition selects random sites for insertion.

Analogous transposable elements have not yet been detected as components of animal cell genomes, but footprints of ancient transposition events are found in the form of direct target repeats flanking dispersed repetitive sequences. In several situations, an RNA sequence has been implicated as the progenitor of a genomic (DNA) sequence. We think that the RNA must have been converted into a duplex DNA copy that was inserted into the genome by a transposition-like event.

The paradigm for this type of event is the ability of retroviruses to insert DNA copies of an RNA viral genome into the chromosomes of a host cell. The mechanism involved is related to bacterial transposition, but a major difference is that the genomic insertions of the element all are derived directly from its extrachromosomal form; that is, once an insertion has occurred, the element probably does not transpose elsewhere.

In conjunction with the ability of plasmids to transfer between bacteria, prokaryotic transposons may confer mobility on some host genes (possibly a useful substitute for the lack of sexual interaction). Along these lines, we should like to know what sort of relationship exists between the transposable elements of different eukaryotic species and whether transposition functions might be retained in passage of an element from one species to another.

Another type of movement within the genome is represented by the "directed" transposition of sequences whose movement from one specific site to another has regulatory functions. In bacteria, there is a clear relationship between the mechanisms of site-specific inversion and transposition. In eukaryotes, directed movement of DNA may occur by other mechanisms, probably not related to the action of transposons.

Yeast displays unidirectional transfer of copies of genes for mating type from "silent" to "active" loci. The process resembles bacterial transposition in leaving the donor locus intact while changing the recipient; however, there is only a single recipient locus, which initiates the transfer instead of being an innocent target.

Another situation in which the content of the genome changes is provided by the immunoglobulin genes, in which recombination is used to construct active genes in the appropriate somatic cells (see Chapter 37). The gamut of changes in the eukaryotic genome thus runs from random transpositions to tissue-specific reconstructions.

YEAST TY ELEMENTS RESEMBLE BACTERIAL TRANSPOSONS

The Ty elements of yeast comprise a family of dispersed repetitive DNA sequences. Each element is 6.3 kb long; the last 330 bp at each end constitute direct repeats, called δ.

Ty is an abbreviation for "transposon yeast;" the existence of a transposition function is suggested by the observation that Ty elements are found at different sites in different strains. We cannot yet estimate the frequency of Ty transposition, but it seems to be less than that of bacterial transposons. Movement of Ty elements has been followed largely by selecting mutations created by insertions into regulatory loci, so most of our information about the results of insertion concerns the effects of Ty on genes adjacent to it.

There is considerable divergence between individual Ty elements. Most elements fall into one of two major classes, called Ty1 and Ty917. Their relationship is illustrated in **Figure 36.1.** The elements share the presence of the delta repeats, a long region at the left end, another region in the center, and a short region adjacent to the right delta. The two regions of major substitution show no homology as judged by restriction mapping or heteroduplex formation.

Individual Ty elements of each type have many changes from the prototype of their class, including base-pair substitutions, insertions, and deletions. There are about 30 copies of the Ty1 type and 6 of the Ty917 type in a typical yeast genome. In addition, there are about 100 independent delta elements, called solo δ's.

The delta sequences also show considerable heterogeneity, although the two repeats of one Ty element are likely to be identical or at least very closely related. We do not know how the identity of the terminal deltas is generated; it could involve gene conversion during or after transposition, or even the generation of both repeats of a "new" Ty element from one of the repeats of the donor. The delta sequences associated with Ty elements show greater conservation of sequence than the solo delta elements, which suggests that recognition of the repeats may be involved in transposition.

A transposition event leaves the same footprint as a bacterial transposition: 5 bp of target DNA are repeated on either side of the inserted Ty element. We do not know the source of whatever enzyme(s) are needed for Ty transposition.

The presence of Ty elements provides regions of portable homology that are targets for recombination events mediated by host systems. The usual tendency of such an event is to mess up the chromosome, either by causing a deletion or inversion (by recombination between two Ty elements on a chromosome), or by causing more dramatic changes when recombination occurs between two Ty elements on different chromosomes. Recombination between Ty elements seems to occur in bursts; when one event is detected, there is an increased probability of finding others. Among

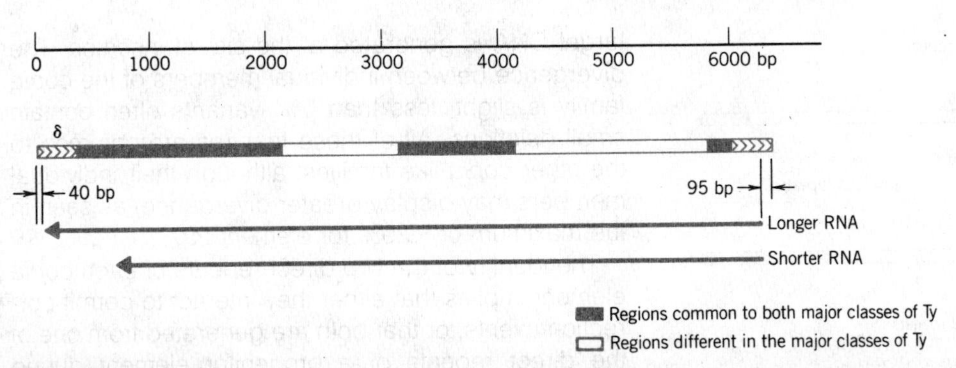

Figure 36.1
Ty elements terminate in short direct repeats and are transcribed into two overlapping RNAs.

the interactions of Ty elements, gene conversion occurs between those at different locations, with the result that one element is "replaced" by the sequence of the other.

Ty elements can excise by homologous recombination between the directly repeated delta sequences. The large number of solo delta elements may be footprints of such events. An excision of this nature may be associated with reversion of a mutation caused by the insertion of Ty; the level of reversion may depend on the exact delta sequences left behind.

The Ty element is transcribed into two poly(A)$^+$ RNA species. As indicated in Figure 36.1, both initiate ~95 bp from the right end of the element. One terminates after 5 kb, within the common region at the left end; the other terminates after 5.7 kb, within 40 bp of the left boundary. This longer RNA therefore both starts and stops within the delta sequence and as a result has repeats at its ends.

A paradox is that both delta elements have the same sequence, yet a promoter is active in the right delta and a terminator is active in the left delta. (A similar feature is found in other transposable elements, including the retroviruses described later.) We do not know whether the Ty RNA is translated into protein.

SEVERAL TYPES OF TRANSPOSABLE ELEMENTS RESIDE IN *D. MELANOGASTER*

The existence of transposable elements in *D. melanogaster* was first inferred from observations analogous to those that identified the first insertion sequences in *E. coli*. Unstable mutations are found that revert to wild type or generate deletions of the flanking material with an endpoint at the original site of mutation. They are caused by several types of transposable sequence.

The best-characterized family is called *copia*. Its name reflects the presence of a large number of closely related sequences that code for abundant mRNAs. The *copia* family is taken as a paradigm for several other types of elements whose sequences are unrelated, but whose structure and general behavior appear to be similar. These are described as "*copia*-like" elements; they are listed in Table 36.1. Together the *copia*-like families account for more than 1% of the DNA of *D. melanogaster*.

The number of copies of the *copia* element depends on the strain of fly; usually it is between 20 and 60. The members of the family are widely dispersed. When the genomes of different strains of flies are compared by restriction analysis, the fragments hybridizing with a *copia*-specific probe show a different (although overlapping) spectrum of sizes in each strain. Hybridization *in situ* with polytene chromosomes shows that *copia* sequences are present at different (again overlapping) series of sites. The variability of the *copia* pattern is indicated by the fact that no one site is occupied by a *copia* sequence in every single strain examined.

These differences have developed over evolutionary periods. Comparisons of strains that have diverged recently (over the past 40 years or so) as the result of their propagation in the laboratory reveal few changes. At present, we have no estimate of the rate of change. However, the nature of the underlying events

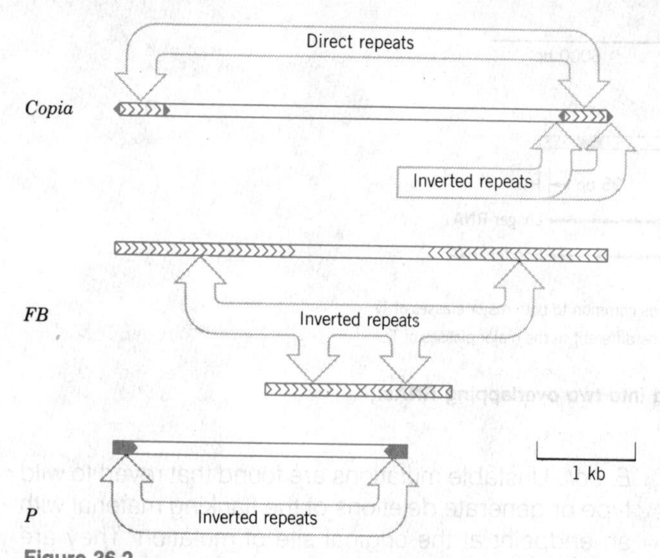

Figure 36.2
Three types of transposable element in *D. melanogaster* have different structures.

is indicated by the result of growing cells in culture. The number of *copia* elements per genome then increases substantially, up to 2–3 times. The additional elements represent insertions of *copia* sequences at new sites. Adaptation to culture in some unknown way transiently increases the rate of transposition to a range of 10^{-3} to 10^{-4} events per generation.

The structure of the *copia* element is illustrated in **Figure 36.2** and detailed in **Table 36.1.** The element is 5000 bp long, with identical direct terminal repeats of 276 bp. Each of the direct repeats itself ends in related inverted repeats. A direct repeat of 5 bp of target DNA is generated at the site of insertion. The divergence between individual members of the *copia* family is slight, less than 5%; variants often contain small deletions. All of these features are common to the other *copia*-like families, although their individual members may display greater divergence, as seen in the maximum of ~20% for element *297*.

The identity of the two direct repeats of each *copia* element implies that either they interact to permit correction events, or that both are generated from one of the direct repeats of a progenitor element during transposition. (We have already described the same feature for yeast Ty elements; it is also characteristic of retroviruses; see later.)

The *copia* elements in the genome are always intact; individual copies of the terminal repeats have not been detected (although we would expect them to be generated if recombination deleted the intervening material). On the other hand, *copia* elements sometimes are found in the form of free circular DNA, a contrast with the absence of free forms of bacterial transposons. The two major types of molecule in the circular *copia* population are 5000 bp and 4700 bp long. The latter has a deletion of one of the terminal repeats (which would be expected if it had originated by excision from the chromosome via reciprocal recombination).

Transcripts of *copia* are found in the form of abundant poly(A)$^+$ mRNAs, representing both full-length and part-length transcripts. The mRNAs have a common 5′ terminus, resulting from initiation in the middle of one of the terminal repeats. The part-length 2 kb species has messenger activity for a 51,000-dalton

Table 36.1
Several classes of transposable elements are found in *D. melanogaster.*

Element	Copies per Genome	Length (bp)	Terminal Direct Repeat	Terminal Inverted Repeat	Target Direct Repeat
copia	20–60	5000	276	None	5
FB	~30	500–5000	None	250–1250	9
P	~50 or 0	500–2900	None	31	8

Elements that resemble *copia* in organization are *412, 297, 17.6, mgd1, mgd3, B104, roo,* and *gypsy*. Their copy numbers vary from 10 to 100 per genome, their lengths from 5500 to 8500 bp, and the direct repeats from 269 to 571 bp. The target repeat lengths are usually 4 or 5 bp.

protein of unknown function. Since the direct repeats are identical, it is possible that both function as promoters, one reading into the element to produce the *copia* RNAs, the other reading into the adjacent DNA to initiate transcription of genes downstream from the site of insertion. (Again, a similar feature is found with retroviruses.)

The members of another family of transposable elements, called *FB* (an abbreviation for foldback), have inverted terminal repeats of variable length. Some FB elements consist solely of juxtaposed inverted repeats; in others the inverted repeats are separated by a region of nonrepetitive DNA.

In spite of the variation in length, the inverted repeats of all members of the FB family are homologous. This feature is explained by their structure, which consists of tandem copies of a simple-sequence DNA, separated by longer stretches of more diverse sequences. Proceeding from the end into the element, the length of the simple-sequence unit increases; initially it is 10 bp, then expands to 20 bp, and finally expands again to 31 bp.

The two copies of the inverted repeat in a single FB element are not identical. The inverted repeats superficially resemble satellite DNA, and it will be interesting to establish the relationship between the repeats of different members of the family.

The structure of the ends poses a puzzle about FB elements; we have no knowledge of what confers their ability to transpose. Sometimes two (nonidentical) FB elements apparently cooperate to transpose a large intervening segment of DNA, possibly in a manner reminiscent of composite bacterial transposons (although the length of the DNA between FB elements can be much greater). Two examples of such units have been found; one is 200 kb long and is called TE (for transposable element). It is possible that any sequence of DNA flanked on either end by FB elements could behave as such a unit.

Another type of transposable element is represented by the much smaller P elements, whose maximum length is 2.9 kb. They terminate in short inverted repeats and transpose only in certain conditions (see later).

A potential relic of a transposable element is identified by the organization of the rRNA genes. About two thirds of the genes for 28S rRNA are interrupted in *D. melanogaster*. The interruptions fall into two classes, located at positions some 50–80 bp apart in the gene. Neither type of interruption can be classed as an intron, because the interrupted genes are not transcribed. The connection between the presence of the interruption and the failure of these genes to initiate at the promoter (which apparently is identical with that of the noninterrupted genes) is provocative but obscure.

Type I insertions are found in rRNA genes on the X chromosome. The insertions are homologous, but vary in length from 0.5 kb to 5.0 kb; the shorter insertions may have been produced by internal deletions of an original 5 kb insert. Their relative structures are illustrated in **Figure 36.3.** The 0.5 or 1.0 kb insertions display a duplication of the flanking coding sequence of 14 or 11 bp, respectively. The same 14 bp target site duplication is present in the type I interrupted genes of the related fly *D. virilis*. Did type I insertions originate in a transposition event? The footprint of the target site duplication seems to have been lost from the 5 kb insertion, which instead of a duplication has a deletion of 9 bp of the coding sequence on the left side of the insertion.

Sequences homologous to the type I insertion occur in other locations besides the rDNA, which is consistent with the idea that they represent an ancient trans-

Figure 36.3

Type I rDNA insertions (red segments) are homologous sequences of various lengths, probably derived by internal deletions. At the point of interruption in the coding sequence, the 5 kb insertion has a 9 bp deletion of rDNA, the 1 kb insertion has a 14 bp duplication of rDNA, and the 0.5 kb insertion has an 11 bp duplication of rDNA.

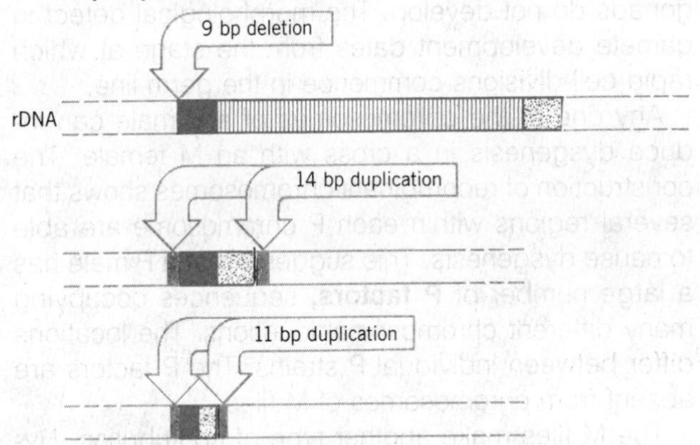

poson. They often occur in the form of tandem repeats interspersed with nonhomologous DNA.

Type II insertions are found in the rRNA genes of both X and Y chromosomes. They vary in length from 1.5 to 4.0 kb. Although also complex in structure, their sequences are not related to the type I insertion. They present no evidence for any ancestral transposition event.

THE ROLE OF TRANSPOSABLE ELEMENTS IN HYBRID DYSGENESIS

Certain strains of *D. melanogaster* encounter difficulties in interbreeding. When flies from two of these strains are crossed, the progeny may display "dysgenic traits," a series of defects inimical to propagation of the organism, including mutations, chromosomal aberrations, distorted segregation at meiosis, and sterility. The appearance of these correlated defects is called **hybrid dysgenesis.**

Two systems responsible for hybrid dysgenesis have been identified in *D. melanogaster*. In the first, flies are divided into the types I (inducer) and R (reactive). Reduced fertility is seen in crosses of I males with R females, but not in the reverse direction. In the second system, flies are divided into the two types P (paternal contributing) and M (maternal contributing). **Figure 36.4** illustrates the asymmetry of the system; a cross between a P male and an M female causes dysgenesis, but the reverse cross does not.

Dysgenesis is principally a phenomenon of the germ cells. In crosses involving the P-M system, the F1 hybrid flies have normal somatic tissues. However, their gonads do not develop. The morphological defect in gamete development dates from the stage at which rapid cell divisions commence in the germ line.

Any one of the chromosomes of a P male can induce dysgenesis in a cross with an M female. The construction of recombinant chromosomes shows that several regions within each P chromosome are able to cause dysgenesis. This suggests that a P male has a large number of **P factors,** sequences occupying many different chromosomal locations. The locations differ between individual P strains. The P factors are absent from chromosomes of M flies.

The M flies make another type of contribution. Hybrid dysgenesis is triggered only when the P factors become exposed to cytoplasm inherited from an M mother. The maternal contribution is called the M cytotype; its basis is not understood, but does not seem to be a simple maternal effect, because it is inherited for more than one generation when there is an M female in the line of descent. In rare lines descended from P male × M female crosses, the P factors convert the line to P cytotype—that is, to the state in which P factors are not active.

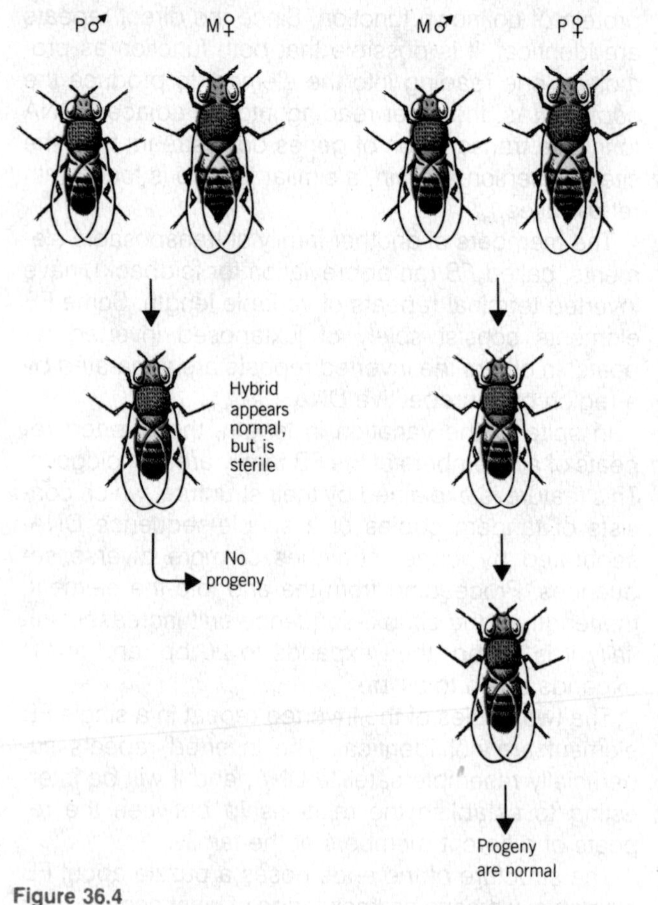

Figure 36.4
Hybrid dysgenesis is asymmetrical; it is induced by P male × M female crosses, but not by M male × P female crosses.

The events responsible for the induction of mutations in dysgenesis have been examined by mapping the DNA of *w* mutants found among the dysgenic hybrids. All the mutations result from the insertion of DNA into the *w* locus. The P insertions remain stable when the chromosome is perpetuated in P cytotype, but become unstable, as detected by their excision, on exposure to M cytotype. In addition to the P insertions, insertions of *copia* elements also are found in dysgenic flies.

The P insertions vary in length but are homologous in sequence. The longest P elements are ~2.9 kb long and have several open reading frames. The shorter elements arise, apparently rather frequently, by internal deletions of a full-length P factor. All P elements possess direct terminal repeats, as indicated in Figure 36.2 and detailed in Table 36.1. Insertion of a P element generates an 8 bp duplication at the target site.

Our current model for the dysgenic interaction supposes that the intact P factors code for a transposase that is inactive in P cytotype, but becomes active in M cytotype. At least some of the shorter P elements may have lost the capacity to produce the transposase, but may be activated in *trans* by the enzyme coded by a complete P element. The ends of the P factor may indicate that it is a suitable substrate for transposition. The P-M dysgenic interaction may also activate the movement of *copia* (and other transposable elements); these insertions, however, remain stable irrespective of the subsequent cytotypes to which the chromosome is exposed.

In situ hybridization identifies 30–50 copies of the P factor in each P strain; the factors are absent from M strains. Chromosomes from P-M hybrid dysgenic flies have P factors inserted at many new sites. The chromosome breaks typical of hybrid dysgenesis occur at hotspots that are the sites of residence of P factors. The average rate of transposition of P elements to M chromosomes is different in male and female dysgenic hybrids, but is of the order of one event per generation. A chromosome that has acquired P elements becomes dysgenic when it is crossed into M cytotype; the capacity of a chromosome to induce sterility is related to the number of P elements it has gained. Independence of the abilities to transpose and to induce sterility is suggested by the existence of the Q strain, which behaves like P strains in all ways except that it does not induce sterility in the hybrids.

Strains of *D. melanogaster* descended from flies caught in the wild more than 30 years ago are always M. Strains descended from flies caught in the past 10 years are almost always P. Does this mean that the P element family has invaded wild populations of *D. melanogaster* in recent years? P elements are indeed highly invasive when introduced into a new population; the source of the invading element would have to be another species. On the other hand, the P element may have been lost from populations maintained in the laboratory over protracted periods.

Because hybrid dysgenesis reduces interbreeding, it is a step on the path to speciation. Suppose that a dysgenic system is created by a transposable element in some geographic location. Another element may create a different system in some other location. Flies in the two areas will be dysgenic for two (or possibly more) systems. If this renders them intersterile and the populations become genetically isolated, further separation may occur. Multiple dysgenic systems may therefore lead to inability to mate—and to speciation.

THE RETROVIRUS LIFE CYCLE INVOLVES TRANSPOSITION-LIKE EVENTS

Retroviruses have genomes of single-stranded RNA that are replicated through a double-stranded DNA intermediate. The class of retroviruses about which we have the most information are those that give rise to the **C-type** particles in birds and mammals. Two copies of the RNA genome are packaged into each virion, making the individual virus particle effectively diploid. When a cell is simultaneously infected by two different C-type viruses, it is possible to generate heterozygous virus particles carrying one genome of each type. The diploidy may be important in allowing the virus to acquire cellular sequences.

A typical retroviral sequence contains three "genes," the term here identifying coding regions, each of which actually gives rise to multiple proteins by processing reactions. The order of the genes is *gag-pol-env*, as indicated in the upper part of Figure 36.5. The viral RNA is translated as a normal messenger—it is capped at the 5′ end and polyadenylated at the 3′ end—into the *gag* polyprotein. One spliced version of the RNA (lacking the *gag* termination codon) is translated into a *gag-pol* protein. Splicing also generates a shorter **subgenomic** messenger that is translated into the *env* polyprotein, a precursor for two proteins.

The *gag* gene gives rise to the protein components of the nucleoprotein core of the virion. The *env* gene codes for components of the envelope of the particle, which also sequesters components from the cellular cytoplasmic membrane. The *pol* gene specifies the enzyme reverse transcriptase, which consists of two subunits, one comprising a processed fragment of the other. As its name implies, the enzyme is responsible for converting the RNA genome into a complementary DNA strand. It also catalyzes subsequent stages in the production of duplex DNA; it has a DNA polymerase activity, behaves like RNAase H (can degrade the RNA part of an RNA-DNA hybrid), and is an endonuclease.

Reverse transcriptase is carried with the genome in the viral particle. The enzyme converts the RNA into a linear duplex in the cytoplasm of the infected cell. The linear DNA makes its way (by unknown means) to the nucleus, where it is converted into circular forms. One or more DNA copies become integrated into the host genome. The integrated **proviral DNA** is transcribed by the host machinery to produce viral RNAs, which serve both as mRNAs and as genomes for packaging into virions. Integration is a normal part of the life cycle and is necessary for transcription.

The structures of the DNA forms of the virus are compared with the RNA in **Figure 36.5.** The viral RNA has direct repeats at its ends. These **R** segments vary in different virus strains from 10–80 nucleotides. Following the R segment at the 5′ end of the virus is the **U5** region of 80–100 nucleotides, whose name indicates that it is unique to the 5′ end. Preceding the R segment at the 3′ terminus is the **U3** segment of 170–1250 nucleotides, which is unique to the 3′ end.

The ends of the linear DNA have additional sequences. A U3 segment is added to the 5′ end; a U5 segment is added to the 3′ end. As a result, each end of the DNA has the sequence U5-R-U3; this is called the **long terminal repeat (LTR).** Its generation involves the reaction illustrated in **Figure 36.6** in which the *reverse transcriptase switches templates*, carrying the nascent DNA with it to the new template. The figure shows the generation of one LTR; a similar event is required to generate another at the other end. The LTRs at each end are identical. The 3′ end of U5 consists of a short inverted repeat relative to the 5′ end of U3, so the LTR itself ends in short inverted repeats.

Thus the organization of the retroviral DNA resembles that of transposable elements like *copia* (see Figure 36.2).

The retroviral life cycle is illustrated diagrammatically in **Figure 36.7.** The stages between the production of linear DNA and its integration to generate provirus are not well understood. After its entry into the nucleus, the linear DNA gives rise to two major forms of monomeric circles. (The number of circular molecules ranges from 0.1 to 5 copies per cell in different systems.)

The more abundant form of circle contains a single copy of the LTR, as though it had originated by a homologous recombination between the LTRs. (We do not know whether this actually is the mechanism for its production, but solitary LTRs that could be relics of the excision are present in some cellular genomes.)

The less abundant form contains two copies of the LTR in tandem; sometimes they are joined directly, as though by blunt-end ligation; sometimes nucleotides are missing from one or both LTRs at the sites of junction. The circular molecule with tandem LTR elements seems to be the form that gives rise to the integrated retrovirus. Perhaps the "integrase" recognizes the unique sequence created by the junction of the two LTR elements. (We note again the similarities with the circular forms of *copia* found in *D. melanogaster*, although these are not known to give rise to genomic elements.)

The viral DNA integrates into the host genome at randomly selected sites. A successfully infected cell gains between one and ten copies of the provirus. At each site of insertion, a short direct repeat of target DNA is generated. The length of the target repeat depends on the particular virus; it may be 4, 5, or 6 bp.

The occurrence of the direct repeats suggests that integration takes place by a mechanism similar to that involved in bacterial transposition. The sequence of the integrated retrovirus falls short of each end of unintegrated linear DNA by 2 bp. Thus the integrated viral DNA has lost 2 bp from the left end of the 5′ terminal U3 and has lost 2 bp from the right end of the 3′ terminal U5.

The U3 region of each LTR carries a promoter. The promoter in the left LTR is responsible for initiating transcription of the provirus. Sometimes (probably rather rarely), the promoter in the right LTR sponsors tran-

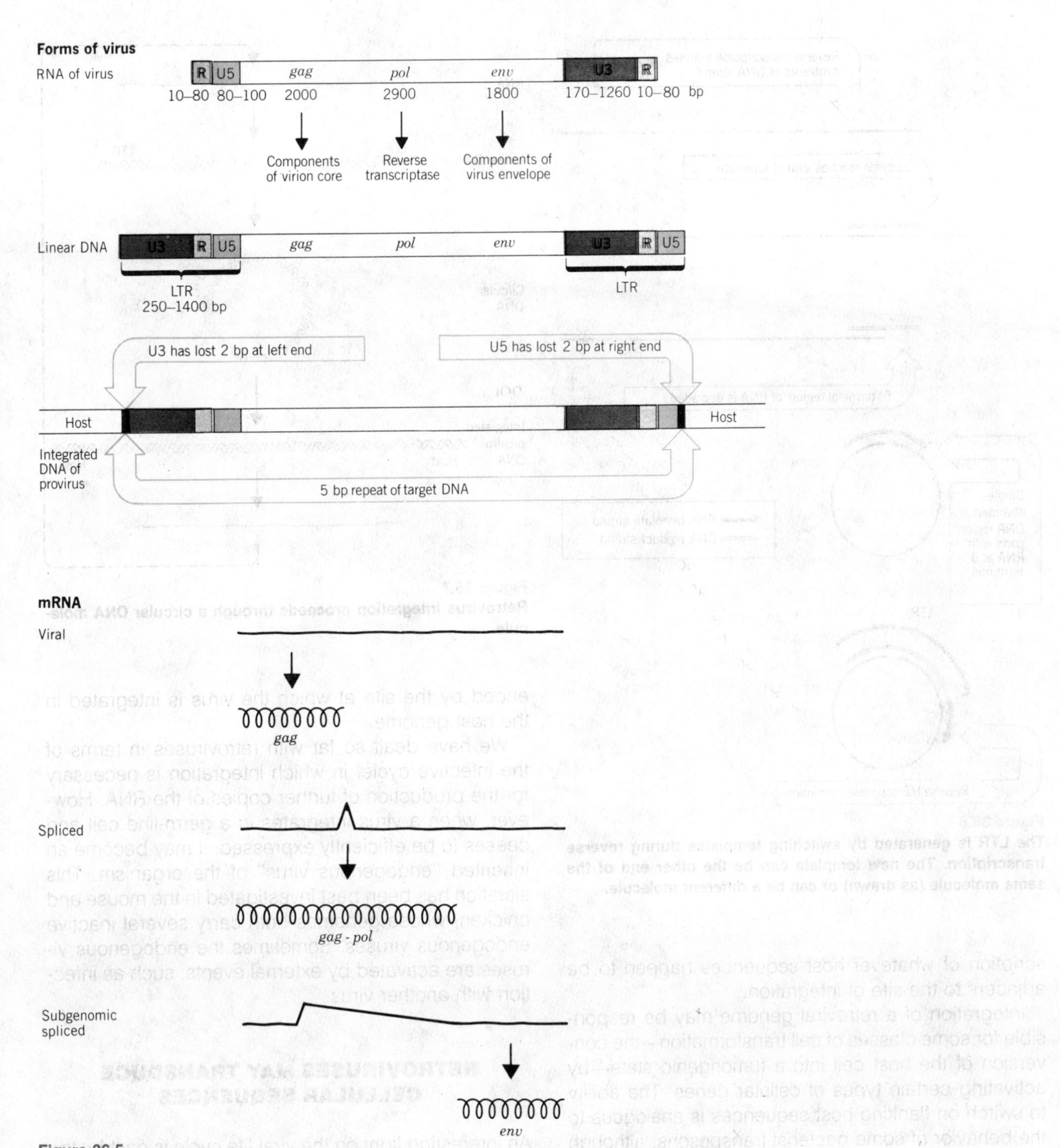

Figure 36.5
Retrovirus RNA ends in direct repeats (R), the free linear DNA ends in LTRs, and the provirus ends in LTRs shortened by two bases each.

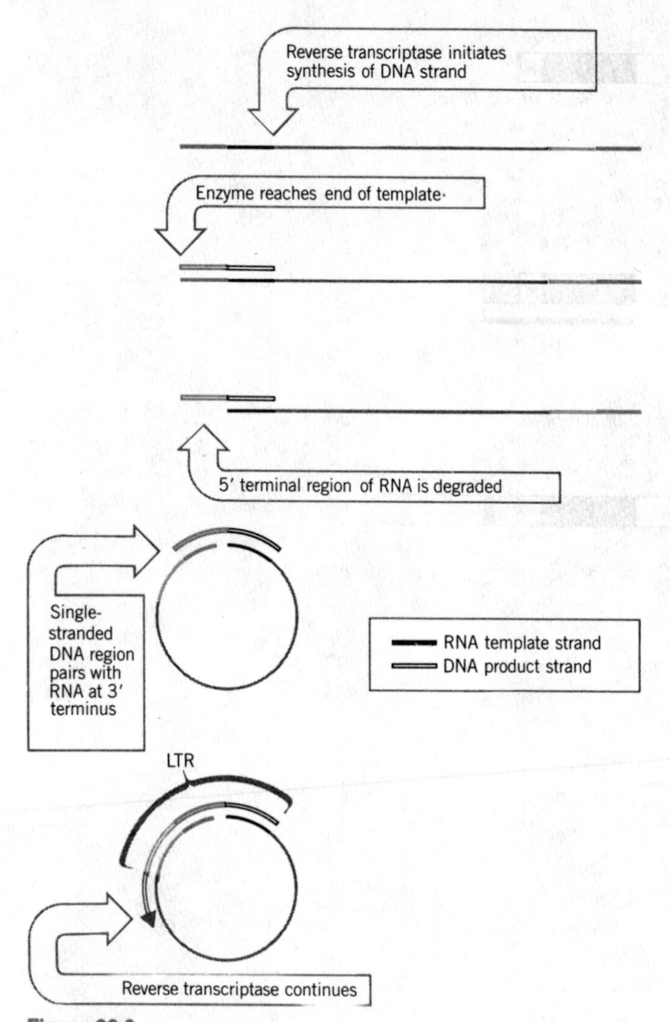

Figure 36.6
The LTR is generated by switching templates during reverse transcription. The new template can be the other end of the same molecule (as drawn) or can be a different molecule.

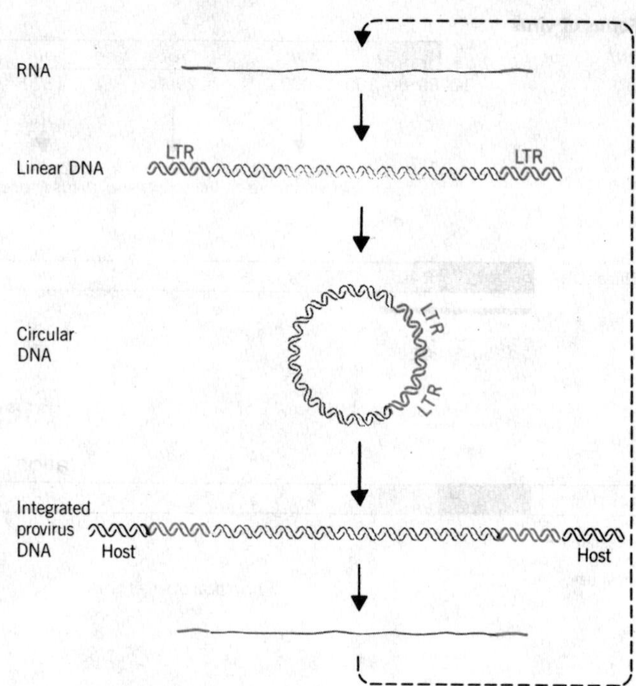

Figure 36.7
Retrovirus integration proceeds through a circular DNA molecule.

scription of whatever host sequences happen to be adjacent to the site of integration.

Integration of a retroviral genome may be responsible for some classes of cell transformation—the conversion of the host cell into a tumorigenic state—by activating certain types of cellular genes. The ability to switch on flanking host sequences is analogous to the behavior of some bacterial transposons, although at least in some cases it may be due to the provision of enhancer rather than promoter sequences. However, the activity of a provirus genome may be influenced by the site at which the virus is integrated in the host genome.

We have dealt so far with retroviruses in terms of the infective cycle, in which integration is necessary for the production of further copies of the RNA. However, when a virus integrates in a germ-line cell and ceases to be efficiently expressed, it may become an inherited "endogenous virus" of the organism. This situation has been best investigated in the mouse and chicken, whose genomes both carry several inactive endogenous viruses. Sometimes the endogenous viruses are activated by external events, such as infection with another virus.

RETROVIRUSES MAY TRANSDUCE CELLULAR SEQUENCES

An interesting light on the viral life cycle is cast by the occurrence of **transducing viruses,** variants that have acquired cellular sequences in the form illustrated in **Figure 36.8.** Part of the viral sequence—in this case,

Figure 36.8

Replication-defective transforming viruses have a cellular sequence substituted for part of the viral sequence. The length of the cellular sequence and the viral region that is replaced are characteristic for each virus.

the *env* gene—has been replaced by the *v-onc* gene. The resulting virus is **replication defective;** it cannot sustain an infective cycle by itself. However, it can be perpetuated in the company of a **helper virus** that provides the missing viral functions.

Onc is an abbreviation for oncogenesis, the ability to transform cultured cells so that the usual regulation of growth is released to allow unrestricted division. Both viral and cellular *onc* genes may be responsible for creating tumorigenic cells.

The *onc* genes carried by retroviruses are called *v-onc,* and each confers upon a virus the ability to transform a certain type of host cell. Loci with homologous sequences found in the host genome are called *c-onc* genes. In their normal state, they play some (usually unknown) role in cellular function. On occasion, however, a mutation may activate a *c-onc* gene, so that it causes the cell to become tumorigenic. A *c-onc* gene may also be activated by integration nearby of a retroviral genome, as described in the previous section.

The *onc* genes constitute a large class; individual members are given names derived from the name of the virus in which they were first found (*src* from Rous sarcoma virus causes sarcoma, *fes* is from feline sarcoma virus, and so on).

How are the *onc* genes acquired by the retroviruses? A revealing feature is the discrepancy in the structures of *c-onc* and *v-onc* genes. The *c-onc* genes usually are interrupted by introns and are expressed at low levels in the cell. The *v-onc* genes are uninterrupted and are expressed at a high level as part of the viral transcription unit. These structures suggest that *the* v-onc *genes originate from spliced RNA copies of the* c-onc *genes.*

A current model for the formation of transforming viruses is illustrated in **Figure 36.9.** A retrovirus has integrated near a *c-onc* gene. A deletion occurs to fuse the provirus to the *c-onc* gene; then transcription generates a joint RNA, containing viral sequences at one end and cellular *onc* sequences at the other end. Splicing occurs to remove the introns in both the viral and cellular parts of the RNA. The RNA has the appropriate signals for packaging into the virion; virions will be generated if the cell also contains another, intact copy of the provirus. Then some of the diploid virus particles may contain one fused RNA and one viral RNA.

A recombination between these sequences could generate the transforming genome, in which the viral repeats are present at *both* ends. (Recombination occurs at a high frequency during the retroviral infective cycle; presumably it employs DNA intermediates or additional strand-transfer reactions. We do not know anything about its demands for homology in the substrates, but we assume that the nonhomologous reaction between a viral genome and the cellular part of the fused RNA proceeds by the same mechanism responsible for viral recombination.)

What is the relationship between retroviruses and transposable elements? A substantial homology of sequence exists between the LTRs of the avian leukosis-sarcoma virus (AL-SV) and the terminal direct repeats of the *Drosophila copia*-like elements *297* and *17.6.* Could the *copia*-like class of elements have evolved by immobilizing a retrovirus?

The common features of the entire retroviral class suggest that it may be derived from a single ancestor. Primordial IS-like elements could have surrounded a host gene for a nucleic acid polymerase; the resulting unit would have the form *LTR-pol-LTR.* It might evolve into a retrovirus by acquiring further nucleic acid polymerizing activities, enabling it to handle both DNA and RNA substrates, followed by the incorporation of genes whose products allowed packaging of the RNA. Other functions, such as transforming genes, might be incorporated later. (There is no reason to suppose that the mechanism involved in acquisition of cellular functions is unique for *onc* genes; but these viruses may have a selective advantage because of their effects on the growth of the cells they infect.)

RNA-DEPENDENT TRANSPOSITIONS MAY HAVE OCCURRED IN THE CELL

Some pseudogenes display external and internal features suggesting that they may have originated in RNA

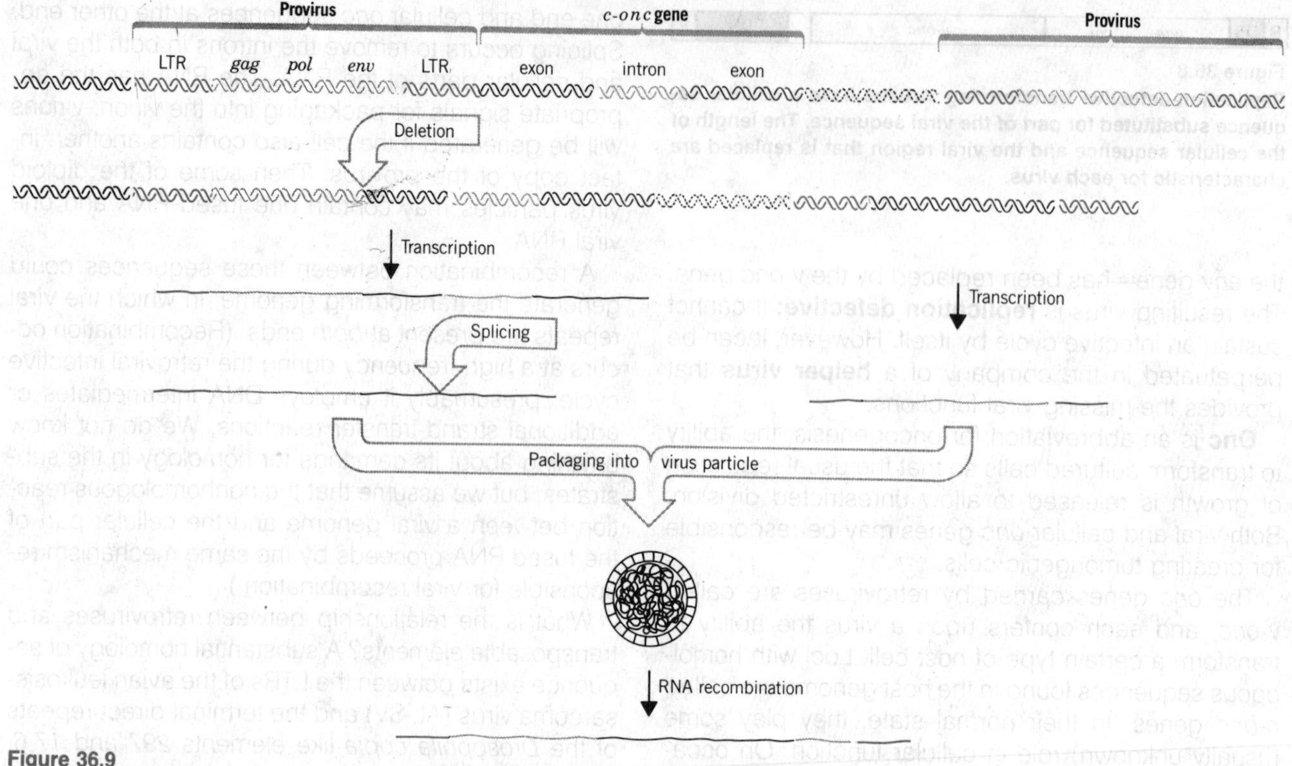

Figure 36.9
Replication-defective viruses may be generated through integration and deletion of a viral genome to generate a fused viral-cellular transcript that is packaged with a normal RNA genome. Nonhomologous recombination is necessary to generate the replication-defective transforming genome.

sequences, although in these cases we can only speculate on how a DNA copy may have been generated. The characteristic features of such a pseudogene are compared in **Figure 36.10** with the features of the original gene and the mRNA. The figure shows *all* the relevant diagnostic features, only some of which may be found in any individual example.

The pseudogene may start at the point equivalent to the 5′ terminus of the RNA, which would be expected only if the DNA had originated from the RNA. Several pseudogenes consist of precisely joined exon sequences; we know of no mechanism to recognize introns in DNA, so this feature argues for an RNA-mediated stage. The pseudogene may end in a short stretch of A-T base pairs, presumably derived from the poly(A) tail of the RNA. On either side of the pseudogene is a short direct repeat, presumed to have been generated by a transposition-like event.

Supporting the idea that these pseudogenes originated by a mechanism different from those responsible for generating pseudogenes found near the active members of gene clusters, the "processed" pseudogenes reside at locations unrelated to their presumed sites of origin.

As Figure 36.10 implies, if a pseudogene is derived from an mRNA sequence, its homology at the 5′ end cannot extend upstream from the initiation site. However, a mouse α globin processed pseudogene does have some upstream homology with the active gene. The processing model could be retained by supposing that initiation occurred farther upstream to generate a longer RNA (either because of an aberrant initiation event or because the promoter was then located differently).

The processed pseudogenes differ in an important respect from the retroviruses: the presumptively trans-

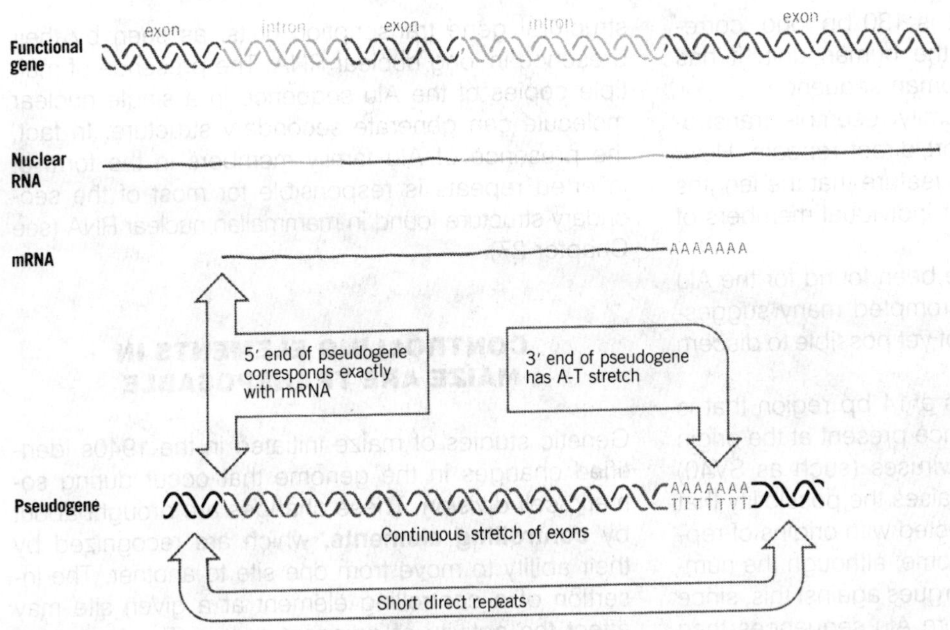

Figure 36.10
Pseudogenes could arise by reverse transcription of RNA to give duplex DNAs that become integrated into the genome.

posed sequences do not themselves carry any information that might be used to sponsor a transposition event (or carry out the preceding reverse transcription of the RNA). Could the process have been mediated by a retrovirus? Was it accomplished by an aberrant cellular system? Perhaps the ends of the transposed sequence fortuitously resembled sequences at the ends of a transposon. Are transposition events currently occurring in these genomes or are we seeing only the footprints of ancient systems?

THE Alu FAMILY

In at least some mammalian genomes, an appreciable proportion of the interspersed moderately repetitive component comprises members of a single family. Its short length and high degree of repetition make it comparable to the simple sequence DNA, except that the individual members of the family are dispersed around the genome instead of being confined to tandem clusters.

In the human genome, a large part of the moderately repetitive DNA exists as sequences of about 300 bp that are interspersed with nonrepetitive DNA. The duplex DNA corresponding to the moderately repetitive sequence component can be isolated by renaturation at intermediate Cot, followed by degradation of the adjacent regions of nonrepetitive DNA that remain unpaired. At least half of the renatured duplex material is cleaved by the restriction enzyme AluI at a single site, located 170 bp along the sequence.

The cleaved sequences all may be members of a single family, known as the **Alu family** after the means of its identification. There are roughly 300,000 members in the haploid genome (equivalent to one member for every 6 kb of DNA). The individual Alu sequences are widely dispersed. A related sequence family is present in the mouse (where the 50,000 members are called the B1 family), in the Chinese hamster (where it is called the Alu-equivalent family), and in other mammals.

The individual members of the Alu family are related rather than identical. The human family seems to have originated by a 130 bp tandem duplication, with an unrelated sequence of 31 bp inserted in the right half of the dimer. The individual sequences have an average homology with the consensus sequence of 87%.

The mouse B1 repeating unit is 130 bp long, corresponding to a monomer of the human unit. It has 70–80% homology with the human sequence.

The members of the Alu family resemble transposons in being flanked by short direct repeats. However, they display the curious feature that the lengths of the repeats are different for individual members of the family.

A variety of properties have been found for the Alu family, and its ubiquity has prompted many suggestions on its function, but it is not yet possible to discern its true role.

Part of the Alu sequence is a 14 bp region that is almost identical with a sequence present at the origin of replication in the papova viruses (such as SV40) and in hepatitis B virus. This raises the possibility that the Alu family could be connected with origins of replication for the eukaryotic genome, although the number of members of the family argues against this, since there are about ten times more Alu sequences than the expected number of replication origins (see Chapter 31).

Another feature is that at least some members of the family can be transcribed *in vitro* into a small RNA by the enzyme RNA polymerase III (which is responsible for transcribing small nuclear RNAs, tRNAs, and 5S RNA). In the Chinese hamster, some (although not all) members of the Alu-equivalent family appear to be transcribed *in vitro*. Transcription units of this sort are found in the vicinity of other transcription units. For example, there are two such units in the region of the human β-globin gene cluster.

Several small RNAs are components of eukaryotic cells (see Chapter 27). The 7S RNA sequence appears to be conserved in higher eukaryotes and also is incorporated into some retrovirus virions. Its sequence has been determined in human and rat cells, where it takes the same general form, corresponding to the insertion of a sequence of 140 bases into an Alu monomeric sequence at a point 100 bases from the 5′ end. Thus the RNA is homologous with the Alu sequence for its first 100 bases and last 40 bases. The central inserted sequence is the same in man and the rat, arguing that the evolution of the coding sequence from an Alu member occurred before the mammalian radiation.

Members of the Alu family may be included within structural gene transcription units, as seen by their presence in long nuclear RNA. The presence of multiple copies of the Alu sequence in a single nuclear molecule can generate secondary structure. In fact, the presence of Alu family members in the form of inverted repeats is responsible for most of the secondary structure found in mammalian nuclear RNA (see Chapter 27).

CONTROLLING ELEMENTS IN MAIZE ARE TRANSPOSABLE

Genetic studies of maize initiated in the 1940s identified changes in the genome that occur during somatic cell division. These changes are brought about by **controlling elements,** which are recognized by their ability to move from one site to another. The insertion of a controlling element at a given site may affect the activity of adjacent genes. The frequency and timing of events associated with the controlling elements are subject to developmental regulation; and their actions may be tissue-specific.

We can now recognize the controlling elements as transposable elements with several functions analogous to those described for other systems. Insertion may generate an unstable allele at a locus. Originally described as "mutable alleles," the terminology still is reflected by the use of *m* to describe such alleles. Loss of the controlling element itself, or of its ability to transpose, converts a mutable allele to a stable allele. Deletions, duplications, inversions, and translocations all may occur at the sites where controlling elements are present. Chromosome breakage is a common consequence of the presence of some elements.

The maize genome contains several families of controlling elements. There are extensive differences in the numbers, types, and sites of presence of the elements in different maize strains. The members of each family may be divided into two classes. **Autonomous** elements have the ability to excise and transpose; their insertion gives rise to unstable (*m*) alleles. **Nonautonomous** elements become unstable *only* when an autonomous member of the same family is present elsewhere in the genome. When complemented in *trans* by an autonomous element, a nonautonomous element may display the usual range of activities associated with autonomous elements. Nonautonomous elements may be derived from autonomous elements by loss of *trans*-acting functions needed for transposition.

Autonomous elements are subject to "changes of state," heritable but relatively unstable alterations in their properties. These changes can influence the timing and frequency of transposition. Some alterations appear to cause autonomous elements to lose their autonomy. *Cis*-acting defects may occur in nonautonomous elements that render them impervious to autonomous elements; thus a nonautonomous element may become permanently stable because it can no longer be activated to transpose.

The families of controlling elements have been defined by the interactions between autonomous and nonautonomous elements. A nonautonomous element can be activated in *trans* only by certain autonomous elements; these are considered to be members of the same family. The three best-characterized families of controlling elements, *Ac-Ds, Spm,* and *Dt,* are summarized in **Figure 36.11.**

Controlling elements may have multiple functions. For example; the dual properties of members of the autonomous *Spm* family are indicated by its name: suppressor-mutator. Both properties may be exercised when *Spm* is introduced into a genome to activate a nonautonomous element already present at a given locus.

"Suppressor" describes the ability to inhibit expression of a structural gene at the locus already occupied by the nonautonomous element. "Mutator" describes the ability to revert (transpose) so that the gene regains full activity. Both properties are held by the same element, but they can be distinguished by $sp^+ m^-$ mutants that have lost one but not the other property. Since the ability of a nonautonomous element to suppress adjacent functions is controlled in *trans* by an autonomous element, the effect cannot result merely from insertion of the nonautonomous element, but must have a specific regulatory basis.

Two features of maize have helped to follow the results of transposition events. Controlling elements often insert near genes that have visible but nonlethal effects on the phenotype. And because maize displays clonal development, the occurrence and timing

Figure 36.11
Each controlling element family has both autonomous and nonautonomous members.
Autonomous elements are capable of transposition and also have other activities (for example, influencing gene expression). Nonautonomous elements are deficient at least in transposition. The names originally given to pairs of autonomous and nonautonomous elements are classified by families. There may be more than one element in each classification; for example, *Spm-s* and *Spm-w* are different versions of *Spm*.

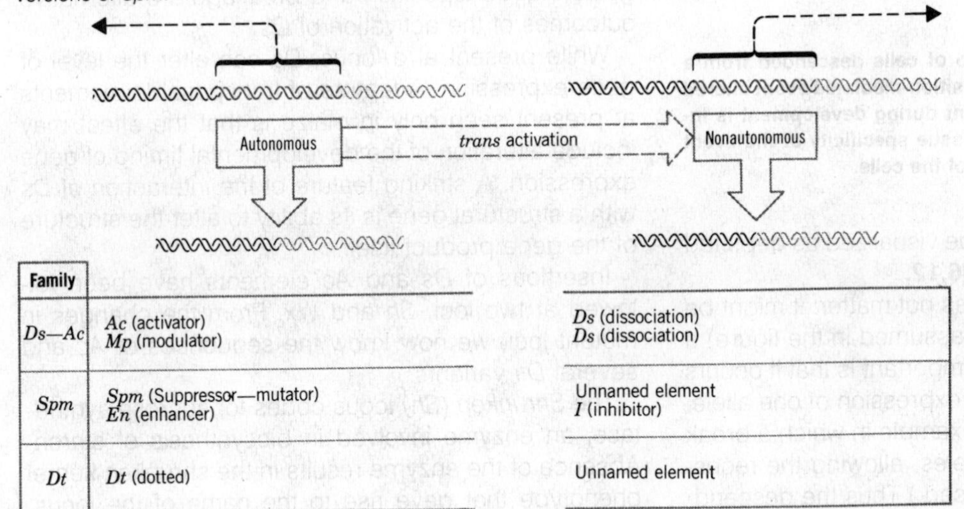

Family		
Ds—Ac	*Ac* (activator) *Mp* (modulator)	*Ds* (dissociation) *Ds* (dissociation)
Spm	*Spm* (Suppressor—mutator) *En* (enhancer)	Unnamed element *I* (inhibitor)
Dt	*Dt* (dotted)	Unnamed element

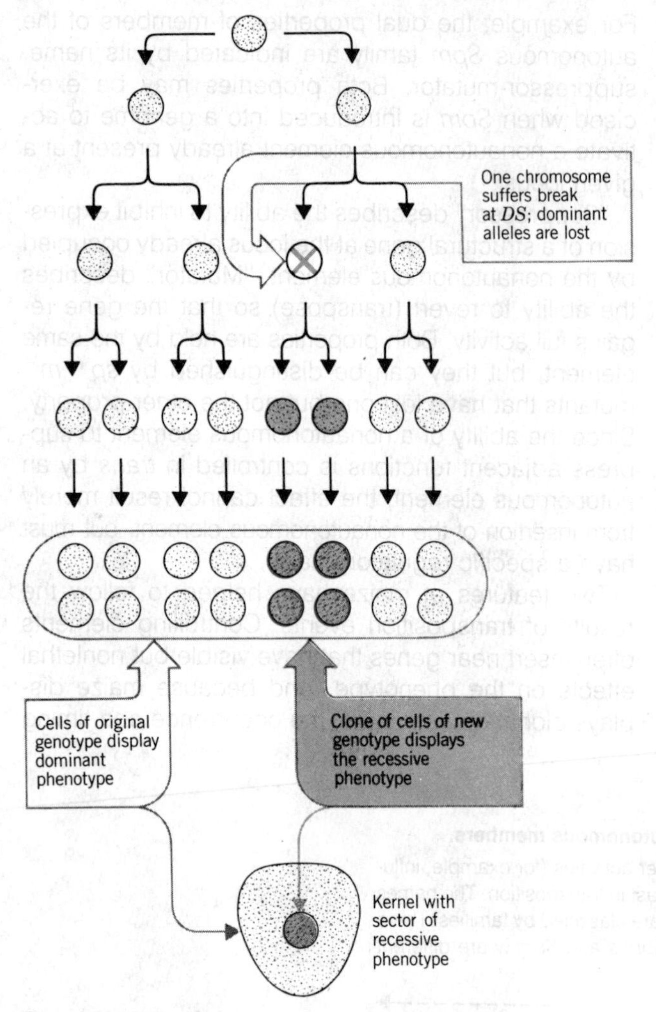

Figure 36.12
Clonal analysis identifies a group of cells descended from a single ancestor in which a transposition-mediated event altered the phenotype. Timing of the event during development is indicated by the number of cells; tissue specificity of the event may be indicated by the location of the cells.

of a transposition event can be visualized as depicted diagrammatically in **Figure 36.12.**

The nature of the event does not matter: it might be an insertion, excision, or (as assumed in the figure) a chromosome break. What is important is that it occurs in a heterozygote to alter the expression of one allele. (The next section shows an example in which a break leads to loss of dominant alleles, allowing the recessive phenotype to be expressed.) Thus the descendants of a cell that has suffered an element-dependent event display a new phenotype, while the descendants of cells not affected by the event continue to display the original phenotype.

Mitotic descendants of a given cell remain in the same location and thus give rise to a **sector** of tissue. A change in phenotype during somatic development is called **variegation;** it is revealed by a sector of the new phenotype residing within the tissue of the original phenotype. The size of the sector depends on the number of divisions in the lineage giving rise to it; so the size of the area of the new phenotype is determined by the timing of the change in genotype. The earlier its occurrence in the cell lineage, the greater the number of descendants and thus the size of patch in the mature tissue.

Ds MAY TRANSPOSE OR CAUSE CHROMOSOME BREAKAGE

The presence of the nonautonomous element *Ds* has several possible consequences for an adjacent locus. Upon activation by an autonomous *Ac* element, *Ds* may either transpose to a new site or cause chromosome breakage. Transposition is accompanied by excision from the donor site, so it converts an unstable allele to a stable allele (not necessarily of wild type). An inverse relationship between reversion to stability and the occurrence of chromosome breakage suggests that transposition and breakage are alternative outcomes of the activation of *Ds*.

While present at a locus, *Ds* can alter the level of gene expression. A feature of transposable elements at present seen only in maize is that the effect may include alteration of the developmental timing of gene expression. A striking feature of the interaction of *Ds* with a structural gene is its ability to alter the structure of the gene product itself.

Insertions of *Ds* and *Ac* elements have been followed at two loci, *Sh* and *Wx*. From the changes in mutant loci, we now know the sequences of *Ac* and several *Ds* variants.

The *Shrunken (Sh)* locus codes for sucrose synthetase, an enzyme involved in biosynthesis of starch. Absence of the enzyme results in the shrunken kernel phenotype that gave rise to the name of the locus.

Sh-m loci in which *Ds* is present have a different restriction map in the 5′ region of the gene, corresponding to insertion of additional DNA, sometimes also associated with a duplication of material from the locus. The exact change in sequence is different for individual *sh-m* alleles.

The presence of *Ds* alters *sh-m* expression. In some mutants, the *sh-m* gene is barely expressed. In other mutants, the locus is transcribed into abundant aberrant RNAs that are shorter than *Sh* mRNA and consist in part of *Sh* sequences and in part of new sequences. They are translated into a protein immunologically related to sucrose synthetase, but having an altered electrophoretic mobility. As the structures of the various *Sh* and *sh-m* loci become elucidated, we shall be able to see which sites within or adjacent to the structural gene are influenced by the presence of *Ds* and how this affects gene expression.

The *Waxy (Wx)* locus determines the amylose contents of pollen and endosperm tissue. It codes for a 58,000 dalton enzyme, UDP-glucose starch transferase, that is bound by starch granules. Insertions of both *Ac* and *Ds* have been found. By comparing the sequences of these insertions and those at the *Sh* locus, we have been able to deduce the structures of *Ac* and *Ds*. They are illustrated in **Figure 36.13.**

The autonomous *Ac* element is 4563 bp long and contains three major open reading frames. Frames 1 and 2, read in the same direction, seem likely to code for proteins, as judged by their triplet constitution. The coding function of frame 3, read in the opposite direction, is more doubtful. Together the reading frames account for most of the sequence of *Ac*. The element itself ends in inverted repeats of 11 bp; and a target sequence of 8 bp is duplicated at the site of insertion.

Figure 36.13
The Ac element has three open reading frames; Ds elements have internal deletions.

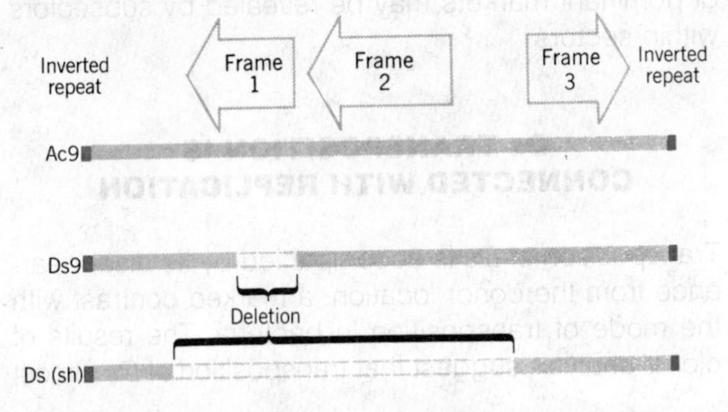

Ds elements vary in both length and sequence, but are related to *Ac*. They end in the same 11 bp inverted repeats. The element *Ds9* has a deletion of 194 bp from reading frame 1. The *Ds6* element has a length of 2 kb, representing 1 kb from each end of *Ac*, so that a major part of frames 1 and 2 is missing. A complex double *Ds* element has one *Ds6* sequence inserted in reverse orientation into another.

Nonautonomous elements evidently take various forms, but share in common the features that they lack some internal sequences, but possess the terminal inverted repeats (and possibly other sequence features). These structures support the general view that nonautonomous elements are derived from autonomous elements by deletions (or other changes) that inactivate the *trans*-acting transposase, but leave the sites (including the termini) on which the transposase acts. The *Ds* elements thus resemble the defective P elements in *D. melanogaster*.

The *Ds* element was originally identified by its ability to provide a site for chromosome breakage upon activation by *Ac*. The consequences are illustrated in **Figure 36.14.** Consider a heterozygote in which *Ds* lies on one homologue between the centromere and a series of dominant markers *(C-I, Bz, Wx)* whose effects can be visualized by the color of the cells or by appropriate staining. (*C-I* is a regulator that causes colorless aleurone; *Bz* glycosylates the anthocyanid pigments; *Wx* is starchy and stains blue with I_2-KI.) The other homologue lacks *Ds* and has recessive markers *(C, bz, wx)*.

Breakage at *Ds* generates an **acentric fragment** carrying the dominant markers. Because of its lack of a centromere, this fragment is lost at mitosis. Thus the descendant cells have only the recessive markers carried by the intact chromosome. This gives the type of situation whose results are depicted in Figure 36.12.

Breakage at *Ds* leads to the formation of two unusual chromosomes. These are generated by joining the broken ends of the products of replication. One is a U-shaped acentric fragment consisting of the joined sister chromatids for the region distal to *Ds* (on the left as drawn in Figure 36.14). The other is a U-shaped

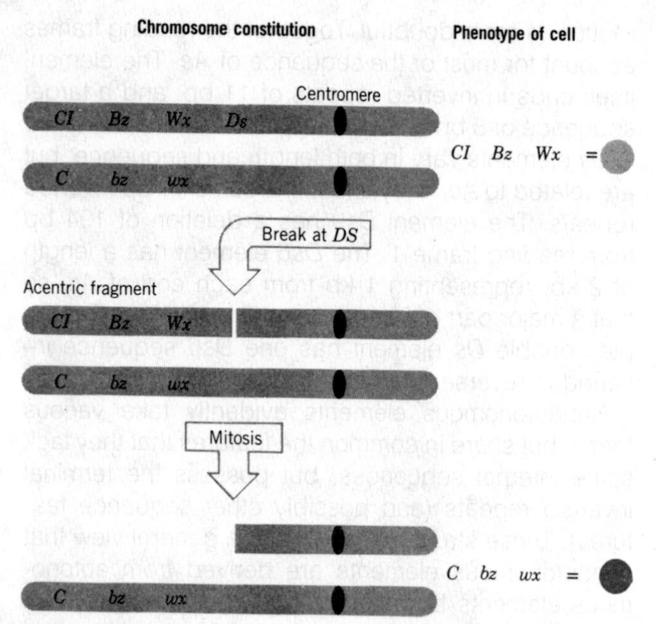

Figure 36.14

A break at a controlling element causes loss of an acentric fragment; if the fragment carries the dominant markers of a heterozygote, its loss changes the phenotype.

dicentric chromosome comprising the sister chromatids proximal to *Ds* (on its right in the figure). The latter structure leads to the classic **breakage-fusion-bridge** cycle illustrated in **Figure 36.15.**

Consider the fate of the dicentric chromosome when it attempts to segregate on the mitotic spindle. Because it has two centromeres, it is pulled by each centromere toward an opposite pole. The tension breaks the chromosome at a random site between the centromeres. In the example of the figure, breakage occurs between loci *A* and *B*, with the result that one daughter chromosome has a duplication of *A*, while the other has a deletion. If *A* is a dominant marker, the cells with the duplication will retain **A** phenotype, but cells with the deletion will display the recessive **a** phenotype.

The breakage-fusion-bridge cycle continues through further cell generations, allowing genetic changes to continue in the descendants. For example, consider the deletion chromosome that has lost *A*. In the next cycle, a break may occur between *B* and *C*, so that the descendants are divided into those with a duplication of *B* and those with a deletion. Successive losses

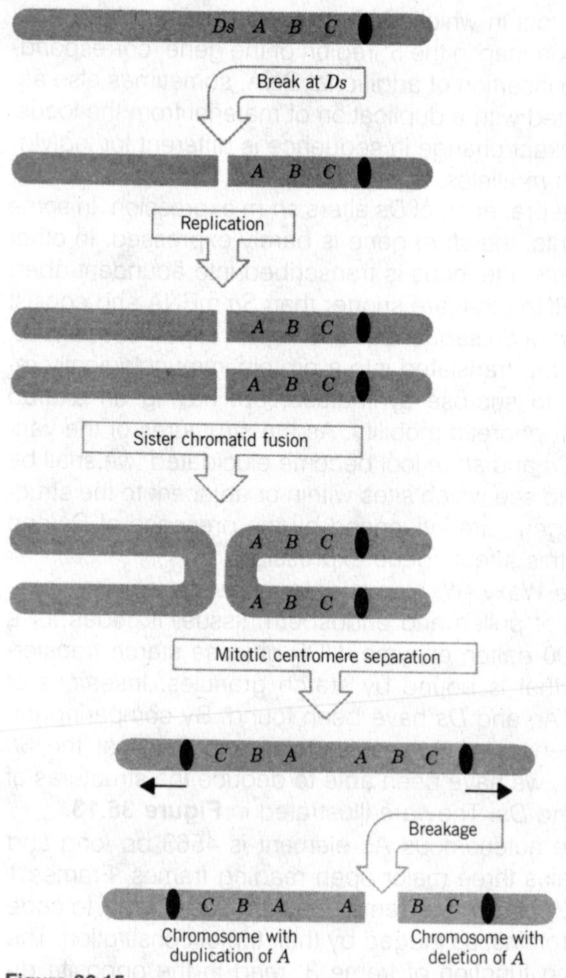

Figure 36.15

***Ds* provides a site to initiate the chromatid fusion-bridge-breakage cycle (on activation by *Ac*). The products can be followed by clonal analysis.**

of dominant markers may be revealed by subsectors within sectors.

Ds TRANSPOSITION IS CONNECTED WITH REPLICATION

Transposition of *Ds* is accompanied by its disappearance from the donor location, a marked contrast with the mode of transposition in bacteria. The results of clonal analysis suggest that transposition of *Ds* almost

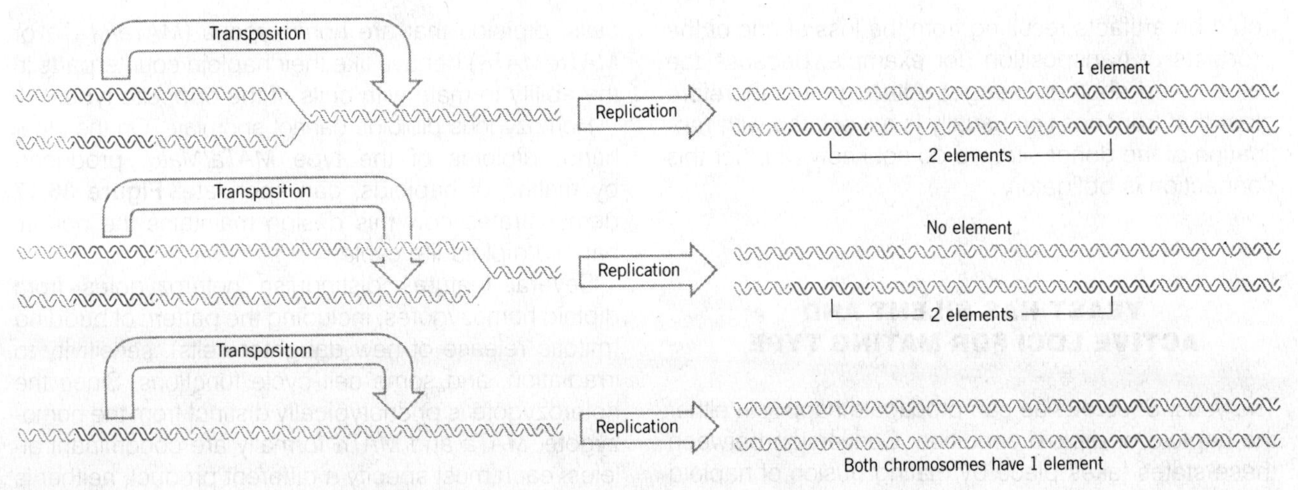

Figure 36.16

Maize transposition involves movement of an element from the donor site after it has been replicated; the nature of the products depends on whether the recipient site has been replicated prior to the transposition.

Top: If the recipient site has not been replicated, the element is duplicated after transposition, so each daughter chromosome gains an element. Only one of the daughter chromosomes has lost the element. Thus one daughter chromosome has an element only at the recipient site; the other has elements at both the donor and recipient sites.

Center: If the recipient site has been replicated, only one daughter chromosome gains an element at the recipient position. Usually an element moves from a donor site on one chromatid to a recipient site on the *other* chromatid, so that both chromatids have a change in genotype as shown. (If the element moved from a donor to a recipient site on the *same* chromatid, the genetic result would be indistinguishable from that shown in the bottom drawing.)

Bottom: If transposition were to occur before replication of the donor site, both the progeny chromosomes would have the element at the recipient site.

always occurs after the donor element has been replicated. The most revealing information is provided by transpositions from a donor to a recipient site on the same chromosome. Studies of this sort have been performed with *Mp*, an element of the *Ac* type.

Two possible outcomes of the transposition are distinguished by whether the recipient site has replicated. They are illustrated in the top two rows of **Figure 36.16.** If the recipient site has not replicated before transposition, the element will be replicated in its new site. Thus it is present at the recipient site on both daughter chromosomes. If the recipient site has previously replicated, only the chromatid physically gaining the element will display a change at the recipient locus. In either case, the element is lost from the donor site of one chromatid and remains at the donor site of the other chromatid.

In the first two situations illustrated in the figure, cell division produces daughter cells whose genotypes are different from each other as well as from the parent cell. Clonal analysis reveals their descendants in the form of **twin sectors:** the appearance of a sector of one new genotype is accompanied by the appearance of another sector representing the other new genotype. Note that twin sectors are the inevitable result of transposition to an unreplicated recipient site. When the recipient site has been replicated, they result only if the target lies on the opposite chromatid from the donor.

The predominance of twin sectors (80% of all maize transpositions) demonstrates that these are the customary events. The minority of transpositions generating single sectors could represent insertions at recipient sites on the same chromatid as the donors, could result from transposition not connected with replication (shown in the lowest line of the figure), or could be artifacts resulting from the loss of one of the products of transposition (for example, because the cell entered another lineage). Although it is therefore clear that transposition usually is associated with replication of the donor site, we do not know whether this connection is obligatory.

YEAST HAS SILENT AND ACTIVE LOCI FOR MATING TYPE

The yeast *S. cerevisiae* can propagate happily in either the haploid or diploid condition. Conversion between these states takes place by mating (fusion of haploid spores to give a diploid) and by sporulation (meiosis of diploids to give haploid spores). The ability to engage in these activities is determined by the **mating type** of the strain.

The properties of the two mating types are summarized in **Table 36.2.** We may view them as resting on the teleological proposition that there is no point in mating unless the haploids are of different genetic types; and there is no point sporulating unless the diploid is heterozygous and thus can generate recombinants.

The two mating types are determined by the alleles *MATa* and *MATα* of the mating type locus. Haploid cells *must* be either *MATa* or *Matα*. Cells of opposite type can mate; cells of the same type cannot.

Recognition of cells of opposite mating type is accomplished by the secretion of **pheromones.** *MATα* cells secrete the small polypeptide **α** factor; *MATa* cells secrete **a** factor. Each factor acts on cells of the opposite mating type to coordinate cell and nuclear fusion.

Although usually we deal with the mating of haploid cells, diploids that are homozygous (*MATa/MATa* or *MATα/MATα*) behave like their haploid counterparts in the ability to mate with cells of the oppcsite type.

Homozygous diploids cannot sporulate. On the other hand, diploids of the type *MATa/Matα*, produced by mating of haploids, can sporulate. **Figure 36.17** demonstrates how this design maintains the normal haploid/diploid life cycle.

Several features distinguish heterozygotes from diploid homozygotes, including the pattern of budding (mitotic release of new daughter cells), sensitivity to irradiation, and some cell cycle functions. Since the heterozygote is phenotypically distinct from the homozygote, *MATa* and *MATα* formally are codominant alleles: each must specify a different product; neither is a defective version of the other.

A remarkable feature is the ability of some yeast strains to **switch** their mating types. These **homothallic** strains carry a dominant allele *HO* and *change their mating type frequently,* as often as once every generation. **Heterothallic** strains with the recessive allele *ho* have a stable mating type, subject to change with a frequency $\sim 10^{-6}$.

Expression of the *HO* gene is itself under mating type control, since it does not function in *MATa/MATα* diploids—teleologically, we may think that there is no need for switching when *both MAT* alleles are expressed anyway. When *HO* is present in a yeast population, irrespective of the initial mating type, in very few generations there are large numbers of cells of both mating types, leading to the formation of *MATa/MATα* diploids that take over the population. The production of stable diploids from a haploid population can be viewed as the raison d'être for switching.

The existence of switching suggests that all cells contain the *potential* information needed to be either *MATa* or *MATα*, but express only one type. This con-

Table 36.2
Mating type controls several activities.

Property	*MATa* or *MATa/MATa*	*MATα* or *MATα/MATα*	*MATα/MATα*
Mating	Yes	Yes	No
Sporulation	No	No	Yes
Secretion	**a** factor	**α** factor	None

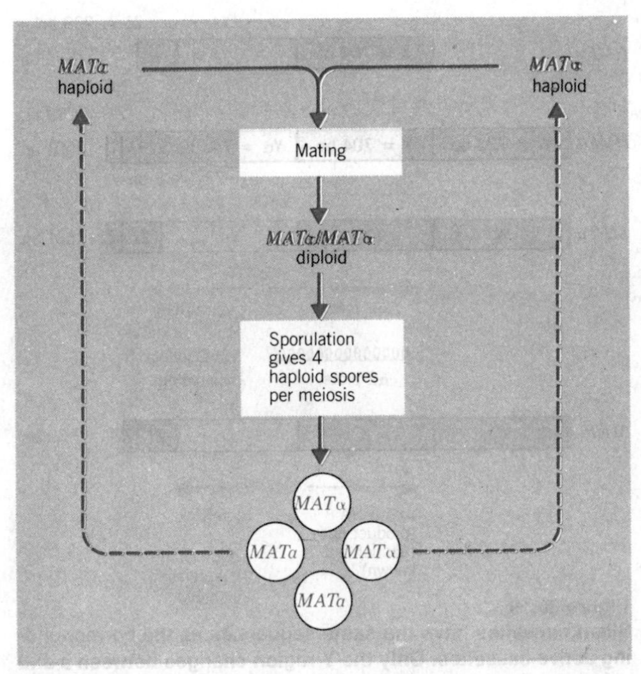

Figure 36.17
The yeast life cycle proceeds through mating of *MATa* and *MATα* haploids to give heterozygous diploids that sporulate to generate haploid spores.

clusion is confirmed by analysis of mutants in either type of *MAT* allele. Switching in *HO* strains allows the mutant allele to be replaced by either a wild-type *MATa* or *MATα* gene.

Where does the wild-type information come from? Two additional loci are needed for switching. *HMLα* is needed for switching to give a *MATα* type; *HMRa* is needed for switching to give a *MATa* type. These loci lie on the same chromosome that carries *MAT; HML* is far to the left, *HMR* is far to the right.

The **cassette model** for mating type proposes that *MAT* has an **active cassette** of either type α or type *a*. *HML* and *HMR* have **silent cassettes** that are physically substituted for the cassette at MAT by the switching process. Usually *HML* carries an α cassette, while *HMR* carries an *a* cassette. The cassettes constitute transposable coding information.

The process of switching is illustrated in **Figure 36.18.** Switching is nonreciprocal, because the copy at *HML* or *HMR replaces* the allele at *MAT*. We know this because a mutation at *MAT* is lost permanently when it is replaced by switching—it does not exchange with the copy that replaces it.

The copies present at *HML* or *HMR* also can be mutated. In this case, switching introduces a mutant allele into the *MAT* locus. The mutant copy at *HML* or *HMR* remains there through an indefinite number of switches. This immediately identifies a resemblance with replicative transposition, in which the donor element generates a new copy at the recipient site, while itself remaining inviolate. Mating type switching corresponds to a directed transposition, since there is only one recipient *(MAT)* and two donors *(HML* and *HMR)*.

As illustrated in Figure 36.18, switching usually changes the mating type. It involves replacement of *MATa* by the copy at *HMLα* or replacement of *MATα* by the copy at *HMRa*. However, studies with mutant loci show that *HMLα* occasionally replaces *MATα*, and *HMRa* occasionally replaces *MATa*. Such switches pass unobserved unless a mutation is present at one of the loci to mark it genetically.

Figure 36.18
Changes of mating type occur when silent cassettes replace active cassettes of opposite genotype; when transpositions occur between cassettes of the same type, the mating type remains unaltered.

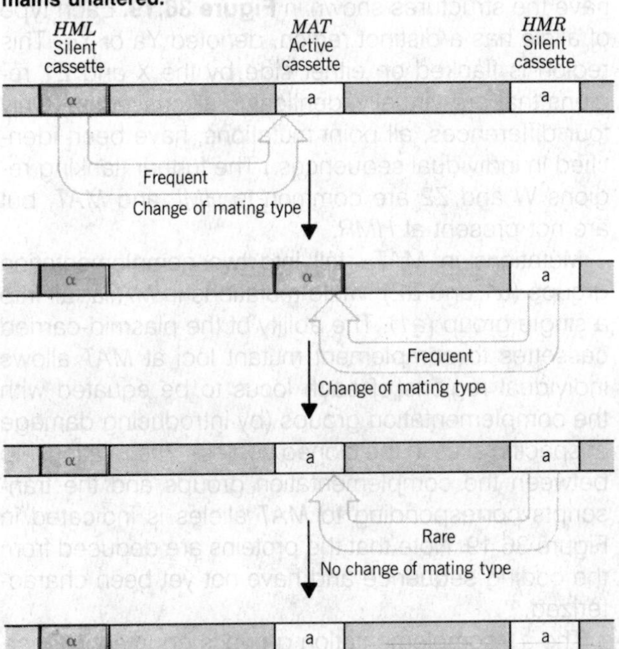

In 80–90% of switches, the *MAT* allele is replaced by one of opposite type, an effect apparently determined by the phenotype of the cell. Cells of **a** phenotype preferentially choose *HML* as donor; cells of **α** phenotype preferentially choose *HMR*. It is possible to obtain yeast strains in which the usual orientation of the silent cassettes is reversed. When their genotypes are *HMLa* and *HMRα*, 92% of the switches are homologous, because the choice of donors remains unaltered.

An interesting feature of the switching event is its dependence on cell lineage. Switching is detected only in the products of a division; *both daughter cells have the same mating type*, switched from that of the parent. This suggests a connection with replication in the sense that switching must occur *before* the MAT locus is replicated. (Compare this with the connection between transposition and replication in maize shown in Figure 36.16.)

SILENT AND ACTIVE CASSETTES HAVE THE SAME SEQUENCES

Both the silent and active mating type loci have been cloned onto plasmids that replicate in yeast. The genes have the structures shown in **Figure 36.19.** Each type of allele has a distinct region, denoted Ya or Yα. This region is flanked on either side by the X and Z1 regions that are virtually identical in all cassettes. (Only four differences, all point mutations, have been identified in individual sequences.) The further flanking regions W and Z2 are common to *HML* and *MAT*, but are not present at *HMR*.

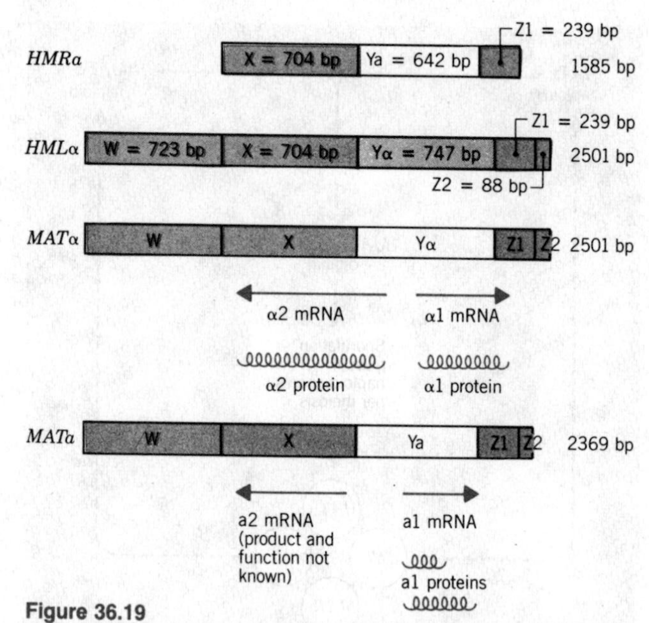

Figure 36.19
Silent cassettes have the same sequences as the corresponding active cassettes. Only the Y region changes between *a* and α types. *HMR* lacks the W and Z2 sequences common to *HML* and *MAT*.

Mutations in *MATα* fall into two complementation groups (α1 and α2), while mutations in *MATa* fall into a single group *(a1)*. The ability of the plasmid-carried cassettes to complement mutant loci at *MAT* allows individual regions of each locus to be equated with the complementation groups (by introducing damage at specific sites in the cloned copies). The relationship between the complementation groups and the transcripts corresponding to *MAT* alleles is indicated in Figure 36.19. Note that the proteins are deduced from the coding sequence and have not yet been characterized.

The *a1* complementation group is equated with the a1 transcript, which is coded entirely within the region of Ya and therefore is unique to the *a*-type locus. The *a1* sequence has two potential products. The shorter terminates at a UGA codon; however, we know that readthrough must occur to give the longer protein, because mutations beyond the termination codon prevent *a1* function. We do not know whether the shorter protein plays a different role from the longer. No function has been detected for the a2 transcript, which is transcribed from a promoter at least part of which lies in Ya. The transcript lies within the common region X and could be translated into either of two short polypeptides.

Each of the transcripts generated at *MATα* corresponds to one of the known complementation groups. The transcripts initiate divergently within Yα and each extends into the adjacent common region. The α1 transcript codes for a protein largely coded by Yα. The α2 transcript can be translated from an initiation codon in Yα, but most of the sequence corresponds to X.

The functions of the products, as deduced from the deficiencies of mutants in each complementation group, are illustrated in **Figure 36.20.** In haploids, **a** mating functions are expressed constitutively. The function of

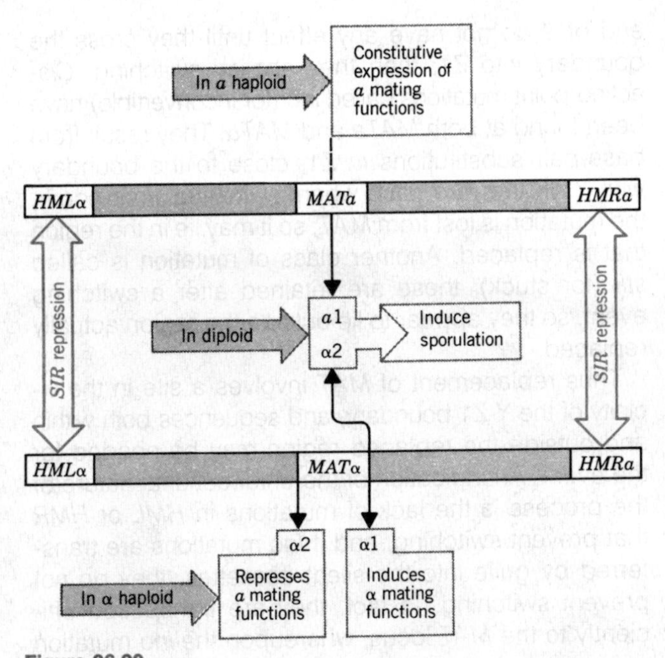

Figure 36.20

The *a*1 and α2 functions cooperate to induce sporulation in diploids. In haploids, the α2 function represses *a* mating functions; α1 induces α mating functions.

a1 is needed, in cooperation with α2, to induce sporulation in diploids. The α2 group also has responsibility for repressing **a** mating functions in haploids. The α1 product induces **α** mating functions in haploids.

The transcription map in Figure 36.19 reveals an intriguing feature. Transcription of both *MATa* and *MAT*α initiates within the Y region. Only the MAT locus is expressed; *yet the same Y region is present in the corresponding nontranscribed cassette, HML* or *HMR*. This implies that regulation of expression is not accomplished by direct recognition of some site overlapping with the promoter. A site *outside the cassettes* must distinguish *HML* and *HMR* from *MAT*.

Can we find the basis for the control of cassette activity by identifying other loci that act on the cassettes? Some are revealed by mutations that prevent expression of the allele occupying *MAT*. Other mutations allow the usually silent cassettes at *HML* and *HMR* to be expressed. (When this results in expression of both **a** and **α** functions, the cells behave like *MATa/MAT*α diploids.)

Four complementation groups have been identified in which mutations lead to expression of *HML* and *HMR*. They are called *SIR* (silent information regulator). The four wild-type *SIR* loci are needed to maintain *HML* and *HMR* in the repressed state; mutation in any one of these loci to give a *sir*$^-$ allele has two effects. Both *HML* and *HMR* can be transcribed. And both the silent cassettes become targets for replacement by switching. *So the same regulatory event is involved in repressing a silent cassette and in preventing it from being a recipient for transposition.*

Why are there four different *sir* loci? We do not know whether they code for subunits of a single regulatory protein or represent individual functions that interact with one another or individually with the target loci.

The different activities of the silent and active cassettes may result from the structure of chromatin. The use of probes representing the regions adjacent to each of the cassettes allows chromatin structure to be visualized by the end-labeling technique described in Chapter 30. All cassettes possess sites hypersensitive to DNAase I, but there are significant differences between the active and silent cassettes. Also, there are changes in the nucleosome ladders and in the degree of supercoiling, which may be associated with gene activation.

The afflicted hypersensitive sites are summarized in **Figure 36.21.** The general pattern of hypersensitivity is similar in *MATa* and *MAT*α. In each case, there is a site at the 3′ end of the leftward transcript (a2 or α2). There is either a single (*a*) or double (α) site in the region between the two transcripts. And there is a doublet of hypersensitive sites around the Y-Z boundary.

Figure 36.21

Most of the DNAase I hypersensitive sites found in *MAT* are not present in *HML* or *HMR*; but they are activated in *sir*$^-$ mutants. Some permanent hypersensitive sites are found in the regions flanking *HML* and *HMR*.

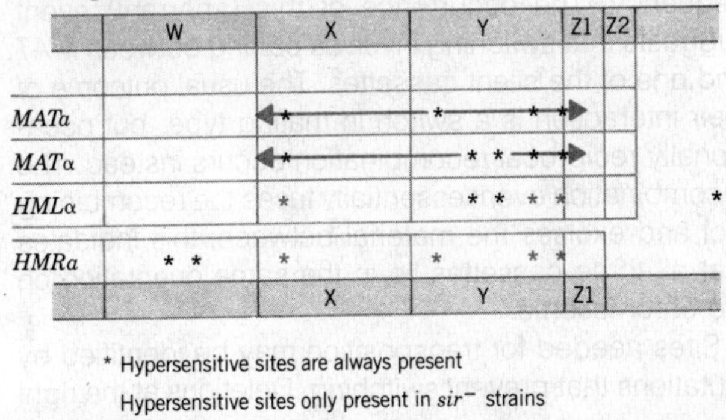

W X Y Z1 Z2

MATa
MATα
HMLα
HMRa

X Y Z1

* Hypersensitive sites are always present

Hypersensitive sites only present in *sir*$^-$ strains

All of these sites are absent from *HML* and *HMR* except for the doublet between α1 and α2. However, the introduction of a *sir*⁻ mutation restores all the missing hypersensitive sites. Thus activation of the cassettes is associated with the presence of hypersensitive sites. The various sites may have different functions. The sites within Y may be necessary (although not sufficient) to initiate transcription. The function of the sites within X is unknown. The doublet at the X-Y boundary may be important in initiating the transposition event (see next section).

Some sites are found in the flanking DNA near *HML* and *HMR*. Their presence does not depend on *SIR* function; in fact, these sites could identify the silent cassettes as targets for SIR action. Deletion mapping shows that the flanking site on the left of *HML* is essential for the response to *SIR;* the flanking site on the right also is important. When these sites are recognized by *sir,* a signal must be transmitted to prevent formation of the hypersensitive sites within the cassettes. In lieu of this action, the various hypersensitive sites allow the locus to be both expressed and subject to switching.

How can the *sir* functions act at a distance? One possibility is that a structural effect is transmitted along the DNA. For example, the degree of supercoiling could influence sites within the cassettes, or a pattern of nucleosome phasing could determine whether a site is available.

UNIDIRECTIONAL TRANSPOSITION IS INITIATED BY THE RECIPIENT MAT LOCUS

In populations undergoing switching, recombination between *MAT* and either *HML* or *HMR* occurs at a low frequency. The occurrence of this (aberrant) event suggests that switching involves pairing between *MAT* and one of the silent cassettes. The usual outcome of their interaction is a switch in mating type, but occasionally reciprocal recombination occurs instead. The recombination event essentially fuses the recombining loci and excises the material between; this indicates that all three cassettes lie in the same orientation on the chromosome.

Sites needed for transposition may be identified by mutations that prevent switching. Deletions at the right end of Y do not have any effect until they cross the boundary into Z1; then they abolish switching. *Cis*-acting point mutations called *inc* (for inconvertible) have been found at both *MATa* and *MATα*. They result from base-pair substitutions in Z1, close to the boundary with Y. In the rare case when switching does occur, the mutation is lost from *MAT*, so it may lie in the region that is replaced. Another class of mutation is called *stk* (for stuck); these are retained after a switching event, so they appear to lie outside the region actually replaced.

Thus replacement of *MAT* involves a site in the vicinity of the Y-Z1 boundary, and sequences both within and outside the replaced region may be needed for the event. An indication of the unidirectional nature of the process is the lack of mutations in *HML* or *HMR* that prevent switching; and if *inc* mutations are transferred by guile into the silent cassettes, they do not prevent switching. In fact, they are transposed efficiently to the *MAT* locus, whereupon the *inc* mutation reappears.

The Y-Z boundary is the site of a change in DNA that may be connected with transposition. In populations of cells undergoing switching, 1–3% of the DNA of the *MAT* locus has a double-stranded cut at this site. The cut lies within 10 bp of the boundary and coincides with the DNAase hypersensitive site.

We can now suggest a possible series of events for switching. The hypersensitive site at the Y-Z1 boundary of *MAT* may be accessible because it lacks a nucleosome. It is recognized by an endonuclease controlled by the *HO* locus. In fact, this is equivalent to suggesting that hypersensitivity to DNAase I *in vitro* reflects a natural sensitivity to the *HO* endonuclease *in vivo*!

The *HO* endonuclease—also called YZ endo—makes a staggered break in the Z1 region, 3 bp from the YZ junction. It does not cut mutant *MAT* loci that cannot switch. Deletions that leave less than 11 bp remaining of the target sequence are not cleaved by the enzyme, so the recognition site is relatively large for a nuclease. Probably it occurs nowhere else in the genome.

Synthesis of the *HO* endonuclease is under cell cycle control. The timing of its production may explain the characteristic lineage of switching. The gene is expressed only during the G1 phase of a mother cell, which explains why both progeny have the new mating type.

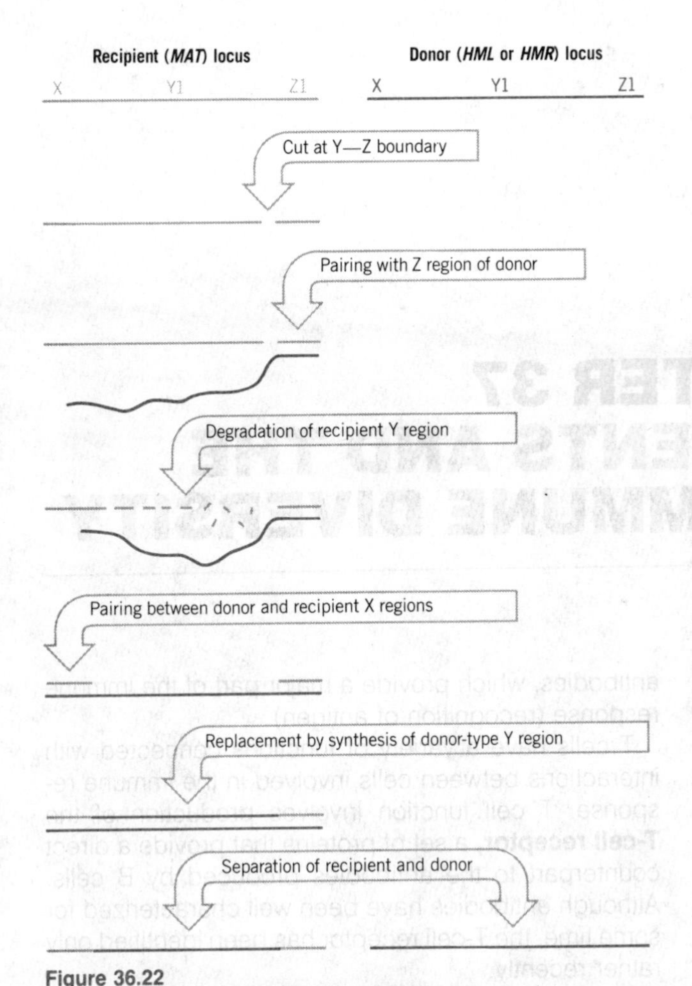

Figure 36.22

Cassette substitution is initiated by the recipient locus and may involve pairing on either side of the Y region with the donor locus. The sequence that is removed and replaced may extend into the X region. The figure shows the reaction in terms of a single recipient and single donor strand; actually it involves duplex regions of DNA.

The reaction triggered by the cleavage is illustrated schematically in terms of single strands in **Figure 36.22.** (Of course, it must be more complex in terms of double strands, and several pathways are possible.) Suppose that the free Z end of *MAT* invades either the *HML* or *HMR* locus and pairs with the Z region. The Y region of *MAT* is degraded until a region with homology to X is exposed. At this point, *MAT* is paired with *HML* or *HMR* at both the left side (X) and the right side (Z1). The Y region of *HML* or *HMR* is copied to replace the region lost from *MAT* (which might extend beyond the limits of Y itself). The paired loci separate. (The order of events could be different.) The model is related to the scheme for general recombination drawn in Figure 34.10, and the stages following the initial cut require the enzymes involved in general recombination.

Two features are notable. Unlike other transpositions that we have described, the switching event involves pairing at both ends of the transposed segment, with degradation and replacement of the material between. And the process is initiated by *MAT*, the *locus that is to be replaced.* In this sense, the description of *HML* and *HMR* as donor loci refers to their ultimate role, but not to the mechanism of the process, which is the reverse of the usual initiation of transposition by the donor. The sequences of the loci do not identify any short direct repeats in the relevant regions, which argues that the mechanism of joining may be different from the use of staggered breaks found in other transposition events.

FURTHER READING

Eukaryotic transposable elements have been extensively reviewed in *Mobile Genetic Elements* (ed. Shapiro, Academic Press, New York, 1983). Chapters of particular interest: **Fink** on Ty elements of yeast; **Rubin** on the tour de force of characterizing *D. melanogaster* transposons and mapping the *white* locus; **Varmus** provided a powerful analysis of retroviruses as both viruses and transposable elements; **Fedoroff** brought the topic of maize-controlling elements into the molecular era; and **Haber** provided a comprehensive summary of yeast mating type. *Drosophila* P elements have been summarized by **Engels** (*Ann. Rev. Genet.* **17,** 315–344, 1983). The determination of mating type has been reviewed from a biological perspective by **Nasmyth** (*Ann. Rev. Genet.* **17,** 439–500, 1983). The rDNA of *D. melanogaster* was discussed briefly by **Glover** (*Cell,* **26,** 297–298, 1981).

CHAPTER 37
REARRANGEMENTS AND THE GENERATION OF IMMUNE DIVERSITY

The **immune response** of the mammal is exercised by an apparently limitless repertoire of **antibodies,** each synthesized in response to provocation by a different **antigen.** Usually the antigen is a protein (or protein-attached moiety) of foreign origin that has entered the bloodstream of the animal—for example, it may be the coat protein of an infecting virus. Exposure to an antigen initiates production of an antibody that *specifically recognizes the antigen.*

Antibody production occurs in the B lymphocytes and enables them (in conjunction with T lymphocytes and macrophages, the other white blood cells) to remove the offending antigen. Once an organism has synthesized a particular antibody, the manufacturing capacity remains available to combat any future reexposure.

One of the mysterious features of this process has been the animal's ability to produce an appropriate antibody whenever it is exposed to a new antigen. How can the organism be prepared to produce antibody proteins each designed specifically to recognize an antigen whose structure cannot be anticipated?

The two classes of lymphocytes are named after the tissues in which they originate. B cells are produced by the bone marrow, while T cells are produced by the thymus. The major function of B cells is to produce antibodies, which provide a major part of the immune response (recognition of antigen).

T cells have a variety of functions connected with interactions between cells involved in the immune response. T cell function involves production of the **T-cell receptor,** a set of proteins that provide a direct counterpart to the antibodies produced by B cells. Although antibodies have been well characterized for some time, the T-cell receptor has been identified only rather recently.

Many reactions of the immune system are influenced by another set of proteins, the **histocompatibility antigens.** Every individual has a characteristic set of histocompatibility antigens. Their name reflects their importance in the graft reaction; a graft of tissue from a different individual is rejected because of the difference in histocompatibility antigens between the donor tissue and recipient individual, an issue of major medical importance. Its set of histocompatibility antigens defines a cell as "self" or "nonself" with regard to another cell. At the cellular level, the histocompatibility antigens are involved in the reactions between lymphocytes.

Each of these three sets of proteins—antibodies, T-cell receptors, histocompatibility antigens—is diverse. Examining a large number of individuals, we find many variants of each protein. Each protein is coded by a large family of genes; and in the case of antibodies and the T-cell receptor, the diversity of the population owes much to rearrangements of the DNA that occur in the relevant lymphocytes.

Antibodies and T-cell receptors are related in structure, and their genes are related in organization. The histocompatibility antigens also share some common features with the antibodies. In dealing with the genetic organization of the immune system, we are therefore concerned with a series of related gene families, perhaps a super-family that evolved from some common ancestor representing a primitive immune response.

The importance of the immune response for the mammal is indicated by the large number of genes contained in these families; and they offer us the opportunity to characterize the massive gene clusters. Eventually we should be able to analyze the entire immune response in terms of the properties of the gene products, and perhaps to account for the evolution of the system.

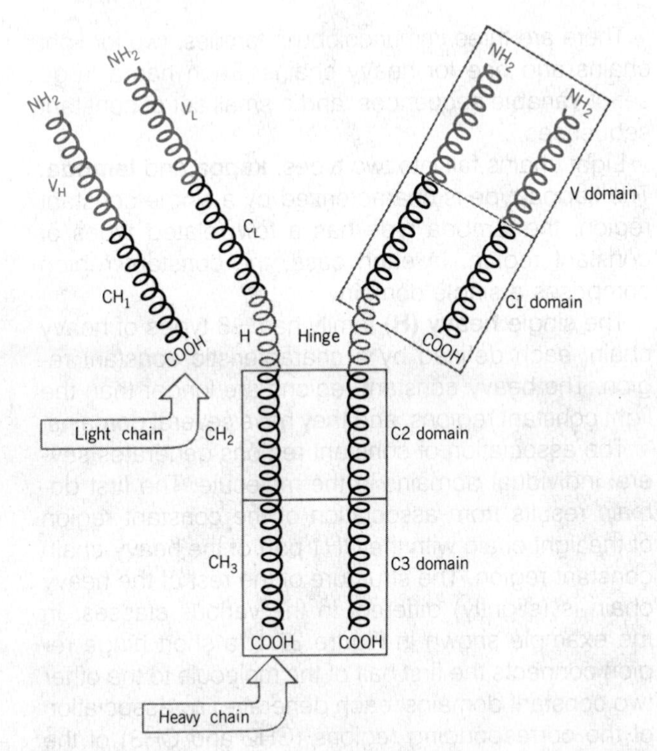

Figure 37.1
Heavy and light chains combine to generate an immunoglobulin with several discrete domains.

ORGANIZATION OF IMMUNOGLOBULINS

For practical purposes, we usually reckon that the organism has the ability to produce $>10^6$ different antibodies. Each antibody is a tetrameric **immunoglobulin** consisting of two identical **light (L) chains** and two identical **heavy (H) chains.** The theory of combinatorial association supposes that any type of light chain can associate with any type of heavy chain. To produce 1,000,000 potential antibodies therefore requires >1000 types of light chain and >1000 types of heavy chain.

The structure of the immunoglobulin tetramer is illustrated in **Figure 37.1.** Each protein chain consists of two principal regions: the N-terminal **variable (V) region;** and the C-terminal **constant (C) region.** They were defined originally by comparing the amino acid sequences of different immunoglobulin chains. As the names suggest, the variable regions show considerable changes in sequence from one protein to the next, while the constant regions show substantial homology. The corresponding regions of the light and heavy chains associate to generate distinct domains in the immunoglobulin protein.

The variable domain is generated by association between the variable region of the light chain and the variable region of the heavy chain. The V domain is responsible for recognizing the antigen. Production of V domains of appropriate specificities creates the ability to respond to diverse antigens. The total number of variable regions for either light or heavy-chain proteins is measured in hundreds. Thus the protein displays the maximum versatility in the region whose target differs in each individual antibody-antigen reaction.

Within the variable region, some parts are more variable than others. The most variable stretches are called **hypervariable (HV) regions.** The relatively less variable stretches are called **framework (FR) regions.** A variable region can be written as an alternating series of these types of subregion: FR1-HV1-FR2-HV2-FR3-HV3-FR4. The hypervariable segments are shorter than the framework segments.

There are three immunoglobulin families, two for light chains and one for heavy chains. Each has a large set of variable sequences, and a small set of constant sequences.

Light chains fall into two types: **kappa** and **lambda.** The kappa type is characterized by a single constant region; the lambda type has a few related types of constant region. In each case, the constant region comprises a single domain.

The single **heavy (H)** family has ~8 types of heavy chain, each defined by a characteristic constant region. The heavy constant regions are longer than the light constant regions, and they have several domains.

The association of constant regions generates several individual domains in the molecule. The first domain results from association of the constant region of the light chain with the CH1 part of the heavy-chain constant region. The structure of the rest of the heavy chain is (slightly) different in the various classes. In the example shown in Figure 37.1, a short hinge region connects the first half of the molecule to the other two constant domains, each generated by association of the corresponding regions (CH2 and CH3) of the heavy chains.

The constant domains mediate the effector functions responsible for executing the immune response. Thus the regions of the molecule with constant functions are conserved. Actually, different classes of immunoglobulins have related but not necessarily identical effector functions; the type of heavy-chain constant region determines the type of effector function.

Comparing the characteristics of the variable and constant regions, we see a central dilemma in immunoglobulin gene structure. How does the genome code for a set of proteins in which any individual polypeptide chain must have one of fewer than 10 possible constant regions, but can have any one of more than 1000 possible variable regions? It turns out that the number of coding sequences for each type of region reflects its variability. There are many genes coding for V regions, only a few genes coding for C regions.

In this context, *"gene" means a sequence of DNA coding for a discrete part of the final immunoglobulin polypeptide* (heavy or light chain). Thus **V genes** code for variable regions and **C genes** code for constant regions, although *neither type of gene is expressed as an independent unit.* To construct a unit that can be expressed in the form of an authentic light or heavy chain, a V gene must be joined physically to a C gene. In this system, two "genes" code for one polypeptide.

For each of the three immunoglobulin chain families—kappa, lambda, and heavy—any one of many V genes usually may be joined to any one of the few C genes. The process occurs by **somatic recombination** of the appropriate DNA sequences *in the B lymphocyte in which the antibody is expressed.* In most cases, the number of V genes is very high, more than 100, organized in a huge gene cluster. (We note that the mouse lambda type is an exception, consisting of only a few genes.) The variety in choice of V genes is responsible for some part of the diversity of immunoglobulins. However, not all diversity is coded in the genome; some is generated by changes that occur during the process of constructing a functional gene.

The construction of a functional immunoglobulin gene might seem superficially to be a Lamarckian process, representing a change in the genome responding to a particular feature of the phenotype (the antigen). At birth, the organism does not possess the functional gene for producing a particular antibody. The subsequent construction of the gene allows the antibody to be synthesized so that it is available to react with the antigen.

The feedback from antigen to the production of antibody is a provocative phenomenon. However, it may be explained without resort to Lamarckian heresy by the **clonal selection theory** summarized in **Figure 37.2.**

Recombination of V and C genes to give functional loci occurs in a population of immature B lymphocytes. *Each cell produces only a single antibody;* it undertakes only one productive rearrangement of light-chain genes and one of heavy-chain genes. The antibody is different in each cell, because different V genes and C genes have been joined in each reconstruction.

On exposure to antigen, a cell whose antibody is able to bind the antigen will be stimulated to divide, probably by virtue of some feedback from the surface of the cell, where the antibody-antigen reaction occurs. Now large numbers of this lymphocyte are produced, its antibody is secreted in large quantities, and it may even come to dominate the antibody population. We note that the entire process occurs in somatic

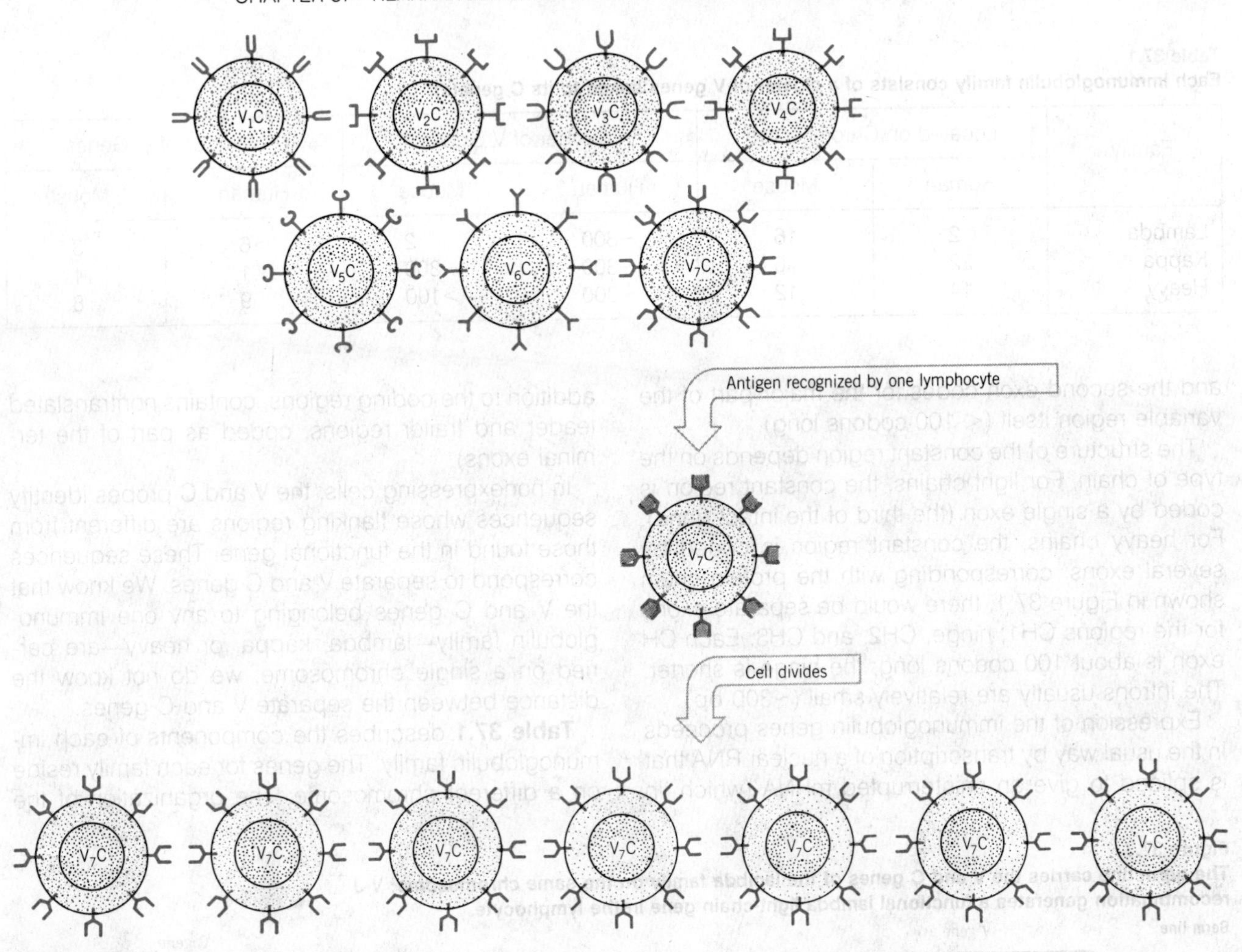

Figure 37.2
Reaction with an antigen leads to clonal expansion of the lymphocyte with the corresponding antibody.

cells and does not affect the germ line; so the response to an antigen is not inherited.

IMMUNOGLOBULIN GENES ARE ASSEMBLED FROM THEIR PARTS

Probes for immunoglobulin genes have been obtained from myelomas or plasmacytomas—tumor cell lines derived from immunoglobulin-expressing ancestors. Each tumor cell line usually synthesizes a single antibody. The protein is a major product of the cell, and its mRNA may occupy up to 10% of the poly(A)$^+$ mRNA.

Cloned probes representing the V and C regions can be used to identify the corresponding genomic sequences. The arrangement of these sequences is *different in the cells producing the immunoglobulins* from all other somatic cells or germ cells. Most of our information is about the immunoglobulin genes of mouse and man.

In a cell expressing an antibody, each immunoglobulin chain is coded by a single (interrupted) gene. The exons of the gene correspond precisely with the functional domains of the protein.

In the variable region, the first exon codes for the signal sequence (involved in membrane attachment),

Table 37.1
Each immunoglobulin family consists of a cluster of V genes linked to its C gene(s).

Family	Located on Chromosome		Number of V Genes		Number of C Genes	
	Human	Mouse	Human	Mouse	Human	Mouse
Lambda	2	16	~300	2	>6	3
Kappa	22	6	~300	~300	1	1
Heavy	14	12	~300	>100	9	8

and the second exon codes for the major part of the variable region itself (< 100 codons long).

The structure of the constant region depends on the type of chain. For light chains, the constant region is coded by a single exon (the third of the intact gene). For heavy chains, the constant region is coded by several exons; corresponding with the protein chain shown in Figure 37.1, there would be separate exons for the regions CH1, hinge, CH2, and CH3. Each CH exon is about 100 codons long; the hinge is shorter. The introns usually are relatively small (~300 bp).

Expression of the immunoglobulin genes proceeds in the usual way by transcription of a nuclear RNA that is spliced to give an uninterrupted mRNA (which, in addition to the coding regions, contains nontranslated leader and trailer regions, coded as part of the terminal exons).

In nonexpressing cells, the V and C probes identify sequences whose flanking regions are different from those found in the functional gene. These sequences correspond to separate V and C genes. We know that the V and C genes belonging to any one immunoglobulin family—lambda, kappa, or heavy—are carried on a single chromosome; we do not know the distance between the separate V and C genes.

Table 37.1 describes the components of each immunoglobulin family. The genes for each family reside on a different chromosome. The organization of the

Figure 37.3
The germ line carries the V and C genes of the lambda family on the same chromosome; V-J recombination generates a functional lambda light-chain gene in the lymphocyte.

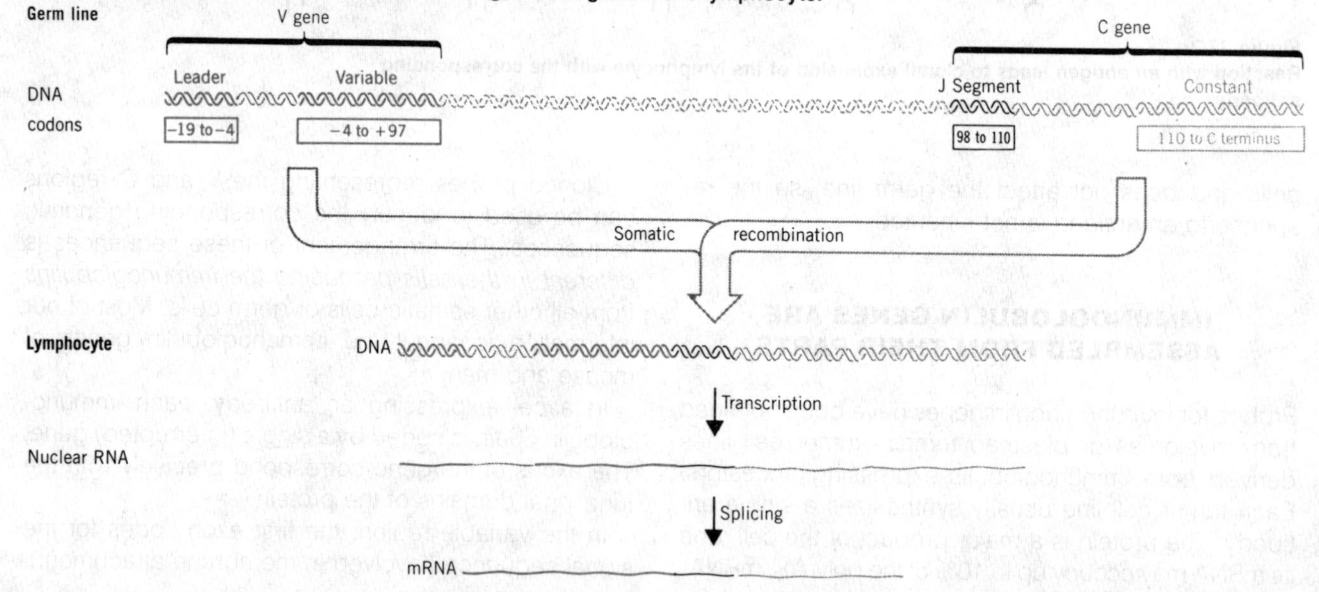

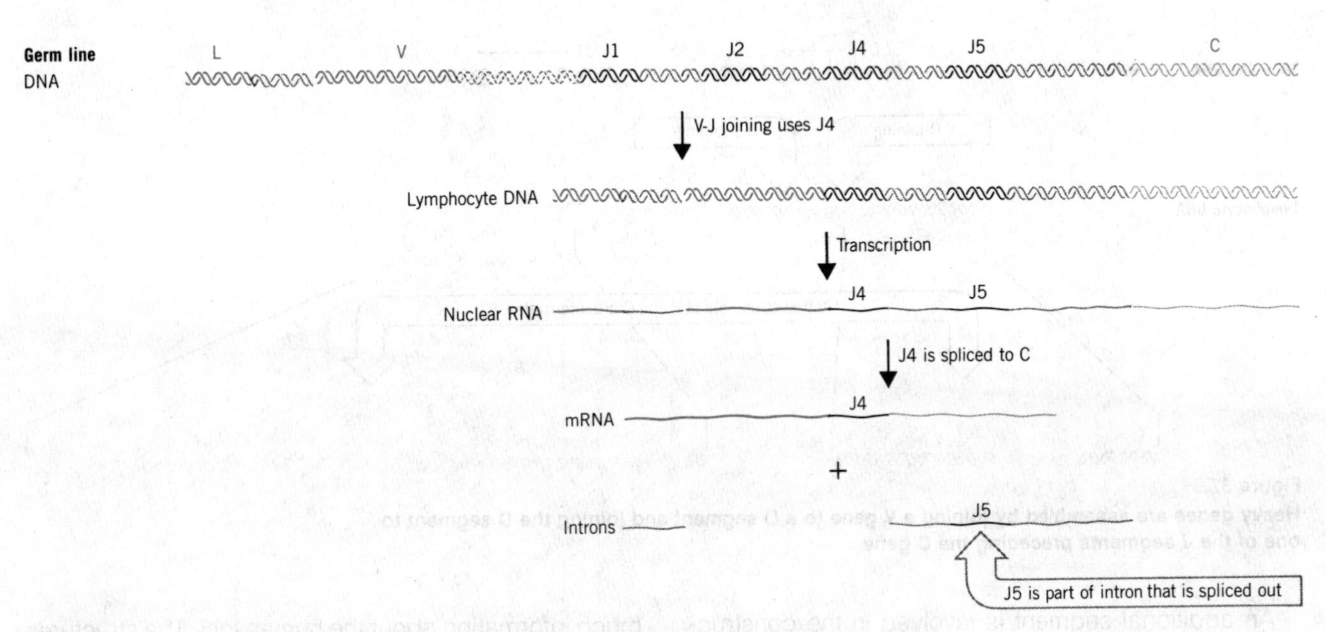

Figure 37.4
The kappa C gene is preceded by multiple J segments in the germ line; V-J joining in the lymphocyte may recognize any one of the J segments, which is then spliced to the C region during RNA processing.

separate V and C genes is described as the **germ-line** pattern (it is also found in all somatic cells of lineages other than the immune system).

A lambda light chain is assembled from two parts, as illustrated in **Figure 37.3.** The V gene consists of the leader exon (L) separated by a single intron from the variable (V) segment. The C gene consists of the J segment separated by a single intron from the constant (C) exon.

The name of the **J segment** is an abbreviation for joining, since it identifies the region to which the V segment becomes connected. The J segment is short and actually codes for the last few (13) amino acids of the variable region, as defined by amino acid sequences. In the intact gene, the V-J segment therefore constitutes a single exon coding for the entire variable region.

A V_λ gene may be joined to any one of a few C_λ genes, each of which has the same J-C dipartite structure. In view of this multipartite structure, we shall try to avoid confusion about the nature of the DNA sequences by referring to V, J, or C "segments" of the genome or "regions" of the polypeptide.

A kappa light chain also is assembled from two parts, but there is a difference in the structure of the C region. A group of five J segments is spread over a region of 500–700 bp, separated by an intron of 2–3 kb from the single C_κ exon. In the mouse, the central J segment is nonfunctional (ψJ3). A V_κ segment may be joined to any one of the J segments. The consequences of the kappa joining reaction are illustrated in **Figure 37.4.**

Whichever J segment is used becomes the terminal part of the intact variable exon. Any J segments on the left of the recombining J segment are lost (J1 and J2 have been lost in the figure). Any J segment on the right of the recombining J segment is treated as part of the intron between the variable and constant exons (J5 is included in the region that is spliced out in the figure).

All functional J segments possess a signal at the left boundary that makes it possible to recombine with the V segment; and they possess a signal at the right boundary that can be used for splicing to the C exon. Whichever J segment is recognized in DNA recombination uses its splicing signal in RNA processing.

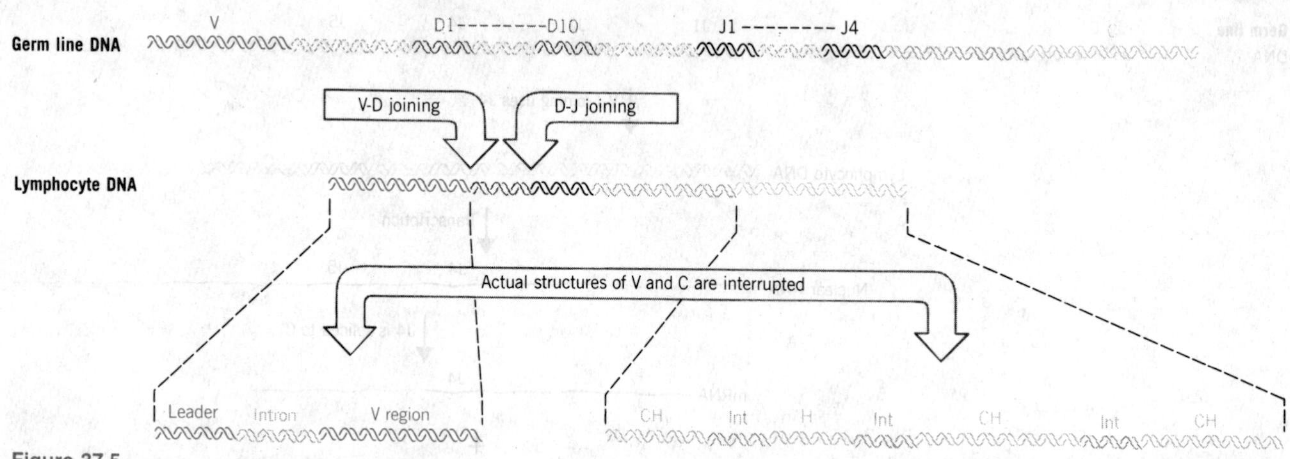

Figure 37.5
Heavy genes are assembled by joining a V gene to a D segment and joining the D segment to one of the J segments preceding the C gene.

An additional segment is involved in the construction of a heavy-chain gene, as illustrated in **Figure 37.5.** The **D** (for diversity) segment was discovered by the presence in the protein of a few (2–13) amino acids between the sequences coded by the V segment and the J segment. An array of D segments lies on the chromosome between the V_H segments and the J_H segments. A V_H segment is recombined with one of the D segments; and the D segment is recombined with one of four J_H segments (which vary in length between 4 and 6 codons). The reconstruction leads to expression of the adjacent C_H segment (which consists of several exons).

The organization of the D segments is not entirely clear, but many seem to lie in a tandem array. The mouse heavy-chain locus contains about ten D segments of variable length. Some unknown mechanism must ensure that the *same* D segment is involved in the V-D joining and D-J joining reactions.

THE DIVERSITY OF GERM-LINE INFORMATION

Now we must examine the different types of V and C genes to see how much diversity can be accounted for by the variety of the coding regions carried in the germ line. Immunoglobulin genes are most completely characterized in the mouse, and we have almost as much information about the human loci. The structures of the two light Ig gene families of the mouse genome are summarized in **Figure 37.6.**

In the mouse, almost all the light chains are of the kappa type; only about 5% are of the lambda type. In man, about 60% of the light chains are kappa and about 40% are lambda. A given lymphocyte generates either a kappa or a lambda light chain to associate with the heavy chain.

The structures of the kappa loci in mouse and man are similar. We have seen that several 13-codon long J segments precede the single C_κ gene. Many different variable sequences can be joined to the J-C locus. What is the origin of the V_κ diversity? Somatic recombination involving (randomly selected) pairs of individual V and J segments creates a variety of intact (V-J) coding regions; and somatic mutation may generate new V_κ segments by introducing changes into the germ-line segments.

Because of the varying degrees of divergence between germ-line genes, it is difficult to make a direct estimate of the number of V_κ genes in the germ line. An individual probe may react with several genes; when one of these genes is used as a probe, it may react with several others; and so on. This is the classic situation of a repetitive gene family, some members of which are closely related and some of which are distantly related. There are probably about 300 V_κ germ-line genes.

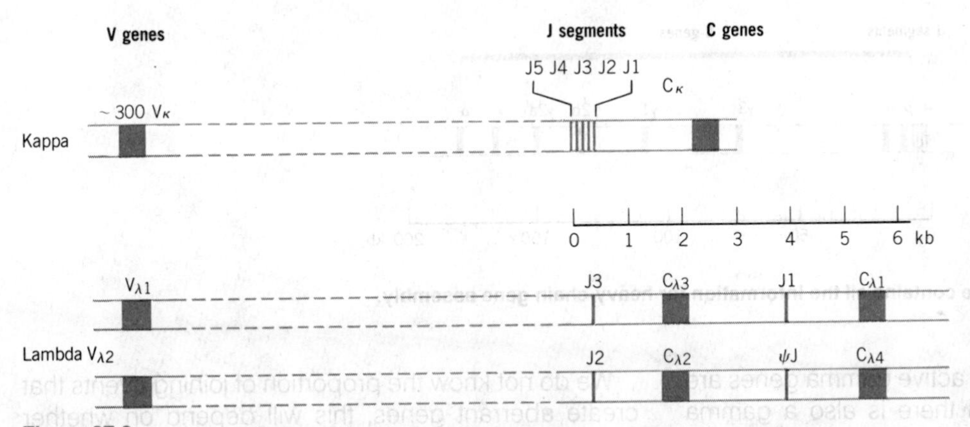

Figure 37.6
The kappa and lambda light-chain families have different complexities in the mouse.

By combining any one of 300 V_κ genes with any one of 4 J segments, the mouse genome has the potential to produce some 1200 kappa chains. However, when closely related variants are examined, often there are more proteins than can be accounted for by the number of corresponding V_κ genes. The new members are created by somatic changes in individual germ-line genes (not by recombination between V_κ genes). We shall consider the relevant mechanisms in later sections.

We do not have much information about the relative organization of the germ-line genes. We assume that the V_κ genes occupy a large cluster on the chromosome, an unknown distance upstream from the constant region. It will not be surprising if the more closely related genes form subclusters, generated by duplication and divergence of individual ancestral members.

The lambda genes show an interesting difference between mouse and man. The diversity of the murine V_λ regions is enormously reduced compared with that of the human V_λ regions (which are comparable to kappa). The reason is that the mouse genome contains only two V_λ genes. We assume that at some time in the past, the mouse suffered a catastrophic deletion of most of its germ-line V_λ genes. This may explain why most of its light chains are of the kappa type.

The mouse lambda genes fall into two groups, presumably generated by an ancestral duplication. Each of the V_λ genes is linked to two constant regions, each of which comprises a J-C structure. The $V_{\lambda 1}$ gene can be linked to either J-$C_{\lambda 3}$ or J-$C_{\lambda 1}$. The V gene therefore has a choice of J segments for joining. (The difference from the kappa gene is that each J segment is associated with a different C gene instead of with the same one.) The $V_{\lambda 2}$ gene usually joins only with J-$C_{\lambda 2}$. The $C_{\lambda 4}$ region is not used, because its J segment has an aberrant structure. Occasionally $V_{\lambda 2}$ joins with J-$C_{\lambda 1}$ or J-$C_{\lambda 3}$. We do not know the physical relationship between the two lambda groups; presumably they lie near one another. In man, not only are there more V_λ genes, but at least six C_λ genes have been identified in a 25 kb cluster.

The single locus for heavy-chain production consists of several discrete sections, whose structure in the mouse genome is summarized in **Figure 37.7.** There are probably about 100 V_H genes. At an unknown distance from them lies the cluster of D segments. Probably not much farther along comes the cluster of J segments. Over the next 170 kb lie all the C_H genes. By combining any one of 100 V_H genes, 10 D segments, and 4 J segments, the mouse genome potentially can produce 4000 variable regions to accompany any C_H gene.

The eight C_H genes of the mouse all are functional; usually only a single one is expressed in any given cell. The cluster has diverged by the usual processes of duplication, divergence, and reassortment of exons by means of unequal crossing-over or gene conversion. The four gamma genes appear to be the most closely related on the basis of protein structure.

There are more C_H genes in man. The mu and delta

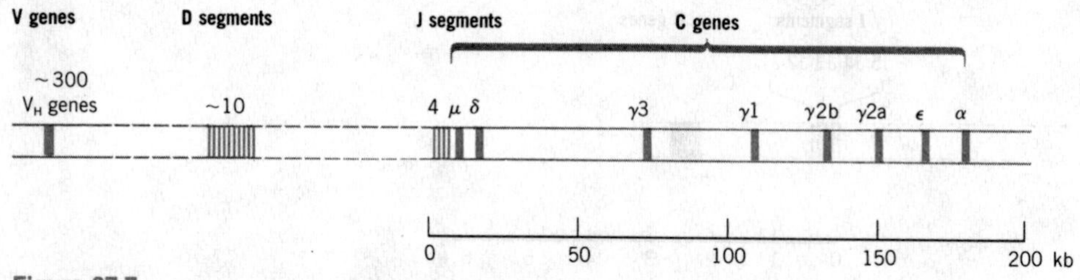

Figure 37.7
A single gene cluster in the mouse contains all the information for heavy-chain gene assembly.

genes remain single, the four active gamma genes are numbered 1 through 4, and there is also a gamma pseudogene; there are three epsilon genes of which at least one is a pseudogene, and there are two alpha genes.

JOINING REACTIONS GENERATE ADDITIONAL DIVERSITY

The joining reaction itself is a source of somatic variation, able to change the amino acid coded at the V-J junction. If the joining reaction is not fixed in position, but can occur between different base pairs, several different amino acids can be generated at the site of each potential V-J recombination. An example for kappa joining is illustrated in **Figure 37.8.**

The use of five potential frames for recombination generates three different amino acids at position 96, including one (arginine) *not coded in the germ line.* Since other V_κ and J_κ segments have different codons at these positions, great diversity becomes possible at the point of junction. It is interesting that the amino acid at position 96 forms part of the antigen-binding site and also is involved in making contacts between the light and heavy chains.

Even greater diversity becomes possible if recombination is not restricted to positions in the *corresponding* reading frames, but can occur between any pair of points in the relevant regions of V and J. There is evidence for the occurrence of such events. Some delete a codon from the somatic gene. Others join the V-J region so that the J segment is out of phase and is translated in the wrong reading frame. The resulting gene is aberrant, since its expression is terminated by a nonsense codon in the incorrect frame.

We do not know the proportion of joining events that create aberrant genes; this will depend on whether there is preference for particular joining reactions. However, we may think of the formation of aberrant genes as comprising the price the cell must pay for the increased diversity that it gains by being able to adjust the site of the recombination event.

Similar although even greater diversity is generated in the joining reactions that involve the D segment of the heavy chain. The same problem remains of gen-

Figure 37.8
Kappa V-J joining takes place in the sequences coding for amino acids 95 and 96. Use of alternative sites for recombination between aligned V and J segments creates new codons. Similar events occur in the V-J lambda and V-D-J heavy joining reactions.

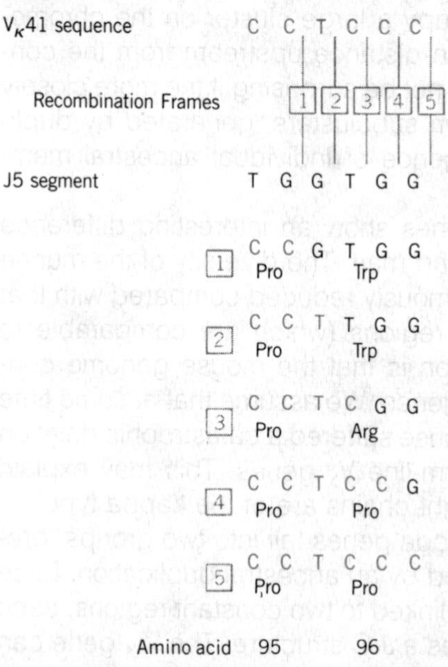

erating nonproductive genes by recombinations that place J and C out of phase.

A puzzle about D segments is that some heavy chains have amino acid residues that do not correspond to any of the known D segments. Two ideas have been proposed to explain this observation.

One possibility is that new D segments are generated in the somatic cell by D-D recombination. This recombination would involve a different mechanism from that involved in V-D-J joining.

Another possibility is that additional nucleotides are incorporated into the V-D or D-J or both junctions during the recombination process. A speculation is that their insertion could occur via the activity of the enzyme deoxynucleoside transferase (known to be an active component of lymphocytes) at a free 3′ end generated during the joining process. At all events, the existence of new sequences identifies another mechanism for generating diversity.

RECOMBINATION OF V AND C GENES GENERATES DELETIONS AND REARRANGEMENTS

Assembly of the parts of both light- and heavy-chain genes seems to involve similar events (although the number of parts is different). The employment of the same mechanism is indicated by the presence of the same consensus sequences at the boundaries of all germ-line segments that participate in joining reactions. Each consensus sequence consists of a palindromic heptamer separated by either 12 or 23 bp from a nanomer.

Figure 37.9 illustrates the relationship between the consensus sequences at the mouse Ig loci. At the kappa locus, each V_κ gene is followed by a consensus sequence with a 12 bp spacing. Each J_κ segment is preceded by a consensus sequence with a 23 bp spacing. The V and J consensus sequences are inverted in orientation. The reverse arrangement is found at the lambda locus; each V_λ gene is followed by a consensus sequence with 23 bp spacing, while each J_λ gene is preceded by a consensus of the 12 bp spacer type.

Comparing these two situations suggests a rule for the joining reaction: *a consensus sequence with one type of spacing can be joined only to a consensus sequence with the other type of spacing.* Since the consensus sequences at V and J segments can lie in either order, the different spacings do not impart any directional information, but serve to prevent one V gene from recombining with another or one J segment from recombining with another.

Figure 37.9
Consensus sequences are present in inverted orientation at each pair of recombining sites. One member of each pair has a spacing of 12 bp between its components; the other has 23 bp spacing.

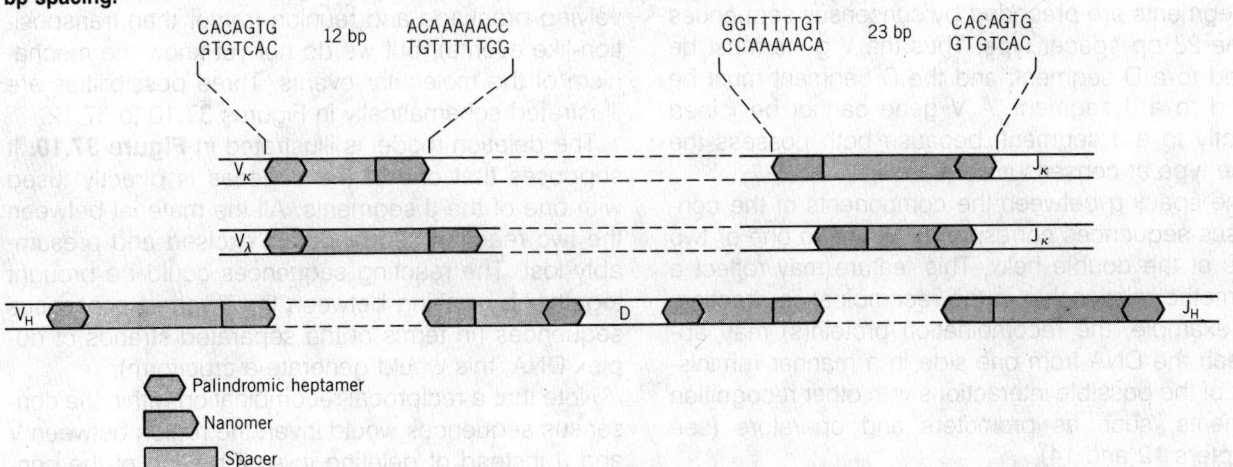

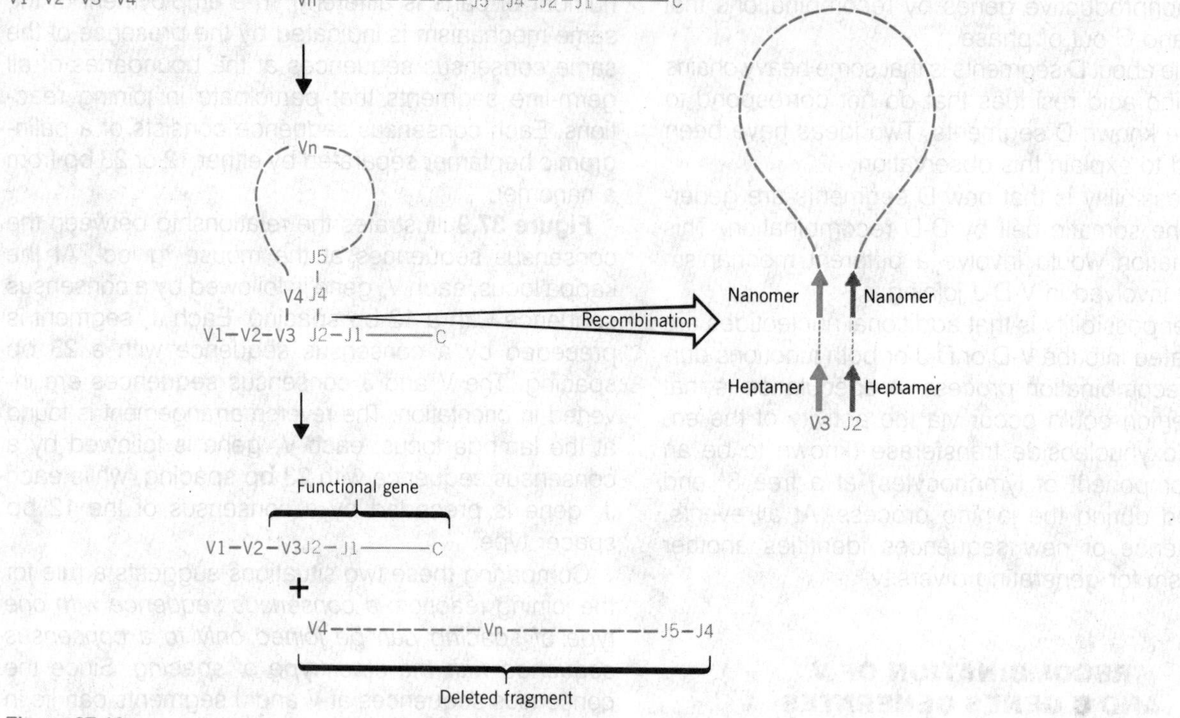

Figure 37.10
The consensus sequences could be used to bring the V and J regions into juxtaposition so that a breakage and reunion excises the material between them.

This concept is borne out by the structure of the components of the heavy genes. Each V_H gene is followed by a consensus sequence of the 23 bp spacer type. The D segments are flanked on either side by consensus sequences of the 12 bp spacer type. The J_H segments are preceded by consensus sequences of the 23 bp spacer type. Thus the V gene must be joined to a D segment; and the D segment must be joined to a J segment. A V gene cannot be joined directly to a J segment, because both possess the same type of consensus sequence.

The spacing between the components of the consensus sequences corresponds almost to one or two turns of the double helix. This feature may reflect a geometric relationship in the recombination reaction. For example, the recombination protein(s) may approach the DNA from one side in a manner reminiscent of the possible interactions with other recognition elements, such as promoters and operators (see Chapters 12 and 14).

In using the term "recombination" to describe the joining of the components of immunoglobulin genes, we do not imply that the reaction involves a reciprocal recombination between homologous sequences. It does involve a physical rearrangement of sequences, involving breakage and reunion (rather than transposition-like events), but we do not yet know the mechanism of the molecular events. Three possibilities are illustrated schematically in Figures 37.10 to 37.12.

The deletion model is illustrated in **Figure 37.10.** It supposes that one of the V genes is directly fused with one of the J segments. All the material between the two reacting sequences is excised and presumably lost. The reacting sequences could be brought together by pairing between the inverted consensus sequences (in terms of the separated strands of duplex DNA, this would generate a cruciform).

Note that a reciprocal recombination within the consensus sequences would invert the region between V and J instead of deleting it; so if pairing of the con-

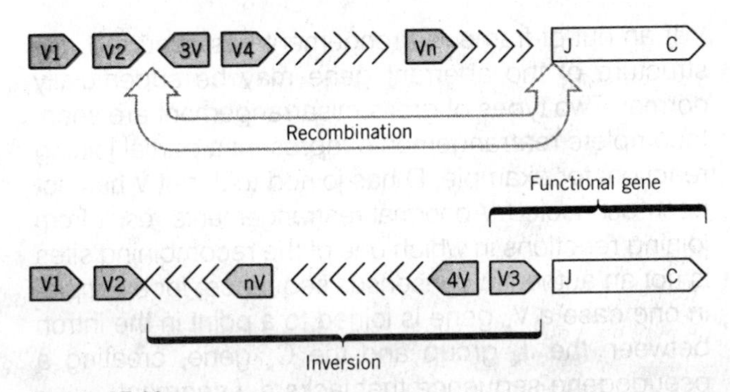

Figure 37.11
A joining reaction involving a V gene in inverse orientation to the J segment could generate a functional V-J-C gene by inversion of the region between V and J.

sensus sequences leads to a deletion, we must be dealing with a different type of breakage and reunion. One possibility is that the loop created by the pairing could be removed by cleavage at the base, with the ends of the V and J regions subsequently ligated together.

Several inversion models are possible, and one form is illustrated in **Figure 37.11.** We have supposed so far that all the V genes lie in the same orientation as the J segments, although we do not yet have information linking the V-coding regions to the J segments. If a V gene were present in reverse orientation, a recombination reaction could generate a functional gene by inverting the intervening material to fuse V to J. If some V genes are in the same orientation as J but others are inverted, some V-J joinings might be accompanied by deletions while others are accompanied by inversions.

The unequal crossing-over model is illustrated in **Figure 37.12.** It involves an unequal recombination between sister chromatids (replicated copies) of the gene. (The model begs the question of what role the consensus sequences may play.) The recombination event produces one chromatid that has a functional gene, accompanied by deletion of the material between the recombining V and J segments. The other chromatid has a duplication of the region between V and J; on either side of the aberrant duplicated region are V and C genes that may remain available for further recombinations. The two chromatids segregate to daughter cells at mitosis.

The structure of the active gene is the same in both the deletion and unequal crossing-over models. The models can be distinguished by the fate of the progeny of the cell in which the event occurred. In the deletion model, all descendants have the deletion. In the unequal crossing-over model, half of the descendants have the deletion, but the other half have the potentially reactive duplication. Pedigree analysis of cells in which the recombination reaction is occurring should be able to resolve the actual situation. The in-

Figure 37.12
Unequal crossing-over between sister chromatids could generate one recombinant with a functional gene and a deletion of material between V and J, while the other recombinant has a duplication of the region between V and J.

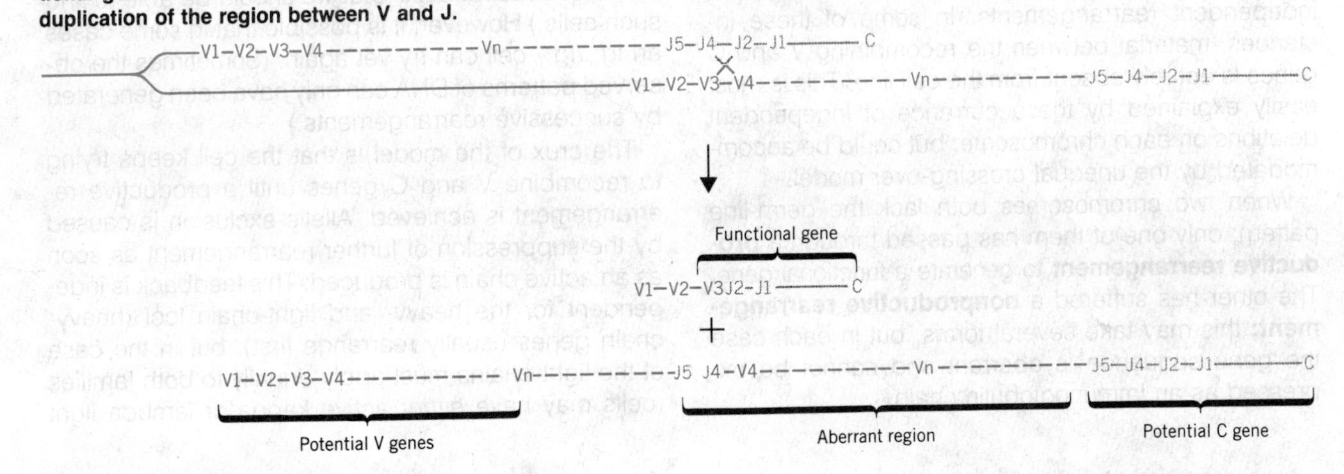

version model is readily distinguished from both of these models by the retention in the inverted segment of the material between the V and C genes.

SOME POSSIBLE CAUSES OF ALLELIC EXCLUSION

Each V-J or V-D-J recombination is a unique event. A single productive rearrangement of each type occurs in a given lymphocyte, to produce one light and one heavy chain gene. Because each event involves the genes of only *one* of the homologous chromosomes, the alleles on the other chromosome are not expressed in the same cell. This phenomenon is called **allelic exclusion.**

The occurrence of allelic exclusion complicates the analysis of somatic recombination. A probe reacting with a region that has rearranged on one homologue will also detect the allelic sequences on the other homologue. We may therefore be compelled to analyze the different fates of the two chromosomes together. However, the usual pattern displayed by a rearranged active gene can be interpreted in terms of a deletion of the material between the recombining V and C loci.

Sometimes an active cell displays *both* the rearranged and germ-line patterns of organization for a given gene. We assume then that joining has occurred on one chromosome, while the other chromosome has remained unaltered. The rearrangement could have occurred by either deletion or unequal crossing-over.

In some lines, both chromosomes have suffered independent rearrangements. In some of these instances, material between the recombining V and C genes is entirely absent from the cell line. This is most easily explained by the occurrence of independent deletions on each chromosome, but could be accommodated by the unequal crossing-over model.

When two chromosomes both lack the germ-line pattern, only one of them has passed through a **productive rearrangement** to generate a functional gene. The other has suffered a **nonproductive rearrangement;** this may take several forms, but in each case the gene sequence is aberrant and cannot be expressed as an immunoglobulin chain.

If an out-of-frame rearrangement has occurred, the structure of the aberrant gene may be superficially normal. Two types of gross misarrangement are seen. Incomplete rearrangements represent a partial joining reaction; for example, D has joined to J, but V has not been connected. Abnormal rearrangements result from joining reactions in which one of the recombining sites is not an authentic consensus sequence; for example, in one case a V_κ gene is joined to a point in the intron between the J_κ group and the C_κ gene, creating a pseudogene sequence that lacks a J segment.

Nonproductive rearrangements are common in myelomas and plasmacytomas. The assumption that their presence is not an aberration of the tumor cells, but reflects the usual pattern of rearrangement, suggests a stochastic model for rearrangements. Each cell starts with two loci of the unrearranged germ-line configuration Ig^0. Either of these loci may be rearranged to generate a productive gene Ig^+ or a nonproductive gene Ig^-. If the rearrangement is productive, the synthesis of an active chain provides a trigger to prevent rearrangement of the other allele. The active cell has the configuration Ig^0/Ig^+.

If the rearrangement is nonproductive, it creates a cell with the configuration Ig^0/Ig^-. There is no impediment to rearrangement of the remaining germ-line allele. If this rearrangement is productive, the expressing cell has the configuration Ig^+/Ig^-. Again, the presence of an active chain suppresses the possibility of further rearrangements.

Two successive nonproductive rearrangements produce the cell Ig^-/Ig^-. (Indeed, one prediction of the stochastic model is that we should be able to find such cells.) However, it is possible that in some cases an Ig^-/Ig^- cell can try yet again. (Sometimes the observed patterns of DNA can only have been generated by successive rearrangements.)

The crux of the model is that the cell keeps trying to recombine V and C genes until a productive rearrangement is achieved. Allelic exclusion is caused by the suppression of further rearrangement as soon as an active chain is produced. This feedback is independent for the heavy- and light-chain loci (heavy-chain genes usually rearrange first), but in the case of the light chains must apply equally to both families (cells may have *either* active kappa or lambda light chains). It is likely that the cell rearranges its kappa genes first, and only tries to rearrange lambda if both kappa attempts are unsuccessful.

FURTHER DNA RECOMBINATION CAUSES CLASS SWITCHING

The class of immunoglobulin is defined by the type of C_H region. IgM has the C_μ sequence, IgD has the C_δ sequence, IgG has one of the C_γ sequences, IgE has the C_ϵ sequence, and IgA has the C_α sequence. A lymphocyte generally produces only a single class of immunoglobulin at any one time, but the class may change during the cell lineage. This is accomplished by a substitution in the type of C_H region that is expressed; the change in expression is called **class switching.**

Switching involves only the C_H gene; the same V_H gene continues to be expressed. Thus a given V_H gene may be expressed successively in combination with more than one C_H gene. The same light chain continues to be expressed throughout the lineage of the cell. Class switching therefore allows the type of effector response (mediated by the C_H region) to change, while maintaining a constant antigen-recognition facility (mediated by the V regions).

All lymphocytes start productive life as immature cells engaged in synthesis of IgM. Cells expressing IgM have the germ-line arrangement of C_H genes shown in Figure 37.7. The V-D-J joining reaction is sufficient to trigger expression of the C_H gene.

Changes in the expression of C_H genes are made in two ways. Some occur at the level of RNA processing (see next section). The majority occur via further DNA recombinations, involving a system different from that concerned with V-D-J joining (and able to operate only at a subsequent point in time).

Cells expressing later C_H genes usually have deletions of C_μ and the other genes preceding the expressed C_H gene. Thus class switching may be accomplished by a recombination to bring a new C_H gene into juxtaposition with the expressed V-D-J unit. The sequence of the switched V-D-J-C_H units shows that the sites of switching lie upstream of the C_H genes themselves. The switching sites are called **S regions. Figure 37.13** shows a (hypothetical) example of two successive switches.

In the first switch, expression of C_μ is succeeded by expression of $C_{\gamma1}$. The $C_{\gamma1}$ gene is brought into the expressed position by recombination between the sites S_μ and $S_{\gamma1}$, deleting the material between. The S_μ site lies between V-D-J and the C_μ gene. The $S_{\gamma1}$ site lies upstream from the $C_{\gamma1}$ gene. The region be-

Figure 37.13
Class switching of heavy genes may occur by recombination between switch regions (S), deleting the material between the recombining S sites. Successive switches may occur.

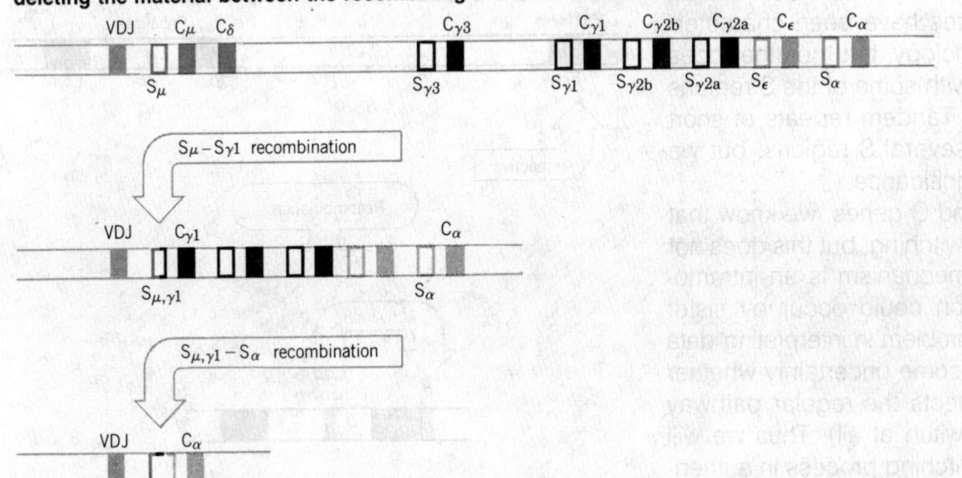

tween V-D-J and the $C_{\gamma 1}$ gene is removed as an intron during processing of the RNA.

The linear deletion model imposes a restriction on the heavy-gene locus: *once a class switch has been made, it becomes impossible to express any C_H gene that used to reside between C_μ and the new C_H gene.* In the example of Figure 37.13, cells expressing $C_{\gamma 1}$ should be unable to give rise to cells expressing $C_{\gamma 3}$, which has been deleted.

However, it should in principle be possible to undertake another switch to any C_H gene downstream from the expressed gene. The figure shows a second switch to C_α expression, accomplished by recombination between S_α and the switch region $S_{\mu,\gamma 1}$, generated by the original switch.

We assume that all of the C_H genes have S regions upstream from the coding sequences. We do not know whether there are any restrictions on the use of S regions. Sequential switches do occur, but we do not know whether they are optional or an obligatory means to proceed to later C_H genes. We should like to know whether IgM can switch directly to *any* other class.

We know that switch sites are not uniquely defined, because different cells expressing the same C_H gene prove to have recombined at different points. When enough of these S sites have been sequenced, we may be able to define the limits of the regions within which switching can occur. In the meantime, we can scrutinize the sequences in the area for common features that might be involved.

The S regions lie about 2 kb upstream from the C_H genes. Three distinct S_μ sites have been characterized; two of them show homology, but the other does not. There are homologies with some of the S regions of other genes, but not all. Tandem repeats of short sequences are present in several S regions, but we have yet to discern their significance.

As with the joining of V and C genes, we know that deletions occur during C_H switching; but this does not prove that the underlying mechanism is an intramolecular deletion; the reaction could occur by sister chromatid exchanges. Our problem in interpreting data obtained with myelomas is some uncertainty whether the observed switching reflects the regular pathway (most tumor cells do not switch at all). Thus we will need to characterize the switching process in authentic lymphocytes.

REARRANGEMENT IS RESPONSIBLE FOR ACTIVATING Ig GENES

What is the connection between joining of V and C genes and their activation? Unrearranged V genes are not represented in RNA. But when a V gene is joined productively to a C_κ gene, the resulting unit is transcribed. However, since the sequence upstream from a V gene is not altered by the joining reaction, *the promoter must be the same in unrearranged, nonproductively rearranged, and productively rearranged genes.*

If the V promoter is activated by its relocation to the C region, the effect must depend on sequences downstream. What role might they play? The best characterized case is the heavy locus, where an enhancer is present in the major intron of the first C_H gene. The enhancer is tissue specific; it is active only in B cells. Its existence suggests the model illustrated in **Figure 37.14,** in which the V gene promoter is activated as soon as it is brought into reach of the enhancer. It seems likely that the same mechanism is used in the other immunoglobulin loci.

Does the enhancer continue to function throughout switching of heavy genes? Because it is located upstream from the first switching site (S_μ), it remains able

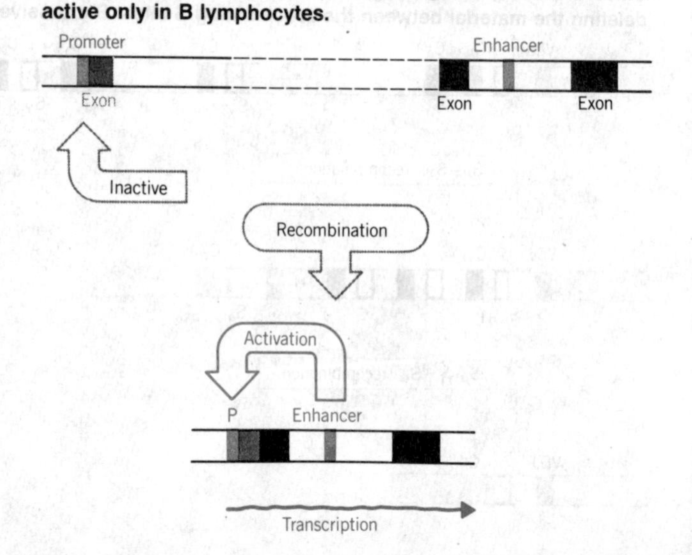

Figure 37.14
A V gene promoter is inactive until it is brought within range of an enhancer in the intron of the C gene. The enhancer is active only in B lymphocytes.

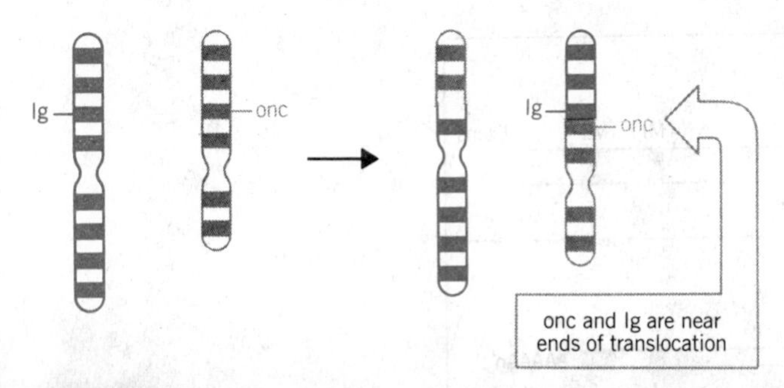

Figure 37.15
A chromosomal translocation is a reciprocal event that exchanges parts of two chromosomes. When an *onc* gene from one chromosome is translocated into an Ig locus, it may become active.

to sponsor transcription through the series of switching events.

A fascinating connection has been found between the Ig loci and the occurrence of certain tumors. Specific chromosomal translocations are often associated with plasmacytomas in the mouse and with Burkitt lymphomas in man. These tumors represent aberrant B lymphocytes. The common feature in both species is that an oncogene on one chromosome is brought into the proximity of an Ig locus on another chromosome (the chromosomes carrying the Ig loci are summarized in Table 37.1). The nature of the translocation event is illustrated in **Figure 37.15.**

In the mouse, most of the chromosomal translocations involve the joining of part of chromosome 15 to the region of chromosome 12 that carries the Ig H locus; some involve joining of the same part of chromosome 15 to the part of chromosome 6 that carries the kappa locus.

In man, the translocations usually involve chromosome 8 and chromosome 14, which carries the Ig H locus; about 10% involve chromosome 8 and either chromosome 2 (kappa locus) or chromosome 22 (lambda locus).

The same oncogene is involved in man and mouse, *c-myc*, located on human chromosome 8 and murine chromosome 15. When *c-myc* is translocated to the Ig locus, it becomes activated. Its activation is one of the events responsible for converting the cell into a tumorigenic state.

Translocation takes a variety of forms at the molecular level, involving different sites on both chromosomes. It seems likely that several different mechanisms can activate the *c-myc* gene in its new location.

The basic cause of the translocation event may lie in a malfunction of the system responsible for recombining Ig genes. The majority of translocations involve the Ig H locus, and the translocation often seems to involve a switching site, so it may represent an aberration of the class switching system. For some reason, at present unknown, the recombining enzymes may act on sites in the vicinity of the *c-myc* gene, located on a different chromosome, as well as on the sites around the Ig loci themselves.

Although the *c-myc*-Ig translocations are now the best characterized, they are far from being the only example of the phenomenon. Other translocations involve different *onc* genes and different recipient sites. Translocation to a new locus may be one of several mechanisms by which an oncogene is activated.

EARLY HEAVY-CHAIN EXPRESSION CAN BE CHANGED BY RNA PROCESSING

The period of IgM synthesis that begins lymphocyte development falls into two parts, during which different versions of the mu constant region are synthesized. As a stem cell differentiates to a pre-B lymphocyte, an accompanying light chain is synthesized, and the IgM molecule (μ_2L_2) appears at the surface of the cell. This form of IgM contains the μ_m version of the constant region (*m* indicates that IgM is located in the membrane). When the B lymphocyte differentiates further into a plasma cell, the μ_s version of the constant region is expressed. (The IgM actually is secreted as a pentamer IgM_5y, in which J is a joining polypeptide that forms disulfide linkages with mu chains.)

The μ_m and μ_s versions of the mu heavy chain are identical for most of their length. However, the μ_m chain is longer than the μ_s chain; the difference results from a change in the amino acid sequence at the C-terminal end. The μ_m chain ends in a 41-residue hydrophobic sequence that may be responsible for securing its place in the membrane. This sequence is replaced by a 20-residue hydrophilic sequence in μ_s; the substitution may allow the mu heavy chain to pass through the membrane.

The μ_m and μ_s chains are coded by different mRNAs.

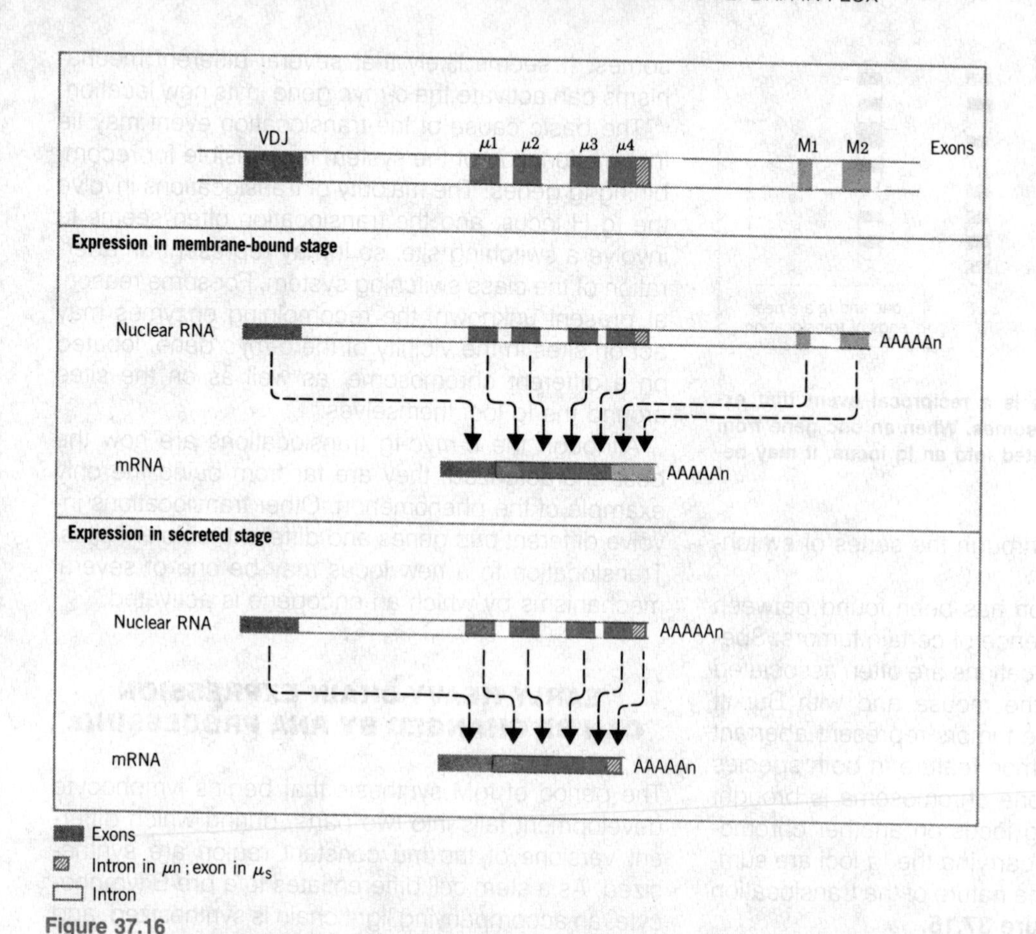

Figure 37.16

The site of termination or cleavage and polyadenylation may control the use of splicing junctions so that alternate forms of the heavy gene are expressed.

The two mRNAs share an identical sequence up to the end of the last constant domain. Then they differ; μ_m has 41 more codons followed by a nontranslated trailer; and μ_s has 20 codons followed by a different nontranslated trailer.

The genome sequence shows that the terminal regions of μ_m and μ_s are coded in different exons. The relationship between the structure of the gene and the mRNAs is illustrated in **Figure 37.16.**

At the membrane-bound stage, the constant region is produced by splicing together six exons. The first four code for the four domains of the constant region. The last two, M1 and M2, code for the 41-residue hydrophobic C-terminal region and its nontranslated trailer.

At the secreted stage, the constant region is generated by joining only the first four exons. The last of these exons extends farther than it did at the previous (membrane) stage; it brings in the last 20 codons and its nontranslated trailer.

The difference between the two mRNAs therefore hinges on whether a splicing junction *within* the exon for the last constant domain is spliced to M1 and M2 (shortening the exon and generating the coding sequence for μ_m) or is ignored (lengthening the exon and generating the coding sequence for μ_s). How is the use of this splicing site controlled?

A plausible model is illustrated in Figure 37.16. At the membrane-bound stage, the nuclear RNA has a polyadenylated end after M2 (probably by virtue of cleavage at this site, possibly by termination). Because the nuclear RNA contains the splicing site at the beginning of M1, it uses the splicing site within the last constant exon. At the secreted stage, the nuclear RNA ends at an earlier site, after the last constant exon. Because no subsequent exons are present in the nuclear RNA, the splicing site within the exon cannot be utilized.

By placing the onus for regulation on the choice of a site for cleavage and polyadenylation (or for termination), we avoid the dilemma of needing to choose between splicing alternatives for a single RNA. To confirm this model, we will need to characterize the nuclear RNA.

A similar transition from membrane to secreted forms is found with other constant regions. The conservation of exon structures suggests that the mechanism is the same. The M1 and M2 exons of three gamma chains are closely related, coding for a sequence of 66 amino acids, the first 41 of which show homology with the mu membrane sequence. The conserved sequence may provide a signal recognized by a receptor protein that anchors the *m* form of the immunoglobulin in the membrane. Perhaps the anchorage is related to the need to initiate cell proliferation in response to the initial recognition of an antigen.

An exception to the rule that only one immunoglobulin type is synthesized by any one cell is presented by the simultaneous production of IgM and IgD in mature B lymphocytes. The two immunoglobulins are identical except for the substitution between mu and delta constant regions in the heavy chain. It seems likely that this is the outcome of alternative pathways for RNA processing. The delta constant region is close to mu; if transcription sometimes continues through the region, the VDJ exon could be spliced to the series of delta exons.

This situation is reminiscent of the situation involved in late adenovirus expression, and we may invoke the same type of model illustrated previously in Figure 20.20; again, the generation of different polyadenylated ends for the nuclear RNA controls the selection of coding regions for splicing to the initial exon.

SOMATIC MUTATION GENERATES ADDITIONAL DIVERSITY

Comparisons between the sequences of expressed immunoglobulin genes and the corresponding V genes of the germ line show that new sequences appear in the expressed population. We have seen that some of this additional diversity results from shifts in the site of recombination during assembly of the V-J light-chain exons or the V-D-J heavy-chain exons. However, some changes occur at preceding locations within the variable domain; they represent **somatic mutations** induced specifically in the active lymphocyte.

With kappa light chains or heavy chains, many potential germ-line V genes could be used to construct the variable domain. A probe representing an expressed V gene can be used to identify all the corresponding fragments in the germ line. Their sequences should identify the complete repertoire available to the organism. Any expressed gene whose sequence is different must have been generated by somatic changes.

The main experimental problem in this analysis lies in ensuring that every potential contributor in the germ-line V genes actually has been identified. This problem is overcome by the fortuitous simplicity of the mouse lambda chain system. A survey of several myelomas producing λ_1 chains showed that many have the sequence of the single germ-line gene. But others have new sequences that must have been generated by mutation of the germ-line gene.

To determine the frequency of somatic mutation in other cases, we need to examine a large number of cells in which the same V gene is expressed. A practical procedure for identifying such a group is to characterize the immunoglobulins of a series of cells, all of which express an immune response to a particular antigen.

(Antigens used for this purpose are small molecules whose discrete structure is likely to provoke a consistent response, unlike a large protein, parts of which may provoke potentially different antibodies. The small molecules are called haptens and are conjugated with a nonreactive protein to form the antigen. The cells are obtained by immunizing mice with the antigen, obtaining the reactive lymphocytes, and sometimes fusing these lymphocytes with a myeloma cell to generate a **hybridoma** that continues to express the desired antibody indefinitely.)

A survey of 19 different cell lines producing antibodies directed against the hapten phosphorylcholine showed that ten have the same V_H sequence (expressed in conjunction with one of the mu, gamma, or alpha constant regions). The V_H sequence can be identified in the germ line as T15, one of four V_H genes hybridizing with a probe for the expressed V_H sequence. The other nine expressed genes differ from each other and from all four germ-line members of the family. They are more closely related to the T15 germ-line sequence than to any of the others, and their flanking sequences are the same as those around T15. This suggests that they have arisen from the T15 member by somatic mutation.

The sequences of these expressed genes vary from the germ-line sequence in all regions of the variable domain, including frameworks as well as hypervariable regions. Sequence changes are found in the downstream flanking regions as well as in the coding regions; they do not seem to extend far upstream. All the variation is due to substitutions of individual nucleotide pairs. The variation is different in each case. One gene shows 3.8% divergence from the germ line; another shows 1.4% divergence. This translates to 8 and 3 amino acid substitutions, respectively. Only a minority of the mutations affect the amino acid sequence, since many lie in third-base coding positions as well as in nontranslated regions.

The large proportion of ineffectual mutations suggests that somatic mutation occurs more or less at random in a region including the V gene and extending beyond it. In the heavy-chain genes, class switching may provide the trigger for mutation. The V_H genes associated with mu constant chains are rarely mutated; the (same) V_H genes associated with gamma or alpha constant chains may be mutated.

Random mutations have unpredictable effects on protein function; some may inactivate the protein, others may confer high specificity for a particular antigen. Thus a critical feature of the process may be selection among the lymphocyte population for those cells bearing antibodies in which chance mutation has created a suitable V domain to bind whatever antigen is at hand.

This does not entirely solve our dilemma about the generation of antibody diversity. Consider the large number of V genes, not to mention the additional diversity generated by alternative joinings to D and J segments. We see in the case of phosphorylcholine that in many separate antigenic provocations, the same V_H gene is used to generate the antibody. For this particular V_H gene always to be available, in either original or mutated sequence, an immense range of lymphocytes must have suffered rearrangements while waiting for an antigen. Does this mean that so many rearrangements occur that every potential V gene actually is available in expressed form?

T-CELL RECEPTOR IS RELATED TO IMMUNOGLOBULINS

The immune response requires a T cell to recognize a cell "presenting" a foreign antigen. To do so, the T cell simultaneously recognizes the foreign antigen and a histocompatibility antigen carried by the presenting cell.

The T-cell receptor is responsible for this dual recognition. In a manner analogous to immunoglobulins, on one hand it must recognize a foreign antigen of unpredictable structure, while on the other it exercises its effector function (recognition of histocompatibility). A common view has been that nature might well solve the problem of antigen recognition by B cells and T cells in the same way, in which case we might expect the organization of the T-cell receptor to resemble the immunoglobulins in the use of variable and constant regions.

Identified recently, the receptor is a glycoprotein of about 80,000 daltons, consisting of an α-chain and a β-chain, each about 40,000 daltons. The chains are held together by disulfide bonds. Peptide mapping suggests that the receptors of different T cell lines share part of their sequences in common, but differ in others (like the immunoglobulins). The receptor is a surface protein, probably anchored in the membrane.

Genes coding for the T-cell receptor have been identified by finding T cell mRNAs whose corresponding sequences in the genome have rearranged in T cells but not in B cells. Their organization appears similar to that of immunoglobulins. In one case, the N-terminal region is related to the variable Ig regions and has a leader followed by a ~120 residue variable region. The C terminal region is related to the constant Ig regions and has a single constant domain followed by transmembrane and cytoplasmic portions.

Germ-line genes corresponding to the β-chain have components that resemble the Ig heavy chain. Compared with the mRNA, the genome has a leader exon, a variable exon, a missing region, a J segment, and several constant exons. The missing region may represent a D segment.

We do not yet have any estimate for the number of V genes. There are two constant genes ($C_{\beta1}$ and $C_{\beta2}$), each preceded by a cluster of several J segments.

At the 3′ side of the V segment and the 5′ side of the J segment are conserved sequences resembling those involved in reconstruction of immunoglobulin genes. The germ-line sequences differ from the mRNA by point substitutions in the leader, variable and J segments, suggesting that somatic mutation may be involved in generating T-cell receptor variability.

The nature of the events involved in constructing the T cell receptor will become clear as more genes are characterized. But already the similarities between the Ig loci and the T-cell receptor sequences suggest strongly that similar mechanisms operate in B cells on the Ig loci and in T cells on the receptor loci. If the same enzymes are involved in constructing expressed genes from germ-line segments (as suggested by the conservation of the adjacent recognition sequences), we may ask how the process is controlled so that Ig loci are rearranged in B cells, while T-cell receptors are rearranged in T cells.

COMPLEXITY OF MAJOR HISTOCOMPATIBILITY LOCI

The major histocompatibility locus occupies a small segment of a single chromosome in the mouse (where it is called the *H2* locus) and in man (called the *HLA* locus). Within this segment are many genes coding for functions concerned with the immune response. At those individual gene loci whose products have been identified, many alleles have been found in the population; the locus is described as highly **polymorphic,** meaning that individual genomes are likely to be different from one another.

The several types of function identified in *H2* and the adjacent region are summarized on the map of **Figure 37.17.** The classic *H2* region occupies 0.3 map units and includes three classes of genes. The adjacent region extends for another map unit; within it are genes related to those present in the *H2* region. In molecular terms, this "small segment" of the chromosome is sizable; the 1.3 map units together potentially represent >2500 kb of DNA. The size of the human locus is comparable; it contains similar functions, although not in the identical order.

Class I genes code for **transplantation antigens,** proteins present on every cell of the mammal. As their name suggests, these proteins are responsible for the rejection of foreign tissue, which is recognized as such by virtue of its particular array of transplantation antigens. In the immune system, their presence on certain (cytotoxic) T lymphocytes plays a role in killing target cells. The K, D, L, and R regions all code for proteins of this type; these functions are defined serologically (by their antigenic properties). Each mouse strain has one of several possible alleles for each of these functions.

The class II genes code for proteins found on the surfaces of both B and T lymphocytes as well as macrophages. These proteins are involved in the communication between cells that is necessary to execute the immune response. The I region of the class II genes is divided into subregions A, B, J, E, and C (not shown on the map).

The class III genes code for **complement proteins.** This part of the locus is also known as the S region; S stands for serum, indicating that the proteins are components of the serum. Their role is to interact with antibody-antigen complexes to cause the lysis of cells.

To the right of the *H2* region lie the *Qa* and *Tla* loci, which code for proteins found on hematopoietic cells. The proteins are described as differentiation antigens, because each is found only on a particular subset of the blood cells, presumably related to their function. Since these proteins are structurally related to the class I H2 proteins, we may regard these genes as extending the region of the chromosome devoted to functions concerned with the development of lymphocytes and macrophages.

The transplantation antigens are transmembrane proteins consisting of two chains with the structure illustrated in **Figure 37.18.**

One chain is the 12,000-dalton β2 microglobulin (coded by a single gene located on another chromosome). This component is needed for the dimeric protein to reside at the cell surface.

The transmembrane component is a 45,000-dalton chain coded at the histocompatibility locus. The protein has three **external domains** (each about 90 amino acids long, one of which interacts with β2 microglobulin), a **transmembrane region** of about 40 residues,

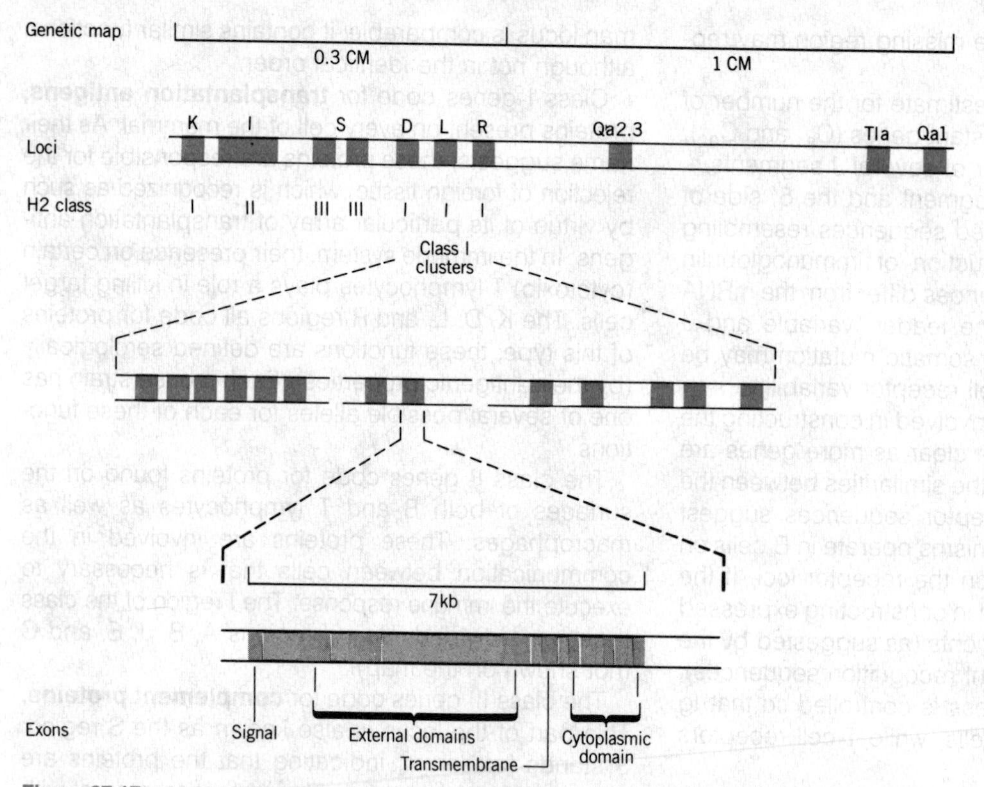

Figure 37.17

The histocompatibility locus contains several genetic regions; class I genes are organized in large structures; the structure of the individual gene corresponds with the structure of the protein.

and a short **cytoplasmic domain** of about 30 residues that resides within the cell.

The organization of class I genes coincides with the protein structure. The first exon codes for a signal sequence (cleaved from the protein during membrane passage). The next three exons code for each of the external domains. The fifth exon codes for the transmembrane domain. And the last three rather small exons together code for the cytoplasmic domain. The only difference in the genes for human transplantation antigens is that their cytoplasmic domain is coded by only two exons.

The class I mouse genes reside in clusters. A total of 36 genes lying over some 850 kb of DNA have been identified so far (in the form of individual clusters that have yet to be connected). The largest individual group has 7 genes spread over 190 kb. All of the genes in each cluster are oriented in the same direction; adjacent genes tend to be more closely related, which suggests their origin by ancestral tandem duplications.

Differences in the restriction patterns of the DNA of different strains of mice can be used to relate the locations of the individual clusters with the genetic loci. (Essentially this involves using the restriction differences as markers in genetic mapping.) By this means, most of the class I genes have been mapped to the *Qa* and *Tla* loci; only 5 of the genes map in the *H2* cluster.

This result implies that class I genes are present at both ends of the vast cluster; the class II and class III genes either evolved within the cluster or were translocated to their present positions. Most of the polymorphism in individual genes occurs in those of the *H2* locus. Variation in the number of genes between different mouse strains seems to occur largely in the *Qa* and *Tla* loci.

What is responsible for generating the high degree

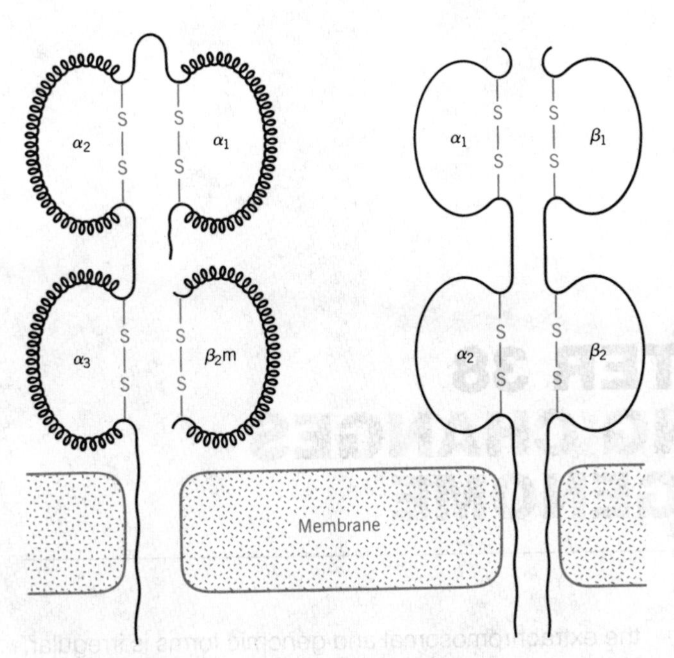

Figure 37.18
Class I and class II histocompatibility antigens have a similar structure. Class I antigens consists of a single (α) polypeptide, with three external domains, that interacts with β_2 microglobulin. Class II antigens consist of two (α and β) polypeptides, whose interaction generates a similar overall structure.

of polymorphism in these genes? Most of the sequence variation between alleles occurs in the first and second external domains, sometimes taking the form of a cluster of base substitutions in a small region. One mechanism involved in their generation may be gene conversion between class I genes.

Pseudogenes are present as well as functional genes; at present we have some way to go before estimating the total number of active genes in the region.

The exon coding for the third external domain of the class I genes is highly conserved relative to the other exons. The conserved domain probably represents the region that interacts with β2 microglobulin, which explains the need for constancy of structure. This domain also exhibits homologies with the constant region domains of immunoglobulins.

Like the class I genes, the class II and class III genes also are interrupted, with the exons related to protein domains. There are fewer genes in these classes, ~10 each.

The gene for β2 microglobulin has been identified. It has four exons, the first coding for a signal sequence, the second for the bulk of the protein (from amino acids 3 to 95), the third for the last four amino acids and some of the nontranslated trailer, and the last for the rest of the trailer.

The length of β2 microglobulin is similar to that of an immunoglobulin gene; there are certain similarities in amino acid constitution; and there are some (limited) homologies of nucleotide sequence between β2 microglobulin and Ig constant domains or type I gene third external domains. It is possible that all the groups of genes that we have discussed in this chapter are descended from a common ancestor that coded for a primitive domain.

FURTHER READING

An overall view of the mechanisms involved in generating Ig diversity has been given by **Tonegawa** (*Nature* **302,** 575–581, 1983). A deletion model for V–J joining was presented by **Max, Seidman, & Leder** (*Proc. Nat. Acad. Sci. USA* **76,** 3450–3454, 1979); a model based on sister chromatid exchange was proposed by **Van Ness et al.** (*Proc. Nat. Acad. Sci. USA* **79,** 262–266, 1982). The structure of the heavy chain genes was elucidated by **Sakano et al.** (*Nature* **286,** 676–683, 1980); the role of consensus sequences in V-(D)-J joining was analyzed by **Sakano et al.** (*Nature* **286,** 676–683, 1980) and **Early et al.** (*Cell* **19,** 981–992, 1980). The nature of rearrangements and allelic exclusion has been reviewed by **Early & Hood** (*Cell* **24,** 1–3, 1981). The connection between translocations involving the Ig loci and tumor formation has been reviewed by **Klein** (*Cell* **32,** 311–315, 1983) and **Perry** (*Cell* **33,** 647–649, 1983). Class switching was reviewed by **Davis, Kim, & Hood** (*Cell* **22,** 1–2, 1980) and **Shimizu & Honjo** (*Cell* **36,** 301–303, 1984). Alternate processing of RNA was reported by **Early et al.** (*Cell* **20,** 313–319, 1980) and **Maki et al.** (*Cell* **24,** 353–365, 1981). Somatic mutation has been reviewed by **Baltimore** (*Cell* **26,** 295–296, 1981). T-cell receptor sequences have been reported by **Hedrick et al.** (*Nature* **308,** 153–158, 1984) and **Siu et al.** (*Cell* **37,** 393–401, 1984). The organization of histocompatibility genes was reviewed by **Steinmetz & Hood** (*Science* **222,** 727–733, 1983).

CHAPTER 38
ENGINEERING CHANGES IN THE GENOME

Pressing on the discovery of transposition and other rearrangements of DNA is the knowledge that DNA sequences are surprisingly adjustable. Sequences may be moved within a genome, modified, or even lost, as a natural event; or they may be introduced into cells by experimental means.

We have seen that reorganization of particular sequences occurs in the immune system; although this is the only example so far characterized in higher animals, examples of rearrangement or loss of specific sequences are legion in the lower eukaryotes. Usually these changes involve somatic cells; the germ line remains inviolate.

Alterations in the relative proportions of components of the genome during somatic development occur to allow insect larvae to increase the number of copies of certain genes. The occasional amplification of genes in cultured mammalian cells is indicated by our ability to select variant cells that have an increased copy number of some gene. Initiated within the genome, the amplification event can create additional copies of the gene that survive in either intrachromosomal or extrachromosomal form.

Extraneous DNA can be introduced into eukaryotic cells, where it survives and may be expressed. In some cases it remains extrachromosomal; in others it is integrated into the genome. The relationship between the extrachromosomal and genomic forms is irregular, depending on chance and to some degree unpredictable events, rather than resembling the interchange between free and integrated forms of bacterial plasmids.

Yet, however accomplished, the process may lead to stable change in the genome; following its injection into animal eggs, DNA may even be incorporated into the genome and inherited thereafter as a normal component, sometimes continuing to function. The ability to introduce specific genes that function in an appropriate manner could become a major medical technique for curing genetic diseases.

Considerable manipulation of DNA sequences therefore is achieved both in authentic situations and by experimental fiat. We are only just beginning to work out the mechanisms that permit the cell to respond to selective pressure by changing its bank of sequences or that allow it to accommodate the intrusion of additional sequences.

TISSUE-SPECIFIC VARIATIONS OCCUR IN THE *DROSOPHILA* GENOME

The content of the genome remains unchanged in most somatic cells. However, we now know of some situations in which changes are made in the relative *proportions* of certain sequences; the best-characterized examples occur during larval development in insects. The adjustment takes the form of underreplicating or overreplicating specific sequences.

Underreplication occurs during development of the polytene tissues of the fly. In these tissues, the giant chromosomes described in Chapter 28 are formed by multiple successive duplications of the original (synapsed) diploid set.

About a quarter of the genome of *D. melanogaster* is contained in the heterochromatic regions of the chromosomes, which aggregate to form a chromocenter (a large part of which consists of satellite DNA sequences). The relative amount of heterochromatin is much less in polytene cells compared with diploid cells. The explanation is that this part of the genome has failed to duplicate during the polytenization of the euchromatic DNA sequences. Measurements of satellite DNA content suggest that these sequences duplicate no or very few times, compared with the 9 duplications of euchromatic DNA in the salivary gland.

The rRNA genes also are underreplicated in polytene tissues. The region of rDNA passes through only 6–7 duplications (actually reaching the same final level whether only one or both of the nucleolar organizers are present. This type of dosage control may be specific for rDNA.)

The genome of *D. melanogaster* can therefore be divided into three types of region according to the control of replication in polytene tissues. A natural assumption is that the origins for replication in each type of region control the number of initiation events. Origins in satellite DNA may fail to be recognized; origins in rDNA stop functioning in response to some feedback from the number of rRNA genes; and general euchromatic origins continue initiating to the bitter end.

Differential amplification of particular protein-coding sequences also occurs in insects. In the salivary gland of *Rhyncosciara* and other Sciarids, **DNA puffs** are generated. Their appearance is superficially similar to the puffs that represent active bands of Dipteran polytene chromosomes, but the Sciarid puffs contain locally amplified DNA as well as RNA. The amplified sequences may have undergone up to 4 additional initiations of replication, increasing the copy number 16-fold.

Insight into possible mechanisms for differential amplification is provided by the state of the chorion genes in the development of *D. melanogaster*. The proteins that make up the chorion (eggshell) are synthesized and secreted by the polyploid ovarian follicle cells. Insect chorion genes tend to be clustered, and two groups have been identified in *D. melanogaster*. Prior to their expression in the follicle cell, chorion genes on the X chromosome are amplified by up to 16-fold (4 additional doublings), while the genes on chromosome III are amplified up to 60 times (6 additional doublings).

In each case, amplification extends for a distance of about 45–50 kb on either side of the chorion genes. The level of maximum amplification represents a plateau of about 20 kb surrounding the chorion genes. The extent of amplification shows a gradient of decline on either side of the plateau.

What is responsible for the amplification? The gradient of decline suggests that the endpoints of the amplified regions in individual molecules are heterogeneous. This could be explained by the model illustrated in **Figure 38.1.**

Multiple initiations of bidirectional replication occur at an origin in the center of the region. The replication forks progress for distances varying from 10 to 50 kb. This model views the entire amplified region as the replicon, present in multiple but only partially replicated copies. (The relationship between underreplicated and more-replicated regions in salivary gland chromosomes could be explained by a similar model, in which the junctions between regions are represented by static replication forks.)

Does the amplified region contain a *cis*-acting origin responsible for replication? In this case, amplification should be abolished for any chromosome whose origin is mutated. In a heterozygote, the failure should apply only to the mutated chromosome. And if the origin is translocated elsewhere, it should sponsor the amplification of whatever replicon it finds itself in.

The properties of the *ocelliless* mutant conform with these predictions. This mutation causes complex changes in the phenotype; it is an inversion of about three bands, containing the region of the chorion genes on the X chromosome. The chorion genes lie within 3 kb of the left end of the inversion.

The inversion causes a major change in the pattern

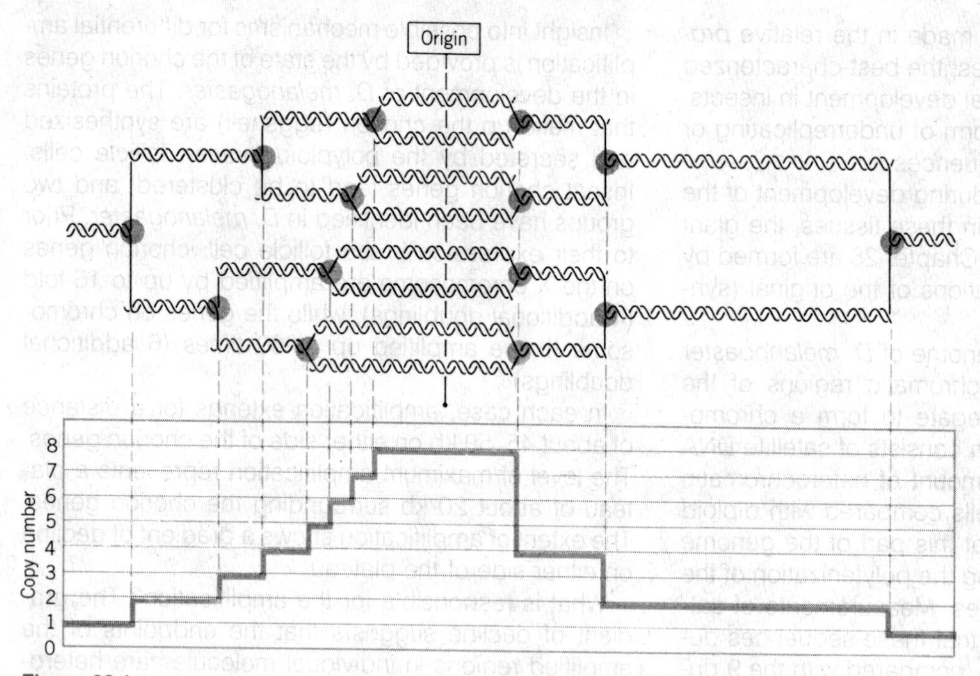

Figure 38.1
Amplification of a local region could be accomplished by multiple initiations within a single replicon.

Each replicated region is joined to the flanking unreplicated regions by static replication forks. If the forks cease to move at imprecisely determined positions, they generate a gradual gradient, as indicated on the left. If several forks terminate coordinately at a fixed site, they generate sudden decreases in the level of amplification, as indicated on the right.

of amplification. **Figure 38.2** compares the regions amplified in wild-type and *ocelliless* files. Of the original amplified region, the 40 kb on the left side of the inversion breakpoint fail to be amplified at all. The 50 kb on the other side are amplified, although at a reduced level, reaching only about half the usual maximum. (The reason for the reduction in level is unknown.) Amplification now spreads into new regions, as seen by the increase in copy number of sequences on the right that are not amplified in the wild type.

These results suggest that an origin able to sponsor bidirectional amplification lies in the vicinity of the chorion genes. The extent of the amplified region remains similar whether the origin lies in its usual position in the wild type or is transferred to a new position by the *ocelliless* inversion.

The origin at which amplification occurs is clearly more complex in nature than the normal origins of replicons. The amplification origin must respond to some *trans*-acting factor in a tissue- and time-dependent manner. It engages in multiple cycles of initiation, which is different from the single round of replication initiated in each replicon in a normal cell cycle. The amplification origin appears to be fairly large, >1 kb, as seen by the ability only of fragments larger than this to sponsor amplification when incorporated into other loci (by P element-mediated transposition).

The classic description of the type of effect produced by the inversion is encompassed by the phenomenon of the **position effect.** This term describes the influence of location on expression of a gene, as seen by the change in activity resulting from a translocation.

A well-known example is the inactivity that results when a gene is translocated from its wild-type position in euchromatin to a new location close to heterochro-

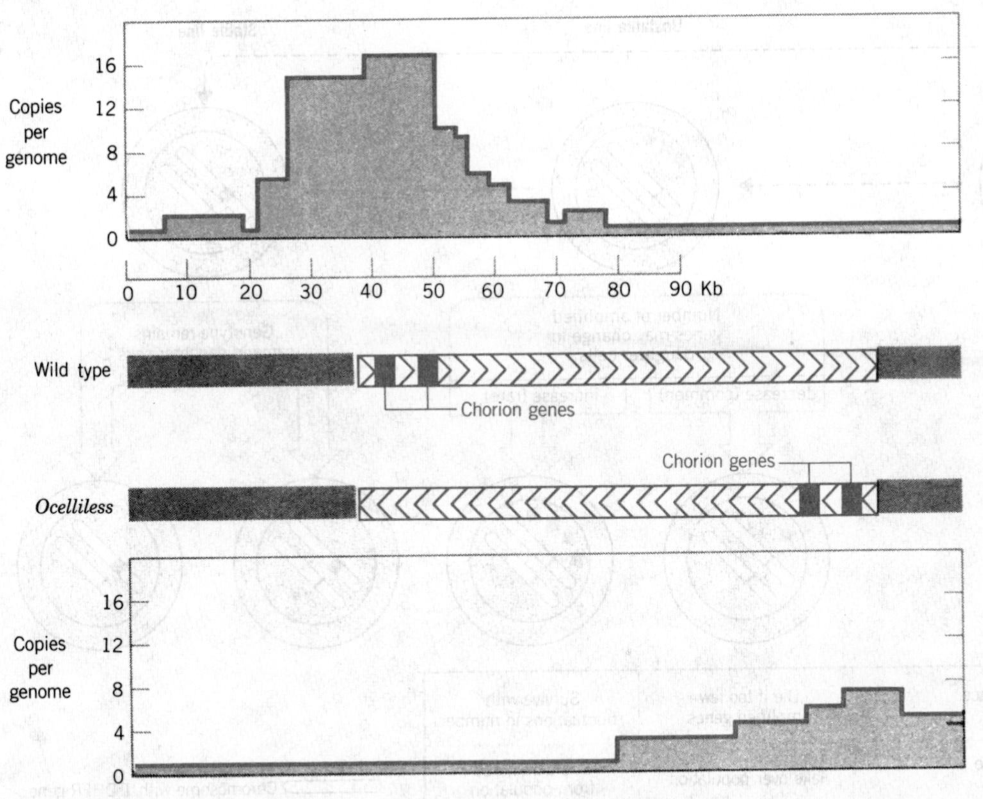

Figure 38.2
The *ocelliless* inversion changes the region of DNA that is amplified.
The left end of the inversion lies within 3 kb of the left boundary of the chorion genes; the right end is not known, but for the purposes of illustration the inversion is assumed to be about 85 kb long.

matin. Although the gene itself is unchanged, its activity is influenced by its surroundings.

In the case of *ocelliless,* the inversion causes a failure in the amplification of some genes, while others are amplified instead. The dosage effects produce complex changes in phenotype, without any alteration of the genes themselves.

SELECTION OF AMPLIFIED GENOMIC SEQUENCES

The eukaryotic genome has the capacity to accommodate additional sequences of either exogenous or endogenous origin. Whether added to cells or originating in the chromosomes, DNA sequences may give rise to multiple copies that survive as a tandem array in either extrachromosomal or chromosomal location. In extrachromosomal form, the additional material is inherited in an irregular way (it does not segregate evenly at division in the manner of an authentic plasmid). In the chromosomal form, however, the material has been integrated into the resident genome and becomes a component of the genotype.

The **amplification** of endogenous sequences is provoked by selecting cells for resistance to certain agents. The same general pattern of flexibility is seen in both transfection and amplification, which very likely involve overlapping mechanisms. We know nothing about the enzymatic activities involved in this manipulation, but we suppose that they are engaged in related activities in the normal course of perpetuating DNA.

The best-characterized example of amplification re-

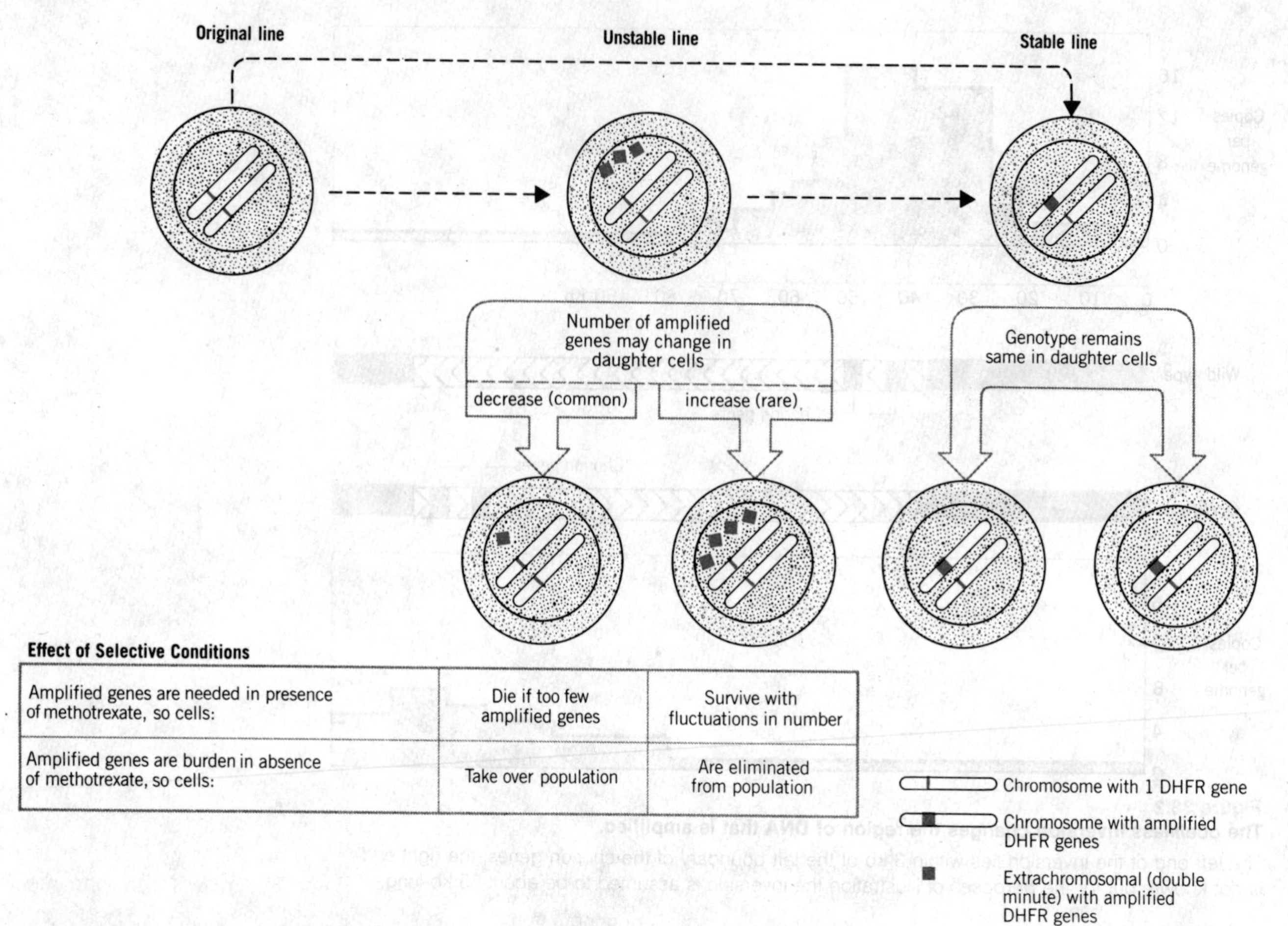

Amplified genes are needed in presence of methotrexate, so cells:	Die if too few amplified genes	Survive with fluctuations in number
Amplified genes are burden in absence of methotrexate, so cells:	Take over population	Are eliminated from population

Figure 38.3
The *dhfr* gene can be amplified to give additional copies that are extrachromosomal (unstable) or chromosomal (stable).

sults from the addition of **methotrexate** (mtx) to any one of several cultured cell lines. The reagent blocks folate metabolism. Resistance to it may be conferred by mutations that change the activity of the enzyme dihydrofolate reductase (DHFR). As an alternative to change in the enzyme itself, the amount of enzyme may be increased. The cause of this increase is an amplification of the number of *dhfr* structural genes. Amplification occurs at a frequency greater than the spontaneous mutation rate, generally ranging from 10^{-4} -10^{-6}.

Although amplification represents a relatively rarely used mechanism for acquiring resistance to a toxic reagent, it is not unique to DHFR. Resistance to an inhibitor of the enzyme aspartyltranscarbamylase also is conferred by an amplification of the corresponding structural gene. Cells resistant to the heavy metal cadmium have an increased copy number of the gene coding for the detoxifying protein metallothionein.

A common feature in these (and other) systems is that highly resistant cells are not obtained in a single step, but instead appear when the cells are adapted to gradually increasing doses of the toxic reagent. Thus gene amplification may require several stages.

The number of *dhfr* genes in a cell line resistant to methotrexate varies from 40 to 400, depending on the stringency of the selection and the individual cell line. The *mtx*^{r} lines fall into two classes, distinguished by

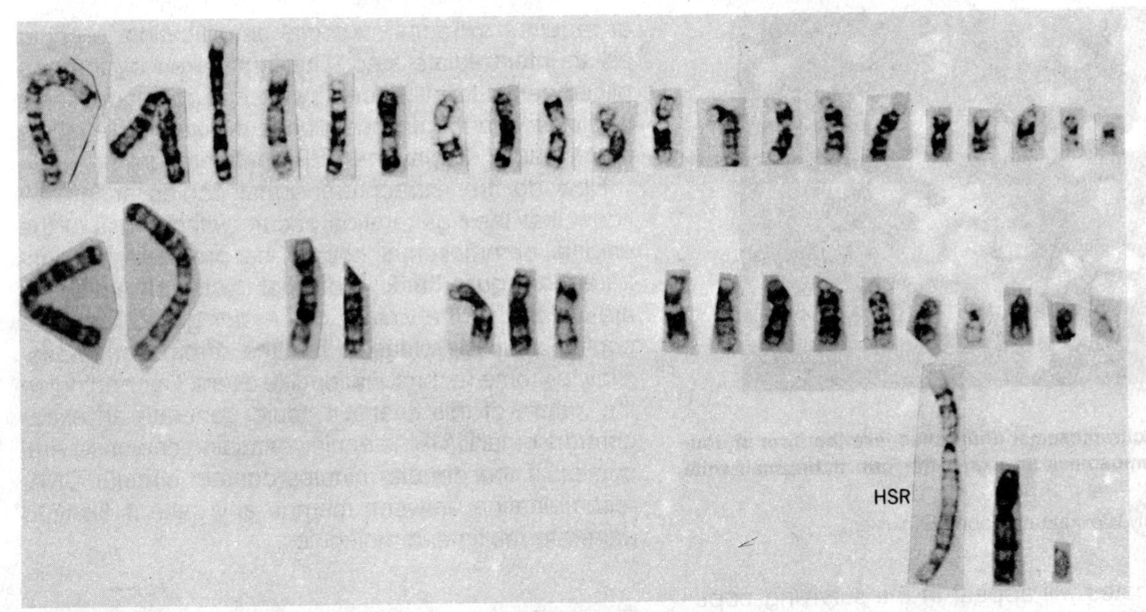

Figure 38.4
Amplified copies of the *dhfr* gene produce a homogeneously staining region (HSR) in the chromosome.
Photograph kindly provided by Robert Schimke.

their response when cells are grown in the absence of methotrexate, removing the former pressure for high levels of DHFR activity.

In **stable** lines, the amplified genes are retained. In **unstable** lines, the amplified genes are at least partially lost when the selective pressure is released. The cause of this difference is the condition of the amplified *dhfr* genes in the two lines. The situation responsible for this state of affairs is summarized in **Figure 38.3.**

In a stable *mtx*ʳ line, the amplified *dhfr* genes are chromosomal, occupying the usual site of *dhfr* on one of the chromosomes. Usually the other chromosome retains its normal single copy of *dhfr*. Thus amplification generally occurs at only one of the two *dhfr* alleles; and increased resistance to methotrexate is accomplished by further increases in the degree of amplification at this locus. (Sometimes the chromosome with the amplified *dhfr* locus itself is duplicated, presumably through nondisjunction at mitosis.)

Gene amplification has a visible effect on the chromosome; the locus can be visualized in the form of a **homogeneously staining region (HSR).** An example is shown in **Figure 38.4.** The HSR takes its name from the presence of an additional region that lacks any chromosome bands after treatments such as G-banding (see, for example, Figure 28.10). This change suggests that some region of the chromosome between bands has undergone an expansion.

In unstable cell lines, no change is seen in the chromosomes carrying *dhfr*. However, large numbers of elements called **double-minute chromosomes** are visible. An example is shown in **Figure 38.5.** In a typical cell line, each double minute carries 2–4 *dhfr* genes. The double minutes appear to be self-replicating; but they lack centromeres. As a result, they do not attach to the mitotic spindle and therefore segregate erratically, frequently being lost from the daughter cells. Notwithstanding their name, the actual status of the double minutes is regarded as extrachromosomal.

The irregular inheritance of the double minutes explains the instability of methotrexate resistance in these lines. Double minutes will be lost continuously during cell divisions; and in the presence of methotrexate, cells with reduced numbers of *dhfr* genes will die. Only those cells that have retained a sufficient number

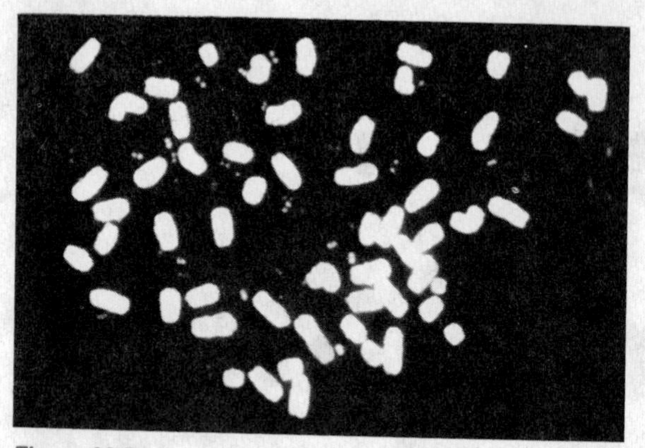

Figure 38.5

Amplified extrachromosomal *dhfr* genes take the form of double-minute chromosomes, as seen in the form of the small white dots.

Photograph kindly provided by Robert Schimke.

of double minutes will appear in the surviving population.

The presence of the double minutes reduces the rate at which the cells proliferate. Thus when the selective pressure is removed, cells lacking the amplified genes have an advantage; they generate progeny more rapidly and soon take over the population. This explains why the amplified state is retained in the cell line only so long as cells are grown in the presence of methotrexate.

Because of the erratic segregation of the double minutes, increases in the copy number can occur relatively quickly as cells are selected at each division for progeny that have gained more than their fair share of the *dhfr* genes. Cells with greater numbers of copies are found in response to increased levels of methotrexate. The behavior of the double minutes explains the stepwise evolution of the mtx^r condition and the incessant fluctuation in the level of *dhfr* genes in unstable lines.

Both stable and unstable lines are found after long periods of selection for methotrexate resistance. What is the initial step in gene amplification? After short periods of selection, most or all of the resistant cells are unstable. The formation of extrachromosomal copies clearly is a more frequent event than amplification within the chromosome.

We do not know whether intrachromosomal amplification simply proceeds less often as a de novo step or requires extrachromosomal amplification to occur as an intermediate step. The form taken by the amplified genes is influenced by the cell genotype; some cell lines tend to generate double minutes, while others more readily display the HSR configuration.

How do the extrachromosomal copies arise? We know that their generation occurs without loss of the original chromosomal copy. One possibility is illustrated in **Figure 38.6.** Additional cycles of replication are initiated in the vicinity of the *dhfr* gene. The extra copies could be released from the chromosome, possibly by some recombination-like event. Depending on the nature of this event, it could generate an extrachromosomal DNA molecule containing one or several copies. If the double minutes contain circular DNA, recombination between them in any case is likely to generate multimeric molecules.

Figure 38.6

Extrachromosomal copies of DNA could be generated by reinitiation of replication followed by nonhomologous recombination between the replicas.

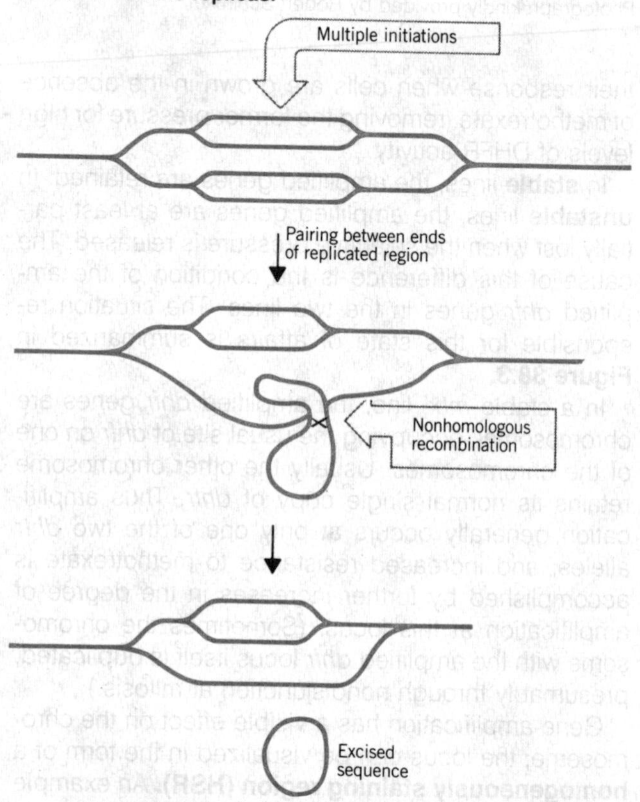

The origin of amplified chromosomal copies may be more difficult to explain. The HSR consists of a large number of tandemly repeated units. One problem in supposing that they arise by integration of double-minute sequences is that the number of chromosomal repeats is many times greater than the number of repeats in the individual double minutes. The large number of integrated copies is also an impediment to constructing models for a strictly intrachromosomal amplification event.

The amplified region is much longer than the *dhfr* gene itself. The gene has a length of ~31 kb, but the average length of the repeated unit is 500–1000 kb in the chromosomal HSR. The amount of DNA contained in a double minute seems to lie in a range from 100 to 1000 kb.

The extent of the amplified region is different in each cell line. We might speculate that it always contains the origin of the replicon that contains the amplified genes, but has variable termini.

The same amplification events presumably occur in the absence of methotrexate, but the products are lost from the cell population in the absence of any selective pressure. Why is this phenomenon seen with so few genes? Perhaps it occurs only with genes that lie in an appropriate location relative to a replicon or when appropriate repetitive sequences occur at locations that permit recombination to excise the region.

Some information about the events involved in perpetuating the double minutes is given by an unstable cell line whose amplified genes code for a mutant DHFR enzyme. The mutant enzyme is not present in the original (diploid) cell line (so the mutation must have arisen at some point during the amplification process). Despite variations in the number of amplified genes, these cells display *only* the mutant enzyme. Thus the wild-type chromosomal genes cannot be continuously generating large numbers of double minutes anew, because these amplified copies would produce normal enzyme.

Once amplified extrachromosomal genes have arisen, therefore, *changes in the state of the cell are mediated through these genes and not through the original chromosomal copies.* When methotrexate is removed, the cell line loses its double minutes in the usual way. On reexposure to the reagent, *normal* genes are amplified to give a new population of double minutes. This shows that none of the extrachromosomal copies of the mutant gene had integrated into the chromosome.

Another striking implication of these results is that the double minutes of the mutant line carried *only* mutant genes—so if there is more than one *dhfr* gene per double minute, all must be of the mutant type. This suggests that multicopy double minutes can be generated from individual extrachromosomal genes.

EXOGENOUS SEQUENCES CAN BE INTRODUCED BY TRANSFECTION

The procedure for introducing exogenous donor DNA into recipient cells is called **transfection.** Transfection experiments began with the addition of preparations of metaphase chromosomes to cell suspensions. The chromosomes are taken up rather inefficiently by the cells and give rise to unstable variants at a low frequency. Intact chromosomes rarely survive the procedure; the recipient cell usually gains a fragment of a donor chromosome (which is unstable because it lacks a centromere). Rare cases of stable lines may have resulted from integration of donor material into a resident chromosome.

Similar results are obtained when purified DNA is added to a recipient cell preparation. However, with purified DNA it is possible to add particular sequences instead of relying on random fragmentation of chromosomes. Transfection with DNA yields stable as well as unstable lines, with the former relatively predominant. (These experiments are directly analogous to those performed in bacterial transformation, as indicated in Figure 1.9, but are described as transfection because of the historical use of transformation to describe changes that allow unrestrained growth of eukaryotic cells.)

The low frequencies of transfection make it necessary to use donor markers whose presence in the recipient cells can be selected for. Note that the transfected sequence is expressed. Most transfection experiments have used markers representing readily assayed enzymatic functions, but in principle, any marker that can be selected can be assayed. This allows the isolation of genes responsible for morphological phenomena. For example, as we have already mentioned, transfected cells can be selected for acquisition of the transformed (tumorigenic) phenotype. Then we can identify the DNA responsible for conferring the phenotype. This type of protocol has led to the isolation of several cellular *onc* genes.

Cotransfection with more than one marker has proved informative about the events involved in transfection and has extended the range of questions that we can ask with this technique. When tk^- cells are transfected with a DNA preparation containing both a purified tk^+ gene and the φX174 genome, *all the* tk^+ *transformants have both donor sequences*. This is a useful observation, because it allows unselected markers to be introduced routinely by cotransfection with a selected marker.

The arrangement of *tk* and φX174 sequences is different in each transfected line, but remains the same during propagation of that line. Often multiple copies of the donor sequences are present, the number varying with the individual line. Revertants lose the φX174 sequences together with *tk* sequences. Amplification of transfected sequences under selective pressure results in the increase of copy number of all donor sequences pari passu. Thus the two types of donor sequence become physically linked during transfection and suffer the same fate thereafter.

To perform a transfection experiment, the mass of DNA added to the recipient cells is increased by including an excess of "carrier DNA," a preparation of some other DNA (often from salmon sperm). Transfected cells prove to have sequences of the carrier DNA flanking the selected sequences on either side. Transfection therefore appears to be mediated by a large unit, consisting of a linked array of all sequences present in the donor preparation.

Since revertants for the selected marker lose all of this material, it seems likely that the *transfected cell gains only a single large unit.* The unit may be formed by a concatemeric linkage of donor sequences in a reaction that is rapid relative to the other events involved in transfection. This transfecting package may be of the order of 1000 kb in length.

Because of the size of the donor unit, we cannot tell from blotting experiments whether it is physically linked to recipient chromosomal DNA (the relevant end fragments are present in too small a relative proportion). From the characteristics of the transfected lines, it seems possible that the first stage is the establishment of an unstable extrachromosomal unit, followed by the acquisition of stability via integration.

In situ hybridization can be used to show that transfected cells have donor material integrated into the resident chromosomes. Any given cell line has only a single site of integration; but the site is different in each line. Probably the selection of a site for integration is a random event; sometimes it is associated with a gross chromosomal rearrangement.

Some interesting results obtained by cotransfection with two selectable markers are summarized in **Figure 38.7.** An $APRT^-$ TK^- recipient cell line was transfected with DNA carrying the *aprt* and *tk* genes. The

Figure 38.7
When transfected genes are amplified, the entire cluster responds coordinately to further selective pressure.

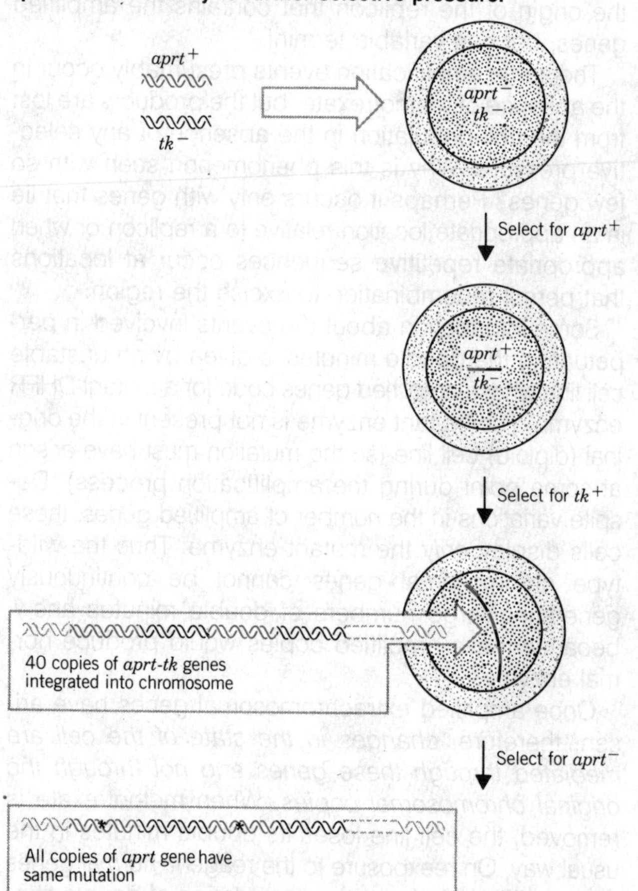

aprt gene was the active allele *aprt*$^+$, which codes for the enzyme adenosine phosphoribosyltransferase. The *tk* gene was an inactive allele, retaining the coding region for the enzyme thymidine kinase, but lacking the promoter. The transfected cells were selected for APRT$^+$ function, but both donor sequences were acquired.

Now these APRT$^+$ TK$^-$ cells were selected for their ability to provide *tk*$^+$ function. Its acquisition depends on two events. The *tk* must be transcribed from some site substituting for its promoter; and because this achieves only a low level of expression, the number of copies must be amplified. The selected cells become APRT$^+$ TK$^+$ and they have about 40 copies of the *aprt-tk* transfected unit, integrated at a single chromosomal site.

Finally these cells were selected for the APRT$^-$ TK$^+$ phenotype. Thus they are required to eliminate the expression of the multiple copies of the *aprt*$^+$ gene without affecting the expression of the *tk* gene. The interesting feature is that *all* copies of the *aprt* gene in these cells acquired the same negative mutation. This result implies the existence of some mechanism that allows the entire gene cluster to be regenerated from a single member or corrected to reflect the sequence of one member; apparently this happens within a single cell generation.

TRANSFECTED DNA CAN ENTER THE GERM LINE

An exciting development of transfection techniques is their application to introduce genes into animals. An animal that gains new genetic information from the addition of foreign DNA is described as **transgenic.**

The first questions we ask about any transgenic animal are how many copies it has of the foreign material, where these copies are located, and whether they are present in the germ line and inherited in a Mendelian manner.

An important issue that can be addressed by experiments with transgenic animals concerns the independence of genes and the effects of the region within which they reside. If we take a gene, including the flanking sequences that contain its known regulatory elements, will it be expressed in the same way as usual irrespective of its location in the genome? In other words, do the regulatory elements function independently, or is gene expression (also) controlled by some other effect, for example, location in an appropriate chromosomal domain?

The answers to these questions are by no means certain as yet, and may depend on the organism and vary with the particular gene. We now have data on several strains of transgenic mice and fruit flies.

An approach that has proved successful with the mouse is summarized in **Figure 38.8.** Plasmids carrying the gene of interest are injected into the germinal vesicle (nucleus) of the oocyte or into the pronucleus of the fertilized egg. The egg is implanted into a pseudopregnant mouse. After birth, the recipient mouse can be examined to see whether it has indeed gained the foreign DNA, and, if so, whether it is expressed.

One series of experiments with the mouse has used a "fusion gene" that has the promoter for metallothionein linked to the thymidine kinase coding sequence. The MT promoter is derived from a natural mouse gene and can be tested for its ability to respond to the customary induction by heavy metals or glucocorticoids; the response can be assayed by measuring thymidine kinase activity.

In a series of 69 injected eggs, 10 of the resulting mice carried the injected sequence as shown by the analysis of kidney DNA. Usually, multiple copies of the plasmid appear to have been integrated in tandem into a single chromosomal site. The number of copies varies from 1 to 150. They are inherited by the progeny of the injected mouse as expected of a Mendelian locus.

Seven of the mice expressed the thymidine kinase gene in liver and kidney (tissues in which MT usually is active). Expression did not occur in brain, a tissue in which the gene usually does not function. Expression is induced by the addition of heavy metals, although not by glucocorticoids.

In the progeny of the injected mice, expression of the donor gene is extremely variable; it may be extinguished entirely, reduced somewhat, or even increased. Even in the original parents, the level of gene expression does not correlate with the number of tandemly integrated genes. Probably only a few of the genes are active. (Also note that if all the additional genes were active, it would be surprising if the normal

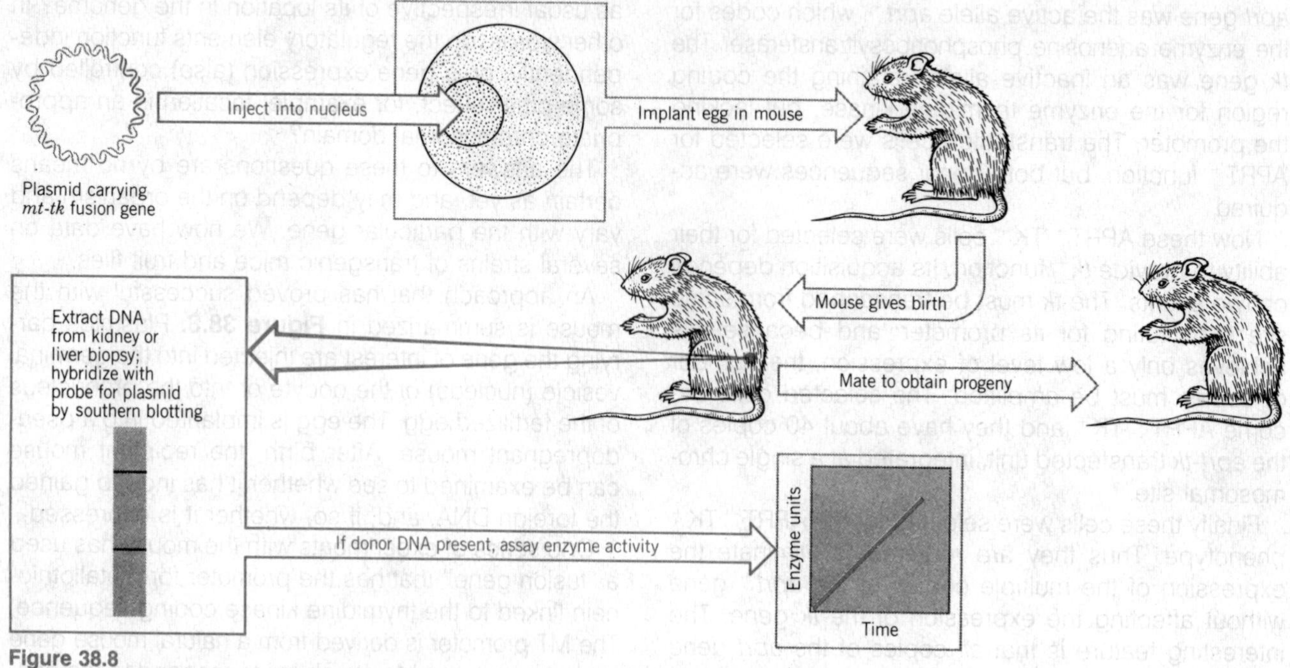

Figure 38.8
Transfection can introduce DNA into the germ lines of animals.

regulatory response were retained, since the large number of promoters might well dilute out the regulatory molecules.)

What could be responsible for the variation in gene expression? One possibility that has often been discussed for transfected genes, and that applies also to integrated retroviral genomes, is that the site of integration may be important. Perhaps a gene is expressed if it integrates within an active domain, but not if it integrates in another area of chromatin. Another possibility is the occurrence of epigenetic modification. Some of the transfected genes in the mouse are methylated. (Since the donor DNA lacked methyl groups, this observation implies the existence of an enzyme able to engage in de novo methylation.) Perhaps changes in the pattern of methylation might be responsible for changes in activity. Alternatively, the genes that happened to be active in the parents may have been deleted or amplified in the progeny.

Another gene that has been used to create transgenic mice is chicken transferrin. The transferrin gene tends to show a partial preference for expression in the appropriate tissue (liver), raising the possibility that regulatory signals have been conserved between chicken and mouse.

In a series of experiments with rabbit β-globin, there was no specificity for expression in blood cells, but one mouse line transcribed the gene in skeletal muscle and another did so in testis. In both cases, the pattern was inherited. This raises the possibility that the gene was expressed because it fell within the orbit of some other regulatory network. Given the unlikelihood of insertion immediately next to a promoter, this suggests the possibility that overall chromosome structure influences expression of the gene.

A particularly striking example of the effects of an injected gene is provided by a strain of transgenic mice derived from eggs injected with a fusion consisting of the MT promoter linked to the rat growth hormone structural gene. Growth hormone levels in some of the transgenic mice were several hundred times greater than normal. The mice grew to nearly twice the size of normal mice, as can be seen from **Figure 38.9.**

The introduction of oncogene sequences can lead to tumor formation. Transgenic mice containing the

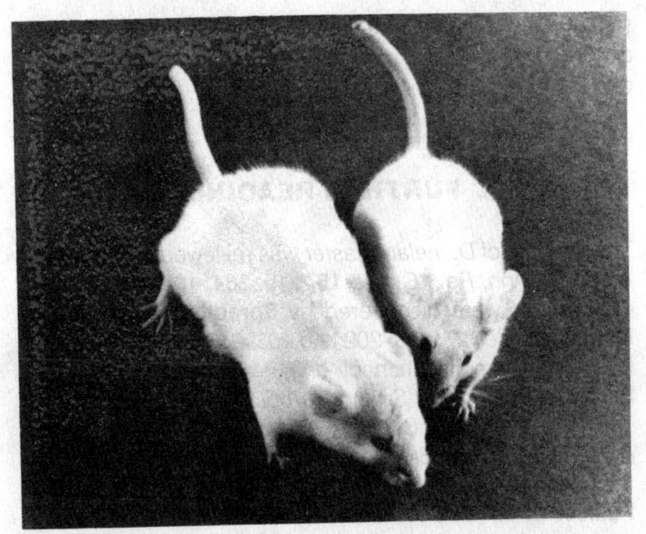

Figure 38.9
A transgenic mouse with an active rat growth hormone gene (left) is twice the size of a normal mouse (right).
Photograph kindly provided by Ralph Brinster.

SV40 early coding region and regulatory elements express the viral genes for large T and small t antigen only in some tissues, most often brain, thymus, and kidney. These mice usually die before reaching 6 months, as the result of developing a tumor in the brain; sometimes tumors are found also in thymus and kidney. In the tumor lines, the SV40 region is often amplified, by up to 20-fold.

Although SV40 is known to cause tumors in hamsters, it had not previously been thought to be oncogenic in mice. In the transgenic strain, however, the SV40 T/t antigens behave as the products of an integrated oncogene. These mice may provide an extremely useful model for investigating the origin of this type of cancer. Other oncogenes may be used to generate mice developing other cancers, thus making possible a range of model systems.

Some significant differences are seen when analogous experiments are performed with *D. melanogaster*. A system for adding DNA to the fly genome has been developed by taking advantage of the P element. The protocol is illustrated in **Figure 38.10.**

A defective P element carrying the gene of interest is injected together with an intact P element into preblastoderm embryos. The intact P element provides a transposase that recognizes not only its own ends but also those of the defective element. As a result, either or both elements may be inserted into the genome.

Only the sequences between the ends of the P DNA are inserted; the sequences on either side are not part of the transposable element. An advantage of this technique is that only a single element is inserted in any one event, so the transgenic flies usually carry only one copy of the foreign gene, a great aid in analyzing its behavior.

Several genes that have been introduced in this way all show the same behavior. They are expressed only in the appropriate tissues and at the proper times during development. These genes have been integrated into various random locations; and, even more strik-

Figure 38.10
Injecting *D. melanogaster* embryos with an active P element together with a defective P element carrying a gene creates transgenic flies that have a single, normally expressed copy of the gene.

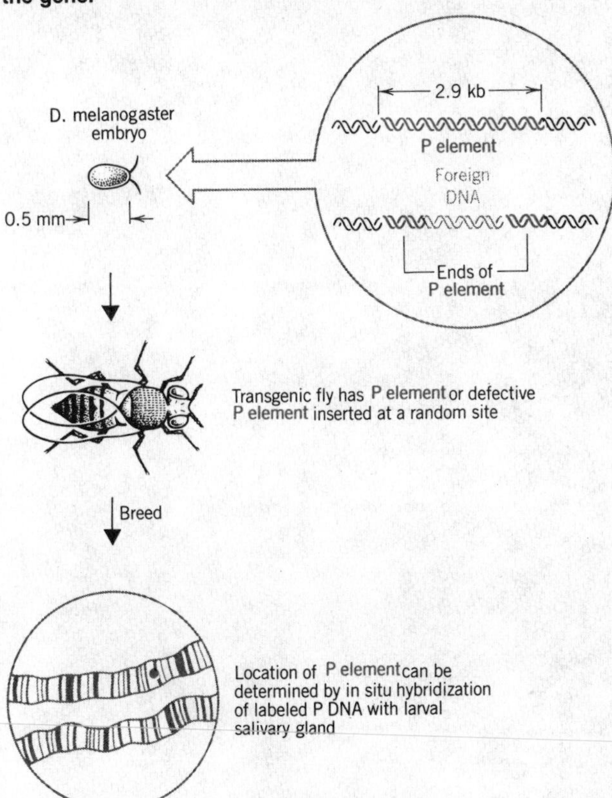

ingly, these features remain true for a gene integrated into heterochromatin. So in *D. melanogaster*, all the information needed to regulate gene expression may be contained within the gene locus itself, and can be relatively impervious to external influence.

With these experiments, we see the possibility of extending from cultured cells to animals the option of examining the regulatory features. The ability to introduce DNA into the genotype allows us to make changes in it, to add new genes that may have had particular modifications introduced *in vitro*. Thus it may become possible to delineate the features responsible for tissue-specific gene expression. Ultimately it may be possible to replace defective genes in the genotype.

We have come a long way from the concept that the gene is an isolated and fixed locus coding for a single protein. Genes may reside in huge clusters containing related sequences. They may be expressed in terms of alternate versions of a protein. They may be reconstructed by physical rearrangement of sequences during somatic development. New genes may even be added to the germ line by introducing DNA into eggs. The shibboleths of genetics continue to tumble.

FURTHER READING

The genome of *D. melanogaster* was reviewed by **Spradling & Rubin** (*Ann. Rev. Genet.* **15,** 219–264, 1981); the effects of *ocelliless* were discovered by **Spradling, Spradling, & Mahowald** (*Cell* **27,** 193–202 and 203–210, 1981). The events involved in amplification of *dhfr* have been reviewed by **Schimke et al.** (*Cold Spring Harbor Symp. Quant. Biol* **45,** 785–797, 1981) and (*Cell* **37,** 705–713, 1984) and by **Stark & Wahl** (*Ann. Rev. Biochem.* **53,** 447–491, 1984). The uses and insights offered by transfection have been reviewed by **Pellicer et al.** (*Science* **209,** 1414–1422, 1980). A series of experiments on injecting genes into mice have been reported by **Palmiter et al.** (*Cell* **29,** 701–710, 1982, *Nature* **300,** 611–615, 1982, *Cell* **37,** 367–379, 1984). P-element mediated transfection was developed by **Spradling & Rubin** (*Science* **218,** 341–353, 1982).

GLOSSARY

Abundance of an mRNA is the average number of molecules per cell.

Abundant mRNAs consist of a small number of individual species, each present in a large number of copies per cell.

Acceptor splicing site is the junction between the right end of an intron and the left end of an exon.

Acentric fragment of a chromosome (generated by breakage) lacks a centromere and is lost at cell division.

Allele is one of several alternate forms of a gene occupying a given locus on the chromosome.

Allelic exclusion describes the expression in any particular lymphocyte of only one of the alleles coding for the ex- [illegible]

Allosteric control refers to the ability of an interaction at one site of a protein to influence the activity of another site.

Alu family is a set of dispersed and related sequences, each ~300 bp long, in the human genome.

Alu-equivalent family is a set of sequences in any mammalian genome, related to the human Alu family.

α-Amanitin is a bicyclic octapeptide derived from the mushroom *Amanita phalloides* that inhibits transcription by certain eukaryotic RNA polymerases, especially RNA polymerase II.

Amber codon is the nucleotide triplet UAG, one of three *nonsense* codons that cause termination of protein synthesis.

Amber mutation describes any change in DNA that creates an amber codon at a site previously occupied by a codon representing an amino acid in some protein.

Amber suppressors are mutant genes that code for tRNAs whose anticodons have been altered so that they can respond to UAG codons as well as or instead of to their previous codons.

Aminoacyl-tRNA synthetases are enzymes responsible for covalently linking amino acids to the 2′- or 3′-OH position of tRNA.

Amplification refers to the production of additional copies of a chromosomal sequence, found as either intrachromosomal or extrachromosomal DNA.

[illegible] to direct the synthesis of RNA and is complementary to it.

Anticodon is a triplet of nucleotides in a constant position in the structure of tRNA that is complementary to the codon(s) in mRNA to which the tRNA responds.

Antigen is any molecule whose entry into an organism provokes synthesis of an antibody (immunoglobulin).

Antitermination proteins allow RNA polymerase to transcribe through certain terminator sites.

AP endonucleases make incisions in DNA on the 5′ side of either apurinic or apyrimidinic sites.

Apoinducer is protein that binds to DNA to switch on transcription by RNA polymerase.

att **sites** are the loci on a phage and the bacterial chromosome whose recombination integrates the phage into, or excises it from, the bacterial chromosome.

Attenuation is the regulation of termination of transcription involved in controlling expression of some bacterial operons.

Attenuator is the terminator sequence at which attenuation occurs.

Autogenous control describes the action of a gene product that either inhibits (negative autogenous control) or activates (positive autogenous control) expression of the gene coding for it.

Autoradiography detects radioactively labeled molecules by their effect in creating an image on photographic film.

Autosomes are all chromosomes except the sex chromosomes; a diploid cell has two copies of each autosome.

Backcross involves crossing an unknown genotype to a recessive homozygote so that the phenotypes of the progeny correspond directly to the chromosomes carried by the parent of unknown genotype.

Back mutation reverses the effect of a mutation that had inactivated a gene; thus it restores wild phenotype.

Bacteriophages are viruses that infect bacteria; often abbreviated as *phages*.

Balbiani ring is an extremely large puff at a band of a polytene chromosome.

Bands of *polytene chromosomes* are visible as dense regions that contain the majority of DNA; *bands of normal chromosomes* are relatively much larger and are produced as regions that retain a stain on certain chemical treatments.

Base pair (bp) is a partnership of A with T or of C with G in a DNA double helix; other pairs are possible in RNA under some circumstances.

Bidirectional replication is accomplished when two replication forks move away from the same origin in opposite directions.

Bivalent is the structure containing all four chromatids (two representing each homologue) at the start of meiosis.

Blunt-end ligation is a reaction that joins two DNA duplex molecules directly at their ends.

Branch migration describes the ability of a DNA strand partially paired with its complement in a duplex to extend its pairing by displacing the resident strand with which it is homologous.

Breakage and reunion is a mode of genetic recombination in which two DNA duplex molecules are broken at corresponding points and then rejoined crosswise.

C banding is a technique for generating stained regions around centromeres.

C genes code for the constant regions of immunoglobulin protein chains.

C value is the total amount of DNA in a haploid genome.

CAAT box is part of a conserved sequence at about 75 bp upstream from the startpoint of eukaryotic transcription units; it may be involved in binding RNA polymerase II.

Cap is the structure at the 5′ end of eukaryotic mRNA introduced after transcription by linking the 5′ end of a guanine nucleotide to the terminal base of the mRNA and methylating at least the additional G; the structure is $^{7Me}G^{5'}ppp^{5'}Np$. . . .

CAP (CRP) is a positive regulator protein activated by cyclic AMP, needed for RNA polymerase to initiate transcription of certain (catabolite-sensitive) operons of *E. coli*.

Capsid is the external protein coat of a virus particle.

Catabolite repression describes the decreased expression of many bacterial operons that results from addition of glucose; caused by decrease in intracellular level of cyclic AMP that inactivates CAP regulator.

cDNA is a single-stranded DNA complementary to an RNA, synthesized from it by *in vitro* reverse transcription.

cDNA clone is a duplex DNA sequence representing an RNA, carried in a cloning vector.

Cell cycle is the period from one division to the next.

Centromere is a constricted region of a chromosome that includes the site of attachment to the mitotic or meiotic spindle.

Chemical complexity is the amount of a DNA component measured by chemical assay.

Chi sequence is an octameric sequence that provides a hotspot for RecA-mediated genetic recombination in *E. coli*.

Chi structure is a joint between two duplex molecules of DNA revealed by cleaving an intermediate of two joined circles to generate linear ends in each circle.

Chiasma (*pl.* chiasmata) is a site at which two homologous chromosomes appear to have exchanged material during meiosis.

Chromatids are the copies of a chromosome produced by replication.

Chromatin is the complex of DNA and protein in the nucleus of the interphase cell, originally recognized by its reaction with stains specific for DNA.

Chromocenter is an aggregate of heterochromatin from different chromosomes.

Chromomere is a densely staining granule visible as an apparent component of the chromosome under certain conditions, especially early in meiosis.

Chromosome is a discrete unit of the genome carrying many genes, consisting of proteins and a very long molecule of DNA, visible as a morphological entity only during the act of cell division.

Chromosome walking describes the sequential isolation of clones carrying overlapping sequences of DNA to span large regions of the chromosome (often in order to reach a particular locus).

***Cis*-acting** locus affects the activity only of DNA sequences on its own molecule of DNA; usually implies that the locus does not code for protein.

***Cis*-acting** protein has the exceptional property of acting only on the molecule of DNA from which it was expressed.

***Cis*-configuration** describes two sites on the same molecule of DNA.

***Cis/trans* test** assays the effect of relative configuration on expression of two mutations. In a double heterozygote, two mutations in the same gene show mutant phenotype in *trans* configuration, wild-type in *cis* configuration.

Cistron is the genetic unit defined by the *cis/trans* test; equivalent to *gene* in comprising a unit of DNA representing a protein.

Class switching is a change in the expression of the C region of an immunoglobulin chain during lymphocyte differentiation.

Clone describes a large number of cells or molecules identical with a single ancestral cell or module.

Closed reading frame contains termination codons that prevent its translation into protein.

Coconversion is the simultaneous correction of two sites during gene conversion.

Coding strand of DNA has the same sequence as mRNA.

Codominant alleles both contribute to the phenotype; neither is dominant over the other.

Codon is a triplet of nucleotides that represents an amino acid or a termination signal.

Coevolution —*see* concerted evolution.

Cognate tRNAs are those recognized by a particular aminoacyl-tRNA synthetase.

Coincidental evolution —*see* concerted evolution.

Cointegrate structure is produced by fusion of two replicons, one originally possessing a transposon, the other lacking it; the cointegrate has copies of the transposon present at both junctions of the replicons, oriented as direct repeats.

Cold-sensitive mutant is defective at low temperature but functional at normal temperature.

Colony hybridization is a technique for using *in situ* hybridization to identify bacteria carrying chimeric vectors whose inserted DNA is homologous with some particular sequence.

Compatibility group of plasmids contains members unable to coexist in same bacterial cell.

Complementation refers to the ability of independent (nonallelic) genes to provide diffusible products that produce wild phenotype when two mutants are tested in *trans* configuration in a heterozygote.

In vitro **complementation assay** consists of identifying a component of a wild-type cell that can confer activity on an extract prepared from some mutant cell; identifies the component rendered inactive by the mutation.

Complementation group is a series of mutations unable to complement when tested in pairwise combinations in *trans;* defines a genetic unit (the cistron) that might better be called a *noncomplementation group*.

Complex locus (of *D. melanogaster*) has genetic properties inconsistent with the function of a gene representing a single protein.

Complexity is the total length of different sequences of DNA present in a given preparation.

Composite transposons have a central region flanked on each side by insertion sequences, either or both of which may enable the entire element to transpose.

Concatemer of DNA consists of a series of unit genomes repeated in tandem.

Concatenated circles of DNA are interlocked like rings on a chain.

Concerted evolution describes the ability of two related genes to evolve together as though constituting a single locus.

Conditional lethal mutations kill a cell or virus under certain (nonpermissive) conditions, but allow it to survive under other (permissive) conditions.

Consensus sequence is an idealized sequence in which each position represents the base most often found when many actual sequences are compared.

Conservative recombination involves breakage and reunion of preexisting strands of DNA without any synthesis of new stretches of DNA.

Constitutive genes are expressed as a function of the interaction of RNA polymerase with the promoter, without additional regulation; sometimes also called household genes in the context of describing functions expressed in all cells at a low level.

Constitutive heterochromatin describes the inert state of permanently nonexpressed sequences, usually satellite DNA.

Constitutive mutations cause genes that usually are regulated to be expressed without regulation.

Controlling elements of maize are transposable units originally identified solely by their genetic properties.

Coordinate regulation refers to the common control of a group of genes.

Cordycepin is 3′ deoxyadenosine, an inhibitor of polyadenylation of RNA.

Core DNA is the 146 bp of DNA contained on a core particle.

Core particle is a digestion product of the nucleosome that retains the histone octamer and has 146 bp of DNA; its structure appears similar to that of the nucleosome itself.

Corepressor is a small molecule that triggers repression of transcription by binding to a regulator protein.

Cosmids are plasmids into which phage lambda *cos* sites have been inserted; as a result, the plasmid DNA can be packaged *in vitro* in the phage coat.

Cot is the product of DNA concentration and time of incubation in a reassociation reaction.

Cot$_{1/2}$ is the Cot required to proceed to half completion of the reaction; it is directly proportional to the unique length of reassociating DNA.

Crossing-over describes the reciprocal exchange of material between chromosomes that occurs during meiosis and is responsible for genetic recombination.

Crossover fixation refers to a possible consequence of unequal crossing-over that allows a mutation in one member of a tandem cluster to spread through the whole cluster (or to be eliminated).

Cruciform is the structure produced at inverted repeats of DNA if the repeated sequence pairs with its complement on the same strand (instead of with its regular partner in the other strand of the duplex).

Cryptic satellite is a satellite DNA sequence not identified as such by a separate peak on a density gradient; that is, it remains present in main-band DNA.

Cyclic AMP is a molecule of AMP in which the phosphate group is joined to both the 3′ and 5′ positions of the ribose; its binding activates the CAP, a positive regulator of prokaryotic transcription.

Cytological hybridization —*see in situ* hybridization.

Cytoplasmic inheritance is a property of genes located in mitochondria or chloroplasts (or possibly other extranuclear organelles).

Cytoplasmic protein synthesis is the translation of mRNAs representing nuclear genes; it occurs via ribosomes attached to the cytoskeleton.

Cytoskeleton is a network of fibers in the cytoplasm of the eukaryotic cell.

D loop is a region within mitochondrial DNA in which a short stretch of RNA is paired with one strand of DNA, displacing the original partner DNA strand in this region. The same term is used also to describe the displacement of a region of one strand of duplex DNA by a single-stranded invader in the reaction catalyzed by RecA protein.

Degeneracy in the genetic code refers to the lack of an effect of many changes in the third base of the codon on the amino acid that is represented.

Deletions constitute the removal of a sequence of DNA, the regions on either side being joined together.

Denaturation of DNA or RNA describes its conversion from the double-stranded to the single-stranded state; separation of the strands is most often accomplished by heating.

Denaturation of protein describes its conversion from the physiological conformation to some other (inactive) conformation.

Derepressed state is the same as *induced* when describing the normal state of a gene; it has the same meaning as *constitutive* in describing the effect of mutation.

Dicentric chromosome is the product of fusing two chromosome fragments, each of which has a centromere. It is unstable and may be broken when the two centromeres are pulled to opposite poles in mitosis.

Diploid set of chromosomes contains two copies of each autosome and two sex chromosomes.

Direct repeats are identical (or related) sequences present in two or more copies in the same orientation in the same molecule of DNA; they are not necessarily adjacent.

Discontinuous replication refers to the synthesis of DNA in short (Okazaki) fragments that are later joined into a continuous strand.

Disjunction describes the movement of members of a chromosome pair to opposite poles during cell division. At mitosis and the second meiotic division, disjunction applies to sister chromatids; at first meiotic division it applies to sister chromatid pairs.

Divergence is the percent difference in nucleotide sequence between two related DNA sequences or in amino acid sequences between two proteins.

Divergent transcription refers to the initiation of transcription at two promoters facing in the opposite direction, so that transcription proceeds away in both directions from a central region.

dna mutants of bacteria cannot synthesize DNA at 42°C but can do so at 37°C.

DNA-driven hybridization involves the reaction of an excess of DNA with RNA.

Domain of a chromosome may refer *either* to a discrete structural entity defined as a region within which supercoiling is independent of other domains; *or* to an extensive region including an expressed gene that has heightened sensitivity to degradation by the enzyme DNAase I.

Domain of a protein is a discrete continuous part of the amino acid sequence that can be equated with a particular function.

Dominant allele determines the phenotype displayed in a heterozygote with another (recessive) allele.

Donor splicing site —*see* left splicing junction.

Down promoter mutations decrease the frequency of initiation of transcription.

Downstream identifies sequences proceeding farther in the direction of expression; for example, the coding region is downstream from the initiation codon.

Early development refers to the period of a phage infection before the start of DNA replication.

Elongation factors are proteins that associate with ribosomes cyclically, for addition of each amino acid to the polypeptide chain.

End labeling describes the addition of a radioactively labeled group to one end (5′ or 3′) of a DNA strand.

End-product inhibition describes the ability of a product of a metabolic pathway to inhibit the activity of an enzyme that catalyzes an early step in the pathway.

Endonucleases cleave bonds within a nucleic acid chain; they may be specific for RNA or for single-stranded or double-stranded DNA.

Enhancer element is a viral sequence of 72 bp repeats that increases the utilization of (some) eukaryotic promoters in *cis*-configuration, but that can function in any location, upstream or downstream, relative to the promoter.

Epigenetic changes influence the phenotype without altering the genotype.

Episome is a plasmid able to integrate into bacterial DNA.

Euchromatin comprises all of the genome in the interphase nucleus except for the heterochromatin.

Evolutionary clock is defined by the rate at which mutations accumulate in a given gene.

Excision of phage or episome or other sequence describes its release from the host chromosome as an autonomous DNA molecule.

Excision-repair systems remove a single-stranded sequence of DNA containing damaged or mispaired bases and replace it in the duplex by synthesizing a sequence complementary to the remaining strand.

Exon is any segment of an interrupted gene that is represented in the mature RNA product.

Exonucleases cleave nucleotides one at a time from the end of a polynucleotide chain; they may be specific for either the 5′ or 3′ end of DNA or RNA.

Extranuclear genes reside in organelles such as mitochondria and chloroplasts outside the nucleus.

Facultative heterochromatin describes the inert state of sequences that also exist in active copies—for example, one mammalian X chromosome in females.

Fast component of a reassociation reaction is the first to renature and contains highly repetitive DNA.

Figure-eight describes two circles of DNA linked together by a recombination event that has not yet been completed.

Filter hybridization is performed by incubating a denatured DNA preparation immobilized on a nitrocellulose filter with a solution of radioactively labeled RNA or DNA.

Foldback DNA consists of inverted repeats that have renatured by intrastrand reassociation of denatured DNA.

Footprinting is a technique for identifying the site on DNA bound by some protein by virtue of the protection of bonds in this region against attack by nucleases.

Forward mutations inactivate a wild-type gene.

Frameshift mutations arise by deletions or insertions that are not a multiple of 3 bp, and that therefore change the frame in which triplets are translated into protein.

G banding is a technique that generates a striated pattern in metaphase chromosomes that distinguishes the members of a haploid set.

G1 is the period of the eukaryotic cell cycle between the last mitosis and the start of DNA replication.

G2 is the period of the eukaryotic cell cycle between the end of DNA replication and the start of the next mitosis.

Gamete is either type of reproductive (germ) cell—sperm or egg—with haploid chromosome content.

Gap in DNA is the absence of one or more nucleotides in one strand of the duplex.

Gene (cistron) is the segment of DNA that is involved in producing a polypeptide chain; it includes regions preceding and following the coding region (leader and trailer) as well as intervening sequences (introns) between individual coding segments (exons).

Gene conversion is the alteration of one strand of a heteroduplex DNA to make it complementary with the other strand at any position(s) where there were mispaired bases.

Genomic (chromosomal) **DNA clones** are sequences of the genome carried by a cloning vector.

Genotype is the genetic constitution of an organism.

Gratuitous inducers resemble authentic inducers of transcription but are not substrates for the induced enzymes.

GT-AG rule describes the presence of these constant dinucleotides at the first two and last two positions of introns of nuclear genes.

Gyrase is a type II topoisomerase of *E. coli* with the ability to introduce negative supercoils into DNA.

Hairpin describes a double-helical region formed by base pairing between adjacent (inverted) complementary sequences in a single strand of RNA or DNA.

Haploid set of chromosomes contains one copy of each autosome and one sex chromosome; the haploid number *n* is characteristic of gametes.

Hapten is a small molecule that acts as an antigen when conjugated to a protein.

Helper virus provides functions absent from a defective virus, enabling the latter to complete the infective cycle during a mixed infection.

Heterochromatin describes regions of the genome that are permanently in a highly condensed condition and are not genetically expressed.

Heteroduplex (hybrid) DNA is generated by base pairing between complementary single strands derived from the different parental duplex molecules; it occurs during genetic recombination.

Heterogametic sex has the chromosome constitution 2A + XY.

Heterogeneous nuclear (hn) RNA comprises transcripts of nuclear genes made by RNA polymerase II; it has a wide size distribution and low stability.

Heteromultimeric proteins consist of nonidentical subunits (coded by different genes).

Heterozygote is an individual with different alleles at some particular locus.

Highly repetitive DNA is the first component to reassociate and is equated with satellite DNA.

Histones are conserved DNA binding proteins of eukaryotes that form the nucleosome, the basic subunit of chromatin.

Hogness (TATA) box is a conserved A-T-rich septamer about 25 bp upstream from the startpoint of nuclear genes; may be involved in positioning RNA polymerase II for initiation.

Homeo box describes the conserved sequence that is part of the coding region of *D. melanogaster* homeotic genes; it is also found in amphibian and mammalian genes expressed in early embryonic development.

Homeotic genes are identified by mutations that convert one body part into another; for example, an insect leg may replace an antenna.

Homogametic sex has the chromosome constitution 2A + XX.

Homologues are chromosomes carrying the same genetic loci; a diploid cell has two copies of each homologue, one derived from each parent.

Homomultimeric protein consists of identical subunits.

Homozygote is an individual with the same allele at corresponding loci on the homologous chromosome.

Hotspot is a site at which the frequency of mutation (or recombination) is very much increased.

Housekeeping (constitutive) genes are those (theoretically) expressed in all cells because they provide basic functions needed for sustenance of all cell types.

Hybrid-arrested translation is a technique that identifies the cDNA corresponding to an mRNA by relying on the ability to base pair with the RNA *in vitro* to inhibit translation.

Hybrid dysgenesis describes the inability of certain strains of *D. melanogaster* to interbreed, because the hybrids are sterile (although otherwise they may be phenotypically normal).

Hybridization is the pairing of complementary RNA and DNA strands to give an RNA-DNA hybrid.

Hybridoma is the cell line produced by fusing a myeloma with a lymphocyte; it continues indefinitely to express the immunoglobulins of both parents.

Hyperchromicity is the increase in optical density that occurs when DNA is denatured.

DNAase I **hypersensitive site** is a short region of chromatin detected by its extreme sensitivity to cleavage by DNAase I and other nucleases; probably comprises an area from which nucleosomes are excluded.

Idling reaction is the production of pppGpp and ppGpp by ribosomes when an uncharged tRNA is present in the A site; triggers the stringent response.

Immunity refers to the ability of a prophage to prevent another phage of the same type from infecting a cell or to the ability of a plasmid to prevent another of the same type from becoming established in a cell. Different mechanisms are involved in the two types of immunity. Also, transposition immunity refers to the ability of certain transposons to prevent others of the same type from transposing to the same DNA molecule.

***In situ* hybridization** is performed by denaturing the DNA of cells squashed on a microscope slide so that reaction is possible with an added single-stranded RNA or DNA; the added preparation is radioactively labeled and its hybridization is followed by autoradiography.

Incompatibility is the inability of certain bacterial plasmids to coexist in the same cell.

Incomplete dominance —*see* codominance.

Indirect end labeling is a technique for examining the organization of DNA by making a cut at a specific site and isolating all fragments containing the sequence adjacent to one side of the cut; it reveals the distance from the cut to the next break(s) in DNA.

Induced mutations result from the addition of a mutagen.

Inducer is a small molecule that triggers gene transcription by binding to a regulator protein.

Induction refers to the ability of bacteria (or yeast) to synthesize certain enzymes only when their substrates are present; applied to gene expression, refers to switching on transcription as a result of interaction of the inducer with the regulator protein.

Induction of prophage describes its excision from the host genome and entry into the lytic (infective) cycle as a result of destruction of the lysogenic repressor.

Initiation factors (IF) are proteins that associate with the small subunit of the ribosome specifically at the stage of initiation of protein synthesis.

Insertions are identified by the presence of an additional stretch of base pairs in DNA.

Integration of viral or another DNA sequence is its insertion into a host genome as a region covalently linked on either side to the host sequences.

Interallelic complementation describes the change in the properties of a heteromultimeric protein brought about by the interaction of subunits coded by two different mutant alleles; the mixed protein may be more or less active than the protein consisting of subunits only of one or the other type.

Interbands are the relatively dispersed regions of polytene chromosomes that lie between the bands.

Intercistronic region is the distance between the termination codon of one gene and the initiation codon of the next gene in a polycistronic transcription unit.

Intermediate component(s) of a reassociation reaction are those reacting between the fast (satellite DNA) and slow (nonrepetitive DNA) components; contain moderately repetitive DNA.

Interphase is the period between mitotic cell divisions; divided into G1, S, and G2.

Intervening sequence is an intron.

Intron is a segment of DNA that is transcribed, but is removed from within the transcript by splicing together the sequences (exons) on either side of it.

Inverted repeats comprise two copies of the same sequence of DNA repeated in opposite orientation on the same molecule. Adjacent inverted repeats constitute a palindrome.

Inverted terminal repeats are the short related or identical sequences present in reverse orientation at the ends of some transposons.

IS is an abbreviation for **insertion sequence,** a small bacterial transposon carrying only the genetic functions involved in transposition.

Isoaccepting tRNAs represent the same amino acid.

Isotype is a group of closely related immunoglobulin chains.

kb is an abbreviation for 1000 base pairs of DNA or 1000 bases of RNA.

Kinetic complexity is the complexity of a DNA component measured by the kinetics of DNA reassociation.

Kinetochore is the structural feature of the chromosome to which microtubules of the mitotic spindle attach.

Lagging strand of DNA must grow overall in the 3′-5′ direction and is synthesized discontinuously in the form of short fragments (5′-3′) that are later connected covalently.

Late period of phage development is the part of infection following the start of DNA replication.

Leader is the nontranslated sequence at the 5′ end of mRNA that precedes the initiation codon.

Leading strand of DNA is synthesized continuously in the 5′-3′ direction.

Leaky mutations allow some residual level of gene expression.

Left splicing junction is the boundary between the right end of an exon and the left end of an intron.

Lethal locus is any gene in which a lethal mutation can be obtained.

Library is a set of cloned fragments together representing the entire genome.

Ligation is the formation of a phosphodiester bond to link two adjacent bases separated by a nick in one strand of a double helix of DNA. (The term can also be applied to blunt-end ligation and to joining of RNA.)

Linkage describes the tendency of genes to be inherited together as a result of their location on the same chromosome; measured by percent recombination between loci.

Linkage group includes all loci that can be connected (directly or indirectly) by linkage relationships; equivalent to a chromosome.

Linker fragment is short synthetic duplex oligonucleotide containing the target site for some restriction enzyme; may be added to ends of a DNA fragment prepared by cleavage with some other enzyme during reconstructions of recombinant DNA.

Linker DNA is all DNA contained on a nucleosome in excess of the 146 bp core DNA.

Linking number is the number of times the two strands of a closed DNA duplex cross over each other.

Liquid (solution) hybridization is a reaction between complementary nucleic acid strands performed in solution.

Locus is the position of a chromosome at which the gene for a particular trait resides; locus may be occupied by any one of the alleles for the gene.

Long-period interspersion is a pattern in the genome in which long stretches of moderately repetitive and nonrepetitive DNA alternate.

Loop is a single-stranded region at the end of a hairpin in RNA (or single-stranded DNA); corresponds to the sequence between inverted repeats in duplex DNA.

LTR is an abbreviation for **long-terminal repeat,** a sequence directly repeated at both ends of a retroviral DNA.

Luxury genes are those coding for specialized functions synthesized (usually) in large amounts in particular cell types.

Lysis describes the death of bacteria at the end of a phage infective cycle when they burst open to release the progeny of an infecting phage.

Lysogen is a bacterium that possesses a repressed prophage as part of its genome.

Lysogenic immunity is the ability of a prophage to prevent another phage genome of the same type from becoming established in the bacterium.

Lysogenic repressor is the protein responsible for preventing a prophage from reentering the lytic cycle.

Lysogeny describes the ability of a phage to survive in a bacterium as a stable prophage component of the bacterial genome.

Lytic infection of bacteria by a phage ends in destruction of bacteria and release of progeny phage.

Main band of genomic DNA consists of a broad peak on a density gradient, excluding any visible satellite DNAs that form separate bands.

Major histocompatibility locus is a large chromosome region containing a giant cluster of genes that code for transplantation antigens and other proteins found on the surfaces of lymphocytes.

Map distance is measured as cM (centiMorgans) = percent recombination (sometimes subject to adjustments).

Marker (DNA) is a fragment of known size used to calibrate an electrophoretic gel.

Marker (genetic) is any allele of interest in an experiment.

Maternal inheritance describes the preferential survival in the progeny of a cross of genetic markers provided by one parent.

Meiosis occurs by two successive divisions (meiosis I and II) that reduce the starting number of 4n chromosomes to 1n in each of four product cells. Products may mature to germ cells (sperm or eggs).

Melting of DNA means its denaturation.

Melting temperature (T_m) is the midpoint of the temperature range over which DNA is denatured.

Micrococcal nuclease is an endonuclease that cleaves DNA; in chromatin, DNA is cleaved preferentially between nucleosomes.

Minichromosome of SV40 or polyoma is the nucleosomal form of the viral circular DNA.

Mitosis is the division of a eukaryotic somatic cell.

Modification of DNA or RNA includes all changes made to the nucleotides after their initial incorporation into the polynucleotide chain.

Modified bases are all those except the usual four from which DNA (T, C, A, G) or RNA (U, C, A, G) are synthesized; they result from postsynthetic changes in the nucleic acid.

Monocistronic mRNA codes for one protein.

Multicopy plasmids are present in bacteria at amounts greater than one per chromosome.

Multiforked chromosome (in bacterium) has more than one replication fork, because a second initiation has occurred before the first cycle of replication has been completed.

Multimeric proteins consist of more than one subunit.

Mutagens increase the rate of mutation by causing changes in DNA.

Mutation describes any change in the sequence of genomic DNA.

Myeloma is a tumor cell line derived from a lymphocyte; usually produces a single type of immunoglobulin.

Negative complementation occurs when interallelic complementation allows a mutant subunit to suppress the activity of a wild-type subunit in a multimeric protein.

Negative regulators function by switching off transcription or translation.

Negative supercoiling comprises the twisting of a duplex of DNA in space in the opposite sense to the turns of the strands in the double helix.

Neutral substitutions in a protein are those changes of amino acids that do not affect activity.

Nick in duplex DNA is the absence of a phosphodiester bond between two adjacent nucleotides on one strand.

Nick translation describes the ability of *E. coli* DNA polymerase I to use a nick as a starting point from which one strand of a duplex DNA can be degraded and replaced by resynthesis of new material; is used to introduce radioactively labeled nucleotides into DNA *in vitro*.

Nondisjunction describes failure of chromatids (duplicate chromosomes) to go to opposite poles during mitosis or meiosis.

Nonpermissive conditions do not allow conditional lethal mutants to survive.

Nonrepetitive DNA shows reassociation kinetics expected of unique sequences.

Nonsense codon is any one of three triplets (UAG, UAA, UGA) that cause termination of protein synthesis. (UAG is known as amber; UAA as ochre.)

Nonsense mutation is any change in DNA that causes a nonsense (termination) codon to replace a codon representing an amino acid.

Nonsense suppressor is a gene coding for a mutant tRNA able to respond to one or more of the nonsense codons.

Nontranscribed spacer is the region between transcription units in a tandem gene cluster.

Northern blotting is a technique for transferring RNA from an agarose gel to a nitrocellulose filter on which it can be hybridized to a complementary DNA.

Nuclear matrix is a network of fibers surrounding and penetrating the nucleus.

Nucleolar organizer is the region of a chromosome carrying genes coding for rRNA.

Nucleoid is the compact body that contains the genome in a bacterium.

Nucleolus is a discrete region of the nucleus created by the transcription of rRNA genes.

Nucleosome is the basic structural subunit of chromatin, consisting of ~200 bp of DNA and an octamer of histone proteins.

Ochre codon is the triplet UAA, one of three nonsense codons that cause termination of protein synthesis.

Ochre mutation is any change in DNA that creates a UAA codon at a site previously occupied by another codon.

Ochre suppressor is a gene coding for a mutant tRNA able to respond to the UAA codon to allow continuation of protein synthesis.

Okazaki fragments are the short stretches of 1000–2000 bases that are produced during discontinuous replication; they are later joined into a covalently intact strand.

Open reading frame contains a series of triplets coding for amino acids without any termination codons; sequence is (potentially) translatable into protein.

Operator is the site on DNA at which a repressor protein binds to prevent transcription from initiating at the adjacent promoter.

Operon is a complete unit of bacterial gene expression and regulation, including structural genes, regulator gene(s), and control elements in DNA recognized by regulator gene product(s).

Origin (*ori*) is a sequence of DNA at which replication is initiated.

Orphons are isolated individual genes that are related to members of a gene cluster.

Overwinding of DNA is caused by positive supercoiling (which applies further tension in the direction of winding of the two strands about each other in the duplex).

Packing ratio is the ratio of the length of DNA to the unit length of the fiber containing it.

Pairing of chromosomes—*see* synapsis.

Palindrome describes a sequence of DNA that is the same when one strand is read left to right or the other is read right to left; consists of adjacent inverted repeats.

Papovaviruses are a class of animal viruses with small genomes, including SV40 and polyoma.

Paranemic joint describes a region in which two complementary sequences of DNA are associated side by side instead of being intertwined in a double helical structure.

pBR322 is one of the standard plasmid cloning vectors.

Periodicity of DNA is the number of base pairs per turn of the double helix.

Permissive conditions allow conditional lethal mutants to survive.

Petite strains of yeast lack mitochondrial function.

Phage (bacteriophage) is a bacterial virus.

Phase variation describes an alternation in the type of flagella produced by a bacterium.

Phenotype is the appearance or other characteristics of an organism, resulting from the interaction of its genetic constitution with the environment.

Plasmid is an autonomous self-replicating extrachromosomal circular DNA.

Playback experiment describes the retrieval of DNA that has hybridized with RNA to check that it is nonrepetitive by a further reassociation reaction.

Plectonemic winding describes the intertwining of the two strands in the classical double helix of DNA.

Pleiotropic gene affects more than one (apparently unrelated) characteristic of the phenotype.

Point mutations are substitutions of single base pairs.

Polarity refers to the effect of a mutation in one gene in influencing the expression (at transcription or translation) of subsequent genes in the same transcription unit.

Polyadenylation is the addition of a sequence of polyadenylic acid to the 3′ end of a eukaryotic RNA after its transcription.

Polycistronic mRNA includes coding regions representing more than one gene.

Polymorphism refers to the simultaneous occurrence in the population of genomes showing allelic variations (as seen either in alleles producing different phenotypes or—for example—in changes in DNA affecting the restriction pattern).

Polyploid cell has more than two sets of the haploid genome.

Polyprotein is a gene product that is cleaved into several independent proteins.

Polysome (polyribosome) is an mRNA associated with a series of ribosomes engaged in translation.

Polytene chromosomes are generated by successive replications of a chromosome set without separation of the replicas.

Position effect refers to a change in the expression of a gene brought about by its translocation to a new site in the genome; for example, a previously active gene may become inactive if placed near heterochromatin.

Positive regulator proteins are required for the activation of a transcription unit.

Positive supercoiling describes the coiling of the double helix in space in the same direction as the winding of the two strands of the double helix itself.

Postmeiotic segregation describes the segregation of two strands of a duplex DNA that bear different information (created by heteroduplex formation during meiosis) when a subsequent replication allows the strands to separate.

Pribnow box is the consensus sequence TATAATG centered about 10 bp before the startpoint of bacterial genes. It is a part of the promoter especially important in binding RNA polymerase.

Primary transcript is the original unmodified RNA product corresponding to a transcription unit.

Primer is a short sequence (often of RNA) that is paired with one strand of DNA and provides a free 3′-OH end at which a DNA polymerase starts synthesis of a deoxyribonucleotide chain.

Primosome describes the complex of proteins involved in the priming action that initiates synthesis of each Okazaki fragment during discontinuous DNA replication; the primosome may move along DNA to engage in successive priming events.

Prokaryotic organisms (bacteria) lack nuclei.

Processive enzymes continue to act on a particular substrate, that is, do not dissociate between repetitions of the catalytic event.

Promoter is a region of DNA involved in binding of RNA polymerase to initiate transcription.

Proofreading refers to any mechanism for correcting errors in protein or nucleic acid synthesis that involves scrutiny of individual units *after* they have been added to the chain.

Prophage is a phage genome covalently integrated as a part of the bacterial chromosome.

Provirus is a duplex DNA sequence in the eukaryotic chromosome corresponding to the genome of an RNA retrovirus.

Pseudogenes are inactive but stable components of the genome derived by mutation of an ancestral active gene.

Puff is an expansion of a band of polytene chromosome associated with the synthesis of RNA at some locus in the band.

Pulse-chase experiments are performed by incubating cells very briefly with a radioactively labeled precursor (of some pathway or macromolecule); then the fate of the label is followed during a subsequent incubation with a nonlabeled precursor.

Quick-stop *dna* mutants of *E. coli* cease replication immediately when the temperature is increased to 42°C.

R loop is the structure formed when an RNA strand hybridizes with its complementary strand in a DNA duplex, thereby displacing the original strand of DNA in the form of a loop extending over the region of hybridization.

Rapid lysis (*r*) mutants display a change in the pattern of lysis of *E. coli* at the end of an infection by a T-even phage.

Reading frame is one of three possible ways of reading a nucleotide sequence as a series of triplets.

Reassociation of DNA describes the pairing of complementary single strands to form a double helix.

RecA is the product of the *recA* locus of *E. coli;* a protein with dual activities, acting as a protease and also able to exchange single strands of DNA molecules. The protease activity controls the SOS response; the nucleic acid handling facility is involved in recombination-repair pathways.

Recessive allele is obscured in the phenotype of a heterozygote by the dominant allele, often due to inactivity or absence of the product of the recessive allele.

Recessive lethal is an allele that is lethal when the cell is homozygous for it.

Reciprocal recombination is the production of new genotypes with the reverse arrangements of alleles according to maternal and paternal origin.

Recombinant progeny have a different genotype from that of either parent.

Recombinant joint is the point at which two recombining molecules of duplex DNA are connected (the edge of the heteroduplex region).

Recombination nodules (nodes) are dense objects present on the synaptonemal complex; could be involved in crossing-over.

Recombination-repair is a mode of filling a gap in one strand of duplex DNA by retrieving a homologous single strand from another duplex.

Regulatory gene codes for an RNA or protein product whose function is to control the expression of other genes.

Relaxed mutants of *E. coli* do not display the stringent response to starvation for amino acids (or other nutritional deprivation).

Relaxed replication control refers to the ability of some plasmids to continue replicating after bacteria cease dividing.

Release (termination) factors respond to nonsense codons to cause release of the completed polypeptide chain and the ribosome from mRNA.

Renaturation is the reassociation of denatured complementary single strands of a DNA double helix.

Repeating unit in a tandem cluster is the length of the sequence that is repeated; appears circular on a restriction map.

Repetition frequency is the (integral) number of copies of a given sequence present in the haploid genome; equals 1 for nonrepetitive DNA, $\geqslant 2$ for repetitive DNA.

Repetitive DNA behaves in a reassociation reaction as though many (related or identical) sequences are present in a component, allowing any pair of complementary sequences to reassociate.

Replacement sites in a gene are those at which mutations alter the amino acid that is coded.

Replication-defective virus has lost one or more genes essential for completing the infective cycle.

Replication eye is a region in which DNA has been replicated within a longer, unreplicated region.

Replication fork is the point at which strands of parental duplex DNA are separated so that replication can proceed.

Replicon is a unit of genome in which DNA is replicated; contains an origin for initiation of replication.

Replisome is the multiprotein structure that assembles at the bacterial replicating fork to undertake synthesis of DNA. Contains DNA polymerase and other enzymes.

Repression is the ability of bacteria to prevent synthesis of certain enzymes when their products are present; more generally, refers to inhibition of transcription (or translation) by binding of repressor protein to specific site on DNA (or mRNA).

Repressor protein binds to operator on DNA or to RNA to prevent transcription or translation, respectively.

Resolvase is enzyme activity involved in site-specific recombination between two transposons present as direct repeats in a cointegrate structure.

Restriction enzymes recognize specific short sequences of (usually) unmethylated DNA and cleave the duplex (sometimes at target site, sometimes elsewhere, depending on type).

Restriction map is a linear array of sites on DNA cleaved by various restriction enzymes.

Retroregulation describes the ability of a sequence downstream to regulate translation of an mRNA.

Retrovirus is an RNA virus that propagates via conversion into duplex DNA.

Reverse transcription is synthesis of DNA on a template of RNA; accomplished by reverse transcriptase enzyme.

Reversion of mutation is a change in DNA that either reverses the original alteration (true reversion) or compensates for it (second site reversion in the same gene).

Revertants have been derived by reversion of a mutant cell or organism.

Rho factor is a protein involved in assisting *E. coli* RNA polymerase to terminate transcription at certain (rho-dependent) sites.

Rho-independent terminators are sequences of DNA that cause *E. coli* RNA polymerase to terminate in the absence of rho factor.

Rifamycins (including rifampicin) inhibit transcription in bacteria.

Right splicing junction is the boundary between the right end of an intron and the left end of the adjacent exon.

RNA-driven hybridization reactions use an excess of RNA to react with all complementary sequences in a single-stranded preparation of DNA.

Rolling circle is a mode of replication in which a replication fork proceeds around a circular template for an indefinite number of revolutions; the DNA strand newly synthesized in each revolution displaces the strand synthesized in the previous revolution, giving a tail containing a linear series of sequences complementary to the circular template strand.

Rot is the product of RNA concentration and time of incubation in an RNA-driven hybridization reaction.

S phase is the restricted part of the eukaryotic cell cycle during which synthesis of DNA occurs.

S1 nuclease is an enzyme that specifically degrades unpaired (single-stranded) sequences of DNA.

Saltatory replication is a sudden lateral amplification to produce a large number of copies of some sequence.

Satellite DNA consists of many tandem repeats (identical or related) of a short basic repeating unit.

Saturation hybridization experiment has a large excess of one component, causing all complementary sequences in the other component to enter a duplex form.

Scaffold of a chromosome is a proteinaceous structure in the shape of a sister chromatid pair; generated when chromosomes are depleted of histones.

Scarce (complex) mRNA consists of a large number of individual mRNA species, each present in very few copies per cell.

scRNA is any one of several small cytoplasmic RNAs, molecules present in the cytoplasm and (sometimes) nucleus. Of unknown function.

Segmentation genes are concerned with controlling the number or polarity of body segments in insects.

Selection describes the use of particular conditions to allow survival only of cells with a particular phenotype.

Semiconservative replication is accomplished by separation of the strands of parental duplex, each then acting as a template for synthesis of a complementary strand.

Semidiscontinuous replication is mode in which one new strand is synthesized continuously while the other is synthesized discontinuously.

Sex chromosomes are those whose content are different in the two sexes; usually labeled X and Y (or W and Z), one sex has XX (or WW), the other sex has XY (or WZ).

Sex linkage is pattern of inheritance shown by genes carried on a sex chromosome (usually the X).

Shine-Dalgarno sequence is part or all of the polypurine sequence AGGAGG located on mRNA just prior to an AUG initiation codon; is complementary to the sequence at the 3′ end of 16S rRNA; involved in binding of ribosome to mRNA.

Short-period interspersion is a pattern in a genome in which moderately repetitive DNA sequences of ~300 bp alternate with nonrepetitive sequences of ~1000 bp.

Shotgun experiment is cloning of an entire genome in the form of randomly generated fragments.

Sigma factor is the subunit of bacterial RNA polymerase needed for initiation; is the major influence on selection of binding sites (promoters).

Signal hypothesis describes the role of the N-terminal sequence of a secreted protein in attaching nascent polypeptide to membrane; that is, mRNA and ribosome are attached to membrane via the N-terminal end of the protein under synthesis.

Silent mutations do not change the product of a gene.

Silent sites in a gene describe those positions at which mutations do not alter the product.

Simple-sequence DNA equals satellite DNA.

Single-copy plasmids are maintained in bacteria at a ratio of one plasmid for every host chromosome.

Single-strand assimilation describes the ability of RecA protein to cause a single strand of DNA to displace its homologous strand in a duplex; that is, the single strand is assimilated into the duplex.

Single-strand exchange is a reaction in which one of the strands of a duplex of DNA leaves its former partner and instead pairs with the complementary strand in another molecule, displacing its homologue in the second duplex.

Sister chromatids are the copies of a chromosome produced by its replication.

Site-specific recombination occurs between two specific (not necessarily homologous) sequences, as in phage integration/excision or resolution of cointegrate structures during transposition.

Slow component of a reassociation reaction is the last to reassociate; usually consists of nonrepetitive DNA.

Slow-stop *dna* mutants of *E. coli* complete the current round of bacterial replication but cannot initiate another at 42°C.

snRNA (small nuclear RNA) is one of many small RNA species confined to the nucleus; roles are unknown, but U1 snRNA might be involved in splicing.

Solution hybridization is the same as liquid hybridization.

Somatic cells are all the cells of an organism except those of the germ line.

SOS box is the DNA sequence (operator) of ~20 bp recognized by LexA repressor protein.

SOS response in *E. coli* describes the coordinate induction of many enzymes, including repair activities, in response to irradiation or other damage to DNA; results from activation of RecA protease to cleave LexA repressor.

Southern blotting describes the procedure for transferring denatured DNA from an agarose gel to a nitrocellulose filter where it can be hybridized with a complementary nucleic acid.

Spheroplast is a bacterial or yeast cell whose wall has been largely or entirely removed.

Spindle describes the reorganized structure of a eukaryotic cell passing through division; the nucleus has been dissolved and chromosomes are attached to the spindle by microtubules.

Splicing describes the removal of introns and joining of exons in RNA; thus introns are spliced out, while exons are spliced together.

Splicing junctions are the sequences immediately surrounding the exon-intron boundaries.

Spontaneous mutations are those that occur in the absence of any added reagent to increase the mutation rate.

Sporulation is the generation of a spore by a bacterium (by morphological conversion) or by a yeast (as the product of meiosis).

SSB is the single-strand protein of *E. coli*, a protein that binds to single-stranded DNA.

Staggered cuts in duplex DNA are made when two strands are cleaved at different points near each other.

Startpoint (startsite) refers to the position on DNA corresponding to the first base incorporated into RNA.

Stem is the base-paired segment of a hairpin.

Sticky ends are complementary single strands of DNA that protrude from opposite ends of a duplex or from ends of different duplex molecules; can be generated by staggered cuts in duplex DNA.

Strand displacement is a mode of replication of some viruses in which a new DNA strand grows by displacing the previous (homologous) strand of the duplex.

Streptolydigins inhibit the elongation of transcription by bacterial RNA polymerase.

Stringent replication describes the limitation of single-copy plasmids to replication pari passu with the bacterial chromosome.

Stringent response refers to the ability of a bacterium to shut down synthesis of tRNA and ribosomes in a poor-growth medium.

Structural gene codes for any RNA or protein product other than a regulator.

Supercoiling describes the coiling of a closed duplex DNA in space so that it crosses over its own axis.

Superrepressed means the same as uninducible.

Suppression describes the occurrence of changes that eliminate the effects of a mutation without reversing the original change in DNA.

Suppressor (extragenic) is usually a gene coding a mutant tRNA that reads the mutated codon, either in the sense of the original codon or to give an acceptable substitute for the original meaning.

Suppressor (intragenic) is a compensating mutation that restores the original reading frame after a frameshift.

Synapsis describes the association of the two pairs of sister chromatids representing homologous chromosomes that occurs at the start of meiosis; resulting structure is called a bivalent.

Synaptonemal complex describes the morphological structure of synapsed chromosomes.

Syntenic genetic loci lie on the same chromosome.

T_m is the abbreviation for melting temperature.

Tandem repeats are multiple copies of the same sequence lying in series.

TATA (Hogness) box is a conserved A-T-rich septamer found about 25 bp before the startpoint of each eukaryotic RNA polymerase II transcription unit; may be involved in positioning the enzyme for correct initiation.

Telomere is the natural end of a chromosome.

Temperature-sensitive mutation creates a gene product that is functional at low temperature but inactive at higher temperature (the reverse relationship is usually called cold-sensitive).

Terminal redundancy describes the repetition of the same sequence at both ends of (for example) a phage genome.

Termination codon is one of three triplet sequences, UAG (amber), UAA (ochre), or UGA that cause termination of protein synthesis; they are also called nonsense codons.

Terminator is a sequence of DNA, represented at the end of the transcript, that causes RNA polymerase to terminate transcription.

Thalassemia is disease of red blood cells resulting from lack of either α or β globin.

Topoisomerase is an enzyme that can change the linking number of DNA (in steps of 1 by type I; in steps of 2 by type II).

Topological isomers are molecules of DNA that are identical except for a difference in linking number.

Tracer is a radioactively labeled nucleic acid component included in a reassociation reaction in amounts too small to influence the progress of reaction.

Trailer is a nontranslated sequence at the 3′ end of an mRNA following the termination codon.

Trans configuration of two sites refers to their presence on two different molecules of DNA (chromosomes).

Transcribed spacer is the part of an rRNA transcription unit that is transcribed but discarded during maturation; that is, it does not give rise to part of rRNA.

Transcription is synthesis of RNA on a DNA template.

Transcription unit is the distance between sites of initiation and termination by RNA polymerase; may include more than one gene.

Transduction refers to the transfer of a bacterial gene from one bacterium to another by a phage; phage carrying host as well as its own genes is called transducing phage.

Transfection of eukaryotic cells is the acquisition of new genetic markers by incorporation of added DNA.

Transformation of bacteria describes the acquisition of new genetic markers by incorporation of added DNA.

Transformation of eukaryotic cells refers to their conversion to a state of unrestrained growth in culture, resembling or identical with the tumorigenic condition.

Transgenic animals are created by introducing new DNA sequences into the germ line as the result of its addition to the egg.

Transition is a mutation in which one pyrimidine is substituted by the other or in which one purine is substituted for the other.

Translation is synthesis of protein on the mRNA template.

Translocation of a chromosome describes a rearrangement in which part of a chromosome is detached by breakage and then becomes attached to some other chromosome.

Translocation of the ribosome is its movement one codon along mRNA after the addition of each amino acid to the polypeptide chain.

Transplantation antigen is protein coded by a major histocompatibility locus, present on all mammalian cells, involved in interactions between lymphocytes.

Transposase is the enzyme activity involved in insertion of transposon at a new site.

Transposition immunity refers to the ability of certain transposons to prevent others of the same type from transposing to the same DNA molecule.

Transposon is a DNA sequence able to replicate and insert one copy at a new location in the genome.

Transvection describes the ability of a locus to influence activity of an allele on the other homologue only when two chromosomes are synapsed.

Transversion is a mutation in which a purine is replaced by a pyrimidine or vice versa.

True-breeding organisms are homozygous for the trait under consideration.

Twisting number of a DNA is the number of base pairs divided by the number of base pairs per turn of the double helix.

Underwinding of DNA is produced by negative supercoiling (because the double helix is itself coiled in the opposite sense from the intertwining of the strands).

Unequal crossing-over describes a recombination event in which the two recombining sites lie at nonidentical locations in the two parental DNA molecules.

Unidirectional replication refers to the movement of a single replication fork from a given origin.

Uninducible mutants cannot be induced.

Unscheduled DNA synthesis is any DNA synthesis occurring outside the S phase of the eukaryotic cell.

Up promoter mutations increase the frequency of initiation of transcription.

Upstream identifies sequences proceeding in the opposite direction from expression; for example, the bacterial promoter is upstream from the transcription unit, the initiation codon is upstream from the coding region.

V gene is sequence coding for the major part of the variable (N-terminal) region of an immunoglobulin chain.

Variegation of phenotype is produced by a change in genotype during somatic development.

Cloning **vector** is any plasmid or phage into which a foreign DNA may be inserted to be cloned.

Virion is the external protein coat of a virus particle.

Virulent phage mutants are unable to establish lysogeny.

Wobble hypothesis accounts for the ability of a tRNA to recognize more than one codon by unusual (non-G-C, A-T) pairing with the third base of a codon.

Writhing number is the number of times a duplex axis crosses over itself in space.

Zero time-binding DNA enters the duplex form at the start of a reassociation reaction; results from intramolecular reassociation of inverted repeats.

Zygote is produced by fusion of two gametes—that is, it is a fertilized egg.

INDEX